AF682433

Herbs for Cancer Treatment

Bhupendra Koul

Herbs for Cancer Treatment

Springer

Bhupendra Koul
Department of Biotechnology, School
of Bioengineering and Biosciences
Lovely Professional University
Phagwara, Punjab, India

ISBN 978-981-32-9146-1 ISBN 978-981-32-9147-8 (eBook)
https://doi.org/10.1007/978-981-32-9147-8

© Springer Nature Singapore Pte Ltd. 2019
This work is subject to copyright. All rights are reserved by the Publisher, whether the whole or part of the material is concerned, specifically the rights of translation, reprinting, reuse of illustrations, recitation, broadcasting, reproduction on microfilms or in any other physical way, and transmission or information storage and retrieval, electronic adaptation, computer software, or by similar or dissimilar methodology now known or hereafter developed.
The use of general descriptive names, registered names, trademarks, service marks, etc. in this publication does not imply, even in the absence of a specific statement, that such names are exempt from the relevant protective laws and regulations and therefore free for general use.
The publisher, the authors, and the editors are safe to assume that the advice and information in this book are believed to be true and accurate at the date of publication. Neither the publisher nor the authors or the editors give a warranty, expressed or implied, with respect to the material contained herein or for any errors or omissions that may have been made. The publisher remains neutral with regard to jurisdictional claims in published maps and institutional affiliations.

This Springer imprint is published by the registered company Springer Nature Singapore Pte Ltd.
The registered company address is: 152 Beach Road, #21-01/04 Gateway East, Singapore 189721, Singapore

Preface

Cancer is extremely a lethal disease where the body's cells divide to grow in an uncontrollable way. These cells then infect the healthy cells and can spread throughout the body. Cancer is not a single disease; rather, it is a collection of related diseases. It has been estimated that the fastest rate of mortality in the world human population is due to cancer. It affects six million lives every year. According to the American Cancer Society, approximately 0.5% of US population is diagnosed with cancer yearly. Mutation causes cancer which alters the molecular mechanism of genes for controlling the normal cell growth. Proper balance of the gene functioning is crucial for preventing oncogenesis. We shall focus on the different types of cancers such as cancers of the breast, prostate, cervix, lung, bowel or colon and anal region, etc. Our studies shall also cover the molecular basis of cancer. Herbal drugs with anticancer potential have gained great importance, and research work has also excelled in this field since the last decade. It has been estimated that there are altogether 250,000 species of higher plants on Earth, and among them, 35,000–70,000 species are being used to treat various diseases due to the presence of secondary metabolites. Science of Ayurveda asserts new knowledge on therapeutic aspect of cancer by using the herbs. Hence, an attempt is made about the curative management of various cancers described in Ayurveda. This book consists of a review of 149 plant families with a large number of species reported to possess anticancer property and other medicinal properties, so that further in-depth research be done and novel biomolecules be extracted for the treatment of different cancers.

The objectives of the present book is to

- Be informed about the different cancer types and the ill effects of conventional cancer treatment regimes
- Encourage the youth to be open-minded and seek information about alternative therapies for cancer
- Be a starting point for your own research so that you can suggest a combination of cancer treatment(s)
- Be informed about the power of Ayurveda as an alternate therapy for cancer management

The present book entitled *Herbs for Cancer Treatment* encompasses the details of the studies under four major chapters as described below.

Chapter 1 – **Cancer Genomics:** Cancer is probably the most complex of all diseases as it is associated with mutation in the genomes. Mutation is caused by certain reasons like toxic chemicals, poisonous gases, radiations, etc. Cancer-causing genes are called as oncogenes (*Bcl-2, C-myc, foc, jun, HER-2lneu/c-erb-2)*, and the genes which suppress the cancer are known as suppressor genes (*Bax, cx32, cx43, p53*, etc.). During cell cycle, only healthy cells can enter from G phase to cytokinesis phase, and unhealthy cells undergo apoptosis which is under the control of p53 gene ("guardian of DNA"). Mutation in this gene affects the whole cycle, which may result in the formation of tumors. According to the recent studies, mutation in genes like *EGFR, PSA*, and *BRCA* may lead to lung, prostate, and breast cancer, respectively. Our work is based on reliable data collected from various cancer databases like Integrative Oncogenomics Cancer Browser (IntOGen), Mouse Retrovirus-Tagged Cancer Gene Database, CEO Cancer Life Sciences Consortium, COSMIC, NPACT, NATTS, International Cancer Genome Consortium, CAB Abstract, J-GATE, ERIC, ProQuest, INMEDPLAN, and websites such as www.sciencedirect.com, www.ncbi.nlm.nih.gov/pubmed/, www.jstor.org, etc. The objective of this chapter is to state the role of genes in cancer. Moreover, the information presented in this report may help the "to be" oncologists to learn more about the causes and therapies for cancer treatment.

Chapter 2 – **Types of Cancer:** Cancer is an extremely fatal disease, not only in developed but also in developing countries. It affects six million lives every year worldwide. According to the American Cancer Society, 0.5% of US population is diagnosed with cancer yearly, and billions of dollars are invested on cancer research every year, and yet the success rates are not satisfactory. Cancer cells arise due to the imbalance in the body functioning, and they invade and infect the normal cells. Cancer is not a single disease; rather, it is a collection of related diseases. Our work includes explicit information of the different types of cancers such as blood, lung, colorectal, prostrate, skin, breast, endometrial, thyroid, lymphatic, etc. Cancers can be treated by various methods which include surgery, stem cell transplant, precision medicine, radiation, chemotherapy, hormone therapy, immune therapy, and targeted therapy. In chemotherapy and radiotherapy, survival rates are usually very low because of the challenging side effects, but herbal therapy has shown some promising potential as it does not possess any side effects. In some cases, cancer can be treated by single method, but, mostly, several methods are being used to cure the disease. Moreover, the treatment specifically depends upon the stage and type of cancer. It gives an insight into the types of cancers and also may help the readers to find out the exact way of treatment.

Chapter 3 – **Role of Ayurveda in Cancer Treatment**: In the present day, cancer is a burning issue and is caused due to the mutation in the cells, which divide uncontrollably, so as to form abnormal mass of cells inside the body. It can also be caused by various reasons such as heredity, loss of exercise, alcohol, smoking, physical trauma, radiations, etc. Every year, about six million cases of cancer are reported worldwide. Cancer cells possess immense potential to divide and to invade other parts of the body at a tremendous speed. The treatment includes chemotherapy, radiotherapy, surgery, etc. which have their own side effects. Ayurveda is a science of life with historical roots in the Indian subcontinent. Ayurvedic therapies have been practiced in Western countries as well. Ayurveda defines cancer as a chronic imbalance of Tridosha (pathophysiological entities), dhatus (tissue), and Triguna (mental qualities) and can be treated without any side effect. Emphasis has been made on how Ayurvedic medicine or the herbal remedies help to retard the process of cancer by targeting the specific tissue and also improves the quality of patient's life. Moreover, the success rates of herbal therapy have also been discussed.

Chapter 4 – **Plants with Anticancer Potential**: Herbs and trees have been utilized for health and medicinal purposes for hundreds of years. It has been estimated that there are altogether 250,000 species of higher plants on Earth, and among them, 35,000–70,000 species are being used to treat various diseases due to the presence of secondary metabolites (alkaloid, flavonoids, steroids, glycosides, etc.). The use of herbs for treatment of various diseases and disorders has been mentioned in the ancient Unani manuscripts, Egyptian papyrus, and Chinese writings. The present work encompasses explicit information of 149 plant families with several species reported to possess anticancer property. Moreover, the biological properties of bioactive compounds are also covered. We have tried to amalgamate individual information of plant species exhibiting anticancer potential into a tabular form, so that the reader can figure out specific information related to the efficiency and efficacy of the concerned plant in the cancer treatment. In the near future, biologically active compounds derived from these herbs can treat various types of cancers. The information present in this work may also inspire new biomedical applications in different plant species.

Phagwara, Punjab, India — Bhupendra Koul

Contents

About the Author

Bhupendra Koul is currently serving as an Assistant Professor at the School of Bioengineering and Biosciences, Lovely Professional University, Punjab, India. During his Ph.D. at the Plant Transgenics Lab, CSIR-National Botanical Research Institute (CSIR-NBRI) in Lucknow, he worked on the optimization, introduction, and expression of modified full-length and truncated versions of Bt-*cry1Ab* and *1Ac* genes in tomato for developing non-chimeric and stable transgenic lines resistant to two lepidopteran insects (*Helicoverpa armigera* and *Spodoptera litura*) and evaluated the performance of both the versions of *cry1Ab* and *1Ac* genes for the stability and efficacy of insecticidal toxin in transgenic plants. He also evaluated the performance and role of various *cis*-motifs of artificial synthetic promoters for overexpression of genes in tomato and performed comparative in silico analyses of several *cry1A* genes for toxicity to target insects. He has 6 years of research experience and 5 years of teaching experience and received the "Teacher Appreciation Award 2016" from the LPU in the Discipline of Biotechnology, through the MHRD Minister, Government of India. He has designed the full-length synthetic *cry1Ac* gene (NCBI submission accession No., KP195020.1; GI, 768678299) and has published 17 research papers in national and international journals, including *BMC Genomics*, *Plant Physiology and Biochemistry*, *Journal of Ethnopharmacology*, and *South African Journal of Botany*, as well as 5 book chapters and 1 authored book with Springer Nature.

Abbreviations

aCML	atypical chronic myeloid leukemia
AEL	acute eosinophilic leukemia
AITL	angioimmunoblastic T-cell lymphoma
AL	acute leukemia
ALCL	anaplastic large-cell lymphoma
ALL	acute lymphocytic leukemia
AML	acute myeloid leukemia
APAF1	apoptotic protease activating factor
APL	acute promyelocytic leukemia
B-ALL	B-cell acute lymphocytic leukemia
B-CLL	B-cell lymphocytic leukemia
BCL2	B-cell lymphoma 2
B-NHL	B-cell non-Hodgkin lymphoma
BAD	BCL2 antagonist of cell death
BAX	BCL2-associated X protein
Bcl X	B-cell lymphoma factor that inhibits cell death
BID	BH3-interacting domain death agonist
CLL	chronic lymphocytic leukemia
CML	chronic myeloid leukemia
CMML	chronic myelomonocytic leukemia
CNL	chronic neutrophilic leukemia
CNS	central nervous system
CMM1	cutaneous malignant melanoma
D	large deletion
DAXX	death-associated protein 6
DFSP	dermatofibrosarcoma protuberans
DGC	diffuse-type gastric carcinoma
DIPG	diffuse intrinsic pontine glioma
DLBCL	diffuse large B-cell lymphoma
DLCL	diffuse large-cell lymphoma
Dom	dominant
DBD	DNA-binding core domain

E	epithelial
ERKs	extracellular signal-regulated kinases
ETP-ALL	early T-cell precursor acute lymphoblastic leukemia
F	frameshift
FADD	Fas-associated death domain
FAS	CD95 and Apo 1
FASL	Fas ligand
FPC	familial adenomatous polyposis of the colon
GBM	glioblastoma multiforme
GIST	gastrointestinal stromal tumor
HES	hypereosinophilic syndrome
HNPCC	hereditary nonpolyposis colorectal cancer
HNSCC	head and neck squamous cell carcinoma
ICGC	International Cancer Genome Consortium
JMML	juvenile myelomonocytic leukemia
L	leukemia/lymphoma
M	mesenchymal
MALT	mucosa-associated lymphoid tissue lymphoma
MAP	mitogen-activated protein
MCL	mantle cell lymphoma
MDS	myelodysplastic syndrome
MOMP	mitochondrial outer membrane permeabilization
Mis	missense
MLCLS	mediastinal large-cell lymphoma with sclerosis
MM	multiple myeloma
MPN	myeloproliferative neoplasm
NHL	non-Hodgkin lymphoma
NK/T	natural killer T cell
NSCLC	non-small cell lung cancer
OCGs	oncogenes
PMBL	primary mediastinal B-cell lymphoma
pre-B ALL	pre-B-cell acute lymphoblastic leukemia
PCD	programmed cell death
PUMA	p53 upregulated modulator of apoptosis
RCC	renal cell carcinoma
Rec	recessive
RSK	ribosomal S6 kinase
sAML	secondary acute myeloid leukemia
SCC	squamous cell carcinoma
SCCOHT	small cell carcinoma of the ovary of hypercalcemic type
SM-AHD	systemic mastocytosis with associated hematological disorder
SMZL	splenic marginal zone lymphoma

TAD	transcription activation domain
TAZ	transcriptional coactivator with PDZ-binding motif
T-ALL	T-cell acute lymphoblastic leukemia
T-CLL	T-cell chronic lymphocytic leukemia
T-PLL	T-cell prolymphocytic leukemia
TCGA	The Cancer Genome Atlas
TGCT	testicular germ cell tumor
TRAIL	tumor necrosis factor-related apoptosis-inducing ligand
TSG	tumor suppressor gene
WHO	World Health Organization
WM	Waldenström's macroglobulinemia
YAP	Yes-associated protein

Cancer Genomics 1

Abstract

Cancer is probably the most complex of all diseases as it is associated with mutation in the genomes. Mutation is caused by certain reasons like toxic chemicals, poisonous gases, radiations, etc. Cancer-causing genes are called as oncogenes (*Bcl-2*, *C-myc*, *foc*, *jun*, *HER-2/neu/c-erb-2*), and the genes which suppress cancer are known as suppressor genes (*Bax*, *cx32*, *cx43*, *p53*, etc). During cell cycle, only healthy cells can enter from G phase to cytokinesis phase, and unhealthy cells undergo apoptosis which is under the control of p53 gene ("guardian of DNA"). Mutation in this gene affects the whole cycle, which may result in the formation of tumors. According to the recent studies, mutation in genes like *EGFR*, *PSA*, and *BRCA* may lead to lung, prostate, and breast cancer, respectively. Our work is based on reliable data collected from various cancer databases like Integrative Oncogenomics Cancer Browser, Mouse Retrovirus Tagged Cancer Gene Database, the CEO Life Sciences Consortium, COSMIC, IntOGen, International Cancer Genome Consortium, CAB Abstract, J-GATE, ERIC, ProQuest, and INMEDPLAN and websites such as www.jstor.org, www.sciencedirect.com, www.ncbi.nlm.nih.gov/pubmed/, etc. The objective of this chapter is to state the role of genes in various cancers. Moreover, the information presented in this report may help the "to be" oncologists to learn more about the causes and therapies for cancer treatment.

Keywords

Proto-oncogenes · Tumor suppressor genes · Carcinomas · Sarcoma · Melanoma · Leukemia

© Springer Nature Singapore Pte Ltd. 2019
B. Koul, *Herbs for Cancer Treatment*,
https://doi.org/10.1007/978-981-32-9147-8_1

1.1 Introduction

The animal body functions as that of a society or ecosystem whose individual components are cells. A healthy tissue contains numerous cells, and each cell is preordained to die a peaceful and timely death in its old age, through a process called "apoptosis" also known as "programmed cell death." All cells are in communication, behave in a social manner, and may perform several activities as and when required for the good of the organism. These activities include resting, dividing, differentiating, or dying. The communication can occur directly, through cell-to-cell contact, or through exchange of messenger compounds like growth factors or hormones. Through proper communication, appropriate cells proliferate to produce new cells.

A single mutation in a cell and its selection can jeopardize the social harmony of the multicellular society. This mutant cell divides more vigorously than its neighbors, beyond the control of the cellular machinery. This cell becomes cancerous and it prospers at the expense of its neighboring cells. Such a vigorous growth culminates into the damage of the tissue, organ, organ system, and whole body as well. Cancer cell divides to produce daughter cells, which further divides to form a colony of cancer cells. These cells also develop the characteristics of metastasis and the ability to evade the immunity and also induce the formation of new blood vessels which is called angiogenesis and is the third stage in carcinogenesis. Cancer cells lose the touch with their neighboring cells, as compared to the other cells. Uncontrolled proliferation and spread are the two main characteristics of cancer cells. Thus, cancer is the result of a micro evolutionary process that operates or activates on a time scale of months or years in a cellular society.

Indeed cancer is one of the notorious and devastating groups of diseases affecting millions of people each year, across the globe. It has been rated as the second main reason of death in human beings (American cancer society 2018). Intensive studies conducted through the large-scale cancer genome projects such as The Cancer Genome Atlas (TCGA) (Cancer Genome Atlas Research Network 2008), The Wellcome Trust Sanger Institute's Cancer Genome Project (Pleasance et al. 2010), and the International Cancer Genome Consortium (ICGC) (International Cancer Genome Consortium 2010) produced multidimensional genome-wide big data which has tremendously advanced cancer research, in terms of its genomics or genetics (Chin et al. 2011).

Basically, cancer originates from any change between the balancing of cell growth and cell death or programmed cell death. Most commonly cancer cells originate in epithelial tissues or in the organs such as the breast, pancreas, lung, skin, and liver (Sever and Brugge 2015). Certain chemicals, genes, viruses, and radiation exposure are the main factors responsible for cancer. It can lead to death of the patient if the spread is not controlled. Cancer is caused by external factors like infectious organisms, tobacco use, etc. and internal factors like immune conditions, genetic mutations, or hormonal imbalance. Hormone therapy, surgery, chemotherapy, radiation, and targeted therapy are either life prolonging or expensive.

Undoubtedly, these treatments are not free from harmful effects and also include nausea and pain, skin and nail problems, general fatigue, decreased sex drive, hair loss, hearing loss, diarrhea, etc.

There has been a continuous search on various natural sources to develop natural product drugs to combat this multifaceted syndrome. Since time immemorial, medicinal plant extracts and their formulations have been used as therapeutics for many diseases (Discussed in Chap. 5). Few out of the many plant bioactive compounds with anticancer properties have been reported clinically active against various types of cancer cells, and many are under still under clinical trials. Further research in this area may lead to better prevention and treatment of cancer.

1.2 Cancer: Overview

1.2.1 Normal Cell vs Cancer Cell

It is interesting that cancer is a genetic disease (caused because of alteration in specific genes) but not necessarily an inherited disease (genetic defect present in chromosome of a parent is transferred to the offspring). In contrast to the above information, most of the genetic alterations that cause cancers arise in the DNA of a somatic cell. Normal cells are those which divide only when stimulated by the homeostatic machinery of the body. They do not survive if excessively damaged and follow the process of apoptosis (programmed cell death). On the other hand, the cancer cells are less specialized (do not differentiate into other cell types) than normal cells. They do ignore the signals of programmed cell death or apoptosis and continue to divide without a halt (Fig. 1.1). Cancer cells activate the non-cancerous cells to form blood vessels so as to supply nutrients and oxygen to the tumor cells. Cancer cells are also able to surpass the immune system. Unlike normal cells, the cancer cells are invasive and invade new cells and tissue to develop new cancer cell colonies.

The normal cells can be converted to cancer cells by treating them with carcinogenic (mutagenic) chemicals, radiation, or tumor viruses. The agents which cause such type of transformation are called carcinogens. The normal cells grow as monolayer, but the cancer cells grow as clumps. The cancer cells can be separated by dissociating from the malignant tumor. The normal cells undergo an aging process after a finite number of mitotic divisions, but the cancer cells proliferate, undergo uncontrolled cell division, and exhibit immortal characteristic. This characteristic cancer cells are due to the presence of telomerase in them (absent in normal cells). This telomerase enzyme maintains the telomeres at the ends of the chromosomes which allow the cells to continue and divide.

In comparison to the non-cancerous cells, the growth of cancer cells may not exhibit standard diploid chromosome number (aneuploid) and often has a disorganized cytoskeleton. They may also produce certain unusual proteins and display them on their surfaces. If the chromosome number of the normal cells gets disturbed, then the cells are guided (signaling pathway) to the apoptotic

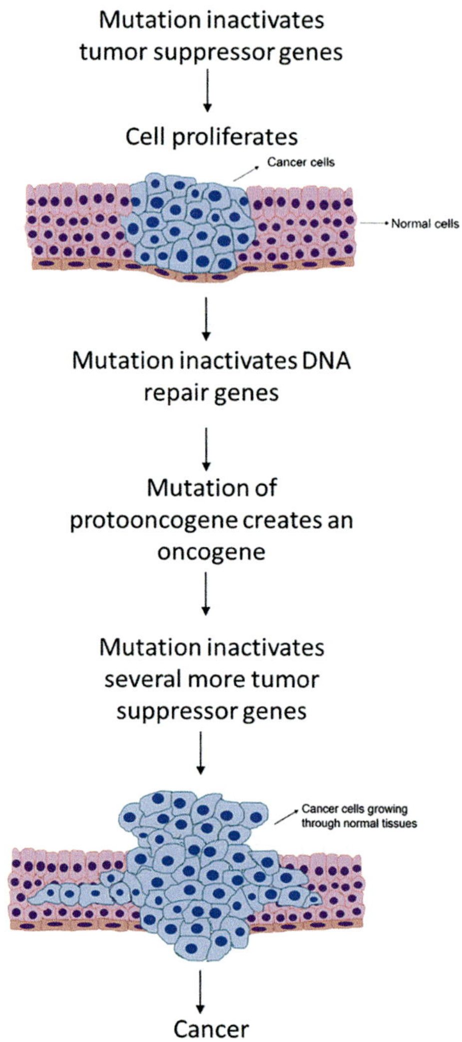

Fig. 1.1 Mutation in a normal lead to cancer

(self-destruction) pathway. On the other hand, the cancer cells do not follow apoptotic response. They depend on glycolysis (anaerobic metabolic pathway) for their energy (ATP) supplies. Interestingly, glycolysis remains the choice even when there is adequate supply of oxygen. In comparison to the normal cells, the cancer cells rapidly uptake glucose to produce ATP. This property reflects the high metabolic requirements of cancer cells. Interestingly, under limited oxygen supplies, cancer cells activate HIF (transcription factor) that induces the formation of new blood vessels and promotes cancer cell migration to spread the tumor.

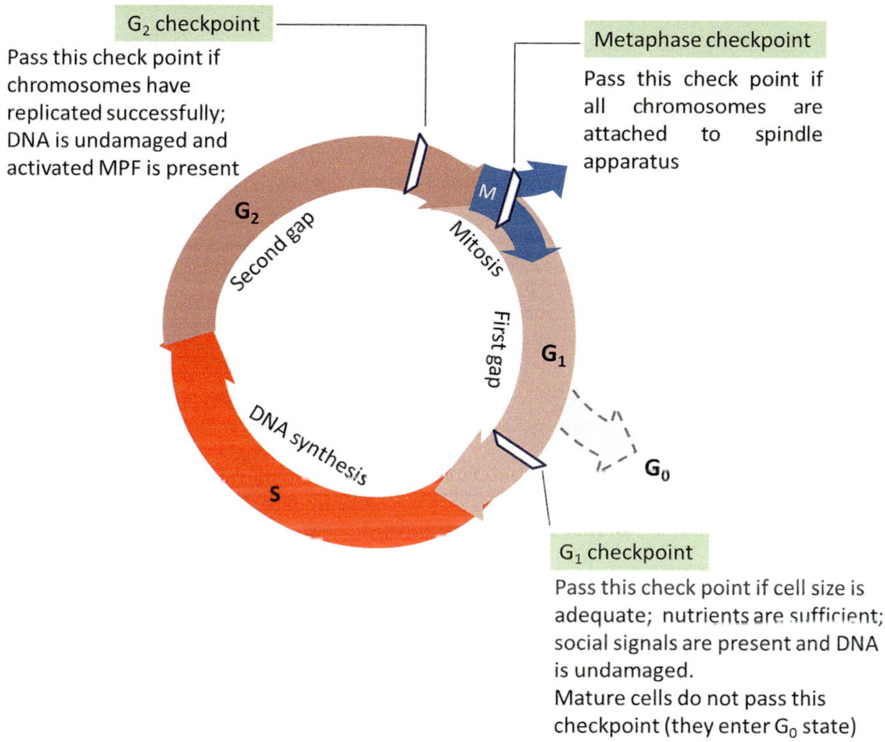

Fig. 1.2 Cell cycle checkpoints for normal cell division

1.2.2 Cell Cycle and Cancer

The cell cycle is composed of periods of growth followed by synthesis of DNA and cell division. The four phases of the cell cycle are G1, S, G2, and M. The transition of G1 and G2 to S and M, respectively, requires chemical signals. However, if the signals are not properly sensed or if the cell is not prepared to respond to the signals, then it becomes cancerous. Therefore, the transitions between each phase are regulated at checkpoints (Fig. 1.2). These checkpoints prevent the progression of cell cycle if the DNA synthesis is not complete or DNA damage is present. Among many there are two main proteins which are associated with the checkpoints: the cyclins and the cyclin-dependent kinases (CDKs). As the name indicates, CDKs require cyclin for phosphorylation of other proteins and thus regulate their activity. Thus, progression or halt through each of the checkpoint requires successive binding and dissociation of cyclin and CDK. An important checkpoint involved in mid G1 phase of the cell cycle is START. It responds to the internal as well as external signals to determine when it is feasible to move into the S phase. If the DNA within a cell is damaged, then the cell's entry into the S phase shall be delayed to allow the repair of the damaged DNA. Cells with a dysfunctional/mutated START checkpoint may become cancerous.

1.2.3 Oncogenes and Tumor Suppressor Genes

As discussed above, cancer occurs by various types of mutation and epigenetic changes in the genes that are concerned with the normal cell proliferation. Those cells with mutation surpass all controls and are selected by a Darwinian process, leading to the formation of a tumor. The products of the genes that are concerned with the development of a tumor may (1) regulate the cell proliferation directly, (2) control apoptosis, and (3) directly involve in damaged DNA repair. Based on the mechanism of action of these cancer-related genes, they can be classified into two categories, i.e., growth-promoting oncogenes (OCGs) and growth inhibitory tumor suppressor genes (TSGs).

1.2.4 Proto-Oncogenes and Oncogenes

Proto-oncogenes code for those proteins which send a signal to the nucleus so as to activate the process of cell division or to promote the process of apoptosis. These proteins act in a stepwise manner which is also known as signal transduction cascade. This cascade consists of membrane receptors for signal molecule and intermediary proteins which transmit the signal through the cytoplasm.

During early developmental stages, proto-oncogenic activity remains high owing to the continuous development of tissues and organs. The proto-oncogenes are classified on the basis of their function inside the cell. The examples include human epidermal growth factor receptor gene (2HER2), this gene codes for a transmembrane tyrosine kinase receptor; RAS: Ras protein family members belong to the GTPase protein family and are responsible for cellular signal transduction; WNT: this gene codes for signal transduction pathway cell surface receptor proteins; MYC: this is a family of proto-oncogenes and regulator genes which codes for transcription factors involved in cell proliferation; ERK: extracellular signal-regulated kinases (ERKs) or mitogen-activated protein (MAP) kinases are widely expressed protein kinase intracellular signaling molecules that are involved in cell division; and TRK: this gene codes for receptors that mediate several signaling cascades for neuronal survival and differentiation.

When the proto-oncogenes get mutated, they become oncogenes and gain the capacity to turn normal cells into cancer cells. These oncogenes then activate the signaling processes which further lead to increased production of growth stimulants. For example, RAS is oncogene family which functions in an on-off way. A particular type of mutation in RAS causes them to remain "on" which lead to the unrestrained cell division and growth. Nearly about 30% of tumors (thyroid, lung, pancreas, colon) occurs due to mutations in RAS.

Another example MYC is a type of proto-oncogene and codes for a transcription factor. Mutations in MYC proto-oncogene convert them into an active oncogene which is responsible for cancer.

The change of a proto-oncogene to an oncogene occurs by various types of mutations in the proto-oncogene (insertion and deletion mutations, point mutations) by rearrangement of genes in the chromosome that moves the proto-oncogene to a new location or by an increase in the number of copies (gene amplification, increased transcription) of the normal proto-oncogene. It is important here to understand that mutations are permanent alterations of the nucleotide sequence of the DNA (~ gene) that codes for a protein that is abnormal or functionally different or having an increased activity. These changes cause the cells which were once normal to divide continuously, bypassing the normal apoptotic pathway which mediates natural cell death. In that situation, the cells which were supposed to die remain inside the body and cause tumors and cancers. Some viruses can also insert their DNA near the proto-oncogene which coverts them into oncogene. Single copy of some oncogenes is appropriate for the expression of uncontrolled growth of cells, because of dominant mutations. Tumor can also be inherited from parent to offspring in some conditions if the oncogenes are present in germ cell lines.

1.2.5 Tumor Suppressor Genes

These genes transcribe and translate to produce proteins which normally hinder cell growth and prevent the formation of tumor. These genes produced such kind of substances which act on the cell membrane, cytoplasm, or nucleus of the cell. Any type of mutations in these genes leads to the abnormal growth and division of cells. Table 1.1 consists of an exhaustive list of alphabetically arranged genes that are associated with cancer. The information in Table 1.1 consists of the data derived from the Catalogue of Somatic Mutations in Cancer (COSMIC) database (COSMIC website: cancer.sanger.ac.uk/cosmic). Retinoblastoma is an example of this type of cancer; if both the parents carry RB tumor suppressor gene, the chances to carry this cancer is about 90% by the offspring (Fig. 1.3). Retinoblastoma can also occur by mutation in RB in the somatic cell of a person.

Certain other examples are hereditary breast cancer, which occurs due to mutations in both copies of BRCA2; familial adenomatous polyposis of the colon (FPC), which occurs due to mutations in both copies of the APC gene; and hereditary breast and ovarian cancer, which occurs due to mutations in both copies of BRCA1. Some environmental factors are also responsible for somatic mutations in tumor suppressor genes.

From the above discussion, it is now clear that the proto-oncogenes and tumor suppressor genes work together. The proto-oncogenes enhance cell growth, while tumor suppressor genes retard the growth of cell, and the controlled cell growth is regulated by proto-oncogenes. The mutations which occur in tumor suppressor genes stop the normal inhibition process of cell growth, while the mutations which occur in proto-oncogenes enhance the growth of cells.

Table 1.1 A list of genes associated with cancer

Gene	Name	Entrez Gene Id	Tumor types (somatic)	Tumor types (germline)
ABI1	abl-Interactor 1	10006	AML	
ABL1	v-abl Abelson murine leukemia viral oncogene homolog 1	25	CML, ALL, T-ALL	
ABL2	c-abl oncogene 2, non-receptor tyrosine kinase	27	AML	
ACKR3	Atypical chemokine receptor 3	57007	Lipoma	
ACSL3	Acyl-CoA synthetase long-chain family member 3	2181	Prostate	
ACVR1	Activin A receptor, type I	90	DIPG	
ACVR2A	Activin A receptor type 2A	92	Large intestine carcinoma, stomach carcinoma, pancreatic carcinoma, biliary tract, esophagus	
AFF1	AF4/FMR2 family, member 1	4299	AL	
AFF3	AF4/FMR2 family, member 3	3899	ALL, T-ALL	
AFF4	AF4/FMR2 family, member 4	27125	ALL	
AKT1	v-akt murine thymoma viral oncogene homolog 1	207	Breast, colorectal, ovarian, NSCLC	
AKT2	v-akt murine thymoma viral oncogene homolog 2	208	Ovarian, pancreatic	
ALK	Anaplastic lymphoma kinase (Ki-1)	238	ALCL, NSCLC, neuroblastoma, inflammatory myofibroblastic tumor, Spitzoid tumor	Neuroblastoma
AMER1	APC membrane recruitment protein 1	139285	Wilms tumor	
APC	Adenomatous polyposis of the colon gene	324	Colorectal, pancreatic, desmoid, hepatoblastoma, glioma, other CNS	Colorectal, pancreatic, desmoid, hepatoblastoma, glioma, other CNS
APOBEC3B	Apolipoprotein B mRNA editing enzyme catalytic subunit 3B	9582		Breast cancer
AR	Androgen receptor	367	Prostate	

ARHGAP26	Rho GTPase activating protein 26	23092	AML, MDS	
ARHGEF12	RHO guanine nucleotide exchange factor (GEF) 12 (LARG)	23365	AML	
ARID1A	AT-rich interactive domain 1A (SWI-like)	8289	Clear cell ovarian carcinoma, RCC, breast	
ARID1B	AT-rich interactive domain 1B	57492	Breast, hepatocellular carcinoma	
ARID2	AT-rich interactive domain 2	196528	Hepatocellular carcinoma	
ARNT	Aryl hydrocarbon receptor nuclear translocator	405	AML	
ASPSCR1	Alveolar soft part sarcoma chromosome region, candidate 1	79058	Alveolar soft part sarcoma	
ASXL1	Additional sex combs like 1	171023	MDS, CMML	
ATF1	Activating transcription factor 1	466	Malignant melanoma of soft parts, angiomatoid fibrous histiocytoma	
ATIC	5-Aminoimidazole-4-carboxamide ribonucleotide formyltransferase/IMP cyclohydrolase	471	ALCL	
ATM	Ataxia telangiectasia mutated	472	T-PLL	Leukemia, lymphoma, medulloblastoma, glioma
ATP1A1	ATPase, Na+/K+ transporting, alpha 1 polypeptide	476	Adrenal aldosterone-producing adenoma	
ATP2B3	ATPase, Ca++ transporting, plasma membrane 3	492	Adrenal aldosterone-producing adenoma	
ATR	ATR serine/threonine kinase	545	Endometrial, gastric, epithelial ovarian, myeloma	Oropharyngeal
ATRX	Alpha thalassemia/mental retardation syndrome X-linked	546	Pancreatic neuroendocrine tumors, pediatric GBM	

(continued)

Table 1.1 (continued)

Gene	Name	Entrez Gene Id	Tumor types (somatic)	Tumor types (germline)
AXIN1	Axin 1	8312	Colorectal, endometrial, prostate, hepatocellular carcinoma, hepatoblastoma, sporadic medulloblastoma	
AXIN2	Axin 2	8313	Colorectal carcinoma, stomach carcinoma, hepatocellular carcinoma	Colorectal carcinoma
B2M	Beta-2-microglobulin	567	DLBCL, melanoma, colorectal adenocarcinoma	
BAP1	BRCA1 associated protein-1 (ubiquitin carboxy-terminal hydrolase)	8314	Uveal melanoma, breast, NSCLC, RCC	Mesothelioma, uveal melanoma
BCL10	B-cell CLL/lymphoma 10	8915	MALT	
BCL11A	B-cell CLL/lymphoma 11A	53335	B-CLL	
BCL11B	B-cell CLL/lymphoma 11B (CTIP2)	64919	T-ALL	
BCL2	B-cell CLL/lymphoma 2	596	NHL, CLL	
BCL3	B-cell CLL/lymphoma 3	602	CLL	
BCL6	B-cell CLL/lymphoma 6	604	NHL, CLL	
BCL7A	B-cell CLL/lymphoma 7A	605	BNHL	
BCL9	B-cell CLL/lymphoma 9	607	B-ALL	
BCL9L	B-cell CLL/lymphoma 9-like	283149	Colorectal cancer, endometrial carcinoma, gastric cancer	
BCOR	BCL6 corepressor	54880	Retinoblastoma, AML, APL (translocation)	
BCORL1	BCL6 corepressor-like 1	63035	AML, HNSCC	
BCR	Breakpoint cluster region	613	CML, ALL, AML	
BIRC3	Baculoviral IAP repeat-containing 3	330	CLL, MALT, MCL, MM	
BLM	Bloom syndrome	641		Leukemia, lymphoma, skin squamous cell, other tumor types
BMPR1A	Bone morphogenetic protein receptor, type IA	657		

BRAF	v-raf murine sarcoma viral oncogene homolog B1	673	Melanoma, colorectal, papillary thyroid, borderline ovarian, NSCLC, cholangiocarcinoma, pilocytic astrocytoma, Spitzoid tumor, pancreas acinar carcinoma, melanocytic nevus, prostate, gastric	
BRCA1	Familial breast/ovarian cancer gene 1	672	Ovarian	Breast, ovarian
BRCA2	Familial breast/ovarian cancer gene 2	675	Breast, ovarian, pancreatic	Breast, ovarian, pancreatic, leukemia (FANCB, FANCD1)
BRD3	Bromodomain containing 3	8019	Lethal midline carcinoma of young people	
BRD4	Bromodomain containing 4	23476	Lethal midline carcinoma of young people	
BRIP1	BRCA1 interacting protein C-terminal helicase 1	83990		AML, leukemia, breast
BTG1	B-cell translocation gene 1, antiproliferative	694	B-CLL	
BTK	Bruton agammaglobulinemia tyrosine kinase	695	CLL, mantle cell lymphoma, WM	
BUB1B	BUB1 budding uninhibited by benzimidazoles 1 homolog beta (yeast)	701		Rhabdomyosarcoma
C2orf44	Chromosome 2 open reading frame 44	80304	Colorectal	
CACNA1D	Calcium channel, voltage-dependent, L type, alpha 1D subunit	776	Adrenal aldosterone-producing adenoma	
CALR	Calreticulin	811	MPN, MDS	
CAMTA1	Calmodulin-binding transcription activator 1	23261	Epithelioid hemangioendothelioma	
CANT1	Calcium-activated nucleotidase 1	124583	Prostate	

(continued)

Table 1.1 (continued)

Gene	Name	Entrez Gene Id	Tumor types (somatic)	Tumor types (germline)
CARD11	Caspase recruitment domain family, member 11	84433	DLBCL	
CARS	Cysteinyl-tRNA synthetase	833	ALCL	
CASC5	Cancer susceptibility candidate 5	57082	AML	
CASP8	Caspase 8, apoptosis-related cysteine peptidase	841	Hepatocellular, oral squamous cell, breast	
CBFA2T3	Core-binding factor, runt domain, alpha subunit 2; translocated to, 3 (MTG-16)	863	AML	
CBFB	Core-binding factor, beta subunit	865	AML	
CBL	Cas-Br-M (murine) ecotropic retroviral transforming	867	AML, JMML, MDS	
CBLB	Cas-Br-M (murine) ecotropic retroviral transforming sequence b	868	AML	
CBLC	Cas-Br-M (murine) ecotropic retroviral transforming sequence c	23624	AML	
CCDC6	Coiled-coil domain containing 6	8030	Papillary thyroid, CML, NSCLC	
CCNB1IP1	Cyclin B1 interacting protein 1, E3 ubiquitin protein ligase	57820	Leiomyoma	
CCND1	Cyclin D1	595	CLL, B-ALL, breast	
CCND2	Cyclin D2	894	NHL, CLL	
CCND3	Cyclin D3	896	MM	
CCNE1	Cyclin E1	898	Serous ovarian	
CD274	CD274 molecule	29126	PMBL, Hodgkin lymphoma	
CD74	CD74 molecule, major histocompatibility complex, class II invariant chain	972	NSCLC	
CD79A	CD79a molecule, immunoglobulin-associated alpha	973	DLBCL, WM	

CD79B	CD79b molecule, immunoglobulin-associated beta	974	DLBCL, WM	
CDC73	Cell division cycle 73	79577	Parathyroid adenoma	Parathyroid adenoma, multiple ossifying jaw fibroma
CDH1	Cadherin 1, type 1, E-cadherin (epithelial) (ECAD)	999	Lobular breast, gastric	Gastric
CDH11	Cadherin 11, type 2, OB-cadherin (osteoblast)	1009	Aneurysmal bone cyst	
CDK12	Cyclin-dependent kinase 12	51755	Serous ovarian	
CDK4	Cyclin-dependent kinase 4	1019		Melanoma
CDK6	Cyclin-dependent kinase 6	1021	ALL	
CDKN1B	Cyclin-dependent kinase inhibitor 1B (p27, Kip1)	1027	Breast, small intestine neuroendocrine tumors	Pituitary, parathyroid
CDKN2A	Cyclin-dependent kinase inhibitor 2A (p16(INK4a)) gene	1029	Melanoma, multiple other tumor types	Melanoma, pancreatic
CDKN2C	Cyclin-dependent kinase inhibitor 2C (p18, inhibits CDK4)	1031	Glioma, MM	
CDX2	Caudal type homeo box transcription factor 2	1045	AML	
CEBPA	CCAAT/enhancer binding protein (C/EBP), alpha	1050	AML, MDS	
CHCHD7	Coiled-coil-helix-coiled-coil-helix domain containing 7	79145	Salivary adenoma	
CHD4	Chromodomain helicase DNA binding protein 4	1108	Uterine serous carcinoma	

(continued)

Table 1.1 (continued)

Gene	Name	Entrez Gene Id	Tumor types (somatic)	Tumor types (germline)
CHEK2	CHK2 checkpoint homolog (*S. pombe*)	11200		Breast
CIC	Capicua homolog	23152	Oligodendroglioma, soft tissue sarcoma	
CIITA	Class II, major histocompatibility complex, transactivator	4261	PMBL, Hodgkin lymphoma	
CLIP1	CAP-GLY domain containing linker protein 1	6249	Spitzoid tumor	
CLTC	Clathrin, heavy polypeptide (Hc)	1213	ALCL, renal	
CLTCL1	Clathrin, heavy polypeptide-like 1	8218	ALCL	
CNBP	CCHC-type zinc finger, nucleic acid-binding protein	7555	Aneurysmal bone cyst	
CNOT3	CCR4-NOT transcription complex subunit 3	4849	T-ALL	
CNTRL	Centriolin	11064	MPN, NHL	
COL1A1	Collagen, type I, alpha 1	1277	DFSP, aneurysmal bone cyst	
COL2A1	Collagen, type II, alpha 1	1280	Chondrosarcoma, enchondroma	
CREB1	cAMP responsive element binding protein 1	1385	Clear cell sarcoma, angiomatoid fibrous histiocytoma	
CREB3L1	cAMP responsive element binding protein 3-like 1	90993	Myxofibrosarcoma	
CREB3L2	cAMP responsive element binding protein 3-like 2	64764	Fibromyxoid sarcoma	
CREBBP	CREB binding protein (CBP)	1387	ALL, AML, DLBCL, B-NHL	
CRLF2	Cytokine receptor-like factor 2	64109	B-ALL, Down syndrome-associated ALL	
CRTC1	CREB-regulated transcription coactivator 1	23373	Salivary gland mucoepidermoid	
CRTC3	CREB-regulated transcription coactivator 3	64784	salivary gland mucoepidermoid	

CSF3R	Colony-stimulating factor 3 receptor (granulocyte)	1441	aCML, CNL, leukemia lymphoma, and MDS associated with severe congenital neutropenia, CMML, de novo AML, MDS	
CTCF	CCCTC-binding factor	10664	Endometrial, breast, head and neck cancer	
CTNNB1	Catenin (cadherin-associated protein), beta 1	1499	Colorectal, ovarian, hepatoblastoma, pleomorphic salivary gland adenoma, other tumor types	
CUX1	Cut-like homeobox 1	1523	Endometrial, melanoma, colorectal, AML, MDS, other tumor types	
CXCR4	C-X-C motif chemokine receptor 4	7852	WM	
CYLD	Familial cylindromatosis gene	1540	Cylindroma	Cylindroma
DAXX	Death domain-associated protein	1616	Pancreatic neuroendocrine tumor, pediatric glioblastoma	
DCTN1	Dynactin 1	1639	Inflammatory myofibroblastic tumor, Spitzoid tumor	
DDB2	Damage-specific DNA binding protein 2	1643		Skin basal cell, skin squamous cell, melanoma
DDIT3	DNA-damage-inducible transcript 3	1649	Liposarcoma	
DDR2	Discoidin domain receptor 2	4921	Squamous cell carcinoma, NSCLC	
DDX10	DEAD (Asp-Glu-Ala-Asp) box polypeptide 10	1662	AML*	
DDX3X	DEAD-box helicase 3, X-linked	1654	CLL, medulloblastoma	
DDX5	DEAD (Asp-Glu-Ala-Asp) box polypeptide 5	1655	prostate	
DDX6	DEAD (Asp-Glu-Ala-Asp) box polypeptide 6	1656	B-NHL	
DEK	DEK oncogene (DNA binding)	7913	AML	

(continued)

Table 1.1 (continued)

Gene	Name	Entrez Gene Id	Tumor types (somatic)	Tumor types (germline)
DICER1	Dicer 1, ribonuclease type III	23405	Sex cord-stromal tumor, TGCT, embryonal rhabdomyosarcoma, pleuropulmonary blastoma, pituitary blastoma, Wilms tumor, thyroid cancer, other tumor types	Pleuropulmonary blastoma
DNAJB1	DnaJ heat shock protein family (Hsp40) member B1	3337	Fibrolamellar hepatocellular carcinoma	
DNM2	Dynamin 2	1785	T-ALL	
DNMT3A	DNA (cytosine-5-)-methyltransferase 3 alpha	1788	AML	
DROSHA	Drosha ribonuclease III	29102	Wilms tumor, NSCLC, bladder carcinoma	
EBF1	Early B-cell factor 1	1879	Lipoma	
EGFR	Epidermal growth factor receptor (erythroblastic leukemia viral (v-erb-b) oncogene homolog, avian)	1956	Glioma, NSCLC	NSCLC
EIF3E	Eukaryotic translation initiation factor 3, subunit E	3646	Colorectal	
EIF4A2	Eukaryotic translation initiation factor 4A, isoform 2	1974	NHL	
ELF4	E74-like factor 4 (ets domain transcription factor)	2000	AML	
ELK4	ELK4, ETS-domain protein (SRF accessory protein 1)	2005	Prostate	
ELL	ELL gene (11-19 lysine-rich leukemia gene)	8178	AL	
EML4	Echinoderm microtubule-associated protein like 4	27436	NSCLC	

EP300	300 kd E1A-Binding protein gene	2033	Colorectal, breast, pancreatic, AML, ALL, DLBCL	
EPAS1	Endothelial PAS domain protein 1	2034	Paraganglioma, pheochromocytoma, central nervous system hemangioblastomas	
EPS15	Epidermal growth factor receptor pathway substrate 15 (AF1p)	2060	ALL	
ERBB2	v-erb-b2 erythroblastic leukemia viral oncogene homolog 2, neuro/glioblastoma derived oncogene homolog (avian)	2064	Breast, ovarian, other tumor types, NSCLC, gastric	
ERBB3	erb-b2 receptor tyrosine kinase 3	2065	Colon, gastric, head and neck, bladder, skin	
ERBB4	erb-b2 receptor tyrosine kinase 4	2066	Melanoma, gastric, NSCLC	
ERC1	ELKS/RAB6-interacting/CAST family member 1	23085	Papillary thyroid, Spitzoid tumor	
ERCC2	Excision repair cross-complementing rodent repair deficiency, complementation group 2 (xeroderma pigmentosum D)	2068		Skin basal cell, skin squamous cell, melanoma
ERCC3	Excision repair cross-complementing rodent repair deficiency, complementation group 3 (xeroderma pigmentosum group B complementing)	2071		Skin basal cell, skin squamous cell, melanoma
ERCC4	Excision repair cross-complementing rodent repair deficiency, complementation group 4	2072		Skin basal cell, skin squamous cell, melanoma
ERCC5	Excision repair cross-complementing rodent repair deficiency, complementation group 5 (xeroderma pigmentosum, complementation group G (Cockayne syndrome))	2073		Skin basal cell, skin squamous cell, melanoma
ERG	v-ets erythroblastosis virus E26 oncogene like (avian)	2078	Ewing sarcoma, prostate, AML	

(continued)

Table 1.1 (continued)

Gene	Name	Entrez Gene Id	Tumor types (somatic)	Tumor types (germline)
ESR1	Estrogen receptor 1	2099	Breast	
ETNK1	Ethanolamine kinase 1	55500	aCML, CMML, SM-AHD, HES	
ETV1	ets variant gene 1	2115	Ewing sarcoma, prostate	
ETV4	ets variant gene 4 (E1A enhancer binding protein, E1AF)	2118	Ewing sarcoma, prostate carcinoma	
ETV5	ets variant gene 5	2119	Prostate	
ETV6	ets variant gene 6 (TEL oncogene)	2120	Congenital fibrosarcoma, multiple different leukemia, and lymphoma tumor types including ALL, secretory breast, MDS	
EWSR1	Ewing sarcoma breakpoint region 1 (EWS)	2130	Ewing sarcoma, desmoplastic small round cell tumor , ALL, clear cell sarcoma, sarcoma, myoepithelioma, mesothelioma	
EXT1	Multiple exostoses type 1 gene	2131		Exostoses, osteosarcoma
EXT2	Multiple exostoses type 2 gene	2132		Exostoses, osteosarcoma
EZH2	Enhancer of zeste homolog 2	2146	DLBCL	
EZR	Ezrin	7430	NSCLC	
FAM46C	Family with sequence similarity 46, member C	54855	MM	
FANCA	Fanconi anemia, complementation group A	2175		AML, leukemia
FANCC	Fanconi anemia, complementation group C	2176		AML, leukemia
FANCD2	Fanconi anemia, complementation group D2	2177		AML, leukemia
FANCE	Fanconi anemia, complementation group E	2178		AML, leukemia
FANCF	Fanconi anemia, complementation group F	2188		AML, leukemia
FANCG	Fanconi anemia, complementation group G	2189		AML, leukemia

FAS	Fas cell surface death receptor	355	TGCT, nasal NK/T lymphoma, skin squamous cell carcinoma-burn scar related	
FAT1	FAT atypical cadherin 1	2195	Oral squamous cell, chemorefractory CLL, head and neck, pancreatic acinar cell carcinoma	Pancreatic
FAT4	FAT atypical cadherin 4	79633	Lymphoma, pancreatic, head and neck, melanoma, hepatocellular carcinoma	
FBXO11	F-box protein 11	80204	DLBCL	
FBXW7	F-box and WD-40 domain protein 7 (archipelago homolog, Drosophila)	55294	Colorectal, endometrial, T-ALL	
FCGR2B	Fc fragment of IgG, low affinity IIb, receptor for (CD32)	2213	ALL	
FCRL4	Fc receptor-like 4	83417	B-NHL	
FES	FES proto-oncogene, tyrosine kinase	2242	HNSCC, ovarian carcinoma	
FEV	FEV protein – (HSRNAFEV)	54738	Ewing sarcoma	
FGFR1	Fibroblast growth factor receptor 1	2260	MPN, NHL, salivary adenoma	
FGFR1OP	FGFR1 oncogene partner (FOP)	11116	MPN, NHL	
FGFR2	Fibroblast growth factor receptor 2	2263	Gastric, NSCLC, endometrial	
FGFR3	Fibroblast growth factor receptor 3	2261	Bladder, MM, T-cell lymphoma	
FGFR4	Fibroblast growth factor receptor 4	2264	Rhabdomyosarcoma	
FH	Fumarate hydratase	2271		Leiomyomatosis, renal
FHIT	Fragile histidine triad gene	2272	Pleomorphic salivary gland adenoma	
FIP1L1	FIP1 like 1 (*S. cerevisiae*)	81608	Idiopathic hypereosinophilic syndrome	
FLCN	Folliculin	201163		Renal, fibrofolliculomas, trichodiscomas
FLI1	Friend leukemia virus integration 1	2313	Ewing sarcoma	
FLT3	fms-related tyrosine kinase 3	2322	AML, ALL	
FLT4	fms-related tyrosine kinase 4	2324	Soft tissue sarcoma	
FOXA1	Forkhead box A1	3169	Breast, prostate	

(continued)

Table 1.1 (continued)

Gene	Name	Entrez Gene Id	Tumor types (somatic)	Tumor types (germline)
FOXL2	Forkhead box L2	668	Granulosa cell tumor of the ovary	
FOXO1	Forkhead box O1	2308	Alveolar rhabdomyosarcoma	
FOXO3	Forkhead box O3	2309	AL	
FOXO4	Forkhead box O4	4303	AL, soft tissue sarcoma	
FOXP1	Forkhead box P1	27086	ALL	
FSTL3	Follistatin-like 3 (secreted glycoprotein)	10272	B-CLL	
FUBP1	Far upstream element (FUSE)-binding protein 1	8880	Oligodendroglioma	
FUS	Fusion, derived from t(12;16) malignant liposarcoma	2521	Liposarcoma, AML, Ewing sarcoma, angiomatoid fibrous histiocytoma, fibromyxoid sarcoma	
GAS7	Growth arrest-specific 7	8522	AML*	
GATA1	GATA binding protein 1 (globin transcription factor 1)	2623	Megakaryoblastic leukemia of Down syndrome	
GATA2	GATA binding protein 2	2624	AML (CML blast transformation)	
GATA3	GATA binding protein 3	2625	Breast	
GNA11	Guanine nucleotide binding protein (G protein), alpha 11 (Gq class)	2767	Uveal melanoma, primary central nervous system melanocytic neoplasms	
GNAQ	Guanine nucleotide binding protein (G protein), q polypeptide	2776	Uveal melanoma, primary central nervous system melanocytic neoplasms	
GNAS	Guanine nucleotide binding protein (G protein), alpha stimulating activity polypeptide 1	2778	Pituitary adenoma, pancreatic intraductal papillary mucinous neoplasm, fibrous dysplasia	
GOLGA5	Golgi autoantigen, golgin subfamily a, 5 (PTC5)	9950	Papillary thyroid, Spitzoid tumor	

GOPC	Golgi associated PDZ and coiled-coil motif containing	57120	Glioblastoma, cholangiocarcinoma, borderline ovarian	
GPC3	Glypican 3	2719		Wilms tumor
GPHN	Gephyrin (GPH)	10243	AL	
GRIN2A	Glutamate receptor, ionotropic, N-methyl D-aspartate 2A	2903	Melanoma, colorectal carcinoma, gastric carcinoma, lung carcinoma	
H3F3A	H3 histone, family 3A	3020	Glioma	
H3F3B	H3 histone, family 3B (H3.3B)	3021	Chondroblastoma	
HERPUD1	Homocysteine-inducible, endoplasmic reticulum stress-inducible, ubiquitin-like domain member 1	9709	Prostate	
HEY1	Hairy/enhancer-of-split related with YRPW motif 1	23462	Mesenchymal chondrosarcoma	
HIF1A	Hypoxia-inducible factor 1 alpha subunit	3091	Endometrioid carcinoma, glioblastoma, colorectal, renal, lung, pancreatic	
HIP1	Huntingtin-interacting protein 1	3092	CMML, NSCLC	
HIST1H3B	Histone cluster 1, H3b	8358	Glioma	
HIST1H4I	Histone 1, H4i (H4FM)	8294	NHL	
HLA-A	Major histocompatibility complex, class I, A	3105	Spitzoid tumor	
HLF	Hepatic leukemia factor	3131	ALL	
HMGA1	High mobility group AT-hook 1	3159	Microfollicular thyroid adenoma, various benign mesenchymal tumors	
HMGA2	High mobility group AT-hook 2 (HMGIC)	8091	Lipoma, leiomyoma, pleomorphic salivary gland adenoma	
HNF1A	HNF1 homeobox A	6927	Hepatic adenoma, hepatocellular	Hepatic adenoma, hepatocellular carcinoma
HNRNPA2B1	Heterogeneous nuclear ribonucleoprotein A2/B1	3181	Prostate	

(continued)

Table 1.1 (continued)

Gene	Name	Entrez Gene Id	Tumor types (somatic)	Tumor types (germline)
HOOK3	Hook homolog 3	84376	Papillary thyroid	
HOXA11	Homeobox A11	3207	CML	
HOXA13	Homeobox A13	3209	AML	
HOXA9	Homeobox A9	3205	AML*	
HOXC11	Homeobox C11	3227	AML	
HOXC13	Homeobox C13	3229	AML	
HOXD11	Homeobox D11	3237	AML	
HOXD13	Homeobox D13	3239	AML*	
HRAS	v-Ha-ras Harvey rat sarcoma viral oncogene homolog	3265	Infrequent sarcomas, rare other tumor types	Rhabdomyosarcoma, ganglioneuroblastoma, bladder
HSP90AA1	Heat shock protein 90kDa alpha (cytosolic), class A member 1	3320	NHL	
HSP90AB1	Heat shock protein 90kDa alpha (cytosolic), class B member 1	3326	NHL	
IDH1	Isocitrate dehydrogenase 1 (NADP+), soluble	3417	Glioblastoma	
IDH2	Isocitrate dehydrogenase 2 (NADP+), mitochondrial	3418	Glioblastoma	
IGH	Immunoglobulin heavy locus	3492	MM, Burkitt lymphoma, NHL, CLL, B-ALL, MALT, MLCLS	
IGK	Immunoglobulin kappa locus	50802	Burkitt lymphoma, B-NHL	
IGL	Immunoglobulin lambda locus	3535	Burkitt lymphoma	
IKBKB	Inhibitor of kappa light polypeptide gene enhancer in B cells, kinase beta	3551	SMZL	
IKZF1	IKAROS family zinc finger 1	10320	ALL, DLBCL	

IL2	Interleukin 2	3558	Intestinal T-cell lymphoma	
IL21R	Interleukin 21 receptor	50615	NHL	
IL6ST	Interleukin 6 signal transducer (gp130, oncostatin M receptor)	3572	Hepatocellular carcinoma	
IL7R	Interleukin 7 receptor	3575	ALL, ETP ALL	
IRF4	Interferon regulatory factor 4	3662	MM	
ITK	IL2-inducible T-cell kinase	3702	Peripheral T-cell lymphoma	
JAK1	Janus kinase 1	3716	ALL	
JAK2	Janus kinase 2	3717	ALL, AML, MPN, CML	
JAK3	Janus kinase 3	3718	Acute megakaryocytic leukemia, ETP ALL	
JUN	jun oncogene	3725	Sarcoma	
KAT6A	K(lysine) acetyltransferase 6A	7994	AML	
KAT6B	K(lysine) acetyltransferase 6B	23522	AML, leiomyoma	
KCNJ5	Potassium inwardly-rectifying channel; subfamily J; member 5	3762	Adrenal adenoma	
KDM5A	Lysine (K)-specific demethylase 5A, JARID1A	5927	AML	
KDM5C	Lysine (K)-specific demethylase 5C (JARID1C)	8242	Clear cell renal carcinoma	
KDM6A	Lysine (K)-specific demethylase 6A, UTX	7403	Renal cell carcinoma, bladder carcinoma, esophageal SCC, MM, medulloblastoma, T-ALL, other tumor types	
KDR	Vascular endothelial growth factor receptor 2	3791	NSCLC, angiosarcoma	melanoma
KDSR	3-ketodihydrosphingosine reductase	2531	B-NHL	
KEAP1	Kelch-like ECH associated protein 1	9817	NSCLC, breast carcinoma	
KIF5B	Kinesin family member 5B	3799	NSCLC, Spitzoid tumor	
KIT	v-kit Hardy-Zuckerman 4 feline sarcoma viral oncogene homolog	3815	GIST, AML, TGCT, mastocytosis, mucosal melanoma	GIST, epithelioma

(continued)

Table 1.1 (continued)

Gene	Name	Entrez Gene Id	Tumor types (somatic)	Tumor types (germline)
KLF4	Kruppel-like factor 4	9314	Meningioma	
KLF6	Kruppel-like factor 6	1316	Prostate, glioma	
KLK2	Kallikrein-related peptidase 2	3817	Prostate	
KMT2A	Lysine (K)-specific methyltransferase 2A	4297	AML, ALL	
KMT2C	Lysine (K)-specific methyltransferase 2C	58508	Medulloblastoma	
KMT2D	Lysine (K)-specific methyltransferase 2D	8085	Medulloblastoma, renal	
KRAS	v-Ki-ras2 Kirsten rat sarcoma 2 viral oncogene homolog	3845	Pancreatic, colorectal, lung, thyroid, AML, other tumor types	
KTN1	Kinectin 1 (kinesin receptor)	3895	Papillary thyroid	
LASP1	LIM and SH3 protein 1	3927	AML	
LCK	Lymphocyte-specific protein tyrosine kinase	3932	T-ALL	
LEF1	Lymphoid enhancer binding factor 1	51176	B-ALL, T-ALL, eyelid sebaceous carcinoma, AML, lymphomas	
LIFR	Leukemia inhibitory factor receptor	3977	Salivary adenoma	
LMNA	Lamin A/C	4000	Spitzoid tumor	
LMO1	LIM domain only 1 (rhombotin 1) (RBTN1)	4004	T-ALL, neuroblastoma	Neuroblastoma
LMO2	LIM domain only 2 (rhombotin-like 1) (RBTN2)	4005	T-ALL	
LPP	LIM domain containing preferred translocation partner in lipoma	4026	Lipoma, leukemia	
LRIG3	Leucine-rich repeats and immunoglobulin-like domains 3	121227	NSCLC	
LRP1B	LDL receptor related protein 1B	53353	CLL, ovarian cancer, esophageal squamous cell carcinoma, urothelial cancer	

LYL1	Lymphoblastic leukemia derived sequence 1	4066	T-ALL	
LZTR1	Leucine-zipper-like transcription regulator 1	8216	Glioblastoma	Glioblastoma, schwannoma
MAF	v-maf musculoaponeurotic fibrosarcoma oncogene homolog	4094	MM	
MAFB	v-maf musculoaponeurotic fibrosarcoma oncogene homolog B (avian)	9935	MM	
MALT1	Mucosa-associated lymphoid tissue lymphoma translocation gene 1	10892	MALT	
MAML2	Mastermind-like 2 (Drosophila)	84441	Salivary gland mucoepidermoid	
MAP2K1	Mitogen-activated protein kinase kinase 1	5604	NSCLC, melanoma, colorectal	
MAP2K2	Mitogen-activated protein kinase kinase 2	5605	NSCLC, melanoma	
MAP2K4	Mitogen-activated protein kinase kinase 4	6416	Pancreatic, breast, colorectal	
MAP3K1	Mitogen-activated protein kinase kinase kinase 1, E3 ubiquitin protein ligase	4214	Luminal A breast	
MAP3K13	Mitogen-activated protein kinase kinase kinase 13	9175	Breast	
MAPK1	Mitogen-activated protein kinase 1	5594	CLL, ovarian mixed germ cell tumor, cervical carcinoma	
MAX	Myc-associated factor X	4149	Pheochromocytoma, endometrioid carcinoma, colon carcinoma	Pheochromocytoma
MDM2	Mdm2 p53 binding protein homolog	4193	Sarcoma, glioma, colorectal, other tumor types	
MDM4	Mdm4 p53 binding protein homolog	4194	Glioblastoma, bladder, retinoblastoma	
MECOM	MDS1 and EVI1 complex locus	2122	AML, CML, MDS	
MED12	Mediator complex subunit 12	9968	Uterine leiomyoma, fibroadenoma, phyllodes tumor	

(continued)

Table 1.1 (continued)

Gene	Name	Entrez Gene Id	Tumor types (somatic)	Tumor types (germline)
MEN1	Multiple endocrine neoplasia type 1 gene	4221	Parathyroid tumors, pancreatic neuroendocrine tumor	Parathyroid adenoma, pituitary adenoma, pancreatic islet cell, carcinoid
MET	met proto-oncogene (hepatocyte growth factor receptor)	4233	Papillary renal, head-neck squamous cell	Papillary renal
MITF	Melanogenesis-associated transcription factor	4286	Melanoma	
MKL1	Megakaryoblastic leukemia (translocation) 1	57591	Acute megakaryocytic leukemia	
MLF1	Myeloid leukemia factor 1	4291	AML	
MLH1	*E.coli* MutL homolog gene	4292	Colorectal, endometrial, ovarian, CNS tumors	Colorectal, endometrial, ovarian, CNS
MLLT1	Myeloid/lymphoid or mixed-lineage leukemia (trithorax homolog, *Drosophila*); translocated to, 1 (ENL)	4298	AL	
MLLT10	Myeloid/lymphoid or mixed-lineage leukemia (trithorax homolog, *Drosophila*); translocated to, 10 (AF10)	8028	AL	
MLLT11	Myeloid/lymphoid or mixed-lineage leukemia (trithorax homolog, *Drosophila*); translocated to, 11	10962	ALL	
MLLT3	Myeloid/lymphoid or mixed-lineage leukemia (trithorax homolog, *Drosophila*); translocated to, 3 (AF9)	4300	ALL	

MLLT4	Myeloid/lymphoid or mixed-lineage leukemia (trithorax homolog, *Drosophila*); translocated to, 4 (AF6)	4301	AL	
MLLT6	Myeloid/lymphoid or mixed-lineage leukemia (trithorax homolog, *Drosophila*); translocated to, 6 (AF17)	4302	AL	
MN1	Meningioma (disrupted in balanced translocation) 1	4330	AML, meningioma	
MPL	Myeloproliferative leukemia virus oncogene, thrombopoietin receptor	4352	MPN	MPN
MSH2	mutS homolog 2 (*E. coli*)	4436	Colorectal, endometrial, ovarian	Colorectal, endometrial, ovarian
MSH6	mutS homolog 6 (*E. coli*)	2956	Colorectal	Colorectal, endometrial, ovarian
MSI2	Musashi homolog 2 (*Drosophila*)	124540	CML	
MSN	Moesin	4478	ALCL	
MTCP1	Mature T-cell proliferation 1	4515	T cell prolymphocytic leukemia	
MTOR	Mechanistic target of rapamycin	2475		
MUC1	Mucin 1, transmembrane	4582	B-NHL	
MUTYH	mutY homolog (E. coli)	4595		Colorectal
MYB	v-myb myeloblastosis viral oncogene homolog	4602	Adenoid cystic carcinoma	
MYC	v-myc myelocytomatosis viral oncogene homolog (avian)	4609	Burkitt lymphoma, amplified in other cancers, B-CLL	
MYCL	v-myc avian myelocytomatosis viral oncogene lung carcinoma derived homolog	4610	Small cell lung carcinoma	
MYCN	v-myc myelocytomatosis viral related oncogene, neuroblastoma derived (avian)	4613	Neuroblastoma	

(continued)

Table 1.1 (continued)

Gene	Name	Entrez Gene Id	Tumor types (somatic)	Tumor types (germline)
MYD88	Myeloid differentiation primary response gene (88)	4615	ABC-DLBCL	
MYH11	Myosin, heavy polypeptide 11, smooth muscle	4629	AML	
MYH9	Myosin, heavy polypeptide 9, non-muscle	4627	ALCL	
MYO5A	Myosin VA (heavy chain 12, myoxin)	4644	Spitzoid tumor	
MYOD1	Myogenic differentiation 1	4654	Embryonal rhabdomyosarcoma	
NAB2	NGFI-A binding protein 2	4665	Solitary fibrous tumor, meningeal hemangiopericytoma	
NBN	Nibrin	4683		NHL, glioma, medulloblastoma, rhabdomyosarcoma
NCOA1	Nuclear receptor coactivator 1	8648	Alveolar rhabdomyosarcoma	
NCOA2	Nuclear receptor coactivator 2 (TIF2)	10499	AML, chondrosarcoma, rhabdomyosarcoma	
NCOA4	Nuclear receptor coactivator 4 – PTC3 (ELE1)	8031	Papillary thyroid	
NCOR1	Nuclear receptor corepressor 1	9611	Breast	
NCOR2	Nuclear receptor corepressor 2	9612	Prostate	
NDRG1	N-myc downstream regulated 1	10397	Prostate	
NF1	Neurofibromatosis type 1 gene	4763	Neurofibroma, glioma	Neurofibroma, glioma
NF2	Neurofibromatosis type 2 gene	4771	Meningioma, acoustic neuroma, renal	Meningioma, acoustic neuroma
NFATC2	Nuclear factor of activated T cells, cytoplasmic, calcineurin-dependent 2	4773	Ewing sarcoma	
NFE2L2	Nuclear factor (erythroid-derived 2)-like 2 (NRF2)	4780	NSCLC, HNSCC	
NFIB	Nuclear factor I/B	4781	Adenoid cystic carcinoma, lipoma	

NFKB2	Nuclear factor of kappa light polypeptide gene enhancer in B cells 2 (p49/p100)	4791	B-NHL	
NFKBIE	Nuclear factor of kappa light polypeptide gene enhancer in B-cell inhibitor, epsilon	4794	CLL, desmoplastic melanoma	
NIN	Ninein (GSK3B interacting protein)	51199	MPN	
NKX2-1	NK2 homeobox 1	7080	NSCLC	
NONO	Non-POU domain containing, octamer-binding	4841	papillary renal	
NOTCH1	Notch homolog 1, translocation-associated (Drosophila) (TAN1)	4851	T-ALL, breast, bladder, skin SCC, lung SCC, head and neck SCC	
NOTCH2	Notch homolog 2	4853	Marginal zone lymphoma, DLBCL, bladder	
NPM1	Nucleophosmin (nucleolar phosphoprotein B23, numatrin)	4869	NHL, APL, AML	
NR4A3	Nuclear receptor subfamily 4, group A, member 3 (NOR1)	8013	Extraskeletal myxoid chondrosarcoma	
NRAS	Neuroblastoma RAS viral (v-ras) oncogene homolog	4893	Melanoma, MM, AML, thyroid	
NRG1	Neuregulin 1	3084	NSCLC	
NSD1	Nuclear receptor binding SET domain protein 1	64324	AML	
NT5C2	5'-Nucleotidase, cytosolic II	22978	Relapse ALL	
NTRK1	Neurotrophic tyrosine kinase, receptor, type 1	4914	Papillary thyroid, Spitzoid tumor	
NTRK3	Neurotrophic tyrosine kinase, receptor, type 3	4916	Congenital fibrosarcoma, secretory breast	
NUMA1	Nuclear mitotic apparatus protein 1	4926	APL	
NUP214	Nucleoporin 214kDa (CAN)	8021	AML, T-ALL	
NUP98	Nucleoporin 98kDa	4928	AML	
NUTM1	NUT midline carcinoma, family member 1	256646	Lethal midline carcinoma	

(continued)

Table 1.1 (continued)

Gene	Name	Entrez Gene Id	Tumor types (somatic)	Tumor types (germline)
NUTM2A	NUT family member 2A	728118	Endometrial stromal sarcoma	
NUTM2B	NUT family member 2B	729262	Endometrial stromal sarcoma	
OLIG2	Oligodendrocyte lineage transcription factor 2 (BHLHB1)	10215	T-ALL	
P2RY8	Purinergic receptor P2Y, G-protein coupled, 8	286530	B-ALL, Down syndrome-associated ALL	
PAFAH1B2	Platelet-activating factor acetylhydrolase, isoform Ib, beta subunit 30kDa	5049	MLCLS	
PALB2	Partner and localizer of BRCA2	79728		Wilms tumor, medulloblastoma, AML ,breast
PAX3	Paired box gene 3	5077	Alveolar rhabdomyosarcoma	
PAX5	Paired box gene 5 (B-cell lineage specific activator protein)	5079	NHL, ALL, B-ALL	
PAX7	Paired box gene 7	5081	Alveolar rhabdomyosarcoma	
PAX8	Paired box gene 8	7849	follicular thyroid	
PBRM1	Polybromo 1	55193	Clear cell renal carcinoma, breast	
PBX1	PreB-cell leukemia transcription factor 1	5087	PreB-ALL, myoepithelioma	
PCM1	Pericentriolar material 1 (PTC4)	5108	Papillary thyroid, CML, MPN	
PDCD1LG2	Programmed cell death 1 ligand 2	80380	PMBL, Hodgkin lymphoma	
PDE4DIP	Phosphodiesterase 4D interacting protein (myomegalin)	9659	MPN	
PDGFB	Platelet-derived growth factor beta polypeptide (simian sarcoma viral (v-sis) oncogene homolog)	5155	DFSP	

PDGFRA	Platelet-derived growth factor, alpha-receptor	5156	GIST, idiopathic hypereosinophilic syndrome, pediatric glioblastoma	GIST
PDGFRB	Platelet-derived growth factor receptor, beta polypeptide	5159	MPN, AML, CMML, CML	
PER1	period homolog 1 (Drosophila)	5187	AML, CMML	
PHF6	PHD finger protein 6	84295	ETP ALL	
PHOX2B	Paired-like homeobox 2b	8929	Neuroblastoma	Neuroblastoma
PICALM	Phosphatidylinositol binding clathrin assembly protein (CALM)	8301	T-ALL, AML	
PIK3CA	Phosphoinositide-3-kinase, catalytic, alpha polypeptide	5290	Colorectal, gastric, glioblastoma, breast	
PIK3R1	Phosphoinositide-3-kinase, regulatory subunit 1 (alpha)	5295	Glioblastoma, ovarian, colorectal	
PIM1	pim-1 oncogene	5292	NHL	
PLAG1	Pleomorphic adenoma gene 1	5324	Salivary adenoma, lipoblastoma	
PLCG1	Phospholipase C, gamma 1	5335	Angiosarcoma	
PML	Promyelocytic leukemia	5371	APL, ALL	
PMS2	PMS2 postmeiotic segregation increased 2 (S. cerevisiae)	5395		Colorectal, endometrial, ovarian, medulloblastoma, glioma
POLE	Polymerase (DNA directed), epsilon, catalytic subunit	5426	Colorectal carcinoma, endometrioid carcinoma, stomach carcinoma, skin cancer	Colorectal cancer susceptibility
POT1	Protection of telomeres 1	25913	CLL	
POU2AF1	POU domain, class 2, associating factor 1 (OBF1)	5450	NHL	
POU5F1	POU domain, class 5, transcription factor 1	5460	Sarcoma	
PPARG	Peroxisome proliferative activated receptor, gamma	5468	Follicular thyroid	

(continued)

Table 1.1 (continued)

Gene	Name	Entrez Gene Id	Tumor types (somatic)	Tumor types (germline)
PPFIBP1	PTPRF interacting protein, binding protein 1 (liprin beta 1)	8496	Spitzoid tumor, inflammatory myofibroblastic tumor	
PPM1D	Protein phosphatase, Mg2+/Mn2+ dependent 1D	8493	Glioma, ovarian cancer, breast cancer	
PPP2R1A	Protein phosphatase 2, regulatory subunit A, alpha	5518	Clear cell ovarian carcinoma	
PPP6C	Protein phosphatase 6, catalytic subunit	5537	Melanoma	
PRCC	Papillary renal cell carcinoma (translocation-associated)	5546	Papillary renal	
PRDM1	PR domain containing 1, with ZNF domain	639	DLBCL	
PRDM16	PR domain containing 16	63976	MDS, AML	
PREX2	Phosphatidylinositol-3,4,5-trisphosphate dependent Rac exchange factor 2	80243	Melanoma, pancreatic ductal adenocarcinoma	
PRF1	Perforin 1 (pore forming protein)	5551		Various leukemia, lymphoma
PRKACA	Protein kinase cAMP-activated catalytic subunit alpha	5566	Fibrolamellar hepatocellular carcinoma, cortisol secreting adrenal adenoma	
PRKAR1A	Protein kinase, cAMP-dependent, regulatory, type I, alpha (tissue specific extinguisher 1)	5573	Papillary thyroid	Myxoma, endocrine, papillary thyroid
PRRX1	Paired related homeobox 1	5396	AML	
PSIP1	PC4 and SFRS1 interacting protein 1 (LEDGF)	11168	AML	
PTCH1	Homolog of Drosophila Patched gene	5727	Skin basal cell, medulloblastoma	Skin basal cell, medulloblastoma
PTEN	Phosphatase and tensin homolog gene	5728	Glioma, prostate, endometrial	Hamartoma, glioma, prostate, endometrial
PTK6	Protein tyrosine kinase 6	5753	Renal, gastric, head and neck, ovarian, others	

PTPN11	Protein tyrosine phosphatase, non-receptor type 11	5781	JMML, AML, MDS	
PTPN13	Protein tyrosine phosphatase, non-receptor type 13	5783	Lung, NSCLC, gastric, peritoneal carcinomatosis	Hepatocellular
PTPRB	Protein tyrosine phosphatase, receptor type, B	5787	Angiosarcoma	
PTPRC	Protein tyrosine phosphatase, receptor type, C	5788	T-ALL	
PTPRK	Protein tyrosine phosphatase, receptor type, K	5796	colorectal	
PTPRT	Protein tyrosine phosphatase, receptor type T	11122	HNSCC, colorectal cancer, gastric cancer, lung cancer, melanoma	
QKI	QKI, KH domain containing, RNA binding	9444	Angiocentric glioma, colorectal cancer	
RABEP1	Rabaptin, RAB GTPase binding effector protein 1	9135	CMML	
RAC1	ras-related C3 botulinum toxin substrate 1 (rho family, small GTP binding protein Rac1)	5879	Melanoma, carcinoma	
RAD21	RAD21 homolog (*S. pombe*)	5885	AML, endometrium, colorectal, lung	
RAD51B	RAD51 paralog B	5890	Lipoma, uterine leiomyoma	
RAF1	v-raf-1 murine leukemia viral oncogene homolog 1	5894	Pilocytic astrocytoma, prostate	
RANBP2	RAN binding protein 2	5903	Inflammatory myofibroblastic tumor	
RAP1GDS1	RAP1, GTP-GDP dissociation stimulator 1	5910	T-ALL	
RARA	Retinoic acid receptor, alpha	5914	APL	
RB1	Retinoblastoma gene	5925	Retinoblastoma, sarcoma, breast, small cell lung carcinoma	Retinoblastoma, sarcoma, breast, small cell lung carcinoma

(continued)

Table 1.1 (continued)

Gene	Name	Entrez Gene Id	Tumor types (somatic)	Tumor types (germline)
RBM10	RNA binding motif protein 10	8241	Lung adenocarcinoma	
RBM15	RNA binding motif protein 15	64783	Acute megakaryocytic leukemia	
RECQL4	RecQ protein-like 4	9401		Osteosarcoma, skin basal cell, skin squamous cell
REL	v-rel reticuloendotheliosis viral oncogene homolog (avian)	5966	Hodgkin lymphoma	
RET	ret proto-oncogene	5979	Medullary thyroid, papillary thyroid, pheochromocytoma, NSCLC, Spitzoid tumor	Medullary thyroid, papillary thyroid, pheochromocytoma
RHOA	ras homolog family member A	387	Diffuse gastric, T cell lymphomas, pediatric Burkitt lymphoma	
RHOH	ras homolog family member H	399	NHL	
RMI2	RecQ mediated genome instability 2	116028	PMBL, Hodgkin lymphoma	
RNF213	Ring finger protein 213	57674	ALCL	
RNF43	Ring finger protein 43	54894	Cholangiocarcinoma, ovary, pancreas	
ROS1	v-ros UR2 sarcoma virus oncogene homolog 1 (avian)	6098	Glioblastoma, NSCLC, Spitzoid tumor, cholangiocarcinoma, borderline ovarian	
RPL10	Ribosomal protein L10	6134	T-ALL	
RPL22	Ribosomal protein L22 (EAP)	6146	AML, CML	
RPL5	Ribosomal protein L5	6125	T-ALL	
RPN1	Ribophorin I	6184	AML	
RSPO2	R-spondin 2	340419	Colorectal	
RSPO3	R-spondin 3	84870	Colorectal	
RUNX1	Runt-related transcription factor 1 (AML1)	861	AML, preB-ALL, T-ALL	
RUNX1T1	Runt-related transcription factor 1; translocated to, 1 (cyclin D-related)	862	AML	

SALL4	Spalt like transcription factor 4	57167	Colorectal cancer, breast cancer, prostate cancer, glioblastoma, melanoma	
SBDS	Shwachman-Bodian-Diamond syndrome protein	51119		AML, MDS
SDC4	Syndecan 4	6385	NSCLC	
SDHA	Succinate dehydrogenase complex, subunit A, flavoprotein (Fp)	6389	GIST, paraganglioma	Paraganglioma
SDHAF2	Succinate dehydrogenase complex assembly factor 2	54949		Paraganglioma
SDHB	Succinate dehydrogenase complex, subunit B, iron sulfur (Ip)	6390		Paraganglioma, pheochromocytoma
SDHC	Succinate dehydrogenase complex, subunit C, integral membrane protein, 15kDa	6391		Paraganglioma, pheochromocytoma
SDHD	Succinate dehydrogenase complex, subunit D, integral membrane protein	6392		Paraganglioma, pheochromocytoma
SET	SET translocation	6418	T-ALL	
SETBP1	SET binding protein 1	26040	aCML, sAML, MDS/MPN-U, CMML, JMML	Neuroepithelial tumors
SETD2	SET domain containing 2	29072	Clear cell renal carcinoma	
SF3B1	Splicing factor 3b, subunit 1, 155kDa	23451	Myelodysplastic syndrome	
SFPQ	Splicing factor proline/glutamine rich(polypyrimidine tract binding protein associated)	6421	Papillary renal	
SFRP4	Secreted frizzled related protein 4	6424	Colorectal cancer, melanoma, SCC, gastric cancer, esophageal SCC	
SH2B3	SH2B adaptor protein 3	10019	MPN, sAML, erythrocytosis, B-ALL	
SH3GL1	SH3-domain GRB2-like 1 (EEN)	6455	AL	
SLC34A2	Solute carrier family 34 (sodium phosphate), member 2	10568	NSCLC	
SLC45A3	Solute carrier family 45, member 3	85414	Prostate	

(continued)

Table 1.1 (continued)

Gene	Name	Entrez Gene Id	Tumor types (somatic)	Tumor types (germline)
SMAD2	SMAD family member 2	4087	Colorectal carcinoma, hepatocellular carcinoma	
SMAD3	SMAD family member 3	4088	Colorectal carcinoma, oral squamous cell carcinoma	
SMAD4	SMAD family member 4	4089	Colorectal, pancreatic, small intestine	Gastrointestinal polyp
SMARCA4	SWI-/SNF-related, matrix-associated, actin-dependent regulator of chromatin, subfamily a, member 4	6597	NSCLC, SCCOHT	
SMARCB1	SWI-/SNF-related, matrix-associated, actin-dependent regulator of chromatin, subfamily b, member 1	6598	Malignant rhabdoid	Malignant rhabdoid
SMARCD1	SWI-/SNF-related, matrix-associated, actin-dependent regulator of chromatin, subfamily d, member 1	6602	Breast	
SMARCE1	SWI-/SNF-related, matrix-associated, actin-dependent regulator of chromatin, subfamily e, member 1	6605		Meningioma
SMO	Smoothened homolog (*Drosophila*)	6608	Skin basal cell	
SND1	Staphylococcal nuclease and tudor domain containing 1	27044	Pancreas acinar carcinoma	
SOCS1	Suppressor of cytokine signaling 1	8651	Hodgkin lymphoma, PMBL	
SOX2	SRY (sex determining region Y)-box 2	6657	NSCLC, esophageal squamous carcinoma	
SPEN	Spen family transcriptional repressor	23013	SMZL, adenoid cystic carcinoma, DLBCL	
SPOP	Speckle type POZ protein	8405	Prostate, endometrial	Prostate cancer
SRC	SRC proto-oncogene, non-receptor tyrosine kinase	6714	Colorectal cancer, endometrial carcinoma	
SRSF2	Serine/arginine-rich splicing factor 2	6427	MDS, CLL	
SRSF3	Serine/arginine-rich splicing factor 3	6428	Follicular lymphoma	

SS18	Synovial sarcoma translocation, chromosome 18	6760	Synovial sarcoma	
SS18L1	Synovial sarcoma translocation gene on chromosome 18-like 1	26039	Synovial sarcoma	
SSX1	Synovial sarcoma, X breakpoint 1	6756	Synovial sarcoma	
SSX2	Synovial sarcoma, X breakpoint 2	6757	Synovial sarcoma	
SSX4	Synovial sarcoma, X breakpoint 4	6759	Synovial sarcoma	
STAG2	Stromal antigen 2	10735	Bladder carcinoma, glioblastoma, melanoma, Ewing's sarcoma, myeloid neoplasms	
STAT3	Signal transducer and activator of transcription 3 (acute-phase response factor)	6774	T-cell large granular lymphocytic leukemia	Pediatric large granular lymphocytic leukemia
STAT5B	Signal transducer and activator of transcription 5B	6777	Large granular lymphocytic leukemia, skin basal cell, APL	
STAT6	Signal transducer and activator of transcription 6, interleukin-4 induced	6778	Solitary fibrous tumor, meningeal hemangiopericytoma	
STIL	SCL/TAL1 interrupting locus	6491	T-ALL	
STK11	Serine/threonine kinase 11 gene (LKB1)	6794	NSCLC, pancreatic	Jejunal hamartoma, ovarian, testicular, pancreatic
STRN	Striatin, calmodulin binding protein	6801	Anaplastic thyroid, papillary thyroid, lung	
SUFU	Suppressor of fused homolog (Drosophila)	51684	Medulloblastoma	Medulloblastoma
SUZ12	Suppressor of zeste 12 homolog (Drosophila)	23512	Endometrial stromal tumor	
SYK	Spleen tyrosine kinase	6850	MDS, peripheral T-cell lymphoma	
TAF15	TAF15 RNA polymerase II, TATA box binding protein (TBP)-associated factor, 68kDa	8148	Extraskeletal myxoid chondrosarcoma, ALL	
TAL1	T-cell acute lymphocytic leukemia 1 (SCL)	6886	Lymphoblastic leukemia/biphasic	
TAL2	T-cell acute lymphocytic leukemia 2	6887	T-ALL	

(continued)

Table 1.1 (continued)

Gene	Name	Entrez Gene Id	Tumor types (somatic)	Tumor types (germline)
TBL1XR1	Transducin (beta)-like 1 X-linked receptor 1	79718	Splenic marginal zone lymphoma, primary central nervous system lymphoma, colorectal carcinoma, gallbladder carcinoma	
TBX3	T-box 3	6926	Breast, large intestine, skin	
TCEA1	Transcription elongation factor A (SII), 1	6917	Salivary adenoma	
TCF12	Transcription factor 12 (HTF4, helix-loop-helix transcription factors 4)	6938	Extraskeletal myxoid chondrosarcoma	
TCF3	Transcription factor 3 (E2A immunoglobulin enhancer binding factors E12/E47)	6929	PreB-ALL	
TCF7L2	Transcription factor 7-like 2	6934	Colorectal	
TCL1A	T-cell leukemia/lymphoma 1A	8115	T-CLL	
TERT	Telomerase reverse transcriptase	7015	Melanoma, glioblastoma, hepatocellular carcinoma, bladder, skin basal cell, skin squamous cell, mesothelioma, medulloblastoma, other tumor types	Melanoma
TET1	tet methylcytosine dioxygenase 1	80312	AML	
TET2	tet oncogene family member 2	54790	MDS	
TFE3	Transcription factor binding to IGHM enhancer 3	7030	Papillary renal, alveolar soft part sarcoma, renal	
TFEB	Transcription factor EB	7942	Renal cell carcinoma (childhood epithelioid)	
TFG	TRK-fused gene	10342	Papillary thyroid, ALCL, NSCLC, extraskeletal myxoid chondrosarcoma	
TGFBR2	Transforming growth factor beta receptor II	7048	Head and neck, colorectal	Colorectal
TLX1	T-cell leukemia, homeobox 1 (HOX11)	3195	T-ALL	

TLX3	T-cell leukemia, homeobox 3 (HOX11L2)	30012	T-ALL	
TMEM127	Transmembrane protein 127	55654		Pheochromocytoma, renal cell carcinoma
TMPRSS2	Transmembrane protease, serine 2	7113	Prostate	
TNFAIP3	Tumor necrosis factor, alpha-induced protein 3	7128	Marginal zone B-cell lymphomas, Hodgkin lymphoma, PMBL	
TNFRSF14	Tumor necrosis factor receptor superfamily, member 14 (herpesvirus entry mediator)	8764	Follicular lymphoma	
TNFRSF17	Tumor necrosis factor receptor superfamily, member 17	608	Intestinal T-cell lymphoma	
TOP1	Topoisomerase (DNA) I	7150	AML*	
TP53	Tumor protein p53	7157	Breast, colorectal, lung, sarcoma, adrenocortical, glioma, Spitzoid tumor, multiple other tumor types	
TP63	Tumor protein p63	8626	HNSCC, DLBCL	NSCLC
TPM3	Tropomyosin 3	7170	Papillary thyroid, ALCL, NSCLC, Spitzoid tumor	
TPM4	Tropomyosin 4	7171	ALCL	
TPR	Translocated promoter region	7175	Papillary thyroid, NSCLC	
TRA	T cell receptor alpha locus	6955	T-ALL	
TRAF7	Tumor necrosis factor receptor-associated factor 7	84231	Meningioma	
TRB	T cell receptor beta locus	6957	T-ALL	
TRD	T cell receptor delta locus	6964	T-cell leukemia	
TRIM24	Tripartite motif containing 24	8805	APL	
TRIM27	Tripartite motif-containing 27	5987	Papillary thyroid	
TRIM33	Tripartite motif-containing 33 (PTC7, TIF1G)	51592	Papillary thyroid	
TRIP11	Thyroid hormone receptor interactor 11	9321	AML	

(continued)

Table 1.1 (continued)

Gene	Name	Entrez Gene Id	Tumor types (somatic)	Tumor types (germline)
TRRAP	Transformation/transcription domain-associated protein	8295	Melanoma	
TSC1	Tuberous sclerosis 1 gene	7248	Renal cell carcinoma, bladder carcinoma	
TSC2	Tuberous sclerosis 2 gene	7249	Pulmonary lymphangioleiomyomatosis (LAM), renal angiomyolipoma, HNSCC	
TSHR	Thyroid stimulating hormone receptor	7253	Toxic thyroid adenoma	Thyroid adenoma
U2AF1	U2 small nuclear RNA auxiliary factor 1	7307	CLL, MDS	
UBR5	Ubiquitin protein ligase E3 component n-recognin 5	51366	Mantle cell lymphoma, gastric, colorectal	
USP6	Ubiquitin-specific peptidase 6 (Tre-2 oncogene)	9098	Aneurysmal bone cyst	
USP8	Ubiquitin-specific peptidase 8	9101	Corticotroph adenoma	
VHL	von Hippel-Lindau syndrome gene	7428	Renal, hemangioma, pheochromocytoma	Renal, hemangioma, pheochromocytoma
WAS	Wiskott-Aldrich syndrome	7454		Lymphoma
WHSC1	Wolf-Hirschhorn syndrome candidate 1(MMSET)	7468	MM	
WHSC1L1	Wolf-Hirschhorn syndrome candidate 1-like 1 (NSD3)	54904	AML	
WIF1	WNT inhibitory factor 1	11197	Pleomorphic salivary gland adenoma	
WRN	Werner syndrome (RECQL2)	7486		Osteosarcoma, meningioma, other tumor types
WT1	Wilms tumor 1 gene	7490	Wilms tumor, desmoplastic small round cell tumor	Wilms tumor
WWTR1	WW domain containing transcription regulator 1	25937	Epithelioid hemangioendothelioma	

XPA	Xeroderma pigmentosum, complementation group A	7507		Skin basal cell, skin squamous cell, melanoma
XPC	Xeroderma pigmentosum, complementation group C	7508		Skin basal cell, skin squamous cell, melanoma
XPO1	Exportin 1 (CRM1 homolog, yeast)	7514	CLL	
YWHAE	Tyrosine 3-monooxygenase/tryptophan 5-monooxygenase activation protein, epsilon polypeptide (14-3-3 epsilon)	7531	Endometrial stromal sarcoma	
ZBTB16	Zinc finger and BTB domain containing 16	7704	APL	
ZFHX3	Zinc finger homeobox 3	463	Endometrial, gastric, prostate	
ZNF198	Zinc finger protein 198	7750	MPN, NHL	
ZNF278	Zinc finger protein 278 (ZSG)	23598	Ewing sarcoma	
ZNF331	Zinc finger protein 331	55422	Follicular thyroid adenoma	
ZNF384	Zinc finger protein 384 (CIZ/NMP4)	171017	ALL	
ZNF521	Zinc finger protein 521	25925	ALL	
ZRSR2	Zinc finger (CCCH type), RNA-binding motif and serine/arginine rich 2	8233	MDS, CLL	

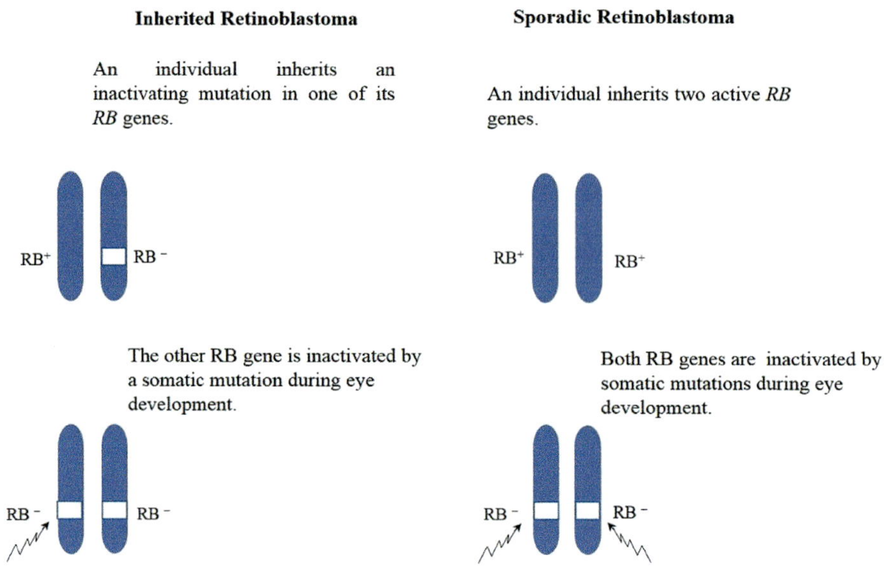

Fig. 1.3 Two mutations are required to eliminate the function of *RB* gene

1.2.6 DNA Repair Genes

The environmental factors, like UV light, chemicals, ionizing radiation, etc., can damage the DNA which led to mutations. DNA repair genes repair DNA, minimize the mutations in cell, and maintain the chromosome structure. When mutation occurs in this gene, the repair process is hampered which lead to the accumulation of other mutations in cell. For example, there are seven XP (xeroderma pigmentosum) DNA repair genes, whose products remove DNA damage caused by UV light or other factors. A defect in XP gene results in individuals who are very sensitive to UV light and have a thousand-fold increase in the incidence of all types of skin cancer. Bloom syndrome is an inherited disorder that leads to increased risk of cancer, lung disease, and diabetes. Mutation in the BLM gene causes a high frequency of chromosome breaks and interchanges, which culminates in the activation of oncogenes.

1.3 Roles of Tumor Suppressor Genes/Proteins

P53 is a 53-kilodalton (393 amino acids) tumor suppressor protein encoded by TP53 gene. Abnormal functioning of the p53 protein due to somatic mutations that deactivate both the copies of its gene is a key step in carcinogenesis. The p53 protein consists of three different domains: (i) N-terminal transcription activation domain (TAD); (ii) central DNA-binding core domain (DBD); and (iii) C-terminal homo-oligomerization domain (OD). Most of the mutations (more than 50% of human tumors) that inactivate TP53 gene occur in the DBD. The level of the p53 protein is low in the normal cells but increases with an increase of cellular stress. The response

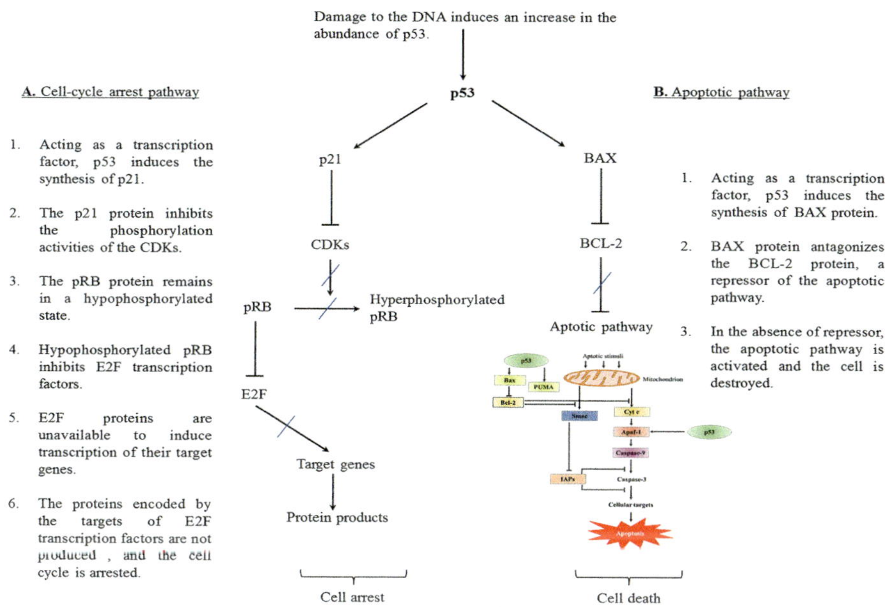

Fig. 1.4 Mechanism of action of *TP53* gene. (**a**) Cell cycle arrest and (**b**) apoptotic pathway

to DNA damage is mediated by an unknown signaling pathway during which phosphorylation of the p53 protein occurs, thereby making it stable and active. The phosphorylated protein stimulates the transcription of those genes whose product arrests the cell cycle. Interestingly, p53 prevents tumor formation by two different processes (Fig. 1.4). In the first, p53 activates the p21 Cdk inhibitor gene in response to DNA damage and stress. p21 protein inhibits cyclin-CDK protein complexes. During stress, it inactivates cyclin-CDK complexes which further mediate cell cycle arrest. During this phase the cell may repair its damaged DNA. Loss of p53 in cells inhibits the p21 gene from being transcribed, which enhanced the activity of multiple Cdks that were normally turned off by p21, resulting in increased cell proliferation. A second way in which p53 prevents tumor formation is by triggering a suicidal response against the damaged cell which lead to induction of apoptosis. Caspase proteinases (activated by apoptogenic factors released from mitochondria, cytochrome c, and Smac/DIABLO) induced the cell death via p53 pathway.

Apoptosome is a complex of high molecular weight which is formed due to the activity of cytochrome c and is comprised of adapter protein caspase 9 and Apaf-1. Once the caspase 9 gets activated, it cleaves and activates the effector caspases, such as caspases 3 and 7, which accomplish the cell death phenomenon. Pro- and anti-apoptotic Bcl-2 family proteins regulate the release of apoptogenic factors from mitochondria which can inhibit or induce permeabilization of outer mitochondrial membrane. The exact mechanism through which p53 signals Bcl-2 family proteins is not yet clear. However, in some cases, pro-apoptotic family members like PUMA, BAX, and NOXA are considered as the transcriptional targets of p53. The elucidation of the p53-dependent pathway, resulting in permeabilization of outer mitochondrial membrane via pro-apoptotic Bcl-2 family proteins, is a key to reveal the mechanism of stress-induced apoptosis. Because p53 acts as an important "brake" for tumor formation and maintains the cell's genetic integrity, it has been nicknamed as the "guardian of the genome." In 60% of cancer cases, P53 is found to be mutated or inactivated. In case of transformed cells, it is found in very large amount. For example, Li-Fraumeni is a familial cancer syndrome which occurs due to the variants of p53. Barrett's adenocarcinoma, which is a disease of lower esophagus, is caused by mutation in p53.

1.3.1 pAPC

pAPC protein was reported in adenomatous *Polyposis coli* and has been responsible for colorectal cancer. It is a 283 kilodalton, long protein which plays a major role in the renewal of intestinal epithelial cells. In a way, the pAPC protein balances cell death and cell proliferation. However, when the function of pAPC stops, the balance gets upset and the cell proliferation surpasses the cell death mechanism. Due to the rapid cell division, many small, benign tumors called as polyps or adenomas form in the intestinal epithelium. This condition is inherited as a rare autosomal dominant state called familial adenomatous polyposis (FAP). It is now discovered that pAPC maintains the balance between cell death and cell proliferation by sequestering β catenin. β catenin binds with several proteins inside the cells. The mutated cells where pAPC loses its function possess high levels of β catenin which binds with certain transcription factors and promotes apoptosis.

1.3.2 PhMSH2

It is a DNA repair protein associated with human cancer. The mutations in the MSH2 gene are related with instability of microsatellite and some cancers, most commonly with the inheritance of HNPCC (hereditary nonpolyposis colorectal cancer), a dominant autosomal condition. During this, small adenomas form, and one of which becomes cancerous. Msh2 or MutS protein homolog 2 is a protein in human being which is encoded by tumor suppressor gene MSH2 and is located on the second chromosome.

MSH2 forms dimer and heterodimer with MSH3 and MSH6 to form MutSβ DNA repair complex and MutSα mismatch repair complex, respectively. MSH2 also participates in DNA repairing processes like homologous recombination, transcription-coupled repair, and base excision repair.

1.3.3 BRCA1/BRCA2

Breast cancer susceptibility genes 1 and 2 are of human origins that produce tumor suppressor proteins. BRCA1 was identified in 1990 and located on chromosome 17, while BRCA2 was identified in 1994 and is located on chromosome 13. These

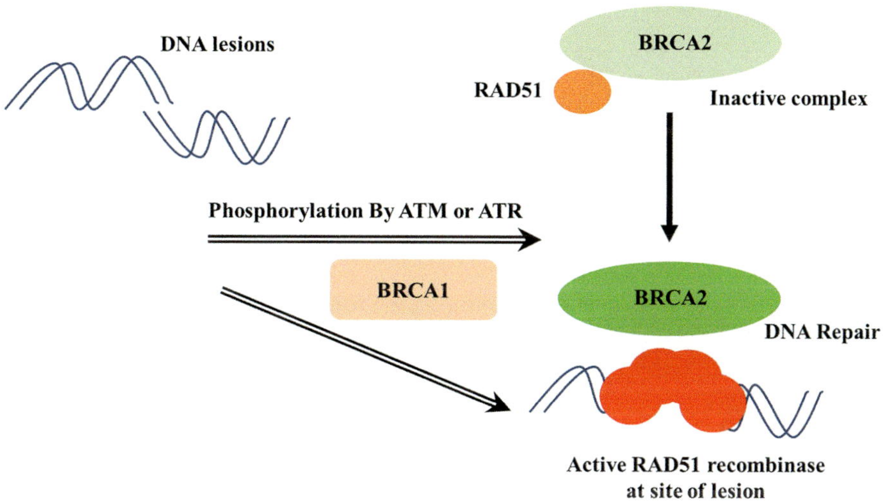

Fig. 1.5 Schematic representation of the role of BRCA2 in DNA double-strand breaks (DSBs) repair (Davies et al. 2001)

BRCA1and BRCA2 proteins ensure genetic stability of the cell by repairing the damaged DNA. For repairing the DNA lesions, an "inactive" complex between BRCA2 and RAD51 in the nucleus undergoes transition after phosphorylation by kinases such as ATM or ATR to form an active complex in which RAD51 becomes "active." This active state promotes the formation of nucleoprotein filaments at the sites of DNA damage (Fig. 1.5). The precise role of BRCA1 in these events is still unclear, but it is expected to have multiple functions in DNA damage responses.

Any mutations, alteration, and malfunction of either of the two genes produce extra genetic changes which can lead to cancer. Mutations in both the genes cause breast cancer and enhance the risk of ovarian, fallopian tube, pancreatic, as well as peritoneal cancer. It has been noticed that 5–10 % of the breast cancers are due to mutations in BRCA1 and BRCA2. Approximately 60 % (3 in 5) of the women who inherit these mutations are liable to suffer from breast cancer. Since we inherit two of each of the genes (BRCA1 and BRCA2) from each parent, we have their two copies. If one of the genes (from a pair) transferred from parent to offspring is mutated, then the breast cancer may not develop as long as the other copy of the gene functions normally. Therefore, every woman who has a single mutated gene may not develop breast cancer.

Moreover, mutation in BRCA1 gene increases the risk of uterine, cervical, and colon cancer, while mutations in BRCA2 enhance the probability of gallbladder, stomach, and bile duct cancer and melanoma. Men with BRCA2 and BRCA1 mutations are having high risk of breast and prostate cancer. Moreover, certain mutations in BRCA1 (also known as FANCS) and BRCA2 (also known as FANCD1) can cause Fanconi anemia.

1.3.4 CDK4

CDK4 genes produce cyclin-dependent kinase 4 protein and are responsible for cell cycle regulation. Certain non-hereditary cancers occur due to the mutations in these genes.

1.3.5 CMM1

CMM1 (cutaneous malignant melanoma) genes are associated with familial malignant cutaneous melanoma.

1.3.6 HER2

Also known as *NEU* or *ERBB2*, it produces a protein found on the cell surface called neuregulin receptor complex. This gene is associated with non-hereditary breast cancer.

1.3.7 MLH1

This gene is involved to fix the error in DNA replication and is also linked with familial hereditary nonpolyposis colon cancer (HNPCC) which is one of the most common genetic disorders of the Western world.

1.3.8 p16

Protein 16 is also called CDKN2. It inhibits the cell division and slows down the cell growth.

This leads to tumor formation in different types of skin tissues.

1.3.9 pRb

These genes are associated with various types of hereditary cancer. The main function of this gene is to inhibit the cell cycle process until a cell is not ready for division. This gene is then phosphorylated and gets activated when a cell is ready to divide. RB1 was used to describe the "two-hit theory" for hereditary cancers, which says that if an individual is carrying a susceptibility gene, then with a random mutation (non-inherited mutation) in the normal gene, the individual can have cancer. Any deletion or alteration of this protein results in the childhood eye cancer (retinoblastoma) which accounts for ~2% of all childhood malignancies. The chances of retinoblastoma to occur genetically and to show clinical symptoms are very low, while in most of the cases, it occurs sporadically.

1.4 Cell Signaling and Cancer

Two major pathways which are responsible for cell signaling and are associated with cancer are PI3K-Akt signaling (Hemmings and Restuccia 2012) and Ras-ERK (Morrison 2012).

1.5 Molecular Mechanism of Cancer

Uncontrolled proliferation of cells is an important characteristic of most of the cancers. From the aforementioned information, we came to know that any mutations in the proto-oncogenes and tumor suppressor genes can hyperactivate the pathways which were normal before. It is interesting to note that the signaling target molecules of cell cycle can also suffer genetic alteration to bring cancer into effect. The genes for cell cycle mediators cyclin D, cyclin E, and CDK4 get enlarged in some cancers, while pRB (G1 inhibitor) and p16 can get mutated or deleted. Similarly, the constitutive hyperactivity of Ras-ERK and PI3K-Akt pathways which regulate cell proliferation can culminate in rapid proliferation (Fig. 1.6).

The Myc transcriptional factor is a major target of Ras-ERK pathway which is phosphorylated by ERK which leads to stabilization due to ubiquitylation suppression. ERK also phosphorylates mitogen and stress-activated kinase (MSK), ribosomal S6 kinase (RSK), and mitogen-activated protein kinase (MAPK)-interacting kinase (MNK) families. These kinases further phosphorylate the transcription factors (e.g., FOS and CREB) that regulate the progression of cell cycle. FOS (oncogene) encodes for a transcription factor AP1 which controls several genes associated with cell proliferation. Myc induces genes that enhance the proliferation of cell. These include G1/S cyclins, CDKs, and E2F family transcription factors that regulate the cell cycle. Myc also represses the expression of various cell cycle inhibitors, hinders the activity of transcription factors which enhance differentiation of cell, and promotes those genes which enhance translation.

ERK activates RSK which further activates mTORC pathway through TSC2 phosphorylation. The RSK also phosphorylates eIF4B, which increases its interaction with eIF3 (translation initiation factor). This mechanism promotes cell growth and cell multiplication. Akt controls the TORC1 and regulates cell growth during cell cycle progression. It mediates the activation of mTORC1 by GTP-bound Rheb. It further phosphorylates eIF4-binding protein and releases the eIF4E cap binding factor to bind the mRNAs and p70 RSK. This mechanism increased protein synthesis, which is crucial for enhanced cell growth during cell cycle progression (Richardson et al. 2004). Akt inhibits the catalytic activity of the kinase GSK3. Inhibition of GSK3 by Akt causes stabilization of cyclin D and Myc which are

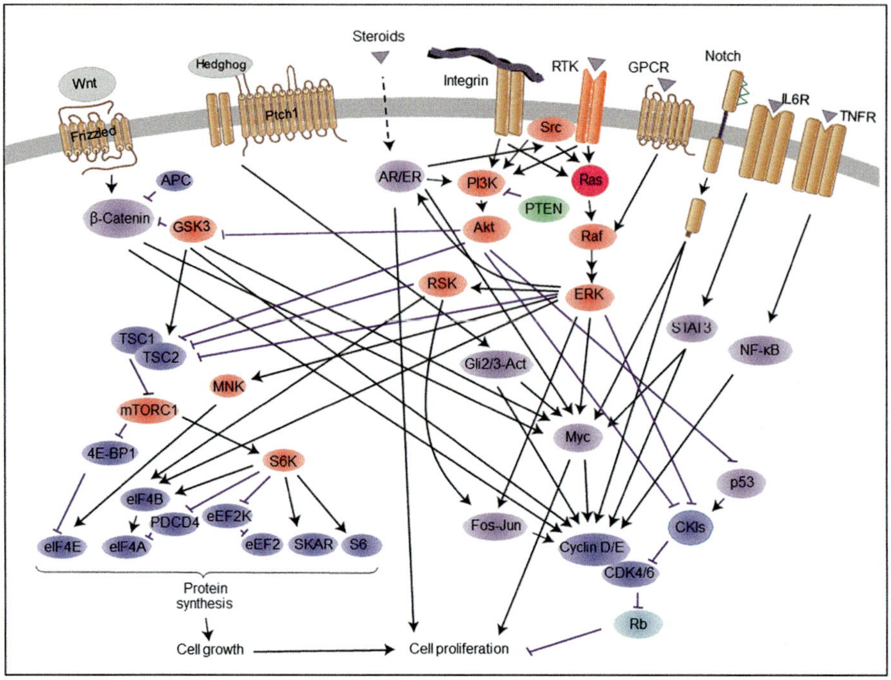

Fig. 1.6 Cell proliferation associated with Ras-ERK and PI3K-Akt pathways. (Redrawn from Sever and Brugge 2015)

important cell cycle regulators (Sears et al. 2000). In addition, Akt inhibits various inhibitors of cell cycle, like CKIs p27 and p21. Their phosphorylation leads to their sequestration in the cytoplasm by 14-3-3 proteins.

Akt also inhibits the FoxO-dependent transcription of cell cycle inhibitors such as p27 and RBL2 (retinoblastoma-like protein 2) (Burgering and Medema 2003). It also phosphorylates and activates MDM2 which mediates degradation of p53, thus releases a key brake on the cell cycle. In the later stage of the cell cycle, Akt may also regulate enzymes involved in the G2/M transition (Xu et al. 2012). Akt is also involved in the induction of b-catenin target genes that regulate proliferation, including those encoding Myc and cyclin D (Fang et al. 2007; Korkaya et al. 2009).

1.6 Regulation of Cell Death

Apart from the aforementioned pathway, several other signaling processes can also drive proliferation of cells in cancer. The role of Myc has already been described. RTK and cytokine signaling activate STAT3 which further promote the synthesis of Myc and cyclin D (Harrison 2012). Notch, Wnt/b-catenin, and hedgehog, all of which are concerned with cancer, also promote the activity of Myc and cyclin D. The transcription factor NF-kB also regulates cyclin D expression upon activation by TNF. The signaling through ERs and the androgen receptor is coupled to and enhanced by Ras-ERK and PI3K-Akt signaling (Renoir et al. 2013). The estrogen signaling mediates cell proliferation through activation of the ERa subtype, which further induces cyclin D and Myc function. An imbalance of function between ERa and ERb or mutations in ERa can dysregulate this pathway (Robinson et al. 2013; Toy et al. 2013).

Another pathway, the Hippo tumor suppressor pathway coordinates cell differentiation, cell proliferation, and cell death so as to control the growth of tissue. The major target of the Hippo core kinases is the mammalian transcriptional activator TAZ and YAP (TAZ, transcriptional coactivator with PDZ-binding motif; YAP, yes-associated protein). In cancer, the Hippo signaling is inactivated, and YAP and TAZ are activated and remain free to move into the nucleus to promote the proliferation of cells. Nuclear YAP/TAZ suppresses or activates transcription factors which regulate the target genes associated with proliferation of cells, metastasis, tissue growth, and organ shape and size. The PI3K-Akt and Ras-ERK pathways may also intersect with Hippo pathway components so as to inhibit its tumor-suppressive activity (Collak et al. 2012).

Cell death is a homeostatic control mechanism to control the cell number and thus prevents rapid cell proliferation and cancer. This mechanism is activated during the early oncogenic stage when the cell proliferation signals get unbalanced. Mutations that disable cell death signaling cause rapid cell proliferation and cancer. Bcl2 is an anti-apoptotic protein; its overexpression in B lymphocytes causes follicular lymphoma, by preventing apoptosis. As already discussed, p53 also controls apoptosis by inducing the transcription of pro-apoptotic regulators and directly binds to the Bax protein (Green and Llambi 2014). The anti-apoptotic proteins like Puma and Bok are frequently deleted in cancer, while Mcl1 and Bcl-xL protein genes get amplified. The pro-apoptotic regulators like Bim and Bad and the anti-apoptotic regulators like Bcl2 and Mcl1, in normal cells, ensure that cells must undergo apoptosis in the absence of appropriate signals supplied by growth factors or the tissue microenvironment. Due to oncogenic mutations in the pathways of Ras-ERK and PI3K-Akt, the hyperactivated signals disturb the balance which culminates in tumor cell survival and multiplication.

It is now well reported that the PI3K-Akt and Ras-ERK pathways regulate cell death in several ways (Fig. 1.7) (Zhang et al. 2011) which are (a) Akt phosphorylates forkhead family transcription factors like FoxO3A, which enhance cytoplasmic sequestration through 14-3-3 proteins. This activity blocks the induction of FasL and TRAIL (death ligands) and Bim (pro-apoptotic Bcl2 family protein); (b) Akt and RSK also phosphorylate Bad (pro-apoptotic Bcl2 family protein), for sequestration by 14-3-3 proteins; (c) Akt phosphorylates and activates XIAP (apoptosis inhibitor); (d) Akt also activates NF-kB, which regulates anti-apoptotic proteins like Bcl2, BCLxl, and Mcl1 and the intracellular death receptor inhibitor FLIP (Shen and Tergaonkar 2009); and (e) Akt-induced ubiquitylation and further degradation of p53 suppress p53-induced apoptosis (Ogawara et al. 2002). ERK phosphorylates IkBa and Bim (NF-kB inhibitor) and targets them for degradation. Caspase 9 scaffolding protein APAF is phosphorylated by ribosomal S6 kinase (RSK) and inhibits the cytochrome c to form nucleate apoptosome (Kim et al. 2012).

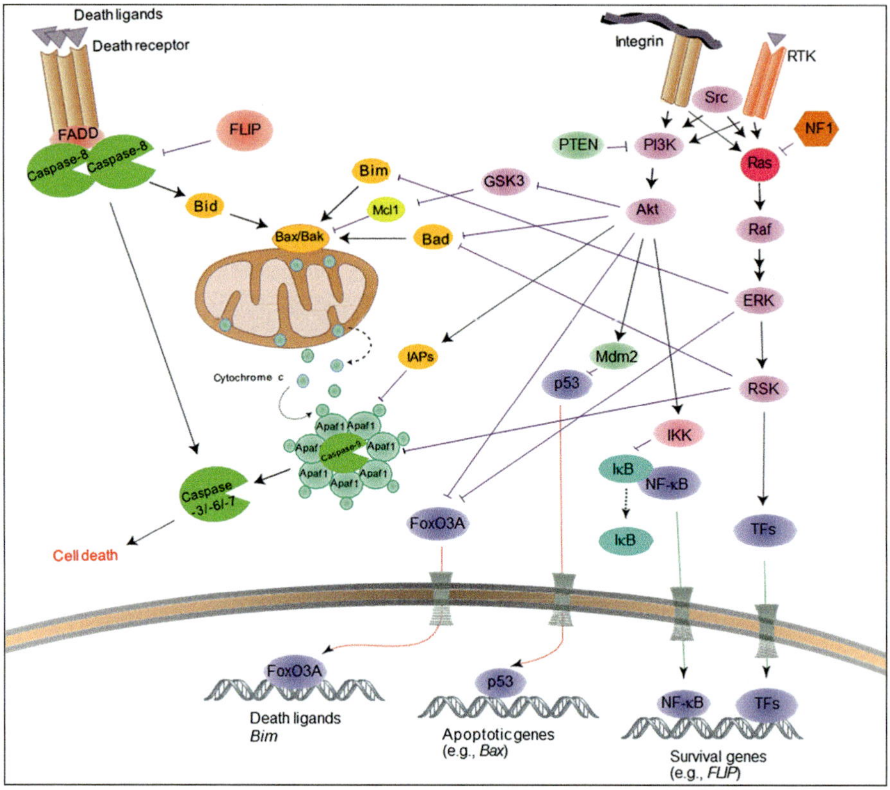

Fig. 1.7 Signaling in cancer: Ras-ERK and PI3K-Akt pathways related to cell death. (Redrawn from Sever and Brugge 2015)

1.7 Conclusions

Cancer is a state of dysregulation of the normal cellular signaling process that culminates in excessive cell proliferation. It may be caused by genetic or epigenetic changes. During the last 20 years, adequate scientific data has now been accumulated regarding the various mechanisms of cancer. It is now time to apply the knowledge that is generated to design the strategies for its cure. The strategies include pharmacologic and antibody-based inhibitors to target the mutated signaling process. The complexity of signaling network is such that it is challenging to develop a trustworthy anticancer drug that may show reproducible effects. Moreover, the heterogeneity of the tumor cells also leads to variations in drug sensitivity. The inhibitors of the non-receptor tyrosine kinase Abl and RTK ErbB2 have improved the life expectancy of the patients suffering from chronic myelogenous leukemia and breast cancer. Inhibitors that target B-Raf, EGFR, and the kinase ALK cause drastic reduction of tumor size. This further extends the survival chances of the patients suffering from melanoma and non-small-cell lung carcinoma. However, the rate of recurrence is also high because of drug resistance (Holohan et al. 2013; Lito et al. 2013). The RaS-ERK and Akt-PI3K pathways which get mutated can be a promising target to cancer treatment. A combination of the available treatments that target tumor cells can increase the efficacy of targeted therapies. Moreover, the existing cancer treatment regimes can be supplemented with Ayurvedic formulations to analyze the synergistic response of the drugs to treat cancers. Therefore, in-depth studies and analyses of the nature of resistance mechanisms of the tumor cells are required.

References

Burgering BM, Medema RH (2003) Decisions on life and death: FOXO Forkhead transcription factors are in command when PKB/Akt is off duty. J Leukoc Biol 73:689–701

Chin L, Andersen JN, Futreal PA (2011) Cancer genomics: from discovery science to personalized medicine. Nat Med 17(3):297

Collak FK, Yagiz K, Luthringer DJ, Erkaya B, Cinar B (2012) Threonine- 120 phosphorylation regulated by phosphoinositide-3-kinase/Akt and mammalian target of rapamycin pathway signaling limits the antitumor activity of mammalian sterile 20-like kinase 1. J Biol Chem 287:23698–23709

Davies AA, Masson JY, McIlwraith MJ, Stasiak AZ, Stasiak A, Venkitaraman AR, West SC (2001) Role of BRCA2 in control of the RAD51 recombination and DNA repair protein. Mol Cell 7(2):273–282

Fang D, Hawke D, Zheng Y, Xia Y, Meisenhelder J, Nika H, Mills GB, Kobayashi R, Hunter T, Lu Z (2007) Phosphorylation of b-catenin by AKT promotes b-catenin transcriptional activity. J Biol Chem 282:11221–11229

Green DR, Llambi F (2014) Cell death signaling. Cold Spring Harb Perspect Biol:a006080

Harrison DA (2012) The Jak/STAT pathway. Cold Spring Harb Perspect Biol 4:a011205

Hemmings BA, Restuccia DF (2012) PI3K-PKB/Akt pathway. Cold Spring Harb Perspect Biol 4:a011189

Holohan C, Van Schaeybroeck S, Longley DB, Johnston PG (2013) Cancer drug resistance: an evolving paradigm. Nat Rev Cancer 13:714–726

Kim J, Parrish AB, Kurokawa M, Matsuura K, Freel CD, Andersen JL, Johnson CE, Kornbluth S (2012) Rsk-mediated phosphorylation and 14-3-31 binding of Apaf-1 suppresses cytochrome c-induced apoptosis. EMBO J 31:1279–1292

Korkaya H, Paulson A, Charafe-Jauffret E, Ginestier C, BrownM DJ, Clouthier SG, Wicha MS (2009) Regulation of mammary stem/ progenitor cells by PTEN/Akt/b-catenin signaling. PLoS Biol 7:e1000121

Lito P, Rosen N, Solit DB (2013) Tumor adaptation and resistance to RAF inhibitors. Nat Med 19:1401–1409

Morrison DK (2012) MAP kinase pathways. Cold Spring Harb Perspect Biol 4:a011254

Ogawara Y, Kishishita S, Obata T, Isazawa Y, Suzuki T, Tanaka K, Masuyama N, Gotoh Y (2002) Akt enhances Mdm2-mediated ubiquitination and degradation of p53. J Biol Chem 277:21843–21850

Pleasance ED, Cheetham RK, Stephens PJ, McBride DJ, Humphray SJ, Greenman CD, Ye K (2010) A comprehensive catalogue of somatic mutations from a human cancer genome. Nature 463(7278):191

Renoir JM, Marsaud V, Lazennec G (2013) Estrogen receptor signaling as a target for novel breast cancer therapeutics. Biochem Pharmacol 85:449–465

Richardson CJ, Schalm SS, Blenis J (2004) PI3-kinase and TOR: PIKTORing cell growth. Semin Cell Dev Biol 15:147–159

Robinson DR, Wu YM, Vats P, Su F, Lonigro RJ, Cao X, Kalyana-Sundaram S, Wang R, Ning Y, Hodges L (2013) Activating ESR1 mutations in hormone-resistant metastatic breast cancer. Nat Genet 45:1446–1451

Sears R, Nuckolls F, Haura E, Taya Y, Tamai K, Nevins JR (2000) Multiple Ras-dependent phosphorylation pathways regulateMyc protein stability. Genes Dev 14:2501–2514

Sever R, Brugge JS (2015) Signal transduction in cancer. Cold Spring Harb Perspect Med 5(4):a006098

Shen HM, Tergaonkar V (2009) NFkB signaling in carcinogenesis and as a potential molecular target for cancer therapy. Apoptosis 14:348–363

Toy W, Shen Y, Won H, Green B, Sakr RA, Will M, Li Z, Gala K, Fanning S, King TA (2013) ESR1 ligand-binding domain mutations in hormone-resistant breast cancer. Nat Genet 45:1439–1445

Xu X, Hou Y, Yin X, Bao L, Tang A, Song L, Li F, Tsang S, Wu K, Wu H (2012) Single-cell exome sequencing reveals singlenucleotide mutation characteristics of a kidney tumor. Cell 148:886–895

Zhang X, Tang N, Hadden TJ, Rishi AK (2011) Akt, FoxO and regulation of apoptosis. Biochim Biophys Acta 1813:1978–1986

Types of Cancer

2

Abstract

Cancer is an extremely fatal disease, not only in developed but also in developing countries as it affects six million lives each year worldwide. According to the American Cancer Society each year 0.5% of US population is diagnosed with cancer. Annually, billions of dollars are invested on cancer research, yet the success rates are not satisfactory. Cancer cells arise due to the imbalance in the body functions and they invade and infect the normal cells. Cancer is not one disease; rather it is a group of various diseases. The present chapter includes explicit information on cancer types such as cancer of blood, lungs, colon and rectum, prostate, skin, breast, uterus, thyroid, lymphatic system, etc. Cancers are being treated by various methods which include chemotherapy, precision medicine, radiation therapy, surgery, stem cell transplant, hormone therapy, immune therapy, and targeted therapy. In some cases, cancer can be treated by single method, but mostly several methods are used to cure the disease. Moreover, the treatment specifically depends upon the stage and type of cancer. In chemotherapy and radiotherapy, survival rates are usually very low because of their undesirable side effects on human health. On the contrary, herbal therapy alone and in combination with the routine cancer treatment regimes has shown some promising potential, as it is free from side effects.

Keywords

Lymphoma · Carcinoma · Sarcoma · B cell · T cell · Metastasis

2.1 Introduction

The term cancer has its origin from a Greek word "Karakinos" by a physician Hippocrates (460–370 B.C), to describe carcinoma tumors. Some of the earliest evidence of human bone cancer were reported in mummies of ancient Egypt and in

© Springer Nature Singapore Pte Ltd. 2019

B. Koul, *Herbs for Cancer Treatment*,
https://doi.org/10.1007/978-981-32-9147-8_2

ancient manuscripts of 1600 B.C. The inscriptions reveal that the surface tumors were surgically removed in a similar way as they are removed today. In the 1980s groundbreaking evidences on carcinogenicity came into limelight. According to the International Agency for Research on Cancer (IARC), carcinogenicity in humans is caused by human T-cell lymphotropic virus, human immunodeficiency virus, hepatitis B virus, hepatitis C virus, human papillomavirus, Epstein-Barr virus, and human herpesvirus 8 (Blackadar 2016). Being an assemblage of diseases, cancer is characterized by the rapid growth and spread of abnormal cells. It can be called as a genetic disease as it is related to alterations within specific genes. However, in most of the cases, it is not found to be an inherited disease.

Our body is made up of 100,000,000,000,000 cells (Cancer Research, UK). Basically, different types of cells in the body do their own work, but they are similar. They have their own control centers, which are called as the nucleus. Inside the nucleus, chromosome harbors genes. Genes contain long strings of DNA that codes messages and send signals to the cells and make them do their own given work. Proteins and RNA control the cell and decide its duty, the timing to divide, what to do, and when to die. Mutations happen in genes when they divide, in which the gene may be damaged or lost or copied twice. When mutation occurs, it means that the cell no longer understands their given instruction and proliferates rapidly. Cancer happens due to transformation of the cell in which it bypasses apoptosis (Weinberg 2014). Cells produce signals to control how long will they live and how often will they divide. If any of the signals goes wrong or it is missing, then the cell starts to proliferate and form a lump called tumor. Cancer starts with a primary tumor (Weinberg 2014). For proliferation, the cancer cells obtain macromolecules from the microenvironment and break down in the lysosome to produce substrates for ATP generation and anabolism (Finicle et al. 2018). Oncogenes and tumor suppressor genes (TSGs) are the genes that are associated with cancer. These genes act antagonistically to regulate many normal cellular activities, such as cell survival metabolism and proliferation. Mutations in these genes onsets the progression of chronic and life-threatening diseases, especially cancer (Weinberg 2014).

The types of cancer are usually named after the organ or tissue where they arise (Fig. 2.1) or by the type of cell that formed them. According to the recent stats, men are more prone to cancers of prostate, lung, colon and rectum, and urinary bladder, whereas women are more affected by cancers of breast, lung, colon and rectum, uterus, and thyroid. It is rather alarming that prostate cancer in men and breast

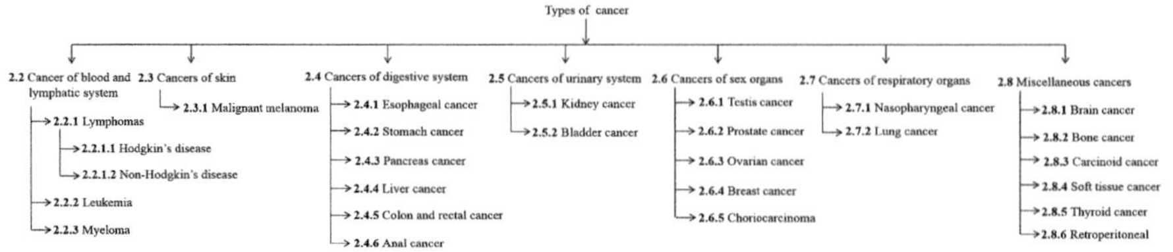

Fig. 2.1 Different types of cancer based on the organ affected

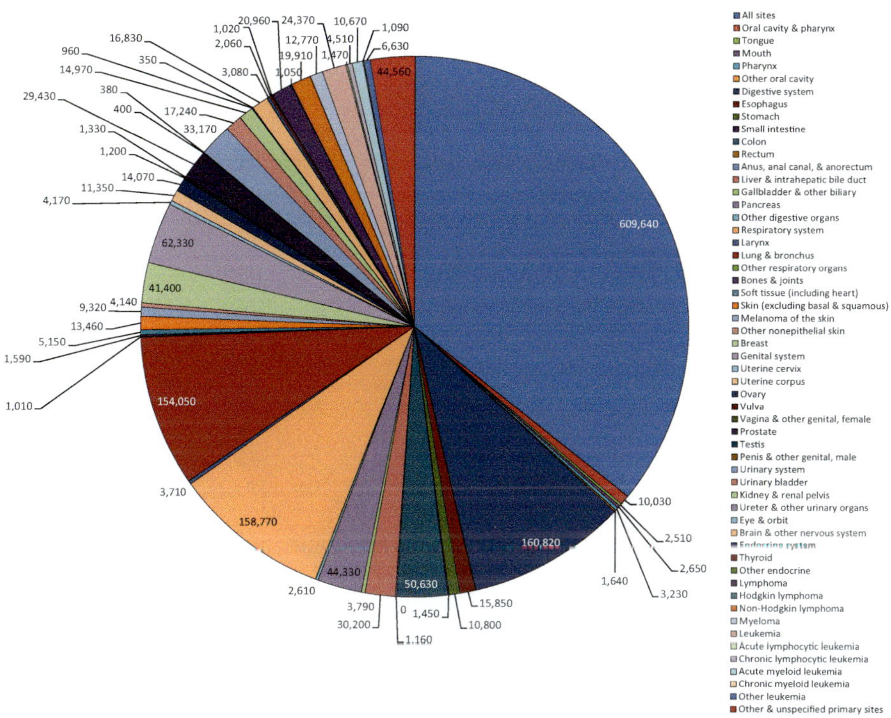

Fig. 2.2 Estimated deaths due to cancer in the US (cancer facts and figures 2018)

cancer in women are responsible for cancer-related deaths in men and women, respectively (Siegel et al. 2016). Cancer-related mortality in children is caused mostly due to cancer of blood, brain, and lymph nodes (Yoo and Shin 2003; Siegel et al. 2016).

In the United States, cancer has a major impact on the society as it occurs in large groups of people (Fig. 2.2). In 2018, 1,735,350 new cases of cancer (Fig. 2.3) are estimated to be diagnosed, and 609,640 people may die from cancer (National Cancer Society, USA 2018). India ranks third in cancer cases after China and the United States, and more than one million new cases are reported each year. Here, 25% of cancer-mediated death in males occur due to oral cavity and lung cancer, while 25% of cancer-mediated deaths in females is caused by oral cavity and breast cancer (Fig. 2.4). Several women die of cervical cancer every 8 min (India.com lifestyle 2018). Moreover, the possibility of cancer-mediated death before the age of 75 in males is 7.34%, while in females it is 6.28%. In 2010, the total cancer patients in India were 979,786, out of which 462,408 were males and 517,378 were females. This clearly shows that there has been a hike in cancer cases over time (Ali et al. 2011). A recent survey by the Punjab government revealed that Malwa region has the highest number (compared to other states) of cancer patients – 107 in 100,000 population, which makes Punjab the "cancer capital of India."

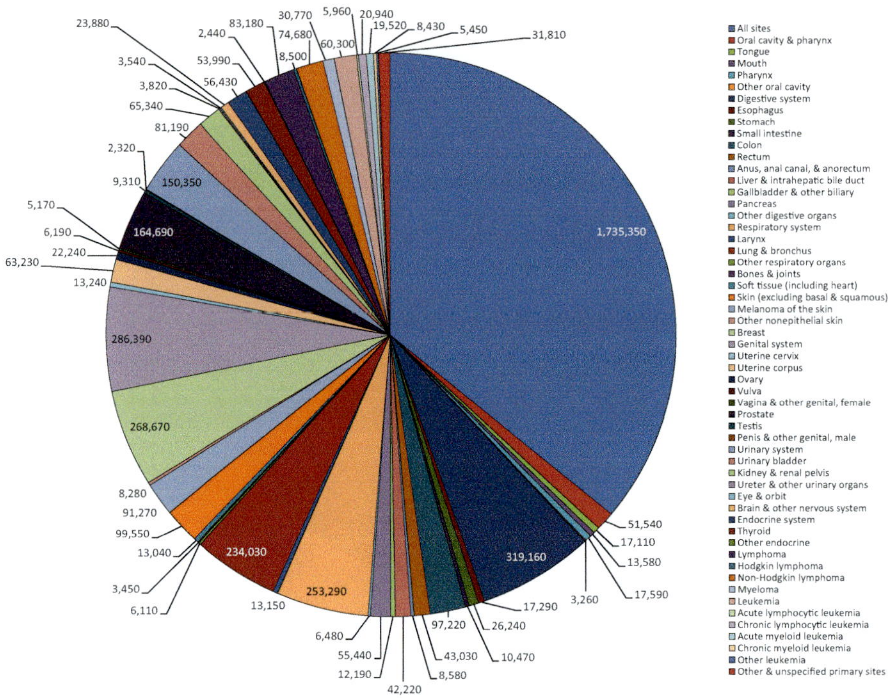

Fig. 2.3 Estimated new cases of cancer in the US (cancer facts and figures 2018)

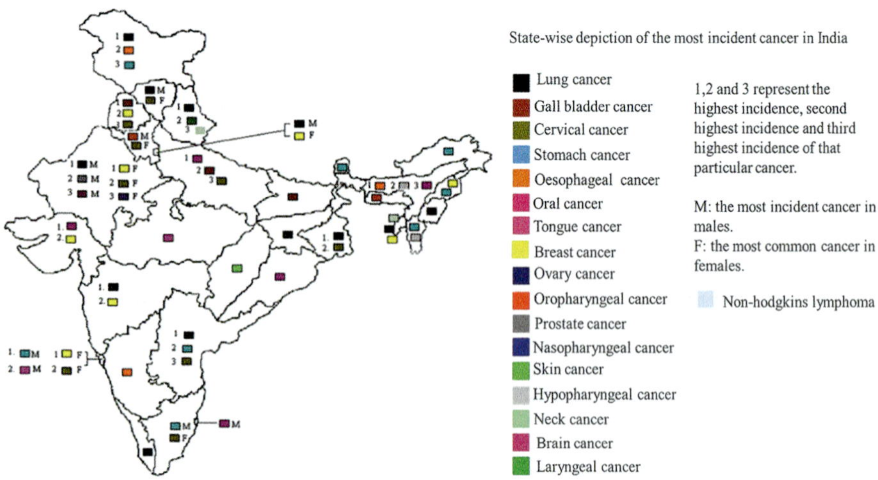

Fig. 2.4 State-wise depiction of the most incident cancer in India (Ali et al. 2011)

Several treatments are available to cope with cancers and to prolong the life of the patient. The treatment regime depends upon the organ/tissue affected, cancer stage, and age and health status of the individual. Some might undergo only one treatment, while others might need dual treatments, such as surgery with chemotherapy or radiotherapy and some might need immunotherapy, hormone therapy, or target therapy (National Cancer Institute 2018). There are many molecular-based treatments, but herbal therapies have also been introduced. By further meticulous researches and in-depth understanding of cancer and its causes, we can find effective treatments for different types of cancers and their recurrence. In this chapter, we will deal with different types of cancers, their symptoms, diagnosis, staging, and the possible available treatments.

2.2 Cancer of Blood and Lymphatic System

Blood cancers or hematologic cancers are a group of malignancies that affect the bone marrow (where blood is produced), blood, and lymphatic system (Ali and Sultana 2012). The lymphatic system is a part of the immune system, and it consists of thin tubes (lymph vessels or lymphatic vessels) and lymph nodes (lymph glands) which run all over the body. It includes lymph nodes, lymphatic vessels, tonsils, thymus, spleen, and digestive tract lymphoid tissue. The lymph vessels carry lymph (a colorless liquid) and branch throughout the body like the blood vessels (arteries and veins). The lymph nodes are connected by lymphatic channels. Lymph flows through the lymph nodes, as well as through the spleen, tonsils, bone marrow, and thymus gland. It contains lymphocytes (type of WBCs) that fight with infectious and damaged or abnormal cells. The white blood cells attack any bacteria or viruses that are present. This activity causes swelling and tenderness of a localized infection. These enlarged and often merged collections of lymph nodes are known as "swollen glands," and this condition is known as lymphadenopathy. Cancer cells which break from the affected site get stuck at these lymph nodes. Blood cancer such as leukemia and myeloma starts in the bone marrow, while lymphoma starts from the lymphatic system. Blood cancers are of three different types: (i) lymphoma, (ii) leukemia, and (iii) myeloma.

2.2.1 Lymphomas

Lymphocytes are the cells of the immune system which recognize abnormal cells and infectious organisms and protect the body against them. The two major subtypes of lymphocytes are B lymphocytes (B cells) and T lymphocytes (T cells). B cells produce antibodies that circulate through the blood and lymph and attach to infectious organisms or abnormal cells. Antibodies activate the other cells of the immune system so as to recognize and destroy the pathogens. This phenomenon is known as humoral immunity. T cells when activated kill the pathogens directly and also prevent the immune system from inappropriate overactivity or underactivity. It

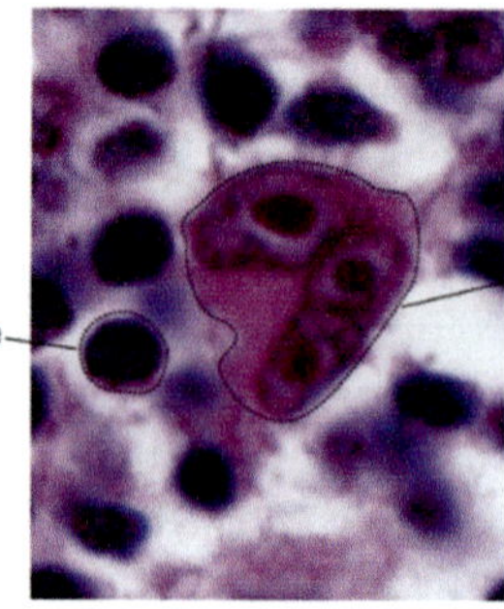

Fig. 2.5 Difference between normal lymphocyte and Reed-Sternberg cell

is interesting that after fighting off the invader, some of the B and T lymphocytes become memory cells which remember the invader and remain prepared for a second encounter.

Lymphoma is a type of cancer wherein a malignant transformation of either B or T cells or their subtypes occurs. Lymphoma is the most common type of blood cancer (seventh most common cancer in adults and third most common in children) in the United States that develops in the lymph nodes and tissues of the lymphatic system. As the abnormal cells multiply, they form a mass called as a tumor which deprives the surrounding tissues of the oxygen and nutrients needed for normal function. Lymphomas are usually confined to lymph nodes and other lymphatic tissues but can also spread anywhere in the body (extra nodal disease).

The classification of lymphomas is based upon both the microscopic appearance and genetic and molecular markers. The two major classes of lymphomas are [i] Hodgkin's lymphoma or HL (previously called Hodgkin's disease) and [ii] non-Hodgkin's lymphomas or NHLs (all other lymphomas). Both types of lymphomas may affect the same place, with same appearance and symptoms, but exhibit distinct cell surface markers.

2.2.1.1 Hodgkin's Disease [HL]

It was first discovered by Dr. Thomas Hodgkin in 1832. In Hodgkin's disease, Reed-Sternberg [RS] cells are observed in biopsies (Fig. 2.5). Reed-Sternberg cells are large cells of diameter > 45 μm and are binucleated, each with large acidophilic central nucleoli surrounded by a clear "halo," i.e., owl's eye appearance. Some recent studies indicate that RS cells of HL are lymphocytic in nature and in most cases originate from B cell (Laura 2016). HL mostly affects the young generation and patients with immune deficiency (Laura 2016). It may originate from germinal center B-cell or Reed-Sternberg cells.

Germinal centers or GCs are specific sites within secondary lymphoid organs, lymph nodes, and the spleen where mature B cells proliferate, differentiate, and mutate their genes and change the class of the antibodies during a normal immune response to an infection. They develop speedily after the activation of follicular B cell by T-dependent antigen (Zotos and Taclinton 2012).

Symptoms of HL

The symptoms of HL include:

(a) Asymptomatic lymph node enlargement (lymphadenopathy), especially in the neck, armpit, and around the groin.
(b) Mediastinal complains such as chest pain, cough, and dyspnea (trouble breathing) occur due to various tumors in mediastinum (central compartment of the thoracic cavity).
(c) Pel-Ebstein fever also known as Ebstein-Cardarelli fever occurs. This fever persists for days to weeks followed by afebrile intervals and then reoccurrence. It is a rarely found symptom noticed in patients with HL.
(d) Nocturnal hyperhidrosis occurs when there is excessive sweating during sleeping.
(e) Weight loss and fatigue may be due to lack of nutrition or due to cancer and viral infection such as HIV infection, depression, bowel diseases, and overactive thyroid.

Stages of HL

Table 2.1 and Fig. 2.6 show different stages of Hodgkin's lymphoma.

Table 2.1 Different stages of Hodgkin's lymphoma

Stage	Explanation
Stage I	Characterized by the occurrence of cancer in one lymph node region
Stage II	Characterized by the spread of cancer to two or more lymph nodes on one side of the diaphragm. It may involve an organ and lymph nodes near to it
Stage III	Characterized by the spread of cancer to both sides of the diaphragm. In addition, an extra lymphatic organ may be involved
Stage IV	The lymphoma spreads to organs (liver, bone marrow, or lungs) beyond the lymph nodes

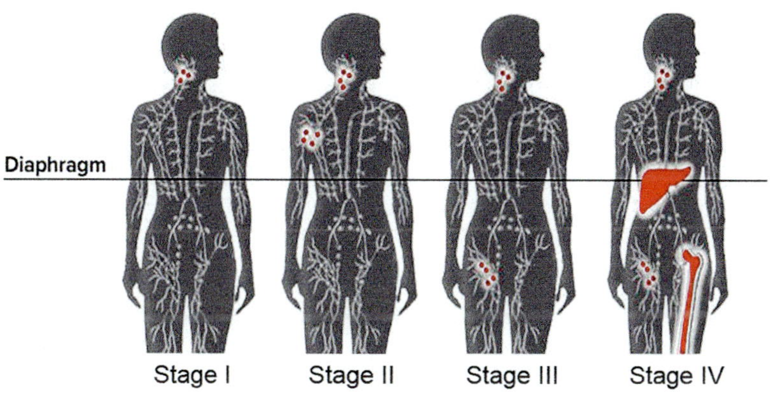

Fig. 2.6 Stages of Hodgkin's lymphoma disease

Recurrent Lymphoma

Lymphoma may relapse years later after the first treatment. It might return in the region where it first started or in another part of the body. In such a situation, another round of tests (similar to the previous) is required to diagnose the extent of the recurrence.

Diagnosis/Tests for HL

The diagnosis/tests for HL include:

(i) *Physical examination and medical history:* The doctor analyses the medical history of the symptoms, family history, and possible risk factors. The doctors are cautious about lymph nodes because they expand abnormally during infections in the body.

(ii) *Lymph node biopsy*: The two major types of biopsies are (a) *excisional biopsy (most common)* which involves removal of the entire lymph node and (b) *incisional biopsy* which involves removal of a small part of a tumor or node. The abovementioned are the most common types of biopsies. Another type of biopsy is "needle biopsy." However, it is less invasive than the aforementioned two biopsies but cannot extract enough sample to diagnose Hodgkin's lymphoma and its types. The two major types of needle biopsies are (a) *fine needle aspiration (FNA)* which involves the use of a fine, hollow needle to extract small volume of fluid and tiny fragments of tissue and (b) *core needle biopsy* which involves the use of a large needle to remove a slightly larger piece of tissue. If the enlarged node is superficially present, the biopsy sample can be extracted by feeling the node. But, if the node or tumor is deeply seated, then the doctor can guide the needle using a computed tomography (CT) or ultrasound.

(iii) *Bone marrow aspiration and biopsy:* If Hodgkin's lymphoma has been diagnosed, its spread in the bone marrow is analyzed through bone marrow aspiration. The bone marrow samples are usually collected from the back of the pelvic (hip) bone. This process is performed under anesthesia but the patients do feel brief pain. The aspiration process is followed by bone marrow biopsy wherein a small piece of bone and marrow is removed with a needle. This procedure is also not free from pain.

(iv) *Imaging tests*: X-rays, ultrasound waves, magnetic fields, or radioactive particles can be used to determine the exact stage of Hodgkin's lymphoma, to analyze the effect of treatments, and to look for signs of cancer reoccurrence.

(v) *Computed tomography scan (CT scan):* A CT scan can be used to generate detailed cross-sectional images of the enlarged lymph nodes or organs to locate Hodgkin's lymphoma in the neck, chest, abdomen, and pelvis. CT scan can also aid in guiding a biopsy needle to a suspicious area.

(vi) *Magnetic resonance imaging (MRI):* MRI can be used to generate detailed images of soft tissues in the body. Unlike the CT scans, it uses computer-aided radiofrequency waves and strong magnetic field to locate the spread of cancer to the spinal cord or brain.

(vii) *Positron-emission tomography (PET):* It uses small amounts of radioactive material (radioactive form of sugar which collects in cancer cells), a special camera, and a computer to help investigate the body areas exhibiting radioactivity. PET scan is usually repeated after a few courses of chemotherapy to see that the lymph nodes no longer take up radioactive sugars. The image generated is not detailed as that of a CT or MRI scan but can render useful information regarding Hodgkin's lymphoma. A combination of PET/CT scan gives accurate information because the lymphoma prone areas which look normal on a CT scan can be better spotted by a PET scan. Thus, a doctor can compare areas of higher radioactivity on the PET scan and a more detailed appearance of that area on the CT scan.

(viii) *Blood tests:* Blood tests such as complete blood count (CBC) and erythrocyte sedimentation rate (ESR) are usually performed when the doctor suspects Hodgkin's lymphoma. They are also done to check liver and kidney function. If the lymphoma spreads to the bone marrow (site of blood cell production), then the patient suffers from anemia, which is indicated by a low RBC count. An increase in the number of WBCs and a high ESR are observed during Hodgkin's lymphoma.

(ix) *Electrocardiogram and pulmonary function test*: If chemotherapeutic drug(s) is expected to affect the heart and the lungs, then echocardiogram or a multigated acquisition (MUGA) scan is done, to examine the heart activity. A MUGA scan uses a radioactive tracer (radionuclide) and a specific camera to assess the heart functioning. Similarly, pulmonary function tests (PFTs) involve the analysis of lung function.

(x) *Immunophenotyping:* The biopsy samples (cells from a lymph node, blood, or bone marrow) are observed under a microscope for the presence of Hodgkins' lymphoma (HL) cells (Reed-Sternberg cells). The pathologists perform immunohistochemistry test to look for proteins such as CD15 and CD30 which are displayed on the surface of Reed-Sternberg cells. Similarly, tests for the presence of other surface proteins are also performed depending on the type of lymphoma. After correct diagnosis, the stage of the Hodgkin's disease (HD) can be determined (Fig. 2.5).

Treatment of HL

Some combination chemotherapy treatments that are used for HL are as follows:

(i) ESHAP (etoposide, methylprednisolone, cisplatin, cytarabine), (ii) DHAP (dexamethasone, cisplatin, cytosine arabinoside), (iii) DICE (dexamethasone, ifosfamide, cisplatin, etoposide), (iv) ICE (ifosfamide, carboplatin, etoposide), and (v) IGEV (ifosfamide, gemcitabine, etoposide, vinblastine). Some new immunotoxin and immunomodulatory drugs (IMiD) presently under clinical trials are XMABT 2513, MDX- 1401, SGN30, HGN35, and Revlimid (lenalidomide).

2.2.1.2 Non-Hodgkin's Lymphoma [NHL]

In non-Hodgkin's lymphoma, B cells and T cells (white blood cells with special roles in immunity) are affected. In non-Hodgkin's lymphoma, B lymphocytes become abnormal (Reed-Sternberg cells). NHL usually begins when a lymphocyte (usually a B cell) becomes abnormal and divides rapidly to form a tumor. In the United States, B-cell lymphomas are much more common than the T-cell lymphomas. The symptoms of NHL include occurrence of swollen, painless lymph nodes in the neck, armpits, or groin; unexplained weight loss; fever and shivering; night sweats; coughing, troubled breathing, or chest pain; weakness; and a feeling of fullness in the abdomen.

Staging of NHL (Table 2.2)

Table 2.2 Stages of NHL

Stage	Explanation
Stage I	This stage indicates that the lymphoma is localized to only one lymph node
Stage II	This stage indicates that two or more groups of lymph nodes occur on the same side, above, or below the diaphragm. It includes nodes in the underarm and neck area
Stage III	This stage indicates that the lymphoma is present in lymph node areas, on both sides, above, and below the diaphragm
Stage IV	This stage indicates that the lymphoma has spread to the organs (liver, lungs, or bone marrow)

Diagnosis of NHL

If any swollen lymph node or any symptoms of NHL are seen, then it is advisable to consult a medical practitioner. We may have the following examination and tests done:

(i) *Physical examination:* if any swollen lymph node is seen in the neck, underarms, and groin, the doctor also examines for a swollen spleen or liver.
(ii) *Blood tests:* a complete blood count is done to check the number of white blood cells. Since the lymphoma causes a hike in the level of lactate dehydrogenase (LDH), so its level is also analyzed.
(iii) *Chest X-rays:* it is usually done to check for swollen lymph nodes or any other signs of NHL in the chest.
(iv) *Biopsy:* excisional biopsy (entire lymph node removed) or incisional biopsy (some part of lymph node tested) may be performed.

There are several differences between HL and NHL (Table 2.3).

There are two main types of NHL:

1. Diffuse large B-cell lymphoma (DLBCL).
2. Indolent follicular lymphoma.

1. **Diffuse Large B-Cell Lymphoma [DLBCL]**

DLBCL affects the B cell and can occur between adolescence and old age. It is more common in men around the age of 60. DLBCL is an extra nodal disease (begins in body parts, e.g., stomach or bowel). Through gene expression profiling (GEP), DLBCL has been segregated into two distinct molecular subtypes: (i) germinal center B cell (GCB) and (ii) activated B cell (ABC). The early sign of DLBCL is usually a painless swelling in the neck, armpit, or groin. General symptoms (B symptoms) of DLBCL include loss of appetite, tiredness, night sweats, high fevers, and weight loss. Four stages have been reported in DLBCL (Table 2.4).

Table 2.3 Difference between and HL and NHL

Hodgkin's disease (HL)	Non-Hodgkin's disease (NHL)
HL develops from a specific abnormal B lymphocyte lineage	NHL may be derived from either abnormal B or T cells and are characterized by the presence of unique genetic markers
HL has five subtypes	NHL has 30 subtypes
HL is most common in two age groups: Young adults 16–34 years of age and in older people 55 years and above	NHL occurs in older people
HL is not much common	NHL is far more common

Table 2.4 Different stages of DLBCL

Stage	Explanation
Stage I	It indicates that only one group of the lymph nodes in an area is affected
Stage II	It indicates that two or more groups of lymph nodes present above or below the diaphragm are affected
Stage III	It indicates that lymphoma is occurring in the lymph nodes both above and below the diaphragm
Stage IV	It indicates that lymphoma has invaded other organs such as bone marrow, liver, etc.

Diagnosis and Treatment of DLBCL

The most common diagnosis of DLBCL is a lymph biopsy. Some other tests may also be done. These include (i) bone marrow tests, (ii) blood tests, (iii) X-rays, and (iv) lumbar puncture to check the presence of lymphoma cells in CNS. During the early stages of DLBCL (stage 1 or 2), the patients are treated with a short-course chemotherapy or chemo-immunotherapy followed by radiotherapy, while the advanced stage (stage 3 or 4) is usually treated with a longer course of chemotherapy. Table 2.5 summarizes the treatment regime for DLBCL. The R-CHOP

Table 2.5 Treatment regime for DLBCL

DLBCL	Treatment type	Drug(s) combination		Duration of treatment
Early stage (stages 1, 2)	Chemotherapy	CHOP	Cyclophosphamide, hydroxydaunorubicin, and vincristine	2–4 cycles
	Chemo-immunotherapy	R-CHOP	Rituximab (Rituxan®), cyclophosphamide, hydroxydaunorubicin, vincristine, prednisolone	
	Radiotherapy			
Advanced stage (stages 3, 4)	Strong chemotherapy	R-CODOX-M	Rituximab, cyclophosphamide, doxorubicin, methotrexate	6–8 cycles
		R-IVAC	Rituximab, ifosfamide, vincristine, cytarabine	
		DA-EPOCH-R	Dose-adjusted etoposide, prednisolone, cyclophosphamide, hydroxydaunorubicin, rituximab	
	Mild chemotherapy for patients with health issues	R-miniCHOP	Rituximab, cyclophosphamide, hydroxydaunorubicin, vincristine, prednisolone	6–8 cycles
		R-GCVP	Rituximab, gemcitabine, cyclophosphamide, vincristine, prednisolone	
		R-CEOP	Rituximab, cyclophosphamide, etoposide, prednisolone	
Relapsed and refractory	Strong chemotherapy	R-GDP	Rituximab with gemcitabine, dexamethasone, and cisplatin	
		R-DHAP	Rituximab with dexamethasone, cytarabine, and cisplatin	
		R-ICE	Rituximab with ifosfamide, carboplatin, and etoposide	

Table 2.6 New targeted treatments for DLBCL

Drug(s)	Trade name	Category/mode of action
Ibrutinib	Imbruvica®	Cell signal blockers
Bortezomib	Velcade®	Proteasome inhibitors
Obinutuzumab	Gazyva®	Newer antibodies against CD20
Brentuximab vedotin	Adcetris®	Antibody-drug conjugates
Lenalidomide	Revlimid®	Immunomodulatory drugs
Nivolumab and pembrolizumab	–	Immune checkpoint inhibitors
Pixantrone	Pixuvri®	Newer chemotherapy drugs
CAR T-cell therapy	–	Engineered immune cells recognize and attack the lymphoma cells

(rituximab + cyclophosphamide + doxorubicin + vincristine + prednisolone) chemotherapy regime is usually adopted for treating diffuse large B-cell lymphoma. More promising alternative regimes are also under phase III trials (Table 2.6).

2. **Indolent Follicular Lymphoma (IFL)**

Indolent follicular lymphoma (IFL) is an NHL which affects the white cells that helps our body to fight from infections. In IFL, the affected blood cells migrate to different locations of the body such as the bone marrow, lymph nodes, etc. and form tumors. Since the IFL grows slowly, the patient can live long with it. The symptoms of IFL include painless swelling of the lymph nodes in the neck, armpits, groin, and stomach, breathing trouble, night sweats, general fatigue, and weight loss. The IFL may be caused by exposure to radiation or carcinogenic chemicals. Patients who suffer from immune deficiency disease(s) (HIV, rheumatoid arthritis, lupus, or celiac disease) are more susceptible to IFL.

Diagnosis and Treatment of IFL

The diagnostic tests include bone marrow test, CT scan, and PET scan. The treatment of IFL includes radiation therapy, monoclonal antibodies, chemotherapy, radio-immunotherapy [treatment by Y90 ibritumomab tiuxetan (Zevalin)], and stem cell treatment.

Waldenstrom's Disease

It is an NHL that grows slowly and is also identified as Waldenstrom's macroglobulinemia (WM) or lymphoplasmacytic lymphoma. The American Cancer Society (ACS) observes 1100–1500 cases of Waldenstrom's disease each year, in the United States (Advani et al. 2019). WM is a lymphoplasmacytic lymphoma with immunoglobulin M (IgM) monoclonal protein (Gertz 2019). It is believed that WM originates from B lymphocytes that proliferate in the germinal centers of lymph nodes and accumulate all the genetic changes that occur in these centers (Coimbra et al. 2014). The symptoms of this disease include general body fatigue, numbness and weakness, sudden weight loss, diarrhea, enlargement of lymph nodes or spleen, peripheral neuropathy (pain in hands or feet), abdominal swelling, shortness of breath, skin lesions, changes in color of the fingertips on exposure to cold, changes in vision (blurry vision or double vision), and heavy night sweats. These symptoms are often termed as B symptoms. Those who do not experience the B symptoms are categorized as A.

Diagnosis and Treatment of Waldenstrom's Disease
To diagnose WM, doctors may check for any swelling in the spleen, liver, or lymph nodes. Blood test may be done to determine the IgM level. A bone marrow biopsy, CT scan of the bones or soft tissues, and X-rays of bones or soft tissue may also be done. There is no cure for WM but its treatment can be effective in controlling the aforementioned symptoms. However, the treatment of this disease usually depends on the symptoms and their severity. The available treatments include:

(i) *Chemotherapy*: One can get this treatment as a pill or intravenously. Chemotherapy for WM is planned to damage the atypical cells producing excess IgM (Advani et al. 2019).
(ii) *Plasmapheresis/plasma exchange*: It is a process of removal of excess IgM from the plasma, followed by the addition of donor plasma and returning it to the body.
(iii) *Biotherapy:* It involves biological therapies which boost the immune system to cope with cancer.
(iv) *Surgery:* It is also known as "splenectomy" (removal of spleen). This may reduce or eliminate the aforementioned symptoms for many years.

2.2.2 Leukemia

There are two main types of WBCs in our body: lymphoid cells and myeloid cells. Leukemia is a cancer of these blood cells or the blood-forming tissues including bone marrow. Thus, it is a neoplastic proliferation of one particular cell type: granulocytes, monocytes, lymphocytes, or infrequently RBCs (Jabbour and Kantarjian 2018; Flis and Chojnacki 2019). Leukemia is characterized based upon (i) the rate of progression and (ii) the category of blood cell involved (myeloid or lymphoid). Based on these characteristics, leukemias are of two types: (i) **acute** leukemia which occurs when most of the abnormal blood cells (myeloblasts, red blood cells, or platelets) stay immature and (ii) **chronic** leukemia which occurs when some cells among the normal cells remain immature. Hence, the four main subtypes of leukemia (Table 2.7) are (a) acute myeloid leukemia (AML), (b) chronic myeloid leukemia (CML), (c) acute lymphocytic leukemia (ALL), (d) chronic lymphocytic leukemia (CLL), and (e) hairy cell leukemia (HCL). The country-wise incidence of leukemia cases (per 100,000 individuals) has been reported to be: Denmark, 29; Luxembourg, 28.7; Italy, 28.4; Belgium, 27; France, 26.8; Croatia, 26.6; Latvia, 26.4; Germany, 26; New Zealand, 26; and Australia, 25.3 (Pokharel 2012).

Table 2.7 Different types of leukemia and their cause(s)

Type of leukemia	Description	Causes
Acute myelogenous leukemia (AML)	This cancer develops from inside the bone marrow involving immature cells that would have turned white blood cells. It is the most common type of leukemia, and the disease progresses rapidly	It is caused by exposure to harmful chemicals and rays, blood disorders or weakened immune system
Chronic myelogenous leukemia (CML)	Slow buildup of the immature WBCs which hampers the function of the bone marrow	Due to the presence of an abnormal chromosome called Philadelphia chromosome; radiation exposure may also be a cause
Acute lymphocytic lymphoma (ALL)	Production of large number of immature lymphocytes hinders the of bone marrow function	The direct cause is not reported; exposure to toxins like benzene and radiation, chemotherapy, and chromosomal abnormality can increase the risk of ALL
Chronic lymphocytic leukemia (CLL)	Slow increase in lymphocytes affects the lymph nodes and the spleen which ultimately causes the bone marrow to stop their function	Not known
Hairy cell leukemia (HCL)	Rare category of cancer which affects WBCs The cells look hairy under a microscope	Not known

1. **Acute Myeloid Leukemia (AML)**

Acute myeloid leukemia (AML) or acute myelogenous leukemia or acute myeloblastic leukemia is a type of cancer that grows fast in the blood and bone marrow. Being the most common type of acute leukemia, it arises when the bone marrow starts to synthesize blasts (immature blood cells, incapable to fight infection). These cells increase in number and begin to compete with the normal WBCs, RBCs, and platelets. AML has eight different subtypes: (i) myeloblastic (M0), (ii) myeloblastic (M1), (iii) myeloblastic (M2), (iv) promyelocytic (M3), (v) myelomonocytic (M4), (vi) monocytic (M5), (vii) erythroleukemia (M6), and (viii) megakaryocytic (M7). According to the cancer treatment centers of America (2017), the specific symptoms of AML include fever, cold, shiver, swollen lymph nodes (due to infection), lightness of skin color, etc.

2. **Chronic Myelogenous Leukemia (CML)**

Chronic myeloid leukemia (CML) occurs in the blood-forming cells (myeloid cells) found in the bone marrow. Due to CML (chronic granulocytic, chronic myelocytic, or chronic myelogenous leukemia), a hike is observed in the number of WBCs such as neutrophils or granulocytes that normally fight infection (Jabbour and Kantarjian 2018; Flis and Chojnacki 2019). In CML, a part of the chromosome 9 binds to the other sections of chromosome 22 forming the Philadelphia (Ph) chromosome. The Ph chromosome is made up of two genes called *BCR* and *ABL* which form the fusion gene called *BCR-ABL*. It is found only in the blood-forming cells, not in other organs of the body. The *BCR-ABL* gene causes myeloid cells to synthesize abnormally activated tyrosine kinase (enzyme: fusion protein) which allows WBCs to grow uncontrollably. The symptoms of CML include breathing trouble, general fatigue, fever, weakness, weight loss, night sweats, feeling of fullness, and enlarged spleen.

3. **Acute Lymphocytic Leukemia (ALL)**

Acute lymphoblastic leukemia (ALL) or acute lymphoid leukemia is a malignant proliferation of lymphoid cells and accounts for three-fourth of all cases of child leukemia (Miranda et al. 2016). This cancer develops in the bone marrow and the leukemia cells invade the other parts of the body, including the lymph nodes, spleen, liver, testicles, and CNS. People of any age can develop ALL, but it is most prevalent in young children and adults who are more than 50 years old. The symptoms of ALL include paleness of skin, fatigue, easy bruising, abdominal pain, weight loss, bleeding, fever, back or breathing problem, recurrent infections, swollen lymph nodes, enlarged liver or spleen, long-lasting menstruation (in women), dizziness, headache, blurred vision, nausea, and vomiting.

4. **Chronic Lymphocytic Leukemia (CLL)**

Chronic lymphocytic leukemia (CLL) is a cancer of the lymphocytes and is usually detected in the last stages. In patients with CLL, the abnormal cells compete and prevent the production of the healthy blood cells, including RBCs, other types of WBCs such as neutrophils or granulocytes, and platelets. The patient with CLL shall face anemia (due to the deficiency of RBCs), frequent infections (due to the deficiency of RBCs), and slow-healing wounds (due to the deficiency of platelets). There are two general types of CLL which are based on whether the disease affects B cells or T cells: (i) B-cell CLL and (ii) T-cell prolymphocytic leukemia. More than 95% of the patients with CLL have the B-cell-type leukemia. The T-cell type of CLL is also called T-cell prolymphocytic leukemia (T-PLL). About 1% of the patients with CLL have the T-cell-type leukemia. These two types of CLL have several subtypes that differ genetically. The symptoms of CLL include painless swelling of lymph nodes (in the neck, armpits, or groin). Due to the enlarged spleen, the patient complains of the discomfort or fullness in the upper left part of the abdomen. "B symptoms" that include fever, chills, night sweats, weight loss, frequent infections, abnormal bleeding, breathing problem, general fatigue, early satiety, and malaise are also reported in patients suffering from CLL.

5. **Hairy Cell Leukemia (HCL)**

Hairy cell leukemia (HCL) is a slow-growing form of leukemia. The abnormal lymphocytes exhibit hairlike projections when seed under a microscope. Therefore,

Table 2.8 Different stages of leukemia

Stage	Description
ALL	ALL stage is based on the cell type and maturity
AML	AML is staged according to the French American British (FAB) system, which considers the number of healthy blood cells, size and number, the changes in the chromosome of the leukemia cells, and other genetic abnormalities, in leukemia cells
CLL	It is staged using the rai system, which considers three main factors: The number of lymphocytes in the blood; the degree of enlargement of the lymph node, spleen, or liver; and the presence of anemia or thrombocytopenia
CML	It is staged on the basis of the number of diseased cells observed in blood and bone marrow tests

the disease is called "hairy cell" leukemia. These cells multiply and aggregate in the bone marrow, blood, and spleen. Since they are abnormal, they do not protect the body from disease and infection and outnumber the healthy cells. The hairy cells build up in the spleen and bone marrow and may not cause characteristic lymph node swelling. The symptoms like easy bruising, delayed wound healing, decreased appetite and feeling of fullness, unexplained weight loss and fatigue, persistent abdominal pain (due to the enlarged spleen), frequent infections, fever, chills, and night sweats occur.

Staging of Leukemia

The different stages of leukemia are enlisted in Table 2.8.

Treatment of Leukemia

The treatment of leukemia includes:

(i) *Chemotherapy*: chemotherapy is a routine treatment for leukemia which uses chemicals (as pills or intravenous) to kill leukemia cells. Usually, acute leukemia is treated by chemotherapy.

(ii) *Targeted therapy*: on one hand the biological therapy works by using treatments that help our immune system attack leukemia cells, while the targeted therapy uses drugs that targets specific components within the cancer cells. For example, the drug imatinib halts the action of a protein within the leukemia cells to control the CML disease.

(iii) *Radiation therapy*: radiation therapy uses X-rays or other high-energy beams to damage leukemia cells and stop their growth. Radiation therapy may be used before a stem cell transplant.

(iv) *Stem cell transplant*: a stem cell transplant is a procedure to replace the diseased bone marrow with healthy bone marrow. Before a stem cell transplant, high doses of chemotherapy or radiation therapy are used to treat the diseased bone marrow. The patient may receive stem cells from a donor, or in some cases the patient himself may be able to use his own stem cells.

2.2.3 Myeloma/Multiple Myeloma/Plasma Cell Myeloma

The blood cells in the human body are formed in the bone marrow. B lymphocytes or the plasma cells are produced here. These cells produce antibodies and form a crucial component of the body's immune system. Myeloma (type of plasma cell neoplasm), also known as multiple myeloma (MM), is a blood cancer that arises from plasma cells [plasma cell myeloma (PCM) or Kahler's disease] (Chng et al. 2007; Cancer Facts and Figures 2009). It is derived from a post-germinal center B cell that has undergone somatic hypermutation and productive immunoglobulin heavy-chain class switching (Lorsbach et al. 2011). Here, the plasma cells multiply rapidly, form tumors on the surface of the solid bones, and thus make the bones functionless (does not produce healthy blood cells). The plasma cells also make an abnormal protein (antibody) [monoclonal immunoglobulin/monoclonal protein (M-protein)/M-spike/ paraprotein] which affects the kidney function negatively. PCM accounts for 15% of the total blood cancers and 2% of all cancers. There are two types of myeloma:

(i) **Hyperdiploid myeloma** – In this the myeloma cells have more chromosomes than normal. This occurs in about 45% of myeloma patients and is less aggressive.
(ii) **Hypodiploid myeloma** – In this the myeloma cells have less chromosomes than normal. This occurs in about 40% of myeloma patients and is more aggressive. The World Health Organization (WHO) has classified the plasma cell neoplasms (Table 2.9) which encompass clonal plasma cell proliferations with a wide range of clinical manifestations and behavior (Chng et al. 2007).

Symptoms and Diagnosis of Myeloma

Since myeloma is manifested in the form of tumors on the bones, the most prevalent symptoms (reported in 70% cases) of myeloma are joint pain, bone fracture, and bone lesions. Increase in the bone resorption leads to abnormally high levels of calcium in the blood which culminates in renal failure. Myeloma also interferes with the ability of the bone marrow to produce healthy blood cells. So, anemia accompanied by low platelet count, paleness of skin, body fatigue, and shortness of breath occurs. Abnormal and easy bleeding and bruising are also reported. If the

Table 2.9 Classification of plasma cell neoplasms

Type of plasma cell neoplasm	Subtype
Monoclonal gammopathy of unknown significance	Asymptomatic myeloma, nonsecretory myeloma, plasma cell leukemia
Plasmacytoma	Solitary osseous plasmacytoma, extramedullary plasmacytoma
Immunoglobulin deposition diseases	Primary amyloidosis, systemic light and heavy-chain deposition diseases
Osteosclerotic myeloma	POEMS [polyneuropathy, organomegaly, endocrinopathy, monoclonal gammopathy, and skin changes]

cancer invades the nervous system, it culminates in neuropathies (body weakness, fatigue, pain, numbness and confusion, loss of bowel and bladder control, and paraplegia). A number of blood tests are performed to determine the complete blood count, albumin, calcium, and total protein. Blood and urine are also tested to check for antibodies and proteins (in case of renal failure). Bone marrow biopsy, bone density monitoring, and bone X-rays are performed to detect the disease.

Treatment of Myeloma

The routine treatment for aggressive myeloma consists of chemotherapy and radiation therapy. Chemotherapy includes the administration of several drugs, especially bisphosphonates which maintain the health of the bones. Steroids are sometimes administered. Radiotherapy is used to treat localized bone pain caused by concentrated tumors. Surgery may be performed to repair bone damage. Both autologous (patient's own stem cells used) and allogenic (stem cells from a donor is used) bone marrow transplant are used in certain cases. Although this treatment option has serious risks, it offers long-term possibilities of cure.

2.3 Cancers of Skin

Skin cancer is one of the most common types of cancer in the United States. There are three major types of skin cancers: (a) basal cell carcinoma (BCC), (b) squamous cell carcinoma (SCC), and (c) melanoma. The squamous and basal cell carcinoma (keratinocyte carcinoma or KC) are both considered non-melanoma skin cancer.

1. **Non-melanoma Skin Cancer**

As the name indicates, non-melanoma skin cancer does not metastasize to different locations of the body. It is of various types: (i) angiosarcoma; (ii) basal cell carcinoma (BCC), found in 75%–80% of all skin cancers; (iii) cutaneous B-cell lymphoma; (iv) cutaneous T-cell lymphoma; (v) dermatofibrosarcoma protuberans (rare type); (vi) Merkel cell carcinoma (rare type); (vii) sebaceous carcinoma; and (viii) squamous cell carcinoma (SCC), found in about 20% of all skin cancers. In a recent study (2012) on US citizens, 5.4 million cases of keratinocyte carcinoma (KC) were diagnosed among 3.3 million people (Cancer Facts and figures 2017). Melanoma alone represents about 1% of all the cases of skin cancers reported worldwide.

2. **Melanoma Skin Cancer**

Melanoma skin tumors are caused due to abnormal (cancerous) pigmented cells (melanocytes). As the cells continue to synthesize melanin, therefore the tumors are usually brown or black in color. If not diagnosed early, malignant melanoma may become dangerous. However, most of the melanomas are diagnosed early before spread so it is likely to be cured (Hartman and Lin 2019). The depth of a melanoma (into the dermis) is directly proportional to the risk of spreading through the lymph or blood vessels. Melanomas are of three types: (i) cutaneous melanoma, (ii) mucosal melanoma, and (iii) ocular melanoma.

Causes of Melanoma

Repeated exposure to sunlight [ultraviolet (UV) radiation] is one of the main causes of melanoma (Wang et al. 2001). Moreover, the whites are more susceptible to the damages caused by sun. The risk of melanoma is more among people who have a large number of dysplastic moles (nevi) or have a family history of both dysplastic nevi and melanoma. Ten percent of the cases of melanoma had a family member (family history) who had suffered from it. Those people (personal history) who in the past (adolescence, teen age, youth) had faced severe, blistering, or sunburn are at risk of developing melanoma. Moreover, people who develop more than two melanomas (BCC or SCC) and have been treated for melanoma or have a weak immune system are also at a higher risk (Abbasi 2008).

Symptoms of Melanoma

The first sign of melanoma is a change in the color, shape, size, and feel of an existing mole. Most of the melanomas appear black or blue-black. Melanoma also may appear as a new, black, "ugly-looking" mole (Bermick 2008). The shape of the melanoma may become asymmetric and the borders be ragged, notched, blurred, or irregular. The color may be uneven or of different shades: black, tan, brown, gray, white, pink, red, or blue.

Stages of Melanoma

According to the NCCN Guidelines for Patients (2018), there are five stages of melanoma as shown in Table 2.10.

Table 2.10 Different stages of melanoma

Stage	Explanation
Stage 0	It indicates that the melanoma is localized to one place and does not invade inside the dermis
Stages I	It indicates that the tumor is thinner than 1 mm, the cells are dividing slowly, and there is no ulceration viewed under the microscope. Stage 1B tumors are thinner than 1.0 mm and have ulceration
Stage II	This stage is divided into three groups – "A, B, and C" which are based on tumor thickness and ulceration status
Stage III	It indicates that the melanoma has reached the nearby lymph vessels, nodes, and skin. It includes tumors of any depth with metastases in lymph nodes and lymph vessels
Stage IV	It indicates that melanoma has reached to one or more distant locations. The tumor may be of any thickness and with any range of spread in lymph vessels and lymph nodes

Diagnosis of Melanoma

A per the Clinical practice guideline (2009), the diagnosis of melanomas involves several tests which includes:

(i) *Blood tests:* in the early stages of the disease, TA-90, a glycoprotein that can stimulate an immune response, is found in immune complexes in the blood. If the blood test detects the presence of TA-90, then it can be predicted if a deadly skin cancer, malignant melanoma, has metastasized or spread.
(ii) *Chest X-rays:* it helps to provide information on the spread of melanoma to the lungs.
(iii) *CT (computed tomography) scan*: in contrast to the regular and normal X-ray, CT scan provides details of the inner organs/softer tissues in the body. An image of the internal body is provided, so that organs like liver and lymph node may be tested.
(iv) *MRI (magnetic resonance imaging):* similar to X-rays and CT scan, MRI also provides detailed information of the body's inner soft organs. But MRI uses radio waves and powerful magnet, to generate the signals for further translation by a computer.
(v) *PET scan* (positron-emission tomography scan): this involves injection of radioactive sugar (glucose) and tracing its accumulation/consumption by the cancerous cells using a high-defined camera, although the information is not as detailed as that of CT and MRI. PET scan is most useful in cases with advanced stages of melanoma to observe the spread of cancer to various parts of the body.

Treatment of Melanoma

1. **Surgery**: Surgical removal of a melanoma is a normal treatment for this disease.
2. **Chemotherapy**: Drugs are used to treat and/or kill cells that are cancerous. Usually chemotherapy is given in cycles that consist of treatment period followed by a recovery period. Some drugs may also be used for the treatment of melanomas:
   (i) *5-flourouracil*: it is an antimetabolite that inhibits DNA synthesis. It is used at a concentration of 5%, as an ointment (Paek 2008).
   (ii) *Carmustine*: carmustine or BiCNU is used in chemotherapy. It is a nitrogen mustard β-chloro-nitrosourea compound that is used as an alkylating agent.
   (iii) *Dacarbazine*: dacarbazine is an antineoplastic chemotherapy drug with alkylating property. It is used in the treatment of various cancers (melanoma, Hodgkin's lymphoma, sarcoma, and islet cell carcinoma of the pancreas).
   (iv) *Cisplatin*: it is a platinum-based chemotherapy drug used for the treatment of various types of cancers including carcinomas, sarcomas, lymphomas, and germ cell tumors.

(v) *Other synthetic drugs*: these include Aldesleukin, Blanoxan, Blastolem, Blenoxane, Bleolem, Bleomycin, Carbazine, CeeNU, Cisplatin, Cosmegen, Dactinomycin, Docetaxel, Droxia, DTIC, DTIC-Dome, Hydrea, Hydroxyurea, Lomustine, Matulane, Mylocel, Natulan, Platinol-AQ, Procarbazine, Proleukin, Roferon-A, Taxotere, Tecnoplatin, Temodal, Temodar, and Temozolomide. Latest drugs used for treatments for melanoma are dacarbazine, cisplastin, vinblastine, carmustine, and tamoxifen.

Some medicinal plants/plant preparations used in the treatment of melanomas are (i) *Rubia cordifolia*, (ii) *Ocimum sanctum*, (iii) *Podophyllum*, (iv) *Semecarpus anacardium*, (v) Abha Guggulu, (vi) *Laksha Guggulu*, (vii) *Kaishore Guggulu*, (viii), *Gandhaka Rasayana*, (ix) Kanchanar Guggulu, (x) Triphala Guggulu, (xi) Tribang Bhasma, and (xii) Shilajatu Vati.

3. **Biological therapy**: It is also a systemic therapy which involves the use of substances that are known as biological response modifiers (BRMs). Although these substances are produced in small amounts in response to infection or disease but the patients with metastatic melanoma or those with high risk of relapse may be treated with interferon-alfa and interleukin-2, after surgery.
4. **Radiation therapy**: Radiation therapy is used to heal some of the symptoms caused by melanoma. It usually prevents melanoma from spreading to the bones, brain, as well as other locations in the body.

2.4 Cancers of the Digestive System

2.4.1 Esophageal Cancer

The esophagus is a tube-shaped organ of alimentary canal, located towards the back of the chest just in front of the spine. It is almost 25.4 cm long and helps in the movement of food and drinks from the throat to stomach. The wall of the esophagus has four main layers. The inner layer that has contact with food is called the mucosa. It is made of three sublayers: (i) the epithelium, (ii) submucosa, and (iii) muscularis mucosae (NCCN 2016). Esophageal cancer mostly develops in the squamous and glandular cells and epithelial cells of the esophageal wall. Cancers of squamous cells are known as squamous cell carcinomas, while those of glandular cells (mucus-secreting cells) are known as adenocarcinomas (Fig. 2.7).

Squamous Cell Carcinoma

It is usually common in people between 60 and 70 years of age and most common in men than women. The factors responsible for squamous cell carcinoma are continued smoking, consumption of alcohol, lye, or high starchy diet devoid of fruits and vegetables (Layke and Lopez 2006).

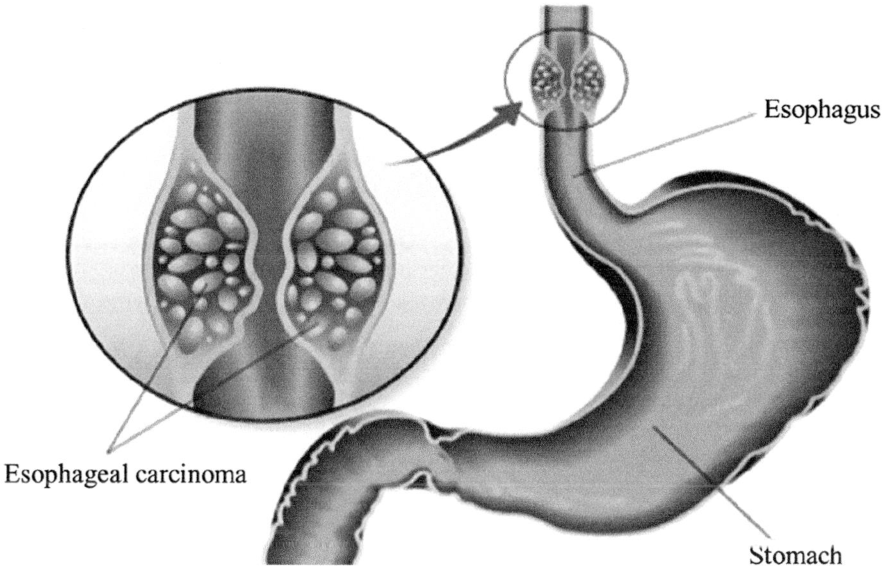

Fig. 2.7 Image showing esophageal cancer

Adenocarcinoma
It is reported in people between 50 and 60 years of age. It is well-documented that there is a connection between adenocarcinoma and Barrett's esophagus, a condition in which normal stratified squamous mucosa is replaced by metaplastic columnar epithelium and appears to arise in response to gastroesophageal reflux disease (GERD). In the United States, there is a rapid growth of problems related to obesity that exhibits a causal relationship with GERD, which increase the risk of developing esophageal adenocarcinoma. The other risk factors for esophageal adenocarcinoma include Zollinger-Ellison syndrome, myotomy-treated achalasia, and scleroderma. They all are associated with esophagitis (Layke and Lopez 2006).

Symptoms and Causes of Esophageal Cancer
The symptoms include troubled swallowing, chest pain, weight loss, severe cough and vomiting, hoarseness, hiccups, and bleeding within the throat. As with most cancers, the cause of esophageal cancer isn't yet known. It may be caused due to mutations in the DNA of the esophageal cells, which cause the cells to multiply rapidly and form tumors. It is interesting to mention that men are three times more susceptible to esophageal cancer than women. It is reported that any activity/agent that causes irritation to the esophageal cells can culminate in development of cancer. Some habits and conditions that can cause irritation include Barrett's esophagus (condition characterized by damaged esophageal lining due to GERD), reflux

disorder [such as gastroesophageal reflux disease (GERD)], achalasia (condition where the muscle at the bottom of the esophagus doesn't relax properly), alcohol consumption, smoking, obesity, and diet without fruits and vegetables.

Stages of Esophageal Cancer

The staging groups are labeled in Roman numerals (0–IV) as shown in Table 2.11. Stage 0 is known as high-grade dysplasia (HGD). The stages are classified using the TNM scores and cancer grade. Here, the esophagus is divided into three sections: (i) *upper*, (ii) *middle*, and (iii) *lower*. The American Joint Committee on Cancer (AJCC) has proposed a modified version of TNM staging of esophageal cancer as shown in Tables 2.11 and 2.12.

Diagnosis of Esophageal Cancer

The diagnosis of the esophagus involves clinical observation, radiological detection, histopathological correlation, and finally confirmation, after immune histochemical analysis. The family history, age, and medical condition of the patient are also considered. Most esophageal cancers are asymptomatic and are only discovered when they are nearly 4 cm in size. Symptomatic esophageal cancer involves gastrointestinal bleeding, feeling of fullness, pain in abdomen, palpable masses, and obstructive signs. It also involves metastasis in the liver (65% cases) and in the omentum (25% cases) (Abbas et al. 2016).

Treatment of Esophageal Cancer

Several treatments are available to treat esophageal cancer such as surgery, chemotherapy, and radiation therapy. These therapies can be used individually or in combination depending upon the age and health status of the patient.

1. **Radiation therapy**: Radiation therapy alone is used for inoperable squamous cell carcinoma in the past. However, it has little effect on relatively radio-insensitive adenocarcinomas. Several, short- and long-term side effects such as gastric irritation and occurrence of structures or perforations have been reported with radiation therapy (Wong and Malthaner 2006).
2. **Chemotherapy**: Chemotherapy alone has been successful in treating tumors in 50% of patients suffering from esophageal cancer (Wong and Malthaner 2006). Cisplatin (Platinol) along with fluorouracil is routinely used chemotherapy drugs. However, recent research suggests that multimodal therapy is more effective than single chemotherapy. Certain newer drugs [e.g., vinorelbine, taxanes] offer promising potential in chemotherapy (Fu et al. 2004).
3. **Surgical resection**: In most cases, surgical resection of esophagus (esophagectomy) may be performed if squamous cell/adenocarcinoma remains localized. With the advancements in restoring esophagogastric continuity, developments in operative techniques, and perioperative management, the mortality rate has reduced to less than 3% and a minimum survival for 5 years (Fu et al. 2004).

Table 2.11 Staging of squamous cell carcinoma and adenocarcinoma

Anatomic stage/prognostic groups										
Squamous cell carcinoma						Adenocarcinoma				
Stage	T	N	M	Grade	Tumor location	Stage	T	N	M	Grade
0	Tis (HGD)	N0	M0	1, X	Any	0	Tis (HGD)	N0	M0	1,X
IA	T1	N0	M0	1, X	Any	IA	T1	N0	M0	1–2, X
IB	T1	N0	M0	2–3	Any		T1	N0	M0	3
	T2–3	N0	M0	1, X	Lower, X	IB	T2	N0	M0	1–2,X
IIA	T2–3	N0	M0	1, X	Upper, middle	IIA	T2	N0	M0	3
	T2–3	N0	M0	2–3	Lower, X		T3	N0	M0	Any
IIB	T2–3	N0	M0	2–3	Upper, middle	IIB	T1–2	N1	M0	Any
	T1–2	N1	M0	Any	Any	IIIA	T1–2	N2	M0	Any
IIIA	T1–2	N2	M0	Any	Any		T3	N1	M0	Any
	T3	N1	M0	Any	Any		T4A	N0	M0	Any
	T4a	N0	M0	Any	Any	IIIB	T3	N2	M0	Any
IIIB	T3	N1	M0	Any	Any	IIIC	T4A	N1–2	M0	Any
IIIC	T4a	N1–2	M0	Any	Any		T4B	Any	M0	Any
	T4b	Any	M0	Any	Any		Any	N3	M0	Any
	Any	N3	M0	Any	Any	IV	Any	Any	M1	Any
IV	Any	Any	M1	Any	Any	Stage	T	N	M	Grade

The original source for this table is the AJCC cancer staging manual, seventh edition, 2010

Table 2.12 Staging strategy of squamous cell carcinoma and adenocarcinoma

Staging strategy of carcinoma		
T = Tumor	T1:	It indicates that tumor has invaded the lamina propria, muscularis mucosae, or submucosa
	T1a:	It indicates that tumor has invaded the muscularis propria
	T1b:	It indicates that tumor has invaded the submucosa
	T2:	It indicates that tumor has invaded the muscularis propria
	T3:	It indicates that tumor has invaded the adventitia
	T4:	It indicates that tumor has invaded nearby tissues
	T4a:	It indicates that tumor has invaded the outer lining of the lungs, heart, abdomen, or diaphragm
	T4b:	It indicates that tumor has invaded tissues such as the trachea that can't be treated
N=Node	NX:	It indicates that it is unknown if there is cancer in "lymph nodes"
	N0:	It indicates that that there is no cancer within the "lymph nodes" region
	N1:	It indicates that that the spread of the cancer is to one or two "lymph nodes"
	N2:	It indicates that that the spread of cancer is from three to six "lymph nodes"
	N3:	It indicates that that the spread of cancer is more than seven "lymph nodes" and beyond
M = Metastasis	MX:	It indicates that it is unknown if cancer has spread to distant sites
	M0:	It indicates that there is no growth to distant sites
	M1:	It indicates that the cancer has spread to distant sites

4. **ESD (endoscopic submucosal dissection)**: It is the latest and effective endoscopic treatment that removes a tumor in one go. ESD technique is usually performed under general anesthesia and involves 2–4 h (NCCN 2016). ESD technique involves ablation which destroys very small tumors without damaging nearby tissues. The most recommended types of ablation are as follows:
   (a) *Cryoablation*: it involves killing the cancer cells by freezing them. For this liq.N_2 is passed through an endoscope at the affected site.
   (b) *Radiofrequency ablation*: this technique uses heat from the electrodes, which is passed through an endoscope, to kill the cancer cells.
   (c) *Photodynamic ablation*: it kills cancer cells using a laser that activates a cancer-killing drug in the tumor, which is injected into a vein (NCCN 2016).

2.4.2 Stomach Cancer

The stomach is one of the four organs that make up the GI tract; it is a large, bean-shaped sac. The wall has four main layers: (i) mucosa, (ii) submucosa, (iii) muscularis propria, and (iv) serosa. The inner layer that remains in contact with food is known as mucosa. It is made of three sublayers: (a) the epithelium, (b) lamina propria, and (c) muscularis mucosae (NCCN 2016). Epithelium makes sticky, thick liquid called mucus that protects the stomach. The lamina propria consists of connective tissue, tiny lymph vessels, and glands. The third sublayer consists of a thin layer of muscle known as muscularis mucosa. The second layer that lines the wall of the stomach is called the submucosa. It consists of connective tissue, blood, nerve cells, and large lymph vessels. The third layer is called the muscularis propria which consists of muscle fibers, and these muscles help to move food through the stomach. The fourth layer is serosa or the visceral peritoneum and forms the outermost layer of the stomach wall. It consists of a thin layer of connective tissue, called the subserosa, which contains cells that produce lubricating fluid, to facilitate the stomach to move smoothly against other organs.

Stomach cancer (gastric cancer) begins when its normal cells grow out of control and culminate into a tumor. The cancer can begin in any of the inside surface of the stomach and may also affect the nearby lymph nodes including other organs like the liver, bones, etc. Mostly, the stomach cancers are a type of adenocarcinoma, but few other types like gastric sarcoma, lymphoma, and carcinoid tumors are also reported.

Symptoms and Causes of Stomach Cancer

The symptoms exhibited by the advanced cancer of the stomach include nausea and vomiting, frequent heartburn, loss of appetite, sudden weight loss, constant bloating, and early satiety while having food, bloody stools, jaundice, excessive fatigue, and stomach pain after meals.

Cause of Stomach Cancer

These are several general and specific causes of stomach cancer. It is common in individuals near the age of 50 and above. Individuals that have undergone stomach surgery, have polyps or growths on the lining of the stomach, have a family history of stomach cancer, and have been suffering from *Helicobacter pylori* bacterial infection are more susceptible to stomach cancer. Daily routine habits such as smoking; low intake of fruits, vegetables, and fiber; and eating smoked, salted, and pickled foods increase the risk of stomach cancer.

Stages of Stomach Cancer (NCCN 2016)

Like esophageal cancer TNM staging is applicable in case of stomach cancer also (Table 2.13).

Diagnosis and Treatment of Stomach Cancer

Stomach cancer can be diagnosed and confirmed through MRI scan, ultrasound, X-ray, and CT scan. When the cancer is localized to the stomach (stages 0 and I), surgery is done to remove the affected. If it reaches to the outer wall of the stomach,

Table 2.13 Stages of stomach cancer (NCCN 2016)

Anatomic stage			
Stage 0	Tis	N0	M0
Stage IA	T1	N0	M0
Stage IB	T2	N0	M0
	T1	N1	M0
Stage IIA	T3	N0	M0
	T2	N1	M0
	T1	N2	M0
Stage IIB	T4a	N0	M0
	T3	N1	M0
	T2	N2	M0
	T1	N3	M0
Stage IIIA	T4a	N1	M0
	T3	N2	M0
	T2	N3	M0
Stage IIIB	T4b	N0	M0
	T4b	N1	M0
	T4a	N2	M0
	T3	N3	M0
Stage IIIC	T4b	N2	M0
	T4b	N3	M0
	T4a	N3	M0
Stage IV	Any T	Any N	M1

chemotherapy plus surgery or radiation therapy may be deployed. Usually, advanced-stage stomach cancer is treated with chemotherapy. The common treatments are as follows:

(i) *Surgery*: the three main surgical procedures that are used to treat stomach cancer are (a) endoscopic mucosal resection, (b) partial gastrectomy, and (c) total gastrectomy. Endoscopic mucosal resection is applicable in cases of early-stage cancer and involves the use of a camera inserted through the mouth. In partial or subtotal gastrectomy, a portion of the stomach is removed along with neighboring tissues and nearby lymph nodes. The remaining sections are then joined together. Total gastrectomy involves complete removal of the entire stomach along with neighboring tissues and nearby lymph nodes, followed by connecting the small intestine to the esophagus.

(ii) *Radiation therapy*: another treatment for stomach cancer is radiation therapy, which are of two types: (a) external beam radiation therapy and (b) internal radiation therapy. In external beam radiation therapy, a high-energy X-ray machine is used to direct radiation at the tumor. Internal radiation therapy destroys cancer cells with small implants that are placed directly into the tumor. The secondary symptoms such as bleeding, pain, and difficulty in eating can be reduced through radiation therapy.

(iii) *Chemotherapy*: chemotherapy drugs may slow down tumor growth and reduce pain.
(iv) *Targeted therapy*: targeted therapy, as the name indicates, stops the action of target molecules [e.g., vascular endothelial growth factor (VEGF) or human epidermal growth factor receptor 2 (HER2)]. It is safer than chemotherapy.

2.4.3 Pancreatic Cancer

The pancreas is situated at the back of the stomach. The head of the pancreas is located at the right side of the abdomen, while the tail is situated towards the left side, immediately after the spleen. It is 15.24 cm in length and below 5.08 cm in breadth and is connected to the duodenum through a small tube called the pancreatic duct. There are two main types of pancreatic cells: (i) exocrine (makes pancreatic enzymes for digestion of food) and (ii) endocrine (produces hormones glucagon and insulin that regulate sugar level). Pancreatic cells (exocrine and endocrine) produce tumors of different types which can be differentiated on the basis of risk factor and symptoms. They are:

(a) *Exocrine pancreatic cancers*: They are the most common type of pancreatic cancer (Fig. 2.8). Of all the types of exocrine cancers in the world, 95% are adenocarcinomas. These cancers usually develop/start in the ducts of the pancreas and not from the cells involved in enzyme production. The less common types of exocrine cancer are squamous cell carcinomas, adenosquamous carcinomas, signet ring cell carcinomas, undifferentiated carcinomas, and undifferentiated carcinomas with osteoclastic-like giant cells.

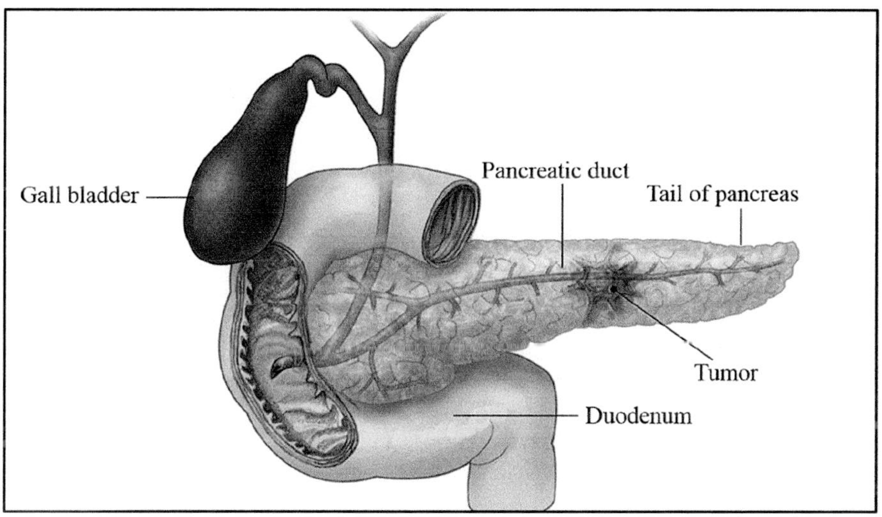

Fig. 2.8 Diagram showing a tumor in the distal part of pancreas

(b) *Ampullary cancer (carcinoma of the ampulla of Vater)*: It relates to the cancer originating in the ampulla of Vater (the region where the pancreatic duct and bile duct meets). The ampullary cancers at initial stage block the bile duct which leads to the accumulation of the bile within the body that results in symptoms such as yellowing of eyes and skin. Therefore, this disease has an excellent prognosis.

(c) *Pancreatic endocrine tumors (neuroendocrine tumors)*: These tumors account for only 5% of the total pancreatic cancers. These are commonly referred to as pancreatic neuroendocrine tumors (NETs) or islet cell tumors. It is difficult to differentiate between the non-cancerous (benign) and cancerous (malignant) tumor because both have similar appearances. Generally, there are many types of pancreatic NETs, and their names are based on the hormone produced by the specific pancreatic cells (Table 2.14). Generally, most of these NETs are either "gastrinomas" or "insulinomas"; the other forms/types are very uncommon.

Symptoms and Causes of Pancreatic Cancer (NCCN 2017)

The symptoms include weight loss accompanied by jaundice, yellowish tinge to the skin and white parts of the eye (due to the increased level of the bilirubin), nausea, floating stools, pain in the belly or back, indigestion, depression or sadness, and lack of interest in any activities. It's not yet scientifically proved what really cause several cancers of pancreas; however, various risk factors have been mentioned, some of which significantly affect DNA of pancreatic cells that leads to abnormal growth of cells culminating in tumors.

Table 2.14 Types of pancreatic cancers

Cell function	Cancer type	Nature
Cells that make gastrin	Gastrinomas	Malignant
Cells that make insulin	Insulinomas	Benign
Cells that make glucagon	Glucagonomas	Benign
Cells that make somatostatin	Somatostatinomas	Malignant
Cells that make vasoactive intestinal peptide (VIP)	Vipomas	Malignant
Cells that make up the pancreatic polypeptide	PPomas	Malignant
Cells of the pancreas	Serous cystic neoplasms (SCNs)/ cystadenomas	Benign
Cells of the pancreas	Mucinous cystic neoplasms (MCNs)/ mucinous cystadenomas	Benign/malignant
Cells of the pancreatic duct	Intraductal papillary mucinous neoplasms (IPMNs)	Benign/malignant
Cells of the pancreas	Solid pseudopapillary neoplasms (SPNs)	Benign/malignant

Staging of Pancreatic Cancer

The staging of pancreatic cancer has been reported to be of two types: (i) the AJCC system and (ii) based on information retrieved from imaging tests. The AJCC system has grouped the pancreatic cancer into five stages (stage 0–stage IV). These stages are based on the growth of the primary tumor and its travel to the various sites within the body. In the AJCC system, cancer may be staged twice: first based on tests before surgery and then based on tests of tissue removed during surgery. Imaging tests provide the key information used to determine the clinical stages of pancreatic cancer. Some doctors plan the treatment by classifying the stages based on the imaging results and other tests done before surgery. In this kind of system, doctors classify pancreatic cancer into four main groups:

(i) *Resectable*: cancer has not spread outside the pancreas and appears to be easily treated through surgery.
(ii) *Borderline resectable*: cancer is localized in the pancreas but approaches to nearby structures.
(iii) *Locally advanced unresectable*: cancer is spread outside the pancreas close to the blood vessel or related sites (tissues).
(iv) *Metastatic*: cancer has spread outside the pancreas to organs and tissues far away in the body.

Diagnosis of Pancreatic Cancer

The techniques like multi-detector-row-computed tomography (MDCI), magnetic resonance imaging (MRI), and endoscopic ultrasound (EUS) have brought advancement in the diagnosis of pancreatic carcinoma (Miura et al. 2006; NCCN 2016). However, the common techniques deployed for the diagnosis of pancreatic cancer include:

(i) *Ultrasonography (US):* it is a noninvasive and cost-effective method of locating chronic pancreatitis, pancreatic adenocarcinoma, and endocrine cell tumors. These pancreatic tumors are revealed as hypoechoic areas and accuracy for diagnosis by US in 50–70%. A newer version of the technique is the contrast-enhanced Doppler US which is more advantageous for the diagnosis of pancreatic tumors. This technique reveals the pancreatic adenocarcinoma as hypovascularized while the endocrine cell tumors as hypervascularized and pancreatitis-associated mass as isovascularized regions. Another highly sensitive (90–95%) and specific technique to diagnose pancreatic carcinoma is contrast-enhanced coded phase-inversion harmonic US. The ultrasonography is involved in the initial screening examination which may be followed by CT and MRI for further accurate diagnosis.
(ii) *Endoscopic US (EUS):* EUS detects small tumors and can also locate lymph node metastases and vascular tumor infiltration with high sensitivity. EUS-guided fine needle aspiration biopsy is a safe and highly accurate method of studying the tissue for diagnosis of pancreatic carcinoma. Thus, EUS is superior to spiral CT, MRI, and PET. However, this technique is not free from

limitations. The technique fails to detect (limited visualization) the metastatic spread to the liver and peritoneum. The technique is also expert dependent.

(iii) *Computed tomography (CT)*: CT is also used as the imaging technique for detection and staging of pancreatic carcinoma. The sensitivity range of helical CT in revealing pancreatic carcinoma is 89–97%. Moreover, multi-detector-row (MD) CT may offer an improvement in the early detection and accurate staging of pancreatic carcinoma. It facilitates the acquisition of superb 3-D images.

(iv) *Endoscopic retrograde cholangiopancreatography (ERCP):* diagnostic ERCP is valuable if a tumor is suspected despite negative results on US and CT or may be used as an additional aid to differentiate between cancer and chronic pancreatitis.

(v) *Magnetic resonance imaging (MRI):* MRI is more reliable for diagnosis of pancreatic carcinoma than CT because it provides soft tissue contrast before the administration of contrast material. It also makes possible the examination of pancreatobiliary system, noninvasively.

(vi) *Positron-emission tomography (PET):* the cells of the malignant tumors exhibit increased glucose utilization due to an increased number of glucose transporter and increased hexokinase and phosphofructokinase activity. PET uses the radio-labeled glucose analog ^{18}F-fluorodeoxyglucose (FDG) to differentiate between benign and malignant lesions. The FDG-PET offers 71–100% sensitivity and 64–90% specificity for detecting malignant pancreatic tumors. However, there are certain limitations of FDG-PET imaging. The chronic and acute pancreatitis can also accumulate FDG and provides false-positive interpretations of PET images.

Treatment of Pancreatic Cancer

Many studies have been conducted to find out the most suitable treatment of pancreatic cancer. Radiation therapy and surgery are the most reliable. However, under certain circumstances, combination of two or more treatments works well.

1. **Surgery:** The surgical procedure of the treatment of pancreatic cancer is based on factors such as the tumor location, tumor size, and tumor staging. For cancers of the head and/or neck of the pancreas, the classic Whipple procedure (pancreatoduodenectomy) is recommended. When cancer invades body and tail, distal/subtotal pancreatectomy is followed, and when the tumor spreads to surrounding tissues, extended resection is advisable.
2. **Radiation therapy:** Radiotherapy is one of the options for treating unresectable pancreatic cancers. Although it can kill cancer cells and keep them from growth and recurrence, it is not free from side effects. These side effects include fatigue, gastrointestinal troubles, skin rashes, and damage to normal tissues. New forms of radiotherapy are known as intensity-modulated radiotherapy (IMRT), stereotactic body radiotherapy (SBRT), and image-guided radiotherapy (IGRT) and are more effective and tolerable (Son et al. 2012). SBRT can deliver a high dose of radiation accurately without harming the surrounding normal tissues.

3. **Chemotherapy:** The use of gemcitabine in combination with other drugs can prolong the life of patient. Certain newer chemo drugs are also under trial. For unresectable pancreatic cancer, drugs such as GEM/erlotinib, GEM/NAB-paclitaxel, FOLFIRINOX, capecitabine/oxaliplatin, and (XELOX)GEM/capecitabine are recommended.
4. **Targeted therapies:** As the name indicates, this therapy includes the use of drugs specifically designed to target specific cancer cells or their neighboring cells:
   (a) *Drugs [such as erlotinib (tarceva)] that specifically target epidermal growth factor (EGFR)*: This therapy can be more efficacious in combination with other recommended treatments. As with the other techniques, these therapies too have ill-effects. The cancers depend on blood vessels for nourishment and growth. Blockage of the blood vessel obstructs the tumors growth. Blockage of the blood vessel obstructs the tumors growth. Clinical trials of anti-angiogenesis drugs for treating pancreatic cancer are in progress.
   (b) *Drugs that target the tumor stroma (supporting tissue)*: Chemo drugs such as PEGPH20 (still in trial) usually break down stroma and make the drug accessible/available to the tumor.
   (c) *Drugs that target cancer stem cells*: Cancer stem cells are a group of cancerous cells which make the chemotherapy less or noneffective by making the tumor more resistant. Some drugs such as BBI-608 and demcizumab are meant to attack the stem cells. However, these drugs are still under trial in conjunction with chemotherapy.
5. **Immune therapy:** The aim of the immune therapy is to provide the body with components that can help fight cancerous cells. An example of immune therapy is the use of monoclonal antibodies. They are developed in response to specific molecule (tumor marker) such as carcinoembryonic antigen (CEA) located on the cancerous pancreatic cells. These treatments are still under trial for use in cancer of the pancreas.
6. **Cancer vaccines**: Various vaccines that increase the response of the immune system to pancreatic cancer are currently under trial. Although these types of treatment carry little chances of side-effect, it still needs stringent trials. Generally, cancer cells manage to use immune system checkpoints and escape the immune system. Advanced drugs that can attack checkpoints shall provide treatment of other types of cancers.
7. **Individualization of therapy:** Many types of drugs tend to be more effective, when there is a mutation in the tumor cells. An example to this is "erlotinib drug" that works effectively when there is alteration in the EGFR gene.

2.4.4 Liver Cancer

The liver is a vital organ of the body which filters and removes waste and toxins from the blood; makes bile which breaks down fats in the digestive process;

synthesizes plasma proteins, such as albumin, clotting factors, bilirubin, cholesterol, hormones, and drugs needed by the body; and stores glycogen, vitamins, and minerals. The occurrence of cancer cells in the liver may be due to some liver damage. For example, an infection from hepatitis B or C virus, liver cirrhosis, fatty liver, family history of liver cancer, or other risk factors such as the advancing age, alcohol intake, and consumption of fatty foods [culminates in nonalcoholic fatty liver disease (NAFLD)].

There are several types of liver cancer. (i) *Primary liver cancer*: when cancer begins in the functional cells of the liver called the hepatocytes, it is called primary liver cancer. Hepatocellular carcinoma (HCC) accounts for 90% of all reported cases of liver cancers. It is a global problem and is the most common cancer diagnosed in men worldwide. In 2013, approximately 30,600 Americans were diagnosed with primary liver cancer, along with the increasing numbers of hepatitis C cases. (ii) *Hepatoblastoma*: young children (below the age of three) are usually affected by this rare type of primary liver cancer. (iii) *Fibrolamellar HCC*: it is a rare type of HCC, but compared to other types of liver cancer, it is more responsive to treatment. It is reported in people between the age group of 20–30 and above and is not usually associated with cirrhosis or infection of hepatitis B or C. The patients suffering from fibrolamellar carcinomas may not exhibit higher levels of alpha fetoprotein (AFP) in their blood. (iv) *Cholangiocarcinoma (bile duct cancer)*: it occurs in the bile ducts that carry bile from the liver to the gallbladder. This type of cancer is rare and accounts for only 10–20% of all liver cancers. As the name indicates, intrahepatic bile duct cancer begins in ducts within the liver, while the extrahepatic bile duct cancer develops in ducts outside the liver. (v) *Angiosarcoma (hemangiosarcoma)*: it is an extremely rare cancer and accounts for about 1% of all liver cancers. Angiosarcoma develops and grows rapidly in the blood vessels of the liver and is therefore diagnosed at an advanced stage. (vi) *Secondary liver cancer*: this cancer begins in another part of the body (except the liver) and then metastasizes to the liver.

Symptoms of Liver Cancer

A hard lump or swelling may be found in the right side of the belly, just below the ribs. Pain in the belly or the right shoulder blade which may be accompanied by (i) *jaundice*, (ii) *nausea*, (iii) *loss of appetite*, (iv) *fatigue*, (v) *bleeding*, and (vi) *enlarged tender liver* (Fig. 2.9).

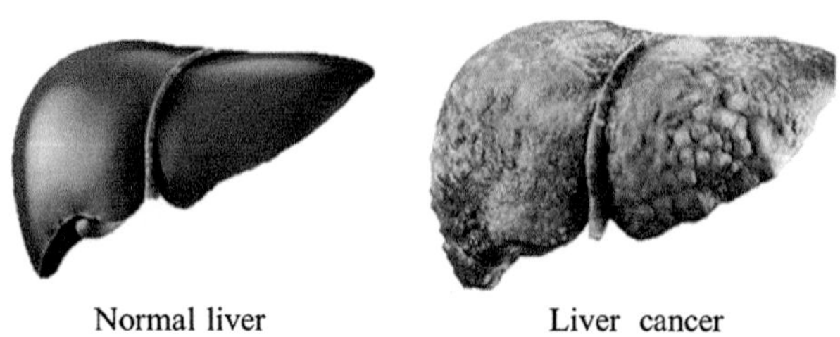

Fig. 2.9 Liver morphology in normal and cancerous condition

Staging of Liver Cancer

Several liver cancer staging systems have been developed that consider a variety of factors including tumor spread, tumor invasion into blood vessels, liver function, and the overall health of the individual. The staging information guides the multidisciplinary team to take a decision whether the patient is eligible for surgical resection, liver transplantation, systemic or targeted therapy, or intrahepatic or regional therapy. The American Joint Committee on Cancer (AJCC)-TNM staging system of liver cancer is mentioned in Table 2.15.

Table 2.15 Staging of liver cancer

AJCC stage	TNM category	Description
IA	T1a, N0, M0	T1a: It indicates the presence of a tumor of 2 cm or smaller size and has not grown into blood vessels
		N0: It indicates that the tumor has not spread to nearby lymph nodes
		M0: It indicates that the tumor has not spread to other locations
IB	T1b, N0, M0	T1b: It indicates the presence of a tumor of more than 2 cm size that has not grown into blood vessels
		N0: It indicates that the tumor has not spread to nearby lymph nodes
		M0: It indicates that the tumor has not spread to other locations
II	T2, N0, M0	T2: It indicates the presence of one or more tumors of 2–5 cm size and has grown into the blood vessels
		N0: It indicates that the tumor has not spread to nearby lymph nodes
		M0: It indicates that the tumor has not spread to other locations
IIIA	T3, N0, M0	T3: It indicates the presence of one or more tumors of 2–5 cm size and at least one having size larger than 5 cm
		N0: It indicates that the tumor has not spread to nearby lymph nodes
		M0: It indicates that the tumor has not spread to other locations
IIIB	T4, N0, M0	T4: It indicates the presence of one tumor of any size that has grown into the portal or hepatic vein
		N0: It indicates that the tumor has not spread to nearby lymph nodes
		M0: It indicates that the tumor has not spread to other locations
IVA	Any T, N1, M0	Any T: It indicates the presence of one tumor or multiple tumors of any size
		N1: It indicates that the tumor has spread to nearby lymph nodes
		M0: It indicates that the tumor has not spread to other locations
IVB	Any T, any N, M1	Any T: It indicates the presence of one tumor or multiple tumors of any size
		Any N: It indicates that the tumor has or has not spread to nearby lymph nodes
		M1: It indicates that the tumor has spread to bones or lungs

Diagnosis of Liver Cancer

The diagnosis of liver cancer involves certain imaging techniques and tests like (i) serum alpha fetoprotein (AFP) test, (ii) laparoscopy, (iii) CT scan, (iv) ultrasound, (v) magnetic resonance imaging (MRI), and (vi) chest X-ray and bone scan tests (refer to Section No.2.8.2).

Treatment of Liver Cancer

1. **Surgery:** Live surgery is performed if the cancer has not spread to the adjoining lymph nodes or other locations of the body. It involves (i) *partial hepatectomy* which is the removal of one or more of the eight liver segments or an entire lobe and (ii) *total hepatectomy and liver transplant* in which the entire liver is replaced by a healthy liver. Transplantation is done when the cancer mass is smaller than 5 cm and localized in the liver.
2. **Regional therapies:** For unresectable liver cancer, other therapies are performed. These treatments are usually performed by a doctor called an interventional radiologist. They are used for patients with small tumors that are not resectable or sometimes when a patient is waiting for a transplant. It includes:
   - (i) *Ablation*: it involves destruction of liver cancer cells. Radiofrequency ablation involves introducing a probe into the tumor to destroy the cancer cells with the heat. Cancer cells may also be subjected to freezing or ethanol.
   - (ii) *Embolization*: it involves blocking the tumor's blood supply.
   - (iii) *Chemoembolization*: the tumor is treated with a high dose of chemotherapy followed by cutting off its blood supply (transcatheter arterial chemoembolization).
   - (iv) *Radioembolization* (selective internal radiation therapy): radioactive beads that are deposited close to the tumor site to destroy the cancer cells.
3. **Systemic therapies:** It aims at treating the unresectable or advanced-stage liver cancers. Sorafenib or Nexavar® is an oral drug which specifically targets and blocks vascular endothelial growth factor (VEGF), a protein required by the tumor to grow its own blood vessels. Thus, this drug is a promising candidate for the treatment of liver cancer.
4. **Radiation:** It involves external beam radiation therapy and internal beam radiation therapy (brachytherapy). The former involves treating the liver with high-powered energy beams, such as X-rays or protons, for 5 days a week for several weeks, while the latter, brachytherapy, involves implanting many rice-sized radioactive seeds in the tumorous tissue using a needle guided by ultrasound images. The radioactive seeds slowly release a low dose of radiation.

2.4.5 Colon and Rectal Cancer

The colon is also called as large bowel or large intestine. Colorectal cancer forms due to anomalous growth (known as polyps) of the epithelial cell lining of the colon or rectum (ACS 2017). These polyps are often benign but may have the tendency to become cancerous. About two thirds of the colorectal polyps have the potential to develop colorectal cancer. Screening and awareness can reduce colorectal cancer-mediated mortality. This is achieved by early diagnosis and removal of polyps before they become cancerous.

Symptoms of Colon and Rectal Cancer
The symptoms include:

(i) Rectal bleeding: rectal bleeding is the bleeding through the anus, as well as the stool becomes maroon colored or black. It may be due to anal fissure, hemorrhoids, diverticulosis, colon cancer, colitis, etc.
(ii) A mass in the abdomen or rectum: it is the abnormal growth in the abdomen, which causes visible swelling changing the shape of the abdomen. It might be accompanied by weight gain and abdominal discomfort such as pain and bloating.
(iii) Change in bowel habit: perianal symptoms, such as lesions or abscesses, bloating constipation, and vomiting.

Causes of Colon and Rectal Cancer
A person can develop colorectal cancer at any age, but the chance of developing it increases after the age of 50. Lack of exercise, sedentary lifestyle, excessive alcohol consumption, smoking, and obesity are associated with a higher risk of colorectal cancer. Moreover, the probability of acquiring colorectal cancer doubles if a direct relative or more than one relative had suffered from the disease. If a person has a history of polyps or inflammatory bowel disease or inherited disorders such as familial adenomatous polyposis (FAP), the risk of developing colorectal cancer is much higher.

Staging of Colon and Rectal Cancer (NCCN 2016)
The different stages of colon cancer are shown in Table 2.16 and Fig. 2.10.

Table 2.16 Stages of colon cancer

Stage	Explanation
Stage 0	It is a noninvasive stage wherein the cancer has not grown beyond the first layer of the colon wall
Stage I	It indicates that the cancer has grown into either the second or third layer of the colon wall. There is no cancer in nearby or distant sites
Stage II	It indicates that the cancer has grown into the fourth layer of the colon wall. There is no cancer in nearby or distant sites
Stage III	It indicates that the cancer has reached the neighboring lymph nodes or presence of tumor deposits (small secondary tumors within the colon)
Stage IV	It indicates that the cancer has spread to the other organs like the liver and lungs

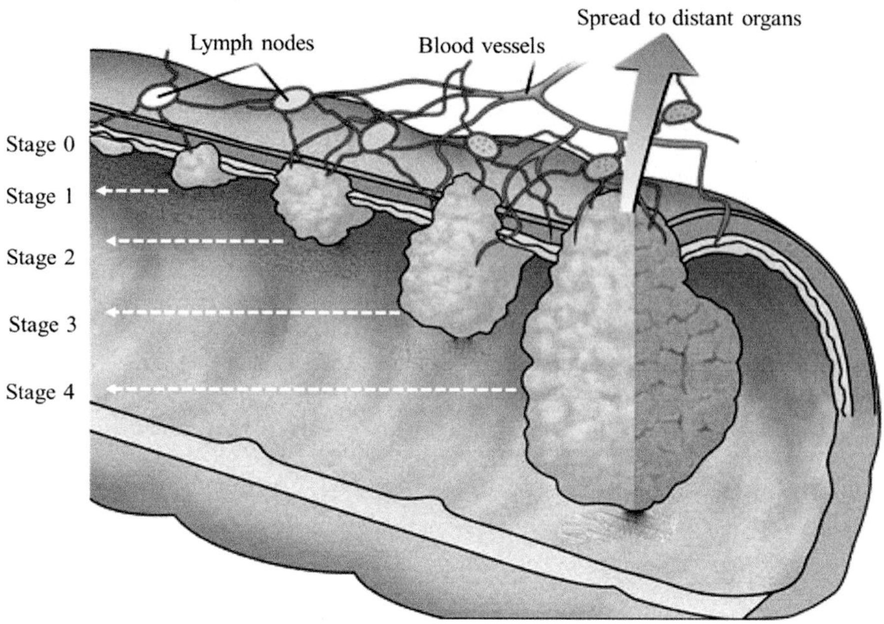

Fig. 2.10 Pictorial representation of the stages of colon cancer

Diagnosis of Colon and Rectal Cancer

There are several ways to diagnose a colon or rectal cancer. These include:

(i) *Colonoscopy*: a colonoscope is used to view entire length of the colon.
(ii) *Sigmoidoscopy*: a sigmoidoscope is deployed to view the lower colon.
(iii) *Double-contrast barium enema*: X-rays of the colon and rectum are performed after a barium meal. Barium lines the colon and allows an outline to be viewed in an X-ray.
(iv) *Biopsy*: during a colonoscopy or sigmoidoscopy, biopsy sample is collected and examined to determine its stage.

Treatment of Colon and Rectal Cancer

1. **Surgery**: The early-stage colorectal tumor is removed by the procedure called as "resection." The area near the cancerous tissue and the lymph nodes is also removed to decrease the danger of the cancer spread. In laparoscopic surgery, the affected area of the colon is removed through keyhole surgery. Patients with advanced-stage disease may also undergo resection particularly when the cancer has metastasized to the liver.
2. **Chemotherapy**: Patients diagnosed with advanced disease usually undergo surgery followed by chemotherapy. This involves a combination of a fluoropyrimi-

dine or fluorouracil with leucovorin and oxaliplatin (known as FOLFOX19) or with irinotecan (known as FOLFIRI). Some patients who cannot undergo surgery due to invasive tumors can undergo neoadjuvant treatment which involves chemotherapy regime before a surgery.

3. **Use of potential protective agents**: Use of nonsteroidal anti-inflammatory drugs (NSAIDs) like aspirin, hormone replacement therapy (in women), and intake of fiber-rich diet are possible protective factors which help in reducing the risk of colorectal cancer.
4. **Biological therapies**: In this therapy metastatic colorectal cancer can be treated with anti-angiogenics which block cancer blood vessel growth. These therapies require chemotherapy also.

2.4.6 Anal Cancer

The anus is the opening at the posterior end of the alimentary canal which allows solid waste to move out of the body. Various cancers are reported in different parts of the anus (Salatı and Al Kadi 2012). Types of anal cancer include:

(i) **Squamous cell carcinomas:** this is the commonest form of anal cancer, wherein the cancer starts in the cells lining the anal margin and the anal canal in about 75% of cases. The initial state of squamous cell carcinoma is called as carcinoma in situ or Bowen's disease.

(ii) **Adenocarcinomas:** almost 15% of anal cancers are adenocarcinoma. This affects glands in the anal area and one type of adenocarcinoma that can occur in the anal area, known as Paget's disease. It can also affect the vulva, breasts, and other areas of the body. Anal adenocarcinomas are usually treated similarly as rectal cancers. Other very rare types of anal cancer are lymphomas and gastrointestinal stromal tumors (GIST).

Symptoms of Anal Cancer

Presence of small lumps seen or felt around the anus, increased number or size of piles, pain in the anal area, difficulty in passing stools, and extreme constipation are common symptoms. Feeling a continuous urge to pass a motion or fecal incontinence, discharge from the back passage, swelling, itching and persistent redness or soreness around the anal area, and one or more lumps in the groin area are related symptoms of anal cancer.

Causes of Anal Cancer

There are several causes of anal cancer:

(i) *Age*: people above the age of 50 are usually reported with anal cancer.

(ii) *African-American race*: compared to others African-Americans have a greater risk of developing colon cancer.

Table 2.17 Different stages of anal cancer

Stage	Explanation
Stage 0	It's referred to as Bowen's disease or anal carcinoma in situ (AIN)
Stage 1	It indicates that the cancer has only affected the anus, the affected area is smaller than 2 cm in size, and the cancer has not begun to spread into the sphincter muscle
Stage 2	It indicates that the size of the cancer is beyond 2 cm but has not reached to the lymph nodes or other locations within the body
Stage 3	It indicates that lymph node and/or its neighboring parts or organs like bladder or vagina are also affected by cancer
Stage 4	It indicates that the cancer has affected the distant organs of the body, like the liver

(iii) *A person's polyp or colorectal cancer*: a person who has already suffered from colon cancer or adenomatous polyps is likely to develop colon cancer in the near future.

(iv) *Inflammatory intestinal conditions*: patients with ulcerative colitis or Crohn's disease can develop colon cancer.

(v) *Inherited syndromes that increase colon cancer risk*: genetic syndromes [(familial adenomatous polyposis and hereditary nonpolyposis (Lynch syndrome)] passed through generations can increase chance of developing colon cancer.

The other causes of anal cancer are diabetes, intake of low fiber, intake of high-fat diet, obesity, sedentary lifestyle, smoking, and alcohol consumption.

Staging of Anal Cancer

The different stages of anal cancer are shown in Table 2.17.

Diagnosis of Anal Cancer

Several techniques that are deployed to diagnose anal cancer are:

(a) *Proctoscopy*: It involves the insertion of a short, illuminated tube together with a lens or a very small camera that can video inner side of the rectum. It can also be used to collect tissue samples from inside the anal canal. Drugs may be used to make the patient sleepy during these tests.

(b) *Anoscopy*: For anoscopy, the proctologist uses a short hollow tube, called an anoscope which may have a light at its end. The doctor lubricates the anoscope with a lubricant and then gently pushes it into the anus and rectum. It enables the doctor to get a clear view of the inner lining of the lower rectum and anus. This exam usually does not involve much discomfort.

(c) *Rigid proctosigmoidoscopy*: The rigid proctosigmoidoscopy uses a longer tube (10 inch longer) which facilitates the proctologist to see the whole of rectum together with the below location of the sigmoid colon.

(d) *Colonoscopy*: It involves the examination of the large intestine and the distal part of the small intestine with a fiber optic.

(e) *Ultrasound scan*: The sound waves are used to generate the inner side image of the organ or the masses. The wand-like transducer is put right into the rectum. This is called a transrectal or endorectal ultrasound. This test can be used to identify the depth and how serious the cancer has developed within the tissues around the anus. The test can be uncomfortable but free from pain or injury.

(f) *Biopsy*: Biopsy of any abnormal lumps or tissue in the anal canal (during endoscopic exam) is done for the confirmation of cancer. If the tumor is very small, the doctor might eliminate the whole of the tumor by the biopsy. Also, biopsies are done to check the spread of cancer to the nearby lymph node. A fine needle aspiration (FNA) is often used to collect small fluid sample from lymph nodes, to be sent for examination. If cancer is detected then the lymph node may be surgically removed;

(g) *Imaging tests*: As the name indicates, the aim of these tests is to create the image of the internal body parts by the use of X-rays, magnetic fields, sound waves, or radioactive substances. Imaging tests might be done to detect the cancer, to check the spread of cancer, to evaluate the success of the treatment, and to check the relapse of the cancer. The imaging tests include the following:

   (i) *Computed tomography (CT) scan*: CT scan is a common test for the people with anal cancer, and it tells if cancer has reached into the lymph nodes or some other regions like the liver, lungs, etc. CT scanner rotates and captures many images of the affected area, which are then combined and visualized on a computer screen. CT-guided needle biopsy may also be conducted wherein a biopsy sample is then taken out from the suspected tissue with the help of CT scanner and sent to a lab for further identification.

   (ii) *Magnetic resonance imaging (MRI)*: This form of scan involves the use of radio waves and strong magnets followed by pattern translation that are obtained into detailed images for studying the tissues in normal and diseased condition. This test may identify if the neighboring lymph node(s) is enlarged, which indicates cancer spread.

   (iii) *Positron-emission tomography (PET) scan*: The cancer cells utilize high quantity of sugar for their activity. PET scan makes use of this property of cancer cells. In PET scan, radioactive sugar fluorodeoxyglucose (FDG) is taken into blood, and a high quantity of radioactive sugar is absorbed by the cancerous cells. A camera creates pictures of the regions of higher radioactivity. Although the resulting images are not detailed as that of a CT or MRI scan, significant information is generated. Nowadays, the same machine can function both as a PET scanner and a CT scanner, through which the doctor may locate the higher radioactivity areas visible on a PET scan with the detailed image of the same region shown by a CT scan.

Treatment of Anal Cancer

The treatment of anal cancer includes three types of standard treatments:

1. **Radiotherapy:** It uses X-rays in small doses to destroy cancer cells. The treatment can be done through external radiotherapy or internal radiotherapy (brachytherapy). However, radiotherapy is not free from side effects. It can cause temporary but significant redness and tenderness of the area treated, which develops gradually after several sessions. Radiotherapy treatment to the anal area can also cause narrowing of the vagina in women.
2. **Chemotherapy:** For an effective treatment, chemotherapy may be used along with radiotherapy (chemoradiation). The chemotherapy drug, mitomycin-C, is used in combination with 5-fluorouracil (5-FU). A drug called capecitabine (Xeloda) may take as an alternative to 5-FU. Usually, in a radiotherapy treatment consisting of 5 weeks, chemotherapy is given during the first and fifth week. The most common ill effects of mitomycin-C include nausea, vomiting, diarrhea, tiredness, low blood cell count, anemia, more susceptibility to infections, mouth ulcers, and soreness. The other ill effects include kidney and liver problems, skin rashes, hair loss or thinning of hair, and fertility issues.
3. **Surgery:** After analyzing the extent of spread and size of the tumor, surgery may be recommended by the specialist. Surgery can eliminate small tumors if they do not respond to any treatment. The local excision involves removing small tumors on the anal margin. This does not affect the sphincter muscle, and the continence is not affected. The abdominoperineal excision involves the removal of the anus and rectum. Non-temporary colostomy may be required, which involves diverting the open end of the bowel to wall of the abdomen to allow waste to be passed out of the stoma (opening on the abdominal wall) into a colostomy bag.

2.5 Cancers of Urinary System

2.5.1 Kidney Cancer

The kidneys (left and right) are situated at the upper region of abdomen, and their shape resembles that of the bean. They have support and protection of the lower rib cage. Adrenal gland is situated at the top of each kidney. The basic role of the kidney in the body is “filtration” whereby unwanted materials such as excess water and salt are eliminated from the blood in the form of waste and are discharged out of the body as urine. Urine passes through ureters and remains stored in the bladder, until urination. Another crucial role of the kidney is synthesis of rennin, a hormone that helps in blood pressure control.

Cancer of the kidney (Fig. 2.11) is a very serious disease and is ranked as seventh frequently occurring cancer in the world. It is also the tenth leading cause of death of males and females in the United States (American Cancer Society 2015). In the year 2015, 14,080 deaths, out of 61,560 cases of kidney cancers, were reported in

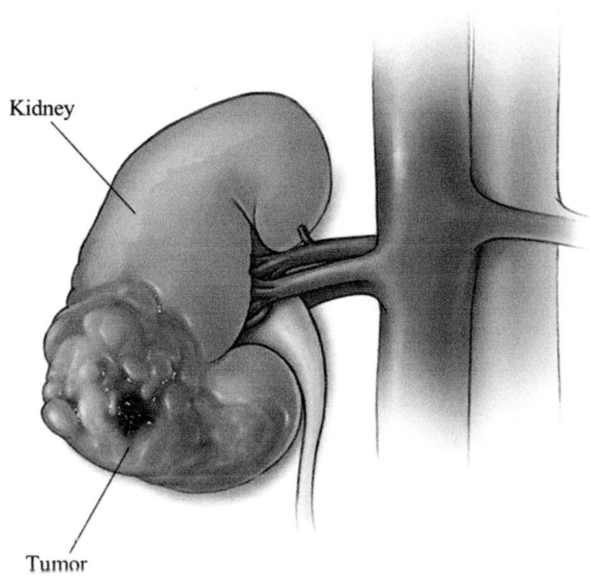

Fig. 2.11 Image showing cancer of the kidney

the United States. Although the 5-year disease-specific survival has enhanced from about 50% in 1975–1977 to 65% in 2000–2005, survival of the patients with an advanced-stage kidney cancer still remains at 10% (Novara et al. 2010; Chaffer and Weinberg 2011; Gandaglia et al. 2014). Although nephrectomy is practiced for non-metastatic renal cell carcinoma, results are distant metastasis in 30% of the patients (Pichler et al. 2011). In addition, the patients often discontinue the drug treatment due to severe side effects including hand-foot syndrome, liver dysfunction, and interstitial pneumonia (Banumathy and Cairns 2010; Suttle et al. 2014; Ivanyi et al. 2014; Lai et al. 2016; Willemsen et al. 2016). Therefore, a robust therapy for kidney cancer is still expected.

Kidney Cancer and Its Types

The different types of kidney cancer are (i) renal cell carcinoma, (ii) urothelial carcinoma, (iii) renal sarcoma, (iv) Wilms' tumor, (v) benign (non-cancerous) kidney tumors, (vi) renal adenoma, (vii) oncocytoma, and (viii) angiomyolipoma (Muglia and Prando 2015).

(i) *Renal cell carcinoma (RCC)*: This cancer is called by other names like "renal cell cancer" and/or "renal cell adenocarcinoma." It is considered the most common cancer of the kidney. RCC accounts for less than 3% of adult solid tumors, and each year, 30,000 new cases in the United States and 20,000 in the European Union are detected. Moreover, compared to women, men are affected twice, with the highest incidence in those between the age of 50 and 70 years (Kirkali et al. 2001). Out of every ten cases of kidney cancer, nine are RCC. Even though the cancer develops inside the kidney as a single tumor, more than one tumor can also be found per kidney or pair. RCC are of various subtypes, depending on the nature of the affected cells, when viewed under electron microscope. Having knowledge of these subtypes will help in selection of most appropriate treatment:

(a) *Clear cell renal cell carcinoma*: This form of cancer occurs in about seven out of every ten patients, thus making it the most common type of RCC. Viewing under electron microscope, the cells of RCC appear clear but in a pale color.
(b) *Papillary renal cell carcinoma*: This disease is the second most important form of RCC, occurring in at least one patient out of every ten RCC patients. Here the tumors take the shape of a papillae. However, the cancers are also called chromophilic due to pinky appearance (after dyes treatment) of the cell when viewed by electronic microscope.
(c) *Chromophobe renal cell carcinoma* (CRCC): Another subtype which occurs in about 5% of the RCC patients is "chromophobe renal cell carcinoma." This type of cancer forms in the cells lining the small tubules (helps filter waste from the blood to make urine) in the kidney. It is frequently diagnosed in stage I or stage II between ages 40 and 50. It has an overall better prognosis than other types of RCC, and its treatment generally involves surgery.
(d) *Rare types of renal cell carcinoma*: Rare subtypes of RCC affecting less than 1% of patients include multilocular cystic renal cell carcinoma, mucinous tubular and spindle cell carcinoma, collecting duct renal cell carcinoma, medullary carcinoma, and renal cell carcinoma associated with neuroblastoma.
(e) *Unclassified renal cell carcinoma*: Sometimes, RCC is not classified because the cancer has more than one subtype. As a result, the nature of the cancer does not fall in any categories.

(ii) *Transitional cell carcinoma* (TCC): Usually, TCC occurs in almost five to ten patients out of every hundred kidney cancer patients. These cancers are also called as "urothelial carcinomas (UC)" and are the most common type of bladder cancer and cancer of the ureter and urethra. TCC develops from the inner lining (of renal sinus, ureter, bladder, urethra, and urachus) of the urinary tract (transitional urothelium).
(iii) *Renal sarcoma*: These types of cancers of the kidney are usually not common (reported in <1% of all kidney cancers). The cancer normally starts from the blood vessels and/or connective tissues.
(iv) *Wilms' tumor (nephroblastoma)*: This type of tumors usually develops in children, and not so common among adults. Based on the microscopic examination of the cancer cells, Wilms' tumor is of two types: (a) favorable histology and (b) anaplastic histology. Contrary to the cells with anaplastic histology, the cells with favorable histology can be successfully treated with radiation therapy and chemotherapy.
(v) *Benign (non-cancerous) kidney tumors*: These tumors are non-cancerous in nature (known as benign) and usually do not disperse to various locations or

sites in patients. However, these tumors may grow further and develop a problem. The tumors are usually removed by surgery or destroyed. The option for treating these tumors is based on its size, number, as well as the general health of the patients.

(vi) *Renal adenoma*: As tumors of kidneys, "renal adenomas" are the most common non-cancerous tumors. Although they are non-cancerous and small and exhibit non-rapid growth, the tumors may resemble RCC when cells are observed under a microscope.

(vii) *Oncocytoma*: Another form of non-cancerous tumors of the kidney are "oncocytomas." These tumors can develop and grow so big, but do not disperse to neighboring locations and are treated by surgery.

(viii) *Angiomyolipoma*: This disease is not so common; it is a condition that occurs among patients with "tuberous sclerosis." This is usually a genetic disorder which affects the organs such as the heart, eyes, skin, etc.

Symptoms of Kidney Cancer

The symptoms include blood in the urine loss of appetite, shortness of breath, extreme fatigue, cough, prolonged fever, lump abdomen, weight loss, anemia, swelling in ankles or legs, and pain in bones (Ljungberg et al. 2014).

Staging of Renal Cancer

After diagnosis, the next step is to determine the stages of the cancer. Staging may be in grade form (Table 2.18) or TNM classification.

Diagnosis of Renal Cancer (www.mayoclinic.com)

The diagnosis of renal cancer includes:

(i) *Blood and urine tests:* Blood test is necessary for individual suffering from cancer of the kidney. Some enzymes appear to be in quantity beyond normal, when the level of calcium is above normal. This may signify that the cancer has reached to the bone. These tests are usually done prior to kidney-related operation. Generally, about 50% of RCC patients urinate blood; however, if an individual is affected by TCC, then the urine cytology (actual number of cancerous cell present in the urine) is required.

Table 2.18 Stages of renal cancer

Stage	Explanation
Stage I	It indicates that the tumor is up to 7 cm in diameter and confined to the kidney
Stage II	It indicates that the tumor is larger than that of stage I but confined to the kidney
Stage III	It indicates that the tumor extends beyond the kidney to the surrounding tissues and may also spread to the surrounding lymph nodes
Stage IV	It indicates that the tumor has spread outside the kidney, to multiple lymph nodes or to other distant parts such as bones, liver, or lungs

(ii) *Imaging tests:* The imaging tests reveal the site, size, as well as the shape of the tumor. They are instrumental in determining the extent of cancer invasion to the lymph nodes and neighboring locations. The imaging tests deployed for kidney cancer are:

(a) *Intravenous pyelogram (IVP):* IVP test is a radiological procedure done to identify problems including cancer of the pelvis or urethra. It is basically a scan wherein a contrast agent (dye) is introduced inside the vein to get a contrast in the pyelogram. The contrast agent later eliminates out of the bladder as part of urine.

(b) *Magnetic resonance imaging (MRI) scan*: MRI scan is recommended for people diagnosed with kidney cancer. It produces better image than that provided by CT scan. It is usually done when the cancer has reached the blood vessels.

(c) *Ultrasound*: This technique helps to determine the mass of the kidney and see if internal environment of the kidney is filled entirely with fluids or it's internally solid. The technique makes use of the echo pattern to identify kidney mass for stones or cancerous or non-cancerous tumor of the kidney.

(d) *Positron-emission tomography (PET) scan*: PET scan is recommended when the cancer is suspected to have traveled to neighboring lymph nodes of the kidney or to an unknown location. This test involves the use of a kind of sugar (radioactive) that is injected into the bloodstream. As already known, compared to the normal cells, the cancerous cells absorb or utilize more sugar, so the entire radioactivity in the patient's body can be imaged using a special camera.

(e) *Angiography*: As the name indicates, it is a scan of the blood vessels of the kidney by the use of a dye (different from that used in IVP test). During this scan, the dye is introduced into "renal artery" to identify the vessels that transport to the tumors of the kidney. However, to avoid further kidney damage, prior to the scan, it must be ensured that the patient's kidneys are in the working condition. (f) *Bone scan*: If during kidney problems a higher level of calcium is recorded in blood, then a bone scan or bone scintigraphy is recommended for identification of the occurrence of bone cancer. It is a nuclear medicine imaging technique of the bone that can diagnose bone inflammation and fractures and location of cancer or metastasis. PET scan may also solve the purpose.

(iii) *Kidney biopsy*: In case when the imaging test does not provide ample information regarding kidney cancer, then a biopsy (sample extraction and analysis) is conducted. It may be a needle core or a fine needle aspiration (FNA) biopsy. If RCC is confirmed, then its grading is done by nuclear grading system proposed by Fuhrman et al. (1982).

Treatment of Renal Cancer

If the localized RCC is diagnosed early, then nephrectomy is the most suitable remedy (Kirkali and Canda 2008). Paladini et al. (2017) reported the first case of simultaneous robotic surgery of kidney cancer and its metastasis in the pancreas, in a 68-year-old woman. They reported the first case of simultaneous robotic treatment of left renal mass and pancreatic metastasis. The robotic surgery also ensures less trauma and rapid postoperative recovery (Paladini et al. 2017). With the same objective of treating kidney tumors, *cryoablation* may be performed, wherein cold gas is released through a hollow needle inserted into the kidney tumor and freeze the cancer cells. Contrary to this *radiofrequency ablation* may be performed. In this treatment a special kind of probe is inserted into the skin and tumor of the kidney, followed by passing of electrical current to the cancer cells, thus generating heat to these cells. In a recent study by Kato et al. (2016), inhibitor molecules against T-lymphokine-activated killer cell-originated protein kinase [TOPK (OTS514)] and maternal embryonic leucine zipper kinase [MELK (OTS167)] can effectively suppress the kidney cancer cell growth. These two kinases play a crucial role in cancer cell proliferation. Obstruction of either or both of these kinases causes downregulation of forkhead box protein M1 (FOXM1) activity (oncogenic transcriptional factor).

2.5.2 Bladder Cancer

Among all known cancers worldwide, bladder cancer is the ninth most common cancer. It is also the second most common urological cancer and accounts for 3% of the new cancers. A strong male predominance is observed with three-quarters of all bladder cancer cases occurring in men (GLOBOCAN 2012). Although bladder cancer is frequent in men as compared to women, in contrast to most cancers, its risk is unrelated to family history (Yaxley 2016). Sixty percent cases of bladder cancer are associated with cigarette smoking. Certain occupational exposures are also associated with bladder cancer, especially to those who work in chemical and textile industries. Radiation exposure also increases risk of bladder cancer. The symptoms of bladder cancer include lower back pain, blood or blood clots in the urine, frequent micturition, and pain or burning sensation during urination (www.cancer.net).

Staging of Bladder Cancer

The TNM staging of bladder cancer (Babjuk et al. 2017) is shown in Table 2.19.

Table 2.19 TNM staging of urinary bladder cancer

T – Primary tumor (1°)	
TX	Assessment of 1° tumors found none
T0	1° tumors not available
Ta	Noninvasive papillary carcinoma
Tis	Carcinoma in situ "flat tumor"
T1	Tumor has invaded subepithelial connective tissue
T2	Tumor has invaded muscle
T2a	Tumor has invaded superficial muscle (inner half)
T2b	Tumor has invaded deep muscle (outer half)
T3	Tumor has invaded perivisceral tissue
T3a	Microscopically
T3b	Macroscopically (extravesical mass)
T4	Tumor has invaded prostate/uterus/vagina/pelvic wall/abdominal wall
T4a	Tumor has invaded prostate, uterus, or vagina
T4b	Tumor has invaded pelvic wall or abdominal wall
N – Lymph nodes	
(NX)	Assessment of lymph nodes (regional) found none
(N0)	Regional lymph node metastasis is not available
(N1)	Metastasis in a single lymph node in the true pelvic
(N2)	Metastasis in multiple lymph nodes in the true pelvic
(N3)	Metastasis in common iliac lymph nodes
M – Distant metastasis	
(MX)	Distant metastasis non-traceable
(M0)	Absence of distant metastasis
(M1)	Metastasis is of a distant type

Diagnosis of Bladder Cancer

The following tests may be done to diagnose bladder cancer (www.cancer.net).

(i) *Urine tests:* The urine sample is diagnosed for the presence of blood and bacteria (urinalysis). The sample is analyzed for the presence of cancer cells.

(ii) *Cystoscopy*: Flexible cystoscopy is a 10–20 min procedure wherein an anesthetic jelly is introduced into the urethra of the patient to numb the area. Then a cystoscope (flexible tube with a light and a camera) is passed into the bladder to examine its inside lining. If an abnormal tissue is observed, a rigid cystoscopy is performed under general anesthesia to collect the biopsy sample. This procedure may also involve transurethral resection of bladder tumor (TURBT) which is done to treat bladder cancer. After the rigid cystoscopy, the patient may experience pain or blood in the urine, for a few days.

(iii) *Imaging tests:*

(a) *CT-urography:* During a computerized tomography (CT) urography, a contrast dye is introduced into the kidneys, ureters, and bladder, through a vein in the hand. X-ray images that are taken provide a detailed view of the cancerous tissue. A variant of this technique is retrograde pyelogram. It is a type of X-ray exam mainly done to get detailed information of the upper urinary tract. During this test, a thin tube (catheter) is inserted into the bladder through the urethra to inject a radiocontrast dye into the ureters so that it reaches the kidneys. The ureters and the kidneys can be visualized through fluoroscopy or radiography.

(b) *Ultrasound*: Ultrasound scans are noninvasive and may take 15–20 min. Usually, an empty bladder scan follows a full-bladder scan. The scan can detect the presence of large tumors only. If the tumors are small and undetectable, then further tests are required.

(c) *CT scan*: The CT scan for the detection of bladder cancer is called as CT intravenous pyelogram (CT-IVP) or a triple-phase abdominopelvic CT. As a part of the procedure, a dye sometimes called the contrast (usually contains iodine) is injected into one of the veins. The dye reaches the kidneys, ureters, and bladder and facilitates the visibility of abnormal areas. The scan is usually done thrice: (i) before the dye is injected, (ii) immediately after the dye is injected, and (iii) a few minutes later.

(d) *Magnetic resonance imaging scan*: This scan uses magnetic waves to receive detailed cross-sectional images of organs in the abdomen. This technique may also help in diagnosis of bladder cancer. A dye may be injected to highlight the organs in the body.

Treatment of Bladder Cancer (Chang et al. 2017)

1. **Neoadjuvant/Adjuvant Chemotherapy**

Cisplatin-based neoadjuvant chemotherapy is generally deployed for bladder cancer therapy.

2. **Radical Cystectomy**

This technique involves removal of the prostate, bladder, and seminal vesicles in males and the bladder, fallopian tubes, uterus, anterior vaginal wall, and ovaries in females.

3. **Urinary Diversion**

It is a procedure (involves surgery) to reroute the urine flow when there is defect in the bladder. A thin flexible tube called catheter is used in the bladder to drain the urine. This process is known as urinary catheterization.

4. **Pelvic Lymphadenectomy**

It is also known as pelvic lymph node dissection (PLND). It is performed at the time of a radical cystectomy or a partial cystectomy. It also provides an information regarding stage of cancer.

Ephrin receptor A2 (EphA2) is a member of the ephrin-ephrin receptor cell-to-cell signaling family of molecules and plays a key role in proliferation, differentiation, and migration. It has been observed that EphA2 is overexpressed in a broad range of cancers, including bladder cancer, and is associated with increased metastasis and poor prognosis. Recently, Kamoun et al. (2018) have proposed to use a novel antibody-directed nanotherapeutic (ADN) approach to target EphA2 for treating bladder cancer. His work characterized the expression of EphA2 in bladder cancer patients, evaluated its prognostic power, and tested an EphA2-targeted ADN, MM-310, in patient-derived xenograft (PDX) models. EphA2 expression in tumors was investigated using a validated immunohistochemistry assay performed in 147 bladder cancer samples, enabling analysis of prevalence and prognostic power. Whole sections of primary and metastatic tumor resections were used to assess EphA2 expression in tumor-associated vessels and tumor cells. Four EphA2+ PDX models were used to evaluate the activity of MM-310 compared to free docetaxel, alone or in combination with gemcitabine. EphA2 was expressed in tumor cells and tumor-associated blood vessels in primary and metastatic lesions. EphA2 overexpression was seen in 80–100% of tumors and correlated with shorter survival. In the PDX models, MM-310 controlled the tumor growth and was more active than free docetaxel at equitoxic dosing. Additionally, the combination of MM-310 and gemcitabine-controlled tumor grew better than each drug alone and outperformed the combination of free docetaxel and gemcitabine in the single PDX model where it was compared. Thus, in bladder cancer models, a docetaxel-based ADN as a monotherapy in combination with gemcitabine targeted EphA2 led to a significant tumor regression (Kamoun et al. 2018).

2.6 Cancers of Sex Organs

2.6.1 Testicular Cancer (TC)

It is the most common malignancy affecting men between the age of 15 and 40 years. Most types of testicular cancers develop in the germ cells and are referred to as germ cell tumors (Huang 2018). It may begin in several locations: the testicles, the retroperitoneum, the mediastinum, and the pineal gland. TC is almost curable if diagnosed at an early stage and even curable in the late stages. There are two main kinds of TC: (i) seminomas and (ii) non-seminomas. Non-seminomas have four subtypes, i.e., (a) choriocarcinoma, (b) embryonal carcinoma, (c) yolk sac tumor, and (d) teratoma.

Symptoms and Risk Factors of Testicular Cancer

The symptoms of testicular cancer are minor ache in the lower abdomen or groin, painless lump or swelling on either side of testes, fluid accumulation in the scrotum, pain or discomfort in the testicles or scrotum, heavier testicle, swelling of one or both legs, breast tenderness or growth (gynecomastia), bloody sputum or phlegm, chest pain, and shortness of breath.

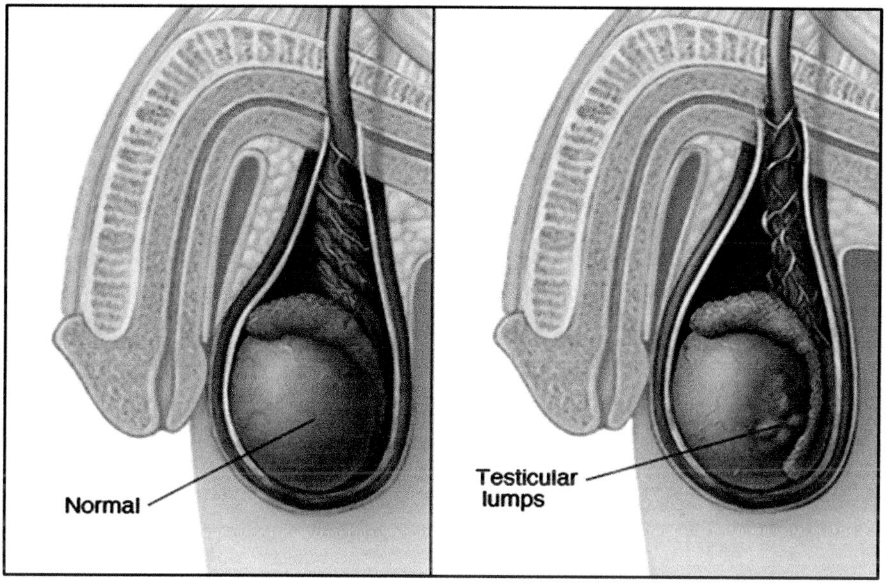

Fig. 2.12 Image showing (**a**) normal testicles and (**b**) testicles with lumps. (Courtesy: MAYO foundation)

Some risk factors that affect TC are:

(a) *Age*: TC is diagnosed in men between the age of 20 and 45 and even in their teens or 60s. Pain, swelling, or lumps in the testicle or groin area may be an indication of testicular cancer (Fig. 2.12).
(b) *Cryptorchidism*: it is a state when one or both testicles do not descend into the scrotum before birth. Men with this condition have high risk for TC. Sometimes a surgical procedure known as orchiopexy is performed as a remedy.
(c) *Family history:* most men with cancer of testicles have no trace of the disease in their history. Klinefelter's syndrome (inherited disease) may also lead to testicular cancer.
(d) *Personal history*: about 3–4% of men who have been cured of cancer in one testicle are more susceptible to cancer in the other testicle.
(e) *Race and ethnicity:* the white men are at a higher risk (about 4–5 times) of developing testicular cancer than black and Asian-American men.

Staging of Testicular Cancer

The following classification is used to analyze the stage of testicular cancer (Albers et al. 2015) as shown in Table 2.20.

Table 2.20 TNM classification for staging of testicular cancer

pT: Primary tumor	
pTX	Primary tumor cannot be assessed
pT0	No evidence of primary tumor (e.g., histological scar in testis)
pTis	Intratubular germ cell neoplasia (testicular intraepithelial neoplasia)
pT1	Tumor limited to testis and epididymis without vascular/lymphatic invasion: tumor may invade tunica albuginea but not tunica vaginalis
pT2	Tumor limited to testis and epididymis with vascular/lymphatic invasion or tumor extending through tunica albuginea with involvement of tunica vaginalis
pT3	Tumor invades spermatic cord with or without vascular/lymphatic invasion
pT4	Tumor invades scrotum with or without vascular/lymphatic invasion
N: Regional lymph nodes clinical	
(NX)	Assessment of the lymph nodes (regional) non-traceable
(N0)	Absence of lymph node (regional) metastasis
(N1)	Metastasis with a lymph node mass 2 cm or less in greatest dimension or multiple lymph nodes, none more than 2 cm in greatest dimension
(N2)	Metastasis with a lymph node mass more than 2 cm but not more than 5 cm in greatest dimension, or multiple lymph nodes, any one mass more than 2 cm but not more than 5 cm in greatest dimension
(N3)	Metastasis with a lymph node mass more than 5 cm in greatest dimension
Pn: Pathological	
pNX	Assessment of the lymph nodes (regional) non-traceable
pN0	Absence of lymph node (regional) metastasis
pN1	Metastasis with a lymph node mass 2 cm or less in greatest dimension and 5 or fewer positive nodes, none more than 2 cm in greatest dimension
pN2	Metastasis with a lymph node mass more than 2 cm but not more than 5 cm in greatest dimension, or more than 5 nodes positive, none more than 5 cm, or evidence or extra nodal extension of tumor
pN3	Metastasis with a lymph node mass more than 5 cm in greatest dimension
M: Distant metastasis	
MX	Distant metastasis non-traceable
M0	Absence of distant metastasis
M1	Metastasis is of distant type
M1a	Non-regional lymph node(s) or lung
M1b	Other sites

Diagnosis and Treatment of Testicular Cancer

The diagnosis of testicular cancer can be mediated through physical examination, ultrasound, blood tests/tumor markers, lactate dehydrogenase markers, placental alkaline phosphatase tumor markers, orchiectomy/surgical pathology tests, X-rays, MRI scan, CT scan, PET scan, and biopsy.

The treatment of testicular cancer includes:

(a) *Radical inguinal orchidectomy:* it is the surgical removal of cancerous testicles. The testicle may be removed for almost all stages of testicular cancer.

(b) *Retroperitoneal lymph node dissection* (RPLND): it is the surgical removal of the retroperitoneal lymph nodes that are located at the back of the abdomen.
(c) *Chemotherapy:* drugs such as bleomycin, carboplatin, cisplatin, etoposide, gemcitabine, ifosfamide, oxaliplatin, paclitaxel, and vinblastine are used as chemotherapeutic agents.
(d) *Radiation therapy:* Using X-rays or related particles, cancerous cells are killed.

2.6.2 Prostate Cancer (NCCN 2016)

It is a very common (second most common) form of cancer reported in men. In the year 2012, 1.1 million cases were diagnosed worldwide, which accounts for 15% of all known cancers (Mottet et al. 2017). The incidence of prostate cancer diagnosis varies widely between different geographical areas. It is reported to be highest in Northern America, Australia/New Zealand, and Western and Northern Europe while low in Eastern and South-Central Asia. The symptoms of prostate cancer include weak or interrupted urine flow or frequent urination (especially at night) requiring pressure to empty the bladder, pain or burning sensation during urination, onset of erectile dysfunction, presence of blood in the urine or in seminal fluid, and discomfort when sitting.

Causes of Prostate Cancer
This cancer is said to occur due to exogenous factors (Leitzmann and Rohrmann 2012) such as:

(a) *Diet*: a diet rich in dairy products increases the chance of prostate cancer.
(b) *Chronic inflammation*: it contributes to prostate cancer.
(c) *Sexual behavior*: a patient suffering from sexually transmitted diseases (gonorrhea and syphilis) may also suffer from prostate cancer. Infection caused due to *Trichomonas vaginalis* may also lead to prostate cancer. In a recent study, a higher ejaculation frequency diminishes the chance of developing prostate cancer.
(d) *Exposure to ultraviolet radiation*: a higher physical activity and outdoor exposure to UV radiation decreases the risk of prostate cancer and vice versa.
(e) *Age:* it is found abundantly in men older than 65.
(f) *Geography:* prostate cancer is most common in Northwestern Europe, North America, Caribbean islands, and Australia.
(g) *Family history:* A person having a close relative (father or brother) who had suffered prostate cancer is at a high risk.

Staging of Prostate Cancer (NCCN 2016)
The different stages of prostate cancer are shown in Table 2.21.

Table 2.21 Staging of prostate cancer

Stage	Explanation
Stage I	This stage indicates tumor is growing slowly
Stage II	This stage indicates that the tumor is present only in the prostate. PSA level is medium or low
Stage III	This stage indicates that the tumor is growing or the cancer is high grade (locally advanced cancer that is likely to spread). PSA level is high
Stage IV	This stage indicates that distant lymph nodes plus other neighboring sites and/or bones have been infected

Diagnosis of Prostate Cancer
The diagnosis of prostate cancer includes:

(i) *Digital rectal examination (DRE)*: the doctor inserts gloved finger (lubricated) inside the rectum to examine the prostate for any abnormalities in the shape or size or texture of the gland. Further tests may be required later.
(ii) *Prostate-specific antigen (PSA) test*: PSA is a protein released by the prostate tissue whose level rises in response to an inflammation, infection, enlargement, or cancer in the prostate.
(iii) *Biomarker tests*: A biomarker may also be called a tumor marker. Biomarker tests for prostate cancer include the 4Kscore and the prostate health index (PHI), which may forecast the chances of developing prostate cancer in a man.
(iv) *PCA3 test:* The prostate cancer gene 3 (PCA3) assay is a confirmatory test for prostate cancer, which looks for the presence of *PCA3* gene product in a man's urine. It is used along with a PSA test to confirm whether a prostate biopsy is needed.
(v) *Transrectal ultrasound (TRUS) and biopsy*: During TRUS a probe is inserted into the rectum to capture images of the prostate. Together with TRUS, several samples of biopsy (12–14 pieces of the tissue) are collected from different areas of the prostate for diagnosis.

To get an idea of the spread of prostate cancer, imaging techniques like CT scan, PET scan, or MRI may be done when required.

Treatment of Prostate Cancer
To treat the prostate cancer, several procedures may be performed like:

(a) *Surgical removal of the prostate gland*: radical prostatectomy involves removing of prostate gland and some neighboring tissue and a few lymph nodes. It can be performed by retropubic surgery or by using a robot. But it may cause urinary incontinence and erectile dysfunction.
(b) *Cryosurgery*: it is used as an alternative to radiation therapy. For cryosurgery, the prostate cells are treated with alternating cycles of freezing (cold gas passed through a needle) and heat to kill the cancer cells and some normal cells in the vicinity.

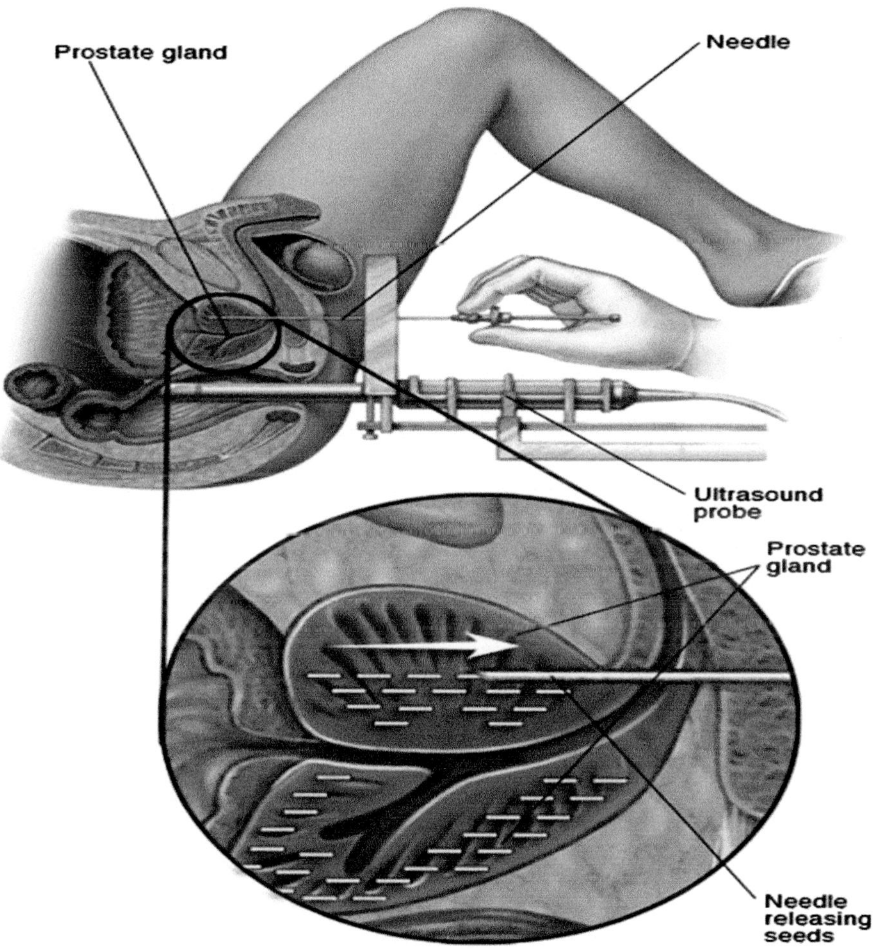

Fig. 2.13 Image showing brachytherapy for treating prostate cancer

(c) *Radiation therapy*: external beam radiation therapy involves treating the prostate with high-powered energy beams, such as X-rays or protons, for 5 days a week for several weeks, while brachytherapy involves implanting many rice-sized radioactive seeds in the prostate tissue using a needle guided by ultrasound images (Fig. 2.13). The radioactive seeds slowly release a low dose of radiation. However, this technique is not free from side effects and causes painful or frequent urination, loose motions, and even erectile dysfunction.

(d) *Androgen deprivation therapy*: the male sex hormones (androgen: testosterone) promote the cancer cells to grow and divide. Thus, strategically, lowering down the testosterone level or preventing it from reaching the cancer cells can kill or slow down the growth of cancer cells. Drugs that are used in hormone therapy are goserelin (Zoladex), leuprolide (Lupron, Eligard), histrelin (Vantas), triptorelin (Trelstar), and ketoconazole and abiraterone (Zytiga). Some anti-androgens that prevent testosterone from reaching the cancer cells include flutamide, nilutamide (Nilandron), bicalutamide (Casodex), and enzalutamide (Xtandi).

(e) *Chemotherapy*: chemotherapy drugs may kill the fast-growing cells, including cancer cells. Chemotherapy may be used as an alternative to hormone therapy.

(f) *Biological therapy*: this therapy makes use of body's immune system to kill cancer cells. Biological therapy with sipuleucel-T (Provenge) can successfully treat advanced and relapsed prostate cancer.

2.6.3 Ovarian Cancer

Ovarian cancer refers to various malignancies (heterogeneous disease) affecting the ovary (Karst and Drapkin 2010). One of three cell types may be involved in ovarian cancer: epithelial cells, germ cells, and stromal cells. They are named accordingly as:

(a) *Epithelial carcinoma*: it accounts for 85–90% of all ovarian cancers. These types of epithelial carcinomas include serous (52%), endometrioid (10%), clear cell (6%), and mucinous carcinomas (6%) (Fig. 2.14).

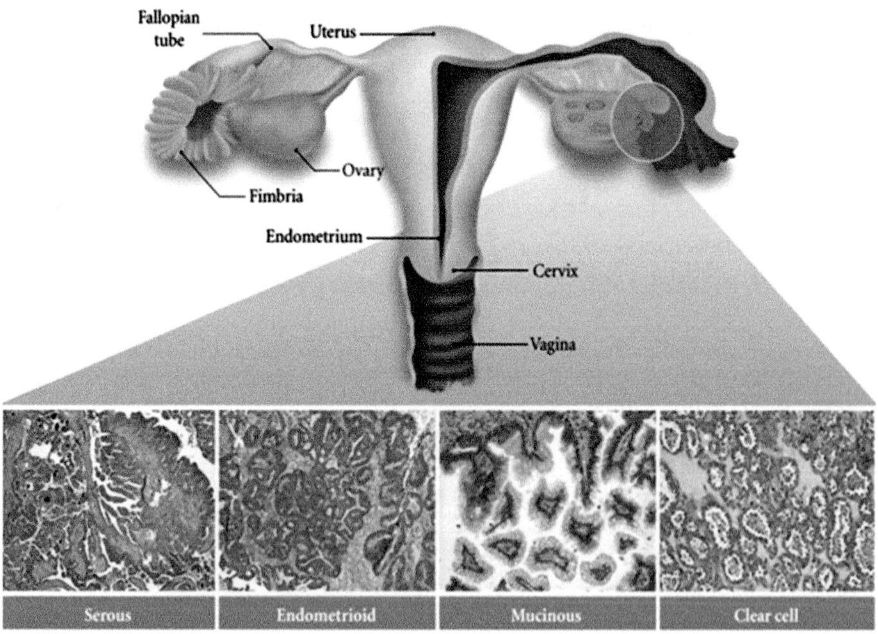

Fig. 2.14 The major histologic subtypes of ovarian carcinoma. Serous carcinomas resemble fallopian tube epithelium, endometrioid carcinomas resemble endometrial glands, and mucinous carcinomas resemble endocervical epithelium. The photographs show specific tumor section

(b) *Germ cell malignancies*: it is an uncommon type of ovarian cancer which develops in the egg-producing cells of the ovaries.
(c) *Stromal malignancies*: this is also a rare form of ovarian cancer which develops in the connective tissue cells and granulosa and theca cells that hold the ovary together. Approximately 40% of all ovarian tumors are nonepithelial in origin, and only 10% of the ovarian cancers progresses to a malignant state (Karst and Drapkin 2010).

Symptoms of Ovarian Cancer

The symptoms include abdominal swelling, abdominal bloating, abdominal and/or pelvic pain, early satiety, indigestion, constipation, back pain, fatigue, menstrual irregularities, and vaginal discharge.

Causes and Risk Factors of Ovarian Cancer

Several factors may cause a woman to develop ovarian cancer. Such factors are:

(a) *Family history*: if a women's mother or sister suffered from ovarian, breast, or uterine cancer, then she is likely to develop ovarian cancer (Permutt and Sellers 2009).
(b) *Age*: women above the age of 50 years may develop ovarian cancer (Roett and Evans 2009).
(c) *Childbirth and menopause*: women who did not give birth faced early menstruation or late menopause and have a potential risk of developing ovarian cancer. However, most ovarian cancers are diagnosed after the menopause (Permutt and Sellers 2009).
(d) *Genetics*: women with mutations in the BRCA1 or BRCA2 genes have 23–54% greater risk of developing ovarian cancer.
(e) *Previous gynecological problems*: women who complain of ovarian cysts or endometriosis can also develop ovarian cancer.
(f) *Lifestyle*: sedentary lifestyle, obesity, and smoking can lead to ovarian cancer (Roett and Evans 2009).

Staging of Ovarian Cancer

Ovarian cancer is staged to determine how far it has spread beyond the ovary. The different stages for ovarian cancer are explained in Table 2.22.

Table 2.22 Stages of ovarian cancer

Stage	Explanation
Stage I	It indicates that the tumor is confined to the ovary
Stage II	It indicates that the tumors have extended beyond the ovary to adjacent pelvic structures such as the fallopian tube or uterus
Stage III	It indicates metastasis to the peritoneum and/or regional lymph nodes
Stage IV	It indicates tumor has metastasized beyond the peritoneum to distant sites

Diagnosis of Ovarian Cancer

The early-stage ovarian cancer can be detected by transvaginal sonography (TVS) and serum biomarker testing:

(i) *Transvaginal sonography (TVS) or transvaginal ultrasound (TVUS):* TVS/TVUS may be used to locate a tumor in the uterus, fallopian tubes, and ovaries. But, it cannot predict the nature (cancerous or benign) of the tumor.
(ii) *Serum biomarker testing:* the use of biomarker measurements as tools for early detection is a promising research innovation currently being applied to a number of human cancers. The amount of cancer antigen 125 (CA125) in the blood is used as a serum biomarker specific for ovarian cancer (Karst and Drapkin 2010). Many women, but not all, with ovarian cancer may have high levels of CA-125. The success of the treatment can be analyzed by the decline in the level of CA125 in the patient's blood.

Treatment of Ovarian Cancer

The treatment of ovarian cancer is based on tumor size, position, degree of spread, and patient's health. Before 2011 the treatment options for ovarian cancer were limited to surgery and chemotherapy. However, in 2011, the European Commission (EC) announced the approval of a biological therapy in combination with standard chemotherapy (carboplatin and paclitaxel) as initial treatment for advanced (stages III and IV) epithelial ovarian, fallopian tube, or primarily peritoneal cancer.

1. **Surgery:** Surgical removal of both the ovaries (bilateral oophorectomy) or removal of uterus (hysterectomy) or surgical reduction of tumor (debulking surgery or cytoreduction) may be done. For advanced-stage tumors, both debulking surgery and chemotherapy are useful (Gubbels et al. 2010).
2. **Chemotherapy:** Patients who cannot undergo surgery are subjected to chemotherapy regime which involves treatment with a combination of drugs (usually carboplatin and paclitaxel). However, it is unfortunate that in most cases reoccurrence of cancer culminates in the death of the patient (Heitz et al. 2006).
3. **TRAIL (tumor necrosis factor-related apoptosis-inducing ligand):** Tumor necrosis factor-related apoptosis-inducing ligand (TRAIL) is a potentially promising technique for the treatment of ovarian cancer which induces apoptosis, in many malignant cells without affecting the normal cells. TRAIL may be used alone or in combination with other chemotherapeutic agents (Khaider et al. 2011).

2.6.4 Breast Cancer

In women, breast cancer is the second most important cancer after skin cancer. As per the recent US cancer statistics, although the death rates due to breast cancer have been decreasing since 1989, still about 41,760 women are expected to die in 2019 from breast cancer. Breast cancer is not confined only to women; it is reported

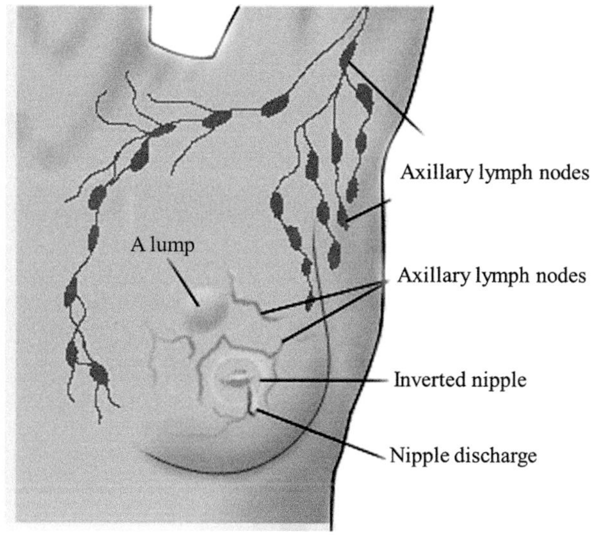

Fig. 2.15 Symptoms of breast cancer

in men also. The symptoms include lump(s) (a hard knot or a thickening) in the breast, change in breast shape accompanied with pain, dimpling of the skin, fluid coming from the nipple (even blood), a newly inverted nipple, or a red or scaly patch on the skin (skin irritation). This cancer most commonly develops in cells lining the milk ducts (*ductal carcinoma*) and the lobules (*lobular carcinoma*) that supply the ducts with milk (Fig. 2.15). The other subtypes include (a) in situ *ductal carcinoma*, which is a type of cancer, located mainly at duct, and develops from preinvasive lesions, and (b) *invasive or infiltrating ductal carcinoma*, which is a type of cancer that spreads outside the duct. Some common types of breast cancer are mucinous, papillary, medullary, metaplastic, tubular, and inflammatory cancer.

Risk Factors of Breast Cancer

The risk factors include:

(a) *Age:* the women above 50 years of age are more susceptible to develop cancer.
(b) *Gender*: women are likely to have cancer of the breast than their counterpart males with 100 times chances.
(c) *Genes*: about 5–10% of breast cancers can be linked to gene mutations. Those females with BRCA1 and BRCA2 gene mutations are more liable to develop breast cancer.
(d) *Inherited genetic predisposition*: if a close female relative has had breast cancer, you are at a risk for developing it.
(e) *Personal history*: women who have their first child after the age of 35 are at an increased risk of breast cancer. Females who had their first period before the age of 12 have an increased risk for developing breast cancer. Women who never became pregnant also have higher chances to develop breast cancer. Similarly, females that did not undergo menopause until after age 55 have

higher possibility to develop breast cancer. Moreover, a woman who had breast cancer in one side of the breast has higher chances of having other breast affected.

(f) E*strogen and progesterone exposure*: women who consume/continue with post-menopausal estrogen and progesterone medications also have a higher risk of developing breast cancer.

(g) *Lifestyle*: drinking excessive amounts of alcohol raises the risk of cancers.

Staging of Breast Cancer

The following classification is used to detect the stage of breast cancer (Table 2.23).

Diagnosis of Breast Cancer

The diagnosis of breast cancer includes:

(i) *Diagnostic mammography*: the X-ray of the breast is known as mammogram. Screening mammography is routinely conducted, but a diagnostic mammography is conducted after a suspicious result on a screening mammogram. It is conducted when a woman is experiencing some symptoms, such as lump or nipple discharge (Cancer.net 2017). Mammograms cannot only spot tumors but also ductal carcinoma in situ (DCIS).

(ii) *Ultrasound:* it is conducted after a suspicious breast self-exam or a screening mammogram. It uses sound wave to create a picture of the breast tissue. The ultrasound result (sonogram) distinguishes between a solid mass (which may be cancerous) and a fluid-filled cyst (i.e., not cancerous) in the breast. In MRI, the MRI machine conducts magnetic energy and radio waves through the breast tissue and scans the tissue for generating detailed images of breast tissues, to distinguish between normal and diseased tissue. MRI is highly sensitive for detection and characterization of occult breast carcinoma (Zujewski et al. 2011).

Table 2.23 Staging of breast cancer (Cserni et al. 2018)

Stage I	T < 2.0 cm	It indicates that the tumors are small (less than or equal to 2 cm) with no known lymph node involvement and no metastasis
	N0	
	M0	
Stage II	T > 2.0 < 5 cm	It indicates that the tumor is slightly larger than that of stage 1 (between 2 and 5 cm). Their lymph node may be affected but no metastasis
	N1	
	M0	
Stage III	T > 5 cm (T3)	It indicates large tumors (greater than 5 cm) with signs of inflammatory breast cancer. Ipsilateral nodes are involved. There is metastasis to ipsilateral internal mammary lymph nodes
	N2	
	N3	
	T4d	
Stage IV	Any T	It indicates metastatic cancer, irrespective of the size of the tumor and the involvement of lymph nodes
	Any N	
	M1	

(iii) *Biopsy:* the biopsy sample may be collected by fine needle aspiration (FNA), core needle biopsy (CNB), or surgical biopsy for examination. The biopsy may involve the use of ultrasound, vacuum-assisted device (VAD), and a marker (thin metal clip) (Zujewski et al. 2011).

(iv) *Mammotome and minimally invasive breast biopsy (MIBB):* it is different from the core needle biopsy wherein a needle is inserted once and multiple samples are collected. This technique uses a vacuum system wherein a probe is inserted in the breast and tissue samples are collected in a chamber (Zujewski et al. 2011).

Treatment of Breast Cancer

Different treatments are recommended on the basis of staging (Anjum et al. 2017 as shown in Table 2.24).

Table 2.24 Stage-dependent treatment of breast cancer

Stage	Explanation	Treatment
Stage 0	Presence of noninvasive breast tumor [e.g., ductal carcinoma in situ (DCIS), – lobular carcinoma in situ (LCIS), and Paget's disease of the nipple]	Hormone therapy medication (tamoxifen) helps prevent cancer cells from growing
Stage I	Stage IA: It indicates that the tumor is smaller than 2 cm or even smaller in size and has not spread to the lymph nodes	Lumpectomy can be sufficient with radiation therapy. For early breast cancer, systematic chemotherapy is used with radiotherapy
	Stage IB: Cancer has spread to the lymph nodes	
Stage II	Stage IIA: It indicates that there is no tumor, and less than four axillary lymph nodes contain cancer cells	Large tumor (>1 cm) and node-positive breast cancer can be treated with chemotherapy and adjuvant endocrine therapy which is followed by hormone receptor status evaluation
	Stage IIA: It indicates that the tumor is less than 2 cm, and less than four axillary lymph nodes contain cancer cells	
	Stage IIA: It indicates that the tumor is between 2 and 5 cm and has not yet spread to the lymph nodes	
	Stage IIB: It indicates that the tumor is between 2 and 5 cm and has spread to less than four axillary lymph nodes	
	Stage IIB: It indicates that the tumor is larger than 5 cm but has not spread to any axillary lymph nodes	

(continued)

Table 2.24 (continued)

Stage	Explanation	Treatment
Stage III	Stage IIIA: It indicates that the tumor is less than 2 cm and cancer has spread to 4–9 lymph nodes	A modified radical mastectomy and adjuvant chemotherapy (to shrink the size of the primary tumor) with radiation therapy are used. Adjuvant endocrine therapy is mostly followed by hormone receptor status evaluation
	Stage IIIA: It indicates that the tumor is larger than 5 cm and cancer clusters are present in the lymph nodes	
	Stage IIIA: It indicates that the tumor is larger than 5 cm and cancer has spread to the lymph nodes near the breast bone or underarm	
	Stage IIIB: It indicates that the tumor is of any size; cancer has spread to the chest wall or breast skin and to the lymph nodes	
Stage IV	It indicates that the cancer has spread to the brain, bones, lung, and liver	There is no cure, and the treatments are available only to increase survival rate. To control symptoms, palliation is important, and to avoid bone fractures, bisphosphonates can be used to treat bone metastasis

2.6.5 Choriocarcinoma

It is a rare type of tumor which grows fast and affects the woman's uterus. The abnormal cells usually develop within tissues that later become the placenta. Gestational choriocarcinoma affects pregnant women during gestation. It starts from the uterus and later spreads (malignant tumor) to other parts of the body through both venous and lymphatic systems (www.webmd.com). Choriocarcinomas are characterized by trophoblastic malignant cells that may produce the pregnancy hormone, human chorionic gonadotropin (hCG) in the absence of an ongoing pregnancy (Yamamoto 2009). They are of two types: (i) gestational and (ii) non-gestational. (i) *Gestational type*: it includes ovarian metastasis from primary uterine choriocarcinoma which occurs in association with the normal pregnancy or abortion, complete hydatidiform mole or partial mole, and primary gestational ovarian choriocarcinoma (Yamamoto 2009; ACS 2017). (ii) *Non-gestational type*: it is a very rare malignant tumor and is a mixture of germ cell tumor and a pure ovarian choriocarcinoma. (Yamamoto 2009). There are several forms of gestational trophoblastic disease (GTD) which includes (a) hydatidiform mole (complete or partial), (b) invasive mole, (c) choriocarcinoma, (d) placental site trophoblastic tumor, and (e) epithelioid trophoblastic tumor (Fig. 2.16).

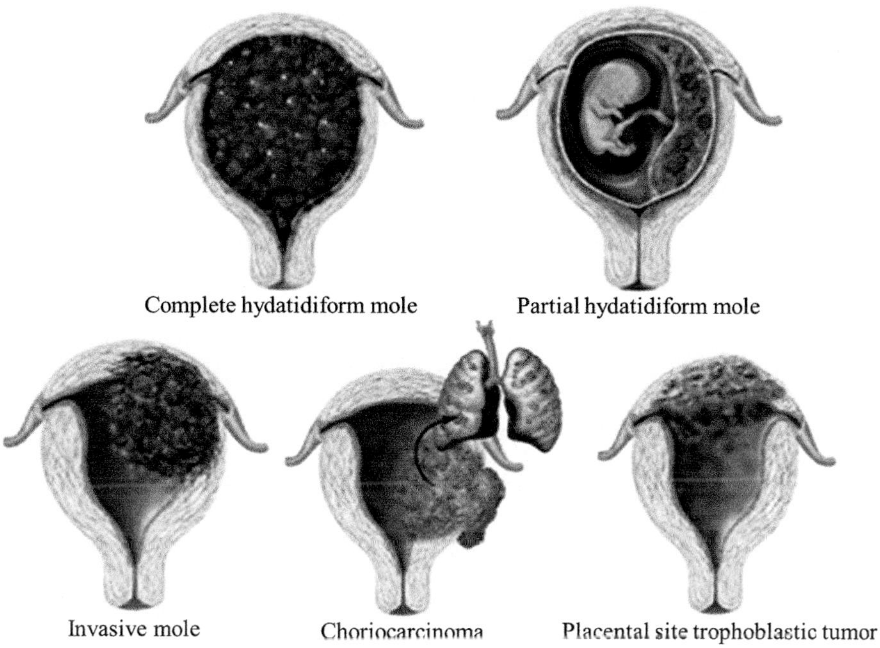

Fig. 2.16 Types of gestational trophoblastic diseases

(a) ***Hydatidiform Mole***

It is the most common form of GTD and also named as molar pregnancy. It occurs when the placenta is formed without/partial/abnormal fetus that is incapable to develop into a baby. Usually hydatidiform mole forms are non-cancerous and are of two types: (a) complete and (b) partial. A complete hydatidiform mole forms when one or two sperms fertilize an empty egg cell (cell devoid of nucleus). Then all genetic material come from the male parent and fetal tissue does not form. Although these moles can be surgically removed, some may become invasive mole or a choriocarcinoma. A partial hydatidiform mole forms when two sperms fertilize a normal egg. In that case the tumor contains some fetal tissue mixed in with trophoblastic tissue. Partial mole usually does not become cancerous and are surgically removed.

(b) ***Invasive Mole***

An invasive mole develops from either partial or complete hydatidiform mole. However, as compared to partial, complete mole is more invasive (Wan et al. 2014). It grows into the muscle layer of the uterus. A woman is more likely to develop an invasive mole if there is at least a 4-month gap between the last menses and the surgery (complete removal of mole), if she is more than 40 years old, if her uterus has become enlarged, and if she has a past record of gestational trophoblastic disease. When an invasive mole (tumor) successfully develops in the lining of the womb, it may bleed within the pelvic cavity that may be so dangerous (Wan et al. 2014). However, even if the tumor is eliminated, it does spread to neighboring locations or organs (lungs are commonly the affected organ in 4% of the cases).

(c) ***Choriocarcinoma***

It is a malignant form of carcinoma which constitutes 5% of all GTDs. It spreads to the muscle layer of the uterus and the adjacent organs. Fifty percent of gestational choriocarcinoma develop as molar pregnancy; 25% develops in women after miscarriage, intentional abortion, and tubal pregnancy. Another 25% develops post-pregnancy and delivery. Choriocarcinomas that are not related to pregnancy are reported both in men (testicles, chest, abdomen) and women (ovaries, chest, abdomen). Therefore, this type of cancer is known as "mixed germ cell tumor."

(d) ***Placental Site Trophoblastic Tumor (PSTT)***

PSTT is a type of rare tumor and constitutes 0.23–3% of GTDs. It usually grows at a location where the uterus lining unites with the placenta. These tumors develop after a normal pregnancy or abortion.

It may also occur even after partial and/or complete removal of a mole. These tumors have the potential to grow into the muscle layer of the uterus. Surgery is recommended for the treatment of PSTTs as they do not respond to chemotherapy.

(e) ***Epithelioid Trophoblastic Tumor (ETT)***

This disease is a very rare form of GTD often very difficult to be diagnosed and treat through chemotherapy and so surgery is usually recommended. Similar to other tumors, such as those of the placenta, the ETT usually develops after full-term pregnancy. However, it requires several years to develop. It is often found to have metastasized when diagnosed.

Symptoms

The symptoms of choriocarcinoma include vaginal bleeding during early pregnancy, spontaneous abortion, and postpartum bleeding. If it is dispersed to neighboring location within the body, then different symptoms may be noticed, such as cough, hemoptysis, troubled breathing, chest pain, headache, dizziness, and increase in thyroid-stimulating hormone (www.cancerwall.com; www.rightdiagnosis.com).

Staging of Choriocarcinoma

The different stages of choriocarcinoma and their explanation are shown in Table 2.25.

Table 2.25 Stages of choriocarcinoma

Stage	Explanation
Stage I	It indicates that choriocarcinoma is limited only to the uterus. The malignant cells are limited inside the womb
Stage II	It indicates spreading of choriocarcinoma to the genital tract
Stage III	It indicates metastasis of choriocarcinoma to the lungs. The metastasis is hematogenous in nature
Stage IV	It indicates metastasis of cancerous cells to various body locations through hematogenous and lymphatic routes

Diagnosis of Choriocarcinoma

The diagnosis of choriocarcinoma includes:

(i) *Blood, urine,* and *spinal fluid test*: It is well-known that the blood human chorionic gonadotropin (HCG) levels go high during a normal pregnancy. Some of the HCG from the blood is excreted out in the urine, and its presence is an indicator of positive pregnancy. However, it is also known that a complete mole releases much higher levels of HCG than a normal placenta which indicates the presence of a mole. It is also interesting to note that not all females suffering from GTD exhibit a higher HCG level than the normal females as seen in patients with PSTT, ETT, and partial moles. Detecting the level of HCG (level going down) also facilitates in analyzing whether or not the treatment of a GTD has been successful. Similarly, measuring the levels of a unique marker, human placental lactogen (hPL), has been useful to follow up patients diagnosed with PSTT. If the levels of HCG remain high without any signs of tumor during routine radiological examination, then it is suspected that GTD might have spread to the nervous system. To confirm this spinal fluid sample is extracted by the procedure of lumbar puncture or spinal tap. The spinal fluid is analyzed for the signs of tumor spread.

(ii) *Pelvic examination*: It is done to examine the presence of tumor(s).

(iii) *Imaging tests*: Ultrasound, chest X-ray, CT scan, MRI, or PET scan is also performed to know the presence and location of the tumor inside the uterus.

(iv) *Liver function test*: The liver is initially affected by choriocarcinoma so analyzing the presence of high liver enzymes may indicate liver metastases.

Treatment of Choriocarcinoma

1. **Surgery:** As already mentioned before, surgery is usually the first treatment used for a molar pregnancy (www.cancer.net 2017). For gestational-type choriocarcinoma, the type of surgery depends on the stage of the tumor. When there is a molar pregnancy, the molar tissue is removed by suction dilation and curettage

(D&C). During D&C, a doctor dilates the cervix and removes the molar tissues inside the uterus and skillfully scrapes the walls of the uterus (to remove the remaining molar tissue), to preserve the woman's fertility. During this procedure, oxytocin is injected to maintain the normal size of the uterus. The reported side effects of this surgery are scarring inside the uterus, vaginal bleeding, cramping, infection, and blood clots.

2. **Hysterectomy:** In most cases, hysterectomy (removal of the woman's uterus and uterine cervix) is not necessary because the molar tissue can be removed by D&C, followed by chemotherapy regime. But, if the beta HCG levels increases after an initial drop, it indicates reoccurrence of tumor cells in the uterus.
3. **Chemotherapy**: Chemotherapy drugs kill the tumor cells by hampering their ability to grow and divide. Compared to PSTT and ETT, molar pregnancy can be effectively treated by chemotherapy. Chemotherapy may be accompanied with surgery. Chemotherapy drugs may be administered intravenously, intramuscularly, or orally (pills). Drugs like dactinomycin (actinomycin D, cosmegen), methotrexate (multiple brand names), cyclophosphamide (Cytoxan, Neosar), etoposide (VePesid, Toposar, Etopophos), cisplatin (Platinol), and vincristine (Oncovin, Vincasar) are used in chemotherapy. The chemotherapy regime depends on the stage and type of the tumor. A low-risk invasive mole or a cancerous one can often be treated successfully with methotrexate alone or in combination with leucovorin. Dactinomycin can be used for patients with unhealthy liver. To treat the high risk, metastatic disease more than one drug called combination chemotherapy is followed. The success of the treatment is routinely gauged by testing the patient's HCG levels. Generally, 3–4 cycles of chemotherapy are given until beta HCG levels become normal. Chemotherapy may develop some side effects like fatigue, nausea, infection, vomiting, mouth sores, hair loss, appetite changes, neuropathy, and ototoxicity. These side effects may subside after the treatment is discontinued.

2.7 Cancers of Respiratory Organs

2.7.1 Nasopharyngeal Cancer (NPC)

Nasopharyngeal cancers are associated with the head and neck. They are rare throughout the world but common in certain geographical areas like southeast Asia, southern China, North Africa, Middle East, and Alaska (Marnouche 2017). NPC originates from the nasopharynx epithelium (Chou et al. 2017) (Fig. 2.17). The WHO has classified NPC on the basis of histology (Table 2.26). The symptoms of NPC are lump(s) in the neck, stuffiness in the nose, hearing problems, earache (due to fluid accumulation in the middle ear), ringing in the ear, sore throats, nosebleeding, unexplained weight loss, and difficulty in opening the mouth (Marnouche 2017). Scientists are not sure what exactly causes nasopharyngeal cancer. NPC has

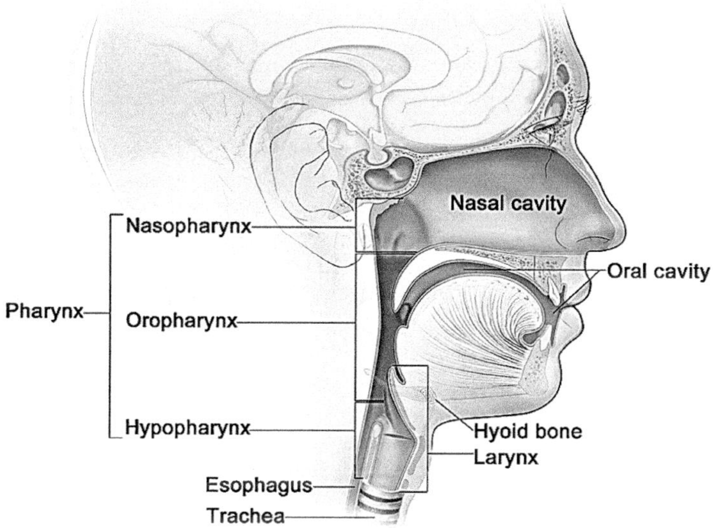

Fig. 2.17 Figure showing anatomy of the pharynx

Table 2.26 Classification of NPC by the WHO

Type	Classification	Characteristics
Type I	Keratinizing squamous carcinoma	It is characterized by well-differentiated cells that produce keratin. These are not associated with Epstein-Barr virus (EBV) infection
Type II	Non-keratinizing squamous carcinoma	It varies in cell differentiation but does not produce keratin. Associated with EBV infection
Type III	Non-keratinizing carcinoma	It is less differentiated, having highly varied cell types (clear cell, spindle cell, anaplastic). Associated with EBV infection

been strongly linked to the EBV. Though EBV infection is common, everyone who has EBV may not develop nasopharyngeal cancer (Chou et al. 2017).

Stages of Nasopharyngeal Cancer

The staging of NPC as framed by the American Joint Committee on Cancer is shown in Table 2.27.

Diagnosis and Treatment of NPC

NPC may be diagnosed by the following methods: (i) physical examination and blood test, (ii) endoscopy, (iii) biopsy, (iv) X-ray, (v) CT scan, (vi) PET scan, (vii) MRI, (ix) hearing tests, and (x) neurological exam.

Table 2.27 Staging of nasopharyngeal cancer

AJCC stages	Stage grouping	Characteristics of various stages
(0)	(Tis, N0, M0)	Tis: Presence of tumor within the wall of the cells within the nasopharynx, but it's not gone deeper
		N0: There is absence of cancer at neighboring lymph nodes
		M0: There is absence of cancer in the distant organs/locations within body
(I)	(T1, N0, M0)	T1: Presence of tumor at nasopharynx, and it may even spread to the throat region behind the mouth or the region of nasal cavity
		N0: There is absence of cancer at neighboring lymph nodes
		M0: There is absence of cancer in the distant organs/locations within body
(II)	(T1 or T0, N1, M0)	T1: Presence of tumor at nasopharynx, and it may even spread to the throat region behind the mouth or the region of nasal cavity
		T0: Absence of tumor in the nasopharynx but the appearance of cancer in the neck region may be observed. EBV-positive
		N1: Single or more than one lymph node at the region of the neck is affected by cancer or the lymph node at the back of the throat
		M0: There is absence of cancer in the distant organs/locations within body
	OR	
	(T2, N0 or N1, M0)	T2: Development of tumor at the throat region (both right and left upper part)
		N0: Neighboring lymph node has no cancer
		N1: One or more of the lymph nodes of the neck region may have the cancer and sometimes spreads to those nodes at the back of the throat. In both cases all the nodes do not exceed beyond 6 cm
		M0: There is absence of cancer in the distant organs/locations within body
(III)	(T1 or T0, N2, M0)	T1: Presence of tumor at the nasopharynx and the tumor may even develop at throat region behind the mouth or the region of nasal cavity
		T0: Tumor may not appear at nasopharynx but the appearance of cancer at the neck region in the lymph nodes with the presence of EBV
		N1: One or more of the lymph nodes (not larger than 6 cm) of both sides of the neck region affected by cancer
		M0: There is absence of cancer in the distant organs/locations within the body
	OR	
	(T2, N2, M0)	T2: Development of tumor at the upper region of the throat (right or left sides)
		N2: One or more of the lymph nodes of the neck region affected by cancer and size of node has not exceeded 6 cm
		M0: Cancer has not spread to the distant parts of the body
	OR	
	(T3, N0 to N2, M0)	T3: Growth of tumors within sinuses and/or in the neighboring bones

(continued)

Table 2.27 (continued)

AJCC stages	Stage grouping	Characteristics of various stages
		N0 to N2: Cancer may be absent or present at the neighboring lymph nodes to both the neck and throat and the size of node has not exceeded 6 cm
		M0: Cancer has not spread to the distant parts of the body
(IVA)	(T4, N0 to N2, M0)	T4: Development of tumor at various locations such as the skull and/or cranial nerves, throat region, salivary gland, or eyes or the neighboring tissues
		N0 to N2: Cancer may be absent or present at the neighboring lymph nodes to both the neck and throat and the size of node has not exceeded 6 cm
		M0: Cancer has not spread to the distant parts of the body
	OR	
	(Any T, N3, M0)	Any T: Tumor has or has not grown into structures outside the nasopharynx
		N3: Cancer travels to lymph nodes usually beyond 6 cm, sometimes even around the above region of the collarbone
		M0: Cancer has not spread to the distant parts of the body
(IVB)	(Any T, any N, M1)	Any T: Tumor has or has not grown into structures outside the nasopharynx
		Any N: Cancer may or may not have reached to the neighboring lymph nodes
		M1: The presence of cancer in the distant parts of the body

The treatments for NPC are:

1. **Radiation therapy:** It involves destruction of the cancerous cells of the body, by the use of an "X-ray" and/or other useful radiations. This procedure involves several sittings. It may be of different types. (i) *External beam radiation therapy*: this process of therapy is one of the most common forms when it comes to use of radiation for NPC treatment; the process is known as the "external beam radiation therapy." An improved version of this known as intensity-modulated radiation therapy (IMRT) is more effective in reducing the damage to healthy cells and offers fewer side effects. (a) *Proton therapy*: proton therapy may be provided for advanced-stage NPC (skull-base tumors) that is located close to parts of the central nervous system, which includes the brain and spinal cord. (b) *Stereotactic radiosurgery*: stereotactic radiosurgery delivers radiation precisely to the tumor that has grown into the base of the skull or a tumor that has recurred at the base of the brain or skull. (ii) *Internal-beam radiation therapy or brachytherapy*: it involves radiation treatment through implants. To treat NPC using brachytherapy, tiny pellets or rods that contain radioactive materials are surgically implanted in or near the cancer site, by a doctor. This approach is most often used to treat the relapse of NPC. It may also be used to treat new tumor(s). The radiation therapy is not free from side effects as it causes redness or skin irritation, hypothyroidism, dry mouth, sore throat, mouth sores, difficulty in swallowing, appetite loss, bone pain, fatigue, nausea, pain, hearing loss, and formation of earwax.

2. **Chemotherapy:** the chemotherapy drugs cause the destruction of cancerous cells. A chemotherapy schedule usually consists of a specific number of cycles given over a set period of time. Both chemotherapy and radiation therapy may be used to treat NPC. It may lead to dryness in the mouth, general fatigue, nausea, vomiting, hair loss, constipation or diarrhea, loss of appetite, and change in sense of taste.
3. **Surgery:** The tumor along with some surrounding healthy tissue may be surgically removed. If the surgical oncologist suspects that lymph nodes were attacked by cancer, then surgical removal of the lymph nodes is also done (neck dissection).

2.7.2 Lung Cancer

Lung cancer is one of the leading causes of cancer-related death in both men and women of high- income or low-middle income countries of the world (Aggarwal et al. 2016). It accounts for 1.6 million cancer-related deaths annually, with an estimated 1.8 million new cases annually, worldwide (Aggarwal et al. 2016). The two types of lung cancer are (i) non-small cell lung cancer and (ii) small cell lung cancer. Non-small cell lung cancer can be characterized into three subtypes: (a) squamous cell carcinoma, (b) adenocarcinoma, and (c) substantial cell lung cancer. Smoking is most firmly connected with small-cell lung cancer and squamous cell carcinoma (Fig. 2.18). Adenocarcinoma is mostly reported in non-smokers (Jaggi 2017).

The symptoms of lung cancer are cough and pain in the chest, change in color of the sputum, blood in sputum, breathing problems, hoarseness of voice, and harsh

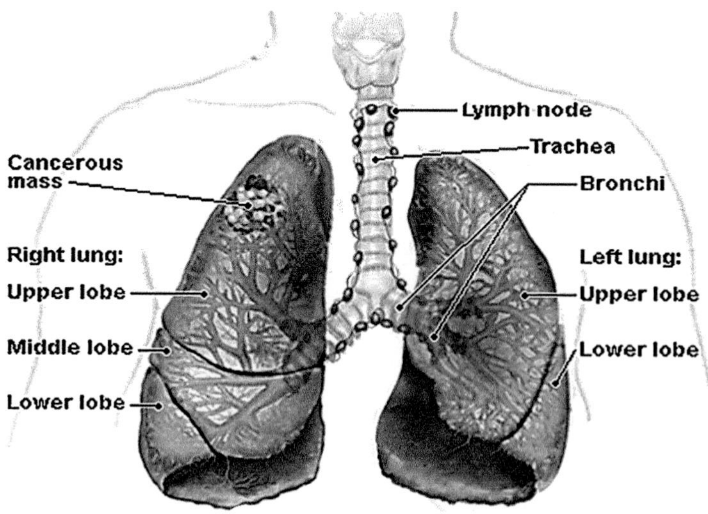

Fig. 2.18 Figure showing human lung cancer

Table 2.28 Staging of lung cancer

Stage I	It indicates that the cancer cells are limited to the lungs and have not spread to nearby lymph nodes
Stage II	It indicates that the cancer cells are present in the lungs and also in the nearby lymph nodes
Stage III	It indicates that the cancer cells are present in the lungs and in the lymph nodes in the middle of the chest
Stage IV	It indicates that the cancer cells are present in the lungs and have spread to the distant organs like the liver

sound with each breath, such as bronchitis and pneumonia. The lung cancer is strongly related to cigarette smoking, because 90% of the cases arise as a result of tobacco consumption/use. The risk associated with lung cancer rises with the pack-years of smoking (packs of cigarettes smoked per day multiplied by the number of years of smoking). It is interesting to note that compared to non-smokers, the passive smokers remain at a higher risk (> 24%). Radon gas and asbestos exposure also contribute to lung cancer (Webmed.com).

Staging of Lung Cancer

A simple classification to determine the stage of lung cancer (Reck and Rabe 2017) is shown in Table 2.28.

Diagnosis of Lung Cancer

The first step for diagnosis is through scanning (CT scan, MRI, or PET scan) followed by biopsy. A biopsy can be done during a bronchoscopy, wherein the doctor inserts a bronchoscope (diagnostic and therapeutic instrument for the air passage) through nostrils or mouths down to his lungs to collect the lung tissue. A liquid biopsy may also be performed wherein the free-floating DNA in the bloodstream is analyzed for mutation and other abnormalities (Jaggi 2017).

Treatment of Lung Cancer

The lung cancer therapy depends on the stage (stages 1–4), grade (low, moderate, or high grade), and type of cancer cells. The routine treatments include:

1. **Surgery:** Surgery is most commonly done for non-small cell lung cancer. If the tumor is localized in one of the lungs or nearby lymph nodes, then surgery is possible. For that, a small segment of the lung may be removed (wedge resection or segmentectomy), or small affected area along with the surrounding tissue may be removed (sub lobar resection), or the whole lobe of the lung may be removed (lobectomy), or the entire lung may be removed (pneumonectomy). All these procedures are performed keeping in mind the extent of tumor spread, the age and the health of the patient, and the chance of mortality.
2. **Radiofrequency ablation (RFA) and microwave ablation (MWA)**: For RFA, high-frequency electrical currents are passed through an electrode, creating a small region of heat. Similarly, in MWA, microwaves are used to create a small

region of heat. This heat destroys the lung cancer cells. Both RFA and MWA may be used to treat lung tumors.

3. **Radiation therapy**: "X-rays" can kill cancerous cells. It has many uses in lung cancer therapy: (a) as pretreatment to shrink the tumor before surgery, (b) as a posttreatment to eliminate any cancer cells that remain in the treated area after surgery, (c) as treatment of cancer of the lung which spreads to the brain and other distant organs. It may be given alone or in combination with chemotherapy. Patients with small localized lung cancer are being treated with stereotactic body radiation therapy (SBRT) which involves treatment of lung tumor with focused radiation beams. A three-dimensional conformal radiation therapy, called the intensity-modulated radiation therapy (IMRT), is a new technique based on a 3-D image of the tumor taken with CT scanning. This method minimizes radiation exposure to flanking normal lung tissue. In brachytherapy, radiation is delivered directly to the affected area by surgically placing the radioactive seeds at the edge of the surgical resection (after resection of primary tumor).
4. **Chemotherapy**: Drugs (single or in combination) that are toxic to cancer cells or slow down the tumor growth are used in chemotherapy. It is given after or before radiotherapy (as injection or as tablets) and surgery in patients who cannot have surgery. These drugs include cisplatin, carboplatin, etoposide (VePesid, vp-16), docetaxel (Taxotere), gemcitabine (Gemzar), pemetrexed (Alimta), paclitaxel (Taxol), vinblastine (Velbe), topotecan (Hycamtin), vinorelbine (Navelbine), and vincristine (Oncovin). Chemotherapy is not free from side effects as it may cause nausea with vomiting and damage to the WBCs. However, latest drugs exhibit less side effects. There are several case studies available which have shown that amalgamation of radiotherapy and chemotherapy give better response with less side effects.
5. **Biological therapy**: Biological therapies are "targeted therapies" or "personalized medicine" that uses the natural immunity of the patient to fight cancer. Molecular testing may allow the doctor to find out detailed information about the tumor such as mutations in the DNA of the tumor and specific proteins in the tumor. The biological therapies that are commonly used to treat lung cancer are alectinib (Alecensa), ceritinib (Zykadia), afatinib (Giotrif), cetuximab (Erbitux), nintedanib (Vargatef), bevacizumab (Avastin), gefitinib (Iressa), crizotinib (Xalkori), erlotinib (Tarceva), and thalidomide.

2.8 Miscellaneous Cancers

2.8.1 Brain Cancer

The brain is a delicate organ protected by cranium (bony skull). It is a soft mass of supportive tissues and nerve cells that connects to the spinal cord (Wuellner 2014). The nervous system is categorized into (i) into central nervous system (CNS) and (ii) peripheral nervous system (PNS). The brain and the spinal cord together form

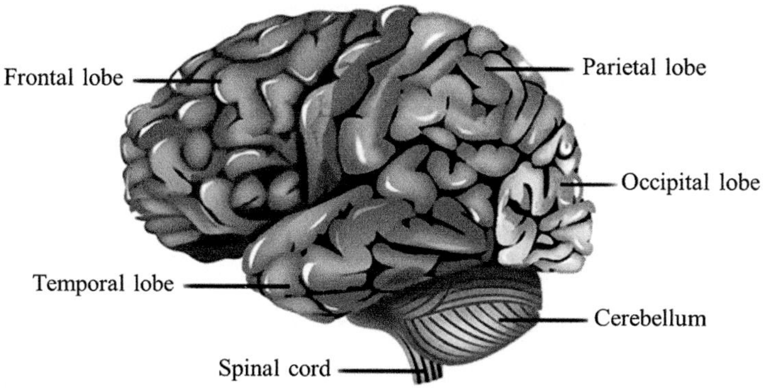

Fig. 2.19 An image showing major parts of the human brain

the CNS, while the spinal nerves which branch from the spinal cord and cranial nerve that branch from the brain form the PNS. Cerebrospinal fluid (CSF) is clear watery fluid that bathes and cushions the brain and spinal cord. The brain is composed of three main parts (Fig. 2.19):

(a) *Cerebrum*: it is the major part of the brain and has right and left hemispheres. It is interesting that the right cerebral hemisphere controls the left side of the body and the left cerebral hemisphere controls the right side of the body. The activities like speech and emotions, thinking, reading, and learning are controlled by the cerebrum.
(b) *Cerebellum*: being the second major part of the brain, it also consists of two hemispheres and a middle portion. The activities like standing, balance for walking, and other complex actions are controlled by it.
(c) *Brain stem*: being the bottommost portion of the brain, it connects cerebrum with the spinal cord. It includes midbrain, pons, medulla oblongata, and reticular formation. The body temperature, blood pressure, breathing, and other basic functions are controlled by the brain stem (Wuellner 2014).

Brain cancer is a disease wherein the cancer cells develop in the brain tissue and grow to form a tumor. However, malignant tumors are cancerous whereas benign tumors are noncancerous (Patil et al. 2013). The brain tumors may be (i) the primary brain tumors and (ii) the secondary brain tumors. The cancer cells which develop in brain tissue itself form the primary brain tumor.

(i) Primary brain tumors may be:
    (a) *Benign brain tumors*: it does not contain cancer cells. It grows slowly and has distinct boundaries. Benign brain tumors can be removed. It spreads rarely to other parts of the body. However, it can press on the sensitive areas of the brain which may cause serious health issues (Moore and Kim 2010).

Table 2.29 Grading of brain tumor

Grade	Explanation
Grade I	It indicates that the tissue is benign and cells are growing normally
Grade II	It indicates that the tissue is malignant, having relatively slow-growing cells with abnormal appearance
Grade III	It indicates that the tissue is malignant with actively dividing abnormal cells
Grade IV	It indicates that the tissue is malignant with most abnormal cells which reproduce rapidly (Wuellner 2014)

(b) *Malignant brain tumor*: it is also called as brain cancer. It grows quickly, has irregular boundaries, and spreads to nearby brain areas. It can also spread within the brain and spinal cord (Laws and Thapar 1993).

(ii) *Secondary brain tumors*: it is also called as metastatic brain tumors. The tumors that spread from the other part of the body to the brain form the secondary brain tumor. Cancer of the lungs, breast, and skin spreads to the brain via the bloodstream (Laws and Thapar 1993).

Malignant tumors that originate in the brain grow faster than benign tumor. The symptom of the brain cancer varies with the form, size, and location of the tumor within the brain (Moore and Kim 2010). The WHO has predicted 17 million deaths and 27 million new cases of brain cancer, worldwide, by the year 2030 (Magalhaes et al. 2016). In India, the incidence of CNS tumor ranges from 5 to 10 per 100,000 individuals. The causes of cancer in India are same as in other parts of the world. Causes for cancer can be either internal, such as inherited mutations, hormones, and immune conditions or external, such as tobacco consumption, diet, radiation, etc. (Yeole 2008).

Tumor Grades

Brain tumors can be graded to manage the treatment strategy and to predict the progress/success of the treatment (Jarmusch et al. 2016). The WHO has graded the brain tumors under the following types as shown in Table 2.29.

There are over 120 different types of brain tumors. Some common brain tumors are shown in Table 2.30.

Causes and Symptoms of Brain Cancer

Specific risk factors for brain tumors are as follows:

(i) *Ionizing radiation*: people who were subjected to radiation therapy may develop a tumor (Moore and Kim 2010).

(ii) *Family history*: very small percentage of brain tumors are linked to heredity (Laws and Thapar 1993).

(iii) *Exposure to infections, viruses, and allergens*: certain allergies and common infections including chicken pox and shingles and viruses have shown to cause brain tumors (Moore and Kim 2010).

Table 2.30 Type of brain tumors

Type of brain tumor	Explanation
Astrocytoma	In this type the tumors arise from astrocytes (star-shaped glial cells). These types of tumors are common in adults. In adults an astrocytoma arises mostly in the cerebrum. It can be of any grades such as:
	Grade I and II: Also called as low-grade glioma
	Grade III: Also called anaplastic astrocytoma or high-grade tumor
	Grade IV: also called as malignant astrocytic glioma or glioblastoma (Laws and Thapar 1993)
Oligodendroglioma	Tumor develops from cells that make the fatty substance which helps to cover and protect nerves
Meningioma	Tumor develops in the meninges, grows slowly, and can be graded as grade I, II, or III
Medulloblastoma	It is a grade IV tumor and occurs most commonly in children (Jarmusch et al. 2016). It usually arises in the cerebellum and is also called as primitive neuroectodermal tumor
Ependymoma	This type of tumor is commonly reported in children. In this type, the tumor arises from cells that line the ventricles or the central canal of the spinal cord
Brain stem glioma	In this type, low-grade or high-grade tumor occurs in the lowest part of the brain

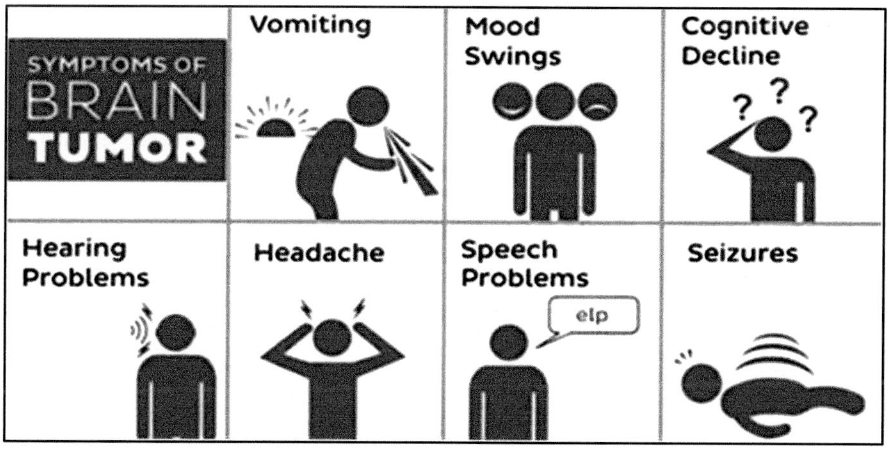

Fig. 2.20 Photograph showing the common symptoms of brain tumor

(iv) *Chemicals in the workplace*: workers of synthetic rubber, polyvinyl chloride, and petrochemical and oil industries are more vulnerable to brain tumors (Patil et al. 2013).

(v) *Cellular telephones/cell phones*: strong evidences regarding the consistent usage of cellular telephones/cell phones in causing brain tumors are still awaited (Laws and Thapar 1993).

The general symptoms of brain cancer are shown in Fig. 2.20.

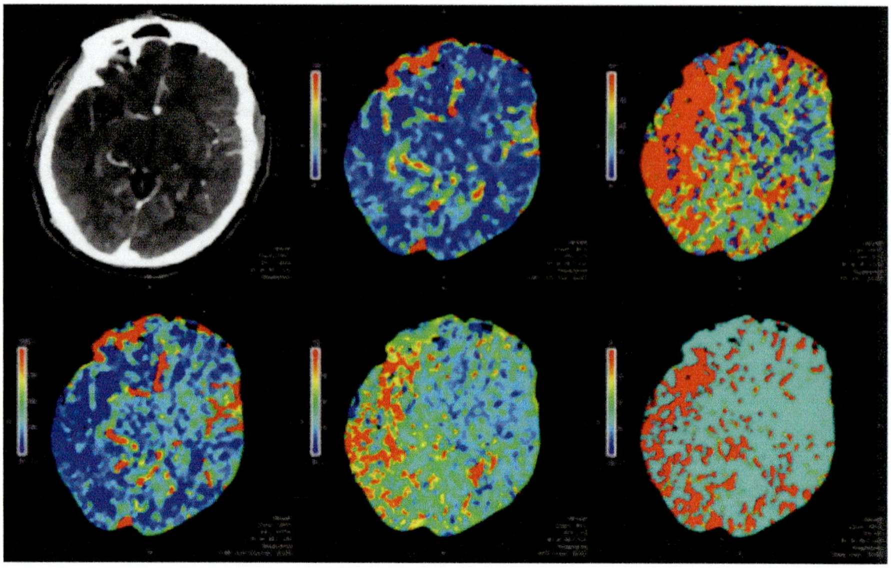

Fig. 2.21 Photograph showing the CT scan of the human brain (Wuellner 2014)

Diagnosis of Brain Cancer

In diagnosis a number of tests are performed to investigate symptoms of brain cancer. They are:

(i) *Physical examination*: In physical examination doctor will assess (neurological examination) the working of nervous system including hearing, vision, and movement. The doctor also studies patient's family medical history (Laws and Thapar 1993).

(ii) *CT* scan: It is a safe noninvasive test that uses an X-ray beam to click pictures of the internal parts of the body. The X-ray takes a single picture, but the CT scan can collect several pictures of the affected part (Fig. 2.21). During this scan, a dye is introduced inside the blood stream which facilitates to make the scanned pictures clear (Wuellner 2014).

(iii) *MRI scan*: MRI scan is also a noninvasive scan that uses a magnetic field and radiofrequency waves to build up very detailed pictures of the affected part. It can make three-dimensional images of the brain (Fig. 2.22). A contrast dye injected into the blood stream facilitates a clear scan. Through this scan, brain lesions and their effects on the surrounding tissues can be analyzed (Wuellner 2014).

(iv) *Biopsy*: When the scans do not provide the clear diagnosis, then biopsy is required to determine the type of brain tumor. The neurosurgeon may make a small opening in the skull and insert a needle to collect sample or may remove a large part of the skull to get the tumor (Wuellner 2014) (Fig. 2.23).

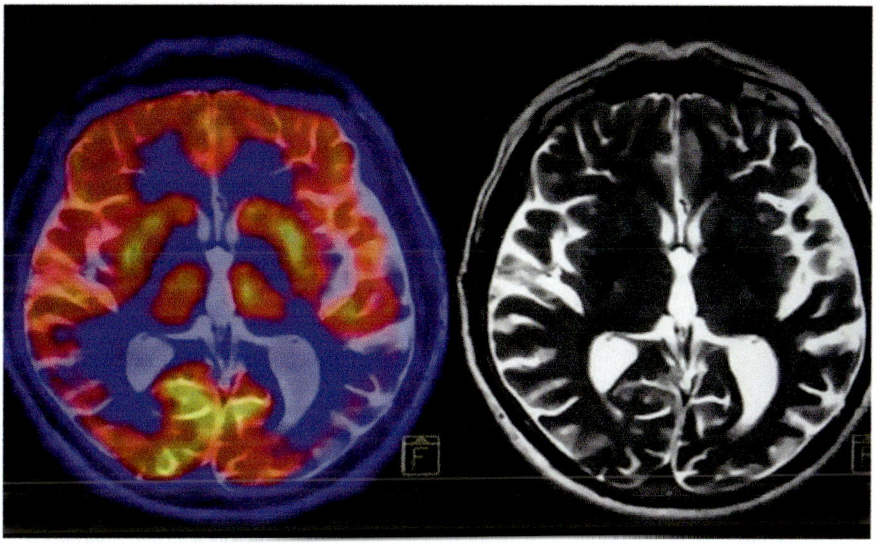

Fig. 2.22 Photograph showing the MRI scan of the human brain (Wuellner 2014)

Treatments of Brain Cancer

The treatment for brain cancer depends on the stage of disease, the symptoms, and the person's general health.

1. **Surgery:** Neurosurgery is the first promising treatment for brain tumors. Some tumors may be removed by gross total resection or partial resection or debulking. Some tumors cannot be removed because it may be attached to the important part of the brain. This type of tumors is called inoperable or unresectable tumor (Moore and Kim 2010). Before craniotomy begins, the patient is given general anesthetic, the scalp of the patient is shaved, and incision is made with a special type of saw to remove a piece of skull bone (Laws and Thapar 1993).
2. **Radiation therapy:** Two types of radiation therapy may be used to treat the brain tumor (Wuellner 2014):
   (a) *External radiation therapy*: a machine outside the body to generate and target the radiation beam at the head. Stereotactic radiosurgery (SRS), fractionated stereotactic radiotherapy (FSR), and whole brain radiotherapy (WBRT) are types of external radiation therapy (Laws and Thapar 1993).
   (b) *Internal radiation therapy*: it is also called as implant radiation or brachytherapy. Here, the radiation is delivered from inside the body by placing radioactive material (seed) directly into the tumor through surgery. It is not commonly used for treating the brain tumors (Jarmusch et al. 2016).
3. **Chemotherapy:** Chemotherapy drug travels through the blood stream and damages or destroys cancer cells. Temozolomide (Temodar) and bevacizumab (Avastin) are the most commonly used drugs to treat brain tumors. The tumor

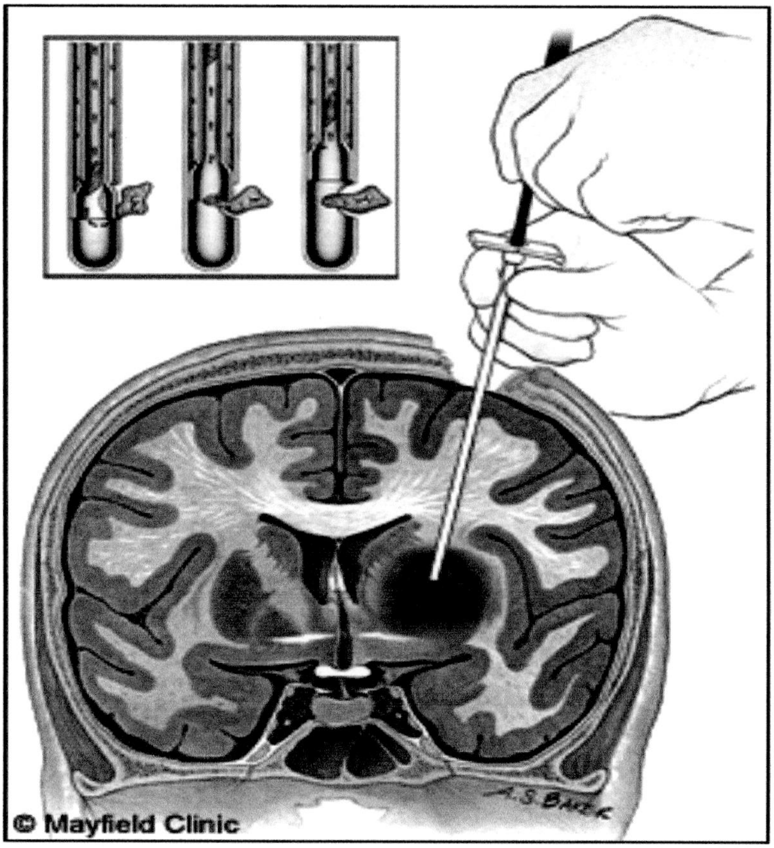

Fig. 2.23 Photograph showing needle biopsy (Wuellner 2014)

bed may also be treated with chemotherapy drugs (BCNU wafer) after removing the tumor (Laws and Thapar 1993).

2.8.2 Bone Cancer

Bone cancer arises when an abnormal mass of tissue or tumor (malignant) forms in a bone which may have the tendency to spread to other parts of the body. The bone cancer may be (i) *1° cancer*, a notorious bone cancer that can also spread from its origin (bone) to the surrounding tissues such as cartilages, and (ii) *2° cancer*, a type of cancer that spreads from another part of the body to the bones. The 1° bone cancers consist of:

(i) *Chondrosarcoma*: this is the second frequently occurring 1° bone cancers. It is the cancer of shoulders, pelvis, and thigh areas of the elderly. It occurs in the tough connective tissue (subchondral tissue) present between the bones.

(ii) *Multiple myeloma*: another most common type of bone cancer that is generally reported in older adults. It occurs when cancer cells proliferate in the bone marrow and affects other bones.
(iii) *Ewing's sarcoma*: it is a rare cancer that affects either the soft tissues associated with the bones or the long bones of the body (bones of arms, legs, and pelvis), in young adults and children.
(iv) *Osteosarcoma:* it is also known as osteogenic sarcoma and generally affects tips of the long bones in the arms and legs, hips, shoulders, or other locations, in adolescents and children (Fig. 2.24).

Causes and Symptoms of Bone Cancer
The exact cause of the bone cancer is not known, but patients with long-term inflammatory diseases (such as Paget's disease) are at a higher risk of developing bone cancer, later in life. The people who have received radiation therapy, people with a close relative who has bone cancer, individuals with hereditary retinoblastoma (a type of eye cancer), individuals with a history of Paget's disease, people with

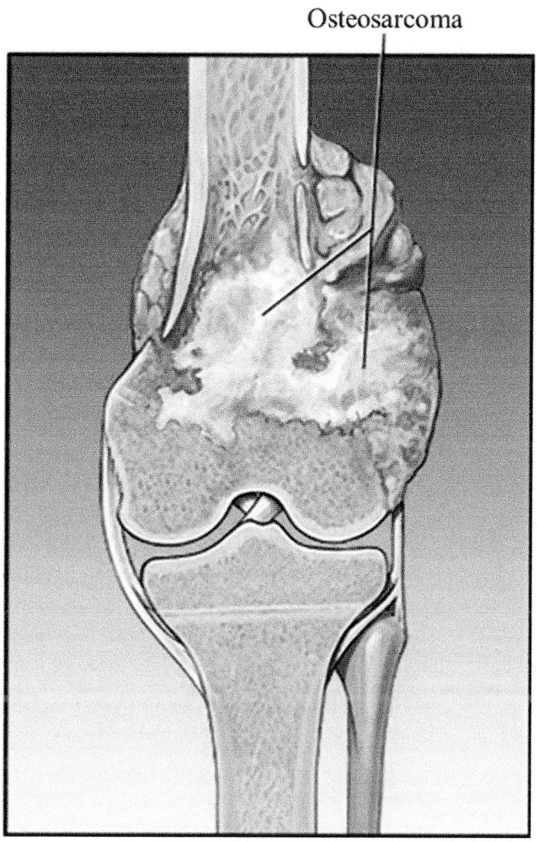

Fig. 2.24 An image showing osteosarcoma

Li-Fraumeni syndrome (a rare genetic condition), and young adults are at a higher risk of developing bone cancer. The symptoms or signs of bone cancer are pain and swelling in the affected bone, joint swelling and stiffness, limping (breaking or fractures in the bone with tumor), fever, feeling of being unwell, weight loss, and anemia.

Diagnosis of Bone Cancer
Based on the signs and/or symptoms and the techniques of imaging and blood testing, the presence of bone cancer can be confirmed. The diagnostic techniques used for the same are:

(i) *Computed tomography (CT) scans*: this type of scan helps in identifying whether cancer has reached to other organs such as lungs, liver, etc. It is also used for "CT-guided needle biopsy."
(ii) *X-rays*: generally, cancer of the bone is detectable on "X-ray scan." The affected bone usually looks ragged when scanned. Tumor in the bone can sometimes spread to the neighboring tissues. Lungs are checked for cancer attack by the X-ray of the chest.
(iii) *Magnetic resonance imaging (MRI) scans*: this scan is considered among the most useful tests in determining the tumor of the bone. It helps in identifying the presence of infection/disease in the brain and spinal cord.
(iv) *Radionuclide bone scans*: It is a nuclear imaging technique in which radioactive material is introduced into the blood stream for detection by a scanner. By this test the blood flow to the bone and bone cell activity can help to diagnose the cause of bone pain, inflammation, degenerative diseases, and even tumors.
(v) *Positron-emission tomography (PET) scans*: this scan involves tracing the location of consumption/absorption of radioactive sugar by the cancerous cells by the help of a camera. This test is important in searching for cancer at any location of the body. It may even reveal whether a tumor is benign or malignant.
(vi) *Biopsy*: the cells/tissue are observed under a specialized microscope, so that it may be revealed if the tumor is malignant or another form of disease. Depending on the nature and location of the tumor, there are many forms of biopsy:
   (a) *Needle biopsy*: the needle biopsy may be "fine needle" (aspiration) or "core needle." In fine needle, syringe and needle are used to get small portion of fluids together with cells out of the tumor. However, for the near-surface tumor, needle is aimed by feeling the tumor. But if it's very deep, the needle is usually guided down to the tumor while observing the direction in the CT scan (CT-guided needle biopsy), whereas larger needles are used to take out portion of small cylindrical part of tissue (diameter 1/16 inch and length 1/2 inch) in core biopsy. According to experts, core needle biopsy is much better than the FNA.
   (b) *Surgical bone biopsy*: in incisional biopsy the surgeon has to operate to take out the tissue sample of the tumor. When the whole of tumor is eliminated, then it's referred to as excisional biopsy.

Staging of Bone Cancer

Cancer of the bone is staged I to IV. Generally, as a rule, the lesser the number, the lower the spread of cancer to other parts of the body and vice versa. The most advanced staging system is that put forward by the American Joint Committee on Cancer (AJCC) for the cancer of bones. The details of the staging system are tabulated in Table 2.31.

Table. 2.31 AJCC staging of bone cancer

The cancer stages (AJCC)	Grouping of the cancer stages	Features of cancer stages
(IA)	T1, N0, M0, G1, or GX	T1: It indicates that size of the cancer is 8 cm or less
		N0: It indicates that the cancer did not reach to neighboring nearby lymph nodes
		M0: It indicates that the cancer did not reach distant sites of the body
		G1: It indicates a grade 1 cancer
		GX: It indicates that the grade is not identifiable
(IB)	T2, N0, M0, G1, or GX	T2: It indicates that the size of the cancer is more than 8 cm
		N0. It indicates that the cancer did not reach to neighboring nearby lymph nodes
		M0: It indicates that the cancer did not reach distant sites of the body
		G1: It indicates a grade 1 cancer
		GX: It indicates that the grade is not identifiable
	OR	
	T3, N0, M0, G1, or GX	T3: It indicates that the location of the cancer is > 1 in a bone
		N0: It indicates that the cancer did not reach to neighboring lymph nodes
		M0: It indicates that the cancer did not reach distant sites of the body
		G1: It indicates a grade 1 cancer
		GX: It indicates that the grade is not identifiable
(IIA)	T1, N0, M0, G2 or G3	T1: It indicates that size of the cancer is 8 cm or less
		N0: It indicates that the cancer did not reach to neighboring lymph nodes
		M0: It indicates that the cancer did not reach distant sites of the body
		G2 or G3: It indicates that the cancer is of grade G2 or G3
(IIB)	T2, N0, M0, G2 or G3	T2: It indicates that the size of the cancer is beyond 8 cm
		N0: It indicates that the cancer did not reach to neighboring lymph nodes

(continued)

Table. 2.31 (continued)

The cancer stages (AJCC)	Grouping of the cancer stages	Features of cancer stages
		M0: It indicates that the cancer did not reach distant sites of the body
		G2 or G3: It indicates that the cancer is of grade G2 or G3
(III)	T3, N0, M0, G2 or G3	T3: It indicates that there are numerous cancer patches per bone
		N0: It indicates that the cancer did not reach to neighboring lymph nodes
		M0: It indicates that the cancer did not reach distant sites of the body
		G2 or G3: It indicates that the cancer is of grade G2 or G3
(IVA)	Any T, N0, M1a, Any G	Any T: It indicates that there are numerous spots of cancer and of various sizes per bone
		N0: It indicates that the cancer did not reach to neighboring lymph nodes
		M1a: It indicates that the cancer has reached the lungs
		Any G: It indicates that the cancer may fall under any grade
(IVB)	Any T, N1, any M, Any G	Any T: It indicates that there are numerous spots of cancer and of various sizes per bone
		N1: It indicates that the cancer does not reach to neighboring lymph nodes
		Any M: It indicates that the cancer has reached or not reached to the organs such as lungs or other bones
		Any G: It indicates that the cancer may fall under any grade category
	OR	
	Any T, any N, M1b, any G	Any T: It indicates that there are numerous spots of cancer and of various sizes per bone
		Any N: It indicates that the cancer has or has not reached to the neighboring lymph nodes
		M1b: It indicates that the cancer has reached to different organs such as the liver, brain, or other bones
		Any G: It indicates that the cancer may fall under any grade category

Treatment of Bone Cancer

The treatment regimes and options depend on the type, stage, grade of cancer, and possible side effects and patient's health. The treatment options include:

1. **Surgery**: The low-grade tumor can be surgically removed but for a high-grade tumor; a combination of treatments such as chemotherapy and/or radiation therapy and/or surgery is suitable. Sometimes it becomes necessary to remove the whole arm or leg along with the tumor; however, a wide excision surgical

technique has reduced the number of bone cancer cases involving amputations. The patients can be treated with "reconstructive surgery" rather than amputation.

2. **Chemotherapy**: Surgery alone is not the solution of osteosarcoma. It has to be supplemented with chemotherapy to increase the survival rates of bone cancer patients. Neoadjuvant chemotherapy or induction chemotherapy is the chemotherapy given prior to surgery. To shrink the high-grade tumors, three to four cycles of chemotherapy prior to surgery is recommended. Moreover, postoperative or adjuvant chemotherapy (chemotherapy after surgery) facilitates the patient to live longer. However, the chemotherapy drugs and treatment depend on the type of sarcoma. The commonly used chemotherapy drugs are cisplatin, cyclophosphamide, doxorubicin, dactinomycin, etoposide, ifosfamide, methotrexate, and vincristine. The negative impact of this therapy is based on dose as well as condition of patient and causes nausea and vomiting, loss of appetite, fatigue, hair loss, risk of infection, and diarrhea. However, after treatment is over, the negative impacts slowly disappear.
3. **Radiation therapy**: When surgery is not possible, then the radiation oncologist commonly performs "external beam radiation therapy" whose schedule usually consists of a specific number of treatments per week. Another therapy, "internal radiation therapy" or "brachytherapy," involves treatment using implants. For bone cancer, radiation therapy is most often used for patients who have a tumor that cannot be removed with surgery. Radiation therapy may also be done prior to surgery to shrink the tumor or post-surgery to destroy the remaining cancer cells. The side effects of this treatment are general fatigue, skin and belly problems, etc.

Recent Advances in Bone Cancer Therapy

In a study conducted by Luger et al. (2001), osteolytic sarcoma cells (2472) were injected into the intramedullary space of the femur of C3H/HeJ mice to understand the mechanism of pain in bone cancer. The mice were administrated daily with osteoprotegerin for 21 days. Movement-evoked pain and bone destruction were assessed routinely (10th, 12th, 14th, 17th, and 21st day). The results indicate that treatment with osteoprotegerin successfully stopped further bone destruction and reduced ongoing and movement-evoked pain.

In another study, Kotz et al. (2002) examined 721 patients (in institution) suffering from primary malignant bone tumors. From year 1965 to 1974, 154 patients were treated, of whom 36% underwent resection, 17% had no surgery, and 46% underwent amputation with overall survival rate of 24%. From year 1975 to 1984, 174 patients were treated, of whom 54% underwent limb salvage procedures, 12% had no surgery, and 34% underwent amputation or resection-replantation with overall survival rate of 46%. From year 1985 to 1994, 393 patients were treated of whom 77% underwent limb salvage procedures, 7% had no surgery, and 15% underwent amputation with overall survival rate of 62%. From this study it can be concluded that advancements in the treatment of 1° malignant bone tumors and an amalgam of amputation with poly-chemotherapy resulted into an increase in the overall survival rate of more than 60%. Therefore, it is suggested that combination of these therapies will prove more beneficial for cancer treatment (Goblirsch et al. 2006).

Recurrence of osteosarcoma at the site of surgical resection, chemotherapy, and radiation treatment is one of the major threats in successful cancer treatment. To deal with this, poly-l-lactic acid materials coated with selenium nanoparticles were generated as these materials decrease the density of osteosarcoma cell and promote the formation of healthy, noncancerous, osteoblast cells (Stolzoff and Webster 2016). Recently, a novel nano-device based in doxorubicin (DOX)-loaded mesoporous silica nanoparticles (MSNs) as nanoplatforms has been developed for the treatment of bone cancer. This nano-device resulted into almost 100% osteosarcoma cell death as the cytotoxicity of this device is eight times higher than that of free drug delivery system. These findings have open up new vistas for the treatment of bone cancer (Martínez-Carmona et al. 2018). Bone defects caused by surgeries are a challenge during bone cancer treatment. Recently, Fe-$CaSiO_3$ composite scaffold material has been developed with 3-D printing technique which has high compressive strength and mechanical support. Such type of materials with high compressive and mechanical strength can function as efficient and versatile biomaterials for the treatment of bone cancer as well as future regeneration of cortical bone defects (Winkler et al. 2018).

2.8.3 Thyroid Cancer

The thyroid is an endocrine gland that plays a crucial role in the metabolic activities and growth processes of the body (Bãrbuş et al. 2017). Thyroid cancer (Fig. 2.25) is the most commonly known endocrine neoplasia, well-known in women as compared to men (Bãrbuş et al. 2017). The reports of thyroid cancer in the United States

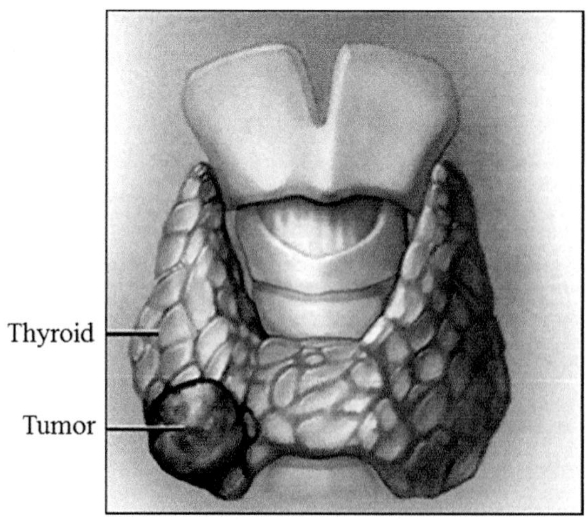

Fig. 2.25 An image showing thyroid tumor

are alarming. In the year 1975, 4.9 cases per 100, 000 persons were reported which grew up to 14.3 cases per 100,000 persons in the year 2014 (Lin et al. 2017). The increasing incidences of thyroid cancer are radiation, environmental factors, genetic factors, etc. Most importantly, better accessibility to early detection of any changes in the thyroid gland is low (Bãrbuş et al. 2017). The general classes of thyroid cancer are (i) follicular, (ii) papillary, (iii) anaplastic, and (iv) medullary thyroid cancer. The symptoms that are often observed are a lump or swelling in the neck which might grow quickly, pain around the neck that may reach the ears, hoarseness of voice, trouble in swallowing, and trouble in breathing and cough.

Staging of Thyroid Cancer

The stages of thyroid cancer framed by the AJCC are mentioned in Table 2.32.

Table 2.32 Staging of thyroid cancer

Stages of cancer (AJCC)	Patients' age during diagnosis (years)	Grouping of cancer stage	Features of various stages
(I)	Below 55	Any T	Any T: It indicates that the cancer is of any size
		Any N	Any N: It indicates that the cancer did not reach neighboring lymph nodes
		M0	M0: It indicates that the cancer did not reach the other locations
	OR		
	55 and above	T1	T1: It indicates that the cancer size is smaller than 2 cm and located in the thyroid
		N0 or NX	N0: It indicates that the cancer has reached to neighboring lymph nodes
		M0	M0: It indicates that the cancer did not reach the other locations
	OR		
	55 or above	T2	T2: It indicates that size of the cancer exceeds 2 cm, but below 4 cm, and restricted to the thyroid
		N0 or NX	N0: It indicates that cancer has reached to neighboring lymph nodes
		M0	M0: It indicates that the cancer did not reach the other locations
(II)	Below 55	Any T	Any T: It indicates that the size of cancer is of any type
		Any N	Any N: It indicates that sometimes cancer may even reach to neighboring lymph nodes but may as well not
		M1	M1: It indicates that cancer reached to various organs like the liver, etc.

(continued)

Table 2.32 (continued)

Stages of cancer (AJCC)	Patients' age during diagnosis (years)	Grouping of cancer stage	Features of various stages
	OR		
	55 or above	T1	T1: It indicates that the size of cancer is smaller than 2 cm and restricted to thyroid
		N1	N1: It indicates that cancer has reached the neighboring lymph nodes
		M0	M0: It indicates that the cancer did not reach the other locations
	OR		
	55 or above	T2	T2: It indicates that the size of the cancer is larger than 2 cm but below 4 cm and is restricted to thyroid
		N1	N1: It indicates that cancer has reached to neighboring lymph nodes
		M0	M0: It indicates that the cancer did not reach the other locations
	OR		
	55 or above	T3a or T3b	T3a: It indicates that size of the cancer is more than 4 cm and restricted to the thyroid
			T3b: It indicates that the cancer has invaded the strap muscles of the thyroid
		Any N	Any N: It indicates that that the cancer may (or may not) have reached to neighboring lymph nodes
		M0	M0: It indicates that the cancer did not reach the other locations
(III)	55 or above	T4a	T4a: It indicates that size of the cancer is of any type and its growth exceed to neighboring tissues around the neck, trachea, and esophagus, among others
		Any N	Any N: It indicates that that the cancer may (or may not) have reached to neighboring lymph nodes
		M0	M0: It indicates that the cancer did not reach the other locations
(IVA)	55 or above	T4b	T4b: It indicates that size of cancer is of any type; it has spread from the thyroid gland and reached up to the spine and vessels of the blood
		Any N	Any N: It indicates that that the cancer may (or may not) have reached to neighboring lymph nodes
		M0	M0: It indicates that the cancer did not reach the other locations
(IVB)	55 or above	Any T	Any T: It indicates that the size of cancer is of any type
		Any N	Any N: It indicates that the cancer may (or may not) have reached to neighboring lymph nodes
		M1	M1: It indicates that the cancer has reached to various organs like the liver, etc.

1° Tumor inaccessible TX

No trace of 1° Tumor T0.

Diagnosis of Thyroid Cancer

A diagnostic algorithm shown in Fig. 2.26 clearly explains the steps involved in the analysis of thyroid cancer. The diagnostic features of thyroid cancer depend upon the size and sonography patterns of suspected thyroid nodules (Schmidbauer et al. 2017). Thyroid nodules larger than 1 cm are subjected to scintigraphic examination. The thyroid-stimulating hormone (TSH) level is analyzed by routinely performing technetium 99 m pertechnetate (99mTcO4−) thyroid scan, so that the patient's risk of fine needle aspiration (FNA) is reduced. A radionuclide scan should only be done when TSH level is subnormal.

Treatment of Thyroid Cancer

There are five standard treatments for thyroid cancer:

(i) *Surgery*: it is the most common treatment that consists of (a) *lobectomy*, the removal of the lobe where the thyroid is found; (b) *near-total thyroidectomy*, the removal of all lobes except the small part of the thyroid; (c) *total thyroidectomy*, the complete removal of the thyroid; and (d) *tracheostomy*, the surgery to create a hole or an opening called as stoma into the windpipe to help the patient breath.

(ii) *Radiation therapy*: They are of two kinds: (a) *external radiation therapy*, in which it uses an external device to generate radiations for treating the internally affected (cancerous) area of the body, and (b) *internal radiation therapy*, in which a radioactive substance is placed inside the body near the cancer for its treatment.

(iii) *Chemotherapy*: it uses drugs (single or in combination) to inhibit/kill the cancer cells. These drugs may be injected intravenously, intramuscularly, or orally, and their dose depends upon stage of the thyroid cancer.

(iv) *Thyroid hormone therapy*: it involves blocking the production of thyroid-stimulating hormone (TSH) as it increases the chance of thyroid cancer cell proliferation or cancer relapse.

(v) *Targeted therapy*: as the name indicates, this treatment ensures identification and attacks specific infected cancer cells without damaging normal cells (e.g., tyrosine kinase inhibitor therapy).

2.8.4 Soft Tissue Cancer

Soft tissues refer to nonepithelial extra-skeletal tissues (Gogoi et al. 2017). Soft tissue cancers are extremely rare types of cancer that contributes to less than 1% of all known cancers. They can occur at any age and may be benign or malignant (Gogoi et al. 2017). Based upon the cell types, they are of more than 50 different forms

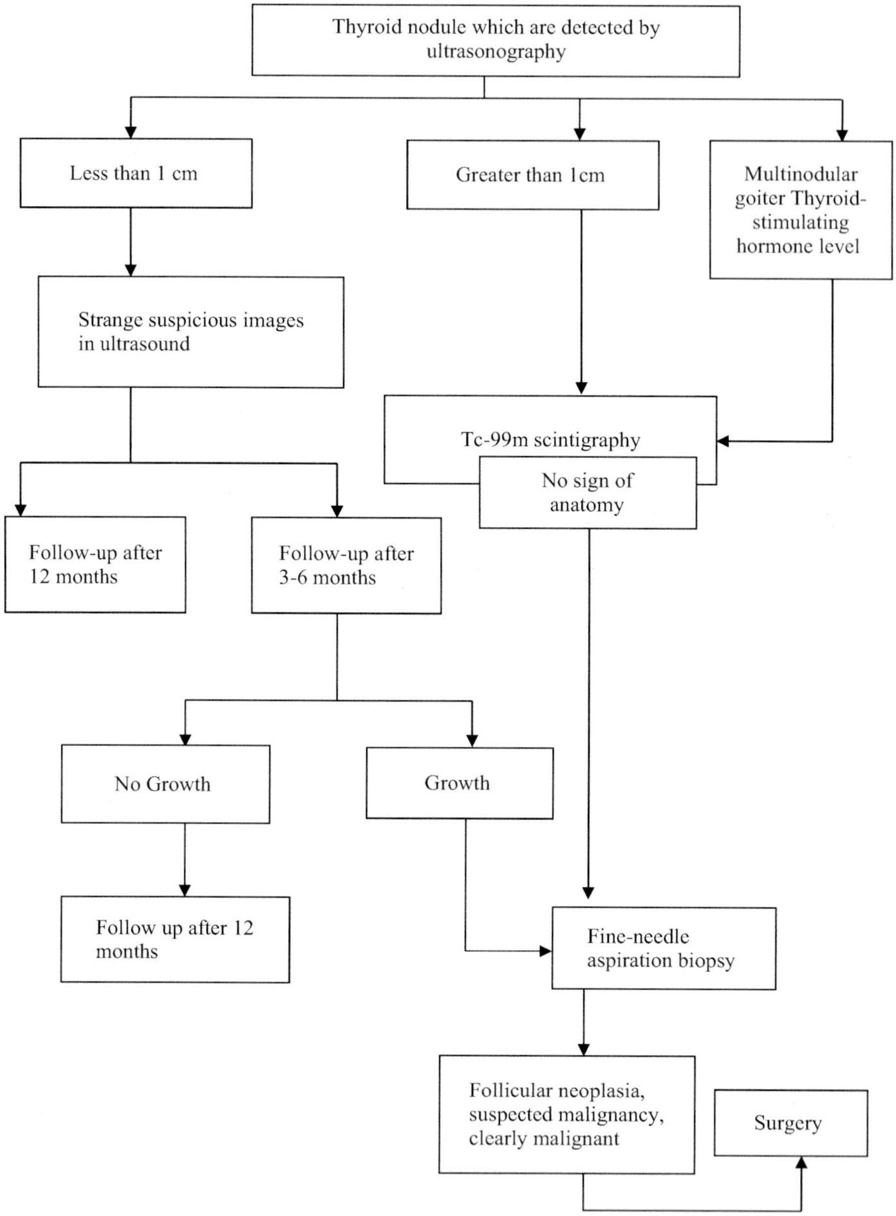

Fig. 2.26 Flow-chart for the assessment of thyroid nodules (Schmidbauer et al. 2017)

(Maguire et al. 2017). WHO has reported more than 100 distinct histological subtypes including angiosarcoma, synovial sarcoma, rhabdomyosarcoma, liposarcoma, leiomyosarcoma, and Ewing's sarcoma (Hoang et al. 2018).

Symptoms and Diagnosis of Soft Tissue Cancer

The indications of soft tissue cancer are lump(s) growing in any part of the body, persistent abdominal pain, blood in vomiting and stool, and black or tarry color of the stool due to bleeding inside the stomach and bowels. The diagnosis of soft tissue cancer includes (i) *medical history* (if a person in the family had soft tissue cancer, then there is a possible risk for the others also) and (ii) *imaging tests* (these tests use X-rays, sound waves, radioactive substances, and magnetic fields to create an image of the inner part of the body suspected to have cancer and also to analyze the extent of cancer spread). These are:

(a) *X-ray*: an X-ray scan of the chest helps to diagnose if the lungs were attacked by sarcoma.
(b) *CT scans*: it makes detailed cross-sectional images of our body. It might use biopsy needle if it has to examine a tumor inside the body.
(c) *MRI*: this technique works on the normal magnetic properties of the human body to produce informative images.
(d) *Ultrasound*: this process uses sound waves together with echoes produced, to capture images of the parts of the organs inside our body. Very special equipment known as transducer sends out the sound, and the echoes bounce away from the organ.
(e) *PET scan*: in this scan, radioactive sugar (fluorodeoxyglucose) is introduced intravenously. The cancer cells consume more sugar as compared to normal tissue, and a scanner can spot the radioactive deposits and thus tell the location of cancer.

Grading and Staging of Soft Tissue Cancer

The grade of a sarcoma helps predict how rapidly it will grow and spread. It helps to determine treatment options (Maguire et al. 2017). Table 2.33 shows detailed staging of soft tissue cancer.

Treatment of Soft Tissue Cancer

1. **Surgery**: Among the soft tissue sarcomas, 15% of the tumors are those of the retroperitoneum. It is very challenging for surgical resection of the tumors with clear margins. But for low-graded tumors, resection with wide margins is the standard procedure (Maguire et al. 2017). During surgery if the nearby lymph node is suspected to have cancer, then they are removed. Neoadjuvant treatment (chemotherapy or radiation therapy or both: chemoradiation) and adjuvant treatment may be deployed prior and post-surgery, respectively.

Table 2.33 Staging of soft tissue cancer of trunk and extremities

AJCC stage	Stage grouping	Description
IA	T1, N0, M0, G1 or GX	T1: It indicates that the cancer is less than or equal to 5 cm
		N0: It indicates that the cancer has not spread to nearby lymph nodes
		M0: It indicates that the cancer has not spread to other locations
		G1 or GX: It indicates that the cancer is of grade 1 or the grade cannot be determined
IB	(T2, T3, T4), N0, M0, G1 or GX	T2: It indicates that the cancer is larger than 5 cm but not more than 10 cm
		T3: It indicates that the cancer is larger than 10 cm but not more than 15 cm
		T4: It indicates that the cancer is larger than 15 cm
		N0: It indicates that the cancer has not spread to nearby lymph nodes (N0)
		M0: It indicates that the cancer has not spread to other locations
		G1 or GX: It indicates that the cancer is of grade 1 or the grade cannot be determined
II	T1, N0, M0, G2 or G3	T1: It indicates that the cancer is 5 cm or smaller
		N0: It indicates that the cancer has not spread to nearby lymph nodes
		M0: It indicates that the cancer has not spread to other locations
		G2 or G3: It indicates that the cancer is of grade 2 or grade 3
IIIA	T2, N0, M0, G2 or G3	T2: It indicates that the cancer is larger than 5 cm but not more than 10 cm
		N0: It indicates that the cancer has not spread to nearby lymph nodes
		M0: It indicates that the cancer has not spread to other locations
		G2 or G3: It indicates that the cancer is of grade 2 or grade 3
IIIB	T3 or T4, N0, M0, G2 or G3	T3 or T4: It indicates that the cancer is larger than 10 cm but not more than 15 cm or larger than 15 cm
		N0: It indicates that the cancer has not spread to nearby lymph nodes
		M0: It indicates that the cancer has not spread to other locations
		G2 or G3: It indicates that the cancer is of grade 2 or grade 3
IV	Any T, N1, M0, any G	Any T: It indicates that the cancer is of any size
		N1: It indicates that the cancer has spread to nearby lymph nodes
		M0: It indicates that the cancer has not spread to other locations
		Any G: It indicates that the cancer is of any grade
	Any T, any N, M1, any G	OR
		Any T: It indicates that the cancer is of any size
		Any N: It indicates that the cancer has spread to any nearby lymph node(s)
		M1: It indicates that the cancer has spread to other locations
		Any G: It indicates that the cancer is of any grade

2. **Radiotherapy:** The radiation therapy includes:
   (i) *External beam radiation*: usually, intensity-modulated radiation therapy (IMRT) is given 5 days per week, to treat sarcomas. With IMRT radiation can be easily focused on the cancer which reduces the damage to normal cells.
   (ii) *Proton beam radiation*: this technique uses streams of protons to treat the cancer. Technically, it has some advantages over IMRT, but for treating soft tissue sarcoma, it is of little use.
   (iii) *Intraoperative radiation therapy (IORT)*: in this technique after removing the cancer, the wound is subjected to single large dose of radiation to restrict any residual cancerous cells from spreading.
   (iv) *Brachytherapy*: it may also be called internal radiation therapy and can be used along with external radiation therapy. To treat the soft tissue sarcoma, small pellets of radioactive materials are kept in catheters to be placed near the cancer cells.
   (v) The radiation therapy causes certain side effects which include skin irritations, bone weakness, nausea, vomiting, difficulty in swallowing, general fatigue, and lung problems (difficulty in breathing).
3. **Chemotherapy**: The most commonly used drugs for soft tissue cancer are ifosfamide (Ifex) and doxorubicin (Adriamycin) (NCI 2018). When ifosfamide is used, a drug mesna (non-chemo drug) is also administered to protect the bladder from ifosfamide-mediated toxicity. Other chemo drugs are also used including epirubicin, dacarbazine (DTIC), docetaxel (Taxotere), temozolomide (Temodar), vinorelbine (Navelbine), gemcitabine (Gemzar), eribulin (Halaven), and trabectedin (Yondelis). The drugs may also be given in combinations: MAID (ifosfamide, mesna, dacarbazine, and adriamycin) or AIM (ifosfamide, adriamycin, and mesna) (NCI 2018). The common ill effects of chemotherapy include nausea and vomiting, loss of appetite, mouth sores, hair loss, fatigue, and low blood count. The severe effects include nerve damage (neuropathy), damage to ovaries and testicles, and even infertility.
4. **Targeted therapy:** Targeted therapy may be used for the treatment of advanced soft tissue sarcomas that did not respond to chemotherapy. Pazopanib drug blocks tyrosine kinases (cell enzymes) which are concerned with cell survival and growth. Thus, it may help to slow down the tumor growth and reduce the side effects in patients with sarcomas for which surgery could not be done.

2.8.5 Retroperitoneal Sarcoma (RPS)

Retroperitoneal sarcomas (cancer of connective tissue) occur in the retroperitoneum which is an area behind the peritoneum and covers the abdominal organs. The RPS is an infrequent disease and contributes to 0.1–0.2% of all malignancies. Liposarcoma and leiomyosarcoma are the main types of sarcoma, while the other types include solitary fibrous tumor, Ewing's sarcoma, synovial sarcoma, malignant nerve sheath tumor, and pleomorphic sarcoma. The symptoms of retroperitoneal sarcoma include a lump in the abdomen causing dull back pain, increase in abdominal girth, bleeding in stools, early satiety, weight loss, and hernia or anemia. The symptoms are diagnosed using a series of tests that may include (i) *sonography*, (ii) *MRI*, (iii) *CT scan*, (iv) *histopathology*, and (v) *blood test*.

Grading and Staging of Retroperitoneal Sarcoma

The AJCC staging of cancer is the modified and informative version of TNM staging. The staging of retroperitoneal carcinoma is shown in Table 2.34.

Table 2.34 Grading of retroperitoneum sarcomas of trunk and extremities

AJCC stage	Sarcoma stage groups	Stage description
IA	T1, N0, M0, G1, or GX	T1: It indicates that the cancer is 5 cm or smaller
		N0: It indicates that the cancer has not spread to nearby lymph nodes
		M0: It indicates that cancer has not yet spread to the other locations
		G1: It indicates that the cancer is of grade 1
		GX: It indicates that the grade cannot be determined
IB	(T2, T3, T4), N0, M0, G1 or GX	T2: It indicates that the cancer is larger than 5 cm but not more than 10 cm
		T3: It indicates that the cancer is larger than 10 cm but not more than 15 cm
		T4: It indicates that the cancer is larger than 15 cm
		N0: It indicates that the cancer has not spread to nearby lymph nodes
		M0: It indicates that cancer has not yet spread to the other locations
		G1: It indicates that the cancer is of grade 1 or the grade cannot be determined
II	T1, N0, M0, G2 or G3	T1: It indicates that the cancer is 5 cm or smaller
		N0: It indicates that the cancer has not spread to nearby lymph nodes
		M0: It indicates that the cancer has not yet spread to the other locations
		G2 or G3: It indicates that the cancer is of grade 2 or grade 3
IIIA	T2, N0, M0, G2 or G3	T2: It indicates that the cancer is larger than 5 cm but not more than 10 cm
		N0: It indicates that the cancer has not spread to nearby lymph nodes
		M0: It indicates that the cancer has not yet spread to the other locations
		G2 or G3: It indicates that the cancer is of grade 2 or grade 3

(continued)

Table 2.34 (continued)

AJCC stage	Sarcoma stage groups	Stage description
IIIB	T3 or T4, N0, M0, G2 or G3	T3 or T4: It indicates that the cancer is larger than 10 cm but not more than 15 cm or the cancer is larger than 15 cm
		N0: It indicates that the cancer has not spread to nearby lymph nodes
		M0: It indicates that cancer has not yet spread to the other locations
		G2 or G3: It indicates that the cancer is of grade 2 or grade 3
	Any T, N1, M0, any G	Any T: It indicates that the cancer is of any size
		N1: It indicates that the cancer has spread to nearby lymph nodes
		M0: It indicates that the cancer has not yet spread to the other locations
		Any G: It indicates that cancer can be of any grade
IV	Any T, any N, M1, any G	Any T: It indicates that the cancer is of any size
		N1: It indicates that the cancer has spread to nearby lymph nodes
		M1: It indicates that the cancer has spread to the other locations such as the lungs
		Any G: It indicates that cancer can be of any grade

Treatment of Retroperitoneal Sarcoma

The treatment of retroperitoneal sarcoma includes:

1. **Surgery**: It aims at removing the tumor along with its margins. En bloc surgery (removing tumor along with part(s) of any organ close to it) may be performed to remove whole of the cancerous tissue.
2. **Radiotherapy**: Treatment high-energy radiation may be performed before or after the surgery to completely eradicate the cancer. Radiotherapy remains the only option if there is risk of surgery on patient's health/survival.
3. **Chemotherapy**: The chemotherapeutic drugs destroy the cancer cells. Ewing's sarcoma, synovial sarcoma, and rhabdomyosarcoma are sensitive to chemotherapy, and so full treatment may not be possible through it. However, surgical removal of leiomyosarcomas, followed by chemotherapy, may prove beneficial. On the other hand, chemotherapy is not used to treat liposarcoma because they are not sensitive to chemo drugs.

2.9 Conclusions

Cancer is a group of diseases wherein the cells grow abnormally and uncontrollably and have a tendency to spread and damage other parts of the body. There are more than 100 types of cancer that affect different parts of the body. There is always a tussle between oncogenes and tumor suppressor genes as these act antagonistically to regulate many normal cellular processes, such as cell proliferation, metabolism, and survival. Strategies for an early detection of cancer are necessary as they are often detected at late stage, which reduce the patient's survival rate. So, advanced diagnostic and painless noninvasive techniques are key components to overcome cancer. In most of the cancers, tissue biopsy, sonography, CT scan, MRI scan, and PET scan are routinely performed to localize the cancer and formulate the cancer treatment regime, which generally includes chemotherapy, radiation therapy, chemoradiation, targeted therapy, and surgery. Although these treatments can kill the cancer cells, they can also be detrimental for the normal cells. The short-term or long-lasting side effects of these therapies often make the patient's life miserable and vulnerable to further infections, and the chances of survival remain unpredictable in higher stages of cancer. It is true that there is not one magic bullet to treat all types of cancers. Each type needs to be diagnosed before the final stage and needs specific treatment or a combination of treatments which further depends upon the location and stage of the cancer and the age and health status of the patient. It should not be like "operation (treatment) successful-patient died," instead "operation successful-patient survived," because every living being deserves a healthy life.

The information on the use of new molecular-based therapies, targeted therapies, immunological treatments, or an amalgam of therapies (e.g., chemotherapy plus Ayurveda) are still not adequate. To make it successful, we need deeper understanding of the physiology and biochemistry of cancer cells. Moreover, we need to protect our environment and change our lifestyle to prevent the occurrence of cancer. Thus, with meticulous studies, clinical trials, and analyses of cancers at cellular and molecular level, we can surely find a cure to various cancers.

References

Abbas J, Dar WR, Latief M, Farooq S, Parry MA, Ziaulhaq P, Sofi NU, Dar I (2016) Gastrointestinal stromal tumors: a review. J Mol Imag Dynamic 10(4172):2155–9937

Abbasi NR (2008) Utility of lesion diameter in the clinical diagnosis of cutaneous melanoma. Arch Dermatol 144:538–540

Advani P, Paulus A, Ailawadhi S (2019) Updates in prognostication and treatment of Waldenström's macroglobulinemia. Hematology/oncology and stem cell therapy (in press)

Aggarwal A, Lewison G, Idir S, Peters M, Aldige C, Boerckel W, Boyle P, Trimble EL, Roe P, Sethi T, Fox J, Sullivan R (2016) The state of lung cancer research: a global analysis. J Thorac Oncol 11(7):1040–1050

Albers P, Albrecht W, Algaba F (2015) Guidelines on testicular cancer: 2015 update. Eur Urol 68(6):1054–1068

Ali MZ, Sultana S (2012) Blood cancer. Bangladesh Research Institute 221:4303–4602

Ali I, Wani WA, Saleem K (2011) Cancer scenario in India with future perspectives. Cancer Ther 8:56–70

American Cancer Society. (2015) Cancer facts and figures

American Cancer Society, 2017, USA

Anjum F, Ravi N, Msood MA (2017) Breast cancer therapy: a mini review. MOJ Drug Des Dev Ther 1(2):00006

Babjuk M, Böhle A, Burger M et al (2017) EAU guidelines on non-muscle-invasive urothelial carcinoma of the bladder: update 2016. Eur Urol 71:447–461

Banumathy G, Cairns P (2010) Signaling pathways in renal cell carcinoma. Cancer Biol Ther 10:658–664

Bărbuş E, Peştean C, Larg MI, Piciu D (2017) Quality of life in thyroid cancer patients: a literature review. Clujul Med 90(2):147–153

Bermick (2008) Are tanning beds "safe". Human studies of melanoma. Pigment Cell Melanoma Res 21(5):517–519

Blackadar CB (2016) Historical review of the causes of cancer. World J Clin Oncol 7(1):54–86

Chaffer CL, Weinberg RA (2011) A perspective on cancer cell metastasis. Science 331:1559–1564

Chng WJ, Glebov O, Bergsagel PL (2007) Genetic events in the pathogenesis of multiple myeloma. Best Pract Res Clin Haemotal 20(4):571–596

Chou J, Lin Y, Kim J, You L, Xu Z (2017) Nasopharyngeal carcinoma– a review of the molecular mechanisms of tumorigenesis. Head Neck 30(7):946–963

Coimbra S, Neves R, Lima M, Belo L, Silva AS (2014) Waldenstrom's macroglobulinemia review. Rev Assoc Med Bras 60(5):490–499

Cserni G, Chmielik E, Cserni B, Tot T (2018) The new TNM-based staging of breast cancer. Virchows Arch 472(5):697–703

Finicle BT, Jayashankar V, Edinger AL (2018) Nutrient scavenging in cancer. Nat Rev Cancer 18:619–633

Flis S, Chojnacki T (2019) Chronic myelogenous leukemia, a still unsolved problem: pitfalls and new therapeutic possibilities. Drug Des Devel Ther 13:825–843

Fu JH, Rong TH, Li XD, Hu Y, Qu W, Hu YH, Li Q (2004) Chemotherapy followed by surgery in treatment of locally advanced esophageal carcinoma: a phase II trial. Ai Zheng 23(11):1473–1476

Fuhrman SA, Lasky LC, Limas C (1982) Prognostic significance of morphogenic parameters in renal cell carcinoma. Am J Surg Pathol 6(7):655–663

Gandaglia G, Ravi P, Abdollah F, Abd-El-Barr AE, Becker A, Popa I, Briganti A, Karakiewicz PI, Trinh QD, Jewett MA, Sun M (2014) Contemporary incidence and mortality rates of kidney cancer in the United States. Can Urol Assoc J 8:247–252

Gertz MA (2019) Waldenström macroglobulinemia: 2019 update on diagnosis, risk stratification, and management. Am J Hematol 94(2):266–276

GLOBOCAN (2012) v1.0, cancer incidence and mortality worldwide: IARC Cancer Base no. 11. International Agency for Research on Cancer Web site. http://globocan.iarc.f

Goblirsch MJ, Zwolak PP, Clohisy DR (2006) Biology of bone Cancer pain. Clin Cancer Res 12:6231s–6235s

Gogoi G, Borgohain M, Saikia P, Patel B, Hazauka RK, Brahma RC, Gogoi B, Saikia N, Rabha J (2017) Histomorphological study of soft tissue tumours and review of literature of rarer types. Int Clin Pathol J 4:00113–00123

Gubbels JAA, Claussen N, Kapur AK, Connor JP, Patankar MS (2010) The detection, treatment and biology of epithelial ovarian cancer. J Ovarian Res 3(8):1757–2215

Hartman RI, Lin JY (2019) Cutaneous melanoma-a review in detection, staging, and management. Hematol Oncol Clin North Am 33(1):25–38

Heitz APM, Odicino F, Maisonneuva P, Quinn MA, Benedet JL, Creasmam WT, Ngan HYS, Pecorelli S, Beller U (2006) Carcinoma of the ovary. Intituto Europeo di Oncologia 95(1):60033–60037

Hoang NT, Acevedo LA, Mann MJ, Tolani B (2018) A review of soft-tissue sarcomas: translation of biological advances into treatment measures. Cancer Manag Res 10:1089–1114

http://cancerwall.com

Huang S (2018) Physical activity and risk of testicular cancer-a systematic review. BMC Cancer 18(1):189

Ivanyi P, Fuehner T, Adam M, Eichelberg C, Herrmann E, Merseburger AS, Ganser A, Grunwald V (2014) Interstitial lung disease during targeted therapy in metastatic renal cell carcinoma: a case series from three centres. Med Oncol 31:147

Jabbour E, Kantarjian H (2018) Chronic myeloid leukemia: 2018 update on diagnosis, therapy and monitoring. Am J Hematol 93:442–459

Jaggi P (2017) A review article on lung cancer diagnosis and treatment. JMAHS 6(1):2318–9865
Jarmusch AK, Pirro V, Baird Z, Hattab EM, Cooks G (2016) Lipid and metabolite profiles of human brain tumors by desorption electrospray ionization. MS PNAS 113(6):1486–1419
Kamoun W, Elden S, Christine P, Lia L, Jason C, Irawati K, Richard H (2018) Targeting EphA2 in bladder cancer using a novel antibody-directed nanotherapeutic. Am Assoc Cancer Res 78(13):5771–5771
Karst AM, Drapkin R (2010) Ovarian cancer pathogenesis: a model of evolution. J Oncol 10:93237
Kato T, Inoue H, Imoto S, Tamada Y, Miyamoto T, Matsuo Y, Nakamura Y, Park JH (2016) Oncogenic roles of TOPK and MELK and effective growth suppression by small molecular inhibitors in kidney cancer cells. Oncotarget 7(14):17652
Khaider NG, Lane D, Matte I, Rancourt C, Piche A (2011) Targeted ovarian cancer treatment: the TRAILs of resistance. AMJ Cancer Res 2(1):75–92
Kirkali Z, Canda AE (2008) Open partial nephrectomy in the management of small renal masses. Adv Urol 309760
Kirkali Z, Tuzel E, Mungan MU (2001) Recent advances in kidney cancer and metastatic disease. BJU Int 88(8):818–824
Kotz R, Dominkus M, Zettl T, Ritschl P, Windhager R, Gadner H, Zielinski C, Salzer-Kuntschik M (2002) Advances in bone tumour treatment in 30 years with respect to survival and limb salvage. A single institution experience. Int Orthop 26(4):197–202
Lai JS, Beaumont JL, Diaz J, Khan S, Cella D (2016) Validation of a short questionnaire to measure symptoms and functional limitations associated with hand-foot syndrome and mucositis in patients with metastatic renal cell carcinoma. Cancer 122:287–295
Laura H (2016) Hodgkin's disease and lymphoma. Am Cancer Soc 32(7):34–42
Laws ER Jr, Thapar K (1993) Brain tumors. CA Cancer J Clin 43:263–271
Layke JC, Lopez PP (2006) Esophageal cancer: a review and update. Am Fam Physician 73:2187–2194
Leitzmann MF, Rohrmann S (2012) Risk factors for the onset of prostatic cancer: age, location, and behavioral correlates. Clin Epidemiol 4:1–11
Lin JS, Bowles EJ, Williams SB, Morrison CC (2017) Screening for thyroid cancer: updated evidence report and systematic review for the US preventive services task force. JAMA 31:888–903
Ljungberg B, Bensalah K, Bex A et al (2014) Guidelines on renal cell carcinoma. European Association of Urology. Available at https://uroweb.org/wp-content/uploads/10-Renal-Cell-Carcinoma_2017_web.pdf
Lorsbach RB, His ED, Dogan A, Fend F (2011) Plasma cell myeloma and related neoplasm. Am J Clin Pathol 136(2):168–182
Luger NM, Honore P, Sabino MA, Schwei MJ, Rogers SD, Mach DB, Clohisy DR, Mantyh PW (2001) Osteoprotegerin diminishes advanced bone cancer pain. Cancer Res 61(10):4038–4047
Magalhaes KCS, Vaz JPM, Gontijo PAM, Carvalho GTC, Christo PP, Simoes RT, Silva KR (2016) Profile of patients with brain tumors and tumors and the role of nursing care. Rev Bras Enferm 69(1):138–143
Maguire R, Andelkovic V, Chauhan H, Nolan GJ (2017) Advances in the management of soft tissue sarcomas- focus on emerging therapies. Clin Oncol 2:1189–1196
Marnouche E (2017) Diagnosis, therapeutic and evolutionary characteristics of nasopharyngeal Cancer in Morocco. J Cancer Sci Ther 9:439–444
Martínez-Carmona M, Lozano D, Baeza A, Colilla M, Vallet-Regí M (2018) A novel visible light responsive nanosystem for cancer treatment. Nanoscale 9(41):15967–15973
Miranda MB, Leusekes M, Kraus MP, Hanftein B, Fabarin A, BAerloches GM (2016) Secondary malignancies in chronic myeloid leukemia patients after imatinib based treatment: long term observation. Leukemia 30(6):1255–1262
Miura F, Takada T, Amano H, Yoshida M, Furui S, Takeshita K (2006) Diagnosis of pancreatic cancer. HPB (Oxford) 8(5):337–342
Moore K, Kim L (2010) Primary brain tumors: characteristics practical diagnostic and treatment approaches. Molecular mechanism of pathogenesis and current therapeutic strategies 10:42–75

Mottet N, Bellmunt J, Bolla M, Briers E, Cumberbatch MG, Santis MD, Fossati N, Gross T, Henry AM, Joniau S, Lam TB, Mason MD, Matveev VB, Moldovan PC, van den Bergh RCN, Van den Broeck T, van der Poel HG, van der Kwast TH, Rouvière O, Schoots IG, Wiegel T, Cornford P (2017) EAU-ESTRO-SIOG guidelines on prostate cancer. Part 1: screening, diagnosis, and local treatment with curative intent. Eur Urol 71(4):618–629

Muglia VF, Prando A (2015) Renal cell carcinoma: histological classification and correlation with imaging findings. Radiol Bras 48(3):166–174

National Comprehensive Cancer Network. Guidelines for patients version 1, 2016

National Comprehensive Cancer Network. Guidelines for patients 2017

Novara G, Ficarra V, Antonelli A, Artibani W, Bertini R, Carini M, Cosciani Cunico S, Imbimbo C, Longo N, Martignoni G, Martorana G, Minervini A, Mirone V (2010) Validation of the 2009 TNM version in a large multi-institutional cohort of patients treated for renal cell carcinoma: are further improvements needed? Eur Urol 58:588–595

Paek SC (2008) Cutaneous melanomas. Fitzpatrick's Dermatology in General Medicine 109(1):100–108

Paladini A, Boni A, Quadrini F, Cochetti G, Mearini E (2017) First report of simultaneous robotic treatment of kidney cancer and enucleation of pancreatic metastasis. Eur Uro 16(6):2385

Patil M, Prabhu S, Patil S, Patil S (2013) Brain tumor identification using K–means clustering. IJETT 4(3):354–357

Permutt WJ, Sellers TA (2009) Epidemiology of ovarian cancer. Methods Mol Biol 472:413–437

Pichler M, Hutterer GC, Chromecki TF, Jesche J, Kampel-Kettner K, Rehak P, Pummer K, Zigeuner R (2011) External validation of the Leibovich prognosis score for nonmetastatic clear cell renal cell carcinoma at a single European center applying routine pathology. J Urol 186:1773–1777

Pokharel M (2012) Leukemia a review article. Leukemia IJARP 2(3):397–407

Reck M, Rabe KF (2017) Precision diagnosis and treatment for advanced non–small cell lung cancer. N Engl J Med 377:849–861

Roett MA, Evans P (2009) Ovarian cancer: an overview. Am Fam Physician 80(6):609–616

Salati SA, Al Kadi A (2012) Anal cancer. Int J Cancer Health Sci 6(2):206–230

Schmidbauer B, Menhart K, Hellwig D, Grosse J (2017) Differentiated thyroid Cancer-treatment: state of the art. Int J Mol Sci 18(6):1292

Siegel RL, Miller KD, Jemal A (2016) Cancer statistics. CA Cancer J Clin 66:7–30

Son SH, Song JH, Choi BO et al (2012) The technical feasibility of an image-guided intensity-modulated radiotherapy (IGIMRT) to perform a hypofractionated schedule in terms of toxicity and local control for patients with locally advanced or recurrent pancreatic cancer. Radiat Oncol 7, Article 203

Stolzoff M, Webster TJ (2016) Reducing bone cancer cell functions using selenium nanocomposites. J Biomed Mater Res 104:476–482

Suttle AB, Ball HA, Molimard M, Hutson TE, Carpenter C, Rajagopalan D, Lin Y, Swann S, Amado R, Pandite L (2014) Relationships between pazopanib exposure and clinical safety and efficacy in patients with advanced renal cell carcinoma. Br J Cancer 111:1909–1916

Wan J, Li XM, Gu J (2014) Primary Choriocarcinoma of the fallopian tube: a case report and literature review. Eur J Gynaecol Oncol 35(5):604–607

Wang SQ, Setlow R, Berwick M, Polsky D, Marghoob AA, Kopf AW (2001) Ultraviolet a and melanoma: a review. J Am Acad Dermatol 44:837–846

Weinberg RA (2014) The biology of cancer, 2nd edn

Willemsen AE, Grutters JC, Gerritsen WR, van Erp NP, van Herpen CM, Tol J (2016) mTOR inhibitor-induced interstitial lung disease in cancer patients: comprehensive review and a practical management algorithm. Int J Cancer 138:2312–2321

Winkler T, Sass FA, Duda GN, Schmidt-Bleek K (2018) A review of biomaterials in bone defect healing, remaining shortcomings and future opportunities for bone tissue engineering: the unsolved challenge. Bone Joint Res 7(3):232–243

Wong R, Malthaner R (2006) Combined chemotherapy and radiotherapy. Cochrane Database Syst Rev 1:CD002092

Wuellner L (2014) Understanding brain tumours. Cancer Council NSW 54(12):134–164

www.cancer.net (2017)
www.cancer.org National Cancer Institute, 2018
www.medicinenet.com
www.rightdiagnosis.com
www.webmd.com
Yamamoto E (2009) Ovary: choriocarcinoma. Atlas genet. Cytogenet Oncol Haematol 13(9):683–685
Yaxley JP (2016) Bladder cancer. J Family Med Prim Care 5(3):533–538
Yeole BB (2008) Trends in the brain cancer incidence in India. Asian Pac J Cancer Prev 9(2):267–270
Yoo KY, Shin HR (2003) Cancer epidemiology and prevention. Korean J Epidemiol 25:1–15
Zotos D, Taclinton D (2012) Determining B-cell fate. Trends Immunol 33(6):280–288
Zujewski JA, Harlan LC, Morrell DM, Stevens JL (2011) Ductal carcinoma in situ: trends in treatment over time in the US. Breast Cancer Res Treat 127(1):251–257

Role of Ayurveda in Cancer Treatment

3

Abstract

In the present day, cancer is a burning issue and is caused due to the mutation in the cells, which divide uncontrollably, so as to form abnormal mass of cells inside the body. It can also be caused by various reasons such as heredity, loss of exercise, alcohol, smoking, physical trauma, radiations, etc. Every year about 6 million cases of cancer are reported worldwide. Cancer cells possess immense potential to divide and to invade other parts of the body at a tremendous speed. Cancer treatment includes chemotherapy, radiotherapy, surgery, etc. which have their own side effects. Ayurveda is a science of life with historical roots in the Indian subcontinent. Ayurvedic treatments have also been practiced in Western countries as well. Ayurveda defines cancer as a chronic imbalance of Tridosha (pathophysiological entities), dhatus (tissue), and Triguna (mental qualities) and can be treated without any side effect. In this chapter, emphasis has been made on how Ayurvedic medicine or the herbal remedies help to retard the process of cancer by targeting the specific tissue to improve the quality of patient's life. The success rates of herbal therapy have also been discussed.

Keywords

Cancer · Ayurveda · Herbal medicine · Phytochemicals

3.1 Introduction

Cancer is one among the most dreadful diseases of the twenty-first era, and about 6 million cases are reported each year (Reddy et al. 2003). After the cardiovascular diseases, the second major causes of deaths are lung, colon, prostate, and breast cancer (Hartwell 1967; Srivastava et al. 2005). Mother Nature has blessed us with an immense treasure of valuable medicinal flora to cure number of diseases. The utilization of characteristic herbs as a panacea to several maladies can be traced

© Springer Nature Singapore Pte Ltd. 2019

B. Koul, *Herbs for Cancer Treatment*,
https://doi.org/10.1007/978-981-32-9147-8_3

back to 1500 BC (Boik 1995). The huge distinction among Western and Ayurvedic cure of malignancy is that Ayurveda does not include any harsh treatments (Sastry 2001). Ayurvedic treatments were observed to have the capacity to treat such kind of unending maladies in a better way, which were earlier not agreeable to Western therapeutic practices (Balachandran and Govindarajan 2005). Ayurvedic arrangement of solution was very much established on the fundamental standards of nature, and its components are extensive investigation of human physiology. Ayurveda portrays good health as the perfect condition of physical, mental, social, and profound segments of an individual (Pandey 2002).

The objective of Ayurvedic treatment is considerably more on empowering the self-healing capacities of body without any side effects by utilizing the medicinal herbs to cleanse and bolster body tissues for recuperation. The first step is to control the Tridosha and Triguna (Prasad 1987). Hundreds of years back in India, herbal medicines have been used successfully, and in the previous few decades, it has been consumed throughout the world. In 2004, the pharmaceutical market was worth US $550 billion, and by the year 2009, the prediction was US $900 billion. The current requirement for medicinal plants is US $14 billion per year, and it would be increased to US $5 trillion till 2050. After knowing the dangerous side effects of synthetic drugs and their high costs, people start using herbal drugs in developing countries as well as in developed countries (Aneesh et al. 2009). Traditional Indian drugs with its advancement through hundreds of years have constantly attracted professionals and specialists for its potential to cure cancer on a research-based foundation. It has been assessed that 60% of the popularized (affirmed) drugs for tumor are extracted from common sources (Farnsworth et al. 1985; Cragg et al. 1997). Herbal extracts possess tremendous potential to cure a disease and are normally utilized as a part of Ayurveda. The benefit of home-prepared decoctions is that it works in a systematic way by supporting different organs of the body (Treadway 1998). Ayurvedic grouping of neoplasms relies on different clinical manifestations in connection to Tridoshas. Group I includes diseases that can be named as clear threat, which incorporates arbuda and granthi, e.g., mamsarbuda (melanoma) and raktarbuda (leukemia), mukharbuda (oral tumor), and so forth. Group II includes diseases that can be considered as tumor, for example, serious ulcers such as tridosaj gulmas (tumors of the stomach like carcinomas of the liver or lymphomas and stomach). Group III includes diseases with the likelihood of danger, e.g., Visarpa (erysipelas), asadhya kamala (hopeless jaundice), and nadi vrana (sinusitis) (Prasad 1987; Singh 2002).

AYUSH is the acronym of the medical systems that are being practiced in India such as Ayurveda, yoga and naturopathy, Unani, Siddha, and homeopathy. These systems are based on definite medical philosophies and represent a way of healthy living with established concepts on prevention of diseases and promotion of health. The basic approach of all these systems on health, disease, and treatment is holistic. Because of this, there is a resurgence of interest on AYUSH systems.

Charaka and Sushruta Samhita (Sharma 1981; Bhishagratha 1991) are the two famous Ayurvedic works which portray cancer as fiery or non-provocative inflammations and specify them as either granthi (minor neoplasm) or arbuda (major neoplasm). Ayurvedic writing characterizes three body control frameworks, viz., the sensory system (Vata or air), the venous structure (Pitta or fire), and the blood vessel structure (Kapha or water) which jointly play out the typical capacity of the body. Sushruta describes six sorts of tumors. The initial four tumors, Paitika granthi (Pitta dosha), Medas granthi (affected fat tissue), Vatika granthi (Vata dosha), and Kaphaja granthi (Kapha dosha), are benign and can be successfully treated during their early stages with the correct treatments. The last two tumors, Rakta arbuda (blood tumor) and Mamsa arbuda (muscle tumor), are malignant. Normally, the cells divide and die, and the dead cells are replaced by new healthy cell. If dead cells are not replaced, they may form abnormal growths called tumor. Tumors can be classified into two types, i.e., benign tumors and malignant tumor.

3.1.1 Benign Tumors

The non-cancerous growths in the body which don't spread to other parts are termed as benign tumors. The exact cause of a benign tumor is still unknown and can form anywhere in the body. Benign tumors are reparable when identified in an early state, but whenever these tumors are in the circulatory system, it becomes harder to cure. There are several types of benign tumors that can develop in various body parts such as adenomas and fibromas.

Adenomas These are benign tumors arising in the epithelial tissue which is a thin layer of tissue which covers the glands and organs. A common type of adenoma is known as polyp (lump-like structure) in the colon and might also grow in the pituitary, adrenal, and thyroid gland and liver. Adenomas can be taken out from the body through surgery; otherwise they may become malignant.

Fibromas (or Fibroids) These are tumors of connective tissue that can develop in any part of body, most commonly in the uterus. It may lead to bladder problems, heavy vaginal bleeding, pelvic pain, or pressure in the pelvic region of the body.

3.1.2 Malignant Tumors

Malignant tumors are cancerous, in which cells grow, behave, and function differently than normal cells. These cancerous cells break and spread in different directions inside the body and invade other nearby tissues, lymphatic vessels or blood vessels. They become life-threatening by interfering with normal body functions. Malignant tumors can reappear after they are removed. There are several types of benign tumors that can develop in various body parts such as Rakta arbuda and Mamsa arbuda.

Rakta arbuda This happens when troubled doshas slow down veins in their working, hindering the blood course too. This stimulates the unpleasant influences, like gathering of toxic compounds in the blood (ama). These toxins (e.g., cancer-causing agents) can be the reason for a tumor. At the point when tumors occur in the veins, they spread through the blood and lymphatic structures, causing death. The vast spreading of tumors will make recuperation amazingly troublesome. The side-effects of blood tumor are slight liquid or discharge collection (incendiary signs) around the tumor, rapid development with little metastases, and release of vitiated blood. This is extremely hard to cure.

Mamsa arbuda This happens due to the muscle destruction. At the point of punching, muscles demonstrate a swelling which is stony hard, painless, and gleaming, called as the tumor. Mamsa arbuda is amazingly hard to cure similarly as that of Rakta arbuda. This type is more regular in non-vegans.

3.2 Sushruta Describes Three Forms of Metastasizing

These forms of metastasizing are *Raktarbuda* (when a tumor is enclosed by littler tumors), *Adhyarbuda* (advancement of another tumor over the main tumor), and *Dviarbuda* (when two tumors emerge at the same time). In the Sushruta Samhita, the essential drug describes the presence of benign and malignant tumors in an equivalent way. Charak clarifies that undesirable tumors in the stomach area, digestion tracts, neck, and indispensable organs are extremely hard to treat. In the present age, it is possible to expel the tumor surgically, but generally there is high risk of repeating chance. Both (Sushruta and Charak) works for the treatment of wounds with particular herbs.

3.3 Astanga Hrdayam

The "Ashtanga Hridayam" (heart or the essence of the eight branches of Ayurveda) encompasses the primary ancient literatures of Ayurveda. It continues to be the root source of Ayurvedic philosophy and protocols which offers health and social insurance to everyone (Patel 2003). This therapeutic work was composed a couple of hundreds of years after Charak and Sushruta. Thus, it gives and outline of the two past works. Tumors and their growth are likewise examined in this work and represent nine types of benign. Among these nine tumors, six can form into malignant tumors (three doshas and three tissues). The enlargement of a benign tumor (granthi) is represented as aggravated doshas, of which kapha is influenced the most. This tumor enters the fatty tissues and the muscles where it forms a round, hard, light, and somewhat nodular structure.

The current science has made significant progress in understanding the mechanism of cancer disease and learning principles in order to prevent or treat it (Garodia et al. 2007). As opposed to the basic explanatory and advanced Western solution,

Ayurveda follows a comprehensive approach encompassing wide understanding of the components of life, health and illness (Singh 2002; Aggarwal et al. 2011). Although the technique of Ayurveda is all complete and general, yet the practices are still limited.

3.4 Modern Medical Point of View

3.4.1 Definition

Malignancy refers to the condition wherein the cells divide relentlessly forming solid tumors. The division and expansion of the abnormal cells is so dynamic and uncontrolled that it hinders the growth of neighboring normal cells. The abnormal cells have a tendency to invade fresh tissue through circulation and spread all through the body.

3.5 Stages of Tumor Development

3.5.1 Development of a Benign Tumor

3.5.1.1 First Stage: Tumor Initiation Stage

Tumor initiation is the first phase in the development of tumor. A substance which reacts with the normal cell DNA and changes its genetic makeup is called as mutagen. The mutagen that initiates the cell to develop tumor is called as initiator. If initiators are nonreactive to DNA first they get changed with the drug metabolizing enzymes inside the body and then only cause mutation. Initiators are specific to the species and tissue type. Initiation is a permanent change; thus daughter cells also carry the mutation.

3.5.1.2 Second Stage: Promotion

The nonreactive compounds which can stimulate the development of tumor are known as promoters. They do not bind covalently to DNA but may bind to the receptor of cell surface so as to affect the intracellular pathway that causes the cell proliferation. There are two types of promoters: specific and non-specific. Specific promoters interact in or on the receptor of target cell or specific tissue. Non-specific promoters interfere in the expression of genes, without the presence of receptor. There are specific promoters for specific species and tissue type.

3.5.2 Development of a Malignant Tumor

3.5.2.1 Third Stage: Progression

The conversion of benign to neoplasm and to malignant tumor is called as progression. It is connected with the karyotypic change that refers to all types of tumors caused by abnormal number of chromosomes (aneuploid).

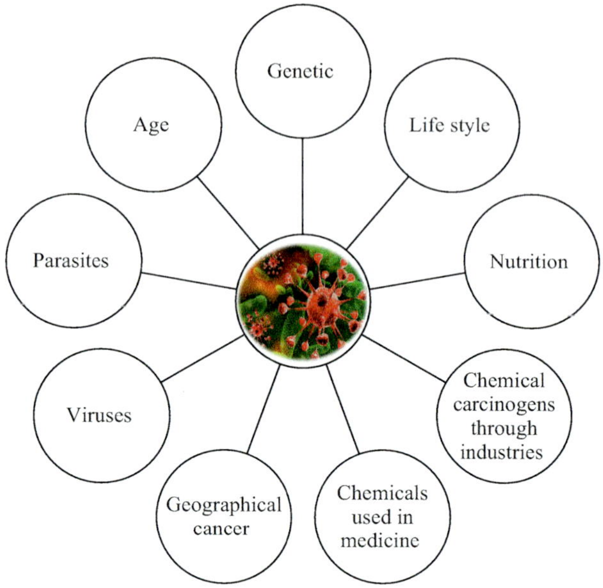

Fig. 3.1 Factors responsible for cancer

3.6 Triguna and Cancer

Dynamic ancient Indian system introduced a concept called Triguna that performs a vital role in the applied research on the good health and spirituality. This concept includes that the psyche comprises of three gunas (energy): sattva (quality of truth, reality, purity, goodness, and virtue), rajas (quality of enthusiasm), and tamas (quality of dullness and inertia).

According to this concept, these three types of energies are present in everyone, and they explain the differences in the patient's behavior as well as his good health.

A disease like malignancy not only irritates the body, but it additionally influences the psyche also. At certain conditions a person can break off, cannot recover, and may be lost. This irregularity of the brain may lead to hyperactivity, laziness, detachment, or even sorrow. Psychotherapy is considered as a part of the treatment in Ayurveda because well-being can be accomplished by balancing/synchronizing the activities of body, brain, and soul.

The main factors responsible for the occurrence of cancer are shown in Fig. 3.1.

3.7 Causes of Cancer

Genetic Cause Various recent researches have revealed that some cancers are genetically driven, such as skin, breast, and colon cancer. Persons with Down syndrome have been reported to have more risk to develop leukemia as compared to normal persons.

Lifestyle Tobacco smoking is one of the major causes of lung, neck, mouth, and larynx cancer.

Radiations Excessive sun burning, tanning, and UV exposure also enhance the risk of cancer. X-rays can easily penetrate into the body and are used in nuclear power plants, space researches, and petrochemical industries. Long-time exposure to these rays can also increase the risk of blood cancer or acute leukemia. Apart from these, excessive sexual activities, excessive physical effort, stress, and smoky diet contribute to the occurrence of cancer.

Nutrition Smoked or preserved foods contain such kind of substances which may lead to stomach cancer, when eaten in excess. Foods containing less than 30% calories from fat reduce the risk of breast and lung cancer. Intake of stale foods increases the risk of cancers. Excessive intake of alcohol increases the risk of esophagus, throat, and mouth cancer. Many vitamins do inhibit the growth of tumors, but clinical research is still in progress. The risk of bowel cancer is considered to increase in persons having deficiency in vitamins A and D.

Chemical Carcinogens Prolonged inhalation of chemical carcinogenic substances like asbestos and nickel increases the risk of lung cancer and pleural cancer. Other potential carcinogenic substances such as aromatic amines, vinyl chloride, benzene, arsenic, chromates, etc. do increase the risk of cancers. Some chemicals used in medicine like DES (diethylstilbestrol) (a synthetic estrogen), oxymetholone, and Thorotrast (radiographic contrast agent) develop the risk of cancer.

Geographic Area The risk of certain types of cancer is also associated to the geographic area of residence. For instance, no cases of gastric cancer are reported in Japan. But the risk increased and eventually equalled the rest of the American population in Japanese people who have shifted to the United States, after some generations. Papillomaviruses, Hepatitis B virus, Epstein-Barr virus, and HIV cause risk of cervical cancer, liver cancer, nose and pharynx cancer, and blood cancer, respectively. Some species of parasites like *Schistosoma* and *Clonorchis* are known to cause bladder cancer, pancreatic cancer, and bile duct cancer, respectively. As per the research data, most forms of cancer are developed in people over 60 years of age, although cervical cancer is reported among younger age group also.

3.8 Ayurvedic Treatment

Ayurveda-based cancer treatment involves therapeutic methods such as (i) purification and elimination of vitiated doshas (sodhana chikitsa using Panchakarma procedures), (ii) pacification of doshas (somana chikitsa), (iii) correction of metabolic defects (dhatwagni chikitsa), (iv) immunotherapy (rasayana prayoga), (v) anticancerous drugs (vyadhipratyanika chikitsa), (vi) symptomatic treatment (lakshanika chikitsa), (vii) surgical treatment (sastra chikitsa), (viii) health maintenance (prakritisthapani chikitsa), (ix) cure of disease (roganashani chikitsa), and (x) spiritual therapy (naishthiki chikitsa) which are extensively used in cancer management (Thatte and Dhahanukar 1991; Sonata 1986). However, surgical removal of tumorous tissue is only the last option. Ayurvedic treatment acts as an additional or a co-therapy along with chemotherapy or radiotherapy and can minimize the side effects of these therapies and is helpful in postsurgical care. The general methods of Ayurveda for cancer treatment include:

- Avoidance of food and lifestyle that cause imbalance in Tridosha (Vata, Pitta, Kapha)
- Recovery of healthy digestive power (jatharagni)
- Elimination of toxins through Panchakarma
- Revitalization through rasayanas

There is no one specific reason of cancer. As mentioned before, genetic factors, lifestyle, and geographical factors play a decisive role in the occurrence of cancers. There are general suggestions to prevent cancers:

3.8.1 Medical Advice

It is recommended to get experts' advice if you find any following change in your body like lumpiness in the neck or anywhere else in the body, coughs that do not go away, sores or ulcers that do not heal, changes in toilet habits, new moles or skin spots, unexplained weight loss, etc. Expert should be the leading Ayurveda specialists trained especially for the treatment of cancer. He or she should basically set up the plan for caring a cancer patient and then to motivate the patient for herbal therapy. Specialist needs to be skillful so as to find a way to help the patient and his entire family through the cancer journey. A cancer patient deserves a specialist who supports the good balance of confidence and belief.

3.8.2 Ideal Lifestyle

One should maintain the lifestyle as follows so as to avoid cancer: maintain body as slim as possible; perform active physical exercise (running, cycling, skipping, etc.) and mental exercise (yoga meditation, etc.) for 30 min daily; avoid high calorie food, consume fresh fruits, vegetables, whole grains, and legumes, and low consumption of red meat (pork, beef, lamb, etc.); avoid processed meat, alcohol, and supplements; and breastfeed should be necessary for a mother till 6 months.

3.9 Cancer and Panchakarma

Panchakarma is a Sanskrit word which refers to "five treatments" for cleansing and removing the waste materials accumulated in the body due to improper nutrition, diseases, and environmental exposure. Panchakarma treatments (oleation: use of oil for massage; fomentation: sweat producing treatment) are prescribed to enhance the body's self-healing capacity to remove the vitiated doshas and toxic substances from the body. These therapies include "Vaman," "Virechan," "Basti," "Nasya," and "Raktamokshan." Vaman therapy treats asthma, hyperacidity, weight gain, etc.; Virechan treats colitis, celiac, herpes zoster, jaundice, infection, etc.; Basti treats constipation, arthritis, piles, etc.; Nasya treats sleep disorder, headache, cerebral pain, hair issues, sinusitis, neurological disorders, and chronic rhinitis; and Raktamokshan treats skin lesions (abscesses and pigmentation) and skin infections (psoriasis and dermatitis).

However, the treatments are highly individual specific depending on the needs of an individual, imbalanced dosha, age group, digestive potential, immunity status, and several other factors. A specially trained therapist analyzes an individual's case and prescribes all or few among the five therapies. After chemotherapy, Panchakarma will be effective for recuperation of both the body and the brain. Malignancy is a ceaseless awkwardness of the Tridosha, dhatus, and Triguna. For curing it, Ayurvedic treatment focuses on re-establishing the balance of the body, mind, and soul. Few diseases are definitely curable, but still some require specialized treatments.

Ayurvedic treatment is empowered by treating the cause of the change of the Tridosha and Triguna. For both, there are herbs, herbal products, and particular treatments available (Table 3.3). Whenever radiation and chemotherapy are the final options, Ayurveda can support between and after these therapies. Ayurveda offers concrete rules about how to cooperate with our surroundings, about our self-improvement (mental and physical), way of life, sustenance, and how to adjust to the rhythms of nature.

3.10 Herbal Preparations for Cancer Treatment

In Ayurvedic treatments the extracts which are rich in nutraceuticals and have pharmaceutical values are used because of their several benefits to human health (Satyavati et al. 1976; Park and Pezzuto 2002; Sinha et al. 2003). These Ayurvedic formulations can be used in various forms such as swallowing; consuming with milk, ghee, tea, and honey; chewing; applying over the skin; etc. (Ramawat and Goyal 2008). Till date, more than 25,000 herbs or plant-based formulations are used in Indian traditional medicine system. Nearly about 75 formulations are available in the Indian market for human health and vitality. All of these formulations contain *Asparagus racemosus* (81.5%), *A. adscendens* (48%), *Withania somnifera* (100%), *Curculigo orchioides* (15 %), etc. (Ramawat et al. 1998). Triphala (*Terminalia chebula*, *T. bellerica*, and *Embelica officinalis*) containing potent anticancer agents have been found in more than 219 formulations (Kamboj 2000). Some of the bioactive compounds that are commercially being used for cancer therapy are listed in Table 3.1. Some plant extracts are known for their therapeutic enhancement potential and act as adjuvants during cancer treatment regimes (Table 3.2).

3.11 Success Rates of Herbal Cancer Treatment

Cancer is a severe metabolic syndrome and is one of the prominent causes of death irrespective of the advancements in its diagnosis, cure, and prevention processes (He et al. 2016; Qin et al. 2017; Zhang et al. 2017). According to a report of the World Health Organization (WHO 2012), nearly about 8.2 million deaths and 14.1 million new cases of cancer have been reported during year 2012, and it has been estimated that in the coming 20 years, this number would be increased by 70% (Thakore et al. 2012; Ferlay et al. 2015; Faria et al. 2017). Cure of cancer demands a particular strategy such as chemotherapy, surgery, immunotherapy, vaccinations, photodynamic therapy, radiotherapy, stem cell transformation, etc. No doubt, these treatments are effective and can increase the life span of cancer patient to some extent; besides that it has various unbearable side effects such as toxicity and non-specificity. Besides that, plants are the reservoirs for various phytochemicals (flavonoids, saponins, coumarins, alkaloids, terpenes, lignans, glycosides, etc.) having the anticancer or antitumor properties. These phytochemicals are safe, cost-effective, less toxic, selective, eco-friendly, and fast in comparison to other traditional methods for cancer treatment (Iqbal et al. 2017). More than 80% of world's population depends upon traditional herbal medicines for their well-being, and 60% of the clinically approved anticancer drugs are also the derivatives of medicinal flora (Khan 2014, Cragg et al. 2016). A number of case studies have been conducted which indicated the anticancer potential of herbal preparations. Traditional Chinese medicines (TCM) are most commonly used worldwide for the treatment of cancer. PC-SPES are the patented dietary supplements which are manufactured by Botanic Lab, Brea, California, USA. PC stands for prostate cancer, and SPES is derived from Latin word which means "hope." These are commercially available capsules for treatment of prostate cancer. The original ingredients of one capsule is the 320 mg of mixture of eight different Chinese herbs, i.e., 32.0 mg of dyer's woad or glastum (*Isatis tinctoria* or *Isatis indigotica*) leaves, 19.2 mg of saw palmetto (*Serenoa repens*) berry, 3.2 mg of Chinese liquorice (*Glycyrrhiza uralensis*) roots, 25.6 mg of *Panax pseudoginseng* roots, 51.2 mg of Baikal skullcap (*Scutellaria baicalensis* or *Scutellaria lateriflora*) roots, 35.2 mg of *Rabdosia rubescens* or *Isodon rubescens leaves*, 99.2 mg of Lingzhi or Reishi mushroom (*Ganoderma lucidum*) stem, and 25.6 mg of *Dendranthema morifolium* or *Chrysanthemum morifolium* flowers (Lee 2005).

Various reports are available regarding the successful use of PC-SPES for the cure of prostate cancer. It has been reported that PSA (prostate-specific antigen) level decreased significantly from 8.8 to 0.1 ng/ml (Moyad et al. 1999), 100 to 24 ng/ml, and 386 to 114 ng/ml (De la Taille et al. 2000) in the patients having prostate cancer. About 28–98% of cancer patients in Asia (Abdullah et al. 2003; Lam et al.

Table 3.1 List of plant derivatives used in cancer therapy

Plant	Semisynthetic analogs of plant derivatives	Structure	Image	Experiments on various cancer cells	Mechanism of action	Reference(s)
Catharanthus roseus (Apocynaceae)	Vindesine	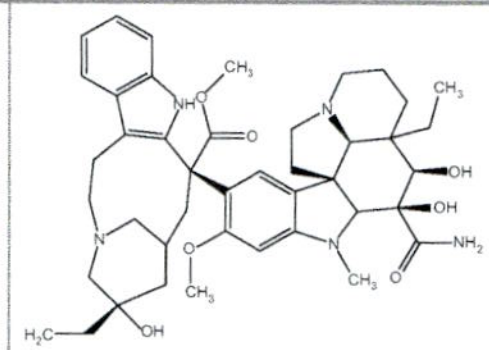		Advanced testicular cancer, leukemia, lymphomas, lung cancer, Kaposi's sarcoma, and breast cancer	Mitotic block	Cragg and Newman (2005)
	Vinorelbine	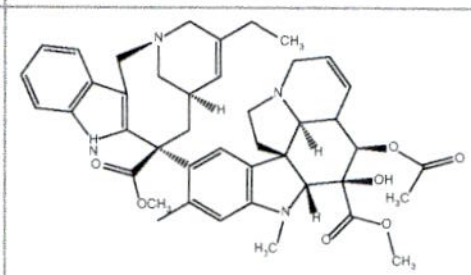				
	Vinflunine	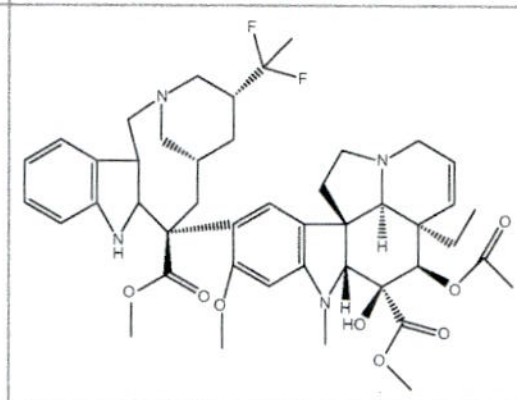		Reduced toxicity in animal models, bladder cancer	Mitotic block	Okouneva et al. (2003), Simoens et al. (2008)

(continued)

Table 3.1 (continued)

Plant	Semisynthetic analogs of plant derivatives	Structure	Image	Experiments on various cancer cells	Mechanism of action	Reference(s)
Podophyllum peltatum and *Podophyllum emodi* (Berberidaceae)	Etoposide	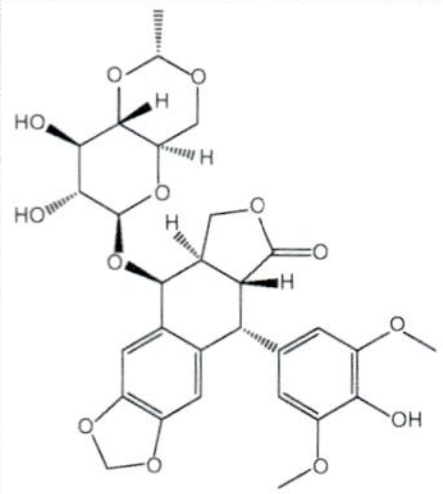		Bronchial lymphomas and testicular cancers		Shoeb (2006)
	Teniposide	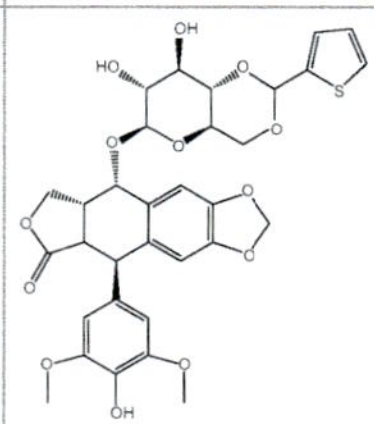				
Taxus brevifolia, *Taxus baccata* (Taxaceae)	Taxol	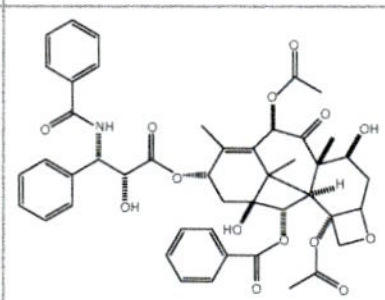		Breast, metastatic, ovarian, prostate, and lung cancer and lymphoid malignancies	Antimitotic	Kingston (2007)

	Taxotere	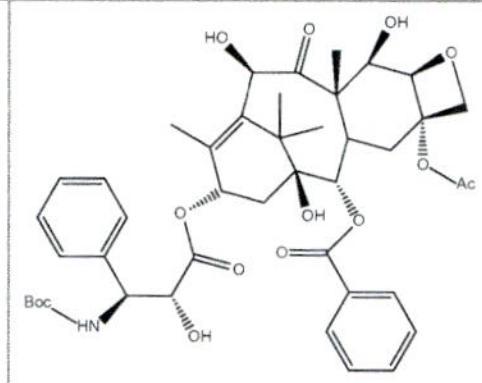		Used in paclitaxel-resistant patients	Antimitotic	Hait et al. (2007)
Camptotheca acuminate (Nyssaceae)	Topotecan	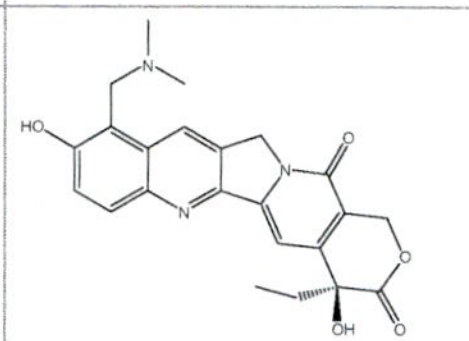		Small cell lung cancer and epithelial ovarian cancer	DNA topoisomerase I inhibition	Creemers et al. (1996)
	Irinotecan	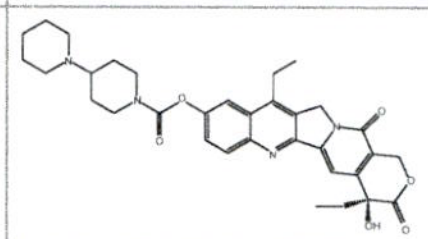		Colorectal and metastatic cancer	DNA topoisomerase I inhibition	Fuchs et al. (2006)

(continued)

Table 3.1 (continued)

Plant	Semisynthetic analogs of plant derivatives	Structure	Image	Experiments on various cancer cells	Mechanism of action	Reference(s)
	Exatecan	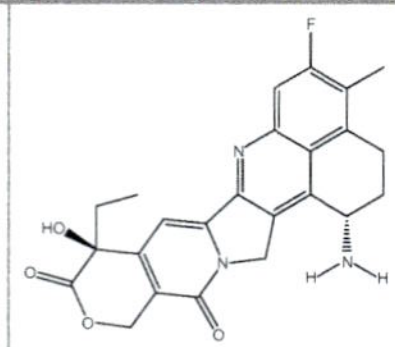		Potential antitumor activity both in vitro and in vivo	DNA topoisomerase I inhibition	Mineko et al. (2000)
	LE–SN–38	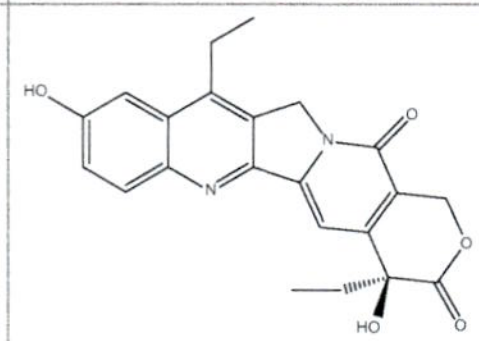		Various cancer cell lines	DNA topoisomerase I inhibition	Zhang et al. (2004)
Berberis amurensis (Berberidaceae)	Berbamine	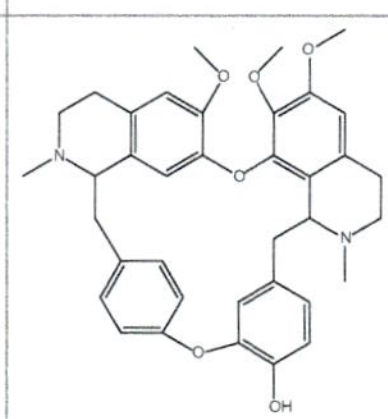		Chronic myeloid leukemia	Caspase-3-dependent apoptosis	Xu et al. (2006), Xie et al. (2009)

Hydrastis canadensis L., *Berberineeris* sp., *Arcangelisia* flava (Ranunculaceae)	Berberine	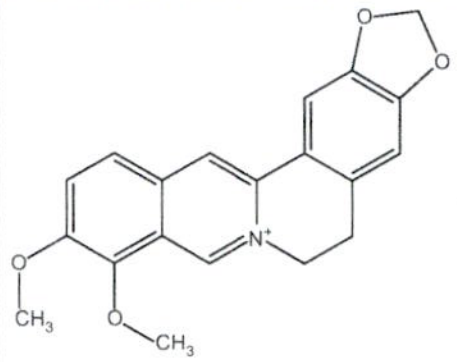		Osteosarcoma, lung, liver, prostate, and breast cancer	Not known	Patil et al. (2010), Wang et al. (2011)
Tabebuia avellanedae (Bignoniaceae)	Betalapachone	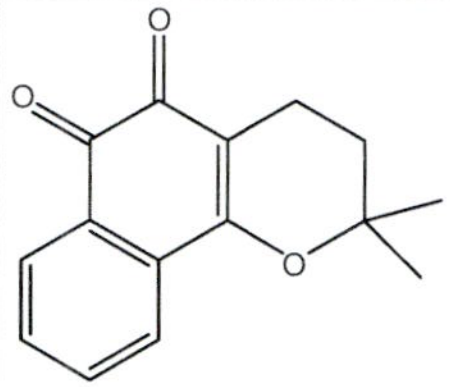		Promyelocytic leukemia, breast, prostate, lung, and pancreatic cancer	Inhibition of topoisomerase I and II	Li et al. (2000), De Almeida (2009)
Betula alba (Bignoniaceae)	Betulinic acid	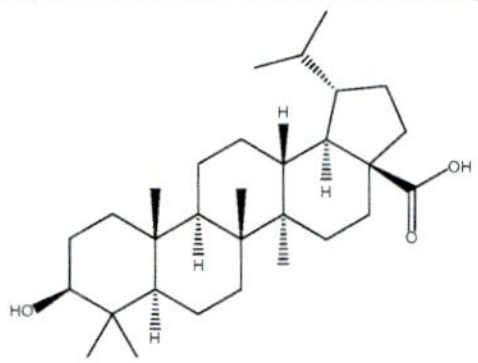		Exhibits anticancer activity in humans	Triggers mitochondrial pathway of apoptosis	Fulda (2008)

(continued)

Table 3.1 (continued)

Plant	Semisynthetic analogs of plant derivatives	Structure	Image	Experiments on various cancer cells	Mechanism of action	Reference(s)
Colchicum autumnale and *Gloriosa superba* L. (Colchicaceae)	Colchicine	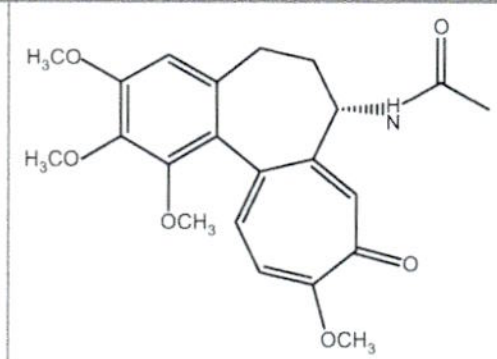		Leukemic and solid tumors	Antimitotic	Dubey et al. (2008)
Combretum Caffrum Kuntze (Combretaceae)	Combretastatin A-4	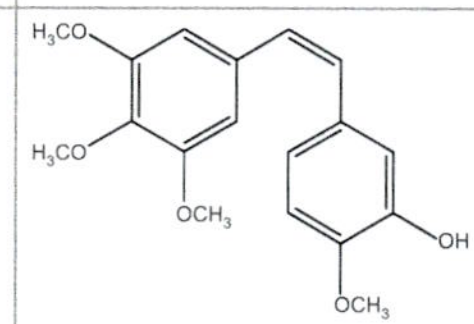		Phase II clinical trials	Tubulin structure disruption	Thomson et al. (2006), Ley et al. (2007)
Cucurbitaceae species	Cucurbitacin	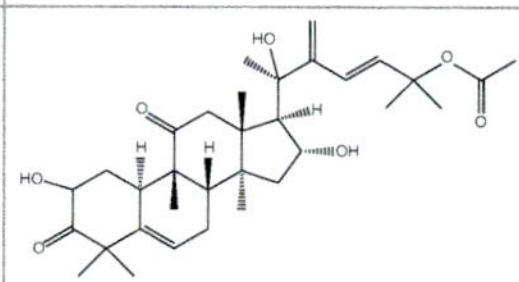		Various cancer cell lines	Inhibits signal transducer/JAK 2 activity and activates STAT3 pathway	Molavi et al. (2008), Bernard and Olayinka (2010)

Curcuma longa (Zingiberaceae)	Curcumin	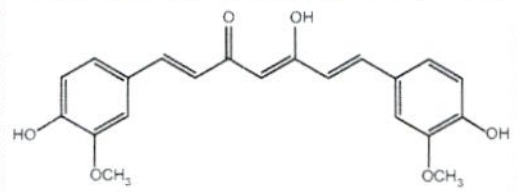		Colorectal cancer, multiple myeloma and pancreatic cancer	Exact mechanism of action is still unknown	Goel et al. (2008), Sa et al. (2010)
Wikstroemia indica (Thymelaeaceae)	Daphnoretin	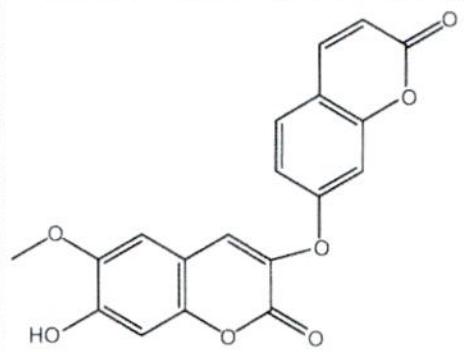		Ehrlich ascites carcinomas and human hepatoma Hep3B cells	(a) Suppression of protein and DNA synthesis (b) Suppresses hepatitis B surface antigen expression	Diogo et al. (2009), Lu et al. (2011)
Lupinus species, *Vicia faba*, *Glycine max*, *Psoralea corylifolia* (Leguminosae)	Daidzein	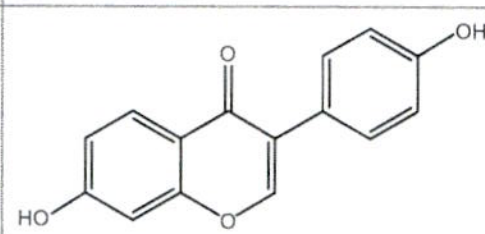		Genistein inhibits breast and ovarian cancers and also chemically induced cancers of lung, colon, stomach, bladder, prostate, and blood	Inhibits 3A 4- mediated metabolism and oxidative metabolism	Kaufman et al. (1997), Dixon and Ferreira (2002), Moon et al. (2006)

(continued)

Table 3.1 (continued)

Plant	Semisynthetic analogs of plant derivatives	Structure	Image	Experiments on various cancer cells	Mechanism of action	Reference(s)
	Genistein	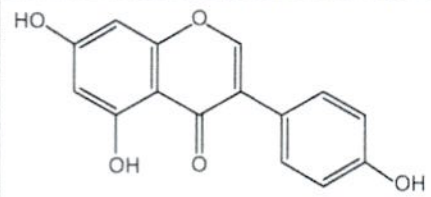				
Ochrosia borbonica, *Excavatia coccinea*, *Ochrosia elliptica* (Apocynaceae)	Ellipticine	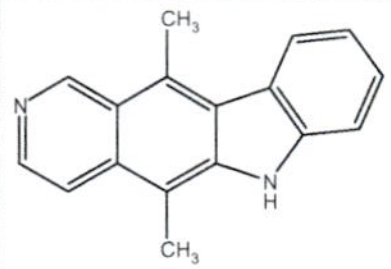		Various cancer cell types	DNA intercalation and inhibition of topoisomerase II	Kuo et al. (2006)
Rhizome of *Rheum emodi* (rhubarb) (Polygonaceae)	Emodin	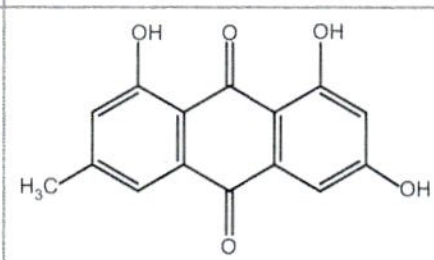		Lung, liver, ovarian, and blood cancer	Apoptosis of cancer cells by several pathways	Huang et al. (2009)

Amoora rohituka and *Dysoxylum binectariferum* (Meliaceae)	Flavopiridol	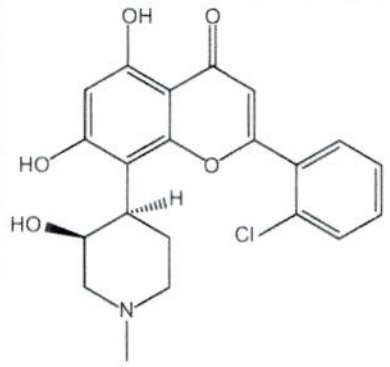		Hodgkin's lymphoma, colorectal and renal cell carcinoma, chronic lymphocytic leukemia, non-small cell lung cancer and, also, solid tumors	Inhibits cell cycle progression at G1 or G2 phase	Mans et al. (2000)
Cephalotaxus harringtonia, *C. hainanensis*, and *C. sinensis* (Cephalotaxaceae)	Harringtonine	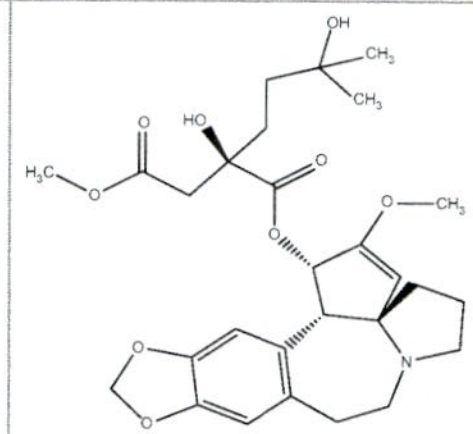		Acute myeloid leukemia and chronic myeloid leukemia	Inhibition of protein synthesis and chain elongation during translation	Cragg and Newman (2005), Efferth et al. (2007)
	Homoharring-tonine	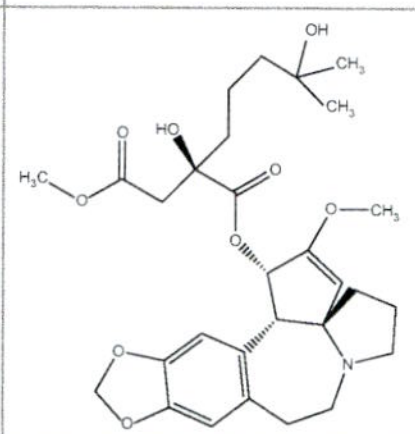				

(continued)

Table 3.1 (continued)

Plant	Semisynthetic analogs of plant derivatives	Structure	Image	Experiments on various cancer cells	Mechanism of action	Reference(s)
Chinese herb, Dang gui	Indirubin	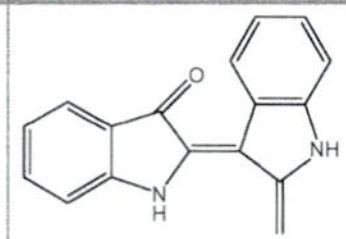		Chronic myeloid leukemia	Inhibits cyclin-dependent kinases	Nam et al. (2005)
Euphorbia peplus L. (Euphorbiaceae)	Ingenol-3-angelate	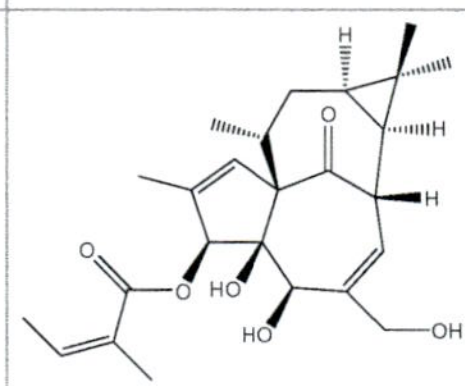		Actinic keratosis and basal cell carcinoma	Causes necrosis of tumor by the activation of PKC	Hampson et al. (2005)
Ipomoea batatas (Convolvulaceae)	4-Ipomeanol	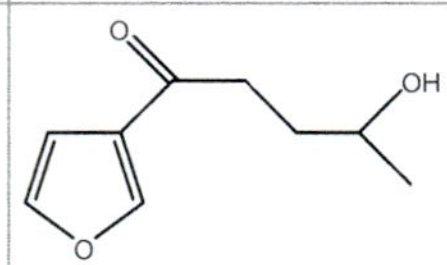		Lung-specific cancer in animal models	Cytochrome P-450-mediated conversion into DNA binding metabolites	Ancuceanu and Istudor (2004)

Iridaceaelatea pallasii and *Iris kumaoensis* (Iridaceae)	Irisquinone	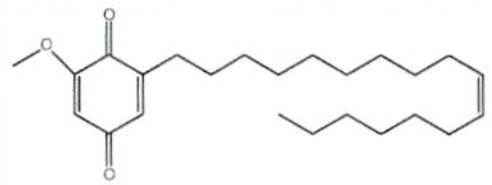		Good activity in transplantable rodent tumors	Acts as a chemosensitizer	Hazra et al. (2004)
Plant isoflavone, genistein	Phenoxodiol	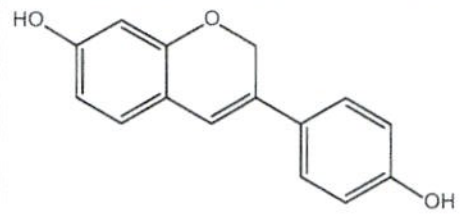		Prostate, ovarian, and cervical cancer	Inhibit plasma membrane electron transport and cell proliferation	Herst et al. (2009)
Saponins of ginseng	Pandimex	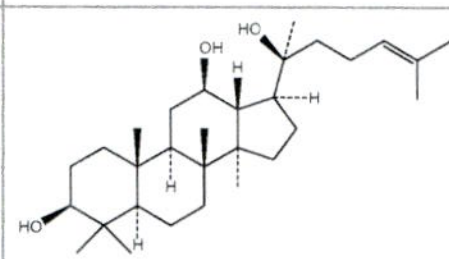		Solid tumors, advanced cancers of the colon, rectum, breast, lung, and pancreas	Cell cycle arrest and acts as P-glycoprotein blocker	Pan et al. (2010)

(continued)

Table 3.1 (continued)

Plant	Semisynthetic analogs of plant derivatives	Structure	Image	Experiments on various cancer cells	Mechanism of action	Reference(s)
Many plant species like mints, cherries, lavenders, and many others	Perillyl alcohol	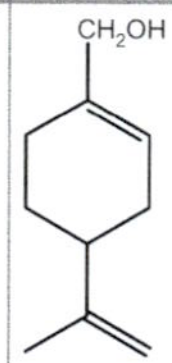		Prostate, breast, non-small cell lung and colon cancer	Exact mechanism is yet to be identified	Bardon et al. (2002); Yeruva et al. (2007); Pan et al. (2010)
Erythroxylum pervillei (Erythroxylaceae)	Pervilleines	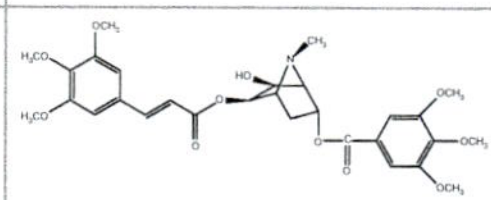		Facilitates chemotherapy	Inhibitors of P-glycoprotein	Mi et al. (2001), Mi et al. (2002), Mi et al. (2003)
Salvia prionitis Hance (Lamiaceae)	Salvicine	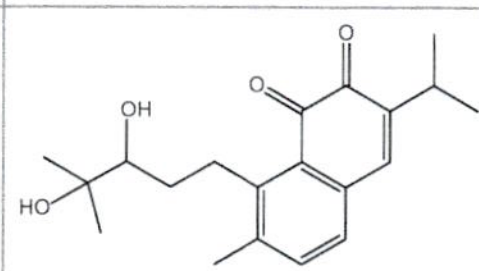		Malignant tumors	Inhibition of topoisomerase II	Deng et al. (2011)

Centaurea schischkinii (Asteraceae)	Schischkinnin	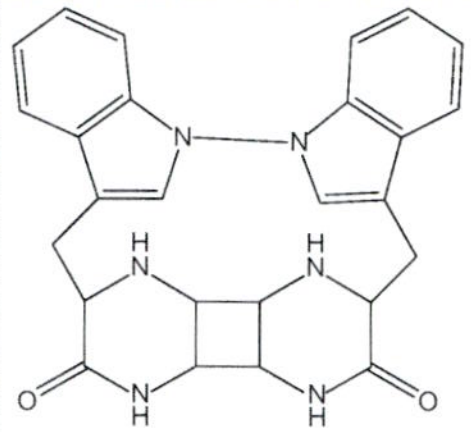		Colon cancer lines in vitro	Not known	Shoeb et al. (2005)
Centaurea montana (Asteraceae)	Montamine	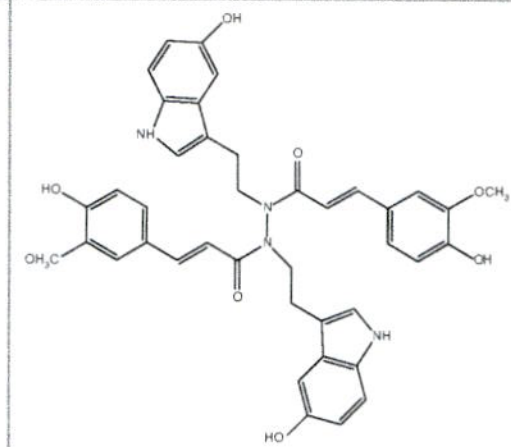		$CaCO_2$ colon cancer cell line in vitro	Not known	Shoeb et al. (2006)
Aglaia foveolata Panell (Meliaceae)	Silvestrol	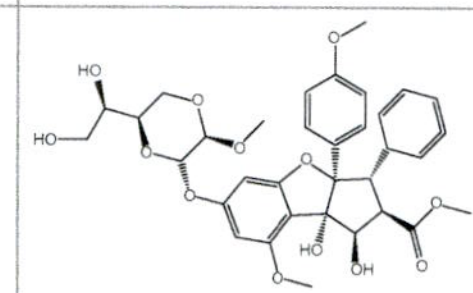		Prostate, breast, and lung cancers	Apoptosome/ mitochondrial pathway was involved in triggering extrinsic pathway of programmed cell death of tumor cells	Kim et al. (2007), Kinghom et al. (2009)
Tripterygium wilfordii Hook F. (Celastraceae)	PG490-88	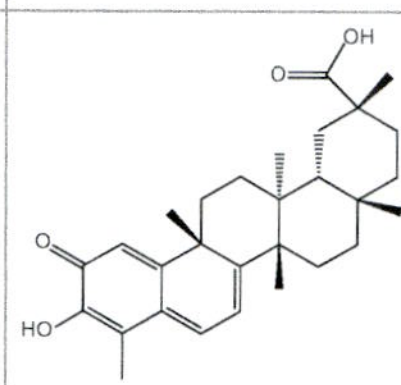		Prostate cancer	Enhances the antitumor effects of cytotoxic and chemotherapeutic agents, thereby induces apoptosis	Liu (2011)

Table 3.2 Therapeutic enhancement potential of Ayurvedic herbs on cancer chemotherapy/radiation

Herb	Image	Chemotherapy/Ayurvedic herb intervention studies	Reference(s)
Allium sativum (Amaryllidaceae)		S-allylmercaptocysteine (SAMC), a water-soluble derivative of garlic, repressed progression of cell cycle and proliferation and in two human colon cancer cell lines, i.e., HT-29 and SW-480, similar to the effects of sulindac sulfide (SS), a well-known colon cancer chemopreventive drug. Combination of SS with SAMC increased the growth inhibitory and apoptotic effects of SS, which suggests the future use of SAMC alone or in combination with certain other chemopreventive agents	Shirin et al. (2001)
Aloe vera (Xanthorrhoeaceae)		In a clinical trial involving patients with advanced solid tumors, for whom no other standard effective therapy was available, co-administration of pineal indole melatonin (MLT) and extracts of aloe vera produced some therapeutic benefits, in terms of disease stabilization and survival as compared to alone MLT treatment	Lissoni et al. (1998); Olsen et al. (2001)
Alstonia scholaris (Apocynaceae)		Pre-treatment with the extract of *Alstonia scholaris* enhanced the radiation effect by enhancement of cell killing in HeLa and KB cells, followed by HL60, MCF-7, and HePG2 cells. The combination of plant extract with cyclophosphamide has been reported to be highly effective against cyclophosphamide and increased the survival rate of patient	Jagetia and Baliga (2003a, b)

(continued)

Table 3.2 (continued)

Herb	Image	Chemotherapy/Ayurvedic herb intervention studies	Reference(s)
Curcuma longa (Zingiberaceae)		Curcuma showed radiation sensitizing effect in K-562, HeLa, and IM-9 cell lines when used in a combination with radiation therapy. The active compound curcumin increased the anticancer potential of cisplatin and also reduces its nephrotoxicity	Sriganth and Premalatha (1999); Baatout et al. (2004)
Heliotropium indicum (Boraginaceae)		An alkaloid indicine N-oxide isolated from the plant has been used in persons having ovarian carcinoma and skin melanoma or in the persons who have undergone radiation or chemotherapy	Ohnuma et al. (1982)
Moringa oleifera (Moringaceae)		Leaf extract of the plant is used in pre-treatment so as to protect the chromosomes of bone marrow in mice, and this could be useful to overcome side effects of radiation therapy	Rao et al. (2001)
Nigella sativa (Ranunculaceae)		Thymoquinone isolated from oil of *Nigella sativa* significantly increased the therapeutic effects of ifosfamide by improving its antitumor effects and by reducing its nephrotoxicity. Mice treated with the combination of ifosfamide and thymoquinone showed less loss of body weight in comparison to single ifosfamide therapy	Badary (1999)
Ocimum sanctum (Lamiaceae)		Two compounds, vicenin and orientin, which are soluble in water, extracted from leaves of the plant protect the human lymphocytes from radiation lethality and clastogenic effect of radiation under in vivo conditions	Vrinda and Uma Devi (2001)

(continued)

Table 3.2 (continued)

Herb	Image	Chemotherapy/Ayurvedic herb intervention studies	Reference(s)
Taxus baccata (Taxaceae)		A combination of three compounds, i.e., carboplatin, ifosfamide, and taxol (active constituent of *Taxus baccata*), was found to be safe and easy to deliver on a person with advanced stage IIIB–IV non-small-cell lung cancer. Another combination of taxol and herceptin significantly enhanced the response rate and survival rate in a patients having breast cancer	Lang (2002); Zaniboni et al. (2003)
Withania somnifera (Solanaceae)		*W. somnifera* causes significant reversal of neutropenia of paclitaxel in mice when given 4 days before the treatment and given for 12 continuous days. It can also be used in a combination with the anticancer drugs used in chemotherapy. The extract obtained from this plant inhibits the growth of tumor in Ehrlich ascites carcinoma mice and also increase their rate of survival. The active phytochemical withaferin A extracted from the plant exhibit antitumor properties without any side effects	Devi (1996); Sharada et al. (1996); Gupta et al. (2001)

2010) and 25–47% of North America (Maskarinec et al. 2000; Ferro et al. 2007) have been reported to use herbal formulations as a part of their cancer treatment. Various traditional Chinese herbal formulations such as Kang-Fu-Zhi-Tong (Xu et al. 2007), PHY906 (Saif et al. 2010), Bojungikki-tang (Jeong et al. 2010), Xiao-Chai-Hu-Tang, Huang-Lian-Jie-Du-Tang, and Yin-Chen-Wu-Ling-San (Liu et al. 2011) are commercially available to reduce the effects of cancer such as pain, fatigue, and respiratory and gastrointestinal infections.

Various herbal formulations have been prepared by Divyajyot Ayurvedic Research Foundation, Gujarat, India, for successful treatment of different types of cancers. These formulations include *Bacopa monnieri* (mandukaparni) and kastoori bhairav rasa for brain cancer; *Cassia occidentalis* (kasamarda) and mahalaxmi vilas rasa for oropharyngeal cancers; *Piper longum* (Pippali) and hirak rasayan for lung cancer; *Rubia cordifolia* (manjishtha) and kaishore gugglu, gandhak rasayan, and samira panaga rasa for skin cancer; chinchabhallataka and gojivha for breast cancer; madhu malini vasant rasa and aabha gugglu for bone cancer; *Hemidesmus indicus* (Anantmula) and suvarna vasant malti rasa for blood cancer; arogyavardhini and bhumyamalaki for liver cancer; makardhvaja and triphala for male genital cancers; *Saraca asoca* (Ashoka) and vaikranta bhasma for female genital cancers; *Moringa oleifera* (shigru) and panchamrut purpti for intestinal cancer; and *Phyllanthus emblica* (amlaki), *P. nurare* (bhoy-amli), *Tephrosia purpurea* (sarphunkha), *Cassia fistula* (amaltas), *Asparagus racemosus* (shatavari), *Aloe vera*, and banga bhasma for stomach cancers.

Nowadays, Tibetan medicine system has gained interest in worldwide because of their successful cancer treatments. This system is based upon the three diagnostic approaches, i.e., touch, observations, and investigative questioning. Minimal published literature is available regarding the clinical trials of Tibetan medicines for cancer treatment. Furthermore a collaborative research of Tibetan and other pharmaceutical sciences should be carried out in order to expand the area of herbal formulations. "Sangdak Dharyaken," "Gawa Chudruk," "Dashel Dhuetsema," "Khyunga Nila," "Khyunga," and "Rinchen Drangjor Chenmo" are the main Tibetan formulations which are used for the treatment of cancer. A person suffering from blood cancer was treated with Tibetan medicine "Rinchen Drangjor Chenmo" and "Rinchen Mangjor Chenmo." Within 3 months of the treatment, his hemoglobin levels (from 10.5 to 15.0) and WBC count (from 149,000 to 8300) were found normal. Another person suffering from gastric carcinoma was treated with Tibetan formulations "Rinchen Tsodru Dashel" and "Rinchen Mangjor Chenmo" on a weekly basis. After 29 months of the treatment, no evidence of cancer was shown in abdominal computed tomography (CT) scans (Bauer-Wu et al. 2014). Some clinical trials seem to show that immunomodulatory herbs may help people to live longer, reduce side effects, and help to prevent cancer coming back, especially when combined with conventional treatment (Table 3.3).

3.12 Mechanism of Action of Herbal Medicines

More than 5000 individual phytochemicals have been reported till date, and this number is increasing continuously because of the introduction of new techniques for phytochemical isolation (Yehya et al. 2017). These phytochemicals can be classified as alkaloids, organosulfur, carotenoids, phenolics, and nitrogen-containing compounds (Asif et al. 2016). These phytochemicals possess various properties such as antioxidant, anti-angiogenic, stimulates the enzyme activities of various hormones, protect the non-tumorous cells from radiations, and interfere with DNA replication (American Thoracic Society 2000). These chemicals can also be used in combination with the other antineoplastic drugs so as to enhance their anticancer activity (Asif et al. 2016; Lachumy et al. 2013; Lu et al. 2016) (Fig. 3.2).

Alkaloids are an important class of phytochemicals containing nitrogen and contribute significantly as anticancer compounds. Compounds like vincristine, camptothecin, and vinblastine belong to alkaloid family (Li et al. 2007). Alkaloids

Table 3.3 Immunomodulatory herbs, compounds, and their products

Herbs/formulation	Ingredients	Benefits
Chyawanprash awaleha 	*Agaru (Aquilaria agallocha), Agnimantha (Premna corymbosa), Bala (Sida cordifolia), Bilwa (Aegle marmelos), Brihati (Solanum indicum), Chandana (Santalum album), Draksha (dry grapes), Ela (Elettaria cardamomum), Gambhari (Gmelina arborea), Gokshura (Tribulus terrestris), Hareetaki (Terminalia chebula), Jeevaka (Microstylis muscifera), Jeevanti (Leptadenia reticulata), Kakanasika (Martynia diandra), Kakoli (Lilium polyphyllum), Kantakari (Solanum xanthocarpum), Karkata shrungi (Pistacia integerrima), Mashaparni (Teramnus labialis), Meda (Polygonatum cirrhifolium), Mudgaparni (Vigna trilobata), Musta (Cyperus rotundus), Patala (Stereospermum suaveolens), Pippali (Long pepper), Prishnaparni (Uraria picta), Punarnava (Boerhaavia diffusa), Pushkara (Inula racemosa), Ruddhi (Habenaria intermedia), Shalaparni (Desmodium gangeticum), Shati (Hedychium spicatum), Shyonaka (Oroxylum indicum), Taamalaki (Phyllanthus fraternus), Utpala (Nymphaea stellata), Vidari (Ipomea digitata), Vrishabhaka (Microstylis wallichii), Vrushamula (Adhatoda vasica)*	Chyawanprash (CA) is a polyherbal Ayurvedic *rasayan*, used as a health supplement, acts as adjuvant in the treatment of pulmonary TB, provides excellent protection against burning effects of radiation, reduces the tumor weight inside the tissue, and balances Vata and Kapha. This is mainly due to the antioxidant effects of *E. officinalis* contained in the preparation
Amrit prash 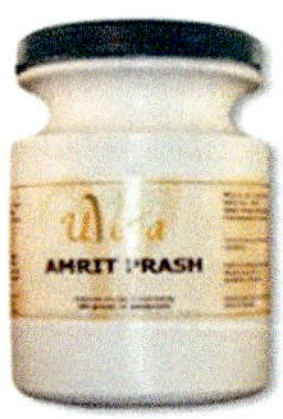	Herbal paste	Improves overall body functions
		Balances Tridosha (Vata, Pitta, Kapha)

(continued)

Table 3.3 (continued)

Herbs/formulation	Ingredients	Benefits
Ashwagandha 	*Withania somnifera* (Indian ginseng)	Reduces Vata/Kapha (with excess Pitta), used for recovery from excess fatigue, tissue rejuvenation, sexual debility, old age, memory loss, emaciation of children, muscular exhaustion, nervous exhaustion, and tissue deficiency
Bala 	*Sida cordifolia*	Balances Tridosha (Vata, Pitta, Kapha), provides stamina or inner strength, used as a cardiac stimulant, diuretic, decongestant, anti-asthmatic, and anti-depressant
Bhringaraj 	*Eclipta erecta*, *Eclipta alba*, *Eclipta punctata*, *Verbesina prostrate*	Balances Tridosha (Vata, Pitta, Kapha), prevents aging, maintains and rejuvenates memory, sight, hearing, teeth, bones and hair, kidneys, and liver
Dashmool 	Traditional herbal formula	Balances Tridosha, antioxidant, anticancer Strengthens the body, enhances the production of tissue
Guduchi 	*Tinospora cordifolia*	Balances Tridosha (Vata, Pitta, Kapha), protects white blood cells, builds up the body's own defense, increases memory, prevents liver infection and digestive disorders, recommended before and after chemotherapy

(continued)

Table 3.3 (continued)

Herbs/formulation	Ingredients	Benefits
Kounch 	*Mucuna pruriens*	Balances Vata, Pitta, and Kapha; strengthens and revitalizes muscles; stimulates nervous system; prevents painful menstruation, and stimulates the male genital system
Lavan Bhaskar 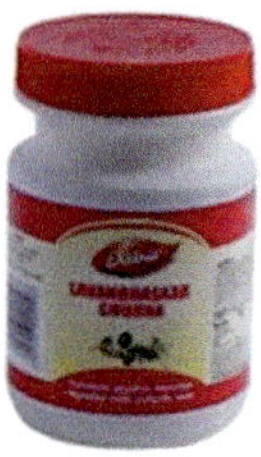	Traditional herbal formula	Reduces Vata and Kapha, increases Pitta, stimulates appetite, improves digestion (constipation, piles, colic, loss of appetite, etc.)
Neem 	*Melia azadirachta*	Pacifies Vata and Kapha, but aggravates Pitta; used as a cooling herb; purifies the blood and cleanses the liver; exhibits antibacterial, antiviral, antiparasitic activity; used against tumors, overweight, and infections
Pippali 	*Piper longum*	Pacifies Vata and Kapha, aggravates Pitta, stimulates digestion, useful in respiratory discomfort (including asthma) and cough, tranquilizing (pain killer) action, increases circulation, specifically to the lungs, against tumors
Shatavari 	*Asparagus racemosus*	Reduces Vata and Pitta, aggravates Kapha, supports the female organs, prevents sexual debility, helps menopausal conditions, supports healthy immune system digestion, treats stomach ulcers, production of tissue, fertility and a healthy libido, inflammation, and chronic fevers

(continued)

Table 3.3 (continued)

Herbs/formulation	Ingredients	Benefits
Shilajit 	Asphaltum (mineral)	Balances Tridosha (Vata, Pitta, Kapha); rejuvenative; promotes strength and immunity, especially in the urinary tract; and a primary rejuvenative for the prostate
Triphala 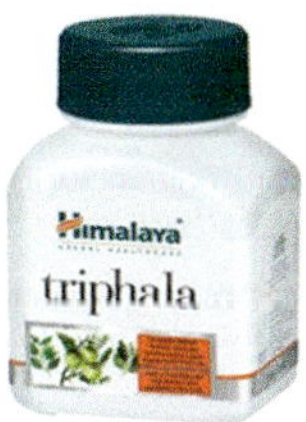	Traditional herbal formula	Balances Tridosha (Vata, Pitta, and excess Kapha), appetite, and digestion, increases red blood cells, removes undesirable fats, prevents diabetes, used for natural internal cleaning, nourishes and rejuvenates the tissues, and acts as antioxidant
Tulsi 	*Ocimum sanctum* (Holy Basil)	Pacifies Vata and Kapha, increases Pitta; useful in management of digestive fire cardiac tonic skin diseases, respiratory discomfort, and cough; acts as anti-inflammatory, antioxidant, and wound healer; reduces chronic stress and blood sugar levels in diabetes (type II); improves memory and lowers cholesterol; benefits the immune system; preventive against tumors; and especially beneficial in the first stage of many types of tumors
Vidari-Kanda 	*Ipomoea digitata*	Rejuvenative for Vata and Pitta and increases Kapha; nourishes and rejuvenates the body; supports proper function of the male and female reproductive systems; promotes healthy muscle and nerve tissue

(continued)

Table 3.3 (continued)

Herbs/formulation	Ingredients	Benefits
Yashtimadhu 	*Glycyrrhiza glabra* – liquorice	Pacifies Vata and Pitta; aggravates Kapha; has soothing effect on inflamed mucous membranes of the throat, lungs, stomach, and intestines; regulates the menstrual cycle; improves eyesight, strength, sexual power, and complexion; strengthens hair; beneficial to all kinds of inflammations; boiling Yashtimadhu in milk prevents inhibited absorption of calcium and potassium
Yogaraj guggul  	Traditional herbal formula	Reduces Vata and Kapha; aggravates Pitta; prevents joint pain; lowers cholesterol and excess fat; and clears toxins from joints and muscles

have been reported to inhibit the process of angiogenesis in cancer by blocking the VEGF-induced proliferation, migration, and tubular 6 structure formation in endothelial cell in tumor cells (Lu et al. 2012). It has also been demonstrated that alkaloids can suppress or inhibit the activity of STAT3 (signal transducer and activator of transcription) proteins and DNA binding activity of tumor cells (Chen et al. 2012).

Terpenes are a large family of phytochemicals which consists of nearly about 25,000 derivatives. Terpenoids can be classified as monoterpenoids, diterpenoids, triterpenoids, and tetraterpenoids on the basis of their structure. Various mechanisms have been demonstrated due to which terpenes suppress angiogenesis and metastasis of tumor cells (American Thoracic Society 2000). Artemisinin (sesquiterpenoids) have been reported to inhibit the angiogenic process by downregulating the VEGFR-1 and VEGFR-2 (vascular endothelial growth factor receptors 1 and 2) and CD34 (cluster of differentiation-34) genes in HUVECs (human umbilical vein endothelial cells) (Abdolmaleki et al. 2016) (Fig. 3.3).

A triterpenoid obtained from *Tripterygium wilfordii* Hook has been reported to suppress the angiogenesis of tumor cells by inhibiting the activities of proteins ICAM-1(intercellular adhesion molecule-1), MMP-9 (matrix metallopeptidase-9), and VEGF (vascular endothelial growth factors) that are known for angiogenic and metastasis activities (Huang et al. 2012; Wang et al. 2015).

Nitrogen-containing and organosulfur compounds also play a key role in the cure of various diseases including respiratory, cardiovascular, and hypertension disorders (Stabler et al. 2012; Wiatrak 2003). These compounds can be categorized as

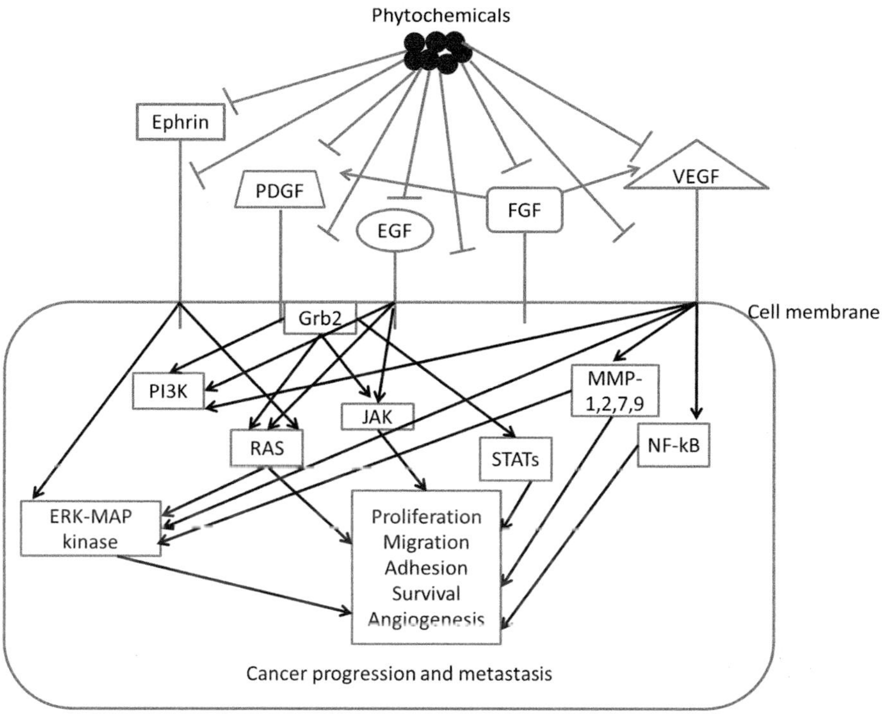

Fig 3.2 Mechanism of the anticancer activity of phytochemicals

S-methyl cysteine sulfoxide and glucosinolates. Several compounds of this family such as glucosinolates, gluconasturtiin (2-phenylethyl), glucoraphanin (4-methylsulfinylbutyl), isothiocyanates, and glucobrassicin (3-indolymethyl) have been reported to show significant anticarcinogenic activities (Stoewsand 1995). A compound named R-(−)-b-omethylsynephrine, isolated from family Rutaceae, demonstrated significant anti-angiogenic activity by inhibiting the expression of VEGF in HUVECs which lead to the inhibition of capillary formation in tumor cells (Kim et al. 2010). Condensed tannins, i.e., proanthocyanidins, have also been reported to exhibit anti-angiogenic activities in breast, colon, and prostate cancer by inhibiting the VEGF signaling (Huang et al. 2012).

Polyphenols are an important group of chemical compounds which have been used as antioxidant, anti-angiogenic, anti-atherosclerotic, antimutagenic, and antiviral agent (Aboul-Enein et al. 2013; Bhattacharjee and Chatterjee 2013; Lall et al. 2015). Caffeic acid is one among the most important compounds of polyphenol family and is found in coffee, cereals, fruits, and vegetables (Liao et al. 2013). This compound has been reported to inhibit the angiogenesis in mouse model (having human renal carcinomas cells) by suppressing the activity of STAT3-mediated VEGF and HIF-1 (hypoxia-inducible factors 1) expression (Jung et al. 2007). Another compound named scopoletin (6-methoxy-7-hydroxycoumarin) extracted

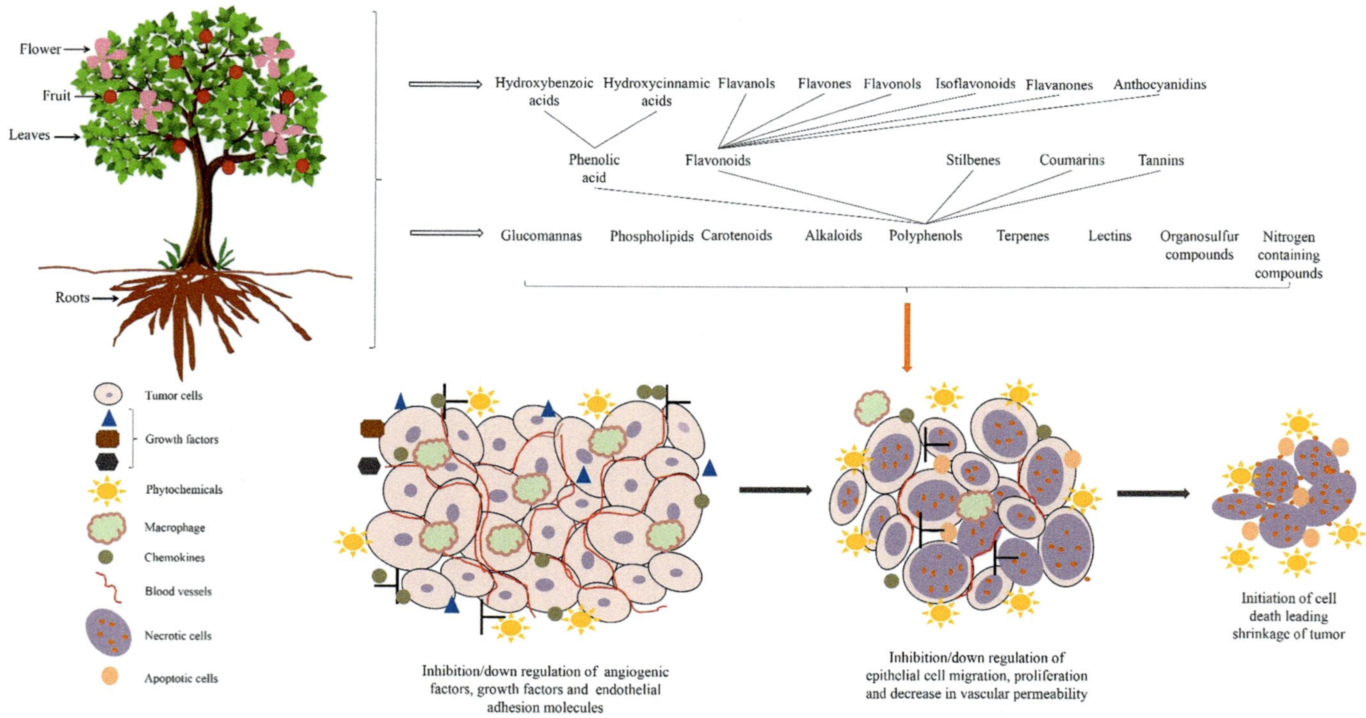

Fig. 3.3 Anti-angiogenic and antitumor potential of phytochemicals

from stems of *Eyrie obtusifolia* Benth exhibited significant anti-angiogenic activity among tumor cells by inhibiting the HUVECs proliferation, migration, and tubule formation through FGF-2 (basic fibroblast growth factor). It has also been reported to suppress the expression of certain proteins by blocking or inhibiting the signaling of NF-κB (nuclear factor-κB) (Pan et al. 2011). Another study suggested that anti-angiogenic activity of scopoletin compound extracted from *Nicotiana glauca* might be due to the blockage or inhibition of VEGF-A, ERK1 (extracellular signal-regulated kinases 1), and FGF-2 or the downregulation of tumor angiogenesis factors (Tabana et al. 2016).

3.13 Conclusions

Among all the benefits of Ayurvedic medicines, relief from cancer symptoms is especially valuable. At one side, it is true that Ayurvedic medicines are capable of preventing, if not fully curing all the cancer types. Herbal products contain multiple body-friendly bioactive chemicals which heal the body indirectly by strengthening the immune system to fight against cancer. Health-promoting and self-healing benefits of Ayurvedic treatments make these treatments attractive. These medicines are cost-effective and reduce unnecessary side effects. Active compounds of these herbal preparations can operate in a synergetic manner with other routine treatments which can reduce their adverse side effects. Personal attention, empowerment, and participation in healing process are the primary requirements for these treatments to be successful.

Moreover, these herbal medicines are supplied as formulations. Although the nature and property of individual herbal powder are known, much researches are required to know as to how these combinations/formulations potentiate and synergize the anticancer activity. Once the mechanism is known, then these formulations can be used to treat various types of cancers in a more effective manner. It has also been found that the anticancer preparations which are available commercially can vary in terms of their preparations. Furthermore the clinical efficiency and toxicity of these natural products are greatly unknown. Therefore, some of the Ayurvedic medicines and methods of cancer treatment need new observations, hypothesis-driven testing, standardization, innovation, and peer review.

On the other side, it is also true that the Ayurvedic formulations heal the human body slowly as compared to the allopathic medicines. Although the latter bring quick relief to the patient, they are not free from side effects. There is no question of comparison between herbal medicines and modern medicines as both have different mechanism and efficacy. But, this is true that Ayurvedic formulations against cancer are safe and effective if they are used at the right time, that is, in the first stage of cancer. As mentioned before, the mode of action of Ayurvedic medicine is to make the body strong enough to cope with or fight with cancers. In addition to it, they aim at healing the illnesses with changes in diet and daily activities, keeping in view of the age and body requirements of the patient. Interestingly, the synergism of Ayurvedic therapeutic regimes (chemotherapy, radiotherapy, etc.) excellently fights cancerous tissue. There are successful reports of integrated cancer management using a combination of herbal and modern medicines. However, these aspects require more intensive researches along with the analysis of feedbacks from treated patients after 4–5 years of recovery. In a nutshell, the science of Ayurveda encourages the healthy as well as the diseased person to opt herbal medicines and vegetarian lifestyle devoid of alcohol and tobacco consumption and to meditate so as to relax the body to potentiate it to heal itself.

References

Abdolmaleki Z, Arab HA, Amanpour S, Muhammadnejad S (2016) Anti-angiogenic effects of ethanolic extract of *Artemisia sieberi* compared to its active substance, artemisinin. Rev Bras 26(3):326–333

Abdullah AS, Lau Y, Chow LW (2003) Pattern of alternative medicine usage among the Chinese breast cancer patients: implication for service integration. Am J Chin Med 31(4):649–658

Aboul-Enein H, Berczynski P, Kruk I (2013) Phenolic compounds: the role of redox regulation in neurodegenerative disease and cancer. Mini-Rev Med Chem 13(3):385–398

Aggarwal B, Kannappan R, Park B, Prasad S, Reuter S, Yadav V (2011) Identification of novel anti-inflammatory agents from Ayurvedic medicine for prevention of chronic diseases: "reverse pharmacology" and "bedside to bench" approach. Curr Drug Targets 12(11):1595–1653

American Thoracic Society (2000) Idiopathic pulmonary fibrosis: diagnosis and treatment. International consensus statement. American Thoracic Society (ATS), and the European Respiratory Society (ERS). Am J Respir Crit Care Med 161

Ancuceanu RV, Istudor V (2004) Pharmacologically active natural compounds for lung cancer. Altern Med Rev 9:402–419

Aneesh TP, Mohamed H, Sonal SM, Manjusree DTV (2009) International market scenario of traditional Indian herbal drugs – India declining. Int J of Green Pham 28:543–547

Asif M, Yehya AHS, Al-Mansoub MA, Revadigar V, Ezzat MO, Khadeer Ahamed MB, Oon CE, Murugaiyah V, Abdul Majid AS, Abdul Majid AMS (2016) Anticancer attributes of *Illicium verum* essential oils against colon cancer. South Afr J Bot 103:156–161

Baatout S, Derradji H, Jacquet P, Ooms D, Michaux A, Mergeay M (2004) Effect of curcuma on radiation-induced apoptosis in human cancer cells. Int J Oncol 24:321–329

Badary OA (1999) Thymoquinone attenuates ifosfamide-induced Fanconi syndrome in rats and enhances its antitumour activity in mice. J Ethnopharmacol 67:135–142

Balachandran P, Govindarajan R (2005) Cancer-an ayurvedic perspective. Pharmacol Res 51(1):19–30

Bardon S, Foussard V, Fournel S, Loubat A (2002) Monoterpenes inhibit proliferation of human colon cancer cells by modulating cell cycle-related protein expression. Cancer Lett 181:187–194

Bauer-Wu S, Lhundup T, Tidwell T, Lhadon T, Ozawa-de Silva C, Dolma J, Yeshi T (2014) Tibetan medicine for cancer: an overview and review of case studies. Integr Cancer Ther 13(6):502–512

Bernard SA, Olayinka OA (2010) Search for a novel antioxidant, antiinflammatory/analgesic or antiproliferative drug:cucurbitacins hold the ace. J Med Plant Res 4:2821–2826

Bhattacharjee B, Chatterjee J (2013) Identification of proapoptopic, anti-inflammatory, antiproliferative, anti-invasive and anti-angiogenic targets of essential oils in cardamom by dual reverse virtual screening and binding pose analysis. Asian Pac J Cancer Prev 14(6):3735–3742

Bhishagratha KL (1991) Sushruta samhita. Choukhamba Orientalia, Varanasi

Boik J (1995) Cancer and natural medicine: a textbook of basic science and clinical research, p315 cells. Int J Oncol 24:321–329

Chen ML, Lin YH, Yang CM, Hu ML (2012) Lycopene inhibits angiogenesis both *in vitro* and *in vivo* by inhibiting MMP-2/uPA system through VEGFR2-mediated PI3K-Akt and ERK/p38 signaling pathways. Mol Nutr Food Res 56(6):889–899

Cragg GM, Newman DJ (2005) Plants as a source of anti-cancer agents. J Ethnopharmacol 100:72–79

Cragg GM, Newman DJ, Snader KM (1997) Natural products in drug discovery and development. J Nat Prod 60(1):52–60

Cragg GM, Newman DJ, Yang SS (2016) Natural product extracts of plant and marine origin having antileukemia potential. The NCI experience. J Nat Prod 69:488–498

Creemers GJ, Bolis G, Gore M, Scarfone G, Lacave AJ, Guastalla JP, Despax R, Favalli G, Kreinberg R, VanBelle S, Hudson I, Verweij J, Ten Bokkel Huinink WW (1996) Topotecan an active drug in the second-line treatment of epithelial ovarian cancer: results of a large European phase II study. J Clin Oncol 14:3056–3061

De Almeida ER (2009) Preclinical and clinical studies of lapachol and betalapachone. Open Nat Prod J 2:42–47

Deng F, Lu JJ, Liu HY, Lin LP, Ding J, Zhang JS (2011) Synthesis and antitumor activity of novel salvicine analogues. Chin Chem Lett 22:25–28

Devi PU (1996) *Withania somnifera* Dunal (Ashwagandha): potential plant source of a promising drug for cancer chemotherapy and radio sensitisation. Indian J Exp Biol 34:927–932

Diogo CV, Felix L, Vilela S, Burgeiro A, Barbosa IA, Carvalho MJM, Oliveira PJ, Peixoto FP (2009) Mitochondrial toxicity of the phyotochemicalsdaphnetoxin and daphnoretin -relevance for possible anti-cancer application. Toxicol in Vitro 23:772–779

Dixon RA, Ferreira D (2002) Molecules of interest: Genistein. Phytochemistry 60:205–211

Dubey KK, Ray AR, Behera BK (2008) Production of demethylated colchicines through microbial transformation and scale–up process development. Process Biochem 43:251–257

Efferth T, Li PCH, Konkimalla VSB, Kaina B (2007) From traditional chinese medicine to rational cancer therapy. Trends Mol Med 13:353–361

Faria SS, Morris CFM, Silva AR, Fonseca MP, Forget P, Castro MS (2017) A timely shift from shotgun to targeted proteomics and how it can be groundbreaking for cancer research. Front Oncol 7:13

Farnsworth NR, Akerele O, Bingel AS, Soejarto DD, Guo Z (1985) Medicinal Plants in therapy. Bull World Health Organ 63(6):965–981

Ferlay J, Soerjomataram I, Dikshit R, Eser S, Mathers C, Rebelo M (2015) Cancer incidence and mortality worldwide: sources, methods and major patterns in GLOBOCA. Int J Cancer 136:359–386

Ferro MA, Leis A, Doll R, Chiu L, Chung M, Barroetavena MC (2007) The impact of acculturation on the use of traditional Chinese medicine in newly diagnosed Chinese cancer patients. Support Care Cancer 15(8):985–992

Fuchs C, Mitchell EP, Hoff PM (2006) Irinotecan in the treatment of colorectal cancer. Cancer Treat Rev 32:491–503

Fulda S (2008) Betulinic acid for cancer treatment and prevention. Int J Mol Sci 9:1096–1107

Garodia P, Ichikawa H, Malani N, Sethi G, Aggarwal B (2007) From ancient medicine to modern medicine: ayurvedic concepts of health and their role in inflammation and cancer. J Soc Integr Oncol 5(1):25–37

Goel A, Kunnumakkara AB, Aggarwal BB (2008) Curcumin as "Curecumin": from kitchen to clinic. Biochem Pharmacol 75:787–809

Gupta YK, Sharma SS, Rai K, Katiyar CK (2001) Reversal of paclitaxel induced neutropenia by *Withania somnifera* in mice. Indian J Physiol Pharmacol 45:253–257

Hait WN, Rubin E, Alli E, Goodin S (2007) Tubulin targeting agents. Update on Cancer Ther 2:1–18

Hampson P, Wang K, Lord JM (2005) Treatment of actinic keratoses, acute myeloid leukemia therapy, treatment of basal cell carcinoma, protein kinase C activator. Drugs Future 30:1003

Hartwell JL (1967) Plants used against cancer. A survey. Lloydia 30:379–436
Hazra B, Das Sarma M, Sanyal U (2004) Separation methods of quinonoid constituents of plants used in oriental traditional medicines. J Chromatogr B 812:259–275
He L, Gu J, Lim LY, Yuan ZX, Mo J (2016) Nanomedicine-mediated therapies to target breast cancer stem cells. Front Pharmacol 7:313
Herst PM, Davis JE, Neeson P, Berridge MV, Ritchie DS (2009) The anticancer drug phenoxodiol kills primary myeloid and lymphoid leukemic blasts and rapidly proliferating T cells. Haematol J 94:928–934
Huang Z, Chen G, Shi P (2009) Effects of emodin on the gene expression profiling of human breast carcinoma cells. Cancer Detect Prev 32:286–291
Huang M, Lu JJ, Huang MQ, Bao JL, Chen XP, Wang YT (2012) Terpenoids: natural products for cancer therapy. Expert Opin Investig Drugs 21(12):1801–1818
Iqbal J, Abbasi BA, Mahmood T, Kanwal S, Ali B, Khalil AT (2017) Plant-derived anticancer agents: a green anticancer approach. Asian Pac J Trop Biomed:1–23
Jagetia GC, Baliga MS (2003a) Modulation of antineoplastic activity of cyclophosphamide by *Alstonia scholaris* in the Ehrlich ascites carcinoma-bearing mice. J Exp Ther Oncol 3:272–282
Jagetia GC, Baliga MS (2003b) Treatment with *Alstonia scholaris* enhances radiosensitivity in vitro and in vivo. Cancer Biother Radiopharm 18:917–929
Jeong JS, Ryu BH, Kim JS, Park JW, Choi WC, Yoon SW (2010) Bojungikki-tang for cancer-related fatigue: a pilot randomized clinical trial. Int Cancer Ther 9(4):331–338
Jung JE, Kim HS, Lee CS, Park DH, Kim YN, Lee MJ, Lee JW, Park JW, Kim MS, Ye SK (2007) Caffeic acid and its synthetic derivative CADPE suppress tumor angiogenesis by blocking STAT3-mediated VEGF expression in human renal carcinoma cells. Carcinogenesis 28(8):1780–1787
Kamboj VP (2000) Herbal medicine. Curr Sci 78:35–38
Kaufman PB, Duke JA, Brielmann H, Boik J, Hoyt JE (1997) A comparative survey of leguminous plants as sources of the isoflavones genistein and daidzein:implications for human nutrition and health. J Altern Complement Med 3:7–12
Khan H (2014) Medicinal plants in light of history: recognized therapeutic modality. J Evid Based Complement Alternat Med 19:216–219
Kim S, Hwang BY, Su BN, Chai H, Mi Q, Kinghorn AD, Wild R, Swanson SM (2007) Silvestrol a potential anticancer rocaglate derivative from Aglaiafoveolata induces apoptosis in LNCaP cells through the mitochondrial/apoptosome pathway without activation of executioner caspase-3 or -7. Anticancer Res 27:2175–2183
Kim NH, Pham NB, Quinn RJ, Kwon HJ (2010) R-(−)-β-O-methylsynephrine, a natural product, inhibits VEGF-induced angiogenesis in vitro and in vivo. Biochem Biophys Res Commun 399(1):20–23
Kingston DGI (2007) The shape of things to come: structural and synthetic studies of taxol and related compounds. Phytochemistry 68:1844–1854
Kuo YC, Kuo PL, Hsu YL, Cho CY, Lin CC (2006) Ellipticine induces apoptosis through p53-dependent pathway in human hepatocellular carcinoma HepG2 cells. Life Sci 78:2550–2557
Lachumy SJ, Oon CE, Deivanai S, Saravanan D, Vijayarathna S, Choong YS, Yeng C, Latha LY, Sasidharan S (2013) Herbal remedies for combating irradiation: a green antiirradiation approach. Asian Pac J Cancer Prev 14(10):5553–5565
Lall RK, Syed DN, Adhami VM, Khan MI, Mukhtar H (2015) Dietary polyphenols in prevention and treatment of prostate cancer. Int J Mol Sci 16(2):3350–3376
Lam W, Bussom S, Guan F, Jiang Z, Zhang W, Gullen EA, Liu SH, Cheng YC (2010) The four-herb Chinese medicine PHY906 reduces chemotherapy-induced gastrointestinal toxicity. Sci Trans Med 18(2):45–59
Lang I (2002) Possibilities and results with Herceptin-Taxol combination in the treatment of breast cancer. Magy Onkol 46:195–196
Lee CO (2005) Complementary and alternative medicine patients are talking about: PC-SPES. Clin J Oncol Nurs 9(1):113–114

Ley CD, Horsmany MR, Kristjansen PEG (2007) Early effects of combretastatin-A4 disodium phosphate on tumor perfusion and interstitial fluid pressure. Neoplasia 9:108–112

Li Y, Li CJ, Yu D, Pardee AB (2000) Potent induction of apoptosis by lapachone in human multiple myeloma cell lines and patient cells. Mol Med 6:1008–1015

Li W, Shao Y, Hu L, Zhang X, Chen Y, Tong L, Li C, Shen X, Ding J (2007) BM6, a new semi-synthetic vinca alkaloid, exhibits its potent in vivo anti-tumor activities via its high binding affinity for tubulin and improved pharmacokinetic profiles. Cancer Biol Ther 6(5):787–794

Liao CC, Ou TT, Wu CH, Wang CJ (2013) Prevention of diet-induced hyperlipidemia and obesity by caffeic acid in C57BL/6 mice through regulation of hepatic lipogenesis gene expression. J Agric Food Chem 61(46):11082–11088

Lissoni P, Giani L, Zerbini S, Trabattoni P, Rovelli F (1998) Biotherapy with the pineal immuno-modulating hormone melatonin versus melatonin plus aloevera in untreatable advanced solid neoplasms. Nat Immun 16:27–33

Liu Q (2011) Triptolide and its expanding multiple pharmacological functions. Int J Immunopharmacol 11:377–383

Liu ML, Chien LY, Tai CJ, Lin KC, Tai CJ (2011) Effectiveness of traditional chinese medicine for liver protection and chemotherapy completion among cancer patients. Evid Based Complement Alternat Med 2011:291843

Lu CL, Li YM, Fu GQ, Yang L, Jiang JG, Zhu L, Lin FL, Chen J, Lin QS (2011) Extraction optimisation of daphnoretin from root bark of Wikstroemiaindica and its antitumour activity tests. Food Chem 124:1500–1506

Lu JJ, Bao JL, Chen XP, Huang M, Wang YT (2012) Alkaloids isolated from natural herbs as the anticancer agents. Evid Based Complement Alternat Med:2012

Lu K, Bhat M, Basu S (2016) Plants and their active compounds: natural molecules to target angiogenesis. Angiogenesis 19(3):287–295

Mans DRA, Rocha AB, Schwartsmann G (2000) Anti-cancer drug discovery and development in Brazil: targeted plant collection as a rational strategy to acquire candidate anti – cancer compounds. Oncologist 5:185–198

Maskarinec G, Shumay DM, Kakai H, Gotay CC (2000) Ethnic differences in complementary and alternative medicine use among cancer patients. J Altern Complement Med 6(6):531–538

Mi Q, Cui B, Silva GL, Lantvit D, Lim E, Chai H, You M, Hollingshead MG, Mayo JG, Kinghorn AD, Pezzuto JM, Pervilleine A (2001) A novel tropane alkaloid that reverses the multidrug-resistance phenotype. Cancer Res 61:4030–4037

Mi Q, Cui B, Silva GL, Lantvit D, Lim E, Chai H, Hollingshead MG, Mayo JG, Kinghorn AD, Pezzuto JM (2002) Pervilleines B and C new tropane alkaloid aromatic esters that reverse the multidrug-resistance in the hollow fiber assay. Cancer Lett 184:13–20

Mi Q, Cui B, Lantvit D, Reyes-Lim E, Chai H, Pezzuto JM, Kinghorn AD, Swanson SM (2003) Pervilleine F a new tropane alkaloid aromatic ester that reverses multidrug resistance. Anticancer Res 23:3607–3615

Mineko I, Michio I, Ikuo M, Megumi M, Setsuko I, Akiko T, Akio E (2000) Growth inhibitory effect of a new camptothecinanalog DX-8951f on various drug-resistant sublines including BCRP-mediated camptothecin derivative-resistant variants derived from the human lung cancer cell line PC-6. Anti Cancer Drugs 11:353–362

Molavi O, Ma Z, Mahmud A, Alshamsan A (2008) Polymeric micelles for the solubilization and delivery of state inhibitor cucurbitacins in solid tumors. Int J Pharm 347:118–127

Moon YJ, Wang X, Morris ME (2006) Dietary flavonoids: effects on xenobiotic and carcinogen metabolism. Toxicol in Vitro 20:187–210

Moyad MA, Pienta KJ, Montie JE (1999) Use of PC-SPES, a commercially available supplement for prostate cancer, in a patient with hormone-naive disease. Urology 54:319–323

Nam S, Buettner R, Turkson J, Kim D, Cheng JQ, Muehlbeyer S, Hippe F, Vatter S, Merz KH, Eisenbrand G, Jove R (2005) Indirubin derivatives inhibit State signaling and induce apoptosis in human cancer cells. Proc Natl Acad Sci 102:5998–6003

Ohnuma T, Sridhar KS, Ratner LH, Holland JF (1982) Phase I study of indicine N-oxide in patients with advanced cancer. Cancer Treat Rep 66:1509–1515

Okouneva T, Hill BT, Wilson L, Jordan MA (2003) The effects of vinflunine, vinorelbine and vinblastine on centromere dynamics. Mol Cancer Ther 2:427–436

Olsen DL, Bradley C, Johnson M, Macias JL, Love V, Markoe A (2001) The effect of *Aloe vera* gel/mild soap versus mild soap alone in preventing skin reactions in patients undergoing radiation therapy. Oncol Nurs Forum 28:543–547

Pan L, Chai H, Kinghom AD (2010) The continuing search for antitumor agents from higher plants. Phytochem Lett 3:1–8

Pan R, Gao X, Lu D, Xu X, Xia Y, Dai Y (2011) Prevention of FGF-2-induced angiogenesis by scopoletin, a coumarin compound isolated from Erycibe obtusifolia Benth, and its mechanism of action. Int Immunopharmacol 11(12):2007–2016

Pandey G (2002) Anticancer herbal drugs of India with special reference to Ayurveda. Sri Satguru Publications, New Delhi, pp 18–121

Park EJ, Pezzuto JM (2002) Botanicals in cancer chemoprevention. Cancer Metastasis Rev 21:231–255

Patel P (2003) Global resurgence and international recognition of Ayurveda

Patil JB, Kim J, Jayaprakasha GK (2010) Berberine induces apoptosis in breast cancer cells (MCF-7) through mitochondrial-dependent pathway. Eur J Pharmacol 645:70–78

Prasad GC (1987) Studies on cancer in Ayurveda and its management. JRAS 3(3):147–167

Qin W, Huang G, Chen Z, Zhang Y (2017) Nanomaterials in targeting cancer stem cells for cancer therapy. Front Pharmacol 8:1

Ramawat KG, Goyal S (2008) The Indian herbal drugs scenario in global perspectives. In: Bioactive molecules and medicinal plants. Springer, Berlin, pp 325–347

Ramawat KG, Jain S, Suri SS, Arora DK (1998) Aphrodisiac plants of Aravalli Hills with special reference to safed musli. Role of biotechnology in medicinal and aromatic plants 1:210–223

Rao AV, Devi PU, Kamath R (2001) In vivo radioprotective effect of *Moringa oleifera* leaves. Indian J Exp Biol 39:858–863

Reddy L, Odhav B, Bhoola KD (2003) Natural products for cancer prevention: a global perspective. Pharmacol Ther 99(1):1–13

Sa G, Das T, Banerjee S, Chakraborty J (2010) Curcumin:from exotic spice to modern anticancer drug. Al Ameen J Med Sci 3:21–37

Saif MW, Lansigan F, Ruta S, Lamb L, Mezes M, Elligers K, Grant N, Jiang ZL, Liu SH, Cheng YC (2010) Phase I study of the botanical formulation PHY906 with capecitabine in advanced pancreatic and other gastrointestinal malignancies. Phytomedicine 17(3–4):161–169

Sastry J (2001) Introduction to oncology cancer in Ayurveda. Chaukhambha orientalia, Varanasi

Satyavati GV, Raina MK, Sharma M (1976) Medicinal plants of India. Indian Council of Medical Research, New Delhi

Sharada AC, Solomon FE, Devi PU, Udupa N, Srinivasan KK (1996) Antitumour and radiosensitising effects of withaferin A on mouse Ehrlich ascites carcinoma in vivo. Acta Oncol 35:95–100

Sharma PV (1981) Charaka samhita. Choukhamba Orientalia, Varanasi

Shirin H, Pinto JT, Kawabata Y, Soh JW, Delohery T, Moss SF, Murty V, Rivlin RS, Holt PR, Weinstein IB (2001) Antiproliferative effects of S-allylmercaptocysteine on colon cancer cells when tested alone or in combination with sulindac sulphide. Cancer Res 61:725–731

Shoeb M, Celik S, Jaspars M, Kumarasamy Y, MacManus S, Nahar L, Kong TLP, Sarker SD (2005) Isolation structure elucidation and bioactivity of schischkiniin a unique indole alkaloid from the seeds of Centaurea schischkinii. Tetrahedron 61:9001–9006

Shoeb M, MacManus SM, Jaspars M, Trevidadu J, Nahar L, Thoo-Lin PK, Sarker SD (2006) Montamine, a unique dimericindole alkaloid, from the seeds of *Centaurea montana* (Asteraceae), and its in-vitro cytotoxic activity against the CaCo2 colon cancer cells. Tetrahedron 62:11172–11177

Simoens C, Lardon F, Pauwels B, de Pooter CMJ, Lambrechts HAJ, Pattyn GGO, Breillout F, Vermorken JB (2008) Comparative study of the radiosensitising and cell cycle effects of vinflunine and vinorelbine, in-vitro. BMC Cancer 8:65

Singh RH (2002) An assessment of the ayurvedic concept of cancer and a new paradigm of anticancer treatment in Ayurveda. J Alternat Complement Med 8(5):609–614

Sinha R, Anderson DE, McDonald SS, Greenwald P (2003) Cancer risk and diet in India. J Postgrad Med 49:222–228

Sonata S (1986) The efficacy of ayurveda drugs on cancer (Arbuda), Workshop on Cancer Souvenir. Central Research Institute for Siddha, Chennai, p 81

Sriganth NPI, Premalatha B (1999) Dietary curcumin with cisplatin administration modulates tumour marker enzymes indices in experimental fibrosarcoma. Pharmacol Res 39:175–179

Srivastava V, Negi AS, Kumar JK, Gupta MM, Khanuja SP (2005) Plant-based anticancer molecules: a chemical and biological profile of some important leads. Bioorg Med Chem 13(21):5892–5908

Stabler SN, Tejani AM, Huynh F, Fowkes C (2012) Garlic for the prevention of cardiovascular morbidity and mortality in hypertensive patients. Cochrane Database Syst Rev 8

Stoewsand G (1995) Bioactive organosulfur phytochemicals in *Brassica oleracea* vegetables-a review. Food Chem Toxicol 33(6):537–543

Tabana YM, Hassan LEA, Ahamed MBK, Dahham SS, Iqbal MA, Saeed MAA, Khan MSS, Sandai D, Majid ASA, Oon CE, Majid AMSA (2016) Scopoletin, an active principle of tree tobacco (*Nicotiana glauca*) inhibits human tumor vascularization in xenograft models and modulates ERK1, VEGF-A, and FGF-2 in computer model. Microvasc Res 107:17–33

Thakore P, Mani RK, Kavitha SJ (2012) A brief review of plants having anti-cancer property. Int J Pharm Res Dev 3:129–136

Thatte U, Dhahanukar S (1991) Ayurveda, the natural alternative. Sci Today 2001:12–18

Thomson P, Naylor MA, Everett SA, Stratford MRL, Lewis G, Hill S, Patel KB, Wardman P, Davis PD (2006) Synthesis and biological properties of bioreductively targeted nitrothienylpro drugs of combretastatin A-4. Mol Cancer Ther 5:2886–2894

Treadway S (1998) An ayurvedic herbal approach to a healthy liver. Clin Nutr Res 6(16):1–3

Vrinda B, Uma Devi P (2001) Radiation protection of human lymphocyte chromosomes in vitro by orientin and vicenin. Mutat Res 498:39–46

Wang F, Gao Y, Gao L, Xing T (2011) Study on the electrochemical behaviour of the anticancer herbal drug berberine and its analytical application. J Chin Chem Soc 58:450–456

Wang Z, Dabrosin C, Yin X, Fuster MM, Arreola A, Rathmell WK (2015) Broad targeting of angiogenesis for cancer prevention and therapy. Semin Cancer Biol 35:224–243

Wiatrak BJ (2003) Overview of recurrent respiratory papillomatosis. Curr Opin Otolaryngol Head Neck Surg 11(6):433–441

Xie J, Ma T, Gu Y, Zhang X, Qiu X, Zhang L, Xu R, Yu Y (2009) Berbamine derivatives: a novel class of compounds for anti–leukemia activity. Eur J Med Chem 44:3293–3298

Xu R, Dong Q, Yu Y, Zhao X, Gan X, Wu D, Lu Q, Xu X, Yu XF (2006) Berbamine:a novel inhibitor of bcr/abl fusion gene with potent anti-leukemia activity. Leuk Res 30:17–23

Xu L, Lao LX, Ge A, Yu S, Li J, Mansky PJ (2007) Review Chinese herbal medicine for cancer pain. Integr Cancer Ther 6(3):208–234

Yehya AH, Asif M, Tan YJ, Sasidharan S, Majid AMA, Oon CE (2017) Broad spectrum targeting of tumor vasculature by medicinal plants: an updated review. J Herb Med 9:1–13

Yeruva L, Pierre KJ, Elegbede A, Wang RC, Carper SW (2007) Perillyl alcohol and perillic acid induced cell cycle arrest and apoptosis in non–small cell lung cancer cells. Cancer Lett 257:216–226

Zaniboni A, Ardizzoni A, De Marinis F, Portalone L, Boni C, Meriggi F, Cafferata MA, Ariganello O, Torri V, Neumaier CE, Rosso R (2003) Phase II study of Taxol combined with ifosfamide and carboplatin in the treatment of stage IIIb-IV non-small-cell lung cancer. Am J Clin Oncol 26:84–88

Zhang JA, Xuan T, Parmar M, Ma L, Ugwu S, Ali S, Ahmad I (2004) Development and characterization of a novel liposome-based formulation of SN-38. Int J Pharm 270:93–107

Zhang LQ, Lv RW, Qu XD, Chen XJ, Lu HS, Wang Y (2017) Aloesin suppresses cell growth and metastasis in ovarian cancer SKOV3 cells through the inhibition of the MAPK signaling pathway. Anal Cell Pathol 2017:1–6

Plants with Anticancer Potential

4

Abstract

Plants have been utilized for health and medicinal benefits for hundreds of years due to their multiple beneficial attributes such as anticancer, antitumor, antioxidant, antimicrobial, antibacterial, anti-ulcer, anti-arthritic, etc. It has been estimated that there are altogether 250,000 species of higher plants on Earth and among them 35,000–70,000 species are being used to treat various diseases due to the presence of secondary metabolites (alkaloids, flavonoids, steroids, glycosides, saponins, etc.). Cancer is a worldwide leading cause of morbidity and mortality. To cure this at right time, herbal drugs are more beneficial than synthetic drugs, because the synthetic medicines can cause heavy damage to normal cells while destroying the tumor cells. The present work consists of a review of 149 plant families harboring 838 species reported to possess anticancer property. Moreover, other biological properties of the bioactive compounds are also covered. This work is based on reliable data collected from multifarious databases such as CAB abstract, MEDLINE, EMBASE, J GATE, ERIC, Proquest, INMEDPLAN, NATTS, NPACT, The Plant List, JSTOR, Google Scholar, Springer, Elsevier, and websites such as www.sciencedirect.com, www.jstor.org, www.eflora.org, and www.pfaf.org. The complete data regarding plant names, synonyms, common names, botanical description, medicinal properties, and bioactive compounds present in the plant parts is compiled. In near future, these bioactive compounds can be deployed singly or in combination with routine chemotherapy and radiotherapy to treat various types of cancers, after proper standardization, dose optimization, and stringent clinical trials.

Keywords

Medicinal plants · Botanical description · Anticancer · Antitumor · Antioxidant

© Springer Nature Singapore Pte Ltd. 2019
B. Koul, *Herbs for Cancer Treatment*,
https://doi.org/10.1007/978-981-32-9147-8_4

4.1 Introduction

From the earliest times, herbs have been prized for their role in culinary preparations, pain-relieving, blood-purifying, purgative, analgesic, immunity-boosting, antiaging, memory-enhancing, and wound-healing potential. Nowadays, we still rely largely on the curative and taste enhancer properties of the plants. Plant-based traditional medicine system is playing a vital role for healthcare in both developing and developed countries. According to the World Health Organization (WHO), 80% of the people living in rural areas depend on medicinal herbs for their primary healthcare. More than 70% of Indian population (1.1 billion) still uses non-allopathic system of medicine to cure several diseases (Vaidya and Devasagayam 2007; Srinivasan et al. 2014).

Cancer is one of the major causes of death in the world, and it is the second most leading cause of mortality after cardiovascular diseases. Cancer starts due to genetic mutations in DNA which is caused by the deformation of a natural cell (Greenwell and Rahman 2015). This abnormal cell then multiplies in an abnormal way by asexual reproduction, that is, it bypasses the signals related to regulation of cell growth around it and obtains invasive characteristics and causes changes in surrounding tissues (Jain and Barton 2018). According to the World Health Organization, each year 14 million people suffer from cancer, and 8 million die due to cancer, worldwide. According to the National Cancer Registry Programme of the Indian Council of Medical Research (ICMR), more than 1300 Indians die every day due to cancer. Between 2012 and 2014, the mortality rate due to cancer increased by approximately 6%. In 2012, 4,78,180 deaths were recorded out of 29,34,314 cancer cases reported. In 2013, 465,169 deaths were recorded out of 3,016,628 cancer cases reported. In 2014, 491,598 deaths were recorded out of 2,820,179 cases reported. As per the data obtained from Population Cancer Registry of the ICMR, the incidence and mortality of cancer is highest in the northeastern region of the country. Breast cancer is the most common cancer, followed by lung and prostate cancer (Jemal et al. 2011; Siegel et al. 2017).

Herbal medicines have a vital role in the prevention and treatment of cancer, and medicinal herbs are easily procurable, cost-effective, safe, and free from side effects (Simon 2018). Scientists across the world are concentrating on the herbal medicines to boost immune cells of the body against cancer. By understanding the complex synergistic interaction of various constituents of anticancer herbs, the herbal formulations can be designed to attack the cancerous cells without harming the normal cells of the body (Priya et al. 2015). In addition, to control/kill the cancer cells, the science of Ayurveda aims at strengthening the immune system by fostering the healthy cells to cope with cancer (Poornima and Efferth 2016). As India is the largest producer of herbal plants and Ayurvedic medicines in the world, so it is called as the "botanical garden of the world." In India, 17,000–18,000 species of flowering plants are found, of which 6000–7000 are estimated to have medicinal properties. The evidences gathered from Indian, Chinese, Greek, Egyptian, Papyrus, Roman, Syrian, and Unani manuscript texts prove that the usage of herbal drugs is safe and time tested (Samal 2016). India has a rich history of very long and safe usage of several medicinal herbs under the recognized alternative systems of medicine like Ayurveda, yoga, Unani, Siddha, homeopathy, and naturopathy.

Nowadays, various methods are used for cancer treatment such as chemotherapy, but in this method, because of improper selection of medicines, a high percentage of healthy cells are also lost along with the cancerous cells. In order to prepare anticancer medicines from natural resources like plants and plant parts, proper testing of cytotoxic compounds and screening of raw extracts is of utmost necessary (Shabani 2016).

AYUSH is the acronym of the medical streams that are being practiced in India such as "Ayurveda," "Yoga" and "Naturopathy," "Unani," "Siddha," and "Homeopathy." These systems are based on definite medical philosophies and represent a way of healthy living with established concepts on prevention of diseases and promotion of health in a holistic manner. Ayurveda, the traditional Indian holistic medicine, uses plant-derived products for the treatment of various types of diseases including cancer (Mukherjee et al. 2007) as the plants are reservoir of bioactive compounds. Nowadays, Ayurvedic medicines are accepted globally. Plant-based treatment can cure the patients from side effects of Western medicines such as dose-mediated toxicity and drug resistance. Moreover, the phytochemicals extracted from plants are of low cost and safe when used in a dose-dependent manner.

As already mentioned, herbal drugs are prepared using various parts of the plants such as roots, stems, leaves, fruits, and seeds. The phytochemicals can be extracted and used in the form of decoction, syrups, juices, and pills. The constituents extracted from the herbs belong to the chemical class acids, alcohols, aliphatic hydrocarbons, alkyl aldehyde, benzofuran, benzopyran, chalcone, coumarins, coumestrol, dihydrofuran, esters, fatty acids, flavonoids, furanocoumarin, isoflavone, meroterpene, phenolic cinnamate, phenols, phenylpropene, sterols, terpenes, rose ketones, saponins, glycosides, etc. (Iqbal et al. 2017). The anticancer activities of the different classes of bioactive compounds are *isoflavone*: reduction of the risk of breast cancer; alkaloids, inhibition of cancer cell growth; *coumarins*: inhibition of cancer cell growth; *flavonoids and polyphenols*: anticancer and antiproliferation effects; *terpenoids*: MCF-7 cell apoptosis; and *quinine*: inhibition of the growth of cancerous cells (Yin et al. 2013). A large volume of clinical studies has reported that medicinal herbs and their bioactive compounds are being recognized increasingly for complementary treatment of cancer. It is reported that herbal medicines are more beneficial for survival, immune modulation, and quality of life of cancer patients and can enhance the healing effect of the existing cancer treatment regimes (Kainsa et al. 2012).

There are several plant families which are a source of a large number of anticancer plants including Acanthaceae (Saha and Paul 2017), Euphorbiaceae (Chen et al. 2016a, b, c), Fabaceae (Bonigala et al. 2017), Apocynaceae (Baliga 2012), Asteraceae (Mallmann et al. 2018), Liliaceae (Jena et al. 2012), Annonaceae, Rosaceae (Tuasha et al. 2018), Rubiaceae (Ye et al. 2013), Rutaceae (Harahap et al. 2018), Verbenaceae (Patel et al. 2014), and many more. The present work can be considered as an offline dictionary of plants with known anticancer properties and those which can be potentially beneficial after stringent experimentation.

With the economic point of view, the herbal medicines with the long history of their use and reliability are gaining popularity in the global market as registered drugs, healthcare products, cosmetics, and dietary supplements. The acceptance of the use of herbal medicines is increasing not only in Asian countries, but also in the Western countries with a rate of 6.4% per annum (Adinarayana et al. 2012). It has been estimated that the current rate of plant loss is nearly hundred to thousand times higher as compared to the natural extinction rate. Thus, "Mother Earth" is losing its one medicinal plant species in every 2 years (Pimm et al. 1995). According to a survey of the International Union for Conservation of Nature and the World Wildlife Fund, about 50,000–80,000 species of flowering plant extracts have been used as medicine worldwide. Among these, about 15,000 species are extinct due to their habitat destruction or overharvesting (Bentley 2010). In such conditions, conservation of herbs is necessary so that the upcoming generations can also use these herbs (Pan et al. 2014). The increasing risk of extinction can be prevented by conservation processes based on both in situ and ex situ strategies (Chen et al. 2016a, b, c).

4.2 Methodology

Plants belonging to 149 families were extensively searched for their anticancer properties and for the information on common names, synonyms, botanical description, and distribution from 12 reliable databases including "The Plant List" (www.theplantlist.org), "eFlora of India" (http://efloraindia.nic.in), "eFlora of South Africa" (ipt.sanbi.org.za), "eFlora of China" (http://www.efloras.org), "JSTOR" (https://www.jstor.org), "Missouri Botanical Garden" (http://www.missouribotanicalgarden.org), "Plants for a Future" (https://www.pfaf.org), "Biodiversity of India" (http://www.biodiversityofindia.org/), "Useful Tropical Plants" (http://tropical.theferns.info/), "Flowers of India" (http://www.flowersofindia.net), BioNET-EAFRINET (http://keys.lucidcentral.org), and "Kew Botanical Garden" (https://www.kew.org/). The information on medicinal properties and experimental data generated using these plants was collected from various research papers and review articles from various databases such as PubMed, Web of Science, Google Scholar, CAB Abstracts, MEDLINE, EMBASE, INMEDPLAN, NATTS, and web links like www.springer.com, www.elsevier.com, www.onlinelibrary.wiley.com, www.academic.oup.com, www.tandfoline.com, etc. The chemical structures were searched in ChemSpider and PubChem databases and redrawn using the ChemDraw® software (version 8.0).

4.3 Results and Discussion

A record of 149 plant families harboring various plant species with known (traditional/experimental) anticancer/antitumor potential has been tabulated in Tables 4.1, 4.2, 4.3, 4.4, 4.5, 4.6, 4.7, 4.8, 4.9, 4.10, 4.11, 4.12, 4.13, 4.14, 4.15, 4.16, 4.17, 4.18, 4.19, 4.20, 4.21, 4.22, 4.23, 4.24, 4.25, 4.26, 4.27, 4.28, 4.29, 4.30, 4.31, 4.32, 4.33, 4.34, 4.35, 4.36, 4.37, 4.38, 4.39, 4.40, 4.41, 4.42, 4.43, 4.44, 4.45, 4.46, 4.47, 4.48, 4.49, 4.50, 4.51, 4.52, 4.53, 4.54, 4.55, 4.56, 4.57, 4.58, 4.59, 4.60, 4.61, 4.62, 4.63, 4.64, 4.65, 4.66, 4.67, 4.68, 4.69, 4.70, 4.71, 4.72, 4.73, 4.74, 4.75, 4.76, 4.77, 4.78, 4.79, 4.80, 4.81, 4.82, 4.83, 4.84, 4.85, 4.86, 4.87, 4.88, 4.89, 4.90, 4.91, 4.92, 4.93, 4.94, 4.95, 4.96, 4.97, 4.98, 4.99, 4.100, 4.101, 4.102, 4.103, 4.104, 4.105, 4.106, 4.107, 4.108, 4.109, 4.110, 4.111, 4.112, 4.113, 4.114, 4.115, 4.116, 4.117, 4.118, 4.119, 4.120, 4.121, 4.122, 4.123, 4.124, 4.125, 4.126, 4.127, 4.128, 4.129, 4.130, 4.131, 4.132, 4.133, 4.134, 4.135, 4.136, 4.137, 4.138, 4.139, 4.140, 4.141, 4.142, 4.143, 4.144, 4.145, 4.146, 4.147, 4.148, and 4.149. The outcome of my research has revealed selected plant families which possess significant anticancer potential. These plant families include Acanthaceae, Anacardiaceae, Annonaceae, Apocynaceae, Araceae, Asteraceae, Berberidaceae, Euphorbiaceae, Fabaceae, Liliaceae, Rosaceae, Rubiaceae, Rutaceae, and Verbenaceae. These families encompass 838 species of medicinal plants including *Taxus baccata* L., *Miquelia dentata* Bedd., *Acanthus ilicifolius* L., *Adhatoda vasica* Nees, *Acer pseudoplatanus* L., *Actinidia chinensis* Planch., *Dracaena draco* (L.), *Achyranthes aspera* var. *australis*, *Amorphophallus paeoniifolius* (Dennst.), *Arisaema lobatum* Engl., *Calotropis gigantea* R.Br., *Abrus precatorius* L., *Caesalpinia coriaria* (Jacq.) Willd., *Caragana cuneata* Moench, *Dalbergia candenatensis* (Dennst.) Prain., and *Dalbergia odorifera* T. Chen. It is interesting to note that a total of 801 anticancer compounds have been reported from these plants' species. Further researches may be conducted to improve the activity and efficacy of these compounds for cancer prevention and care. It is true that there is no single herbal medicine to treat all types of cancers. But, a combination of the aforementioned plants and their parts (leaves, stem, roots, bark, sap, etc.) can be consumed as dried powder or decoction after dose standardization, to boost the body's immunity and to prolong or fight cancer or synergize the potency of the routine chemotherapy and radiotherapy treatments, for cancer prevention and cure. These plant products may also heal the body from the side effects caused during cancer treatment regime (chemotherapy, radiotherapy, chemoradiation, etc.). In the future, meticulous phytochemical and pharmacological studies on these plant species and their sustainable conservation through biotechnological advancements may yield reliable molecules for production of novel anticancer drugs, for better healthcare. In the end, I would request the readers to believe in the power of "Ayurveda" (The Science of Life) to treat cancers.

Table 4.1 Anticancer plants of family Asteraceae

Family	Botanical name	Image	Plant part used	Bioactive compound(s)	Structure	Reference(s)
Asteraceae	*Acanthospermum australe* (Loefl.) Kuntze		Entire plant	Acanthostral	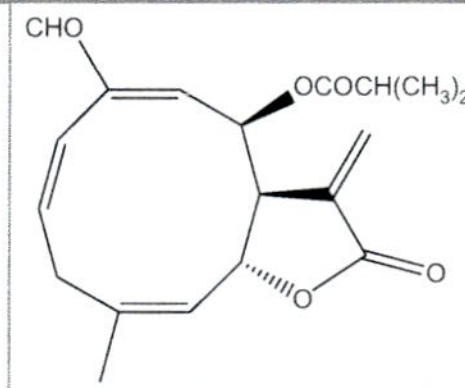	Garcia-Barriga (1975) and Matsunaga et al. (1996)
	Synonym(s): *Acanthospermum brasilum* Schrank, *Acanthospermum hirsutum* DC., *Acanthospermum xanthioides* (Kunth) DC., *Acanthospermum xanthioides* var. *obtusifolium* DC. *Acanthospermum xanthioides* var. *xanthioides*, *Centrospermum xanthioides* Kunth, *Echinodium prostratum* Poit., *Melampodium australe* Loefl., *Orcya adhaerens* Vell., *Orcya adhaerescens* Vell. **Common name(s):** Paraguayan starbur, Ihi kukae hipa, Paraguay bur, Paraquay starbur, Pipili, Sheepbur, Spiny-bur. **Botanical description:** *Acanthospermum australe* is native to South America, within Brazil. The plant is annual or short-term perennial, stems 10–60 (–120) cm, mostly procumbent. Leaves cauline. Leaves rhombic-ovate to triangular, 1.5–3.5 cm long, 1–3 cm wide, with conspicuous glands on both surfaces, margins irregularly serrate above the middle, base cuneate, petioles 0.3–1.5 cm long. Heads 4–6 mm in diameter, generally solitary, with 3–8 white-cream-yellow ray florets and 3–8 yellowish disk florets functionally staminate. **Medicinal use(s):** The plants are depurative, pectoral, anti-inflammatory, hemostatic. They are used in the treatment of coughs, pulmonary infections, and murine leukemia.					

	Acanthospermum hispidum DC.		Leaf	Guaianolides	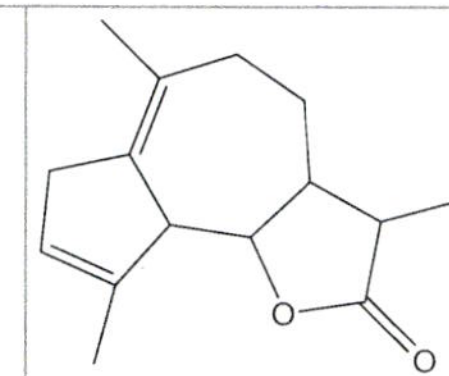	Cartagena et al. (2000)
	Synonym(s): *Acanthospermum humile* var. *hispidum* (DC.) Kuntze. **Common name(s):** Goat's head. **Botanical description:** *Acanthospermum hispidum* is a native plant of Central and South America. It is an annual herb, coarse, erect, to 1.3 m tall. Stems poorly branched, hispid. Leaves sessile or shortly petiolate; blade oblong or obovate; 10–100 × 5–40 mm; both surfaces hispid and gland-dotted; margin subentire and repand-dentate or coarsely dentate. **Medicinal use(s):** *Acanthospermum hispidum* is used to treat the swelling of the brain (encephalitis), fever, boils, and sores. Honeysuckle is also used for urinary disorders, headache, diabetes, rheumatoid arthritis, and gastrointestinal cancer.			Cisgermacranolides	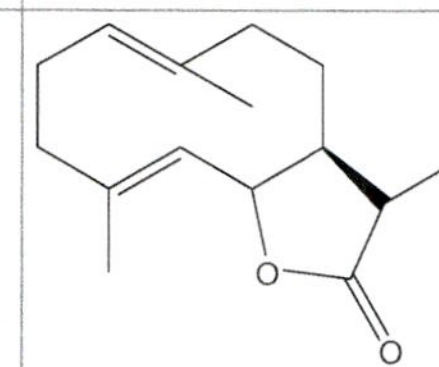	
				Melampolides	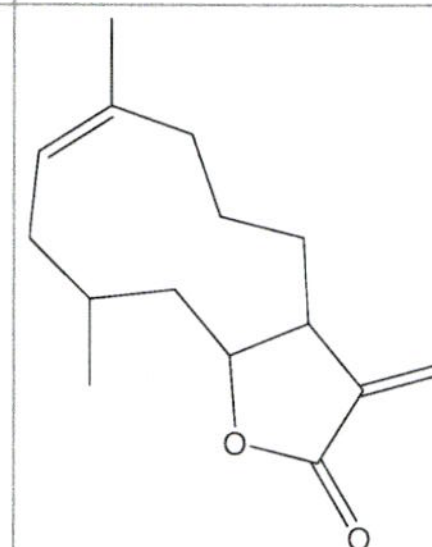	

(continued)

Table 4.1 (continued)

Family	Botanical name	Image	Plant part used	Bioactive compound(s)	Structure	Reference(s)
	Achillea millefolium Ledeb.		Apical bud	Apigenin	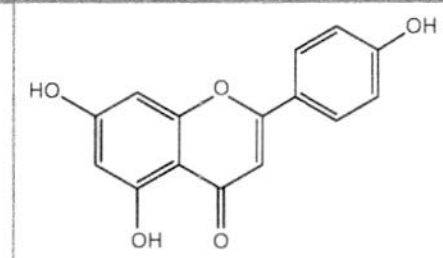	Csupor-Loffler et al. (2009)
				Luteolin	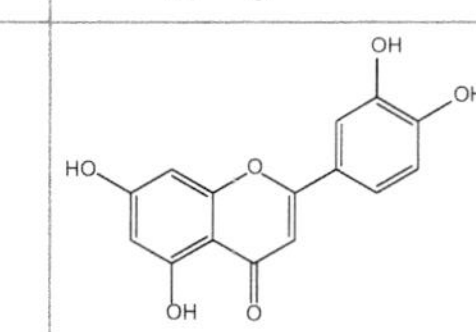	
				Centaureidin	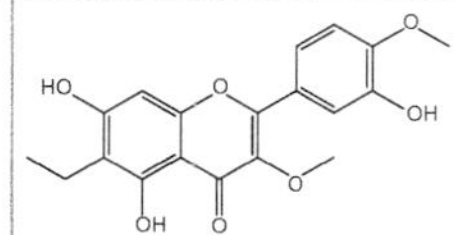	
	Synonym(s): *Achillea millefolium* L., *Achillea millefolium* subsp. *alpestris* (Wimm. & Grab.) Gremli, *Achillea millefolium* subsp. *ceretanica* (Sennen) Sennen. **Common name(s):** Death flower. **Botanical description:** *Achillea millefolium* is native of North America and is an herb, perennial, 40–100 cm tall, with long rhizomes; stems erect, unbranched, or branched in upper part, often with short sterile branches at leaf axils above middle, striate, usually white villous. Leaves sessile; leaf blade lanceolate, oblong-lanceolate, exterior gland-dotted, apex 5-lobed. Achenes greenish and oblong. **Medicinal use(s):** *Achillea millefolium* is used for digestive disorders including pain and swelling (inflammation) of the small intestine (enteritis) and dysentery; upper respiratory tract infections including colds, influenza, swine flu, and pneumonia. This plant is used to treat gastrointestinal cancer.			Casticin	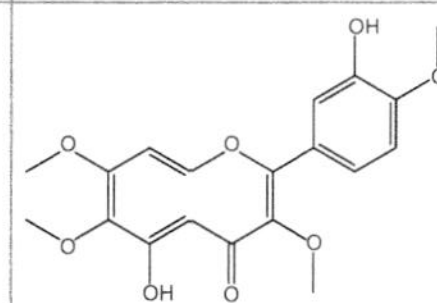	
				Artemetine	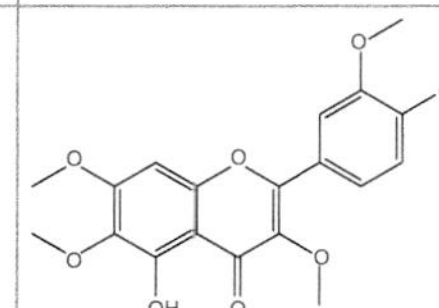	

	Achyrocline lehmannii Hieron		Entire plant	Not available	Not available	Garcia-Barriga (1975)
	Synonym(s): *Achyrocline alata* (Kunth) DC., *Achyrocline albicans* Griseb., *Achyrocline anabelae* Deble., *Achyrocline arrojadoana* Mattf. **Common name(s):** Not available. **Botanical description:** *Achyrocline lehmannii* is a genus of flowering plant in the Asteraceae described as a genus in 1838. It is native to Latin America and Africa. **Medicinal use(s):** Plant extract is used to treat dermatitis, ringworm, scabies, herpes, asthma, and esophagus cancer.					
	Achyrocline satureioides (Lam.) DC.		Entire plant	Luteolin	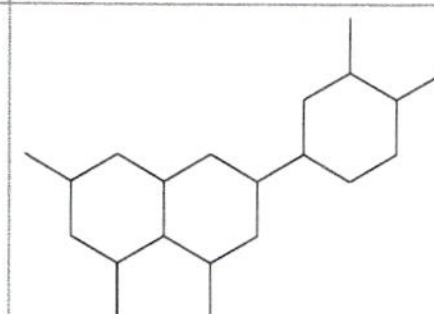	Morton (1975) and de Souza et al. (2018)
				3-O-methyl-quercetin	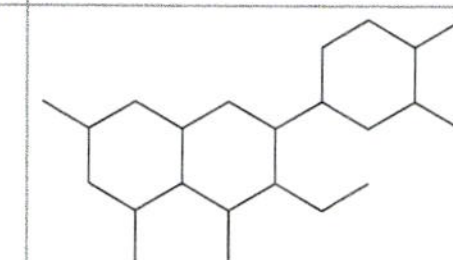	

(continued)

Table 4.1 (continued)

Family	Botanical name	Image	Plant part used	Bioactive compound(s)	Structure	Reference(s)
	Synonym(s): *Achyrocline satureioides* (Lam.) DC. **Common name(s):** Marcela. **Botanical description:** *Achyrocline satureioides* is a medicinal plant found in South America. **Medicinal use(s):** This plant is famous for its biological activities such as anti-inflammatory, hepatoprotective, antioxidant, immunomodulatory, antimicrobial, trypanocidal, and photoprotective. It is also used to cure central nervous system tumors.			Achyrobichalcone	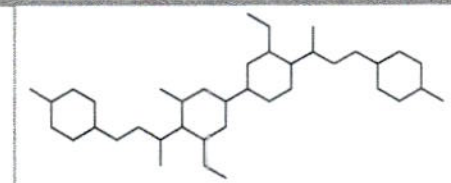	
	Ageratum houstonianum Mill.		Leaf	1, 4-Cyclohexylphenyl	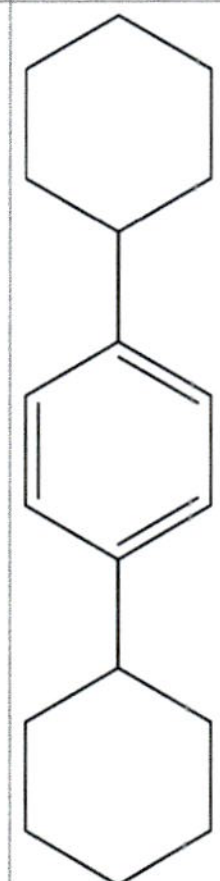	Salie et al. (1996) and Verma et al. (2014)

	Synonym(s): *Ageratum houstonianum* Mill., *Ageratum houstonianum* var. *muticescens* B.L.Rob. **Common name(s):** Blue billygoat weed, Floss flower. **Botanical description:** *Ageratum houstonianum* is an annual herb of North America, 30–70 (–100) cm tall. Stems erect, simple, or branched from the middle to the lower part; stems and branches purple-red, green, or straw-colored, white tomentose, or thinly lanate. Leaves broadly ovate or triangular-ovate; petiole 0.7–3 cm; median stem leaves 2–6 × 1.5–3.5 cm, or length equal to width; upper and axillary leaves smaller; all leaves basally 3-veined or inconspicuously 5-veined. **Medicinal use(s):** *Ageratum houstonianum* is used medicinally. People in Central America (Ecuador) use the plant as an antiphlogistic, to relieve swelling and pain in the throat, and as a poultice for brain tumors.					
	Arctotis auriculata Jacq.		Leaf	Bleomycin	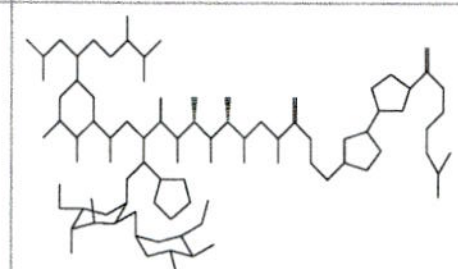	Lao et al. (1984)
	Synonym(s): *Arctotis melanocycla* Willd. ex Harv., *Arctotis melanocycla* Willd., *Arctotis namaquana* Schltr. ex Lewin. **Common name(s):** "African daisy," or "Gousblom" in Afrikaans. **Botanical description:** *Arctotis auriculata* is native of South Africa and is perennial, 0.3–0.4 m with silver trailing foliage and yellow-orange flowers. These plants have daisy-like composite flowers which tend to close in the late afternoon or in dull weather, but numerous cultivars have been developed for garden use which stay open for longer and are available in a wide range of colors. Tender perennials are often grown in temperate regions as half-hardy annuals. **Medicinal use(s)**: This plant is used to treat asthma, pneumonia, infections, parasitic diseases, diabetes mellitus, and liver uterine cancer.			Doxorubicin	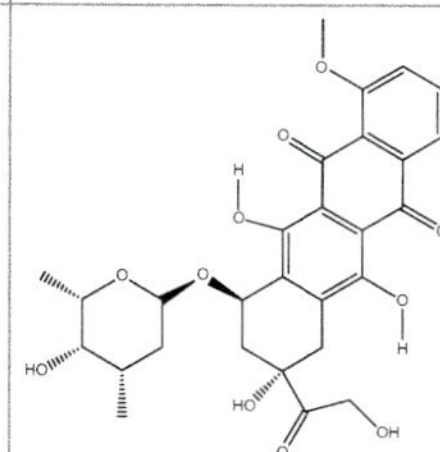	

(continued)

Table 4.1 (continued)

Family	Botanical name	Image	Plant part used	Bioactive compound(s)	Structure	Reference(s)
	Artemisia argyi H. Lév. and Vaniot		Entire plant	Flavones	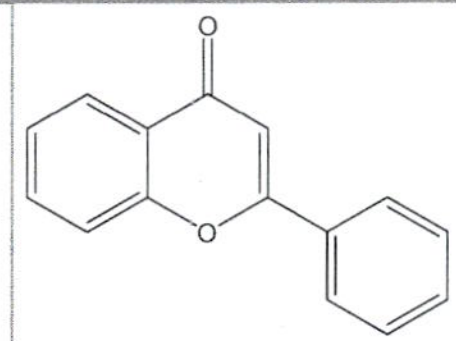	Duke and Ayensu (1985a, b) and Seo et al. (2003)
	Synonym(s): *Artemisia argyi* H. Lév. & Vaniot, *Artemisia argyi* f. *argyi*. **Common name(s):** Silvery wormwood or Chinese mugwort. **Botanical description:** *Artemisia argyi* is a Chinese perennial herb or subshrubs, 80–150 (–250) cm tall, with many lateral roots, shortly branched apically, gray arachnoid pubescent, strongly aromatic. Middle stem leaves: petiole 2–3 mm; leaf blade ovate, triangular-ovate, or subrhombic, 5–8 × 4–7 cm; abaxially densely gray arachnoid-tomentose, adaxially incanous pubescent, and white gland-dotted. **Medicinal use(s):** *Artemisia argyi* is used in making important anti-inflammatory, detoxifying, and hemostatic drugs. It is used to treat many types of cancers like lung, liver, and intestinal cancer.					

	Artemisia annua L.		Stem	Coumarins	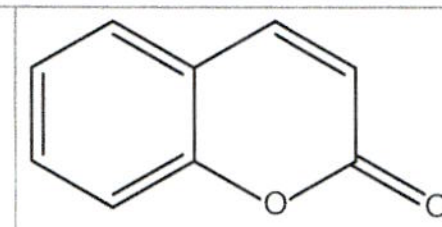	Brisibe et al. (2009)
	Synonym(s): *Artemisia annua* Pall., *Artemisia annua* L. **Common name(s):** Wormwood. **Botanical description:** *Artemisia annua* is a native of China and is widely distributed in North Africa, Asia, Europe, and North America. It is an annual herb, 70–160 (–200) cm tall, much branched, sparsely puberulent, soon glabrous, strongly aromatic. Leaves gland-dotted. Lowermost stem leaves: leaf blade ovate or triangular-ovate, 3–7 × 2–6 cm, 3 (or 4)-pinnatipartite; segments 5–8(–10) pairs. Middle stem leaves: petiole 1–2 cm; leaf blade 2 (or 3)-pinnatisect or pectinatisect; lobules deeply serrate to pectinate; teeth triangular, 1–2 × ca. 0.5 mm; rachis narrowly winged, sparsely serrate or not; midvein prominent adaxially. Uppermost leaves and leaflike bracts 1(or 2)-pinnatipartite. Synflorescence a panicle to 15 × 8 cm, produced from most nodes to form broad, conical compound panicle. **Medicinal use(s):** *Artemisia annua* contains essential oils, artemisinin. It is well-known in traditional Chinese medicine for its anti-inflammatory, antifebrile, and hemostatic properties, and artemisinin has become widely valued as an effective treatment for strains of *Plasmodium* resistant to other antimalarial drugs. The plant is used to treat blood cancer.			Arteannuin B	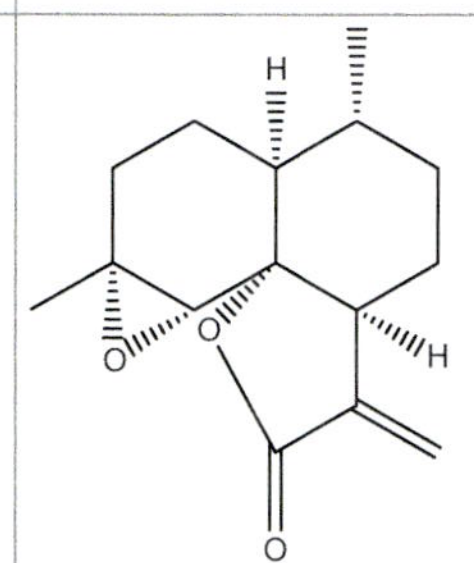	
				Artemisinic acid	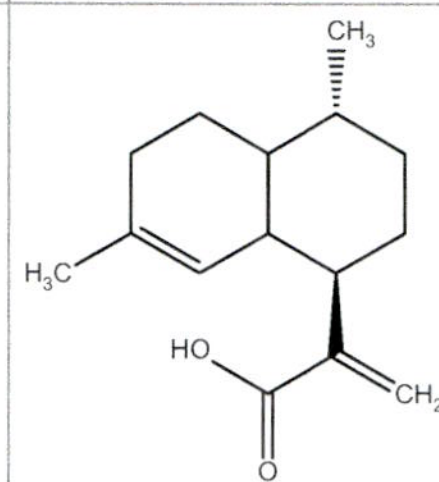	

(continued)

Table 4.1 (continued)

Family	Botanical name	Image	Plant part used	Bioactive compound(s)	Structure	Reference(s)
	Aster amellus L.		Flowers, roots, entire plant	Not available		Cha (1977) and Hussain et al. (1993)
	Synonym(s): *Aster amellus* subsp. *bessarabicus* (Bernh. ex Rchb.) Soo, *Aster amellus* subsp. *ibericus* (Steven) V.E.Avet. **Common name(s):** The European Michaelmas-daisy, Italian ester. **Botanical description:** *Aster amellus* is present on the European mountains from the Pyrenees and the Alps to the Carpathians. Outside Europe it is located in Western Asia (Turkey), the Caucasus, Siberia, and Central Asia (Kazakhstan). In the language of flowers, the Michaelmas-daisy symbolizes a farewell or a departure. *Aster amellus* reaches on average a height of 20–50 cm (7.9–19.7 in.). The stem is erect and branched, the leaves are dark green. The basal leaves are obovate and petiolated; the cauline ones are alternate and sessile, increasingly narrower and lanceolate. The flowers are lilac. The flowering period extends from July through October. The hermaphroditic flowers are either self-fertilized (autogamy) or pollinated by insects (entomogamy). The seeds are an achene that ripens in October. **Medicinal use(s):** The roots are anti-inflammatory, antitussive, depurative, hemostatic, and pectoral. They are used in the treatment of coughs, pulmonary affections, and malaria. The root juice is used internally in Nepal to treat indigestion and externally to treat boils. Mainly the plant is used in inguinal tumors.					

	Aster tataricus		Entire plant, root	Epifriedelanol	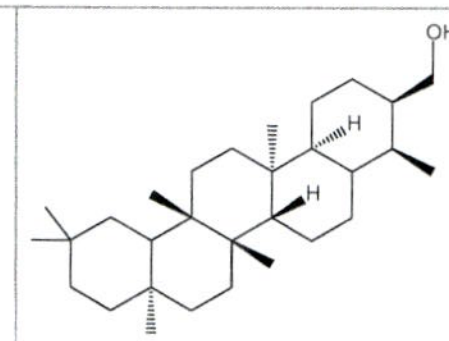	Cha (1977), Han et al. (1984), Duke and Ayensu (1985a, b), Bibi et al. (2011) and Zhou et al. (2010)
	Synonym(s): *Aster fauriei* H. Lév. & Vaniot, *Aster fauriei* Tamamsch. **Common name(s):** Tatarian aster. **Botanical description:** *Aster tataricus* is a perennial herb native of North America and widely distributed in China and Japan. It is 11–150 cm tall; rhizomes becoming woody; stemless rosettes often present. Stems erect, simple, ribbed, glabrate proximally to sparsely strigillose distally, minutely stipitate glandular below leaves. Leaves cauline, much reduced upwards, scabrous, abaxially sparsely minutely stipitate glandular, margin scabrous-ciliate, midvein prominent abaxially, apex acuminate to acute or rounded, mucronate; lowest leaves withered by anthesis, long, slightly winged-petiolate (petiole to 30 cm), petiole base sheathing; blade oblanceolate to ovate. **Medicinal use(s):** The roots of *Aster tataricus* are used medicinally to relieve coughs and reduce sputum. The plant is mostly used to treat breast and liver cancer.					

(continued)

Table 4.1 (continued)

Family	Botanical name	Image	Plant part used	Bioactive compound(s)	Structure	Reference(s)
	Bidens pilosa L.		Whole plant	1,3,6,10-Dodecatetraene, 3,7,11-trimethyl-(Z,E) (Farnesene)	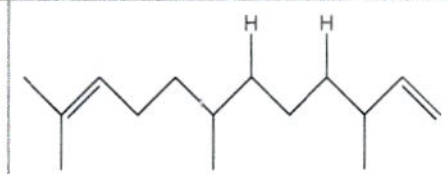	Kviecinski et al. (2008) and Singh et al. (2017)
	Synonym(s): *Bidens pilosa* var. *abyssinica* (Sch.Bip. ex Walp.) Fiori, *Bidens pilosa* f. alausensis (Kunth) Sherff. **Common name(s):** Black-jack. **Botanical description:** *Bidens pilosa* is an annual herb of North America and also found in Anhui, Fujian, Gansu, Guangdong, Guangxi, Guizhou, Hainan, Hebei, Henan, Hubei, Hunan, Jiangxi, Liaoning, Shaanxi, Shandong, Shanxi, Sichuan, Taiwan, Xizang, Yunnan, and Zhejiang. Stems are 30–180 cm tall, glabrous or very sparsely pubescent in upper part. Petiole 10–30 (–70) mm; leaf blade either ovate to lanceolate, 30–70 (–120) × 12–18 (–45) mm, or pinnately 1-lobed, primary lobes 3–7, ovate to lanceolate (10–) 25–80 × (5–)10–40 mm, both surfaces pilosulose to sparsely hirtellous or glabrate, bases truncate to cuneate, ultimate margin serrate or entire, usually ciliate, apices acute to attenuate. **Medicinal use(s):** This plant is used traditionally as a folk remedy for a wide variety of maladies and conditions including dermatitis, abscesses, eczema, sores, warts, corns, ringworm, scabies, herpes, and skin cancer.			1H-3A, 7-Methanozulene, Octahydro-1,4,9,9-tetramethyl (Patchoulane)	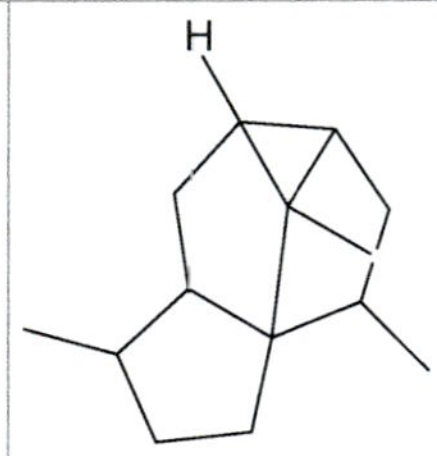	

	Carpesium rosulatum Miq N		Whole Plant	1, 4β,10α-Dihydroxy-guaia-8α,12-olide	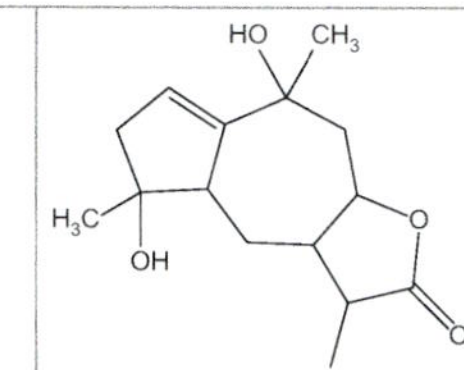	Moon and Zee (2010)
	Synonym(s): No synonyms are recorded for this name. **Common name(s):** Hime gankubi sou. **Botanical description:** *Carpesium rosulatum* (Inuleae, Compositae), uncommon in Korea but distributed in South Korea, has been used in Korean traditional medicine properties. Seeds of *Carpesium* species are explored as a traditional expellent of seatworms in Japan. **Medicinal use(s):** The roots are antipyretic, vermifugic, analgesic, anti-inflammatory, depurative, hemostatic, and pectoral. They are used in the treatment of coughs, pulmonary infections, malaria, and blood cancer.			2, 4β 10α-Dihydroxy-1(2),11(13)-guaiadien-8α 12-olide	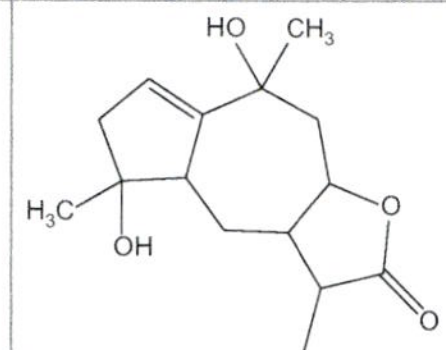	
				3, 3β,8β-Dihydroxy-1α,5α-guaian-10(14)-ene-6α,12-olide	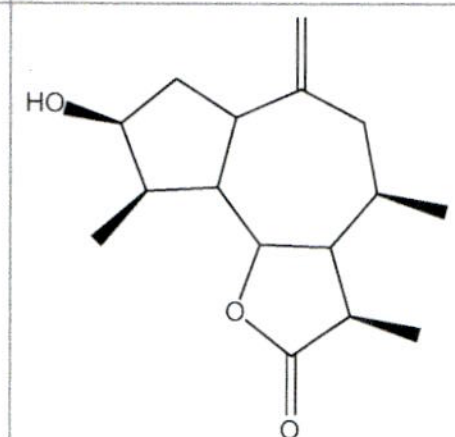	

(continued)

Table 4.1 (continued)

Family	Botanical name	Image	Plant part used	Bioactive compound(s)	Structure	Reference(s)
	Centaurea gigantea Schultz. Bip.		Apical part	Chlorogenic acid	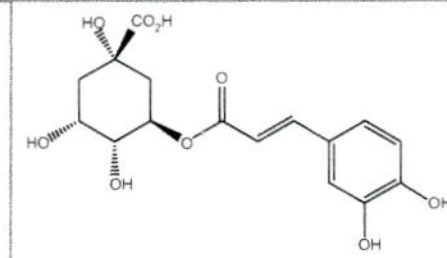	Shoeb et al. (2007a)
	Synonym(s): *Centaurea chaldaeorum Nábělek.* **Common name(s):** Amberboi. **Botanical description:** *Centaurea gigantean* is endemic to Southeastern Anatolia and Turkey and is native of North America. This plant is a biennial with erect stems up to 1–1.80 m, densely adpressed-tomentose leaves, and pale purplish to white flowers. **Medicinal use(s):** Extract of plants has been shown to have a wide variety of potential medicinal properties including anti-HIV-1, antibacterial, and antimycotic. The plant is used to treat breast, colon, and prostate cancer.			2″-(4‴-Hydroxybenzoyl)-isoorientin	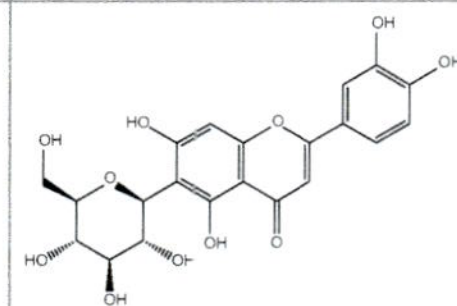	
				Isoorientin	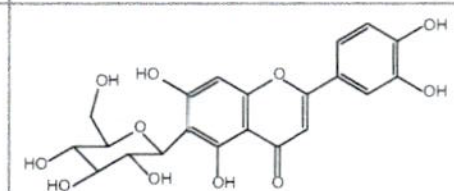	
				Isoquercetin	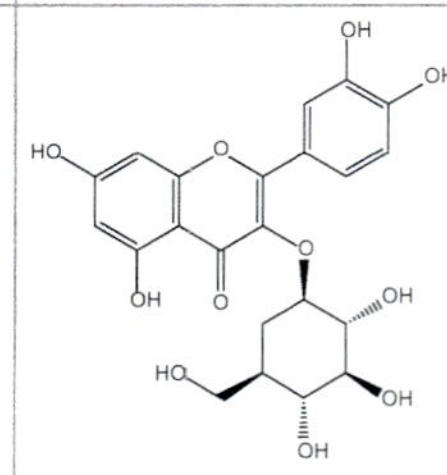	

				Cirsiliol	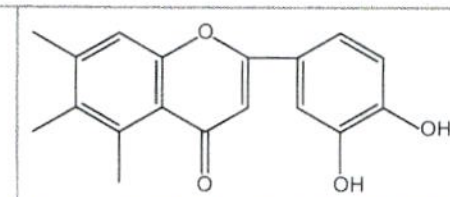	
	Centaurea montana L.		Stem	Montanoside epoxylignans	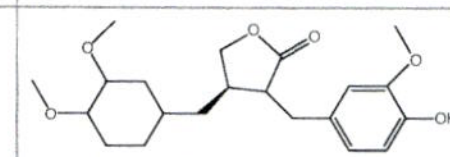	Shoeb et al. (2007b)
				Berchemol	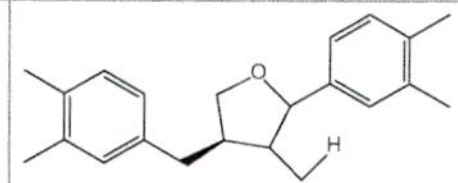	
	Synonym(s): *Centaurea montana* Burm.f., *Centaurea montana* Auct. ex Willk. & Lange. **Common name(s):** Great blue bottle. **Botanical description:** *Centaurea montana* is a very handsome plant, native to the mountains of Europe, now widely cultivated as an ornamental. Europe, from Belgium eastwards to Poland and Serbia. The plant is perennials, 25–80 cm, from rhizomes or stolons. Stems are 1–several, erect, simple, or sparingly branched, villous with septate hairs and thinly arachnoid-tomentose with long, simple hairs. Leaves thinly villous and ± tomentose, glabrate; proximal leaves winged-petiolate. **Medicinal use(s):** The stem of *Centaurea montana* and its tuber are used in skin treatments for rashes, infected grazes, and cracks in the skin and very effective to treat colon and renal cancer.			Flavanone	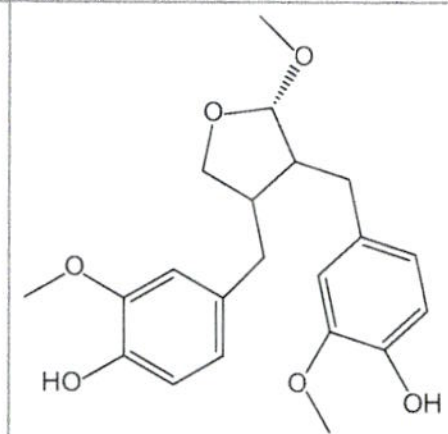	

(continued)

Table 4.1 (continued)

Family	Botanical name	Image	Plant part used	Bioactive compound(s)	Structure	Reference(s)
	Psephellus schischkinii (Tzvelev) Wagenitz		Stem	Mataresinoside	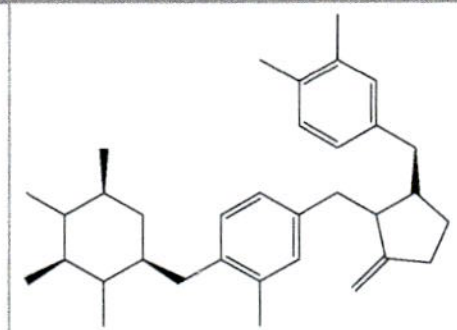	Shoeb et al. (2005)
				Arctiin	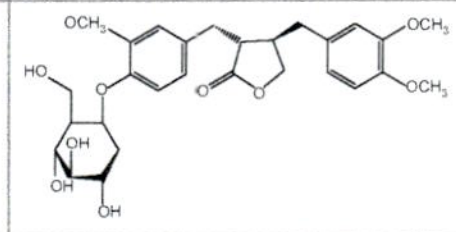	
	Synonym(s): *Centaurea schischkinii* Tzvelev. **Common name(s):** Basket flower. **Botanical description:** These plants are perennial herbs with simple or sparsely branched stems. Leaves are tomentose, undivided, pinnatipartite or lyrate. Involucre 15–30 mm long and 10–30 mm broad and ovoid to sub-globose. Phyllaries in several series, appendages membranous, usually large and semicircular to ovate or rhombic, margin entire, denticulate or regularly ciliate. Flowers rose-purple, the marginal radiant, with staminodes, the central flowers numerous. Achenes 5–8 mm long, pappus (2–)5–13 mm, of scabrous bristles with the inner row of short scales. **Medicinal use(s):** This plant is used rarely in herbal medicine but seems to have good astringent qualities making it useful in baths and facials. *Centaurea schischkinii* are used for aromatherapy and are used to treat brain tumor.			Astragalin	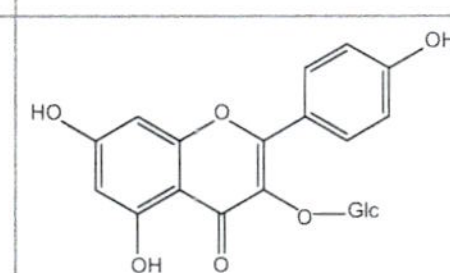	
				Matairesinol	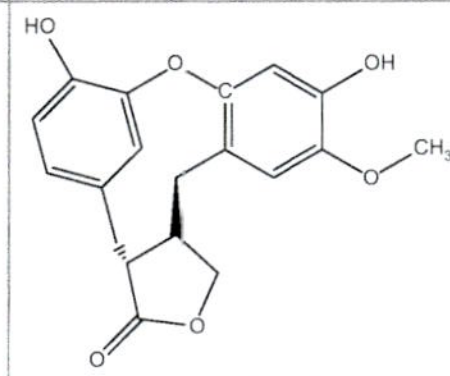	
				Arctigenin	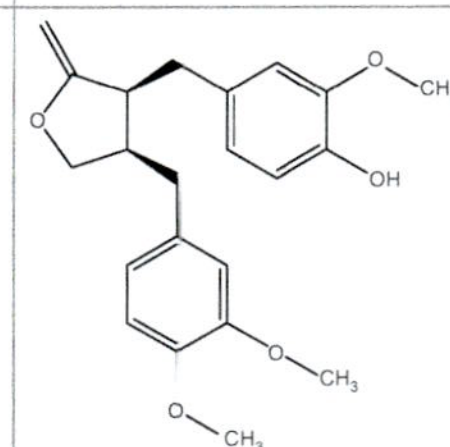	

	Cirsium maackii Maxim		Entire plant	Cirsimaritin	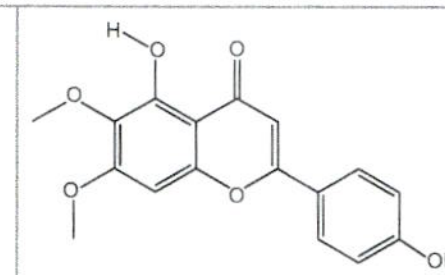	Duke and Ayensu (1985a, b) and Park et al. (2017)
	Synonym(s): *Cirsium maackii* Maxim., *Cirsium maackii* f. *albiflorum* W.Wang & C.Y.Li. **Common name(s):** Plume thistles. **Botanical description:** *Cirsium maackii* is a perennial herbaceous species found in the mountains and fields of Korea, China, and Japan. **Medicinal use(s):** The plant is used in treating a variety of ailments including diarrhea and colds. It is often prescribed as a booster for patients with low energy levels and for hypochondriacs. The plant is used to treat breast cancer.					
	Solanecio angulatus (Vahl) C. Jeffrey		Entire plant	Pyrrolizidine	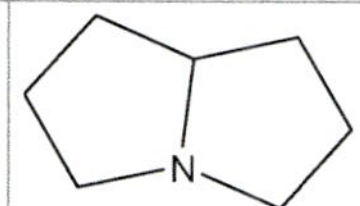	Lin and Lin (1993)
	Synonym(s): *Crassocephalum bojeri* (DC.) Robyns. **Common name(s):** Not available. **Botanical description:** *Solanecio angulatus* is native of Zimbabwe having Liane or climbing shrub; branchlets without conspicuous leaf scars; leaves: lamina 8–14 × 4–8 cm, pinnatifid, glaucous, fleshy, base amplexicaul; margin with 1–4 large runcinate irregularly arranged teeth on each side of the lamina; capitula compact, tubular, in many rounded clusters; and florets yellow. **Medicinal use(s):** Leaves and flowers are good for healing ulcers and for promoting growth and color of hair. The roots are used to make various decoctions for cough, hair loss, and hair graying. The plant is used as treatment for prostate cancer.					

(continued)

Table 4.1 (continued)

Family	Botanical name	Image	Plant part used	Bioactive compound(s)	Structure	Reference(s)
	Echinacea purpurea		Root Leaf	Caftaric acid	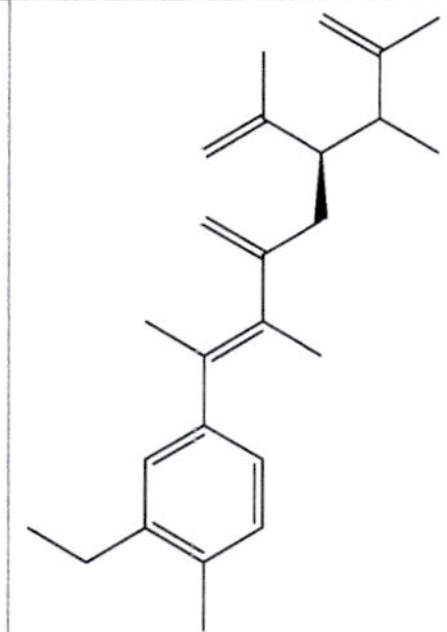	Goey et al. (2013) and Driggins et al. (2017)
				Chlorogenic acid	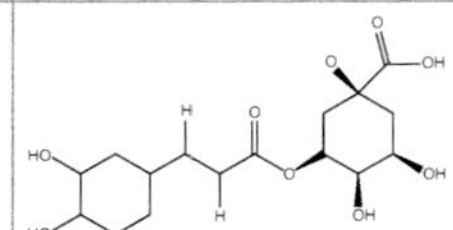 	
	Synonym(s): *Echinacea purpurea* (L.) Moench, *Echinacea purpurea* var. *arkansana* Steyerm. **Common name(s):** Eastern purple coneflower, Hedgehog coneflower, Purple coneflower. **Botanical description:** *Echinacea purpurea* is a North American species of flowering plant in the sunflower family. It is native to Eastern North America and present to some extent in the wild in much of the Eastern, Southeastern, and Midwestern United States as well as in the Canadian Province of Ontario. It is most common in the Ozarks and in the Mississippi/Ohio Valley. **Medicinal use(s):** It is one of the best herbal remedies for treatment of pain related to kidney stones by reducing the size of the stones and allowing them to pass painlessly. It is used to treat lung and kidney cancer.			Chicoric acid	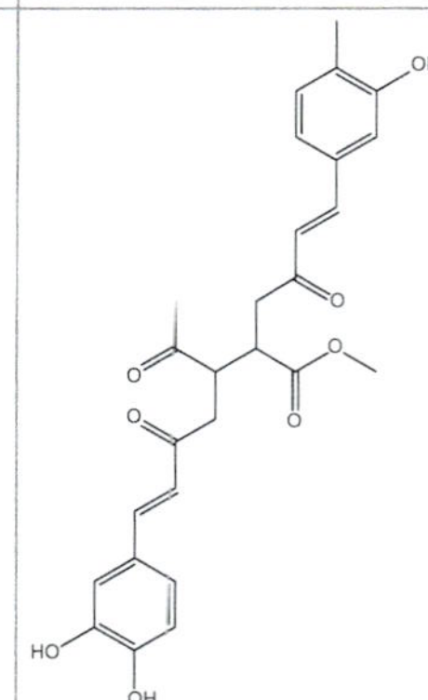 	

Echinops grijsii Hance		Root	5-(4-Hycroxybut-1-ynyl)-2,2′-bithiophene	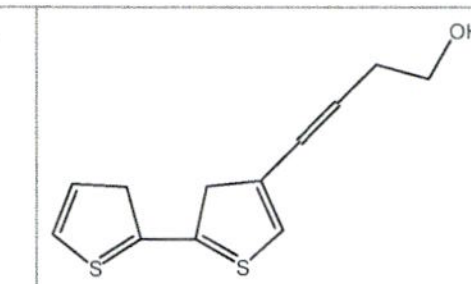	Duke and Ayensu (1985a, b), Lin et al. (1992), Lin and Lin (1993), and Zhang et al. (2009)
Synonym(s): *Echinops aberdaricus* R.E.Fr., *Echinops abstersibilis* Iljin, *Echinops abuzinadianus* Chaudhary, *Echinops acantholepis* Jaub. & Spach. **Common name(s):** Yuzhou loulu. **Botanical description:** The genus *Echinops grijsii* belongs to the family Compositae and comprises over 120 species, of which 17 occur in China. *E. grijsii* is mainly distributed in the southeast of the country. The root of *E. grijsii* is listed in *Chinese Pharmacopoeia*. **Medicinal use(s):** It is used to treat colds, coughs, chronic bronchitis, hectic fever, gonorrhea, and ophthalmia and is used to clear heat, expel miasma, and stimulate milk secretion. Honey produced from this plant is said to be a powerful expectorant and anti-spasmodic. This plant is used for bone tumors and hepatomas.			2-(Penta-1,3-diynyl)-5-(4-hydroxybut-1-ynyl) thiophene	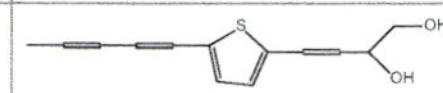	
			Arctinol-b	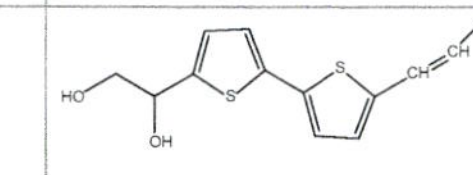	

(continued)

Table 4.1 (continued)

Family	Botanical name	Image	Plant part used	Bioactive compound(s)	Structure	Reference(s)
	Echinops latifolius Tausch		Root	5-{4-[4-(5-pent-1,3-Diynylthiophen-2-yl)-but-3-ynyloxy]-but-1-ynyl}-2,2′-bithiophene	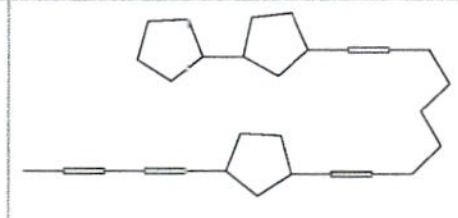	Hsu (1967), Lin et al. (1992), and Wang et al. (2007)
	Synonym(s): *Echinops latifolius* Tausch, *Echinops latifolius* f. *angustilobus* (DC.) Kitag. **Common name(s):** Globe thistles. **Botanical description:** *Echinops latifolius* is native of China and is perennial growing up to 0.5 m (1ft 8in). It is in flower from July to September, and the seeds ripen from August to October. The species is hermaphrodite (has both male and female organs) and is pollinated by insects. Suitable for light (sandy), medium (loamy), and heavy (clay) soils, prefers well-drained soil, and can grow in nutritionally poor soil. Suitable pH: acid, neutral, and basic (alkaline) soils. It cannot grow in the shade. It prefers dry or moist soil and can tolerate drought. **Medicinal use(s):** It is also used in medicine as the powder from dried bulbs is used as water disinfectant and applied on wounds to stop bleeding. The plant is used to treat lung cancers.			5-(4-Hydroxybut-1-one)-2,2′-bithiophene	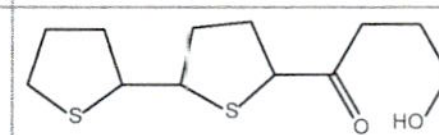	
	Echinops setifer		Whole plant	Echinopsine	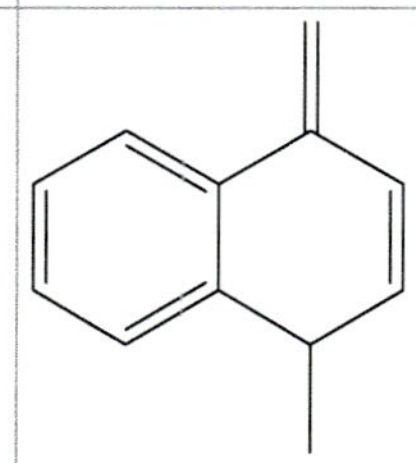	Mohammad (2006), Desai et al. (2008), and Madhuri and Pandey (2009)

	Synonym(s): No synonyms are recorded for this name. **Common name(s):** Arctic glow. **Botanical description:** *Echinops setifer* is a flowering plant in the daisy family Asteraceae. They have spiny foliage and produce blue or white spherical flower heads. They are native to Europe, east to Central Asia, and south to the mountains of tropical Africa. **Medicinal use(s):** It is used traditionally as a folk remedy for a wide variety of maladies and conditions including dermatitis, abscesses, eczema, sores, warts, corns, ringworm, scabies, herpes, asthma, and skin cancer.					
	Eclipta alba L.		Root, leaf, flower	β-Amyrin	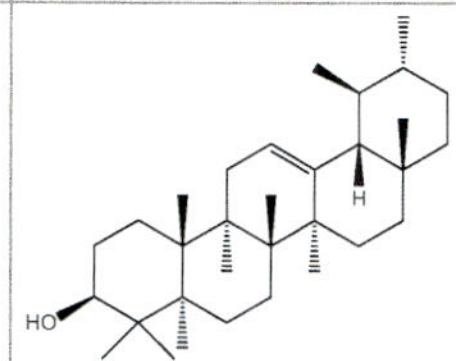	Hsu (1967) and Abdel-Kader et al. (1998)
				Luteolin-7-O-glucoside	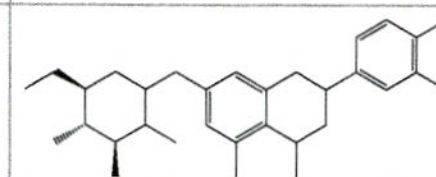	
				Hentriacontarol		
	Synonym(s): *Eclipta alba* (L.) Hassk. **Common name(s):** Bhringraj, Maka bhangra. **Botanical description:** *Eclipta alba* is a small herb native of tropical America, distributed throughout India in wet or moist wastelands, ascending up to 2000 m on the hills. It is an erect or prostrate, much-branched herb with white flowers. **Medicinal use(s):** Flowers are used in cosmetic industries for making perfume and rose water. Buds and petals are used for the removal of gallbladder and kidney stones. Flowers are used to treat asthma and roots to treat bone cancer.			Stigmasterol	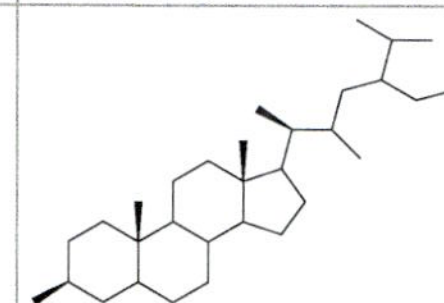	

(continued)

Table 4.1 (continued)

Family	Botanical name	Image	Plant part used	Bioactive compound(s)	Structure	Reference(s)
	Eclipta prostrata L.		Aerial parts Seeds Roots	Wedelolactone	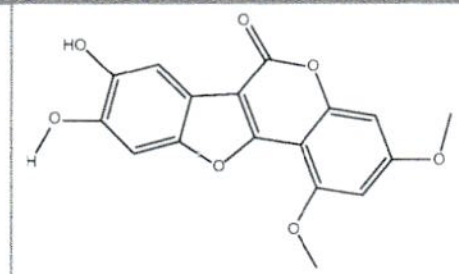	Lirdprapa-mongkol et al. (2008) and Liu et al. (2012)
	Synonym(s): *Eclipta prostrata* (L.) L., *Eclipta prostrata* Lour., *Eclipta prostrata* f. *aureoreticulata* Y.T.Chang. **Common name(s):** Bhringraj, False daisy, Yerba de tago, Karisalankanni. **Botanical description:** *Eclipta prostrata* species grows commonly in moist places in warm temperate to tropical areas worldwide. It is widely distributed throughout India, Nepal, China, Thailand, and Brazil. This plant has cylindrical, grayish roots. The solitary flower heads are 6–8 mm (0.24–0.31 in.) in diameter, with white florets. The achenes are compressed and narrowly winged. **Medicinal use(s):** The ancient Greeks used it to treat ulcers, and snakebites and for consumption. It is used as a cold tea to stop sweating, while the same tea drunk hot produces sweating. The extract is used to treat gastrointestinal cancer.			Eclalbasaponin	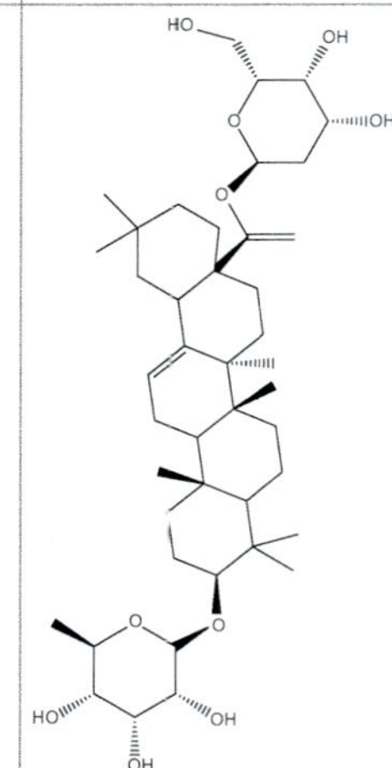	

		Luteolin	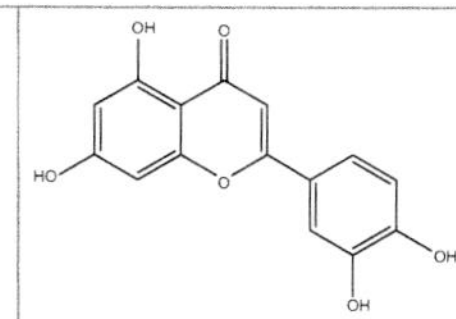	
		Luteolin-7-O-glucoside	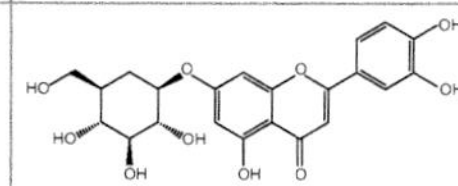	

(continued)

Table 4.1 (continued)

Family	Botanical name	Image	Plant part used	Bioactive compound(s)	Structure	Reference(s)
	Emilia sonchifolia (L.) DC.		Apical part	Pyrrolizidine	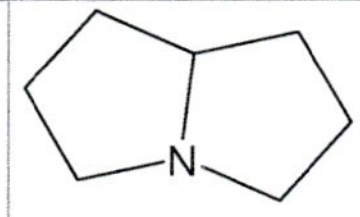	Shylesh et al. (2005) and Pratibha et al. (2014)
	Synonym(s): *Emilia sonchifolia* (L.) DC. ex DC. **Common name(s):** Lilac tassel flower, Cupid's shaving brush. **Botanical description:** *Emilia sonchifolia* is widespread in tropical regions around the world, apparently native to Asia (China, India, Southeast Asia, etc.) and naturalized in Africa, Australia, the Americas, and various oceanic islands. It is a branching, perennial herb up to 40 cm (15.5 in.) tall. Leaves are lyrate-pinnatilobed, up to 10 cm (4 in.) long, sometimes becoming purplish as they get old. One plant can produce several pink or purplish flower heads in Asia. **Medicinal use(s):** It was formerly much used for poultices and reckoned well for sickness of the stomach. A weak infusion of the plant is now sometimes given as a simple and easy purgative and a strong infusion as an emetic. Plant extract is used to treat cervical and pancreatic cancer.			2.4-Diethoxy-6-hydroxy-8-formyl 3oxa bicyclo[4.3.0]-cyclononan-7-ene	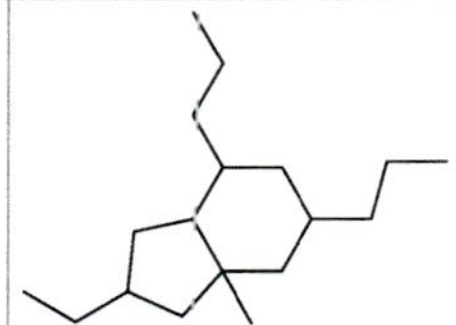	
				Caryophyllene oxide	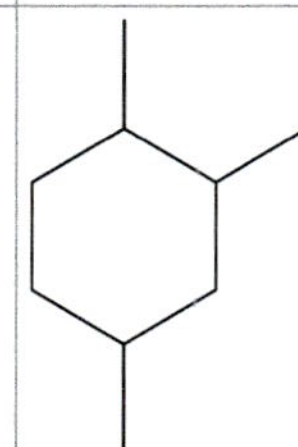	
				5-Fluorouracil	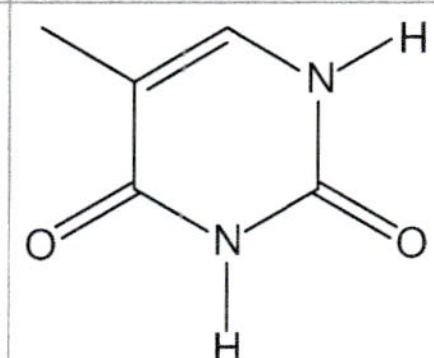	

	Eupatorium cannabinum L.		Whole plant	Pyrrolizidine	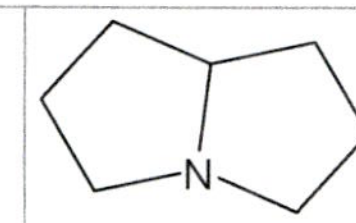	Ribeiro-Varandas et al. (2014)
	Synonym(s): *Eupatorium aboriginum* Greene, *Eupatorium abronium* Klatt, *Eupatorium achillea* Chodat. **Common name(s):** Thoroughworts, Hemp-agrimony, Holy rope. **Botanical description:** *Eupatorium cannabinum* is the only species of that genus native to Europe (with the remainder in Asia or North America). It is a perennial herb up to 1.5 m (4.9 ft.) tall or more and 1.2 m (3.9 ft.) wide. It lives in moist low-lying areas in temperate Eurasia. It is dioecious, with racemes of mauve flower heads which are pollinated by insects from July to early September. The flowers are visited by many types of insects and can be characterized by a generalized pollination syndrome. The flower heads are tiny and fluffy and can be pale dusty pink or whitish. The fruit is an achene about 2 or 3 mm long, borne by a pappus with hairs 3–5 mm long, which is distributed by the wind. The plant over-winters as a hemicryptophyte. **Medicinal use(s):** The flower produces food coloring which is used in poultry in order to intensify the color of the skin and yolk. The plant is also known to have antioxidant activities and used for the treatment of lung cancer.			Bisphenol A	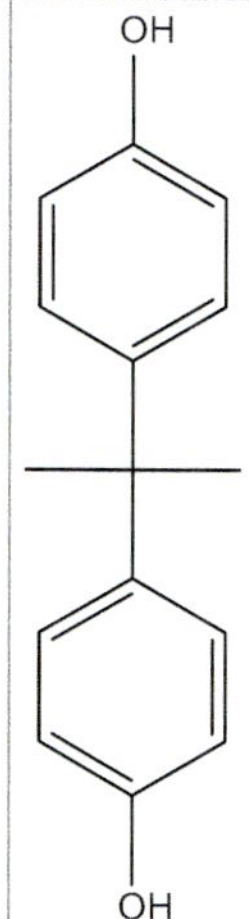	
				Doxorubicin	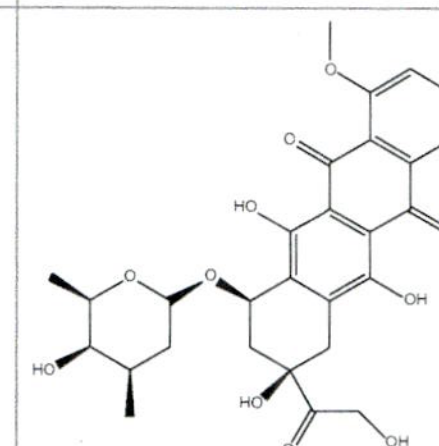	

(continued)

Table 4.1 (continued)

Family	Botanical name	Image	Plant part used	Bioactive compound(s)	Structure	Reference(s)
	Eupatorium formosanum Hayata		Leaf and stem	Eupatolide	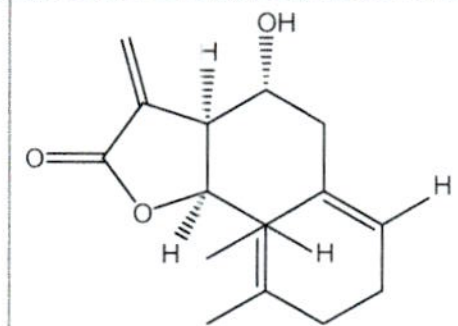	Lee et al. (1972) and Garcia-Barriga (1975)
	Synonym(s): *Eupatorium formosanum* Hayata, *Eupatorium formosanum* var. *formosanum*. **Common name(s):** Bonesets, Snakeroots. **Botanical description:** *Eupatorium formosanum* is a plant species in the family Asteraceae. It is found in Taiwan and the Ryukyu Islands (Japan). It is a perennial herb growing about 2 m (7 ft.) tall. **Medicinal use(s):** The plant is considered as anti-inflammatory, hepatoprotective, antioxidant, immunomodulatory, and antimicrobial. It is also used to treat leukemia.					

Gnaphalium elegans Kunth.		Entire plant, leaf, flowers	5, 7-Dihydroxy-3, 6, 8-trimethoxy flavone (flavone A)	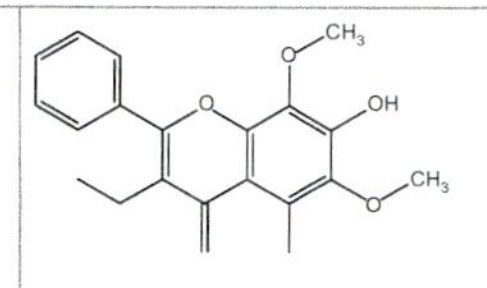	Torrenegra et al. (1980), Duke and Ayensu (1985a, b), and Thomas et al. (2012)
Synonym(s): *Gnaphalium elegans* C.Presl, *Gnaphalium elegans* Salisb. **Common name(s):** Vira-viras. **Botanical description:** *Gnaphalium elegans* is a medicinal plant commonly known in the Andean regions of South America as vira-viras. These plants belong to the tribe Inuleae and to the genera *Gnaphalium*, *Achyrocline*, and *Gamochaeta*. They are annuals or perennials that grow between 2000 and 3200 m above sea level, to an average height of 1.5 m. Because of their morphological characteristics, they can easily be confused with species belonging to different genera. The medicinal use of these plants is not limited to South America, since similar species belonging to these genera grow in various parts of the world and are commonly used for diverse medicinal purposes. **Medicinal use(s):** This plant is used for digestive disorders like pain and swelling, inflammation of the small intestine (enteritis), and dysentery and upper respiratory tract infections including colds, influenza, and swine flu. Another important use of this plant is to treat respiratory cancer.			3, 5-Dihydroxy-6, 7, 8-trimethoxy flavone (flavone B)	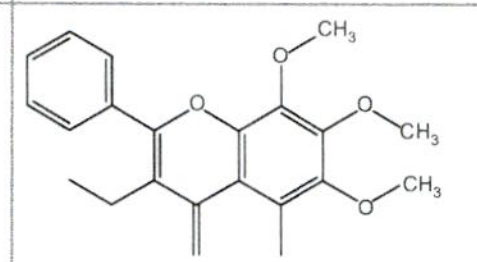	

(continued)

Table 4.1 (continued)

Family	Botanical name	Image	Plant part used	Bioactive compound(s)	Structure	Reference(s)
	Inula britannica L.		Leaf, flowers	1-O-Acetylbritannilactone	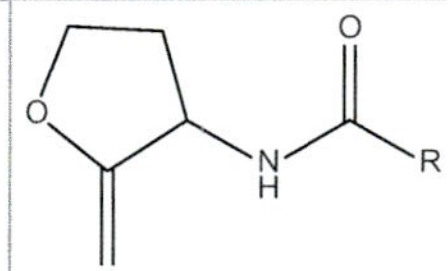	Duke and Ayensu (1985a, b) and Khan et al. (2010)
	Synonym(s): *Inula britannica* L., *Inula britannica* var. *britannica Inula britannica* subsp. *britannica*, *Inula britannica* subsp. *hispanica* (Pau) O.Bolòs & Vigo. **Common name(s):** British yellowhead, Meadow fleabane. **Botanical description:** *Inula britannica* is an Eurasian species of plant in the genus *Inula* within the daisy family. It is widespread across most of Europe and Asia and sparingly naturalized in scattered locations in North America. It is an erect herb up to 75 cm (30 in.) tall, with fine hairs but not the thick woolly coat characterizing some related species. Leaves are lance-shaped, up to 5 cm (2 in.) long. One plant produces a few heads, each on a long-flower stalk. Each contains 50–150 yellow ray flowers and 100–250 yellow disk flowers. **Medicinal use(s):** The plant is used as an emetic, antidiarrheal, cold, and cough remedy, a dermatological and gynecological aid, and a hemostat. It is also used to treat cancer of the pancreas, breast, colon, prostate, and esophagus.			1,6-O,Odiacetylbritannila ctone	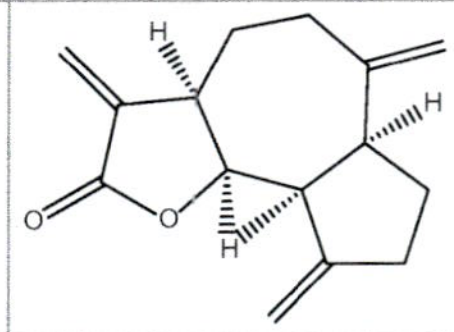	
				6α-O-(2-Methylbutyryl)-britannilactone	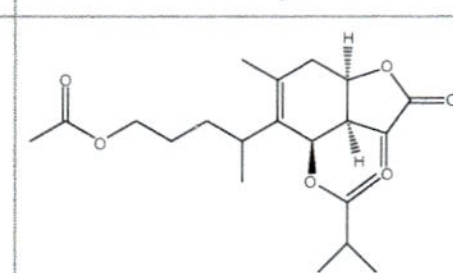	
				Neobritannilactone B	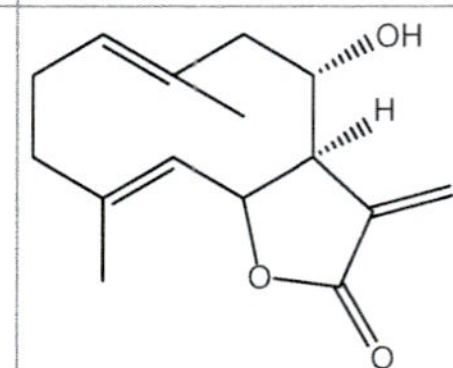	

				Spinacetin	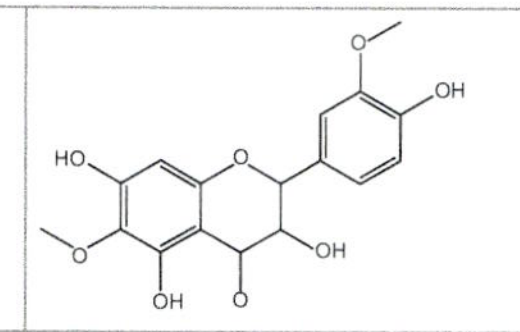	
				Diosmetin B	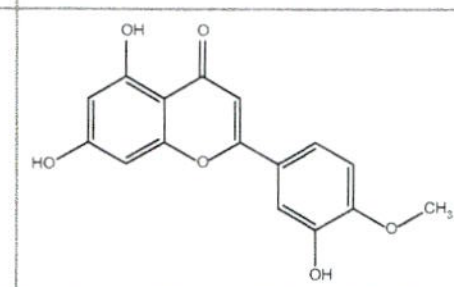	
	Inula falconeri Hook.f.	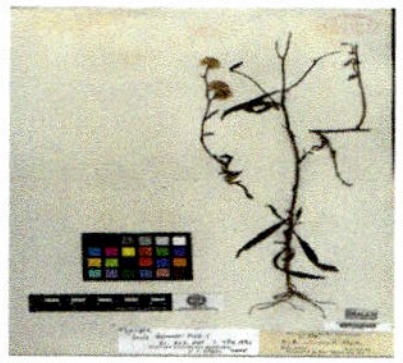	Root	1,6-O,Odiacetylbritannila ctone	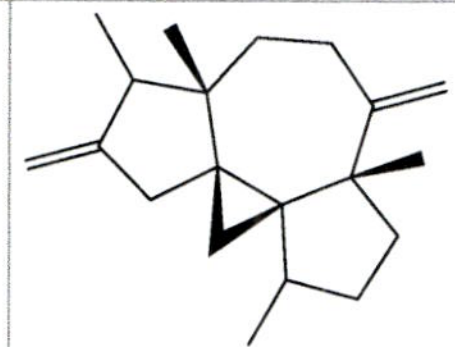	Khan et al. (2010)
	Synonym(s): No synonyms are recorded for this name. **Common name(s):** Inula. **Botanical description:** *Inula falconeri* is native to Europe, Asia, and Africa. It is an annual, herbaceous perennials or subshrubs that vary greatly in size, from small species a few centimeters tall to enormous perennials over 3 m (10 ft.) tall. They carry yellow daisy-like composite flower heads often with narrow ray florets. **Medicinal use(s):** It has antibacterial and anti-inflammatory properties and is used to treat carbuncles, fevers, influenza, and ulcers. It is, however, of cold and yin nature and should not be taken by anyone with a weak and "cold" digestive system. This is used to treat renal cancer.			Eucesmane-type sesquiterpenoid	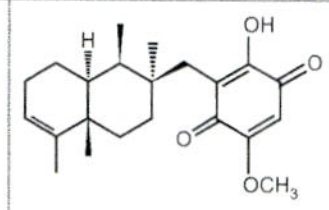	

(continued)

Table 4.1 (continued)

Family	Botanical name	Image	Plant part used	Bioactive compound(s)	Structure	Reference(s)
	Inula linariifolia Turcz.		Flowers	Lactone A	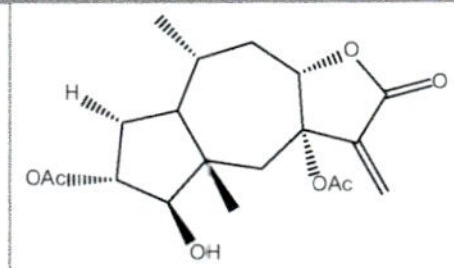	Zani et al. (1995) and Zhang et al. (2016)
	Synonym(s): *Inula britannica* subsp. linariifolia (Turcz.) Kitam., *Inula britannica* var. maximoviczii Regel. **Common name(s):** Not available. **Botanical description:** *Inula linariifolia* is native of Japan and mostly found in Korea, Mongolia, and Russia. It is a perennial herb, 30–80 cm tall; stems simple or branched, covered with white sparse multicellular appressed hairs. Leaves linear-lanceolate, acuminate, and sparsely pilose abaxial. **Medicinal use(s):** *Inula linariifolia* is used to treat the risk of hypertension, glaucoma, heart attack, and anemia. It is also used to treat prostate and brain cancer.					

	Jungia floribunda Less.		Aerial parts	Not available	Not available	Duke and Ayensu (1985a, b)
	Synonym(s): *Jungia affinis* Gardner, *Jungia pyramidalis* D.Don, *Jungia pyramidalis* D.Don. **Common name(s):** Not available. **Botanical description:** *Jungia floribunda* is a flowering plant in the sunflower family. It is native mostly to South America, with one widespread species extending its range into Central America and Southern Mexico. **Medicinal use(s):** It has been reported that the extracts and some compounds isolated from this plant have antitumor, anti-inflammatory, analgesic, antibacterial, antiviral, antifertility, and other pharmacological activities. Especially it is used to treat gastric cancer.					

(continued)

Table 4.1 (continued)

Family	Botanical name	Image	Plant part used	Bioactive compound(s)	Structure	Reference(s)
	Laggera pterodonta (DC.) Benth.		Entire plant	Sabinene	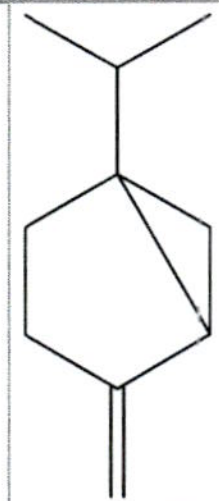	Ma et al. (1997) and Omoregie et al. (2012)
	Synonym(s): *Laggera pterodonta* (DC.) Sch.Bip. ex Oliv., *Laggera pterodonta* (DC.) Benth. **Common name(s):** Not available. **Botanical description:** *Laggera pterodonta* is an annual shrub found growing as common weed in Nigeria. It belongs to the family Asteraceae (Compositae), and the genus consists of about 20 species. The plant is spread throughout the sub-Saharan Africa and the tropical countries of Asia, especially Southeast Asia, in open waste spaces and partially shaded galleried forest. It grows in Senegal, Sierra Leone, Nigeria, and West Cameroons and probably occurs elsewhere in the region. **Medicinal use(s):** The plant extracts have long been used for their antiseptic, disinfectant, or preservative properties. This plant is also used to reduce the risk of breast, liver, and gastric cancer.			p-Cymene	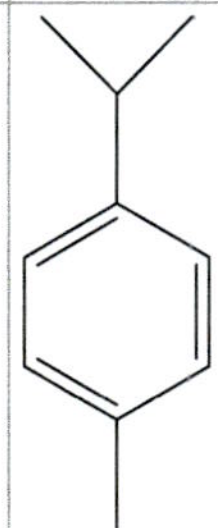	
				4-Terpineol	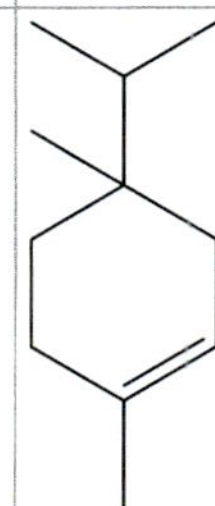	
				7-Methyl-1-undecene		

	Ligularia altaica DC.		Flower	Bisabolene	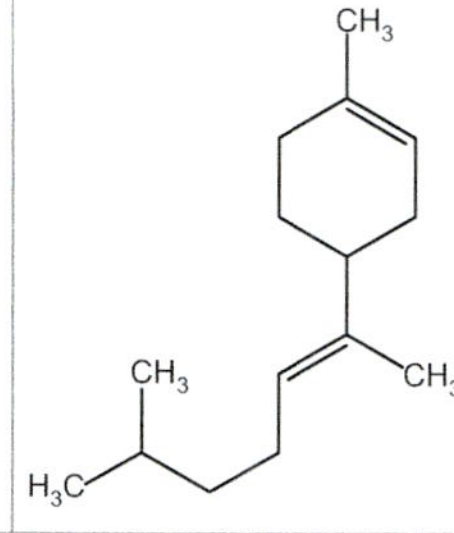	Wang et al. (2010a, b)
	Synonym(s): *Cineraria altaica* C.A.Mey. ex DC., *Ligularia glauca* var. *altaica* (DC.) Krylov, *Senecillis altaica* (DC.) Kitam., *Senecio altaicus* (DC.) Sch.Bip. **Common name(s):** Altainnauhus. **Botanical description:** *Ligularia altaica* is native of Central and Eastern Asia and is found in Kazakhstan, Mongolia, Russia, and Siberia. The plant is grayish green with stem erect, 10–150 cm tall, 4–10 mm in diam. at base, glabrous. Basal leaves petiolate; petiole 13–30 cm, glabrous, base narrowly sheathed, upper narrowly winged. **Medicinal use(s):** *Ligularia altaica* are known to possess a wide spectrum of pharmacologic properties, including antioxidant, antitumor, anti-allergic, anti-inflammatory, antibacterial, antifungal, and antiviral activities. It is used to treat gastric and ovarian cancer.			Altaicalarin	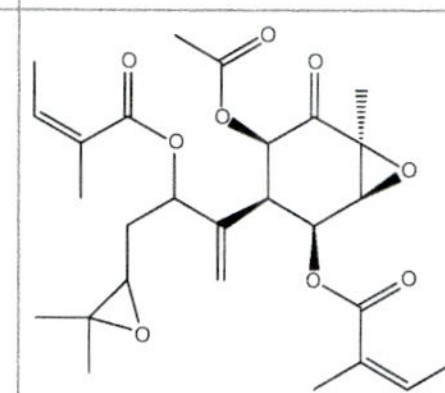	

(continued)

Table 4.1 (continued)

Family	Botanical name	Image	Plant part used	Bioactive compound(s)	Structure	Reference(s)
	Ligularia duciformis (C.Winkl.) Hand.-Mazz		Flower	3β-Acetone-6β-methaxyeremophila-7(11),9(10)-dien-12,8β-oxide	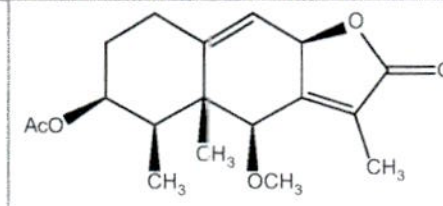	Wang et al. (2008a, b)
	Synonym(s): *Senecillis duciformis* (C.Winkl.) Kitam., *Senecio duciformis* C.Winkl. **Common name(s):** Ligularia. **Botanical description:** The plant is native of China having characteristic features like stem erect, 170 cm tall, 1 cm in diam. at base, glabrous or distally shortly yellow pilose. Basal leaves petiolate; petiole shortly yellow pilose, to 31 cm, base enlarged sheathed. **Medicinal use(s):** *Ligularia duciform* is used to treat a number of ailments, including chest pains, rheumatism, snakebites, diarrhea, eye infection, dyspepsia, and wounds. It has been used as anticancer drug to treat breast, liver, and lung cancer.			3β-Acetone-8α-hydroxy-6β-methoxyeremophila-7(11),9(10)-dien-12,8β-olide	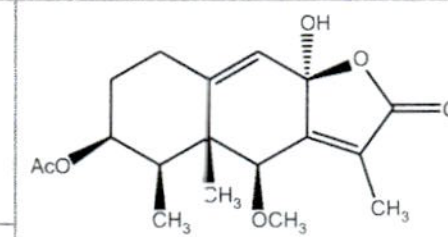	
				3β-Acetoxy-10β-hydroxy-6β, 8β-dimethoxyeremophil-7(11)-en-12,8α -olide	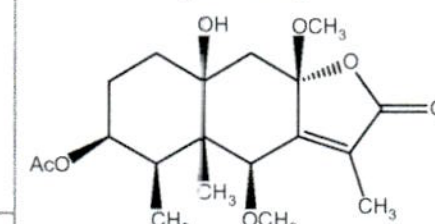 (3)	
				3β-Acetoxy-6β,8β,10β-trihydroxyeremophil-7(11)-en-12,8α -olide	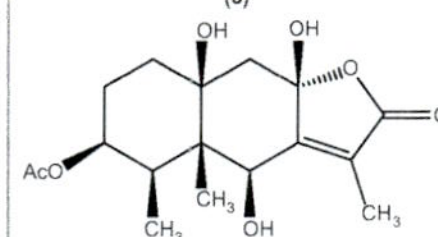	
	Ligularia intermedia Nakai		Root	Eremophilane	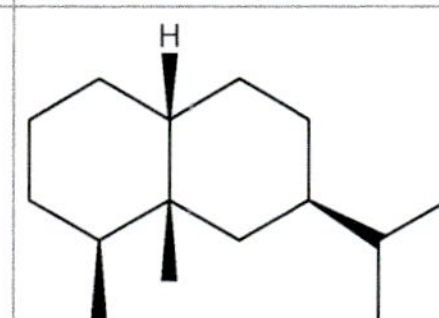	Sarg (1975) and Yuan et al. (2013)

	Synonym(s): *Ligularia intermedia* Nakai, *Ligularia intermedia* var. *intermedia*, *Ligularia intermedia* var. *oligantha* Nakai, *Ligularia intermedia* var. *venusta* Nakai. **Common name(s):** Leopard plant. **Botanical description**: *Ligularia intermedia* is a robust old-world herbaceous perennial plants in the groundsel tribe within the sunflower family. They have yellow or orange composite flower heads with brown or yellow central disk florets and are native to damp habitats mostly in Central and Eastern Asia, with a few species from Europe. There are about 120–140 species in the genus, and over half are endemic to China. The name *Ligularia*, the Latin for "strap," refers to the shape of the ray florets. **Medicinal use(s):** The root extract is very effective in the treatment of coughs, diarrhea, stomach ailments, herpes, sexually associated diseases, and lung caner.					
	Ligularia macrophylla (Ledeb.) DC.		Flower and bud	Eremophilane	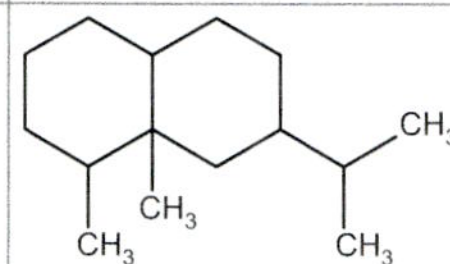	Shen et al. (2005)
	Synonym(s): *Senecillis macrophylla* (Ledeb.) Kitam., *Senecio ledebourii* Sch.Bip. **Common name(s):** Isonauhus. **Botanical description:** *Ligularia macrophylla* is native of Southern Africa and largely found in Kazakhstan, Kyrgyzstan, Pakistan, and Tajikistan. Plants grayish green. Stem erect, 35–110(–180) cm tall, 8–15 mm in diam. at base, proximally glabrous, distally and synflorescence shortly pilose. Basal leaves petiolate; petiole often purplish red, 5–20 cm, glabrous, narrowly winged and base sheathed. **Medicinal use(s):** The medicinal uses include the treatment of breast cancer, coughs, diarrhea, arthritis, and stomach ailments.					

(continued)

Table 4.1 (continued)

Family	Botanical name	Image	Plant part used	Bioactive compound(s)	Structure	Reference(s)
	Ligularia muliensis Hand.-Mazz.		Flower	Bieremoligularolide	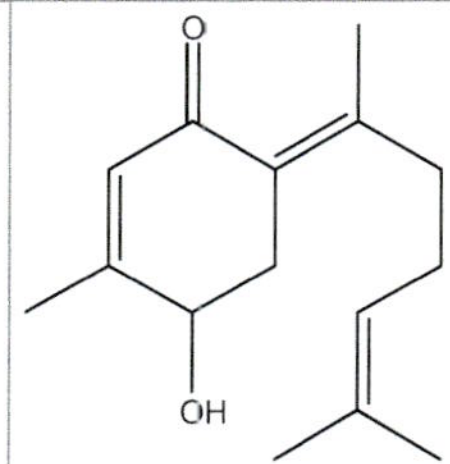	Wu et al. (2006)
	Synonym(s): No synonyms are recorded for this name. **Common name(s):** Big leaf, Golden ray. **Botanical description:** *Ligularia muliensis* is native of Central and Eastern Asia. The stem is erect, 28–32 cm tall, 4–6 mm in diam. at base, proximally glabrous, distally and synflorescence shortly yellowish-brown pilose. Rosette of leaves not seen. Stem leaves oblong or ovate-oblong, 7–15 × 4.5–6 cm, smaller distally, glabrous or abaxially shortly pilose along veins, pinnately veined, base cuneate or rounded, margin denticulate and apex rounded. **Medicinal use(s):** *Ligularia muliensis* is used for digestive disorders including pain and swelling (inflammation) of the small intestine (enteritis), and dysentery; upper respiratory tract infections including colds, influenza, swine flu, and pneumonia; and other viral and bacterial infections. It is also used to treat leukemia.					

	Ligularia virgaurea (Maxim.) Mattf. ex Rehder & Kobuski		Flower	Ceremophilane	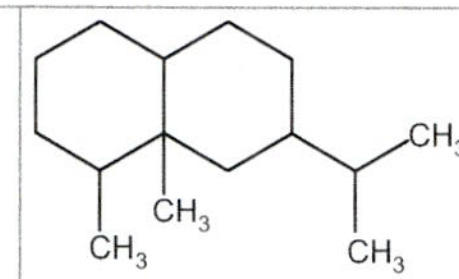	Zhang et al. (2007a)
	Synonym(s): *Ligularia virgaurea* (Maxim.) Mattf. ex Rehder & Kobuski, *Ligularia virgaurea* var. *virgaurea.* **Common name(s):** Gahe. **Botanical description:** *Ligularia virgaurea* is native of South America and Australia and also widely found in Bhutan, India (Sikkim), and Nepal. Plant is grayish green in color with erect stem 15–80 cm tall, 2–9 mm in diam. at base, glabrous. Basal leaves petiolate; petiole winged or only upper winged. **Medicinal use(s):** *Ligularia virgaurea* is an antiseptic, anti-asthmatic, antitumor, anti-ulcer, and fertility-regulating agent. This plant is used to treat cancers of the kidney, lungs, breast, and skin.					

(continued)

Table 4.1 (continued)

Family	Botanical name	Image	Plant part used	Bioactive compound(s)	Structure	Reference(s)
	Ligustrum lucidum Ait		Leaf	Oleanolic acid	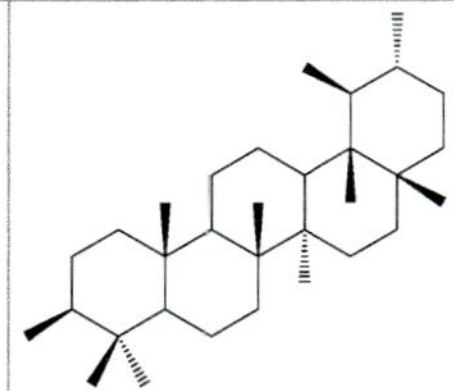	Xia et al. (2011)
	Synonym(s): *Ligustrum lucidum* W.T.Aiton, *Ligustrum lucidum* var. *alivonii* Rehder, *Ligustrum lucidum* var. *aureomarginatum* Rehder, *Ligustrum lucidum* var. *coriaceum* (Carrière) Decne., *Ligustrum lucidum* var. *esquirolii* H.Lév., *Ligustrum lucidum* f. latifolium (W.C.Che). **Common name(s):** Chinese privet wax-leaf privet. **Botanical description:** *Ligustrum lucidum* is native of China and found in southern half of China. The plant is shrubs or trees to 25 m, evergreen or deciduous, glabrous. Branchlets terete. Petiole 1–3 cm; leaf blade ovate to sometimes broadly elliptic or elliptic to lanceolate. **Medicinal use(s):** Plant is used in relieving back pain, as well as the discomforts of arthritis and rheumatism. Indigenous peoples of the Amazon drink decoctions and tinctures of the bark as an herbal tonic. Extracts of the bark are antioxidant and antimutagenic. It is used for the treatment of skin cancer.			Ursolic acid	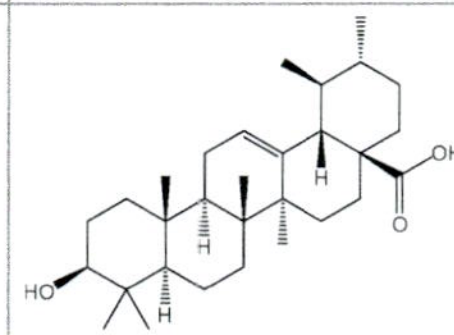	

	Pulicaria crispa (Forsk.) Sch.- Bip.		Aerial parts	2ε-Hydroxyalantolactone	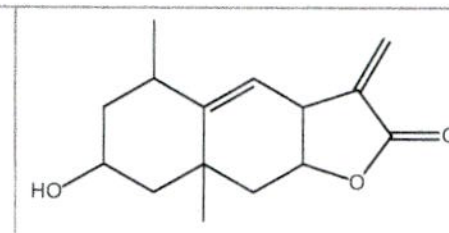	Duke and Ayensu (1985a, b) and Yahya et al. (1988)
	Synonym(s): *Pulicaria crispa* (Forssk.) Oliv., *Pulicaria crispa* Sch.Bip. **Common name(s):** Gethgath, False fleabane. **Botanical description:** *Pulicaria crispa* is a flowering plant in the sunflower family, native to Europe, Asia, and Africa. Commonly found along dry riverbed, fallow fields in sandy soil in Ganganagar, Churu, Jhunjhunu, Nagaur, Jaiselmer, Barmer, Jodhpur, Bikaner, Pali, Udaipur, Dungarpur, Jaipur, Tonk, Bundi, and Bharatpur districts of Rajasthan. The plant is diffused or erect, 24–40 cm high herbs. Stem: ascending, striate. Leaves: auriculate or semi-amplexicaul at base, linear, oblong or oblanceolate. **Medicinal use(s):** This plant is used locally to treat inflammation, is an insect repellent, and is used to treat vaginal tumors.					

(continued)

Table 4.1 (continued)

Family	Botanical name	Image	Plant part used	Bioactive compound(s)	Structure	Reference(s)
	Saussurea lappa DC.		Root	Costunolide	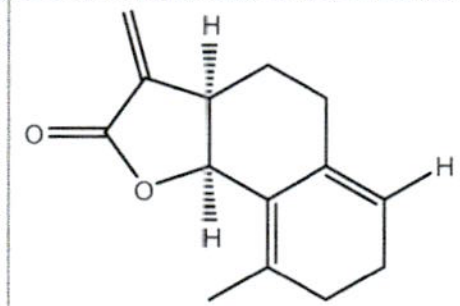	Saxena and Dixit (1993), Ko et al. (2004), and Robinson et al. (2008)
	Synonym(s): *Saussurea lappa* (Decne.) Sch.Bip., *Saussurea lappa* (Decne.) C.B.Clarke. **Common name(s):** Kut root, Costus. **Botanical description:** *Saussurea lappa* is native to India. Essential oils extracted from the root have been used in traditional medicine and in perfumes since ancient times. **Medicinal use(s):** The plant is an anti-inflammatory, analgesic, and antipyretic agent. It is very effective against bone cancer.			Beta-cyclocostunolide	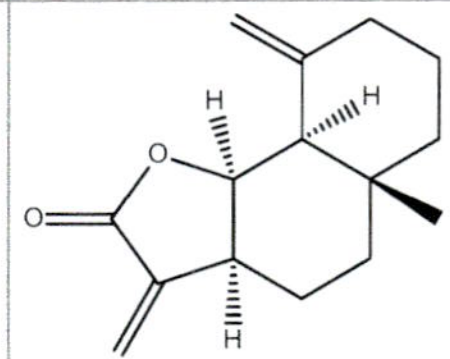	
	Senecio integrifolius (L.) Clairv.			Pyrrolizidine	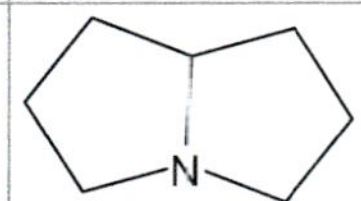	Roeder and Liu (1991)
	Synonym(s): *Senecio integrifolius* Hook. **Common name(s):** Not available. **Botanical description:** *Senecio integrifolius* is a perennial herb usually with one rosette and a single erect flowering stem. The inflorescence is a corymb with 2–6 heads. Flowering time in Southern Sweden is from mid-May to late June, depending on weather conditions. The plant is nectarless, has a strong fragrance, and is visited by several pollen-feeding insect species. **Medicinal use(s):** The bioactive compounds exhibit antibacterial, antiviral, antitumor , and cardioprotective properties.					

	Senecio oryzetorum Diels	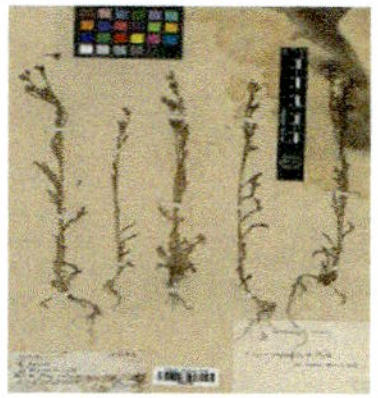		Not available		Duke and Ayensu (1985a, b)
	Synonym(s): No synonyms are recorded for this name. **Common name(s):** Not available. **Botanical description:** *Senecio oryzetorum* is native of China annual herb. Stems erect, 20–30 cm tall, branching from base or in upper part, sparsely pubescent to subglabrous. Leaves oblanceolate-linear to linear-oblong, 3–8 × 1–2 cm, papery, sparsely pubescent to ± glabrous, lower ones attenuate into a petaloid base, middle and upper ones sessile, basally scarcely auriculate, margin somewhat remotely coarsely dentate or pinnatifid with 6–10 linear to shortly linear-oblong, acute, remotely denticulate or entire, ascending-spreading lateral lobes, apex acute or subacute. **Medicinal use(s):** It has many medicinal uses especially in treating breast and intestinal cancer.					

(continued)

Table 4.1 (continued)

Family	Botanical name	Image	Plant part used	Bioactive compound(s)	Structure	Reference(s)
	Silybum marianum (L.) Gaertn		Aerial parts, Seed	Silymarin	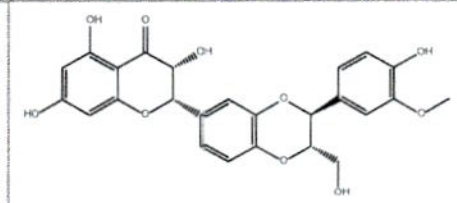	Van Erp (2005) and Bijak (2017)
	Synonym(s): *Carduus lactifolius* Stokes, *Carduus marianus* L., *Carduus versicolor* Salisb., *Carthamus maculatum* (Scop.) Lam., *Mariacantha maculosa* Bub., *Mariana lactea* Hill, *Silybum intermedium* Willk., *Silybum leucanthum* Jord. & Fourr., *Silybum maculatum* (Scop.) Moench, *Silybum mariae* (Cr.) A. Gray, *Silybum pygmaeum* Cass. **Common name(s):** Cardus marianus, Milk thistle, Blessed milk thistle, Marian thistle, Mary thistle, Saint Mary's thistle, Mediterranean milk thistle, Variegated thistle, and Scotch thistle. **Botanical description:** This species is an annual or biennial plant of the Asteraceae family. This fairly typical thistle has red to purple flowers and shiny pale green leaves with white veins. Originally a native of Southern Europe through Asia, it is now found throughout the world. Milk thistles can grow to be 30–200 cm (12–79 in.) tall, and have an overall conical shape. The approximate maximum base diameter is 160 cm (63 in.). The stem is grooved and more or less cottony. The largest specimens have hollow stems. The leaves are oblong to lanceolate. They are either lobate or pinnate, with spiny edges. They are hairless, shiny green, with milk-white veins. **Medicinal use(s):** Plant is famous for its therapeutic properties against prostate, skin, breast, cervical, and liver cancer. It is used to treat kidney damage caused by diabetes, uterine complaints, allergy symptoms, obsessive-compulsive disorder (OCD), Alzheimer's disease, Parkinson's disease, multiple sclerosis (MS), high cholesterol, and menopausal symptoms. It is reported to decrease the side effects of chemotherapy and radiation and also increase the breast milk flow.					

	Spilanthes acmella (L.) Murray		Leaf Stem Flowers	Vanillic acid	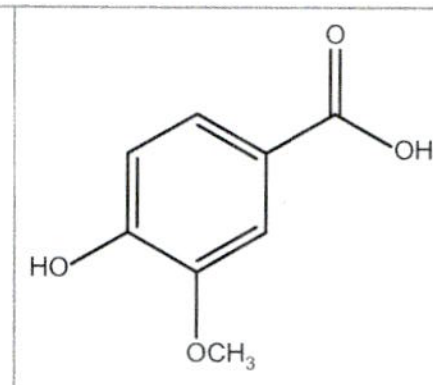	Hsu (1967) and Prachaya-sittikul et al. (2009)
	Synonym(s): *Spilanthes acmella* (L.) L. **Common name(s):** Akarkara, Antitoothache plant. **Botanical description:** *Spilanthes acmella* is a well-known species comprising over 60 species that are widely distributed in tropical and subtropical regions of the world. With high medicinal usage, it has been recognized as an important medicinal plant with an increasing high demand worldwide. **Medicinal use(s):** This plant is an anti-inflammatory, analgesic, and antipyretic agent. It is very effective for the treatment of skin, prostrate, mouth, and liver cancer.			Trans-ferulic acid	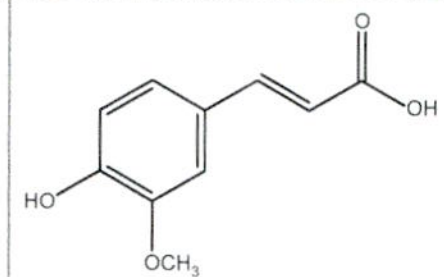	
				Trans-isoferulic acid	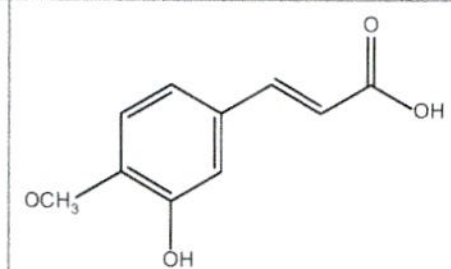	
				Coumarin	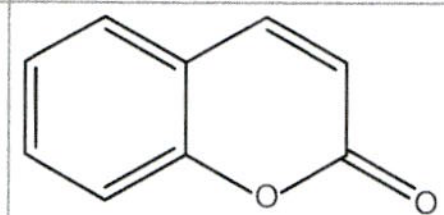	
				Stigmasterol	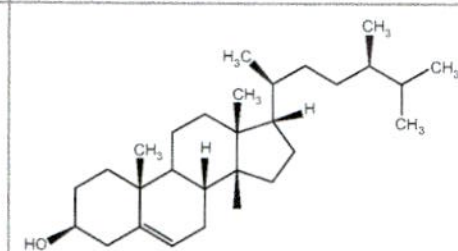	

(continued)

Table 4.1 (continued)

Family	Botanical name	Image	Plant part used	Bioactive compound(s)	Structure	Reference(s)
	Taraxacum coreanum Nakai		Aerial parts	Taraxinic acid	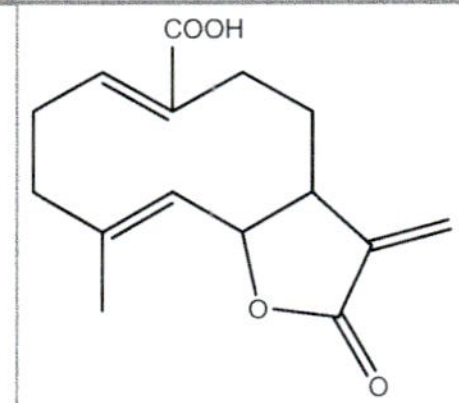	Sigstedt et al. (2008)
	Synonym(s): *Taraxacum pseudo-albidum* Kitag., *Taraxacum pseudo-albidum* var. *lutescens* Kitag. **Common name(s):** Dandelion. **Botanical description:** *Taraxacum coreanum* is native of Korea. The plant is a herb 20–25 cm tall, petiole suffused purplish or ± green, whitish-brown arachnoid above, ± winged; leaf blade mid-green and often ± suffused purplish, ± linear-oblanceolate, 13–20 × (2–) 2.5–3 (–5) cm, sparsely arachnoid, deeply pinnatisect; lateral lobes (3 or) 4–6, usually linear-deltoid to linear-triangular, patent, margin subentire to minutely dentate in outer leaves but dentate and often lobulated in middle and inner ones. **Medicinal use(s):** It has antibacterial and anti-inflammatory properties and is used to dispel heat and remove toxins, including fevers, influenza, and ulcers. This is used to treat lung, breast, ovarian, and renal cancer.					

	Taraxacum mongolicum		Whole plant	Glabellin	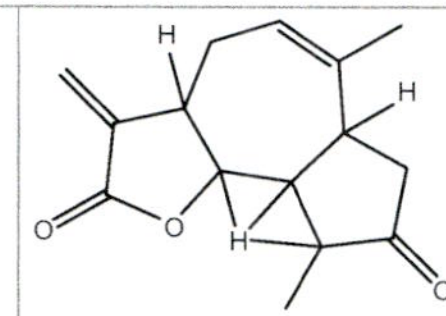	Madhuri and Pandey (2009)
	Synonym(s): *Taraxacum mongolicum* Hand.-Mazz. **Common name(s):** Chinese dandelion. **Botanical description:** *Taraxacum mongolicum* is native to China; it is a herb 8–25 cm, petiole usually narrow; leaf blade mid-green, oblanceolate, 6–15 × 2–3.5 cm, subglabrous to sparsely arachnoid, pinnatilobed, pinnatisect, or rarely undivided. **Medicinal use(s):** *Taraxacum mongolicum* plant is traditionally used in folk medicine for the treatment of arthritis, rheumatism, influenza, arthritis, bronchitis, diarrhea, dysentery, gastrointestinal diseases, hemorrhoids, impotence, inflammation, gastric cancer, and brain and skin tumor.			Arɔorescin	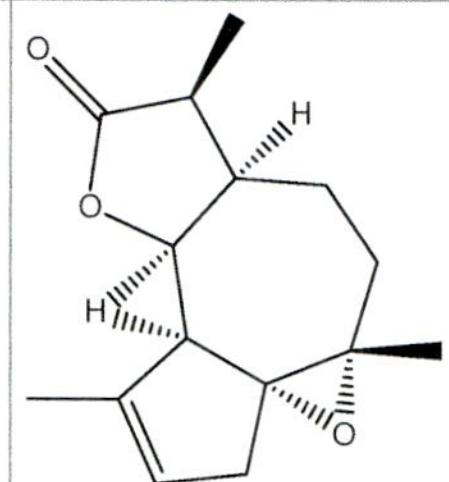	
				Estafiatin	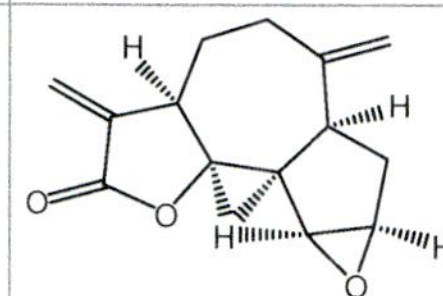	

(continued)

Table 4.1 (continued)

Family	Botanical name	Image	Plant part used	Bioactive compound(s)	Structure	Reference(s)
	Taraxacum officinale agg *Eurasia*		Leaf, root, and flower	Taraxasterol	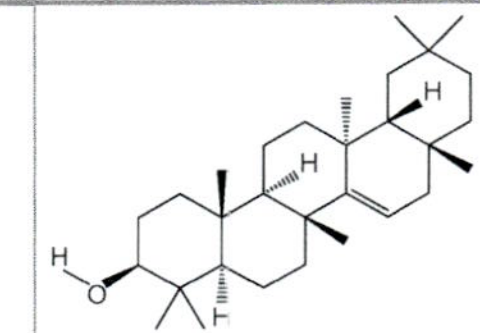	Sigstedt et al. (2008)
	Synonym(s): *Taraxacum officinale* (L.) Weber ex F.H.Wigg., *Taraxacum officinale* var. *albiflorum* Makino, *Taraxacum officinale* subsp. *alpestre* Schinz & Thell., *Taraxacum officinale* subsp. *aquilonare* (Hand.-Mazz.) Schinz & Thell. **Common name(s):** Dandelion. **Botanical description:** Plants of the genus *Taraxacum*, also known as dandelions, are members of the Asteraceae family. These perennial plants are widespread throughout the warmer temperate zones of the Northern Hemisphere and have been used for centuries as a remedy for various ailments by several societies. Dandelions play a pivotal role in traditional Chinese medicine. **Medicinal use(s):** *Taraxacum officinale* plant is used for treatment of breast, uterine, and lung tumors.			Taraxerol	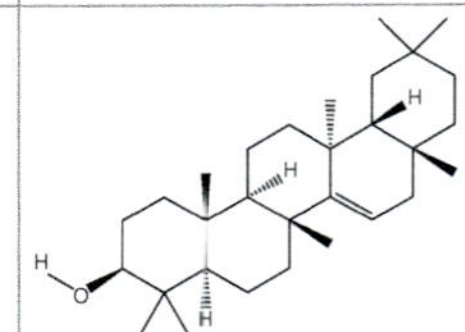	

	Tanacetum parthenium		Leaf Flower	Parthenolide	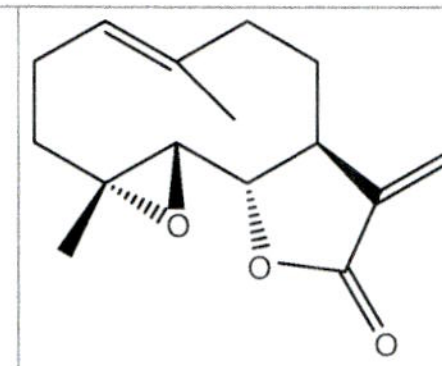	Patel et al. (2000)
	Synonym(s): *Chamaemelum parthenium* (L.) E.H.L.Krause, *Chrysanthemum parthenium* (L.) Pers. **Common name(s):** Feverfew. **Botanical description:** *Tanacetum parthenium* is native to Eurasia, specifically the Balkan Peninsula, Anatolia, and the Caucasus, but cultivation has spread it around the world, and it is now also found in the rest of Europe, North America, and Chile. The plant is an herbaceous perennial that grows into a small bush, up to 70 cm (28 in.) high with pungently scented leaves. The leaves are light yellowish green, variously pinnatifid. The conspicuous daisy-like flowers are up to 20 mm across, borne in lax corymbs. The outer, ray florets have white ligules and the inner, disk florets are yellow and tubular. It spreads rapidly by seed and will cover a wide area after a few years. **Medicinal use(s):** Frequently this plant is used for treatment of breast, uterine, and lung tumors as well as hepatitis and digestive diseases.					

(continued)

Table 4.1 (continued)

Family	Botanical name	Image	Plant part used	Bioactive compound(s)	Structure	Reference(s)
	Vernonia amygdalina Delile.		Leaf	Edotides	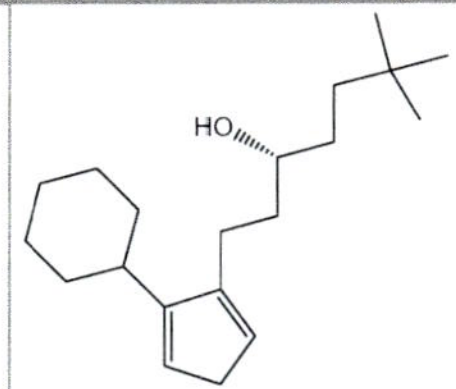	Hernandez et al. (1999)
	Synonym(s): *Vernonia amygdalina* Delile. **Common name(s):** Bitter leaf, Grawa (Amharic). **Botanical description:** *Vernonia amygdalina*, a member of the daisy family, is a small shrub that grows in tropical Africa. *V. amygdalina* typically grows to a height of 2–5 m (6.6–16.4 ft.). The leaves are elliptical and up to 20 cm (7.9 in.) long. Its bark is rough. The cooked leaves are a staple vegetable in soups and stews of various cultures throughout equatorial Africa. **Medicinal use(s):** The stems of this plant have been used in traditional medicine for treatment of gastric cancer.					

	Sphagneticola calendulacea (L.) Pruski		Leaf Seed Flower	Wedelolactone	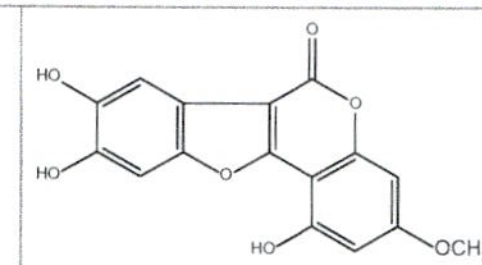	Duke and Ayensu (1985a, b) and Tsai et al. (2009)
	Synonym(s): *Wedelia chinensis* (Osbeck) Merr. **Common name(s):** Pilabhangara. **Botanical description**: *Sphagneticola calendulacea* is native to Taiwan and also found in India, Indonesia, Japan, Myanmar, the Philippines, Sri Lanka, Thailand, and Vietnam. Stems of the plant are prostrate with ascending tips, elongate, rooting from lower nodes, appressed pilose. Leaves sessile or shortly petiolate; blade linear-oblong to lanceolate, 2–10 cm × 6–20 mm, papery, appressed pilose on both surfaces. **Medicinal use(s):** European herbalists authorize the use of *Wedelia chinensis* for fever, boils, eye problems, diabetes, and diarrhea. Moreover it is also used to treat brain tumors.			Luteolin	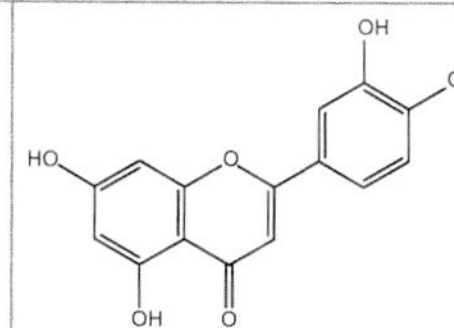	
				Apigenin	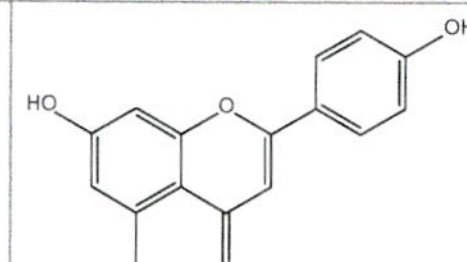	

(continued)

Table 4.1 (continued)

Family	Botanical name	Image	Plant part used	Bioactive compound(s)	Structure	Reference(s)
	Xanthium sibiricum Patrin ex Widder		Entire plant, fruit, root	7,12-Dimethylbenz[a] anthracene	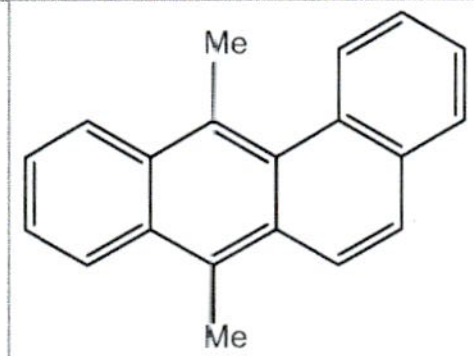	Duke and Ayensu (1985a, b) and Zhang et al. (2006)
	Synonym(s): *Xanthium sibiricum* Patrin ex Widder, *Xanthium sibiricum* var. *subinerme* (C.Winkl.) Widder. **Common name(s):** Clotbur, Marula matangi. **Botanical description:** *Xanthium sibiricum* is an annual species, probably originates in North America, and has been extensively naturalized elsewhere. **Medicinal use(s):** In traditional uses of medicine, these plants have been applied to remedy liver and spleen cancers.					
	Xanthium strumarium L.		Whole plant	Xanthatin	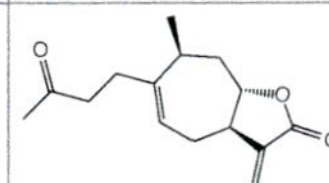	Ramirez-Erosa et al. (2007)
				Xanthinosin	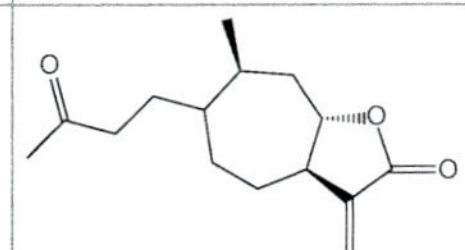	

	Synonym(s): *Xanthium strumarium* L., *Xanthium strumarium* var. *arenarium* (Lasch) Uechtritz, *Xanthium strumarium* subsp. *brasilicum* (Vell.) O.Bolòs & Vigo. **Common name(s):** Cocklebur. **Botanical description:** *Xanthium strumarium* is native to North America. The species is monoecious, with the flowers borne in separate unisexual heads: staminate (male) heads situated above the pistillate (female) heads in the inflorescence. The pistillate heads consist of two pistillate flowers surrounded by a spiny involucre. Upon fruiting, these two flowers ripen into two brown to black achenes, and they are completely enveloped by the involucre, which becomes a bur. **Medicinal use(s):** Native Americans use the roots and herbs of this plant to treat kidney disease, dyspepsia, and heartburn. It can also cure breast and prostate cancer.			

Table 4.2 Anticancer plants of family Acanthaceae

Family	Botanical name	Image	Plant part used	Bioactive compound(s)	Structure	Reference(s)
Acanthaceae	*Acanthus ilicifolius* L.		Leaf, root, and flower part	Betulinic acid	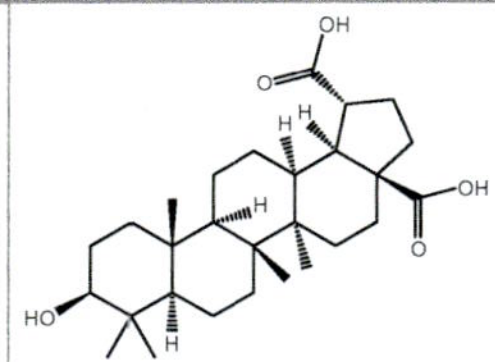	Jongsuwat (1981), Duke and Ayensu (1985a, b), and Khajure and Rathod (2011)
	Synonym(s): *Acanthus doloarius* Blanco, *Acanthus neoguineensis* Engl. **Common name(s)**: Holly-leaved *Acanthus*, Sea holly, Holy mangrove, Hargoza. **Botanical description**: It is a small biennial plant native to India, Sri Lanka, Malaysia, Australia, and Pacific Islands. It has thick fusiform tuberous roots and bear funnel-shaped violet-colored flowers. Its fruit is a 2-cm-long sessile capsule containing about 20 seeds. **Medicinal use(s)**: The plant is used for the treatment of cancer and as a medicine in asthma and rheumatism. The alcoholic extract of this plant was found to be effective against tumor progression. The extract was found to be cytotoxic towards lung fibroblast cells. Oral intake of the extract reduced the tumor volume and increased the life span by 75% in ascites tumor.			Taraxerol	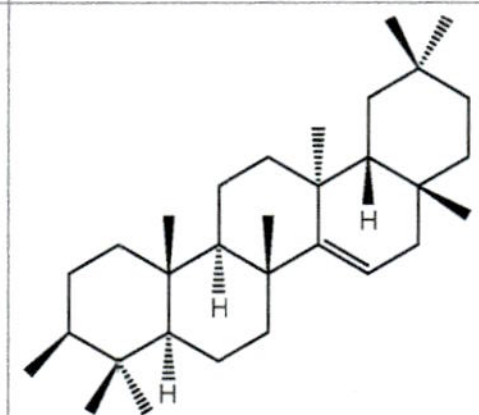	
				β-Sitosterol	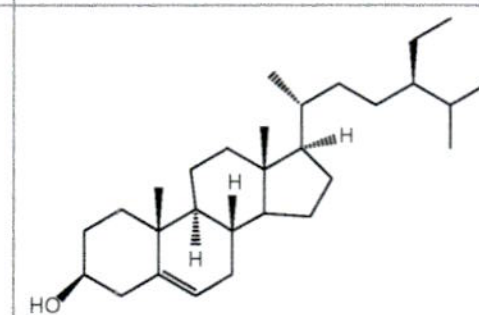	
				Stigmasterol	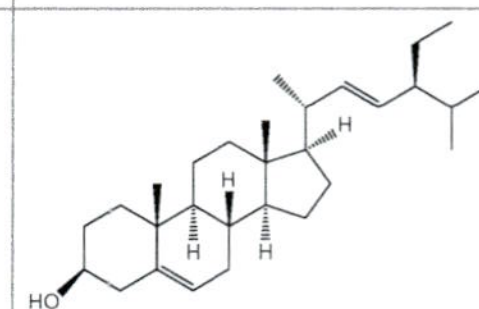	

	Synonym(s): *Justicia adhatoda*. **Common name(s)**: Malabar nut tree and local names in some areas are Ya-Zui-Hua in China, Vasaka (Sanskrit), Arusha (Hindi). **Botanical description**: The plant is native to Asia and is found in Sri Lanka, Nepal, India, Pakistan, Indonesia, Malaysia, and China. It is a small, evergreen, perennial shrub, which reaches an average height of 3 m. Its branches are opposite and ascending. The broad, leathery leaves, which are sometimes used as an insecticide, measure from 10 to 15 cm in length and are about 4 cm in width. **Medicinal use(s)**: It is used to treat human lung carcinoma. It is well-known for preparation of medicine for bronchitis, asthma, and other pulmonary infections and has anti-spasmodic, expectorant, febrifuge, anti-arthritic, antiseptic, antimicrobial, expectorant, sedative, and antituberculosis properties.					
				Deoxyvasicinone	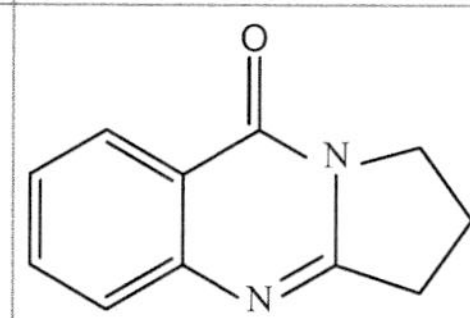	
				Peganine (vasicine)	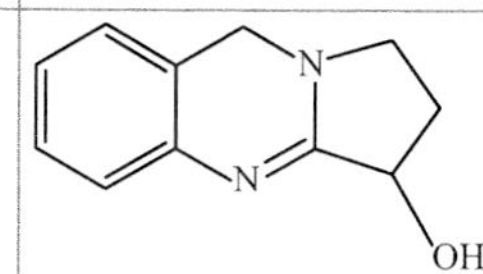	
				Peganidine	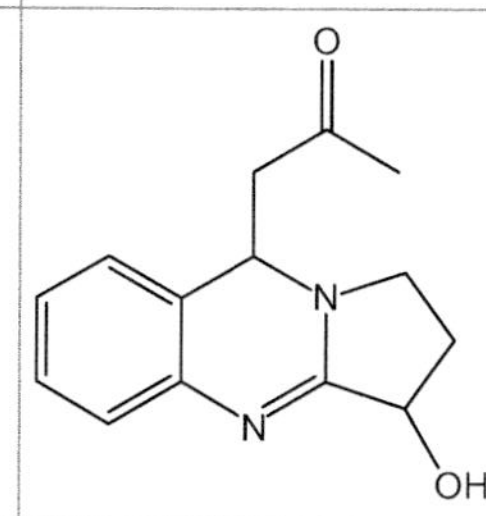	

(continued)

Table 4.2 (continued)

Family	Botanical name	Image	Plant part used	Bioactive compound(s)	Structure	Reference(s)
				Adhatodine	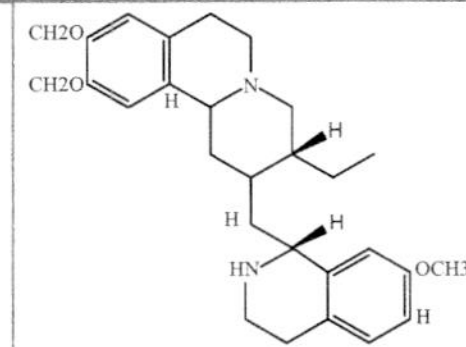	
				Daucosterol	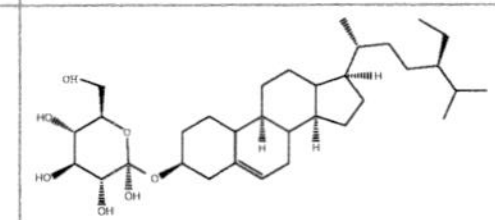	
				Epitaxol	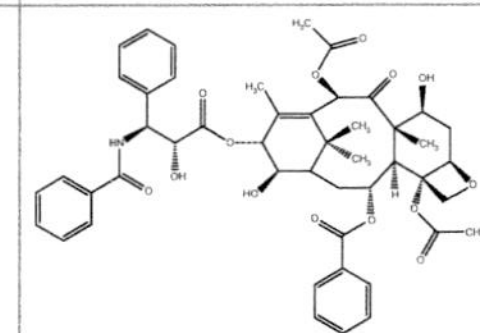	
					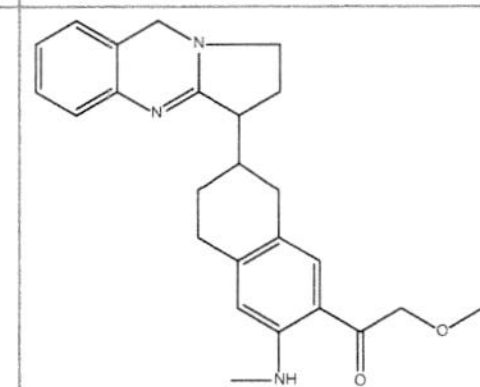	
					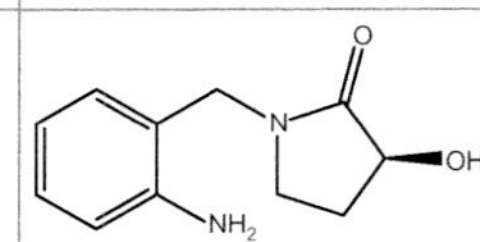	

	Andrographis paniculata (Burm.f.) Wall. ex Nees		Vasicol	Andrographolide	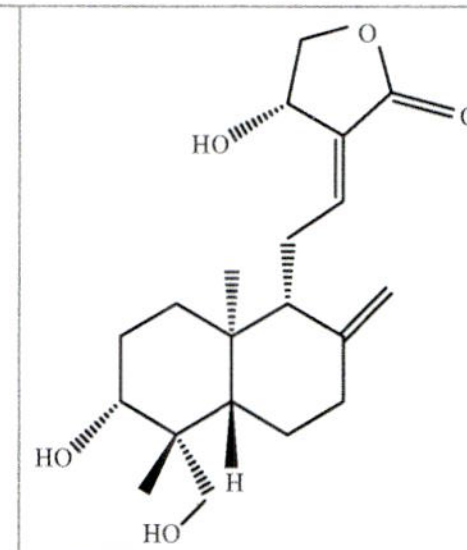	Kumar et al. (2004), Chao and Lin (2010), Rajagopal et al. (2003), Hossain et al. (2014), and Kumoro et al. (2018)
	Synonym(s): *Justicia paniculata* Burm. F. **Common name(s)**: King of bitters, Dark cloud, Green chirayta. **Botanical description**: This plant is native to Taiwan, Mainland China, and India. Grows erect to a height of 30–110 cm in moist, shady places. The slender stem is dark green, squared in cross section with longitudinal furrows and wings along the angles. The lance-shaped leaves have hairless blades. The small flowers are borne in spreading racemes. The fruit is a capsule. It contains many yellow-brown seeds. **Medicinal use(s)**: Used to treat liver disorders, bowel complaints of children, colic pain, common cold, and upper respiratory tract infection and relieves internal heat. It also possesses anti-inflammation, anticancer, immunomodulation, anti-infection, anti-hepatotoxicity, anti-atherosclerotic, antidiabetic, and antioxidant properties.			Neoandrographolide	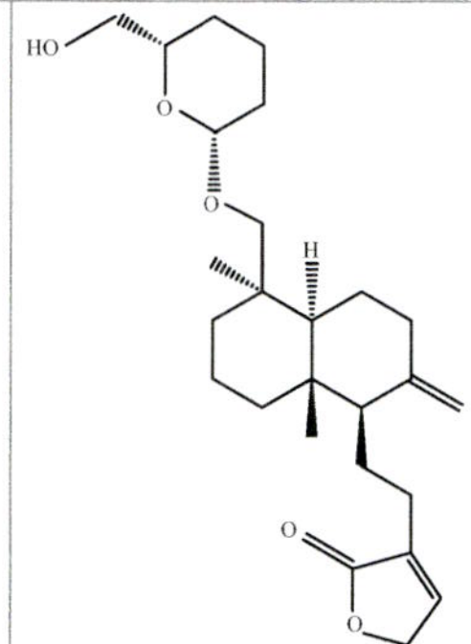	

(continued)

Table 4.2 (continued)

Family	Botanical name	Image	Plant part used	Bioactive compound(s)	Structure	Reference(s)
				14-Deoxyandrographolide	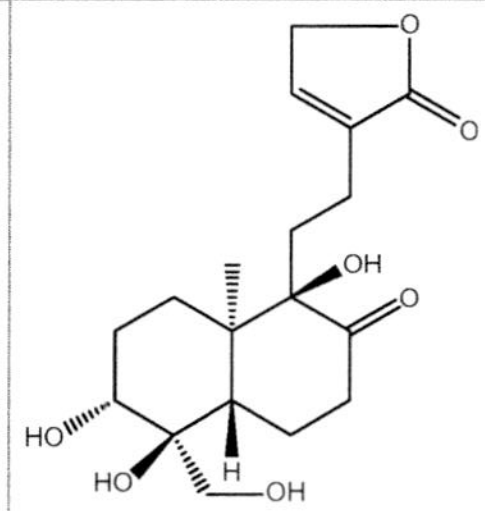	
				14-Deoxy-11,12-didehydroandrographolide	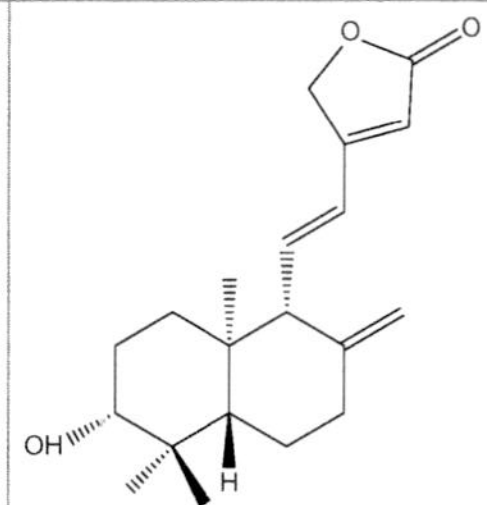	
				Andrograpanin	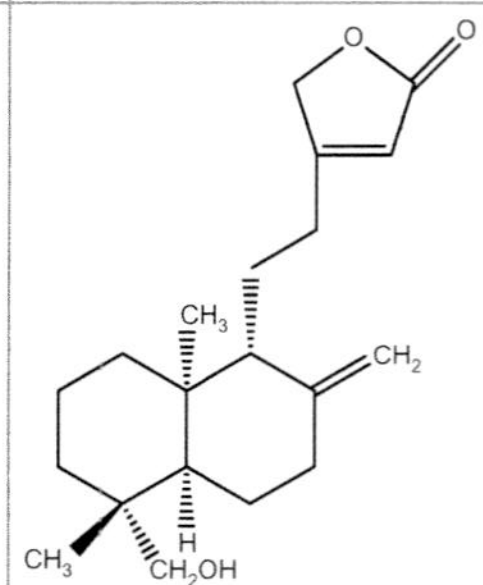	

				Isoandrographolide	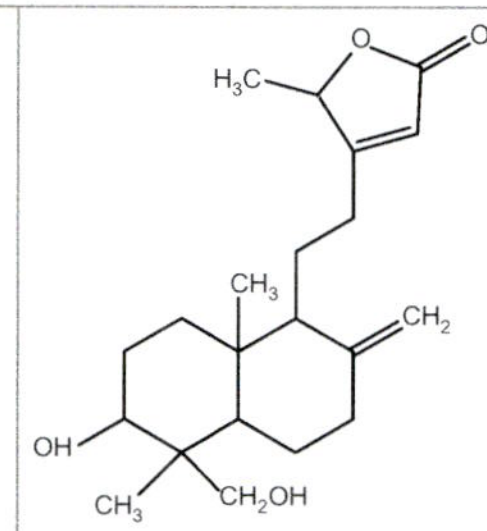	
				14-Acetylandrographolide	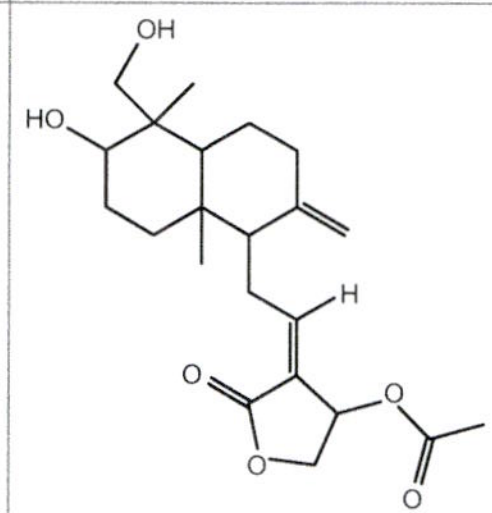	
	Asystasia gangetica (L.) T. Anderson		Leaf and flowers	Luteolin	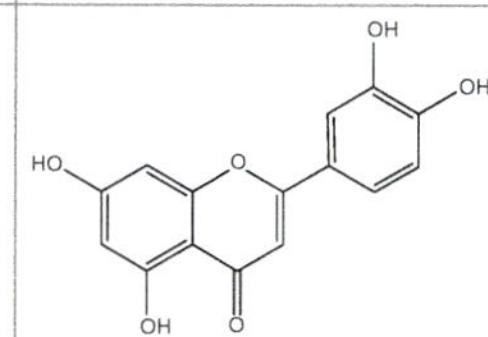	Tamilselvan et al. (2014) and Janakiraman et al. (2014)

(continued)

Table 4.2 (continued)

Family	Botanical name	Image	Plant part used	Bioactive compound(s)	Structure	Reference(s)
	Synonym(s): *Asystasia parvula, Asystasia querimbensis, Asystasia floribunda, Justicia gangetica, Asystasia acuminate, Asystasia coromandeliana.* **Common name(s)**: Chinese violet, Coromandel, or Creeping foxglove, may simply be called Asystasia. **Botanical description**: It is native to tropical Americas and Hawaii and is limited to Asia. It is a spreading herb or groundcover, reaching 600 mm in height or up to 1 m if supported. The stems root easily at the nodes. The leaves are simple and show opposite phyllotaxy. The fruit is an explosive capsule which is green in color but dries to brown after opening. **Medicinal use(s)**: It is used as an astringent, anticancer, antihelmintic, antidiabetic, anti-asthmatic, anti-inflammatory, and vermifuge agent. The leaves provide benefit through a bronchospasmolytic effect of the terpenoid compound. A leaf decoction is used in the treatment of fever aches, epilepsy, stomach pain, and urethral discharge. The pulped leaf is used as suppository for piles.			Quercetin	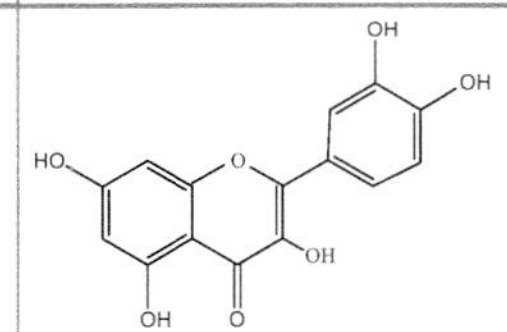	
				Kaempferol	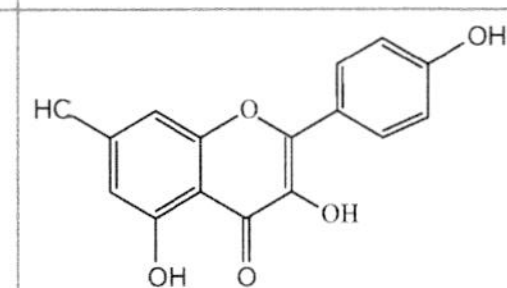	
				Isorhamnetin	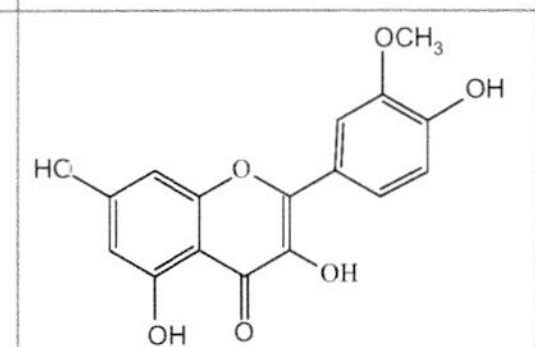	

	Barleria cristata L		Leaf	Desrhamnosylacteoside	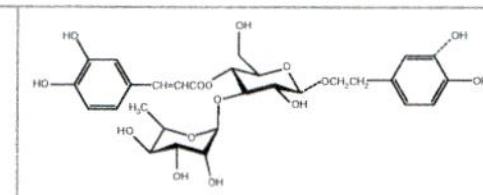	Singh (2012) and Kumar et al. (2018)
	Synonym(s): *Barleria alba* Lodd. **Common name(s)**: Philippine violet, Bluebell barleria, or Crested Philippine violet. **Botanical description**: It is native to a wide area ranging from Southern China to India and Myanmar. It grows as a shrub 60–100 cm tall. The leaves are dark green on the upper surface and pale green on the lower surface, elliptic to narrowly ovate. The flowers are about 5 cm long, funnel-shaped in violet, pink, or white color. The fruits are about 1.5-cm-long ellipsoid capsules, become glabrous and glossy at maturity. **Medicinal use(s)**: It has antioxidant, anti-inflammatory, hepatoprotective, anti-toxicity, hypoglycemic, antidiabetic, tonic, diuretic, and purgative properties.			Acteosice	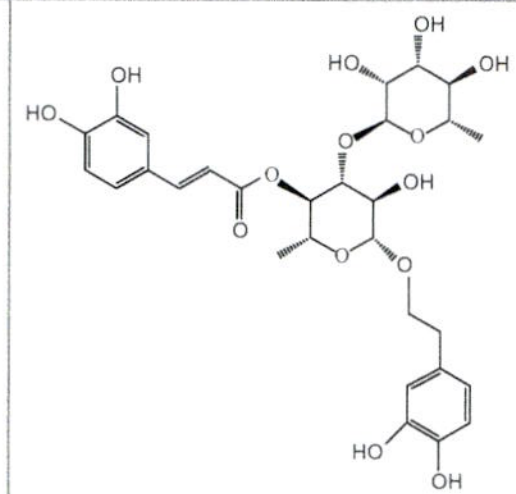	
				Poliumoside	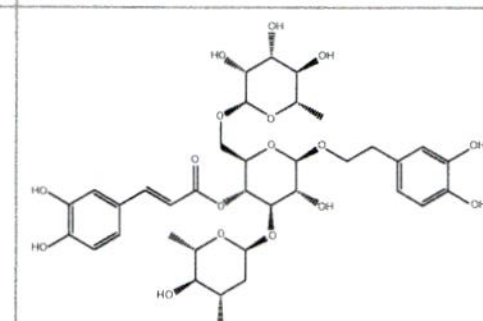	

(continued)

Table 4.2 (continued)

Family	Botanical name	Image	Plant part used	Bioactive compound(s)	Structure	Reference(s)
				Quercetagetin	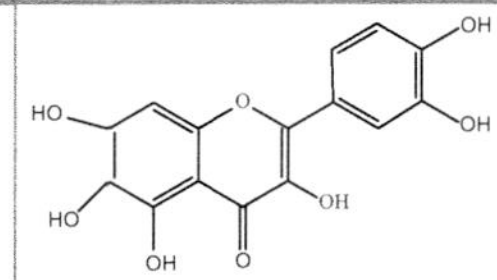	
				Tamarixetin	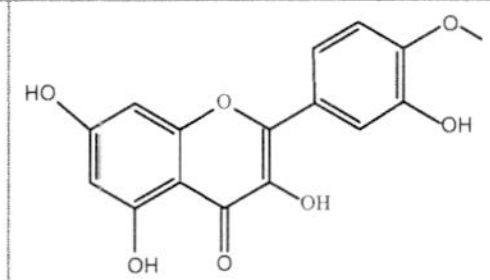	
				Gossypetin	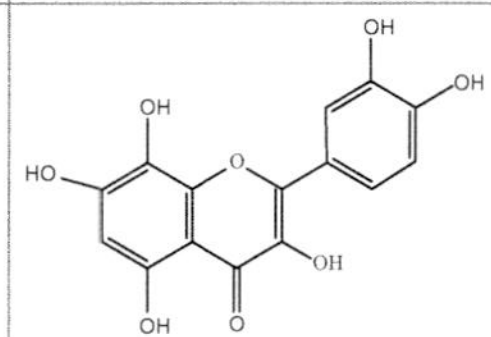	
				Oleanolic acid	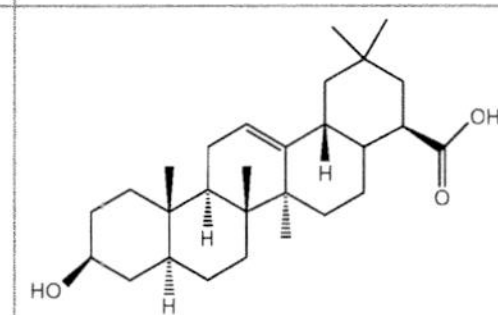	

	Barleria prionitis L.		Leaf, roots and bark	Shanzhiside methyl ester	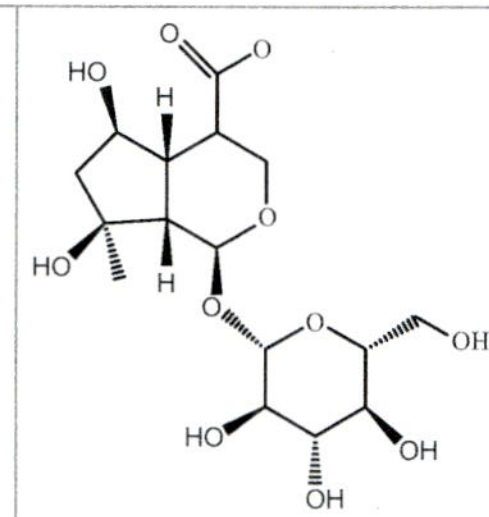	Chopra et al. (1956), Ata et al. (2009), Amoo et al. (2011), and Pandey et al. (2018)
	Synonym(s): *B. hystrix.* **Common name(s)**: Porcupine flower. **Botanical description**: Annual shrub, 1–3 ft. high, native to India, Sri Lanka, and Eastern, Southern, and Central Africa. It is an evergreen shrub growing up to 1.7 m tall from a central tap root. The base of the leaves is armed with three to five sharp, pale-colored spines. **Medicinal use(s)**: It is used in catarrhal affections of children, glandular swellings, boils, fever, toothache, inflammation, and gastrointestinal disorders; bark in whooping cough as an expectorant; the whole plant and especially the roots are used as tonic and diuretic.			6-*O*-*Trans*-*p*-coumaroyl-8-*O*-acetylshanzhiside methyl ester	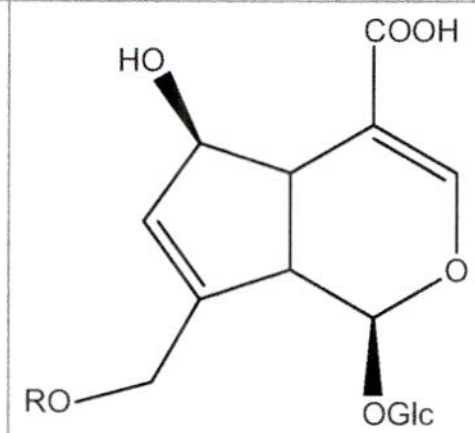	
				Barlerin	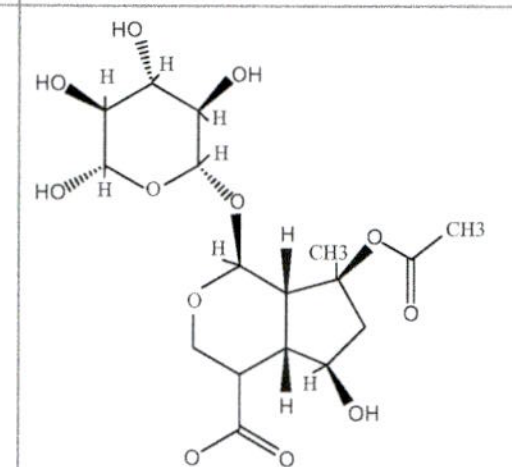	

(continued)

Table 4.2 (continued)

Family	Botanical name	Image	Plant part used	Bioactive compound(s)	Structure	Reference(s)
				Acetylbarlerin	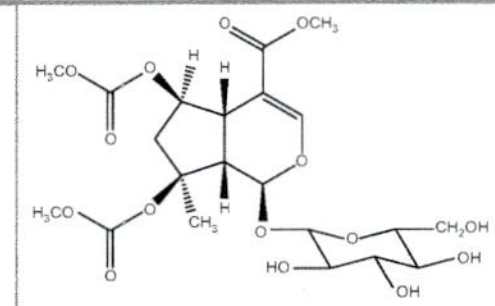	
				Lupulinoside	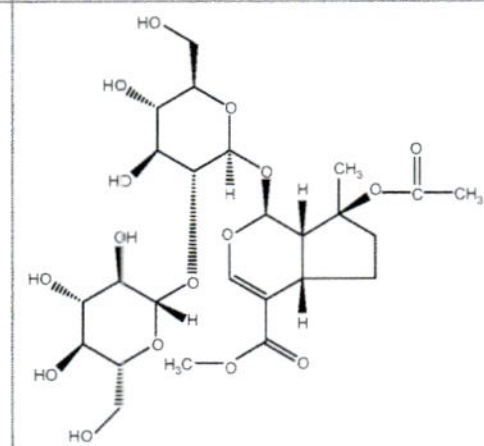	
	Hemigraphis hirta T. Anders		Entire plant	Stigmasterol	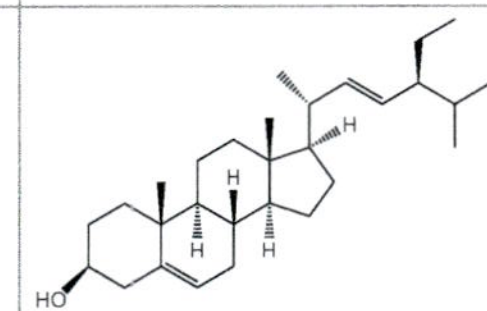	Gangadharan et al. (2013) and Alam et al. (2002)

	Synonym(s): *Ruellia hirta* Vahl., *Ruellia sarmentosa*. **Common name(s)**: Hairy hemigraphis. **Botanical description**: It is native to Tropical Asia. A softly hirsute herb; stem 15–45 cm long, creeping in grass. Leaves small, ovate crenate. Heads 2–6 flowered, axillary. Sepals linear or in fruit subspathulate. **Medicinal use(s)**: It has anticancer properties. It is an antihelmintic agent and is used as treatment for acute wounds, dysentery, stomatitis, glossitis, and abdominal pain.	n-hentriacontane	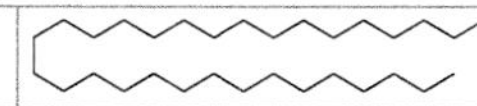	
		Squalene	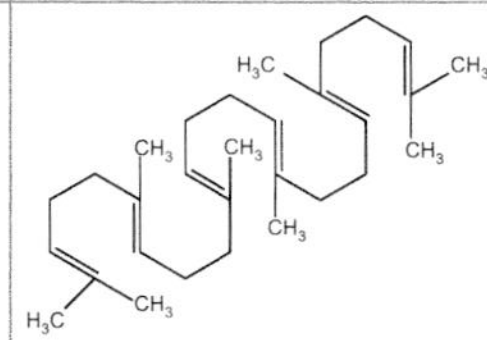	
		Lupeol	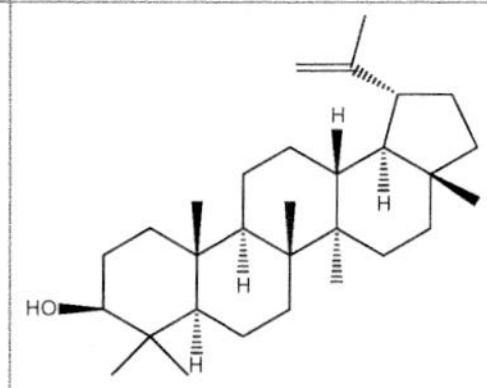	
		β-Sitosterol	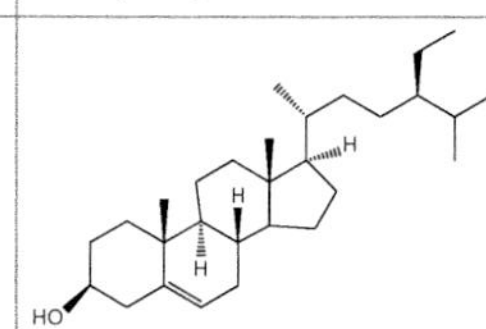	

(continued)

Table 4.2 (continued)

Family	Botanical name	Image	Plant part used	Bioactive compound(s)	Structure	Reference(s)
	Hygrophila auriculata (Schumach.) Heine		Part not specified	Lupeol	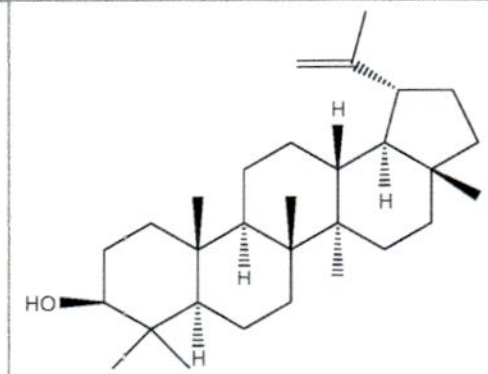	Jain(1970), Chauhan and Dixit (2010), Pattanayak et al. (2013), and Saha and Paul (2017)
	Synonym(s): *Asteracantha longifolia* (L.) Nees, *Barleria auriculata* Schumach, *Barleria longifolia* L., *Hygrophila schulli* M. R. Almeida & S. M. Almeida, *Hygrophila spinosa* T. Anderson. **Common name(s):** Gokulakanta (India) and Neeramulli (Sri Lanka). **Botanical description**: It is an herbaceous medicinal plant growing in marshy places, ponds, ditches, and puddles after rainy season and is native to tropical Asia and Africa. It is an erect, stout, branched, or unbranched, annual herb growing 150 cm or taller. Occasionally the basal part of the plant of the stem is creeping and rooting. **Medicinal properties:** It is used to treat cancer. It is used in chemotherapeutic and antineoplastic drugs. The plant is also used for the treatment of rheumatism, urinary tract infection, inflammation, jaundice, hepatic obstruction, impotence, and abdominal cramps.			Stigmasterol	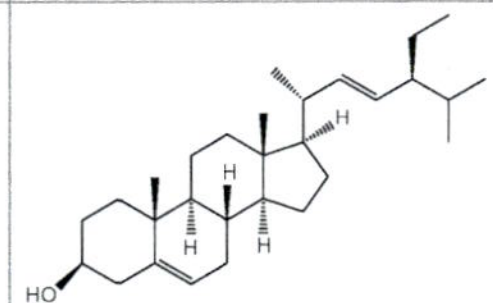	
				Lupenone	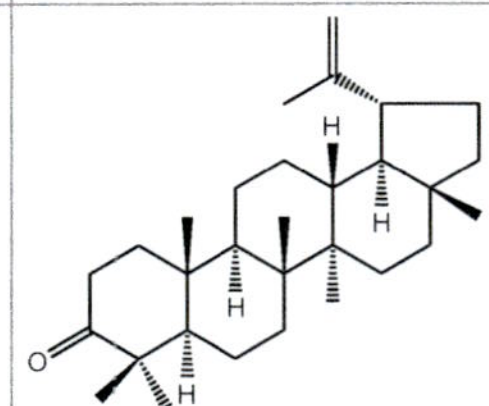	
				25-Oxo-hentriacontanyl acetate	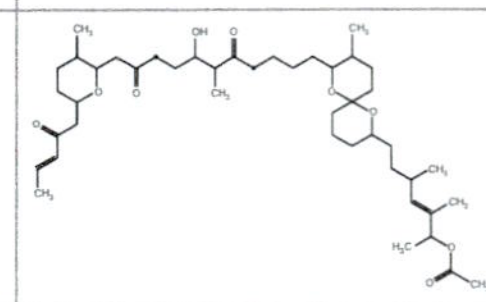	
				Methyl 8-n-hexyltetracosanoate	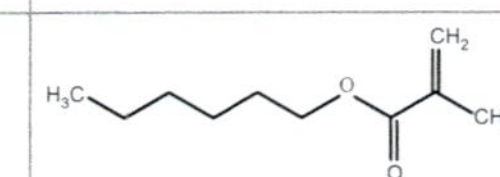	

	Justicia rhodoptera Baker		Apical part	Justicidin	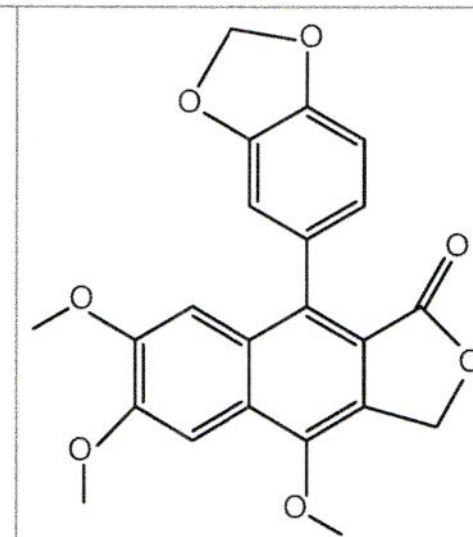	Corrêa and Alcântara (2011, 2012)
	Synonym(s): *Ecbolium rhodopterum*. **Common name(s)**: Not available. **Botanical description**: Native of Madagascar, Taiwan, and Nansei Shoto. It is a perennial plant growing 80–100 cm tall, the decumbent stems rooting at the basal nodes. It has cylindrical corolla tube, four spreading segments, and two stamens, low in tube, short filaments. **Medicinal use(s)**: It possesses anticancer properties. It also shows antitumor activity against different cancer cell lines. It exhibited anticancer activity against P388 lymphocytic leukemia in mice, human ovarian cancer cell line, as well as prevention of some tumor cell growth.					

(continued)

Table 4.2 (continued)

Family	Botanical name	Image	Plant part used	Bioactive compound(s)	Structure	Reference(s)
	Justicia gendarussa Burm. F		Entire plant	β-Sitosterol	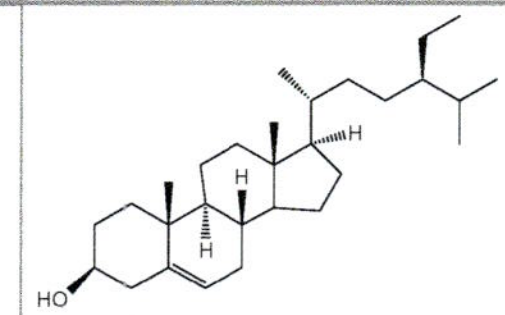	Ayob et al. (2013)
	Synonym(s): *Gendarussa vulgaris.* **Common name(s):** Willow-leaved justicia. **Botanical description**: It is native to Asia. The plant is endemic to India and is a small erect, branched shrub. The shrub has lance-shaped leaves 10–15 cm in length and 4 cm wide. They have oppositely arranged, smooth-edged short petioles. They are of dull brownish-green color when dried. Its trunk has many long, opposite, ascending branches; the bark is yellowish in color. Flowers are usually white in color, and the inflorescence shows large, dense, axillary spikes. The fruits are pubescent and are with club-shaped capsules. **Medicinal use(s)**: It possesses antitumor properties. It is useful in asthma, rheumatism, and colitis of children and is a potential birth control pill for men. It is used as febrifuge and emetic (bark is a good emetic), employed for rheumatic affection. Leaves are insecticidal. Infusion of the leaves is given internally in cephalalgia and hemiplegia and used in chest pain and facial paralysis. The leaves and tender shoots are diaphoretic; decoction is given in chronic rheumatism.			Lupeol	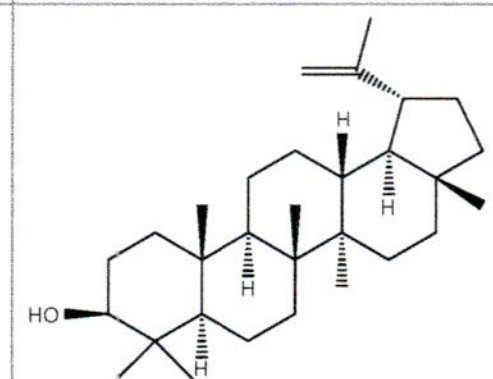	

				Friedelin	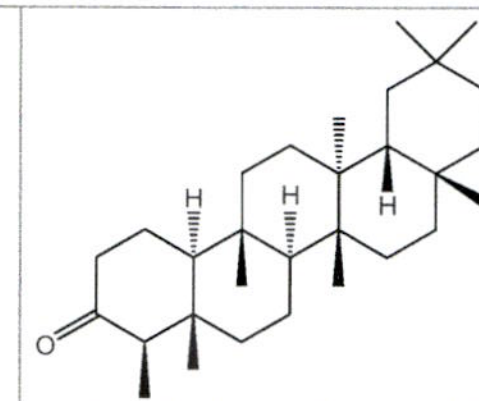	
	Justicia hyssopifolia L.		Leaf	Elenoside	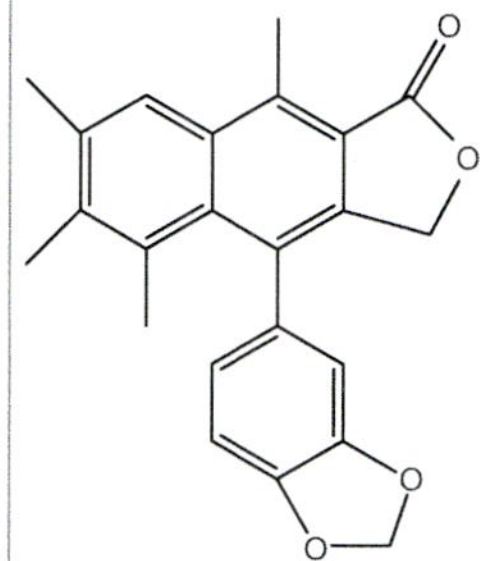	Navarro et al. (2001, 2004), Correa and Alcantara (2011), and Kumar (2016)
	Synonym(s): *Dianthera hyssopifolia* (L.) Medik., *Ecbolium hyssopifolium* (L.) Kuntze, *Gendarussa hyssopifolia* (L.) Webb & Berthel. **Common name(s)**: Water-willow. **Botanical description**: It is native to Tropical America. They are evergreen perennial shrubs with leaves which are often strongly veined, but they are primarily cultivated for their showy tubular flowers in shades of white, cream, yellow, orange, or pink. **Medicinal use(s)**: The plant shows neuropharmacological activity. It is having cytotoxicity, acute toxicity, and pharmacological effect. Elenoside was cytotoxic to human cancer cell lines in a range of concentration from 10^{-5} to 10^{-4}. *M. elenoside* appears to be a sedative with broad-spectrum cytotoxicity.					

(continued)

Table 4.2 (continued)

Family	Botanical name	Image	Plant part used	Bioactive compound(s)	Structure	Reference(s)
	Mendoncia cowanii (S. Moore) Benoist		Root and bark	Meliavolin	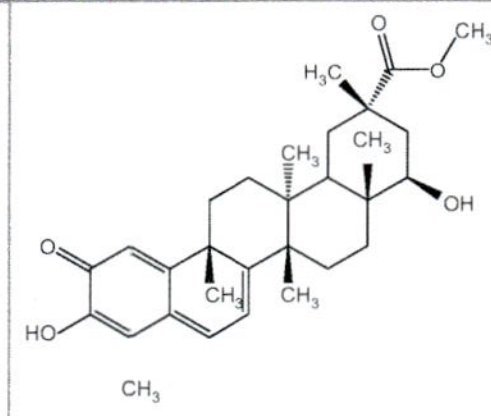	Zeng et al. (1997a, b) and Williams et al. (2006)
	Synonym(s): *Afromendoncia cowanii*. **Common name(s)**: Magnaghi. **Botanical description**: It is native to Madagascar; *Mendoncia cowanii* is a liana that grows in humid and subhumid forests and in vegetation within human settlement. The leaves are eaten, and the fruits are eaten by *Eulemur rubriventer*. This species is endemic to Madagascar where it occurs in the provinces of Antananarivo, Antsiranana, Fianarantsoa, Toamasina, and Toliara between sea level and 1499 masl. This species is threatened by selective logging for subsistence use and by wildfire. **Medicinal use(s)**: It is used as an anticancer agent for cell line of ovarian cancer. The leaves of this species are used for the treatment of syphilis. It is also used as fuel wood.			Meliavolkin	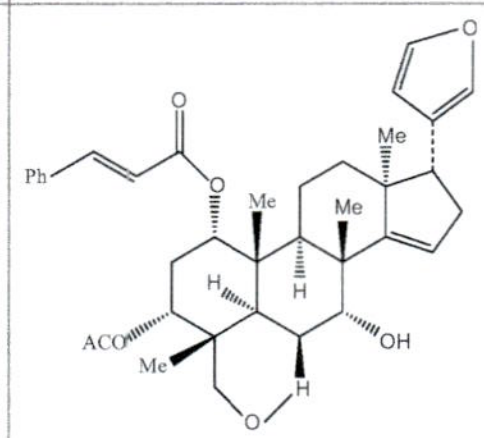	

	Ruellia tuberosa L.		Tuber	Squalene	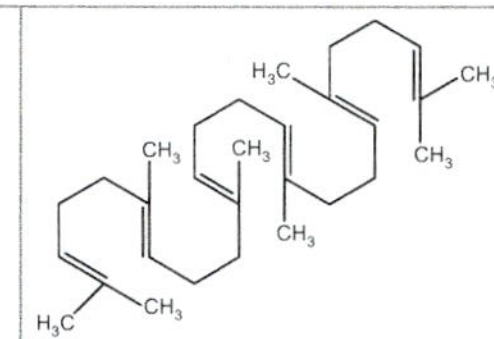	Reddy et al. (1991), Chen et al. (2006), Dey et al. (2013), and Vennapusa et al. (2013)
	Synonym(s): *Cryphiacanthus barbadensis* Nees, *Dipteracanthus clandestinus*, *Ruellia clandestine*, *Ruellia picta*. **Common name(s)**: Minnieroot, Snapdragon root, Fever root, Sheep potato, Iron root, Cracker plant. **Botanical description**: It is native to Central America and India but also found in tropical South and Southeast Asia. It is a small biennial plant with thick fusiform tuberous roots and striking funnel-shaped violet-colored flowers. Its fruit is a 20-seeded, 2-cm-long sessile capsule. It reaches an average height of about 25 cm in moist and shady environments. It grows, however, preferably in grasslands and roadsides often as a weed in cultivated fields and also in xerophilic and ruderal habitats. **Medicinal properties**: Used to treat cancer in the stomach. It is being used as a diuretic, antidiabetic, antipyretic, analgesic, antihypertensive, and gastro-protective agent and to treat gonorrhea. It is also used as natural dye for textiles. It is used in human ovarian cancer cell line.			Cholest-5-ene, 3-bromo-, (3beta)	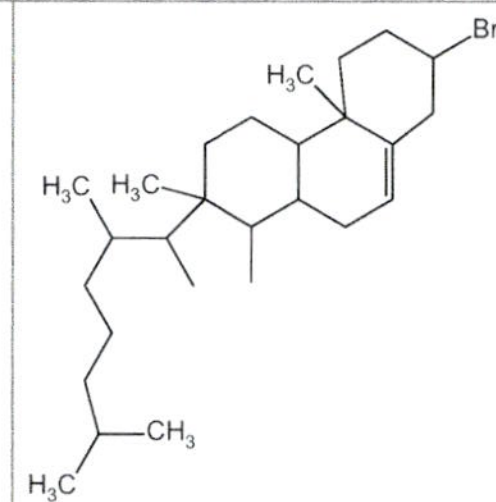	
				Beta-sitosterol	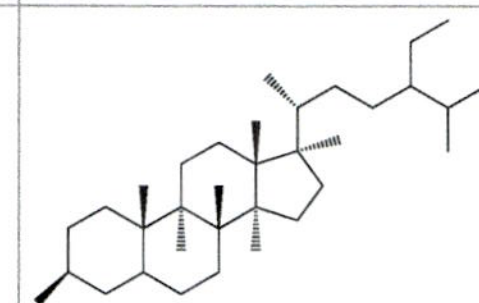	
				Lupeol	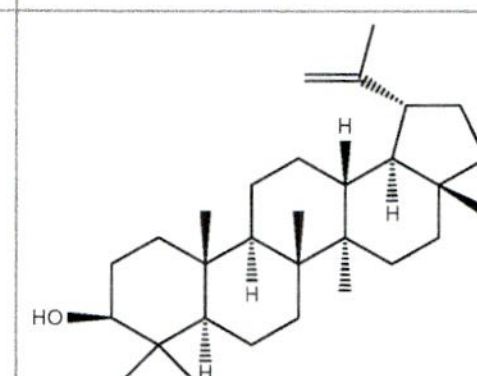	

(continued)

Table 4.2 (continued)

Family	Botanical name	Image	Plant part used	Bioactive compound(s)	Structure	Reference(s)
				α-Sitosterol	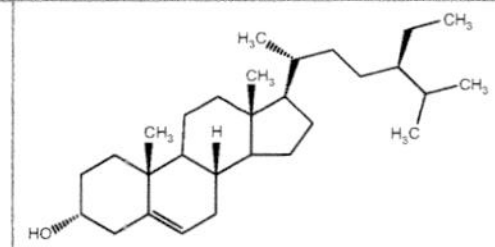	
				Stigmasterol	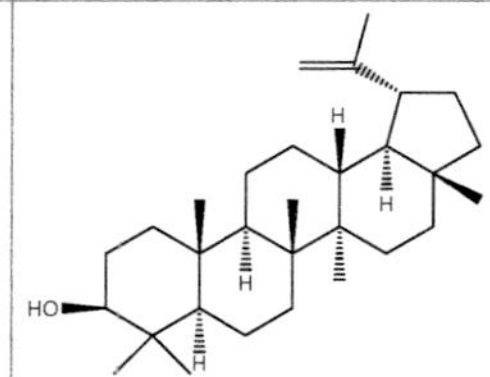	
				2-Methyl- heptacosane	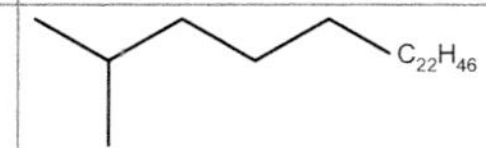	
				3-Bromo- octadecane		
				2-Methyl-nonadecane	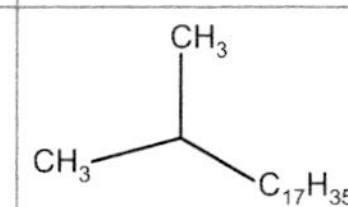	
				2-Methyl- eicosane	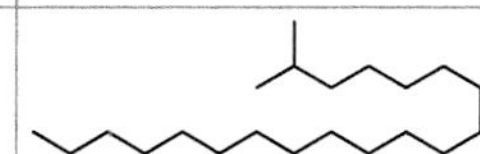	
				Cholest-5-ene	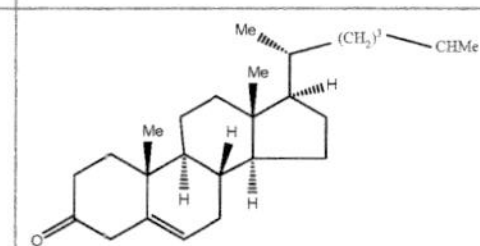	

				Apigenin	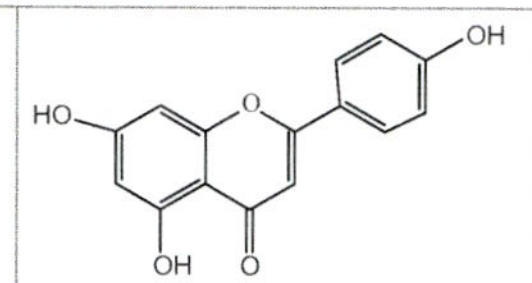	
	Rungia repens (Nees) Lindau		Leaf and roots	Umbelliferone	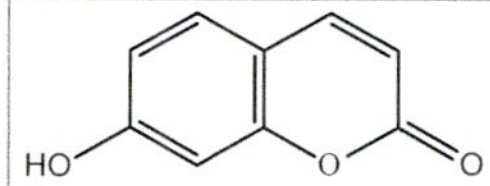	Navarro et al. (2004), Swain et al. (2008), and Corrêa and Alcântara (2012)
	Synonym(s): *Justicia repens*, *Dicliptera repens* (L.) Juss. **Common name(s)**: Creeping rungia, Parpatakah. **Botanical description**: It is native to India and usually grows as a weed in moist habitats such as riverbanks, ditches, channels, and bunds of paddy fields. It is an herb that spreads with angular, zigzag stems, often rooting at lower stem nodes; leaves subsessile. The leaves are lanceolate to oblong-elliptic in shape. Flowers are of different shades such as blue, pink, ivory-white, and yellow. Flowers possess long spikes with broadly elliptic bracts. Fruits or capsules are ovoid-oblong shaped with pale brown seeds. **Medicinal use(s)**: It possesses anti-angiogenic, anti-leishmanial, antifungal, hypolipidemic, anti-asthmatic, antiviral, antineoplastic, antifeedant, insecticidal, cardiotonic, antidepressant, analgesic, antiplatelet, and anti-inflammatory properties.			3′,4′-Dihydroxyflavonol	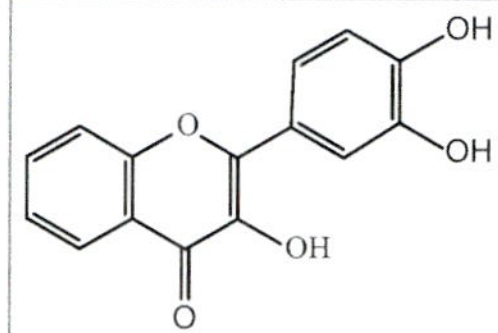	

(continued)

Table 4.2 (continued)

Family	Botanical name	Image	Plant part used	Bioactive compound(s)	Structure	Reference(s)
				Apigenin	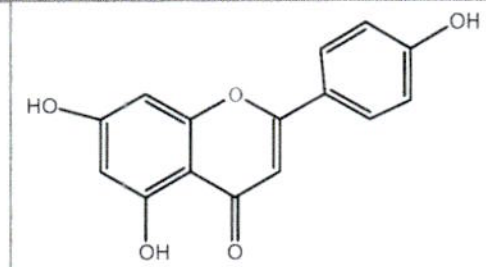	
				Kaempferitrin	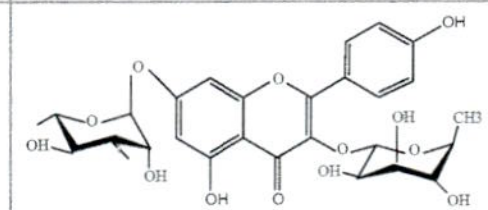	
				Vitexin (flavonoids)	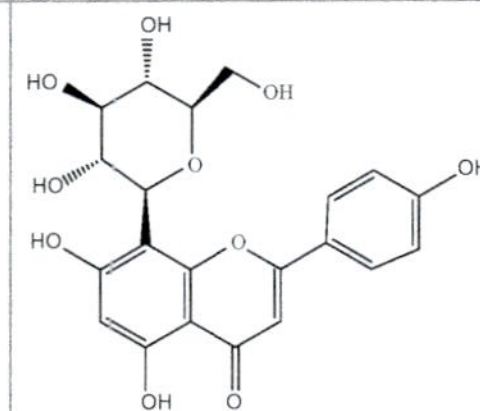	
				Jusbetonin	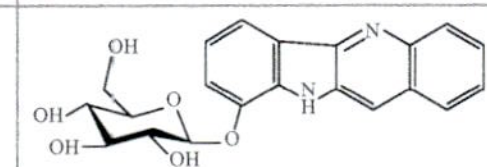	
	Rungia pectinata (L.) Nees		Leaf and root	Serotonin	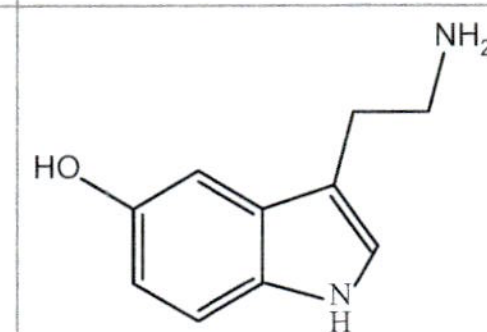	Swain (2008)

	Synonym(s): *R. parviflora* (L.) Nees, *Justicia pectinata* L. **Common name(s)**: Comb rungia. **Botanical description**: It is a branched, procumbent, or erect herb found as a weed throughout the warmer parts of India. It is an herb with hairy stem. Leaves 4–6 cm long, ovate to linear-oblong or obovate. They are elliptic, narrowly winged on one side, flowers are small densely packed; seeds are orbicular, rugose, and brown. **Medicinal use(s)**: The leaf juice is aperient, febrifuge, refrigerant; bruised leaves are applied externally to subside swellings. The juice of leaves is given to children suffering from smallpox. Roots are used as febrifuge.	Histamine	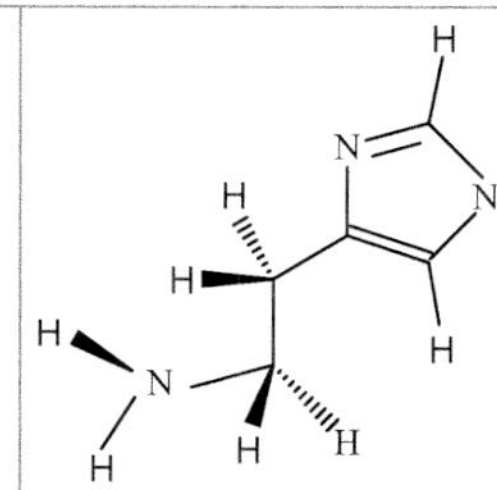	
		Bradykinin	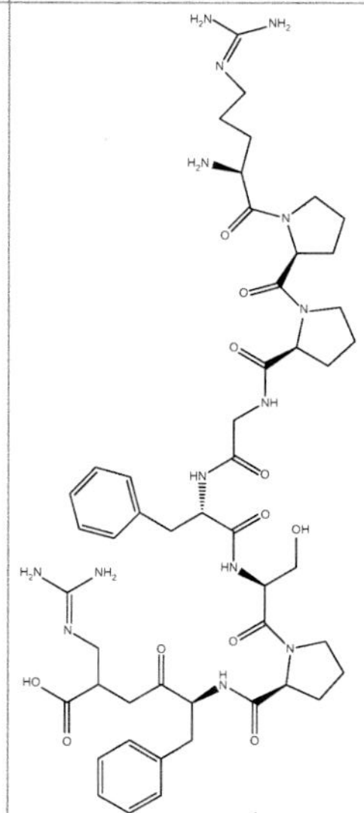	

(continued)

Table 4.2 (continued)

Family	Botanical name	Image	Plant part used	Bioactive compound(s)	Structure	Reference(s)
	Ruellia prostrate (Poir) Nees		Leaf	Hexadecanal	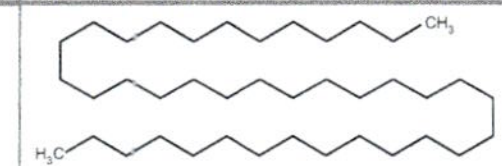	Rai et al. (2015) and Kumar et al. (2015)
	Synonym(s): *Dipteracanthus prostrates*. **Common name(s)**: Bell weed, Prostrate wild petunia, Black weed. **Botanical description**: It is native plant of South Africa, Arabian Peninsula, eastwards to Southern Asia, and the Pacific Islands. It is prostrate perennial herb, with stems often rooting at the nodes. Ovate green leaves, have lower surface conspicuously paler. Leaf stalk is 5–30 mm long. Flowers occur solitary in the leaf axils, each one subtended by oblanceolate to ovate bracts. Sepals are 5 in number. Flowers are violet-blue to occasionally nearly white, the petals slightly spreading. Capsules club-shaped, densely covered with fine hairs. Flowering: August–September. **Medicinal use(s)**: It is used as antidote for snakebite and possesses anticancer, hypoglycemic, anti-inflammatory, anti-ulcer, and antioxidant activities.			Hexatriacontane		
				Spinacene	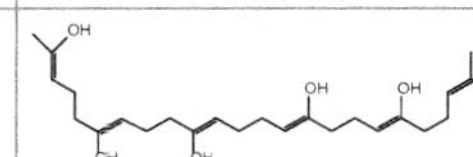	
				Dotriacontane	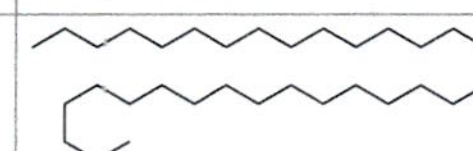	

				Stigmasterol	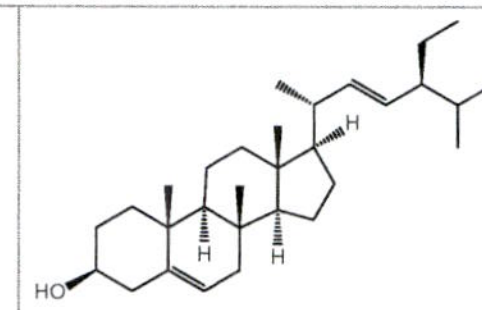	
	Thunbergia laurifolia Lindl.		Entire plant	8-*Epi*-grandifloric acid	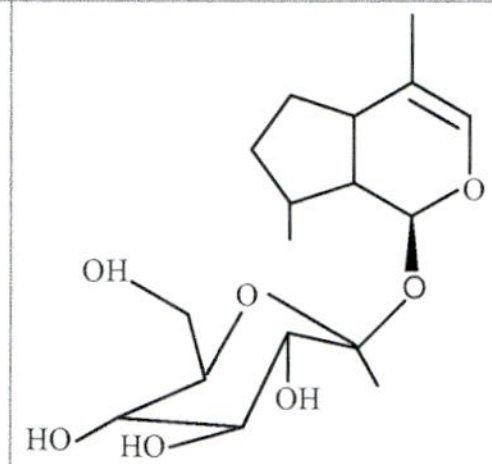	Wasuwat (1967), Kanchanapoom et al. (2002), Wonkchalee et al. (2012), and Jetawattana et al. (2015)
	Synonym(s): *Thunbergia grandiflora* var. laurifolia (Lindl.) Benoist. **Common name(s)**: Blue trumpet vine, Laurel clock vine. **Botanical description**: It is a perennial shrub native to tropical regions of Africa, Madagascar, and Southern Asia and India. The plant is characterized by its heart-shaped serrate leaf margin and non-scented flowers which are yellowish from inner side and white on outside. The plant flowers almost continuously throughout year. Carpenter bees aid in its pollination. **Medicinal properties**: It is used to treat human breast cancer. The leaf juice is used to cure deafness, in Malaysia. In Thailand, leaves are used as an antipyretic as well as an antidote to detoxify the poisons.			3′-*O*-β-Glucopyranosyl-stilbericoside	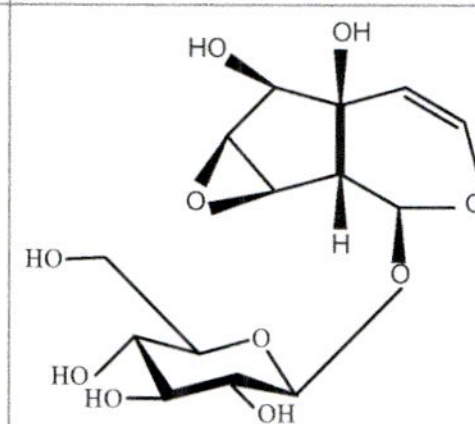	

(continued)

Table 4.2 (continued)

Family	Botanical name	Image	Plant part used	Bioactive compound(s)	Structure	Reference(s)
				Benzyl β-glucopyranoside	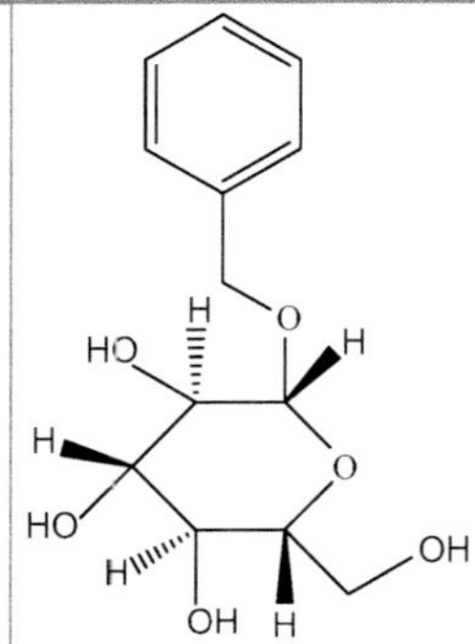	
				Benzyl β-(2′-*O*-β-glucopyranosyl) glucopyranoside	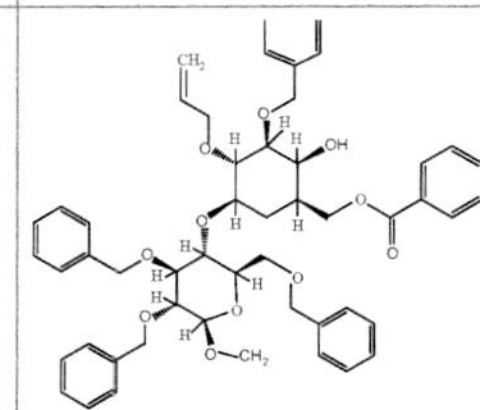	
				Grandifloric acid	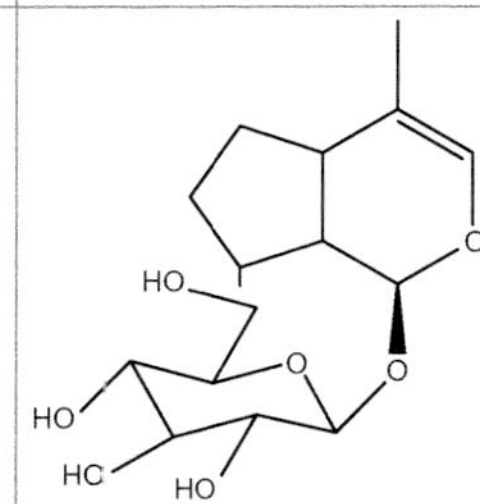	

		Compound	Structure	
		(*E*)-2-hexenyl β-glucopyranoside	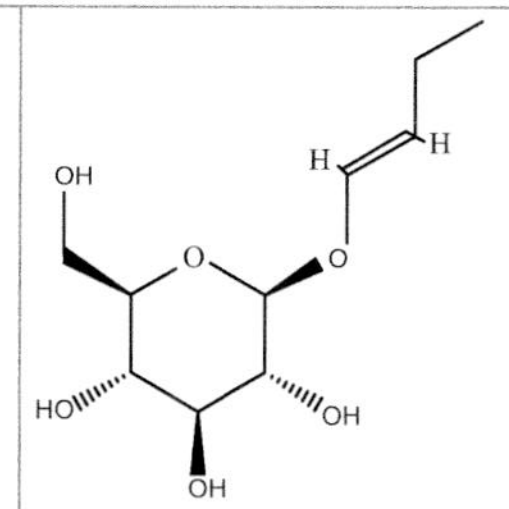	
		Hexanol β-glucopyranoside	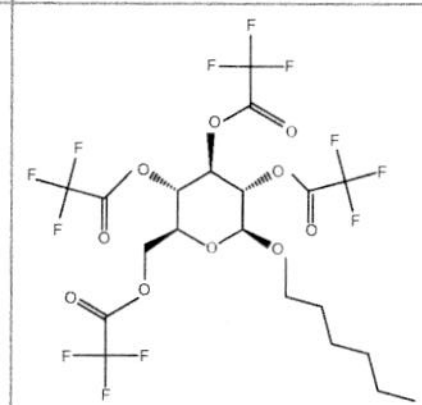	
		6,8-Di-*C*-glucopyranosylapigenin	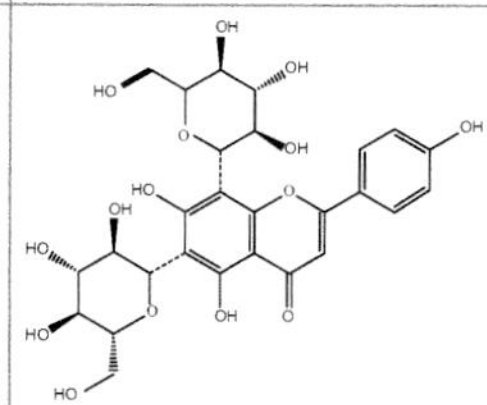	

(continued)

Table 4.2 (continued)

Family	Botanical name	Image	Plant part used	Bioactive compound(s)	Structure	Reference(s)
	Eremomastax speciosa (Hochst.) Cufod.		Entire plant	Iridoids	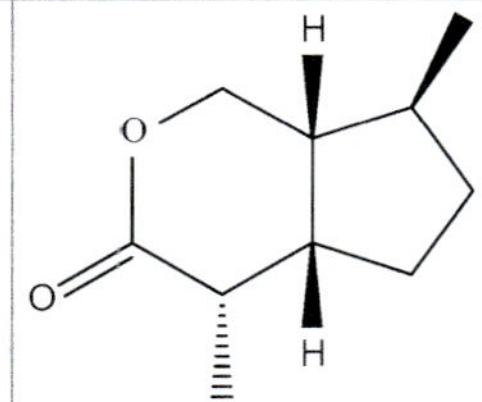	Fongod et al. (2013) and Sagnia et al. (2014)
	Synonym(s): *Paulo-Wilhelmia speciosa* Hochst., *Paulo-wilhelmia glabra* Lindau, *Paulo-wilhelmia pubescens* Lindau. **Common name(s)**: Blood plant, Golden seal, African blood tonic plant. **Botanical description**: The plant is widely distributed in tropical Africa. It is a robust, polymorphous shrub that grows 2 m high and has a characteristic quadrangular stem and violate underside of the leaf. **Medicinal use(s)**: It possesses anticancer properties. The leaf extract is used for the treatment of both male and female infertilities, malaria, kidney pain, scabies, anemia, diabetes, nerve pain, irregular menstruation, menstrual pains, gonorrhea, ulcers, diarrhea, appendicitis, dry burns, and poisoning. It also increases and purifies the blood.			Anthraquinone	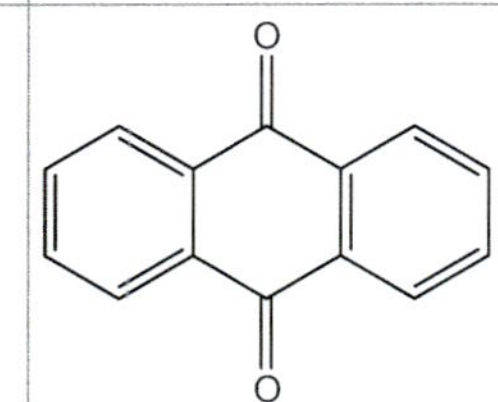	
				Kaempferol	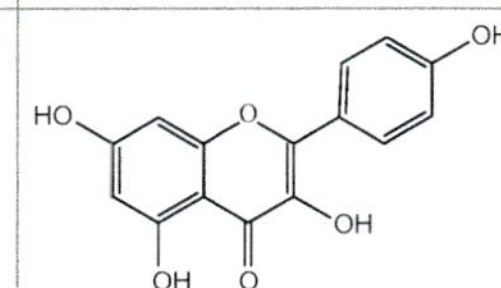	
				Quercetin	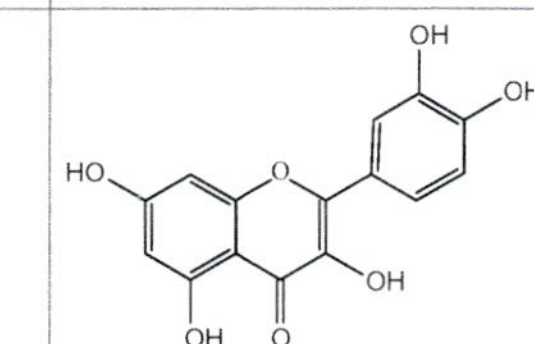	

Table 4.3 Anticancer plants of family Aceraceae

Family	Botanical name	Image	Plant part used	Bioactive compounc(s)	Structure	Reference(s)
Aceraceae	*Acer maximowiczianum*		Bark	Aceroside	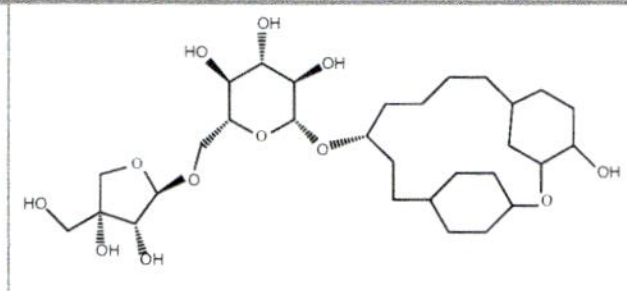	Potvin et al. (2011) and Omar (2013)
	Synonym(s): *Acer nikoense*. **Common name(s)**: Nikko maple. **Botanical description**: It is native to east Asia, eastern China (Anhui, Hubei, Jiangxi, Zhejiang), and Central and Southern Japan. It is a deciduous tree with a round-topped crown; it can grow 12–15 m tall, occasionally reaching 25 m. The stem can be 30–45 cm in diameter, exceptionally 70 cm. The tree is usually smaller in cultivation, becoming a small bushy tree or vase-shaped shrub. **Medicinal use(s)**: It is used to treat various types of cancers and cancer cells (induces cancer cell death). It seems to have free radical scavenging activity and enhance the cytotoxic activity of vitamin C on cancer cells. It also exhibits antitumor, anti-inflammatory, antidiabetic, hepatoprotective, and anti-obesity activities.			Acerogenin	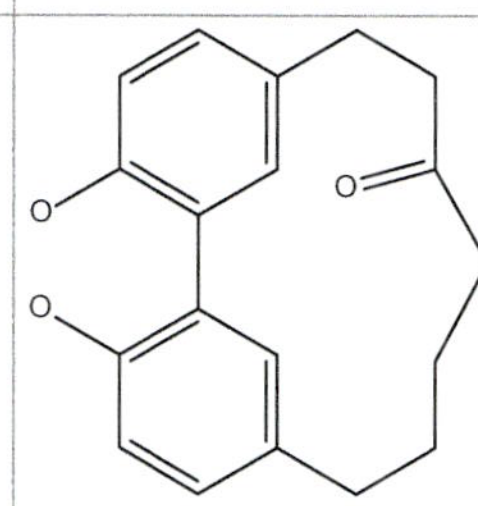	

(continued)

Table 4.3 (continued)

Family	Botanical name	Image	Plant part used	Bioactive compound(s)	Structure	Reference(s)
				(−)-centrolobol	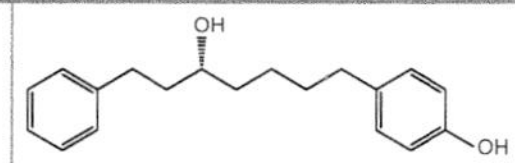	
				(+)-rhododendrol	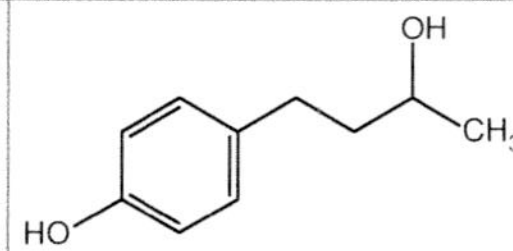	
				(+)-catechin	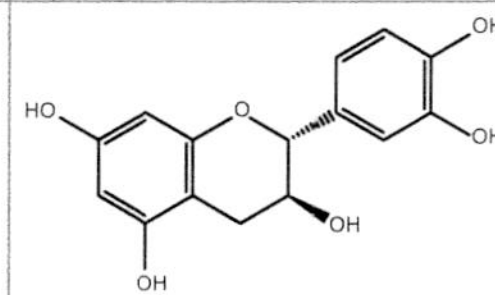	
				Phenylalanine	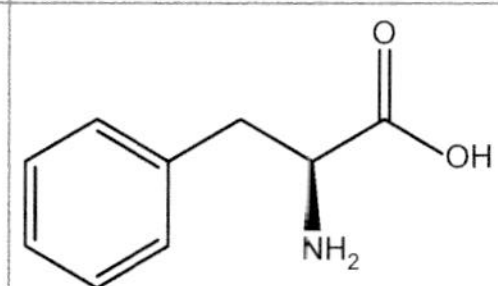	

	Acer rubrum L.		Bark, stem	Maplexin	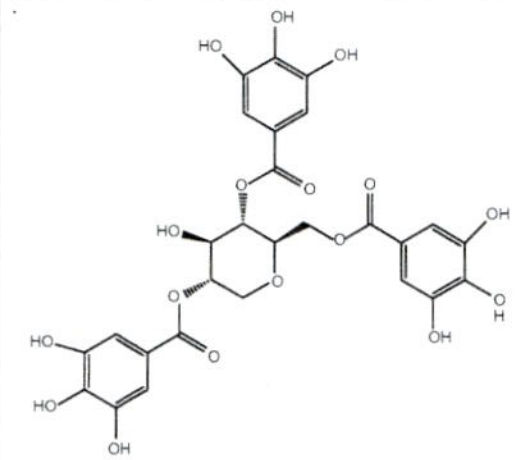	Omar (2013) and Ma (2014)
	Synonym(s): *Acer carolinianum* Walter, *Acer coccineum* F.Michx., *Acer drummondii* Hook. & Arn. ex Nutt., *Acer fulgens* Dippel, *Acer glaucum* Marshall, *Acer hypoleucum* K.Koch, *Acer microphyllum* Pax, *Acer sanguineum* Spach. **Common name(s)**: Red maple, Scarlet maple, Swamp maple. **Botanical description**: It is native plant of South Asia, Malaysia, Indonesia, and the Philippines. It is medium-sized, deciduous tree that is native to eastern North America. It typically grows 40–60 m tall with a rounded to oval crown. Leaves (2–5 cm long) have three principal triangular lobes (sometimes five lobes with the two lower lobes being largely suppressed). Lobes have toothed margins and pointed tips. Upper surface of leaves is dark green, and the lower surface is gray green. Flowers on a given tree are primarily male or female or monoecious and appear in late winter to early spring (March–April) before the leaves. Fruit is a two-winged samara. **Medicinal use(s)**: It possesses cytotoxic property and is used for the treatment of cancer. It is effective against human colon tumorigenic cells (HCT-116) and breast cancer cells (MCF-7). It is used for the abdominal cramps and other menstrual problems. The bark is used for treating sore eyes and as an astringent in several remedies. The bark strips are used for making red dye.			Gallic acid	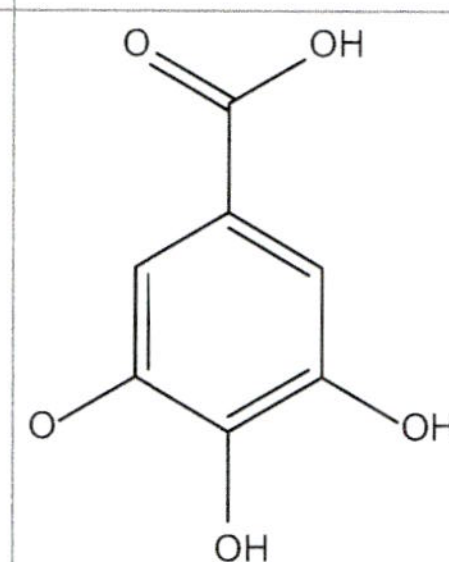	

(continued)

Table 4.3 (continued)

Family	Botanical name	Image	Plant part used	Bioactive compound(s)	Structure	Reference(s)
				Nymphaeoside	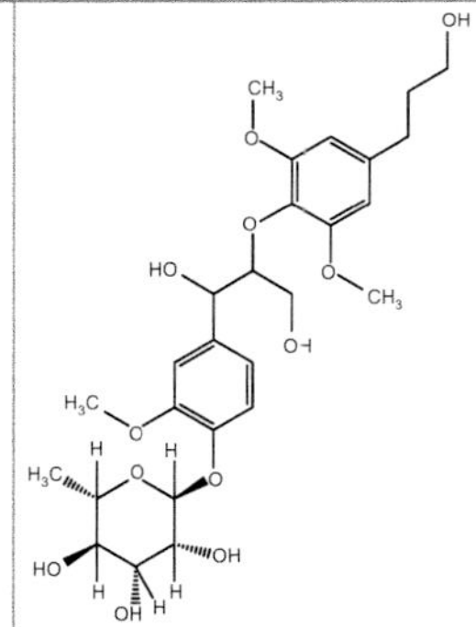	
				Methyl vanillate	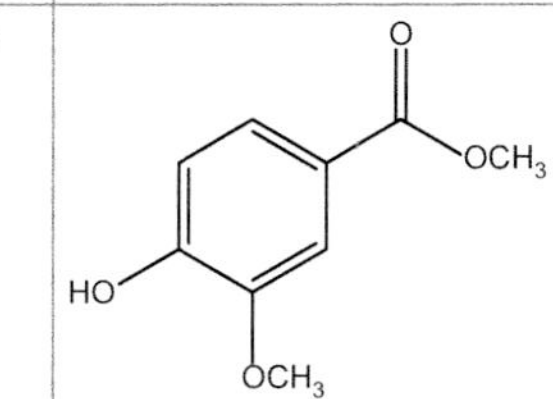	

	Acer pseudoplatanus L.		Bark, fruit	Phloroglucinol	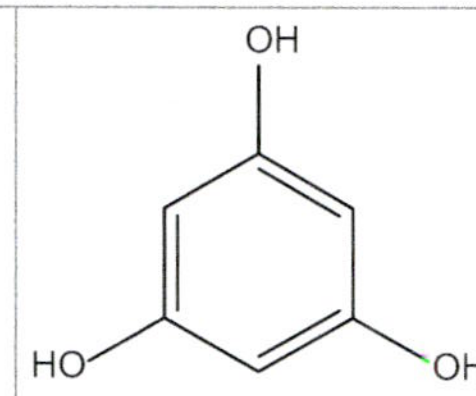	Culpeper (1950) and Bi et al. (2016)
	Synonym(s): *Acer abchasicum* Rupr., *Acer atropurpureum* Dippel, *Acer dittrichii* Ortm., *Acer majus* Gray, *Acer melliodorum* Opiz, *Acer procerum* Salisb., *Acer villosum* C.Presl, *Acer wondracekii* Opiz, *Acer worleei* Dippel. **Common name(s)**: False plane-tree, Great maple, Scottish maple, Mock-plane, Sycamore, or Celtic maple. **Botanical description**: It is a species of maple native to Central Europe and Southwestern Asia, from France eastwards to Ukraine, Northern Turkey, and Caucasus. It is a large deciduous tree reaching to a height between 20 and 35 m. The bark of tree is brown to pinkish in color and very rough in case of mature plants but smooth and gray in case of young plants. **Medicinal use(s)**: Its bark is used to treat tumors, and fruits are used to dissolve tumor. The bark of the tree has been used to make a wash for skin problems and eyewash for sore eyes. The inner bark containing the sweet sap can be used as dressing for wounds.			Graecunins	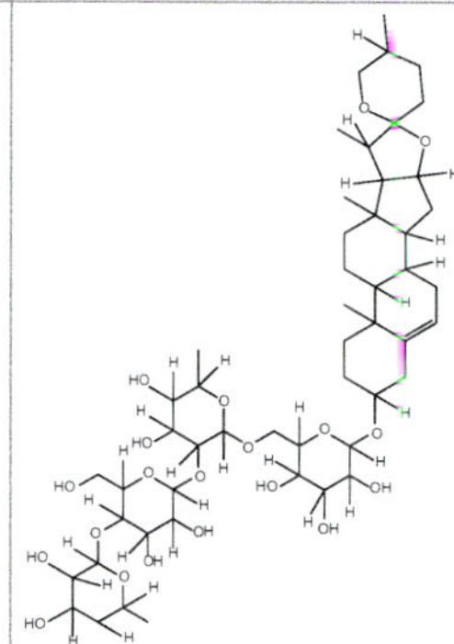	
				L noleic acid	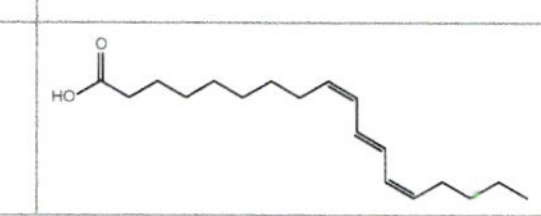	

Table 4.4 Anticancer plants of family Agavaceae

Family	Botanical name	Image	Plant part used	Bioactive compound(s)	Structure	Reference(s)
Agavaceae	*Agave americana*		Leaf	Steroidal saponin (hecogenin)	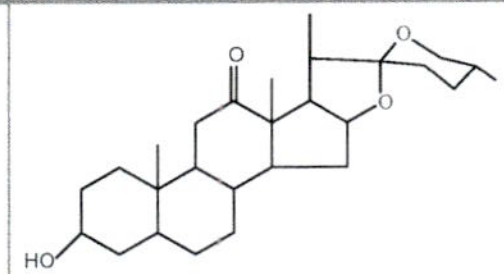	Anajwala et al. (2010) and Khade et al. (2011)
	Synonym(s): *Agave altissima* Zumagl, *Agave americana* subsp. *americana*, *Agave americana* var. americana, *Agave americana* var. marginata Trel., *Agave communis* Gaterau. **Common name(s)**: Agave, American century plant, Century plant or Maguey. **Botanical description**: It is native to southwestern North America and naturalized in the Mediterranean. It is an evergreen perennial plant growing up to 7.5 m (24 ft.) by 2.5 m (8 ft.) at a slow rate. The flowers are hermaphrodite (have both male and female organs) and are pollinated by lepidopterans and bats. It is suitable in light (sandy) and medium (loamy) soils and prefers well-drained soil, and suitable pH is acid, neutral, and basic (alkaline) soils. It cannot grow in the shade. It prefers dry or moist soil and can tolerate drought. **Medicinal use(s)**: It is used to treat breast, liver, and esophagus cancer. It possesses anticancer, antitumor, antibacterial, antifungal, and anti-inflammatory properties. It is used internally to treat ulcers, stomach inflammation, tuberculosis, jaundice and other liver diseases, syphilis, and menstrual problems. Additionally, it is used as a treatment for high fever by inducing sweat. An infusion of the chopped leaf is used as a purgative, and the juice of the leaves is applied to bruises.					

	Dracaena draco (L.) L.		Fruit sap	Draconin	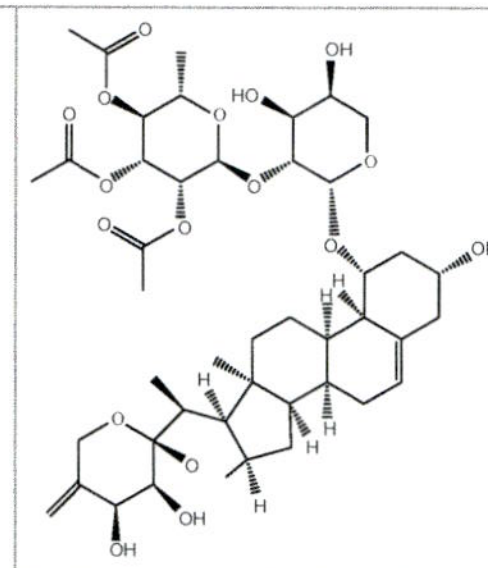	Darias et al. (1989), Gupta (2007), and Jura-Morawiec and Tulik (2016)
	Synonym(s): *Asparagus draco* L., *Dracaena boerhaavei* Ten., *Stoerkia cranziana* Berens, *Palma draco* (L.) Mill. **Common name(s)**: Dragon tree. **Botanical description**: It is native to N. Africa – Morocco to the Canaries, Madeira, and Cape Verde Islands. It is a very slow-growing, evergreen tree that can eventually reach a height of around 10 m. It grows without branching in the early years of its life until it first produces a terminal inflorescence. It produces a domed, spreading head of regular branches. The tree is often harvested from the wild and also sometimes cultivated, for its resin, called dragon's blood. **Medicinal use(s)**: It is used to treat stomach cancer, colon cancer, lung cancer, and microtubular damages in these cell lines. It has several therapeutic uses: hemostatic, anti-diarrheic, anti-ulcer, antimicrobial, antiviral, wound healing, antitumor, anti-inflammatory, antioxidant, antitumor, anti-rheumatic, antifungal, antileukemic, anti-allergic, and antimutagenic. The resin, the so-called dragon's blood, is used as a folk medicine and fumigating agent. It is also used to treat diabetes, sore throat, piles, several types of skin disorder, and several types of respiratory problems.			Dracorubin	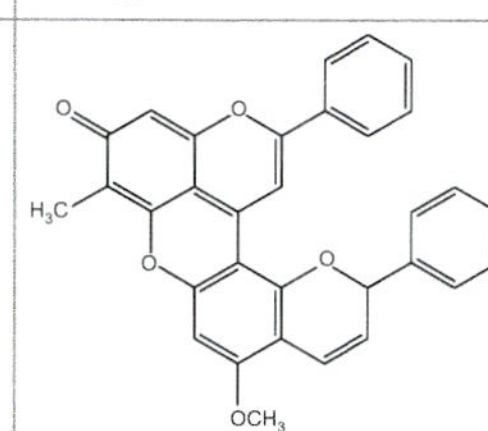	

(continued)

Table 4.4 (continued)

Family	Botanical name	Image	Plant part used	Bioactive compound(s)	Structure	Reference(s)
				Dracorhodin	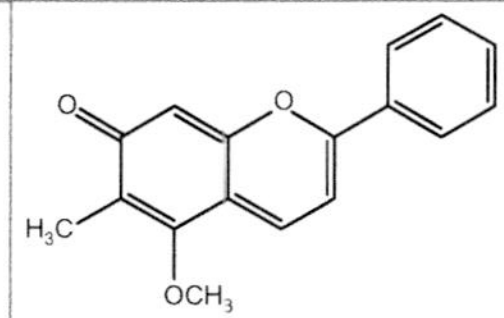	
				Pyrogallol (draco-alban)	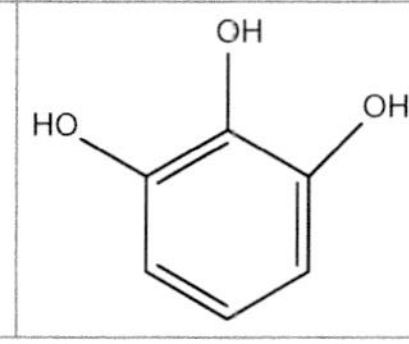	

Table 4.5 Anticancer plants of family Actinidiaceae

Family	Botanical name	Image	Plant part used	Bioactive compound(s)	Structure	Reference(s)
Actinidiaceae	*Actinidia arguta* (Seib. and Zucc.) Planch. Ex Miq.		Entire plant	Steroidal sapinogens (neotogogenin)	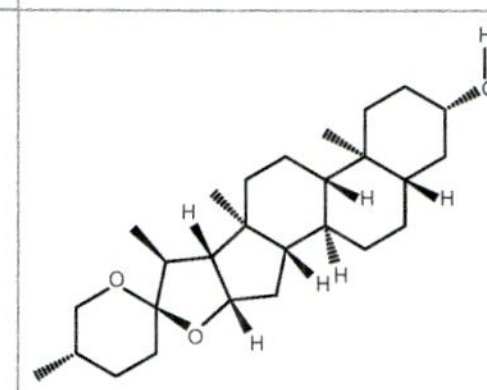	Han et al. (1984) and Zhang et al. (2007a, b)
	Synonym(s): *Actinidia cordifolia*, *Actinidia platyphylla*. **Common name(s)**: Hardy kiwi, Kokuwa, Tara vine. **Botanical description**: It is a perennial vine or climber native to Japan, Korea, Northern China, and Russia. The fruits produced by plants are small and known as hardy kiwifruits or grape kiwi. The fruits resemble like kiwifruits both in size and taste and are grape sized and green in color. The plants can survive at low temperature of −34 °C. **Medicinal use(s)**: It is used to treat stomach cancer. The fruit of *Actinidia arguta* is highly nutritious. It is rich in vitamin C and in other vitamins and minerals, especially folate and potassium. It also possesses antioxidant, antimicrobial, antiseptic, and anti-inflammatory properties.			Leutic acid	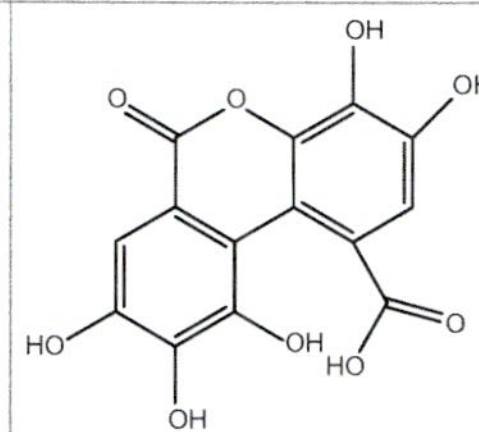	
				Phenylalanine	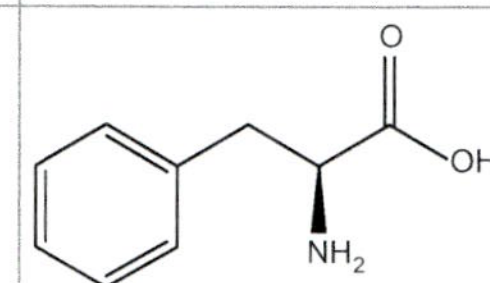	

(continued)

Table 4.5 (continued)

Family	Botanical name	Image	Plant part used	Bioactive compound(s)	Structure	Reference(s)
	Actinidia chinensis Planch.		Entire plant	Campesterol	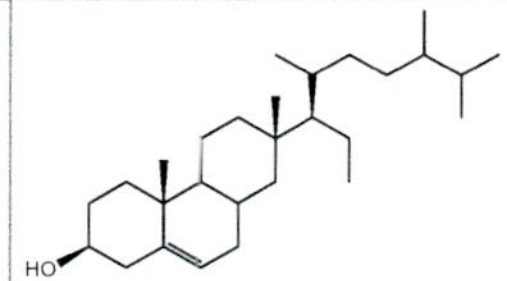	Duke and Ayensu (1985a, b), He et al. (2017), and Lv et al. (2018)
	Synonym(s): *Actinidia deliciosa*, *Actinidia arguta*, *Actinidia callosa*, *Actinidia coriacea*. **Common name(s)**: Kiwifruit. **Botanical description**: It is a deciduous climber growing up to 7.5 m (24 ft. 7 in.) tall. It is native to China where it is used as traditional medicinal plant. It usually grows in thick oak forests and bushlands. The fruit of the plant is edible and of walnut size. **Medicinal use(s)**: It is used to treat cancer. The fruits, leaves, and stems of the plant are used as diuretic, febrifuge, and sedative. They are used in treatment of stones in the urinary tract, rheumatoid arthralgia, and cancers of the liver and esophagus.			Fenugreek (Tannin)	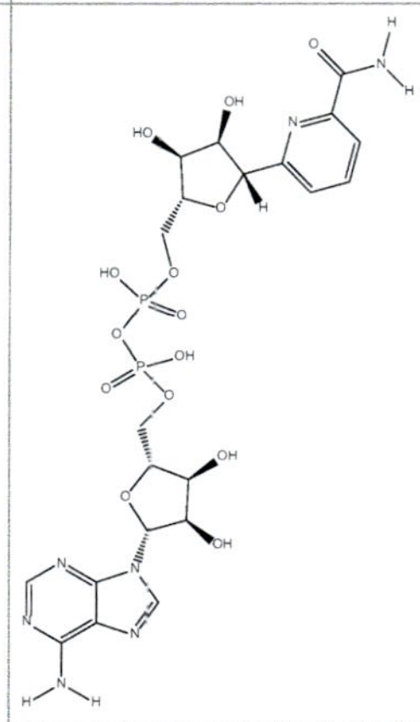	

	Actinidia polygama		Whole plant	Ascorbic acid	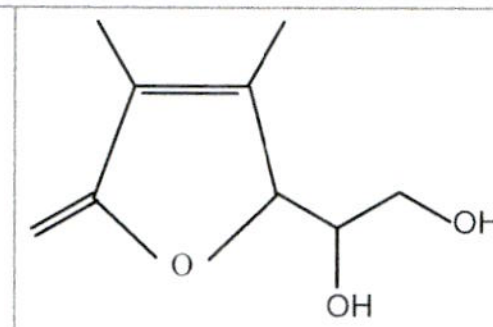	Liao et al. (2012b)
	Synonym(s): *A. volubilis*. (Sieb.&Zucc.) Planch. *Trochostigma polygama* Sieb.& Zucc. **Common name(s)**: Silver vine, Cat powder. **Botanical description**: It grows in the mountainous areas of Japan and China at elevations between 500 and 1900 m. It can reach up to 5–6 m high at maturity. It is a deciduous climber and is frost tender. The petiole leaves are silver and white in color and 6–13 cm long and 4–9 cm wide. The flowering season lasts from late June to early July, in which the plant bears white flowers about 2.5 cm in diameter. The longevity of an individual flower is 2–3 days, when the plant also starts to develop small, yellow to yellow-red, egg-shaped, fleshy, and multi-seeded fruits, which mature from September to October. The fruit is about 1.5 cm wide and 3.0–4.0 cm long. **Medicinal use(s)**: It is used to treat cancer and has anti-inflammatory, antimicrobial, and antitumor properties. It has been used for a wide range of health problems, including rheumatism, cystitis, arthritic pain, hypertension, kidney diseases, stroke, and cardiac ailments, and as heart tonic and circulatory stimulant and to reduce cholesterol and protect the liver. It is also used for treating diuresis, alleviation of pain, hypertension, genital troubles, and bronchitis.			Betulinic acid	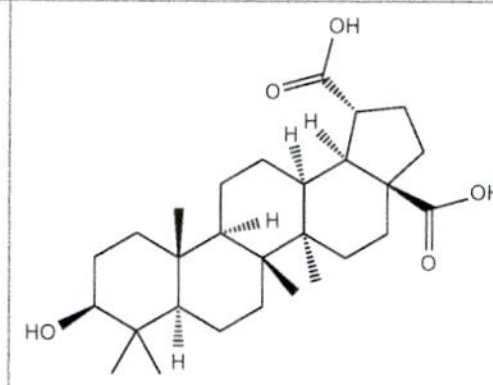	
				Zeaxanthin (flavonoids)	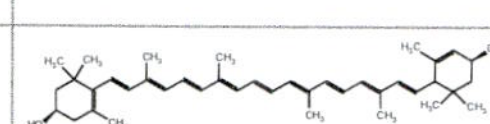	

Table 4.6 Anticancer plants of family Amaranthaceae

Family	Botanical name	Image	Plant part used	Bioactive compound(s)	Structure	Reference(s)
Amaranthaceae	*Achyranthes aspera* var. *australis*		Leaf	Quinic acid	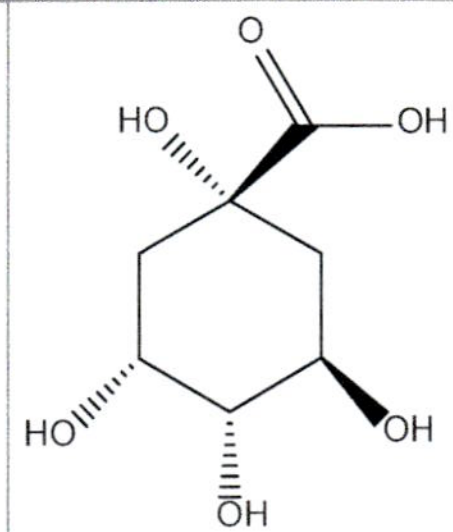	Arora and Tandon (2014), Subbarayan et al. (2010), and Narayan and Kumar (2013)
	Synonym(s): *Achyranthes acuminata* E.Mey. ex Cooke & Wright., *Achyranthes aspera* var. *australis* (R.Br.) Domin., *Achyranthes aspera* var. *obtusifolia* Suess. **Common name(s)**: Chaff flower, Prickly chaff flower, Devil's horsewhip, *Apamarga*. **Botanical description**: It is native to Australia. Erect or procumbent, annual or perennial herb of about 1–2 m in height, often with a woody base. Stems are angular, ribbed, simple, or branched from the base, often with tinged purple color, branches terete or absolutely quadrangular, striate, pubescent; leaves are thick, ovate – elliptic or obovate – rounded, finely and softly pubescent on both sides, entire, petiolate; petiole 6–20 mm long, flowers greenish white, numerous in axillary or terminal spikes up to 75 cm long; seeds subcylindric, truncate at the apex, rounded at the base, and reddish brown in color.			Chrysin	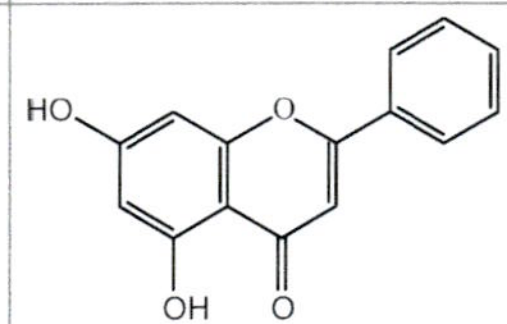	

	Medicinal use(s): It is used to treat breast, pancreatic, and cervical cancer. The plant is used in the cure of asthma, pneumonia, night blindness, and cough. It has pungent, anti-phlegmatic, antioxidant, anticancer, antioxidant, antiproliferative, antiperiodic, diuretic, purgative and laxative properties and is useful in edema, dropsy and piles, boils and eruptions of the skin, etc. Infusion of the root is a mild astringent in bowel complaints and is used as external application for bites of poisonous snakes and reptiles and cutaneous diseases and to cure various types of gastric disorders. Leaves and seeds are emetic, carminative, digestive; resolve swelling and aids in expelling phlegm. Ash of the plant is applied externally for ulcers and warts. The crushed leaves are rubbed on aching back. The paste of the roots in water is used in ophthalmia and opacities of the cornea. Paste of fresh leaves is used for allaying pain from bite of wasps. The plant is useful in liver complaints, rheumatism, scabies, and other skin diseases. It also possesses tranquillizing properties.	Kaempferol	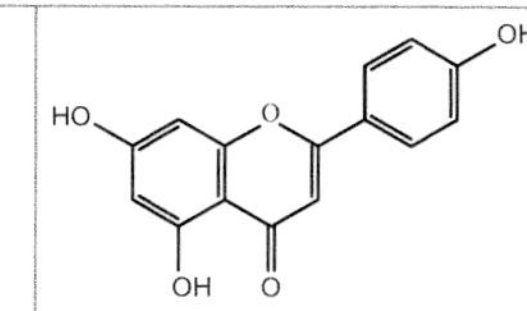	
		Quercetin	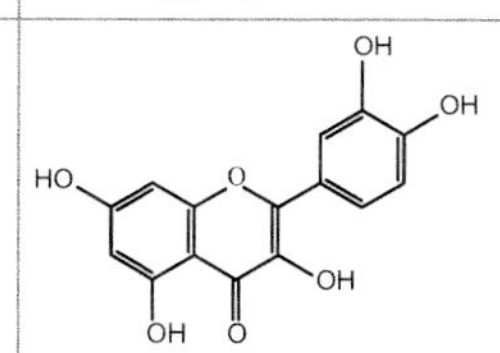	
		Chlorogenic acid	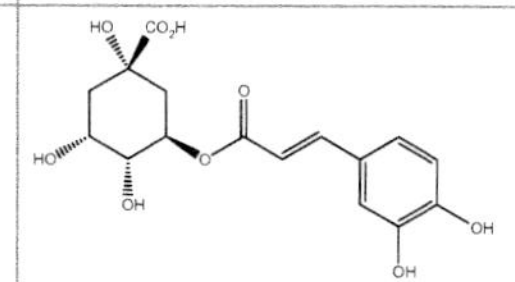	

(continued)

Table 4.6 (continued)

Family	Botanical name	Image	Plant part used	Bioactive compound(s)	Structure	Reference(s)
				Achyranthine	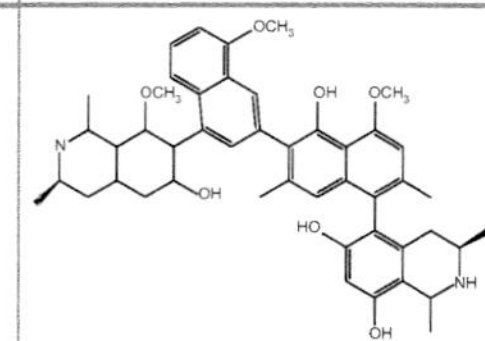	
				Tritriacontane		
				Hexatriacontane	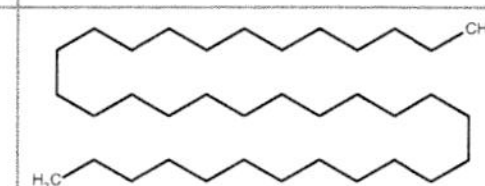	
				Betaine	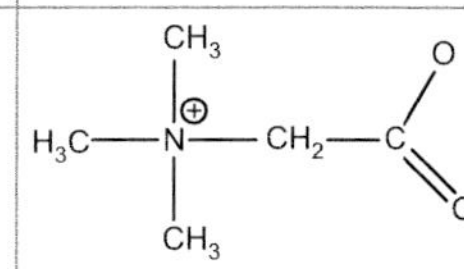	
				Pentatriacontane		
				6-pentatriacontanone	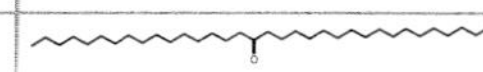	

	Achyranthes bidentata Blume		Entire plant	Kaempferol	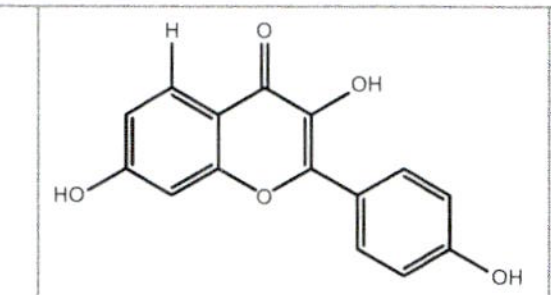	Zhang et al. (2012a, b) and Jiang et al. (2017)
	Synonym(s): *Achyranthes longifolia* Makino. **Common name(s)**: Ox knee. **Botanical description:** Herbs perennial, 70–120 cm tall. Stem green or tinged purple, angulate or quadrangular, appressed or spreading pubescent, or nearly glabrous; branches opposite. Petioles 0.5–3 cm long and hairy; leaf blade is elliptic or elliptic-lanceolate, rarely oblanceolate, 4.5–12 × 2–7.5 cm, both surfaces annexed or spreading pubescent, base cuneate or broadly cuneate, caudate. Spikes terminal or axillary, 3–5 cm; rachis 1–2 cm, white hairy. Flowers dense, ca. 5 mm. Bracts reflexed after anthesis, broadly ovate, 2–3 mm, apex acuminate; bracteoles 2.5–3 mm, spiny, base 2-parted, apex curved. Tepals shiny, lanceolate, 3–5 mm, with a midvein, apex acute. Stamens 2–2.5 mm; pseudo staminodes slightly serrulate, apex rounded. Utricles yellowish brown, shiny, oblong, 2.0–2.5 mm, smooth. Seeds are light brown and oblong. **Medicinal use(s)**: It is used to treat human colon cancer cells. It is a traditional Chinese medicine and inhibits collagen-induced arthritis and acute edema; inhibits excess recruitment of neutrophils to injured tissues.			Beta-carotene	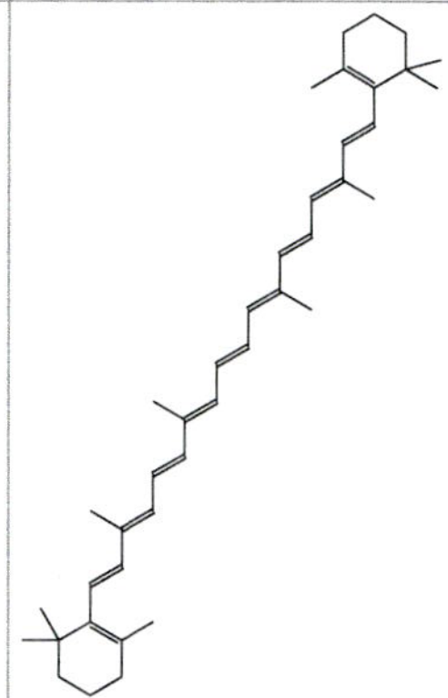	

(continued)

Table 4.6 (continued)

Family	Botanical name	Image	Plant part used	Bioactive compound(s)	Structure	Reference(s)
				(25S)-inokosterone	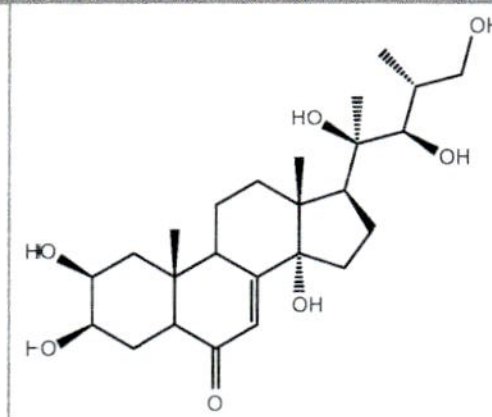	
				20-hydroxyecdysone	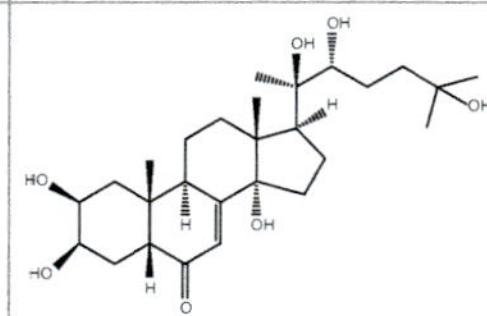	
	Achyranthes japonica (Mic) Nikkai		Root	Tyramine	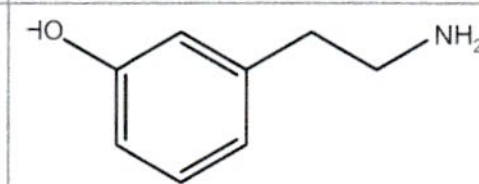	Lee et al. (2016b) and Shim et al. (2016)

	Synonym(s): *Achyranthes japonica* var. *hachijoensis* Honda., *Achyranthes japonica* var. *katsuudakemontana* Tawada. **Common name(s)**: Japanese chaff flower. **Botanical description**: A perennial plant growing up to 50 cm (20 in.)–100 cm (39 in.) tall with thickened roots. Stems are glabrous or slightly pubescent and shape quadrangular and branched. Its nodes are dilated. The leaves opposite and shape elliptic or oval and slightly pubescent and are petiolate. The leaves are 10 cm (3.9 in.)–20 cm (7.9 in.) long and 4 cm (1.6 in.)–10 cm (3.9 in.) wide. Flowers bloom in August to September and inflorescence spikes in axils and at terminals of the stem. From there, tiny green flowers bloom. There are five stamens and a pistil in each flower and unites into one at the bottom. **Medicinal use(s)**: It is used to treat human breast cancer cells. The root of the plant is used in Korea to treat edema, rheumatism, and delayed menses and as a contraceptive and an abortifacient. The root has been shown to have anticancer, analgesic, anti-allergic, anti-inflammatory, anti-spasmodic, diuretic, hypotensive, and uterine stimulant properties. In addition, it contains protocatechuic acid, which has antioxidant properties, and also inhibits the aggregation of platelets.	Diethylene glycol	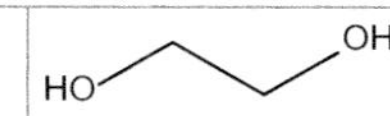	
		Miconazole	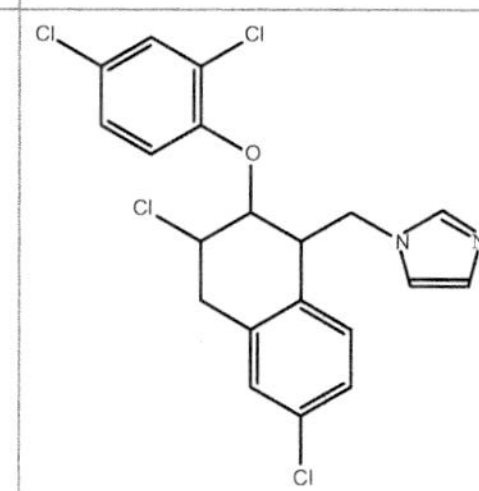	

(continued)

Table 4.6 (continued)

Family	Botanical name	Image	Plant part used	Bioactive compound(s)	Structure	Reference(s)
				Beta-D-glucopyranose	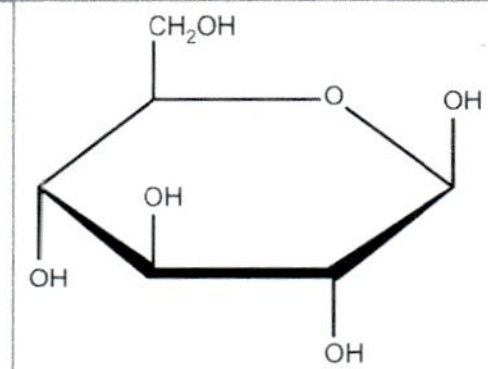	
	Achyranthes longifolia Makino		Entire plant	Trigofoenoside G (saponin)	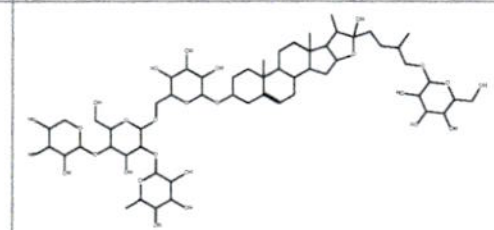	Duke and Ayensu (1985a, b) and Jiang et al. (2017)
	Synonym(s): *Achyranthes bidentata* Blume. **Common name(s)**: Huainiuxi (China). **Botanical description**: *It is found in tropical Africa, Indian subcontinent, Indochina, and Malaysia.* It is a perennial herb, growing along the hillside in Taiwan, Yunnan, Japan, China, Thailand, and Vietnam. The stem of plant is green or tinged purple with leaves having hairy petiole and leaf blade lanceolate type. Flowers are dense. **Medicinal properties**: Used to treat nipple cancer**.** It is used for treatment of arthritis. Besides this, it has also anti-rheumatic, antibacterial, and anti-spasmodic properties.					

	Deeringia amaranthoides (Lam.) Merr.		Fruits	Gallic acid	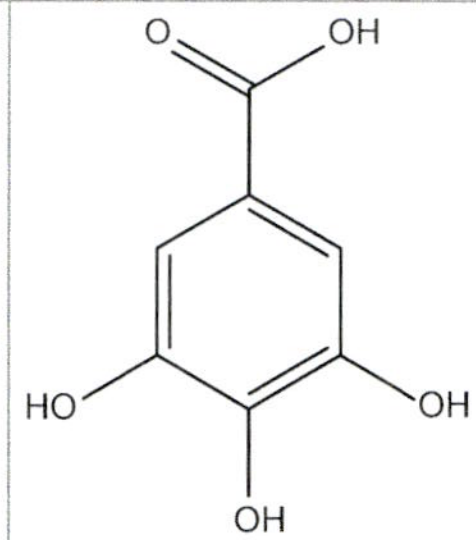	Sati et al. (1990)
	Synonym(s): *Achyranthes amaranthoides* Lam., *Deeringia baccata* (Retz.) Moq., *Deeringia celosioides* R. Br., *Celosia baccata* Retz. **Common name(s)**: Ditiran, Aribugbug (Ilk), Kalalori, Latman, Wali (India) and Bayam besar, Pancaluhur (Indonesia). **Botanical description**: *Found in Western Australia*. Ditiran is a slightly hairy or nearly smooth vine, reaching a length of 5–6 m, with drooping branches. Leaves are entire, ovate to oblong-ovate, 7–14 cm long, with pointed tips and subtruncate bases. Flowers are numerous, shortly pedicelled, greenish white, with reflexed segments, and about 1.5 mm long. Stamens are white and erect. Fruit is ovoid, fleshy and red about 4 mm long. **Medicinal properties**: It is used to treat breast and prostate cancer. In the Philippines the leaves of the very plant are used for dysentery. Leaves are applied to sores. In the Northwest Himalayas, roots are used as sternutatory; least as antiseptic, applied on boils and sores. In India, fruits are used as anticancer; roots are used to treat jaundice.			Saponin(s)	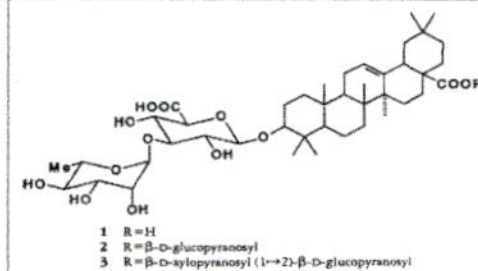	

(continued)

Table 4.6 (continued)

Family	Botanical name	Image	Plant part used	Bioactive compound(s)	Structure	Reference(s)
	Hebanthe eriantha (Poir.) Pedersen		Root	Pfaffic acid	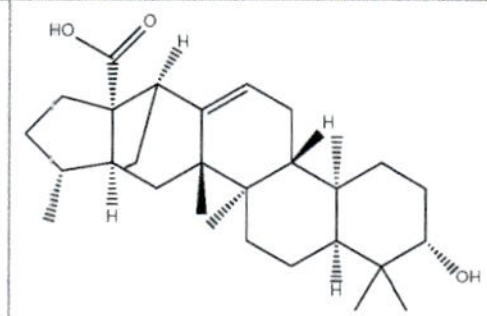	Khan (2005)
	Synonym(s): *Gomphrena eriantha* (Poir.) Moq.; *Gomphrena paniculata* (Mart.) Moq.; *Hebanthe paniculata* Mart.; *Iresine erianthos* Poir.; *Iresine paniculata* (Mart.) Spreng.; *Iresine tenuis* Suess.; *Pfaffia eriantha* (Poir.) Kuntze; *Pfaffia paniculata* (Mart.) Kuntze; *Xeraea paniculata* (Mart.) Kuntze. **Common name(s)**: Para toda, Suma. **Botanical description**: Large, rambling, shrubby ground vine with an intricate, deep and extensive root system. It is indigenous to the Amazon basin and other tropical parts of (southern) Brazil, Ecuador, Panama, Paraguay, Peru, and Venezuela. **Medicinal use(s)**: It is used for the treatment of breast cancer. Suma root is employed as a cellular oxygenator and consumed to stimulate appetite and circulation, increase estrogen production, balance blood sugar levels, enhance the immune system, strengthen the muscular system, and enhance memory. Also, in North American herbal medicine, suma root is used as an adaptogenic and regenerative tonic regulating many systems of the body; as an immunostimulant to treat exhaustion and chronic fatigue, impotence, arthritis, anemia, diabetes, cancer, tumors, mononucleosis, high blood pressure, PMS menopause, hormonal disorders, and many types of stresses. In Ecuador, suma is considered a tonic and "normalizer" for the cardiovascular system, the central nervous system, the reproductive system, and the digestive system; it is used to treat hormonal disorders, sexual dysfunction and sterility, arteriosclerosis, diabetes, circulatory and digestive disorders, rheumatism, and bronchitis.			Nortriterpenes	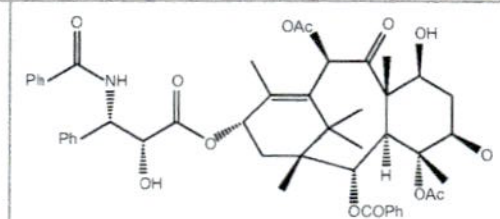	

				Beta-ecdysterone	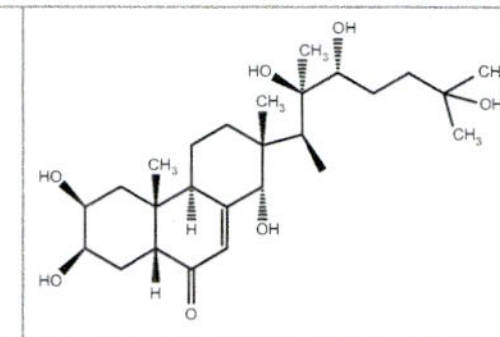	
				Pantothenic acid	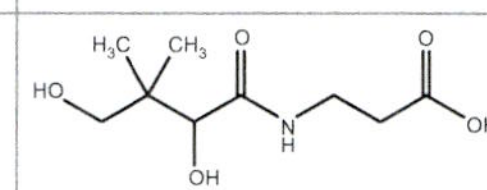	
	Pfaffia paniculata (Mart.) Kuntze		Root	Yuccagenin (steroidal saponin)	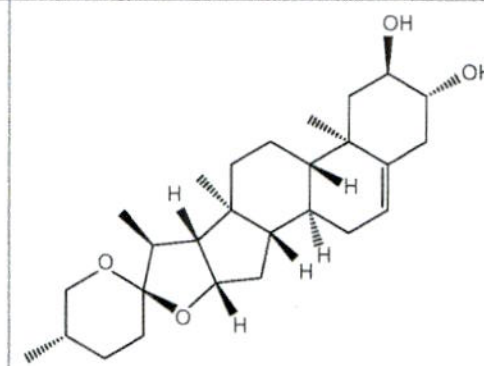	Takemoto et al. (1983) and Nagamine et al. (2009)
	Synonym(s): *Rinorea paniculata*. **Common name(s)**: Suma. **Botanical description**: It is a native plant to South America. It is a large shrubby, ground vine native to Amazon basin and other tropical parts of South America. **Medicinal properties**: It is used to treat human breast cell line, MCF-7 cells, and leukemia. It has been used historically for various conditions like treatment of menstrual disorders, body building agent, and sexual enhancement agent and as a general tonic.					

Table 4.7 Anticancer plants of family Amaryllidaceae

Amaryllidaceae	*Lycoris radiata* (L'Her.) Herbert		Entire plant	Lycoramine	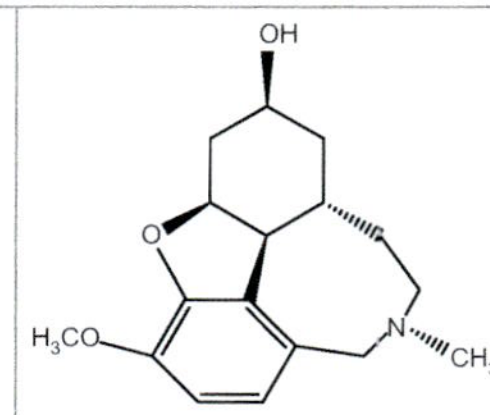	Huang et al. (1983) Feng et al. (2011), Kang et al. (2012), and Chen et al. (2016a, b, c)
	Synonym(s): *Amaryllis radiata* L'Hér, *Lycoris terracianii* Dammann, *Nerine japonica* Miq., *Nerine radiata* (L'Hér.) Sweet, *Orexis radiata* (L'Hér.) Salisb. **Common name(s)**: Red spider lily, Red magic lily, Hurricane lily. **Botanical description**: It is a bulbous perennial and a native of Asia. Originally from China, Korea, and Nepal, it was introduced in Japan and from there to the United States and elsewhere. It is considered naturalized in the Seychelles and in the Ryukyu Islands. The plant flowers late in summer or autumn often in response to heavy rainfall. It normally flowers before the leaves fully appear, on stems 30–70 cm (12–28 in.) tall. The leaves are parallel-sided, 0.5–1.0 cm (0.20–0.39 in.) wide with a paler central stripe. The red flowers are arranged in umbels. Individual flowers are irregular, with narrow segments which curve backwards and long projecting stamens. **Medicinal use(s)**: It is used to treat ovarian cancer. The root is used in the treatment of swellings, ulcers, and nervous afflictions of children. The bulb is emetic and expectorant; a decoction is used to counteract poison and is also applied to ulcers and swellings. The bulb can be made into a plaster and is then applied to burns and scalds.			Lycobetaine	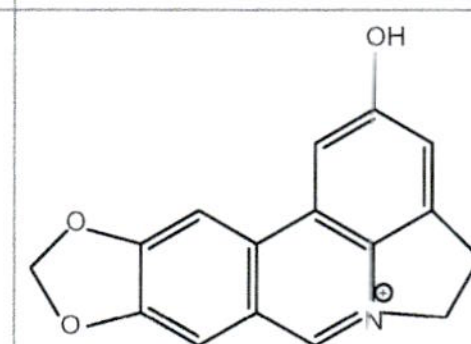	

		Galantamine	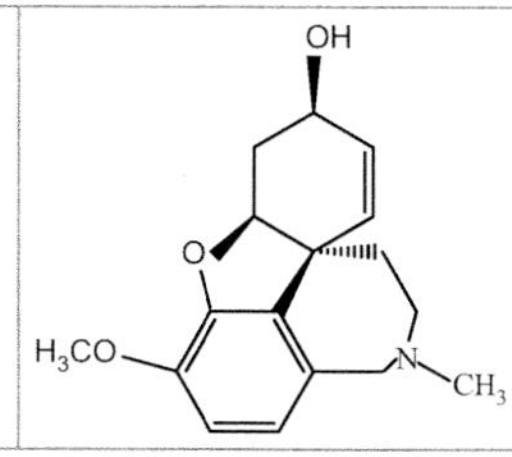	

Table 4.8 Anticancer plants of family Anacardiaceae

Family	Botanical name	Image	Plant part used	Bioactive compound(s)	Structure	Reference(s)
Anacardiaceae	*Anacardium giganteum* Hancock ex Engl.		Bark, fruit	Arachidic acid	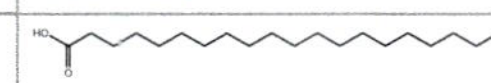	Wijisekera (1978) and Shepherd (2012)
	Synonym(s): *American ginseng, Aralia quinquefolia, Cassuvium Lam.* **Common name(s)**: Caja acu. **Botanical description**: It is native of South America – Northern Brazil and the Guianas. It is found in forests of the mid- and lower Amazon regions, growing both on the flood plains and in areas not subject to inundation. It is a large evergreen tree with a large, spreading crown; it can grow 20–40 m tall. The straight, cylindrical bole can be 50–90 cm in diameter. The edible fruits and seeds are gathered from the wild for local use, and the plant is also sometimes cultivated as a food crop. **Medicinal use(s)**: It is used to treat cancer. This plant has been found to possess strong anti-inflammatory, anti-atherogenic, antimicrobial, antioxidant, anti-reproductive, and anticarcinogenic properties. The bark and fruit are used in the treatment of diarrhea (the bark is scraped and boiled and the liquid drunk as an antidiarrheal), cancerous ulcers, and syphilis. A decoction of the bark is used to treat venereal diseases, iron deficiency, vomiting, and sore throat.			Dihydroflavonol	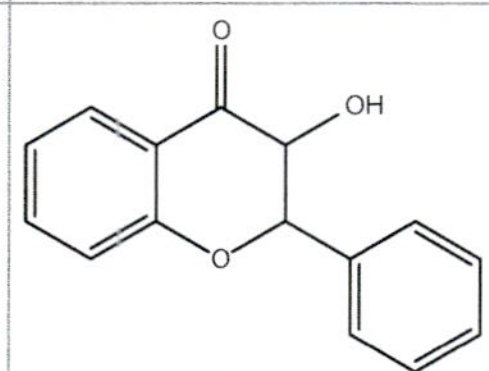	
				m-Phenylenediamine	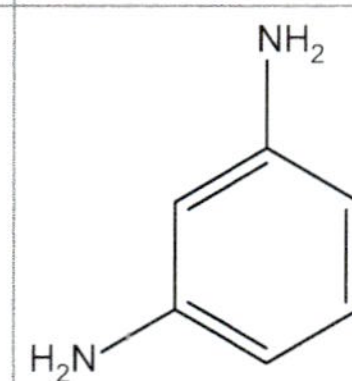	

	Lannea coromandelica (Houtt.) Merr.		Bark and leaf	Eucalyptol	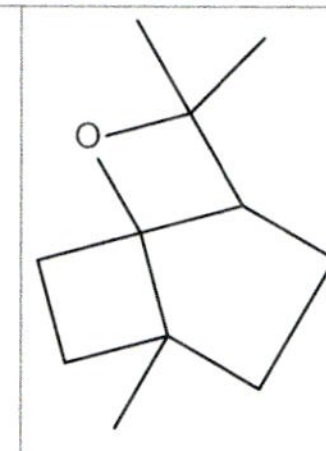	Jain et al. (2013), Akter et al. (2014), and Alam et al. (2017a, b)
	Synonym(s): *Calesium grande* Kuntze, *Dialium coromandelicum* Houtt , *Lannea grandis* Engl. **Common name(s)**: Indian ash tree, Moya, Wodier. **Botanical description**: It is native to East Asia – Southern China, Indian subcontinent, Myanmar, Thailand, Cambodia, Laos, and Vietnam. It is a deciduous tree usually growing 5–10 m tall but with some specimens up to 20 m tall with a bore of 45 cm in diameter. It grows in warm, humid climate. The flowers are minute, yellowish, tinged red, and sessile in long panicles at the ends of leafless branches, and fruits are drupes smooth, glabrous type. **Medicinal use(s)**: It is used to treat cervical cancer. It also used in the treatment of tumor, ulcers, sprains, bruises, skin diseases, and dysentery. Both the leaves and the stem bark have painkilling and astringent property and thus are mixed with toothpowder for use in toothache. The leaves are chopped as fodder and fed to the livestock during inflammatory and digestive problems.			Alpha- Pinene	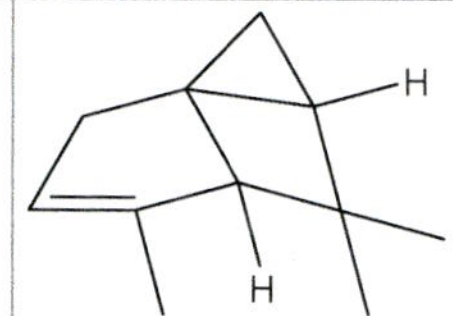	
				Beta-pinene	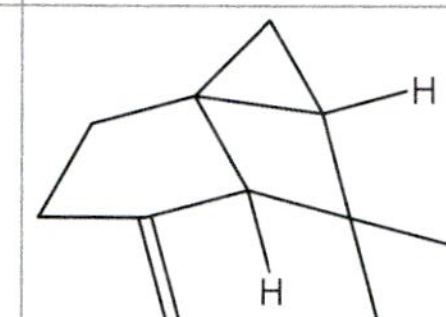	

(continued)

Table 4.8 (continued)

Family	Botanical name	Image	Plant part used	Bioactive compound(s)	Structure	Reference(s)
				Quercitrin	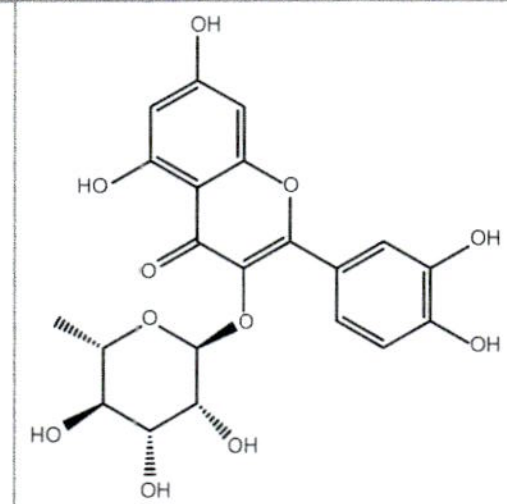	
				(−)-Carvone	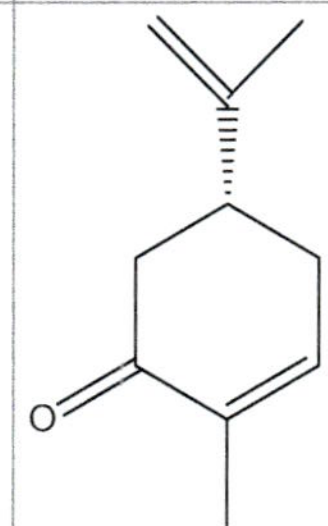	
	Lannea stuhlmannii (Engl.) Engl.		Root, stem, bark	Vincristine	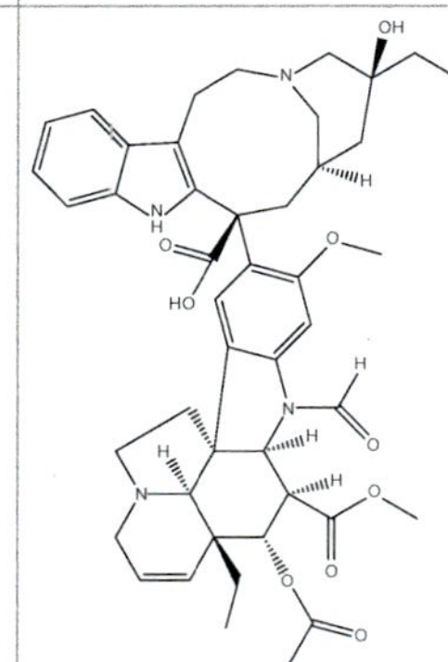	Chhabra et al. (1987, 1991) and Graham et al. (2000)

	Synonym(s): *Lannea stuhlmannii* (Engl.) Engl. **Common name(s)**: *Lannea kirkii, Lannea ambigua.* **Botanical description**: It is native to Eastern Africa – Sudan, Ethiopia, Somalia, Uganda, Kenya, Rwanda, and Tanzania. It is a deciduous shrub or more frequently an irregularly branched tree with a spreading rather open crown and drooping branches; it can grow from 3 to 22 m tall. It is found in the habitat in river valleys, forests, woodlands of several types (including coastal), savannahs, and termite mounds. **Medicinal use(s)**: It is used to treat human bladder carcinoma, human colon adenocarcinoma, human skin carcinoma, leukemia, lymphoma, and breast, lung, and pediatric solid cancers. The bark is used for making a tea that is used as a blood tonic for treating anemia. A decoction of the bark is used for treating diarrhea, stomachache, and headache.	Paclitaxel	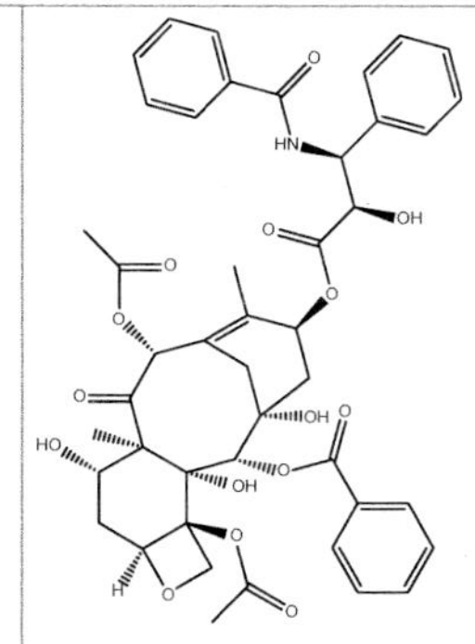	
		Camptothecin	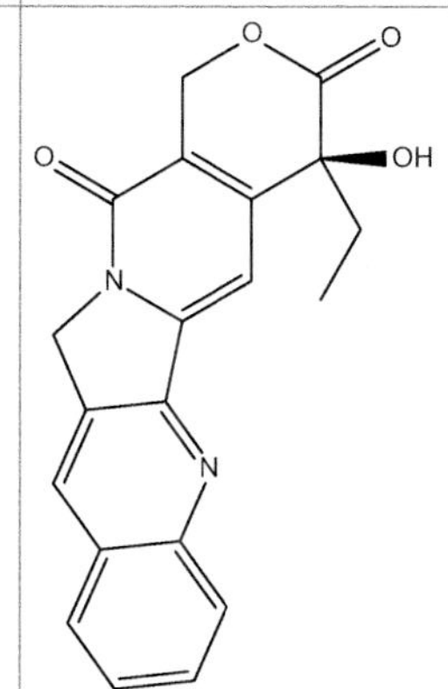	

(continued)

Table 4.8 (continued)

Family	Botanical name	Image	Plant part used	Bioactive compound(s)	Structure	Reference(s)
				Topotecan	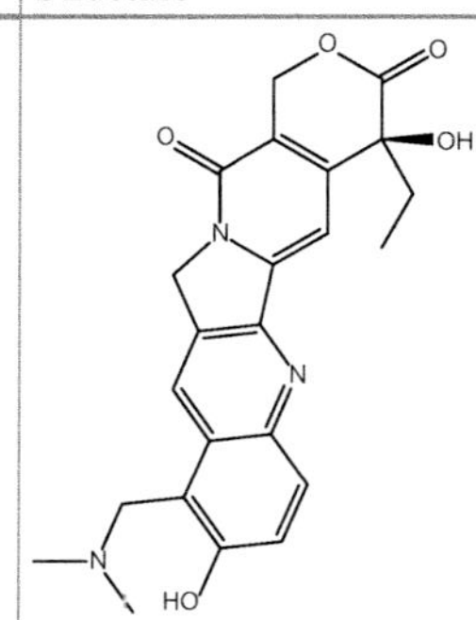	
				Irinotecan	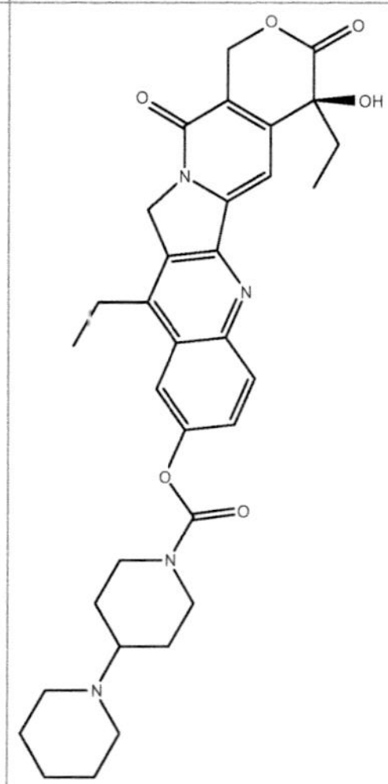	

	Metopium browneí (Jacq.) Urb.		Bark	Pyrocatechcl	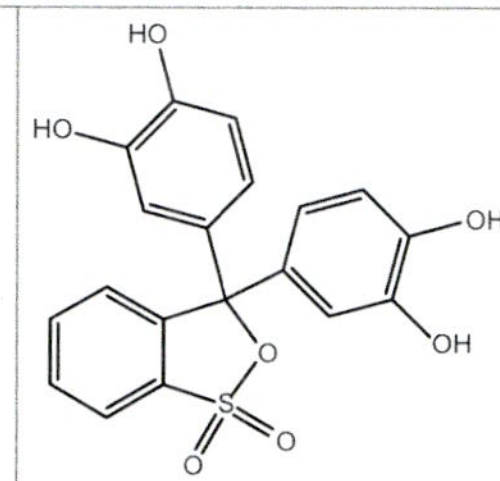	Flores and Ricalde (1996) and Cornelius (2015)
	Synonym(s): *Cotinus metopium* (L.) M.Gomez, *Rhus metopium*, *Terebinthus brownei* jacq. **Common name(s)**: Black poisonwood, Chechem. **Botanical description**: This plant is found in Southern Mexico. A plant with highly irritating sap (obtained from its bark) which can cause dermatitis on one's skin when touched. It is found in the Dominican Republic, Jamaica, Cuba, Belize, and from Yucatan peninsula to Veracruz in the Gulf of Mexico. The plant can reach a height between 8 and 25 m. The wood of the plant is a valuable source of lumber in Central America and West Indies. **Medicinal use(s)**: It is used to treat skin cancer. It has antifungal properties. The plant possesses several medicinal properties like antiviral, cathartic, diaphoretic, anti-rheumatic, and sedative. It is also used to treat warts.			Humilinolide	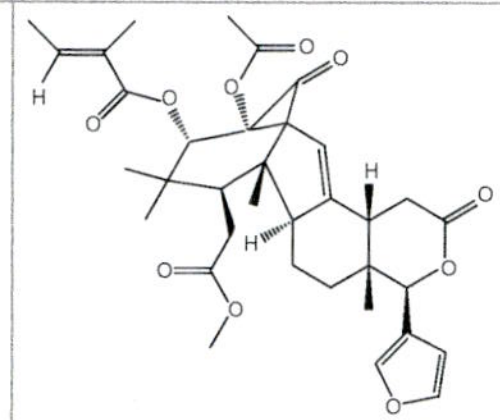	

(continued)

Table 4.8 (continued)

Family	Botanical name	Image	Plant part used	Bioactive compound(s)	Structure	Reference(s)
	Rhus chinensis Mill.		Leaf	Bruceoside B	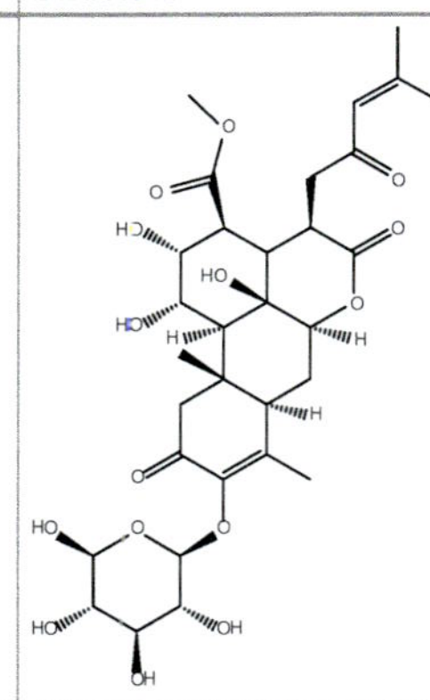	Deiab et al. (2015), Yim et al. (2016), Cai et al. (2003), and Djakpo and Yao (2010)
	Synonym(s): *Rhus amela* D.Don, *Rhus chinensis* var. *chinensis*, *Rhus javanica* var. *chinensis* (Mill.) T.Yamaz., *Rhus semialata* Murray, *Schinus indicus* Burm.f. **Common name(s)**: Chinese sumac, Nutgall tree, Chinese gall. **Botanical description**: It is native of East Asia – China, Japan, Korea, India, Bhutan, Malaysia, and Thailand. It is a deciduous tree growing upto 6 m (19 ft. 8 in.). It flowers in August, and the seeds ripen in October. The flowers are dioecious (individual flowers are either male or female, but only one sex is to be found on any one plant so both male and female plants must be grown if seed is required) and are pollinated by bees. The plant is not self-fertile. It is suitable for light (sandy), medium (loamy), and heavy (clay) soils and prefers well-drained soil. Suitable pH is acid, neutral, and basic (alkaline) soils. It cannot grow in the shade. It prefers moist soil. **Medicinal properties**: It is used to treat human colon cancer. A decoction of the root of the plant is drunk as a treatment for malaria. A decoction of root, combined with leaf sap, is used as laxative and Abortifacient.			Brucein C	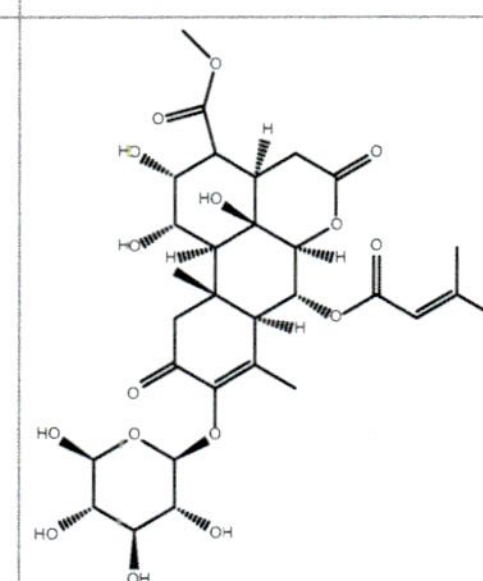	

				Brucein D	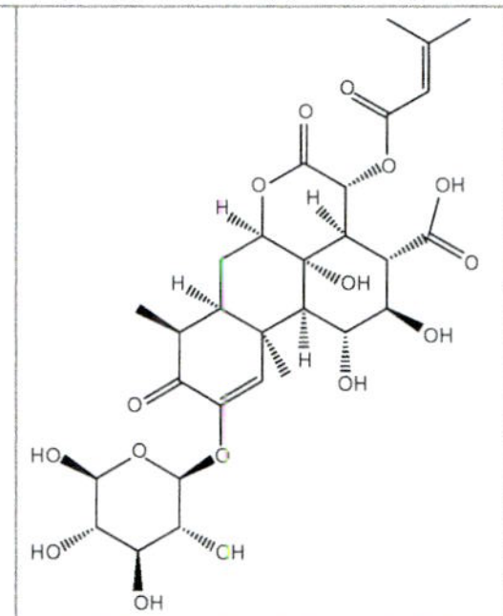	
	Rhus longipes Engl.		Root	Anacardic acid	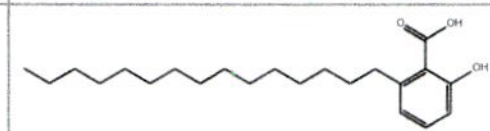	Chhabra et al. (1984) and Olorunnisola et al. (2017)

(continued)

Table 4.8 (continued)

Family	Botanical name	Image	Plant part used	Bioactive compound(s)	Structure	Reference(s)
	Synonym(s): *Searsia longipes* (Engl.) Moffett, *Toxicodendron longipes Kuntze.* **Common name(s)**: Large-leaved rhus. **Botanical description**: It is native of China, Japan, and Korea. It is a medium-sized, semi-deciduous trifoliate tree which attains a height of 10 m. It is found in Africa from Burkina Faso to South Africa as well as in islands of the Indian Ocean. In Zimbabwe, it is found at an altitude between 1000 and 1680 m above sea level. Flowers of this species are small, yellowish green, in axillary or terminal inflorescences. **Medicinal use(s)**: It is used to treat liver cancer. It possesses strong antiviral, antibacterial, anticancer, hepatoprotective, antidiarrheal, and antioxidant activities. It has potential chemotherapeutic properties against lymphoma. A decoction is used in the treatment of hemoptysis, inflammations, laryngitis, snakebite, stomachache, and traumatic fractures. The stem bark is astringent and anthelmintic. The fruit is used in the treatment of colic. The seed is used in the treatment of coughs, dysentery, fever, jaundice, malaria, and rheumatism. The root bark is used as a cholagogue.			Gallic acid	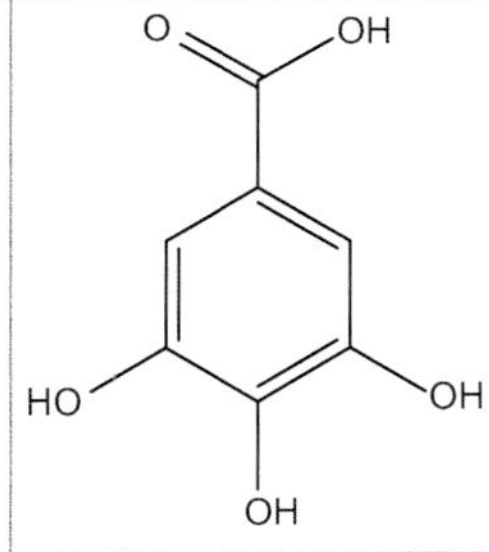	
				Alpha glucosidase	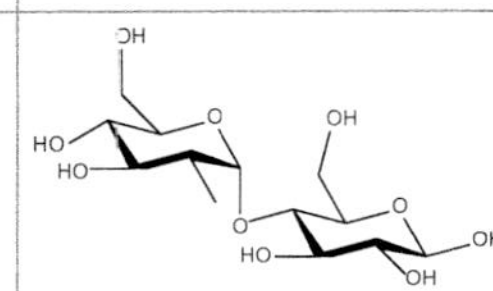	

	Rhus vernicifera DC.		Resin	Butein	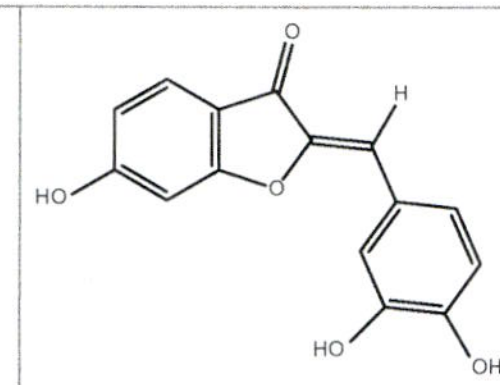	Cha (1977) and Choi et al. (2016)
	Synonym(s): *Toxicodendron verniciflluum* (Stokes) F.A. Barkley. **Common name(s)**: Lacquer tree. **Botanical description**: It is native to Japan, Korea, and Central China. It is a deciduous tree growing up to 15 m (49 ft.) by 10 m (32 ft.). It flowers in July, and the seeds ripen from September to November. The flowers are dioecious and are pollinated by bees. The plant is not self-fertile. It is suitable for light (sandy), medium (loamy), and heavy (clay) soils and prefers well-drained soil. The suitable pH is acid, neutral, and basic (alkaline) soils. It cannot grow in the shade and prefers moist soil. **Medicinal use(s)**: It has potential to use as a chemotherapeutic agent in human gastric cancer. It is also used for the treatment of cancers of the lungs, breast, kidneys, nose, and liver and ascites. The leaves are used in the treatment of wasting diseases and internal parasites. The seed is hemostatic and is used in the treatment of dysentery. A resin from the plant is emmenagogue, hemolytic, stimulant, tonic, and vermifuge. Some caution is advised in the use of the leaves and stems of this plant.			Sulfuretin	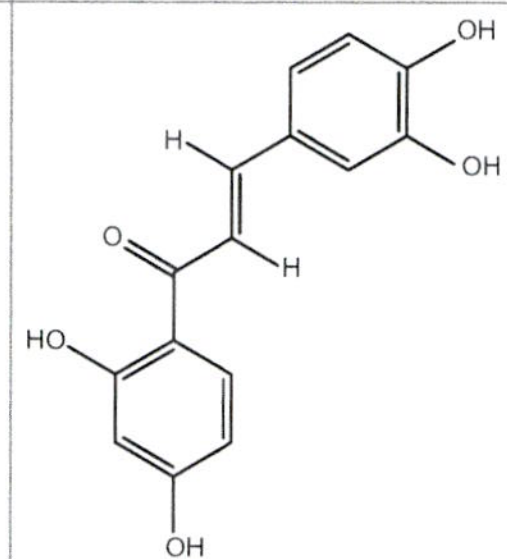	
				Quercetin	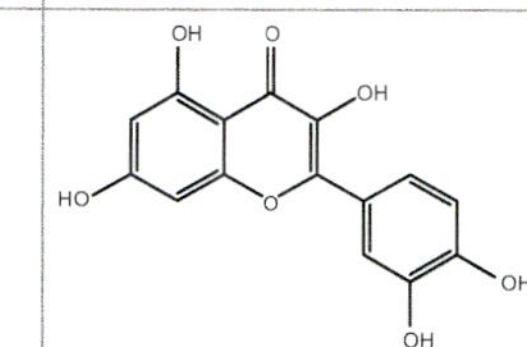	

(continued)

Table 4.8 (continued)

Family	Botanical name	Image	Plant part used	Bioactive compound(s)	Structure	Reference(s)
				Fustin	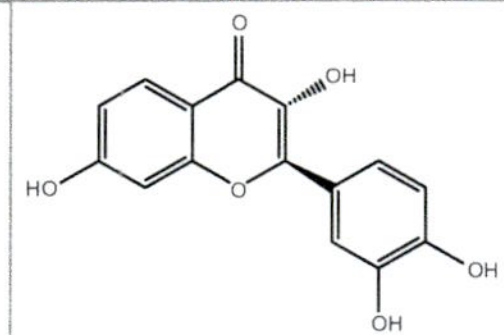	
	Schinus lentiscifolius Marchand		Leaf	α-Terpineol	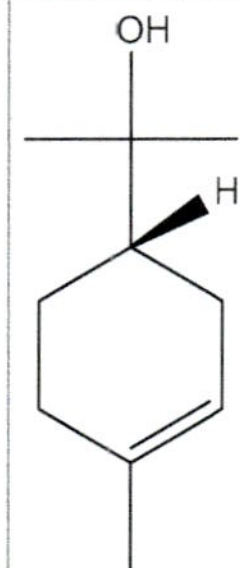	Graziano et al. (1970), Rossini et al. (1996), and Díaz et al. (2008)

	Synonym(s): *Schinus weinmanniifolia*, *Schinus chebataroffi* Herter. **Common name(s)**: Pepper tree. **Botanical description**: It is native to Southern and western South America – Argentina, Uruguay, Paraguay, Brazil, Bolivia, Peru, and Ecuador. It is a fast-growing, evergreen tree with a wide, multi-branched crown of weeping foliage, and it generally grows 6–8 m tall, though on good sites it can reach a height of up to 15 m. The short, cylindrical bole can be 25–35 cm in diameter. It is found in dry regions in the Andes; it is found as a bush in dry lands but reaches tree size in dry riverbeds with accessible underground water. **Medicinal use(s)**: It has antitumor, anti-inflammatory, antiviral, and antibacterial properties. The bark is astringent, diuretic, emmenagogue, tonic, and vulnerary. It is useful in the treatment of diarrhea, inflammations, and tumors.					
	Schinus lentiscifolius (Gill. ex Lindl.) Engler.		Bark	α-phellandrene	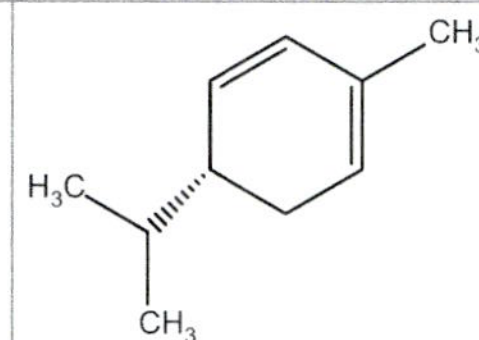	Bendaoud et al. (2010) and Solis-Quispe et al. (2015)

(continued)

Table 4.8 (continued)

Family	Botanical name	Image	Plant part used	Bioactive compound(s)	Structure	Reference(s)
	Synonym(s): *S. angustifolius, S. areira, S. huigan.* **Common name(s)**: Brazilian pepper tree, California pepper tree, Escobilla, Mulli, False pepper, Rose pepper. **Botanical description**: It is the native pant of Southern and Central America. It is also native to Brazil and called as Brazilian pepper tree. It can also be found in the semitropical and tropical regions of the United States and Africa. It is a shrubby tree with narrow, spiky leaves. It grows up to 4–10 m tall. It produces an abundance of small flowers formed. This tropical tree has high oil and essential oil content that produces a spicy, aromatic scent. The berries have a peppery flavor. **Medicinal properties**: It is used to treat human cancer cell lines. The dried bark and leaves are employed for heart problems (hypertension and irregular heartbeat), infection of all sorts, and menstrual disorder with excessive bleeding. It helps to cure tumors and general inflammation. A liquid prepared with the bark is used internally as a stimulant, tonic, and astringent and externally for rheumatism, gout, and syphilis.			β-Phellandrene	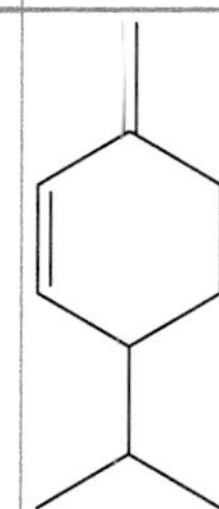	
				α-Pinene	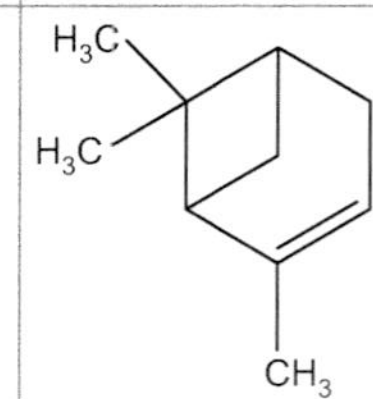	

	Toxicodendron verniciﬂuum (Stokes) F.A. Barkley.		Bark	Rhuspolyphenols G-I	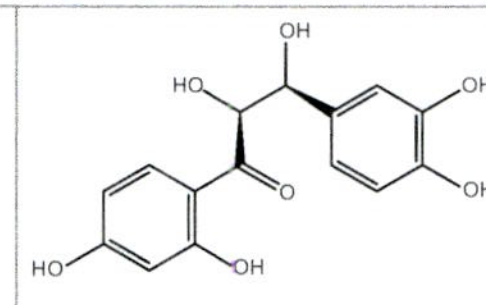	Kim et al. (2015a, b, c)
	Synonym(s): *Rhus verniciflua*. **Common name(s)**: Chinese lacquer tree, Urushi tea, Varnish tree. **Botanical description**: It is a native plant of eastern North America – Maine to Vermont and Ontario, south to Florida, Missouri, and Louisiana. This shrub is 10–25′ tall, consisting of a trunk up to 6″ across and a relatively open crown. Trunk bark is light gray and relatively smooth, except for scattered lenticels that resemble small bumps. Branch bark is also light gray and smoother, while twigs are orange, tan, or brown and usually glabrous (less often pubescent); twigs have numerous small lenticels and prominent terminal buds that are purplish red. Alternate compound leaves occur along the twigs; these leaves are 6–14″ long and odd-pinnate with 7–13 leaflets. **Medicinal use(s)**: It is used as a food supplement and in traditional herbal medicine to treat inflammatory diseases and human breast cancers. The plant has been used in the treatment of fevers, ague, ulcerated bladder, asthma, and wasting diseases. The plant has been used as a wash to treat foul ulcers.			Fisetin	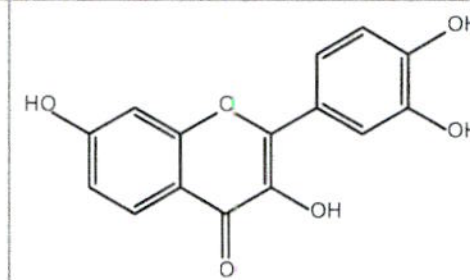	

(continued)

Table 4.8 (continued)

Family	Botanical name	Image	Plant part used	Bioactive compound(s)	Structure	Reference(s)
				Butein	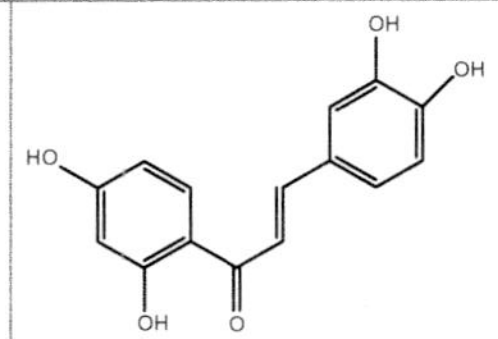	
				Epitaxifolin	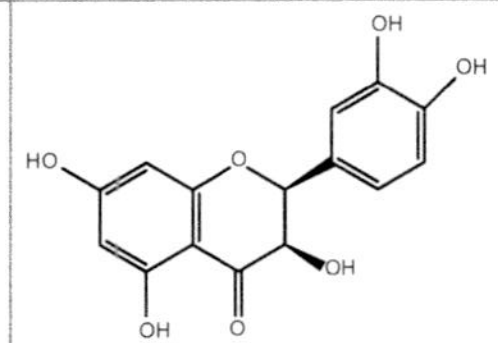	
				(2S,3R)-fustin	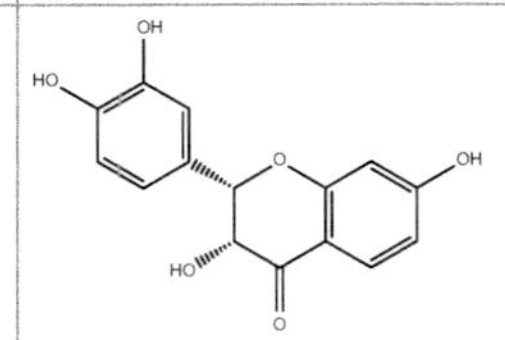	
				Sulfuretin	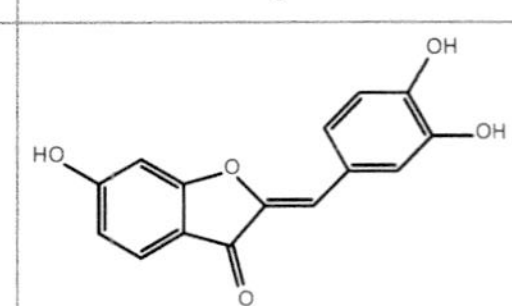	
				(2-benzyl-2,3′,4′,6-tetrahydroxybenzo[b]furan-3(2H)	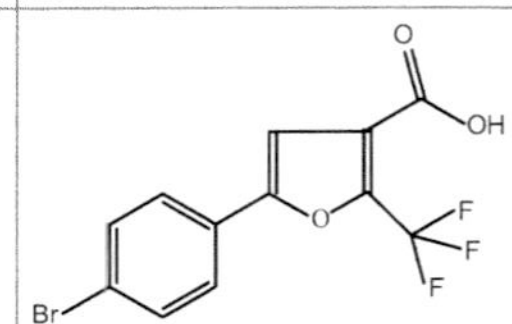	

	Mangifera pajang Kosterm		Methyl gallate	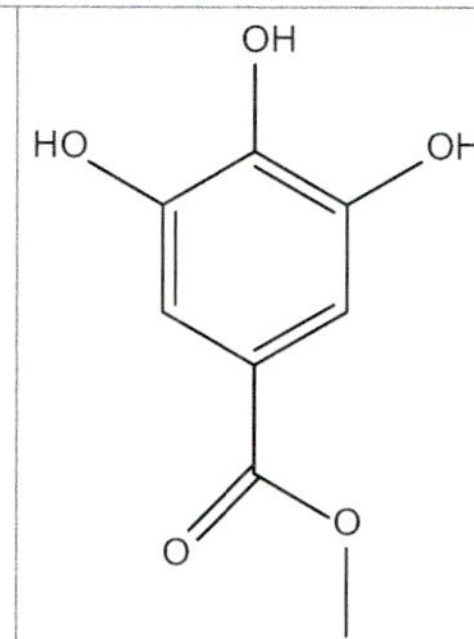	Ahmad et al. (2015)
	Synonym(s): *Phanrangia poilanei* Tardieu. **Common name(s)**: Bambangan, Mango, Mongoro. **Botanical description**: It is native plant of Southeast Asia – Malaysia, Brunei, and Indonesia. It is a deciduous tree and can grow 15–33 m tall. The unbuttressed bole can be 30–70 cm in diameter. It is found in the habitat of primary lowland dipterocarp forest. The tree is rarely found in the wild but is more commonly cultivated. **Medicinal use(s)**: It is used to treat human breast cancer, human cervical cancer (HeLa), and human colon cancer cells. It is used for treating several diseases, including cough, fever, asthma, diarrhea, indigestion, and skin diseases. The extracts from different parts of the plant including fruit have antifungal, antimicrobial, anti-atherosclerotic, anti-hypercholesterolemic, antileukemic, anti-clastogenic, and antiproliferative properties.				

(continued)

Table 4.8 (continued)

Family	Botanical name	Image	Plant part used	Bioactive compound(s)	Structure	Reference(s)
	Semecarpus anacardium		Fruit (nut) India bhilwa	3-(8′(Z),11′(Z)-pentadecadienyl) catechol (SA-3C)	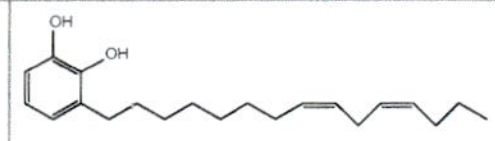	Nair et al. (2009), Ahmad et al. (2015), and Tangah et al. (2017)
	Synonym(s): *Anacardium orientale*. **Common name(s)**: Dhobi nut tree, Indian nut tree, Geeru veej, Marking nut. **Botanical description**: It is native to India. It is a moderate-sized deciduous tree with large stiff leaves. Leaves are 7–24 in. long, 2–12 in. wide, oblong-obovate, rounded at the tip. Leaf base is rounded, heart-shaped, or narrowed into the stalk, leathery in texture. The flowers are small, borne in panicles shorter than the leaves. Fruit is a drupe 1 in. long, ovoid or oblong, smooth and shining, black when ripe, seated on a fleshy cup. **Medicinal use(s)**: It shows inhibitory action on human breast cancer cells. It is an effective and promising chemotherapy agent. It also has strong antioxidant, anticarcinogenic, and antimicrobial activities. It is best used to cure various skin-related disorders such as skin rash, itching, and swelling. It is very good for relieving constipation, digestive disorders, abdominal distention, ascites, piles, and various types of worm infestations.					

Table 4.9 Anticancer plants of family Annonaceae

Family	Botanical name	Image	Plant part used	Bioactive compound(s)	Structure	Reference(s)
Annonaceae	*Annona cherimola* Mill.		Seed	Annocherine	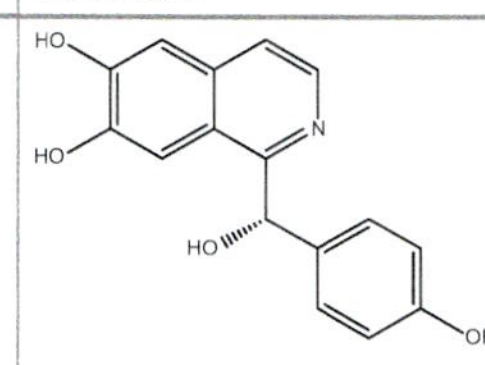	Garcia-Barriga (1974) and Arunjyothi et al. (2011)
	Synonym(s): *Annona pubescens* Salisb., *Annona tripetala* Aiton. **Common name(s):** Cherimoya, Custard apple tree. **Botanical description**: It is native to Andes of South America. It is known for its sweet fruits, hence also named as custard apple. It is a woody, deciduous, but evergreen spreading tree that grows up to 5 m. Leaves are single, alternate, 5–25 cm in length. Flowers are pale green in color and nearly about 1.2 in. long. Fruit is conical or heart-shaped, 10–20 cm and on ripening weigh about 150 gm. **Medicinal properties**: It is used to treat malignant tumors. Fruits are rich in carbohydrates like sugars and are low in protein and fat. Therefore, it has high nutritional value and a high antioxidant property. Also due to its fiber content, it is used to avoid constipation. The leaves of the plant have antidiabetic and antimicrobial activity. The seeds are used for treatment of tumors and also have anti-parasitic activity.			Cherianoine	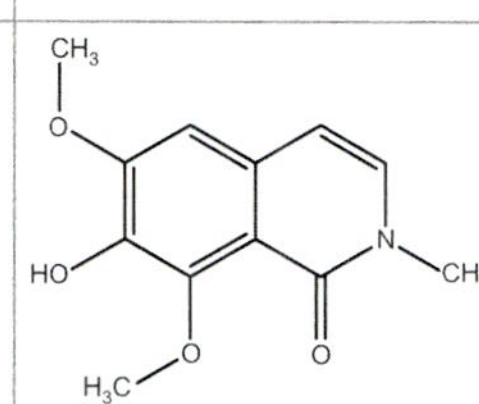	

(continued)

Table 4.9 (continued)

Family	Botanical name	Image	Plant part used	Bioactive compound(s)	Structure	Reference(s)
				Ncis-caffeoyltyramine	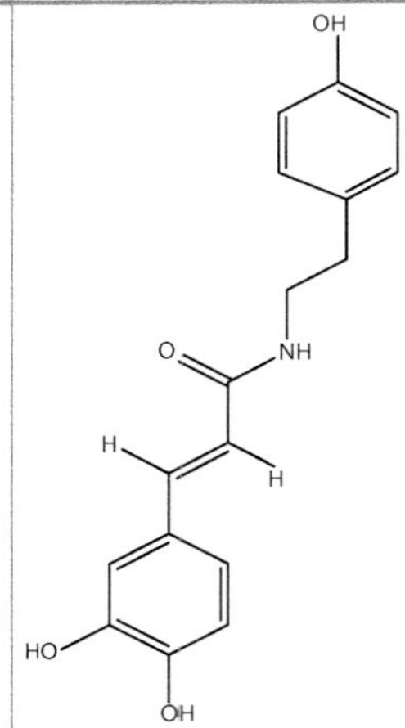	
				Annonaine	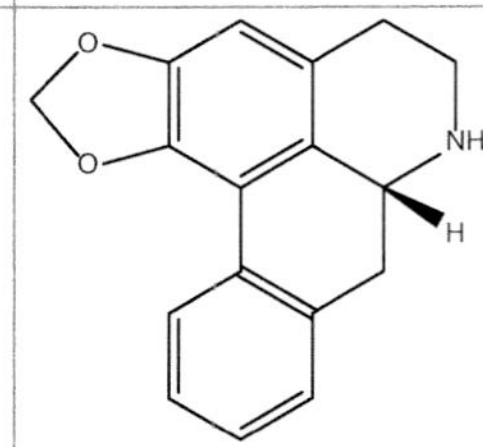	
				Cherimoline	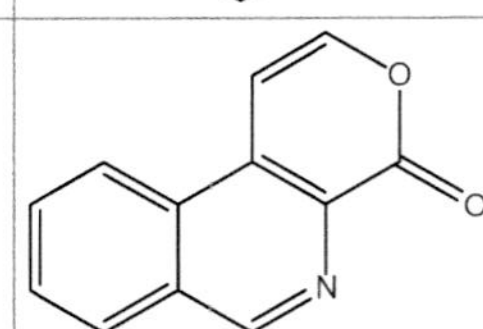	

	Annona dioica A.St.-Hil.		Leaf	Kaempferol	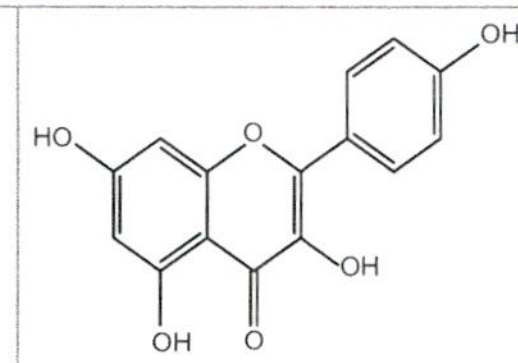	Vega et al. (2007) and Quílez et al. (2018)
	Synonym(s): *Aberemoa dioica* (A.St.-Hil.) Barb. Rodr., *Annona cuyabaensis* Barb.Rodr., *Annona dioica* var. *mattogrossensis* R.E.Fr. **Common name(s)**: Custard apple. **Botanical description**: It is native to South America – Southern, Central, and Eastern Brazil. It is a small, often single-stemmed shrub, growing 0.5–2 m tall. It is widely distributed in tropical and subtropical regions of the world. It is found in the Savannah and Pantanal, at elevations from 80 to 1000 m. **Medicinal use(s)**: It is used to treat liver, colon, prostate, pancreatic, and breast cancer. It also possesses antitumor, anti-inflammatory, antibacterial, and anti-rheumatic properties. Seeds are used to treat diarrhea, and plasters prepared with the fruits are considered to have emollient properties.					
	Annona reticulata L.		Leaf	Solamin	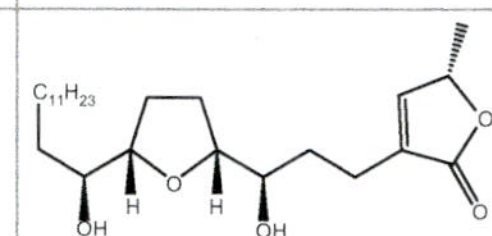	

(continued)

Table 4.9 (continued)

Family	Botanical name	Image	Plant part used	Bioactive compound(s)	Structure	Reference(s)
	Synonym(s): *Annona humboldtiana, Annona humboldtii.* **Common name(s)**: Netted custard apple, Bullock's heart, Bull's heart, Wild-sweetsop. **Botanical description**: It is native to Central America – Panama to Mexico and the Caribbean. It is a fast-growing, deciduous tree with a rounded or spreading crown. It can reach up to 7 m with a bole up to 30 cm in diameter. It is mainly grown in gardens in various areas of the tropics for its edible fruit. It is found in moist or dry thickets and forest, often in second growth, common in cultivation, chiefly at elevations of 1200 m or less in Guatemala. **Medicinal use(s)**: It is used to treat mouth, lung, throat, vocal cord, rectum, colon, esophagus, and stomach cancers. It also possesses anti-inflammatory, antitumor, antimalarial, antioxidant, anti-obesity, antidiabetic, and antifungal properties. Unripe fruits and the bark are rich in tannin. Fragments of the root bark are put around the gums to relieve toothache. The root decoction is taken as a febrifuge.			Diosgenin	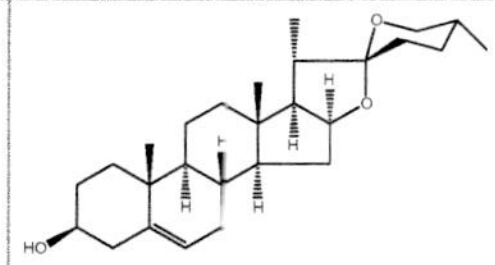	
				Fenugreekine	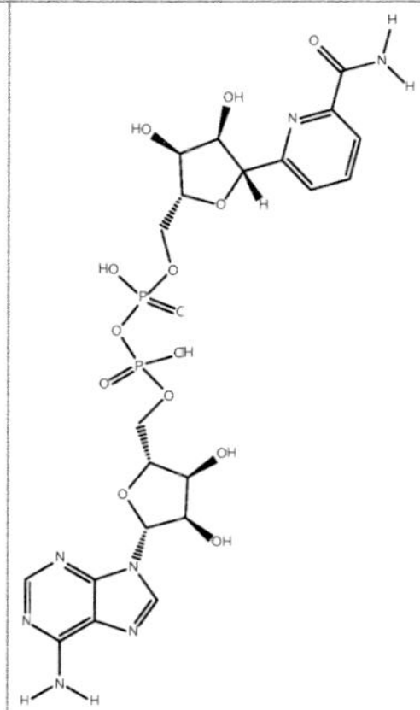	
				Annoreticuin-9-one	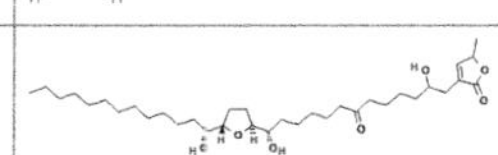	
				Squamone	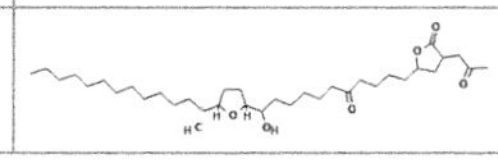	

		Annomonicin	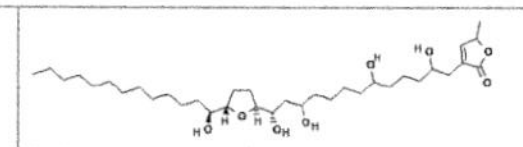	
		Rolliniastatin	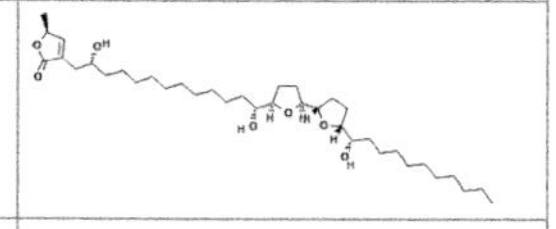	
		Annoreticuin	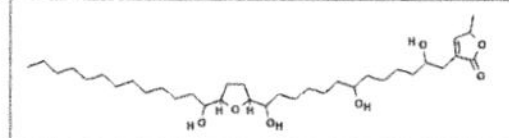	
		Bullatacin	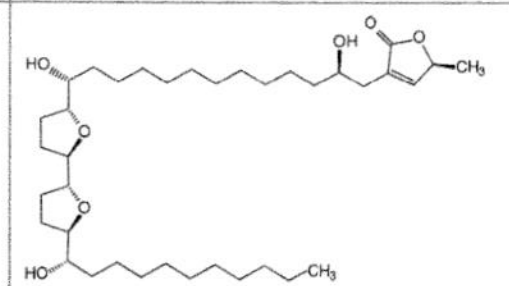	
		Squamocin	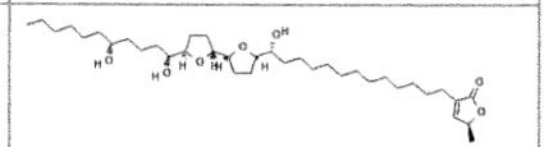	
		cis-bullatacinone	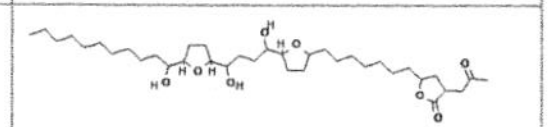	
		cis-murisolinone	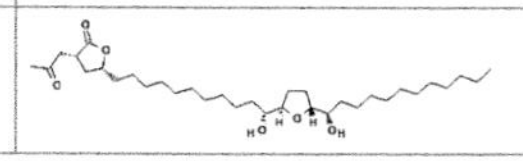	

(continued)

Table 4.9 (continued)

Family	Botanical name	Image	Plant part used	Bioactive compound(s)	Structure	Reference(s)
				Copaene	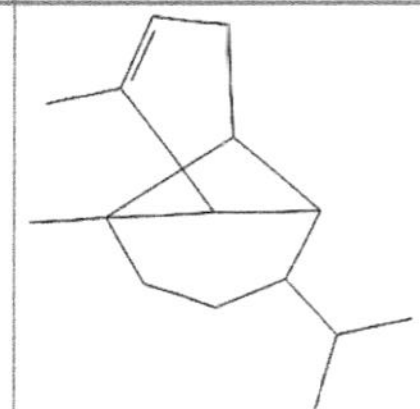	
				Patchoulane	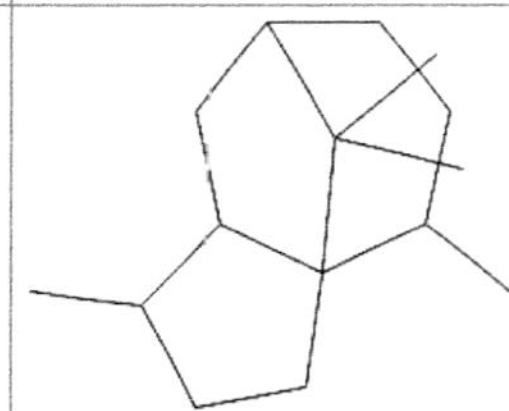	
				Kaur-16-en-19-oic acid	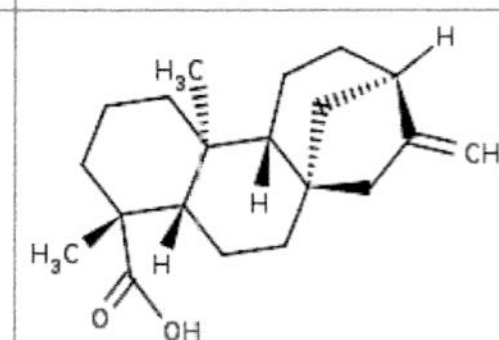	

	Annona crassiflora Mart.		Seeds, Leaves	Araticulin	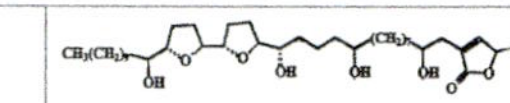	Santos et al. (1996) and Silva et al. (2019)
	Synonym(s): *Annona macrocarpa* Barb.Rodr., *Annona rodriguesii* Barb.Rodr. **Common name(s)**: Araticum, Marolo. **Botanical description**: *Annona crassiflora* is native of 'cerrado' area in Brazil. Araticum is a deciduous tree (4–8 m tall) with a dense, elongated crown. The corky bole is usually crooked and can be 20–30 cm in diameter. The wood is light in weight and soft. The fruit is about 16 cm long × 20 cm wide; flesh is fibrous, succulent and sweet, with a strong, pleasant aroma. **Medicinal use(s)**: *Annona crassiflora* is widely used in human medicine as for treatment of various diseases such as diarrhea, rheumatism and syphilis. The fruit flesh is used as a laxative. The seeds are used in the treatment of snakebite. The bioactive principle 'acetogenins' present in seeds and leaves exhibit cytotoxic, antiparasitic and antitumorigenic activities.			Crassiflorin	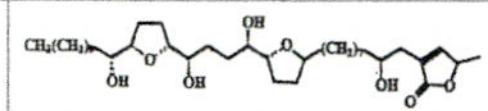	
				Annonin I	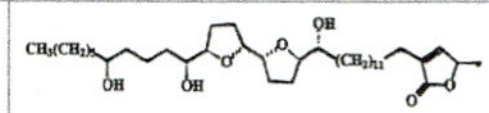	

(continued)

Table 4.9 (continued)

Family	Botanical name	Image	Plant part used	Bioactive compound(s)	Structure	Reference(s)
	Annona senegalensis Pers.		Entire plant	Heneicosane	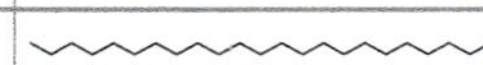	Durodola, (1975a, b), Adesogan and Durodola (1976), Mustapha (2013), and Obiora Nwonuma (2015)
	Synonym(s): *Annona arenaria* Thonn., *Annona chrysophylla* Bojer. **Common name(s)**: African custard apple or Wild custard apple. **Botanical description**: It is native to western tropical Africa – Senegal to Kenya, south to Zimbabwe. It is usually a multi-stemmed deciduous shrub growing 2–6 m tall, but in favorable conditions, especially in East Africa, it can become treelike (11 m tall). It grows naturally in semiarid to humid tropical areas of Africa at elevations ranging from sea level to 2400 m. **Medicinal properties**: It is used to treat mouth, lung, throat, vocal cord, rectum, colon, esophagus, cervical, and stomach cancer. It possesses anti-inflammatory, antitumor, antimalarial, antioxidant, anti-obesity, antidiabetic, and antifungal properties. The leaves of plant are used for treatment of pneumonia. Bark of plant can be used for treating a wide array of ailments, including worms of the intestines or flesh, diarrhea, gastroenteritis, lung infections, toothaches, and even snakebites. The roots are used for treating indigestion, dizziness, and various other venereal diseases. Certain other plant parts are used for treating skin and eye disorders.					

				Pentadecane		
				9, 17-Octadecandienal		
				Octadecanoic acid	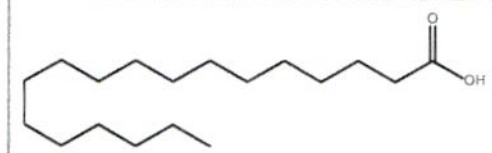	
				13octadecadien-1-ol		
	Annona montana Macfad.		Leaves, Seeds	Montanacin F	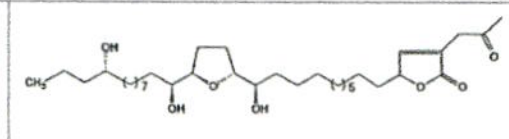	Wu et al. (1993c), Liaw et al. (2005), and Bailon-Moscoso et al. (2016)

(continued)

Table 4.9 (continued)

Family	Botanical name	Image	Plant part used	Bioactive compound(s)	Structure	Reference(s)
	Synonym(s): *Annona marcgravii* Mart., *Annona montana* f. marcgravii (Mart.) Porto, *Annona pisonis* Mart., *Annona sphaerocarpa* Splitg. **Common name(s)**: Mountain soursop. **Botanical description**: *Annona montana* is an edible, fibrous fruits, native to Central America, the Amazon, and islands in the Caribbean. It grows wild at altitudes from 0 to 650 m. The tree has a spreading crown and glossy leaves. It is slightly hard and bears fruit throughout the year. The fruits are nearly round, with dark green skin covered with many short fleshy spines, and are about 15 cm long. Yellow, fibrous pulp – which is aromatic – is sour and bitter, containing many light-brown, plump seeds. **Medicinal properties**: The natural products present in leaves and seeds of *Annona montana* exhibit anticancer, cytotoxic, antiparasitic, insecticidal, and immunosuppressive properties.			Annoretine [3-hydroxy-4-methoxy-N-methyl-tetrahydropyrido(4,3-a) phenanthrene]	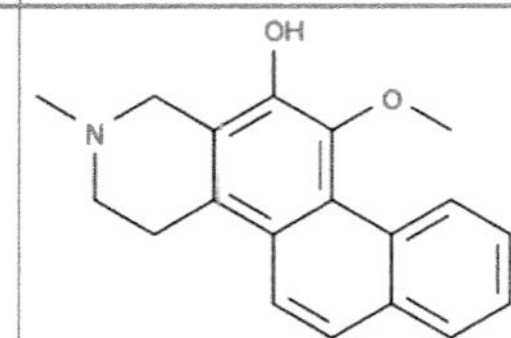	
				Argentinine	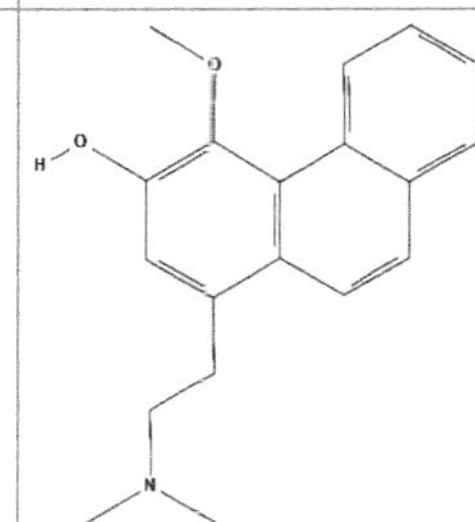	

		Liriodenir.e	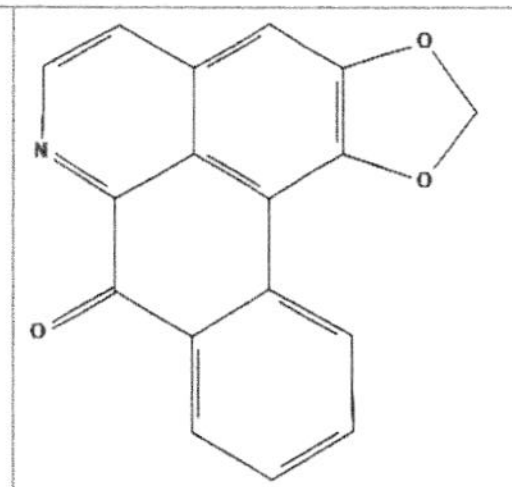	
		Montalicin G	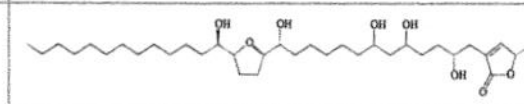	
		Montalicin H	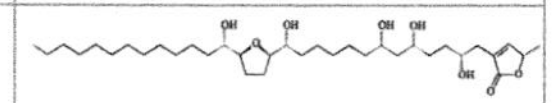	
		Monlicin A	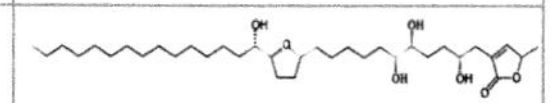	
		Monlicin B	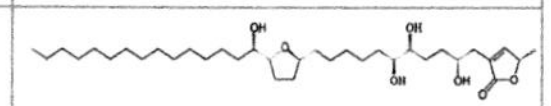	
		(+)-monhexocin	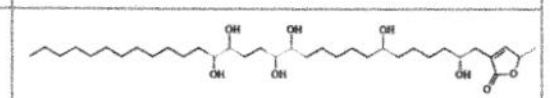	
		Murisolin	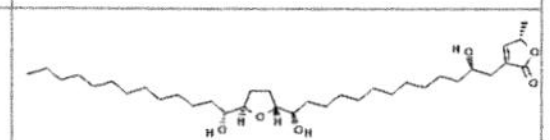	
		4-deoxyann-montacin	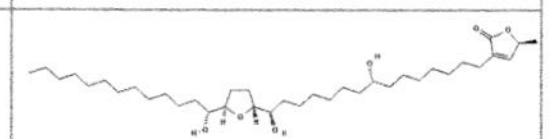	
		Muricatacin	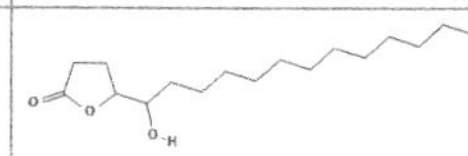	

(continued)

Table 4.9 (continued)

Family	Botanical name	Image	Plant part used	Bioactive compound(s)	Structure	Reference(s)
	Duguetia hadrantha (Diels) R.E. Fr (*Hadrantha*)		Bark	Hadranthine B	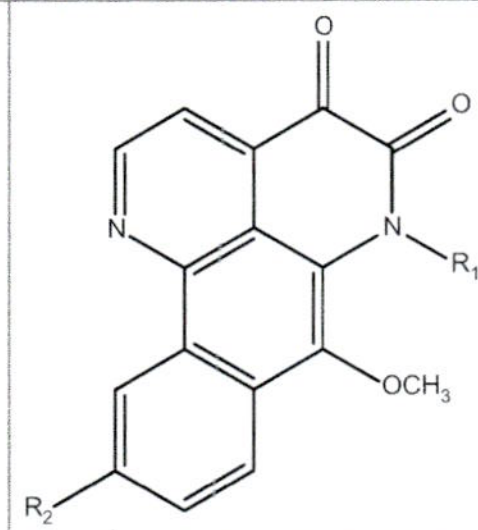	Muhammad et al. (2001)
	Synonym(s): *Aberemoa hadrantha* Diels ex Pilg. **Common name(s)**: Not available. **Botanical description**: It is native to northwest South America – Colombia and Venezuela to Panama. It is a shrub or a tree growing up to 25 m tall. It is found in tropical and subtropical regions. **Medicinal use(s)**: It possesses anticancer, antimalarial, antifungal, anti-inflammatory, and antibacterial properties. It is traditionally used to treat malaria and other fevers.			Sampangine	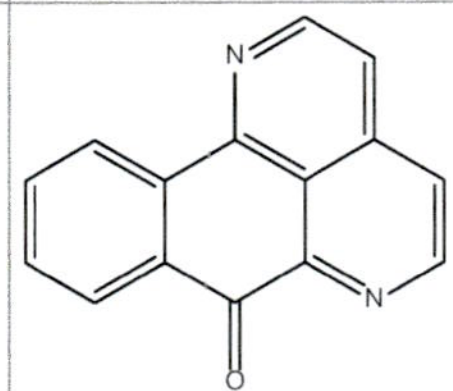	
				3-Methoxysampangine	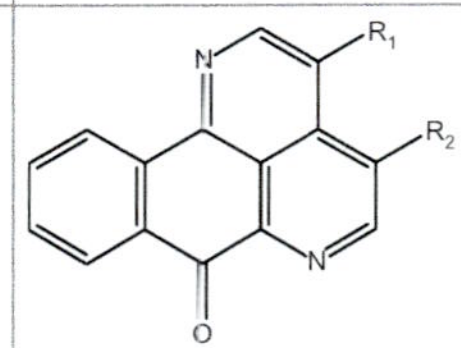	

	Cyathostemma argenteum (Blume) J. Sincl.		Root	Yuccagenin	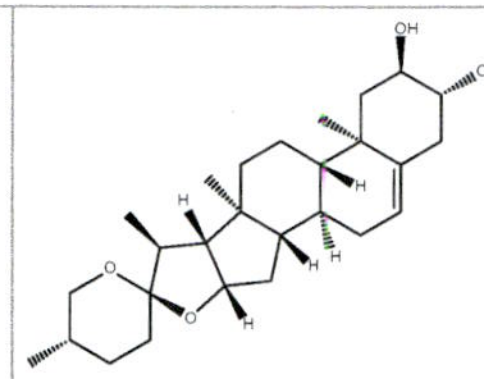	Khamis et al. (1997, 2004)
	Synonym(s): *Cyphostemma abercornense* Wild & R.B.Drumm., *Cyphostemma adamii* Desc., *Cyphostemma adenanthum* (Fresen.) Desc., *Cyphostemma adenocarpum* (Gilg&M.Brandt) Desc. **Common name(s)**: Caimito de mono. **Botanical description**: It is native to S. America, Paraguay, Brazil, Bolivia, Peru, Ecuador, Colombia, Venezuela, and the Guianas; C. America, Panama to Honduras; and the Caribbean, Cuba to Trinidad. It is an evergreen tree with a low, spreading crown; it can grow 10–25 m tall. The short bole can be 10–20 cm in diameter. It is found on moist limestone and lower mountain forests. **Medicinal use(s)**: It is used to treat breast cancer. It also possesses antimicrobial, antibacterial, anti-inflammatory, and antidiarrheal activities. It is used to treat vomiting, stomach problem, pneumonia, indigestion, and several types of infections. The heated bark is used as a medication for sores.					

(continued)

Table 4.9 (continued)

Family	Botanical name	Image	Plant part used	Bioactive compound(s)	Structure	Reference(s)
	Fissistigma oldhamii (Hemsl.) Merr.		Stem	Aristololactam	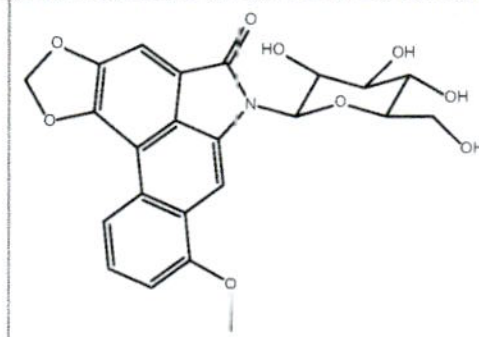	Wu et al. (1994) and Ge et al. (2013)
	Synonym(s): *Melodorum oldhamii* Hemsl. **Common name(s)**: Chinese loquat, Custard apple. **Botanical description**: It is native to Guangdong, Hainan, Jiangxi, Taiwan, Yunnan, and Zhejiang. It is a climber reaching up to 8 m tall. Branchlets fulvous pubescent. Petiole ca. 1 cm, pubescent; leaf blade obovate-elliptic to oblong, 6–13 × 2–5 cm, leathery, abaxially pubescent to glabrescent, adaxially glabrous, secondary veins 10–20 on each side of midvein and adaxially flat, base broadly cuneate to rounded, apex bluntly acute, rounded, or retuse. Inflorescences in glomerulus, 1–8-flowered; peduncle ca. 2.5 cm. Flowers ca. 1.5 × 1–1.7 cm. Sepals broadly triangular, 3–5 (–7) × 5–6 mm, brown tomentose, apex acute. Petals pale yellow to golden; outer petals ovate-elliptic, 21–24 × 11–12 mm, thicker, outside pubescent, inside glabrous; inner petals ovate-lanceolate, ca. 20 × 6 mm, outside puberulent, inside concave and puberulent, margin ciliate. Stamens oblong, ca. 2 cm; connectives slightly curved. Carpels 24–26, golden hairy; ovules 10 per carpel; styles white; stigma apex 2-cleft. Monocarp stipes 2.5–4 cm; monocarps globose, 1.5–1.8 cm in diameter, densely fulvous tomentose. Seeds 4 per monocarp, orbicular. **Medicinal properties**: Widely used in traditional Chinese medicine to treat rheumatoid arthritis. Roots are used as medicine for traumatic injury, arthritis, and inflammation. The alkaloids obtained from stem of the very plant are used to treat tumor and inhibit T- and B-cell proliferation. Used as an antitumor agent.			Fissistigamide	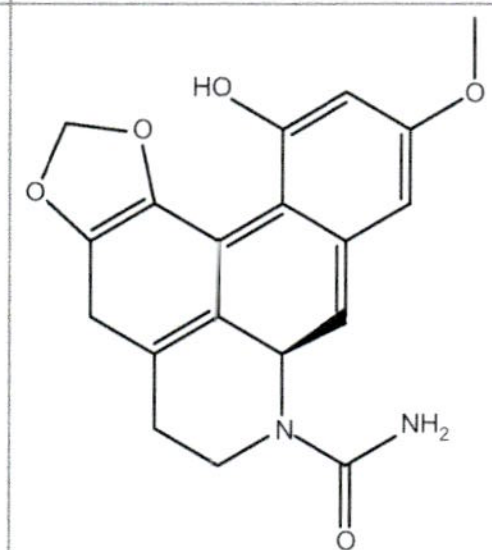	
				Fissistigmine	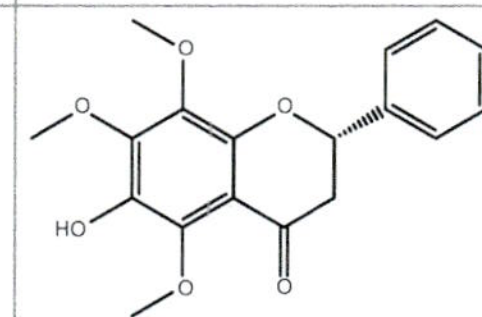	

	Mitrella kentii Miq. (Kiawi)		Bark	Terpenylated dihydrochalcone	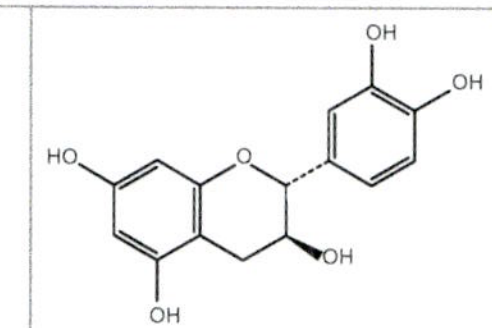	Benosman et al. (1997) and Aziz et al. (2016)
	Synonym(s): *Fissistigma kentii* (Blume) Merr., *Orophea borneensis* Miq., Unona kentii Blume. **Common name(s)**: Not available. **Botanical description**: It is native to Borneo, Jawa, and Malaya. It is a tree-climbing liana. Stamens oblong, ca. 2 cm; connectives slightly curved. Carpels 24–26, golden hairy; ovules 10 per carpel; styles white; stigma apex 2-cleft. Monocarp stipes 2.5–4 cm; monocarps globose, 1.5–1.8 cm in diameter, densely fulvous tomentose. Seeds 4 per monocarp, orbicular. **Medicinal use(s)**: It is used for the treatment of cancer and gastric ulcer. It shows anti-inflammatory, antiacid, and antibacterial activities. It is also used for treatment of fever.			2′,6′- Ihydroxy-4′-methoxydihydrochalcone	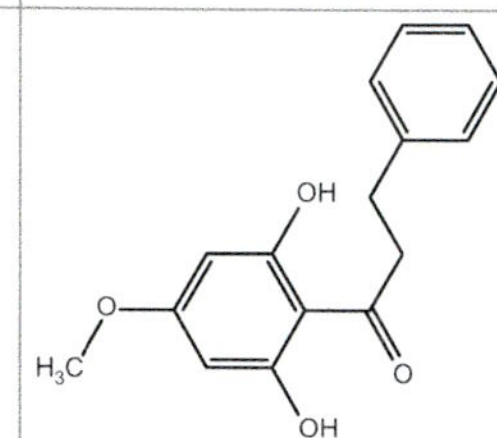	
				(+)-catechin	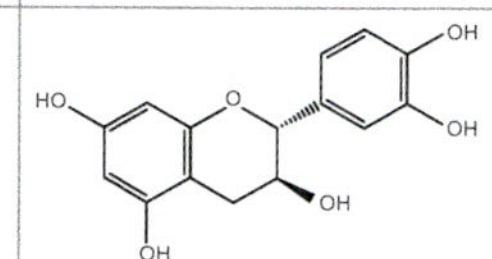	

(continued)

Table 4.9 (continued)

Family	Botanical name	Image	Plant part used	Bioactive compound(s)	Structure	Reference(s)
	Xylopia laevigata		Leaves	γ-Muurolene	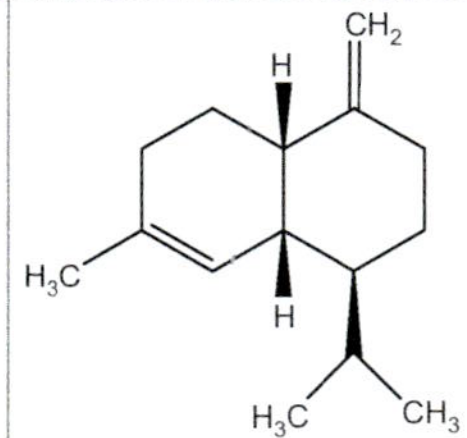	Quintans et al. (2013)
	Synonym(s): *Annona laevigata* Mart. **Common name(s):** Not available. **Botanical description**: It is a native to the lowland rainforest and moist fringe forests in the savannah zones of Africa. It can grow up to 20 m high. It is present in rainforests, especially near the coast. It also grows in riverine and fringing forest and as a pioneer species in arid savannah regions. **Medicinal use(s)**: It is used for the treatment of cancer and tumors. It also possesses antidiarrheal, analgesic, antimicrobial, insecticidal, anti-parasitic, antioxidant, and anti-inflammatory properties.			δ-Cadinene	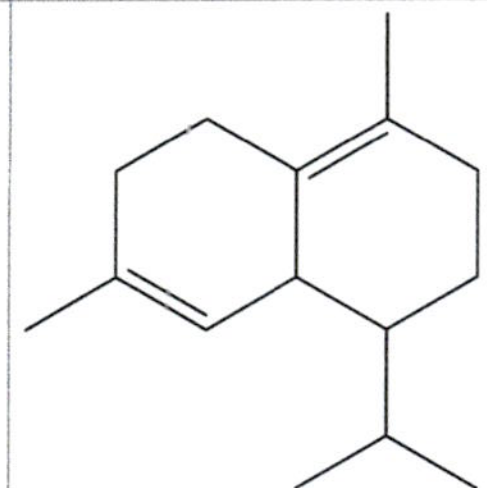	
				Germacrene B	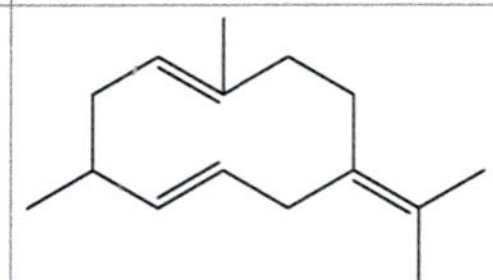	

		α-Copaene	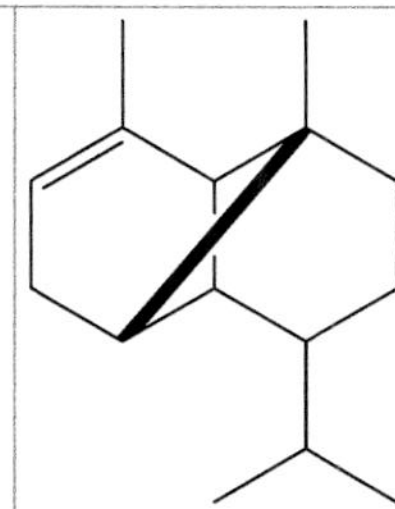	
		Germacrene D	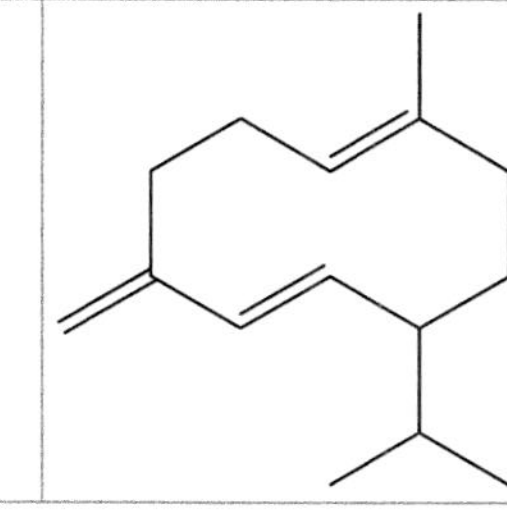	

(continued)

Table 4.9 (continued)

Family	Botanical name	Image	Plant part used	Bioactive compound(s)	Structure	Reference(s)
				Bicyclogermacrene	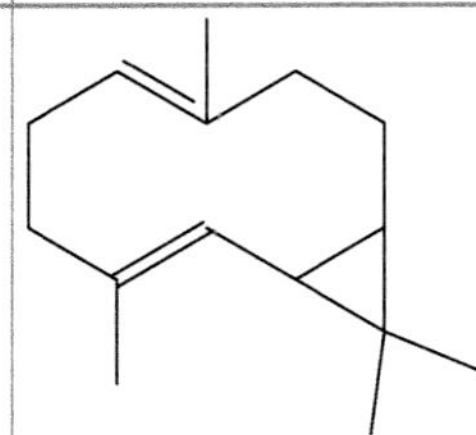	
				(E)-caryophyllene	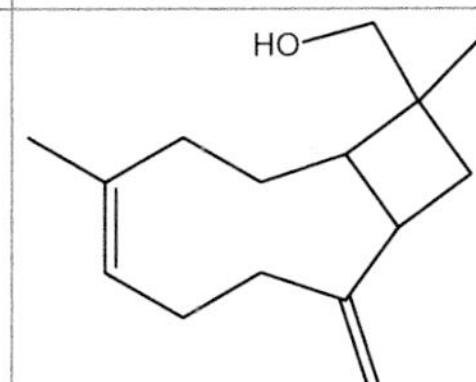	

	Uvaria grandiflora (Lesch. ex DC.) Roxb.		Stem	Zeylenol	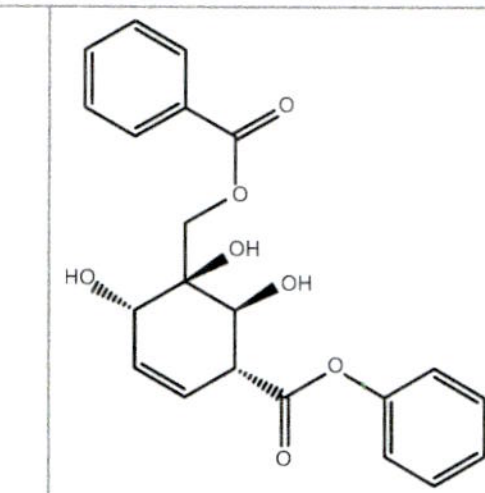	Seangphakdee et al. (2013)
	Synonym(s): *Unona setigera* Blanco., *Uvaria cardinalis* Elmer., *Uvaria flava* Teijsm. & Binn., *Uvaria purpurea* Blume. **Common name(s)**: Large flower *Uvaria*, Red hot poker, Pisangakar, Akarlarak. **Botanical description**: It is native to Southeast Asia – Southern China, India, Sri Lanka, Myanmar, Thailand, Malaysia, Vietnam, Indonesia, and the Philippines. It is a sturdy climbing shrub with large leaves. The stems can be up to 10 m long. Open forests, thickets at elevations from 400 to 1000 m in Southern China. **Medicinal use(s)**: It is used for the treatment of cancer cells in human breast. It has anti-inflammatory and anticancer properties. Leaves and roots are used in traditional medicine to treat stomachache, abdominal pains, and skin diseases.					

(continued)

Table 4.9 (continued)

Family	Botanical name	Image	Plant part used	Bioactive compound(s)	Structure	Reference(s)
	Uvaria dac		Bark	(+)-Grandifloracin	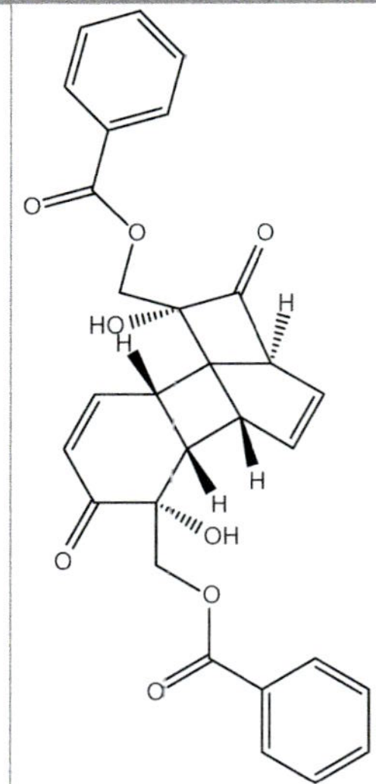	Ueda et al. (2014)
	Synonym(s): *Armenteria Thouars* ex Baill., *Marenteria Thouars.*, *Narum Adans.*, *Pyragma Noronha.* **Common name(s)**: Bush banana. **Botanical description**: It is native to the eastern part of Southern Africa south of the Zambezi, South Africa. These are climbing shrubs or small trees. The flowers are borne singly, in pairs, or in small clusters. There are six petals in two whorls and many stamens. **Medicinal use(s)**: It is used for the treatment of pancreatic cancer and bacterial and fungal infections.					

	Oenanthe javanica ssp. *stolonifera* (Roxb. ex DC.) Murata		Whole plant	Ardisiacrispin	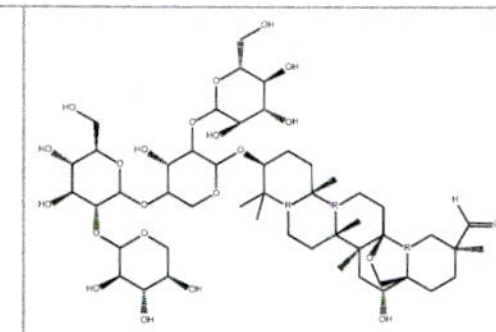	Kwon et al. (2006a, b)
	Synonym(s): *Oenanthe javanica* (Blume) DC., *Oenanthe javanica* subsp. *javanica, Oenanthe javanica subsp. linearis* (*Wall. ex DC.*) *Murata.* **Common name(s)**: Water celery, Water dropwort, Water parsley, Selom, Java water dropwort. **Botanical description**: It is native to East Asia – China, Japan, and Korea – and Australia. It is a perennial growing up to 1 m tall. It flowers from June to August, and the seeds ripen from August to October. The flowers are hermaphrodite (have both male and female organs) and are pollinated by insects. The plant is self-fertile. It is suitable for light (sandy), medium (loamy), and heavy (clay) soils, and suitable pH is acid, neutral, and basic (alkaline) soils. It cannot grow in the shade. It prefers wet soil and can grow in water. **Medicinal use(s)**: It is used to treat liver, small intestine, kidney, lung, skin, and colon cancer. It possesses antitumor, anticancer, antioxidant, antimutagenic, and anti-genotoxic activity. A decoction is used in the treatment of epidemic influenza, fever and discomfort, jaundice, hematuria, and metrorrhagia. The seed contains 3.5% essential oil. This is effective at large dilutions against pathogenic fungi. The leaves are chewed with wild ginger and traditional ash salt as an antidote to poisoning.					

(continued)

Table 4.9 (continued)

Family	Botanical name	Image	Plant part used	Bioactive compound(s)	Structure	Reference(s)
	Centella asiatica L.		Aerial part	Methyl 5-[(E)-9-hydroxy-1-(1-hydroxyhexyl)-2-methoxyundeca-3,10-diene-5,7-diynyloxy] pentanoate (cadiyenol)	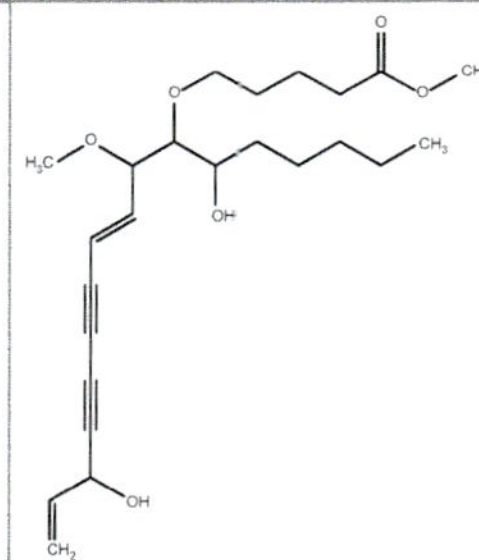	Jamil et al. (2007) and Govindan et al. (2007)
	Synonym(s): *Centella asiatica* var. *asiatica*, *Centella asiatica* var. *crista* Makino, *Centella asiatica* var. *floridana* (J.M.Coult. & Rose) J.M.Coult. & Rose. **Common name(s)**: Gotu kola. **Botanical description**: It is native to South Africa, West Africa, and Asia. It is a low-growing, evergreen perennial plant about 20 cm tall. The plant spreads quickly by the roots, producing long stolons up to 250 cm in length – these roots at the nodes and the plant can soon form large carpets of growth. It is found in the coastal, rocky sunny places in lowland hills; shady, damp, and wet places such as paddy fields; and grass thickets. **Medicinal use(s)**: It is used to treat breast cancer. It also possesses antioxidant, anti-inflammatory, and antitumor properties. It has selective toxicity against tumor cells. It has been suggested that regular use of the herb can rejuvenate the nervous system, and it therefore deserves attention as a possible cure for a wide range of nervous disorders including multiple sclerosis. It is used internally in the treatment of wounds, chronic skin conditions (including leprosy), venereal diseases, malaria, varicose veins, ulcers, etc. It is particularly valued for its effect upon the nervous system, being used in the treatment of conditions such as epilepsy and a range of nervous disorders including senility.					

Table 4.10 Anticancer plants of family Apocynaceae

Family	Botanical name	Image	Plant part used	Bioactive compound(s)	Structure	Reference(s)
Apocynaceae	*Adenium obesum* (Forssk.) Roem. and Schult.		Leaves, stem, bark	3,3′-Bis(0-methy1) quercetin	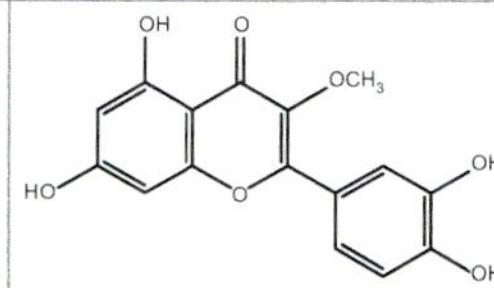	Omino and Kokwaro (1993) and Versiani et al. (2014)
	Synonym(s): *Adenium coetaneum* Stapf., *Adenium honghel* A.D.C., *Adenium somalense* Balf.f., *A. socotranum* Vierh. **Common name(s)**: Sabi star, Kudu, Mock azalea, Impala lily and desert rose, Desert rose, Kudu lily. **Botanical description**: It is native to the Sahel regions, south of Sahara, and tropical-subtropical Eastern Africa and Arabia. It is an evergreen or drought-deciduous succulent shrub that can grow to 1–3 m in height. The stem is stout and with swollen basal caudex. The leaves are spirally arranged on it. The flowers are 2–5 cm long and pentamerous. **Medicinal use(s)**: Leaves are a source of potential anticancer agent. The plant is important in traditional African medicine. It possesses some 30 cardiotoxic glycosides, which act in a similar way as digitalis. In low doses, digitalis is used for treatment of congestive heart failure (CHF) and heart rhythm problems. Latex from plant is applied to decaying teeth and septic wounds. The extracts from roots and bark are used for treatment of skin diseases and skin lumps. Leaves of the plant can be used as an antimicrobial agent and expected that leaves of this plant may be used as therapeutic agents for various diseases.					

(continued)

Table 4.10 (continued)

Family	Botanical name	Image	Plant part used	Bioactive compound(s)	Structure	Reference(s)
	Allamanda cathartica Garden.		Whole plant	Allamandin	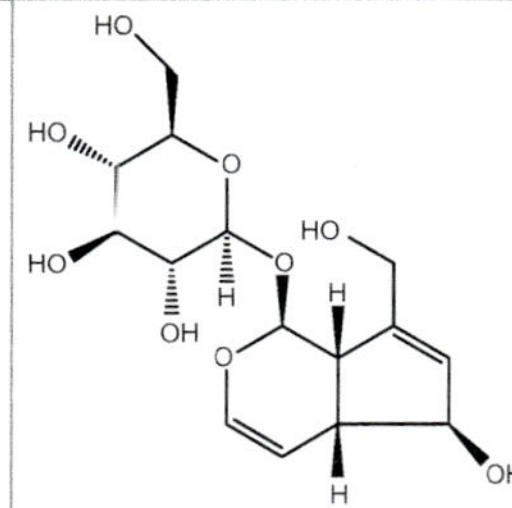	Tiwari et al. (2002)
	Synonym(s): *Echites verticillatus* Sessé & Moc., *Echites salicifolius* Willd. ex Roem. & Schult., *Allamanda williamsii* auct. **Common name(s)**: Golden trumpet vine. **Botanical description**: It is native to South America such as Brazil, Peru, Colombia, Venezuela, and the Guianas; and Central America – Panama, Costa Rica, Nicaragua, and Honduras. It is an evergreen climbing shrub that can reach a freestanding height of 2 m or produce clambering stems 6 m or longer that reach into the crowns of small trees. Older plants often have multiple stems from the root crown and long stems with relatively few branches. **Medicinal use(s)**: It is used to treat malignant tumor and possesses anticancer, antibacterial, antifungal, antimicrobial, and anti-inflammatory properties. An infusion of leaves, in moderate doses, is an excellent cathartic; in larger doses it is purgative and violently emetic. A decoction of the leaves, in small doses, is used as an antidote to poisoning and as a treatment for colic.			Plumericin	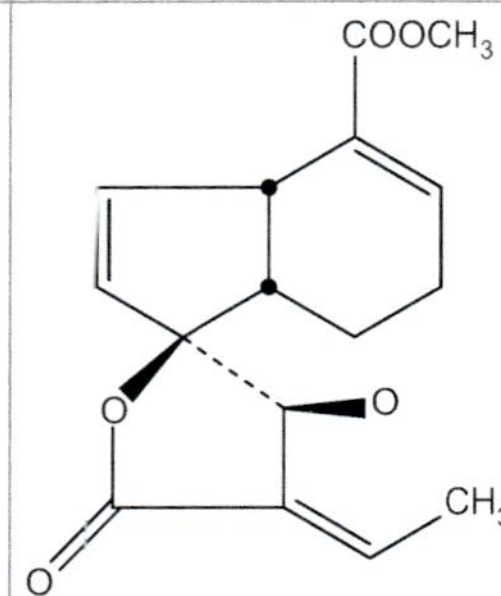	

		Isoplumericin	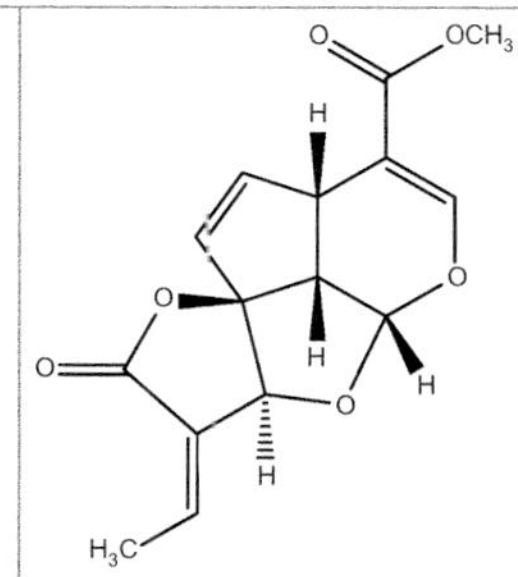	
		Penstemide	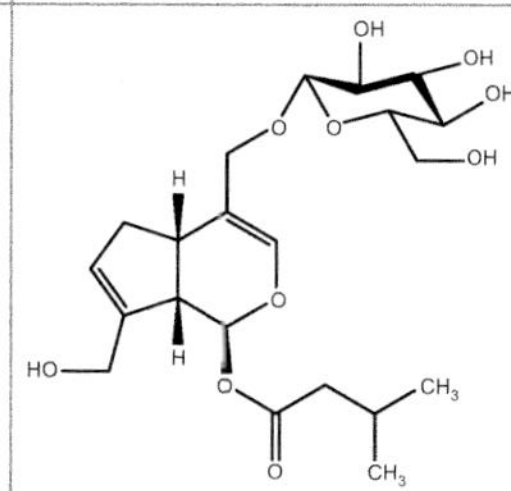	

(continued)

Table 4.10 (continued)

Family	Botanical name	Image	Plant part used	Bioactive compound(s)	Structure	Reference(s)
	Alstonia congensis De wild (*Alstonia*)		Bark	Alstonine	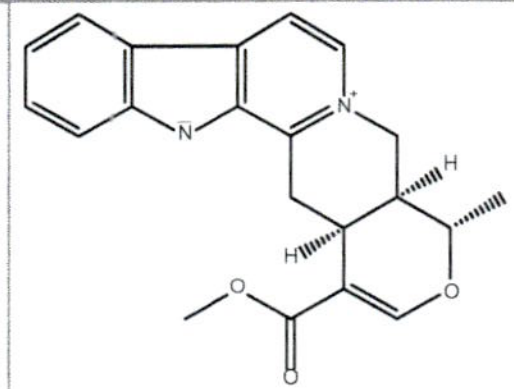	Pratyush et al. (2011)
	Synonym(s): *Alstonia congensis* var. *glabrata* Hutch. & Dalziel., *Alstonia gilletii* De Wild., *Alstonia pedicellata* Pierre ex A.Chev. **Common name(s)**: Stool wood. **Botanical description**: It is native to western tropical Africa – Liberia to Cameroon – and Central African Republic, south to Angola and the DR of the Congo. It is a tall forest tree capable of reaching 30 m tall, bole cylindrical, fluted up to 125 cm in diameter; stout buttress up to 6m tall which spread up to 2 m at the base; outer bark brown or yellowish white; sapwood indistinctly demarcated from heartwood. Growth rings indistinct or absent. The wood is lightweight. **Medicinal use(s)**: It is used for the treatment of cancer, malaria, diabetes, bacterial infections, and fever and for increasing libido. It also possesses antimalarial, anti-ulcer, antidiarrheal, antioxidant, and antimicrobial properties.					

				O-Acetylmacralstonine	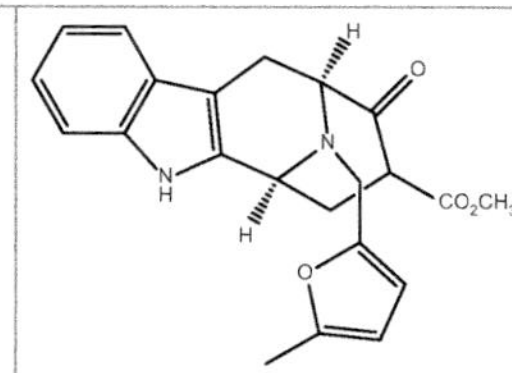	
				Macrocarpamine	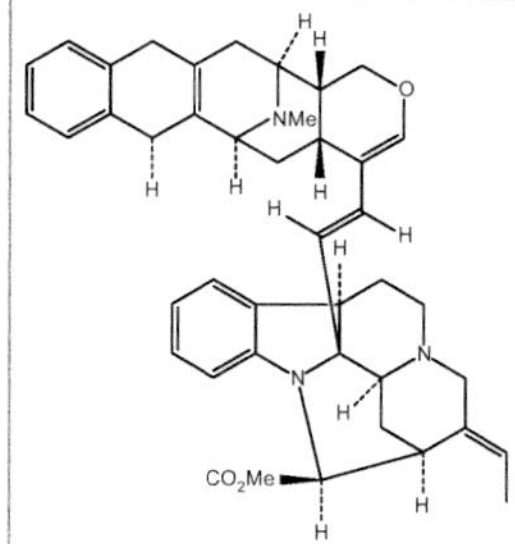	
				Villalstonine	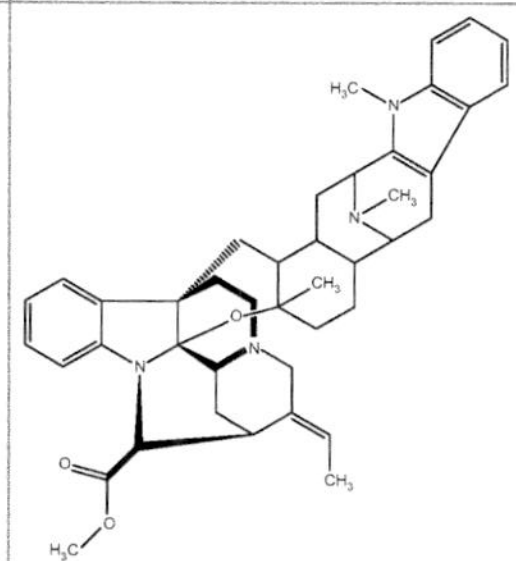	

(continued)

Table 4.10 (continued)

Family	Botanical name	Image	Plant part used	Bioactive compound(s)	Structure	Reference(s)
	Alstonia scholaris R.Br.		Leaves, root, bark, entire plant	Echitamine	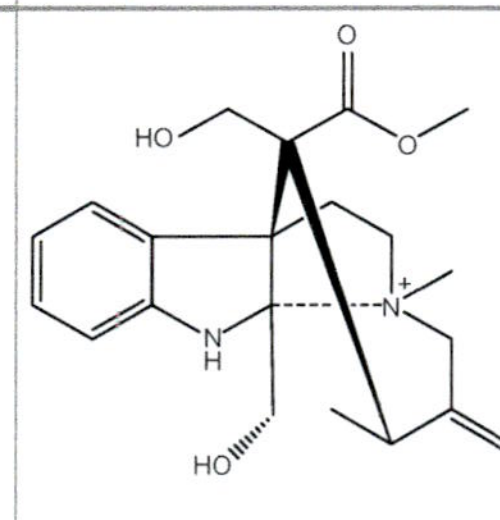	Wasuwat (1967), Gupta (1979), Holdsworth and Wamoi (1982), Anon (1985), Vashi and Patel (1989), Keawpradub et al. (1999), Jahan et al. (2009), Baliga (2010), and Reddy (2016)
	Synonym(s): *Echites scholaris* L. **Common name(s)**: Blackboard tree, Indian devil tree, Milkwood pine, White cheesewood and pulai. **Botanical description**: This evergreen tree is native to the Southeast Asia and Australia, Indian subcontinent, and parts of Indonesia. The leaves are leathery and occur in whorls of 3–10. Petioles are 1–3 cm. The bark of tree is odorless with bitter and milky sap. The flowers bloom in the month of October. **Medicinal properties**: It is used to treat several types of tumors and lung cancer. It is reported to have antimalarial, antibacterial, antioxidant, anticancer, anti-inflammatory, and analgesic activities. The plant is used for treatment of skin disorders, malarial fever, urticaria, chronic dysentery, diarrhea, and snakebite. The milky juice of the tree is applied to treat ulcers. The ethanol extracts of the bark exhibit anticancer activity.			Alstonine	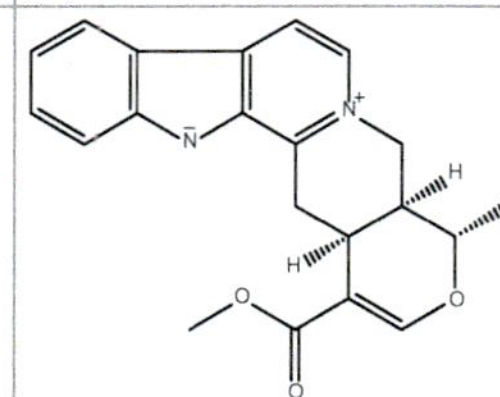	

		Pleiocarpamine	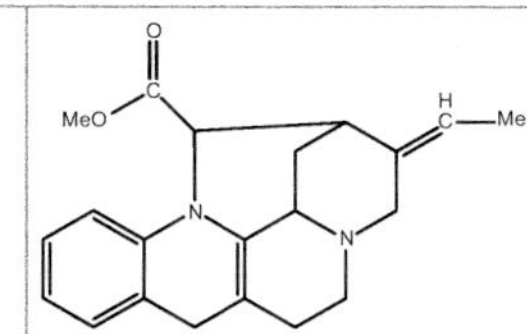	
		Methylmacralstonine	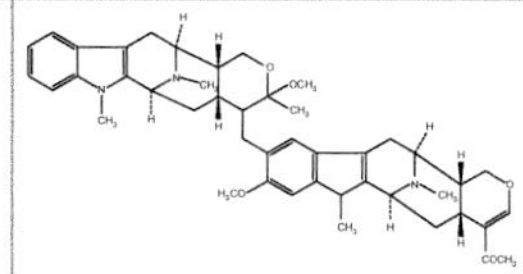	
		Echitamine	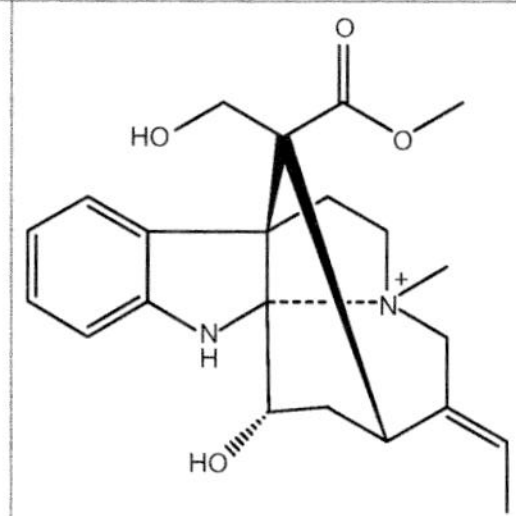	

(continued)

Table 4.10 (continued)

Family	Botanical name	Image	Plant part used	Bioactive compound(s)	Structure	Reference(s)
				Tubotaiwine	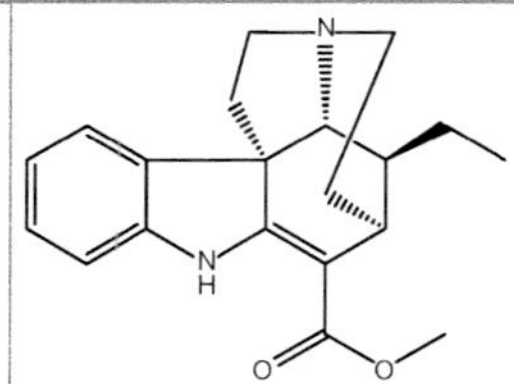	
				Akuammicine	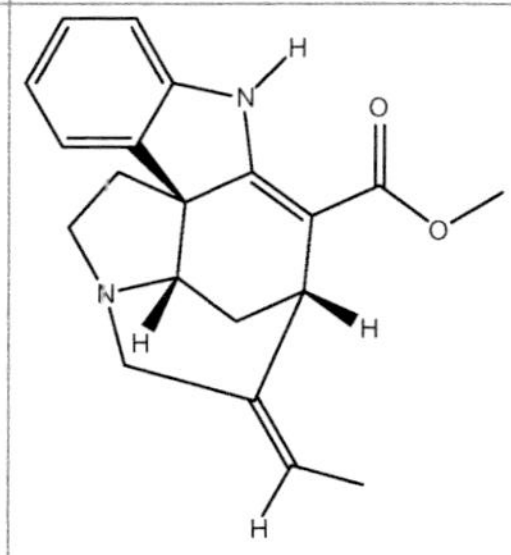	
				Echitamidine-N-oxide-19-O-β-D-glucopyranoside (indole alkaloid)	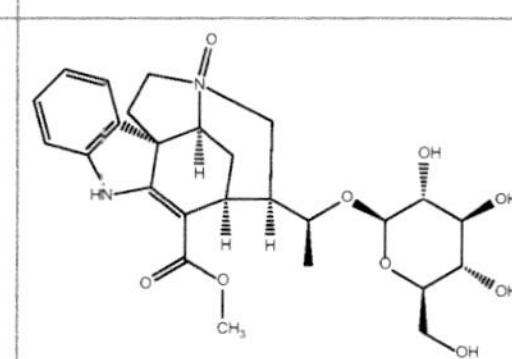	

		Villalstonine	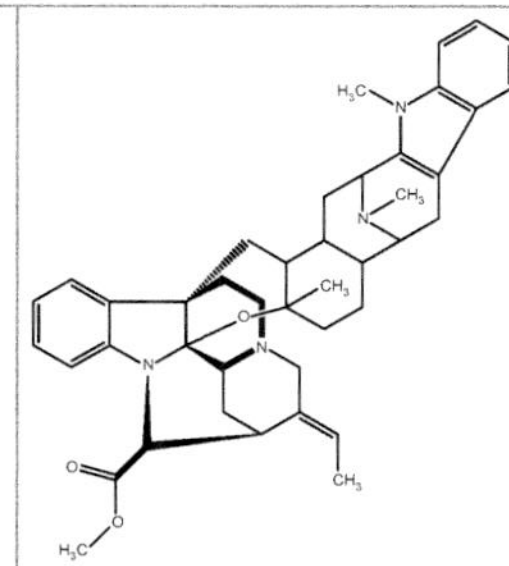	
		Lupeol	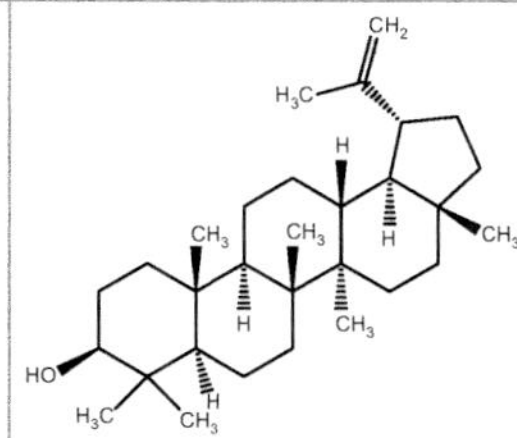	

(continued)

Table 4.10 (continued)

Family	Botanical name	Image	Plant part used	Bioactive compound(s)	Structure	Reference(s)
	Asclepias curassavica L.		Root	Cardenolide	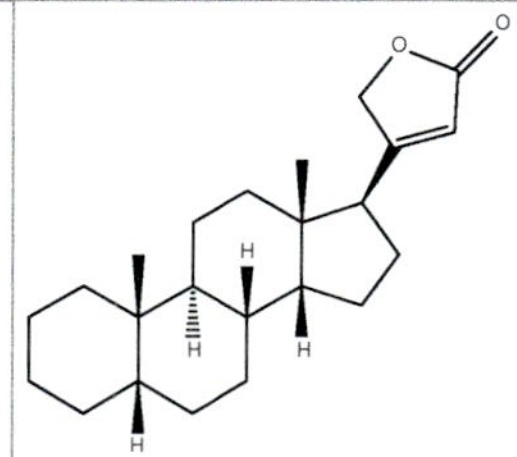	Roy et al. (2005)
	Synonym(s): *Asclepias aurantiaca* Salisb., *Asclepias cubensis* Wender., *Asclepias curassavica* var. *concolor* Krug & Urb., *Asclepias curassavica* f. *flaviflora* Tawada. **Common name(s)**: Milkweed, Mexican butterfly weed. **Botanical description**: It is native to American tropics. The plants are evergreen perennial subshrubs that grow up to 1 m (3.3 ft.) tall and have pale gray stems. The leaves are arranged oppositely on the stems and are lanceolate or oblong-lanceolate-shaped ending in acuminate or acute tips. Like other members of the genus, the sap is milky. The flowers are in cymes with 10–20 flowers each. They have purple or red corollas and corona lobes that are yellow or orange. Flowering occurs nearly year-round. The 5–10 cm long, fusiform-shaped fruits are called follicles. The follicles contain tan to brown seeds that are ovate in shape and 6–7 mm long. The flat seeds have silky hairs which aid dehiscence. **Medicinal use(s)**: It is used in folk medicine in treatment of cancer and malaria. It also possesses anti-asthmatic, antitumor, and antipyretic properties.			Nor-cardenolide	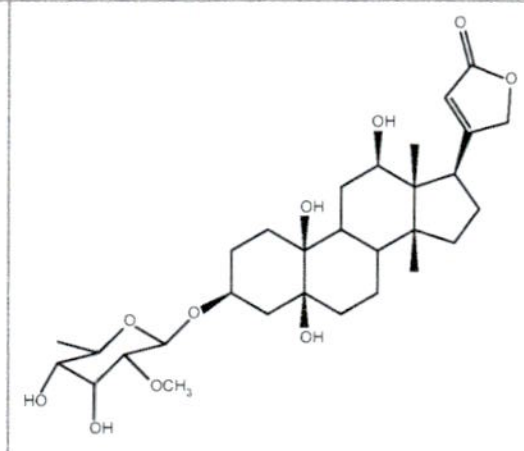	

		Cardenolide genins	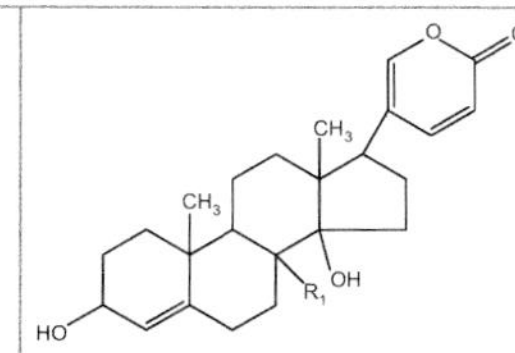	
		Pregnane	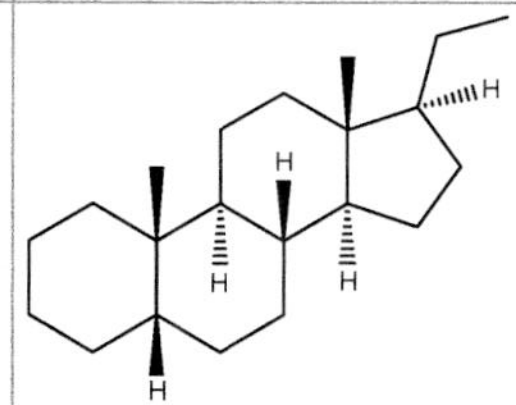	
		Androstane	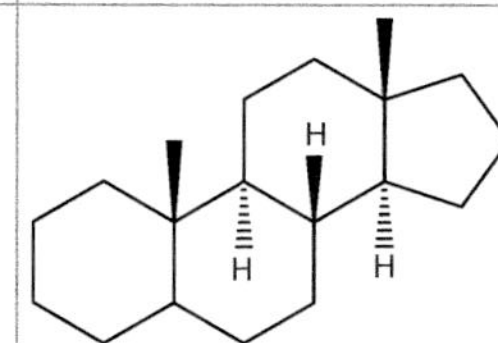	
		3,4-Seco-urs-20(30)-en-3-oic acid	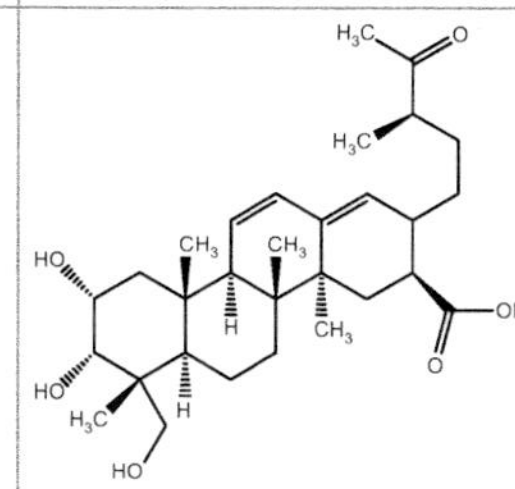	

(continued)

Table 4.10 (continued)

Family	Botanical name	Image	Plant part used	Bioactive compound(s)	Structure	Reference(s)
	Aspidosperma polyneuron Mull. Arg.		Bark and leaf	Pentadecanal	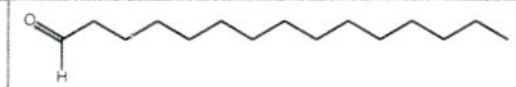	Schultes (1979) and Maffei (2003a, b, c, d, e, f, g, h, i, j, k, l, m)
	Synonym(s): *Thyroma polyneura* (Mull. Arg.) Miers, *Aspidosperma peroba* Saldanha, *Aspidosperma dugandii* Standl. **Common name(s)**: Peroba rosa, Sobro, Peroba mirim, Peroba rajada. **Botanical description**: It is an endangered timber tree native to Brazil, Colombia, Argentina, Peru, and Paraguay. It can grow from 8 to 30 m tall with a bole around 80 cm thick. It is widely used for ornamental purpose. **Medicinal properties**: The leaf is used for treatment of tumors. The bark of tree is bitter in taste and possesses antipyretic and astringent properties.			Kaurene	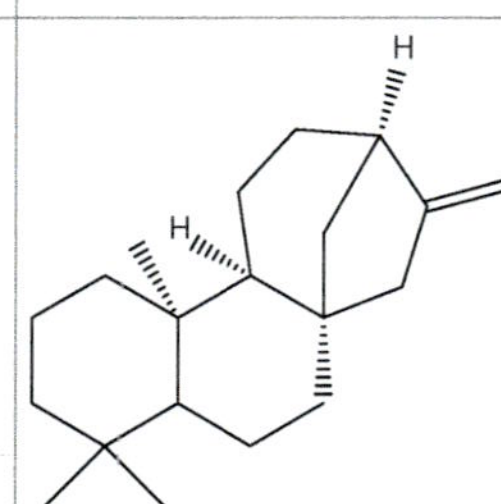	
	Calotropis gigantea R.Br.		Root, whole plant	Cardiac glycoside	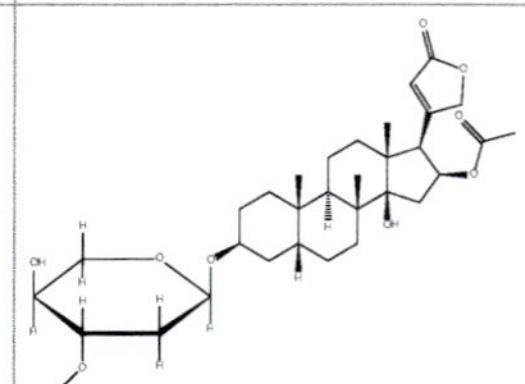	Bhagat et al. (2007) and Bhat and Sharma (2013)

	Synonym(s): *Asclepias gigantea* L., *Calotropis gigantea* (L.) R. Br., *Madorius giganteus* (L.) Kuntze. **Common name(s)**: Milkweed, Swallow-wort. **Botanical description**: It is native to Cambodia, Indonesia, Malaysia, the Philippines, Thailand, Sri Lanka, India, China, Pakistan, Nepal, and tropical Africa. It is a large shrub growing to 4 m tall. It has clusters of waxy flowers that are either white or lavender in color. Flower consists of five pointed petals and a small "crown" rising from the center which holds the stamens. The aestivation found in calotropis is valvate sepals or petals in a whorl just touch one another at the margin, without overlapping. The plant has oval, light green leaves, and milky stem. **Medicinal use(s)**: It possesses anticancer and anti-inflammatory properties. It is used for the treatment of cancer. It is also used for digestive disorders including diarrhea, constipation, and stomach ulcers; for painful conditions including toothache, cramps, and joint pain; and for parasitic infections including elephantiasis and worms.					
	Catharanthus roseus (L.) G.Don.		Flower, whole plant	Vinblastine	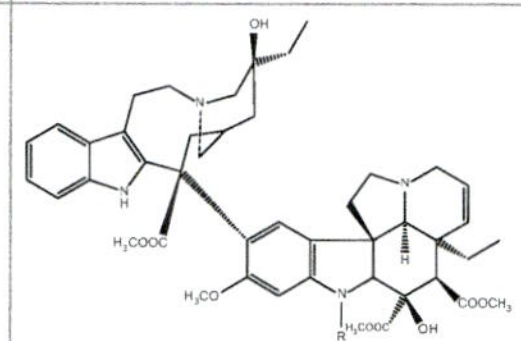	Cragg and Newmann (2005) and Prajakta and Ghosh (2010)

(continued)

Table 4.10 (continued)

Family	Botanical name	Image	Plant part used	Bioactive compound(s)	Structure	Reference(s)
	Synonym(s): *Ammocallis rosea* (L.) Small, *Catharanthus roseus* var. *albus* G. Don, *Catharanthus roseus* var. *roseus*, *Hottonia littoralis* Lour., *Lachnea rosea* (L.) Rchb., *Lochnera rosea* (L.) Rchb. ex Endl., *Lochnera rosea* var. *alba* (G. Don) Hubbard, *Lochnera rosea* var. *flava* Tsiang, *Pervinca rosea* (L.) Gaterau, *Pervinca rosea* (L.) Moench, *Vinca gulielmi-waldemarii* Klotzsch, *Vinca rosea* L., *Vinca rosea* var. *alba* (G. Don) Sweet, *Vinca rosea* var. *albiflora* Bertol. **Common name(s)**: Madagascar periwinkle, Rose periwinkle, or Rosy periwinkle. **Botanical description**: *Catharanthus roseus* is native and endemic to Madagascar. It is an evergreen herbaceous plant that grows 1 m tall. The leaves are oval to oblong, 2.5–9 cm long and 1–3.5 cm broad, glossy green, hairless, with a pale midrib and a short petiole 1–1.8 cm long; they are arranged in opposite pairs. The flowers are white to dark pink with a darker red center, with a basal tube 2.5–3 cm long and a corolla 2–5 cm diameter with five petal-like lobes. The fruit is a pair of follicles 2–4 cm long and 3 mm broad. **Medicinal properties**: *Catharanthus roseus* is used as an anticancer agent. It is most effective for the treatment of lung cancer. It is used for the treatment of leukemia. The medicine prepared from the alkaloid of this plant is used for the treatment of Hodgkin's lymphoma in children. It decreases blood pressure and reduces excitement. It is also proved antidiabetic. It has antibacterial, antiviral, antidiarrheal, anti-inflammatory, antioxidant, anti-acne, anti-eczema, and anti-dermatitis property. It is also useful for the treatment of nose bleeding and gum bleeding as well as treatment of bleeding hemorrhoids. The extract of the plant is also useful for an eye infection and irritation of the eyes. *Catharanthus roseus* regularizes bleeding problem for women; it is useful for the treatment of mouth ulcers; its extract gives comfort during the depression, headache, nausea and fatigue, insect bites, malaria, and asthma. It is diuretic in nature and increases urine.			Vincristine	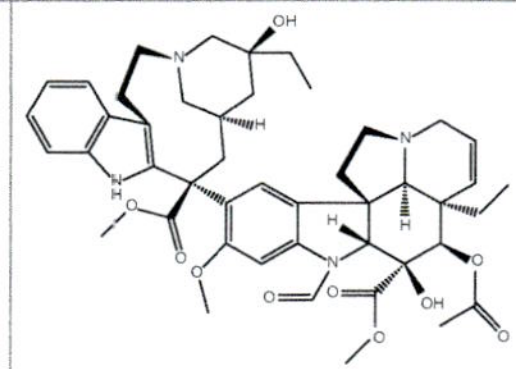	
	Cerbera manghas L.		Leaves, latex	Cerberin	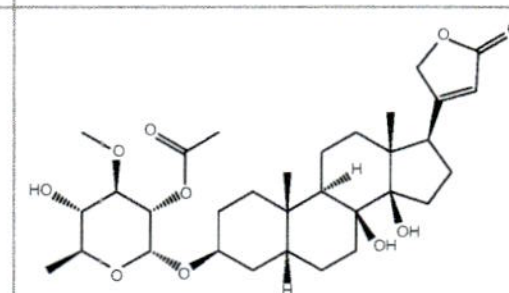	McCuddin (1974), Cheenpracha et al. (2004), Wang et al. (2010c), and Zhao et al. 2011)

	Synonym(s): *Cerbera linnaei* Montr., *Cerbera odollam* var. *mugfordii* F.M. Bailey, *Cerbera venenifera* (Poir.) Steud., *Odollamia manghas* (L.) Raf. **Common name(s)**: Sea mango, Baraibai, Toto, Vasa, Tanguin, Samanta, Kaduru. **Botanical description:** It is a small evergreen coastal tree growing up to 12 m (39 ft.) tall and native to Madagascar, Myanmar, Australia, Andaman and Nicobar Islands, Taiwan, China, Malaysia, Thailand, Sri Lanka, etc. The shiny dark green leaves are in spiral arrangement, ovoid in shape. The flowers are fragrant, possessing a white tubular five-lobed corolla about 3–5 cm in diameter, with pink to red throat. They have five stamens and ovary is positioned above the other flower parts. The fruits are egg-shaped, 5–10 cm long, and turn bright red at maturity. The seeds are known to be deadly poisonous and were used in sentence rituals to poison kings and queens in Madagascar during early times. **Medicinal properties**: It is having antitumor, antioxidant, antifungal, and analgesic properties. Used as a treatment for cancer. Fruit, when fresh, is rubbed on legs for rheumatism, latex obtained from plant is rubefacient on the skin. Fruit combined with *Datura* is used for hydrophobia treatment; the compounds obtained from root (methanol and ethanol extract) have antioxidant and antitumor activity.	Tanghinigenin	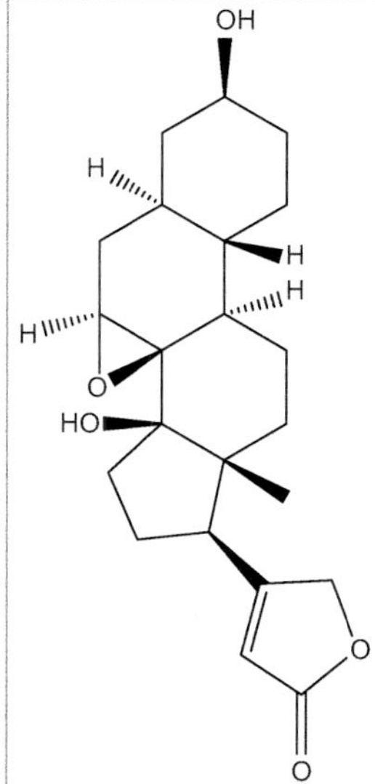	
		Neriifolin	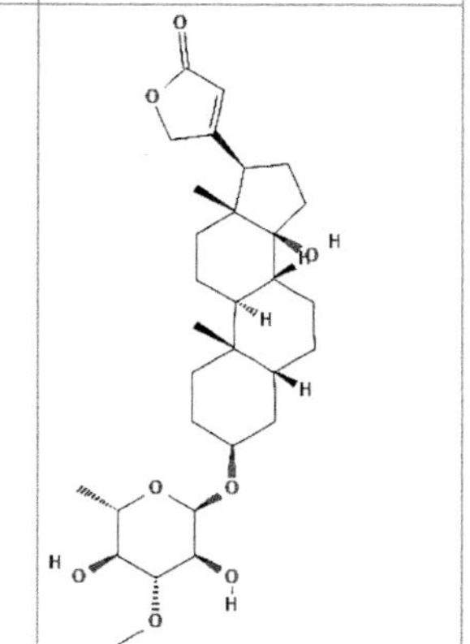	

(continued)

Table 4.10 (continued)

Family	Botanical name	Image	Plant part used	Bioactive compound(s)	Structure	Reference(s)
	Cynanchum auriculatum Royle ex Wight		Root	Caudatin	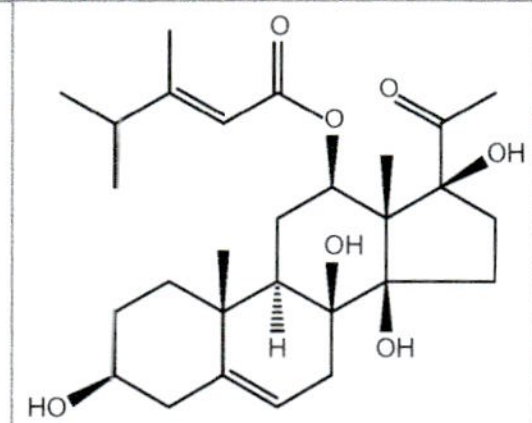	Peng et al. (2008)
	Synonym(s): *Cynanchum auriculatum* var. *sinense* T.Yamaz., *Cynanchum saccatum* W.T.Wang., *Diploglossum auriculatum* (Royle ex Wight) Meisn. **Common name(s)**: Heart-leaf swallow-wort, White he shou wu. **Botanical description**: It is native to Asian temperate and tropical regions; found in China, Nepal, Japan, and the northern parts of Pakistan and India. Its habitat is mountainous shrub land terrain, at elevations from 2800 to 3600 m. It flowers between June and August; fruiting occurs from August to December. **Medicinal use(s)**: It is used for the treatment of lung cancer and gastric cancer. It also possesses antiaging property.			Kidjoranin-3-O-α-diginopyranosyl-(1→4)-β-cymaropyranoside	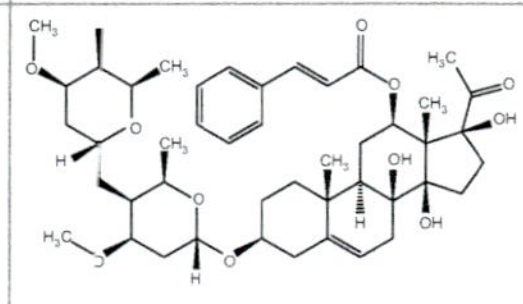	
	Echites yucatanensis (Millspex.) Stand.		Latex (leaf)	Genistein	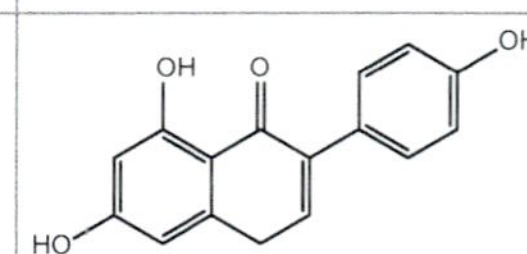	Flores and Ricalde (1996) and Alonso-Castro et al. (2011)

	Synonym(s): *Echites cincinnalis* Woodson, *Echites elegantulus* Woodson, *Echites turriger* Woodson. **Common name(s)**: Not available. **Botanical description**: It is a tree native to Mexico, El Salvador, Honduras, Yucatan, and Belice. It can grow above 30 m tall. The leaves of the plant are oblong, 3–12 cm long. The inflorescence is racemose type. The fruits are near about 30 cm large. **Medicinal properties**: It is considered as a traditional medicinal plant in Mexico. The leaves produce latex which has valuable medicinal use. Some warts can become cancerous, and some skin cancers can look like warts. The latex is used to treat warts and corns.					
	Himatanthus sucuuba (Spruce ex Müll. Arg. (Bellaco Caspi.) Wood		Fruit, seeds, leaf, bark	Lupeol	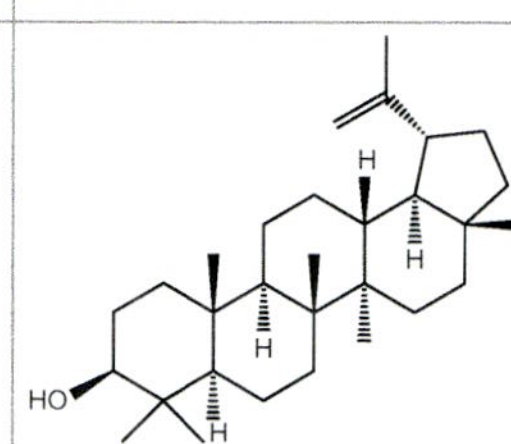	Saleem (2008)
	Synonym(s): *Himatanthus lancifolius* (Mull.Arg.) Woodson, *Plumeria sucuuba* (Spruce ex Mull.Arg.). **Common name(s)**: Agoniada, Agonium, Anaguba, Arapue, Bashi pasha, Bellaco-capsi, Caracuchu, Sucova, Sucoba. **Botanical description**: It is a tropical rainforest tree growing 8–16 m in height with a tall, narrow, pyramidal crown. The trunk is 30–40 cm in diameter with rough mottled bark. The tree produces white perfumed flowers and a 20 cm seed pod with numerous winged seeds inside. The leaves are bright green and about 25–30 cm long. **Medicinal properties**: The plant has also anticancer and antitumor activity. The bark of tree is used for wounds and sores. When a decoction of bark is prepared and internally taken, it is used for rheumatism and body aches and relieves pain. The latex of plant is used for arthritis, pain, and inflammation as well.			Spironolactone	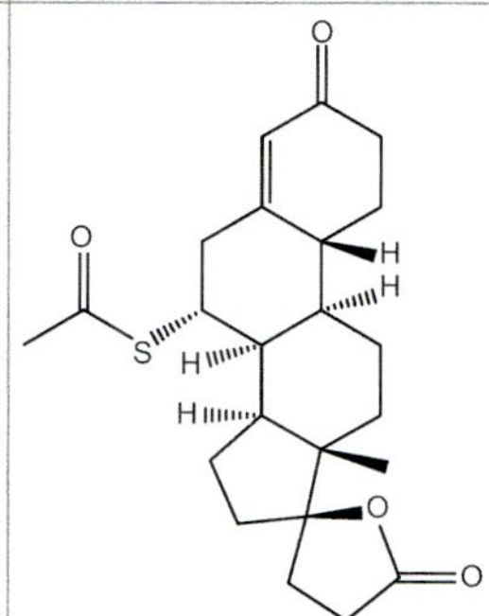	

(continued)

Table 4.10 (continued)

Family	Botanical name	Image	Plant part used	Bioactive compound(s)	Structure	Reference(s)
				Myricetin	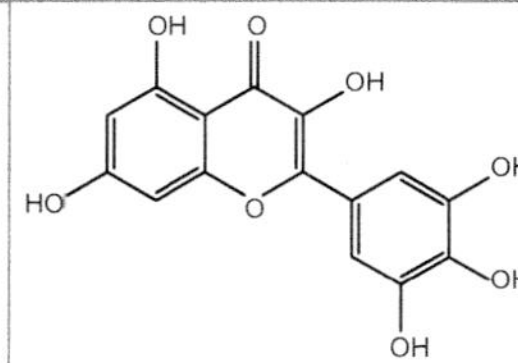	
	Himatanthus obovatus (M. Arg.) W. (agoniada)		Bark	Beta-sitosterol	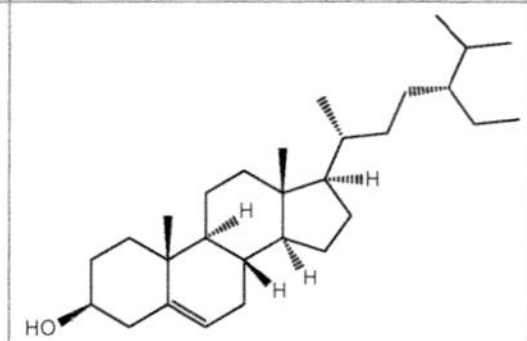	Mousinho et al. (2011)
	Synonym(s): *Himatanthus obovatus* var. *obovatus*., *Plumeria hilariana* Müll. Arg., *Plumeria latifolia* Pilg., *Plumeria obovata* Müll.Arg. **Common name(s):** Sucuba. **Botanical description**: It is native to South America – Brazil, Bolivia, and the Guianas. It is a deciduous tree with an irregular, open crown; it can grow 4–5 m tall. The crooked bole can be 15–25 cm in diameter with a very thick, corky bark. The tree is sometimes harvested from the wild for local use as a medicine and source of wood. It can be grown as a pioneer when restoring native woodland. An ornamental tree, it could be used in landscaping. **Medicinal use(s)**: It is used in the treatment of cancer, tumors, helminths, inflammation, gastric disorders, rheumatism, and bruises.					

	Himatanthus fallax (Mull. Arg.) Plumel (sucuba)		Whole plant	Isoplumericin	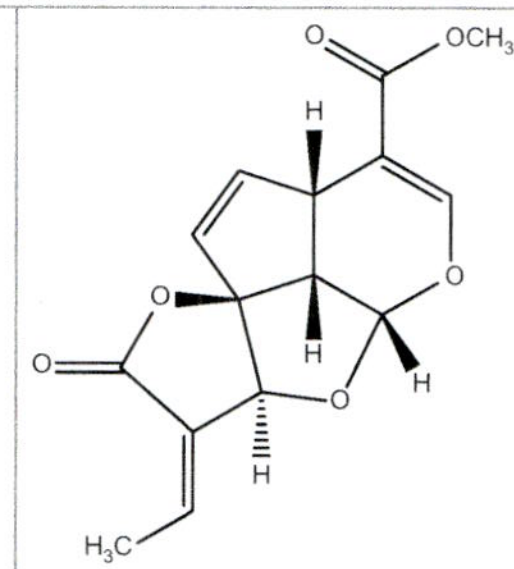	Abdel-Kader et al. (1997) and Soares et al. (2016)
	Synonym(s): *Himatanthus fallax* (Müll.Arg.) Plumel., *Himatanthus semilunatus* Markgr., *Himatanthus phagedaenicus* (Mart.) Woodson., *Himatanthus bracteatus* (A.DC.) Woodson., *Himatanthus tarapotensis* (K.Schum. ex Markgr.) Plumel. **Common name(s)**: Janaguba. **Botanical description**: It is native to South America, Brazil, and North America. It is perennial and grows up to 16 m. **Medicinal use(s)**: It is used for the treatment of cancer. It also possess antimicrobial, antioxidant, anti-inflammatory, antitumor, and anti-leishmanial activity.			Plumericin	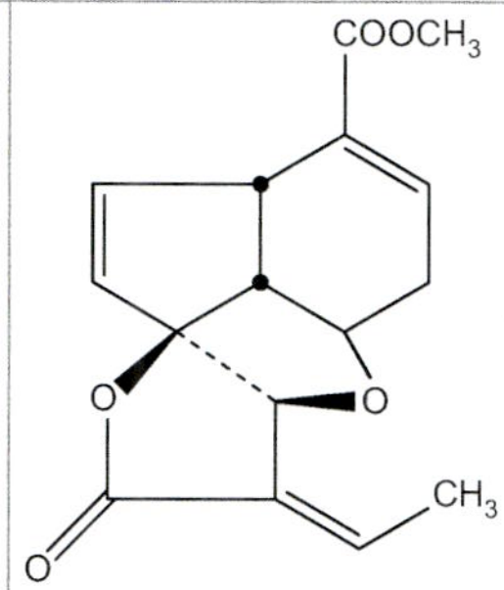	

(continued)

Table 4.10 (continued)

Family	Botanical name	Image	Plant part used	Bioactive compound(s)	Structure	Reference(s)
				7(R)-methoxy-8-epi-matairesinol	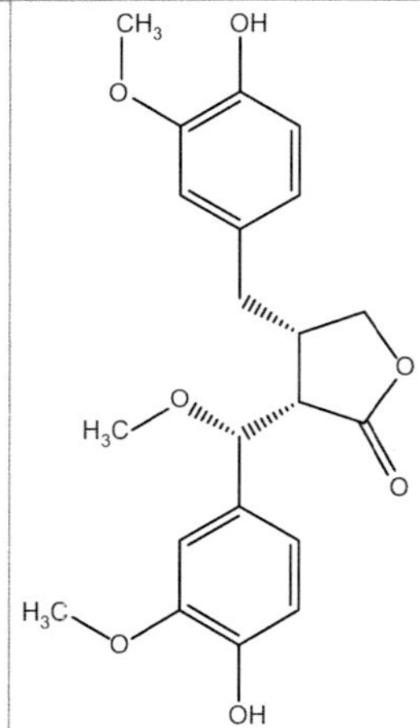	
				Plumieride	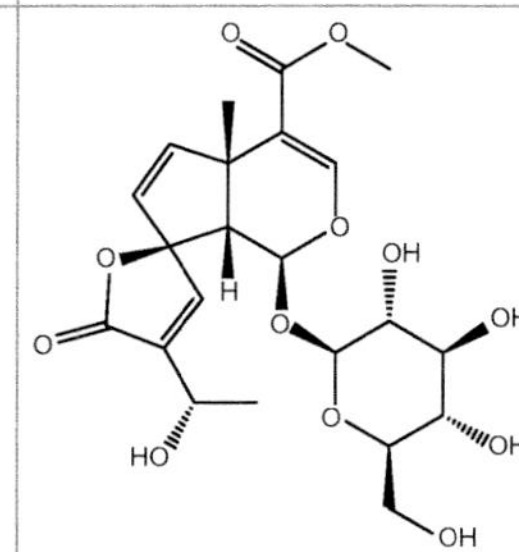	

					Matairesinol	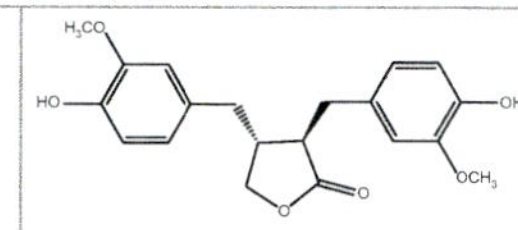	
					Pinoresinol	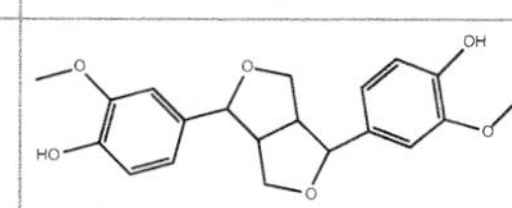	
	Himatanthus drasticus (Mart.) Plume			Bark and latex	Lupeol	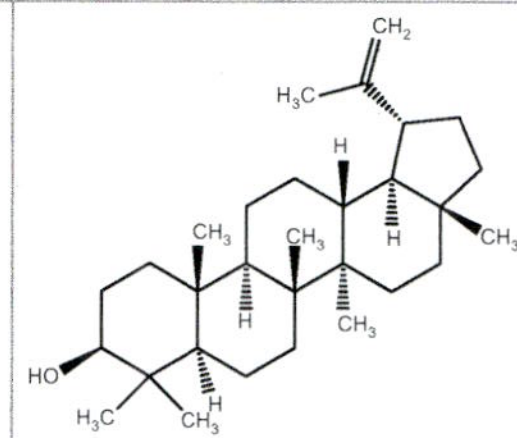	Mousinho et al. (2011)
	Synonym(s): *Plumeria drastica* Mart. **Common name(s)**: Janaguba, Bashi pasha. **Botanical description**: It is native to South America – Brazil and the Guianas. It is a deciduous shrub or small tree with a dense, more or less globose crown; it can grow 3–7 m tall. The bole can be 25–35 cm in diameter with a thick, corky bark. **Medicinal use(s)**: It is used in the treatment of cancer. Research has shown that an ethanolic extract of the leaves has very low toxicity and is effective in the treatment of some forms of cancer. It is used in the treatment of hernias and sloughing ulcers.						

(continued)

Table 4.10 (continued)

Family	Botanical name	Image	Plant part used	Bioactive compound(s)	Structure	Reference(s)
	Himatanthus articulatus (Vahl) Woodson		Bark	Lupeol	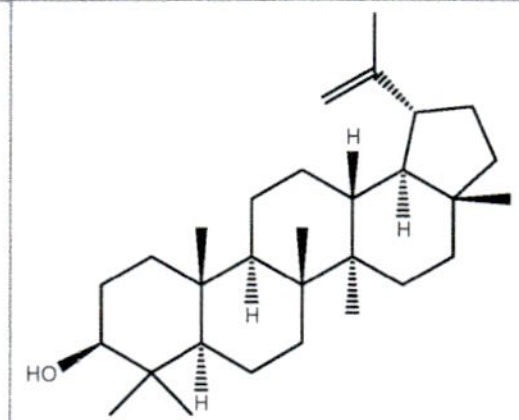	Medeiros et al. (2007), Chaturvedi et al. (2008), Jin et al. (2008), and Reboucas et al. (2011)
	Synonym(s): *Himatanthus rigida* Hoffmanns. ex Schult., *Himatanthus rigidus* Willd. **Common name(s):** Sucuba. **Botanical description**: It is native to South America, Brazil, Panama, east to French Guiana, and south to Bolivia. It is a deciduous shrub or small tree with a dense, more or less globose crown; it can grow 3–7 m tall. The bole can be 25–35 cm in diameter with a thick, corky barky. **Medicinal use(s)**: It used to treat cancer, ulcers, tumors, inflammations, syphilis, and malaria. It is also anti-genotoxic and antinociceptive.			Gallic acid	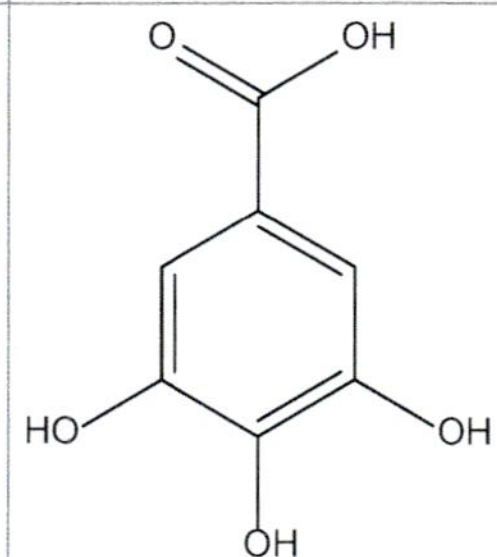	
				α-Amyrin	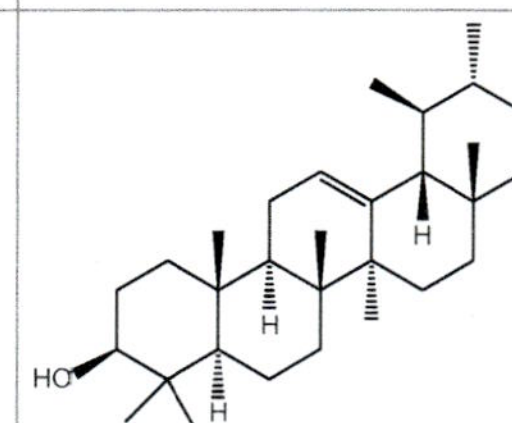	

					Aucubin (iridoid glycoside)	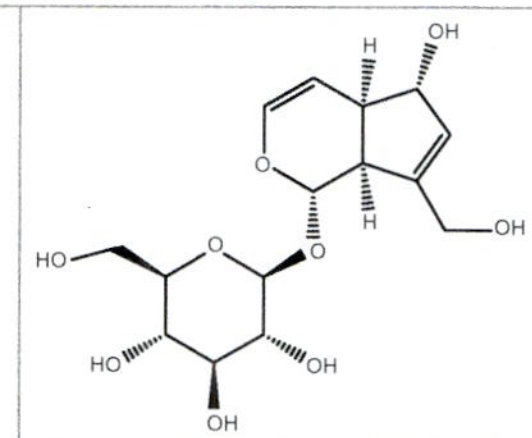	
	Ochrosia elliptica			Trunk, bark	Ellipticine (5, 11-dimethyl-6H-pyrido [4, 3-b] carbazole)	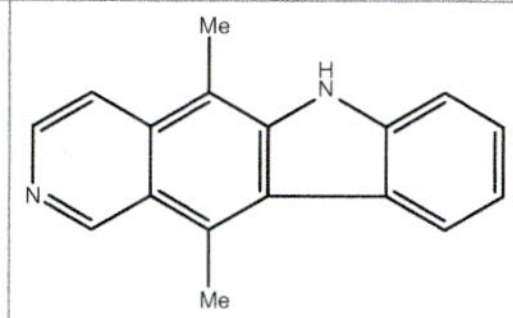	Nirmala et al. (2011)
	Synonym(s): *Bleekeria calocarpa* Hassk., *Bleekeria elliptica* (Labill.) Koidz., *Bleekeria kalocarpa* Hassk., *Excavatia elliptica* (Labill.) Markgr. **Common name(s)**: Yellowwood, Elliptic yellowwood. **Botanical description**: It is native to Northeastern Australia – Queensland to the Southwest Pacific (New Caledonia, Vanuatu), Lord Howe Island, and New Caledonia. The tree bears an inedible red-colored fruit with a white sticky flesh. It has been introduced to Florida where it is considered invasive and also to the Bahamas. *O. elliptica* is an evergreen shrub or small tree usually growing 4–6 m tall but sometimes reaching 12 m. **Medicinal use(s)**: It is used for the treatment of cancer and malaria. Its derivatives exhibited anticancer properties against several experimental neoplasms as well as towards some human cancers.						

(continued)

Table 4.10 (continued)

Family	Botanical name	Image	Plant part used	Bioactive compound(s)	Structure	Reference(s)
	Plumeria rubra L		Stem and bark	Allamcin	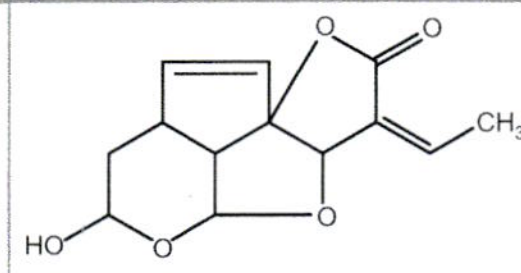	Rekha and Jayakar et al. (2011) and Choudhary et al. (2014)
	Synonym(s): *Plumeria acuminata* W.T.Aiton., *Plumeria acutifolia* Poir., *Plumeria angustifolia* A.DC., *Plumeria arborea* Noronha. **Common name(s)**: Frangipani, Graveyard flower. **Botanical description**: It is native to Mexico, Central America, Colombia, and Venezuela; it has been widely cultivated in subtropical and tropical climates worldwide and is a popular garden and park plant, as well as being used in temples and cemeteries. It has been widely cultivated in subtropical and tropical climates worldwide and is a popular garden and park plant, as well as being used in temples and cemeteries. It grows as a spreading tree to 7–8 m high and wide and is flushed with fragrant flowers of shades of pink, white, and yellow over the summer and autumn. It is also grown in Hawaii to an altitude of 2000 m. **Medicinal use(s)**: It is used for the treatment of cancer. It contains several medically active constituents and has been shown to be uterine stimulant, antifungal, antibacterial, antitumor, antiviral, analgesic, anti-spasmodic, and hypoglycemic. The plant contains fulvoplumierin, which has antibiotic activity and inhibits the growth of *Mycobacterium tuberculosis*. It is also used to treat toothache.			Allamandin	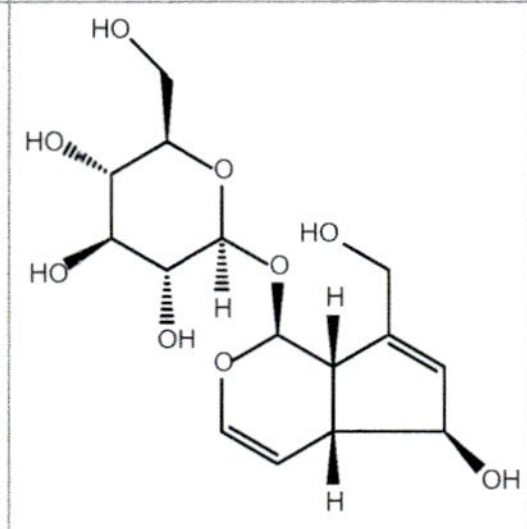	

					Fulvoplumierin	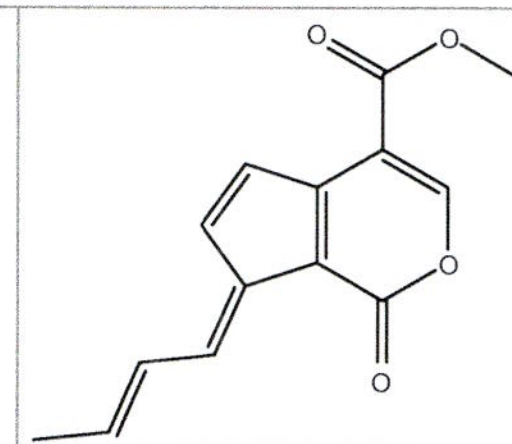	
					2,5-Dimethoxy-p-benzoquinone	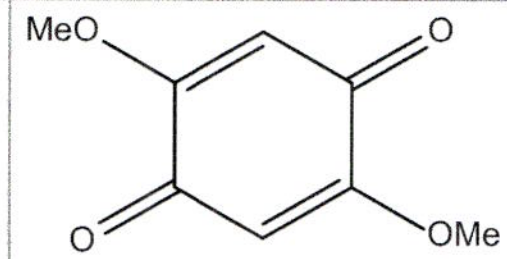	
	Rauvolfia vomitoria Afzel.			Root	Reserpine	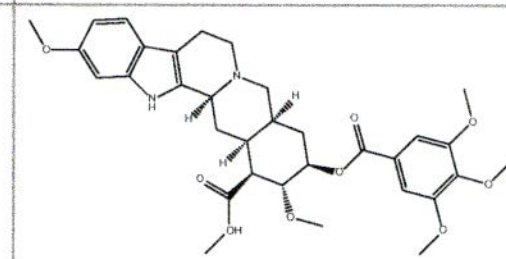	Vasileva (1969) and Yu et al. (2013a)
	Synonym(s): *Hylacium owariense* Afzel., *Rauvolfia senegambiae* A. DC. **Common name(s)**: Swizzle stick, Poison devil's-pepper, Asofeyeje (Yoruba). **Botanical description**: It is a shrub or small tree up to 8 m. Older parts of the plant contain no latex. The branches are whorled and the nodes enlarged and lumpy. Leaves in threes, elliptic-acuminate to broadly lanceolate. Flowers are minute, sweet-scented, branches of inflorescences are distinctly puberulous with hardly any free corolla lobes. Fruits are fleshy and red in color. **Medicinal use(s)**: It possesses anticancer properties. It is used to treat hypertension, insanity, snakebite, and cholera and also used for relief of central nervous system disorder including anxiety, excitement, etc.						

(continued)

Table 4.10 (continued)

Family	Botanical name	Image	Plant part used	Bioactive compound(s)	Structure	Reference(s)
	Rhazya stricta Decne.		Leaf, entire plant	Ebureine	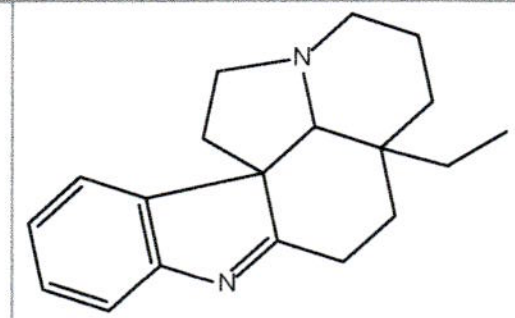	Ahmad et al. (1971), Jewers et al. (1980), and Elkady (2013)
	Synonym(s): No synonyms are recorded for this name. **Common name(s):** Harmal, Sihar, Hisawarg. **Botanical description:** It is a native poisonous plant of Afghanistan, India, Pakistan, Iraq, Yemen, Oman, Iran, and Saudi Arabia. The plant is an evergreen dwarf shrub of Apocynaceae family. **Medicinal properties**: The plant is used as a traditional medicinal plant and has antidiabetic properties. Besides in Pakistan and India, it is widely used as for treatment of tumors.					

	Tabernaemontana amygdalaefolia Jacq.		Latex, leaf	Tabernaemontanine	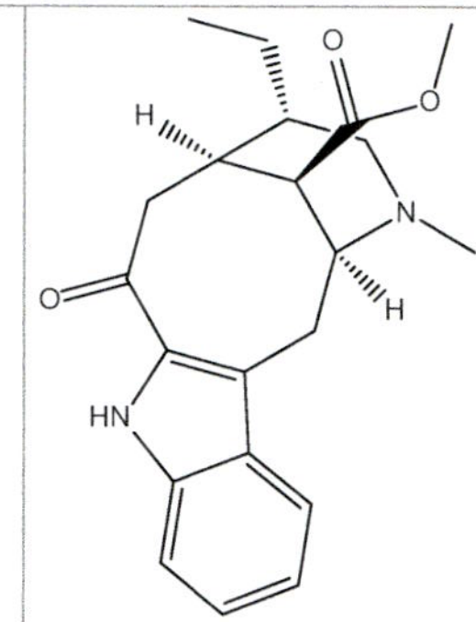	Garcia-Barriga (1975), Schultes (1979), Graham et al. (2000), Dudai et al. (2005), Singh et al. (2011), and Paterna et al. (2017).
	Synonym(s): *Tabernaemontana divaricata.* **Common name(s):** Pinwheel flower, Crape jasmine, East India rosebay, and Nero's crown. **Botanical description:** It is native to India and Southeast Asia. The plant generally grows to a height of 5–6 ft. (1.5–1.8 m) and is dichotomously branched. The large shiny leaves are deep green and about 6 in. (15 cm) in length and 2 in. (5.1 cm) in width. The waxy blossoms are found in small clusters on the stem tips. The (single) flowers have the characteristic "pinwheel" shape also seen in other genera in the family Apocynaceae such as *Vinca* and *Nerium*. Both single- and double-flowered forms are cultivated, the flowers of both forms being white. The plant blooms in spring but flowers appear sporadically all year. The flowers of the single form are unscented, but the double-flowered form has a pleasing fragrance. **Medicinal use(s):** It is used for the treatment of cancer. It possesses antioxidant, antitumor, anti-infection, and analgesic properties.			Coronaridine	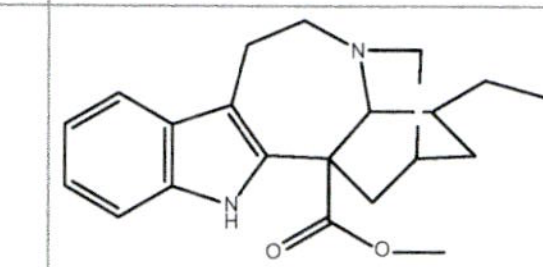	

(continued)

Table 4.10 (continued)

Family	Botanical name	Image	Plant part used	Bioactive compound(s)	Structure	Reference(s)
	Tabernaemontana elegans ramlowii Hieron		Leaf	Corynanthe	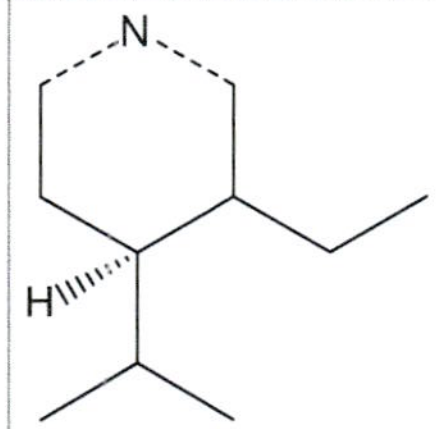	Mansoor et al. (2009)
	Synonym(s): *Conopharyngia elegans* (Stapf) Stapf., *Leptopharyngia elegans* (Stapf) Boiteau. **Common name(s)**: Toad tree. **Botanical description**: It is native to East Africa – Somalia, Kenya, Tanzania, Malawi, Mozambique, Zimbabwe, South Africa, and Swaziland. It is a shrub or a tree growing 1.5–15 m tall. The bole can be 5–30 cm in diameter. The fruit, which is considered a delicacy in parts of Africa, is gathered from the wild for local use, while the plant also has a range of local medicinal uses. *Tabernaemontana elegans* is an unusual but attractive garden tree with its sweet-smelling flowers and unusual fruits covered in brown warts, giving it the look of a toad's skin. **Medicinal use(s)**: The powdered root bark or fruits are used to treat cancer. The seeds, stem bark, and roots are used for treating heart diseases. It is used in treatments of muscular and nervous asthenia, respiratory depression, and type III poliovirus (HPV-3).			Tabernaelegantine (bisindole alkaloids)	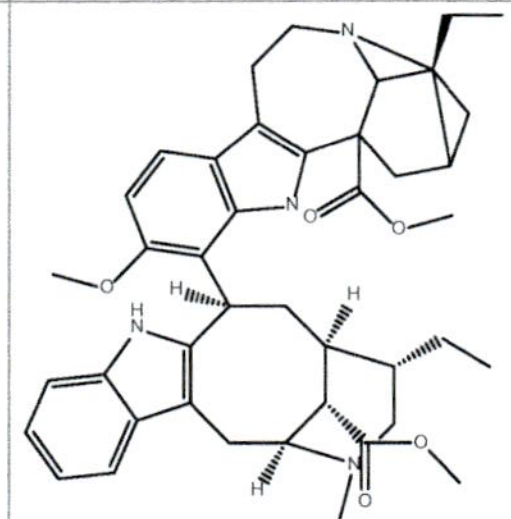	

		Tabernaelegantinine (indole alkaloid)	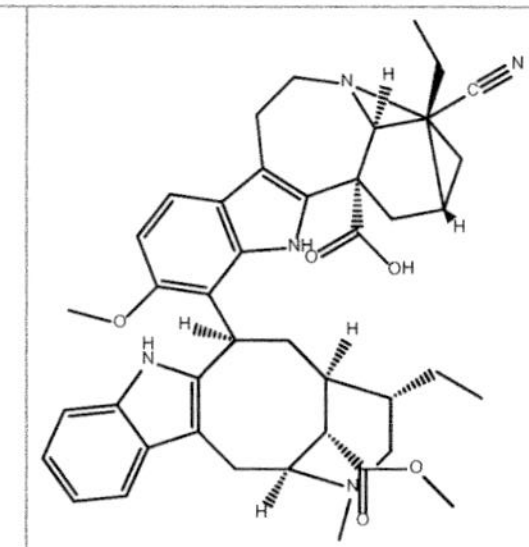	
		Voacangine	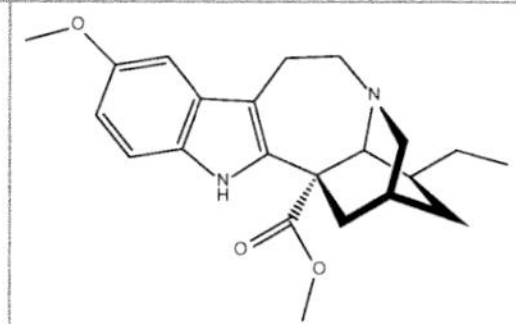	
		Dregamin	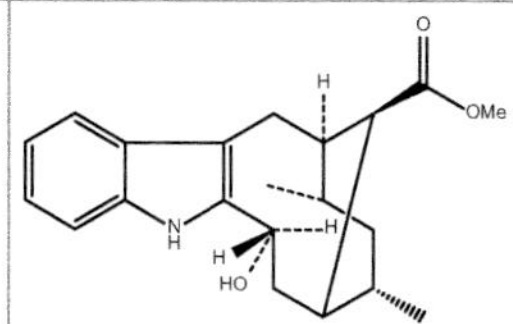	
		Vobasine	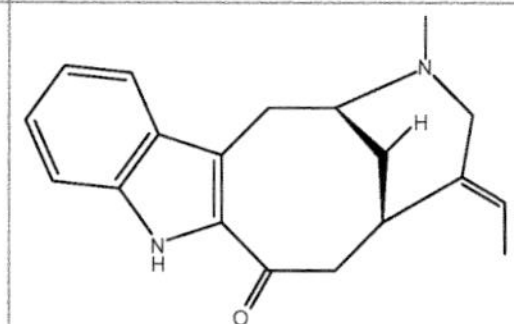	

(continued)

Table 4.10 (continued)

Family	Botanical name	Image	Plant part used	Bioactive compound(s)	Structure	Reference(s)
	Tabernaemontana catharinensis (Engl.) Tiegh.		Bud and leaf	Voacangine (indole alkaloids)	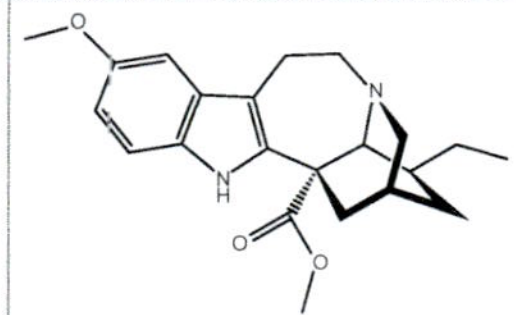	Pereira et al. (2006)
	Synonym(s): *Peschiera acuminata* (Müll.Arg.) Miers., *Peschiera affinis* Miers., *Peschiera albidiflora* Miers., *Peschiera australis* (Müll.Arg.) Miers. **Common name(s)**: Yellow oleander, Banana bush. **Botanical description**: It is native to S. America – Argentina; Uruguay; Paraguay; Southern, Eastern, and Central Brazil; and Bolivia. It is a semi-deciduous tree with an open, more or less globose crown; it can grow 3–8 m tall. The crooked bole can be 25–35 cm in diameter with a thick, corky bark. The tree is sometimes harvested from the wild for local use of its wood. A pioneer species, it can be used for restoring native woodland. An ornamental plant, it can be used in landscaping. **Medicinal use(s)**: It is used for the treatment of cancer and cardiovascular diseases including atherosclerosis and thrombosis.			Coronaridine	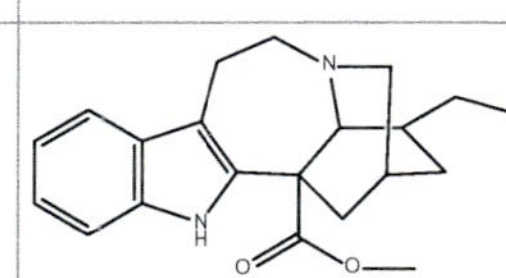	

	Thevetia ahouai (L.) A.DC.		Leaf and stem	Neriifolin (cardenolide glycosides)	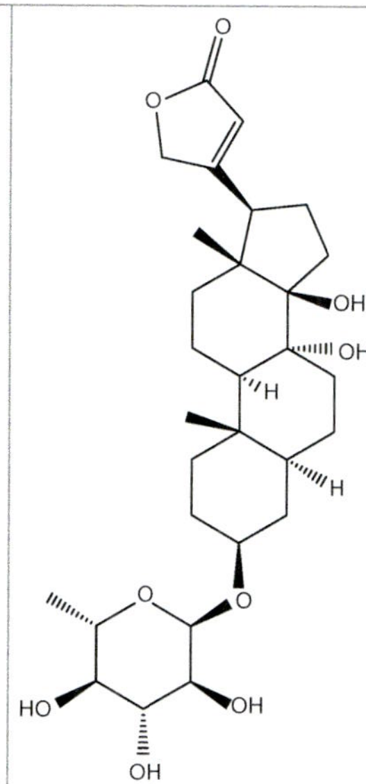	Decosterd et al. (1994), Flores and Ricalde (1996), and Noolvi et al. (2011)
	Synonym(s): *Ahouai nitida* (Kunth) Pichon, *Cerbera ahouai* L., *Cerbera nitida* Kunth, *Plumeriopsis ahouai* (L.) Rusby & Woodson, *Thevetia calophylla* Miers, *Thevetia nitida* (Kunth) A.DC. **Common name(s)**: Dog's tongue, Dog balls, Grandfather's balls. **Botanical description**: It is native to northern South America, Colombia and Venezuela, and Central America, Panama to Mexico. *Thevetia ahouai* is usually an evergreen shrub growing 1–3 m tall, though sometimes it becomes a small tree. Wet to dry thickets, sometimes in secondary growth or in open pine forest, at elevations up to 600 m. **Medicinal use(s)**: It is used for the treatment of different types of cancers and pathogenic infections of viruses, parasites and bacteria. Neriifolin is one of the cardenolide that is effective against non-small cell lung cancer. The fruits are used as a purgative in the treatment of rheumatism and dropsy.					

(continued)

Table 4.10 (continued)

Family	Botanical name	Image	Plant part used	Bioactive compound(s)	Structure	Reference(s)
	Thevetia gaumeri Hemsley		Leaf and stem	Cardiac glycoside	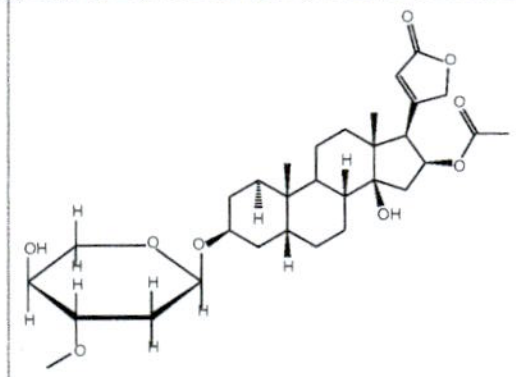	Flores and Ricalde (1996), Atta-ur-Rahman and Choudhary (2017), and RamosSilva (2017)
	Synonym(s): *Thevetia gaumeri* Hemsl., *Thevetia spathulata* Millsp., *Thevetia steerei* Woodson. **Common name(s)**: Lucky nut. **Botanical description**: It is native to northern South America, Colombia and Venezuela, and Central America, Panama to Mexico. It is usually an evergreen shrub growing 1–3 m tall, though sometimes it becomes a small tree. Wet to dry thickets, sometimes in secondary growth or in open pine forest, at elevations up to 600 m. **Medicinal use(s)**: It is used to treat cancer. It also has antifungal, antibacterial, and anti-termite properties.			Thevetiaflavone	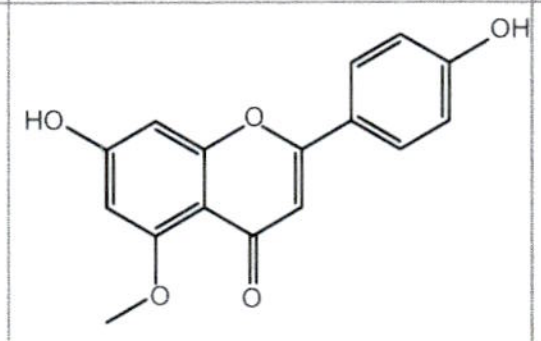	
	Thevetia peruviana (Pers.) K. Schum.		Bark	Neriifolin (1),9 thevefolin	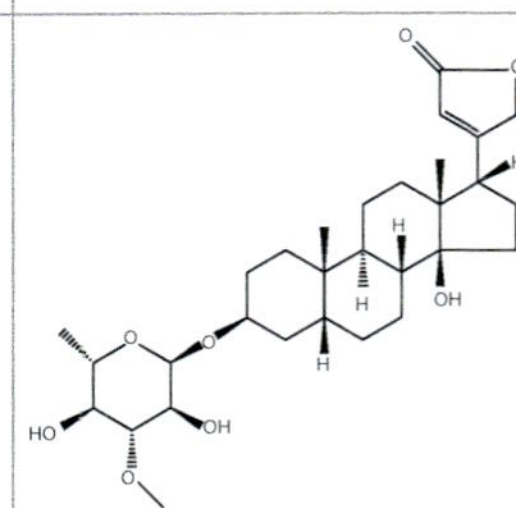	Miyagawa et al. (2009)

	Synonym(s): *Cascabela thevetia* (L.) Lippoid. **Common name(s)**: Yellow oleander. **Botanical description**: Evergreen tropical shrub or small tree; leaves are willow-like, linear-lanceolate, and glossy green in color. Its stem is green turning silver/gray as it ages. Flowers bloom from summer to fall. The long funnel-shaped sometimes fragrant yellow flowers are in terminal clusters. Its fruit is deep red-black in color. **Medicinal importance**: It has antifungal, antibacterial, and anti-termite properties.			Ruvoside	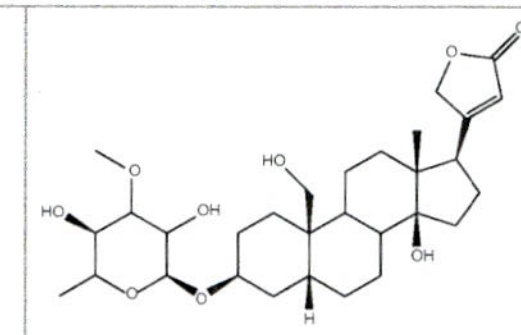	
	Trachyspermum ammi Sprague		Seeds	Thymol	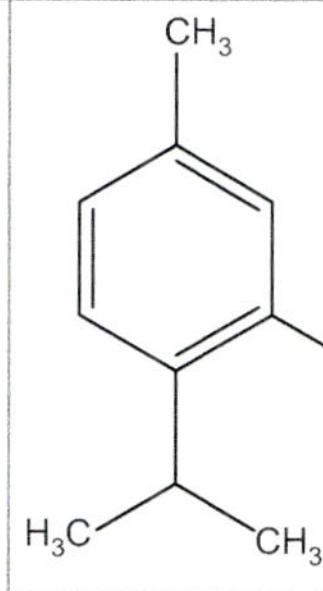	Abdel-Hameed et al. (2014)
	Synonym(s): *Carum copticum* (L.) Benth. & Hook. F. **Common name(s):** Ajwain. **Botanical description**: Ajwain is an erect, hairless or minutely pubescent, branched annual herb. The stems are grooved. The leaves are rather distant, 2–3 pinnately divided into narrow linear segments. Flowers are borne in terminal or seemingly lateral stalked, compound umbels, white and small. The fruits are ovoid, aromatic, grayish brown. The mericarps, which are the components of the fruit, are compressed, with distinct ridges and tubercular surface, 1-seeded. This is what is used as the spice ajwain, in cooking. **Medicinal use(s)**: It has shown having significant anticancer activity against MCF-7 cell lines. It is also traditionally known to aid digestion and relieve abdominal discomfort due to indigestion and antiseptic.			Carvacrol	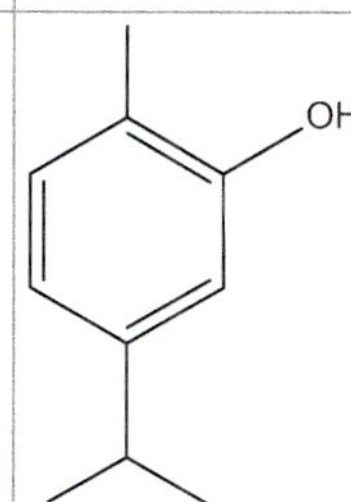	

(continued)

Table 4.10 (continued)

Family	Botanical name	Image	Plant part used	Bioactive compound(s)	Structure	Reference(s)
				p-Cymene	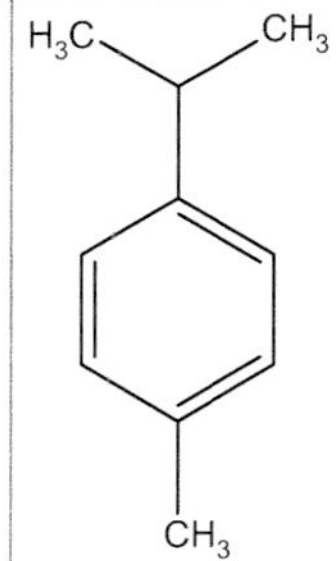	
				γ-Terpinene	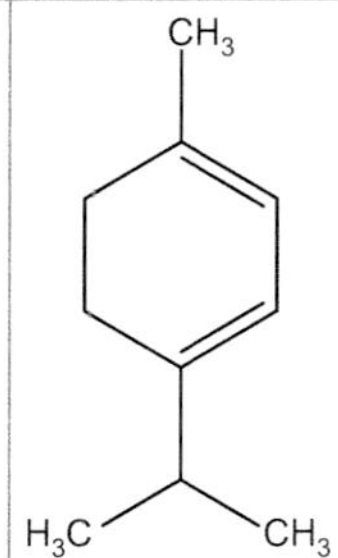	

	Trachelospermum jasminoides (Lindl.) Lem.		Stamen and leaf	Trachelogenin	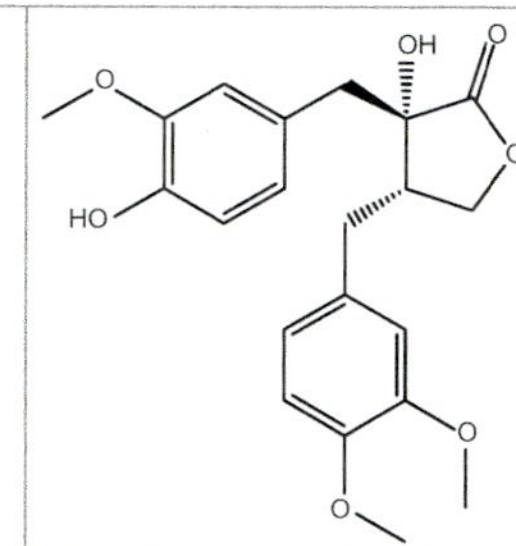	Duke and Ayensu (1985a, b), Jing et al. (2011), and Salama et al. (2015)
	Synonym(s): *Arisaema brevipes* Engl., *A. davidianum* Engl. **Common name(s)**: Himalayan cobra lily, Star jasmine. **Botanical description**: It is tuberous perennial, up to 1 m tall. It has a single leaf (rarely two), with several narrow leaflets, tapering to a threadlike tail. The flowers appear in May–July and have a green or brown and cream striped spathe (about 5 cm long) with a long, narrow point and a whitish, club-shaped spadix. The flowers are followed by a cluster of red berries. **Medicinal importance**: Root possesses antibacterial, antifungal, antiphlogistic, anti-rheumatic, anticancer, antitumor, and anti-spasmodic properties. The dried root is used internally in the treatment of coughs with profuse phlegm, tumors, cervical cancer, epilepsy, tetanus, and complaints involving muscular spasms. The fresh root is applied externally as poultice to ulcers and carcinomatous growth.			Apigenin	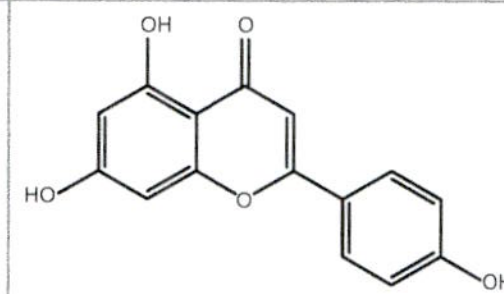	

Table 4.11 Anticancer plants of family Araceae

Family	Botanical name	Image	Plant part used	Bioactive compound(s)	Structure	Reference(s)
Araceae	*Acorus gramineus* Sol. Aiton		Root, entire plant	Acoramol	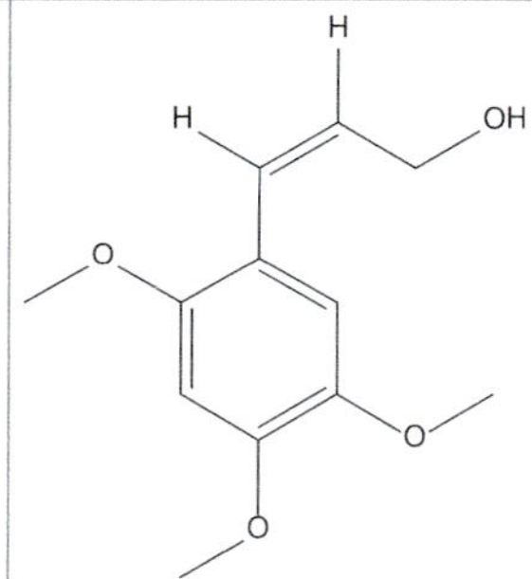	Duke and Ayensu (1985a, b) and Kim et al. (2015d)
	Synonym(s): *Acorus calamus* L., *Acorus casia* Bertol., *Acorus belangeri* Schott., *Acorus flexuosus* Raf. **Common name(s)**: Dwarf sedge, Japanese rush, Grass-leaved sweet rush, Japanese sweet flag, Dwarf sweet flag. **Botanical description**: It is native to East Asia – China, Japan, and India. It is an evergreen perennial growing up to 0.3–0.2 m tall. It flowers from June to July, and the seeds ripen from July to August. The flowers are hermaphrodite (have both male and female organs) and are pollinated by insects. It prefers moist or wet soil and can grow in water. **Medicinal use(s)**: It is used in the treatment of esophageal cancer. It also possesses anticonvulsant, antibacterial, antioxidant, insecticidal, radioprotective, glucosidase inhibitory, insulin-sensitizing, anti-epileptic, larvicidal, smooth muscle relaxant, neuroprotective, hypolipidemic, and immunomodulatory properties.			Surinamensinol	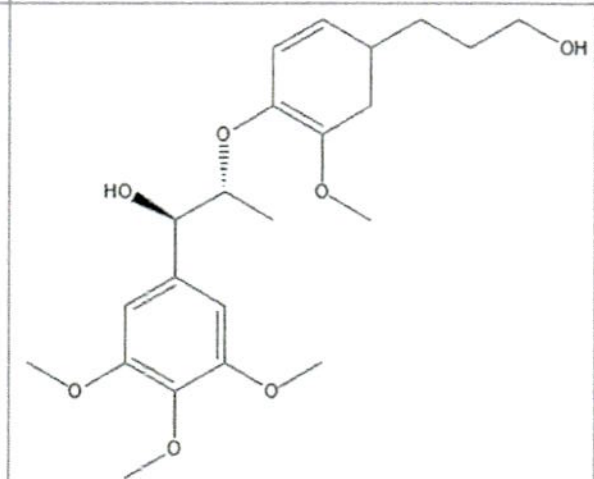	

	Amorphophallus paeoniifolius (Dennst.)		Tuber	Rhamnose Cinnamaldehyde	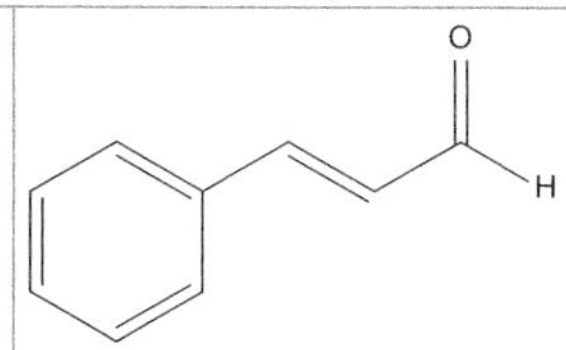	Angayarkanni et al. (2007), Dey et al. (2012), and Ansil et al. (2014)
	Synonym(s): *Amorphophallus campanulatus*. **Common name(s)**: Elephant foot yam, Oal, Gandira, Zaminkand. **Botanical description**: It is a native plant of Southern China, Bangladesh, and India. It is a striking aroid with a flower spike crowned with a bulbous maroon knob and encircled by a fleshy maroon and green-blotched bract. The solitary leaf, which emerges after the flower parts, resembles a small tree. **Medicinal use(s)**: It is used to treat cancer. The tuber is widely used for the treatment of tumor, enlarged spleen, lung swelling, and asthma. It has shown a variety of effects, specifically antiprotease activity, analgesic activity, and cytotoxic activity.			Stigmasterol	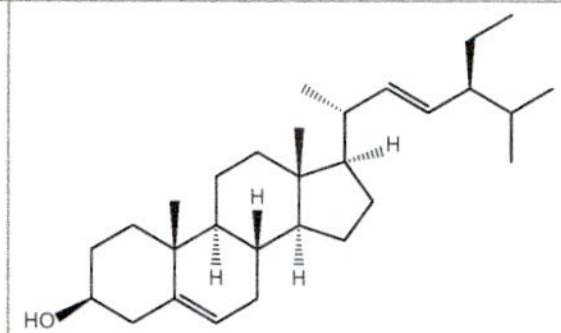	

(continued)

Table 4.11 (continued)

Family	Botanical name	Image	Plant part used	Bioactive compound(s)	Structure	Reference(s)
	Arisaema consanguineum Schott.		Root, entire plant	Asialofetuin	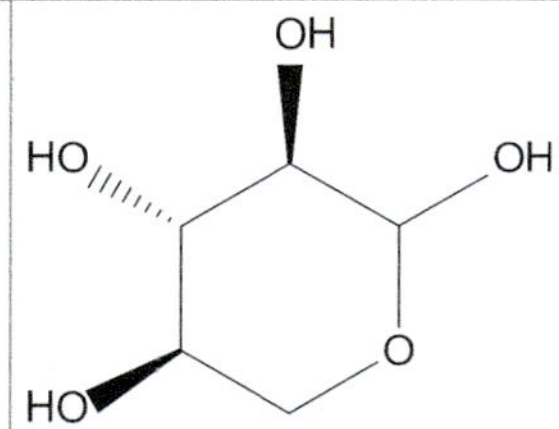	Duke and Ayensu (1985a, b), Shangary et al. (1995), and Cong et al. (2015)
	Synonym(s): *Arisaema brevipes* Engl., *Arisaema consanguineum* f. *alpestre* K.Krause. **Common name(s)**: Tian Nan Xing. **Botanical description**: It is found in the Himalayas, China, North Thailand, and North Vietnam. It is a bulb type of plant with a height of about 1–2 m and a width of 30 cm. Flowering occurs in early summers. 11–20 leaves with long tips radiate out like an umbrella. **Medicinal use(s)**: It is used to treat cervical cancer. The plant has anodyne, anticoagulant, anti-inflammatory, anti-rheumatic, anticancer, antitumor, anti-epilepsy, and anti-tetanus properties and heals muscular spasm.			Betulinic acid	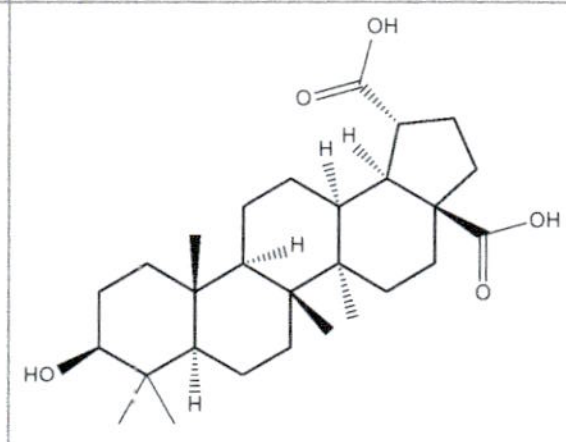	
	Arisaema lobatum Engl.		Corm	Gallic acid (tannin)	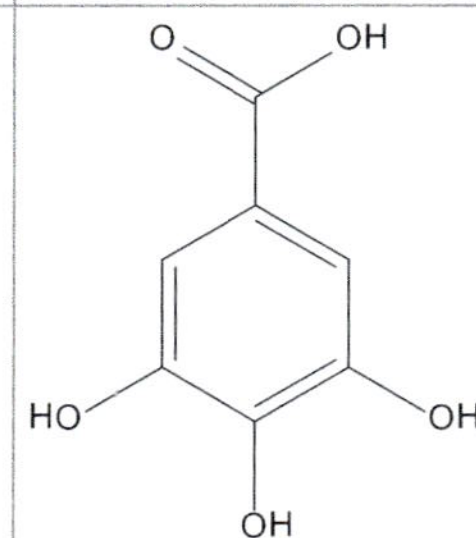	Maffei (2003a, b, c, d, e, f, g, h, i, j, k, l, m)

	Synonym(s): *Arisaema onoticum* Buchet., *A. pictum* N.E.Br.ex Hemsl. **Common name(s)**: Cobra lily. **Botanical description**: It is native of Asia and South Asia. It is a deciduous plant that grows up to 40 cm in height. The shoots when they emerge have attractive snakelike mottling. The spathe is light brown with white lines and white spadix. Deep shiny green three-lobed leaves. It is generally found in the partial shady area. **Medicinal use(s)**: It is used to treat breast cancer. It possesses anticancer, antitumor, antimicrobial, anti-inflammatory, and antifungal properties. It is also use to treat malaria, intestinal parasites, and snake and insect bites.	Phloroglucinol	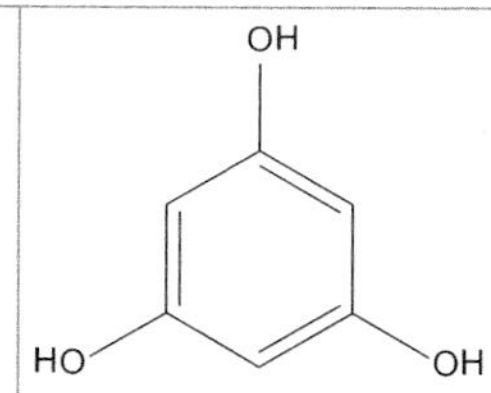	
		Chlorogenic acid	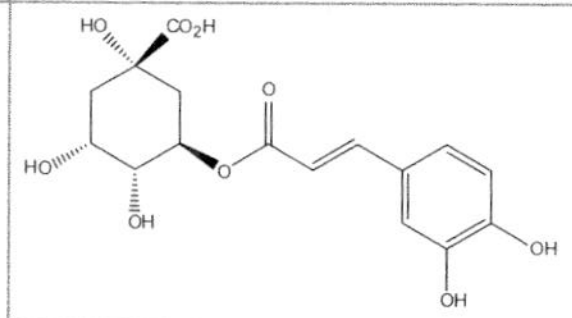	
		Luteic acid	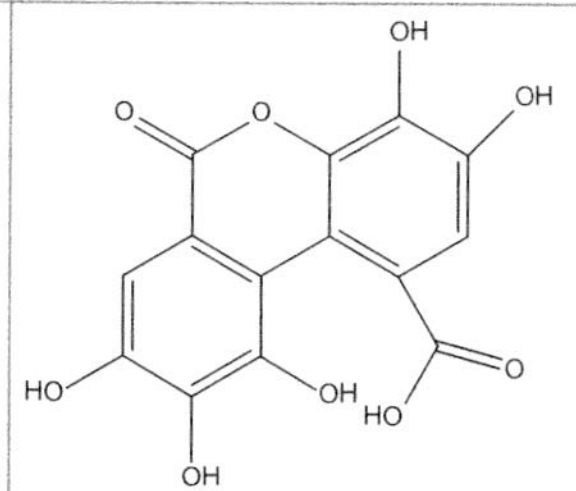	

(continued)

Table 4.11 (continued)

Family	Botanical name	Image	Plant part used	Bioactive compound(s)	Structure	Reference(s)
	Arisaema erubescens (Wall.) Schott		Root, tubers	Trigofoenoside A (saponin)	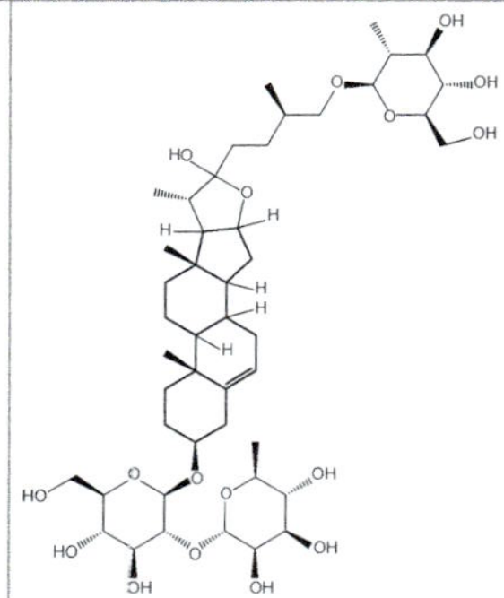	Ducki et al. (1995) and Maffei (2003a, b, c, d, e, f, g, h, i, j, k, l, m)
	Synonym(s): *Arisaema linearifolium* J.T.Yin & Gnsman, *Arum erubescens* Wall. **Common name(s)**: Chinese cobra lily, Blushing cobra lily. **Botanical description**: It is found in Nepal and seen at altitudes of 2000–2600 m in the Himalayas. It reaches 40 cm in height; its spathe varies in color from pale purplish red to purplish pink. The spathe has white stripes on the inside below and ovate long-pointed down-curved blade. Spadix is blunt, slightly longer than the spathe tube. The leaf has 7–14 narrow lancelike, long-pointed radiating leaflets. Flowering occurs in May to June. **Medicinal use(s)**: It is used to treat cervical cancer. It is also used for treating skin diseases with infection and swelling. It possesses antimicrobial, antitumor, and anti-inflammatory properties.			Paeonol	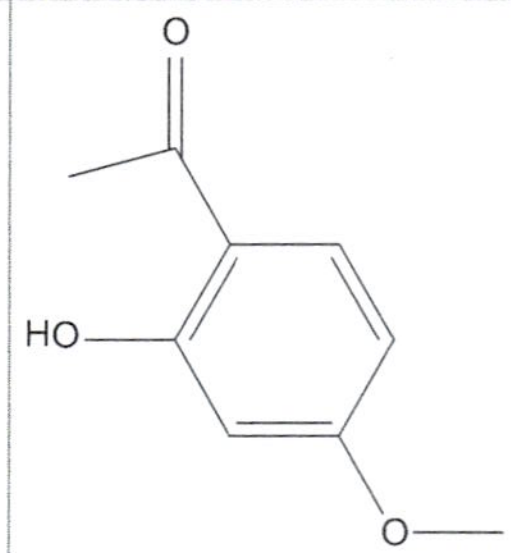	

	Arisaema jacquemontii Blume.		Tuber	Ellagic acid	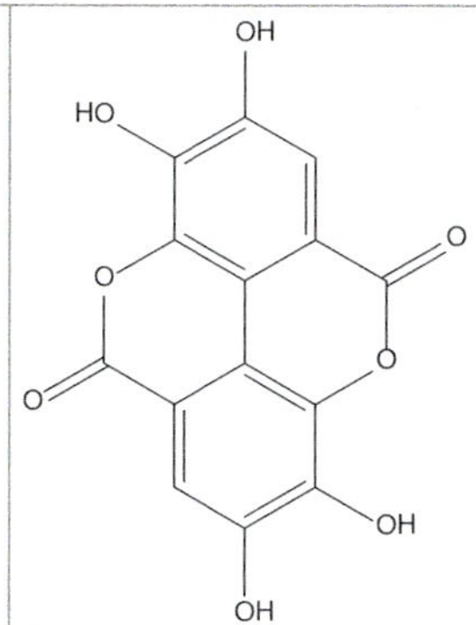	Kaur et al. (2006)
	Synonym(s): A. *brevispathum* bachet, *A. exile* Schott. **Common name(s)**: Cobra lily. **Botanical description**: It is native of Himalayan region, Southern India, and Khasi Hills and can be cultivated in shady areas of temperate regions. It is a tuberous perennial plant, grows about 70 cm tall, and dies back to ground level in the winter or dry season. The fruits are globose berries, which are red in color when mature. This plant has the ability to change sex during their lifetime. It is typically male when small and female or hermaphroditic when large. Sex-changing process depends upon its nutrition and genetics. **Medicinal use(s)**: It is used to treat cancers of the lung, colon, breast, pancreas, and brain. It is reported to have antioxidant, immunomodulating potential, anti-insect, antiproliferative, and anticonvulsant properties.			30-Nor-lanost-5-ene-3β-ol	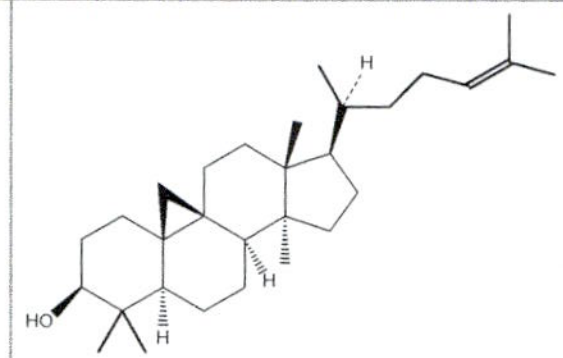	

(continued)

Table 4.11 (continued)

Family	Botanical name	Image	Plant part used	Bioactive compound(s)	Structure	Reference(s)
	Arisaema lobatum Engl.		Corm	Yuccagenin	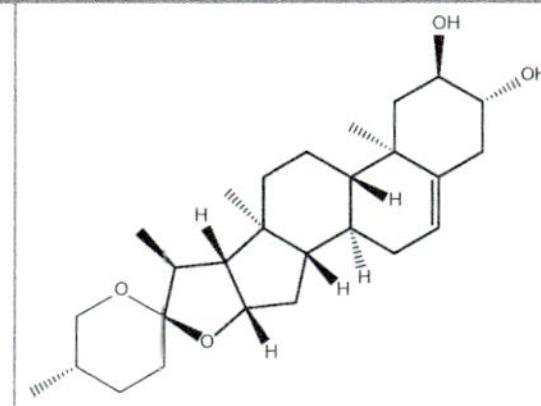	Duke and Ayensu (1985a, b) and Maffei (2003a, b, c, d, e, f, g, h, i, j, k, l, m)
	Synonym(s): *Arisaema abei* Seriz., *Arisaema album N.E.Br.*, *Arisaema anomalum* Hemsl. **Common name(s)**: Jack-in-the-pulpit. **Botanical description**: It is native to China, India, and Nepal. It is a dioecious plant. The plant grows to 150 cm tall to the top of the spadix. It has a distinct horseshoe leaf and a twisted spadix. The plant is typically male when small, and female or hermaphroditic when large, with a single plant capable of changing sex based on nutrition and genetics and perhaps changing sex several times during its life span. **Medicinal use(s)**: It is used as a folkloric antitumor drug. It has been used for treating bronchitis, asthma, chronic catarrh, flatulence, and rheumatic problems. A decoction has been used for treating sore throat. Powdered root is mixed with honey or syrup for internal use and also used as an ointment externally.					

	Arisaema tortuosum Schott		Tuber	Diphenyl-1-picrylhydrazyl	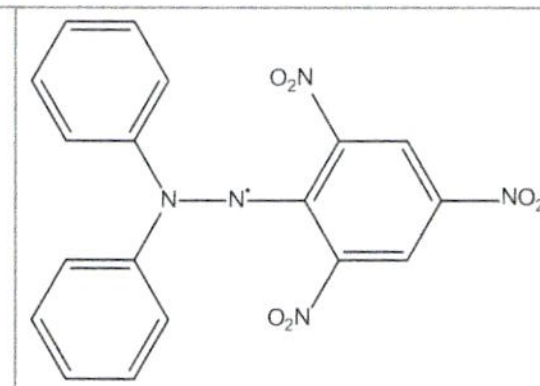	Dhuna et al. (2005), Lan et al. (2007), and Nile and Park (2013)
	Synonym(s): *A. curvatum* (Roxb.) Kunth., *Arum tortuosum*. **Common name(s)**: Whipcord cobra lily. **Botanical description**: It is native to East Asia – the Himalayas from Shimla to Sikkim and Bhutan. It is a perennial growing up to 0.6 m by 0.3 m tall. It flowers from May to June. The flowers are monoecious (individual flowers are either male or female, but both sexes can be found on the same plant) and are pollinated by flies. The plant is not self-fertile. It can grow in full shade (deep woodland), semi-shade (light woodland), or no shade. It prefers moist soil. **Medicinal use(s)**: It is used to treat breast cancer. The tubers have good aesthetic, anticancer, anti-hepatotoxic, antimicrobial, and antioxidant properties. Its proteins comprise of pro-inflammatory components that show anticancer activity against human cancer cell lines.			3-Ethylbenzothiazoline-6-sulfonic acid	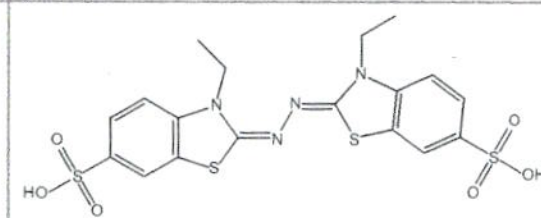	

(continued)

Table 4.11 (continued)

Family	Botanical name	Image	Plant part used	Bioactive compound(s)	Structure	Reference(s)
	Epipremnum pinnatum (L.) Engl.		Leaves	Camptothecin	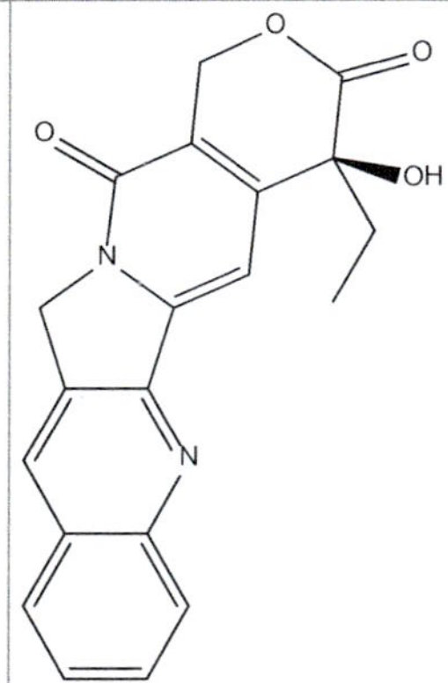	Lan et al. (2007)
	Synonym(s): *Epipremnum aureum*, *Epipremnum angustilobum* K.Krause, *E. crassifolium* Engl. **Common name(s)**: Golden pothos, Golden pothos vine, Pothos, Centipede tongavine, Taro vine, Devil's ivy. **Botanical description**: The plants are lianas. The stems are tree climbing and often have aerial roots. The leaves of older, tree-climbing plants are much larger, deeply pinnately lobed. **Medicinal description**: Possesses anticancer properties. An infusion of the leaves is consumed to effectively treat rheumatism and acts as a general tonic and an anticancer agent. A decoction of the leaves is taken as a treatment for malaria, chest pain, and diabetes and to alleviate toothache.					

	Pistia stratiotes Linn.		Leaves	β-Elemere	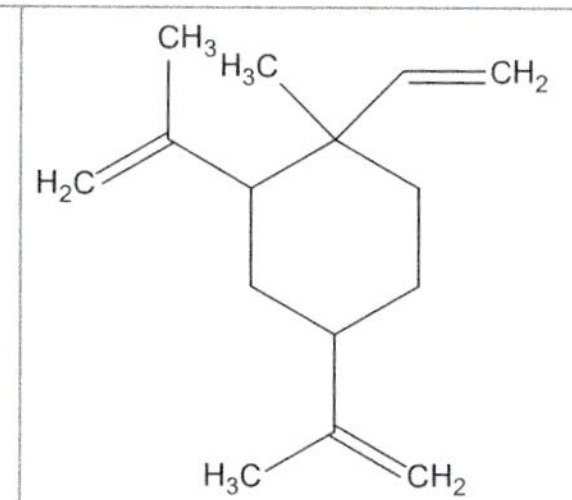	Chan and Wong (2015) and Mathew (2014)
	Synonym(s): *Pistia stratiotes* var. *cuneata* Engl., *Pistia stratiotes* var. *linguiformis* Engl., *Pistia stratiotes* var. *obcordata* (Schleid.) Engl., *Pistia stratiotes* var. *spathulata* (Michx.) Engl. **Common name(s)**: Water cabbage. **Botanical distribution**: It is a perennial monocotyledon with thick, soft leaves that form a rosette. It floats on the surface of the water, its roots hanging submersed beneath floating leaves. The leaves can be up to 14 cm long and have no stem. The flowers are dioecious and are hidden in the middle of the plant among the leaves. Small green berries form after successful fertilization. The plant can also undergo asexual reproduction. Mother and daughter plants are connected by a short stolon, forming dense mats. **Medicinal importance**: The leaf extract is used to treat gout, fevers, tuberculosis, dysentery, eczema, leprosy, ulcer, piles, syphilis, and parasitic worms. The ash of water lettuce is used for curing tinea. *P. stratiotes* works as antioxidant, antibacterial, bronchodilator, antitumor, antifungal, diuretic, antiprotease, emollient, antidiabetic, and antimicrobial.			5,6-Dihydroxyindole	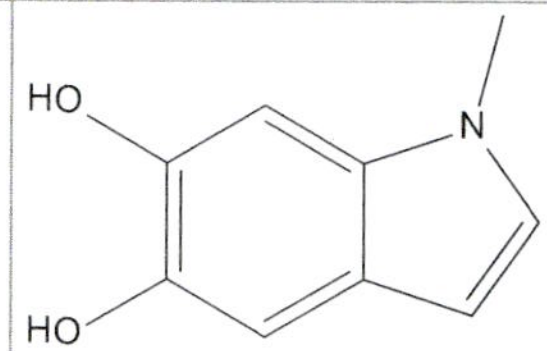	

(continued)

Table 4.11 (continued)

Family	Botanical name	Image	Plant part used	Bioactive compound(s)	Structure	Reference(s)
	Rhaphidophora korthalsii Schott		Leaf	5,6-Dihydroxyindole	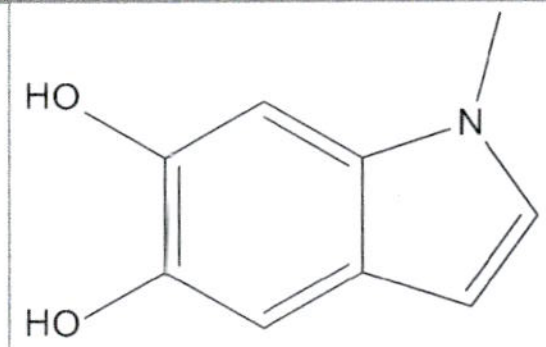	Wong and Tan (1996) and Yeap et al. (2013)
	Synonym(s): *Rhaphidophora palawanensis* Merr. *Pothos bifarius* Wall. ex Hook.f., *P. celatocaulis*N.E.Br., *Rhaphidophora celatocaulis* (N.E.Br.) Alderw. **Common name(s)**: Shingle plant. **Botanical distribution**: Rare tropical climber with juvenile blue-green leaves held tightly together, overlapping like shingles. The young leaves grow about 5 in. wide and give large vertical space in shady area, where it can cling to the surface. **Medicinal importance**: Used to treat cancer and skin diseases.					
	Typhonium divaricatum (L.) Decne		Tuber	N-Acetyllactosamine	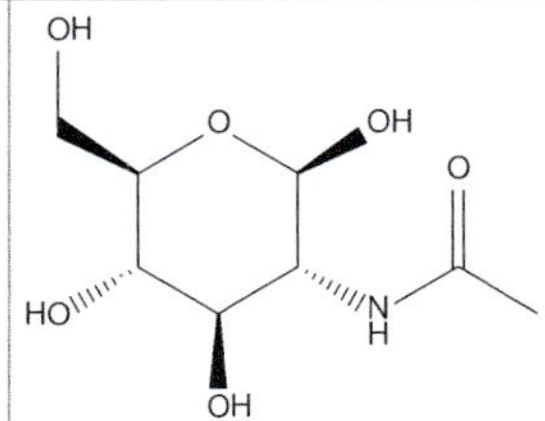	Allen (1995), Luo et al. (2007), and T.K. and Kumar (2015)

	Synonym(s): *Arum diversifolium* Blume., *Arum roxburghii* Thwaites., *Arum trilobatum* Thunb. *Arum maculatum*. **Common name(s)**: Rodent tuber. **Botanical description**: *Typhonium divaricatum* is found in Southeast Asia, growing in wild soggy places such as near streams. The plant at maturity produces flowers ending with a long filament resembling the tail of a rodent. Hence the lay name obtained rodent tuber. *T. divaricatum* is a green plant about one foot tall. Its flower ends in a long filament resembling the tail of a mouse. The plant grows readily in soft, damp, shady areas. It is now grown in household gardens in Malaysia and Singapore for its medicinal value. The plant is highly irritant to the skin and mucous membrane; thus, care should be taken during its preparation for use. **Medicinal use(s)**: It is used for the treatment of breast cancer, colon cancer, cough, and tumor. It possesses anticancer, anti-inflammatory, antibacterial, antiviral, antitumor, and antioxidant properties.					
	Typhonium flagelliforme (Lodd.) Blume		Leaves Root	Pheophorb:de-a	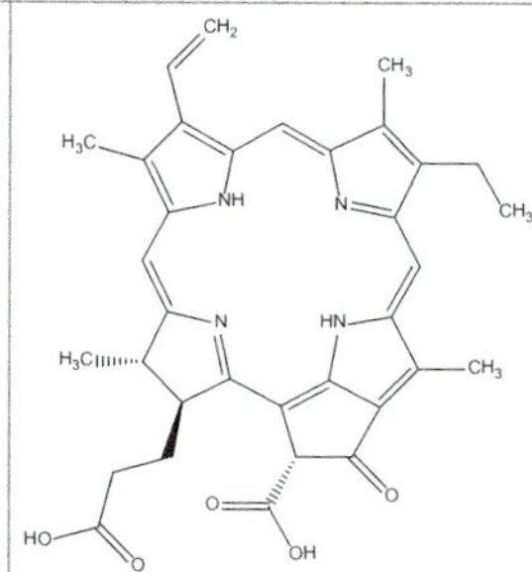	Lai et al. (2010) and Nasir and Bohari (2015)

(continued)

Table 4.11 (continued)

Family	Botanical name	Image	Plant part used	Bioactive compound(s)	Structure	Reference(s)
	Synonym(s): *Arum angulatum* Griff., *Typhonium sylvaticum* Voigt, *Arum divaricatum* L. **Common name(s)**: Karinthakara. **Botanical description**: This plant grows wild in wasteland and is native to the Southeast Asian countries and the southern part of India and Sri Lanka. It naturally grows in tropical forests to a height of 25–30 cm. It has an oblong whitish tuber, triangular leaves, and a spathe which is dilated and round at the base enclosing the yellowish spadix. For mature plants ready to use, the diameter of the bulbs is 1–2 cm. **Medicinal use(s)**: It is effective against the tumor cells of the mammary glands, lung, and liver. It has high antioxidants that are useful in inhibiting the growth of cancer cells. It is also used for treatment of internal injuries and edema.			Pyropheophorbide-a	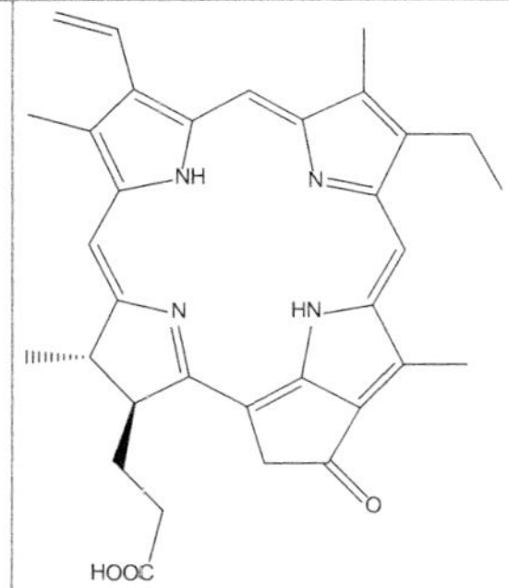	
				Methyl pyropheophorbidea-hexadecanoic acid	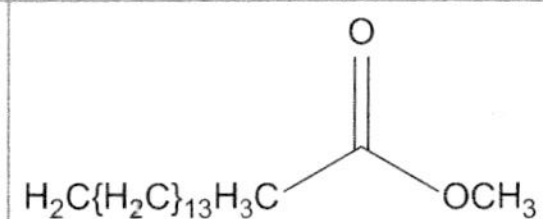	
				Campesterol	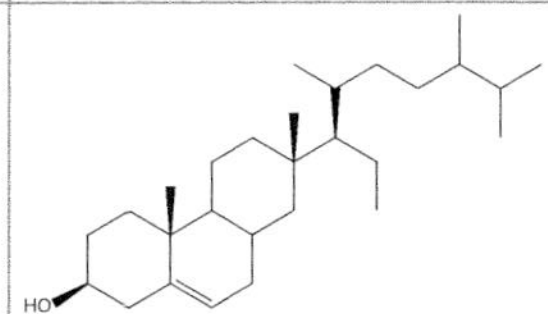	
				Stigmasterol	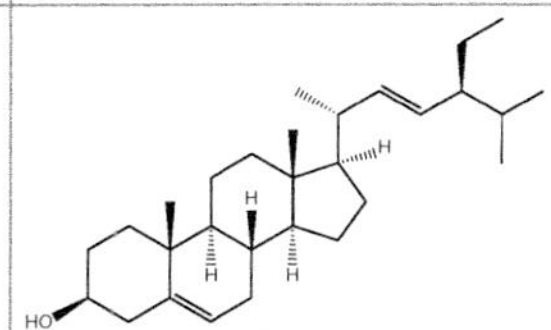	

Aristolochiaceae	*Aristolochia contorta*		Root, fruit	Lysicamine	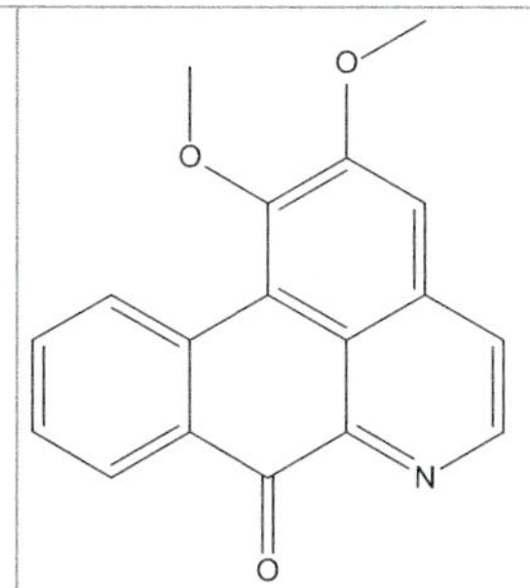	
	Synonym(s): *A. nipponica*. **Common name(s)**: Birthwort, Ma Dou Ling. **Botanical description**: It is native to East Asia – China, Japan, Korea, and Manchuria. It is a perennial herb, growing up to 1.5 m in height. It flowers from May to July, and the seeds ripen from September to October. The flowers are hermaphrodite (have both male and female organs) and are pollinated by flies. It has stout elongated rhizomes. Stem slender, glabrous; leaves alternate, cordate or broadly ovate-cordate, 4–10 cm long, 3.5–8 cm wide, acute or obtuse at tip, cordate at base, entire; petioles 1–7 cm long. **Medicinal use(s)**: The fruit is used in the treatment of cancer, coughs, inflammation of the respiratory organs, hemorrhoids, and hypertension. It is also used to resolve phlegm and lower blood pressure. It has an antibacterial action. The root contains aristolochic acid. This has anticancer properties and can be used in conjunction with chemotherapy and radiotherapy.			Aporphine	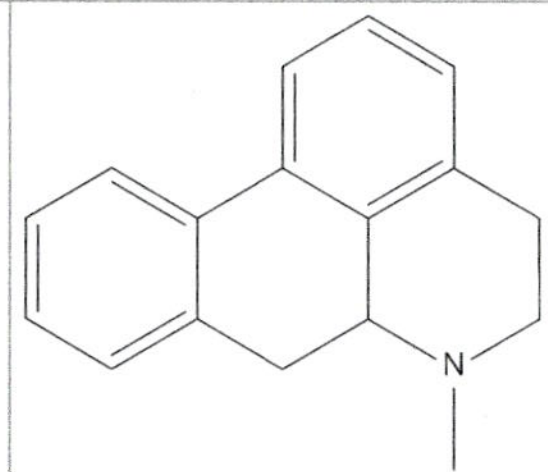	

(continued)

Table 4.11 (continued)

Family	Botanical name	Image	Plant part used	Bioactive compound(s)	Structure	Reference(s)
				Aristolochic acid	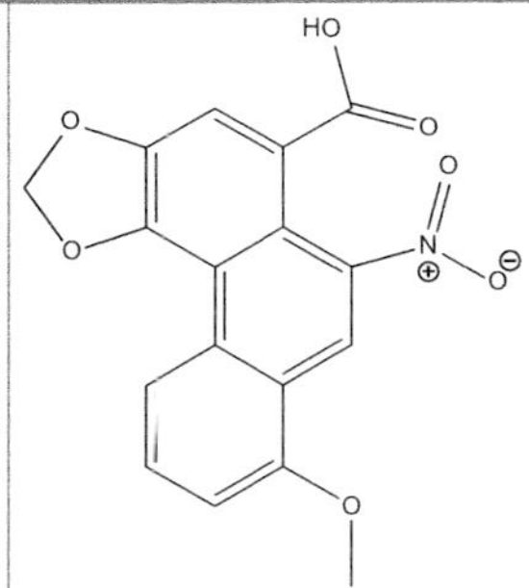	

Table 4.12 Anticancer plants of family Araliaceae

Family	Botanical name	Image	Plant part used	Bioactive compound(s)	Structure	Reference(s)
Araliaceae	*Oplopanax horridus* (Sm.) Torr. & Gray ex. Miq.		Root bark	Oplopantriol A	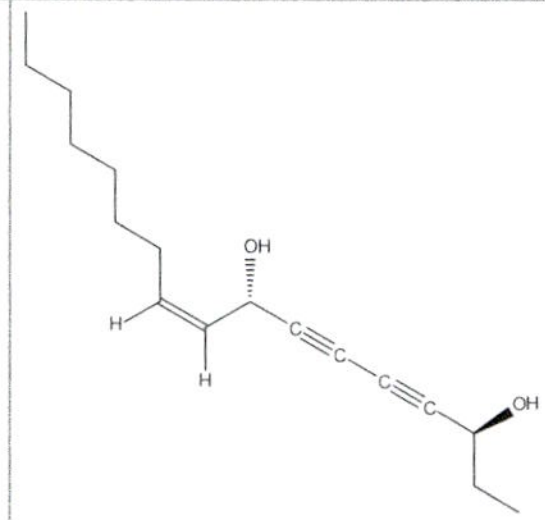	Zhang et al. (2014)
	Synonym(s): *Echinopanax horridus* (Sm.) Decne. & Planch. ex Harms, *Fatsia horrid* (Sm.) Benth. & Hook.f., *Panax horridus* Sm., *Horsfieldia horrida* (Sm.) Seem. **Common name(s)**: Devil's club. **Botanical description**: It is native to western Northern America. It is a deciduous shrub growing up to 2 m by 2 m. It flowers from June to August. The flowers are hermaphrodite. It is suitable for light (sandy), medium (loamy), and heavy (clay) soils, and suitable pH is acid, neutral, and basic (alkaline) soils. It can grow in full shade (deep woodland) or semi-shade (light woodland). It prefers moist soil and can tolerate maritime exposure. **Medicinal use(s)**: It is used to treat pancreatic, colorectal, and colon cancer. It has antitumor and antiproliferative effects. An infusion is used in the treatment of coughs and colds, bronchitis, tuberculosis, stomach problems, etc. A decoction is drunk for the treatment of rheumatism and is also applied externally as a wash on the affected joints. A poultice of the bark has been used to relieve pain in various parts of the body.					

(continued)

Table 4.12 (continued)

Family	Botanical name	Image	Plant part used	Bioactive compound(s)	Structure	Reference(s)
	Panax ginseng C.A.Mey.		Roots	Ginsenosides RG1	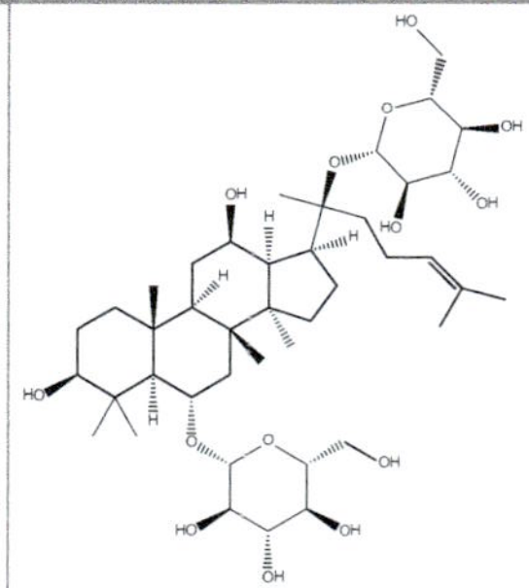	Chang et al. (2003)
	Synonym(s): *Aralia ginseng* (C.A.Mey) Baill., *P. verus* Oken., *P. quinquefolius* var. *ginseng* (C.A.Mey) Regel & Maack. **Common name(s)**: Korean ginseng. **Botanical description**: It is a perennial slow-growing plant with a large spindle-shaped fleshy root and a smooth erect stem, 1–2 ft. high; root spindle-shaped or forked. At the top of the stem are 3 large leaves palmately divided into 4–5 (occasionally 3–7) sharp-toothed oblong-lance-shaped leaflets. In June to July, the leaf axil bears an umbel of yellow-green, scented flowers. Fruits two-seeded red berries. **Medicinal importance**: It is used for the treatment of anxiety, asthma, cancer, diabetes, depression, and erectile dysfunction.					

	Aralia nudicaulis L.		Parillin	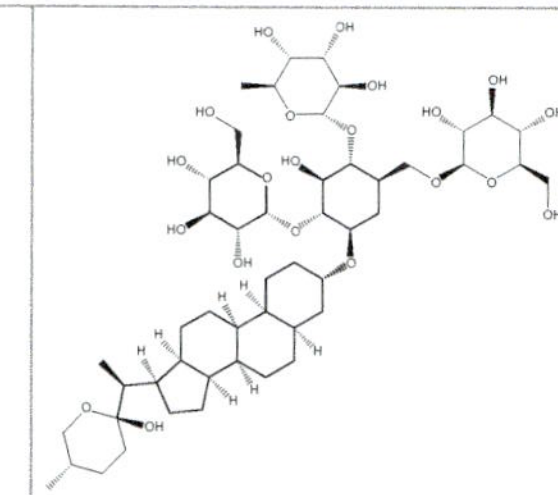	Li (2010)
	Synonym(s): *Aralia nudicaulis* f. *abortiva* Dans., *Aralia nudicaulis* f. *depauperata* Vict., *Aralia nudicaulis* var. *elongata* Nash. **Common name(s)**: Wild sarsaparilla. **Botanical description**: It is native to North America – Newfoundland to Manitoba, south to North Carolina and Missouri. It is a perennial growing up to 0.4 m by 0.3 m. It flowers in June, and the seeds ripen from August to September. The flowers are hermaphrodite (have both male and female organs) and are pollinated by bees. It is suitable for light (sandy), medium (loamy), and heavy (clay) soils and can grow in nutritionally poor soil. It prefers moist soil and can grow in full shade (deep woodland) or semi-shade (light woodland). **Medicinal importance**: It is used to treat cancer and has antifungal, anti-inflammatory effects. It has showed in vitro activity against human colon and leukemia cancer cell. It helps in treating hypofunction of the adrenal cortex and reduces hyperlipidemia associated with liver stress.		Sitosterol stigmasterin	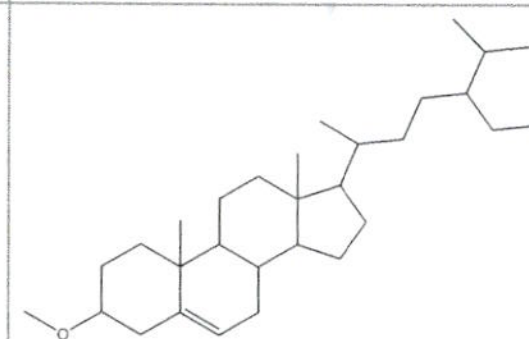	

(continued)

Table 4.12 (continued)

Family	Botanical name	Image	Plant part used	Bioactive compound(s)	Structure	Reference(s)
	Eleutherococcus senticosus Maxim.		Rhizomes	Syringin	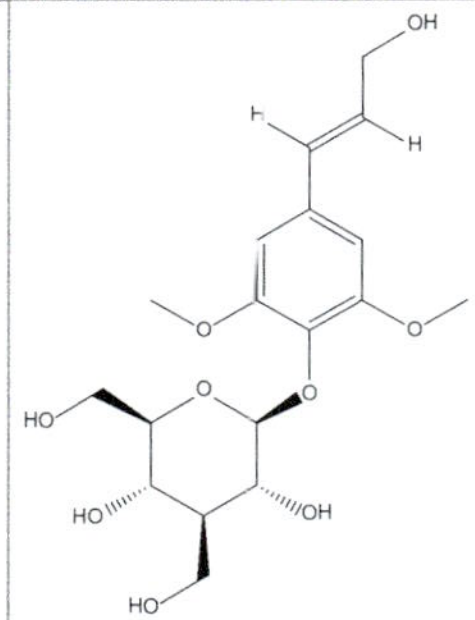	Davydov and Krikorian (2000)
	Synonym(s): *Acanthopanax asperatus* Franch. & Sav., *Acanthopanax senticosus* (Rupr. & Maxim.) Harms, *Eleutherococcus asperatus* (Franch. & Sav.) Koidz., *Hedera senticosa* Rupr. & Maxim. **Common name(s)**: Siberian ginseng. **Botanical description**: It is native to East Asia – China, Japan, and Siberia. It is a deciduous shrub growing up to 2 m at a slow rate. It flowers in July. The flowers are hermaphrodite (have both male and female organs) and are pollinated by insects. It is suitable for light (sandy), medium (loamy), and heavy (clay) soils and can grow in nutritionally poor soil. It can grow in semi-shade (light woodland) or no shade. It prefers moist soil. It can tolerate atmospheric pollution. **Medicinal use(s)**: It is used for treating the side effects of cancer-related chemotherapy. It has been shown to have neuroprotective effects against breast (mammary gland) carcinoma, stomach carcinoma, oral cavity carcinoma, skin melanoma, and ovarian carcinoma. It is used for conditions of the heart and blood vessels such as high blood pressure, low blood pressure, hardening of the arteries (atherosclerosis), and rheumatic heart disease.			Eleutheroside E	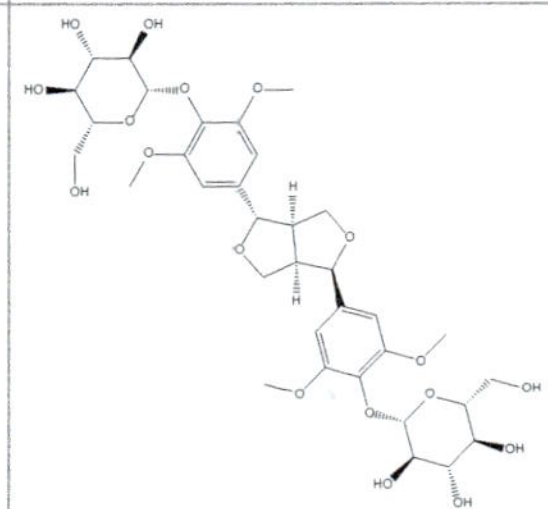	

	Panax notoginseng (Burkill) F.H.Chen		Root	Ardisiacrispin	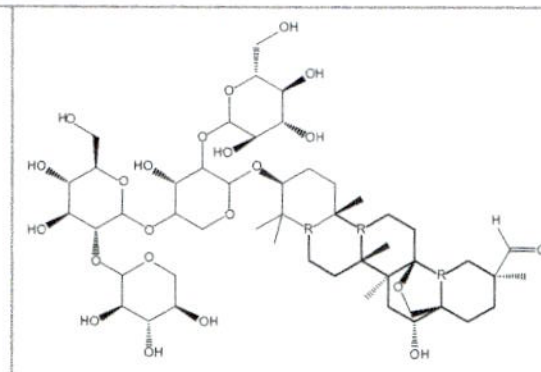	Du et al. (2007) and Sun et al. (2010a, b)
	Synonym(s): *Aralia quinquefolia* var. *notoginseng* Burkill, *Panax pseudoginseng* var. *notoginseng* (Burkill) G.Hoo & C.L.Tseng. **Common name(s)**: Tienchi ginseng. **Botanical description**: It is native to Asia; it is found in forests and shrubberies in China and Burma and up to 2100–4300 m in Nepal and the Himalayas. The perennial plant grows to about 1.2 m high at a slow rate. The flowers are hermaphrodite (have both male and female organs). It can grow in full shade (deep woodland) or semi-shade (light woodland). It requires moist soil. **Medicinal use(s)**: It is used to treat prostate, cervical, lung, and human colorectal cancer. It is also used to relieve pain and to reduce swelling, cholesterol, and blood pressure. It is also used for chest pain (angina), strokes, dizziness, and sore throat.					

(continued)

Table 4.12 (continued)

Family	Botanical name	Image	Plant part used	Bioactive compound(s)	Structure	Reference(s)
	Panax quinquefolium L.		Root	Ginsenosides RG1	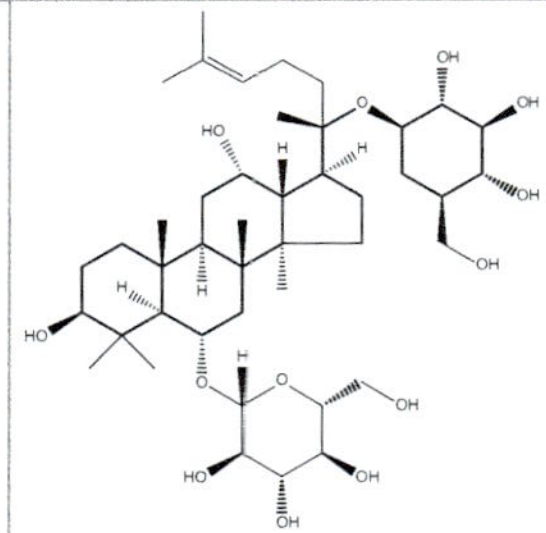	Wang et al. (2008a, b) and Barton et al. (2009)
	Synonym(s): *Aralia quinquefolia* (L.) Decne. & Planch., *Ginseng quinquefolium* (L.) Alph.Wood, *Panax americanus* (Raf.) Raf, *Panax americanus* var. *elatus* Raf. **Common name(s)**: American ginseng. **Botanical description**: It is native to eastern North America – Maine to Georgia, west to Oklahoma and Minnesota. It is a perennial growing up to 0.3–0.5 m at a slow rate. It flowers in June. The flowers are hermaphrodite (have both male and female organs). It is suitable for light (sandy), medium (loamy), and heavy (clay) soils. It prefers moist soil and can grow in full shade (deep woodland) or semi-shade (light woodland). **Medicinal use(s)**: It is used to treat breast cancer. It is also used for the treatment of low iron in the blood (anemia), diabetes, insulin resistance related to HIV treatments, cancer-related fatigue, high blood pressure, trouble sleeping (insomnia), nerve pain, erectile dysfunction (ED), fever, hangover symptoms, attention deficit hyperactivity disorder (ADHD), blood and bleeding disorders, breast cancer, dizziness, headaches, convulsions, fibromyalgia, "hardening of the arteries" (atherosclerosis), memory loss, rheumatoid arthritis, schizophrenia, menopausal symptoms, complications during pregnancy or childbirth, and nervous exhaustion (neurasthenia), improving athletic performance, improving mental performance, and aiding antiaging.					

Table 4.13 Anticancer plants of family Aristolochiaceae

Family	Botanical name	Image	Plant part used	Bioactive compound(s)	Structure	Reference(s)
Aristolochiaceae	*Aristolochia contorta* Bunge		Root, fruit	Lysicamine	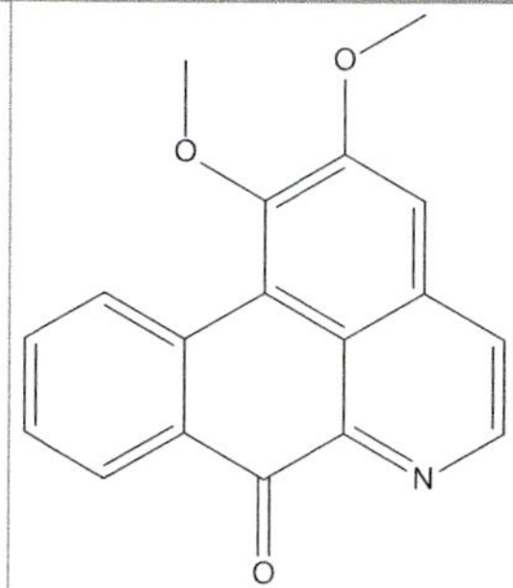	Yang et al. (2014)
	Synonym(s): *Aristolochia nipponica* Makino. **Common name(s)**: Birthwort, Ma Dou Ling. **Botanical description**: It is native to East Asia – China, Japan, Korea, and Manchuria. It is a perennial growing to 1.5 m. It flowers from May to July, and the seeds ripen from September to October. The flowers are hermaphrodite (have both male and female organs) and are pollinated by flies. It is suitable for light (sandy), medium (loamy), and heavy (clay) soils and prefers well-drained moist soil. It can grow in semi-shade (light woodland) or no shade. **Medicinal use(s)**: It is used to treat liver cancer and can be used in conjunction with chemotherapy and radiotherapy. Aristolochic acid can also be used in the treatment of acute and serious infections such as TB, hepatitis, liver cirrhosis, and infantile pneumonia. It also increases the cellular immunity and phagocytosis function of the phagocytic cells. The fruit and its capsule are anti-asthmatic, antiseptic, antitussive, and expectorant. Aristolochic acid is often found as an ingredient in proprietary medicines; however, any internal use of plants in this genus cannot be recommended without the guidance of a skilled therapist.			Aristolochic acid	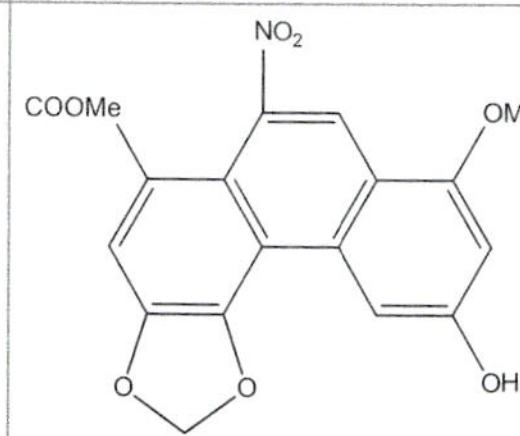	

(continued)

Table 4.13 (continued)

Family	Botanical name	Image	Plant part used	Bioactive compound(s)	Structure	Reference(s)
	Aristolochia mollissima Hance		Entire plant	Oxaaporphine	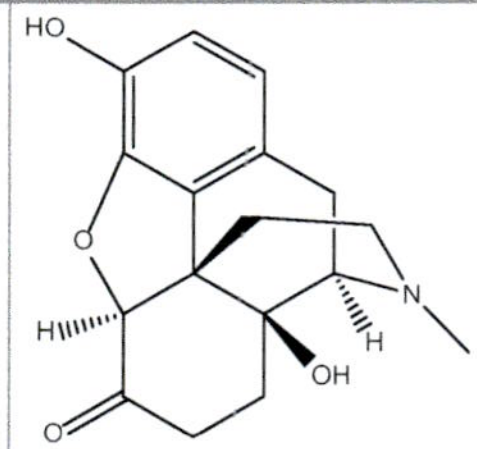	Bayala and Simpore and Bayala (2014) and Duke and Ayensu (1985a, b)
	Synonym(s): *Aristolochia albida* Dich. **Common name(s)**: Hairy birthwort. **Botanical description**: It spreads from the Himalayas to Sri Lanka through Southeast Asia (includes Myanmar, Indonesia, Indochina, and Thailand) and China to Oceania (includes the whole of Malaysia, the Solomon Islands, and Queensland in Australia). It is an evergreen climbing plant growing up to 20 m tall. This is a climber that can reach up to 20 m long. The branches are slightly furrowed with a diameter of up to 5 mm. The 6–20 (–27) cm × 4–10 (–16) cm leaves are usually egg-shaped to egg-shaped-oblong. They are heart-shaped at the base with rounded auricles. It is sparsely short-haired to sub-hairless underneath with 2 pairs of basal veins. There are 3–5 pairs of secondary veins and loosely reticulate or crossbar-like tertiary veins. The flowers are in raceme-like or resembling an indeterminate branches raceme-like inflorescence, 1-lipped whole floral leaves, with faint venation that is either pale yellowish or greenish to purplish or dark reddish brown. The fruit is nearly globular and slightly pear-shaped or oblong that up to 4 cm long. The seeds are winged. **Medicinal use(s)**: Used for urothelial cancer. Despite serious safety concerns, *Aristolochia* is used to prevent seizures, increase sexual desire, boost the immune system, and start menstruation. It is also used to treat snakebite, intestinal pain, gallbladder pain, arthritis, gout, achy joints (rheumatism), eczema, and wounds.					

	Aristolochia tagala Cham.		Leaf	Aristolochic acid	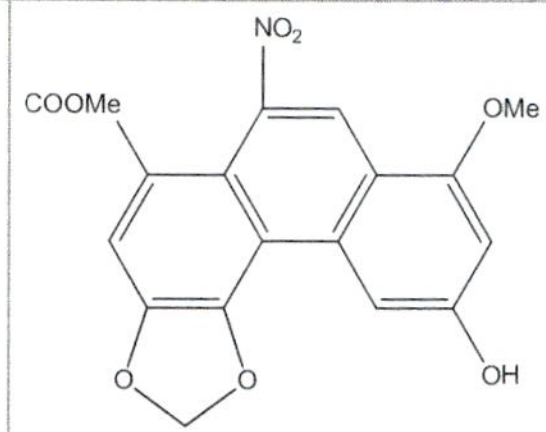 	Masilungan et al. (1971) and Kma et al. (2014)
	Synonym(s): *Aristolochia acuminata* Lam. **Common name(s)**: Indian birthwort. **Botanical description**: It is found in the Himalayas to Sri Lanka through Southeast Asia (includes Myanmar, Indonesia, Indochina, and Thailand) and China to Oceania (includes the whole of Malaysia, the Solomon Islands, and Queensland in Australia). It is an evergreen climbing plant growing up to 20 m tall. This is a climber that can reach up to 20 m long. The branches are slightly furrowed with a diameter up to 5 mm. The leaves are usually egg-shaped to egg-shaped-oblong. They are heart-shaped at the base with rounded auricles. It is sparsely short-haired to sub-hairless underneath with 2 pairs of basal veins. There are 3–5 pairs of secondary veins and loosely reticulate or crossbar-like tertiary veins. The flowers are in raceme-like or resembling an indeterminate branches raceme-like inflorescence, 1-lipped whole floral leaves, with faint venation that is either pale yellowish or greenish to purplish or dark reddish brown. The fruit is nearly globular and slightly pear-shaped or oblong that up to 4 cm long. The seeds are winged. **Medicinal use(s)**: It is used for the treatment of liver cancer and for treatment of snakebites, fever, stomach ache, bilious disorder, swollen limb, menstruation problem, rheumatoid pain, etc. Although scientific validation is still required for most of its therapeutic efficacies, there are reports that different parts of this plant act as cytotoxic, antimicrobial, antiproliferative, analgesic, antioxidant, and insecticidal agent.					

Table 4.14 Anticancer plants of family Asclepiadoideae

Family	Botanical name	Image	Plant part used	Bioactive compound(s)	Structure	Reference(s)
Asclepiadoideae	*Asclepias speciosa* Torrey		Latex	α-Amyrin	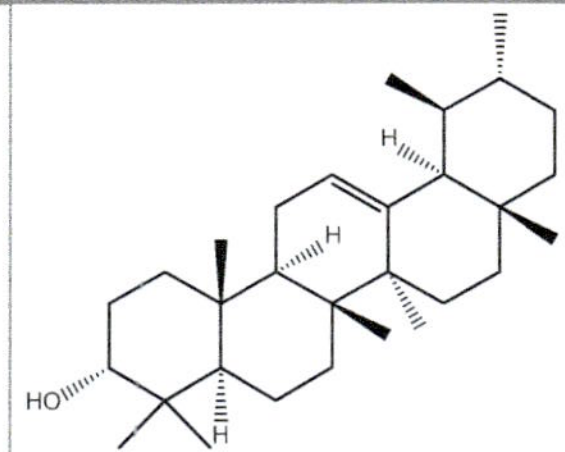	Train et al. (1957) and Eowyndbh (2012)
	Synonym(s): *Asclepias giffordii* Eastw., *Asclepias douglasii* Hook. **Common name(s)**: Showy butterfly weed, Greek milkweed. **Botanical description**: It is native to USA-NV. Showy milkweed is a native herbaceous perennial from widespread rhizomes, which produce stems that grow to 1½–5 ft. tall in summer. The gray-green leaves are opposite, 4–7 in. long, oval, and covered in velvety hairs. Stems and foliage exude milky latex sap when cut. Flowers are in loose clusters at the top of the stems and are rose-purple, aging to yellow. **Medicinal use(s)**: It is used to treat breast cancer. Latex has antiseptic properties, used to cure warts, skin sores, cuts, and ringworms. Root is used to treat stomachache. Used to remove corns.			β-Amyrin	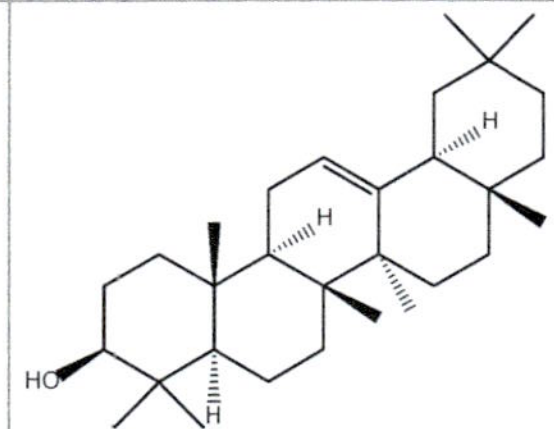	

<table>
<tr><td></td><td>Cynanchum mongolicum (Maxim.) Kom.</td><td></td><td>Entire plant, roots</td><td>Ellagic acid</td><td>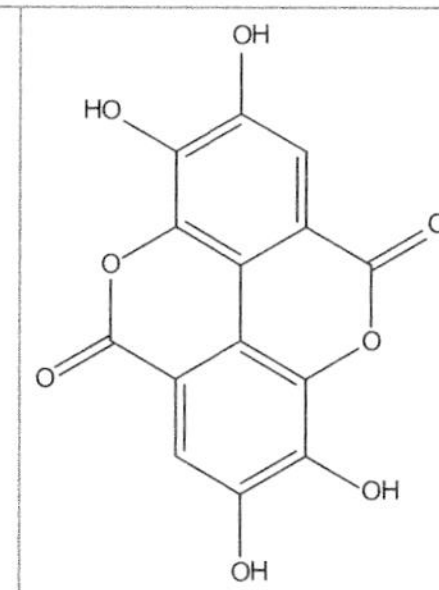</td><td>Konda et al. (1990) and Lou et al. (1991)</td></tr>
<tr><td></td><td colspan="3">Synonym(s): Vincetoxicum mongolicum Maxim., Cynanchum hancockianum (Maxim.) Iljinski.
Common name(s): Hua bei bai qian (North China White).
Botanical description: It is native to North America and China. These are found throughout the tropics and subtropics and also grow in temperate region. These plants are perennial herbs or subshrubs; stems erect, to 50 cm tall, occasionally puberulent along 1 side, glabrescent; the leaves are usually oppositely arranged; and the inflorescences and flowers come in various shapes. Leaves opposite or whorled; petiole 0–5 mm; leaf blade narrowly oblong-elliptic to lanceolate or rhomboid, 3–10 × 0.5–3 cm, leathery or thick papery, base cuneate, margin sometimes ciliate, apex acuminate, or flowering occurs during May–August and fruiting from June to November.
Medicinal use(s): It is used to treat breast cancer, tumor, and leukemia. It possesses anticancer, antioxidant, antibacterial, antitumor, and anti-inflammatory properties.</td><td></td><td></td><td></td></tr>
</table>

(continued)

Table 4.14 (continued)

Family	Botanical name	Image	Plant part used	Bioactive compound(s)	Structure	Reference(s)
	Cynanchum taiwanianum Yamazaki/ *Cynanchum boudieri*		Root	Taiwanoside D	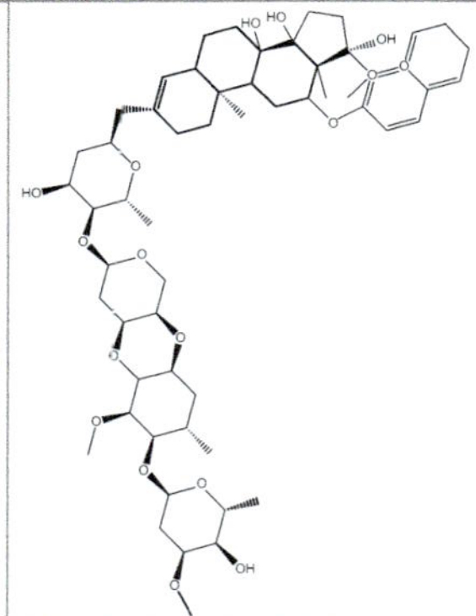	Lin et al. (1997a, b) and Lee et al. (2011a, b)
	Synonym(s): *Cynanchum amphiboleum* C.K.Schneid., *Cynanchum auriculatum* var. *amamianum* (Hatus.) T.Yamaz., *Cynanchum macrorhizon* Carrière. **Common name(s)**: Dog-strangling vine. **Botanical description**: It is native to East Asia – Northern China, Japan, and Korea. It is a perennial growing up to 2.5 m. It flowers from July to August. The flowers are hermaphrodite and are pollinated by insects. It is suitable for light, medium, and heavy soils. It is habituated to open grassy slopes on mountains and hills in Korea. Forest edges and lowlands to elevations of 1600 m in Japan. **Medicinal use(s)**: It is used to treat breast and lung cancer. It possesses anticancer, anti-inflammatory, antioxidant, antibacterial, and antimicrobial properties.					

	Heterostemma brownii Hay	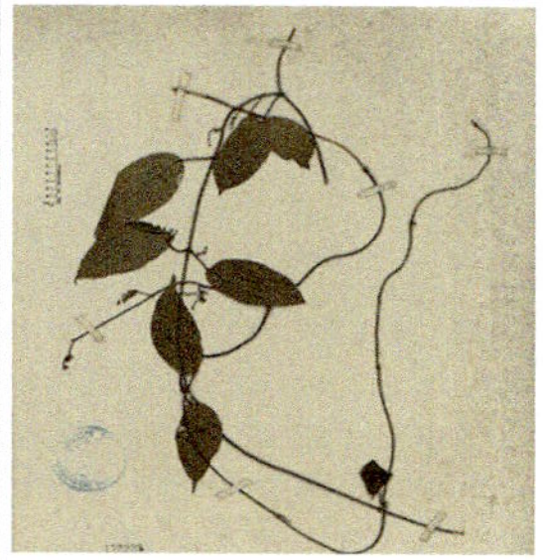	Aerial parts	7,9-Dimethyl-2-(N,N-dimethylamino) guaninium chloride (Heteromines D)	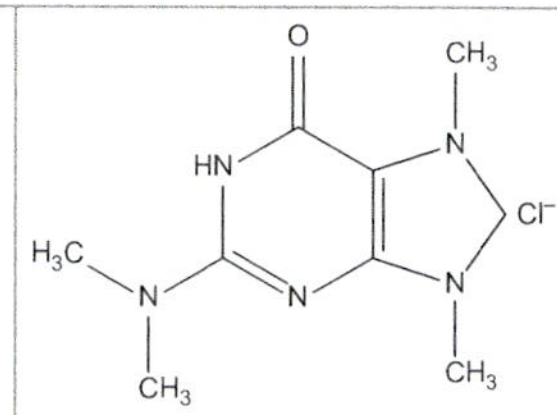	Lin et al. (1997a, b) and Kour (2014)
	Synonym(s): *Heterostemma acuminatum* Decne., *Heterostemma brownii* Hayata., *Heterostemma esquirolii* (H.Lév.) Tsiang. **Common name(s)**: Brown wine. **Botanical description**: It is native to India, China, Southeast Asia, Australia, and certain islands in the Pacific. Stems slender, 5 m in height, glabrous, pubescent along 2 lines when young. Petiole 2–5 cm, flat, pubescent; leaf blade broadly ovate to oblong-ovate, 7–15 × 5–8 cm, papery, puberulent when young but soon glabrescent, base rounded to broadly cuneate, rarely subcordate, apex acuminate; basal veins 3–5, winged when young, lateral veins 3 or 4 pairs. Inflorescences umbel-like becoming raceme-like, 2–6 cm, 10–20-flowered; peduncle 1.5–4.5 cm, puberulent. Pedicel 1–1.5 cm. Sepals ovate, corolla yellow, rotate to shallowly bowl-shaped, ca. 1 cm in diam.; tube 4–5 mm; lobes ovate-triangular, as long as tube. Corona lobes ligulate-oblong, base attenuate, apex obtuse, as long as corona tube. Seeds are broadly ovate, ca. 1.5 × 1 cm. **Medicinal use(s)**: It is used to treat uterine cancer and tumors. It possesses anticancer, antitumor, antibacterial, anti-inflammatory, antimicrobial, and antioxidant properties.			(Heteromines F) 6-methoxy-4-(N-methylamino)-2-(N,N-dimethylamino)-5-(N-methylformamido) pyrimidine	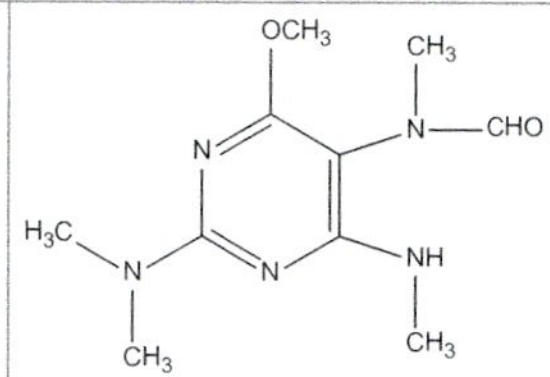	

(continued)

Table 4.14 (continued)

Family	Botanical name	Image	Plant part used	Bioactive compound(s)	Structure	Reference(s)
	Janakia arayalpathra J. Joseph and V. Chandrasekaran		Root	Apigenin	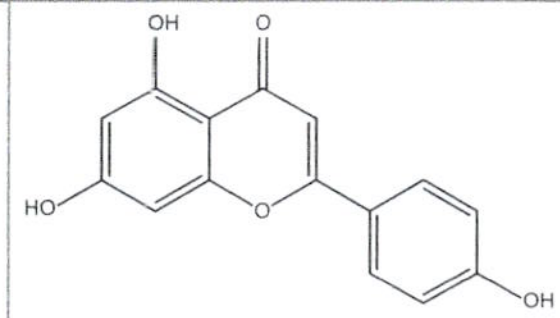	Subramonian et al. (1996) and Subramoniam et al. (2010)
	Synonym(s): *Janakia arayalpathra*. **Common name(s)**: Amruthapala. **Botanical description**: It is native to India. It is a threatened plant found in Western Ghats. It is a woody shrub with root tuberous, strongly smelling. Stems, leaf stacks, and leaves are reddish brown. Some flowers are borne on auxiliary cymes carried on slender, 2–3 cm long stalk. Calyx is bell-shaped. It is endemic to southern forests of the Western Ghats regions of the Kerala, distributed at an elevation of 800–1200 m and growing on the crevices of rocks. **Medicinal properties**: It is used to treat breast cancer. The tuberous roots of this plant are highly aromatic, and the native of Kani tribe use it as an effective remedy for peptic ulcer, for cancer-like afflictions, and as rejuvenating tonic. Recent pharmacological investigation of the root extract of the plant revealed immunomodulatory and anticancer properties.					

	Leptadenia hastata (Pers.) Decne.		Bark	Berberine	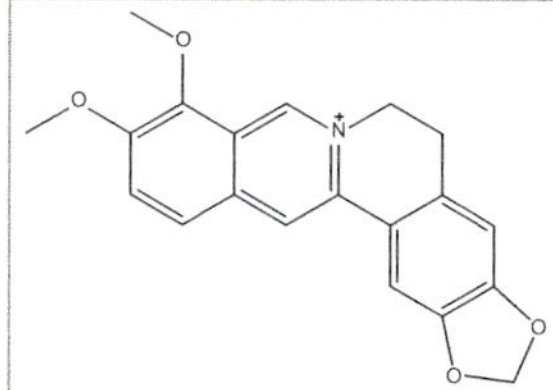	Aquino et al. (1995) and Thomas (2012)
	Synonym(s): *Cynanchum lanceolatum* Poir, *Cynanchum lancifolium* Schumach. & Thonn, *Leptadenia lancifolia* Decne. **Common name(s)**: *Fula-pulaar*. **Botanical description**: It is native to Senegal, tropical Africa, Uganda, and Kenya. It is many stemmed climber, producing soft stems but becoming woody at the base. This plant is able to grow in harsh condition. It is often seen and used as famine food. **Medicinal properties**: It is used to treat prostate cancer. Used as an antitumor drug. The latex of this plant consists of triterpene lupeol, and its derivatives possess anti-inflammatory property. A root macerate is consumed to treat anuria or constipation. The powdered roots in water are taken as a stomachic. The whole plant is used to treat urethral discharge, abdominal complaints.					

(continued)

Table 4.14 (continued)

Family	Botanical name	Image	Plant part used	Bioactive compound(s)	Structure	Reference(s)
	Marsdenia tenacissima (Roxb.) Wight		Entire plant, stem	Carvacrol	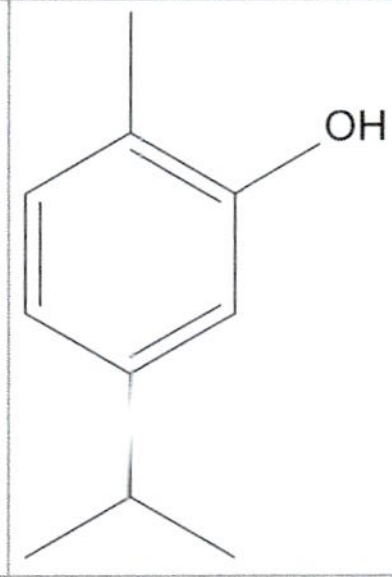	Yang et al. (1981) and Han et al. (2015)
	Synonym(s): *Asclepias tenacissima* Roxb. **Common name(s)**: Murva or moorva. **Botanical description**: It is native to China and India. It is a perennial climber, bearing green flowers and found in tropical hilly tracts of peninsular India and Vindhya ranges as well as lower Himalayan tracts. It is found growing in tropical and subtropical dry and moist deciduous forests having annual rainfall between 1000 and 1500 mm. It grows in moist places in nature and is a shade-loving plant. It is large, extensive twining shrub with very stout, bluntly angular, densely velvety, and hairy branches. Stem and branches are slightly stout and densely tomentose. **Medicinal properties**: It is used as a folkloric antitumor drug and for treatment of lung cancer. It is one of the most controversial drugs in Ayurveda. It is very rich in medicinal properties like antiscorbutic, curing or preventing scurvy; antioxidant, neutralizing the oxidant effect of free radical and other substances; antibacterial, active against bacteria; anti-obesity, reducing or controlling obesity; anthelmintic; and anti-parasitic, expelling parasitic worms and other internal parasites from the body.					

	Raphionacme hirsuta (E. Mey.) R.A.Dyer		Bulb	Campesterol	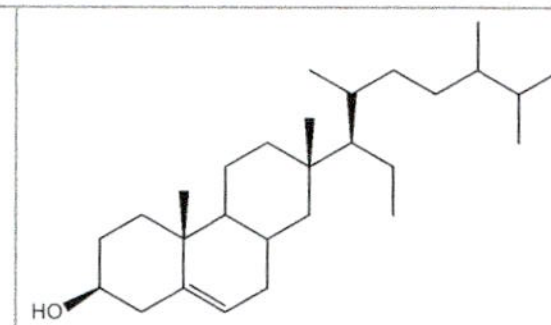	Watt and Breyer-Brandwijk (1962) and Charlson (1980)
	Synonym(s): *Brachystelma hirsutum* E.Mey., *Apoxyanthera pubescens* Hochst., *Mafekingia parquetiana* Baill., *Raphionacme divaricata* Harv. **Common name(s)**: Water roots, False gentian. **Botanical description**: It is native of South Africa, Zimbabwe, and Swaziland. It is a perennial caudiciform asclepiad with herbaceous, annual spreading steams up to 30 m long. The aerial portion seems ridiculously small in proportion with the root tuber. Leaves are broadly elliptic, it is almost glabrous. This species has very wide ecological amplitude and is in found different habitats. **Medicinal use(s)**: It is used to treat esophageal cancer. It is used as an anticancer remedy and also used for the treatment of tumors. Sliced tuber is applied to tumors. The tuber is mixed with hot water and sugar and is used for brewing highly intoxicating liquor. An ethanolic extract of the tubers showed antitumor activity in the P388 lymphocytic leukemia test system.					

(continued)

Table 4.14 (continued)

Family	Botanical name	Image	Plant part used	Bioactive compound(s)	Structure	Reference(s)
	Sarcostemma australe R.Br.		Latex	Podofilox	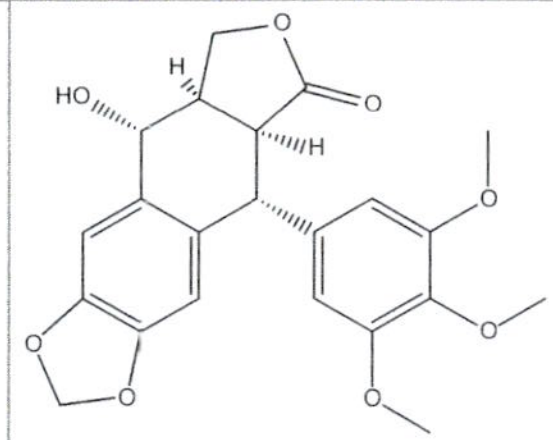	Reid and Betts (1979) and Savigni (2010)
	Synonym(s): *Sarcostemma viminale*, *Sarcostemma aphyllum*. **Common name(s)**: Caustic creeper, Caustic vine. **Botanical description**: It is native plant of Western Australia. It is an erected, tangent, sprawling, succulent shrub, of height 0.3–0.5 m. It grows in sandy red lateritic soils over sandstone or limestone. **Medicinal use(s)**: It is used to treat skin cancer. This plant is having antitumor properties and also has mild psychotropic properties. It is helpful as a weight loss medicine. It is also effective to drive mosquitoes away.					
	Tylophora floribunda Miq.		Entire plant	Allamcin	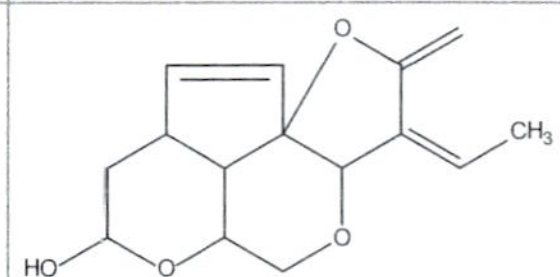	Duke and Ayensu, (1985a, b), Xu (2016)

	Synonym(s): *Tylophora anthopotamica* (Hand.-Mazz.) Tsiang & H.D. Zhang., *Tylophora arenicola* Merr., *Tylophora astephanoides* Tsiang & P.T. Li. **Common name(s)**: Indian Ipecac, Indian ipecacuanha. **Botanical description**: It is a native plant of China, Japan, and Korea. It is a climbing shrub, producing slender stem up to 3 m long that twines into other plant to give support to each other. **Medicinal properties**: It is used to treat metastatic cancer. It is used as an anticancer agent; its roots are used for the treatment of infertile convulsions, diphtheria, traumatic injury, toothache, and snakebite.					
	Tylophora indica (Indian Ipecac)		Root, leaf	Tylophorine	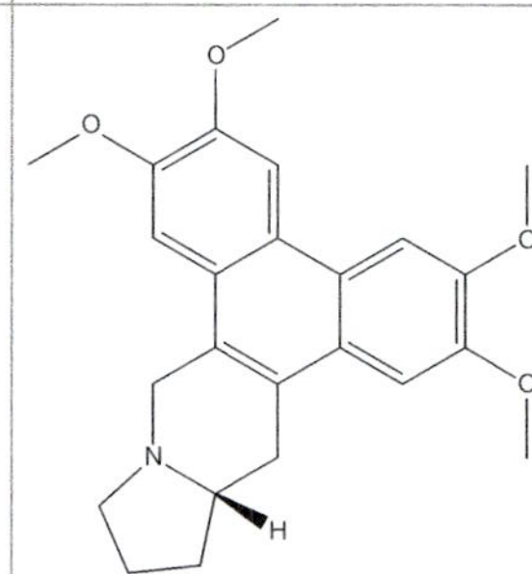	Zj et al. (2011) and Wu et al. (2009)
	Synonym(s): *Cynanchum indicum* Burm.f., *Tylophora indica* var. *glabra* (Decne.) H. Huber. **Common name(s)**: Aadumuttad, Aantmool. **Botanical description**: It is native to India and China. It is a perennial shrub which grows in tropical climate. It grows up to 1.5 m. **Medicinal properties**: It is used to treat lung cancer lines. It is having anti-allergic properties. Best use for snakebites. It is also having antitumor properties, which prevent tumor formation or growth. It inhibits the growth of cancer cells. It is profitable for countering Crohn's diseases. It is a very effective antipyretic agent that reduces fever.					

Table 4.15 Anticancer plants of family Asparagaceae

Family	Botanical name	Image	Plant part used	Bioactive compound(s)	Structure	Reference(s)
Asparagaceae	*Asparagus officinalis* L.		Rhizomes	Shatavarin	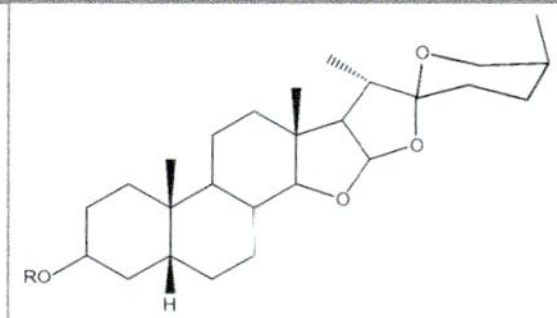	Bousserouel et al. (2013)
	Synonym(s): *Asparagus officinalis* var. *altilis* L. **Common name(s)**: Garden asparagus, Shatavari. **Botanical description**: Herbaceous plant, up to 6 ft. tall, branched, branches thin and drooping, leaves have fernlike look somewhat like pinus needles, flowers whitish green, single or paired flowers from leaf axils. **Medicinal use(s)**: It is used to treat breast cancer, tumor, prostate cancer, and leukemia. It possesses anticancer, antitumor, anti-inflammatory, antioxidant, antibacterial, and antimicrobial properties.			Asparagine	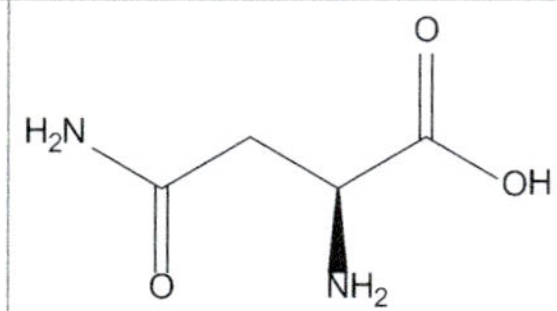	

	Agave americana L.		Leaves	Hecogenin (tetraglycoside)	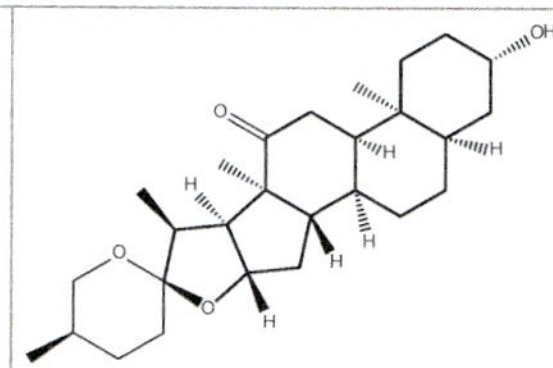	Anajwala et al. (2010) and Yokosuka et al. (2000)
	Synonym(s): *Agave altissima* Zumagl., *Agave communis* Gaterau. **Common name(s)**: Agave, Century plant. **Botanical description**: It has a spreading rosette, about 4 m wide, of gray-green leaves up to 6 ft. long, each with a spiny margin and a heavy spike at the tip. Its common name derives from its habit of only occasionally flowering, but when it does, the spike with a cyme of big yellow flowers may reach up to 25 ft. in height. The plant dies after flowering but produces suckers or adventitious shoots from the base, which continues its growth. **Medicinal use(s)**: It is used for the treatment of breast cancer. It purifies the blood, checks anemia, prevents the formation of pus, promotes the flow of urine, stimulates the bowels, is a mild laxative, checks alopecia, checks internal wounds and swellings, soothes irritated eyes, checks tuberculosis of the lungs, prevents jaundices, and treats infected liver, kidney diseases, and liver problems.					

(continued)

Table 4.15 (continued)

Family	Botanical name	Image	Plant part used	Bioactive compound(s)	Structure	Reference(s)
	Ruscus aculeatus L.		Root Rhizome	Aspartic acid	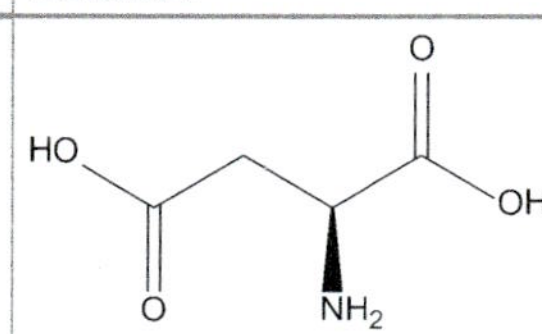	Young (2016)
	Synonym(s): *Oxymyrsine pungens* Bubani, *Ruscus flexuosus* Mill., *Ruscus laxus* Sm., *Ruscus parasiticus* Gueldenst., *Ruscus ponticus* Woronow. **Common name(s)**: Butcher's-broom. **Botanical description**: It is a native plant of Western and Southern Europe and occurs in woodland and hedgerows. It is an evergreen shrub, with flat shoots known as cladodes that give the appearance of stiff, spine-tipped leaves. Small greenish flowers appear in spring and are borne singly in the center of the cladodes. The female flowers are followed by a red berry, and the seeds are bird-distributed, but the plant also spreads vegetatively by means of rhizomes. It is tolerant to deep shade. **Medicinal use(s)**: It is used to treat breast cancer, tumor, prostate cancer, and leukemia. It possesses anticancer, antitumor, anti-inflammatory, antioxidant, antibacterial, and antimicrobial properties.			Ruscogenin diglycoside	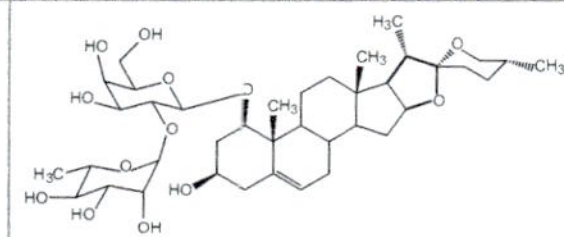	
				26-Glucosyloxy-furostanol saponin	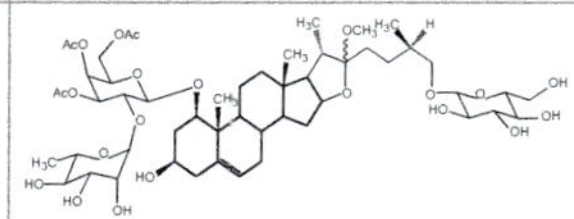	

Table 4.16 Anticancer plants of family Avicenniaceae

Family	Botanical name	Image	Plant part used	Bioactive compound(s)	Structure	Reference(s)
Avicenniaceae	*Avicennia germinans* (L.) L.		Leaf	3- Chlorodeoxylapachol3-chlorodeoxylapachol	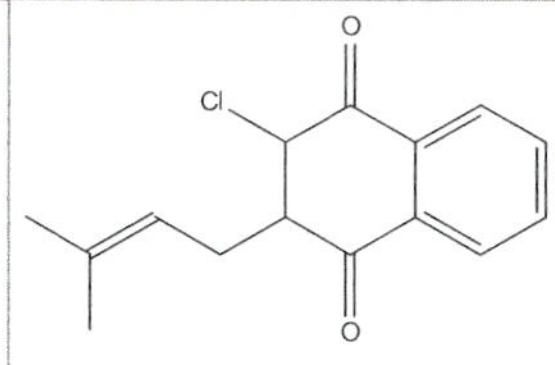	Jones et al. (2005)
	Synonym(s): *Avicennia africana* P.Beauv., *Avicennia elliptica* Thunb., *Avicennia nitida* Jacq., *Avicennia floridana* Gand., *Avicennia lamarckiana* C.Presl, *Avicennia meyeri* Miq., *Avicennia oblongifolia* Nutt. ex Chapm., *Avicennia tomentosa* Jacq., *Bontia germinans* L., *Hilairanthus nitidus* (Jacq.) Tiegh., *Hilairanthus tomentosus* (Jacq.) Tiegh. **Common name(s)**: Black mangrove. **Botanical description**: It grows in tropical and subtropical regions of America on both Atlantic and Pacific coasts and on the Atlantic coast of Tropical Africa. It is a woody shrub that grows in salt marshes. It is adaptable to harsh environment. Its height will vary from 4 to 6 ft.; leaves are 1–5 in. long, elliptical, thick, leathery, dark green in color. **Medicinal properties**: An extract of leafy twigs of this plant showed cytotoxic activity in several human cancer cell lines, with naphthoquinone 3-chlorodeoxylapaehol as an active compound. The bark resin is used in traditional medicine to treat tumors, diarrhea, hemorrhage, hemorrhoids, wounds, and sore throat.			Naphthoquinone	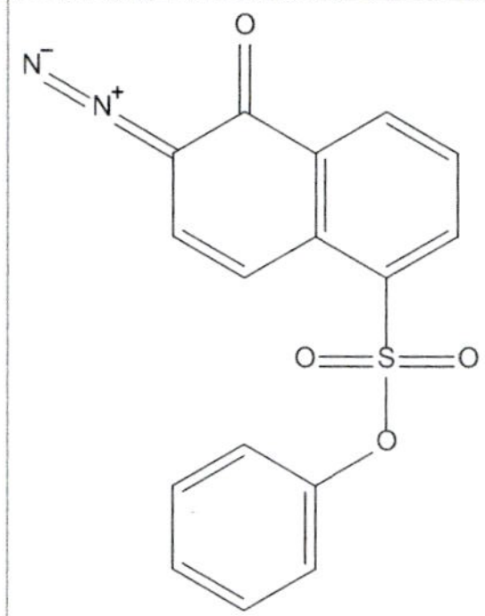	

Table 4.17 Anticancer plants of family Balanophoraceae

Family	Botanical name	Image	Plant part used	Bioactive compound(s)	Structure	Reference(s)
Balanophoraceae	*Sarcophyte sanguinea* subsp. *piriei* (Hutch.) B. Hansen		Entire plant	Diinsininol	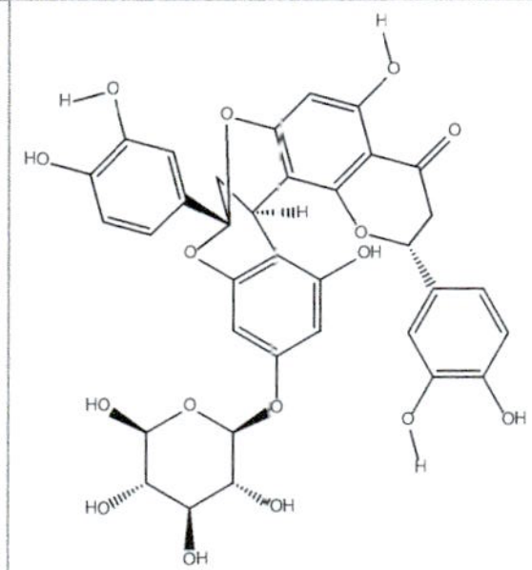	Chhabra et al. (1987) and Schmelzer (2012)
	Synonym(s): *Sarcophyte piriei* Hutch. **Common name(s)**: Not available. **Botanical description**: This plant is native to Tanzania, Kenya, Zambia, Malawi, and Mozambique. It is a dioecious, perennial, holoparasitic herb and reaches a height up to 40 cm. It has a fruity smell. Male inflorescence is up to 40 cm in length, female ones lower, fleshy colored to purplish red except for the pale bracts. Male flowers with tepals of about twice as long as the stamens. **Medicinal use(s)**: It is used to treat pharyngeal tumor. In Tanzania a decoction of the whole plant is consumed to treat cancer. The tuber is used to treat bruises, toothache, sore throat, and abdominal pain. It is also used as a remedy for diarrhea and menstrual disorders.			Diinsinin	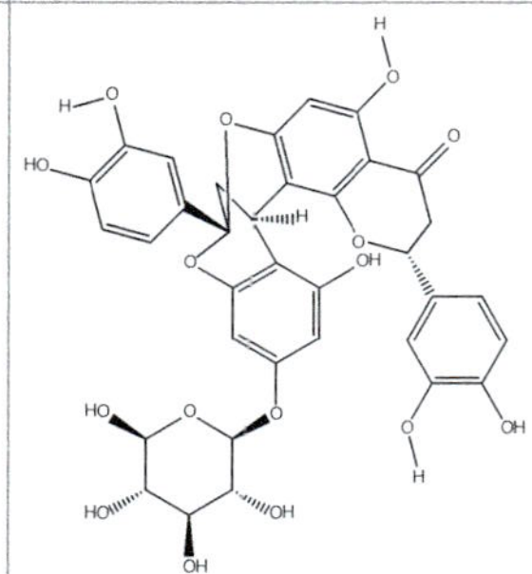	

		Eriodictyol	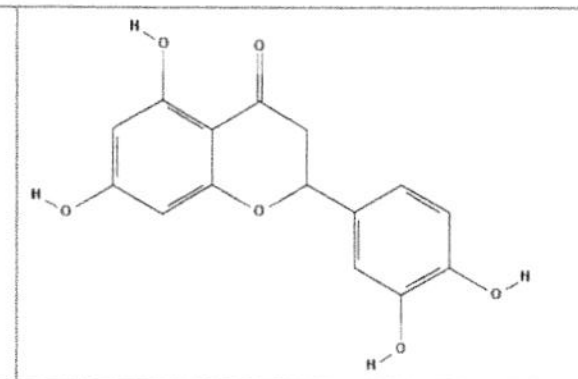	
		Naringenin	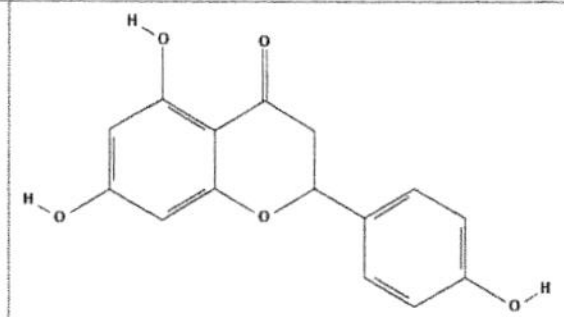	
		Naringenin 5-glucoside	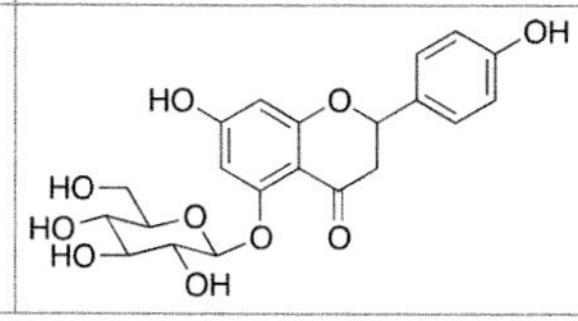	

(continued)

Table 4.17 (continued)

Family	Botanical name	Image	Plant part used	Bioactive compound(s)	Structure	Reference(s)
				Triandrin	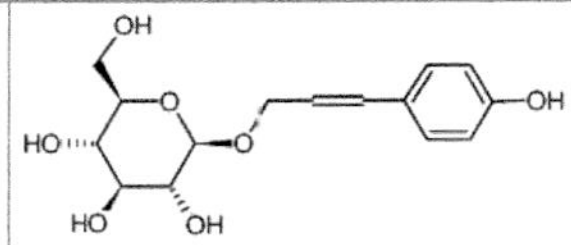	
				D-pinitol	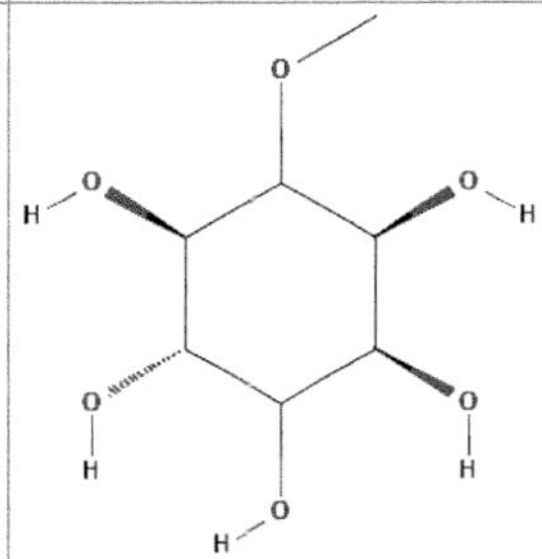	
				trans-p-coumaraldehyde	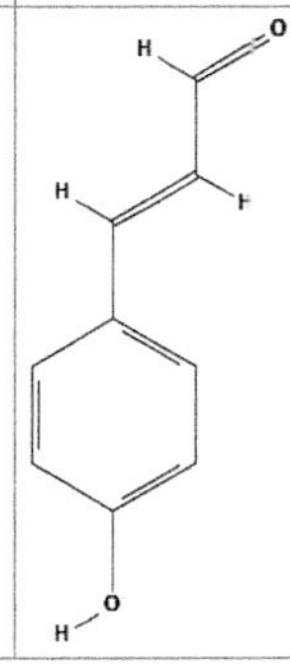	

Table 4.18 Anticancer plants of family Begoniaceae

Family	Botanical name	Image	Plant part used	Bioactive compound(s)	Structure	Reference(s)
Begoniaceae	*Begonia longifolia* Blume		Entire plant	Allamcin	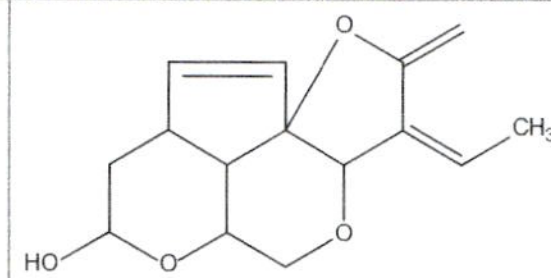	Duke and Ayensu (1985a, b)
	Synonym(s): *Begonia crassirostris* Irmsch., *Begonia inflata* C.B.Clarke, *Begonia tricornis* Ridl., *Begonia trisulcata* (A.DC.) Warb., *Diploclinium longifolium* (Blume) Miq. **Common name(s)**: Hong Tian Kui. **Botanical description**: It is a native plant of China. This plant is a perennial herb of 90–150 cm in height. Stem is erect, glabrous, usually unbranched; leaves are having slender petioles, alternate, the lower ones up to 10 cm; flowers are white in color, male flower has 3 perianth segments, and female flower has 6 perianth segments. It is generally found in streams and wet places. Flowering occurs from April to May and fruiting in July. **Medicinal use(s)**: The herb is useful in treating breast cancer. It fights against the free radicals present inside the body and cures amenorrhea. It shows its action on the upper respiratory tract and, thus, helps in treating asthma. It acts as a liver tonic and also helps in treating rheumatic arthritis.					

(continued)

Table 4.18 (continued)

Family	Botanical name	Image	Plant part used	Bioactive compound(s)	Structure	Reference(s)
	Begonia glabra Aubl.		Entire plant	Cyanidin 3-(6″-(*Z*)-p-coumarylsophoroside)	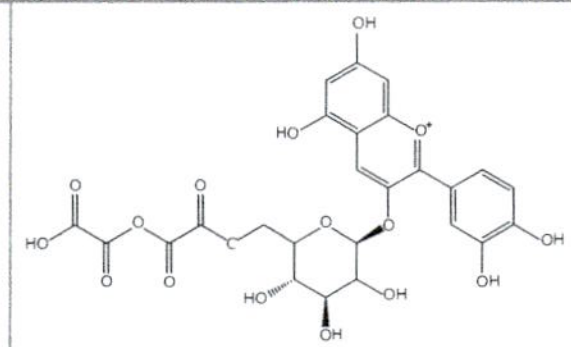	Gupta et al. (1996) and Bhanot et al. (2011)
	Synonym(s): *Begonia elliptica* Kunth, *Begonia hoegeana* Regel & Schmidt, *Begonia locellata* A.DC., *Begonia moritziana* Kunth & C.D.Bouché, *Begonia physalifolia* Liebm., *Wageneria deflexa* Klotzsch, *Wageneria glabra* (Aubl.) Klotzsch, *Wageneria lucida* Klotzsch, *Wageneria montana* Klotzsch. **Common name(s)**: Tes Begonia, Shin Hai Ten Ki, Tea Begonia, Shinhaitenki. **Botanical description**: It is a native plant of India and China. It is a flowering, perennial plant which grows in tropic and subtropical climates. It attains a height of 30 cm. It grows in the shady banks and rock edges in wetter areas. **Medicinal use(s)**: It is used to cure breast cancer and tumors. The juice of this plant is drunk to relieve headaches. The crushed leaves are used as poultice on sore nipples. The root juice is used as an eye wash to treat conjunctivitis. It is also consumed in the treatment of peptic ulcers.					

Table 4.19 Anticancer plants of family Berberidaceae

Family	Botanical name	Image	Plant part used	Bioactive compound(s)	Structure	Reference(s)
Berberidaceae	*Berberis amurensis* Maxim.		Entire plant	Berberine	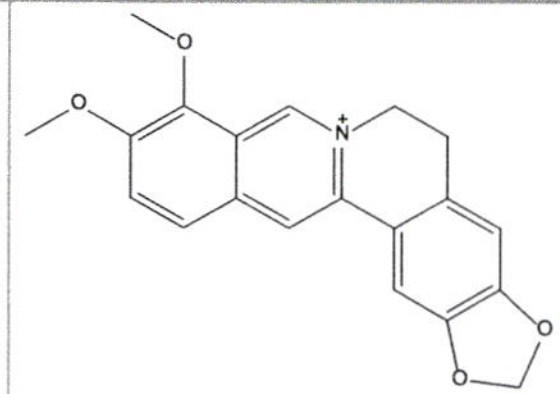	Duke and Ayensu (1985a, b) and Wang et al. (2009)
	Synonym(s): No synonyms are recorded for this name. **Common name(s)**: Dar Hald, Kilmora, Kingora, Rasont. **Botanical description**: It is native plant of China and India. It is a perennial shrub of 3 m in height and grows in temperate climate. It grows on the fringes, in hedges, in the undergrowth of coniferous-deciduous forests, in forests along streams, along the banks of mountain rivers on rocky places, and on sea coasts, with small curtains. Occurs under the canopy of the forest but does not bear fruit. Leaves are large, elliptical, up to 12 cm in length and 5 cm in width, with finely toothed edges, whole, collected in bundles of 2–7 on short shoots, green with a bluish tint. Flowers are pale yellow, bisexual, up to 2 cm in diameter, collected in a brush up to 6 cm long by 10–25 in. each, have a honey smell. Flowering occurs from May to June (3 weeks). Fruits are oblong berries, bright red, up to 10 mm in length and 4–5 mm in width. **Medicinal use(s)**: It halts expansion of breast cancer cells. It is beneficial in the treatment of several diseases like arthritis, rheumatism, typhoid, spleen enlargement, and colon and duodenal ulcers. It is beneficial in the treatment of menopause-related troubles like heavy menses or no menses. It eases constipation. It also possesses antifungal, antibacterial, anti-inflammatory, antioxidant, and antidiarrheal properties.			Berbamine	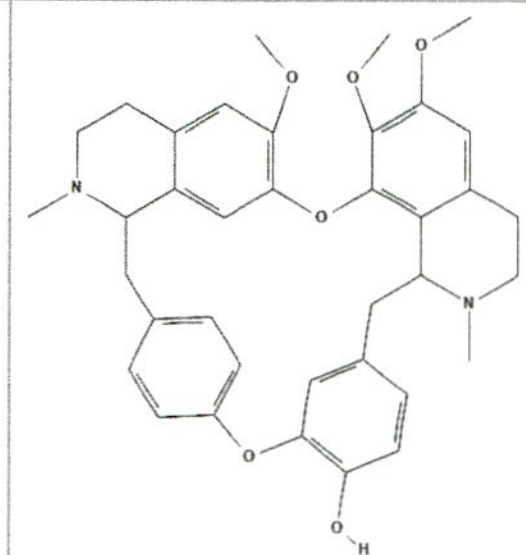	

(continued)

Table 4.19 (continued)

Family	Botanical name	Image	Plant part used	Bioactive compound(s)	Structure	Reference(s)
	Berberis aristata DC. (barberry)		Bark Fruit Root	Rutin	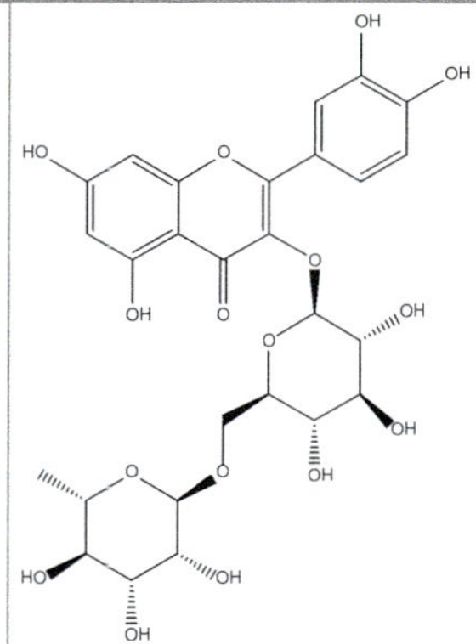	Das et al. (2009)
	Synonym(s): *Berberis bussmul* K.Koch ex Miq., *Berberis coccinea* K.Koch, *Berberis coerulescens* G.Nicholson, *Berberis elegans* K.Koch, Berberis gracilis Lindl., *Berberis gracillima* K.Koch ex Miq., *Berberis macrophylla* K.Koch, Berberis serratifolia K.Koch, Berberis umbellata Lindl., *Berberis undulata* K.Koch. **Common name(s)**: Indian barberry, Chutro or turmeric tree. **Botanical description**: It is an evergreen shrub and is found in tropic and subtropical regions of Asia. It is an erect spiny shrub, ranging between 2 and 3 m in height. It is a woody plant, with bark that appears yellow to brown from the outside and deep yellow to brown from the inside. The leaves are arranged in tufts of five to eight, and appear dark green in color on dorsal surface and light green on the ventral surface. The leaves are simple with pinnate venation. **Medicinal use(s)**: It is a very effective herb with anticancer (human colon cancer) properties. Root powder of this herb triggers the extra formation of insulin and protects the body from harmful effects of diabetes.					

	Dysosma auranticocaulis (Hand.Mazz.) Hu		Root	Ruvoside	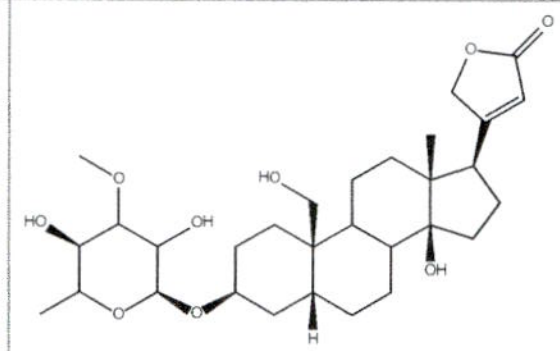	Duke and Ayensu (1985a, b)
	Synonym(s): *Dysosma mairei* (Gagnep.) M.Hiroe, *Podophyllum aurantiocaule* Hand.-Mazz., *Podophyllum mairei* Gagnep., *Podophyllum sikkimense* R.Chatterjee & Mukerjee. **Common name(s)**: Not available. **Botanical description**: It is a native plant of China and found in the damp shady spots in the alpine regions. It is a perennial herb with adventitious roots; stem is erect, light green in color topped by hairs. Leaves grow up to the apex, are peltate rounded, ovate with main rib radiating from the center. **Medicinal use(s)**: It is used to treat breast and prostate cancer. It stimulates the circulation of blood and reduces swelling and detoxifies blood. It cures tumors and is also used in traumatic injuries.					

(continued)

Table 4.19 (continued)

Family	Botanical name	Image	Plant part used	Bioactive compound(s)	Structure	Reference(s)
	Dysosma pleiantha (Hance) Woodson		Root	Apigenin	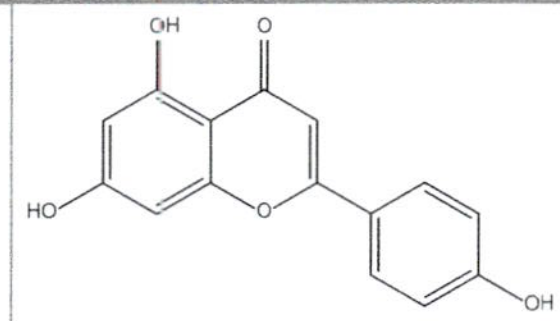	Kao et al. (1992) and Karuppaiya and Tsay (2016)
	Synonym(s): *Dysosma chengii* (J.J.Chien) M.Hiroe, *Dysosma hispida* (K.S.Hao) M.Hiroe, *Podophyllum chengii* S.S. Chien, *Podophyllum hispidum* K.S. Hao, *Podophyllum onzoi* Hayata, *Podophyllum pleianthum* Hance. **Common name(s)**: Woodson. **Botanical description**: The plant originated from China and is native of Taiwan. It grows as perennial, rhizomatous wildflowers on the damp and humus-rich floors of deciduous forest. The single umbrella-shaped leaves grow on an erect stem that usually stands 12–24 in.; the leaves may be completely green or mottled and flecked with purple color. They have an entire or deeply serrated edge. The fruit is a dark redberry. **Medicinal use(s)**: It is used to treat condylomata. From this plant several natural chemicals have been isolated from this herb; those present in podophyllum species have been chemically modified to produce clinically useful drugs such as etoposide and teniposide. Etoposide is of particular value in the treatment of testicular cancer and small-cell lung cancer and is now produced from podophyllotoxin.					

	Epimedium hunanense (Hand.-Mazz.) Hand. -Mazz.		Aerial parts	Betulinic acid	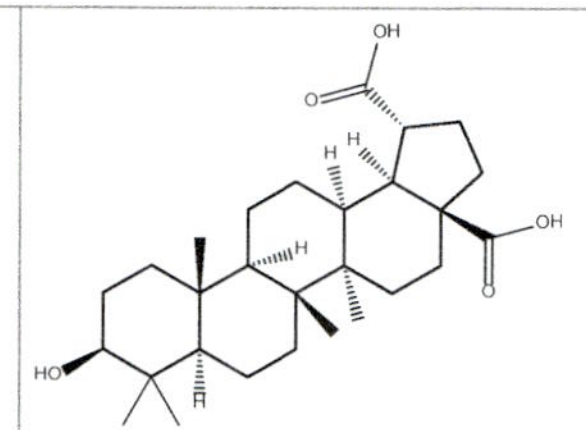	Liang et al. (1997)
	Synonym(s): *Epimedium kunawarense* S.Clay, *Epimedium davidii* var. *hunanense* Hand.-Mazz. **Common name(s)**: Hunan fairy wings. **Botanical description**: It is a native plant of China. It is a deciduous or evergreen, hardy perennial flowering plant. It grows from an underground rhizome. The leaflets usually have spiny margins; the inflorescence is an open raceme or panicle. **Medicinal use(s)**: It is used to treat cervical cancer. It possesses anticoagulant and antiviral properties. It is also used for leukopenia and has various types of components for lowering cholesterol and for treating infertility and thyroid, kidney, and liver disorder.					

(continued)

Table 4.19 (continued)

Family	Botanical name	Image	Plant part used	Bioactive compound(s)	Structure	Reference(s)
	Berberis aquifolium Pursh		Stem, bark, root	Bis-benzylisoquinoline/ protoberberine alkaloids	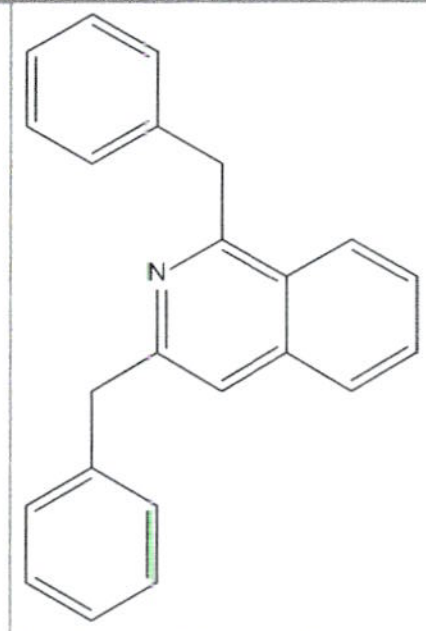	Cernakova et al. (2002)
	Synonym(s): *Mahonia aquifolium* (Pursh) Nutt., *Odostemon aquifolium* (Pursh) Rydb., *Berberis fasciculata* Schult. & Schult.f., *Berberis fascicularis* Sims. **Common name(s)**: Oregon grape, Hollyleaved barberry, Oregon holly grape, Oregon holly. **Botanical description**: It is found in western North America. It is a fruit-bearing evergreen plant that grows 2–6 ft. high; it bears spiny leaves which turn from green to a decorative bronze color. The berries are edible, blue-purple in color and rich in vitamin C. **Medicinal use(s)**: It is used for the treatment of hepatic cancer. It cures psoriasis (fungal infections and skin problems), improves blood flow in the liver, and stimulates intestinal secretions and bile flow. It is used to treat gall bladder disease, hemorrhages, and several forms of cancers.					

	Mahonia bealei (Fortune) Pynaert		Leaf	Berberine	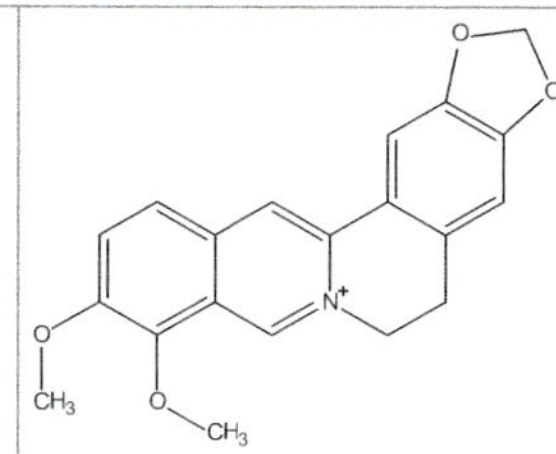	Duke and Ayensu (1985a, b) and Hua et al. (2010)
	Synonym(s): *Berberis bealei* Fortune, Berberis japonica Lindl., *Mahonia bealei* (Fortune) Carrière. **Common name(s)**: Leatherleaf mahonia. **Botanical description**: It is a shrub native to China and Korea. It grows up to 8 m tall; leaves are 50 cm long; flowers are borne in an erect raceme up to 30 cm long; and fruits are a berry which is egg-shaped and dark purple up to 1 mm long. **Medicinal use(s)**: It is used for gastric cancer and mammary carcinoma. It shows antitumor activity. A decoction of the stems and roots is antiphlogistic, anti-rheumatic, depurative, and febrifuge. It is also used for the treatment of pulmonary tuberculosis and recurring fevers.					

(continued)

Table 4.19 (continued)

Family	Botanical name	Image	Plant part used	Bioactive compound(s)	Structure	Reference(s)
	Mahonia fortunei (Lindl.) Fedde.		Leaf	Alkaloid berberine	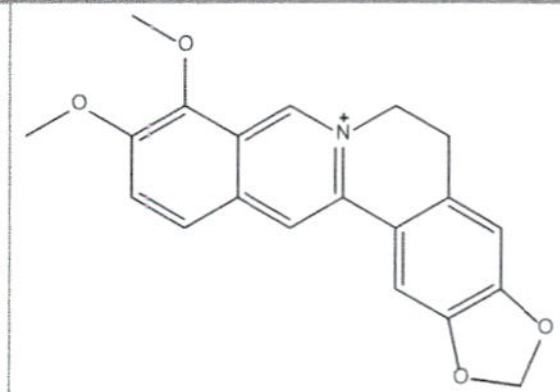	Duke and Ayensu (1985a, b)
	Synonym(s): *Berberis fortunei var. szechuanica* (Ahrendt) Laferr., *Mahonia fortunei* var. *szechuanica* Ahrendt. **Common name(s)**: Chinese mahonia. **Botanical description**: It is the native plant of Eastern Asia and Western-Southern China. It is an evergreen plant, usually unbranched, multi-stemmed shrub growing up to 2 m tall. The plant is harvested from the wild for local use as medicines. This plant mainly grows in the temperate region; it grows best in dumped area where natural humus-rich soil is present. **Medicinal use(s)**: The leaves are used for lung cancer treatment. The stems and roots are said to have properties of clearing heat and dry dampness, purging fire, and removing toxins. It is also used in the treatment of dampness-heat diarrhea, dysentery, jaundice, red urine, red painful swelling eyes, toothache caused by stomach fire, sore and deep-rooted boils, swelling, and abscesses.					

	Mahonia japonica (Thunb.) DC.		Leaf	Dihycroberberine alkalcid	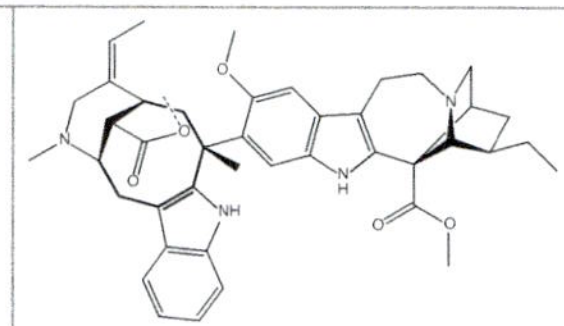	Duke and Ayensu (1985a, b)
	Synonym(s): *Ilex japonica* Thunb., *Mahonia tikushiensis* Hayata. **Common name(s):** Japanese holly-grape. **Botanical description**: It is a native plant of China, Taiwan, and some parts of Asia. It is an evergreen shrub growing up to 2–3 m tall. Flowering occurs from February to April, and the seeds ripen from April to June. The flowers are hermaphrodite. It can grow in full shade and prefers the moist soil. **Medicinal use(s)**: The leaf tonic is used for lung cancer treatment. It also shows antitumor properties. The root and stem of this plant are used as anti-rheumatic, antitussive, expectorant, and febrifuge. It is used in the treatment of bone fractures and bacterial dysentery.					

(continued)

Table 4.19 (continued)

Family	Botanical name	Image	Plant part used	Bioactive compound(s)	Structure	Reference(s)
	Nandina domestica Thunb.		Fruit	Almacin	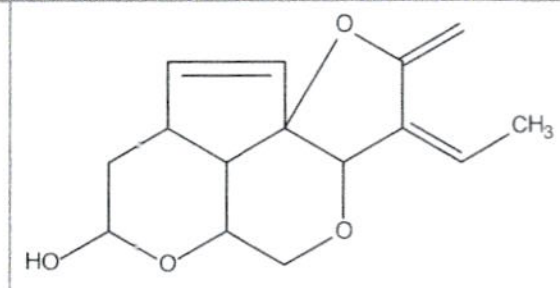	Shoji et al. (1984)
	Synonym(s): *Nandina domestica* var. *linearifolia* C.Y. Wu ex S.Y. Bao. **Common name(s)**: Nandina, Heavenly bamboo. **Botanical description**: It is a flowering plant that is found in the Eastern Asia from the Himalayas to Japan. It is an erect evergreen shrub which grows up to 2 m tall by 1.5 m wide; it bears numerous unbranched stems; the young leaves in spring are brightly colored (pink to red or purplish) again before falling; the flower is white, borne in early summers; and the fruit is bright redberries that ripen in late autumn and persist through the winter. **Medicinal use(s)**: It is used to treat breast cancer and pharyngeal tumors. The roots and stems are antitussive, astringent, febrifuge, stomachic, and tonic. Decoction is used to treat fevers in influenza, whooping cough, and pain in bones. The roots are anti-rheumatic.					

<table>
<tr><td></td><td>Sinopodophyllum hexandrum (Royle) T.S.Ying</td><td></td><td>Root and rhizome</td><td>Podophyllotoxin</td><td>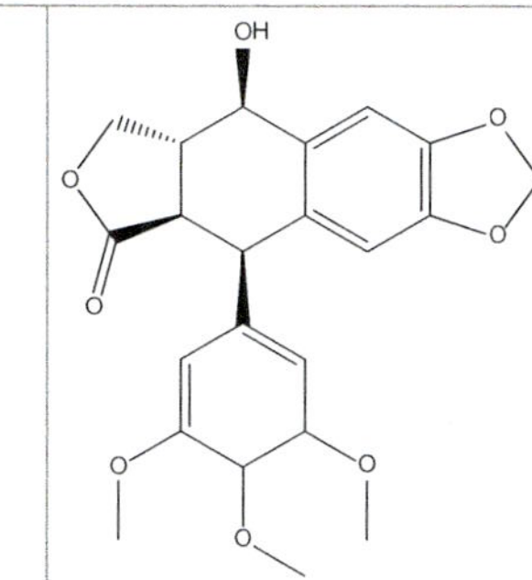</td><td>Giri and Narasu (2000)</td></tr>
<tr><td></td><td colspan="3">Synonym(s): Dysosma emodi (Wall. ex Royle) M.Hiroe, Podophyllum emodi Wall. ex Hook.f. & Thomson, Podophyllum hexandrum Royle, Sinopodophyllum emodi (Wall. ex Hook.f. & Thomson) T.S.Ying.
Common name(s): Mayapple.
Botanical description: It is native plant of the Himalayas and abundant in Kashmir Valley. It is a perennial that grows up to 0.5 m tall and 0.3 m wide. It flowers from May to June, and the seeds ripen from July to August. The flowers are hermaphrodite. It prefers moist soil, grows in full shade, and is suitable for light sandy and medium loamy soil.
Medicinal use(s): It is used to treat precancerous white patches on the tongue and mouth. It is especially used in case of ovarian cancer.</td><td></td><td></td><td></td></tr>
</table>

Table 4.20 Anticancer plants of family Betulaceae

Family	Botanical name	Image	Plant part used	Bioactive compound(s)	Structure	Reference(s)
Betulaceae	*Alnus japonica* (Thunb.) Steud.		Wood	Betulinic acid	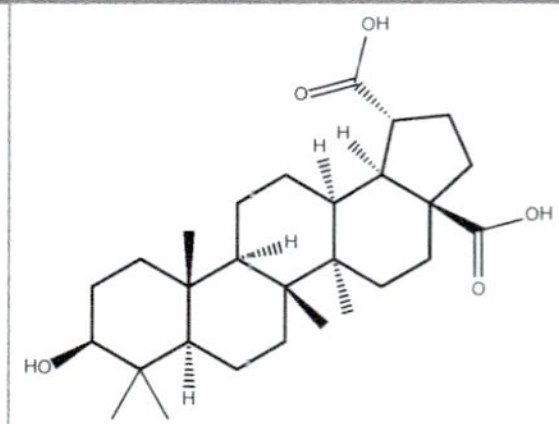	Lee et al. (1992), Rhan et al. (1992), Kim et al. (2005), and Uto et al. (2015)
	Synonym(s): *Alnus reginosa* Nakai, *Betula japonica* Thunb. **Common name(s)**: Japanese alder. **Botanical description:** It is a native plant of South Korea, China, and Japan. It is shrub or tree with a dense crown, usually grows 3–10 m tall; the bole is usually short, deciduous in most of the range; the tree has adopted an evergreen habit. It grows in wet soils along with streams and in the swamps. The tree is a source of fuel. **Medicinal use(s)**: It is used to treat ovarian cancer. It consists of antitumor, anti-inflammatory, antibacterial, and antimicrobial compounds. The stem bark is used in the treatment of cancer, gastric disorders, hepatitis, and fatty liver.			Hirsutanone	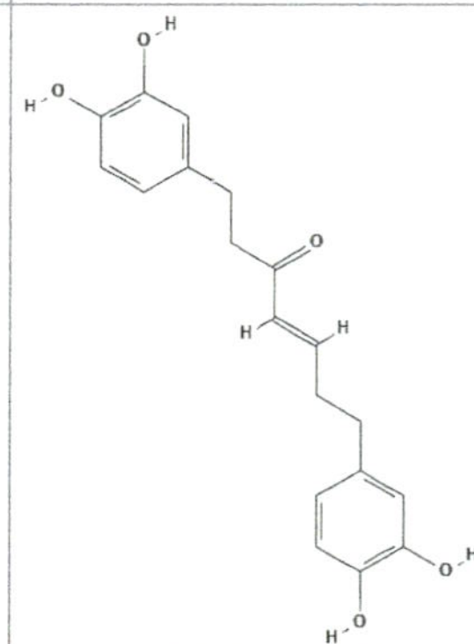	
				Oregonin	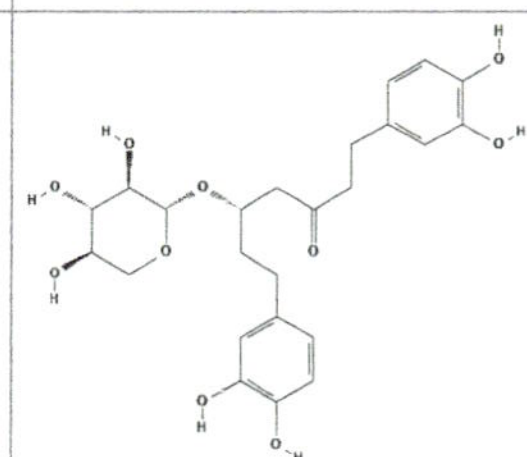	

	Betula platyphylla Sukaczev		Bark, root bark	Diarylheptanoid glycoside	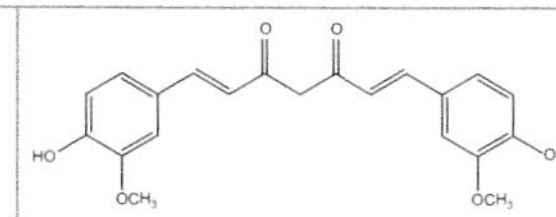	Duke and Ayensu (1985a, b) and MiJu et al. (2004)
	Synonym(s): *Betula ajanensis* Kom., *Betula tauschii* (Regel) Koidz., *Betula latifolia* Kom. **Common name(s)**: White birch, Asian white birch. **Botanical description**: It is a native plant of China, South Korea, Japan, and Siberia. It is a deciduous tree growing up to 20 m at fast rate. It flowers from April to May, and the seeds ripen in August. The flowers are monoecious and pollinated by winds. It cannot grow in shade. **Medicinal use(s)**: It is used for the treatment of breast and prostate cancer. It possesses antifungal and anticancer properties. The root bark shows antitumor properties. It contains several medicinally active constituents including triterpenoids and flavonoids.					

(continued)

Table 4.20 (continued)

Family	Botanical name	Image	Plant part used	Bioactive compound(s)	Structure	Reference(s)
	Betula platyphylla subsp. *mandshurica* (Regel) Kitag.		Bark, root bark	1,1-Diphenyl-2 picrylhydrazyl	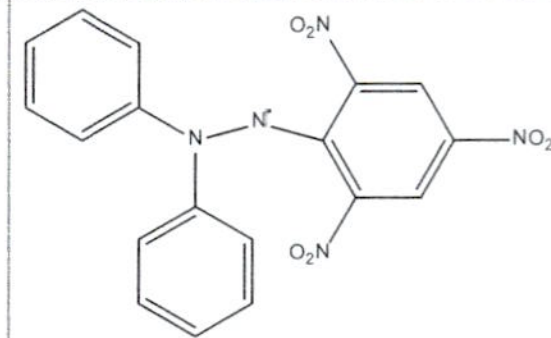	Ju et al. (2004)
	Synonym(s): *Betula mandshurica* (Regel) Nakai, *Betula platyphylla* var. *japonica* (Miq.) Hara. **Common name(s)**: White birch. **Botanical description**: It is native plant of Japan, China, Korea, and Siberia. This grows in conical shape, up to 20 m in height, with white bark, yellow-green male catkins or female catkins, and has dark, green leaves. **Medicinal use(s)**: It is used to treat human colon cancer. The stem and root bark possess anticancer, antitumor, and antifungal properties. It is also used in the treatment of conditions such as internal diseases and inflammation.					

Table 4.21 Anticancer plants of family Bignoniaceae

Family	Botanical name	Image	Plant part used	Bioactive compound(s)	Structure	Reference(s)
Bignoniaceae	*Kigelia africana* (Lam.) Benth.		Leaves, root, bark	β-Sitosterol	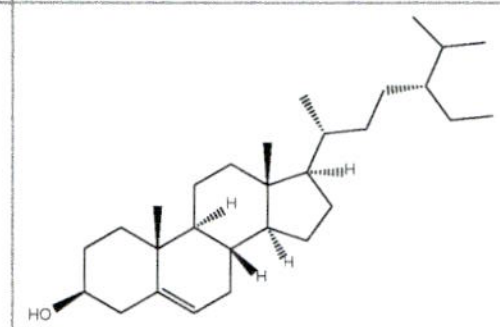	Msonthi and Magombo (1983) and Gabriel and Olubunmi (2009)
	Synonym(s): *Bignonia africana* Lam., *Kigelia abyssinica* A.Rich., *Crescentia pinnata* Jacq., *Kigelia aethiopica* Decne., *Kigelia aethiopum* (Fenzl) Dandy, *Kigelia erytraeae* Mattei, *Kigelia ikbaliae* De Wild., *Kigelia pinnata* (Jacq.) DC., *Kigelia somalensis* Mattei, *Kigelia talbotii* Hutch. & Dalziel, *Sotor aethiopum* Fenzl, *Tanaecium pinnatum* (Jacq.) Willd., *Tecoma africana* (Lam.) G.Don. **Common name(s)**: Sausage tree. **Botanical description**: It is a genus of flowering plant and is native of Africa. It is an evergreen tree that grows up to 20 m tall. The bark is gray and smooth. The fruit is a woody berry from 12 to 39 in. **Medicinal use(s)**: It is used to treat skin and breast cancer. It is used for treating sores, skin ulcer, and skin cancer. Due to strong antimicrobial, antibacterial, antifungal, and antiviral properties, the extract of leaves and bark has become popular in lotions and cosmetics.					

(continued)

Table 4.21 (continued)

Family	Botanical name	Image	Plant part used	Bioactive compound(s)	Structure	Reference(s)
	Stereospermum chelonoides (L.f.) DC.		Root	Lapachol	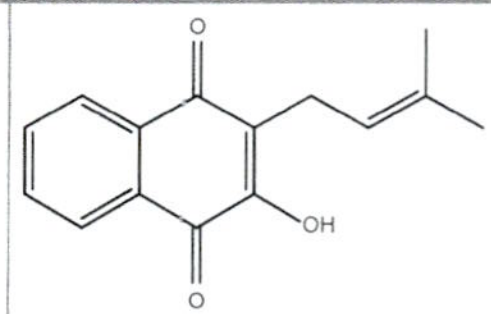	Rao et al. (1968) and Rahmatullah et al. (2010)
	Synonym(s): *Bignonia chelonoides* L.f., *Bignonia gratissima* K.D.Koenig ex DC., *Heterophragma chelonoides* (L.f.) Dalzell & A.Gibson, *Heterophragma suaveolens* (Roxb.) Dalzell & A.Gibson, *Hieranthes fragrans* Raf., *Spathodea suaveolens* (Roxb.) Benth. & Hook.f., *Tecoma suaveolens* (Roxb.) G.Don, *Stereospermum suaveolens* (Roxb.) DC. **Common name(s)**: Padhal, Podal, Parul. **Botanical description**: It is a native plant of India. It is about 6–9 cm long, 1–1.5 cm thick. It is cylindrical in shape; it has rough dye to vertical fissures, cracks, ridges, and transverse fine lenticels, internally dark brown, lamellation of stratification due to the presence of concentric band of fiber. **Medicinal use(s)**: It is used to treat various human cancer cell lines. It is having antitumor potential in the root parts of the plant. It is helpful in the treatment of anorexia and improves the appetite. It is having anti-inflammatory properties; it is also a good blood purifier and removes toxins from the body.					

	Tabebuia cassinoides (Lam.) DC.		Bark	Quinine	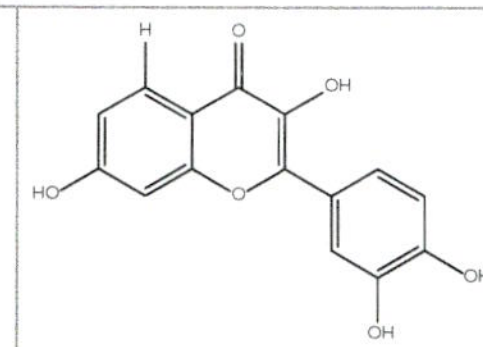 	Mohammad (2006), Desai et al. (2008), and Madhuri and Pandey (2009)
	Synonym(s): *Bignonia cassinoides* Lam., *Bignonia obtusifolia* Lam., *Bignonia tabebuya* Vell., *Bignonia uliginosa* Gomes, *Catalpa cassinoides* (Lam.) Spreng., *Proterpia obtusifolia* (Lam.) Raf., Spathodea *magnolioides* Cham., *Tabebuia magnolioides* (Cham.) Miers, *Tabebuia uliginosa* (Gomes) DC., *Tecoma uliginosa* Mart. ex DC. **Common name(s)**: Amapa, Cortez, Greenheart. **Botanical description**: It is found in tropical America and some of the lesser Antilles. The tree grows on variety of sites, from the ridge top to riverbanks and marsh forests. It grows up to 140–150 ft. in height with a trunk of 6 ft. in diameter. **Medicinal use(s)**: It is used to treat prostate cancer and human tumor cell lines. It has several chemical compounds like alkaloids and flavonoids to fight against malaria and some type of tumor cells. It strengthens the immune system, cleanses the body, and stimulates the production of red blood cells.					

(continued)

Table 4.21 (continued)

Family	Botanical name	Image	Plant part used	Bioactive compound(s)	Structure	Reference(s)
	Tabebuia rosea (Bertol.) Bertero ex A.DC.		Bark	Cinnamaldehyde	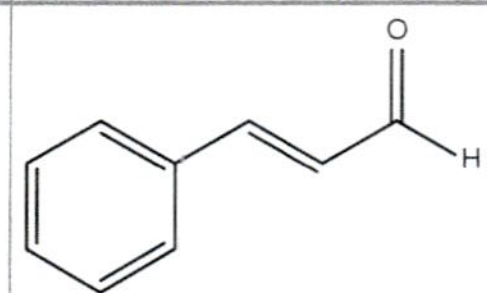	Gentry (1980) and Kaur et al. (2011)
	Synonym(s): *Couralia rosea* (Bertol.) Donn.Sm., *Sparattosperma roseum* (Bertol.) Miers, *Tabebuia mexicana* (Mart. ex DC.) Hemsl., *Tabebuia punctatissima* (Kraenzl.) Standl., *Tecoma mexicana* Mart. ex DC., *Tecoma punctatissima* Kraenzl., *Tecoma rosea* Bertol. **Common name(s)**: Pink pouli, Rosy trumpet tree. **Botanical description**: It is native of Mexico. It is a neotropical flowering tree that grows up to 30 m and can reach a diameter at the breast height up to 100 cm. The tree crown is irregular and wide, with irregular, stratified ramification. **Medicinal use(s)**: It is used for uterine cancer. A decoction of the flower, leaves, and roots has been used to reduce fevers and pain, cause sweating, and treat tonsil inflammation and various other disorders.					

	Handroanthus serratifolius (Vahl) S.O.Grose		Bark	Cardenolide	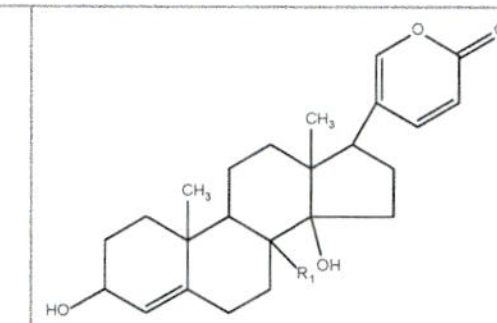	Gentry (1980) and Silva et al. (2012)
	Synonym(s): *Bignonia araliacea* Cham., *Bignonia flavescens* Vell., *Bignonia serratifolia* Vahl, *Gelseminum araliaceum* (Cham.) Kuntze, *Tabebuia serratifolia* (Vahl) G.Nicholson, *Gelseminum speciosum* (DC. ex Mart.) Kuntze, *Handroanthus araliaceus* (Cham.) Mattos, *Handroanthus atractocarpus* (Bureau & K.Schum.) Mattos, *Handroanthus flavescens* (Vell.) Mattos, *Tabebuia araliacea* (Cham.) Morong & Britton, *Tabebuia serratifolia* (Vahl) G. Nicholson, *Tecoma araliacea* (Cham.) DC., *Tecoma atractocarpa* Bureau & K.Schum., *Tecoma conspicua* DC., Tecoma patrisiana DC., *Tecoma serratifolia* (Vahl) G.Don, *Tecoma speciosa* DC. ex Mart., *Vitex moronensis* Moldenke. **Common name(s)**: Suriname greenheart, Yellow poui, Amapa. **Botanical description**: It is a native plant of Central and South America. It grows in the rainforest or tropical rainforest. It bears yellow-colored flowers. It grows up to 150 ft. in height and 4–7 ft. in diameter. **Medicinal use(s)**: It is used to treat terminal types of cancer. It is reported that this plant has antifungal, anticancer, and antitumor properties. It also cures cough, malaria, and anemia and also purifies the blood.			Lapachol	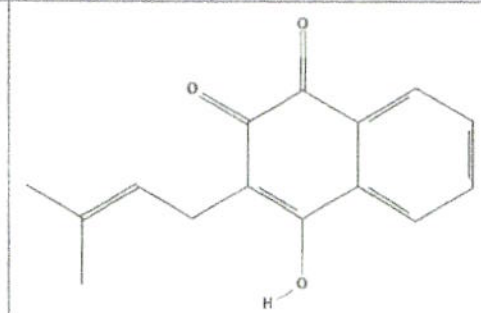	

(continued)

Table 4.21 (continued)

Family	Botanical name	Image	Plant part used	Bioactive compound(s)	Structure	Reference(s)
	Zeyheria tuberculosa (Vell.) Bureau ex Verl.		Trunkwood	Lapachol	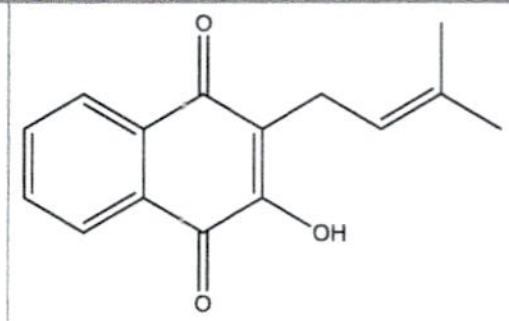	De et al. (1976) and Bastos et al. (2009)
	Synonym(s): *Bignonia tuberculosa* Vell., *Jacaranda tuberculosa* (Vell.) Steud., *Zeyheria kuntzei* K.Schum. **Common name(s)**: Culhoes de bode, Ipe-branco, Buxo de boi, Jopo de mono. **Botanical description**: It is a native plant of Brazil – Minas Gerais – South America, and Bolivia. It is a semi-deciduous shrub or a tree that can grow up to 15–23 m tall. The straight, cylindrical bole can be 40–60 cm in diameter with a thick layer of bark up to 5 cm wide. The plant is occasionally cultivated. **Medicinal use(s)**: It is useful for treating cancer and skin diseases. It also possesses antimicrobial properties.			α-lapachone	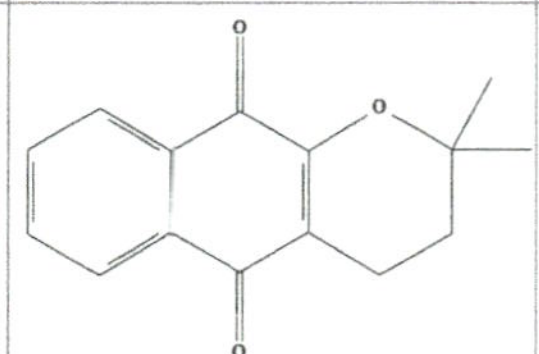	
				dehydro-α-lapachone	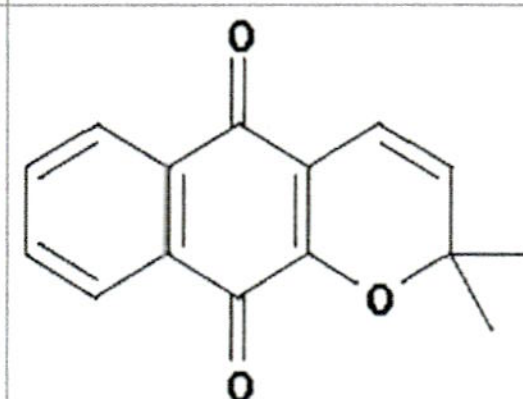	

		Zeyherol	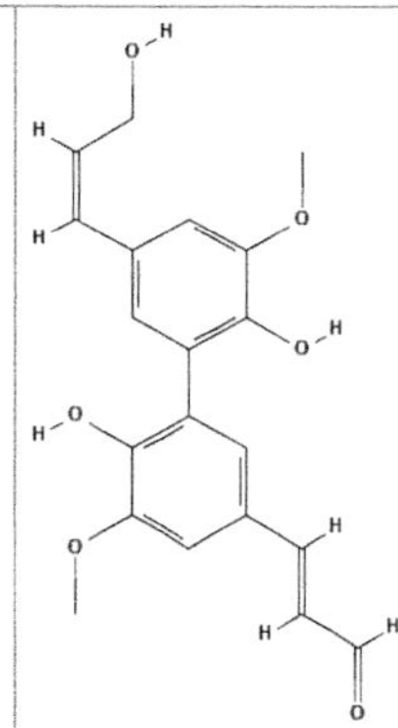	
		β-amyrin	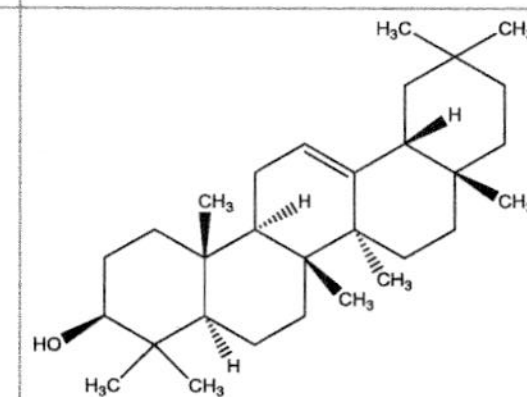	
		β-sitosterol	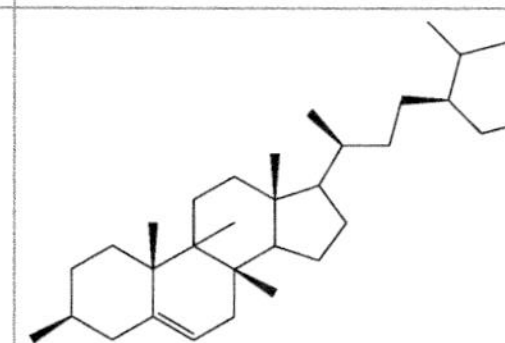	

Table 4.22 Anticancer plants of family Boraginaceae

Family	Botanical name	Image	Plant part used	Bioactive compound(s)	Structure	Reference(s)
Boraginaceae	*Arnebia euchroma* (Royle) I.M. Johnst.		Leaf	Naphthoquinone pigment-LIII	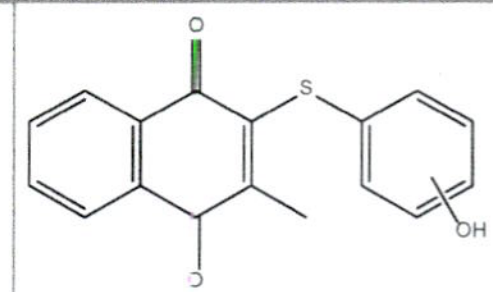	Lu and Liao (1990) and Xiong et al. (2009)
	Synonym(s): *Lithospermum euchromon* Royle, *Macrotomia euchromon* Paulsen. **Common name(s)**: Pink arnebia. **Botanical description**: It is a native plant of East Asia, the Himalayas from Nepal to Afghanistan. It is a perennial plant growing up to 0.3 m. It flowers from June to August. The seeds ripen from July to September. The flowers are hermaphrodite and are pollinated by insects. It cannot grow in shade. It prefers dry or moist soil. **Medicinal use(s)**: It is used to treat lung cancer, breast cancer, and human hepatocellular carcinoma cell line Bel-7402. The root is antipyretic, contraceptive, emollient, and vulnerary. It is used in the treatment of mild constipation. It inhibits the growth of cancer cell lines on the chorionic membrane. It has antitumor anti-inflammatory, antimicrobial, antibacterial, and antipyretic properties.			Shikonin	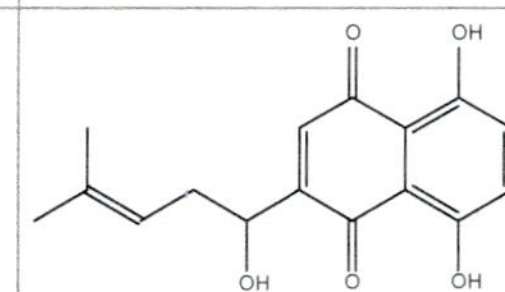	

	Echium arenarium Guss.		Leaf	Thymol	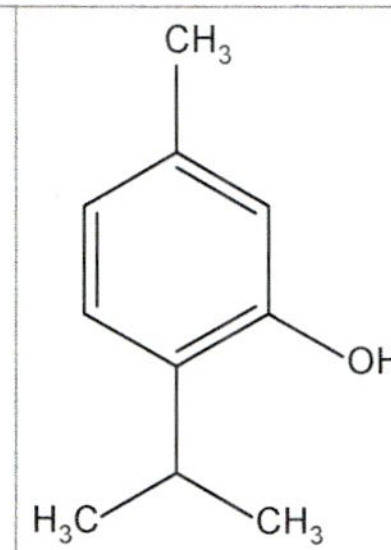	Yousif et al. (1983) and Maffei (2003a, b, c, d, e, f, g, h, i, j, k, l, m)
	Synonym(s): *Echium delileanum* Lojac., *Echium diffusum* Guss., *Echium prostratum* Sieber ex A.DC., *Echium sieberi* (A.DC.) Lojac. **Common name(s)**: Coastal viper's bugloss. **Botanical description**: It is a native plant of Sudan, North Africa, mainland Europe, and the Macaronesian islands. It is a flowering garden plant. It flowers in the month of August. The flowers are hermaphrodite (have both male and female organs) and are pollinated by insects. It is suitable for light (sandy) and medium (loamy) soils and prefers well-drained soil, and suitable pH is acid, neutral, and basic (alkaline) soils. It cannot grow in the shade. It prefers dry or moist soil. **Medicinal use(s)**: It is used to treat lung and breast cancer. It is used to treat tumors and abscesses. It contains high level of alpha linolenic acid, gamma linolenic acid, and stearidonic acid and cures skin problems. It possesses antifebrile, antidepressant, anxiolytic, anti-inflammatory, and antioxidant properties.					

(continued)

Table 4.22 (continued)

Family	Botanical name	Image	Plant part used	Bioactive compound(s)	Structure	Reference(s)
	Lithospermum ruderale Douglas ex Lehm.		Leaf	Quercetin	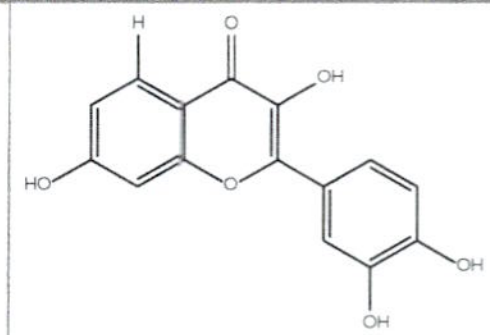	Huang et al. (2009b)
	Synonym(s): No synonyms are recorded for this name. **Common name(s)**: Stoneseed. **Botanical description**: It is a native plant of Western Canada and the Western United States. It is a flowering plant. Bunches of flowers with leaflike bracts appear towards the top of the stem among the leaves. **Medicinal use(s)**: It is used to treat lung cancer. It possesses anti-inflammatory, anti-rheumatic, and anti-infection properties. The infusions of roots are used as diuretics in the treatment of kidney disorders. It decreases the weight of pituitary and thymus glands.			Shikonin	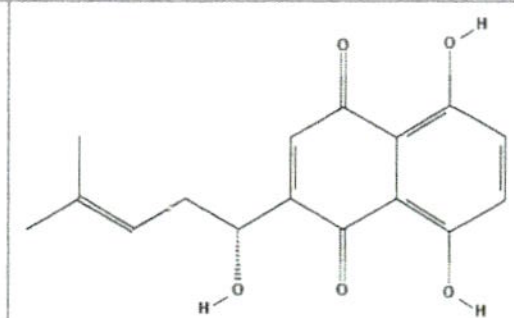	

Table 4.23 Anticancer plants of family Buddlejaceae

Family	Botanical name	Image	Plant part used	Bioactive compound(s)	Structure	Reference(s)
Buddlejaceae	*Buddleja globosa* Hope		Leaf	Campesterol	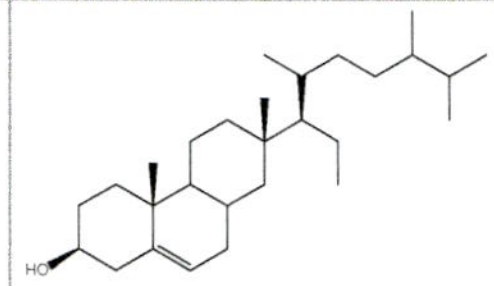	Houghton (1984) and Carmona et al. (2016)
	Synonym(s): *Buddleja capitata* Jacq., *Buddleja connata* Ruiz & Pav., *Buddleja globifera* Mirb. **Common name(s)**: Orange ball tree. **Botanical description**: It is an evergreen shrub native of South America. It grows up to 5 mm at altitudes up to 2000 m. It flowers from June to July. The flowers are monoecious (individual flowers are either male or female, but both sexes can be found on the same plant) and are pollinated by bees, Lepidoptera. It cannot grow in shade. The plant can tolerate maritime exposure. **Medicinal use(s)**: It is used to treat human colon, lung, and breast cancer and myeloid leukemia. It possesses antitumor, antiproliferative, anticancer, antimutagenic, and anti-inflammatory properties. It is also used to treat warts, wounds, burns, and external and internal ulcers.			Verbascoside	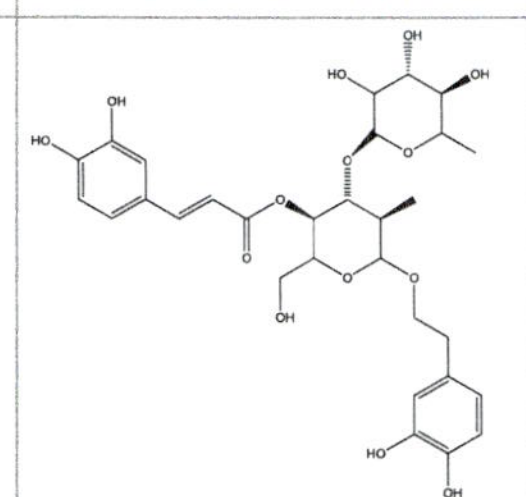	
				Luteolin	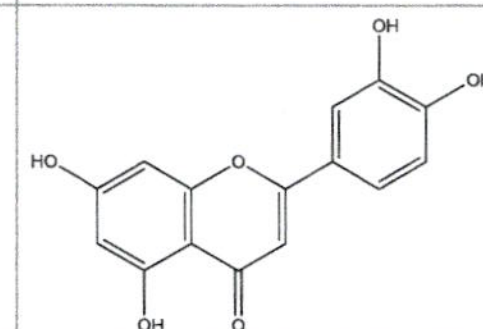	

(continued)

Table 4.23 (continued)

Family	Botanical name	Image	Plant part used	Bioactive compound(s)	Structure	Reference(s)
	Buddleja incana Ruiz & Pav.		Flower and leaf	Carvacrol	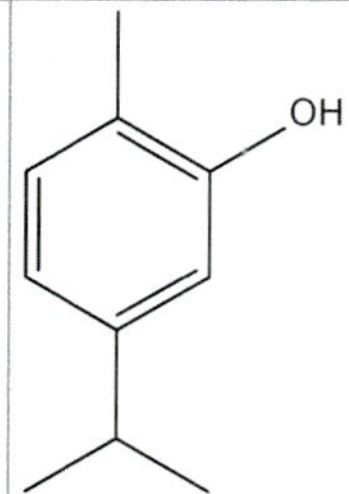	Houghton (1984) and Rehecho et al. (2011)
	Synonym(s): *Buddleja rondeletiiflora* Benth., *Buddleja rugosa* Kunth. **Common name(s)**: Alamo, Chanchunga, Kiswara, Kollia, Kurawara, Quishuar, Quisoar, Quissuar, Quisuar. **Botanical description**: It is a native plant of Bolivia and Colombo. It is a dioecious tree or shrub, 4–15 m tall; the trunk is greater than 50 cm. Paniculate inflorescences have 2–3 order of leafy-bracted branches bearing heads 1–1.5 in. diameter, each with 15–40 flowers, the corollas 3–4 mm long. **Medicinal use(s)**: It is used to treat human colon, lung, and breast cancer and myeloid leukemia cells. It is used to treat warts; it is also used in toothache and as a diuretic. It possesses anti-inflammatory, antibacterial, and antihypertensive properties.					

	Buddleja officinalis Maxim.		Flowers	Benzoic acid	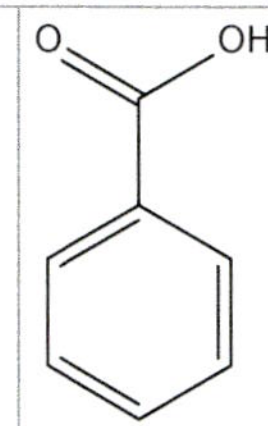	Houghton (1984) and Jung et al. (2017)
	Synonym(s): *Buddleja acutifolia* C.H.Whight, *Buddleja delavayi* L.F.Gagnep., *Buddleja truncate* L.F.Gagnep., *Buddleja lavandulacea* Kraenzl. **Common name(s)**: Buddleia flower. **Botanical description**: It is originated in East Asia and grows in China. It is a shrub which grows up to 2 m tall. It grows in cold climate and is deciduous. **Medicinal use(s)**: It is used for the treatment of lung, breast, colon, and various cancer cell lines. It is effective for blurred vision, chemosis, eye bloodshot, liver diseases, and weak eyesight. It possesses anticancer, antitumor, anti-inflammatory, antioxidant, antibacterial, and antimicrobial properties.					

Table 4.24 Anticancer plants of family Burseraceae

Family	Botanical name	Image	Plant part used	Bioactive compound(s)	Structure	Reference(s)
Burseraceae	*Boswellia serrata* Roxb. ex Colebr.		Frankincense	Boswellic acid	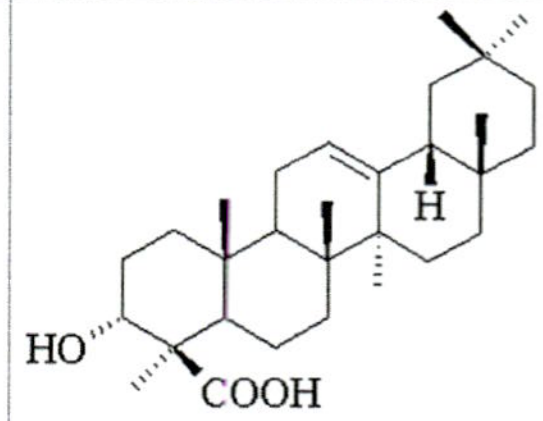	Flavin (2007), Ranjbarnejad et al. (2017), and Sharma and Jana (2019)
			Root	Sabinene	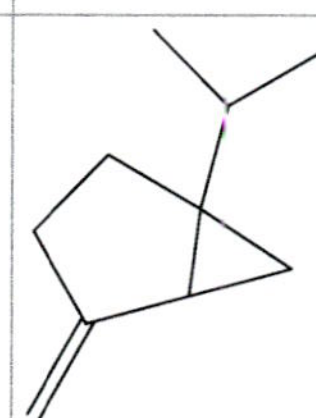	
	Synonym(s): *Boswellia balsamifera* Spreng., *Boswellia glabra* Roxb., *Boswellia thurifera* Roxb. ex Fleming, Chloroxylon dupada Buch.-Ham. **Common name(s)**: Indian olibanum, Indian frankincense, Salaiguggul, Dhuna-goch. **Botanical description**: It is a deciduous tree endemic to India, found on the dry hills and slopes on the gravelly soil. It is a medium-sized tree, 3–5 cm tall with ash color papery bark. Leaves are crowded at the end of branches. **Medicinal use(s)**: It used to treat breast cancer cell lines. It possesses anti-inflammatory, anticancer, antiproliferative, antineoplastic, antiaging, antibacterial, and cytotoxic properties. It cures osteoarthritis and possesses anti-wrinkle properties. It consists of numerous active constituents in the gum resin.			Terpinyl acetate	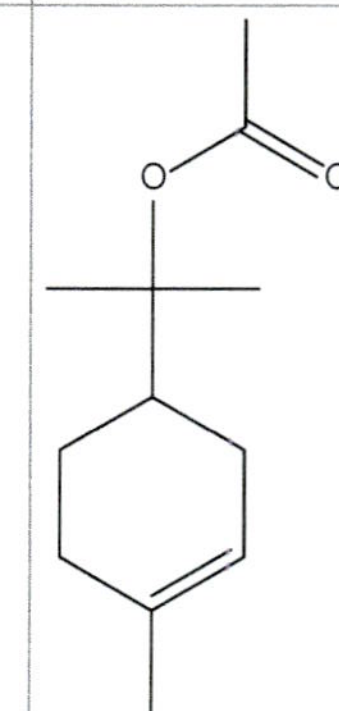	

Table 4.25 Anticancer plants of family Caprifoliaceae

Family	Botanical name	Image	Plant part used	Bioactive compound(s)	Structure	References
Caprifoliaceae	*Lonicera etrusca* G. Santi		Seed	Melatonin	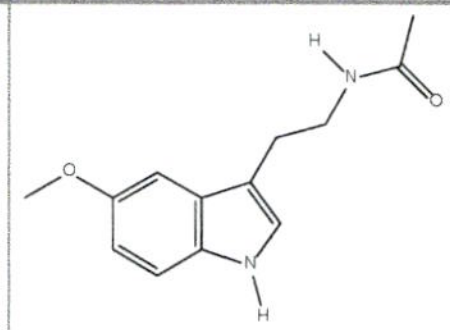	Rivera and Obon (1995) and Çalişkan et al. (2017)
	Synonym(s): *Caprifolium cyrenaicum* Kuntze, *Caprifolium dimorphum* (Tausch) Kuntze, *Caprifolium etruscum* Schult, *Caprifolium germanicum* Quer, and *Lonicera dimorpha* Tausch. **Common name(s)**: Etruscan honeysuckle. **Botanical description**: *Lonicera etrusca* is a species of honeysuckle known by the common name Etruscan honeysuckle. It is native to Europe, and it is known elsewhere, including the Pacific Northwest of North America, as an introduced species where it has escaped cultivation. It is kept in gardens as an ornamental plant. It is a deciduous perennial climber which can reach up to 6 m in length. It is lined with oval leaves which are several centimeters long and bears dense spikes of flowers with pairs of fused leaves at the bases. Each flower has an elongated tubular corolla up to 5 cm long divided partway into two lips. The flower is light yellow to pale reddish pink. The stamens and style protrude from the flower's mouth. The fruit is a bright red rounded berry. **Medicinal use(s)**: Honeysuckle is used for digestive disorders including pain and swelling (inflammation) of the small intestine (enteritis) and dysentery; upper respiratory tract infections including colds, influenza, swine flu, and pneumonia; other viral and bacterial infections; swelling of the brain (encephalitis); fever; boils; and sores. Honeysuckle is also used for urinary disorders, headache, diabetes, rheumatoid arthritis, and cancer. Some people use it to promote sweating, as a laxative, to counteract poisoning, and for birth control. Honeysuckle is sometimes applied to the skin for inflammation and itching and to kill germs. It is also used to care the crown gall tumor.					

(continued)

Table 4.25 (continued)

Family	Botanical name	Image	Plant part used	Bioactive compound(s)	Structure	References
	Sambucus racemosa L.		Bark	Not found		Forlines et al. (1992)
	Synonym(s): *Sambucus pubens*, *Sambucus microbotrys* and *Sambucus callicarpa*. **Common name(s)**: Red elderberry and Red-berried elder. **Botanical description**: *Sambucus racemosa* is often a treelike shrub growing 2–6 m (7–20 ft.) tall. The stems are soft with a pithy center. Each individual leaf is composed of 5–7 leaflike leaflets, each of which is up to 16 cm (6 1/4 in.) long, lance-shaped to narrowly oval, and irregularly serrated along the edges. The leaflets have a strong disagreeable odor when crushed. The inflorescence is a vaguely cone-shaped panicle of several cymes of flowers blooming from the ends of stem branches. The flower buds are pink when closed, and the open flowers are white, cream, or yellowish. Each flower has small, recurved petals and a star-shaped axis of five white stamens tipped in yellow anthers. The flowers are fragrant and visited by hummingbirds and butterflies. The fruit is a bright red or sometimes purple drupe containing 3–5 seeds. **Medicinal use(s)**: It is used as an emetic and antidiarrheal, for cold and cough remedy, as dermatological and gynecological aid, and as a hemostat. It is also used to treat breast cancer.					

	Lonicera caerulea L.		Fruit	Cyanidin 3-O-glucoside	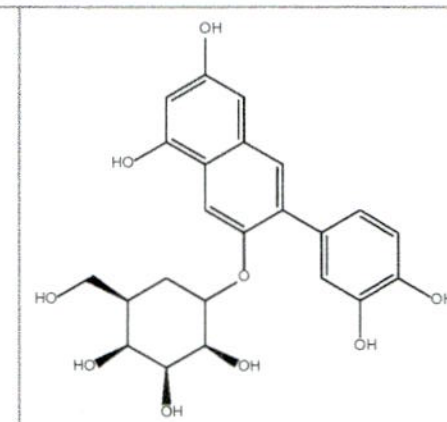	Svarcova et al. (2007) and Luo et al. (2017)
	Synonym(s): *Caprifolium caeruleum* (L.) Lam., *Euchylia caerulea* (L.) Dulac, *Isika coerulea* (L.) Medik and *Xylosteon caeruleum* (L.) Dum.Cours. **Common name(s)**: Blue-berried honeysuckle, Fly honeysuckle, Haskap berry, Sweetberry honeysuckle, and Blue honeysuckle. **Botanical description**: Blue-berried honeysuckle is a deciduous shrub growing up to 1.5–2 m tall. The leaves are opposite, oval, 3–8 cm long and 1–3 cm broad, and grayish green, with a slightly waxy texture. The flowers are yellowish white and 12–16 mm long, with five equal lobes; they are produced in pairs on the shoots. The fruit is an edible, blue berry, somewhat rectangular in shape weighing 1.3–2.2 gm, and about 1 cm in diameter. **Medicinal use(s):** It is believed that haskap has been used in folk medicine to reduce the risk of hypertension, glaucoma, heart attack, anemia, malaria, osteoporosis, and gastrointestinal disease. It is also used to treat prostate cancer.					

(continued)

Table 4.25 (continued)

Family	Botanical name	Image	Plant part used	Bioactive compound(s)	Structure	References
	Weigela subsessilis L.H.Bailey		Leaf	Corosolic acid	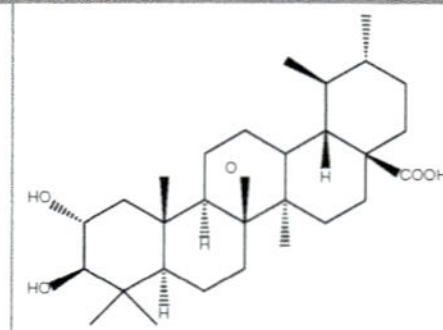	Lee et al. (2010)
	Synonym(s): *Diervilla subsessilis* Nakai. **Common name(s)**: Korean weigela. **Botanical description**: *Weigela subsessilis* is apparently endemic to the Korean peninsula. *W. subsessilis* is a deciduous, broad-leaved shrub, which grows from 2 to 3 m tall. The species usually coexists with azaleas or royal azaleas or sometimes clusters itself. It prefers sandy soil, which helps them survive also in barren land. The leaves of *W. subsessilis* have an opposite leaf arrangement and have a wide egg-shaped body with a sharp tip. The width is up to 2 in., while the length is up to 3 in.. The leaves have hair on each side, and the ones on the bottom has spread hair on the leaf veins. It usually does not have petioles, and the edges are slightly toothed. The fruits ripe in November to October, which is a capsule fruit having a short hair and is 10–15 mm long. The seeds in the fruit have developed wings. **Medicinal use(s)**: It has anti-inflammatory, antihyperlipidemic, antioxidant, and antiviral properties. It is used to treat gastric, lung, and breast cancer.					

	Lonicera macranthoides Thunb.		Leaf	Macranthoside B (hederagenin saponin)	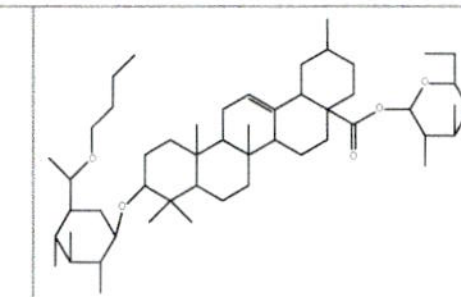	Yu et al. (2009)
	Synonym(s): *Caprifolium macranthum* Kuntze, *Caprifolium nepalense* G.Don ex Loudon, *Lonicera calvescens*, *Lonicera esquirolii H.Lév*, *Lonicera fulvotomentosa*, *Lonicera hirtiflora*, *Lonicera inodora*, and *Lonicera strigosiflora*. **Common name(s)**: Long-flower honeysuckle, Twinberry. **Botanical description**: The canopy of long-flower honeysuckle is dome-shaped. An average height for 5-year-old plants is 1.7 m with a canopy diameter of 4.0 m. Leaves are simple and opposite, ovate to oblong with a round base and acuminate apex. The upper leaf surface is azure green, and the underside is yellowish green. The mean length of mature leaves is 9.3 cm with the widest width of 5.7 cm. The petiole length averages 0.7 cm, and new branches appear reddish in color. **Medicinal use(s):** It has anti-inflammatory, antibacterial, antiviral, and antioxidant effects, inhibits platelet activation, and is used to treat hepatocarcinoma.					

(continued)

Table 4.25 (continued)

Family	Botanical name	Image	Plant part used	Bioactive compound(s)	Structure	References
	Lonicera japonica L.		Flower	Dimethyllanaraflavone	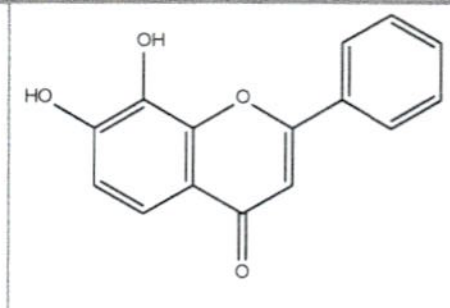	Chen et al. (2004) and Pradhan et al (2009)
	Synonym(s): *Caprifolium japonicum* (Thunb.) Dum.Cours., *Caprifolium roseum* Lam., *Lonicera brachypoda* Siebold, *Lonicera brachypoda* Siebold, *Lonicera fauriei*, and *Lonicera shintenensis* Hayata. **Common name(s)**: Japanese honeysuckle and Jinyinhua in Chinese. **Botanical description**: It is a twining vine able to climb up to 10 m high or more in trees, with opposite, simple oval leaves 3–8 cm long and 2–3 cm broad. The flowers are double-tongued, opening white and fading to yellow, and sweetly scented. The fruit is a globose dark blueberry 5–8 mm diameter containing numerous seeds. The Japanese honeysuckle flower is of high medicinal value in traditional Chinese medicine, where it is called ren dong teng, which means "gold silver flower". **Medicinal use(s):** It has antibacterial and anti-inflammatory properties and is used to dispel heat and remove toxins, including carbuncles, fevers, influenza, and ulcers. It is, however, of cold and yin nature and should not be taken by anyone with a weak and "cold" digestive system. This is used to treat lung, breast, ovarian, and renal cancer.			Agathisflavone 7″-methylagathisflavone	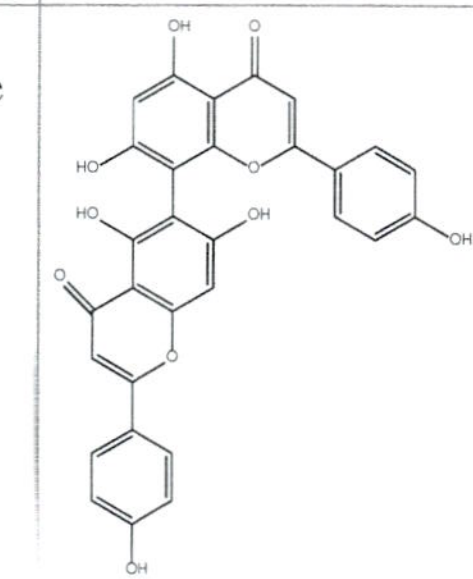	

Table 4.26 Anticancer plants of family Celastraceae

Family	Botanical name	Image	Plant part used	Bioactive compound(s)	Structure	References
Celastraceae	*Celastrus orbiculatus* Thunb.		Root, entire plant	β-Dihydroagarofuran	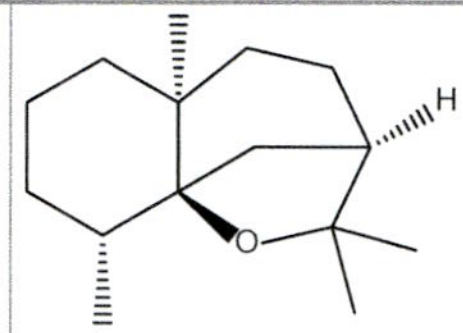	Gupta et al. (1996) and Kim et al. (1998)
	Synonym(s): *Catha articulata* G.Don, *Celastrus insularis* Koidz., *Celastrus jeholensis* Nakai ex Nakai & Kitag., *Celastrus lancifolius* Nakai, *Celastrus orbiculatus*, *Celastrus stephanotiifolius* (Makino) Makino, *Celastrus strigillosus* Nakai, *Celastrus tatarinowii* Rupr., and *Celastrus versicolor* Nakai. **Common name(s)**: Asian bittersweet and Asiatic bittersweet. **Botanical description**: Asian bittersweet was introduced from its native East Asia in 1860 and now grows in much of the Eastern United States. This aggressive vine tolerates both high sun and deep shade and can quickly overtop and girdle trees. This species has long been regarded as an attractive ornamental. Its abundant clusters of yellow seeds surrounded by a fleshy red aril are often used to make festive wreaths and flower arrangements. Seeds are spread when the plants are later thrown away. Birds also relish the fruits and disperse the seeds far and wide. The plant is a liana (i.e., a woody plant with a vine-like growth form); the leaf blade is simple (i.e., lobed or unlobed but not separated into leaflets); there is one leaf per node along the stem; the edge of the leaf blade has teeth; the plant does not have spines, prickles, or thorns; leaf blades are 40–70 mm in length and 35–55 mm in width; the fruit is dry and splits open when ripe; there are three or more scales on the winter bud,; and they overlap like shingles, with one edge covered and the other edge exposed. **Medicinal use(s)**: It has been reported that the extracts and some compounds isolated from COT have antitumor, anti-inflammatory, analgesic, antibacterial, antiviral, antifertility, and other pharmacological activities. Especially it is used to treat gastric cancer.					

(continued)

Table 4.26 (continued)

Family	Botanical name	Image	Plant part used	Bioactive compound(s)	Structure	References
	Elaeodendron transvaalense (Burtt Davy) R.H.Archer		Stem and bark	Taraxastanonol	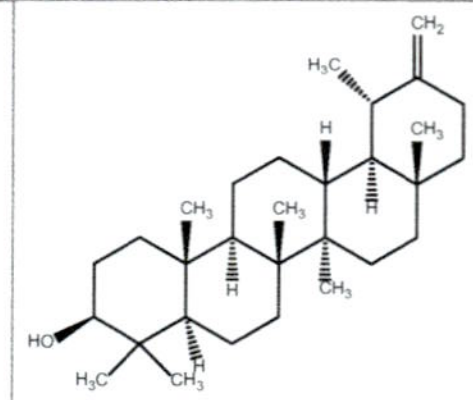	Tshikalange and Hussein (2010)
	Synonym(s): *Cassine transvaalensis* (Burtt Davy) Codd, *Crocoxylon transvaalense* (Burtt Davy) N. Robson, and *Pseudocassine transvaalensis* (Burtt Davy) Bredell. **Common name(s)**: Bushveld saffron (Eng.), Bosveldsaffraan, Lepelhout (Afr.), iNgwavuma, uMgugudo (Zulu), Shimapana (Tsonga), Monamane (Northern Sotho). **Botanical description**: Bushveld saffron is a small to medium-sized, bushy tree that grows up to 6–8 m. In other areas the tree may reach 18 m. It has a conspicuously pale gray, smooth bark that is sometimes finely fissured horizontally. Its dwarf spur-branchlets are characterized by a cluster of leaves at the tips. The leaves are often arranged in threes but can alternate or are arranged spirally on longer stems. They are narrow and linear to narrowly elliptic, light green to dull gray-green. With the entire to finely toothed margin, the leaves are characterized by conspicuous net veins on both sides. The flowers that are borne in summer (December to April) are small, greenish white and are borne in clusters of 20–30 heads, with slender stalks. Flowers give rise to yellow, edible fruits that are somewhat elongated, broadly tapering to both end, up to 20 mm long, and are formed in autumn. **Medicinal use(s)**: *Elaeodendron transvaalense* is used to treat coughs, diarrhea, stomach ailments, herpes, and sexually associated diseases. Other medicinal uses of *E. transvaalense* include the treatment of arthritis, breast cancer, coughs, diarrhea, and stomach ailments. Traditional healers prescribe it presently to people who are suffering from HIV/AIDS.			Sitosterol	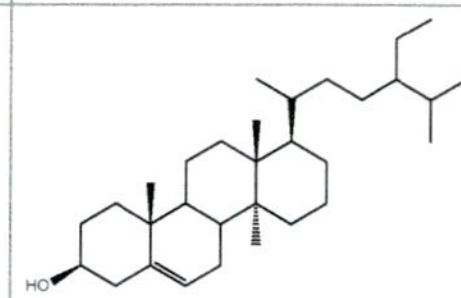	

				4′-O-Methylepi-gallocatechin	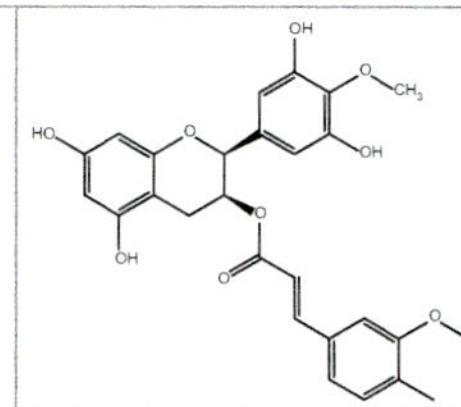	
	Euonymus alatus		Whole plant	Ferulic acid	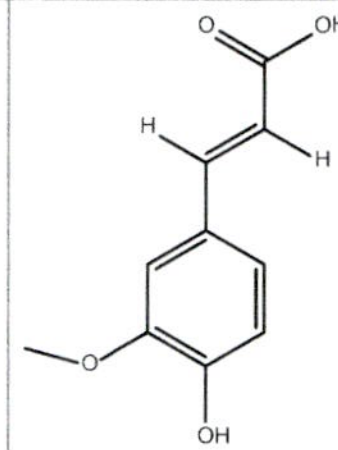	Mohammad (2006), Desai et al. (2008), Madhuri and Pandey (2009), and Jeong et al. (2011)
	Synonym(s): *Euonymus arakianus* Koidz., *Euonymus ellipticus* (Chen H.Wang) C.Y Cheng, *Euonymus kawachianus* Nakai, *Euonymus nakamurae* Makino, *Euonymus rotundatus* (Makino) Nakai, *Euonymus sacrosanctus* Koidz., *Euonymus striatus*, *Euonymus striatus* (Thunb.) Loes., *Euonymus striatus* var. *rotundatus* Makino, *Euonymus subtriflorus* Blume, *Euonymus thunbergianus* Blume, and *Euonymus verrucosus* var. *tchefouensis* Debeaux. **Common name(s)**: Winged spindle, Winged euonymus, or Burning bush. **Botanical description**: This shrub will mature over time to 15–20′ tall but is often pruned shorter. Elliptic to obovate, crenulate to serrulate, green leaves (to 3″ long) turn bright red in fall. Fall color can be spectacular. Small, yellowish-green flowers appear in May but are not showy. Small fruits (1/3″ red capsules) ripen in fall. Fruit capsules split open when ripe to reveal the tiny seeds (each encased in a fleshy orange-red aril). Seeds are attractive to certain birds that eat and distribute them. Greenish-brown stems have distinctive corky ridges ("wings" as used in the common name). Corky-winged stems are more noticeable in winter after leaf drop. Winged euonymus has escaped plantings and naturalized in at least 21 eastern and midwestern states. In some areas, it is now considered to be a threat to native plants because of its ability to establish itself in woodlands, forests, fields, roadsides, and disturbed areas where, if conditions are favorable, it will outcompete native plants to form dense thickets. **Medicinal use(s)**: The stems of *Euonymus alatus* (Thunb.) Siebold have been used in traditional medicine for treatment of gastric cancer.					

(continued)

Table 4.26 (continued)

Family	Botanical name	Image	Plant part used	Bioactive compound(s)	Structure	References
	Maytenus woodsonii Lundell		Root and bark	Isoblepharodol	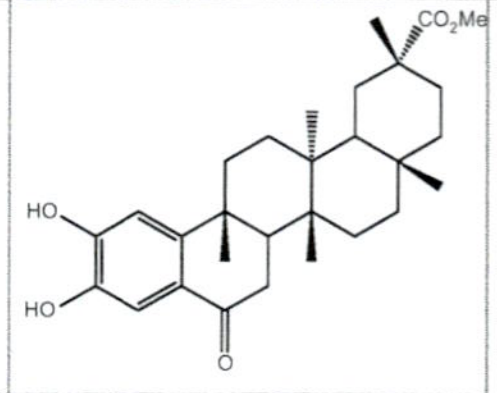	De-Leon et al. (2009) and Gupta (1995)
	Synonym(s): *Maytenus blepharodes* Lundell, Maytenus vulcanicola Standl. **Common name(s)**: Not available. **Botanical description**: *Maytenus woodsonii* Lundell is a species that grows in Panama, Costa Rica, and Colombia. **Medicinal use(s)**: The plant has a long history in traditional medicine since they produce an extraordinary variety of bioactive metabolites of medical interest such as triterpenoid, quinonemethides, or phenolic triterpenes. Phenolic-type antimicrobial agents have long been used for their antiseptic, disinfectant, or preservative properties. This plant is also used to reduce the risk of breast, liver, and gastric cancer.			7-Oxoblepharodol	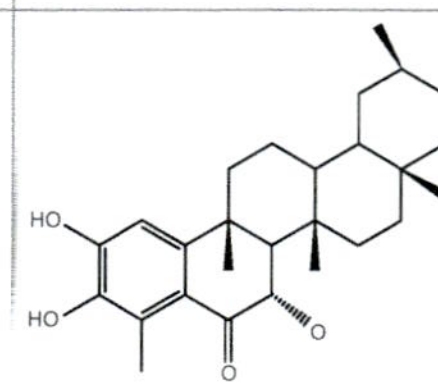	
				Blepharotriol	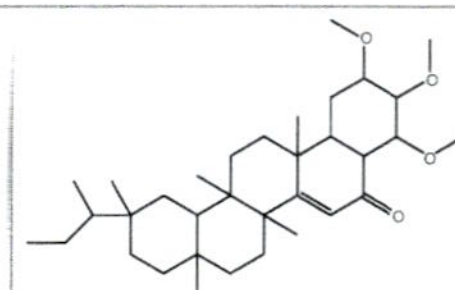	
				6-Deoxoblepharodol	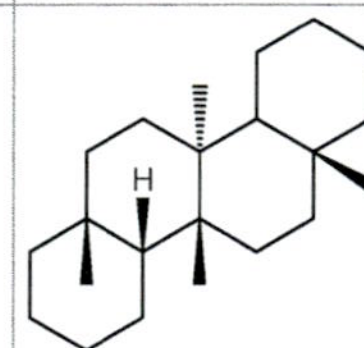	

		Zeylasteral	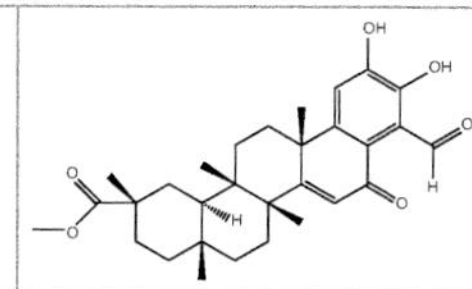 	
		Demethylzeylasteral	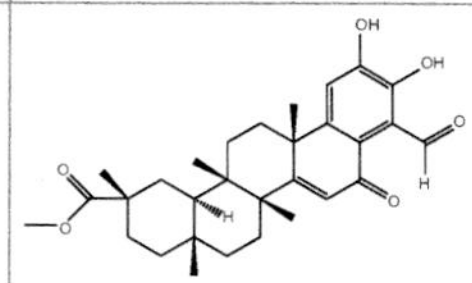 	
		Zeylasterone	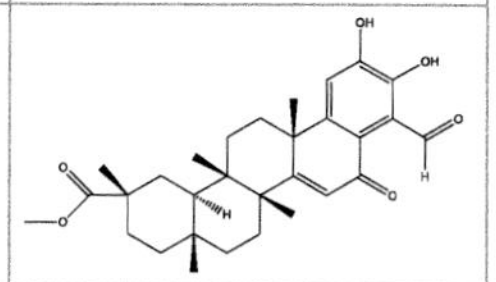 	
		Demethylzeylasterone	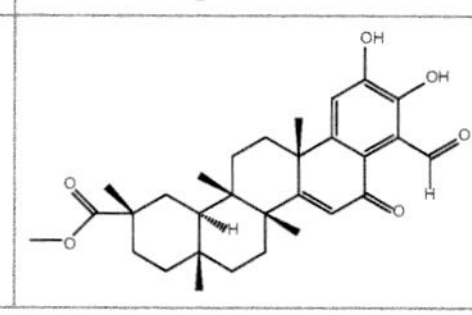 	

(continued)

Table 4.26 (continued)

Family	Botanical name	Image	Plant part used	Bioactive compound(s)	Structure	References
	Gymnosporia cassinoides (L'Hér.) Masf.		Fruit juice	Xuxuarines	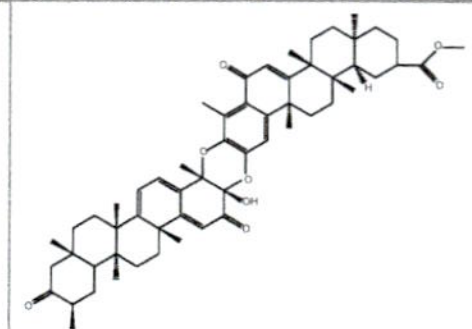	Shirota et al. (1995)
	Synonym(s): *Celastrus cassinoides* L'Herit., *Maytenus canariensis* (Loesl.) Kunk. & Sund., *Maytenus dryandri* var. canariensis Loes. **Common name(s)**: Kanaren-Maytenus, Peralillo. **Botanical description**: *Gymnosporia cassinoides* is native of the Canary Islands. This is a small, highly branching and very knotty tree that reaches 6–8 m in height, although in very exposed and windy areas it may only be shrub-sized. Normally it has a rather small and globose crown. The trunk is somewhat irregular and has dark gray bark with fine, transverse folds. The leaves are simple, perennial, alternate, and slightly leathery and have an irregularly serrated or crenate margin. The blades are shiny green, hairless, and inversely ovate in shape (obovate), although they can sometimes be rounded. They are 4–8 cm long and 2–4 cm wide. In the autumn, numerous creamy white hermaphroditic flowers begin to bloom in short clusters next to the leaf stalks. The fruits, which at first glance look like pears, are pale green capsules that appear fleshy at first. As they ripen, they become hard, brown or reddish in color, and three parts, or valves, open to release blackish-red seeds surrounded by a white aril or seed coat. **Medicinal use(s)**: The plant is used to treat a number of ailments, including chest pains, rheumatism, snakebites, diarrhea, eye infection, dyspepsia, and wounds. It has been used as anticancer drug to treat breast, liver, and lung cancer.					

	Gymnosporia diversifolia Maxim.		Whole plant	Maytansine	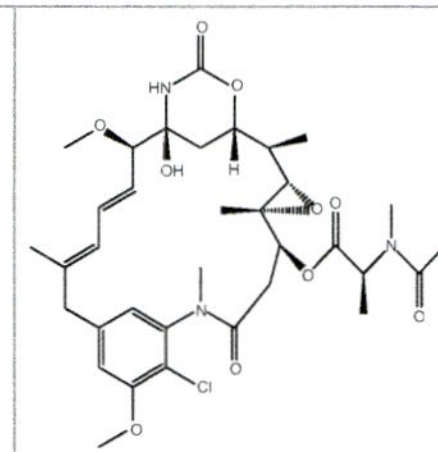	Nakao et al. (2004)
	Synonym(s): *Celastrus diversifolia* (A. Gray) Hemsley, *Maytenus diversifolia* (Maxim.) Ding Hou, *Maytenus garanbiensis* C. E. Chang. **Common name(s)**: Not available. **Botanical description**: It is a spinescent plant found in Philippines, E-Thailand, Vietnam, Taiwan, Ryukyu island, China and Malaysia. **Medicinal use(s)**: The plant extract is very effective in the treatment of coughs, diarrhea, stomach ailments, herpes, and sexually associated diseases. Other medicinal uses include the treatment of breast cancer, lymphocytic leukemia, diarrhea, and arthritis.			Maytanprine	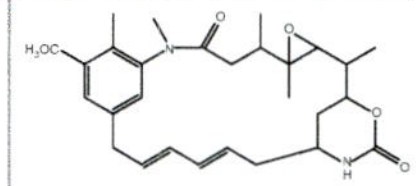	
				Maytenfolic acid	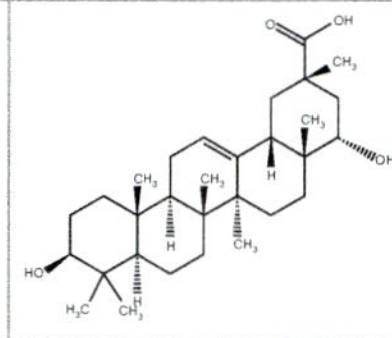	
				Friedelin	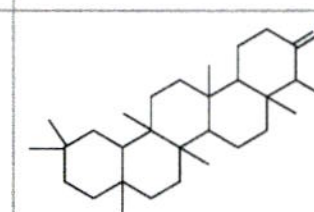	
				Maytensifolin-A	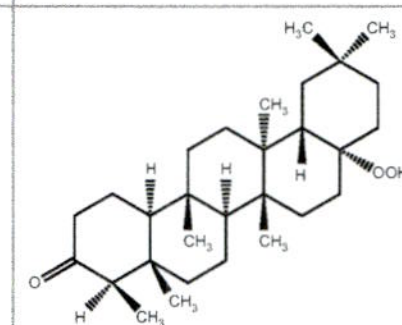	

(continued)

Table 4.26 (continued)

Family	Botanical name	Image	Plant part used	Bioactive compound(s)	Structure	References
				Maytensifolin-B	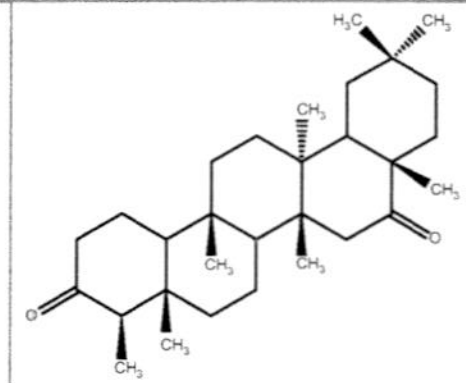	
				Canophyllal	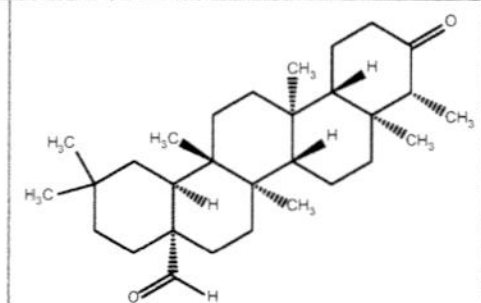	
				β-Amyrin	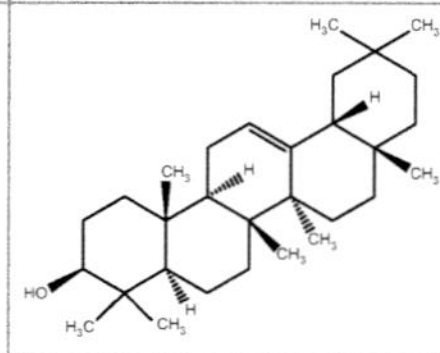	
				Pachysonol	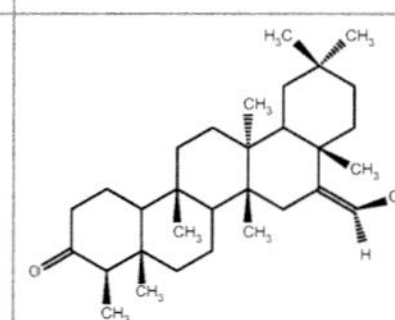	
				Maytenfoliol	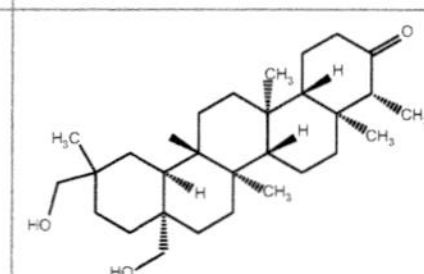	

	Monteverdia ebenifolia (Reissek) Biral		Stem bark	Dulcitol (galactitol)	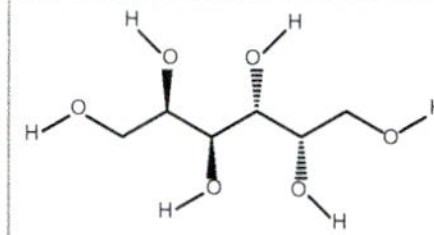	Anon (1985) and Shirota et al. (1998)
	Synonym(s): *Maytenus ebenifolia* Reiss. **Common name(s)**: Not available. **Botanical description**: The plant is native to Bolivia, Brazil North and Peru. **Medicinal use(s)**: The plant is used as an antidiarrheal, cold, and cough remedy and a hemostat. It is also used to treat breast and lung cancer.					
	Gymnosporia emarginata (Willd.) Thwaites		Entire plant	Oppositine A	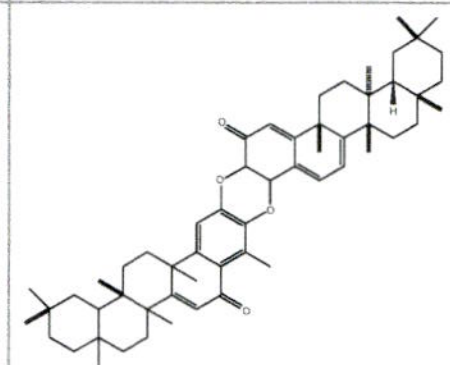	Shirota et al. (1996), Sagwan et al. (2011), and Whitson et al. (2006)

(continued)

Table 4.26 (continued)

Family	Botanical name	Image	Plant part used	Bioactive compound(s)	Structure	References
	Synonym(s): *Maytenus emarginata* (Willd.) Ding Hou. **Common name(s):** Red spike-thorn. **Botanical description**: *Gymnosporia emarginata* is a small shrub, 1–3 m tall; branches gray-brown, shallowly longitudinally fluted, unarmed. Petiole sturdy, 3–5 mm; leaf blade obovate, 3.5–5 × 1.7–3.3 cm, leathery, base cuneate, margin remotely crenate, sometimes glandular spotted in sinus of teeth, apex rounded or subtruncate; lateral veins 4–6 pairs, quite tenuous. Cymes axillary, 1 or 2 × branched; peduncle 1–1.5 cm. Flowers white; calyx lobes deltoid; petals oblong-obovate or oblong, entire, obtuse. Stamens inserted abaxially on disk margin; anther broadly ovoid. Ovary 3-celled; style very short; stigma 3. Capsule subobovoid, trilocular, 7–8 mm in diam., dehiscing into 3 broadly rounded valves, ca. 5 mm. Seeds ellipsoid, red, ca. 3 mm, with small aril at base. **Medicinal use(s)**: Red spike-thorn is used for digestive disorders including pain and swelling (inflammation) of the small intestine (enteritis) and dysentery; upper respiratory tract infections including colds, influenza, swine flu, and pneumonia; and other viral and bacterial infections. It is also used to cure the crown gall tumor.						
	Monteverdia ilicifolia (Mart. ex Reissek) Biral		Root and bark	Pristimerin	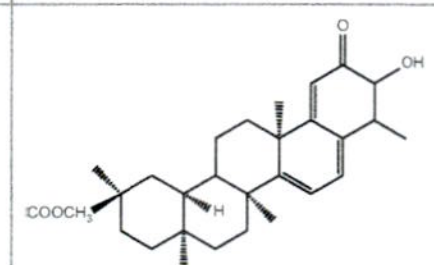	da Costa et al. (2008)
	Synonym(s): *Maytenus hassleri* Briq., *Maytenus ilicifolia* var. boliviana Loes., *Maytenus muelleri* Schwacke, *Maytenus officinalis* D. J. Mabberley, *Maytenus pilcomayensis* Briq. **Common name(s)**: Congorosa, sombra de toro. **Botanical description**: It is distributed in Argentina, S-Brazil, Paraguay, Uruguay, WC-Brazil, SE-Brazil and Bolivia. **Medicinal use(s)**: *Monteverdia ilicifolia* is an antiseptic, anti-asthmatic, fertility-regulating, antitumor, and anti-ulcer agent. This plant is used to treat cancers of the kidney, lungs, breast, and skin.						

<table>
<tr><td rowspan="3"></td><td>Monteverdia krukovii (A. C. Sm.) Biral</td><td></td><td>Stem bark</td><td>Oleanane</td><td>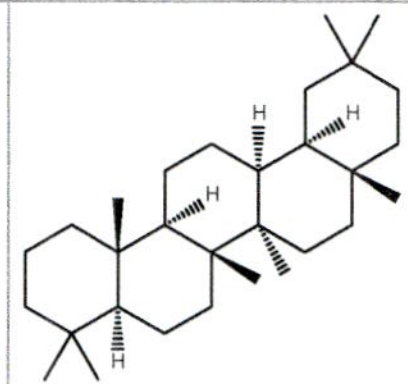</td><td rowspan="3">Shirota et al. (1996), Chavez et al. (1998), and Morit et al. (2008)</td></tr>
<tr><td colspan="3" rowspan="2">Synonym(s): Maytenus krukovii A. C. Sm.
Common name(s): Chuchuhuasi.
Botanical description: Maytenus krukovii is native to the Amazon rainforest; it grows in Bolivia, Colombia, Ecuador, and Peru. With a maximum recorded height of about 30 m (98 ft.) and leaves that span up to 30 cm (12 in.) wide, this large tree contributes significantly to the forest canopy. In the Quechua languages, the tree is called chuchuhuasi (alternately spelled chuchuasi) or chuchuhuasha (alternately spelled chucchu huashu and sometimes shortened to chuchasha). A person can chew the bark, but it tastes very bitter.
Medicinal use(s): Bark of Maytenus krukovii is used in relieving back pain, as well as the discomforts of arthritis and rheumatism. Indigenous peoples of the Amazon drink decoctions and tinctures of the bark as an herbal tonic. Extracts of the bark have antioxidant and antimutagenic properties. It is used for the treatment of skin cancer.</td><td>Ursane</td><td>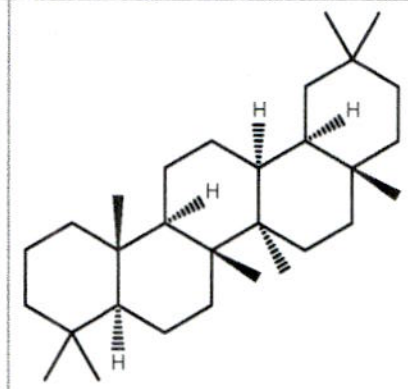</td></tr>
<tr><td>Tingenone</td><td>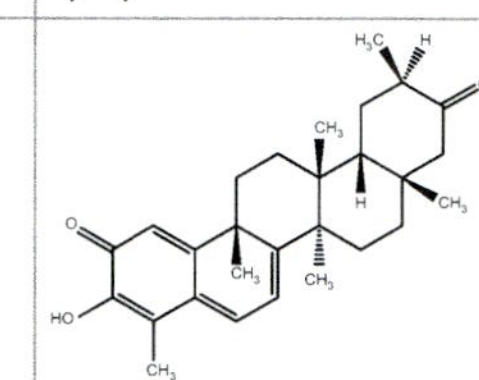</td></tr>
</table>

(continued)

Table 4.26 (continued)

Family	Botanical name	Image	Plant part used	Bioactive compound(s)	Structure	References
				Isoxuxuarines Aα	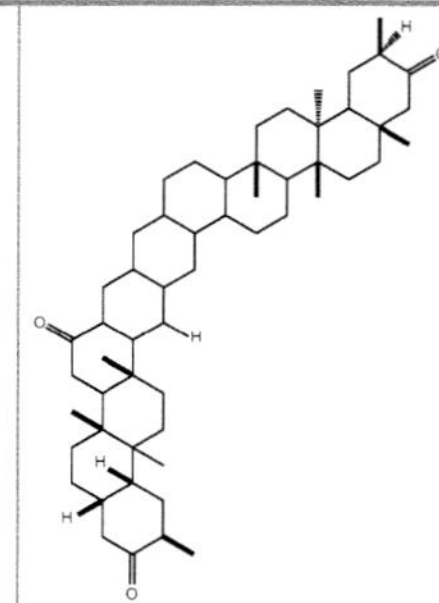	
				Isoxuxuarines Aβ	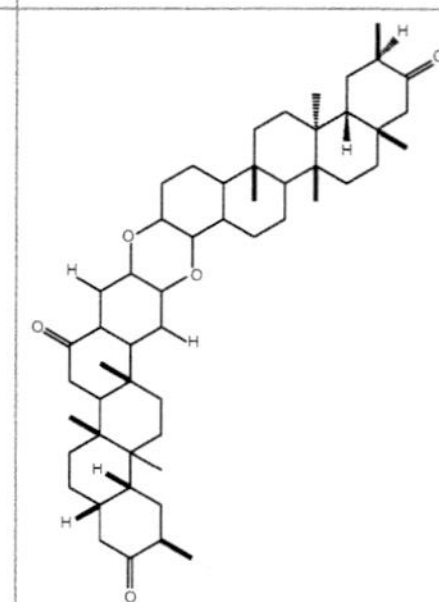	

		Compound	Structure	
		7,8-Dihydroisoxu-xuarine	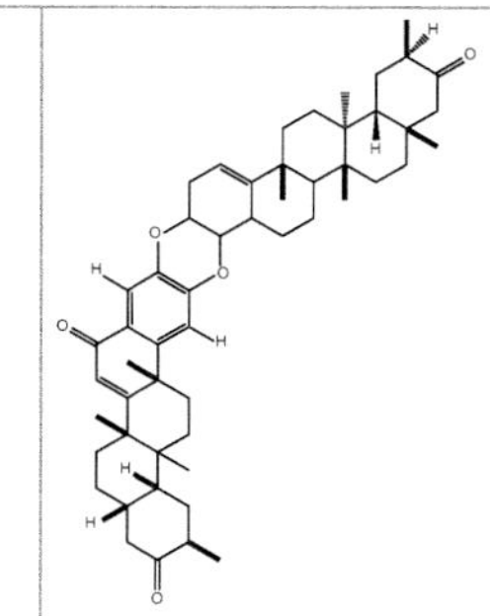	
		7,8-Dihydroxuxuarine	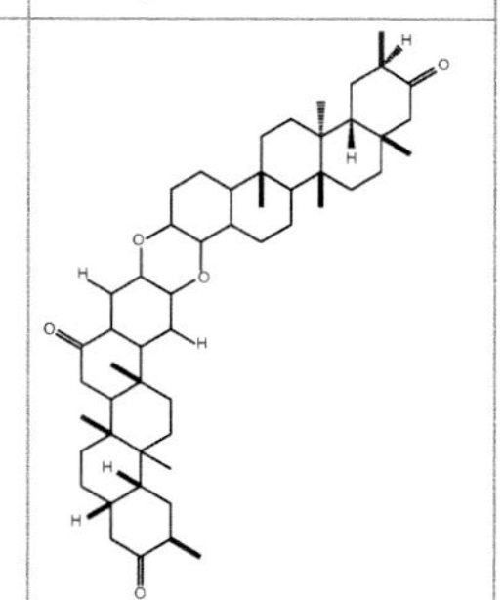	
		Xuxuarine	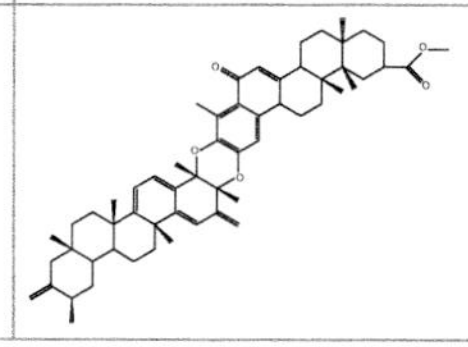	

(continued)

Table 4.26 (continued)

Family	Botanical name	Image	Plant part used	Bioactive compound(s)	Structure	References
				22b-Hydroxytin-genone	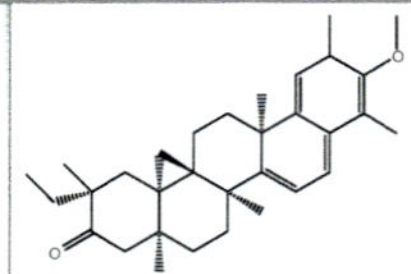	
				Pristimerin	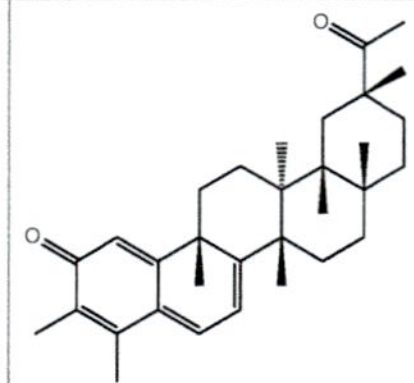	
				Celastrol	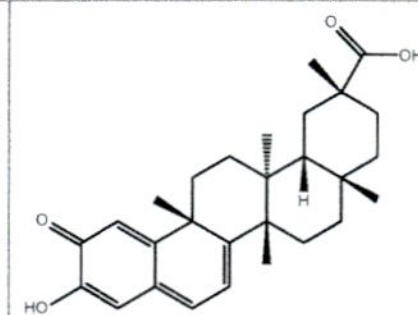	
	Monteverdia macrocarpa (Ruiz & Pav.) Biral		Stem bark	Friedelin	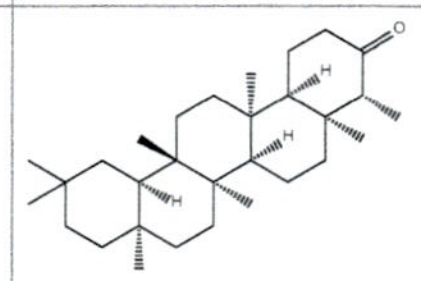	Kokwaro (1976) and Chavez et al. (1998)
				Canophyllol	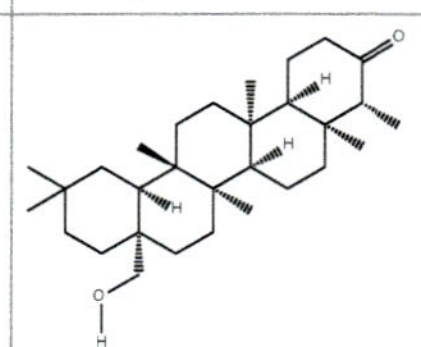	

	Synonym(s): Celastrus macrocarpus Ruiz & Pav., Haenkea macrocarpa (Ruiz & Pav.) Steudel, Haenkea multiflora Ruiz & Pav., Maytenus macrocarpa (Ruiz & Pav.) Briq., Maytenus multiflora (Ruiz & Pav.) Loes., Maytenus tarapotensis Briq. **Common name(s)**: Chuchuhuasi. **Botanical description**: Maytenus macrocarpa is a native plant from the Peruvian Amazon rainforest. Also distributed in Colombia, Venezuela, Ecuador and N-Brazil. **Medicinal use(s)**: Investigations have proved that *Monteverdia macrocarpa* is an antiinflammatory, analgesic, and antipyretic agent. It is very effective in treating skin cancer.					
	Gymnosporia obscura (A. Rich.) Loes.		Leaf	β-Amyrin	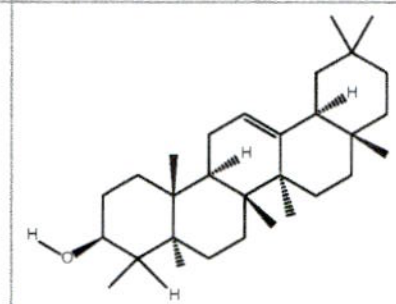	Chhabra et al. (1991)
	Synonym(s): *Celastrus obscurus* A. Rich., *Gymnosporia obscura* Engl., *Gymnosporia serrata* var. obscura (A. Rich.) Fiori, *Maytenus obscura* (A. Rich.) Cuf. **Common name(s)**: Not available. **Botanical description**: *Gymnosporia obscura* is a native plant of East Africa. **Medicinal use(s)**: *Gymnosporia obscura* is an antiseptic, anti-asthmatic, fertility-regulating, antitumor, and anti-ulcer agent. The leaf decoction is taken to fight pains during pregnancy. Pregnant women take powdered pods dissolved in water to relieve stomachache. They are also used to treat gonorrhea and bilharzia. Seeds are used to cure stomach ulcers. Flower buds are crushed and applied to the eye to treat eye problems. This plant is used to treat cancers of the kidney, lungs, breast, and skin.					
	Gymnosporia senegalensis (Lam.) Loes.		Bark, root bark	Phenyldilactone Maysedilactone	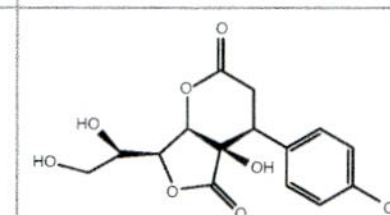	Ulubelen and Cole (1965) and Wilson and Mariam (1979)
				Epigallocatechin	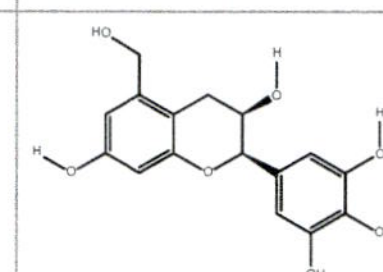	

(continued)

Table 4.26 (continued)

Family	Botanical name	Image	Plant part used	Bioactive compound(s)	Structure	References
	Synonym(s): Catha decolor Webb, *Catha montana* (Roth) G. Don, *Catha senegalensis* (Lam.) G. Don, *Celastrus coriaceus* Guillemin & Perrotet, *Celastrus decolor* Delile, *Celastrus montanus* Roth, *Celastrus obovatus* Hochst. ex Oliv., *Celastrus pallidus* Wall., *Celastrus phyllacanthus* L'Herit., *Celastrus saharae* Batt., *Celastrus senegalensis* Lam., *Gymnosporia baumii* Loes., *Gymnosporia benguelensis* Loes., *Gymnosporia crenulata* Engl., *Gymnosporia eminiana* Loes., *Gymnosporia eremoecusa* Loes., *Gymnosporia europaea* Masf., *Gymnosporia intermedia* Chiov., *Gymnosporia montana* (Roth) Benth. *Gymnosporia saharae* (Battand.) Loes. ex Engl., *Gymnosporia senegalensis* var. angustifolia Engl. & Loes. *Gymnosporia senegalensis* var. inermis (A. Rich.) Loes., *Gymnosporia senegalensis* var. spinosa Engl., *Maytenus baumii* (Loes.) Exell & Mendonca, *Maytenus senegalensis* (Lam.) Exell. **Common name(s)**: Red spike-thorn. **Botanical description**: *Gymnosporia senegalensis* is a shrub or a tree growing up to a height of 15 m. Young branches are often spiny. Leaves are petiolate and alternate, usually glaucous, and coriaceous with pale green lamina. The flowers are dioecious, small in axillary, dichotomous or fasciculate cymes on short branchlets, often forming terminal, elongated panicles. Geographically, *Gymnosporia senegalensis* has a wide distribution range, covering North Africa, Somalia to Senegal, and South Africa, including Madagascar. The plant is also found growing in Afghanistan and India. It grows within wide altitude ranges from 400 to 2400 m. The plant is widespread in the savannah regions of tropical Africa occupying a wide variety of habitats, from deciduous woodland, thickets, shrub and wooded grassland, riverbanks, and swamp margins. **Medicinal use(s)**: Senegalensis plant is traditionally used in folk medicine for the treatment of a number of diseases and disorders including arthritis, rheumatism, skin tumor, influenza, arthritis, bronchitis, diarrhea, dysentery, gastrointestinal diseases, hemorrhoids, impotence, inflammation, menstrual disorders, nausea, osteoarthritis, pain, rheumatism, tumors, virility, eye infections, nausea, snakebites, and severe headache and also as an aphrodisiac. The plant extract is used to treat gastric cancer and brain tumors.					

Monteverdia rigida (Mart.) Biral. (spike-thorn)		Whole plant	Betulinic acid	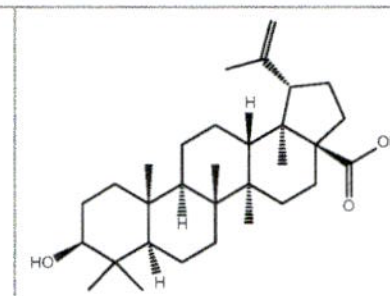	Martucciello et al. (2009)
Synonym(s): *Maytenus rigida* Mart. **Common name(s)**: Not available. **Botanical description**: The plant grows in NE-Brazil. **Medicinal use(s)**: It has anti-inflammatory, antihyperlipidemic, antioxidant, and antiviral properties. It is used to treat gastric, lung, and breast cancer.					
Menyanthes trifoliata L.		Whole plant	Menthiafolin	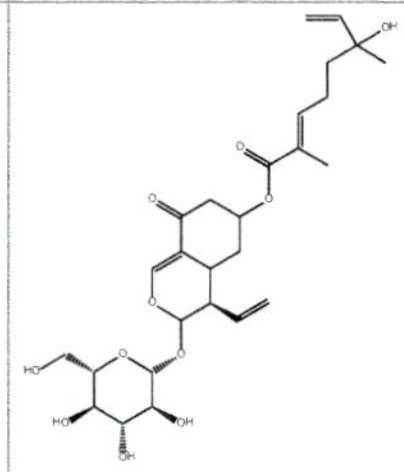	Kuduk-Jaworska et al. (2004)
Synonym(s): *Menyanthes americana* Sweet, *Menyanthes latifolia* Rafin., *Menyanthes palustris* S.F. Gray, *Menyanthes paradoxa* Fries, *Menyanthes tridentata* Rafin., *Menyanthes trifoliata* f. brevistyla L. V. Aver'yanov, *Menyanthes trifoliata* subsp. verna (Raf.) Gervais & M. Parent, *Menyanthes trifolium* Neck., *Menyanthes verna* Rafin. **Common name(s)**: Buckbean, Fieberklee. **Botanical description**: Rhizomes are usually found in mud, sometimes floating. Petiole erect, 12–20 (–30) cm; leaf blade base vaginate; leaflets elliptic, 2.5–4 (–8) cm, base cuneate, margin entire or crenulate, apex obtuse, midvein distinct. Inflorescences many flowered; racemes including scape 30–35 cm; bracts 5–7 mm, margin entire, apex obtuse. Pedicel spreading, 1–1.8 cm. Calyx 4–5 mm; lobes ovate, apex obtuse. Corolla white, tubular, 1.4–1.7 cm, outside glabrous, inside long fimbriate pilose; lobes elliptic-lanceolate, 7.5–10 mm, apex obtuse. Filaments linear, 5.5–6.5 mm; anthers sagittate, 1.8–2 mm. Styles linear, short styles 6–7 mm, long styles 1–1.2 cm; stigma lobes oblong. Capsules globose, 6–7 mm in diam. Seeds orbicular, 2–2.5 mm in dia. Swamps, growing in mud and in open water; 400–3600 m. **Medicinal use(s)**: This plant is an anti-inflammatory, analgesic, and antipyretic agent. It is very effective in the treatment of skin, prostate, and liver cancer.			Loganin	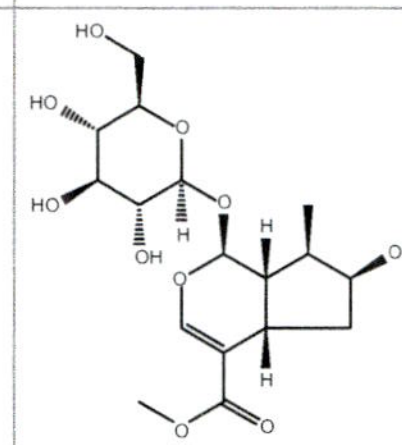	

(continued)

Table 4.26 (continued)

Family	Botanical name	Image	Plant part used	Bioactive compound(s)	Structure	References
				Foliamenthin	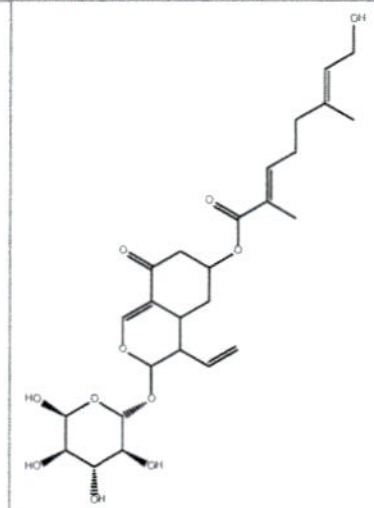	
	Elaeodendron xylocarpum (Vent.) DC.		Entire plant	Maytansine	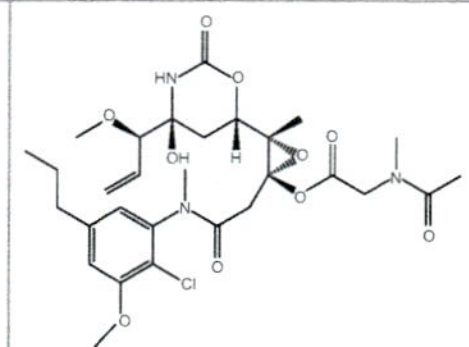	Shen and Zhou (1992) and Massimo (2003)
	Synonym(s): *Cassine rotundata* (DC.) Kuntze, *Cassine xylocarpa* Vent., *Elaeodendron rotundatum* DC., *Elaeodendron trichotomum* (Turcz.) Lundell, *Maytenus trichotoma* Turcz. **Common name(s)**: Marble tree, Spoon tree. **Botanical description**: The plant is grows in Mexico, Panama, Venezuela, Belize, Guatemala, Honduras, El Salvador, Nicaragua, Trinidad, Tobago. **Medicinal use(s)**: *Elaeodendron xylocarpum* has antibacterial and anti-inflammatory properties and is used to dispel heat and remove toxins and treat fevers, influenza, and ulcers. This is used to treat lung, breast, ovarian, and renal cancer.			Maytanprine	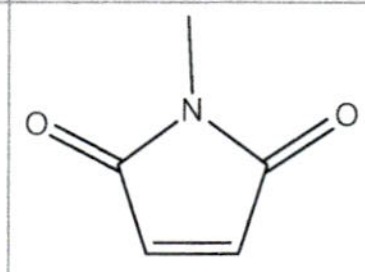	
	Tripterygium wilfordii Hook. F		Stamen and bark	Celastrol	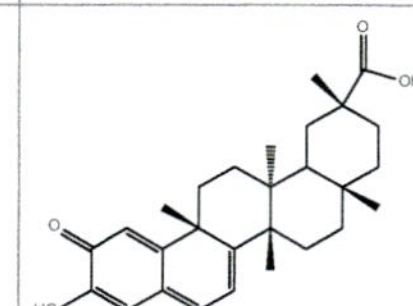	Takaishi et al. (1991), Yang et al (2011a, b), Anon (1976), and Shamona (1997)

	Synonym(s): *Tripterygium regelii* Sprague & Takeda. **Common name(s)**: Seven steps to death, Regel's wingnut. **Botanical description**: Deciduous subshrubs, or scandent and scrambling, or sometimes semiwoody vines, 2–6 m; branching distally, slender or scrambling, minutely pubescent when young or in open sunlight, becoming glabrous when old or in shade, slightly angled, verrucose or not. Petiole 1–2 cm; stipules linear, caducous; leaf blade usually ovate or rounded-ovate, sometimes oblong or elliptic-ovate, (4.6–) 8.6–12.5 (–18.4) × (3.1–) 5.7–8.9 (1–2.3) cm, papery, herbaceous (in shade or low light) to leathery (in direct sunlight or very dry conditions), glabrous or sparsely scurfy tomentose with reddish brown hairs, abaxially ± farinose, base broadly cuneate, or rounded to cordate, margin entire or less often crenulate, apex short to long acuminate or acute, tip often blunt. Thyrses large, composed of several to dozens of thyrselets, (4.5–) 12.5–23.6 (–38) × (2.3–) 4.7–9.3 (–15) cm; flowers 1 normal and 1 an unfertilized male. Flowers whitish, greenish, or yellow green, small, 4–6 × 4–6 mm; calyx 5-lobed, hemispheric, ca. 1 mm; petals 5, oblong to subovate, slightly narrowed to base, 2–2.5 mm, apex rounded. Stamens 5, inserted at margin of cup-shaped and compressed disk. Disk bright green, fleshy, ca. 2 mm in diam. Ovary superior, 3-edged, combined with disk at base, incompletely 3-locular, with 3 prominent ca. 1 mm lobes, short style at apex; stigma capitate, bright purple. Samara usually green or greenish brown when mature, sometimes pink or pinkish purple. **Medicinal use(s)**: *Tripterygium wilfordii* plant is traditionally used in folk medicine for the treatment of arthritis, rheumatism, influenza, arthritis, bronchitis, diarrhea, dysentery, gastrointestinal diseases, hemorrhoids, impotence, inflammation, gastric cancer, and brain and skin tumor.	Triptolide	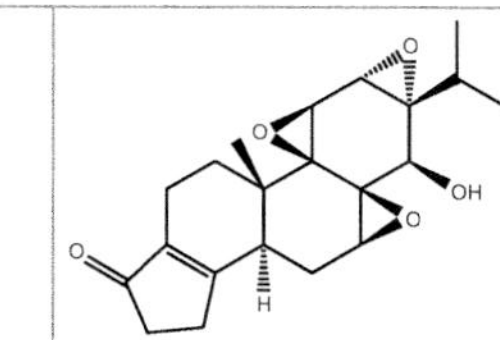	

Table 4.27 Anticancer plants of family Clusiaceae

Family	Botanical name	Image	Plant part used	Bioactive compound(s)	Structure	References
Clusiaceae/ Guttiferae	*Garcinia bracteata*		Leaf	Prenylxanthones	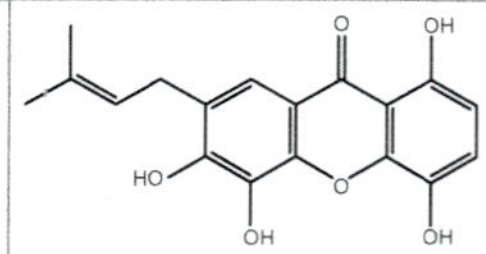	Thoison et al. (2000)
				1-O-Methylneo-bractatin	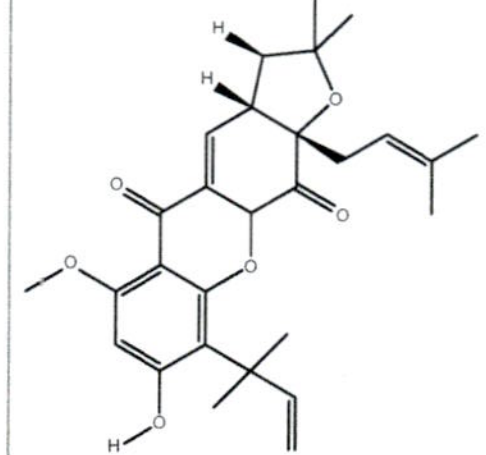	
	Synonym(s): No synonyms are recorded for this name. **Common name(s)**: Wake daily. **Botanical description**: Trees can attain a height of 8 m. Branchlets greenish, robust, striate, glabrous. Petiole robust, 1–1.5 cm; leaf blade abaxially greenish, ovate, ovate-elliptic, or oblong, 8–14 (–18) × 4–8 cm, leathery, midvein raised abaxially, impressed adaxially; secondary veins conspicuous, dense, 20–30 pairs, near margin anastomosing; tertiary veins few and inconspicuous, base broadly cuneate or subrounded, margin cartilaginous, involute, apex acuminate or shortly acuminate, rarely obtuse. Plant dioecious, flowers in 2–7-flowered umbels; umbels usually axillary but male ones occasionally terminal; peduncles (1–) 2–3 cm, with a distal pair of foliar bracts; foliar bracts ovate, large or small, leathery; pedicels 0.6–1.3 cm, 4-bracteolate at base; bracteoles broadly ovate or ovate, ca. 1.5 mm; sepals and petals gradually deflexed after anthesis. Male flowers with staminodes; fertile stamens ca. 40; filaments fleshy, connate into a cup and enveloping pistillode; anthers free, 4-celled, cells longitudinally dehiscent. Female flowers: staminodes ca. 20; filaments membranous, connate into a disk, enveloping ovary base; ovary cylindrical but dilated at the middle, 1-loculed; stigma peltate, smooth, irregularly lobed. Infructescence usually 1-fruited. Fruit ovoid, usually oblique at apex, 2.2–4 cm, to 3 cm in diam. when mature, ± smooth, finely striate, stipitate, wilted tepals persistent. **Medicinal use(s)**: The stems of this plant have been used in traditional medicine for treatment of gastric cancer.					

	Garcinia cowa Roxb.		Tree	Acylphloroglucinol	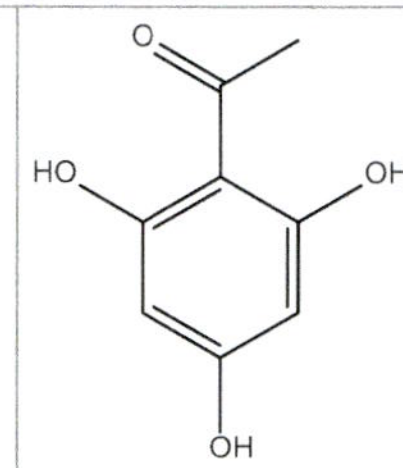	Xu et al. (2010)
	Synonym(s): *Cambogia crassifolia* Blanco, *Garcinia cambogia* Roxb., *Garcinia cornea* Roxb. ex Sm., *Garcinia dioica* Sm., *Garcinia roxburghii* Wight, *Garcinia umbellifera* Wall., *Garcinia wallichii* Choisy, *Oxycarpus gangetica* Buch.-Ham. **Common name(s)**: Brindal berry. **Botanical description**: Trees are 8–12 m tall, 15–20 cm in diam. Bark dark brown. Branches many, borne towards top of trunk, horizontal but usually distally pendulous, slender; twigs dark brown, striate. Petiole 0.8–1.5 (–2) cm; leaf blade lanceolate or oblong-lanceolate, 6–14 × 2–5 cm, papery, midvein raised abaxially, impressed adaxially; secondary veins 12–18 pairs, near margin joining together; tertiary veins conspicuous on both surfaces, base cuneate, sometimes slightly decurrent, margin cartilaginous, involute, apex acuminate or long acuminate, rarely acute or obtuse. Plant dioecious. Male flowers 3–8, terminal or axillary, in an umbel; umbel shortly pedunculate or rarely sessile, 4-bracteate at base; bracts subulate; pedicels 4–8 mm, slender; petals yellow, ca. 2 × as long as sepals; stamen fascicles 4, connate, forming a central capitate 4-sided mass of 40–50 anthers; filaments ± absent, at most short, anthers 4-celled, cells longitudinally dehiscent; pistillode absent. Female flowers usually solitary, axillary, larger than male; pedicels robust, 2–3 mm; staminodes united in lower half and enveloping ovary base; filaments long or short, usually shorter than ovary; ovary ovoid, 4–8-loculed; stigma radiately 4–8-lobed, papillate, 6–7 mm high. Mature fruit opaquely yellow brown, ovoid-globose, oblique, 5–6 × 4–5 cm in diam., 4–8-sulcate, usually apiculate. Seeds 2–4, narrow, fusiform, slightly curved. **Medicinal use(s)**: *Garcinia cowa* is used to treat coughs, diarrhea, stomach problems, herpes, and sexually associated diseases. Other medicinal uses are to treat arthritis and breast cancer.			Garcicowins	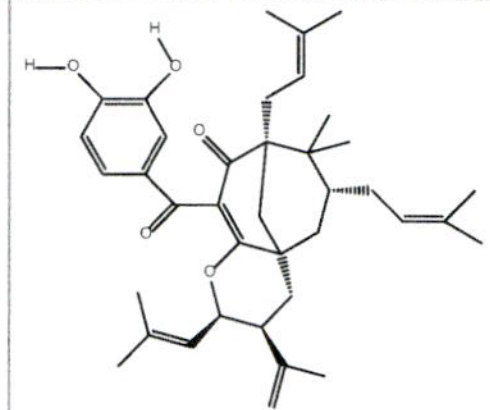	

(continued)

Table 4.27 (continued)

Family	Botanical name	Image	Plant part used	Bioactive compound(s)	Structure	References
	Garcinia gaudichaudii Planch. & Triana		Fruits	Gaudichaudiones A–H	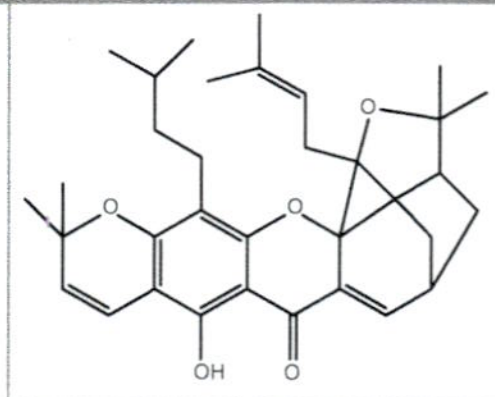	Cao et al. (1998)
				Morellic acid	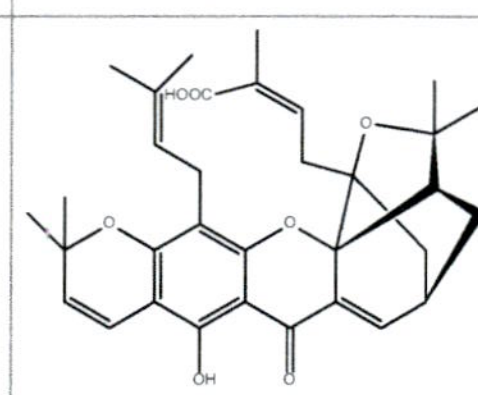	
				Forbesione	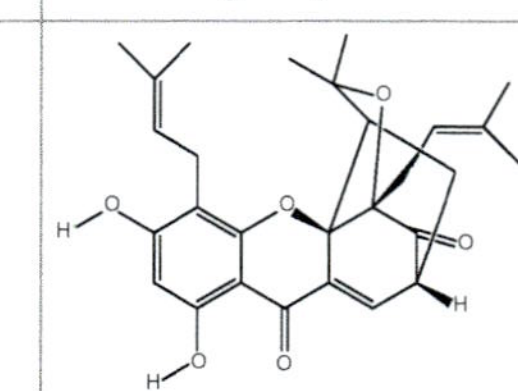	

	Synonym(s): No synonyms are recorded for this name. **Common name(s)**: Not available. **Botanical description**: It is a tree reported from Vietnam that grows up to 12 m tall. Bark is smooth, dark gray in color; blaze is white. Leaves are simple, opposite, decussate; petiole 0.6–1.5 cm long, canaliculate, sheathing at the base, glabrous; lamina is 6.5–15 by 3.5–6 cm, usually elliptic, sometimes narrow obovate; apex acute to acuminate, base attenuate; coriaceious or subcoriaceious, glabrous; secondary nerves are 6–8 pairs; tertiary nerves are obscure. Flowers show inflorescence and are dioecious; male flowers in fascicles, axillary; female flowers larger than male, solitary, axillary. Fruit is globose to slightly elongated, 2 cm in diameter, yellow when ripe. **Medicinal use(s)**: Garcinia gaudichaudii are known to possess a wide spectrum of pharmacologic properties, including antioxidant, antitumor, anti-allergic, antiinflammatory, antibacterial, antifungal, and antiviral activities. This plant is used to treat lung and intestinal cancer.					
	Garcinia hanburyi L.		Leaf	Xanthones	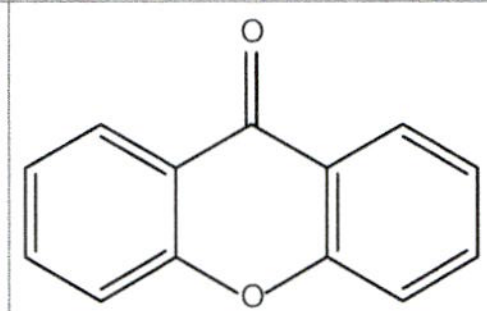	Tao et al. (2009)
	Synonym(s): *Garcinia morella* var. *pedicellata* T.Hanb. **Common name(s)**: Rong Thong, Gamboge. **Botanical description**: *Garcinia hanburyi* Hook f. (is widely distributed throughout Southeast Asia). **Medicinal use(s)**: *Garcinia hanburyi* are widely used for different types of inflammatory diseases. It is used as an antipyretic agent and is employed as a topical anti-inflammatory agent. It is also used to treat arthritis, inflammation of the respiratory tract, and cancer.					

(continued)

Table 4.27 (continued)

Family	Botanical name	Image	Plant part used	Bioactive compound(s)	Structure	References
	Garcinia mangostana L.		Flower	Xanthones	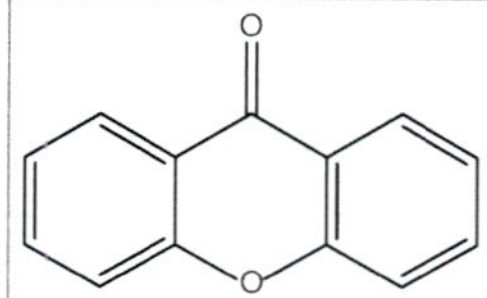	Suksamrarn et al. (2006)
	Synonym(s): *Mangostana garcinia* Gaertn. **Common name(s)**: Purple *mangosteen.* **Botanical description**: Trees small, 12–20 m tall. Branches many, dense, decussate; twigs distinctly angled. Petiole robust, ca. 2 cm, densely and transversely wrinkled when dry; leaf blade shiny, elliptic or elliptic-oblong, 14–25 × 5–10 cm, thickly leathery, midvein raised on both surfaces; secondary veins dense, to 40–50 pairs, joining just within leaf margin, base broadly cuneate or subrounded, margin involute, apex shortly acuminate. Plant dioecious. Male flowers rare, 2–9, clustered at apex of branchlet; pedicels short; stamen fascicles 4, anthers 2-celled, cells longitudinally dehiscent; pistillode conic. Female flowers solitary or paired at apex of branchlet, slightly larger than male ones, 4.5–5 cm in diam.; pedicels ca. 1.2 cm; ovary 5–8-loculed; style nearly absent; stigma 5- or 6-lobed. Mature fruit purple red, sometimes yellow-brown spotted, globose, 5–8 cm in diam., smooth. Seeds 4 or 5 or more, pulp white, juicy, fleshy. **Medicinal use(s)**: The terminal shoot is astringent and diuretic. It is used in the treatment of rheumatism. The seeds are used to inhibit the growth of malignant tumors. Seeds and leaves are used in the treatment of liver cancer.					

	Hypericum japonicum subsp. japonicum		Entire plant	Saroaspidin A	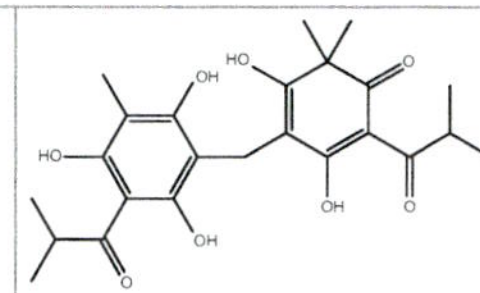	Ishiguro et al. (1987), Ishiguro et al. (1994a, b), Zuo et al. (2012), Zhuang et al. (2015), and Puthur et al. (2018)
	Synonym(s): *Ascyrum humifusum* Labill., *Brathys caespitosa* Bl., *Hypericum calyculatum* Jacquem. ex Dyer, *Reseda cochinchinensis* Lour., *Sarothra laxa* (Bl.) Y. Kimura etc. **Common name(s)**: Not available. **Botanical description**: It is a native plant of East Asia, China, Japan, Indian subcontinent, Myanmar, Thailand, Laos, Vietnam, Malaysia, Indonesia, the Philippines, New Guinea, Australia, and New Zealand. It is a very variable, annual to short-lived perennial plant with erect to decumbent or prostrate stems (2–50 cm tall). **Medicinal use(s)**: *Hypericum japonicum* is widely used as an antipyretic, anti viral and anti-inflammatory agent. It is also used to treat arthritis, respiratory tract, and prostate and liver cancer. It is also used for the treatment of asthma, dysentery, acute hepatitis, liver pain, appendicitis, and boils and is also used as a styptic. It is applied externally to treat wounds, leech and snakebites, swellings, abscesses, scrofula, and fungal skin diseases.			Xanthone	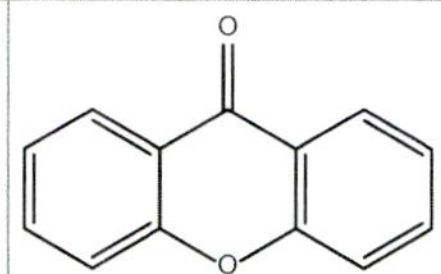	

Table 4.28 Anticancer plants of family Combretaceae

Family	Botanical name	Image	Plant part used	Bioactive compound(s)	Structure	References
Combretaceae	*Anogeissus latifolia* (Roxb. ex DC.) Wall. ex Guillem. & Perr.		Leaves, gum	3,3′-Di-O-methylellagic Acid-4′-β-D-xyloside	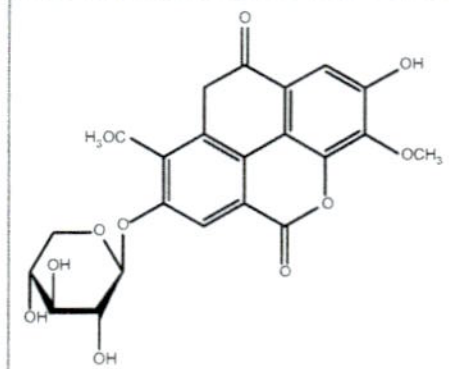	Pettit et al. (1982)
	Synonym(s): *Anogeissus latifolia* var. *glabra* C.B.Clarke, *Anogeissus latifolia* var. *tomentosa* Haines, *Anogeissus latifolia* var. *villosa* C.B.Clarke. **Common name(s)**: Axlewood (English), Bakli, Dhau, Dhawa, Dhawra, or Dhaora (Hindi), Takhian-nu (Thai), and Raam (Vietnamese). **Botanical description**: *Anogeissus latifolia* is a species of small- to medium-sized tree native to India, Nepal, Myanmar, and Sri Lanka. It is one of the most useful trees in India. Its leaves contain large amounts of gallotannins and are used in India for tanning. The tree is the source of Indian gum, also known as ghatti gum, which is used for calico printing among other uses. **Medicinal use(s)**: *Anogeissus latifolia* are known to possess pharmacologic properties, including antioxidant, antitumor, anti-allergic, anti-inflammatory, antibacterial, antifungal, and antiviral. It is used to treat gastric, ovarian, and prostate cancer.			3,4,3′-Tri-O-methylflavellagic acid-4′-β-D-glucoside	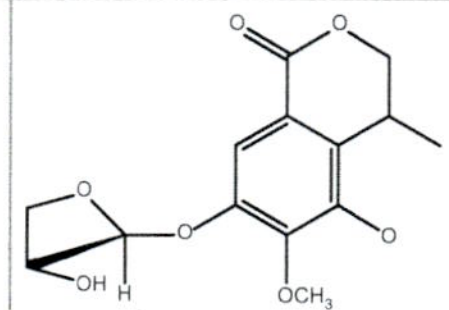	

	Combretum caffrum (Eckl. and Zeyh.) Kuntze		Root, bark	Combretastatins	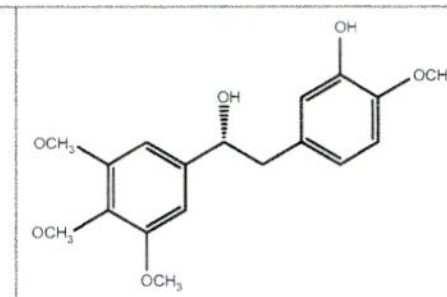	Abela and Santos (1974)
	Synonym(s): *Combretum salicifolium* E.Mey. ex Hook., *Dodonaea caffra* Eckl. & Zeyh., *Dodonaea conglomerata* Eckl. & Zeyh., *Dodonaea dubia* Eckl. & Zeyh. **Common name(s)**: Cape bushwillow, Rooiblaar, Rooiblad, Umdubu, Vaderlandswilgerboom, Vaderlandswilg. **Botanical description**: *Combretum caffrum* is an Eastern Cape South African bushwillow tree. **Medicinal use(s)**: It plays a critical role in normal physiological processes such as wound healing and also in a number of pathological processes, for instance, diabetes retinopathy, arthritis, and the growth of solid brain tumors.					
	Getonia floribunda Roxb.		Leaf	6″-Demethoxyneocalycopterone	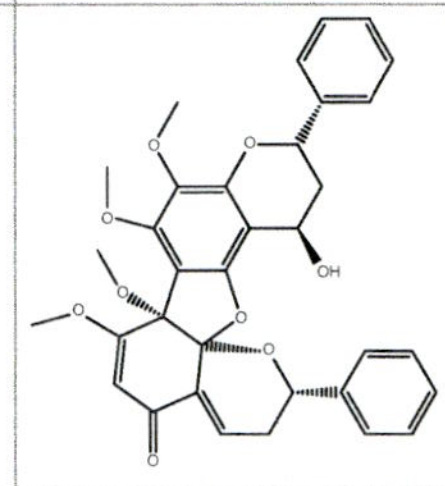	Bhat et al. (2011)

(continued)

Table 4.28 (continued)

Family	Botanical name	Image	Plant part used	Bioactive compound(s)	Structure	References
	Synonym(s): *Calycopteris floribunda* Lam. **Common name(s)**: Ukshi, Kokkarai, Minnarakoti. **Botanical description**: *Calycopteris floribunda* is a large climbing shrub which is 5–10 m long, with vines that are about 5–10 cm in diameter, and the stem and leaves are said to have medicinal properties. Ukshi is found extensively in the low-lying tropical evergreen forests of the Western Ghats and rarely in Eastern Ghats of coastal Andhra. These are also found in "Kavus" or the Sacred Groves of Kerala. Plant is also grown in central and southern parts of India. It bears a gray bark and tenuous branches with thick fluff on the surface. The keratinous leaves, ovoid or oval, are 5–12 cm long. New branches are hairy and rust colored. The flowers that occur in dense clusters are found at the end of the branches. The bracts of the small flowers are ovoid or oval, with thick fluffs on the surface. Petals are absent and the 10 stamens are arranged in 2 cycles. The fruit inception bears 1 ventricle and 3 pendulous ovules inside. The fluffy sham-winged fruit, which is about 8 mm long, has 5 edges and 5 persistent calices which enlarge into the fluffy wing 10–14 mm in length. The sepals are prominent, hairy, and green. **Medicinal use(s)**: The leaves are bitter and used as astringent, laxative, anthelmintic, depurative, diaphoretic, and febrifuge. They are useful to get rid of intestinal worms and to treat colic, leprosy, malarial fever, dysentery, ulcers, and vomiting. Moreover, the fruits are useful in treating jaundice, ulcers, pruritus, and skin cancer.			Calyflorenone	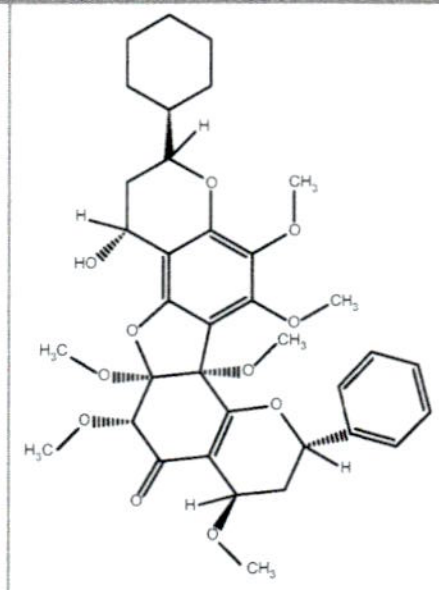	
	Poivrea coccinea (Sonn.) Thouars		Root	Podophyllotoxin	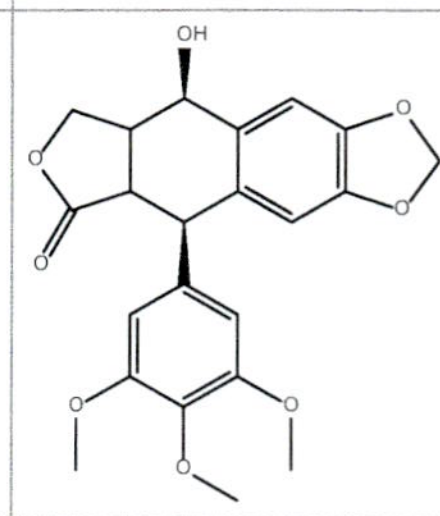	George et al. (2010)

	Synonym(s): *Poivrea coccinea* (Sonn.) DC., *Poivrea coccinea* (Sonn.) Thouars, *Poivrea coccinea* var. *macrophylla* Tul., *Poivrea coccinea* var. *rufipes* (Tul.) H. Perrier. **Common name(s)**: Jungle geranium. **Botanical description**: A native of Madagascar and Mauritius, cultivated in the plains of Pakistan for its beautiful flowers. Usually propagated by cuttings, it does not produce fruits in Pakistan. An elegant looking climbing glabrous shrub. Leaves dark green, 8–13 × 2.5–3.5 cm, elliptic-lanceolate to oblong-lanceolate, acute with a distinct central vein, petiole 1–1.75 (–2) cm long. Racemes terminal and axillary, 5–15 cm long. Flowers red, 1.25–1.5 cm across, with 1–2 mm long pedicel; bract linear-lanceolate 2–3 mm long. Calyx tube 4–5 mm long, 5 angled alternating with calyx teeth, limb 3–4 mm long. Campanulate, teeth triangular, 2 × 2 mm, minutely puberulous outside, villous inside. Petals 5, attached on the hypanthium elliptic-oblong. 5–6 × 2 mm. Stamens 10, exserted, inserted on the throat of the hypanthium, 12–15 mm long, anthers dorsifixed, c. 1 mm long, ovary linear-oblong, 2–4 mm long. Stigma simple, style subulate, filiform, 10–12 mm long. Fruit oblong-elliptic, with 5 papery wings. **Medicinal use(s):** A leafy stem decoction is taken to treat fever and liver complaints, including hepatitis, as it stimulates the secretion of bile from the gallbladder. It is also taken as a diuretic. Crushed leaves are applied to bites of toxic spiders. The seeds are commonly used as an anthelmintic and are especially given to children, followed by a laxative. A decoction of the roots or fruits is also taken as an anthelmintic. The seeds are also given to cattle to get rid of intestinal worms. In the Comoros the seeds are chewed to treat roundworm infection. A root decoction is taken to treat an enlarged spleen.	Deoxypodophyllotoxin	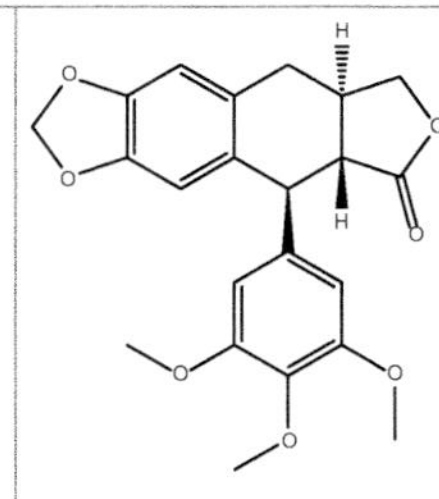	
		4′-Demethylpodophyllotoxin	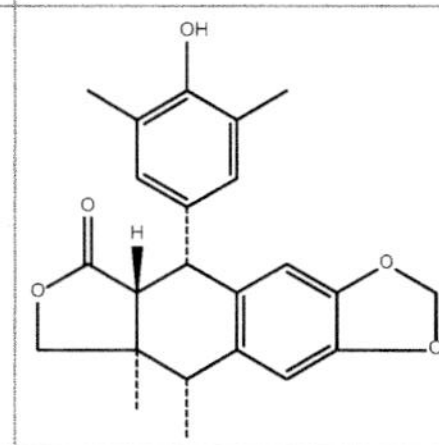	

(continued)

Table 4.28 (continued)

Family	Botanical name	Image	Plant part used	Bioactive compound(s)	Structure	References
				B peltatin	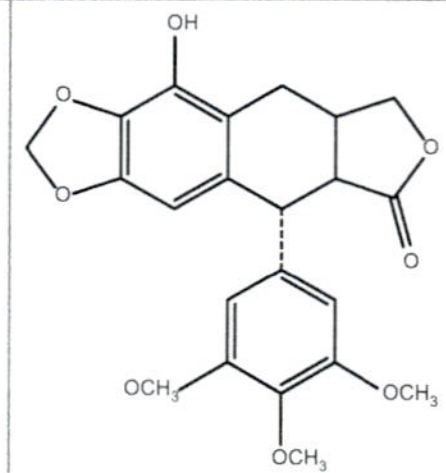	
	Quisqualis indica L.		Leaf	25-O-Acetyl-23,24-dihydro-cucurbitacin F	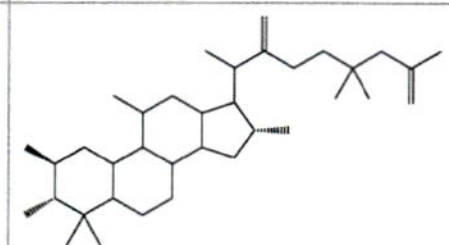	George et al. (2010)
	Synonym(s): *Quisqualis indica* L., *Quisqualis indica* Blanco, *Quisqualis indica* var. *oxypetala* Kurz,, *Quisqualis indica* var. *pierrei* (Gagnep.) O. Lecompte, *Quisqualis indica* var. *villosa* C.B.Clarke. **Common name(s)**: Burma creeper, Rangoon creeper. **Botanical description**: The bark and leaves are used for treating hemorrhages and women diseases such menorrhagia and menstrual pains; leaves are emollient; flowers, when used as an infusion, are laxative and are used in the treatment of coughs and colds, hemorrhages, whooping cough, tuberculosis, and lung cancer. **Medicinal use(s):** It shows various activities such as anti-inflammatory activity, antipyretic activity, immunomodulatory activity, antistaphylococcal activity, anthelmintic activity, antiseptic activity, etc.					

	Terminalia arjuna (Roxb.) W. &A.		Leaf	Arjunin	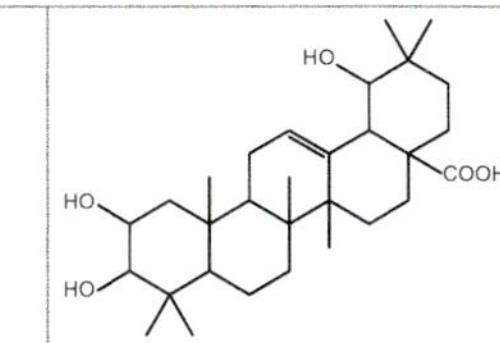	Kadil and Nassar (1998)
	Synonym(s): *Quisqualis indica* L., *Quisqualis indica* Blanco, *Quisqualis indica* var. *oxypetala* Kurz, *Quisqualis indica* var. *pierrei* (Gagnep.) O. Lecompte,, *Quisqualis indica* var. *villosa* C.B.Clarke. **Common name(s)**: Arjuna, Arjun tree. **Botanical description**: This plant is native to India and Ceylon and cultivated in the plains of Pakistan as a shade tree. A large tree, up to 6–15 (–25) m tall with smooth, pale greenish to whitish gray bark. Leaves usually subopposite, oblong-elliptic or somewhat suborbicular, 7–18 (–25) cm long, 4–6 cm broad, obtuse, rarely subacute with rounded or cordate base, glabrous to subglabrous above, partially pubescent beneath, entire somewhat crenate or serrate in the upper half or throughout, petiole 5–10 mm long with 2 (–1) rounded glands at the apex. Inflorescence axillary or terminal paniculate spikes, 3–6 cm long, 1 cm broad. Bracteole small, deciduous, linear-lanceolate. Flowers yellowish white, sessile. Hypanthium broadly campanulate, 4–5 mm long, teeth triangular c. 15 mm long, glabrous. Stamens much exserted. Ovary glabrous; disk barbate. Fruit ovoid-oblong, 2.5–5 cm long, brown with 5 projecting wings, wings striated with 5 ascending veins. **Medicinal use(s):** The plant is used for the treatment of coughs and colds, hemorrhages, whooping cough, tuberculosis, and lung and intestinal cancer.					

(continued)

Table 4.28 (continued)

Family	Botanical name	Image	Plant part used	Bioactive compound(s)	Structure	References
	Terminalia brownii Fresen		Root, Stem, Bark	Betulinic acid	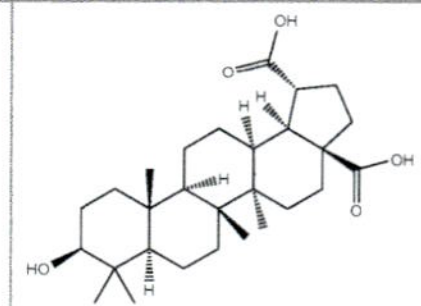	Mbwambo et al. (2007)
				β-Sitosterol	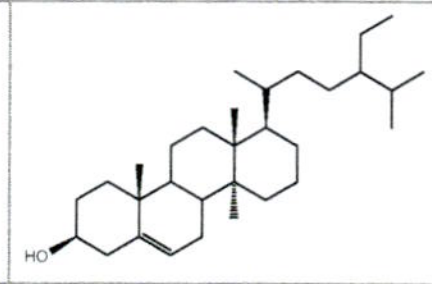	
				Stigmasterol	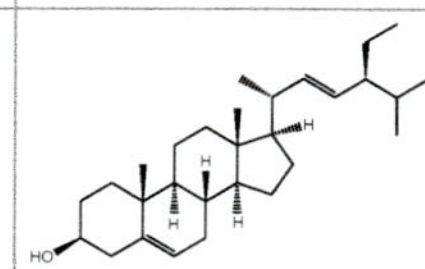	
	Synonym(s): No synonyms are recorded for this name. **Common name(s)**: Darot, Abalo. **Botanical description**: This plant is native of the Republic of the Congo, Kenya, Nigeria, Sudan, Ethiopia, and Somalia. It is a leafy deciduous tree with an attractive somewhat layered appearance, usually 4–15 (25) m high with a rounded, flat topped, spreading crown, and a straight bole. Leaves broadly elliptic to obovate, wider towards the apex and 6–16 × 2.5–8 cm. Fruit winged, smooth, greenish when young, purplish red to brown when mature, broadly elliptic to ovate, apex obtuse to rounded, emarginate, base acute to obtuse, 3.5 × 4.2(5) × 2.5 sometimes up to 7.5 cm long. The tree is grown for shade, shelter, windbreak an as an ornamental. Leaf fall is heavy, making excellent mulch. The tree is widely recommended for agroforestry; despite its rather dense shade, crops do well under its canopy. **Medicinal use(s):** Root, stem, and bark are used to treat cough, jaundice, malaria, tuberculosis, epilepsy, urinogenital problems, syphilis, gonorrhea, and breast and ovarian cancer. It is also an anthelmintic agent.			Monogynol A	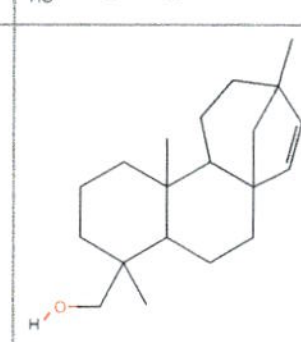	

	Terminalia chebula Retz.		Flower	Casuarinin	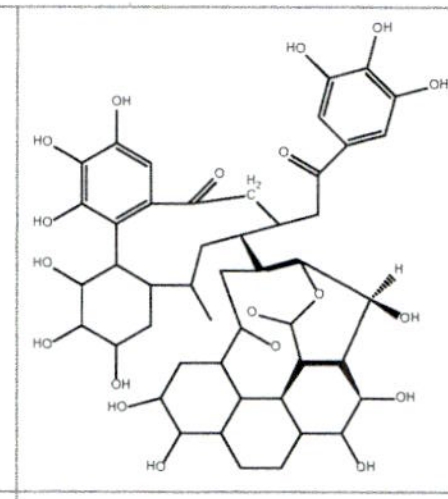	Cheng et al. (2003)
	Synonym(s): *Terminalia chebula* Retz., *Terminalia chebula* var. *chebula*, *Terminalia chebula* var. *tomentella* (Kurz) C.B.Clarke. **Common name(s)**: Chembula, Haritaki. **Botanical description**: This plant is native to India, Ceylon, Burma, Malayan Peninsula, and Siam and cultivated in Pakistan. A medium-sized to large-sized tree, up to 25–30 cm tall with many spreading branches with pale greenish-gray and smooth bark. Leaves alternate to subopposite, 10–15 (–25) × 3.75–7.5 (–8.5) cm, elliptic-oblong or oblong, acute to obtuse, penninerved, covered with silky hairs when young, becoming glabrescent; rounded or cordate at base, usually unequal; petiole 1.25–3.5 cm long, pubescent, usually with 2, rarely more glands on the upper side at the apex. Spikes terminal, rarely panicled, 5–15 (–20) cm long with hispid floral axis. Flowers all bisexual, 5 mm across, yellowish; bracteoles linear, acute, pubescent, 5–6 mm long, caducous. Calyx tube 3–3.5 mm long, glabrous outside, hairy within, calyx teeth very short, broadly triangular. Drupe ellipsoid or obovoid, obscurely 5-ribbed, 2–3.75 cm long. **Medicinal use(s)**: *Terminalia chebula* is a rejuvenative, laxative (unripe), astringent (ripe), anthelmintic, nervine, expectorant, tonic, carminative, and appetite stimulant agent. It is used to treat uterine cancer.			Chebulanin	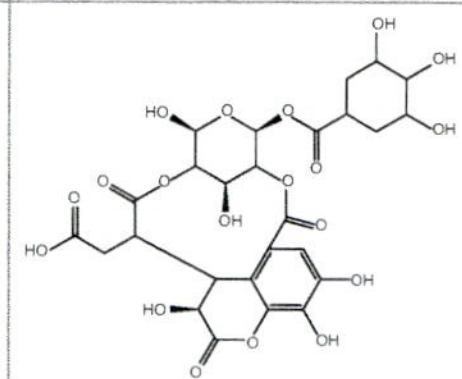	

(continued)

Table 4.28 (continued)

Family	Botanical name	Image	Plant part used	Bioactive compound(s)	Structure	References
				Chebulinic acid	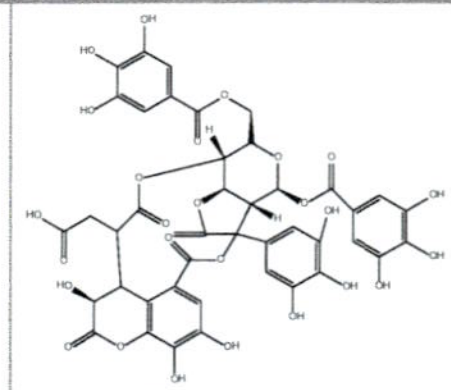	
				1,6-Di-Ogalloyl-b -D-glucose	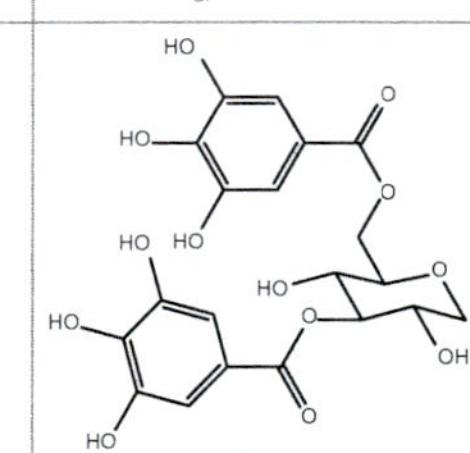	
	Terminalia paniculata (Roxb.) W. & A.		Bark	Doxorubicin	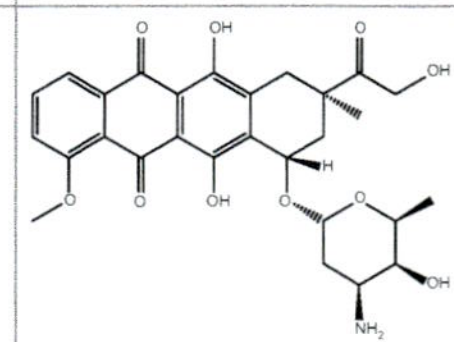	Davey and Atlee (2011)

	Synonym(s): No synonyms are recorded for this name. **Common name(s)**: Kindal tree, Flowering murdah, Kinjal, Pumarutu, Vadamarudu, Pullamaruthu, Putanallamanu, Ulabe, Konkani: Quinzol, Asvakarnah. **Botanical description**: *Terminalia paniculata* is a tree native to Southwest India (including the Western Ghats and Karnataka). Kindal is a tropical tree with a large natural distribution in Western Ghats. The tree has diverging branches. Leaves are nearly opposite, elliptic-oblong, heart-shaped at the base, sharp at the tip. Flowers occur in spiked forming a compound panicle. Fruit has one large and two small wings. It is the fruits, rather than the flowers, which make the tree more colorful. The tree is extensively utilized in pharmaceutical, timber tannin, leather, and silk industries. It is a tall tree. Timber is very useful for ship building and is used as substitute for teak. Fruits are used for tanning and dyeing. In winter, the tree may be partially leafless. Flowering: November–January. **Medicinal use(s):** The plant is used as antibacterial, antifungal, antiprotozoal, antiviral, antidiarrheal, analgesic, antimalarial, antioxidant, anti-inflammatory, and anticancer agent. According to many reports, this plant is used to treat gastric cancer.			

Table 4.29 Anticancer plants of family Crassulaceae

Family	Botanical name	Image	Plant part used	Bioactive compound(s)	Structure	Reference
Crassulaceae	*Kalanchoe pinnata* and *K. daigremontiana* × *butiflora*		Leaf	Bryophyllin A	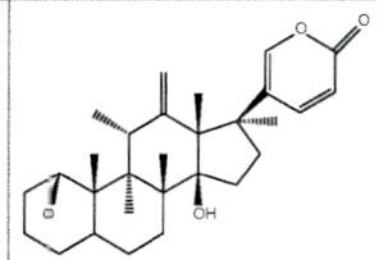	Supratman et al. (2001)
	Synonym(s): *Bryophyllum pinnatum* (Lam.) Oken., *Bryophyllum calycinum* Salisb. **Common name(s)**: *Kalanchoe*. **Botanical description:** Erect, evergreen, usually unbranched, smooth, succulent herb growing from 40 to 200 cm tall. The stems usually become more or less woody, especially near the base. Plants can become more or less leafless when flowering. The plant has a wide range of medicinal uses and is often harvested from the wild. It is commonly grown as an ornamental in the tropics and as a houseplant in colder areas. **Medicinal use(s)**: The leaves are used as astringent, antiseptic, diuretic, and febrifuge. They are also used as a counterirritant against poisonous insect bites. The juice, used on its own or mixed with lard, is used in the treatment of bilious diarrhea, dysentery, cholera, acute nephritis, lithiasis, and phthisis. The leaf juice, combined with salt, is used as a treatment for colds, coughs, grippe, and loose bowels and for teething babies. The fresh leaves are pounded and then applied to burns, are used as poultices on boils and ulcers and as a treatment for headaches and to get rid of ringworm, or are placed on the soles of the feet in order to stop hemorrhages. They are used as a topical application in the treatment of dislocations, equimosis, callosities, etc. Mixed with salt, they are applied as a plaster to the abdomen to relieve enuresis. The leaves are made pliable by keeping over fire and are then applied as a treatment on wounds, bruises, and boils. The dried leaves can be applied as a powder on bad ulcer. The leaf juice is used to treat earache, conjunctivitis, and ophthalmia.					

Table 4.30 Anticancer plants of family Cupressaceae

Family	Botanical name	Image	Plant part used	Bioactive compound(s)	Structure	References
Cupressaceae	*Juniperus phoenicea* subsp. phoenicea		Leaf	Widdrol	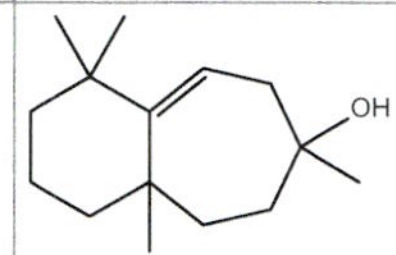	Cirses et al. (2011)
	Synonym(s): *Juniperus phoenicea* subsp. *eumediterranea P.Lebreton & Thivend.* *Juniperus phoenicea* var. *lobelii* Guss., *Juniperus phoenicea* var. *lycia* (L.) St.-Lag., *Juniperus phoenicea* var. *malacocarpa* Endl. **Common name(s)**: Not available. **Botanical description**: *Juniperus phoenicea* is a juniper found throughout the Mediterranean region, from Morocco and Portugal east to Italy, Turkey, and Egypt, south on the mountains of Lebanon, the Palestine region and in Western Saudi Arabia near the Red Sea, and also on Madeira and the Canary Islands. It mostly grows at low altitudes close to the coast but reaches 2400 m (7900 ft.) in altitude in the south of its range in the Atlas Mountains. It is the vegetable symbol of the island of El Hierro. This is a large shrub or small tree reaching 5–8 m (16–26 ft.) tall, with a trunk up to 1–2 m (3 ft. 3 in.–6 ft. 7 in.) in diameter and a rounded or irregular crown. The bark, which can be pealed in strips, is dark grayish brown. The leaves are of two forms, juvenile needlelike leaves 5–14 mm long and 1 mm wide on seedlings, and adult scale leaves 1–2 mm long on older plants with a green to blue-green color; they are arranged in opposite decussate pairs or whorls of three. It is largely monoecious, but some individual plants are dioecious. The female cones are berrylike, 6–14 mm in diameter, orange-brown, occasionally with a pinkish waxy bloom, and contain 3–8 seeds; they are mature in about 18 months and are mainly dispersed by birds. The male cones are 2–4 mm long and shed their pollen in early spring, which is wind-pollinated. **Medicinal use(s)**: The plant extract is used to treat diabetes, diarrhea, rheumatism, and bronchopulmonary disease. They can also be used as a diuretic and can cure lung, mouth, and breast cancer.					

(continued)

Table 4.30 (continued)

Family	Botanical name	Image	Plant part used	Bioactive compound(s)	Structure	References
	Juniperus recurva var. recurva		Part not specified	Podophyllotoxin	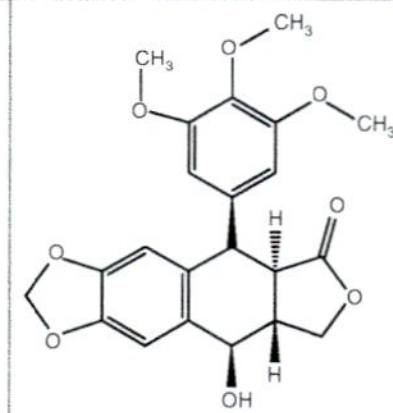	Lechner-Knecht (1982) and Schacter (1996)
	Synonym(s): *Juniperus recurva* var. tenuifolia Spach, *Juniperus recurva* var. typica Patschke, *Juniperus recurva* var. uncinata R.P. Adams, *Juniperus uncinata* (R.P. Adams) R.P. Adams, *Sabina recurva* (Buch.-Ham. ex D. Don) Antoine, *Sabina recurva* var. tenuifolia (Spach) Antoine, *Sabinella recurva* (Buch.-Ham. ex D. Don) Nakai. **Common name(s)**: Not available. **Botanical description**: The plant is native to Afghanistan, Bhutan, N India, Kashmir, N Myanmar, Nepal, Pakistan, and Sikkim. Shrubs or trees monoecious or rarely dioecious; bark light grayish brown or brown; crown conical or broadly pyramidal; branches ascending in apical part of plant and spreading towards base; branchlets pendulous, curved. Leaves in whorls of 3, loosely appressed, greenish white or slightly glaucous adaxially, all needlelike, nearly straight, slightly incurved, 3–10 × ca. 1 mm, concave adaxially, base decurrent, convex with longitudinal grooves at base abaxially, apex sharply pointed. Pollen cones axillary, yellow, ovoid-oblong or ellipsoid-ovoid; microsporophylls 10–16, decussate, each with 3 pollen sacs. Seed cones axillary, slightly glaucous when young, maturing purplish black and not glaucous, ovoid, 6–12 × 5–9 mm, 1-seeded. Seeds ovoid or conical-ovoid, 5–9 × 3–6 mm. **Medicinal use(s)**: The berry oil works best against brain tumors and lung carcinoma, secondarily against liver carcinoma and breast carcinoma, and finally also against cervical carcinoma. The leaf oil is effective against brain tumor, cervix carcinoma, lung carcinoma, liver tumor, and breast carcinoma.					

	Juniperus squamata Buch.-Ham. ex Lamb		Part not specified	Podophyllotoxin	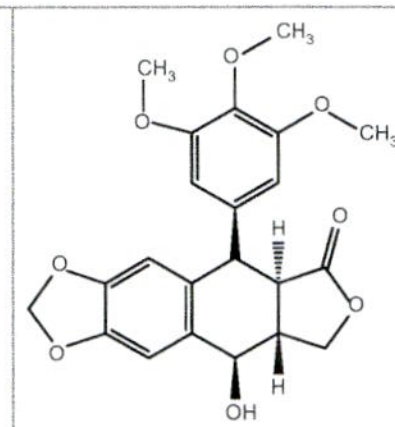	Lechner-Knecht (1982), Renouard et al. (2011), and Kusari et al. (2011)
	Synonym(s): *Juniperus squamata* subsp. *fargesii* (Rehder & E.H.Wilson) Silba, *Juniperus squamata* var. *hongxiensis* Y.F.Yu & L.K.Fu, *Juniperus squamata* var. *loderi* (Hornibr.) Hornibr. **Common name(s)**: Flaky juniper or Himalayan juniper, Gao shan bai, Blue carpet. **Botanical description**: *Juniperus squamata* is a species of juniper native to the Himalayas and China, from Northeastern Afghanistan East to Western Yunnan in Southwestern China, and with disjunct populations North to Western Gansu and east to Fujian. It grows at 1600–4900 m in altitude. It represents the provincial tree of Khyber Pakhtunkhwa (unofficial). It is a coniferous evergreen shrub (rarely a small tree) reaching 2–10 m tall (rarely 15 m), with flaky brown bark, and a prostrate to irregularly conical crown. The leaves are broad needlelike, 3–9 mm long, arranged in six ranks in alternating whorls of three, and often strongly glaucous blue-green in color. The cones are berrylike, globose to ovoid, 4–9 mm long and 4–6 mm in diameter, glossy black, and contain one seed; they are mature in about 18 months. The male cones are 3–4 mm long and shed their pollen in early spring. It is largely dioecious, with pollen and seed cones produced on separate plants, but occasionally monoecious. **Medicinal use(s)**: Plant extract is used for treating urinary tract infections (UTIs) and kidney and bladder stones. Other uses include treating snakebite, diabetes, and skin cancer. Some people apply juniper directly to the skin for wounds and for pain in joints and muscles. The essential oil of juniper is inhaled to treat bronchitis and numb pain.					

(continued)

Table 4.30 (continued)

Family	Botanical name	Image	Plant part used	Bioactive compound(s)	Structure	References
	Thuja occidentalis L.		Leaves and young twigs	Thujone	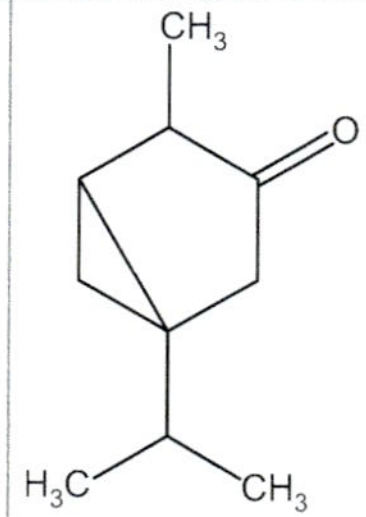	Biswas et al. (2010)
	Synonym(s): *Cupressus arborvitae* Targ. Tozz., *Thuja obtusa* Moench, *Thuja odorata* Marshall, *Thuja procera* Salisb., *Thuja theophrastii* C. Bauhin ex Nieuwl. **Common name(s)**: White-cedar, Swamp cedar, Tree of life, Abendländischer lebensbaum, Obicna americka tuja, Albero della vita, árvore-da-vida, Tuja západná, Tuya amerikanskaya, Bei mei xiang bai, Westerse levensboom, Arbre de vie. **Botanical description**: *Thuja occidentalis* is native to E Canada, NE United States. Trees can grow up to 15 (–38) m tall; trunk up to 0.9 (–1.8) m d.b.h.; bark reddish brown or grayish brown, fibrous, fissured; crown conical. Leaves on both sides of branchlets dull yellowish green; facial leaves (1.5–) 3–5 mm, abaxial gland conspicuous, apex acute; lateral leaves slightly shorter than or as long as facial leaves, apex incurved. Pollen cones reddish, 1–2 mm. Seed cones brown, ellipsoid, (0.6–) 0.9–1.4 cm; fertile cone scales ca. 4. Seeds reddish brown, 4–7 mm including wings. **Medicinal use(s)**: *Thuja occidentalis* are known to possess a wide spectrum of pharmacologic properties, including antioxidant, antitumor, anti-allergic, anti-inflammatory, antibacterial, antifungal, and antiviral.			Isothujone	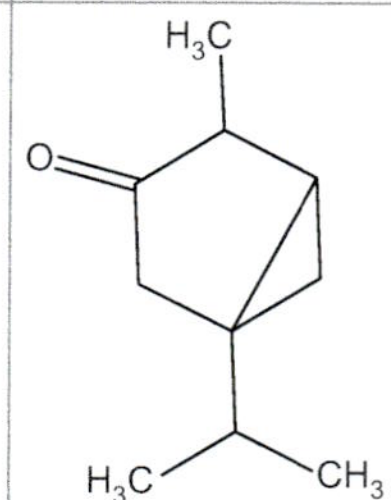	
				Fenchone sabines	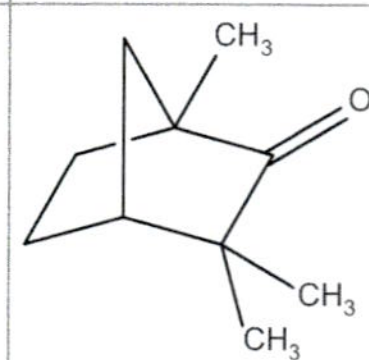	
				p-Coumaric acid	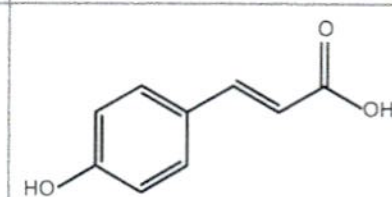	

		Compound	Structure	
		Umbelliferone	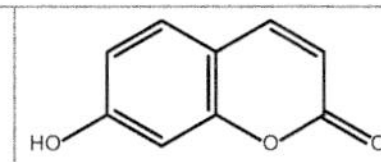	
		Flavonoid(s)	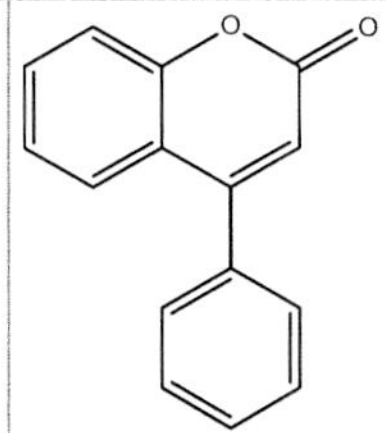	
		(Catechin, gallocatechine)	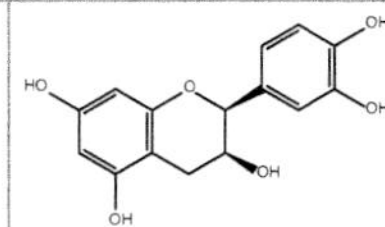	
		Tannic acid	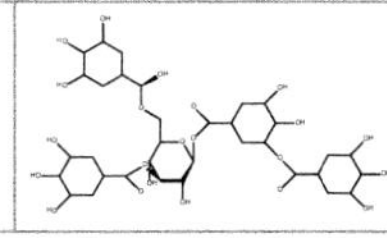	

Table 4.31 Anticancer plants of family Cycadaceae

Cycadaceae	*Cycas revoluta* Thunb.		Leaf	Peptide Cr-ACP	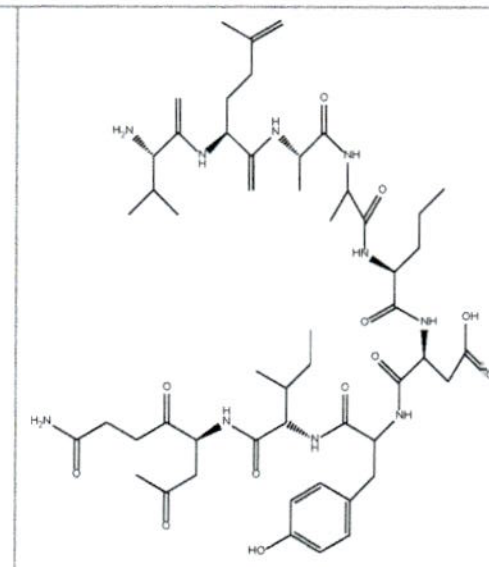	Hsu (1967), Duke and Ayensu (1985a, b), and Mandal et al. (2012)
	Synonym(s): *Cycas inermis Oudem*, *Cycas miquelii Warb*, *Epicycas miquelii* (Warb.) de Laub. **Common name(s)**: Sotetsu (Japanese), Sago palm, King sago, Sago cycad, and Japanese sago palm. **Botanical description**: Cycas revoluta is an evergreen tree growing up to 3.5 m (11 ft.) by 3.5 m (11 ft.) at a slow rate. It is hardy to zone (UK) 9 and is frost tender. It is in leaf all year and in flower from May to July, and the seeds ripen from October to November. The species is dioecious (individual flowers are either male or female, but only one sex is to be found on any one plant, so both male and female plants must be grown if seed is required) and is pollinated by insects and wind. The plant is not self-fertile. It can fix nitrogen. Suitable for light (sandy) and medium (loamy) soils and prefers well-drained soil. Suitable pH: acid, neutral, and basic (alkaline) soils. It can grow in semi-shade (light woodland) or no shade. It prefers dry or moist soil. **Medicinal use(s)**: The leaves are used in the treatment of cancer and hepatoma. The terminal shoot is astringent and diuretic. It is used in the treatment of rheumatism The seeds are used to inhibit the growth of malignant tumors.			Amentoflavone	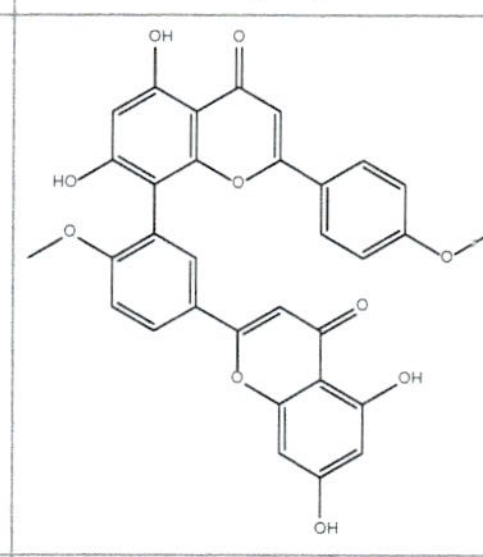	
				Hinokiflavone	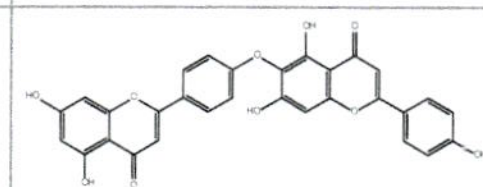	

	Cycas rumphii Miq.		Resin	Isoginkgetin	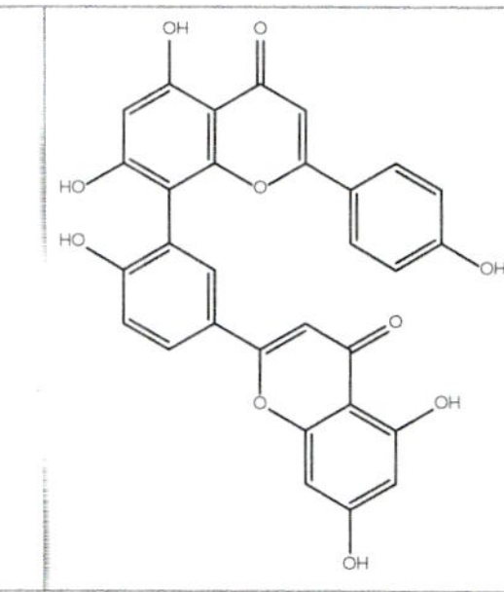	Holdsworth et al. (1983) and Varshney et al. (1973)
	Synonym(s): *Cycas rumphii* subsp. *normanbyana* (F.Muell.) J.Schust. **Common name(s)**: Queen sago or The queen sago palm. **Botanical description**: The cycad is a small tree, growing to about 10 m (33 ft.) in height, with a trunk diameter of up to 40 cm (16 in.). The bark is gray and distinctively fissured into rectangular, or diamond-shaped, segments. The leaves grow from the crown – bright green, glossy, palmlike fronds, 1.5–2.5 m (4.9–8.2 ft.) long, with 150–200 leaflets on each frond. The spiny petiole is 35–60 cm (14–24 in.) long. The male plant's strobilus, or cone, is oblong-ellipsoidal, 30–60 cm (12–24 in.) long, orange in color and fetid in odor. The female's megasporophylls are about 30 cm long, fleshy, brown, and densely hairy, with the fertile area about 35 mm (1.4 in.) wide. The seeds are 45 mm long and 30 mm wide, ripening from green to an orange- or reddish-brown color. **Medicinal use(s)**: *Cycas rumphii* are known to possess a wide spectrum of pharmacologic properties, including antioxidant, antitumor, anti-allergic, anti-inflammatory, antibacterial, antifungal, and antiviral. It is used to treat gastric and ovarian cancer.					

Table 4.32 Anticancer plants of family Calophyllaceae

Family	Botanical name	Image	Plant part used	Bioactive compound(s)	Structure	Reference(s)
Calophyllaceae	*Calophyllum brasiliense* Cambess		Stamen and bark	GUT-70 (tricyclic coumarin)	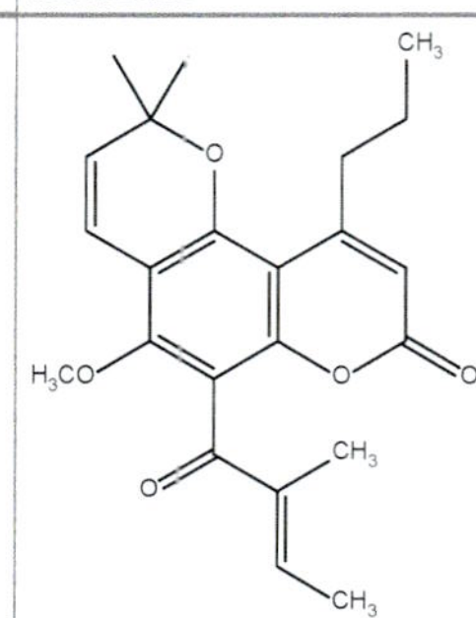	Kimura et al. (2005) and Bernabé-Antonio et al. (2014)
	Synonym(s): *Calophyllum lucidum* Benth., *Calophyllum mariae* Planch. & Triana, *Calophyllum piaroanum* A. Castillo & C. Gil, *Calophyllum revolutum* Rich. ex Vesque. **Common name(s)**: Guanandi. **Botanical description**: It is native to subtropical and tropical regions of Central America, South America, and the Caribbean. It is an evergreen tree growing up to 20–50 m tall, with a trunk up to 1.8 m in diameter, and a dense, rounded crown. The leaves are opposite, 6.3–12.5 cm long and 3.2–6.3 cm broad, elliptic to oblong or obovate, leathery, hairless, glossy green above, paler below, with an entire margin. The flowers are 10–13 mm in diameter, with four white sepals. **Medicinal use(s)**: It is used to treat cancer, ulcer, and gastritis and to avoid prostate damages, skin scarification, and sunburn. It possesses anticancer, anti-inflammatory, and antitumor properties.			5-Methoxy-2,2-dimethyl-6-(2-methyl-1-oxo-2-butenyl)-10-propyl-2H	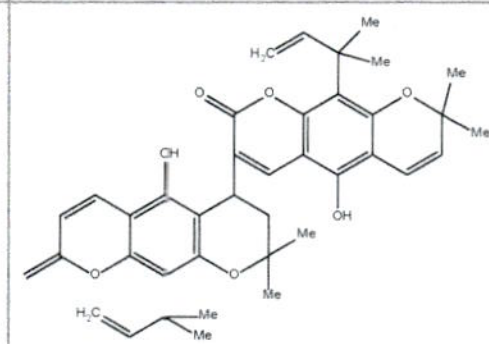	

Table 4.33 Anticancer plants of family Campanulaceae

Family	Botanical name	Image	Plant part used	Bioactive compound(s)	Structure	Reference(s)
Campanulaceae	*Codonopsis lanceolata* (Siebold & Zucc.) Benth. & Hook.f. exTrautv.		Root	Stigmasterol	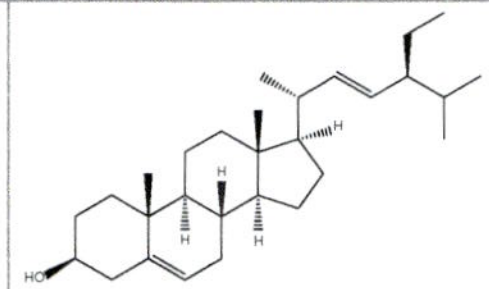	Duke and Ayensu (1985a, b) and Li et al. (2015)
	Synonym(s): *Campanumoea japonica* Siebold ex Merr., *Campanumoea lanceolata* Siebold & Zucc., *Codonopsis bodinieri* H. Lév., *Codonopsis yesoensis* Nakai, *Glosocomia hortensis* Rupr., *Glosocomia lanceolata* (Siebold & Zucc.) Rupr. **Common name(s)**: Deodeok, Lance asiabell. **Botanical description**: It is a flowering plant native to East Asia. It grows up to 1.5 m tall. It has bell-shaped flowers that are purple inside and flowers from August to September, and the seeds ripen from September to October. The flowers are hermaphrodite and are pollinated by bees and wasps. **Medicinal use(s)**: It is used to treat human cancer cell lines, cough, excessive mucus formation in the throat, sore throat, common cold, and flu. It possesses anticancer, antitumor, anti-inflammatory, and antibacterial properties.					

(continued)

Table 4.33 (continued)

Family	Botanical name	Image	Plant part used	Bioactive compound(s)	Structure	Reference(s)
	Codonopsis foetens subsp. nervosa (Chipp) D.Y.Hong		Root	Apigenin	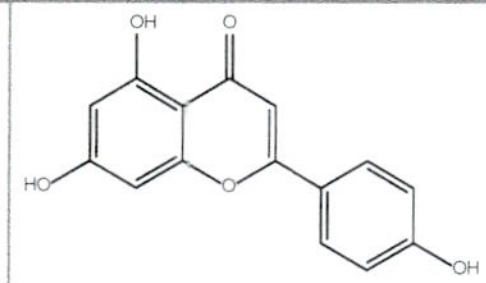	Duke and Ayensu (1985a, b)
	Synonym(s): *Codonopsis macrantha* Nannf., *Codonopsis nervosa* (Chipp) Nannf. **Common name(s)**: Not available. **Botanical description**: It is native to East Asia – the Himalayas in Eastern Tibet. The flowers are hermaphrodite (have both male and female organs) and are pollinated by bees and wasps. Suitable for light (sandy) and medium (loamy) soils and prefers well-drained soil. They are 3–10 cm long; flowering starts from August to October; fruit size 2–5 cm long; seed color brown with wings. **Medicinal use(s)**: It possesses anticancer, antitumor, anti-inflammatory, analgesic, and antibacterial properties. The flowers are used in Tibetan medicine and are said to have a sweet and astringent taste with a cooling potency. They are analgesic and anti-inflammatory, being used in the treatment of pain and swollen joints due to arthritis/gout, stiffening of ureters, and paralysis due to cerebral ischemia.					

	Codonopsis pilosula (Franch.) Nannf.		Root	3H-Benzo[1,2-b;3,4-b′]dipyran-8-one C(23)H(25)O(5)	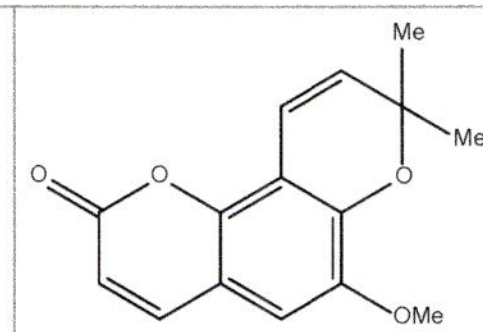	Zhuang et al. (2009)
	Synonym(s): *Campanumoea pilosula* Franch., *Codonopsis glaberrima* Nannf., *Codonopsis microtubulosa* Z. T. Wang & G. J. Xu, *Codonopsis modesta* Nannf., *Codonopsis silvestris* Kom., *Codonopsis volubilis* Nannf. **Common name(s)**: Dang Shen root. **Botanical description**: It is native to Asia, where it grows in forests, meadows, and scrubs. The plant produces twining stems up to 2 m long. It has lateral branches with alternately arranged leaves and small branchlets with oppositely arranged leaves. The ovate leaves are up to 7.3 cm long and are usually coated with short hairs. Solitary flowers occur at the branch tips. **Medicinal use(s)**: It used to treat HIV infection and to protect cancer patients against side effects of radiation treatment. It also boosts the immune system and treats weakness, loss of appetite (anorexia), chronic diarrhea, shortness of breath, noticeable heartbeat (palpitations), asthma, cough, thirst, and diabetes. It possesses anticancer, antitumor, anti-inflammatory, and antibacterial properties.					

(continued)

Table 4.33 (continued)

Family	Botanical name	Image	Plant part used	Bioactive compound(s)	Structure	Reference(s)
	Hippobroma longiflora (L.) G.Don		Entire plant	Lobelin	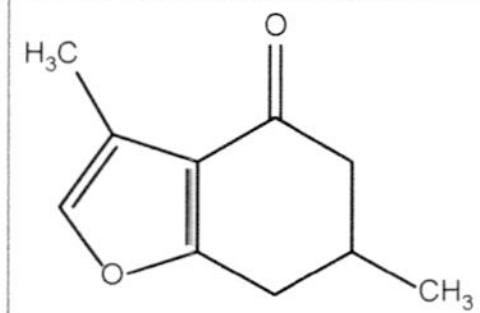	Hsu (1967), Sihotang (2011), and Hapsari et al. (2016)
	Synonym(s): *Isotoma longiflora* (L.) C.Presl, *Isotoma runcinata* Hassk., *Laurentia longiflora* (L.) Peterm., *Lobelia longiflora* L., *Rapuntium longiflorum* (L.) Mill., *Solenopsis longiflora* (L.) M.R. Almeida. **Common name(s)**: Star of Bethlehem. **Botanical description**: It is native to the West Indies but has become naturalized across the American tropics and Oceania. Sepals are about 3 cm long. Flower tube is usually 8–11 cm long, plus the 2–2.5 cm long petals. That explains the other part of its botanical name longiflora, meaning long flowers. **Medicinal use(s)**: It has been used for the treatment of diarrhea, eye problems, toothaches, asthma, epilepsy, and wounds. It possesses anticancer, antitumor, anti-inflammatory, and antibacterial properties.					

	Lobelia nummularia Lam.		Entire plant	Polyacetylene glycosides	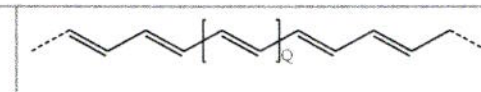	Ho et al. (1995) and Bhanot (2011)
	Synonym(s): *Lobelia arfakensis* Gibbs, *Pratia nummularia* (Lam.) A. Braun & Asch., *Lobelia begoniifolia* Wall., *Lobelia javanica* Thunb., *Lobelia oblique* Buch.-Ham. Ex D. Don, *Lobelia paradoxa* E. Wimm., *Piddingtonia nummularia* (Lam.) A.DC., *Pratia begonifolia* (Wall.) Lindl., *Pratia papuana* S.Moore, *Pratia podenzanae* S.Moore, *Pratia wollastonii* S. Moore, *Pratia zeylanica* Hassk., *Rapuntium nummularia* (Lam.) C. Presl. **Common name(s)**: Roundleaf star creeper. **Botanical description**: It is native to Asia, Australia, and New Zealand. It is small trailing herb occurring in moist places and grasslands. The prostrate stems are 10–60 cm long which root at the nodes. Leaves are either ovate or round, 1–2.5 cm long, or 1–2 cm wide, with prominently toothed margin. Perfectly shaped flowers occur in leaf axils, on 1–2 cm long stalks. Sepals are very narrow, linear, pointed, 2–4 mm long. **Medicinal use(s)**: The plant is used in the treatment of cancer, irregular menstruation, and spermatorrhea. The pounded leaves are sometimes used in the treatment of sprue. The fruits or leaves are used to treat stomachache, difficult urination, scabies, infected eyes, swellings, and boils. It is also used to treat tumors. It possesses anticancer, antitumor, anti-inflammatory, and antibacterial properties.					

Table 4.34 Anticancer plants of family Cannabaceae

Family	Botanical name	Image	Plant part used	Bioactive compound(s)	Structure	Reference(s)
Cannabaceae	*Cannabis sativa* L.		Whole plant	Δ9-Tetrahydrocannabinol	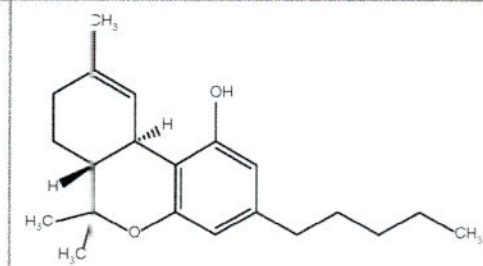	Guzman (2003) and Velasco et al. (2016)
	Synonym(s): *Cannabis americana* Pharm. ex Wehmer, *Cannabis chinensis* Delile, *Cannabis erratica* Siev., *Cannabis foetens* Gilib., *Cannabis generalis* E.H.L.Krause, *Cannabis gigantea* Crevost, *Cannabis indica* Lam., *Cannabis intersita* Soják, *Cannabis kafiristanica* (Vavilov) Chrtek, *Cannabis lupulus* Scop., *Cannabis macrosperma* Stokes, *Cannabis ruderalis* Janisch. **Common name(s)**: Marijuana. **Botanical description**: It is a native plant of Southeast Asia. They are unisexual and plants are most often either male or female. It is a short-day flowering plant, with staminate (male) plants usually taller and less robust than pistillate (female) plants. The flowers of the female plant are arranged in racemes and can produce hundreds of seeds. Male plants shed their pollen and die several weeks prior to seed ripening on the female plant. **Medicinal use(s)**: It is used to treat cancer. It is a multiple-use plant, furnishing fiber, oil, medicine, and narcotics. Modern medicine uses cannabis in glaucoma and alleviating the pains of cancer and chemotherapy. It possesses anticancer, antitumor, anti-inflammatory, and antibacterial properties.					

Ulmaceae	*Celtis philippensis* wightii (Planch.) E. Soepadmo		Twigs	Celtisanin	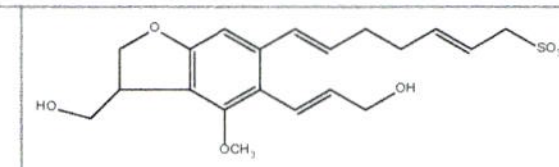	Hwang et al. (2003)
	Synonym(s): *Celtis philippensis* var. consimile (Bl.) Leroy, *Celtis wightii* Planch., *Solenostigma consimile* Bl., *Solenostigma wightii* (Planch.) Bl. **Common name(s)**: Hard celtis. **Botanical description**: *Celtis philippensis* is found in the Asian region. It is 30 m tall, evergreen plant. Bark is grayish white to grayish brown or gray. Branchlets brown or tan, glabrous or pubescent. Winter buds naked, without differentiated scales. Stipules ovate-lanceolate to ovate, 2–8 mm, glabrous or pubescent, enfolding naked terminal bud, apex attenuate. Cymes 1 or 2 per leaf axil, densely flowered, much branched 5 bisexual flowers and 5 or more male flowers. Styles 2, apically broadened and 2-cleft or emarginated. **Medicinal use(s)**: It is used for the treatment of prostate cancer, ovarian cancer, cervical cancer, pancreatic cancer, colon cancer, gastric cancer, skin cancer, stomach cancer, and lung cancer. It also possesses antioxidant, anti-inflammatory, antiproliferative, antifungal, antiviral, and antibacterial properties.			Apigenin	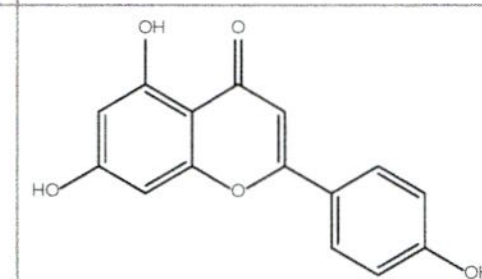	
				Quercetin	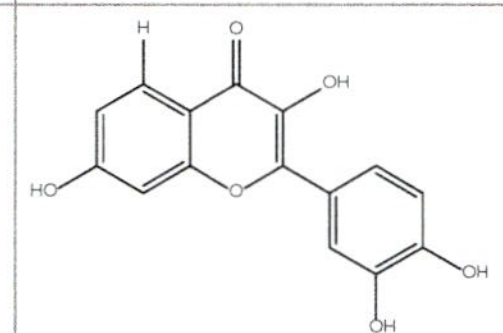	
				Isovitexin	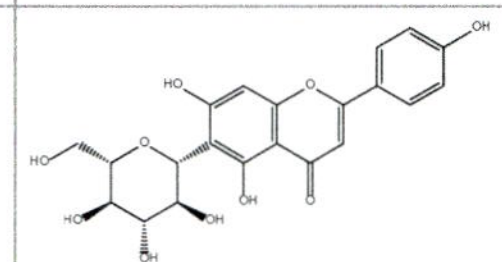	

(continued)

Table 4.34 (continued)

Family	Botanical name	Image	Plant part used	Bioactive compound(s)	Structure	Reference(s)
				Orientin	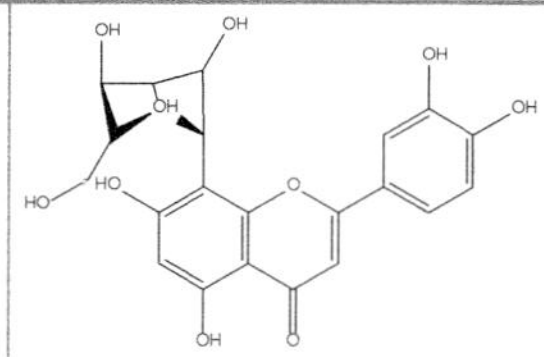	
				Isoorientin	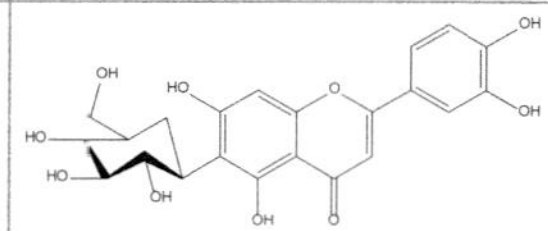	
				Vitexin	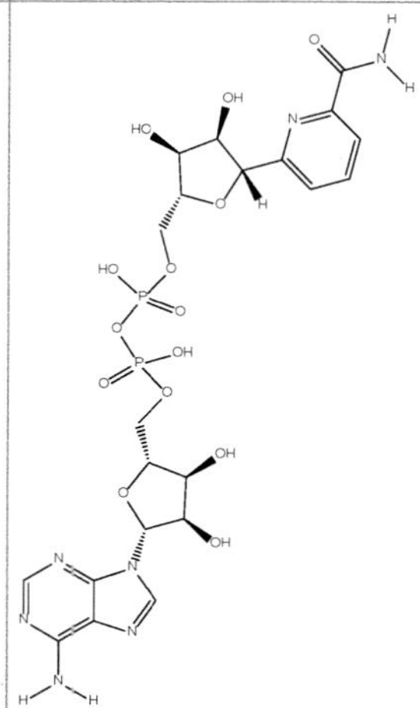	

Table 4.35 Anticancer plants of family Capparaceae

Family	Botanical name	Image	Plant part used	Bioactive compound(s)	Structure	Reference(s)
Capparaceae	*Cadaba glandulosa* Forsk.		Leaf	Stachydrine	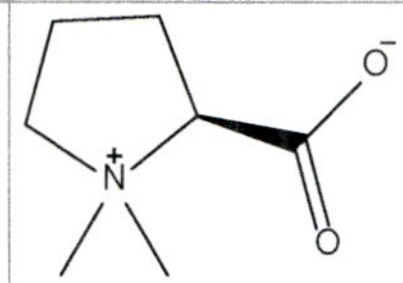	Yousif et al. (1983) and Graham et al. (2000)
	Synonym(s): *Cadaba monopetala* Edgew., *Stroemia glandulosa* (Forssk.) Vahl. **Common name(s)**: Hassaniya. **Botanical description**: These have simple, alternately set leaves. The zygomorphic flowers are solitary or stand in small clusters at the end of short side branches. These flowers consist of four sepals, none or four petals with a narrow claw at base and a wider plate at the top, a tube-shaped nectar producing appendix. **Medicinal use(s)**: It is used to treat cancer and neutralize venom from snakebites. It is used to treat tumors and abscesses. It possesses anticancer, antitumor, antibacterial, and anti-inflammatory properties.					

(continued)

Table 4.35 (continued)

Family	Botanical name	Image	Plant part used	Bioactive compound(s)	Structure	Reference(s)
	Cadaba rotundifolia Forssk.		Leaf	Stachydrine	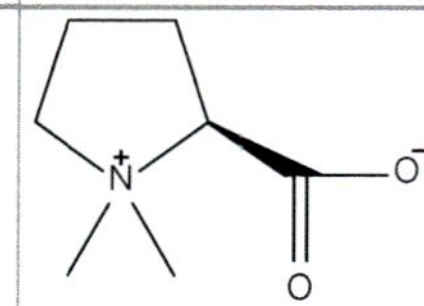	Yousif et al. (1984) and Teklehaymanot (2017)
	Synonym(s): *Stroemia rotundifolia* Vahl. **Common name(s)**: Kadhab. **Botanical description**: It is a much-branched shrub 1–4 m tall; young twigs covered with short glandular hairs; leaves alternate; blade ovate or elliptic, commonly 2.5–4.5 × 2.2–3.8 cm, leathery, apically rounded or retuse, basally subcordate, puberulous; petiole 1–2 cm long, puberulous. **Medicinal use(s)**: It is used to treat cancer and extended flow of menstruation. It is used to massage paralyzed body part. It possesses anticancer, antitumor, antibacterial, and anti-inflammatory properties.			3-Hydroxystashydrine	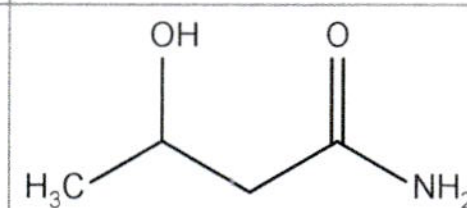	

	Capparis tomentosa Lam.		Entire plant	Beta-carotene	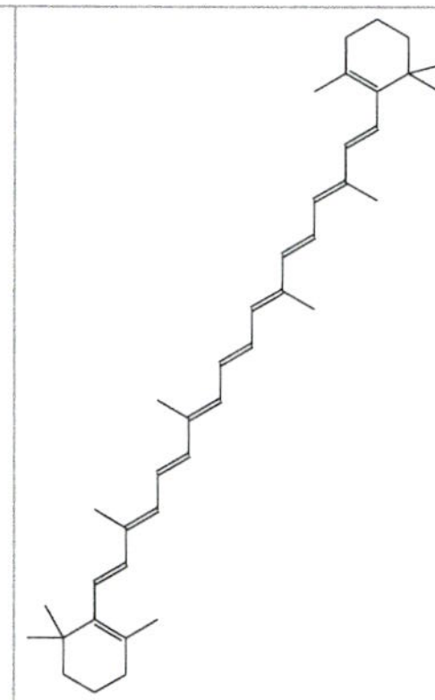	Amico(1977), Hurinanthan et al (2009), and Moshi et al. (2010)
	Synonym(s): *Capparis alexandrae* Chiov., *Capparis biloba* Hutch. & Dalziel, *Capparis corymbifera* E.Mey. ex Harv. & Sond., *Capparis corymbifera* E. Mey. ex Sond., *Capparis floribunda* Wight, *Capparis globifera* Delile, *Capparis hypericoides* Hochst., *Capparis persicifolia* A. Rich., *Capparis puberula* DC., *Capparis subtomentosa* De Wild., *Capparis volkensii* Gilg. **Common name(s)**: Woolly caper bush. **Botanical description**: It is native to Africa. Found in Bushveld and forest from the Eastern Cape of South Africa, through KwaZulu-Natal, Mpumalanga, Limpopo Province, Mozambique, Botswana, Namibia, and into tropical Africa. The northern part of the range extends from Senegal to Eritrea, and this species is also found in the Mascarene Islands. **Medicinal use(s)**: It is used to treat cancer, diarrhea, swelling, and water retention. It is a decorative plant for gardens and can be used for hedging; suitable as a security barrier because of the hooked spines. The fruit may sometimes be eaten by people. It possesses anticancer, antitumor, antibacterial, and anti-inflammatory properties.			Rapamycin	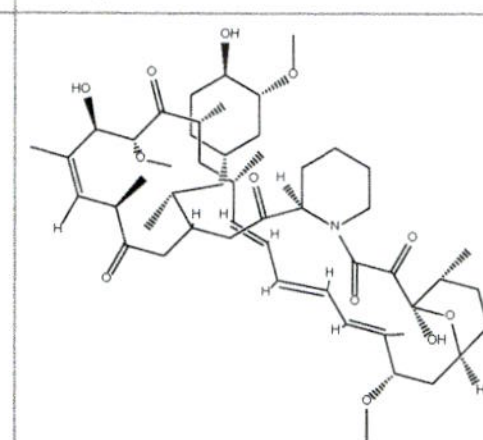	
				Cyclosporine		

(continued)

Table 4.35 (continued)

Family	Botanical name	Image	Plant part used	Bioactive compound(s)	Structure	Reference(s)
	Capparis sikkimensis Kurz		Root	Cappamensin A (2H-1,4-benzoxazin-3(4H)-one, 6-methoxy-2-methyl-4-carbaldehyde),	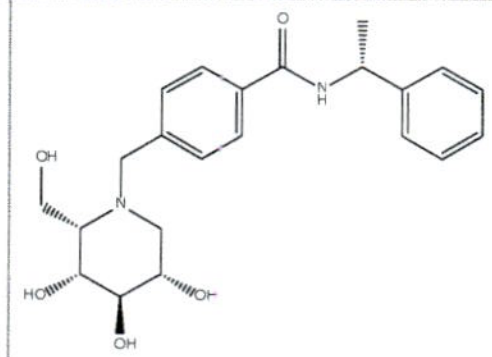	Wu et al. (2003)
	Synonym(s): *Capparis cathcarti* Hemsl. ex Gamble. **Common name(s)**: Not available. **Botanical description:** It is native to Vietnam. Shrubs, vines, or small trees, 4–6 m tall. Branches pale purplish brown when dry, almost terete, pith pale red; new twigs slightly flat, with grooves, densely light reddish brown pubescent with simple trichomes, later glabrescent. Stipular spines to 5 mm, stout, recurved, often aborted or absent on flowering twigs. **Medicinal use(s)**: It possesses anticancer, antitumor, antibacterial, and anti-inflammatory properties.					

Table 4.36 Anticancer plants of family Caricaceae

Family	Botanical name	Image	Plant part used	Bioactive compound(s)	Structure	Reference(s)
Caricaceae	*Carica papaya* L.		Leaf	Fenugreekine	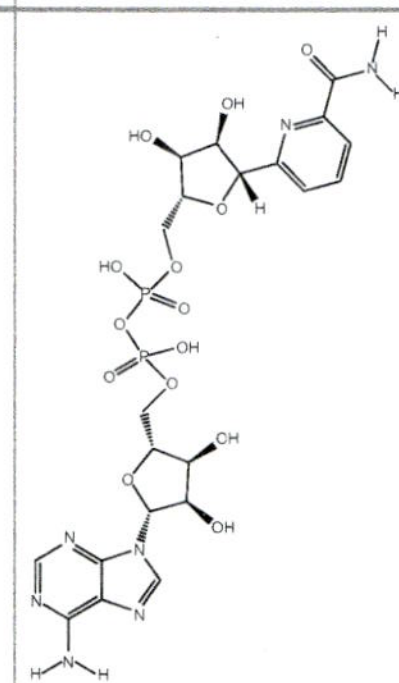	Otsuki et al. (2010) and Chandrasekaran et al. (2018)
	Synonym(s): *Carica citriformis* Jacq., *Papaya carica* Gaertn., *Papaya cimarrona* Sint. ex Kuntze, *Papaya citriformis* (Jacq.) A. DC., *Papaya communis* Noronha, *Papaya cubensis* (Solms) Kuntze, *Papaya cucumerina* Noronha, *Papaya edulis* Bojer, *Papaya edulis* var. *macrocarpa* Bojer, *Papaya hermaphrodita* Blanco, *Papaya peltata* (Hook. & Arn.) Kuntze, *Papaya rochefortii* (Solms) Kuntze, *Papaya sativa* Tuss., *Papaya vulgaris* A. DC., *Vasconcellea peltata* (Hook. & Arn.) A. DC. **Common name(s)**: Pawpaw, Papaya. **Botanical description**: It is native to Mexico and northern South America. It has become naturalized throughout the Caribbean Islands, Florida, Texas, California, Hawaii, and other tropical and subtropical regions of the world. It grows in three sexes: male, female, and hermaphrodite. The male produces only pollen, never fruit. The female produces small inedible fruits unless pollinated. The hermaphrodite can self-pollinate since its flowers contain both male stamens and female ovaries. The papaya is a small, sparsely branched tree, usually with a single stem growing from 5 to 10 m tall, with spirally arranged leaves confined to the top of the trunk. The lower trunk is conspicuously scarred where leaves and fruit were borne. The leaves are large, 50–70 cm in diameter, deeply palmately lobed, with seven lobes. **Medicinal use(s)**: It is used to treat cancer. Papaya leaves are made into tea as a treatment for malaria and to increase the platelets in blood. It works as a tonic to boost the energy during diseases. It possesses anticancer, antitumor, antibacterial, anti-inflammatory, and antimicrobial properties.			Carpaine	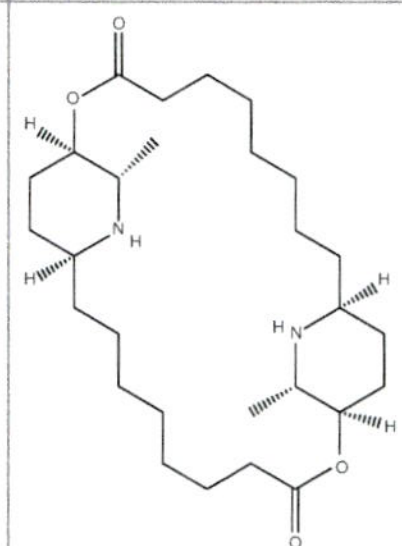	

Table 4.37 Anticancer plants of family Caryophyllaceae

Family	Botanical name	Image	Plant part used	Bioactive compound(s)	Structure	Reference(s)
Caryophyllaceae	*Dianthus zonatus* Fenzl		Flowers	Eugenol	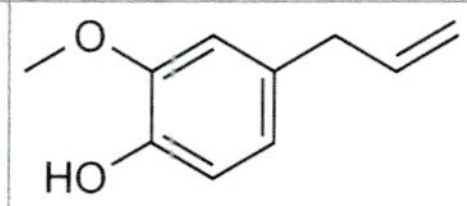	Ishikura (1982) and Arı et al. (2015)
	Synonym(s): *Dianthus hypochlorus* Boiss. & Heldr. **Common name(s)**: Carnation, Karanfil. **Botanical description**: It is native mainly to Europe and Asia. It exhibits a tufted habit (7–30 cm high). Leaves linear, from 2.5 to 8 cm long with thickened margins. Flowers are borne on freely branched inflorescences, 1–2 cm across, deep pink above, yellowish beneath with a darker ring in the throat and dentate petals. The calyx is pale green with a purple tinge on the upper parts. **Medicinal use(s)**: It is used to treat cancer and problems associated with cardiac, urinary, nervous, and gastrointestinal system. It possesses anticancer, antitumor, anti-inflammatory, and antibacterial properties.					

	Drymaria gracilis Cham. and Schltdl.		Leaf and stem	C-Glyccsylflavones	Structure not available	Woo and Woo (1989) and Maffei (2003a, b, c, d, e, f, g, h, i, j, k, l, m)
	Synonym(s): *Drymaria cordata* var. *gracilis* (Cham. & Schltdl.) Rohrb. **Common name(s)**: Not available. **Botanical description**: It is native to Mexico. They are creeping plants that reach up to 13–20 cm in height and with abundant branches. The leaves are kidney-shaped, small and with white flowers. **Medicinal use(s)**: It is used to treat tumors. It possesses anticancer, anti-inflammatory, and antibacterial properties.					
	Melandrium firmum (Siebold and Zucc.) Rohrb.		Entire plant	α-Spinasterol	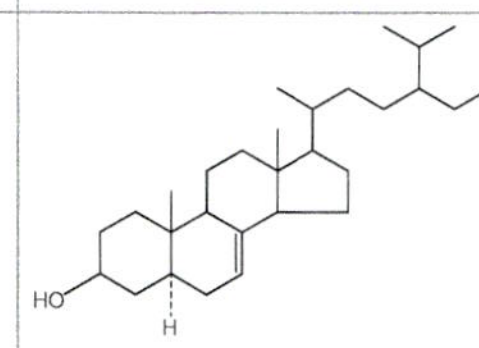	Duke and Ayensu, (1985a, b) and Lee et al. (2014)
	Synonym(s): *Silene firma* Siebold & Zucc. **Common name(s)**: Not available. **Botanical description**: It is native to South Korea. **Medicinal use(s)**: It is used for the treatment of breast cancer. It possesses anticancer, antitumor, anti-inflammatory, and antibacterial properties.					

Table 4.38 Anticancer plants of family Cephalotaxaceae

Family	Botanical name	Image	Plant part used	Bioactive compound(s)	Structure	Reference(s)
Cephalotaxaceae	*Cephalotaxus fortunei* Hook.		Part not specified	Beta-sitosterol	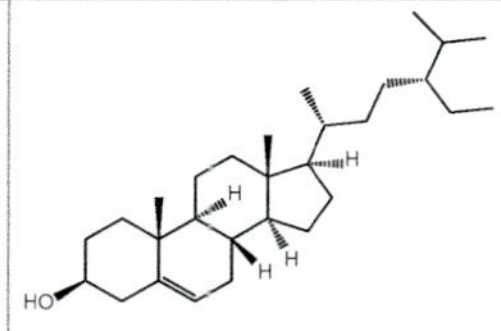	Tai and Hung (1981) and Ni et al. (2018)
	Synonym(s): *Cephalotaxus kaempferi* auct., *Taxus fortunei* (Hook.) C. Lawson. **Common name(s)**: Chinese plum yew. **Botanical description**: It is a native plant to East Asia – East and Central China. It grows in woodlands, especially in limestone regions. Grows in mixed, coniferous, and broad-leaved forests, thickets, and roadsides at elevations of 200–3700 m. It flowers from April to May, and the seeds ripen from October to November. The flowers are dioecious and are wind-pollinated. The plant is not self-fertile. **Medicinal use(s)**: It is used to treat human cancer cell lines. Some substances from the plant have shown anticancer activity. Used for the treatment of tumors. It possesses anticancer, anti-inflammatory, and antibacterial properties.					

Table 4.39 Anticancer plants of family Chloranthaceae

Family	Botanical name	Image	Plant part used	Bioactive compound(s)	Structure	Reference(s)
Chloranthaceae	*Chloranthus glaber* Makino		Aerial parts	Benzyl 2-β-Glucopyranosy-loxybenzoate	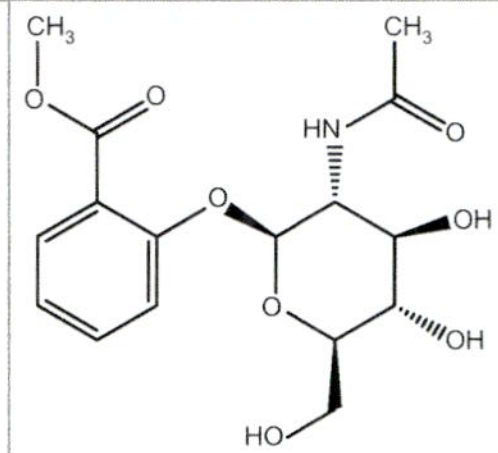	Duke and Ayensu (1985a, b) and Dashtdar et al. (2016)
	Synonym(s): *Ascarina serrata* Blume, *Chloranthus brachystachys* Blume. **Common name(s)**: Nine-knotted flower. **Botanical description**: It is an herb native to Southeast Asia. Leaf blade elliptic or ovate-lanceolate, 6–17 × 2–6 cm, leathery, margin sharply coarsely serrate. **Medicinal use(s)**: The entire plant has anti-stress, antioxidant, anti-inflammatory, detoxifying, blood-activating, and antibacterial effects. It is used to treat cancers. It possesses anticancer, anti-inflammatory, antitumor, and antibacterial properties.					

(continued)

Table 4.39 (continued)

Family	Botanical name	Image	Plant part used	Bioactive compound(s)	Structure	Reference(s)
	Sarcandra glabra (Thunb.) Nakai		Entire plant	Isoastilbin	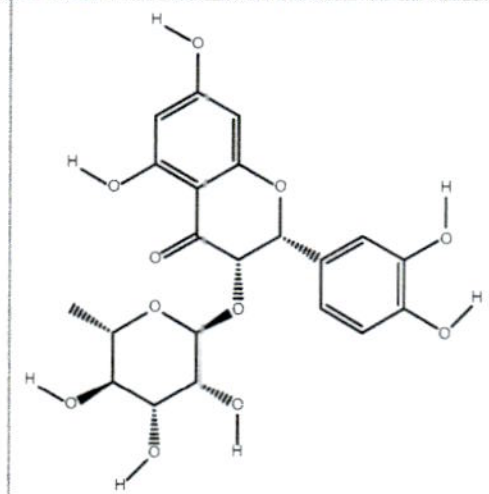	Jain (1970), Takeda et al. (1993), Tsui and Brown (1996), and Li et al. (2007a, b, c)
	Synonym(s): *Ascarina serrata* Blume, *Chloranthus brachystachys* Blume. **Common name(s)**: Nine-knotted flower. **Botanical description**: It is native to East Asia, China, Japan, Korea, India, Sri Lanka, Thailand, Laos, Cambodia, Vietnam, Malaysia, and the Philippines. It is an evergreen perennial plant, the stems becoming woody at least at the base. It grows 50–300 cm tall. Its habitat is at forests, thickets, valleys, ravines, slopes, roadsides, trailsides, grasslands, swamps, streamsides, sandy soil, and elevations from near sea level to 2000 m in China. **Medicinal use(s)**: It is used to treat cancers. It possesses anticancer, anti-inflammatory, antitumor, and antibacterial properties.					

Table 4.40 Anticancer plants of family Commelinaceae

Family	Botanical name	Image	Plant part used	Bioactive compound(s)	Structure	Reference(s)
Commelinaceae	*Cyanotis vaga* (Lour.) Roem. and Schult.		Entire plant	Ecdysterone	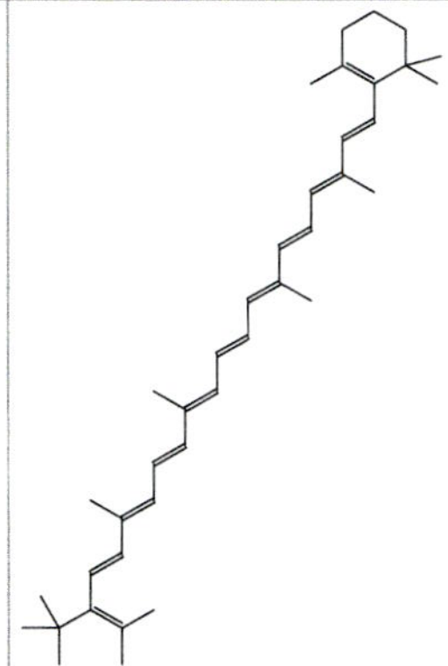	Garcia-Barriga (1975) and Sultan and Raza (2015)
	Synonym(s): *Cyanotis abyssinica* A.Rich., *Cyanotis nobilis* Hassk., *Tradescantia vaga* Lour. **Common name(s)**: Wandering dew-grass. **Botanical description**: It is native to Africa, Southern Asia, and Northern Australia. It is a slender sparsely branched plant, with narrow lance-shaped leaves often with long woolly hairs on the sheaths. Flowers are small dark blue, subtended by long straight or curved bracts. Flowers are about 1 cm across, in clusters in leaf axils or at branch ends. Stamen filaments have long white or blue hairs. Bracts are hairy with enlarged rounded bases. Leaves are 3.0–7.5 cm long, variably hairy. Stem is usually 8–30 cm long, often tufted and rooting at the nodes. **Medicinal use(s)**: It is used for control of breast cancer. It possesses anticancer, antitumor, anti-inflammatory, and antibacterial properties.					

Table 4.41 Anticancer plants of family Convolvulaceae

Family	Botanical name	Image	Plant part used	Bioactive compound(s)	Structure	Reference(s)
Convolvulaceae	*Calystegia sepium* (L.) R.Br.		Entire plant	Calystegines	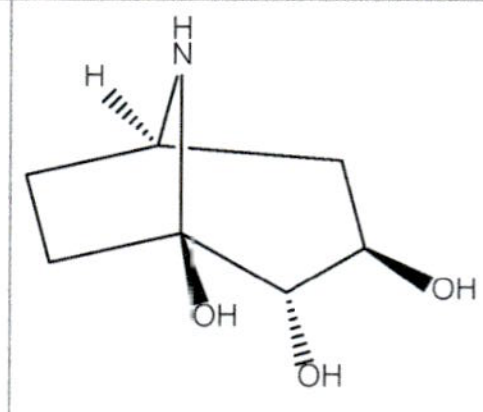	De Feo et al. (1992) and Khan (2017)
	Synonym(s): *Aniseia arifolia* Meisn., *Calystegia rosea* Phil., *Convolvulus sepium* L., *Volvulus sepium* (L.) Junger. **Common name(s)**: Hedge bindweed. **Botanical description**: It is native to North America. It is an herbaceous perennial that twines around other plants, in a counterclockwise direction, to a height of up to 2–4 m, rarely 5 m. The pale matte green leaves are arranged spirally, simple, pointed at the tip and arrowhead shaped, 5–10 cm long and 3–7 cm broad. **Medicinal use(s)**: It possesses anticancer properties. It is used as cholagogue, demulcent, diuretic, febrifuge, purgative, and poultices. It is also used to treat corns.					
	Cuscuta chilensis Ker Gawl.		Entire plant	Quercetin-3-*O*-apiosyl (1→2)-galactoside	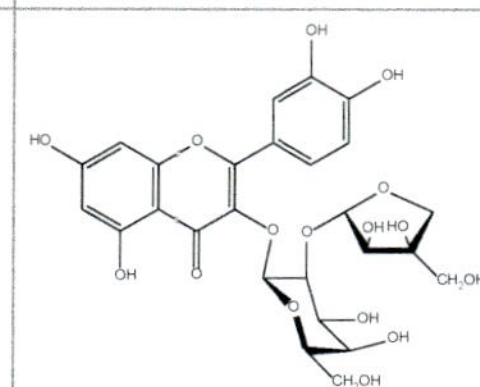	San Martin, (1983), Quattrocchi (2012), and Ibrahim et al. (2008)

	Synonym(s): *Cuscuta aurea* Liebm. **Common name(s)**: Dodder. **Botanical description**: It is native to Southern America. It is a parasitic vine that wraps around other plants for nourishment. Its ripe seeds are used as an annual valuable parasitic herb. **Medicinal use(s)**: It is also used to treat breast cancer and tumors.					
	Cuscuta chinensis Lam.		Entire plant	Trigoneoside Xa	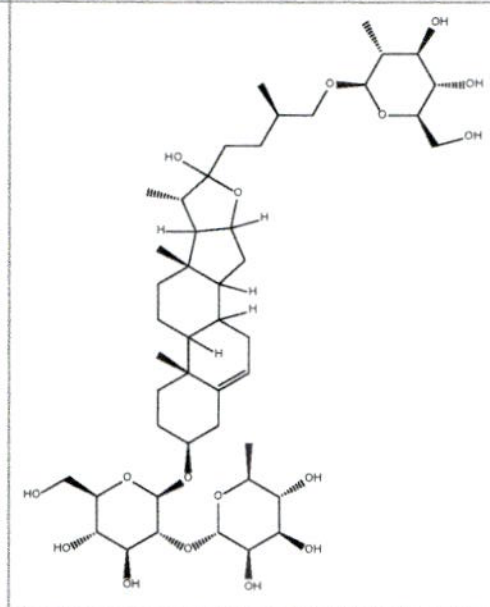	Jain (1970), Nisa et al. (1985), Jafarian et al. (2014), and Riaz et al. (2018)
	Synonym(s): *Cuscuta aurea* Liebm. **Common name(s)**: Dodder. **Botanical description**: It is native to China. It is a parasitic vine that wraps around other plants for nourishment. It creeps up to 1.8 m. **Medicinal use(s)**: It is used to treat human breast carcinoma cell line (MDA-MB-468), human colorectal adenocarcinoma cell line (HT-29), and human uterine cervical carcinoma (HeLa). It is used to treat kidney deficiency and liver failure. It possesses anticancer, antiaging, and anti-inflammatory properties. It is also a pain reliever and an aphrodisiac.			Astragalin	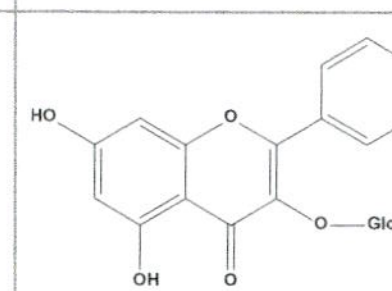	

(continued)

Table 4.41 (continued)

Family	Botanical name	Image	Plant part used	Bioactive compound(s)	Structure	Reference(s)
	Ipomoea carnea Jacq.		Leaf	7-O-ß-D-Glucopyranosyl-dihydroquercetin-3-O-α-D-glucopyranose	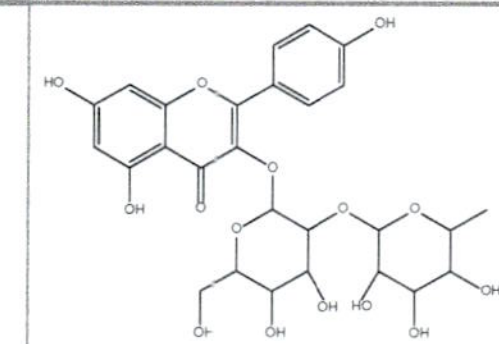	Li (2006) and Uddin et al. (2018)
	Synonym(s): *Ipomoea carnea* subsp. Carnea, *Ipomoea carnea* subsp. fistulosa (Mart. ex Choisy) D. F. Austin. **Common name(s)**: Pink morning glory, Bush morning glory. **Botanical description**: *Ipomoea carnea* is native to South America. The stem (1.25–2.75 m long and 0.5–0.8 cm diameter) is erect, woody, hairy, and more or less cylindrical in shape and greenish in colour. It has alternate leaves. The leaves (10–25 cm long) are light green, heart shaped to lanceolate. The plants bear flowers (4 inch pink flowers all spring and summer long. Flowers are axial, pedicel is green, erect and cylindrical. Its length ranges between 1.5–2.2 cm and diameter ranges between 0.15–0.20 cm. **Medicinal use(s)**: It possesses anticancer, antifungal, antioxidant, antibacterial, anticonvulsant, immunomodulatory, antidiabetic, hepatoprotective, anti-inflammatory, anxiolytic, sedative, and wound healing properties.					

	Ipomoea batatas (L.) Lam.		Flower	Caffeic acid	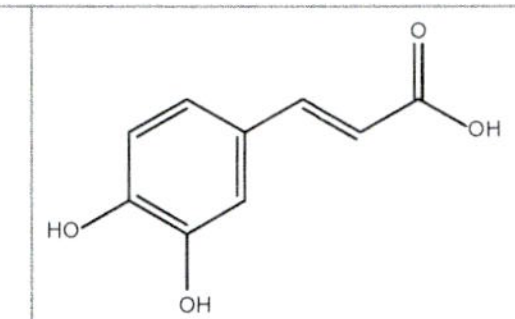	Islam et al. (2006) and Sugata et al. (2015)
	Synonym(s): *Batatas edulis* (Thunb.) Choisy, *Batatas wallii* C. Morren, *Convolvulus apiculata* M. Martens & Galeotti, *Convolvulus attenuatus* M. Martens & Galeotti, *Convolvulus batatas* L., *Convolvulus candicans* Sol. ex Sims, *Convolvulus edulis* Thunb., *Convolvulus esculentus* Salisb., *Convolvulus varius* Vell., *Ipomoea davidsoniae* Standl., *Ipomoea edulis* (Thunb.) Makino, *Ipomoea mucronata* Schery, *Ipomoea purpusii* House, *Ipomoea vulsa* House, *Ipomoea wallii* (C. Morren) Hemsl. **Common name(s)**: Sweet potato. **Botanical description**: The plant is an herbaceous perennial vine, bearing alternate heart-shaped or palmately lobed leaves and medium-sized sympetalous flowers. The edible tuberous root is long and tapered, with a smooth skin whose color ranges between yellow, orange, red, brown, purple, and beige. Its flesh ranges from beige through white, red, pink, violet, yellow, orange, and purple. **Medicinal use(s)**: It possesses anticancer, antidiabetic, and anti-inflammatory activities. It is used to treat stomach, cervical, and prostate cancer.			Cyanidin	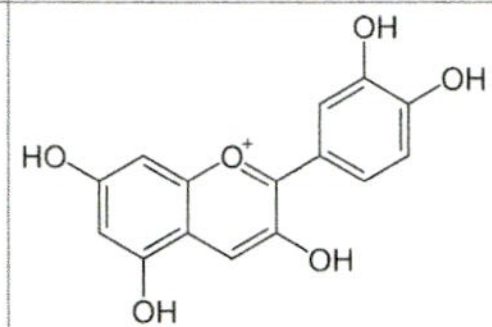	
				Peonidin A	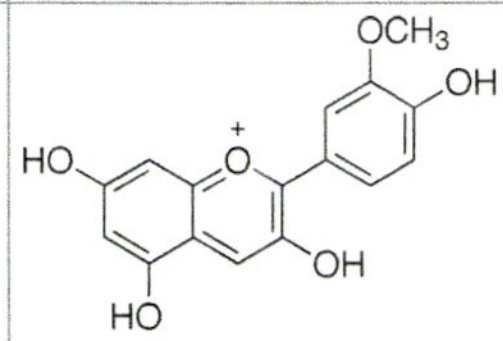	

(continued)

Table 4.41 (continued)

Family	Botanical name	Image	Plant part used	Bioactive compound(s)	Structure	Reference(s)
	Ipomoea cairica (L.) Sweet		Leaf	Myristyl alcohol		Lin et al. (2008), Saied et al. (2011), and Yu et al. (2013b)
	Synonym(s): *Ipomoea palmata* Forssk, *Ipomoea stipulacea* Jacq., *Ipomoea tuberculata* (Desr.) Roem. & Schult. **Common name(s)**: Mile-a-minute vine, Railroad creeper. **Botanical description**: It is native to Eastern Mediterranean; tropical Africa; through tropical Asia to the Philippines, New Guinea, and the Pacific Islands; and South America. It is an evergreen, herbaceous, perennial climbing plant, producing slender stems up to 5 m long from a tuberous rootstock. It bears violet-purple to white flowers. **Medicinal use(s)**: It possesses anticancer, antimicrobial, anti-inflammatory, antibiotic, and antioxidant properties. The entire plant is used for treating external infections. The crushed leaves are taken in a draft to treat for body rashes, especially if accompanied by fever.			Cairicoside(s)	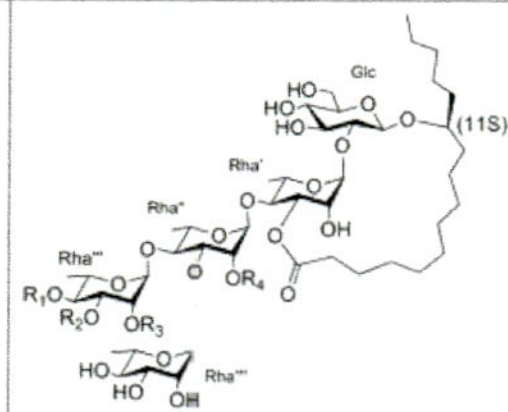	
				Matairesinol	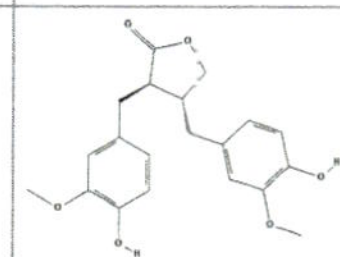	

				Vanillic acid	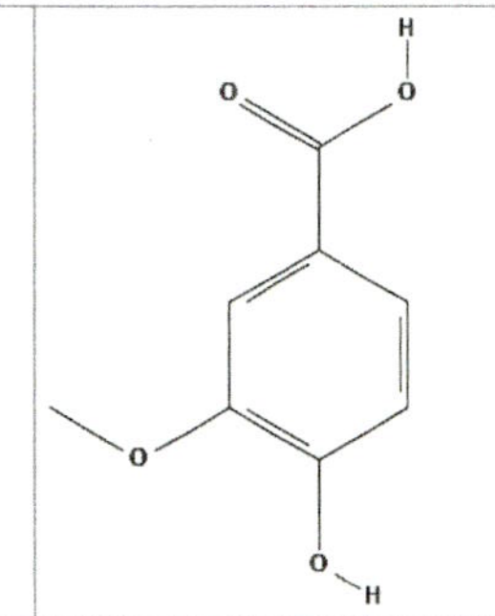	
	Ipomoea squamosa Choisy		Flower and stamens	Ipomoeassin	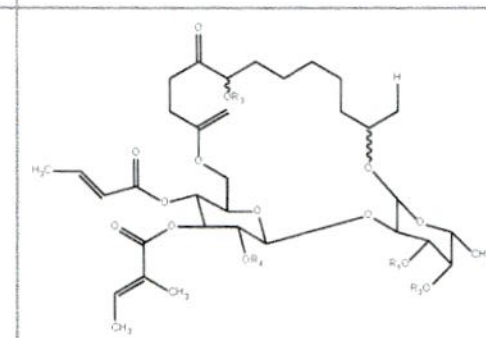	Cao et al. (2005)
	Synonym(s): *Convolvulus mattogrossensis* Kuntze, *Ipomoea callida* House, *Ipomoea mattogrossensis* (Kuntze) K. Schum., *Ipomoea morelii* Duchass. & Walp., *Ipomoea vestalii* Standl., *Ipomoea wilsonii* House. **Common name(s)**: Batatilla. **Botanical description**: It is native to Belize, Bolivia, North Brazil, Northeast Brazil, Southeast Brazil, Central-West Brazil, Colombia, and Costa Rica. **Medicinal use(s)**: It is used to treat ovarian cancer. It is used for the treatment of several diseases, such as diabetes, hypertension, dysentery, constipation, fatigue, arthritis, and rheumatism.					

Table 4.42 Anticancer plants of family Cornaceae

Family	Botanical name	Image	Plant part used	Bioactive compound(s)	Structure	Reference(s)
Cornaceae	*Camptotheca acuminata* Decne		Whole plant	Camptothecin	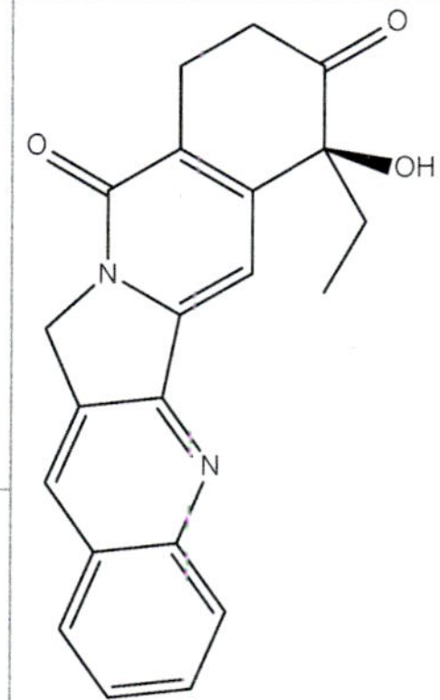	Pasqua et al. (2004)
	Synonym(s): *Camptotheca yunnanensis* Dode., *Cephalanthus esquirolii* H. Lév. **Common name(s)**: Cancer tree, Tree of joy, Happy tree, Tree of life. **Botanical description**: The so-called cancer tree is a deciduous tree native to Southern China and Tibet. It grows to a height of 20–25 m tall and has a light gray bark and oval leaves with heavy pleated veining. The flowers are white to yellowish spherical clusters. **Medicinal use(s)**: It was first used as traditional medicine (ancient Chinese) as a treatment for common colds, psoriasis, liver problems, and digestive problems. Scientific research has found that this plant has cancer-fighting properties; thus, it is called as cancer tree. It contains pentacyclic quinolines camptothecin and 10-hydroxycamptothecin which inhibit DNA topoisomerase I and are very effective against cancer cells. The extract of tender plants is beneficial for brain tumors, liver cancer, cancer in the gastrointestinal tract, leukemia, and other cancers.					

Table 4.43 Anticancer plants of family Cruciferae

Family	Botanical name	Image	Plant part used	Bioactive compound(s)	Structure	Reference(s)
Brassicaceae/ Cruciferae	*Draba litamo* L.Uribe		Entire plant	Cardenolide	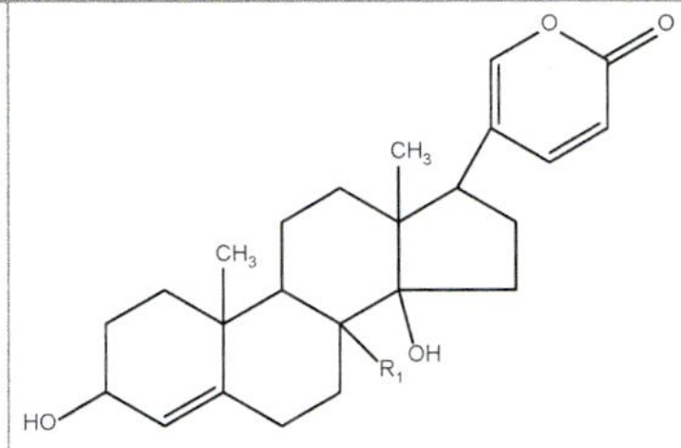	Duke and Ayensu (1985a, b)
	Synonym(s): No synonyms are recorded for this name. **Common name(s)**: Hoary cress, Whitetop. **Botanical description**: It is native plant of Colombia. It is a perennial plant growing up to 1 m. Flowering occurs from May to June. The flowers are hermaphrodite and are pollinated by bees. It can grow in semi-shade and prefers moist soil. **Medicinal use(s)**: It is used to treat breast and skin cancer. It possesses antitumor, anticancer, anti-inflammatory, antimicrobial, and antiscorbutic property. The seeds have been used as a cure for flatulence and fish poison. It also helpful in food poisoning caused by eating suspect fish.					

(continued)

Table 4.43 (continued)

Family	Botanical name	Image	Plant part used	Bioactive compound(s)	Structure	Reference(s)
	Draba nemorosa L.		Entire plant	Vobasine	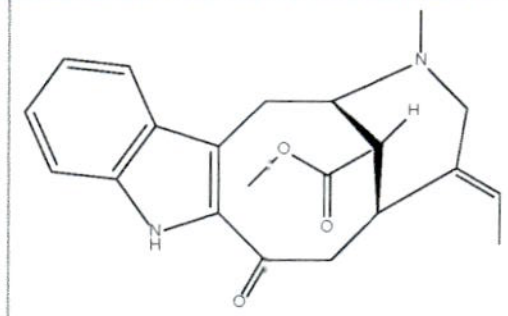	Yu et al. (1994a, b) and Moon et al. (2010)
	Synonym(s): *Draba lutea* Gilib., *Tomostima nemorosa* (L.) Lunell. **Common name(s)**: Woodland draba, Yellow whitlow-wort. **Botanical description**: It is native to East Asia and found mostly in Europe, Japan, and China. It is an annual/biennial plant and the stem attains a height of 0.30 m with elongated cluster of stalked flowers at the top. Flowers are tiny, less than 1/8 in. across, with 4 yellow petals notched at the tip. The green sepals behind the flower are sparsely hairy and more than half as long as the petals. **Medicinal use(s)**: It is used to treat lung and breast cancer. It possesses anticancer, anti-inflammatory, antimicrobial, anti-asthmatic, and antioxidant properties. The seeds are demulcent, deobstruent, diuretic, expectorant, and laxative.			Kaempferol glycoside(s)	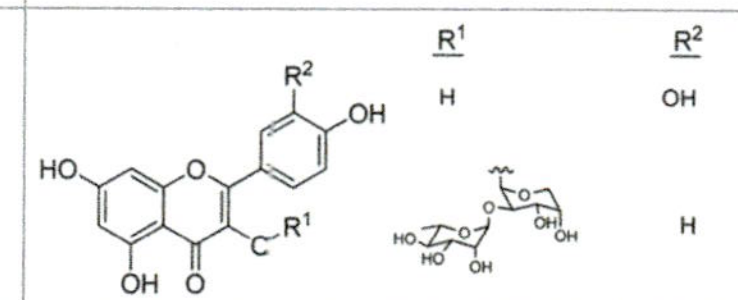	
				Cardenolide glycoside(s)	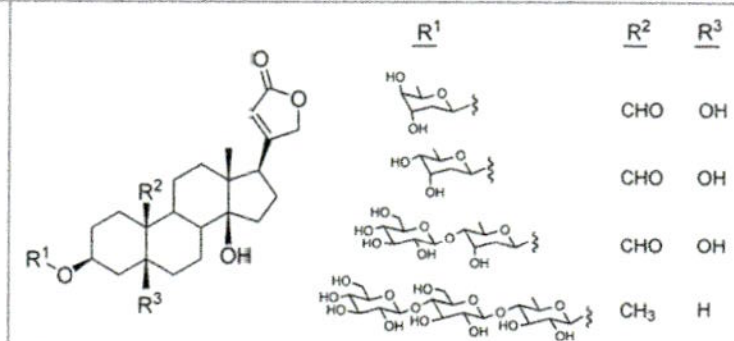	
	Isatis tinctoria L.		Whole plant	Isatin	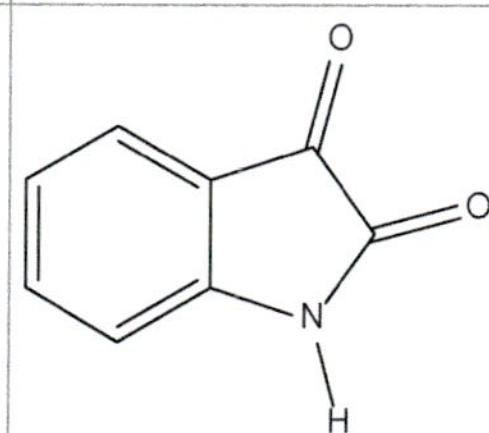	Hamburger (2002)

	Synonym(s): *Isatis bannatica* Link, *Isatis campestris* Stev. ex DC., Isatis *ciesielskii* Zapal., *Isatis japonica* Miq., *Isatis kamienskii* Zapal., *Isatis koelzii* Rech. f., *Isatis macrocarpa* B. Fedtsch. ex Nikitina, *Isatis maeotica* DC., *Isatis maritima* Rupr., *Isatis reticulata* C.A. Mey., *Isatis sibirica* Trautv., *Isatis transsilvanica* Simonk., *Isatis villarsii* Gaudin. **Common name(s)**: Glastum, Dyer's woad, Woad. **Botanical description**: It is a perennial growing up to 1 m by 0.5 m. It flowers from June to August, and the seeds ripen from August to September. The flowers are hermaphrodite (have both male and female organs) and are pollinated by insects. It can grow in very alkaline soils and in semi-shade (light woodland) or no shade. It prefers moist soil. **Medicinal use(s)**: It is used to treat prostate cancer, lung cancer, breast cancer, and skin cancer. It possesses anticancer, anti-inflammatory, antiviral, antibacterial, and anti-ulcer properties. It is used internally in the treatment of a wide range of disorders, including meningitis, encephalitis, mumps, influenza, erysipelas, and heat rash. It is also used in the treatment of fevers, pyogenic inflammation in influenza and meningitis, macula in acute infectious diseases, erysipelas, mumps, and epidemic parotitis.	Tryptanthrin	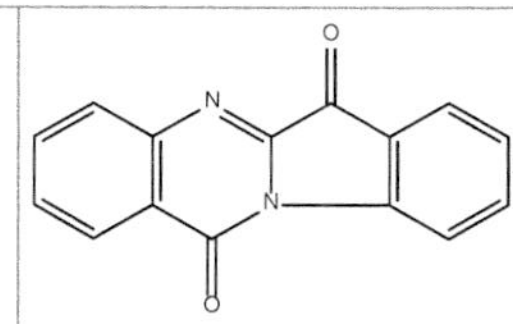	
		Deoxyvasicinone	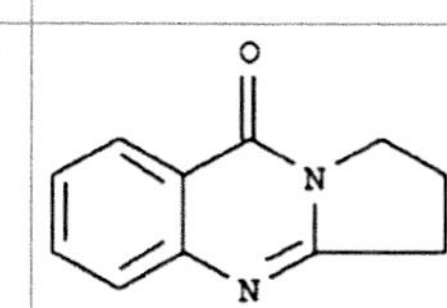	
		Isaindigotone	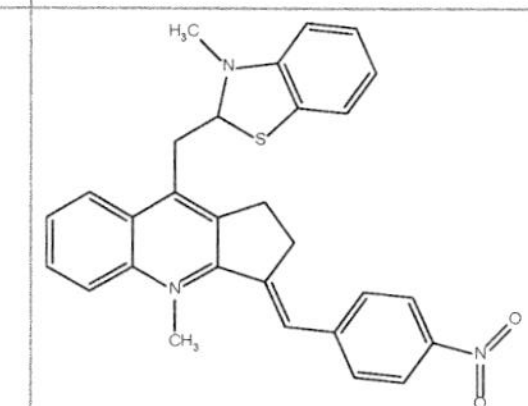	

(continued)

Table 4.43 (continued)

Family	Botanical name	Image	Plant part used	Bioactive compound(s)	Structure	Reference(s)
				Isaindigotidione	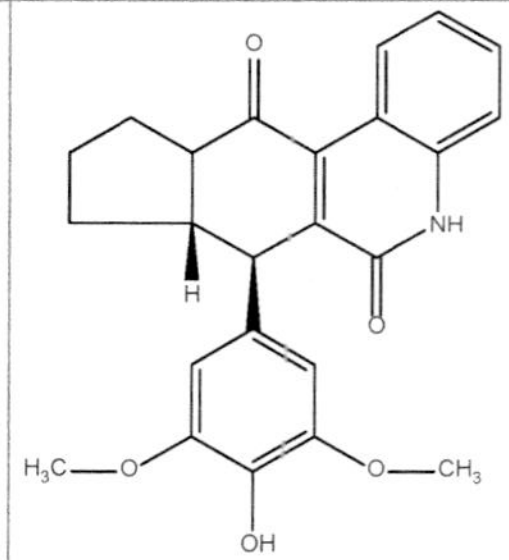	
				Indolinone 8	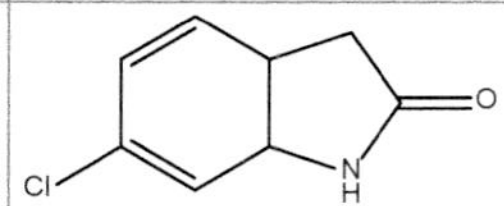	
				Erucin	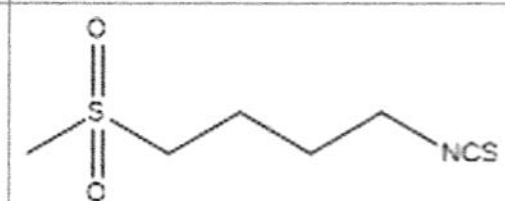	
				Deoxyvasicinone	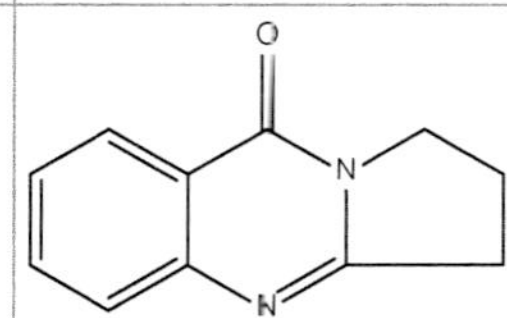	

	Brassica rapa L.		Stem	Sulforaphane β-phenylethyl isothiocyanate	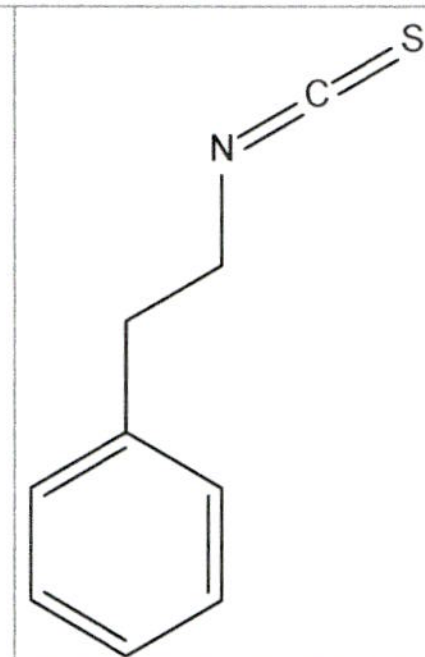	You et al. (2009) and Cartea et al. (2011)
	Synonym(s): *Barbarea derchiensis* S.S. Ying, *Brassica cibaria* Dierb., *Brassica pekinensis* (Lour.) Rupr., *Brassica rapifera* (Metzg.) Dalla Torre & Sarnth., *Brassica rapoasiatica* Sinskaya, *Brassica rapoeuropea* Sinskaya, *Napus campestris* (L.) Schimp. & Spenn., *Napus rapa* (L.) Schimp. & Spenn., *Raphanus campestris* (L.) Crantz, *Raphanus rapa* (L.) Crantz, *Sinapis dichotoma* Roxb., *Sinapis pekinensis* Lour., *Sinapis rapa* (L.) Brot., *Sinapis tuberosa* Poir. **Common name(s)**: Bok choy. **Botanical description**: It is a perennial, growing up to 0.9 m. It flowers from May to August. The flowers are hermaphrodite (have both male and female organs) and are pollinated by bees or self. The plant is self-fertile. It is suitable for light (sandy), medium (loamy), and heavy (clay) soils and prefers well-drained moist soil, and suitable pH is acid, neutral, and basic (alkaline). It can grow in semi-shade (light woodland) or no shade. **Medicinal use(s)**: It is used to treat prostate cancer. It possesses anticancer, antibacterial, anti-inflammatory, antiviral, and antimicrobial properties. The dried root (processed into granules and dissolved in hot water or tea) is called Banlangen Keli and is popular throughout China, where it is used to remove toxic heat, soothe sore throat, and treat influenza, measles, mumps, syphilis, and scarlet fever.			Erucin	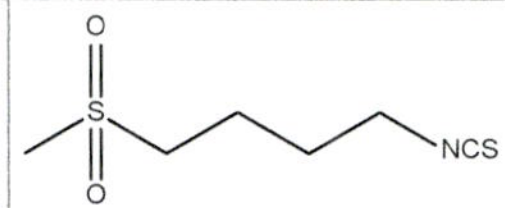	

(continued)

Table 4.43 (continued)

Family	Botanical name	Image	Plant part used	Bioactive compound(s)	Structure	Reference(s)
	Brassica oleracea L.		Stem	Sulforaphane	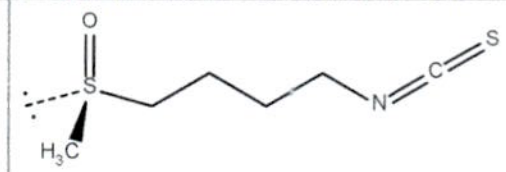	You et al. (2009)
	Synonym(s): *Brassica sabauda* (L.) Lizg., *Brassica subspontanea* Lizg., *Napus oleracea* (L.) K.F. Schimp. & Spenn. **Common name(s)**: Wild cabbage, Broccoli, Tronchuda cabbage, Brussels sprouts, Kohlrabi, Sprouting broccoli. **Botanical description**: It is native to coastal regions of the Mediterranean and Western Europe north to France and Britain. It is a perennial growing up to 1.2 m. It flowers from May to August, and the seeds ripen from July to September. The flowers are hermaphrodite (have both male and female organs) and are pollinated by bees. The plant is self-fertile. It is suitable for light (sandy), medium (loamy), and heavy (clay) soils and prefers well-drained soil, and suitable pH is acid, neutral, and basic (alkaline) soils. It can grow in semi-shade (light woodland) or no shade. It prefers moist soil. The plant can tolerate maritime exposure. **Medicinal use(s)**: It is used to treat lung, stomach, colorectal, and breast cancer. It possesses anticarcinogenic, anti-inflammatory, antibacterial, antiviral, and antimicrobial properties. The leaves are cardiotonic and stomachic. They have been used in the treatment of gout and rheumatism. The leaves can be used as a poultice to cleanse infected wounds – the midrib is removed and the leaf ironed and then placed on the affected area while still hot. The poultice should not be left for too long because it causes blisters. The seeds are anthelmintic, diuretic, laxative, and stomachic.			Erucin	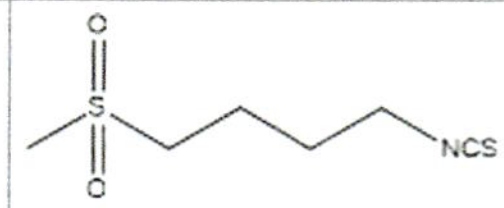	

Table 4.44 Anticancer plants of family Cyatheaceae

Family	Botanical name	Image	Plant part used	Bioactive compound(s)	Structure	Reference(s)
Cyatheaceae	*Cyathea fauriei* (Christ) Copel		Shoots	Vitexin	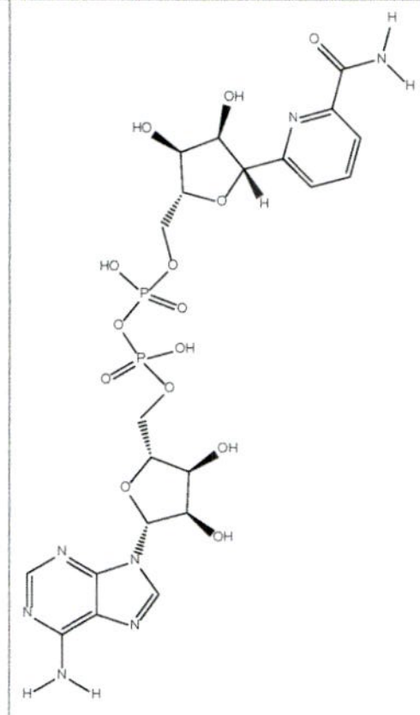	Chiang et al. (1994) and Srinivasan et al. (2014)
	Synonym(s): *Alsophila boninsimensis* Christ in Warb, *Alsophila confucii* Christ, *Alsophila decipiens* J. Scott ex Bedd., *Alsophila fauriei* Christ. **Common name(s)**: Flying spider-monkey tree fern. **Botanical description**: It is widely distributed across Asia including China, Nepal, India, Burma, Myanmar, and Japan. The trunk of this species can grow up to a height of 5 m or more. The stipes are persistent, spiny, and purplish towards the base and covered in brown shiny scales. Fronds are 1–3 m long and three-limbed. The sori, producing the spores, are large and round. **Medicinal use(s)**: It is used to treat tumors. It is used to treat breast, ovarian, lung, gastric, and prostate cancer. Scales from the fronds are used to arrest hemorrhages.					
				Paclitaxel	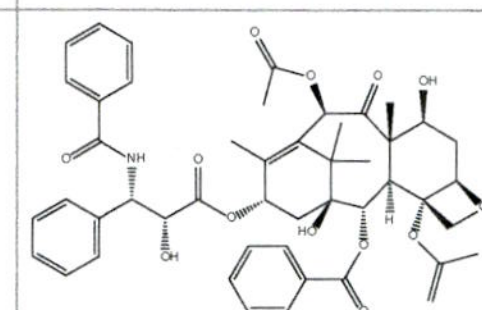	

Table 4.45 Anticancer plants of family Cyperaceae

Family	Botanical name	Image	Plant part used	Bioactive compound(s)	Structure	Reference(s)
Cyperaceae	*Bolboschoenus maritimus* (L.) Palla		Tuber	Resveratrol	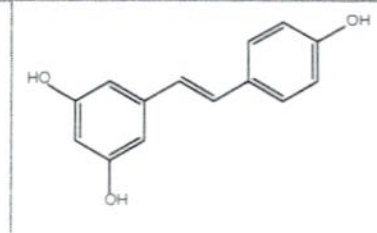	Cha (1977) and Maffei (2003a, b, c, d, e, f, g, h, i, j, k, l, m)
	Synonym(s): *Reigera maritima* (L.) Opiz, *Schoenoplectus maritimus* (L.) Lye, *Scirpus maritimus* L. **Common name(s)**: Sea club rush. **Botanical description**: It is native in seaside wetland habitats worldwide. It is widespread across most of temperate and subtropical Europe, Africa, Asia, North America, South America, and various islands. It is a rhizomatous (with horizontal stems below ground) perennial reaching 1.2 m in height. It has triangular stems that are sheathed by serrated emerald-green leaves. Each individual plant grows from a corm and then puts out a horizontal rhizome from which the next plant grows. **Medicinal use(s)**: It is astringent and diuretic. It is used in the treatment of amenorrhea, dysmenorrhea, abdominal pain or tumors for postpartum females, abdominal distension, and indigestion. It is used for treating abdominal and vaginal tumors. It possesses antiplatelet aggregation, coronary vasodilator, antileukemic, antifungal, and protein-tyrosine kinase inhibitory activities.			Viniferin	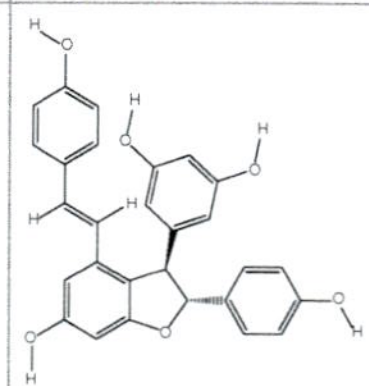	

	Kyllinga brevifolia Rottb.		Part not specified	α-Cyperone	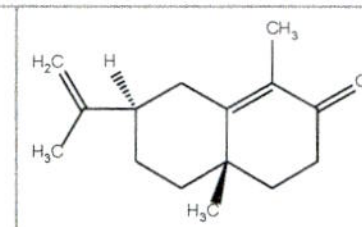	Jain (1970) and Raju et al. (2011)
	Synonym(s): *Cyperus brevifolius* (Rottb.) Hassk., *Cyperus cruciformis* (Schrad. ex Schult.) Endl., *Kyllinga aurata* Nees, *Kyllinga cruciformis* Schrad. ex Schult., *Kyllinga elongata* Kunth, *Kyllinga fuscata* Miq., *Kyllinga gracilis* Kunth, *Kyllinga hohenackeri* Hochst. ex Steud., *Kyllinga honolulu* Steud. ex Jard., *Kyllinga intermedia* R.Br., *Kyllinga intricata* Cherm., *Kyllinga laxa* Schrad. ex Nees, *Kyllinga longiculmis* Miq., *Kyllinga nivea* Pers., *Kyllinga oligostachya* Boeckeler, *Kyllinga pumilio* Steud., *Kyllinga sojauxii* Boeckeler, *Kyllinga sororia* Kunth, *Kyllinga tenuis* Baldwin, *Kyllinga tenuissima* Steud., *Kyllinga tricephala* Salisb., *Mariscus kyllingioides* Steud., *Schoenus capitatus* Crantz. **Common name(s)**: Shortleaf spikesedge, Green kyllinga, Perennial greenhead sedge, and Kyllinga weed. **Botanical description**: It is native to tropical areas in the Americas, but it can be found in warm regions around the world where it is an introduced species. This is a rhizomatous perennial herb growing one to several erect stems to heights up to about half a meter, often much shorter. It produces tiny inflorescences of a few spikelets each which in total are less than a centimeter long. It is sometimes a weed in wet areas such as cultivated land and irrigation ditches. **Medicinal use(s)**: The leaves, stems, and rhizomes are used as analgesic, anti-inflammatory, antitumors, antimalarial, decongestant, diuretic, febrifuge, and sudorific.			β-Selinene	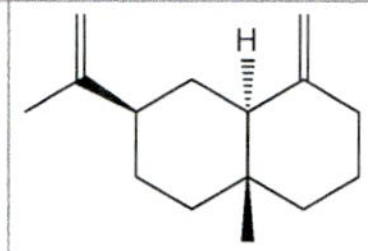	

(continued)

Table 4.45 (continued)

Family	Botanical name	Image	Plant part used	Bioactive compound(s)	Structure	Reference(s)
	Scirpoides holoschoenus (L.) Soják		Inflorescence	Vincristine	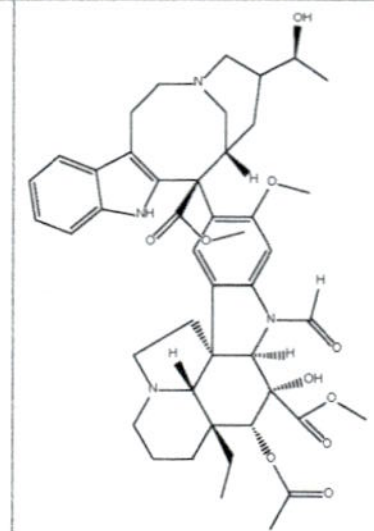	Gonzalez-Tejero et al. (1995) and Bhanot et al. (2011)
	Synonym(s): *Holoschoenus australis* (L.) Rchb., *Holoschoenus diphyllus* Montandon, *Holoschoenus exserens* Rchb., *Holoschoenus filiformis* Rchb., *Holoschoenus globuliferus* (Boiss.) Grossh., *Holoschoenus laxus* Opiz, *Holoschoenus linnaei* Rchb., *Holoschoenus panormitanus* (Parl.) Nyman, *Holoschoenus romanus* (L.) Fritsch, *Holoschoenus tuberculatus* Rchb., *Holoschoenus vulgaris* Link, *Isolepis australis* (L.) G.Don, *Isolepis exserens* (Rchb.) Bluff, Nees & Schauer, *Isolepis fontana* Steud., *Isolepis holoschoenus* (L.) Roem. & Schult., *Isolepis linnaei* (Rchb.) Bluff, Nees & Schauer, *Isolepis paniculata* Gray, *Isolepis panormitana* Parl., *Isolepis poiretii* Roem. & Schult., *Isolepis romana* (L.) Roem. & Schult. Ex G. Don, *Isolepis sphaerophora* Ehrenb. ex Boeckeler, *Scirpoide sromana* (L.) Soják, *Scirpus australis* L., *Scirpus ellychniarius* Molina, *Scirpus holoschoenus* L., *Scirpus longibracteatus* Salzm. ex Rchb., *Scirpus panormitanus* (Parl.) Parl., *Scirpus parlatoris* Biv. ex Parl., *Scirpus polymorphus* Dulac, *Scirpus romanus* L., *Seidlia atrovirens* Opiz. **Common name(s)**: Bolls of reed. **Botanical description**: It is native to Albania, Algeria, Austria, Belarus, Belgium, and Bosnia and Herzegovina. The plant body is devoid of leaves; these have been reduced to pods located at the base, which leave a fibrous reticulum as they age. The cyperaceous lack petals or sepals, but the stamens and styles, which are separated in most species, are surrounded by a small scale called gluma, important for identification. **Medicinal use(s)**: It is used to treat liver cancer and various types of tumors and warts. It possesses anticancer, antioxidant, antimicrobial, antibacterial, and anti-inflammatory properties. It is also used as astringent and diuretic.			Vinblastine	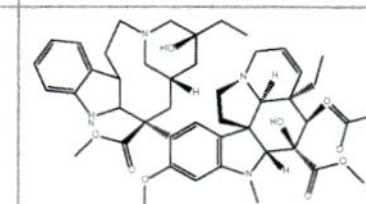	

	Bolboschoenus yagara (Ohwi) Y.C.Yang&M. Zhan		Rhizome	Transresveratrol	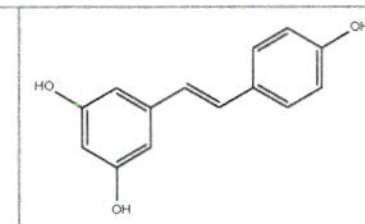	Yang et al. (2010a, b)
	Synonym(s): *Bolboschoenus desoulavii* (Drobow) A.E.Kozhevn., *Scirpus yagara* Ohwi. **Common name(s)**: Ohwi. **Botanical description**: It is native to North America – Newfoundland to Alaska, south to South Carolina, Pennsylvania, Michigan, and Idaho. It is a perennial growing up to 0.6 m. The seeds ripen from July to August. The species is hermaphrodite (has both male and female organs) and are pollinated by wind. It can grow in semi-shade (light woodland) or no shade. It prefers moist or wet soil and can grow in water. **Medicinal use(s)**: It is used to treat ELISA. It possesses anti-inflammatory, anticancer, anti-obesity, and anti-hepatitis properties.			Scirpusin A	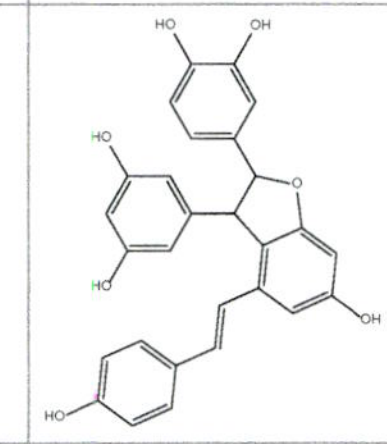	
				Scirpusin B	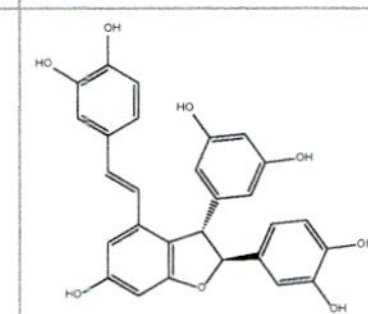	
				P-Hydroxycinnamic acid	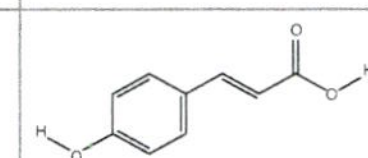	
				Hydroxycinnamic acid	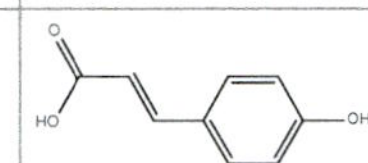	

Table 4.46 Anticancer plants of family Davalliaceae

Family	Botanical name	Image	Plant part used	Bioactive compound(s)	Structure	Reference(s)
Davalliaceae	*Davallia mariesii* T.Moore ex Baker		Protoplasts	Davallioside A	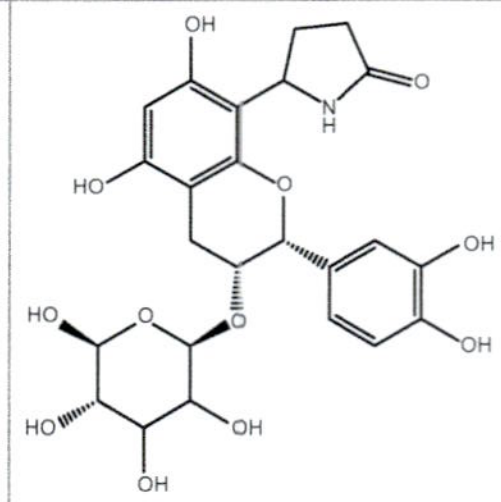	Cui et al. (1990) and Chang et al. (2007)
	Synonym(s): *Davallia stenolepis* Hayata. **Common name(s)**: Squirrel's foot fern, Ball fern, Japan hare's foot fern. **Botanical description**: It is native to East Asia and also found in South Korea. It is an evergreen deciduous fern with finely divided ovate or triangular fronds arising singly from a creeping, conspicuously scaly rhizome. It requires neutral, moist but well-drained loamy soil. It grows up to 0.5 m in height and spread 0.5–1 m. **Medicinal use(s)**: It is used for stomach cancer. The tonic made by this plant is used for the treatment of osteoporosis, arthralgia, and arthritis. It is used to promote healthy pregnancy.			Lutein	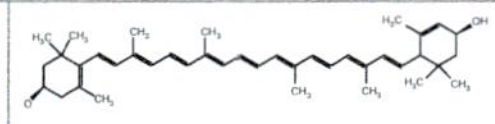	

	Davallia solida (Forst.) Sw.		Glucosylxanthone	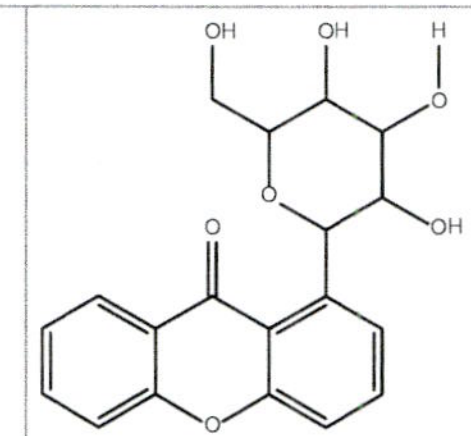	Chen et al. (2008a, b)
	Synonym(s): No synonyms are recorded for this name. **Common name(s)**: Rabbit foot fern, Portuguese lace. **Botanical description**: It is native to the Philippines and also found in the Malay Peninsula to Polynesia. It is an epiphytic, dimorphic fern with stem of 1 cm or more in diameter, sterile leaves are pinnate, subopposite, grow up to 20 cm long. **Medicinal use(s)**: It is used in Chinese medicine for the treatment of ache, inflammation, cancer and bone injuries, asthma, and sore throat. It is also used for fish poisoning.		Mangiferin	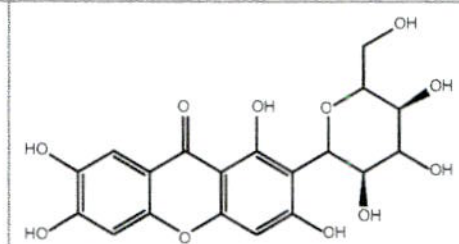	
			4β-Carboxymethyl-(-)-epicatechin	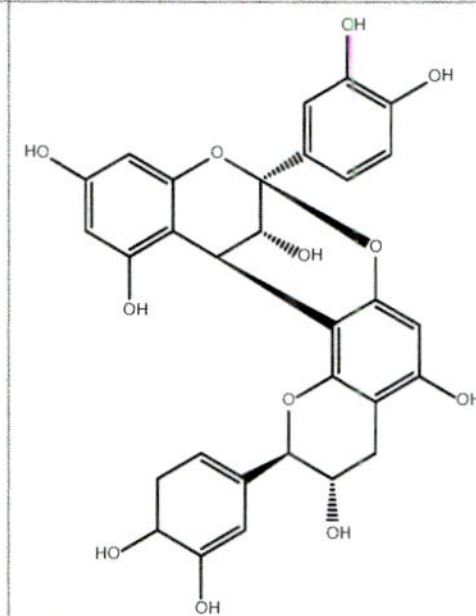	

Table 4.47 Anticancer plants of family Dichapetalaceae

Family	Botanical name	Image	Plant part used	Bioactive compound(s)	Structure	Reference(s)
Dichapetalaceae	*Dichapetalum toxicarium* (G. Don) Baill.		Bark	ω-Fluoro-oleic acid	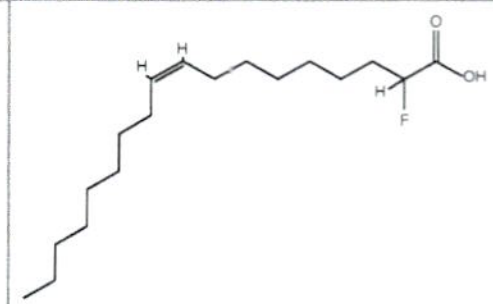	Ward et al. (1964), Vasileva (1969), and Maffei (2003a, b, c, d, e, f, g, h, i, j, k, l, m)
	Synonym(s): *Chailletia toxicaria* G.Don, *Chailletia erecta* G.Don. **Common name(s)**: Broke back. **Botanical description**: It is a native plant to Guinea. It is a shrub tree about 16 cm tall and grows in dry sandy areas in acidic soil. **Medicinal use(s)**: It is also used to treat tumors and used for HIV anti-infective therapy. It possesses anticancer, antibacterial, and anti-inflammatory properties.					

Table 4.48 Anticancer plants of family Dilleniaceae

Family	Botanical name	Image	Plant part used	Bioactive compound(s)	Structure	Reference(s)
Dilleniaceae	*Dillenia indica* L.		Flower	Betulinic acid	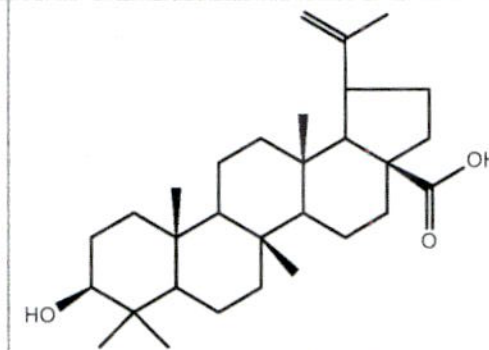	Kumar et al. (2003) and Karan et al (2018)
	Synonym(s): *Dillenia elongata* Miq. **Common name(s)**: Elephant apple, Indian catmon, Chulta. **Botanical description**: It is native to India; it is an evergreen large shrub or small- to medium-sized tree growing up to 15 m tall. The leaves are 15–36 cm long; the flowers are large, up to 5 in. across, with five white petals and numerous yellow stamens. The fruits are pulpy and used in cuisine in berries, jam, and jellies. **Medicinal use(s)**: The fruit is a tonic and laxative. It is used in the treatment of abdominal disorder and mixed with sugar to stop the cough.			Dihydro-isorhamnetin and dillenetin	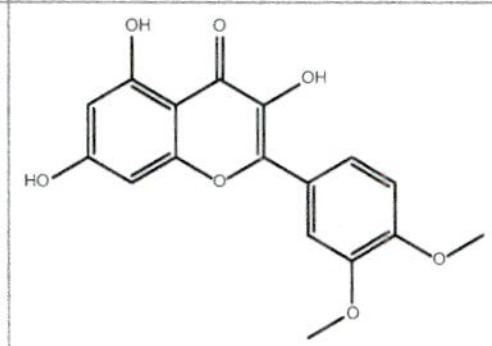	
	Dillenia suffruticosa (Griff.) Martelli		Fruit	Lupeol	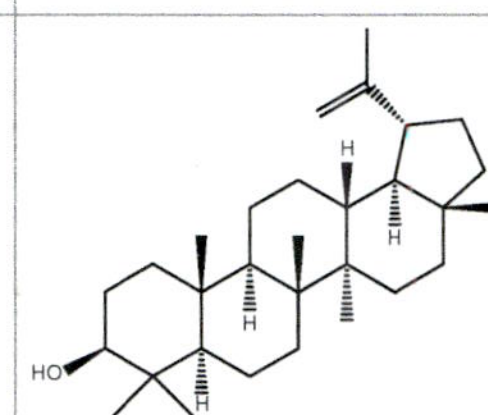	Ahmad and Holdsworth (1995), Yazan et al. (2015), and Tor et al. (2014)

(continued)

Table 4.48 (continued)

Family	Botanical name	Image	Plant part used	Bioactive compound(s)	Structure	Reference(s)
	Synonym(s): *Dillenia burbidgei* (Hook.f.) Martelli, *Wormia burbidgei* Hook.f., *Wormia suffruticosa* Griff. **Common name(s)**: Simpoh ayer. **Botanical description**: It is native plant of Southeast Asia – Malaysia and Indonesia. It is a long shrub or tree growing up to 10 m tall. It grows in a wide range of soils from sandy to clayey and from wet to well-drained. The plant flowers continuously and each flower opens only for a day. **Medicinal use(s)**: It is used to treat cancerous growths. The leaves of the plant and roots are used to treat inflammation, itch, and stomachache and for recovery after delivery.			Kaempferol	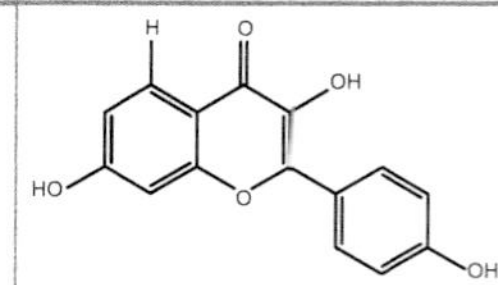	
				Kaempferide	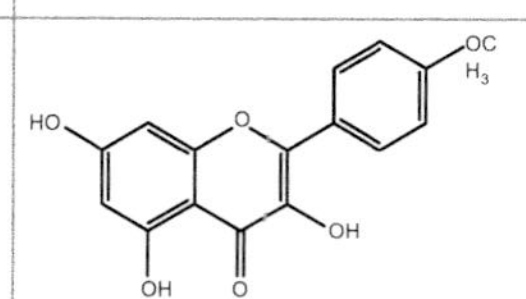	
				Protocatechuic acid	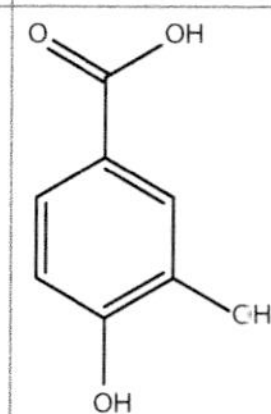	

		Gallic acid	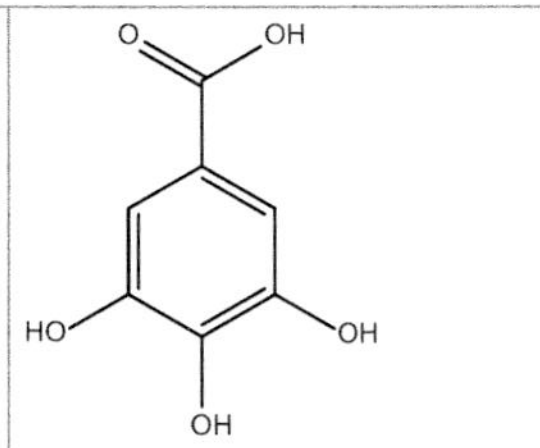	
		3-Epimaslinic acid	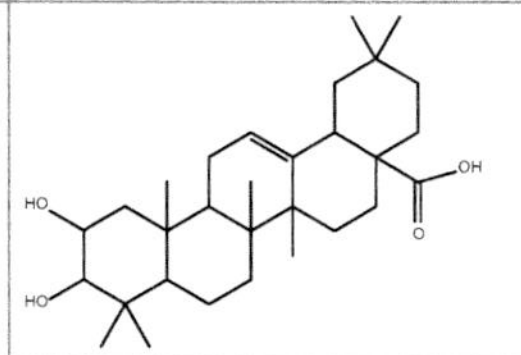	
		β-Sitosterol-3-O-β-D-glucopyranoside	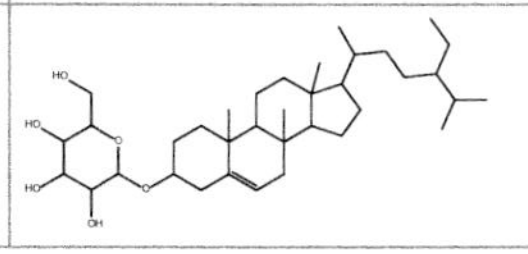	

Table 4.49 Anticancer plants of family Dioscoreaceae

Family	Botanical name	Image	Plant part used	Bioactive compound(s)	Structure	Reference(s)
Dioscoreaceae	*Dioscorea collettii* var. *hypoglauca* (Palib.) S.J.Pei & C.T.Ting		Rhizome	Furostanol	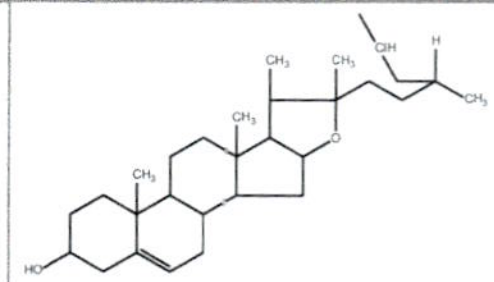	Hu and Yao (2003) and She et al. (2013)
	Synonym(s): *Dioscorea hypoglauca* Palib., *Dioscorea izuensis* Akahori, *Dioscorea kaoi* T.S. Liu & T.C. Huang, *Dioscorea morsei* Prain & Burkill, *Dioscorea undulate* R. Knuth. **Common name(s)**: Wild yam, Izu-dokoro. **Botanical description**: It is the flowering plant native throughout the tropical and warm temperature region of the world. It is the tuberous herbaceous perennial plant which grows up to 2–12 m tall. The leaves are spirally arranged, mostly broad heart-shaped. The fruit is a capsule. **Medicinal use(s)**: It is used to treat cancers of the breast, ovary, uterus, and prostate. It is widely used for the treatment of menstrual cramps and rheumatoid arthritis, stomach cramps, and pain due to gallstone.			Spirostanol	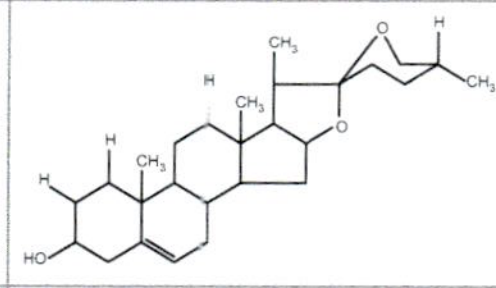	
				Methyl protoneogracillin	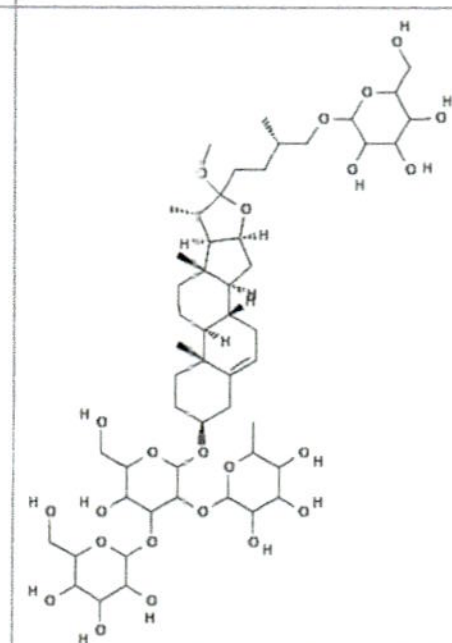	

				Gracillin	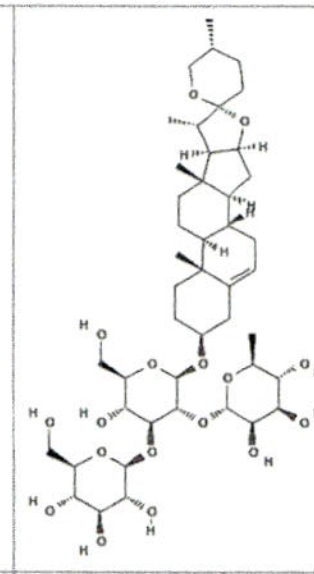	
	Dioscorea composita Hemsl.		Rhizome Root	Furostanol	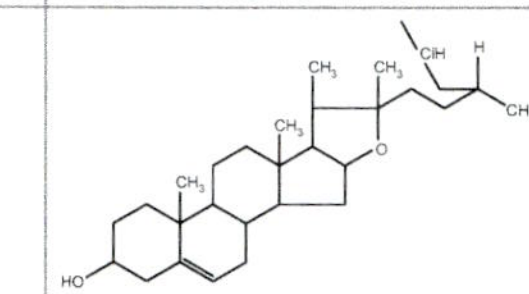	Hu et al. (2007) and Upadhyay et al. (2018)
	Synonym(s): *Dioscorea tepinapensis* Uline ex R.Knuth. **Common name(s)**: Barbasco. **Botanical description**: It is native to Mexico. It is an herbaceous perennial, tropical, climbing plant producing annual stems with tuberous root. **Medicinal use(s)**: It is used to synthesize hormones such as progesterone and cortisone. It is used to develop the combined oral contraceptive pills. It is also used to treat arthritis, diabetes, and neuropathy.			Spirostanol	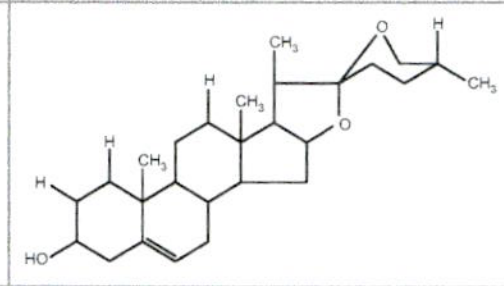	
				Diosgenin	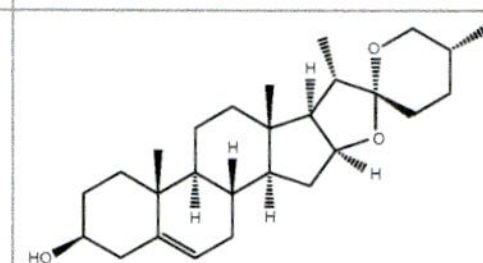	

(continued)

Table 4.49 (continued)

Family	Botanical name	Image	Plant part used	Bioactive compound(s)	Structure	Reference(s)
	Dioscorea japonica Thunb.		Tuber	Furostanol	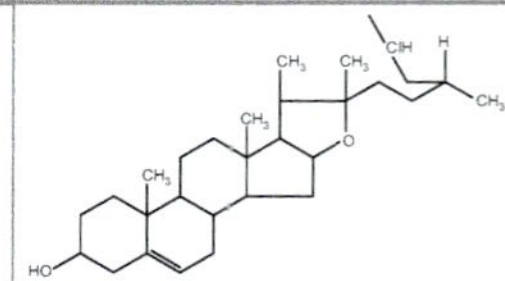	Hu et al. (2007) and Chiang et al. (2018)
	Synonym(s): *Dioscorea belophylloides* Prain & Burkill, *Dioscorea fauriei* R. Knuth, *Dioscorea goeringiana* Kunth, *Dioscorea kiangsiensis* R. Knuth, *Dioscorea neglecta* R.Knuth, *Dioscorea pseudojaponica* Hayata. **Common name(s)**: Cinnamon yam. **Botanical description**: It is native to Taiwan. It is a perennial climber shrub. The flowers are dioecious. The plant is not self-fertile. It is suitable for sandy and loamy soil. **Medicinal use(s)**: It is widely used in the treatment of diarrhea, enteritis, enuresis, and spermatorrhea. The root consists of diosgenin used to manufacture progesterone and steroid drugs.			Spirostanol	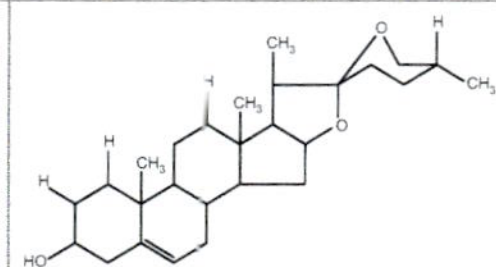	

	Dioscorea villosa L.		Rhizome Roots	Spirostanol glycoside	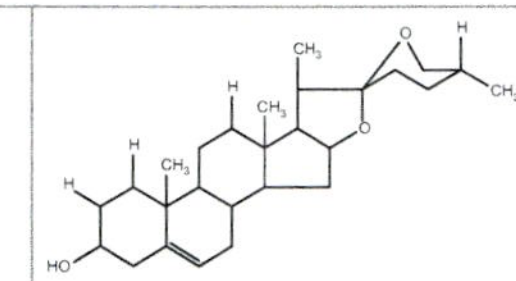	Hu et al. (2007) and Aumsuwan et al. (2016)
	Synonym(s): *Dioscorea cliffortiana* Lam., *Dioscorea glauca* Muhl. ex L.C.Beck, *Dioscorea hexaphylla* Raf., *Dioscorea hirticaulis* Bartlett, *Dioscorea lloydiana* E.H.L. Krause, *Dioscorea longifolia* Raf., *Dioscorea megaptera* Raf., *Dioscorea paniculata* Michx., *Dioscorea pruinosa* Kunth, *Dioscorea quaternata* Walter, *Dioscorea quinata* Walter, *Dioscorea sativa* L., *Dioscorea waltheri* Desf., *Merione villosa* (L.) Salisb. **Common name(s)**: Dios, Wild yam root, Colicroot, Aluka. **Botanical description**: It is native to North and Central America. It is found in tropical and subtropical areas around the world. It is a perennial, twining vine which grows in damp woodlands and thickets. **Medicinal use(s)**: The roots and rhizomes are dried and used for medicinal purposes. It is used as natural progesterone. It has been used to lower blood pressure and cholesterol level in blood. Root is used to treat disorders related to the spleen, kidneys, lungs, and stomach. It is also used to treat irritable bowel syndrome and disorders related to digestion.			Dioscin	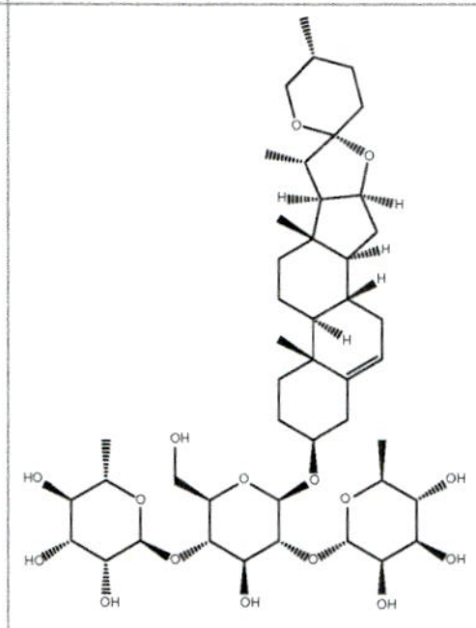	

Table 4.50 Anticancer plants of family Dipsacaceae

Family	Botanical name	Image	Plant part used	Bioactive compound(s)	Structure	Reference(s)
Dipsacaceae	*Dipsacus asper* Wall. ex C.B. Clarke		Root	Akebia saponin D (ASD)	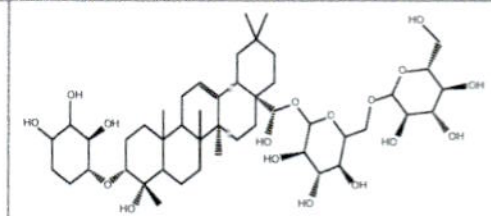	Jeong et al. (1999) and Mazzio and Soliman (2009)
	Synonym(s): No synonyms are recorded for this name. **Common name(s)**: Sichuan teasel. **Botanical description**: It is native to East Asia. It is a biennial flowering plant that grows in tropical climate up to 1.8 m. It is best used for treating liver and kidney problems. **Medicinal use(s)**: It is used to treat breast, colon, lung, and head and neck cancer. It is having anticancer, anti-arthritic, analgesic, and diuretic properties. It is available in the form of capsules, powder, pills, and teas. It is widely used to prevent osteoporosis and cure ruptured tendons and damaged skin like deep cut.					

Table 4.51 Anticancer plants of family Dracaenaceae

Family	Botanical name	Image	Plant part used	Bioactive compound(s)	Structure	Reference(s)
Dracaenaceae	*Dracaena draco* L.		Sap	Draconin A	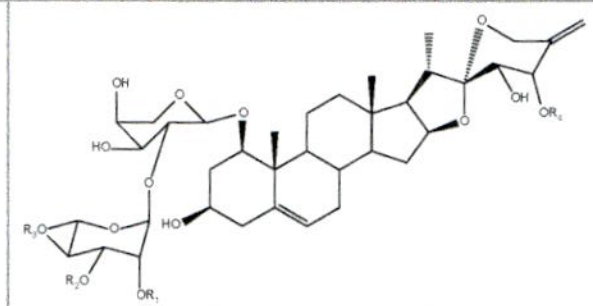 R_1, R_2, R_3 = Ac, R_4 = H	Gonzalez et al. (2003) and Valente et al (2012)
	Synonym(s): *Asparagus draco* L. **Common name(s)**: Dragon tree. **Botanical description**: It is native to the Canary Islands, Cape Verde, and Madeira. It is a monocot with a treelike growth habit. When young it has a single stem. It grows about 10–15 years of age. It grows slowly to a height of 4 ft. **Medicinal use(s)**: The resin, called "dragon's blood," is used to enhance sexual desire and to treat cancer. It is also used as a fumigating agent. Dragon's blood is used externally as a wash to promote healing of wounds and to stop bleeding. It is used internally for chest pains, postpartum bleeding, internal traumas, and menstrual irregularities.			Draconin B	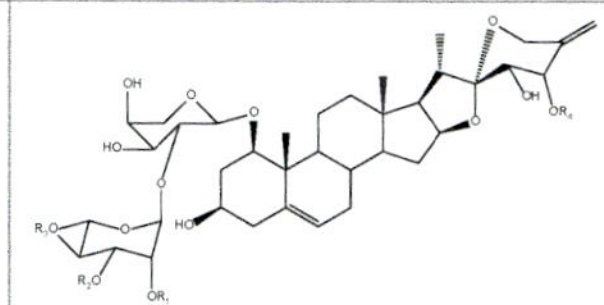 R_1, R_2= Ac, R_3, R_4= H	
				Dracohordin	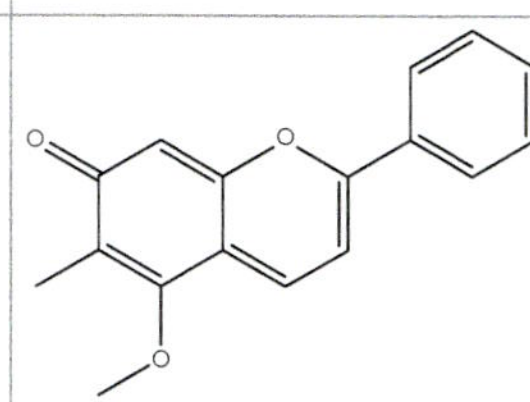	
				3,4-Dihydro-5-methoxy-6-methyl-2-phenyl-2H-1-benzopyran-7-ol	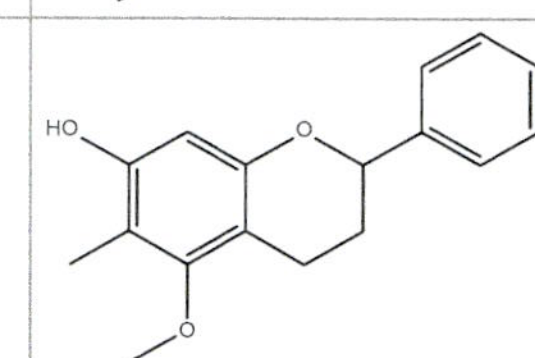	

(continued)

Table 4.51 (continued)

Family	Botanical name	Image	Plant part used	Bioactive compound(s)	Structure	Reference(s)
	Synonym(s): *Asparagus draco* L. **Common name(s)**: Dragon tree. **Botanical description**: It is native to the Canary Islands, Cape Verde, and Madeira. It is a monocot with a treelike growth habit. When young it has a single stem. It grows about 10–15 years of age. It grows slowly to a height of 4 ft. **Medicinal use(s)**: The resin, called "dragon's blood," is used to enhance sexual desire and to treat cancer. It is also used as a fumigating agent. Dragon's blood is used externally as a wash to promote healing of wounds and to stop bleeding. It is used internally for chest pains, postpartum bleeding, internal traumas, and menstrual irregularities.			Trendione	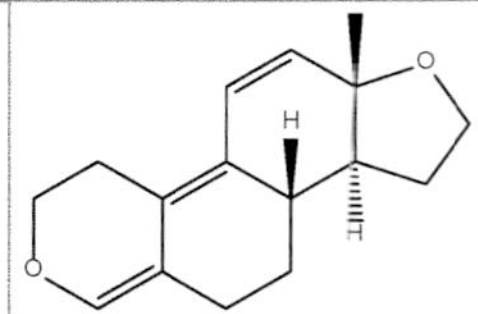	
				Linolenic acid	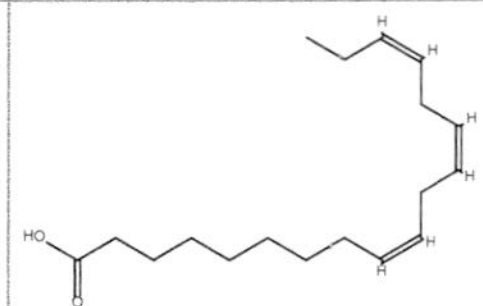	

Table 4.52 Anticancer plants of family Droseraceae

Family	Botanical name	Image	Plant part used	Bioactive compound(s)	Structure	Reference(s)
Droseraceae	*Drosera anglica* Huds.		Leaf	Kaempferol	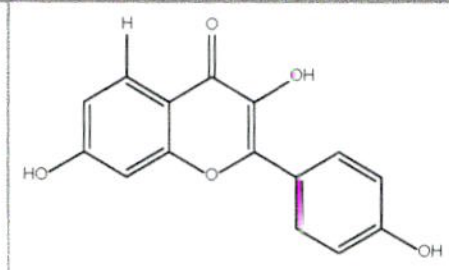	Culpeper (1950) and Maffei (2003a, b, c, d, e, f, g, h, i, j, k, l, m)
	Synonym(s): *Adenopa anglica* (Huds.) Raf., *Drosera kihlmanii* Ikonn., *Drosera linglica* Kusak. Ex R. Gauthier & Gervais, *Drosera longifolia* L., *Drosera woodii* R. Gauthier & Gervais, *Rorella longifolia* (L.) All. **Common name(s)**: Sundew, Kandulessa. **Botanical description**: It is native to India and also found in Africa, Asia, Australia, and Europe. It is a perennial plant that grows in a temperate climate and grows up to 15 cm. **Medicinal use(s)**: It is used to treat cancer. It is a medicinal herb which is effective as antibacterial, antibiotic, antiseptic, and antimicrobial. It is used to treat tuberculosis, whooping cough, windpipe infections, and ulcers.			7-Methyljuglone	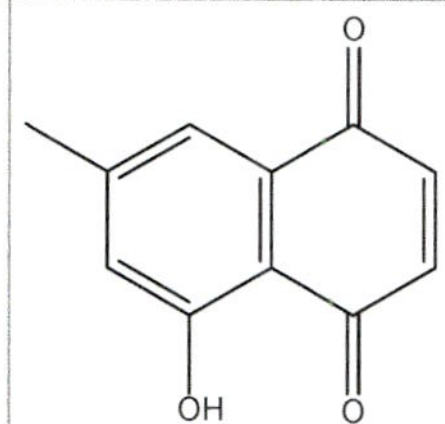	

Table 4.53 Anticancer plants of family Ericaceae

Family	Botanical name	Image	Plant part used	Bioactive compound(s)	Structure	Reference(s)
Ericaceae	*Chimaphila umbellata*		Entire plant	Beta-D-glucopyranoside (arbutin)	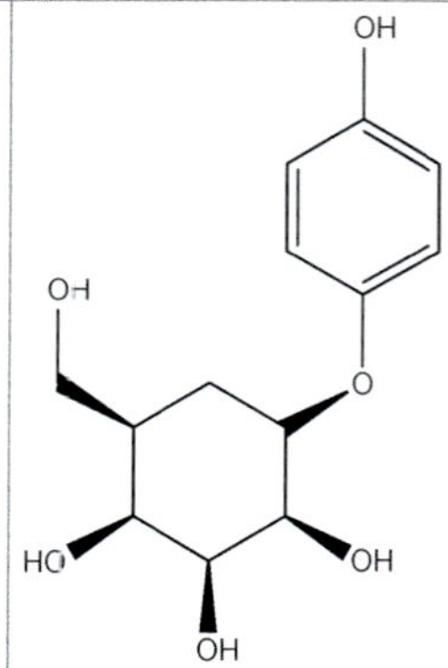	Mohammad (2006), Desai et al. (2008), and Madhuri and Pandey (2009)
	Synonym(s): *Chimaphila acuminata* (Lange) Rydb., *Chimaphila acuta* Rydb., *Chimaphila astyla* Maxim., *Chimaphila corymbosa* Pursh, *Chimaphila costaricens* Andres, *Chimaphila cymosa* J.Presl & C.Presl. **Common name(s):** Pipsissewa, Umbellate wintergreen, Prince's pine. **Botanical description:** *Chimaphila umbellata* is a small perennial flowering plant found in dry woodlands or sandy soils. It is native throughout the cool temperate Northern Hemisphere. It grows 10–35 cm tall and has evergreen shiny, bright green, toothed leaves arranged in opposite pairs or whorls of 3–4 along the stem. Leaves have a shallowly toothed margin, where the teeth have fine hairs at their ends. The flowers are white or pink, produced in a small umbel of 4–8 together. **Medicinal use(s):** Some plateau Indian tribes used a boil of prince's pine to treat tuberculosis. It can be used to treat breast and prostate cancer.					

	Vaccinium bracteatum C.P. Thunb.		Entire plant	Chlorogenic acid	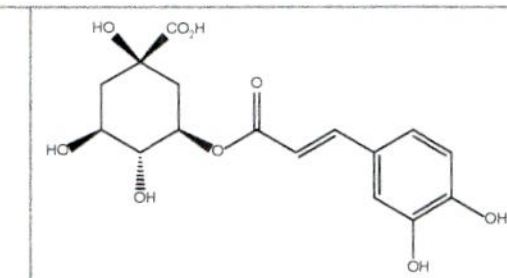	Duke and Ayensu (1985a, b) and Oh et al. (2018)
	Synonym(s): *Vaccinium bracteatum* var. *bracteatum*. **Common name(s):** Sea bilberry. **Botanical description**: *Vaccinium bracteatum* is an evergreen shrub growing up to 1 m (3 ft. 3 in.) native to East Asia, China, Japan, and Korea. It is hardy to zone (United Kingdom). It is in leaf all year. The species is hermaphrodite (has both male and female organs) and is pollinated by insects. Suitable for light (sandy) and medium (loamy) soils and prefers well-drained soil. Suitable pH is acid soils and can grow in very acid soils. It can grow in semi-shade (light woodland) or no shade. It prefers moist soil. **Medicinal use(s):** The plant has been used to treat snakebites, syphilis, and septic wounds. It may be also used to cure skin and blood cancer.			Orientin	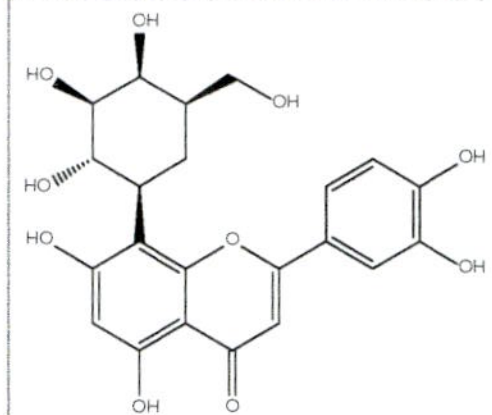	
				Isoorientin	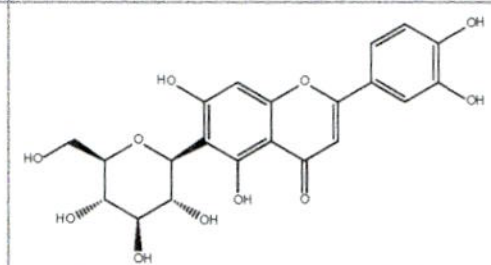	

(continued)

Table 4.53 (continued)

Family	Botanical name	Image	Plant part used	Bioactive compound(s)	Structure	Reference(s)
	Vaccinium dunalianum Wigh		Bud	p-Hydroxyphenyl-6-O-trans-caffeoyl-β-D-glucoside	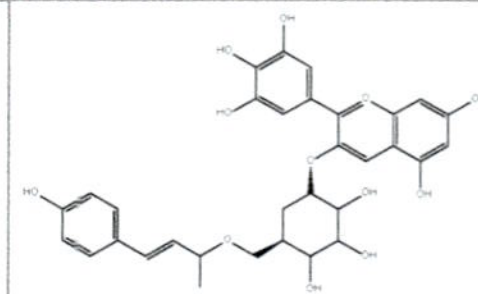	Cheng et al. (2009)
	Synonym(s): *Vaccinium dunalianum* var. *calycinum* Dop, *Vaccinium dunalianum* var. *caudatifolium* (Hayata) H.L. Li, *Vaccinium dunalianum* var. *dunalianum*, *Vaccinium dunalianum* var. *megaphyllum* Sleumer. **Common name(s):** Blueberry. **Botanical description:** *Vaccinium dunalianum* a large growing blueberry native to the temperate rainforests in the mountains of the Sino-Himalayas. Beautiful glossy and reticulated evergreen foliage. Typical blueberry flowers in clusters. **Medicinal use(s):** The plant has been used to treat tonsillitis, periodontitis, viral infections, furunculosis, and nasopharyngeal catarrh. It is also used to treat intestinal cancer.			Caffeic acid	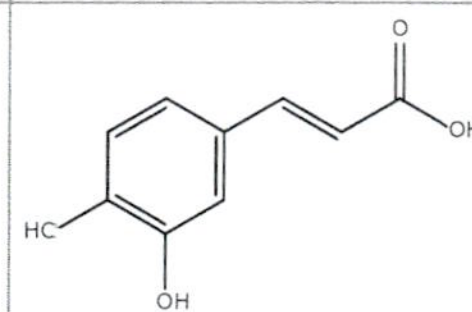	
				Phydroxyl phenyl-β-D-glucoside	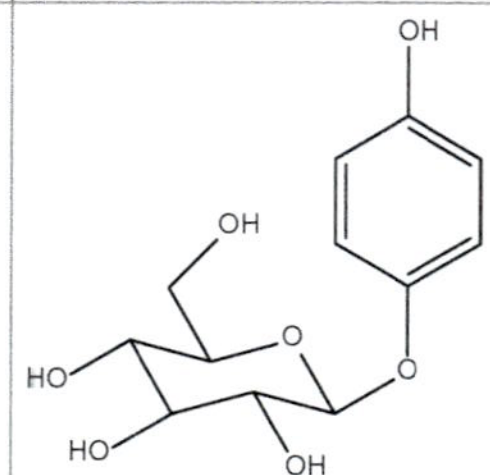	
				Phydroxy benzoic acid	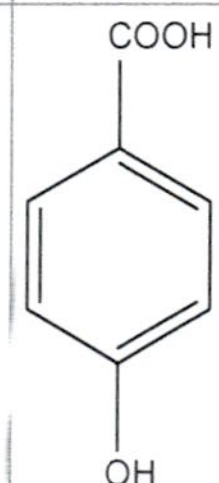	

	Vaccinium macrocarpon Aiton		Flower	cis-3-O-p-hydroxycinnamoyl ursolic acid	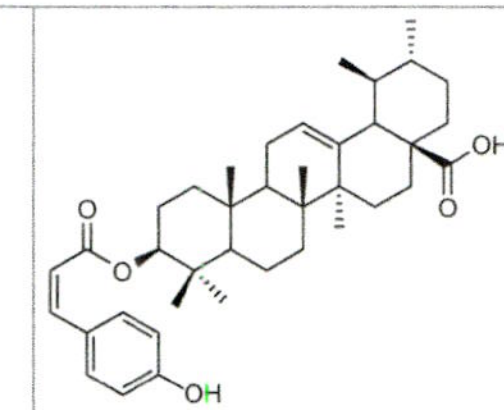	Murphy et al. (2003)
				Phenylboronic acid	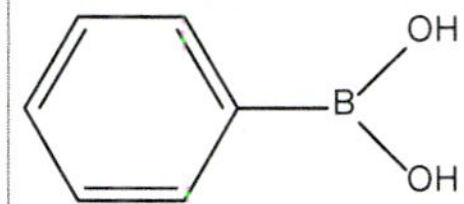	
	Synonym(s): *Oxycoca macrocarpa* (Aiton) Raf., *Oxycoccus macrocarpus* (Aiton) Pers., *Oxycoccus palustris* var. *macrocarpos (Aiton) Pers.*, *Schollera macrocarpa* (Aiton) *Steud.*, *Schollera macrocarpos* (Aiton) Britton. **Common name(s):** Bearberry. **Botanical description:** *Vaccinium macrocarpon* is a North American species of cranberry of the subgenus *Oxycoccus* and genus *Vaccinium*. It is native to Central and Eastern Canada (Ontario to Newfoundland) and the Northeastern and North Central United States (Northeast, Great Lakes Region, and Appalachians as far south as North Carolina and Tennessee). It is also naturalized in parts of Europe and scattered locations in North America along Western Canada (British Columbia) and the Western United States (West Coast). *Vaccinium macrocarpon* is a shrub, often ascending (trailing along the surface of the ground for some distance but then curving upwards). It produces white or pink flowers followed by sour-tasting red or pink berries 9–14 mm (0.35–0.55 in.) across. The species is grown commercially as a cash crop for its edible berries. Many of these are grown in artificial ponds called cranberry bogs. There is some evidence suggesting that the berries or their juice is useful in treating certain urinary tract infections. **Medicinal use(s):** *Vaccinium macrocarpon* play a major role in treating gastrointestinal and brain cancer, heart disease, and vascular and neurodegenerative diseases.			Ursolic acid	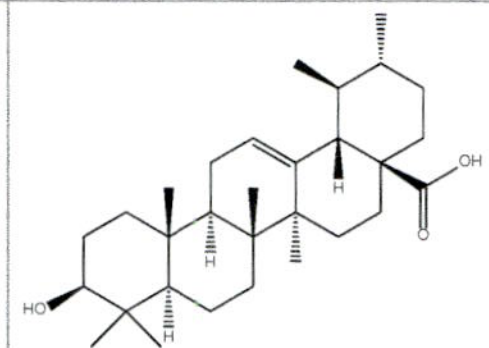	

(continued)

Table 4.53 (continued)

Family	Botanical name	Image	Plant part used	Bioactive compound(s)	Structure	Reference(s)
	Vaccinium myrtillus L.		Bark and fruits	Delphinidin	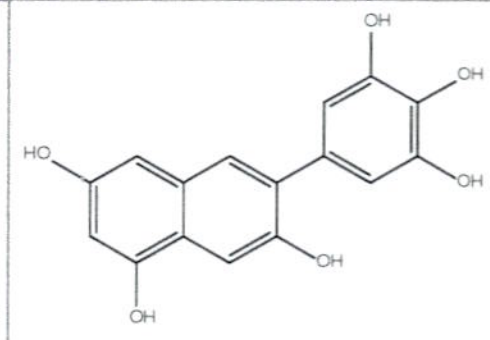	Katsube et al. (2003)
	Synonym(s): *Vaccinium myrtillus* L., *Vaccinium myrtillus* var. *microphyllum* Hook., *Vaccinium myrtillus* var. *oreophilum* (Rydb.) Dorn. **Common name(s):** Dried bilberry, Wimberry, Whortleberry. **Botanical description:** *Vaccinium myrtillus* is found natively in Europe, Northern Asia, Greenland, Iceland, Western Canada, and the Western United States. It occurs in the wild on heathlands and acidic soils. Its berry has been long consumed in the Old World. It is related to the widely cultivated North American blueberry. The flowers are borne singly in leaf axils on 2–3 mm long pedicels. The corolla is pink and shaped like an urn. The leaves are finely toothed and prominently veined on the lower surface. **Medicinal use(s):** The fruit is a rich source of anthocyanins and polyphenols and has a strong antioxidant activity. It is of potential value in the prevention and treatment of many diseases related to aging, including blood, pancreas, and lung cancer.			Malvidin	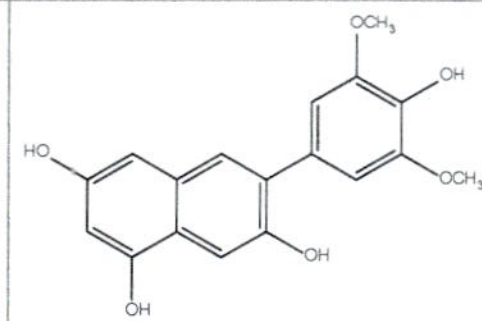	
	Vaccinium stamineum L.		Flower	Resveratrol	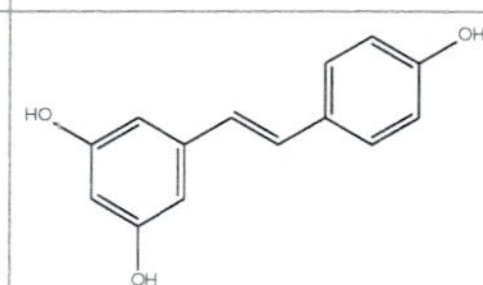	Wang and Ballington et al. (2007)

	Synonym(s): *Vaccinium stamineum* L., *Vaccinium stamineum* var. *affine* (Ashe) Sleumer, *Vaccinium stamineum* var. *austromontanum* (Ashe) Sleumer. **Common name(s):** Deerberry, Tall deerberry, Squaw huckleberry, Highbush huckleberry, Buckberry, Southern gooseberry. **Botanical description:** *Vaccinium stamineum* is a species of flowering plant in the heath family. It is native to North America, including Ontario, the Eastern and Central United States, and parts of Mexico. It is most common in the Southeastern United States. This species is quite variable in morphology [6]. It is a shrub usually growing up to 1.5 m (60 in. or 5 ft.) tall but reaching up to 3 m (10 ft.) at times. It has multiple twisted trunks covered in peeling reddish bark and is highly branched, tapering into thin twigs, some just a millimeter wide. It is deciduous, with alternately arranged leaves. The thin leaf blades are yellow-green, sometimes hairy or waxy in texture, especially on the undersides, and oval in shape with pointed tips and smooth edges. **Medicinal use(s):** *Vaccinium stamineum* is used in the prevention and treatment of common cold, influenza, and respiratory tract infections. The plant has been used to treat lung and blood cancer.	Cyanidin	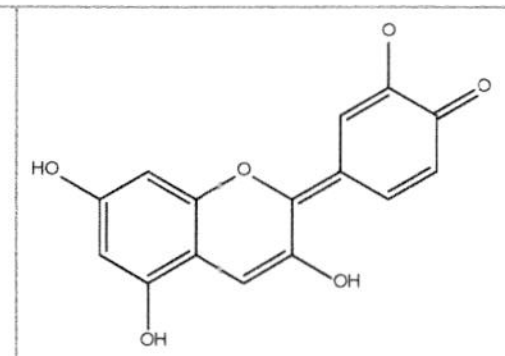	
		Peonidin	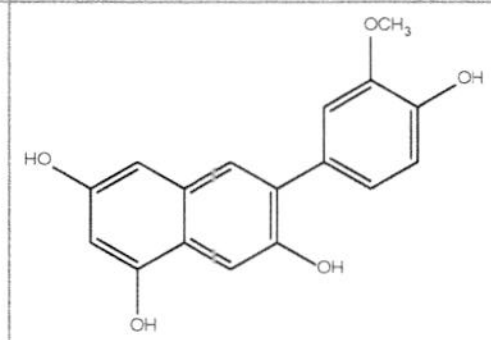	
		Caffeic acid	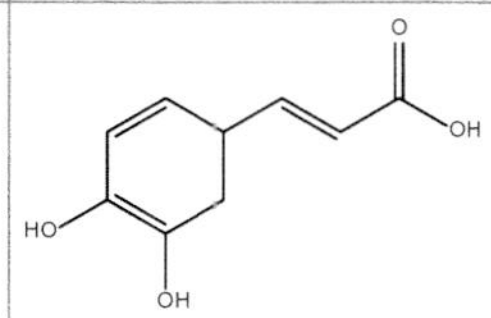	
		P-Coumaric acid	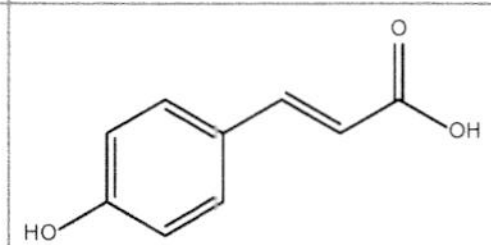	

(continued)

Table 4.53 (continued)

Family	Botanical name	Image	Plant part used	Bioactive compound(s)	Structure	Reference(s)
	Ledum groenlandicum Retzius		Leaf	Ursolic acid	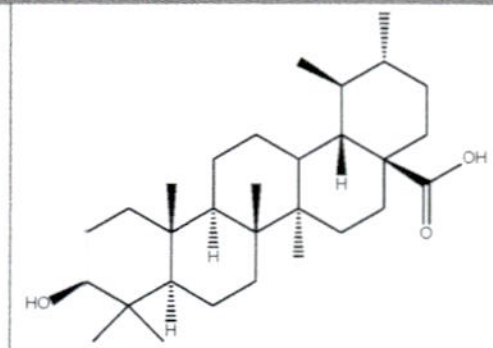	Abdel-Monem et al. (2008)
	Synonym(s): *Ledum groenlandicum* Oeder, *Ledum groenlandicum* subsp. *glandulosum* (Nutt.) Á. Löve & D. Löve. **Common name(s):** Labrador tea. **Botanical description:** Labrador tea is distributed throughout Canada, Alaska, and the northern fringe of the United States. It is an evergreen resinous shrub that grows up to 4.9 ft. (1.5 m) tall. The multiple branches are procumbent or upright and the growth form is rounded. Young stems are coated with curly hairs, while older stems are hairless. Linear leaves are simple and arranged alternately. They measure up to 2 in. (5 cm) long by 2 cm wide and are firm and leathery. Leaf margins are revolute and undersides are woolly. Numerous fragrant flowers are produced in clusters at branch ends. Fruits are 5-parted capsules measuring 4–6 mm long and open at the base releasing a large number of seeds, typically in the fall. The empty capsule may persist for years. **Medicinal use(s):** The leaves of *Ledum groenlandicum* had antioxidant, anti-inflammatory, anti-fatigue, antimicrobial, and antidiabetic activities and protected the retina from light damage. The plant extract is very effective in treating renal and pancreas cancer.					

	Ledum palustre (Led).		Leaf	Rutin	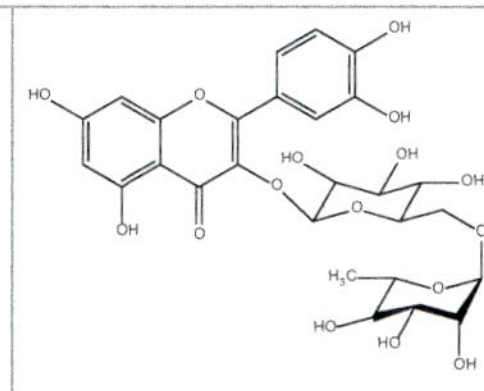	Dufoura et al. (2007)
	Synonym(s): *Ledum palustre* var. *angustum* E.A. Busch, *Ledum palustre* var. *decumbens* Aiton. **Common name(s):** Marsh tea. **Botanical description:** The plant is native of North America; it grows in northern latitudes in Greenland, Canada, and Alaska, in Europe in the northern and central parts, and in Asia South to Northern China, Korea, and Japan. It grows in peaty soils, shrubby areas, moss, and lichen tundra. It is a low shrub growing up to 50 cm (rarely up to 120 cm) tall with evergreen leaves 12–50 mm long and 2–12 mm broad. The flowers are small, with a five-lobed white corolla, and produced several together in a corymb 3–5 cm diameter. They emit strong smell to attract bees and other pollinating insects. **Medicinal use(s):** The plant possesses antioxidant, anti-inflammatory, and immunostimulant properties. It may be used to cure liver and prostate cancer.			Quercitrin	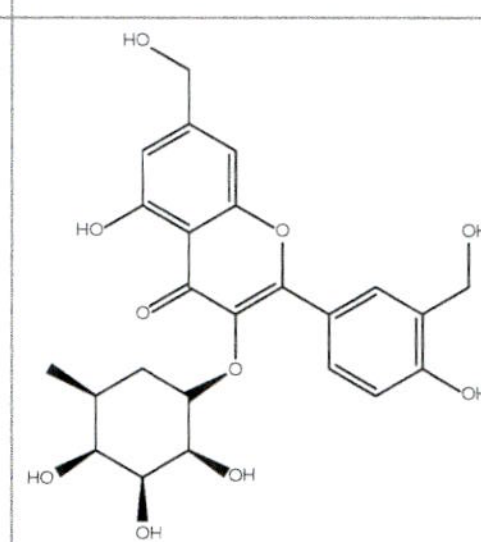	

Table 4.54 Anticancer plants of family Ebenaceae

Family	Botanical name	Image	Plant part used	Bioactive compound(s)	Structure	Reference(s)
Ebenaceae	*Diospyros chloroxylon* Roxb		Whole plant	Beta-sitosterol	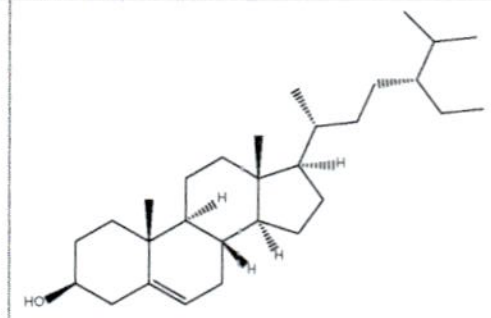	Jain (1970) and Kumari et al. (2018)
	Synonym(s): *Diospyros capitulata* Wight, *Diospyros insculpta* Buch.-Ham., *Diospyros tomentosa* Poir. **Common name(s)**: Kendu, Timburni, Tendu. **Botanical description**: It is a flowering shrub and a native of India. It is the dense black heartwood; it has a very high density and will sink in water. It grows up to 25 m. The bole can be 60 cm in diameter. **Medicinal use(s)**: It is used to treat tumors. The seeds of this plant can be used for intoxification. It also cures mental disorder, nervous breakdown, and palpitation of the heart. The fruit have a cooling and an astringent effect. Dried flower is used for treating skin, urinary, and blood diseases.					

	Euclea crispa (Thunb.) Gurke		Rootbark	Kaempferol	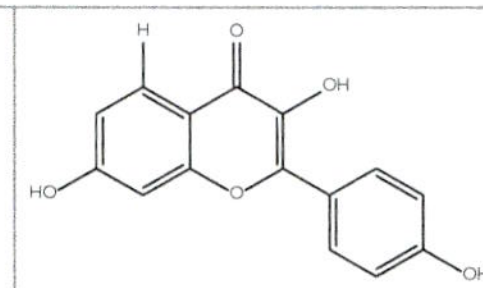	Sibanda et al. (1992), Alayande et al. (2016), and Maroyi (2017)
	Synonym(s): *Euclea antunesii* Gürke, *Euclea bakeriana* Brenan, *Euclea baumii* Gürke, *Euclea crispa* subsp. *crispa*, *Euclea dekindtii* Gürke, *Euclea desertorum* Eckl. & Zeyh., *Euclea guerkei* Hiern, *Euclea lanceolata* E.Mey. ex A.DC. **Common name(s)**: Blue guarri. **Botanical description**: It is found in the regions of Southern Africa, Zambia, and Zimbabwe. It is an evergreen shrub or a small bushy tree growing up to 8 m tall with a bole 5–15 cm in diameter. The shrub grows 30–300 cm tall at a time when it can be rhizomatous suffrutex shrub. **Medicinal use(s)**: It possesses anticancer properties. It is used to treat melanomas. The root is also used for treating various ailments, and the fruit or bark is used as a purgative.					

Table 4.55 Anticancer plants of family Elaeagnaceae

Family	Botanical name	Image	Plant part used	Bioactive compound(s)	Structure	Reference(s)
Elaeagnaceae	*Hippophae rhamnoides* L. ssp. *turkestanica*		Flower	β-Tocopherol	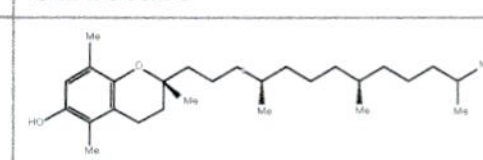	Sadeghi-Aliabadi et al. (2009), Patel et al. (2012), and Olas et al. (2018)
	Synonym(s): No synonyms are recorded for this name. **Common name(s)**: Sea buckthorn, Seaberry. **Botanical description**: It is a native plant of Europe, including Britain, from Norway south and east to Spain and Asia to Japan and the Himalayas. It is a deciduous shrub growing up to 6 m by 2.5 m at a medium rate. It flowers in April, and the seeds ripen from September to October. The flowers are dioecious and are pollinated by wind. The plant is not self-fertile. It can fix nitrogen. **Medicinal use(s)**: It is used as an astringent, poultice, tonic, and vermifuge and for curing cancer and cardiac disorders. A high-quality medicinal oil is made from the fruit and used in the treatment of cardiac disorders; it is also said to be particularly effective when applied to the skin to heal burns, eczema, and radiation injury and is taken internally for the treatment of stomach and intestinal diseases.			β-Tocotrienol	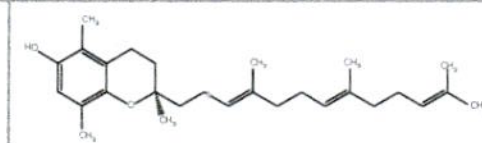	
				δ-Tocotrienol	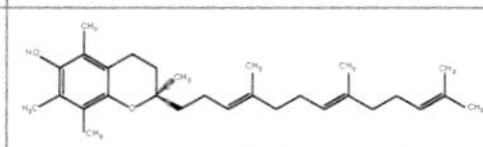	
				β-Tocotrienol	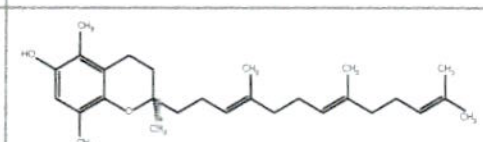	
				γ-Tocopherol	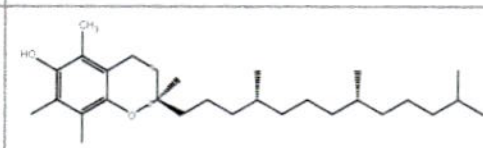	

	Hippophae salicifolia D. Don		Flower, fruit	Tocopherol	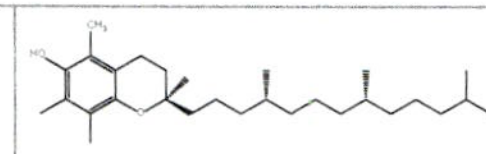	Uniyal (1990), Ravikanth et al. (2003), and Rajchal (2009)
	Synonym(s): *Hippophae rhamnoides* subsp. *salicifolia* (D. Don) Servettaz. **Common name(s)**: Willow-leaved sea buckthorn. **Botanical description**: It is a native plant of East Asia – the Himalayas. It is a deciduous tree growing up to 15 m at a fast rate. It flowers in April, and the seeds ripen from September to October. The flowers are dioecious and are pollinated by wind. The plant is not self-fertile and can fix nitrogen. **Medicinal use(s)**: *Hippophae salicifolia* can cure cancer and cardiac disorders and can be used as a poultice. The fruit is a very rich source of vitamins and minerals, especially vitamins A, C, and E, flavanoids, and other bioactive compounds. It is also a fairly good source of essential fatty acids. It is being investigated as a food that is capable of reducing the incidence of cancer and also as a means of halting or reversing the growth of cancers.					

Table 4.56 Anticancer plants of family Elaeocarpaceae

Elaeocarpaceae	*Aristotelia chilensis* Stuntz		Leaf and stem	Amylodextrin	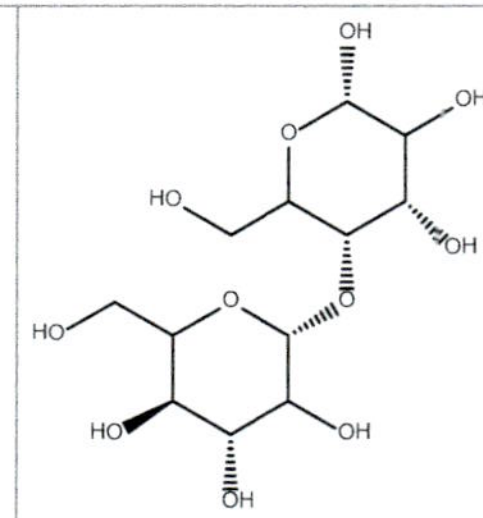	He et al. (1997) and Romanucci et al. (2016)
	Synonym(s): *Cornus chilensis* Molina, *Aristotelia macqui* L'Hér. **Common name(s):** Macqui, Molina. **Botanical description:** It is a native plant of South America – Chile and Argentina. A weedy shrub of cleared forests and waste ground, growing in damp humus-rich soils on lower mountain hillsides by rivers between latitudes 31 and 40° south. It is an evergreen shrub growing up to 3 m by 5 m. It is in flower in May, and the seeds ripen from August to September. The flowers are dioecious and are pollinated by bees and insects. The plant is not self-fertile. It is noted for attracting wildlife. **Medicinal use(s):** It is used to treat colon cancer. It is used as an astringent, a febrifuge, and a poultice. A wine made from the fruit is said to have medicinal properties. It is used in the treatment of throat infections and intestinal tumors. It possesses anticancer, antitumor, antibacterial, and anti-inflammatory properties.					

Table 4.57 Anticancer plants of family Eriocaulaceae

Family	Botanical name	Image	Plant part used	Bioactive compound(s)	Structure	Reference(s)
Eriocaulaceae	*Eriocaulon cinereum* R.Br.		Stem bark	Hispidulin	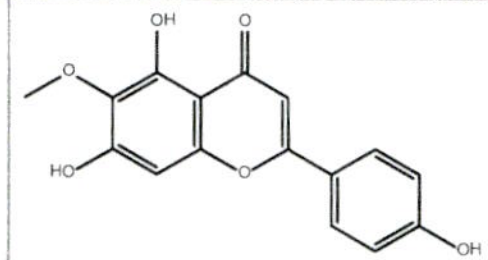	Duke and Ayensu (1985a, b) and Fan et al. (2016)
	Synonym(s): *Eriocaulon amboense* Schinz, *Eriocaulon ciliiflorum* F.Muell., *Eriocaulon sieboldianum* Siebold and Zucc. ex Steud, *Eriocaulon formosanum* Hayata, *Eriocaulon heteranthum* Benth., *Eriocaulon heudelotii* N.E.Br., *Eriocaulon reductum* Ruhland, *Eriocaulon stuhlmannii* N.E.Br., *Eriocaulon taiwanianum* S.S.Ying, *Leucocephala spathacea* Roxb. **Common name(s)**: Pipeworts. **Botanical description**: It is a native plant of Afghanistan, Tien-Shan, Pamir-Alai, Africa, India, Burma, Assam, Sri Lanka, Japan, and China. It is a dwarf plant (3–5 cm tall) with many slender peduncles, more or less of equal length. Our material differs from the typical variety in its larger stature plants up to 11 cm tall, thicker and usually fewer peduncles, fairly unequal in size. **Medicinal use(s)**: It is used to treat cancer. The Aurora kinase family plays critical roles in the regulation of cell division and has attracted great attention in the identification of small-molecule Aurora kinase inhibitors for potential treatment of cancer.					

Table 4.58 Anticancer plants of family Euphorbiaceae

Family	Botanical name	Image	Plant part used	Bioactive compound(s)	Structure	Reference(s)
Euphorbiaceae	*Acalypha californica* Benth		Leaf	Stigmasterol	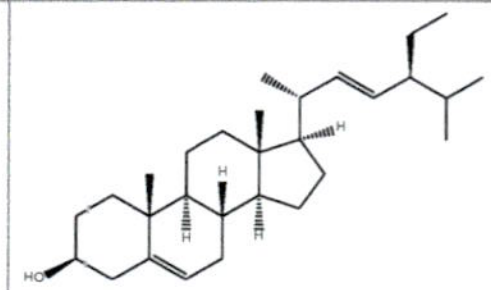	Ishikura (1982) and Rascón-Valenzuela et al. (2015)
	Synonym(s): *Acalypha pringlei* S.Watson, *Acalypha stokesiae* Pax & K. Hoffm., *Ricinocarpus californicus* (Benth.) Kuntze. **Common name(s)**: Copperleaf, Pringle three-seeded mercury. **Botanical description**: It is a native plant of California. The plant bears hairy, juicy, toothed leaves. The flower is made up of a staminate part, which appears as a long spike of tiny red and pink bract and pistillate part at the base of the spike. **Medicinal use(s)**: It possesses anticancer, diuretic, purative, antibacterial, antifungal, and anthelmintic properties. This medicinal plant is used to get rid of intestinal worms and to treat gum problems, stomachaches, hernia, rheumatism, asthma, and skin diseases.			β-Sitosterol	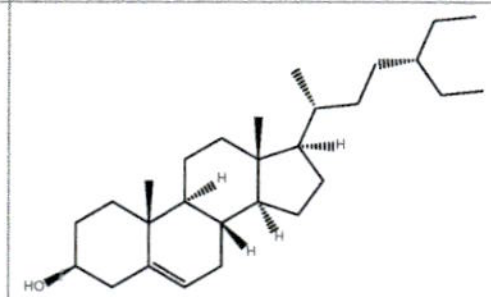	
	Acalypha siamensis Oliv. ex Gage		Leaf	Acalyphaser A	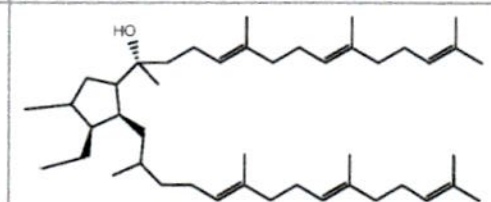	Kambara et al. (2006)

	Synonym(s): *Acalypha evrardii Gagnep.*, Acalypha siamensis var. *denticulata* Airy Shaw. **Common name(s)**: Wild tea, tea leaves. **Botanical description**: It is a native plant of Peninsular Malaysia, Burma (Myanmar), Thailand, Cambodia, Laos, and Vietnam. It is a shrub or small scrambling tree growing up to 4 m tall. It is an evergreen herb. **Medicinal use(s)**: It is used to treat breast cancer. It is used to treat fever, bowel complaints, and kidney diseases. It possesses anticancer, antioxidant, anti-inflammatory, and antipyretic properties.					
	Acalypha wilkesiana Mull. Arg.		Stem	Quercetin (flavonoid)	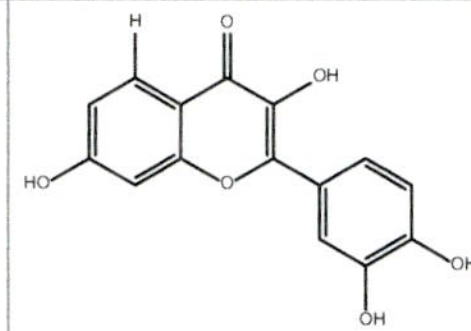	Madziga et al. (2010) and Ikewuchi et al. (2011)
	Synonym(s): *Acalypha compacta* Guilf. ex C.T.White, *Acalypha godseffiana* Mast., *Acalypha hamiltoniana* Bruant, *Acalypha macafeeana* Veitch, *Acalypha musaica* auct., *Acalypha tricolor* Seem., *Acalypha triumphans* L.Linden & Rodigas, *Ricinocarpus wilkesianus* (Müll.Arg.) Kuntze. **Common name(s)**: Acalypha tricolor. **Botanical description**: It is a native plant of Western Pacific, Fiji, and Tuvalu. It is an erect or spreading, evergreen, suckering shrub that can grow 2–4 m tall. It prefers moist soil; the plant can flower and produce fruit all year-round. **Medicinal use(s)**: It is used to treat breast cancer. It is used to treat diarrhea and dysentery. It is having properties of abortifacient, antibacterial, antifungal, and anti-nematodal.			Naringenin (flavanone)	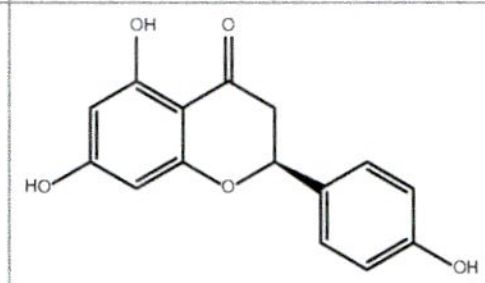	
				Kaempferol	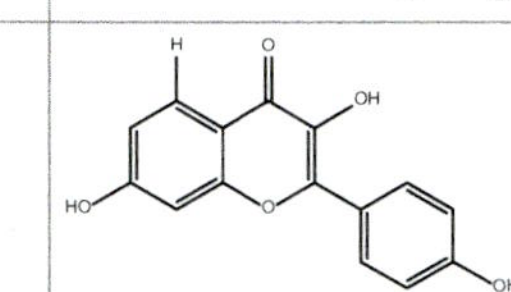	

(continued)

Table 4.58 (continued)

Family	Botanical name	Image	Plant part used	Bioactive compound(s)	Structure	Reference(s)
				Epicatechin (flavonoid)	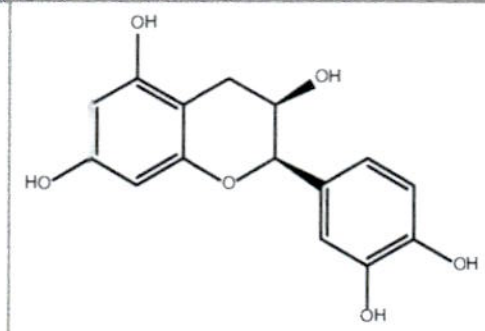	
				β-Sitosterol	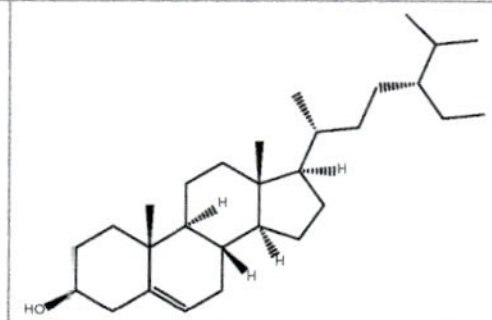	
	Cnidoscolus aconitifolius (Mill.) I.M.Johnst.		Leaf	Campesterol	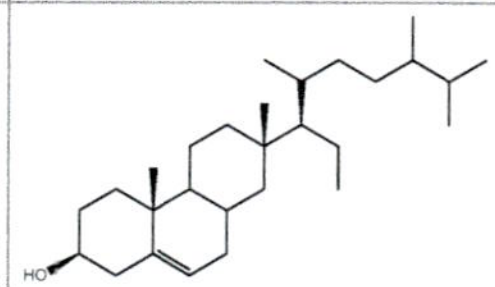	Flores and Ricalde (1996), Kulathuran et al. (2012), and Sanchez-Hernandez et al. (2017)
				Eupafolin	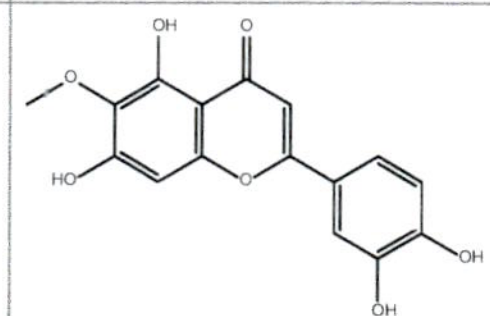	

	Synonym(s): *Cnidoscolus chaya* Lundell, *Cnidoscolus chayamansa* McVaugh, *Cnidoscolus fragrans* (Kunth) Pohl, *Cnidoscolus longipedunculatus* (Brandegee) Pax & K.Hoffm., *Cnidoscolus napifolius* (Desr.) Pohl, *Cnidoscolus palmatus* (Willd.) Pohl, Cnidoscolus *quinquelobatus* (Mill.) Lećn, *Jatropha aconitifolia* Mill., *Jatropha deutziiflora* Croizat, *Jatropha fragrans* Kunth, *Jatropha longipedunculata* Brandegee, *Jatropha napifolia* Desr., *Jatropha palmata* Willd., *Jatropha papaya* Medik., *Jatropha quinqueloba* Sessé, *Jatropha quinquelobata* Mill. **Common name(s)**: Tree spinach, Chaya. **Botanical description**: It is a native plant of Mexico. It is a leafy perennial shrub. It can grow up to 6 m tall. The leaf stem connects to the leaf; the leaf veins are fleshy and cuplike. It has both male and female flowers. **Medicinal use(s)**: It is used to treat cancer. It is a good source of vitamins, proteins, calcium, and iron and also a rich source of antioxidant. It is having antidiabetic, anticancer, antitumor, anti-inflammatory, and antioxidant effect.			Isoquercetin	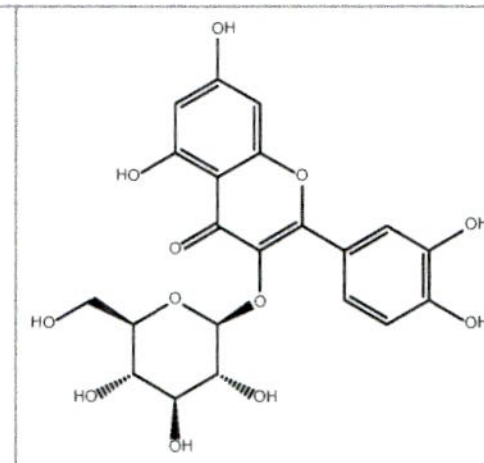	
	Croton argyratus Blume		Leaf and stamen	Styryldehydropyrone	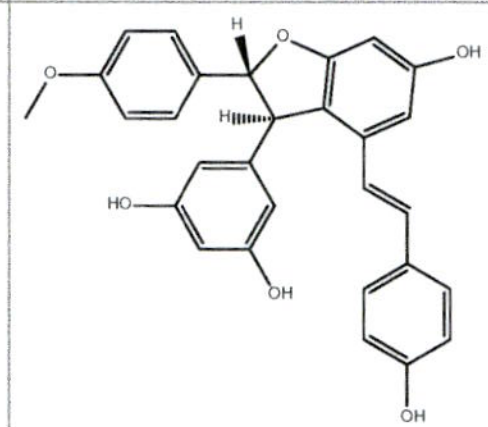	Norizan et al. (2007)

(continued)

Table 4.58 (continued)

Family	Botanical name	Image	Plant part used	Bioactive compound(s)	Structure	Reference(s)
	Synonym(s): *Croton avellaneus* Croizat, *Croton budopensis* Gagnep., *Croton maieuticus* Gagnep., *Croton pilargyros* Croizat, *Croton tawaoensis* Croizat, *Croton zollingeri* Miq., *Oxydectes argyrata* (Blume) Kuntze, *Oxydectes bicolor* (Roxb.) Kuntze. **Common name(s)**: Blume, Croton. **Botanical description**: It is a native plant of Southeast Asia, Malaysia, and Vietnam. It is an evergreen, monoecious, small- to medium-sized tree that grows up to 27 m tall. The cylindrical bole is fairly straight up to 70 cm in diameter. It occurs in well-drained, sandy, clay, and sandy loam soil. **Medicinal use(s)**: It is used to treat cancer and has anticancer properties. It is used to treat diarrhea, ulcers, and fever and used for medicinal baths. It is used as treatment for thrush.			(−)-junceic acid	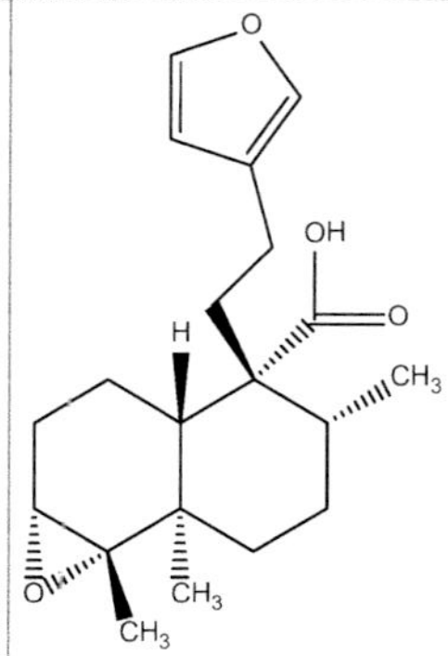	
				Clerodane type (diterpene)	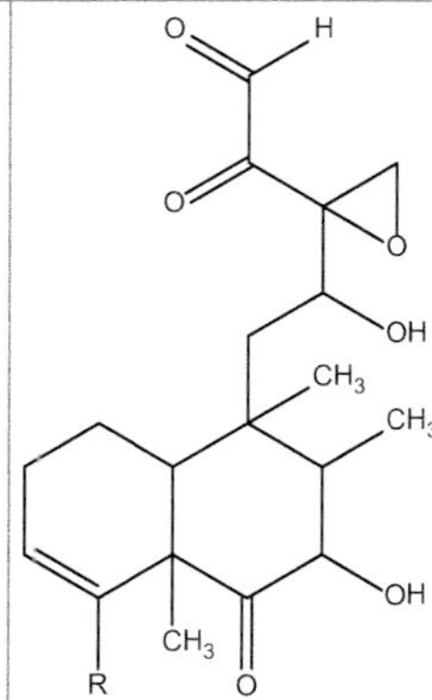	
				Goniothalamin	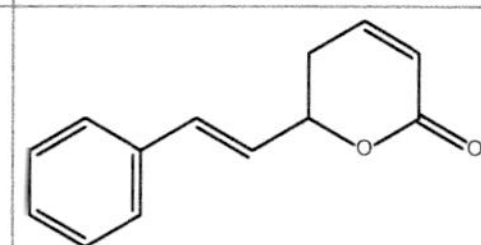	

	Croton bonplandianum Baill.		Leaf	Phorbol 12-myristate 13-acetate diester	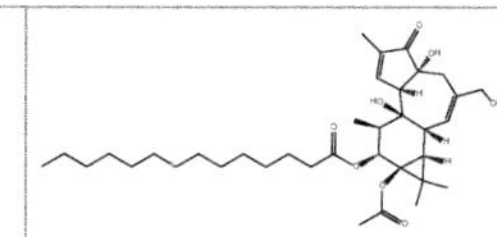	Islam et al. (2011) and Nath et al. (2013)
	Synonym(s): *Croton pauperulus* Müll.Arg., *Croton rivinoides* Chodat, *Croton sparsiflorus* Morong, *Oxydectes bonplandiana* (Baill.) Kuntze, *Oxydectes pauperula* (Müll.Arg.) Kuntze, *Oxydectes sparsiflora* (Morong) Kuntze. **Common name(s)**: Rushfoil. **Botanical description**: It is a native plant of Europe. It is a flowering and a pan tropical plant, with some species extending into temperate area. **Medicinal use(s)**: It possesses anticancer, antitumor, antihyperglycemic, and antioxidant properties. It has organic compound which is antitumor in nature; it is also helpful in treating constipation, heals lesions, and is used as a purgative.			Beta-sitosterol	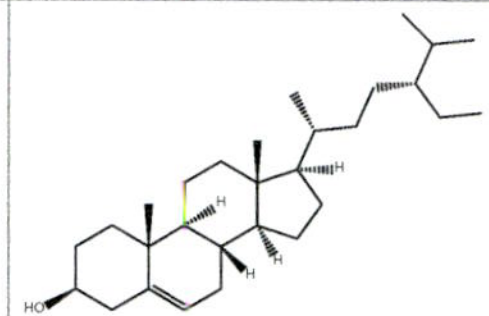	
	Croton draco Schltdl		Aerial parts	Myricitrin	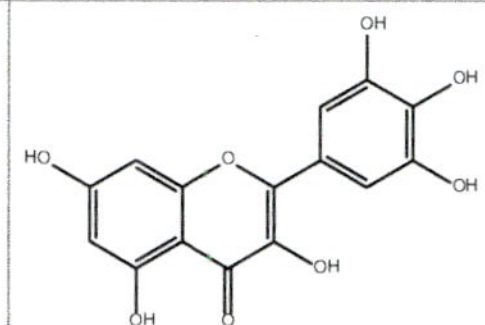	Gupta et al. (1996), Farias et al. (2009), and Gupta et al. (2008)

(continued)

Table 4.58 (continued)

Family	Botanical name	Image	Plant part used	Bioactive compound(s)	Structure	Reference(s)
	Synonym(s): *Croton callistanthus* Croizat, *Croton panamensis* (Klotzsch) Müll.Arg., *Croton steyermarkianus* Croizat, *Croton triumfettoides* Croizat, *Cyclostigma denticulatum* Klotzsch, *Cyclostigma draco* (Schltdl.) Klotzsch, *Cyclostigma panamense* Klotzsch, *Oxydectes draco* (Schltdl.) Kuntze, *Oxydectes panamensis* (Klotzsch) Kuntze. **Common name(s)**: Dragon's blood, Sangre de Drago. **Botanical description**: It is a native plant of South America – Peru, Ecuador, and Colombia. It is a flowering plant with broad and open crown. It grows up to 20 m in height, with a bole diameter of usually 30 cm. It cannot grow in the shade and prefers moist soil. **Medicinal use(s)**: It is used as an antitumor agent. The resin has shown an in vitro effect against stomach cancer and colon cancer; it inhibits the growth of human myelogenous leukemia cell line. The sap has antibacterial, anti-inflammatory, antimicrobial, antifungal, and antitumor properties.			Taspine	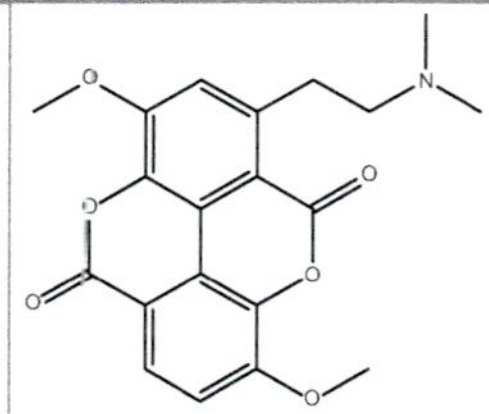	
	Croton erythrochilus Mull.Arg.		Latex	Crotonaldehyde	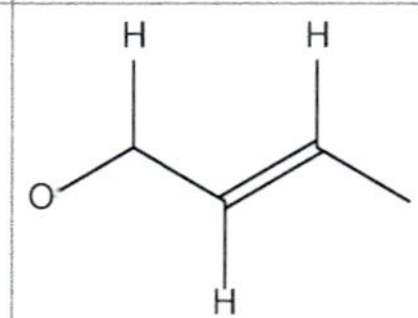	Piacente et al. (1998) and Gupta et al. (2008)

	Synonym(s): *Oxydectes erythrochila* (Müll.Arg.) Kuntze. **Common name(s)**: Garden croton. **Botanical description**: It is native to Indonesia, Malaysia, Australia, and the Western Pacific Ocean islands. It is a perennial plant, grows up to 3 m tall, and has large, thick, leathery, shiny evergreen leaves, alternately arranged, 5–30 cm long and 0.5–8 cm broad. It is found in tropical and subtropical region. **Medicinal use(s)**: It is used for cancer treatment. It possesses antitumor, antiprotease, and anti-angiogenic activities.					
	Croton eluteria (L.) W.Wright		Bark	Prenylbisabolane	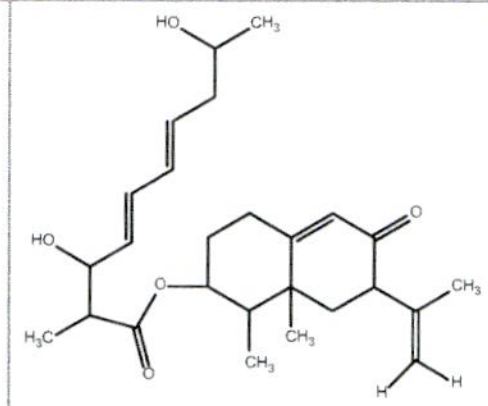	Campagnuolo et al. (2004)
	Synonym(s): *Clutia cascarilla* L., *Clutia decandra* Crantz, Clutia eluteria L., *Croton cascarilla* (L.) L., *Croton homolepidus* Müll.Arg., *Oxydectes cascarilla* (L.) Kuntze, *Oxydectes cascarilloides* Kuntze, *Oxydectes eluteria* (L.) Kuntze, *Oxydectes homolepida* (Müll.Arg.) Kuntze. **Common name(s)**: Cascarilla. **Botanical description**: It is native to the Caribbean – the Bahamas, Cuba, the Dominican Republic, Haiti, and Jamaica. It is a shrub growing from 1.5–12 m tall and found in woodlands on limestone soils in both dry and wet areas at an elevation of 870 m from sea level. **Medicinal use(s)**: It is used to treat cancer and also possesses antitumor property. The bark is used as anti-spasmodic, antitussive, cholagogue, hypotensive agent expectorant, febrifuge, carminative, stimulant, and is a source of a bitter-tonic that aids digestion.			Sesquiterpene lactone LS-1	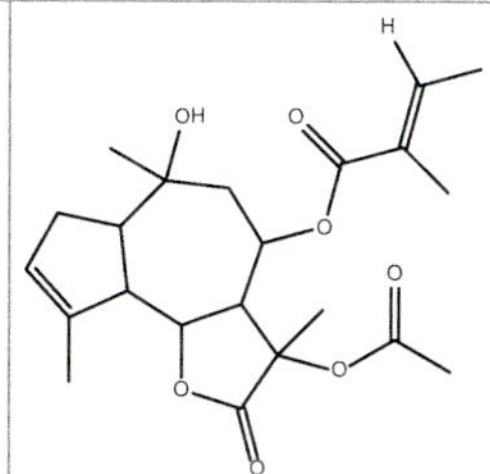	
				Clerodanes	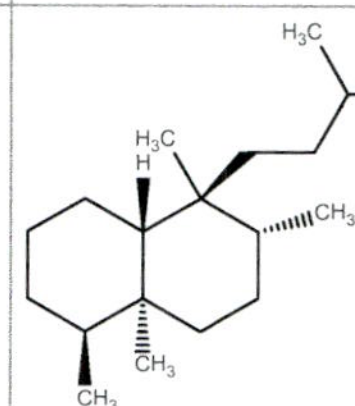	

(continued)

Table 4.58 (continued)

Family	Botanical name	Image	Plant part used	Bioactive compound(s)	Structure	Reference(s)
	Croton flavens L.		Leaf	Viridiflorene	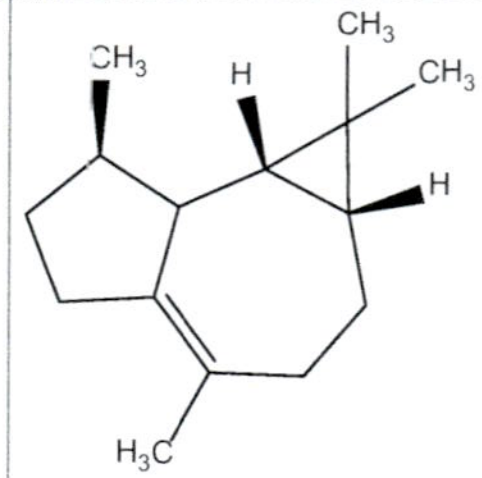	Flores and Ricalde (1996), Sylvestre et al. (2006)
	Synonym(s): *Croton albidus* Müll.Arg., *Croton althaeifolius* Mill., *Croton astroites* Willd., *Croton balsamifer* Jacq., *Croton flocculosus* Geiseler, *Croton lamarckianus* Moldenke, *Croton mauralis* E.H.L.Krause, *Croton mucronatus* Willd., *Croton padifolius* Geiseler, *Croton richardii* Willd., *Croton rigidus* (Müll.Arg.) Britton, *Oxydectes albida* (Müll.Arg.) Kuntze, *Oxydectes althaeifolia* (Mill.) Kuntze, *Oxydectes astroites* (Willd.) Kuntze, *Oxydectes flavens* (L.) Kuntze, *Oxydectes flocculosa* (Geiseler) Kuntze, *Semilta althaeifolia* (Mill.) Raf. **Common name(s)**: Yellow balsam. **Botanical description**: It is native to Southeast Asia. It is a diverse and complex taxonomic group of plants ranging from herbs and shrubs to trees. **Medicinal use(s)**: It is used to treat lung carcinoma cell line A-549 and human adenocarcinoma cell line DLD-1 and was found to be very active against both the tumor cell lines. It is used for callouses and warts.			Alpha-humulene	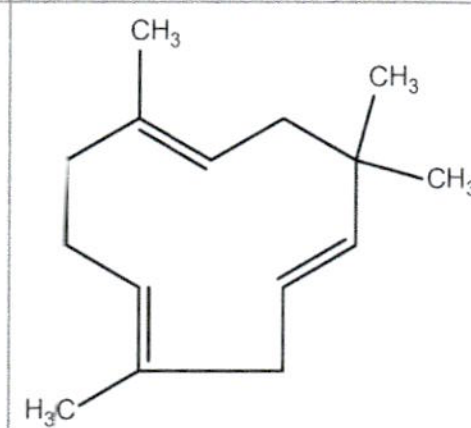	
				Beta-elemene	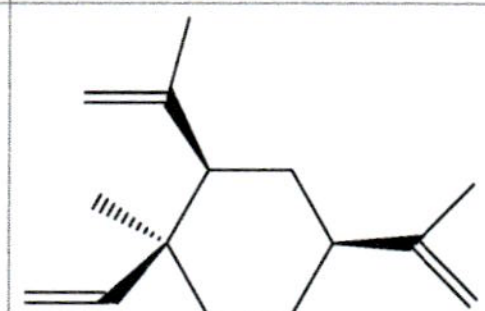	
				Alpha-cadinol	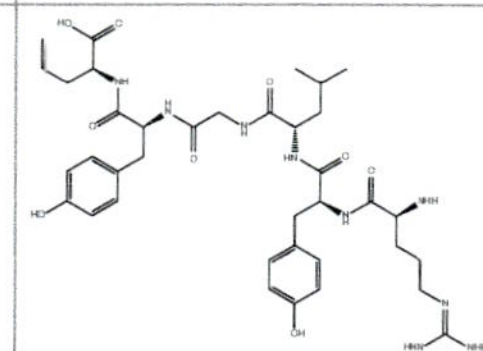	

				Beta-caryophyllene	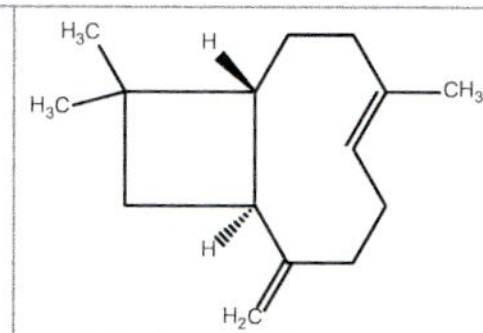	
				Beta-bisabolene	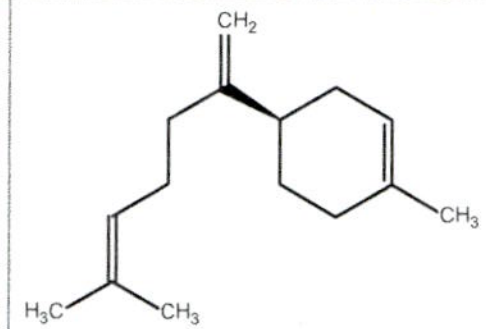	
				Germacrone	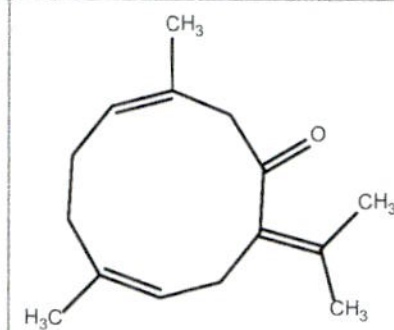	
	Croton lechleri Muell.Arg.		Stamen, latex	Proanthocyanidins	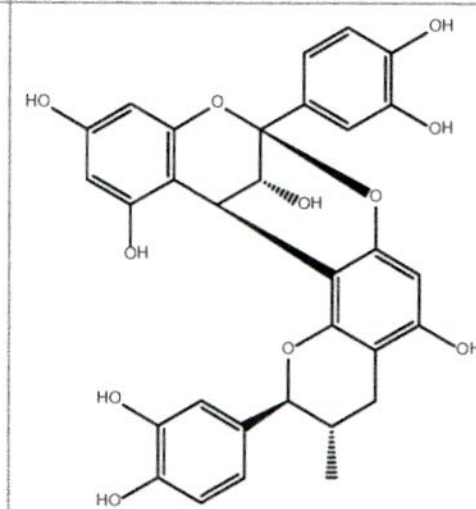	Pieters et al. (1992), Cai et al. (1993), Piacente et al. (1998), and Lopes et al. (2004)

(continued)

Table 4.58 (continued)

Family	Botanical name	Image	Plant part used	Bioactive compound(s)	Structure	Reference(s)
	Synonym(s): *Oxydectes lechleri* (Müll.Arg.) Kuntze. **Common name(s):** Dragon's blood. **Botanical description:** It is native to South America – Peru, Ecuador, and Colombia. It is a flowering plant with a broad but open crown native to northwestern South America. It grows up to 20 m in height with a bole diameter of usually 30 cm. It cannot grow in the shade. It prefers moist soil. The plant can tolerate strong winds but not maritime exposure. **Medicinal use(s):** It is used to treat stomach cancer and colon cancer cells. It possesses anticancer, antioxidant, antibacterial, antifungal, anti-inflammatory, antimicrobial, antipruritic, antitumor, hemostatic, and vulnerary properties.			Epicatechin	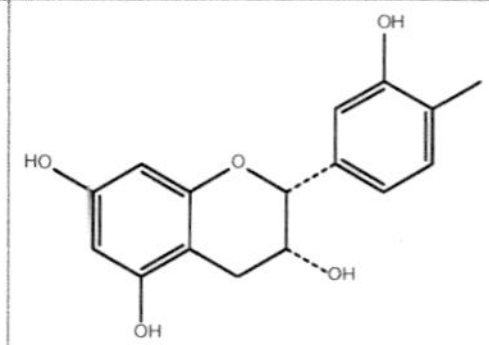	
				Gallocatechin	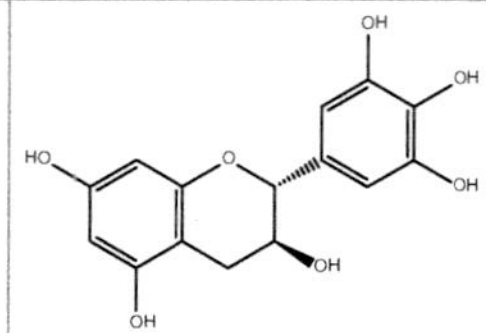	
				Epigallocatechin	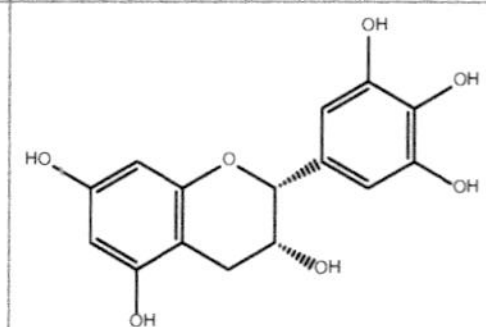	
				Taspine (alkaloid)	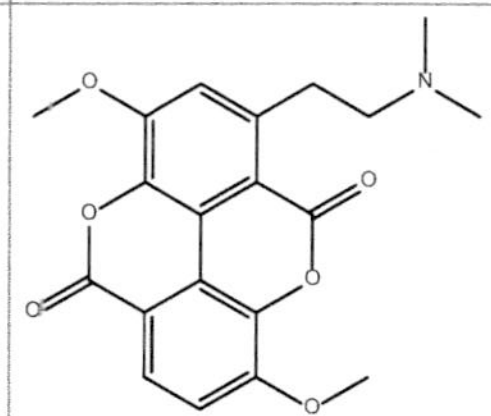	
				Lignan 3,4-O-dimethylcedrusin	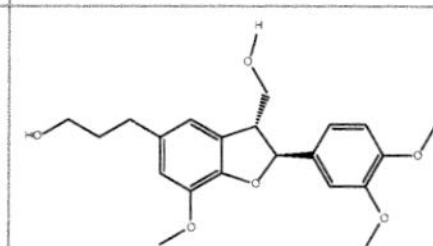	

	Plant	Image	Part used	Compound	Structure	Reference
	Croton insularis Baill.		Aerial parts	Pimarane	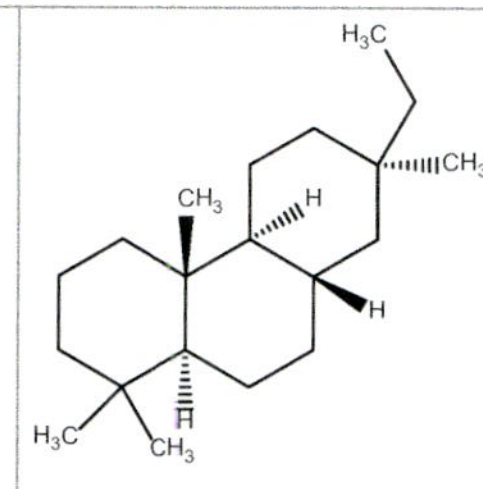	Maslovskaya et al. (2011)
	Synonym(s): *Croton phebalioides* A.Cunn. ex Benth., *Oxydectes insularis* (Baill.) Kuntze. **Common name(s)**: Silver croton. **Botanical description**: It is native to Kurrajong, New South Wales, to Queensland. It is a shrub to small tree growing up to 15 m high, crowns often with occasional leaves turning red or orange before falling. It is found in subtropical, littoral, and dry rainforest, also in eucalypt and brigalow forests. **Medicinal use(s)**: It is used to treat colon-rectal cancer. It possesses anticancer, antitumor, antihypertensive, anti-inflammatory, antimalarial, antimicrobial, anti-spasmodic, anti-ulcer, antiviral, and myorelaxant properties. It is also used to treat constipation, diabetes, digestive problems, dysentery, external wounds, fever, hypercholesterolemia, hypertension, inflammation, intestinal worm infection, malaria, pain, ulcers, and weight loss.			Spruceanol	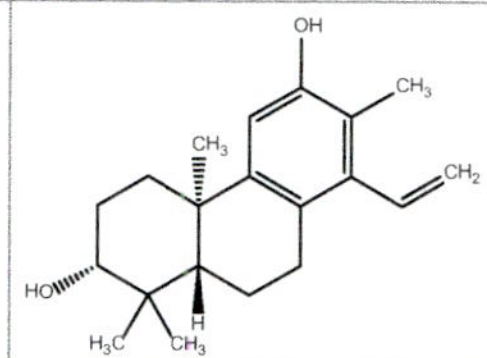	
				Sonderianol	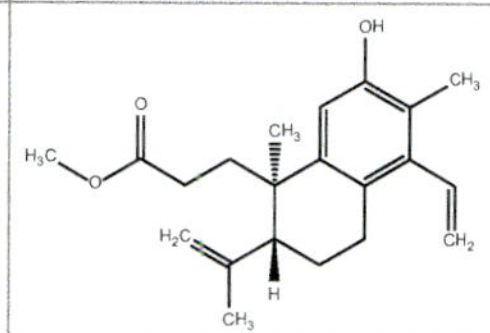	

(continued)

Table 4.58 (continued)

Family	Botanical name	Image	Plant part used	Bioactive compound(s)	Structure	Reference(s)
	Croton persimilis Müll. Arg.		Stem and bark	Furoclerodane	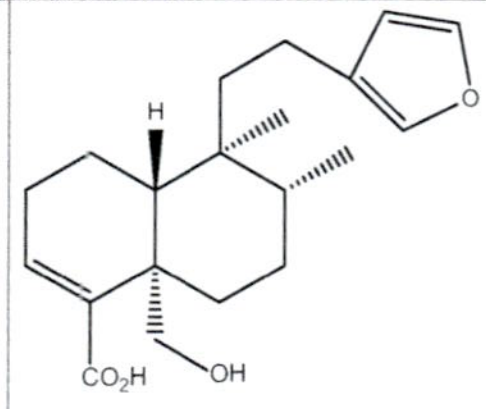	Youngsa-ad et al. (2007)
	Synonym(s): *Croton virbalae* M.R.Almeida, *Oxydectes oblongifolia* Kuntze, *Oxydectes persimilis* (Müll.Arg.) Kuntze. **Common name(s)**: Bol-mangbol chham, Pill-bearing spurge. **Botanical description**: It is native to Chittagong, Chittagong Hill Tracts, and Sylhet and in scrubby and deciduous forests. Also found in village shrubberies of Mymensingh and Dinajpur. It is a small- to middle-sized, deciduous tree. Leaves 12.5–25 cm long, crowded towards the ends of the branchlets, oblong-lanceolate, subacute. **Medicinal use(s):** It is used to treat various human tumor cell lines including HEP-G2, SW620, CHAGO, KATO3, and BT474. Used in reducing chronic enlargement of the liver and in remittent fever. It is externally applied to sprains, bruises, and rheumatic swelling and to the hepatic region in chronic hepatitis.			Nidorellol	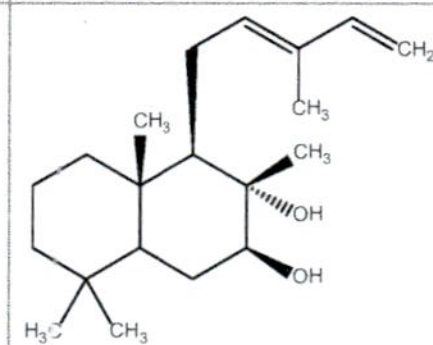	
				Labdane	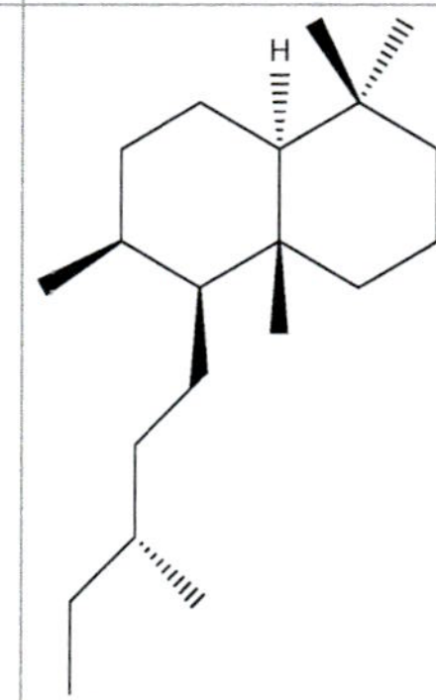	

				Clerodane	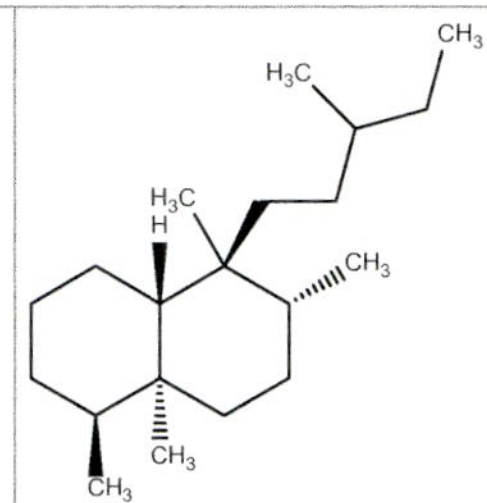	
	Croton palanostigma Mart. ex Klotzsch.		Bark, Resin, Sap	Taspine (3′4′-O-dimethylcedrusin)	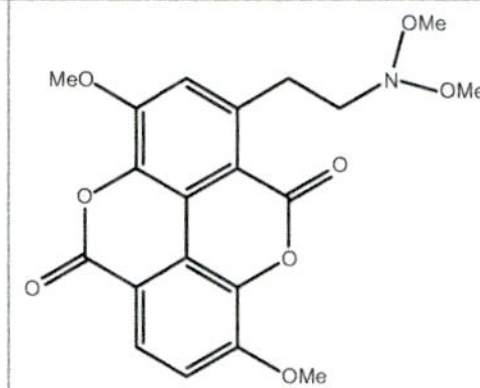	Sandoval et al. (2002b)
	Synonym(s): *Croton benthamianus* Müll.Arg., *Oxydectes palanostigma* (Klotzsch) Kuntze, *Palanostigma martiana* Baill. **Common name(s)**: Sangre de grado, Rushfoil. **Botanical description**: It is native to South America – Northern Brazil, Peru, Ecuador, Colombia, Venezuela, and French Guiana. It is a tree with a spreading crown that grows from 10 to 20 m tall. The straight, cylindrical bole is usually less than 30 cm in diameter and unbranched for around one third of its height. When the trunk of the tree is cut or wounded, a dark red, sappy resin oozes out as if the tree is bleeding. **Medicinal use(s)**: It is used to treat stomach cancer and colon cancer cells. It inhibits the growth of a human myelogenous leukemia cell line and also prevents cells from mutating. The sap has antibacterial, antifungal, anti-inflammatory, antimicrobial, antipruritic, antitumor, hemostatic, and vulnerary properties. It is used internally in the treatment of intestinal and stomach ulcers, upper respiratory viruses, stomach viruses, and HIV.					

(continued)

Table 4.58 (continued)

Family	Botanical name	Image	Plant part used	Bioactive compound(s)	Structure	Reference(s)
	Croton regelianus Muell. Arg.		Leaf	Cymene	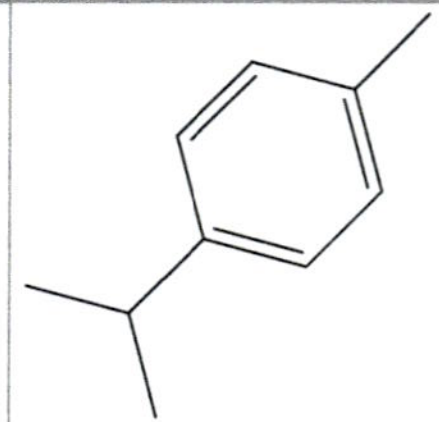	Bezerra et al. (2009)
	Synonym(s): *Oxydectes regeliana* (Müll.Arg.) Kuntze. **Common name(s)**: Gattung. **Botanical description**: It is native to Northeastern Brazil. It is a tree with a spreading crown that grows from 10 to 20 m tall. The straight, cylindrical bole is usually less than 30 cm in diameter and unbranched for around one third of its height. When the trunk of the tree is cut or wounded, a dark red, sappy resin oozes out as if the tree is bleeding. **Medicinal use(s)**: It is used to treat four strains of tumor cells: leukemia, colon, melanoma, and brain human cancer cells. It possesses antitumor, anticancer, and antioxidant properties.			Ascaridole	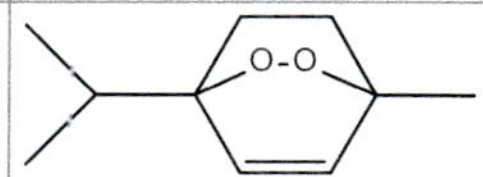	
				Camphor	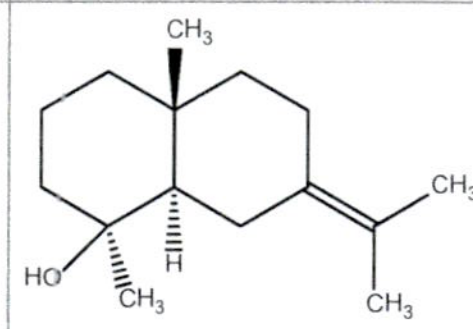	
				Terpinene	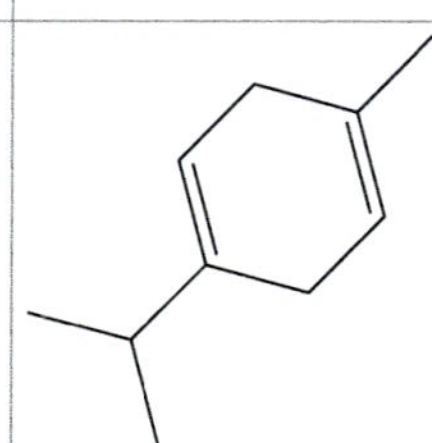	

	Euphorbia helioscopia L.		Leaf, aerial parts, root	Ent-kaurene	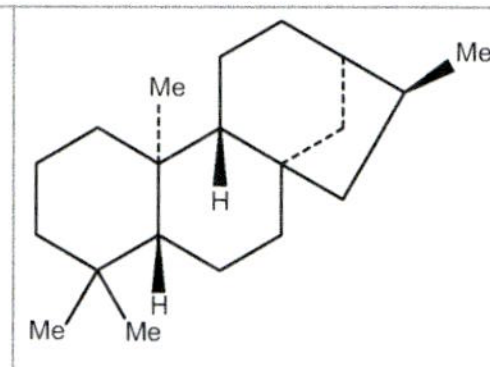	Ahn et al. (1996), Giang et al. (2003), Yu et al. (2004), Hu et al. (2007), and Chen et al. (2008a, b)
	Synonym(s): *Euphorbia dominii* Rohlena, *Euphorbia helioscopioides* Loscos & J.Pardo, *Euphorbion helioscopium* (L.) St.-Lag., *Galarhoeus helioscopius* (L.) Haw., *Tithymalus dominii* (Rohlena) Chrtek & Krísa, *Tithymalus helioscopioides* (Loscos & J.Pardo) Holub, *Tithymalus helioscopius* (L.) Hill, *Tithymalus obovata* Raf. **Common name(s)**: Madwoman's milk, Sun spurge. **Botanical description**: It is native to Europe, including Britain, south to the Mediterranean, and east to Central Asia. It is an annual growing up to 0.4 m. It is in flower from May to October. The species is hermaphrodite (has both male and female organs) and are pollinated by flies. It prefers dry or moist soil. **Medicinal use(s)**: It possesses anticancer, antiperiodic, anti-inflammatory, and anthelmintic properties. The milky sap is applied externally to skin eruptions. The seeds, mixed with roasted pepper, have been used in the treatment of cholera.			Chloroform (E2)	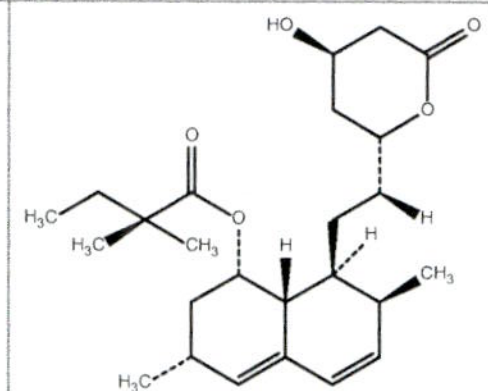	
				Euphornin L	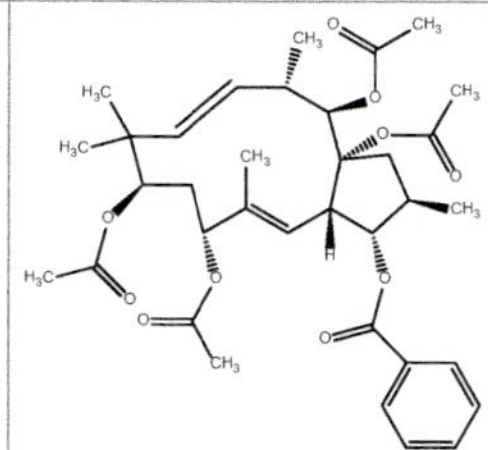	
				Euphoscopin F	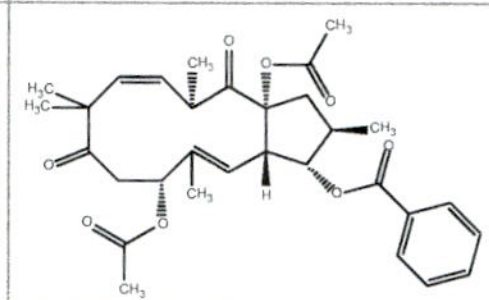	

(continued)

Table 4.58 (continued)

Family	Botanical name	Image	Plant part used	Bioactive compound(s)	Structure	Reference(s)
				Diterpenoids	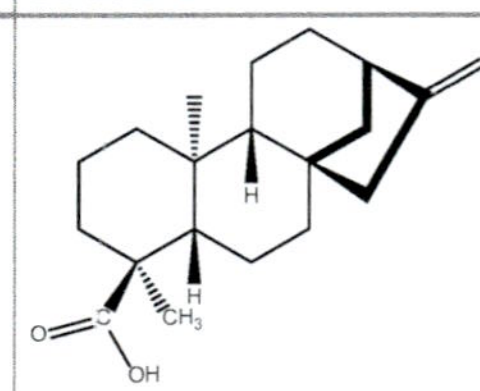	
				Jolkinolide B	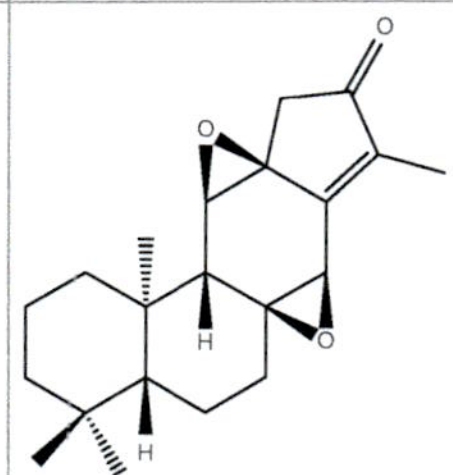	
	Euphorbia guyoniana Boiss. & Reut.		Aerial parts	Guyonianin E	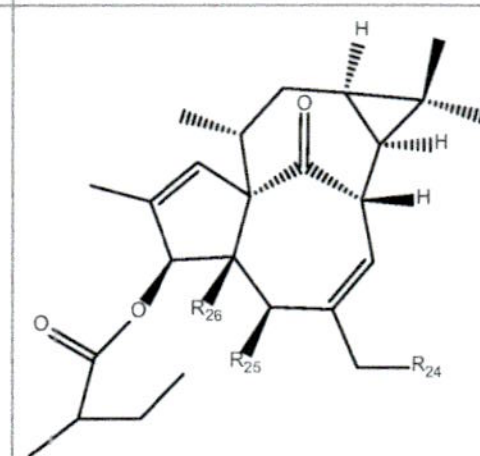	Hegazy et al. (2007)

	Synonym(s): *Tithymalus guyonianus* (Boiss. & Reut.) Klotzsch & Garcke. **Common name(s)**: Guyoniana. **Botanical description**: It is a native plant of America, including Britain, south to the Mediterranean and east to Central Asia. It is common in cultivated ground throughout Britain. It is an annual growing up to 0.4 m. It is in flower from May to October. The flowers are hermaphrodite and are pollinated by flies. It cannot grow in the shade. It prefers dry or moist soil. **Medicinal use(s)**: It possesses anticancer and antifungal properties. It is used for breathing disorders including asthma, bronchitis, and chest congestion. It is also used for mucus in the nose and throat, throat spasms, hay fever, and tumors.			Jatrcphane diterpenes	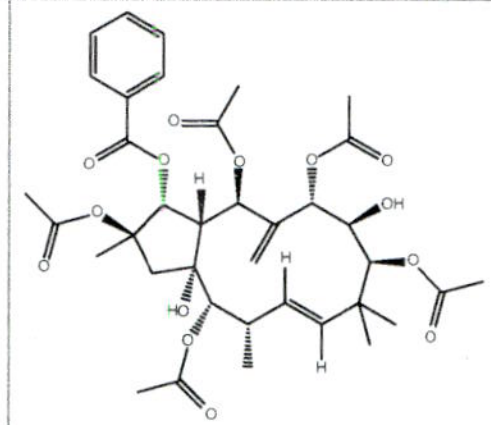	
	Euphorbia heterophylla L.		Whole plant	Akebia saponin D	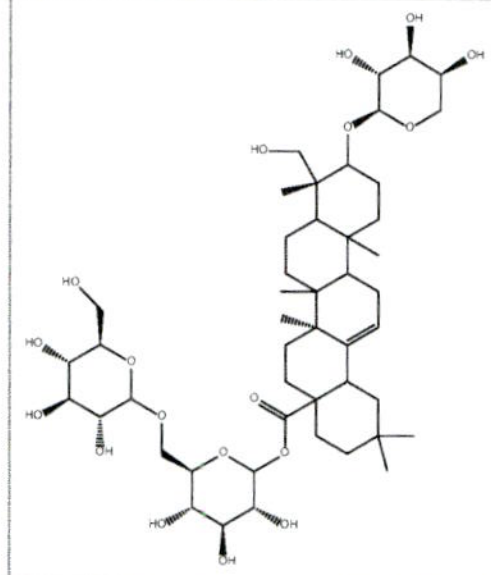	Flores and Ricalde (1996) and Jeong et al. (1999, 2008)

(continued)

Table 4.58 (continued)

Family	Botanical name	Image	Plant part used	Bioactive compound(s)	Structure	Reference(s)
	Synonym(s): *Agaloma angustifolia* Raf., *Cyathophora ciliata* Raf., *Cyathophora heterophyla* (L.) Raf., *Cyathophora picta* Raf., *Euphorbia calyciflora* Sessé & Moc., *Euphorbia elliptica* Lam., *Euphorbia epilobiifolia* W.T.Wang, *Euphorbia frangulifolia* Kunth, *Euphorbia geniculata* Ortega, *Euphorbia morisoniana* Klotzsch, *Euphorbia pandurata* Huber, *Euphorbia prunifolia* Jacq., *Euphorbia taiwaniana* S.S.Ying, *Euphorbia trachyphylla* A.Rich., *Poinsettia frangulifolia* (Kunth) Klotzsch & Garcke, *Poinsettia geniculata* (Ortega) Klotzsch & Garcke, *Poinsettia havanensis* Small, *Poinsettia heterophylla* (L.) Klotzsch & Garcke, *Poinsettia morisoniana* (Klotzsch) Klotzsch & Garcke, *Poinsettia prunifolia* (Jacq.) Klotzsch & Garcke, *Poinsettia ruiziana* Klotzsch & Garcke, *Tithymalus heterophyllus* (L.) Haw., *Tithymalus prunifolius* (Jacq.) Haw. **Common name(s)**: Kaliko splurge. **Botanical description**: It is a native plant of Mexico. It is herbaceous, erect, and 20–200 cm in height. The most common size is 40–60 cm tall. Milky latex is present when most parts of the plant are broken. The stem is branched and cylindrical, with nodes at regular intervals. The surface is smooth and reddish green. Each cyathium comprises a cup-shaped involucre with inconspicuous male flowers, producing a single stamen only, and a female flower, without sepals or petals, producing a 3-lobed, yellowish-green fruit. **Medicinal use(s)**: The latex and preparations of the leaves and root are applied to treat skin tumors. The roots are used as cathartic, emetic, and galactogogue. They are used in small doses in the treatment of gonorrhea and to increase milk production in breast-feeding women. The latex is irritant to the skin and eyes and may be employed as a rubefacient and to remove warts and corns; the latex is also used as an antidote against the irritation caused by the latex of other *Euphorbia* species. A decoction or infusion of the stems and fresh or dried leaves is taken as a purgative and laxative to treat stomachache and constipation and to expel intestinal worms. A leaf infusion is used as a wash to treat skin problems, including fungal diseases and abscesses. The leaf extract is taken to treat body pain.			Fenugreekine (saponins)	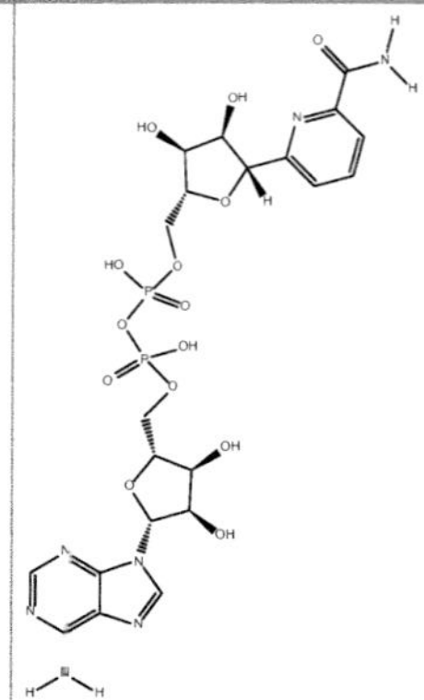	

<table>
<tr><td rowspan="4"></td><td>Euphorbia hirta L.</td><td></td><td>Leaf, Stem, Flower</td><td>Euphorbins B</td><td>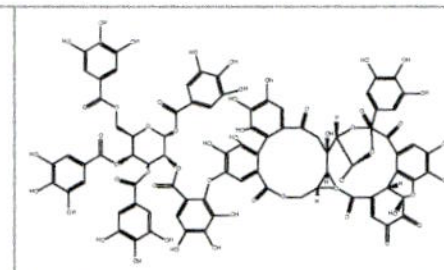</td><td>Hu et al. (2007)</td></tr>
<tr><td></td><td></td><td></td><td>Euphorbins E</td><td>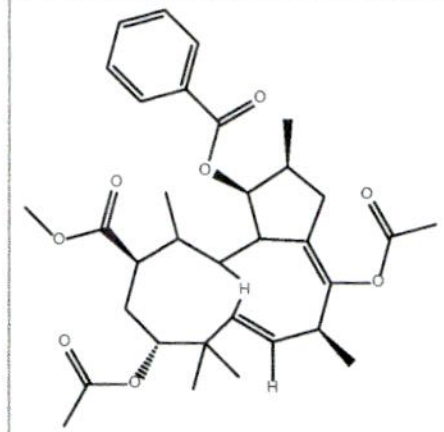</td><td></td></tr>
<tr><td colspan="3" rowspan="2">Synonym(s): Chamaesyce gemella (Lag.) Small, Chamaesyce hirta (L.) Millsp., Chamaesyce karwinskyi (Boiss.) Millsp., Chamaesyce rosei Millsp., Desmonema hirta (L.) Raf., Ditritea hirta (L.) Raf., Euphorbia bancana Miq., Euphorbia capitata Lam., Euphorbia chrysochaeta W.Fitzg., Euphorbia gemella Lag., Euphorbia globulifera Kunth, Euphorbia karwinskyi Boiss., Euphorbia nodiflora Steud., Euphorbia obliterata Jacq.
Common name(s): Asthma plant.
Botanical description: It is found in the Philippines, Australia, South America, and other parts of Asia. Euphorbia hirta is widely found in most parts of India. It is an annual hairy herb with many branches from the base to top, spreading up to 50 cm in height, reddish or purplish in color. The flowers are unisexual and found in axillary cymes at each leaf node. Flowers are purplish to greenish in color, dense, axillary, short-stalked clusters or crowded cymes, about 1 mm in length.
Medicinal use(s): It has many medicinal properties like hypotensive, anticancer, tonic, anxiolytic, analgesic, antimalarial, anti-asthmatic, antidiarrheal, antioxidant, anti-amoebic, anti-inflammatory, antifungal, antibacterial, anti-spasmodic, etc. It is used in the treatment of cancer, diarrhea, dysentery, intestinal infections, asthma, bronchitis, fever, eyelid styes, cough, asthma, bronchial infections, bowel complaints, helmintic infestations, wounds, kidney stones, abscesses, etc. Decoction of dry herbs is used for skin diseases. Decoction of fresh herbs is used as gargle for the treatment of thrush. Root decoction is used for nursing mothers deficient in milk and in the treatment of snakebites, vomiting, chronic diarrheas, and fevers.</td><td>Myricitrin</td><td>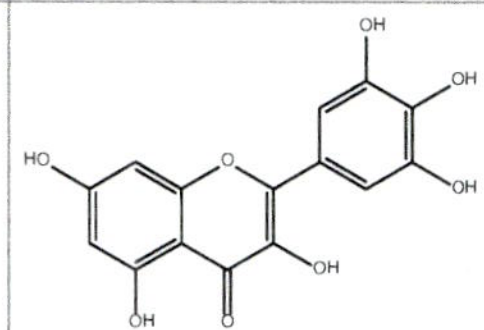</td><td></td></tr>
<tr><td>Quercitol</td><td>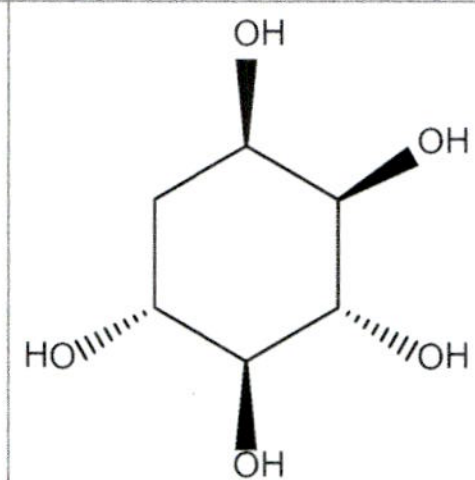</td><td></td></tr>
</table>

(continued)

Table 4.58 (continued)

Family	Botanical name	Image	Plant part used	Bioactive compound(s)	Structure	Reference(s)
				Quercitrin	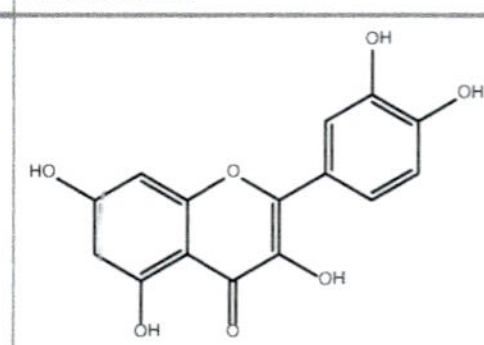	
				Camphol	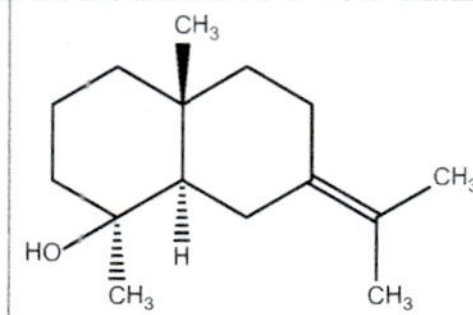	
				Leucocyanidol	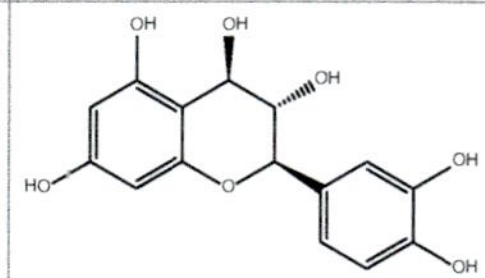	
	Euphorbia kansui Liou ex S.B.Ho		Root	Kansuiphorin	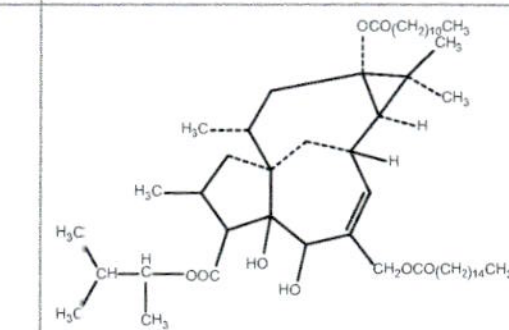	Cha (1977), Wu et al. (1991), Ding and Jia (1992), and Gonzalez et al. (2003)

	Synonym(s): No synonyms are recorded for this name. **Common name(s)**: Euphorbia root, Radix kansui, Radix euphorbiae kansui, Gan sui. **Botanical description**: This herb is mainly distributed in mainland China, including Gansu, Shanxi, Shaanxi, Ningxia, Henan, and other places. Habitats include low hills, slopes, sand, edge of paddy field, roadsides, etc. Its plant is a succulent perennial herb, 25–40 cm high, and the whole plant contains milk. Roots are slender and slightly curved, but some are like a chain of pearls, bar, or elongated oval. Skin is tan. Stems are erect and light purple red. Leaves are alternate, narrow lanceolate or linear-lanceolate, 3–5 cm long, 6–10 mm wide. and with obtuse apex, broadly cuneate base, and entire margin. Cyathium is arranged in cymose, usually 5–9 clustered at the stem end. Flowers are unisexual and without perianth. Capsule is round. Seeds are ovoid and brown. **Medicinal use(s)**: It covers meridians of the spleen, lung, kidney, bladder, large intestine, and small intestine. Primary functions are purgating excessive fluid and relaxing the bowels. Its uses and indications include edema, ascites, accumulation of pathogens in chest, epilepsy, cough, dysuria, and constipation.					
	Euphorbia lathyris L.		Stem	Euphorbia factor L2 (EFL2)	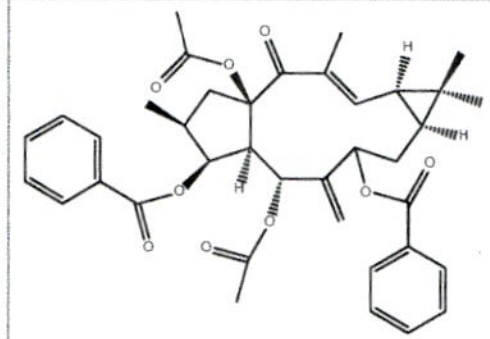	Ayatollahi et al. (2010) and Lin et al. (2017)

(continued)

Table 4.58 (continued)

Family	Botanical name	Image	Plant part used	Bioactive compound(s)	Structure	Reference(s)
	Synonym(s): *Epurga lathyris* (L.) Fourr., *Euphorbia decussata* Salisb., *Euphorbia spongiosa* Ledeb. ex Schrank, *Euphorbion lathyrum* (L.) St.-Lag., *Galarhoeus decussatus* (Salisb.) Gray, *Galarhoeus lathyris* (L.) Haw., *Keraselma lathyris* (L.) Raf., *Tithymalus lathyris* (L.) Hill. **Common name(s)**: Caper spurge, Moleplant. **Botanical description**: It is a native plant of Europe. It is an annual growing up to 1 m. It is in flower from May to June, and the seeds ripen from July to August. The flowers are monoecious and are pollinated by flies. It can grow in semi-shade or no shade. It prefers dry or moist soil. **Medicinal use(s)**: It was used in the past as a violent purgative, while the rubefacient action of the leaves was employed by beggars to raise unsightly sores on their skins to elicit pity and thereby obtain more money. All parts of the plant are emetic and purgative, and the plant is nowadays considered to be far too toxic for it to be used medicinally. The latex in the stems has been used externally as a depilatory and to remove corns, but it is too irritant to be used safely. The seed is used as diuretic, parasitic, and purgative. It has also been used in the treatment of dropsy, edema, tumors, amenorrhea, schistosomiasis, scabies, and snakebites. The fresh seed has an antitumor action, effective against acute lymphocytic and granulocytic leukemia. The plant has anticancer activity. It is also used as antiseptic, cathartic, emetic, and purgative. Use the plant with caution. One seed capsule is said to cause catharsis, several to cause an abortion.						
	Dimocarpus longan Lour.		Flower	Gallic acid	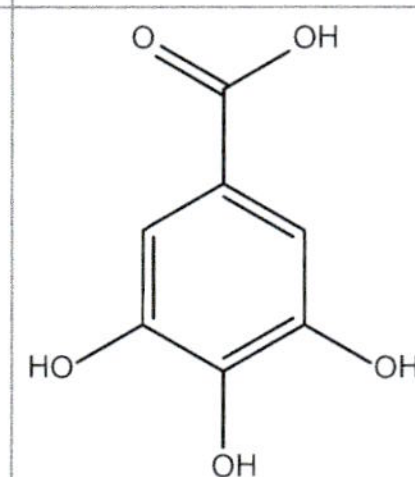	Zahra et al. (2006)

	Synonym(s): *Dimocarpus undulatus* Wight, *Euphoria cinerea* (Turcz.) Radlk., *Euphoria glabra* Blume, *Euphoria gracilis* Radlk., *Euphoria microcarpa* Radlk., *Euphoria nephelioides* Radlk., *Euphoria verruculosa* Salisb., *Nephelium longan* (Lour.) Hook., *Nephelium longana* Cambess. **Common name(s)**: Dragon eye. **Botanical description**: It is a native plant of Asia; it can be pruned to maintain a small tree size. This lychee relative has stiffer leaves and each fruit is encased in a thick brown skin which, like the lychee, separates easily from the aril. The flesh is sweet and is usually eaten fresh but can be canned or dried. The crisp skin is easy to remove, and the white flesh has a texture and sweetness reminiscent of litchi. They generally eat longan fresh at room temperature, but the canned fruit is often served over ice as a dessert. **Medicinal use(s)**: It improves nerve function, helps overcoming irritation, and reduces fatigue. It can also treat insomnia and other sleep disorders. It reduces the risk of cardiovascular disease and stroke. It can fight free radicals in the body and prevent cell damage. It also can reduce the risk of developing several types of cancer. It can be used to prevent anemia. In addition, it proved to be capable of controlling iron levels in the body.			Ellagic acid	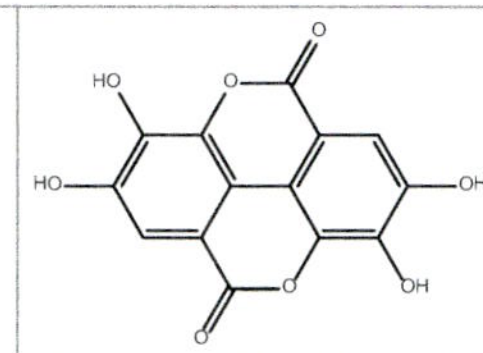	
				Corilagin (an ellagitannin)	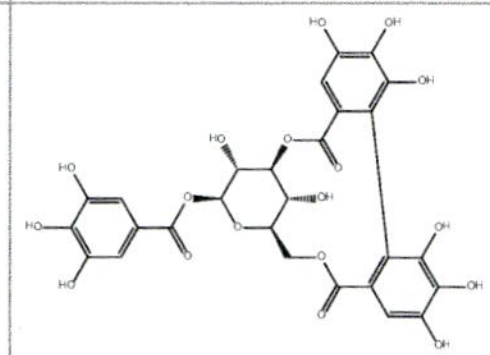	
	Euphorbia macroclada Boiss		Apical part	Dichloromethane	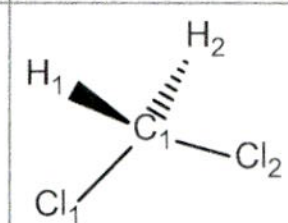	Zhang et al. (2010) and Jafarian et al. (2017)
				Anthraquinone	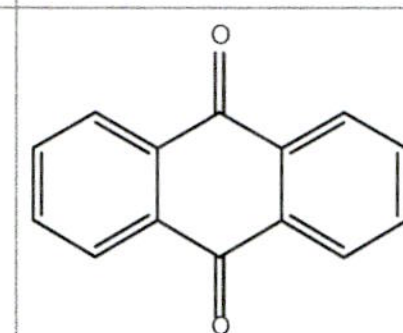	

(continued)

Table 4.58 (continued)

Family	Botanical name	Image	Plant part used	Bioactive compound(s)	Structure	Reference(s)
	Synonym(s): *Euphorbia damascena* Boiss., *Euphorbia lorentii* Hochst., *Euphorbia schizoceras* Boiss., *Euphorbia syspirensis* K.Koch, *Euphorbia tinctoria* Boiss. & A.Huet, *Tithymalus damascenus* (Boiss.) Klotzsch & Garcke, *Tithymalus macrocladus* (Boiss.) Klotzsch & Garcke, *Tithymalus schizoceras* (Boiss.) Klotzsch & Garcke, *Tithymalus syspirensis* (K.Koch) Klotzsch & Garcke. **Common name(s)**: Spurge. **Botanical description**: It is native plant of Iran. The plants are annual or perennial herbs, woody shrubs, or trees with caustic, poisonous milky latex. The roots are fine or thick and fleshy or tuberous. Many species are more or less succulent, thorny, or unarmed. The main stem and mostly also the side arms of the succulent species are thick and fleshy, 15–91 cm tall. The deciduous leaves may be opposite, alternate, or in whorls. In succulent species, the leaves are mostly small and short-lived. **Medicinal use(s)**: It has been used for the treatment of various conditions such as asthma, leukemia, cancer, skin diseases, and intestinal parasitic infections. A number of these species also possess antiviral, antibacterial, antifungal, and cytotoxic properties. The stems and leaves of *E. macroclada* contain some quantities of polyphenols, flavonoids, tannins, alkaloids, saponins, and terpenoid compounds.					

<table>
<tr><td rowspan="3"></td><td>Euphorbia marginata Pursh.</td><td></td><td>Entire plant</td><td>Ingenol mebutate</td><td>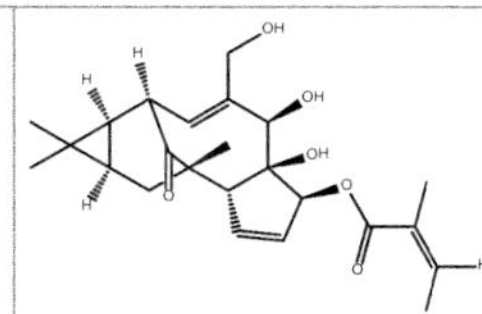</td><td>De Feo et al. (1992) and Seca and Pinto (2018)</td></tr>
<tr><td colspan="3" rowspan="2">Synonym(s): Agaloma marginata (Pursh) Á.Löve & D.Löve, Dichrophyllum marginatum (Pursh) Klotzsch & Garcke, Dichrophyllum variegatum (Sims) Klotzsch & Garcke, Euphorbia bejariensis DC., Euphorbia leucoloma Raf., Euphorbia variegata Sims, Lepadena leucoloma (Raf.) Raf., Lepadena marginata (Pursh) Nieuwl., Tithymalus marginatus (Pursh) Cockerell, Tithymalus variegatus (Sims) Haw.
Common name(s): Snow-on-the-mountain, Smoke-on-the-prairie.
Botanical description: It is native to parts of temperate North America, from Eastern Canada to the Southwestern United States. It is like snow-on-the-mountain and has gray-green leaves along with branches and smaller leaves in terminal whorls with edges trimmed with wide white bands, creating together with the white flowers. It is annual growing to 0.6 m by 0.3 m. It is in flower in September. The flowers are monoidal and are pollinated by insects.
Medicinal use(s): It is used to treat warts, verrucas, and leukorrhea. An infusion of the crushed leaves has been used as a liniment in the treatment of swellings. An infusion of the plant has been used to increase milk flow in nursing mothers.</td><td>Curcumin</td><td>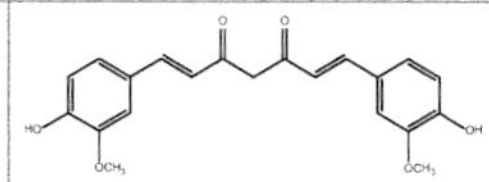</td><td></td></tr>
<tr><td>Paclitaxel</td><td>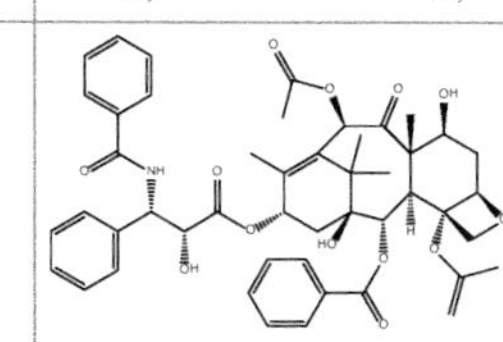</td><td></td></tr>
</table>

(continued)

Table 4.58 (continued)

Family	Botanical name	Image	Plant part used	Bioactive compound(s)	Structure	Reference(s)
	Euphorbia micractina Boiss.		Entire plant	Euphactin D	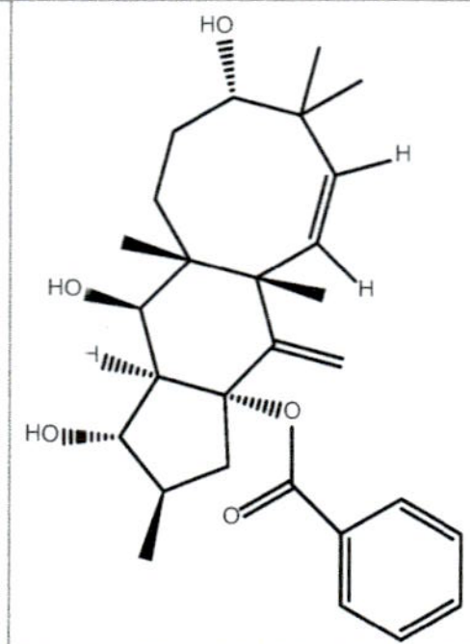	Shi et al. (1994)
	Synonym(s): *Euphorbia tangutica* Prokh., *Euphorbia wangii Oudejans*, *Galarhoeus tanguticus* (Prokh.) Prokh., *Tithymalus micractinus* (Boiss.) Soják, *Tithymalus tanguticus* (Prokh.) Prokh. **Common name(s)**: Not available. **Botanical description**: It is a native plant of China. It is an herb. The parts of the plant that grow above the ground are used to make medicine. It is used for breathing disorders including asthma, bronchitis, and chest congestion. It is also used to help remove mucus in the nose and throat and to treat throat spasms, hay fever, and tumors. Some people use it to cause vomiting. It is also used for treating worm infections, severe diarrhea, gonorrhea, and digestive problems. **Medicinal use(s)**: It is used as an antitumor agent. This plant is also used for the treatment of asthma, abscess, bronchitis, bedsores, cancer, cough, diarrhea, dysentery, eczema, earache, headache, inflammations, jaundice, kidney disease, leprosy, paralysis, skin diseases, scabies, toothache, ulcers, ringworm, and others and as anthelmintic, astringent, and diuretic.			Euphactin C	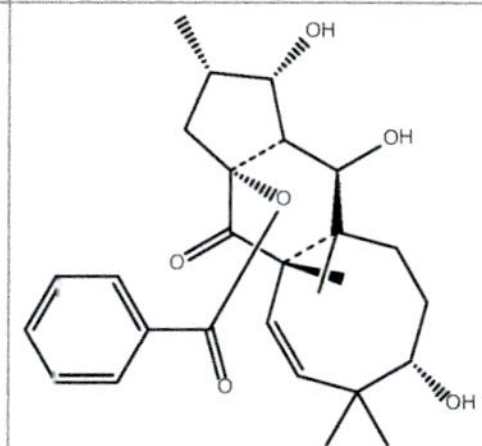	

	Euphorbia neriifolia L.		Leaf	Euphol (triterpenoid sapogenin)	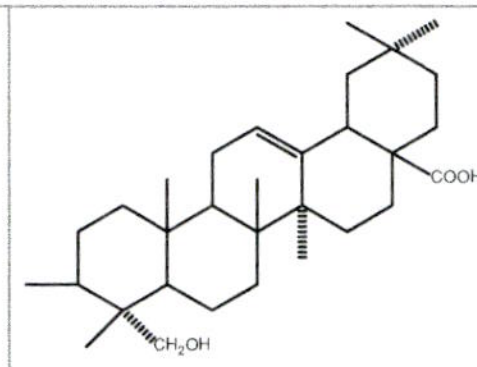	Patil et al. (2009)
	Synonym(s): *Elaeophorbia neriifolia* (L.) A.Chev., *Euphorbia edulis* Lour., *Euphorbia ligularia* Roxb. ex Buch.-Ham., *Tithymalus edulis* (Lour.) H.Karst. **Common name(s)**: Indian spurge tree, Hedge euphorbia. **Botanical description**: It is a native plant of East Asia – South China, India, Myanmar, Vietnam, Malaysia, and New Guinea. It is an erect, branched, prickly, succulent shrub or small tree looking somewhat like a cactus, but with large, persistent leaves on younger parts of the plant, and growing up to 8 m tall. The latex obtained from the plant is used as diuretic, purgative, rubefacient, and vermifuge. **Medicinal use(s)**: It is used in the treatment of asthma. The diluted latex of this plant administered in doses of 10–20 drops thrice a day, has been found to relieve asthma attacks completely. The latex is applied externally on hemorrhoids and is used to relieve sore throat and cracked lips, to treat gonorrhea, and also to remove warts and cutaneous eruptions. The bark has been used as a strong purgative. The root is considered antiseptic. Mixed with black pepper, it is employed in the treatment of snakebites both internally and externally. The leaves are diuretic. The leaves are heated and squeezed, and the sap is taken, sometimes with salt, to treat asthma, wheezing in babies, colds, and stomach upset. The leaves are also used to treat fevers, coughs and colds, and diabetes. And the sap is also used to treat infected nails and to relieve earaches. The plant has been shown to contain terpenes, anthocyanins, alcohols, and steroids.					

(continued)

Table 4.58 (continued)

Family	Botanical name	Image	Plant part used	Bioactive compound(s)	Structure	Reference(s)
	Euphorbia nivulia Buch.-Ham.		Juice of leaves, root bark, stem, and latex	Guyonianin A	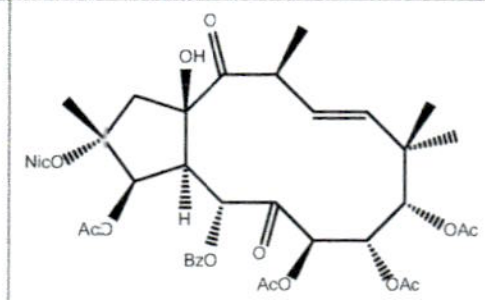	Hegazy et al. (2007)
	Synonym(s): *Euphorbia helicothele* Lem., *Euphorbia nivulia* var. *helicothele* (Lem.) Boiss. **Common name(s)**: Leafy milk hedge. **Botanical description**: It is an open-crowned succulent-branched spiny, glabrous, deciduous tree up to 9 m tall and 1 m in girth. Bark is thick and rough. Branches are terete, 2.5 cm thick, with paws of sharp stipular spines arising from spirally arranged tubercles. Alternately arranged leaves are stalkless or nearly so. Leaf blades obovate-spoon-shaped to inverted- lance-shaped, 10–25 × 3–8 cm, rounded at the tip, tapered to the base, lateral nerves 6–8 pairs, obscure, thick, fleshy, with the midrib prominent beneath. **Medicinal use(s)**: It is used for the treatment of skin cancer and actinic keratosis. It is use in Australian folk medicine. Medicinal uses were identified for >5% of the species in the genus, including descriptions of treatments for a variety of diseases. The most-cited medicinal uses around the world were treatments for digestive system disorders, skin ailments, and, especially in the Southern hemisphere, infections. Consensus ratios indicated that the most-valued medicinal uses of this plant are in the treatment of digestive and respiratory complaints, inflammation, and injuries.			Guyonianin B	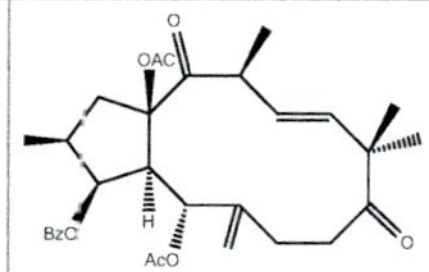	

	Euphorbia inaequilatera Sond.		Part not specified	Kaempferol	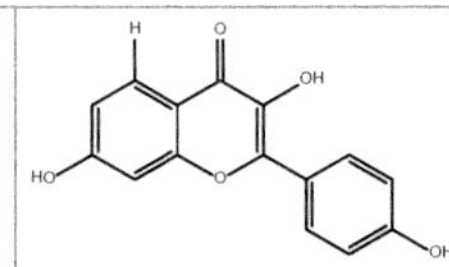	Hirschhorn (1982) and Maffei (2003a, b, c, d, e, f, g, h, i, j, k, l, m)
	Synonym(s): *Alectoroctonum sanguineum* Klotzsch & Garcke, *Anisophyllum inaequilaterum* (Sond.) Klotzsch & Garcke, *Anisophyllum mundii* Klotzsch & Garcke, *Anisophyllum setigerum* Klotzsch & Garcke, *Euphorbia haematodes* Boiss., *Euphorbia inaequalis* N.E.Br., *Euphorbia nelsii* Pax, *Euphorbia parvifolia* E.Mey. ex Boiss. **Common name(s)**: Wartweed. **Botanical description**: It is native to Southeast Asia where it is distributed through Sri Lanka, Cambodia, Laos, Burma, Thailand, and Vietnam. The plants are annual or perennial herbs, woody shrubs, or trees with a caustic, poisonous milky latex. The roots are fine or thick and fleshy or tuberous. Many species are more or less succulent, thorny, or unarmed. The main stem and mostly also the side arms of the succulent species are thick and fleshy, 15–91 cm tall. Leaves may be opposite, alternate, or in whorls. In succulent species, the leaves are mostly small and short-lived. **Medicinal use(s)**: Throughout the ages, the milky sap latex of plants belonging to the spurge family has been regarded as an effective remedy against warts. It has also been used to treat corns and ringworms. If the sap latex is used as a natural treatment for warts, it must be applied with care so it does not come into contact with the surrounding skin.					

(continued)

Table 4.58 (continued)

Family	Botanical name	Image	Plant part used	Bioactive compound(s)	Structure	Reference(s)
	Euphorbia pekinensis Rupr.		Root	Cytotoxic casbane diterpene	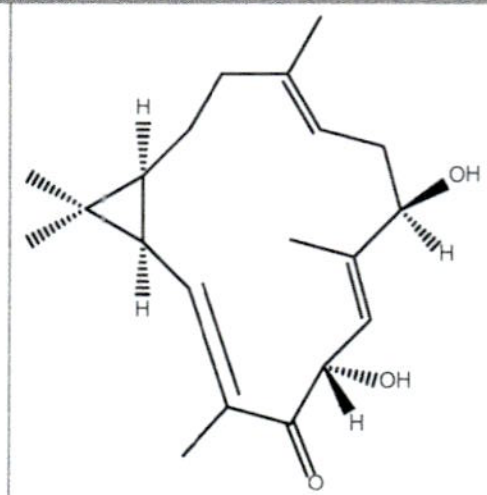	Cha (1977) and Tao et al. (2003)
	Synonym(s): *Euphorbia cavaleriei* H.Lév. & Vaniot, *Euphorbia hurusawae* Oudejans, *Euphorbia imaii* Hurus., *Euphorbia jessonii* Oudejans, *Euphorbia labbei* H.Lév., *Euphorbia lasiocaula* Boiss., *Euphorbia onoi* Franch. & Sav., *Euphorbia sampsonii* Hance, *Euphorbia sinanensis* (Hurus.) T.Kuros. & H. Ohashi, *Euphorbia sinensis* Jesson & Turrill, *Euphorbia subulatifolia* Hurus., *Euphorbia tchen-ngoi* (Soják) Radcl.-Sm., *Galarhoeus lasiocaulus* (Boiss.) Hurus., *Galarhoeus pekinensis* (Rupr.) H.Hara, *Galarhoeus sampsonii* (Hance) Hurus., *Tithymalus pekinensis* (Rupr.) H.Hara, *Tithymalus tchen-ngoi* Soják. **Common name(s)**: Peking spurge. **Botanical description**: It is a native plant of China and India. It is a perennial herb. Usually it reaches up to 30–80 cm high. The whole euphorbia plant contains white sap. Slender roots are in conical shape. Stems are erect, branching on the top, and covered with white pubescence. Single leaf is alternate, almost sessile, oblong or lanceolate, 3–6 cm long, 6 to 12 mm wide, entire at margin, and slightly glaucous below. It is harvested in late fall or early spring. **Medicinal use(s)**: Its ether and hot water extract can stimulate the bowels and thus induce diarrhea. The root is emetic, antibacterial, diuretic, purgative, vasodilator and emmenagogue. It is also used in the treatment of fullness of the chest, oedema, epilepsy, sticky sputum, carbuncle and tubercle.					

	Euphorbia poissonii Pax.		Leaf	12-Deoxyphorbol 13-(9,10-methylene) undecanoate (3),	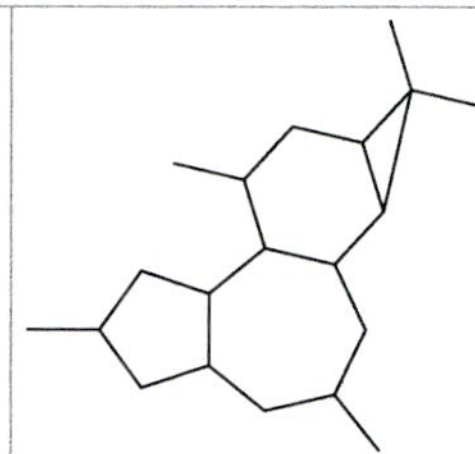	Evans and Kinghorn (1975) and Sundaram et al. (2010)
	Synonym(s): No synonyms are recorded for this name. **Common name(s)**: Euphorbia poissoni, Fula-pulaar. **Botanical description**: It is native to Northern Nigeria. It is having an erect much-branched shrub 1.2–2 m high, with candelabriform branching with one to several columns and topped with large green leaves during the growing season, occasionally sub-spiny. The flowers are short peduncle yellow/green or greenish with red stamens. They grow well in a very draining mineral potting substrate. During the summer, they enjoy average feeding and watering. When dormant, plants are relatively cold tolerant. **Medicinal use(s)**: It is used to treat gastric cancer. It has selective nature of binding and killing pain receptors while leaving other nerve cells intact; resiniferatoxin is currently being researched as a possible treatment for chronic pain.			12-deoxyphorbol	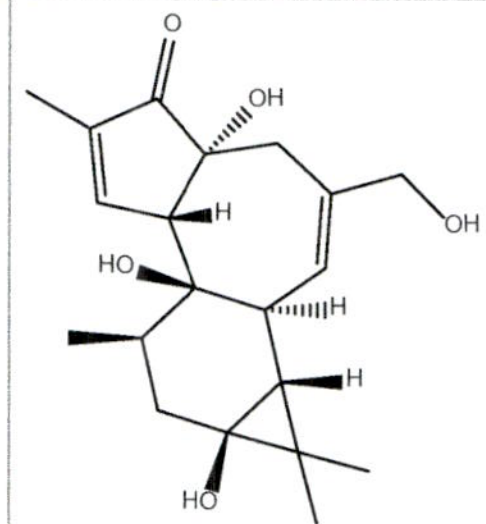	
				Tigliane	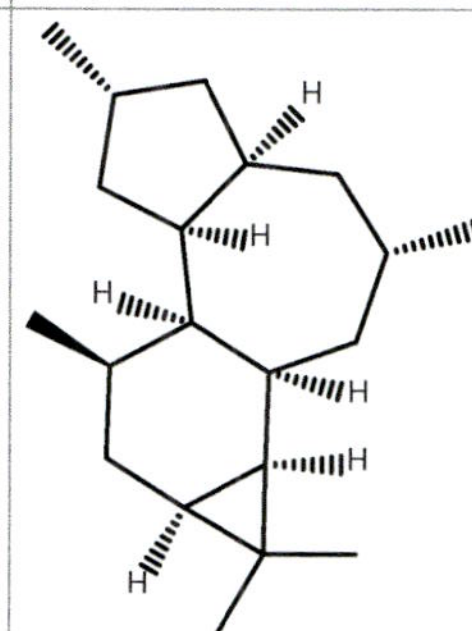	

(continued)

Table 4.58 (continued)

Family	Botanical name	Image	Plant part used	Bioactive compound(s)	Structure	Reference(s)
	Euphorbia prolifera Buch.-Ham. ex D.Don		Latex, root	Myrsinol	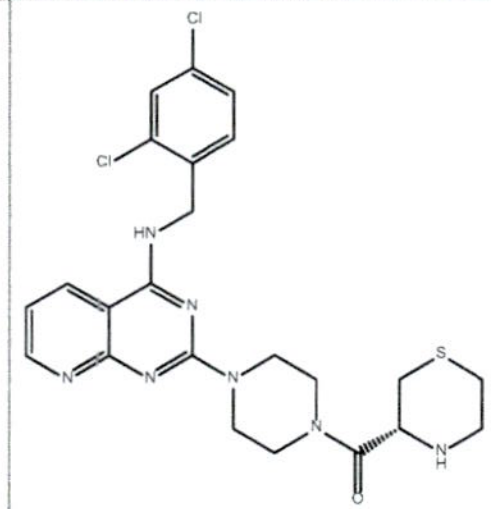	Wu et al. (1991, 1995), Dagang et al. (1994), and Chen et al. (2016a, b, c)
	Synonym(s): *Euphorbia pinus* H.Lév., *Tithymalus proliferus* (Buch.-Ham. ex D.Don) Soják. **Common name(s)**: Milk hedge. **Botanical description**: It is a native plant of Nepal. It is erect glabrous glaucous perennial herb up to 50 cm tall with several subsimple stems arising from a stout rootstock. Stem leaves alternate, subsessile; leaf blades linear, linear-oblong, linear-lanceolate, lanceolate or oblanceolate, 1.5–6 cm × 1–11 mm, acute, subacute, obtuse, rounded or even retuse at the apex, rounded to attenuate at the base, entire, flat or sometimes revolute, coriaceous, palmi- or triplinerved. Pseudumbel, the rays once to 3 times bifid and then often proliferating as densely leafy sterile shoots with linear leaves; axillary rays and/or proliferating shoots often formed below the pseudumbel; pseudumbel leaves 2–5, opposite or whorled, ovate to ovate-lanceolate, otherwise resembling the upper stem leaves; ray leaves opposite, sessile, generally smaller than but otherwise resembling the pseudumbel leaves. **Medicinal use(s)**: The extract has been found effective for treatment of bleeding hemorrhoids due to its contents of flavonoids, phenolics, and phenolic acids. It is having anticancer properties.			Euphorprolitherin	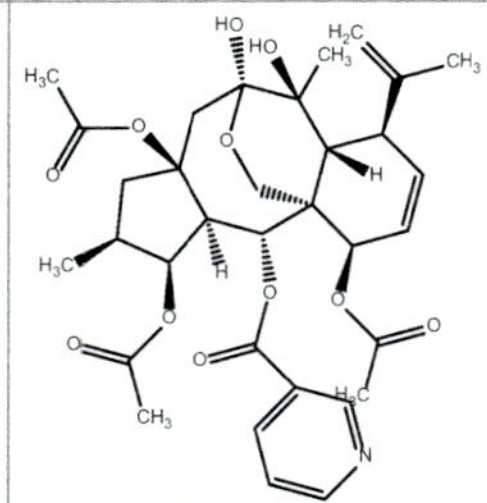	
				Proliferins A	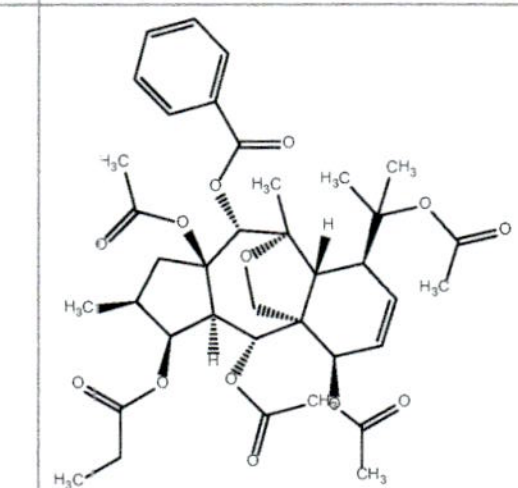	

	Euphorbia pulcherrima Willd. ex Klotzsch		Leaf	Amylodextrin	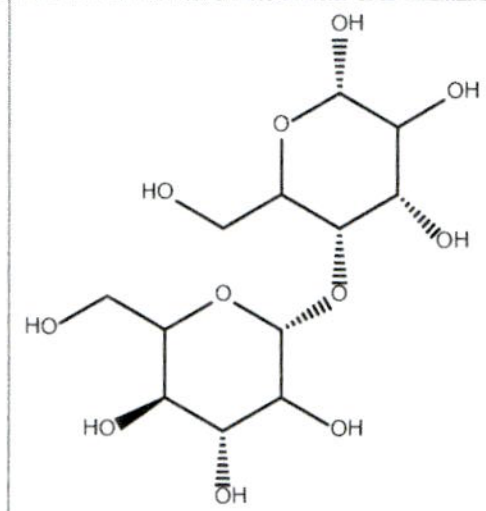	Appendino et al. (2003)
	Synonym(s): *Euphorbia erythrophylla* Bertol., *Euphorbia fastuosa* Sessé & Moc., Pleuradena *coccinea* Raf., *Poinsettia pulcherrima* (Willd. ex Kleitsch) Graham. **Common name(s)**: Poinsettia, Mana angangbi, Christmas star. **Botanical description**: It is a tall, rangy shrub that grows to a height of 10 ft. It has large, dark green, oval leaves that are "toothed" on the sides and pointed at the tips. They emerge from smooth green erect stems. Flowers are borne at the stem tips in winter. The actual flowers are quite small. The bracts are actually modified leaves – in the species, there usually are 8–10 bracts that are 4–7 in. long. **Medicinal use(s)**: The whole plant and its sap are used to make medicine. They also take the latex to kill pain, kill bacteria, and cause vomiting. Some people apply poinsettia latex directly to the skin to remove hair, treat warts, and heal other skin disorders.					
	Euphorbia schimperi C. Presl		Apical part	α-Amyrin	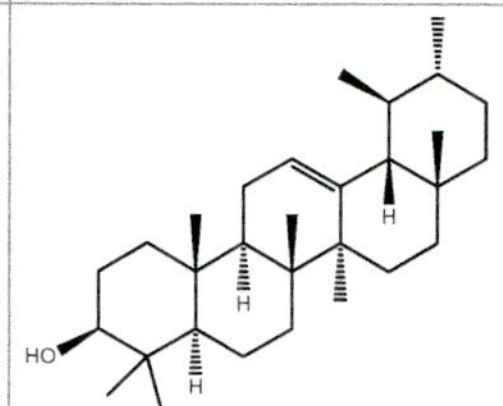	Rangkadilok et al. (2005)

(continued)

Table 4.58 (continued)

Family	Botanical name	Image	Plant part used	Bioactive compound(s)	Structure	Reference(s)
	Synonym(s): *Tirucalia schimperi* (C.Presl) P.V.Heath. **Common name(s)**: Prostrate spurge, Pencil cactus. **Botanical description**: It is a native plant of Nubia Nile Land coast region, Eritrea Nile Land near Acrur, Schweinfurth, and Riva. It is a succulent bush up to 6 ft. high, leafless except on the very young branches, spineless, glabrous. Branchlets alternate, more or less clustered, subparallel or very little divergent, 2 1/2–12 in. long, 1 1/2–3 in. thick, terete, with prominent leaf scars. Leaves not seen, soon deciduous. Umbels terminal, of 3–7 simple rays, 1/5–1/2 in. long, each bearing 1 involucre, glabrous; when young with a whorl of thin green oblong acute bracts at the base of the umbel about as long as the peduncle-like rays and a pair. **Medicinal use(s)**: The diluted latex is drunk to treat infections of tapeworm and other intestinal worms. The latex is applied to wounds in order to promote blood clotting and tissue healing. It is warmed by placing a section of the stem in hot ashes and is then applied as eardrops to treat earache. A leaf decoction, the roasted leaves or the leaf sap mixed with beer, is taken to treat gonorrhea. The roots are a strong purgative and are given to children to chew to treat stomach complaints, whereas adults chew the roots as an emetic to treat epilepsy, poisoning, and snakebites. The ground roots are eaten in a porridge to treat constipation. The stems are used as a tonic, being mixed with soup and given to new mothers after labor. The stems are taken as a recreational stimulating drug by the Maasai people. Stem ash is applied to sores caused by leprosy.			Kaempferol	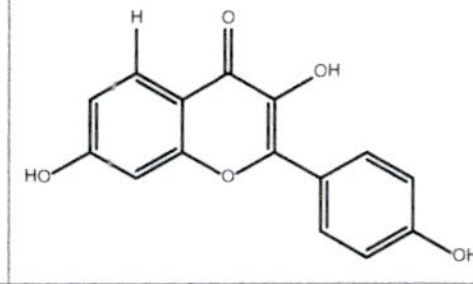	
				Luteolin	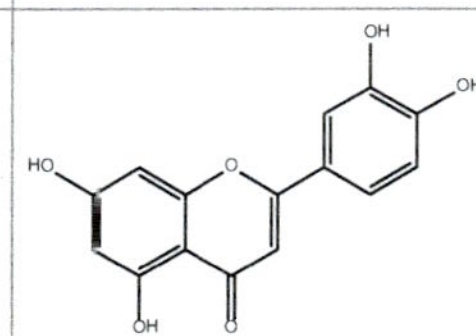	
				Scopoletin	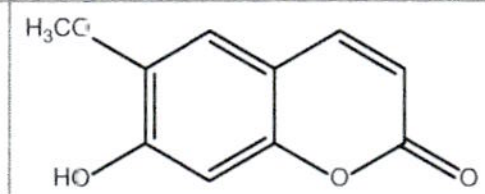	
	Euphorbia serrata L.		Latex	Vincristine	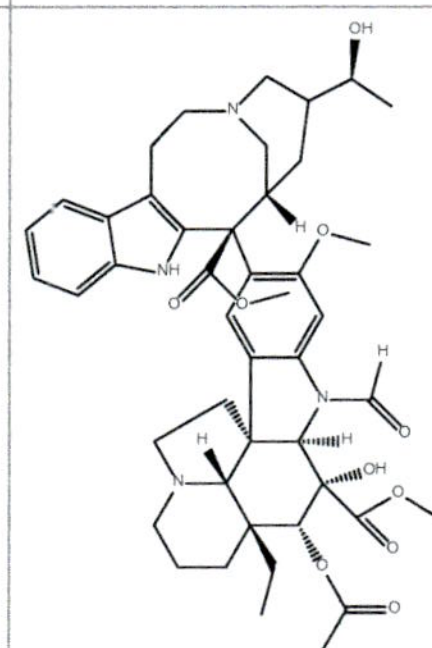	Martinez-Lirola et al. (1996), Maffei (2003a, b, c, d, e, f, g, h, i, j, k, l, m), and Monteiro et al. (2014)

	Synonym(s): *Chylogala serrata* (L.) Fourr., *Euphorbia truncata* (Pers.) Loudon, *Galarhoeus serratus* (L.) Haw., *Tithymalus denticulatus* Moench, *Tithymalus serratus* (L.) Hill. **Common name(s)**: Serrated spurge and Sawtooth spurge. **Botanical description:** It is native to Europe and North Africa, but it is present elsewhere as a weedy introduced species. This is a perennial herb growing anywhere from 20 cm to about half a meter in height. The leaves are long and very narrow on most of the plant, with more oval-shaped leaves towards the tips of the stems. They are finely toothed. At the ends of the branches are inflorescences of tiny flowers. The fruit is a spherical capsule about half a centimeter wide containing tiny gray seeds. **Medicinal use(s)**: The leaves of euphorbia are known to be helpful in treating skin irritations. The milky sap, while poisonous, can be used to clear up warts on the surface of the skin. An extract made from the crushed euphorbia flower can heal eye infections and inflammations like conjunctivitis. The plant is believed to promote healing in cases of dengue fever by facilitating the production of platelets. It can also be used to treat snakebites. It is also known for its anthelmintic properties, and it can be used to get rid of worms and other parasitic organisms. It is also considered to boost breast milk production in lactating mothers. It can be used in the treatment of venereal diseases like gonorrhea. In fact, it has also found use in the treatment of impotency, premature ejaculation, and other sexual disorders. The root of euphorbia can be made into a paste and used for healing stomach pain. However, it should be consumed only in recommended doses, else it can induce vomiting. It is also said to possess antiviral properties, and it has been used in the treatment of dysentery and to alleviate the symptoms of diarrhea. This herb is used as a diuretic and laxative. It is believed to expel water from the body and thus reduce edema and inflammation of the lymph nodes.	Vinblastine	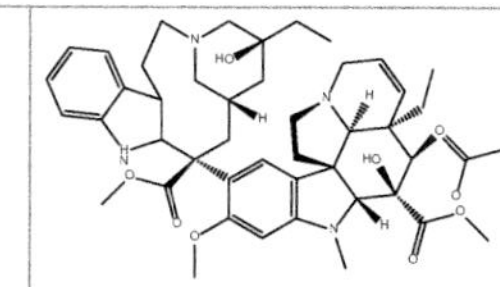	

(continued)

Table 4.58 (continued)

Family	Botanical name	Image	Plant part used	Bioactive compound(s)	Structure	Reference(s)
	Glochidion concolor Mull. Arg.		Leaf	Tannic acid (tannins)	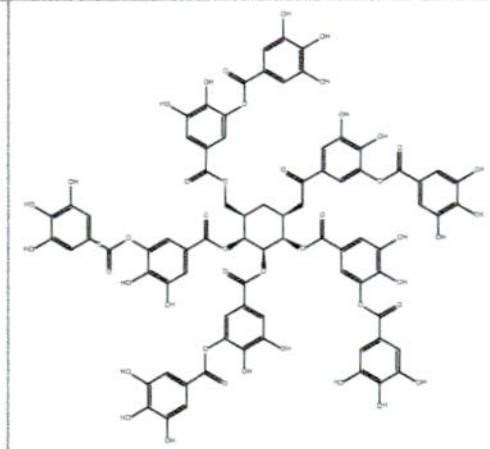	Singh et al. (1984) and Maffei (2003a, b, c, d, e, f, g, h, i, j, k, l, m)
	Synonym(s): *Diasperus concolor* (Müll.Arg.) Kuntze, *Phyllanthus concolor* (Müll.Arg.) Müll.Arg. **Common name(s)**: Not available. **Botanical description**: It is a native plant of the Western Pacific – Wallis-Futuna Island, Fiji, and Tonga. It is a tree with a crown that can be slender or spreading; it usually grows up to 13 m tall, with occasional specimens up to 25 m being recorded. The bole can be up to 50 cm in diameter. The plant remains shrub-like and can be as small as 1 m tall. The tree is harvested from the wild for local use as a medicine and source of wood. The plant is commonly protected when clearing for new gardens and is occasionally grown in home gardens in the Pacific. **Medicinal use(s)**: It is used to treat breast carcinoma. The leaves and bark are used for medicinal purposes. The leaves are eaten to cure stomach troubles, including diarrhea. The plants are used to treat thrush and hemorrhages.					

	Hura polyandra Baill.		Leaf	Carvacrol	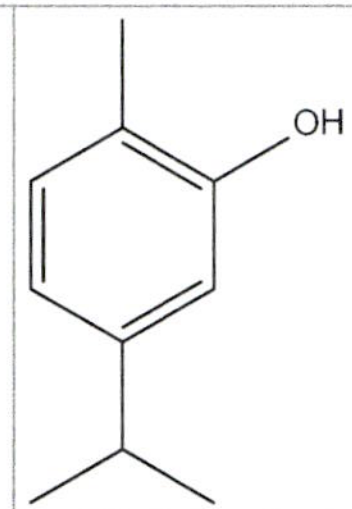 	Flores and Ricalde (1996) and Maffei (2003a, b, c, d, e, f, g, h, i, j, k, l, m)
	Synonym(s): No synonyms are recorded for this name. **Common name(s)**: Arbol del diablo. **Botanical description**: It is a native plant of Central America – Nicaragua, Honduras, El Salvador, Guatemala, and Mexico. It is often 30 m high or more, with a bole a meter or more in diameter. The bole is usually straight and regular and free of branches, usually for a great height. The branches and bole are densely covered with short, very sharp, hard prickles; the crown is broad and spreading. The tree is harvested from the wild for its timber and medicinal uses. This is one of the four or five largest trees of Central America, reaching its best development on the plains or in the foothills of the Pacific Coast, where it often occurs in great abundance and in some regions is a dominant tree. **Medicinal use(s)**: It possesses anticancer properties. The seeds are a violent purgative. The pulverized kernels are sometimes administered in small doses to human beings as a purgative and to expel intestinal parasites.					

(continued)

Table 4.58 (continued)

Family	Botanical name	Image	Plant part used	Bioactive compound(s)	Structure	Reference(s)
	Jatropha curcas L.		Leaf and stamen	Casbane diterpene	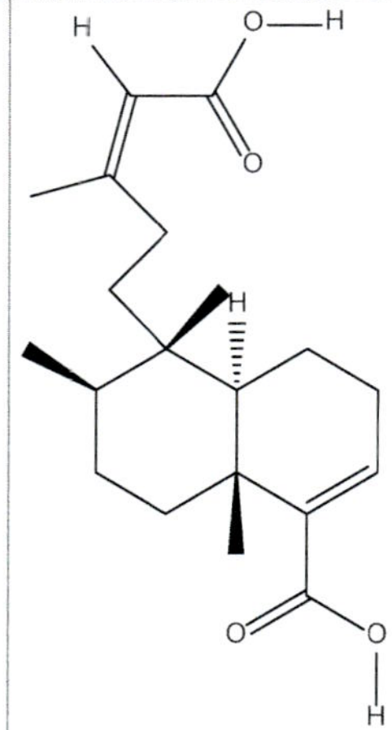	Liang et al. (2009) and Katagi et al. (2016)
	Synonym(s): *Castiglionia lobata* Ruiz & Pav., *Curcas adansonii* Endl., *Curcas drastica* Mart., *Curcas indica* A.Rich., *Curcas lobata* Splitg. ex Lanj., *Curcas purgans* Medik., *Jatropha acerifolia* Salisb., *Jatropha afrocurcas* Pax, *Jatropha edulis* Sessé, *Jatropha yucatanensis* Briq., *Manihot curcas* (L.) Crantz, *Ricinus americanus* Mill., *Ricinus jarak* Thunb. **Common name(s)**: American purging nut, Barbados nut, Barbados purging nut, Bed bug plant, Big purge nut, Black vomit nut, Brazilian stinging nut, Cuban physic nut, Curcas bean, Hell oil, Physic nut, Purge nut, Purging nut, Purging nut tree, Tuba. **Botanical description**: It is native to Texas in the United States as well as Mexico as far south as Oaxaca. It is a flowering plant and dioecious in nature. The arching, succulent stems reach a height of 20–60 cm and have few branches. They are tough and sufficiently flexible to be tied into overhand knots without breaking. The orange rootstock spreads to a length of around 1 m. Leaves are arranged simply, alternately, or fascicularly and are clustered on short shoots extending from the stems. **Medicinal use(s)**: It is used in the treatment of dental issues such as gingivitis, loose teeth, bleeding gums, and toothache. The latex is an astringent and may also be used as a red dye. Leatherstem is sometimes cultivated as an ornamental in xeriscapes or rock gardens.			Isoamericanol A	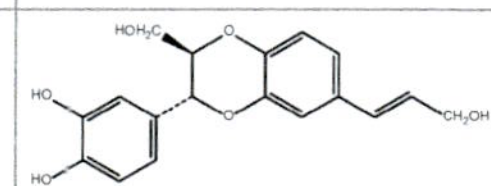	

	Jatropha dioica Sessé	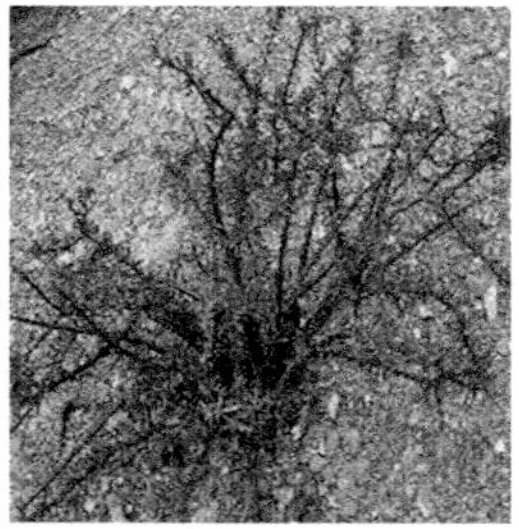	Leaf and root	Citlalitrione	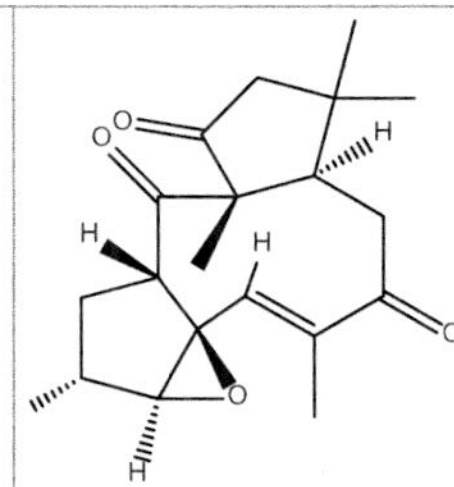	Calixto and Sant'ana (1987), Fatope et al. (1996), Melchor-Martínez et al. (2017), and Oliveira et al. (2013)
	Synonym(s): *Curcas cuneifolia* Baill., *Jatropha cuneifolia* Sessé & Moc., *Jatropha spathulata* (Ortega) Müll.Arg., *Loureira cuneifolia* Cav., *Mozinna sessiliflora* (Hook.) Small, *Mozinna spathulata* Ortega, *Zimapania schiedeanum* Engl. & Pax. **Common name(s)**: Leatherstem. **Botanical description**: It is a small tree or shrub with smooth gray bark, which exudes a white-colored, watery latex when cut. Normally, it grows between 3 and 5 m in height but can attain a height of up to 8 or 10 m in forests, under favorable conditions. It has large green to pale green leaves, alternate to subopposite, three- to five-lobed with a spiral phyllotaxis. **Medicinal use(s)**: It is used to treat cancer. It possesses anticancer, antitumor, anti-inflammatory, and antibacterial properties.			Ellagic acid (tannins)	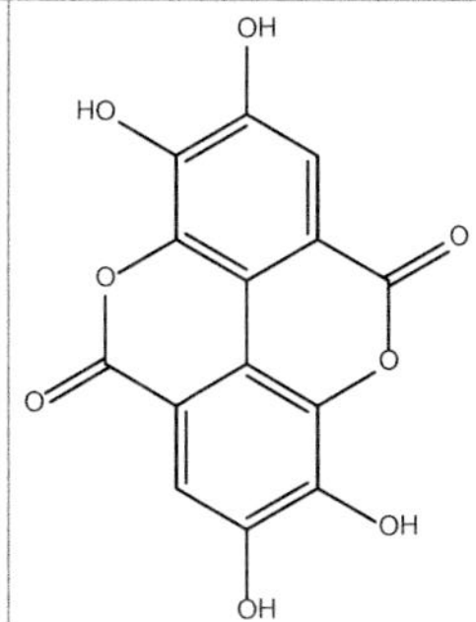	

(continued)

Table 4.58 (continued)

Family	Botanical name	Image	Plant part used	Bioactive compound(s)	Structure	Reference(s)
	Jatropha elliptica (Pohl) Oken		Fruit	2.6-Dimethyl-4-phenyl-pyridine-3,5-dicarboxylic acid diethyl ester	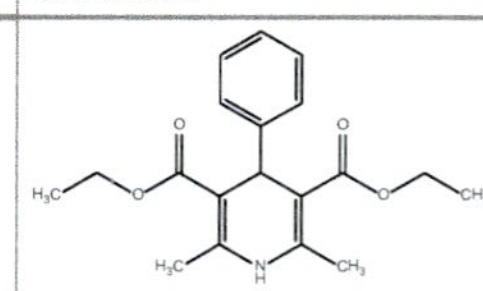	Calixto and Sant'ana (1987), Dos Santos and Sant'Ana (1999), Li et al. (2010), Devappa et al. (2011), and Sabandar et al. (2012)
	Synonym(s): *Adenoropium ellipticum* Pohl, *Adenoropium opiferum* (Mart.) Mart., *Jatropha lacertii* Silva Manso, *Jatropha officinalis* Mart. ex Pohl, *Jatropha opifera* Mart. **Common name(s)**: Physic nut. **Botanical description**: It is a small tree or shrub with smooth gray bark, which exudes a white-colored, watery latex when cut. Normally, it grows between 3 and 5 m in height but can attain a height of up to 8 or 10 m in forests, under favorable conditions. It has large green to pale green leaves, alternate to subopposite, three- to five-lobed with a spiral phyllotaxis. In winter, all the leaves fall and the shrub is leafless. **Medicinal use(s)**: It possesses antimicrobial property and is widely used in folk medicine as a purgative and in the treatment of severe itches, snake bites and syphilis. It is also used for the treatment of ulcer, urinary discharge, neoplasia and abdominal complaints.			Jatrophone	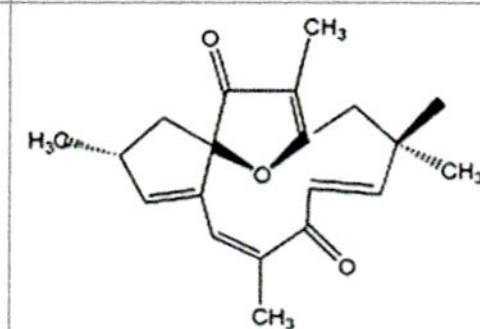	
				Jatropholone A	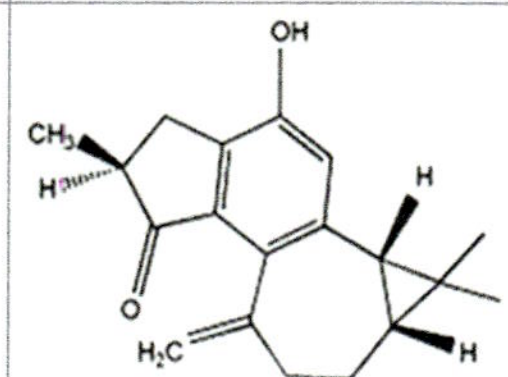	
				Jatropholone B	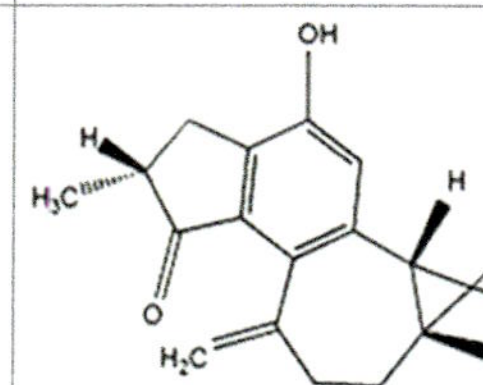	

	Jatropha gossypifolia L.		Leaf	Jatrophone	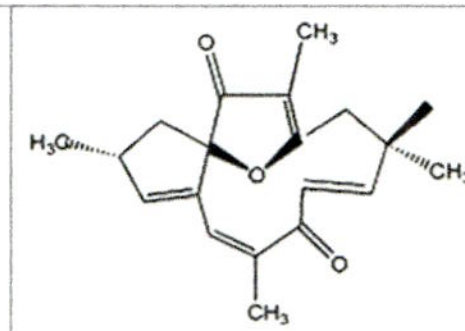	Smith-Kielland et al. (1996), Falodum et al. (2012), and Félix-Silva et al. (2014)
	Synonym(s): *Adenoropium gossypiifolium* (L.) Pohl, *Adenoropium jacquinii* Pohl, *Jatropha jacquinii* Baill., *Jatropha staphysagriifolia* Mill., *Manihot gossypiifolia* (L.) Crantz. **Common name(s)**: American purging nut, Lal bherenda, Tua-tua, Sibidigua, Cotton leaf, Puppet seed plant, Physic nut, Bellyache bush. **Botanical description**: It is native of South America, planted elsewhere in the tropics. These shrubs are purplish green and glandular-hairy on younger parts. Leaves in close spiral, deeply 3–5-lobed, 7–11 × 8–12 cm, orbicular-cordate, margin with gland-tipped hairs, lobes obtusely acuminate, 5-nerved; stipules glandular-hairy; petiole 3–10 cm long, glandular-hairy. Flowers red with yellow center, unisexual in axillary and terminal monoecious corymbose cymes. Bracts linear-lanceolate, glandular-hairy. **Medicinal use(s)**: The oil is applied internally as abortifacient and externally as rubefacient to treat rheumatic conditions and a variety of skin infections, although its use on the skin may also cause an irritative rash. In Senegal the seed oil is also applied to treat leprosy and rabies. The sap has a widespread reputation for healing wounds, as a hemostatic, and for curing skin problems; it is applied externally to treat infected wounds, ulcers, cuts, abrasions, ringworm, eczema, dermatomycosis, scabies, and venereal diseases.			2α-OH Jatrophone	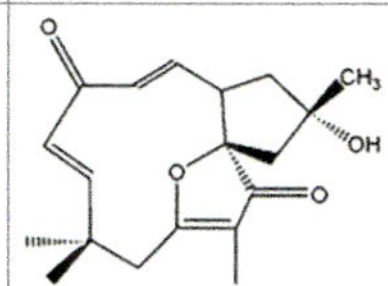	

(continued)

Table 4.58 (continued)

Family	Botanical name	Image	Plant part used	Bioactive compound(s)	Structure	Reference(s)
				2β-OH Jatrophone	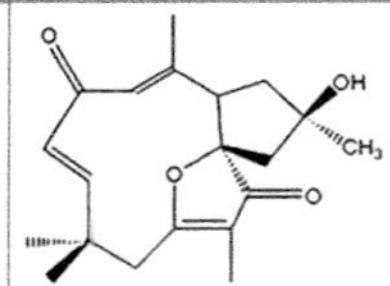	
				2β-OH-5,6-isoJatrophone	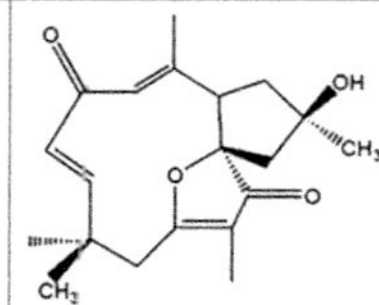	
				9β, 13α-dihydroxyisabellione	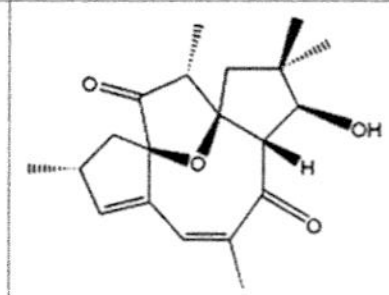	
				Citlalitrione	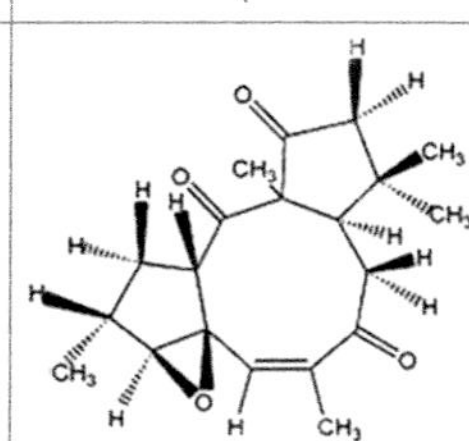	

		Lathyrane-type diterpenes	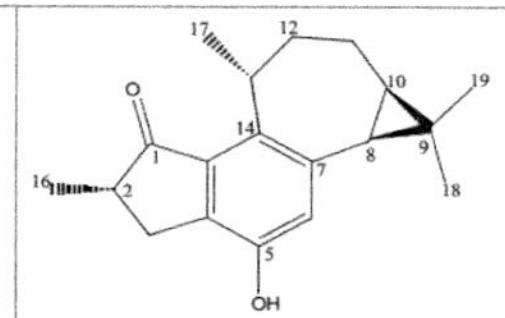	
		Jatropholone-B	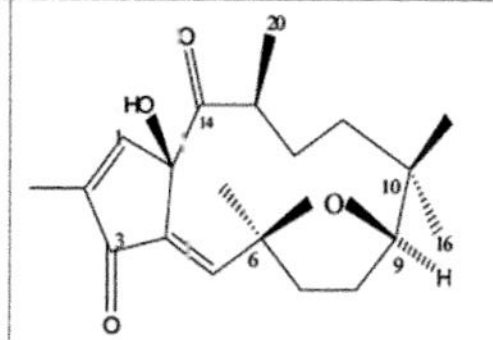	
		(4E)-jatrog-rossidentadione acetate	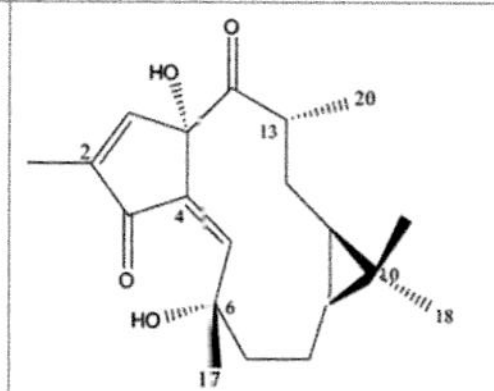	

(continued)

Table 4.58 (continued)

Family	Botanical name	Image	Plant part used	Bioactive compound(s)	Structure	Reference(s)
	Jatropha curcas L.		Leaf and stamen	Carvacrol	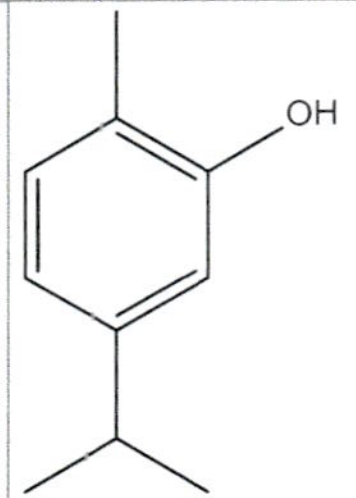	Balaji et al. (2009)
	Synonym(s): No synonyms are recorded for this name. **Common name(s)**: Physic nut, Jatropha, Barbados nut. **Botanical description**: It is a perennial poisonous shrub, up to 5 m high. It is an uncultivated nonfood wild species. The plant, originating in Central America, has been spread to other tropical and subtropical countries as well and is mainly grown in Asia and in Africa. It is used as a living fence to protect gardens and fields from animals. The plant sports large green to pale green leaves, 8–15 cm, broadly ovate, cordate, shallowly 3-lobed, petioles around 5–15 cm. The inflorescence is formed in the leaf axel; flowers are formed terminally, individually, with female flowers usually slightly larger. Flowers greenish yellow. Fruits 3–4 cm, globuse, yellowish green. *Jatropha curcas* seed can be used as biodiesel for any diesel engine without modification. **Medicinal use(s)**: The leaves, seeds, and oil of dravanti are used to treat ulcer, tumor, scabies, wound, hemorrhoid, splenomegaly, skin diseases, rheumatism, and paralysis. Paste of dravanti is applied to clean wounds to promote quick healing. Seeds are severely purgative.			Caniojane	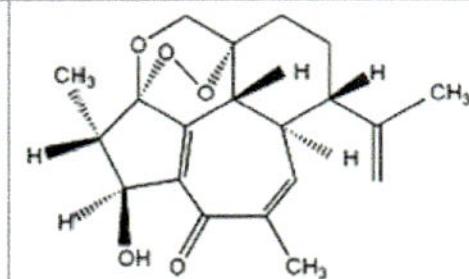	
				Curcasone A	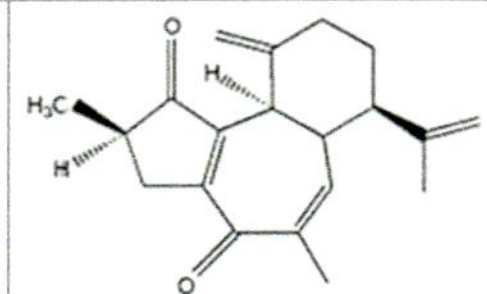	
				Curcasone B	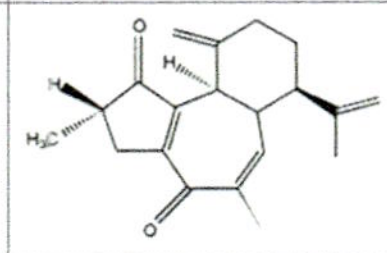	

		Curcasone C	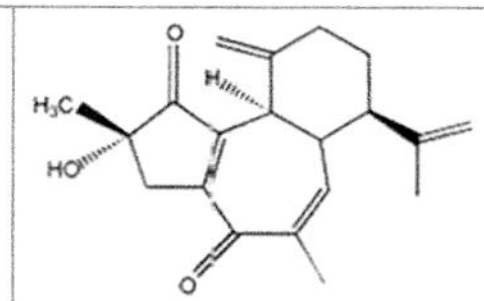	
		Curcasone D	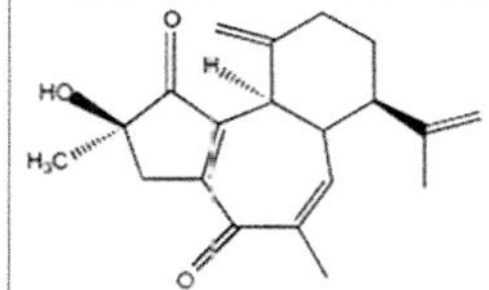	
		Jatropha factor C1	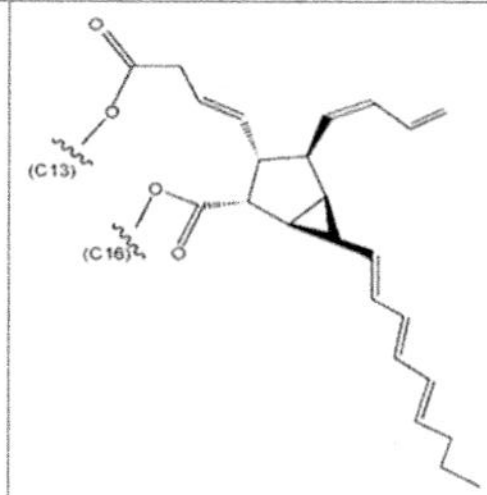	
		Jatrophalactam	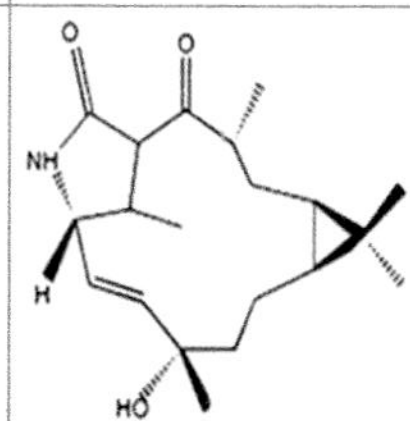	

(continued)

Table 4.58 (continued)

Family	Botanical name	Image	Plant part used	Bioactive compound(s)	Structure	Reference(s)
				Jatropholone A	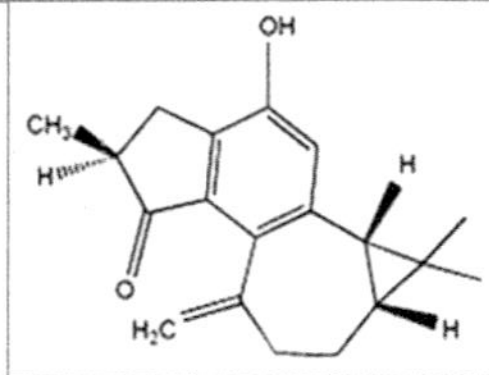	
				Jatropholone B	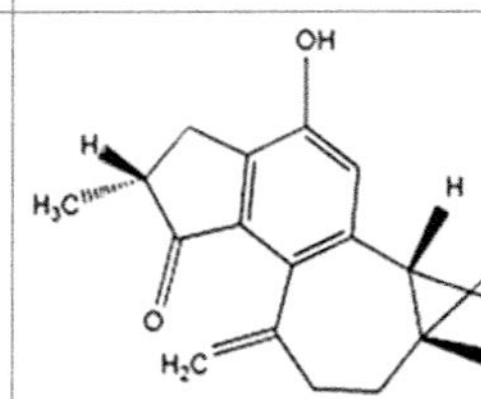	
	Jatropha dioica Sesse ex Cerv. L		Leaf and tree	Fenugreekine (saponins)	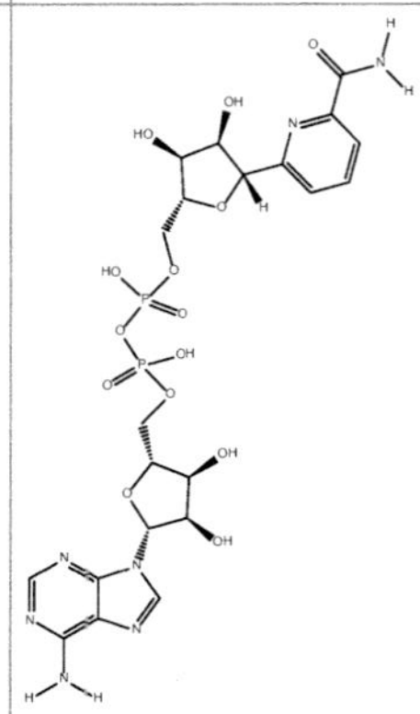	Devappa et al. (2011)

	Synonym(s): *Jatropha spathulata*. **Common name(s)**: Leatherstem. **Botanical description**: It is native to Texas in the United States as well as Mexico as far south as Oaxaca. The arching, succulent stems reach a height of 20–60 cm and have few branches. They are tough and sufficiently flexible to be tied into overhand knots without breaking. The orange rootstock spreads to a length of around 1 m. Leaves are arranged simply, alternately, or fascicularly and are clustered on short shoots extending from the stems. **Medicinal use(s)**: It is traditionally used in the treatment of dental issues such as gingivitis, loose teeth, bleeding gums, and toothache.			Tannic acid (tannins)	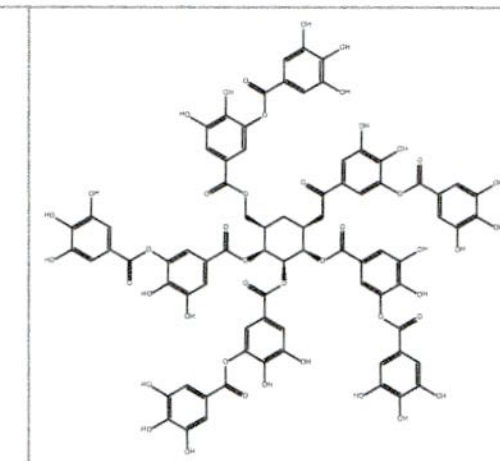	
	Jatropha gossypifolia L.		Leaf	Vetixin (flavonoids)	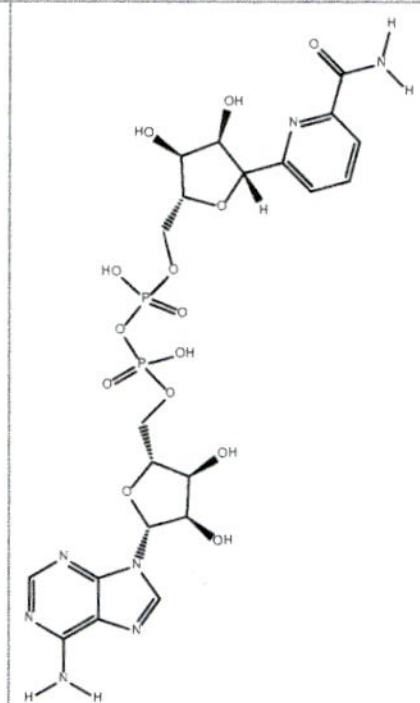	Wu et al. (2019)
	Synonym(s): No synonyms are recorded for this name. **Common name(s)**: Bellyache bush, Red physic nut. **Botanical description**: It is native to South America that is cultivated in tropical countries throughout the world in Africa, India, South America, West Indies, and Central America. It is a small shrub with dark green or more frequently purplish-red dark leaves, with 16–19 cm of length per 10–12.9 cm of width; they are alternate, palmate, and pubescent, with an acuminate apex, cordate base, and serrated margin. **Medicinal use(s)**: It is used as antimicrobial, anti-inflammatory, antidiarrheal, antihypertensive, and anticancer agents.					

(continued)

Table 4.58 (continued)

Family	Botanical name	Image	Plant part used	Bioactive compound(s)	Structure	Reference(s)
	Mallotus japonicus (L.f.) Müll.Arg.		Leaf	Mallotojaponin	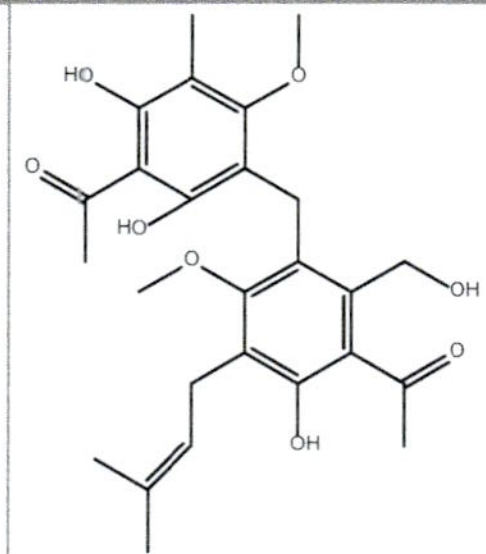	Arisawa et al. (1991) and Taira et al. (2015)
	Synonym(s): *Croton japonicus* L.f., *Rottlera japonica* (L.f.) Spreng. **Common name(s)**: Akamegashiwa, East Asian mallotus. **Botanical description**: It is also found in Japan and Korea. The plant is dioecious. The young shoots are red-colored. **Medicinal use(s)**: The fruit has anthelmintic properties. The large leaves were used to wrap food. The bark is used in the Japanese pharmacopoeia as a decoction against gastric ulcer and duodenal ulcer; young leaves, when boiled, are edible.			Mallotochromanol	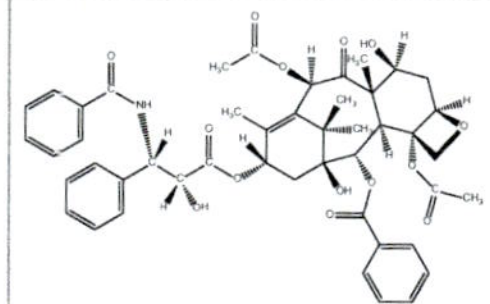	

		Maliotophenone	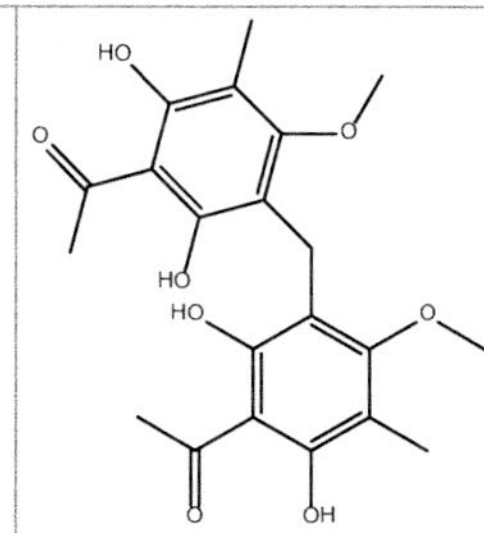	
		Mallctochromene	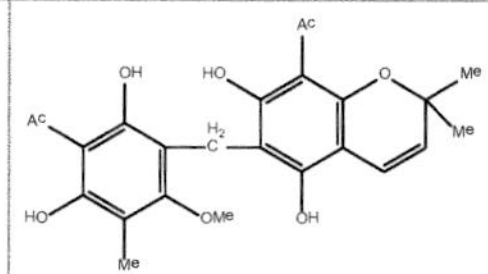	
		Mallotolerin	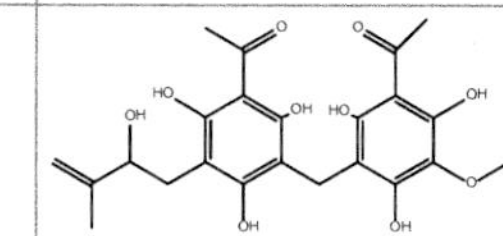	

(continued)

Table 4.58 (continued)

Family	Botanical name	Image	Plant part used	Bioactive compound(s)	Structure	Reference(s)
	Mallotus philippensis (Lam.) Müll. Arg.		Bark	Bergenin	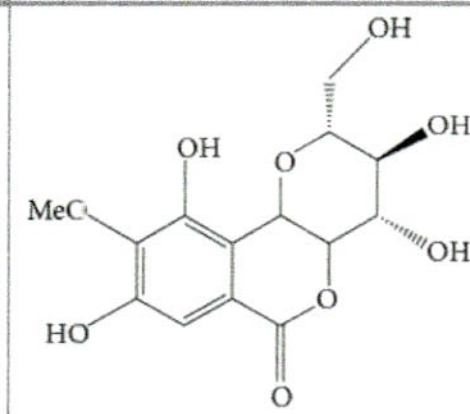	Arfan et al. (2007) and Gangwar et al. (2014)
	Synonym(s): *Aconceveibum trinerve* Miq., *Croton montanus* Willd., *Croton philippensis* Lam., *Echinus philippensis* (Lam.) Baill., *Macaranga stricta* (Rchb.f. & Zoll.) Müll.Arg., *Mallotus bicarpellatus* T.Kuros., *Mallotus reticulatus* Dunn, *Mappa stricta* Rchb.f. & Zoll., *Rottlera affinis* Hassk., *Rottlera aurantiaca* Hook. & Arn., *Rottlera philippensis* (Lam.) Scheff., *Rottlera tinctoria* Roxb., *Tanarius strictus* (Rchb.f. & Zoll.) Kuntze. **Common name(s)**: Kamala tree. **Botanical description**: It is a tree found throughout India. It has been in use as medicinal tree in India for ages. The tree can grow up to 10 m tall. Alternately arranged, ovate or rhombic-ovate leaves are rusty velvety. Male and female flowers occur in different trees. Female flowers are borne in lax spike like racemes at the end of branches or in leaf axils. Male flowers occur three together in the axils of small bracts. Capsule is trigonous-globular, covered with a bright crimson layer of minute, easily detachable reddish powder. Kamala is supposed to be a very useful tree. It is source of Kamala dye which is used in coloring silk and wool. It is used as antioxidant for ghee and vegetable oils. Oil is used as hair fixer and added in ointment. Seed oil is used in paints and varnishes. Seed cake is used as manure. **Medicinal use(s)**: The leaves are bitter and used as cooling agent and appetizer. The fruit has healing property and is used as purgative, anthelmintic, vulnerary, detergent, maturant, carminative, and alexiteric and in the treatment of bronchitis, abdominal diseases, spleen enlargement, etc.			Mallotophilippen C	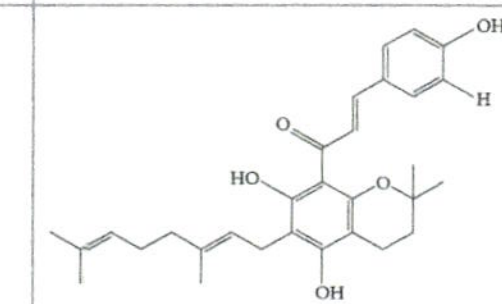	

		Mallotophilippen D	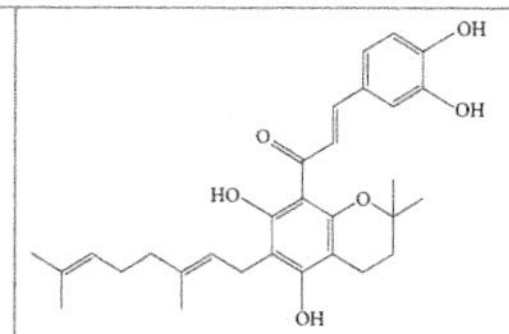	
		Mallotophilippen E	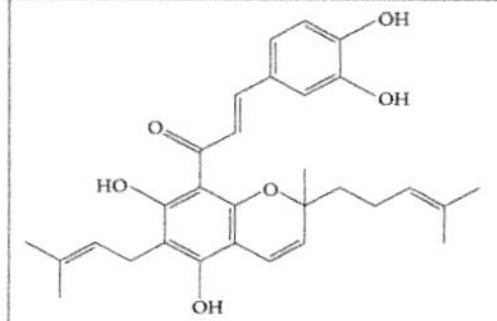	
		Rottlerin	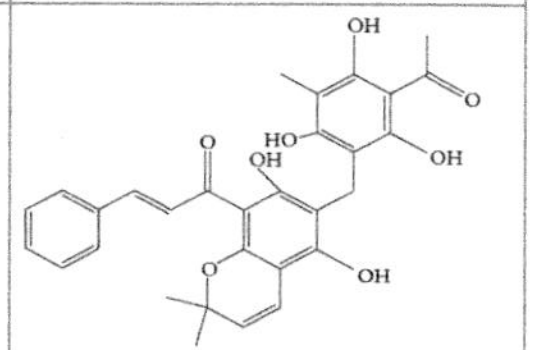	
		Isorottlerin	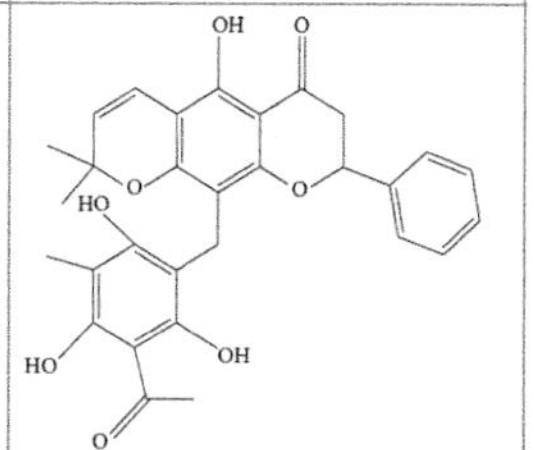	

(continued)

Table 4.58 (continued)

	Mallotus tetracoccus (Roxb.) Kurz.		Leaf	Bis-(2-ethyl hexyl) phthalate	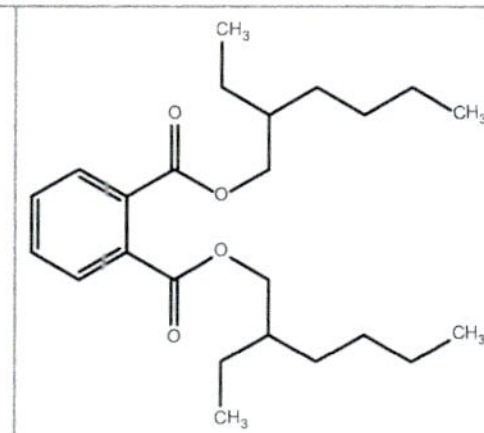	Ramalakshmi and Muthuchelian (2011)
	Synonym(s): *Mallotus ferrugineus* (Roxb.) Müll.Arg., *Rottlera ferruginea* Roxb., *Rottlera tetracocca* Roxb. **Common name(s)**: Pottamaram, Rusty kamala, Uppale mara, Thavatta, Vatta, Vatta kumbil, Vetta kumbil, Mullu polavu, Vatta. **Botanical description**: It is a native plant of Western Ghats of India, Sri Lanka, Myanmar, and Nepal. It is a tree up to 12 m tall. Bark is brown, warty, blaze cream. Young branchlets are flat. Leaves are simple, alternate, spirally arranged, stalk 3–12.5 cm long, swollen at both ends. Leaf blade is 8–25 × 6.5–20 cm, broadly ovate to circular, tip pointed to long pointed. The stalk is attached to the leaf in the middle. Leaves are usually entire or trilobed when young, leathery, and densely white velvety beneath, 3–5 nerved at base. Flowers are unisexual, dioecious, in axillary or terminal panicles, velvety. Capsule is spherical, covered with soft woolly spines and seeds. **Medicinal use(s)**: The plant is known for many Ayurvedic medicinal purposes. The bark and gum of the tree are used for treating fractures and venereal diseases.			3-Methyl-2-(2-oxopropyl) furan	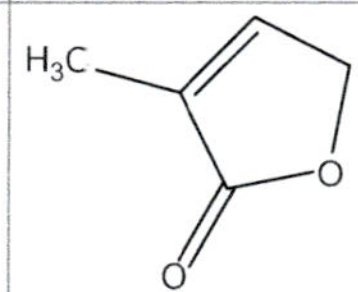	
				E-8-Methyl-9-tetradecen-1-ol-acetate	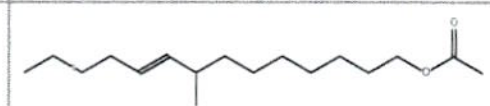	
				Octadecanoic acid	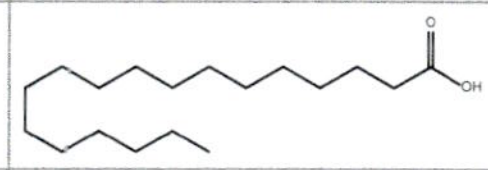	
				Longiborneol	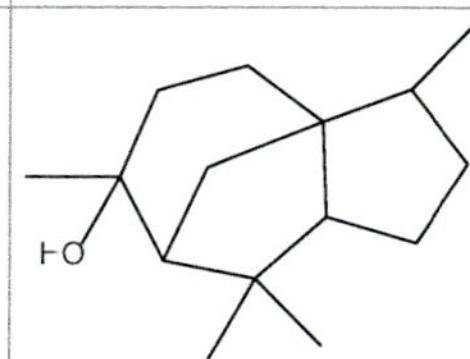	

	Sapium macrocarpum Müll.Arg.		Stem	5-Hydroxymethyl-2-furoic acid	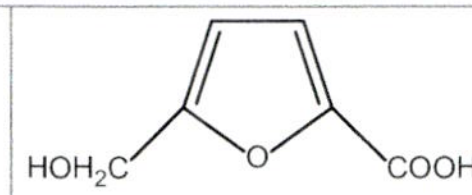 	Flores and Ricalde (1996) and Balick and Arvigo (2015)
	Synonym(s): *Excoecaria macrocarpa* (Müll.Arg.) Müll.Arg., *Sapium bourgeaui* Croizat, *Sapium mexicanum* Hemsl., *Sapium pedicellatum* Huber, *Sapium thelocarpum* K.Schum. & Pittier. **Common name(s)**: Milk tree. **Botanical description**: Trees, 8–35 m tall. Leaves oblong to elliptical or lanceolate, 4–15 cm long and 1.5–4.5 cm wide, acuminate at the apex, tip usually recurved, acute or obtuse to rounded at the base, submargins or indistinctly crenulate, 10–15 pairs of secondary nerves; petioles 1–4 cm long, apical glands paired, cylindrical, ca 1 mm long, deltoid stipules, ca 2 mm long. Terminal inflorescences, solitary, bisexual, mostly 7–15 cm long, oblong bracteal glands, 2.5–3.5 mm long, without covering the bracts; flowers staminate 7–10 per bract, yellowish, chalice 2–2.5 mm long, filaments 2–2.5 mm long; pistillate sessile flowers, sepals 1–3 mm long, styles 2–3 mm long. Capsule more or less pear-shaped, 1.3–2 (–3) cm wide, stipe 3–8 mm long; common in dry or humid forests. **Medicinal use(s)**: It is used for warts treatment. It has radical scavenging and antioxidant activities. The extracts from the aerial parts of the plants were obtained using hexane, acetone, and methanol. The initial qualitative screening of antioxidant.					

(continued)

Table 4.58 (continued)

	Suregada multiflora (A. Juss.) Baill		Wood	Suregadolide C	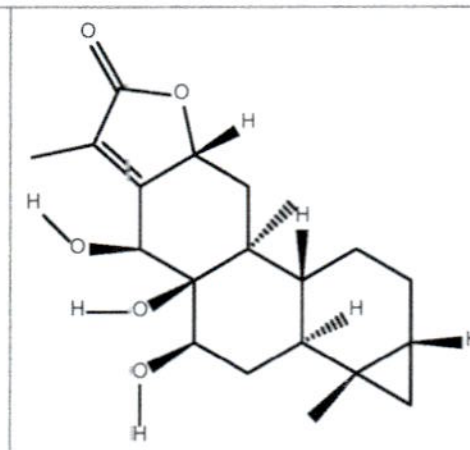	Wasuwat (1967) and Itharat et al. (2004)
	Synonym(s): *Gelonium affine* S.Moore, *Gelonium bifarium* Roxb. ex Willd., *Gelonium fascuculatum* Roxb., *Gelonium multiflorum* A.Juss., *Gelonium obtusum* Miq., *Gelonium oxyphyllum* Miq., *Gelonium sumatranum* S.Moore, *Gelonium tenuifolium* Ridl., *Suregada affinis* (S.Moore) Croizat, *Suregada bifaria* (Roxb. ex Willd.) Baill., *Suregada glabra* Roxb., *Suregada oxyphylla* (Miq.) Kuntze, *Suregada sumatrana* (S.Moore) Croizat, *Suregada tenuifolia* (Ridl.) Croizat. **Common name(s)**: False lime. **Botanical description**: It is a tall shrub or a small tree, growing 2–13 m tall, native to Northeast India. Branches are gray yellow to gray brown, hairless. Leaf stalks are 3–12 mm long. Leaf blade is obovate-elliptic to obovate-lance-shaped or oblong-elliptic, 5–16 cm long, 3–8 cm wide, somewhat leathery, with a pointed tip. Flowers are borne in short-stalked cymes. Flowers are tiny, 5–8 mm across; male and female flowers separate on the same tree. Male flowers have circular sepals, with 30–60 stamens. Female flowers have an annular disk and a spherical ovary. Sepals persistent on the fruit, which is round, 1.1–1.5 cm across, slightly fleshy, 3-seeded. Flowering: March–September. **Medicinal use(s)**: The bark is used for sore and stomach troubles in Khagrachari. It is an important medicinal plant and known as "heavenly fruit" in native range. Phytochemicals possibly effective against HIV and breast cancer. Timber and products. It is cultivated as an ornamental and harvested as timber to be used as rafters and firewood.					

	Tragia involucrata L.		Aerial parts	Bosutinib	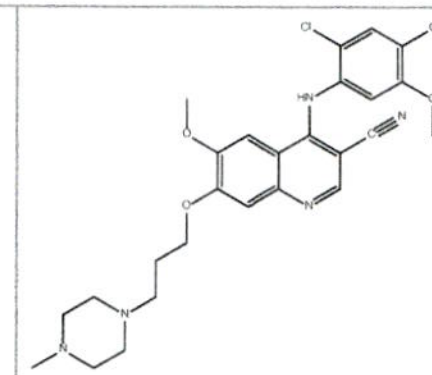	Jain (1970) and Al-Jassabi et al. (2011)
	Synonym(s): *Croton urens* L., *Tragia involucrata* var. *rheediana* Müll.Arg. **Common name(s)**: Indian stinging nettle, Aag paan, Aagya, Laghumedhshing, Aagmavarta, Duhsparsha, Kashagnih. **Botanical description**: It is a native plant of tropical Africa and Asia. It is a perennial twinning herb, covered with stinging hairs. Alternately arranged leaves are oblong-lineolate or ovate, toothed, base heart-shaped or rounded, tip long pointed. Flowers are borne in racemes in leaf axils. Female flowers are few, in lower part of inflorescence; male flowers are many in the upper part. Fruit is a 3-lobed capsule, containing 3 round smooth seeds. **Medicinal use(s)**: Roots are useful in treating pruritic skin eruptions, venereal diseases, diabetes, and guinea worms. Leaves are supposed to be good for cephalalgia.					

Table 4.59 Anticancer plants of family Fabaceae

Family	Botanical name	Image	Plant part used	Bioactive compound(s)	Structure	Reference(s)
Fabaceae	*Abrus precatorius* L.		Leaf, root, and stem	N,N-dimethyl tryptophan metho cation methyl ester	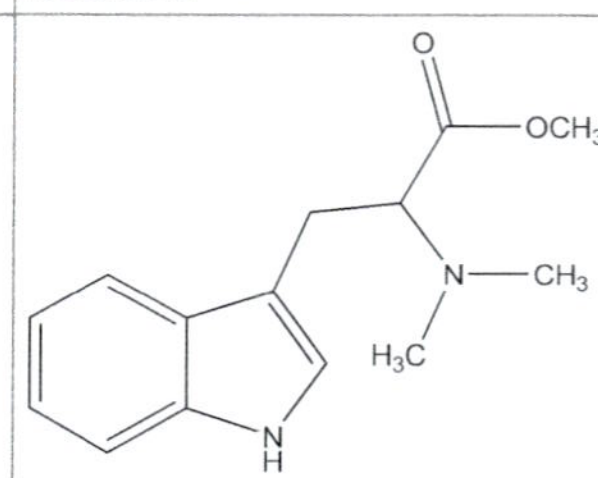	Reddy and Sirsi (1969) and Sofi et al. (2013)
	Synonym(s): *Abrus abrus* (L.) Wright., *Abrus cyaneus* R.Vig., *Abrus maculatus* Noronha, *Abrus minor* Desv., *Abrus pauciflorus* Desv., *Abrus squamulosus* E.Mey., *Abrus tunguensis* Lima, *Glycine abrus* L. **Common name(s)**: Jequirity, Crab's eye. **Botanical description**: It is native to India and Sri Lanka and grows in tropical and subtropical areas of the world. It is an invasive plant in warm temperate to tropical regions. **Medicinal use(s)**: It is used to treat human cancer cell lines, diarrhea, epilepsy, fever, coughs, and cold. It also possesses antitumor, antibacterial, anti-inflammatory, and antioxidant properties.			Picatorine	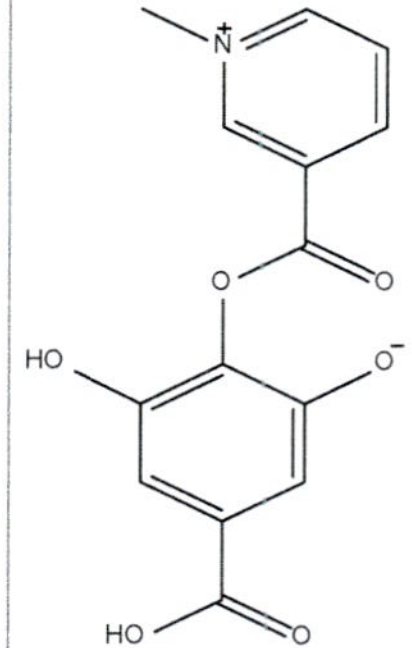	
				Abrine	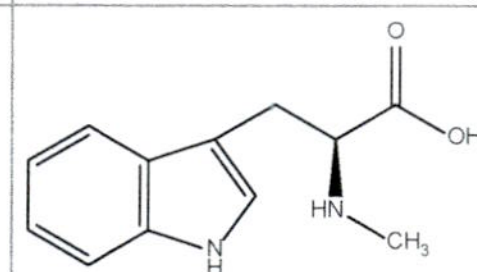	
				Hypaphorine	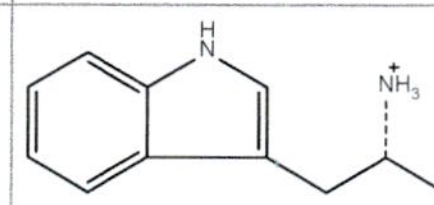	

<table>
<tr><td></td><td></td><td></td><td></td><td>Trigonelline</td><td>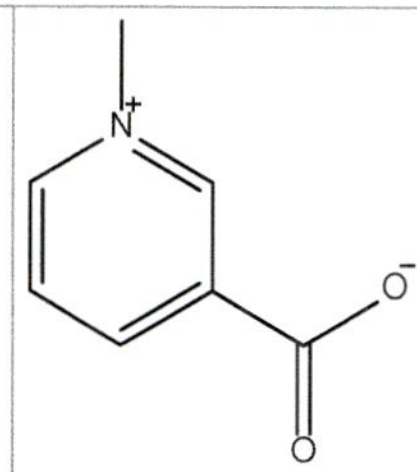</td><td></td></tr>
<tr><td rowspan="2"></td><td>Acacia catechu (L.f.) Willd.</td><td></td><td>Bark, stem</td><td>Catechin</td><td>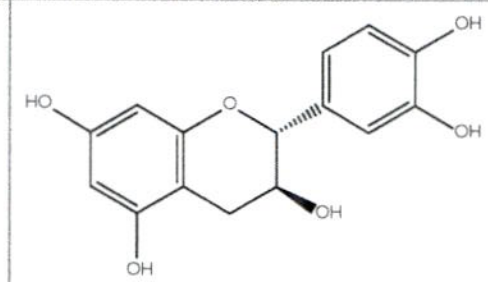</td><td rowspan="2">Cai et al. (2003) and Mandal et al. (2014)</td></tr>
<tr><td colspan="3">Synonym(s): Acacia catechuoides (Roxb.) Benth., Acacia sundra (Roxb.) Bedd., Acacia wallichiana DC., Mimosa catechu L.f., Mimosa catechuoides Roxb.
Common name(s): Cutch tree, Black catechu, Black cutch, Cashoo, Catechu, Wadalee gum.
Botanical description: It is found in Asia, China, India, and the Indian Ocean area. It is native to eastern slopes of the Western Ghats and the Himalayan tracts of India. It is a deciduous, thorny tree which grows up to 15 m (50 ft.) in height.
Medicinal use(s): It is used for the treatment of esophagus cancer, asthma, bronchitis, leukemia, dysentery, and diarrhea. It also possesses antibacterial, antimicrobial, anti-inflammatory, and antioxidant properties.</td><td>Epicatechin</td><td>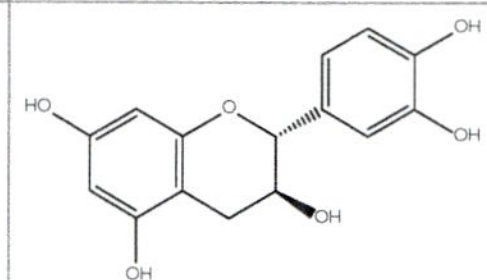</td></tr>
</table>

(continued)

Table 4.59 (continued)

Family	Botanical name	Image	Plant part used	Bioactive compound(s)	Structure	Reference(s)
	Acacia nilotica (L.) Delile		Flower, leaf	Isoquercetin	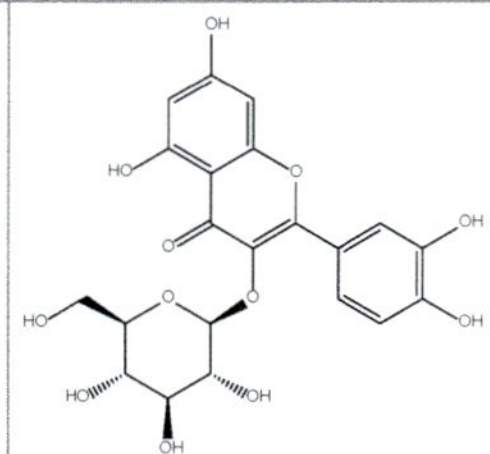	Meena et al. (2006) and Sakthive et al. (2012)
	Synonym(s): *Acacia arabica* (Lam.) Willd., *Acacia scorpioides* W. Wight., *Acacia vera* Willd., *Mimosa arabica* Lam. **Common name(s)**: Lekkerruikpeul, Gum arabic tree, Babul, Thorn mimosa, Egyptian acacia or thorny acacia, *Mimosa nilotica* L., *Mimosa scorpioides* L. **Botanical description**: It is native to West Asia and can reach a height of 10 m, with an average of 4–7 m. The crown is somewhat flattened or rounded, with a moderate density. The branches have a tendency to droop downwards if the crown is roundish. The bark is blackish gray or dark brown and deeply grooved, with longitudinal fissures in mature trees. **Medicinal use(s)**: It is used for the treatment of cancer (ear, eye, and testicles) and diarrhea. It also possesses antibacterial, antimicrobial, anti-inflammatory, and antioxidant properties.			Naringenin	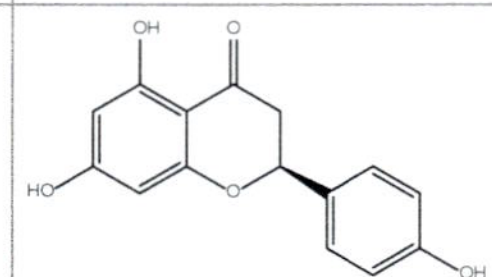	

	Acacia pennatula (Schltdl. & Cham.) Benth.		Whole part	Ardisiacrispin A	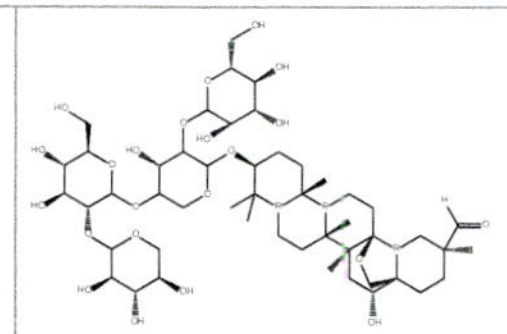	Amin and Mousa (2007) and Bonigala et al. (2017)
	Synonym(s): *Inga pennatula* Schltdl. & Cham., *Pithecellobium minutissimum* M. E. Jones, *Poponax pennatula* (Schltdl. & Cham.) Britton & Ro. **Common name(s)**: Fern leaf acacia, Flagpole tree. **Botanical description**: It is native to Colombia, El Salvador, Mexico, and Venezuela. It is a thorny legume which grows up to a height of 12 m, with an average of 8–10 m. Twigs armed with pairs of short stout stipular spines of 1–1.5 cm long. Leaves are bipinnate having numerous minute leaflets of 1–3 mm long. Flowers are yellow in color and globose heads are present on yellow velvety peduncles. **Medicinal use(s)**: It is used to treat lung cancer and leukemia. It also possesses antioxidant, antibacterial, anti-inflammatory, antitumor, and antimicrobial properties.					
	Acacia victoriae Benth.		Stem	Diosgenin	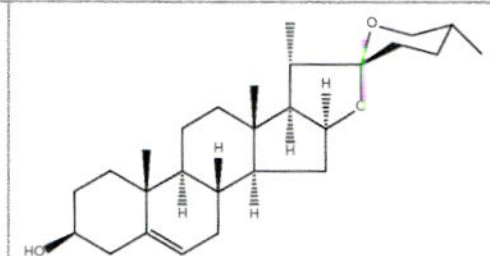	Haridas et al. (2001)

(continued)

Table 4.59 (continued)

Family	Botanical name	Image	Plant part used	Bioactive compound(s)	Structure	Reference(s)
	Synonym(s): *Acacia coronalis* J. M. Black., *Acacia hanniana* Domin, *Acacia sentis* Benth., *Acacia sentis* F.Muell., *Racosperma victoriae* (Benth.) Pedley. **Common name(s)**: Bramble wattle, elegant wattle. **Botanical description**: It is native to arid and semiarid areas of Australia. This plant is generally found in alkaline soils including clayey alluvial, gray cracking clays and saline loams on floodplains, alluvial flats, rocky hillsides, and ridges. It grows into a shrub-like tree with multiple trunks. They reach a height of about 5–6 m and are moderately fast-growing plants. **Medicinal use(s)**: It is used to treat cancer or tumor. It also possesses antioxidant, antibacterial, anti-inflammatory, antitumor, and antimicrobial properties.					
	Acacia xanthophloea Benth.		Fruit	Digitoxin	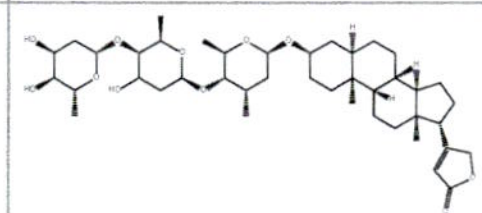	Chhabra et al. (1991) and Kariuki and Njoroge (2011)
	Synonym(s): No synonyms are recorded for this name. **Common name(s)**: Fever tree. **Botanical description**: It is native to Eastern and Southern Africa. It can be found in Botswana, Kenya, Malawi, Mozambique, Somalia, Swaziland, Tanzania, Zambia, and Zimbabwe. It can be used as a landscape tree in other warm climates, outside of its natural range. The trees grow to a height of about 15–25 m (49–82 ft.). **Medicinal use(s)**: It is used to treat lung cancer, diabetes, gingivitis, mouth sores, fever, and eye complaints. It also possesses antioxidant, antibacterial, anti-inflammatory, antitumor, and antimicrobial properties.					

	Albizia julibrissin Durazz.		Stem bark	Coumaric acid	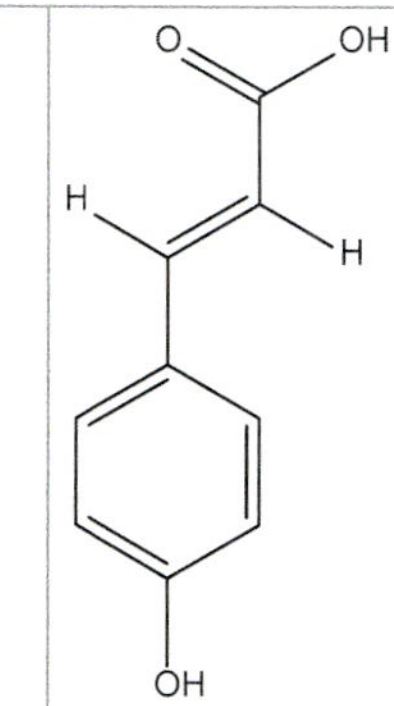	Woo (1985) and Karuppannan et al. (2013)
	Synonym(s): *Acacia julibrissin* (Durazz.) Willd., *Acacia nemu* Willd., *Albizia nemu* (Willd.) Benth., *Feuilleea julibrissin* (Durazz.) Kuntze, *Mimosa julibrissin* (Durazz.) Scop., *Mimosa speciosa* Thunb., *Sericandra julibrissin* (Durazz.) Raf. **Common name(s)**: Mimosa. **Botanical description**: It is native to Iran and Japan. It is a fast-growing, small- to medium-sized, deciduous tree that typically grows in a vase shape up to 20–40 m tall with a spreading, often umbrellalike, crown. **Medicinal use(s)**: It is used to treat lung cancer, tumor, insomnia, and diabetes. It also possesses antioxidant, antibacterial, anti-inflammatory, antitumor, and antimicrobial properties.					
	Albizia schimperiana Oliv.		Stem bark	Allamcin	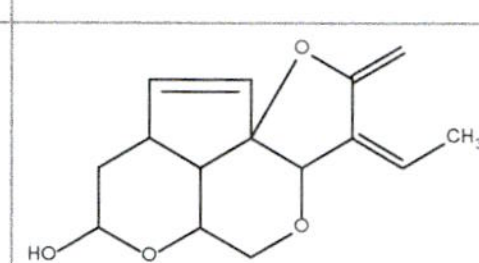	Chhabra et al. (1984) and Karuppannan et al. (2013)

(continued)

Table 4.59 (continued)

Family	Botanical name	Image	Plant part used	Bioactive compound(s)	Structure	Reference(s)
	Synonym(s): *Albizia amanuensis* Baker f. **Common name(s)**: Uachi. **Botanical description**: It is native to east tropical Africa, Eritrea, Ethiopia, Uganda, Kenya, Tanzania, Zambia, Malawi, and Zimbabwe. It is a deciduous tree with a flattened or rounded, often umbrella-shaped, crown, usually growing up to 30 m but occasionally 35 m tall. **Medicinal use(s)**: It is used to treat breast cancer, headache, cough, and warts. It also possesses antioxidant, antibacterial, antitumor, anti-inflammatory, and antimicrobial properties.					
	Arachis hypogaea L.		Stem	Resveratrol (trans-3,5,4′-trihydroxystilbene)	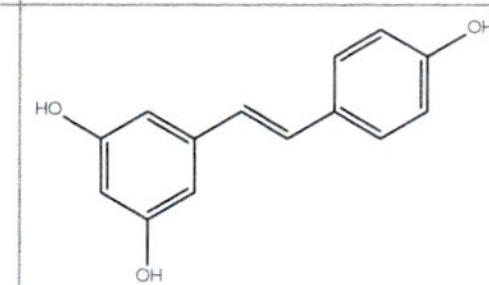	Kim et al. (2009) and Chen et al. (2017)
	Synonym(s): *Arachis nambyquarae* Hoehne. **Common name(s)**: Peanut, Groundnut. **Botanical description**: It is native to South America. It is an annual herbaceous plant that originated and was domesticated in South and Central America and is now grown in tropical and warm temperate regions worldwide for its seeds and oil. The peanut plant has procumbent (trailing) stems and grows to around 0.5 m tall or long. The leaves are alternate and compound, with 4 ovate to oblong leaflets, up to 6 cm long. **Medicinal use(s)**: It is used for the treatment of prostate cancer, diabetes, and tumor. It also possesses antioxidant, anti-inflammatory, antitumor, and antibacterial properties.					

	Astragalus propinquus Schischkin		Whole plant, root	Astragalus saponins	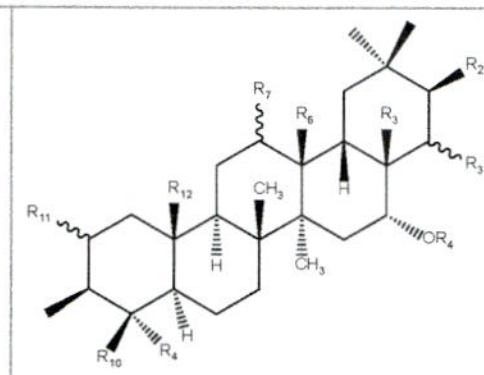	Wong et al. (1992a, b) and Auyeung et al. (2009)
	Synonym(s): *Astragalus membranaceus* (Fisch.) Bunge, *Phacamem branacea* Fisch. **Common name(s)**: Milkvetch. **Botanical description**: It is native from Southwest to Northwest China. It is a perennial plant, approximately 40–80 cm tall. It has hairy stems with leaves made up of 12–18 pairs of leaflets. The root is usually harvested from 4-year-old plants. It thrives in sandy, well-drained soil, with plenty of sunlight. **Medicinal use(s)**: It is used to treat gastrointestinal and liver cancer. It also possesses anticancer, antibacterial, antitumor, anti-inflammatory, and antimicrobial properties.					
	Bauhinia variegate L.		Stem bark, Flowers	Ellipticine	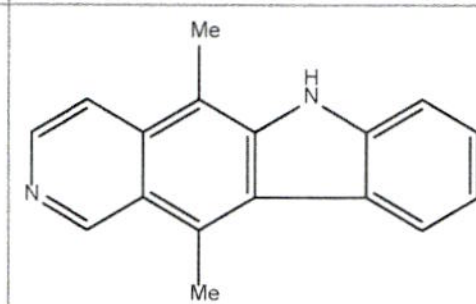	Rajkapoor et al. (2003) and Pandey (2017)

(continued)

Table 4.59 (continued)

Family	Botanical name	Image	Plant part used	Bioactive compound(s)	Structure	Reference(s)
	Synonym(s): *Bauhinia chinensis* (DC.) Vogel., *Bauhinia decora* Uribe., *Phanera variegate* (L.) Benth. **Common name(s)**: Orchid tree, camel's foot tree, kachnar, and mountain ebony. **Botanical description**: It is native to South Asia and Southeast Asia, South China, Burma, India, Nepal, Pakistan, and Sri Lanka. It is a small- to medium-sized tree growing up to 10–12 m tall, deciduous in the dry season. The leaves are 10–20 cm obcordate shaped, long and broad, rounded, and bilobed at the base and apex. The flowers are conspicuous, bright pink or white, 8–12 cm in diameter, with five petals. The fruit is a pod 15–30 cm long, containing several seeds. **Medicinal use(s)**: It is used to treat colon cancer, human cancer cell lines, and skin diseases. It also possesses antibacterial, antitumor, anti-inflammatory, and antimicrobial properties.					
	Caesalpinia coriaria (Jacq.) Willd.		Pod	Digitoxin	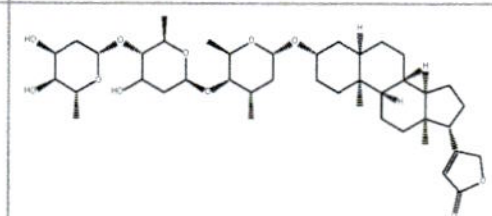	Garcia-Barriga (1974), Morton (1975), and Sánchez-Carranza et al. (2017)
	Synonym(s): *Caesalpinia thomaea* Spreng., *Libidibia coriaria* (Jacq.) Schltdl., *Poinciana coriaria* Jacq. **Common name(s)**: Divi-divi, Cascalote, Guaracabuya, Guatapana, Nacascol, Watapana. **Botanical description**: It is native to the Caribbean, Mexico, Central America, and northern South America. It is a shrub or a small tree with a rounded, spreading crown. It usually grows up to 10 m tall but can be much taller. **Medicinal use(s)**: It is used to treat liver, lung, and rectal cancer and skin problems. It also possesses antibacterial, antitumor, anti-inflammatory, and antimicrobial properties.					

	Caesalpinia ferrea C. Mart.		Stem	Pauferrol A	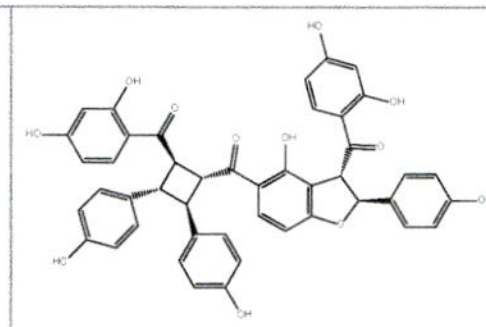	Nozakia et al. (2007) and Freitas et al. (2012)
	Synonym(s): *Caesalpinia ferrea* var. *cearensis* Huber. **Common name(s)**: Leopard tree. **Botanical description**: It is native to Brazil. It is a beautiful tree with white and chocolate flaking bark and pendant, flowing branchlets of compound leaves. The bark of the Brazilian ironwood is the single most attractive feature of the plant. The leaves are double-compound, with broadly oval and rounded leaflets. The flowers are yellow, but unfortunately occur mostly near the top, and hence not easy to observe. **Medicinal use(s)**: It is used to treat human cancer cell lines and inflammatory diseases. It also possesses antibacterial, antitumor, anti-inflammatory, and antimicrobial properties.					

(continued)

Table 4.59 (continued)

Family	Botanical name	Image	Plant part used	Bioactive compound(s)	Structure	Reference(s)
	Caesalpinia volkensii Harms		Stem	Isoquinoline	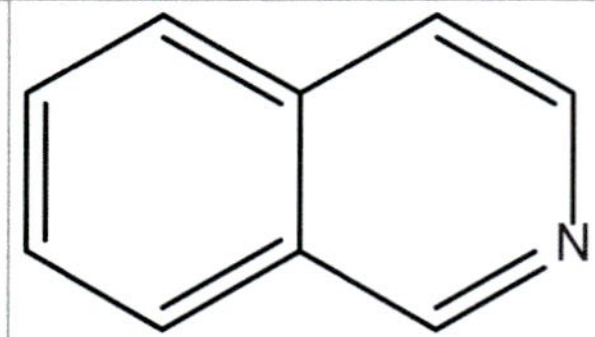	Ogila and Aduol (2017)
	Synonym(s): *Caesalpinia major* sensu Brenan. **Common name(s)**: Not available. **Botanical description**: It is native to eastern tropical Africa, Ethiopia, Kenya, Uganda, and Tanzania. It is a climbing shrub with straggling stems. **Medicinal use(s)**: It is used to treat human cancer cell lines, stomachache, eye problem, stomach ulcers, retinoblastoma, gonorrhea, bilharzia, and malaria. It also possesses antibacterial, antitumor, anti-inflammatory, and antimicrobial properties.					

	Cajanus cajan (L.) Millsp.		Leaf	Hydroxyl stilbenes	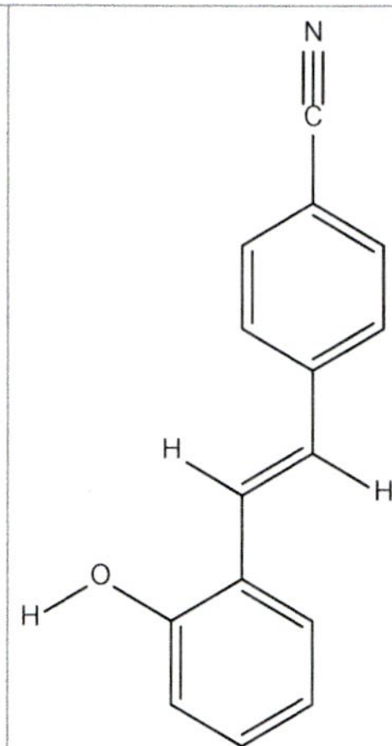	Ashidi et al. (2010)
	Synonym(s): *Cajan indorum* Medik., *Cajanus bicolor* DC., *Cajanus cajan* (L.) Huth, *Cajanus flavus* DC., *Cajanus indicus* Spreng., *Cajanus inodorum* Medik., *Cajanus luteus* Bello, *Cajanus obcordifolius* V. Singh, *Cajanus pseudo-cajan* (Jacq.) Schinz & Guillaumin, *Cajanus striatus* Bojer, *Cytisus guineensis* Schum. & Thonn., *Cytisus pseudocajan* Jacq., *Phaseolus balicus* L. **Common name(s)**: Pigeon pea. **Botanical description**: It is native to Asia, Africa, and Latin America. It is consumed on a large scale in South Asia and is a major source of protein for the population of that subcontinent. **Medicinal use(s)**: It is used to treat human cancer and tumor. It also possesses antibacterial, anti-inflammatory, and antioxidant properties.			Amyrin	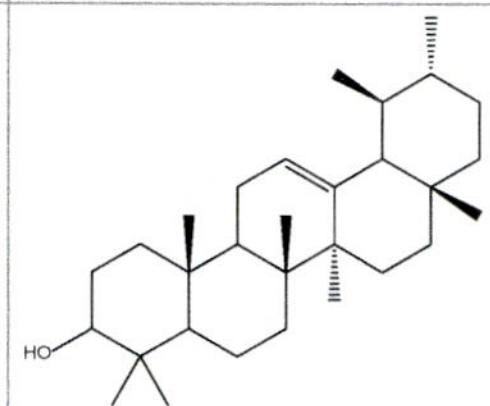	

(continued)

Table 4.59 (continued)

Family	Botanical name	Image	Plant part used	Bioactive compound(s)	Structure	Reference(s)
				Sitosterol	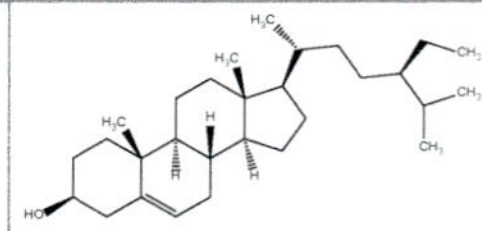	
				Pinostrobin	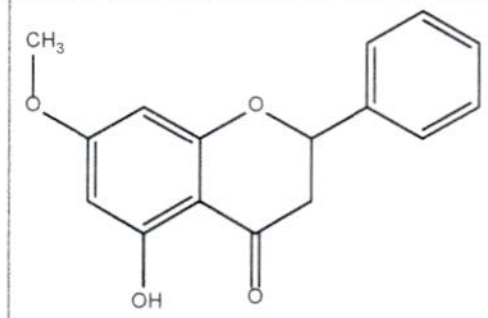	
				Longistylins A	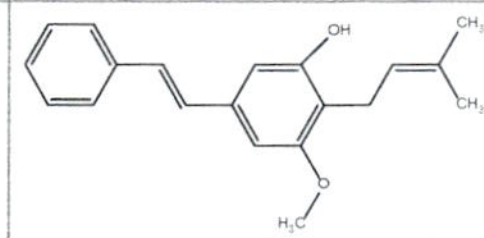	
				Longistylins C	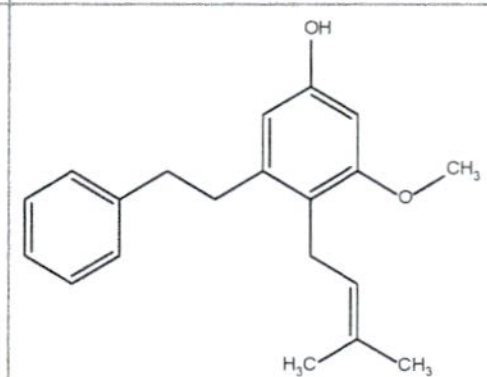	
	Chesneya cuneata (Benth.) Ali		Leaf	Sylvatesmin	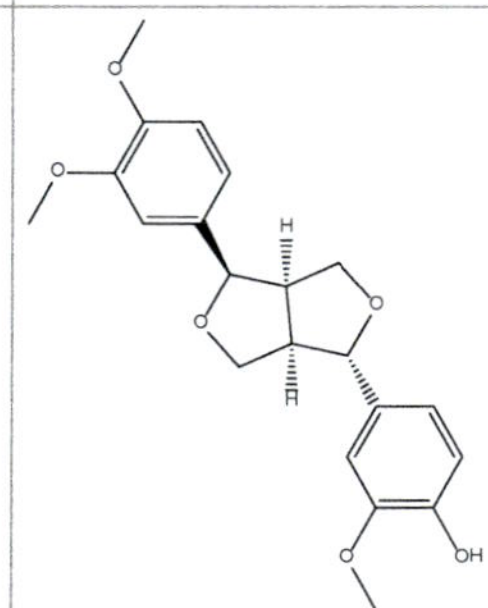	Singh et al. (1996a, b) and Aixia (2014)

<table>
<tr><td></td><td colspan="3">Synonym(s): Calophaca cuneata (Benth.) Kom., Caragana cuneata Moench., Caragana cuneata (Benth.) Baker., Gueldenstaedtia cuneata Benth.
Common name(s): Wedge-leaf chesneya.
Botanical description: It is native to Asia, Africa, and Latin America. It is consumed on a large scale mainly in South Asia and is a major source of protein for the population of that subcontinent.
Medicinal use(s): It is used to treat gastric cancer and tumor. It also possesses anti-inflammatory, antibacterial, and antioxidant properties.</td><td></td><td></td><td></td></tr>
<tr><td></td><td>Senna auriculata (L.) Roxb.</td><td></td><td>Leaf</td><td rowspan="2">Salidroside</td><td rowspan="2">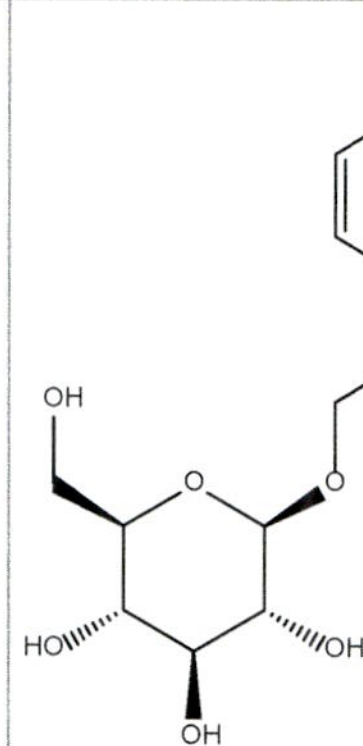</td><td rowspan="2">Prasanna et al. (2009) and Nille and Reddy (2015)</td></tr>
<tr><td></td><td colspan="3">Synonym(s): Cassia auriculata L., Cassia densistipulata Taub.
Common name(s): Avaram senna, Golden shower tree.
Botanical description: It is native to India, Myanmar, and Sri Lanka. It is a shrub with pubescent branchlets. Leaflets are 2 × 1 cm, oblong-obovate or elliptic, shortly acuminate, gland opposite the leaflets, stipitate, petiole 1–1.5 cm long, stipules 1 cm long, lunate and auricled.
Medicinal use(s): It is used to treat human breast, adenocarcinoma, and human larynx cancer. It possesses anti-inflammatory and antioxidant properties.</td></tr>
</table>

(continued)

Table 4.59 (continued)

Family	Botanical name	Image	Plant part used	Bioactive compound(s)	Structure	Reference(s)
	Senna obtusifolia (L.) H.S.Irwin & Barneby		Leaf, seed	Ursolic acid	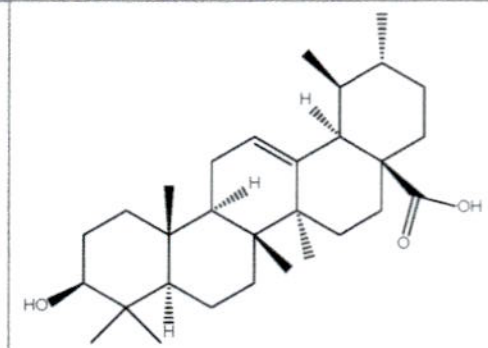	Singh et al. (1979)
	Synonym(s): *Cassia humilis* Collad., *Cassia obtusifolia* L., *Cassia toroides* Roxb., *Cassia toroides* Raf., *Diallobus falcatus* Raf., *Diallobus uniflorus* Raf., *Senna toroides* Roxb. **Common name(s)**: Sicklepod. **Botanical description**: It is native to North, Central, and South America, Asia, Africa, and Oceania. The green leaves of the plant are fermented to produce a high-protein food product called "kawal" which is eaten by many people in Sudan as a meat substitute. Its leaves, seeds, and roots are also used in folk medicine in Asia. The plant's seeds are a commercial source of cassia gum and usually used as a thickener in food additives. Roasted and ground seeds have also been used as a substitute for coffee. **Medicinal use(s)**: It is used to treat human cancer cell lines. It also possesses antitumor, anti-inflammatory, and antioxidant properties.					

	Senna petersiana (Bolle) Lock		Leaf	Petersinone	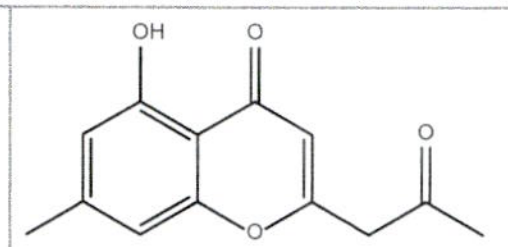	Gamal-Eldeen et al. (2007)
				Sitosterol-3--d-gycoside	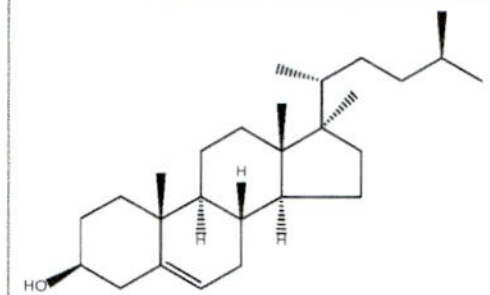	
	Synonym(s): *Cassia petersiana* Bolle. **Common name(s)**: Monkey pod. **Botanical description**: It is native to India, Myanmar, and Sri Lanka. It is a shrub with pubescent branchlets. Leaflets are 2 × 1 cm, oblong-obovate or elliptic, shortly acuminate, stipitate, petiole 1–1.5 cm, stipules 1 cm, lunate and auricled. Corymbs axillary and terminal, peduncle 2 cm, flowers yellow, larger sepals 1.5 × 1 cm. **Medicinal use(s)**: It is used to treat terminal cancer. It also possesses antitumor, anti-inflammatory, and antioxidant properties.					
	Senna italic Mill.		Whole plant	Nerifolin (1)	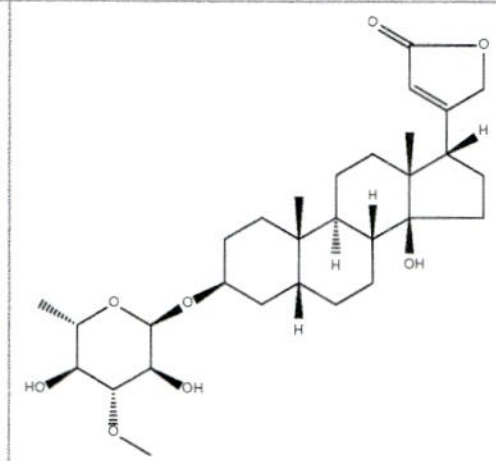	Sermakkani and Thangapandian (2010)

(continued)

Table 4.59 (continued)

Family	Botanical name	Image	Plant part used	Bioactive compound(s)	Structure	Reference(s)
	Synonym(s): *Cassia aschrek* Forssk., *Cassia italic* (Mill.) F. W. Andrews, *Cassia italic* (Mill.) Spreng., *Cassia ligustrina* Mill., *Cassia obovata* Collad., *Cassia obtuse* Roxb., *Cassia porturegalis* Bancr., *Senna obtuse* Roxb. **Common name(s)**: Port royal senna. **Botanical description**: It is native to tropical Africa, Arabia, Iraq, Iran, and Indian subcontinent. It is a deciduous, perennial herb or small shrub growing up to 60 cm tall, with prostrate stems. It is habituated to grassland of the drier regions of tropical Africa, from sea level up to elevations of 1850 m. It is often found close to streams and in disturbed habitats such as roadsides and waste places. **Medicinal use(s)**: It is used for treatments of human cancer cell lines, indigestion, liver complaints, gallbladder disorders, nausea, vomiting, and dysmenorrhea. It also possesses antitumor, anti-inflammatory, and antioxidant properties.					
	Senna tora (L.) Roxb.		Leaf	Procyanidin	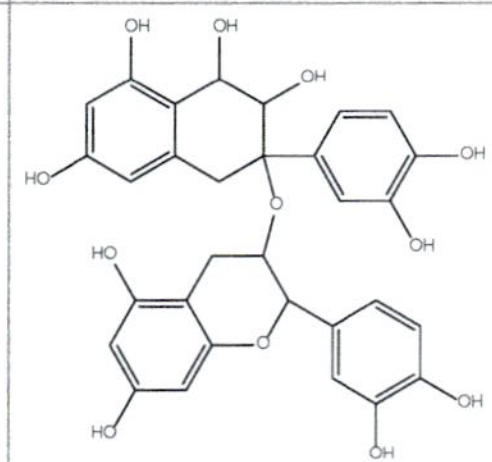	Rejiya et al. (2009)
	Synonym(s): *Cassia borneensis* Miq., *Cassia gallinaria* Collad., *Cassia numilis* Collad., *Cassia tora* L., *Emelista tora* Britton & Rose. **Common name(s)**: Sickle senna, Chakvad. **Botanical description**: It is native to South Asia. It is a small erect hairless shrub, about 1 m tall, commonly found growing wild on roadsides. It is widely spreading with numerous ascending, hairless branches. The compound leaves are arranged spirally and usually have three pairs of symmetrically egg-shaped leaflets up to 2 in. long. **Medicinal use(s)**: It is used to treat colon carcinoma, cell tumors, leprosy, and skin disorders. It also possesses anticancer, anti-inflammatory, and antioxidant properties.					

	Centrosema pubescens Benth.		Flowers	Lutein	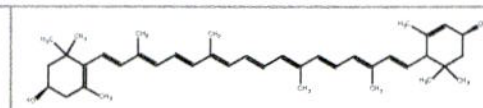	Hastings (1990) and Reddy and Mani (2010)
	Synonym(s): *Bradburya pubescens* (Benth.) Kuntze., *Centrosema ferrugineum* A. Rich., *Centrosema intermedium* A. Rich., *Centrosema salzmannii* Benth., *Centrosema virginianum* (L.) "Benth., p.p.", *Clitoria schiedeana* Schltdl. **Common name(s)**: Butterfly pea. **Botanical description**: It is native to South America and Central America and is a herbaceous, climbing, and perennial herb. The vigorous stems scramble over the ground or twine into other plants for support. The plant has a deeply penetrating taproot system. The plant is considered to be the most productive green manure crop for fertile soils in the humid tropics and is widely grown for this purpose. **Medicinal use(s)**: It is used to treat lung cancer, indigestion, stomachache, wounds, and other inflammatory conditions. It also possesses antitumor, anti-inflammatory, and antioxidant properties.					

(continued)

Table 4.59 (continued)

Family	Botanical name	Image	Plant part used	Bioactive compound(s)	Structure	Reference(s)
	Crotalaria assamica Benth.		Entire plant	Phyllirin	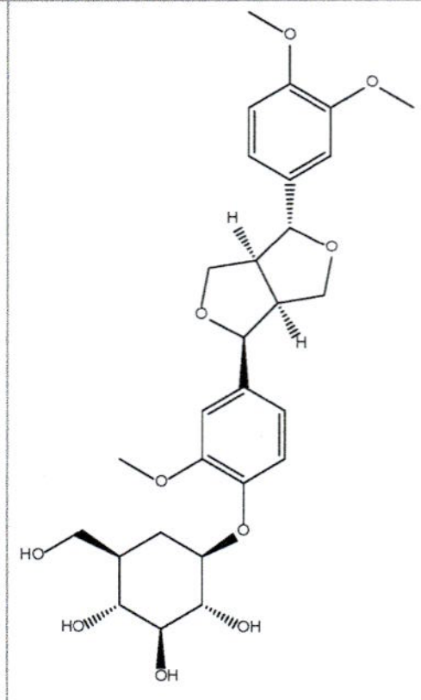	Duke and Ayensu (1985a, b) and Khan et al. (2013)
	Synonym(s): *Crotalaria burmannii* DC., *Crotalaria sericea* Burm.f. **Common name(s)**: Kogane-tanuki-mame. **Botanical description**: It is native to East Asia, China, Japan, Korea, and Manchuria. It is an annual plant growing up to 0.6 m (2 ft.). It flowers from September to October. The flowers are hermaphrodite. It is suitable for light (sandy), medium (loamy), and heavy (clay) soils. It has suitable pH: acid, neutral, and basic (alkaline) soils. It can grow in semi-shade (light woodland) or no shade. It prefers moist soil. It is habituated to grassy places in lowland areas of Central and Southern Japan. **Medicinal use(s)**: It is used to treat breast and non-small cell lung cancer. It possesses anticancer, antitumor, anti-inflammatory, and antioxidant properties.					

	Crotalaria ferruginea Benth.		Entire plant	Gallic acid	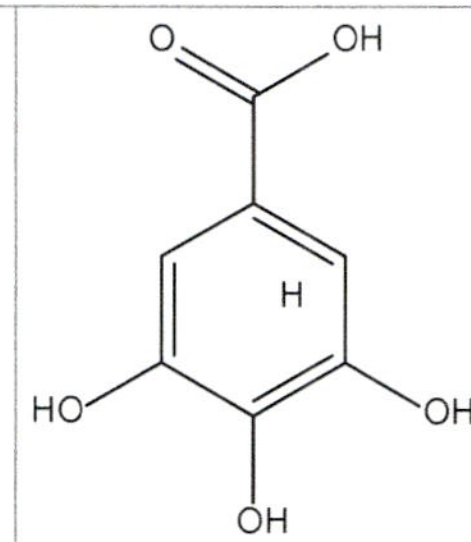	Duke and Ayensu (1985a, b) and Khan et al. (2013)
	Synonym(s): *Crotalaria bodinieri* H. Lev., *Crotalaria lonchophylla* Hand.-Mazz., *Crotalaria pilosissima* Miq., *Crotalaria rufescens* Franch. **Common name(s)**: Rattlepod or rattlebox. **Botanical description**: It is native to East Asia, Southwest China, India, Nepal, Bhutan, Bangladesh, Myanmar, and Thailand. It is an erect to ascending deciduous shrub growing 20–60 cm tall, occasionally to 120 cm. It is habituated to open sunny places at elevations from 1400 to 1900 m in Nepal. It is found in open forests and montane grasslands at elevations from 400 to 2200 m in China. **Medicinal use(s)**: It is used to treat human cancer cell lines, yellow fever, and skin rashes. It also possesses anticancer, anti-inflammatory, and antioxidant properties.					

(continued)

Table 4.59 (continued)

Family	Botanical name	Image	Plant part used	Bioactive compound(s)	Structure	Reference(s)
	Crotalaria sessiliflora L.		Entire plant	Apigenin	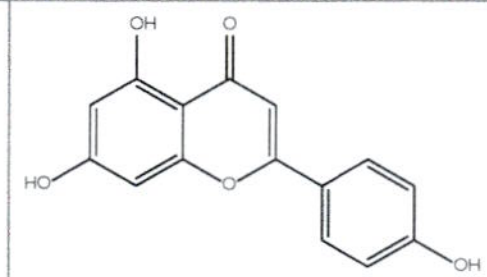	Duke and Ayensu (1985a, b) and Nibret et al. (2009)
	Synonym(s): *Crotalaria anthylloides* Lam., *Crotalaria brevipes* Benth., *Crotalaria nepaulensis* Link. **Common name(s)**: Not available. **Botanical description**: It is native to East Asia, China, Japan, Korea, and Manchuria. It is an annual plant growing up to 0.6 m with hermaphrodite flowers. It is suitable for light (sandy), medium (loamy), and heavy (clay) soils and can grow in semi-shade (light woodland) or no shade areas. It prefers moist soil. **Medicinal use(s)**: It is used for the treatment of skin and cervical carcinomas. It also possesses antitumor, anti-inflammatory, and antioxidant properties.					

Family	Plant	Photo	Part used	Compound	Structure	Reference
Fabaceae	*Dalbergia candenatensis* (Dennst.) Prain.		Whole plant	5-Hydroxybowdichione	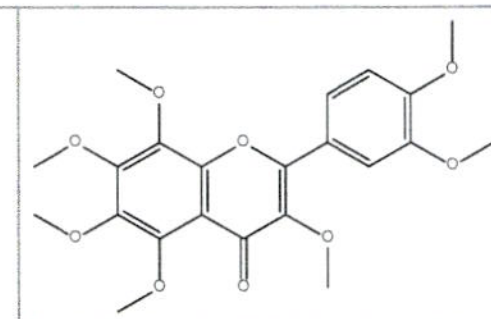	Hamburger and Cordell (1987) and Cheenpracha et al. (2009)
				Mucronulatol	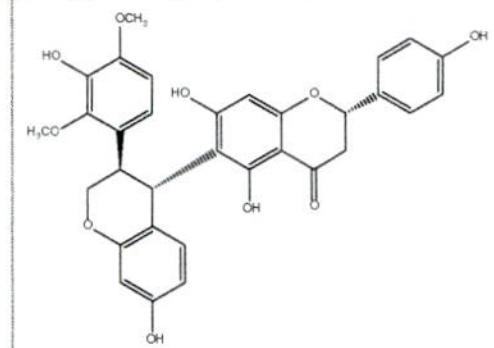	
	Synonym(s): *Amerimnon tortum* (Graham) Kuntze, *Cassia candenatensis* Dennst., *Dalbergia monosperma* Dalzell., *Dalbergia torta* Graham, *Dalbergia torta* A.Gray, *Drepanocarpus monospermus* (Dalzell) Kurz. **Common name(s)**: Takoli, *Dalbergia*. **Botanical description**: It is native to India, Sri Lanka, Nepal, Burma, and Indochina. It is a very conspicuous and beautiful tree when flowering, which appears very profusely during the months of May and June. Large tree with smooth bark and glabrous branches. Leaves are compound and 7.0–15.0 cm long, leaflets 2.5–5.0 cm long, ovate or obovate or elliptic, often emarginate, glabrous, glaucous. Inflorescence is large axillary or terminal panicles flowers unilaterally arranged in inflorescence. **Medicinal use(s)**: It is used for the treatment of colorectal cancer. It also possesses antibacterial, anti-inflammatory, and antioxidant properties.			Claussequinone	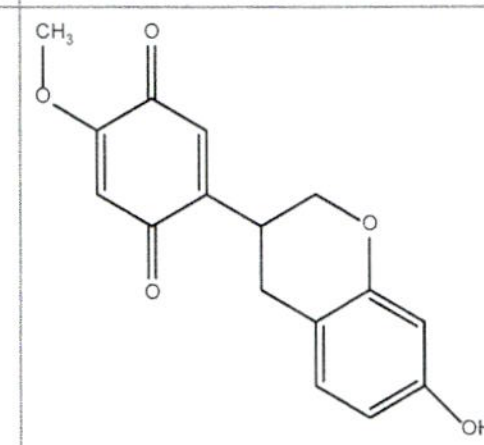	
				Formononetin	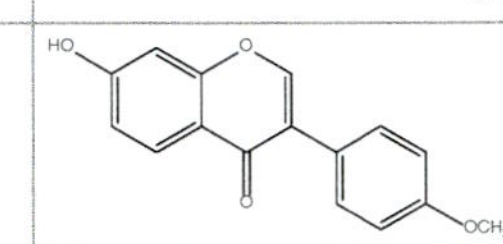	
				Vestitol	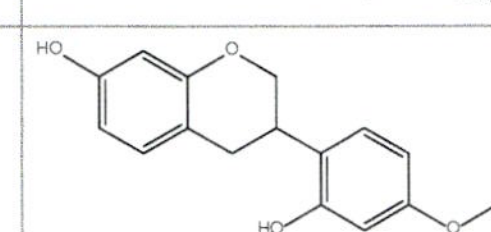	
				Nystatin	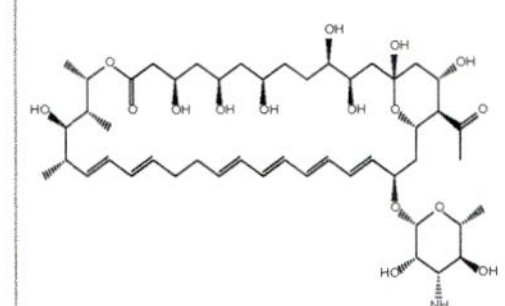	

(continued)

Table 4.59 (continued)

Family	Botanical name	Image	Plant part used	Bioactive compound(s)	Structure	Reference(s)
				Candenatenin B	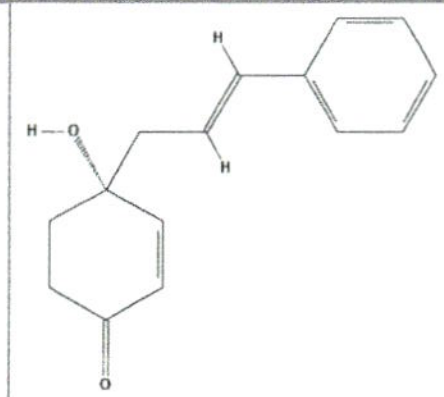	
				Candenatenin C	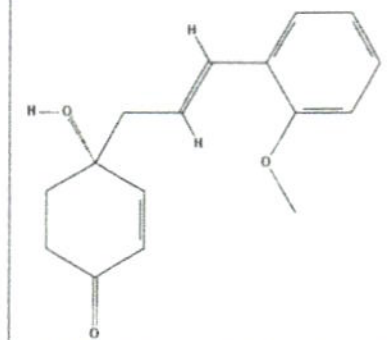	
	Dalbergia odorifera T. Chen		Dried trunk and root	Trihydroxyisoflavanone	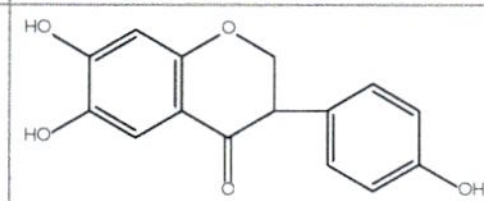	Liu et al. (2005) and Ninh et al. (2017)
	Synonym(s): *Dalbergia hainanensis* auct. non Merr. & Chun. **Common name(s)**: Fragrant rosewood. **Botanical description**: It is native to China, Fujian, Hainan, Zhejiang, and Guangdong. It is a small- or medium-sized deciduous tree and 10–15 m tall. It starts shredding leaves at around December of each year in the Northern Hemisphere. It becomes dormant throughout the winter months; towards the end of the winter or the beginning of spring, the bud begins to swell. **Medicinal use(s)**: It is used to treat cancer, diabetes, blood disorders, ischemia, swelling, necrosis, and rheumatic pain. It possesses antitumor, antibacterial, and anti-inflammatory properties.			Liquiritigenin	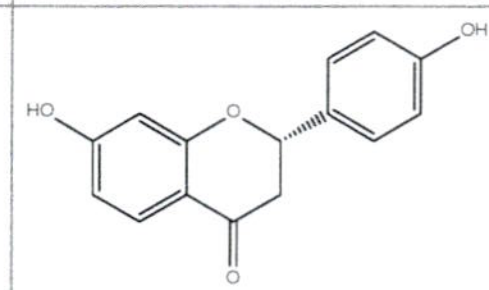	

	Melanettin	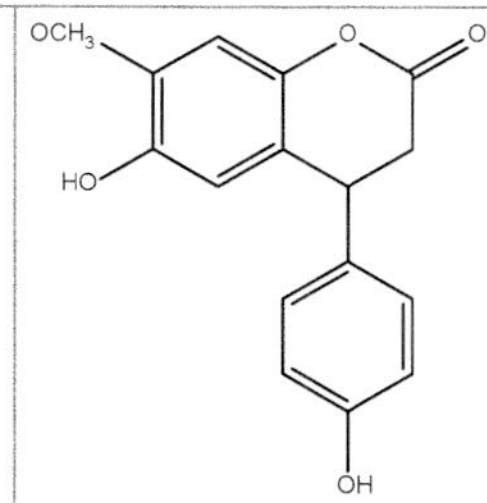	
	Violanone	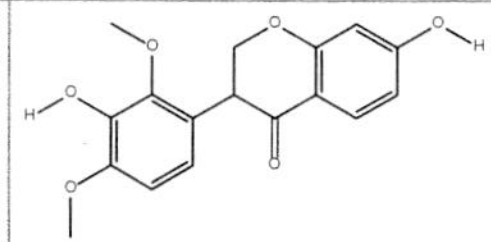	
	Formononetin	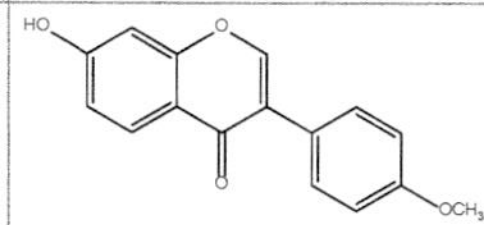	
	Dalbergin	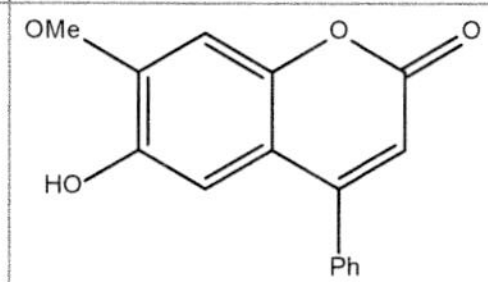	
	Sativanone	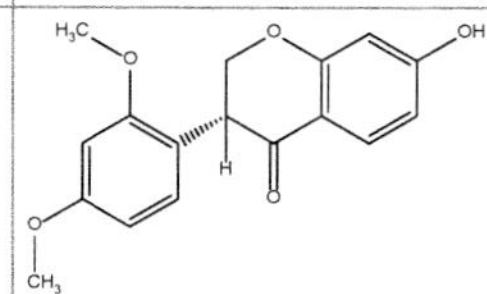	
	Medicarpin	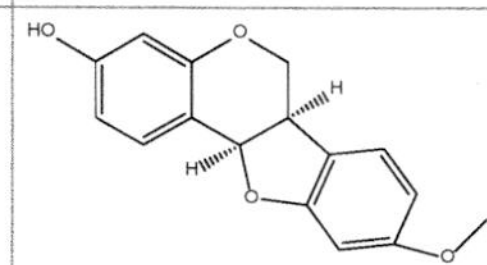	

(continued)

Table 4.59 (continued)

Family	Botanical name	Image	Plant part used	Bioactive compound(s)	Structure	Reference(s)
	Dalea elegans Hook. &Arn.		Root, stem	2-4-Dihidroxy-5-(1-dimethylallyl)-6-prenylpinocembrin) (6PP)	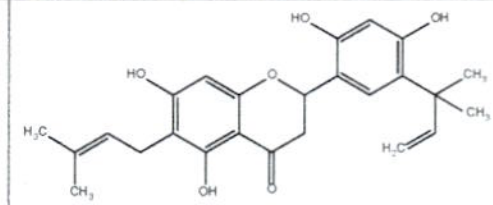	Elingold (2008)
	Synonym(s): *Dalea stenophylla* Griseb., *Parosela elegans* (Gillies ex Hook. &Arn.) J.F. Macbr. **Common name(s)**: Purple prairie clover. **Botanical description**: It is native to Argentina and Bolivia. This species is known from semi-humid thickest and dense pastures. This taxon is known to occur within the protected areas network, but seeds have yet to be collected and stored by a seed bank as a method of ex situ conservation. **Medicinal use(s)**: It possesses anticancer, antitumor, antibacterial, anti-inflammatory, and antioxidant properties.					
	Dalea leporine (Aiton) Bullock		Part not specified	P-Cymene	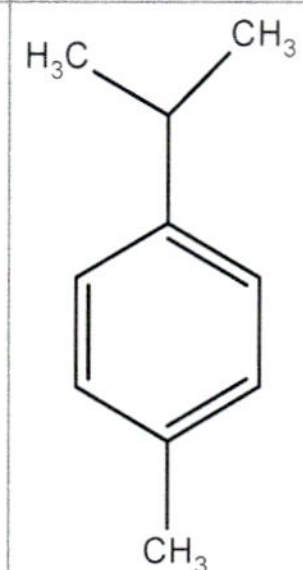	Hastings(1990) and Maffei (2003a, b, c, d, e, f, g, h, i, j, k, l, m)

	Synonym(s): *Dalea alba* Roem., *Dalea alopecuroides* Willd., *Dalea bigelovii* (Rydb.) B. L. Turner, *Dalea lagopus* (Cav.) Willd., *Dalea leporina* (Aiton) Kearney & Peebles, *Dalea linnaei* "Michx., p.p.A", *Dalea oreophila* (Cory) Cory, *Parosela bigelovii* Rydb., *Parosela costaricana* Rydb., *Parosela lagopus* Cav., *Parosela leporina* (Aiton) Rydb., *Petalostemon alopecuroides* (Willd.) Pers., *Petalostemon oreophilum* Cory, *Psoralea alopecuroides* (Willd.) Poir., *Psoralea lagopus* Cav., *Psoralea leporine* Aiton. **Common name(s)**: Foxtail prairie clover. **Botanical description**: It is native to North America but also introduced to New England. The leaves are compound made up of two or more discrete leaflets. The edge of the leaf blade is entire (has no teeth or lobes). The fruit is dry but does not split open when ripe. The fruit length is 2.5–3.0 mm. There are five petals, sepals, or tepals in the flower. **Medicinal use(s)**: It possesses anticancer, antitumor, antibacterial, anti-inflammatory, and antioxidant properties.					
	Dalea lutea (Cav.) Willd.		Whole plant	Not reported	Not reported	Hastings (1990) and Maffei (2003a, b, c, d, e, f, g, h, i, j, k, l, m)
	Synonym(s): *Dalea caudata* (Rydb.) Bullock, *Dalea leucostoma* Schltdl., *Dalea ovalifolia* Ortega, *Dalea plumose* S.Watson, *Parosela caudata* "Rydb., 1920, p.112", *Parosela leucostoma* (Schltdl.) Rose, *Parosela lutea* (Cav.) Cav., *Parosela plumosa* (S.Watson) Rose, *Parosela wardii* Rydb., *Psoralea lutea* Cav. **Common name(s)**: Sierra moonrise. **Botanical description**: It is native to Argentina and Bolivia. This species is known from semi-humid areas. This taxon is known to occur within the protected areas network, but seeds have yet to be collected and stored by a seed bank as a method of ex situ conservation. **Medicinal use(s)**: It possesses anticancer, antitumor, antibacterial, anti-inflammatory, and antioxidant properties.					

(continued)

Table 4.59 (continued)

Family	Botanical name	Image	Plant part used	Bioactive compound(s)	Structure	Reference(s)
	Entada glandulosa Pierre ex Gagnepain		Seed	Rutin	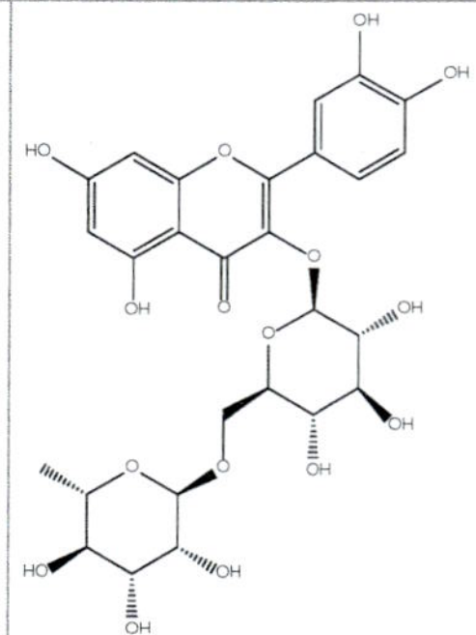	Wasuwat (1967) and Maffei (2003a, b, c, d, e, f, g, h, i, j, k, l, m)
	Synonym(s): *Entada glandulosa* Gagnep. **Common name(s)**: Elephant creeper. **Botanical description**: It is native to Asia and Africa. They have compound leaves and produce exceptionally large seed pods of up to 1.5 m long. Their seeds are buoyant and survive lengthy journeys via rivers and ocean currents. Leaves are cirrhiferous, pinnae 1–2 jugate, leaflets 2–5 jugate (large), ovate, elliptical, or oblong-obovate, obtuse or acuminate or emarginate, often oblique, shining above, glabrous or downy beneath, spikes elongate, solitary or in pairs. Stem climb to the top of lofty trees, very long and ropelike. Petiole end in a simple or branched tendril, pinnae few, 1–2 in. apart, leaflets 1–2 in. long, 1 in. wide. Spikes 2–8 in. long. Legume 2–3 ft. long and 3–4 in. broad. **Medicinal use(s)**: It possesses anticancer, antitumor, antibacterial, anti-inflammatory and antioxidant properties.					

	Entada rheedii Spreng.		Stem, Seed	Phaseoloidin	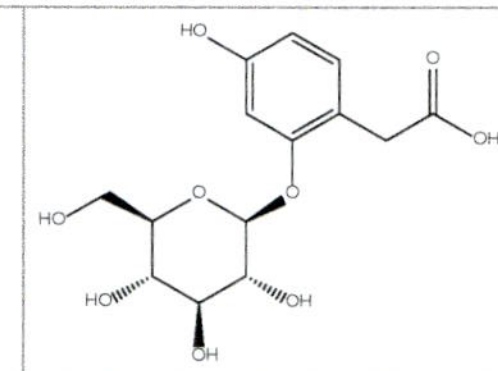	Tapondjou and Pursaethosides (2005), Nzowa et al. (2013), and Shafaat-Al-Mehedi et al. (2015)
	Synonym(s): *Adenanthera gogo* Blanco., *Entada gogo* (Blanco) I. M. Johnst., *Entada monostachya* DC., *Entada pursaetha* DC., *Entada scheffleri* Ridl., *Mimosa entada* L. **Common name(s)**: Vahankarabo. **Botanical description**: It is native to North and South America. Plant is a large woody climber, with twisted stem. Leaves bipinnate, pinnae 2–3 pairs, leaflets 3.8–7.5 cm, oblong or oblong-obovate, rigidly coriaceous. Flowers in panicle or simple spikes, 15–25 cm long, axillary or from the nodes of old leafless branches. Pods are 30–90 cm long, slightly curved and woody. **Medicinal use(s)**: It is used to treat human colon cancer cells. It also possesses antitumor, antibacterial, anti-inflammatory, and antioxidant properties.			Pursaethoside A	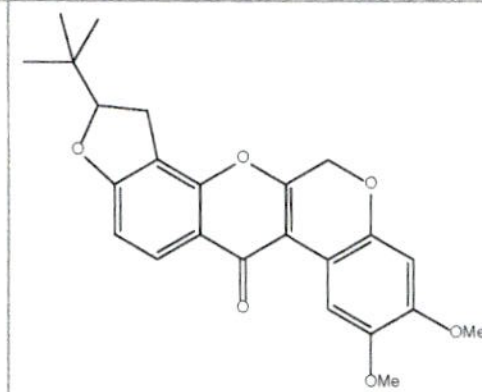	

(continued)

Table 4.59 (continued)

Family	Botanical name	Image	Plant part used	Bioactive compound(s)	Structure	Reference(s)
				Rheediinoside A	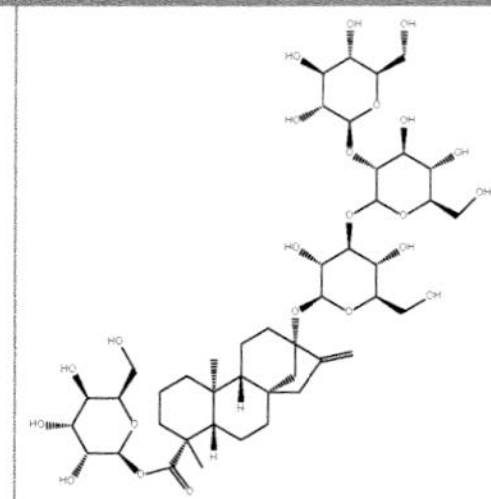	
				Rheediinoside B	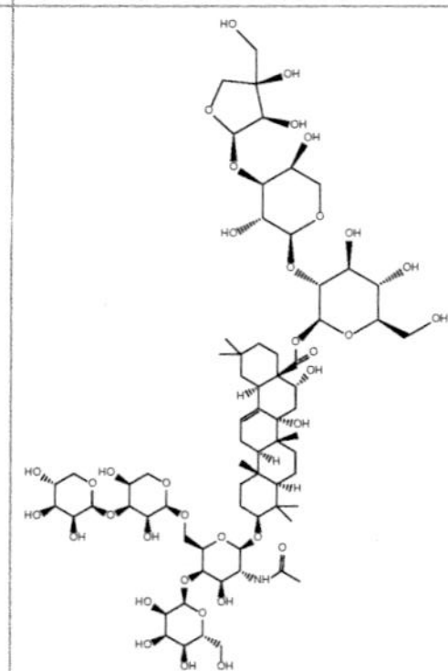	

		Liquiritigenin	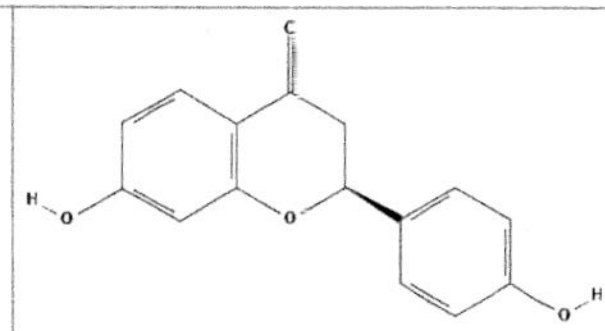	
		Glabridin	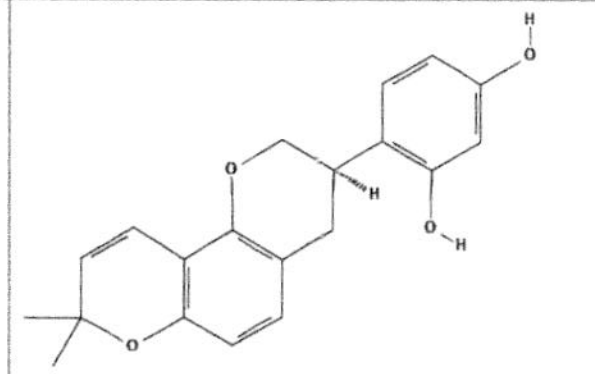	
		4′-O-methylglabridin	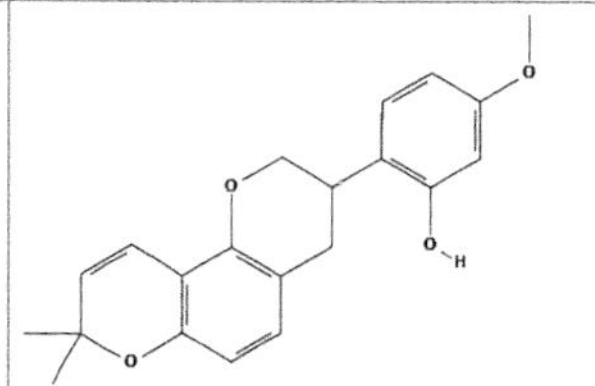	

(continued)

Table 4.59 (continued)

Family	Botanical name	Image	Plant part used	Bioactive compound(s)	Structure	Reference(s)
				Isoliquiritigenin	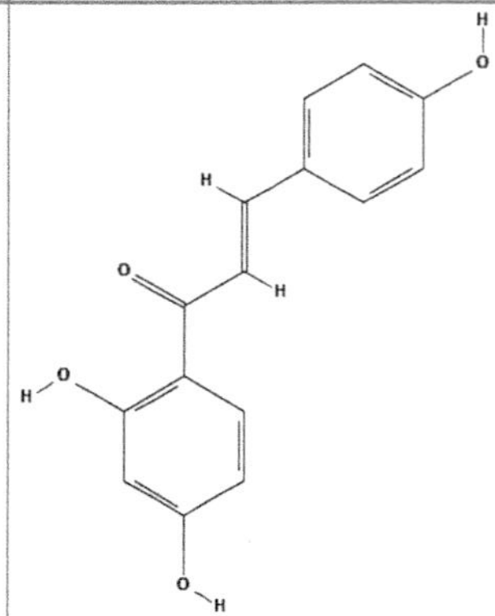	
				Hispaglabridin A	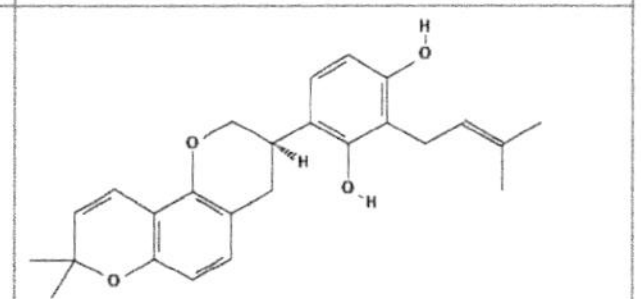	
				Shinflavanone	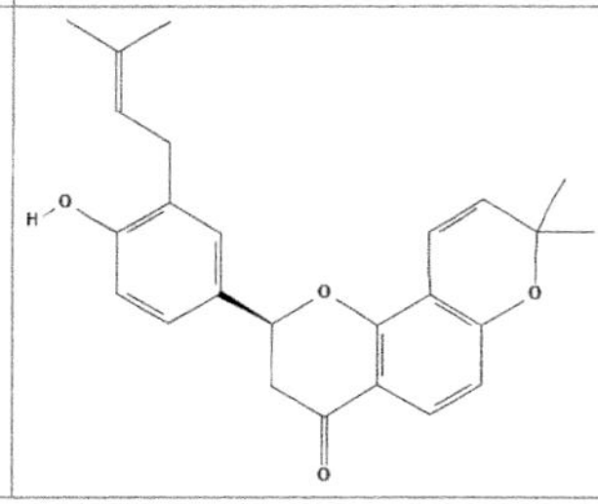	

	Entada gigas (L.) Fawc. & Rendle		Seed	Fulvoplumierin	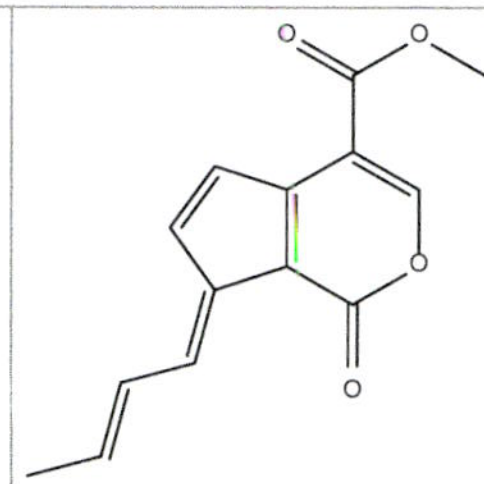	Rahman et al. (2013)
	Synonym(s): *Entada gigalobium* DC., *Entada planoseminata* (De Wild.) G.C.C. Gilbert & Bout, *Entada scandens* (L.) Benth., *Entada umbonata* (De Wild.) G.C.C. Gilbert& Bout, *Mimosa gigas* L., *Mimosa scandens* L. **Common name(s)**: Aathodi. **Botanical description**: It is native to South Asia, Africa, and Australia. It is a large unarmed liana (woody high climber), with stems growing to a size of tree trunks. Leaves with 1–2 pairs of pinnae, leaflets with 3–5 pairs per pinna, oblong, up to 9 × 4 cm, the main spine of the leaf ending in a forked tendril. Flowers are born in long, axillary spikes up to 23 cm long and are cream to pale yellow in color. **Medicinal use(s)**: It is used for the treatment of human cancer cell lines. It also possesses antibacterial, anti-inflammatory, and antioxidant properties.					

(continued)

Table 4.59 (continued)

Family	Botanical name	Image	Plant part used	Bioactive compound(s)	Structure	Reference(s)
	Enterolobium schomburgkii (Benth.) Benth.		Leaf	Corynanthe	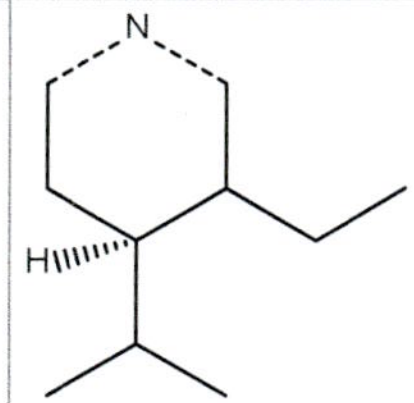	Esposito-Avella et al. (1985) and Maffei (2003a, b, c, d, e, f, g, h, i, j, k, l, m)
	Synonym(s): *Mimosa wilsonii* Standl., *Pithecellobium schomburgkii* Benth., *Feuilleea schomburgkii* (Benth.) Kuntze, *Enterolobium schomburgkii* var. *schomburgkii*. **Common name(s)**: Batibatra. **Botanical description**: It is native to South America (Brazil, Bolivia, Peru, Colombia, Guyana, Surinam) and Central America (Panama, Costa Rica, Nicaragua, Southern Mexico). It is a deciduous tree with a low, wide, dense, and roundish crown. It can grow up to 10–36 m tall. The more or less straight, cylindrical bole can be 60–200 cm in diameter. **Medicinal use(s)**: It possesses anticancer, antitumor, antibacterial, anti-inflammatory, and antioxidant properties.					

	Erythrophleum guineense G.Don		Bark	Taxifolin	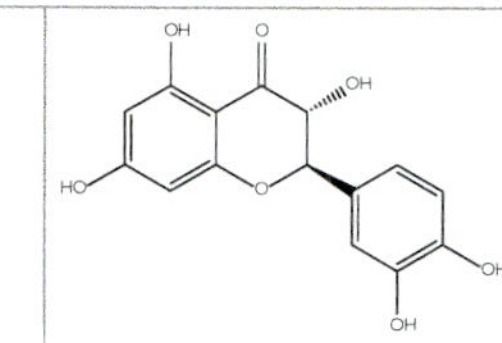	Amico (1977)
	Synonym(s): *Erythrophleum guineense*. **Common name(s)**: Red water tree, Ordeal tree, Sasswood, Sassy bark. **Botanical description**: It is native to west coast of Africa. It is a medium- to large-sized tree, up to 20 m tall. Alternately arranged, pinnate leaves and about 7–13 leaflets at each pinna. Leaflets are large, ovate in shape, about 5 cm long and 2.5 cm wide. Their cream- to yellow-colored flowers are borne on stalked inflorescences. **Medicinal use(s)**: It possesses anticancer, antitumor, antibacterial, anti-inflammatory, and antioxidant properties. It is effective in case of body ache, anthelmintic, dental diseases, dysentery, and internal bleeding. The bark is said to possess astringent and analgesic actions.					
	Gleditsia caspica Desf. (Caspian locust)		Fruit	Gleditsia sapcnins C	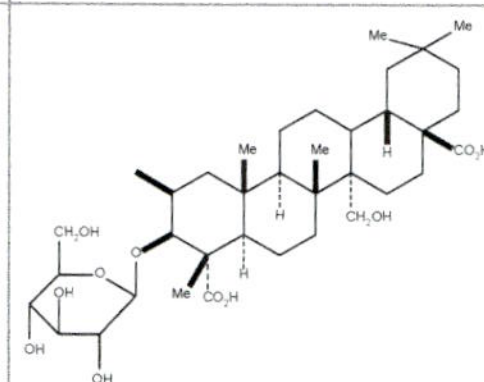	Miyase et al. (2009)

(continued)

Table 4.59 (continued)

Family	Botanical name	Image	Plant part used	Bioactive compound(s)	Structure	Reference(s)
	Synonym(s): *Gleditsia caspia* Desf. **Common name(s)**: Caspian locust. **Botanical description**: It is native to West Asia and North Iran. It is a deciduous tree growing up to 12 m in height. It is in flower from April to May, and the seeds ripen in October. The species is hermaphrodite (has both male and female organs) and is pollinated by insects. It can fix nitrogen. It prefers dry or moist soil and can tolerate drought and atmospheric pollution. **Medicinal use(s)**: It possesses anticancer, antitumor, antibacterial, anti-inflammatory, and antioxidant properties.					
	Gleditsia sinensis Lam.		Fruit	Coronaridine	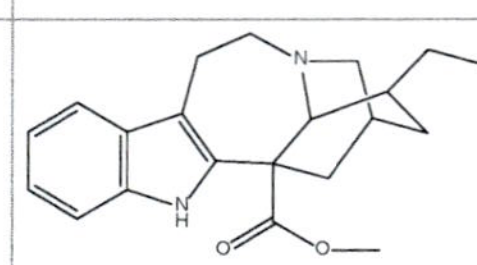	Filly et al. (2005)
	Synonym(s): *Caesalpiniodes caspium* (Desf.) Kuntze, *Gleditsia caspica* var. *subvirescens* hort. ex Loudon, *Gleditsia darwinni* de Vos, *Gleditsia horrida* var. *caspica* C.K. Schneid. **Common name(s)**: Soap bean. **Botanical description**: It is native to West Asia and North Iran. It is a deciduous tree growing up to 12 m in height. It is in flower from April to May, and the seeds ripen in October. The species is hermaphrodite (has both male and female organs) and is pollinated by insects. It can fix nitrogen. It prefers dry or moist soil and can also tolerate drought and atmospheric pollution. **Medicinal use(s)**: It possesses anticancer, antitumor, antibacterial, anti-inflammatory, and antioxidant properties. It is used for the treatment of measles, indigestion, whooping, smallpox, arthrolithiasis, constipation, diarrhea, hematochezia, dysentery, carbuncle, etc.					

	Glycyrrhiza uralensis Fisch.		Stem, root	Kaempferol	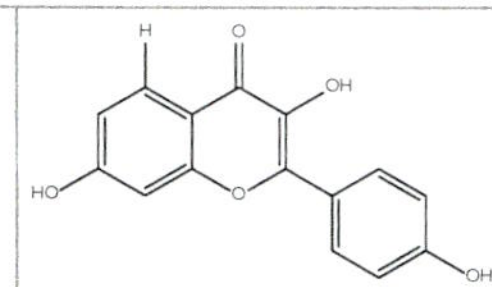	Duke and Ayensu (1985a, b) and Choi et al. (2008)
	Synonym(s): *Glycyrrhiza asperrima* var. *desertorum* Regel, *Glycyrrhiza asperrima* var. *uralensis* Regel, *Glycyrrhiza glandulifera* Ledeb. **Common name(s)**: Liquorice root, Gan cao. **Botanical description**: It is native to East Asia, China, Japan, and Siberia. It is a perennial plant and can grow in semi-shade (light woodland) or no shade. It prefers moist soil. **Medicinal use(s)**: It possesses anticancer, antitumor, antibacterial, anti-inflammatory, and antioxidant properties. It is used for the treatment of Addison's disease, asthma, coughs, and peptic ulcers. Externally, it is also used to treat acne, boils, and sore throats. It should not be prescribed for pregnant women or for people with high blood pressure or kidney disease or anyone taking digoxin-based medications.					
	Gymnocladus chinensis Baill. (Soap tree)		Leaf	Lupeol	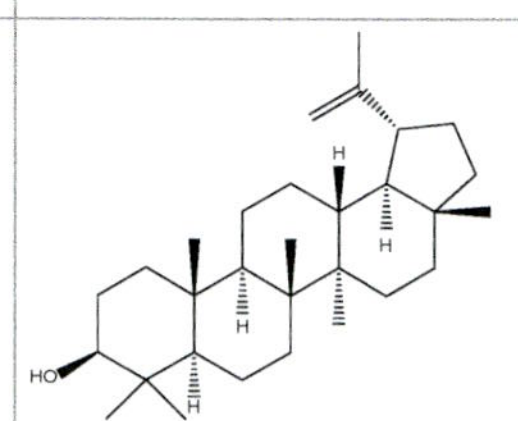	Ma et al. (2007)

(continued)

Table 4.59 (continued)

Family	Botanical name	Image	Plant part used	Bioactive compound(s)	Structure	Reference(s)
	Synonym(s): No synonyms are recorded for this name. **Common name(s)**: Soap tree. **Botanical description**: It is native to East Asia and China. It is a deciduous tree growing up to 12 m in height at a slow rate. The seeds ripen in October. The species is dioecious and not self-fertile and can fix nitrogen. **Medicinal use(s)**: It possesses anticancer, antitumor, antibacterial, anti-inflammatory, and antioxidant properties.			Isoquinoline	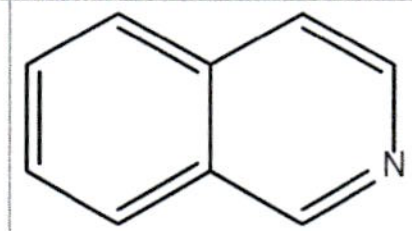	
	Indigofera aspalathoides DC.		Leaves, stems, and seeds	Cianidanol	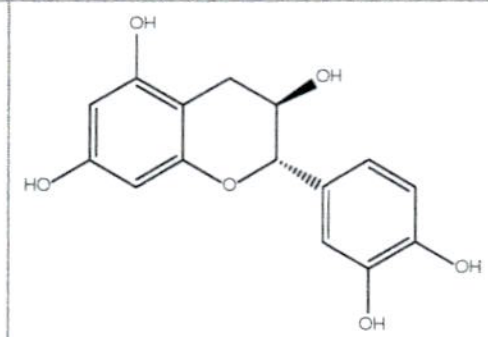	Gupta et al. (2007)
	Synonym(s): *Aspalathus indica* L., *Indigofera aspalathifolia* Roxb. **Common name(s)**: Sivanarvembu. **Botanical description**: It is a deciduous plant native to India and grows in a tropical climate. It grows up to 75 cm having dark brown stem; when young, branches are 0.7–1.5 cm wide and roots are brown colored, woody. The leaf is trifoliate, pale green, oblanceolate, digitate, sessile, and crowded on the young branches. **Medicinal use(s)**: It is used to treat liver tumors. It also possesses anticancer, antibacterial, anti-inflammatory, and antioxidant properties. It is also used to treat abdominal pain, diabetes, spleen, toothache, leprosy, and psoriasis.					

	Indigofera cassioides DC.		Root	Yuccagenin	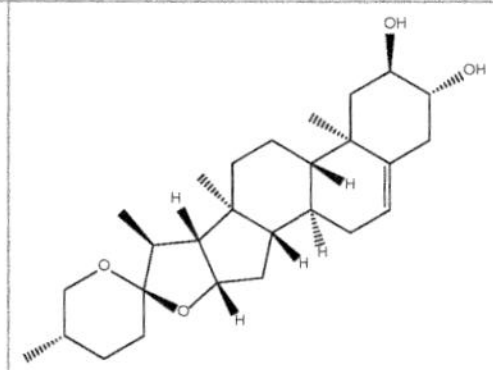	Kumar et al. (2011a, b)
	Synonym(s): *Indigofera arborea* Roxb., *Indigofera byansghatensis* S.N. Biswas, *Indigofera elliptica* Roxb., *Indigofera gibsonii* Graham, *Indigofera leptostachya* DC., *Indigofera pulchella* Roxb., *Indigofera purpurascens* Roxb., *Indigofera violacea* Roxb. **Common name(s)**: Chimnati, Manali, Manneli, Cassia indigo. **Botanical description**: It is a deciduous shrub growing up to 3.5 m and native to the Himalayas. The species is hermaphrodite (has both male and female organs) and is pollinated by insects and can fix nitrogen. **Medicinal use(s)**: It possesses anticancer, antitumor, antibacterial, anti-inflammatory, and antioxidant properties. A decoction of the roots is used for the treatment of coughs. The roots are dried, grounded into a powder, and applied externally for the treatment of pains in the chest.			Quercetin	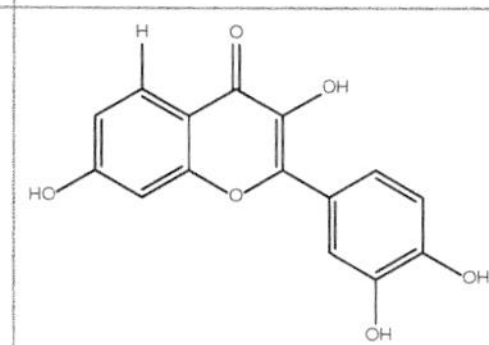	

(continued)

Table 4.59 (continued)

Family	Botanical name	Image	Plant part used	Bioactive compound(s)	Structure	Reference(s)
	Indigofera suffruticosa Mill.		Leaf	Stigmasterol	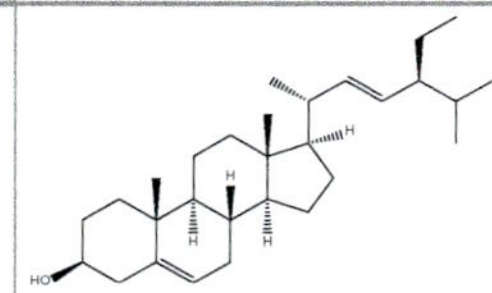	Vieira et al. (2007)
	Synonym(s): *Indigofera angolensis* D. Dietr., *Indigofera anil* L., *Indigofera comezuelo* DC., *Indigofera divaricata* Jacq., *Indigofera drepanocarpa* Bergman, *Indigofera guatimala* Lunan, *Indigofera micrantha* Desv., *Indigofera oligophylla* Lam., *Indigofera tinctoria* Mill., *Indigofera uncinata* G. Don. **Common name(s)**: Wild indigo. **Botanical description**: It is native to subtropical and tropical America, including the Southern United States, the Caribbean, Mexico, and Central America. It is a sparsely branched, stiffly erect, perennial plant growing from 45 to 250 cm tall. It has a woody rootstock from which stems grow that are often woody and persist for more than a year but can also be herbaceous. **Medicinal use(s)**: It is used for the treatment of cancer. It also possesses antitumor, antibacterial, anti-inflammatory, and antioxidant properties.					

	Indigofera tinctoria L.		Leaf, root	Quercetin	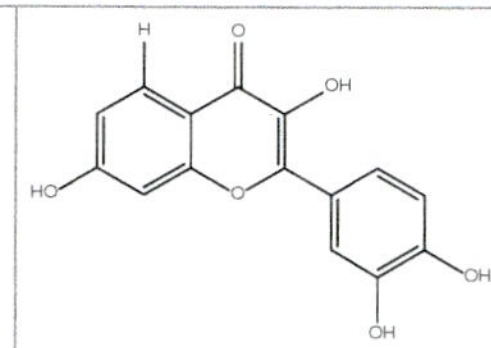 	Magesh et al. (2009)
	Synonym(s): *Indigofera bergii* Vatke, *Indigofera cinerascens* DC., *Indigofera houer* Forssk., *Indigofera indica* Lam., *Indigofera oligophylla* Baker, *Indigofera orthocarpa* (DC.) O. Berg & C.F. Schmidt, *Indigofera sumatrana* Gaertn., *Indigofera tinctoria* Blanco, *Indigofera tulearensis* Drake. **Common name(s)**: Indigo dye, True indigo. **Botanical description**: It is native to China, tropical Asia, and some parts of Africa. It is a deciduous spreading tropical shrub or subshrub of the pea family that typically grows up to 2–3 in. tall. As suggested by the common name, this shrub was the original source of the blue dye known as indigo. It features light green pinnate leaves and short racemes of summer-blooming pink or violet flowers. Flowers are followed by seed pods. **Medicinal use(s)**: It possesses anticancer, antitumor, antibacterial, anti-inflammatory, and antioxidant properties.					

(continued)

Table 4.59 (continued)

Family	Botanical name	Image	Plant part used	Bioactive compound(s)	Structure	Reference(s)
	Lens culinaris Medik.		Seeds	Phillyrin	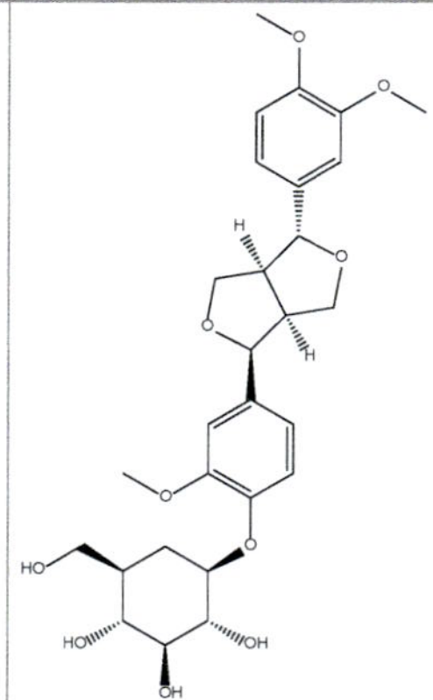	Wang et al. (2000)
	Synonym(s): *Ervum lens* L., *Lens esculenta* Moench, *Lens lens* Huth, *Vicia lens* (L.) Coss. & Germ. **Common name(s)**: Lentil. **Botanical description**: It is native to Western Canada and England. It grows to around 0.5 m in height as a slender bush or twining vine. The leaves are alternate and compound, with 4–7 pinnate (opposite) pairs of leaflets, ending in a tendril or short bristle. The flowers are small and white to bluish, either in solitary or several clusters, and form small fruits, flattened legumes or pods containing one or two lens-shaped seeds. Numerous cultivars vary in the seed size, texture, and color range from green to yellow to orange to red and brown. **Medicinal use(s)**: It is used for the treatment of cancer. It also possesses antitumor, antibacterial, anti-inflammatory, and antioxidant properties.					

	Maackia tenuifolia (Hemsl.) Hand.-Mazz.		Root	Corynanthe	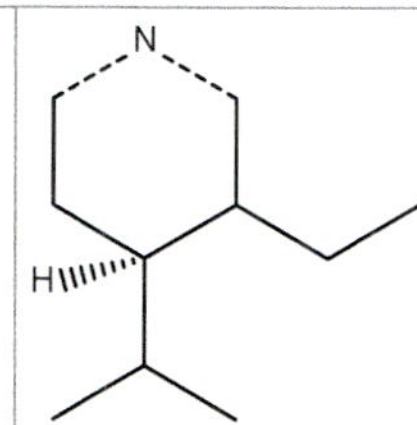	Martin-Cordero et al. (1995) and Levings (2006)
	Synonym(s): *Euchresta tenuifolia* Hemsl, *Maackia honanensis* Bailey. **Common name(s)**: Chinese yellow wood, Amur maackia. **Botanical description**: It is native to East Asia, China, Northern and Central Japan, and Korea. It is a deciduous shrub growing up to 6 m at a slow rate. It is in flower from July to August, and the seeds ripen from October to November. The species is hermaphrodite and is pollinated by insects. It is suitable to grow in light, medium, and heavy soils. **Medicinal use(s)**: It is used for the treatment of cancer and tumor. It also possesses antitumor, antibacterial, anti-inflammatory, and antioxidant properties.					

(continued)

Table 4.59 (continued)

Family	Botanical name	Image	Plant part used	Bioactive compound(s)	Structure	Reference(s)
Fabaceae	*Machaerium aristulatum* (Benth.) Ducke		Stem	Machaerium	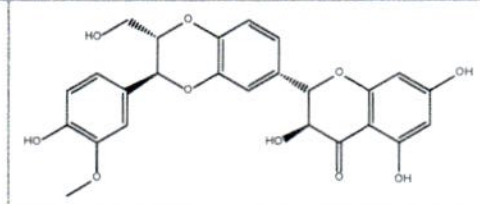	Seo et al. (2001)
				Aristulatum	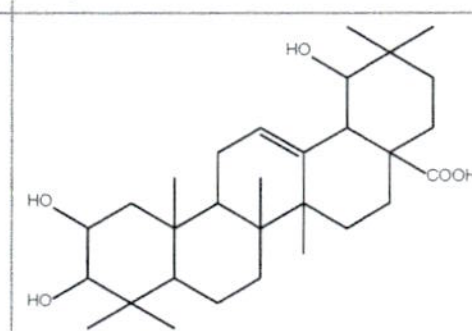	
	Synonym(s): *Drepanocarpus aristulatus* Benth. **Common name(s)**: Una de gavilan. **Botanical description**: It is native to South America, Brazil, Peru, Central America, and Mexico. It is a spiny plant varying in habit from a climbing plant to a shrub or small tree found in moist or dry thickets or forest, often along stream banks, sometimes in pine forest, at elevations up to 1200 m. **Medicinal use(s)**: It is used to treat lung cancer. Its bark is commonly used to treat diarrhea, cough, and cancer. It also possesses antitumor, antiproliferative, antimutagenic, and antioxidant properties.			Pterocarpan	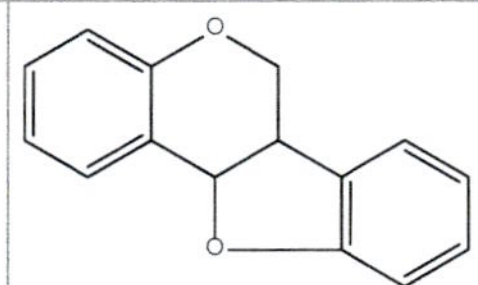	
				Formononetin (flavonoid)	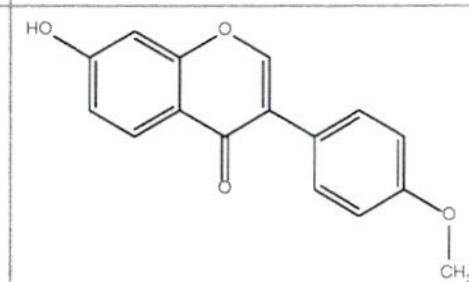	
				Maackiain (isoflavone)	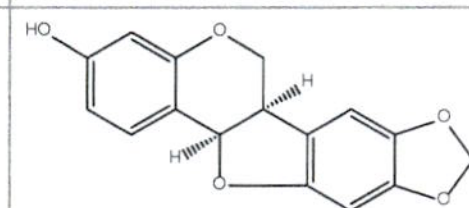	

	Mucuna macrocarpa Wall.		Seed	Berberine	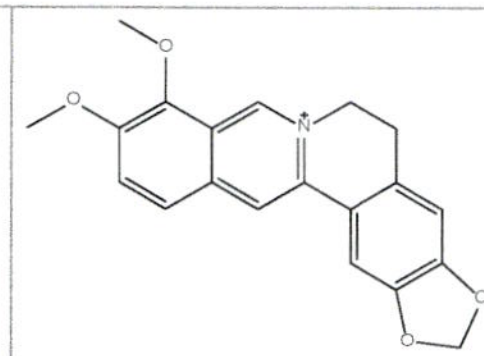	Wasuwat (1967), Cherdshewasart (2009), and Lu et al. (2010, 2012)
	Synonym(s): *Mucuna bodinieri* H.Lev., *Mucuna castanea* Merr., *Mucuna colletii* Lace, *Mucuna ferruginea* Matsum., *Mucuna irukanda* Ohwi, *Mucuna subferrunginea* Hayata, *Mucuna wangii* Hu. **Common name(s)**: Velvet bean. **Botanical description**: It is native to Thailand typically found in the tropical woodlands. The leaves are trifoliate and alternate, and the flowers are large in size with curved destructive petals. **Medicinal use(s)**: It possesses anticancer, antioxidant, antitumor, and antimutagenic properties. It is also used for reducing body pain and joint pain.					
	Ononis hirta Poir.		Apical part	Dregamine	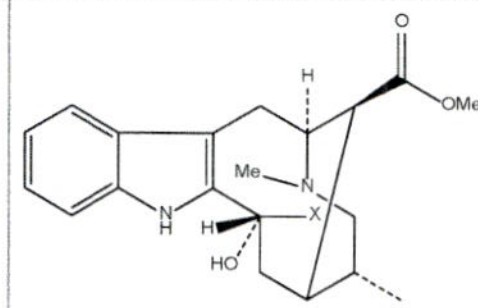	Talib and Mahasneh (2012)

(continued)

Table 4.59 (continued)

Family	Botanical name	Image	Plant part used	Bioactive compound(s)	Structure	Reference(s)
	Synonym(s): *Ononis ellipticifolia* Willk., *Ononis baetica* sensu auct. **Common name(s)**: Restharrow. **Botanical description**: It is a hairy plant and native to South America. It bears several flowers and clustered along with oblong leaves. The leaves appear on either side of the stem. It produces yellow fruits which appear like capsules with reddish seeds. **Medicinal use(s)**: It is used to treat breast cancer. It also possesses antimicrobial and antitumor properties. Traditionally, it was used for treating female disorders, respiratory ailments (cough, bronchitis, and asthma), worm infestations in children, dysentery, jaundice, pimples, gonorrhea, digestive problems, and tumors.					
	Ononis viscosa subsp. *sicula* (Guss.) Hub.-Mor.		Apical part	Cianidanol	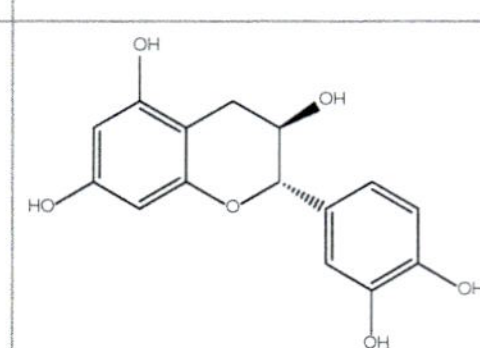	Talib and Mahasneh (2012)
	Synonym(s): *Ononis sicula* Guss. **Common name(s)**: Restharrows. **Botanical description**: It is a small herb native to Europe, Africa, and tropical and temperate Asia. **Medicinal use(s)**: It is used to treat colon, skin, and breast cancer, bladder and kidney problems, and water retention. It also possesses antimicrobial, antiproliferative, and anti-inflammatory properties.					

	Parkinsonia aculeata L.		Twig	Betulinic acid	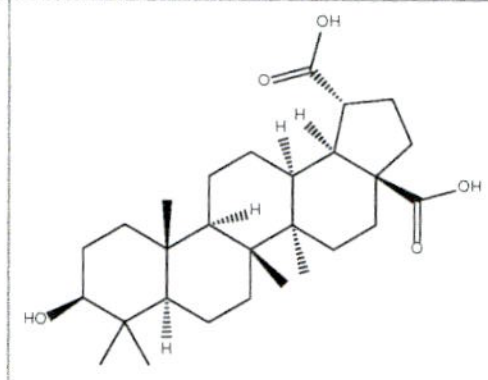	Sharma et al. (1992) and Prameela et al. (2018)
	Synonym(s): *Parkinsonia thornberi* M.E. Jones. **Common name(s)**: Jerusalem thorn. **Botanical description**: It is a small evergreen spiny tree with an open, spreading, globose crown native to South America, Argentina, Uruguay, Paraguay, Brazil, Bolivia, Peru, and Ecuador. It can grow up to 4–10 m tall. **Medicinal use(s)**: It is used to treat gastric cancer. It also possesses antitumor, antimutagenic, antioxidant, and anti-inflammatory properties. Leaf, fruit, and stem decoctions are taken orally or applied externally to treat fever and malaria.					
	Bauhinia thonningii Schum.		Root	Coumaric acid	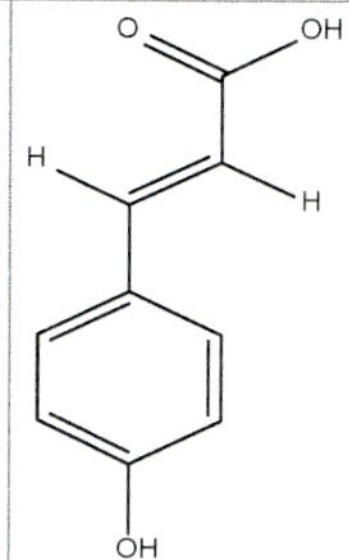	Chhabra et al. (1987), Osunderu (2017)

(continued)

Table 4.59 (continued)

Family	Botanical name	Image	Plant part used	Bioactive compound(s)	Structure	Reference(s)
	Synonym(s): *Bauhinia thonningii* Schumach., *Piliostigma thonningii* (Schum.) Milne-Redh. **Common name(s)**: Camel foot. **Botanical description**: It is native and widespread in tropical Africa and extending to South Africa and Zimbabwe. It has simple leaves, conspicuously 2 lobed with rust-colored indumentum beneath. Flowers are medium to small, unisexual, in racemes or panicles, white or pinkish in color. **Medicinal use(s)**: It is used to treat cancer, dysentery, fever, respiratory ailments, snakebites, hookworm, and skin infections. It also possesses antitumor, antimicrobial, anti-inflammatory, anti-allergic, and antioxidant properties.					
	Piptadeniastrum africanum (Hook.f.) Brenan		Bark, root, leaves	α-Humulene	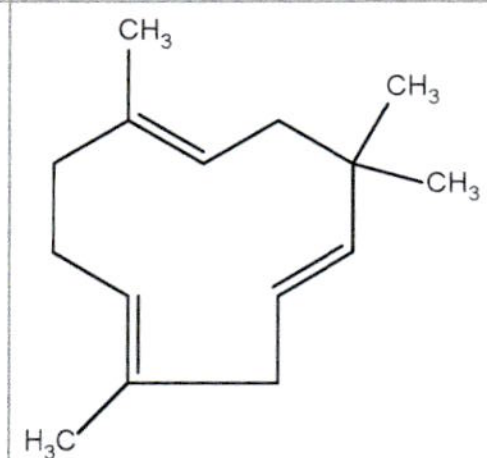	Loizzo et al. (2007) and Marius et al. (2018)
	Synonym(s): *Piptadenia africana* Hook.f., *Piptadeniastrum africanum* var. *africanum*. **Common name(s)**: Mbeli. **Botanical description**: It is native to tropical Africa and Senegal to Southern Sudan. It is a tree that can range in height from only 5 m in the drier areas and 50 m tall in the wetter rainforests. The straight, cylindrical bole, which can be 1.8–3 m in diameter, is up to 20 m high above the buttresses before branching, though it is often slightly sinuous and forked at a lower level. The tree has many thin, high buttresses up to 4.5 m, extending outwards at their base into many surface plank roots. **Medicinal use(s)**: It is used to treat cancer, fever, toothache, pneumonia, edema, skin complaints, and rheumatism, to expel worms, and as a purgative and abortifacient. Root extracts or macerations are used to treat mental disorders and aphrodisia.					

	Pongamia glabra (L.) Pierre		Bark and root	Lonchocarpin	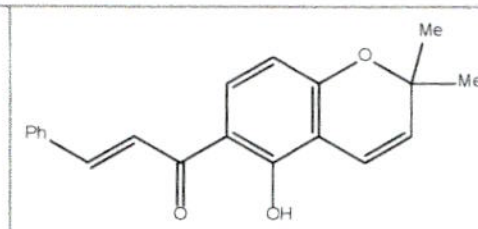	Yin et al. (2006) and Arulvasu et al. (2012)
				Lanceolatin	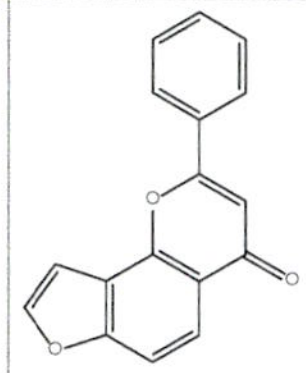	
	Synonym(s): *Cytisus pinnatus* L., *Dalbergia arborea* Willd., *Derris indica* (Lam.) Benn., *Galedupa indica* Lam., *Galedupa pinnata* (L.) Taub., *Galedupa pungum* J.G.Gmel., *Millettia novo-guineensis* Kaneh. & Hatus., *Millettia pinnata* (L.) Panigrahi, *Pongamia glabra* Vent., *Pongamia mitis* (L.) Kurz, *Pongamia xerocarpa* Hassk., *Pterocarpus flavus* Lour., *Robinia* mitis L. **Common name(s)**: Indian beech. **Botanical description**: It is a deciduous tree up to 20 m tall with a large canopy and is native to tropical and temperate Asia and is thought to have originated in India. **Medicinal use(s)**: It is used to treat breast cancer. It also possesses antiproliferative, antifungal, antiseptic, antipruritic, and analgesic properties. Thick brownish oil extracted from the large seeds is used for the treatment of rheumatism.					

(continued)

Table 4.59 (continued)

Family	Botanical name	Image	Plant part used	Bioactive compound(s)	Structure	Reference(s)
	Pongamia pinnata (L.) Pierre		Stem and bark	Tunicatachalcone	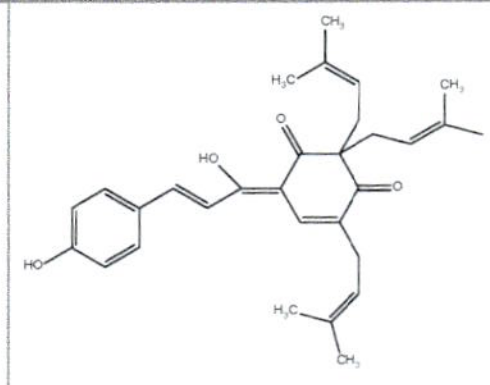	Yin et al. (2006)
	Synonym(s): *Cytisus pinnatus* L., *Dalbergia arborea* Willd., *Derris indica* (Lam.) Benn., *Galedupa indica* Lam., *Galedupa pinnata* (L.) Taub., Galedupa *pungum* J.G.Gmel., *Millettia novo-guineensis* Kaneh. & Hatus., *Millettia pinnata* (L.) Panigrahi, *Pongamia glabra* Vent., *Pongamia mitis* (L.) Kurz, *Pongamia xerocarpa* Hassk., *Pterocarpus flavus* Lour., *Robinia mitis* L. **Common name(s)**: Karanja. **Botanical description**: It is a fast-growing, medium-sized, evergreen or briefly deciduous, glabrous shrub or tree with a broad crown of spreading or drooping branches and is native to East Asia, Indian subcontinent, through Southeast Asia to Northeastern Australia, Fiji, and Japan. It usually grows 15–25 m in height, with a straight or crooked bole of 50–80 cm or more in diameter. **Medicinal use(s)**: It is used to treat breast cancer and epithelial cancer. The leaves are crushed and applied as a poultice for the treatment of parasitic skin diseases and to relieve bleeding hemorrhoids. The fresh stem bark is used as an astringent.					

<table>
<tr><td rowspan="3"></td><td rowspan="2">Cullen corylifolium (L.) Medik.</td><td rowspan="2"></td><td rowspan="2">Stem</td><td>Psoralen</td><td>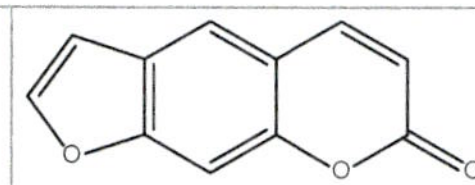</td><td rowspan="3">Wang et al. (2011a, b)</td></tr>
<tr><td rowspan="2">Isopsoralen</td><td rowspan="2">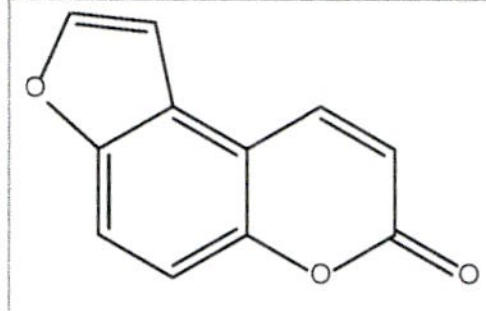</td></tr>
<tr><td colspan="3">Synonym(s): Lotodes corylifolia (L.) Kuntze, Psoralea corylifolia L., Psoralea patersoniae Schonl., Trifolium unifolium Forssk.
Common name(s): Babchi.
Botanical description: It is an annual plant growing up to 0.6 m and is native to Asia, Iran, China, Africa, and the Middle East. It is in flower from July to August, and the seeds ripen from September to October. The flowers are hermaphrodite (have both male and female organs) and are pollinated by insects.
Medicinal use(s): It is used to treat human colorectal, gastric, stomach, oral, and breast cancer cell lines. The seeds are anthelmintic, antibacterial, aphrodisiac, astringent, cardiac, cytotoxic, deobstruent, diaphoretic, diuretic, stimulant, and stomachic. It is used for the treatment of febrile diseases, premature ejaculation, impotence, lower back pains, frequent urination, incontinence, and bed wetting. It is also used externally to treat various skin ailments including leprosy, leucoderma, and hair loss.</td></tr>
</table>

(continued)

Table 4.59 (continued)

Family	Botanical name	Image	Plant part used	Bioactive compound(s)	Structure	Reference(s)
	Pterocarpus erinaceus Poir.		Stem, bark	Lonchocarpin	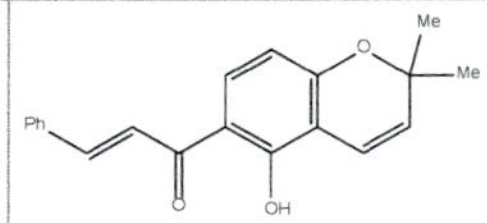	Ayensu (1978) and Noufou et al. (2012)
	Synonym(s): *Lingoum erinaceum* (Poir.) Kuntze, *Pterocarpus adansonii* DC., *Pterocarpus africanus* Hook., in W. Gray. **Common name(s)**: African kino. **Botanical description**: It is a deciduous tree growing to 20 m at a slow rate and is native to west tropical Africa to Central African Republic. It is pollinated by bees and insects. It can also fix nitrogen. **Medicinal use(s)**: It is used to treat cancer and tumors. It possesses antibacterial, antifungal, antimalarial, and antioxidant properties. Decoctions or infusions of the bark or roots are used for the treatment of bronchial infections, toothache, dysentery, menstruation complaints, anemia, gonorrhea, postpartum hemorrhage, ringworm infections, leprosy, wounds, tumors, and ulcers and also used as an antiemetic, purgative, and tonic. Root preparations are administered as an enema to treat venereal diseases. Leaf decoctions are used to treat fever and syphilis and are used as aphrodisiac.					

	Sophora flavescens Aiton		Root	Matrine	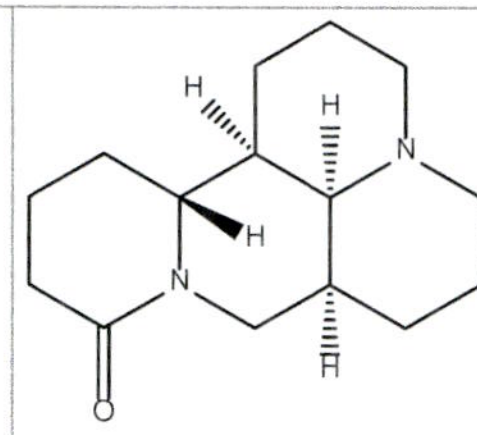	Madhuri and Pandey (2009)
	Synonym(s): *Sophora angustifolia* Siebold & Zucc., *Sophora macrosperma* DC., *Sophora tetragonocarpa* Hayata. **Common name(s)**: Ku shen gen. **Botanical description**: It is an evergreen shrub growing up to 1.5 m by 1 m at a slow rate and is native to East Asia, Russia, and China. It is in flower from July to August, and the seeds ripen in September. The flowers are hermaphrodite (have both male and female organs) and are pollinated by insects. It can also fix nitrogen and cannot grow in the shade. **Medicinal use(s)**: It is used to treat lung cancer, jaundice, dysentery, diarrhea, vaginitis, eczema, pruritus, ringworm, leprosy, syphilis, itching allergic reactions, scabies, and urinary infections. The roots possess anthelmintic, antibacterial, antifungal, and antipruritic properties. It is also used as an astringent, carminative, diuretic, febrifuge, parasiticide, and tonic.					

(continued)

Table 4.59 (continued)

Family	Botanical name	Image	Plant part used	Bioactive compound(s)	Structure	Reference(s)
	Styphnolobium japonicum (L.) Schott		Rhizome, Entire plant	Quercetin	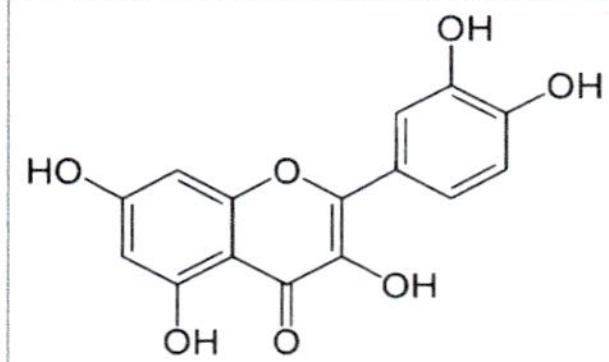	Duke and Ayensu (1985a, b), Shirataki et al. (2004), and Kite et al. (2009)
	Synonym(s): *Sophora japonica* L., *Sophora sinensis* Forrest, Sophora korolkowi Diecks, *Sophora korolkowii* Dieck, *Sophora korolkowi* Dieck ex Koehne. **Common name(s)**: Pagoda tree. **Botanical description**: It is a deciduous tree growing up to 20 m (65 ft.) by 20 m (65 ft.) at a fast rate and is native to East Asia, North China, Japan, and Korea. It is in flower in September, and the seeds ripen in November. The flowers are hermaphrodite (have both male and female organs) and are pollinated by insects. It can also fix nitrogen and cannot grow in the shade. **Medicinal use(s)**: It is used to treat cervical, colon, breast, liver, lung, ovary, stomach, and gastric cancer and hemorrhoids, hematuria, uterine bleeding, constipation, stuffy sensation in the chest, dizziness, epilepsy, red eyes, headache, and hypertension. It also possesses anti-inflammatory, antibacterial, antidiarrheal, anti-spasmodic, anti-cholesterolemic, antifertility, and hypotensive properties. It should be used with caution since it is toxic.			Quercetin 3-O-a-rhamnopyranosyl(1→6)-β-glucopyranoside (rutin)	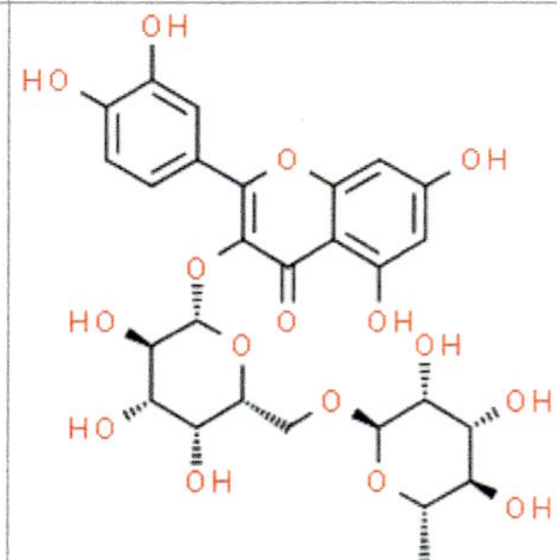	

		Genistein	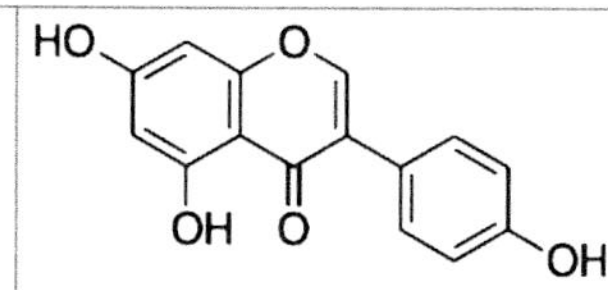	
		Genistein 7,4′-di-O- β –glucopyranoside	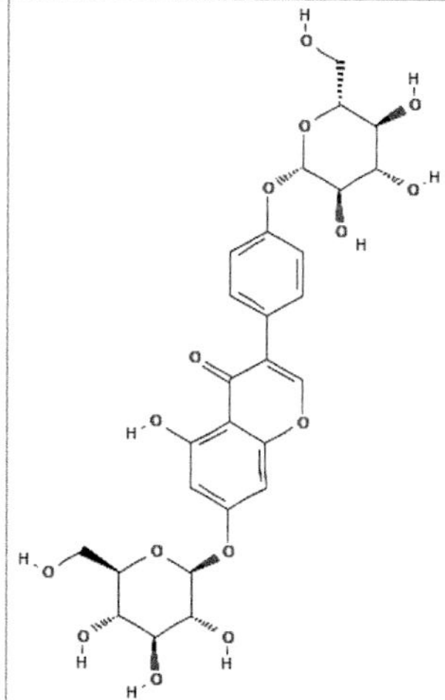	

(continued)

Table 4.59 (continued)

Family	Botanical name	Image	Plant part used	Bioactive compound(s)	Structure	Reference(s)
				Genistein 7-O- β -glucopyranoside (genistin)	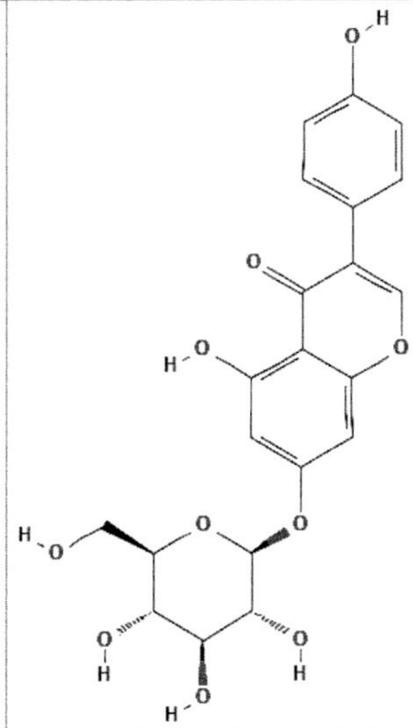	
				Genistein 4′-O- β -glucopyranoside (sophoricoside)	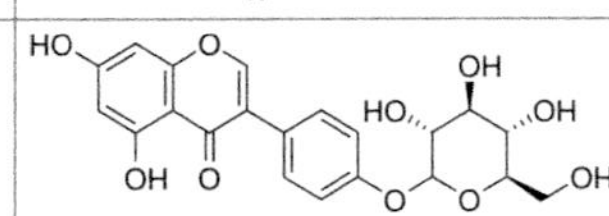	
				Genistein 4′-O-arhamnopyranosyl (1→2)- β -glucopyranoside (sophorabioside)	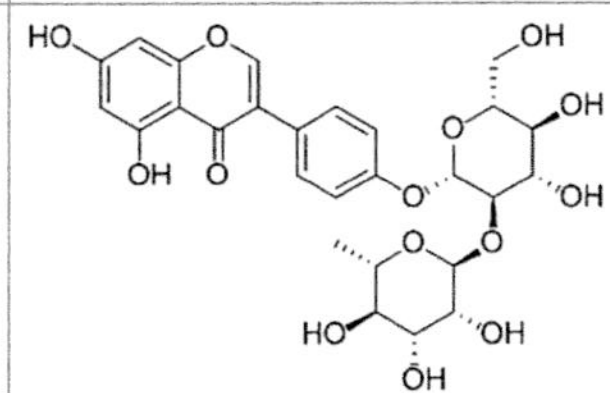	

				Kaempferol	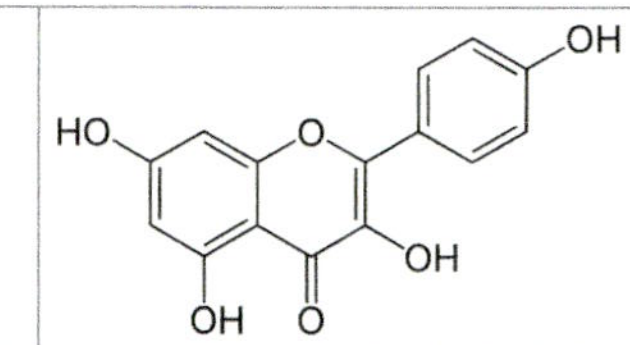	
				Kaempferol 3-O- β -glucopyranosyl(1→2)-β –glucopyranoside	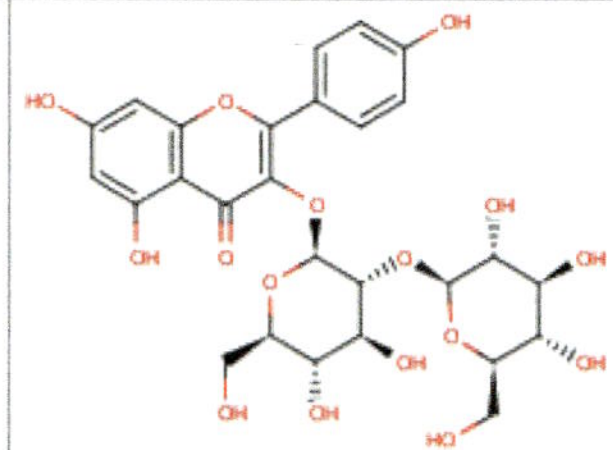	
	Sophora subprostrata		Root	Oxymatrine	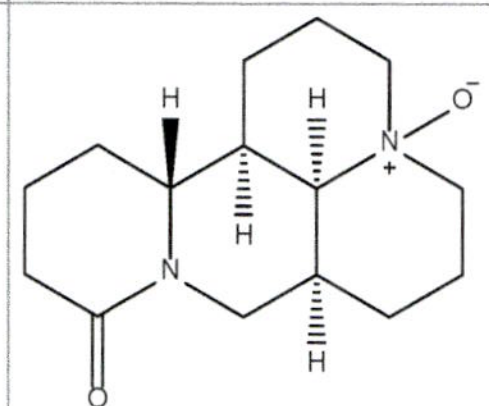	Kyogoku et al. (1975), Sakamoto et al. (1992), and Madhuri and Pandey (2009)
	Synonym(s): *Sophora tonkinensis* Gagnep. **Common name(s)**: Shan dou gen. **Botanical description**: It is a deciduous tree growing up to 6–15 m tall and is native to North Central Pacific and Hawaii. The bole can be up to 60 cm in diameter. **Medicinal use(s)**: It is used to treat breast, nasopharyngeal, and lung cancer. It also possesses antitumor, anti-inflammatory, anti-asthmatic, and anti-ulcer properties. It is also used to treat sore throat, swollen gums, lung-heat cough, and damp-heat jaundice.					

(continued)

Table 4.59 (continued)

Family	Botanical name	Image	Plant part used	Bioactive compound(s)	Structure	Reference(s)
	Sophora tomentosa L.		Root, leaf	Stilbenes	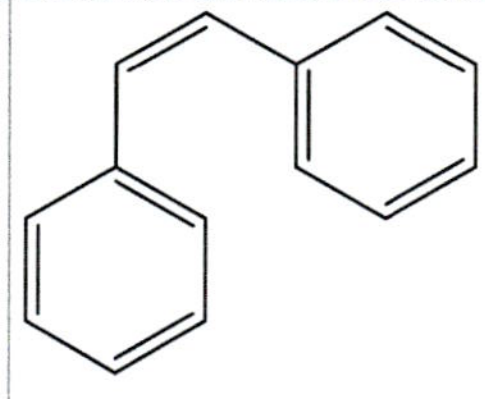	Holdsworth (1991) and Shirataki et al. (2004)
	Synonym(s): *Sophora tomentosa* f. *longifolia* Yakovlev, *Sophora tomentosa* f. *aurea* Yakovlev, *Sorindeia goudotii* Briq. **Common name(s)**: Necklace pod. **Botanical description**: It is an evergreen shrub or small tree that can grow up to 7 m tall and is native to pantropical and coastal regions. The bole can be up to 45 cm in diameter and free of branches for up to 3 m. It is occasionally grown as an ornamental, especially in gardens close to the sea. **Medicinal use(s)**: The bitter seeds are used as astringent, febrifuge, and stomachic; they also possess dangerous emetocathartic properties. The pulverized seeds are used for the treatment of dysentery and cholera and as an antidote against poisonous marine animals. Smaller doses are applied as an antidote in food poisoning and stings of poisonous fish, as well as to treat hemoptysis, painful urination, and gonorrhea.					

<table>
<tr><td rowspan="2"></td><td>Sutherlandia frutescens (L.) R.Br.</td><td></td><td>Whole plant</td><td rowspan="2">Lupeol</td><td rowspan="2">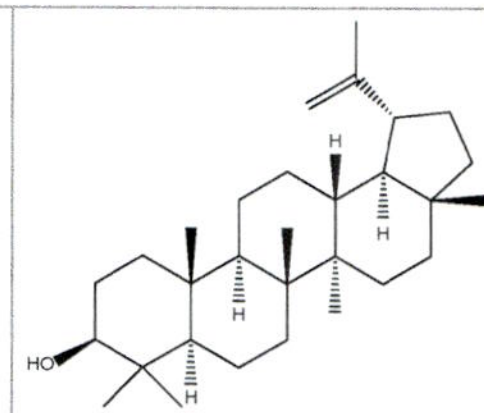</td><td rowspan="2">Chinkwo (2005)</td></tr>
<tr><td colspan="3">Synonym(s): Colutea frutescens L.
Common name(s): Cancer bush.
Botanical description: It is an annual, small, soft-wooded shrublet, 0.5–1 m in height and is native to South Africa, West Cape, Namibia, and Botswana. The leaves are pinnately compound, and the leaflets are 4–10 mm long, gray green in color, giving the bush a silvery appearance.
Medicinal use(s): It is used to treat human breast adenocarcinoma and human nontumorigenic epithelial mammary gland cells and kidney cell lines. It is also used in the treatment of viral hepatitis, bronchitis, flu, asthma, rheumatoid arthritis, peptic ulcers, liver infections, urinary tract infections, stress, and anxiety. It supports the immune system against chronic diseases and immunodeficiency diseases. It also possesses antioxidant, antiviral, anti-stress, antidiabetic, antimutagenic, anti-inflammatory, antiproliferative, antibacterial, and antithrombotic properties.</td></tr>
</table>

(continued)

Table 4.59 (continued)

Family	Botanical name	Image	Plant part used	Bioactive compound(s)	Structure	Reference(s)
	Trifolium pratense L.		Aerial parts, flower tops	Biochanin A	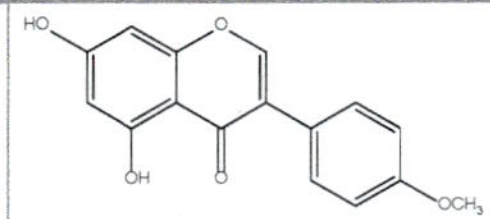	Mohammad (2006), Desai et al. (2008), and Madhuri and Pandey (2009)
	Synonym(s): *Trifolium borysthenicum* Gruner, *Trifolium bracteatum* Schousb., *Trifolium lenkoranicum* (Grossh.) Roskov, *Trifolium ukrainicum* Opperman. **Common name(s)**: Red clover. **Botanical description**: It is a perennial growing up to 0.6 m by 0.6 m at a medium rate and native to Europe, Britain, South or East Norway, Spain, and West Asia. It is in flower from May to September, and the seeds ripen from July to October. The flowers are hermaphrodite and pollinated by bees. It can also fix nitrogen. It can grow in shade and prefers moist soil. The plant can tolerate strong winds but not maritime exposure. **Medicinal use(s)**: It is used to treat ovarian and breast cancer. It also possesses anti-inflammatory, antitumor, anti-spasmodic, anti-scrophulatic, and anti-psoriatic properties. It is also used to treat skin infections, and leaves are estrogenic and are helpful for the treatment of menopausal complaints.					

	Trigonella foenum-graecum L.		Stem and leaf	Graecunin G	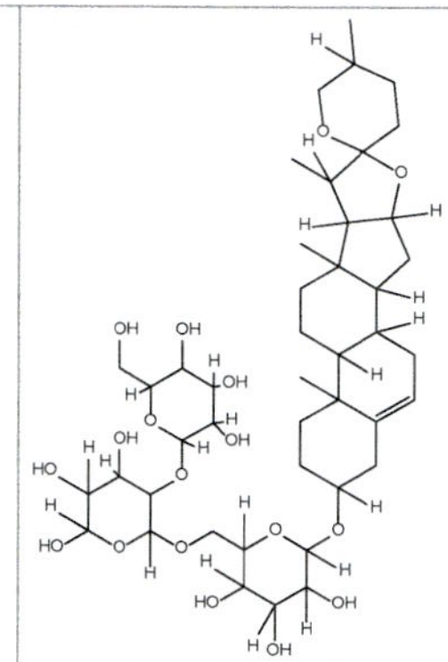	Bhatia et al. (2006)
	Synonym(s): *Trigonella tibetana* (Alef.) Vassilcz. **Common name(s)**: Fenugreek. **Botanical description**: It is an annual herb 30–60 cm in height and is native to Southeastern Europe, West Asia-India, China, Myanmar, Bhutan, Sri Lanka, and Nepal. The leaves are light green and pinnately trifoliate. The flowers are white or yellowish white and axillary. Fruits are legumes 5–7.5 cm long, narrow, curved, tapering with a slender point and containing 10–20 deeply furrowed seeds per pod. **Medicinal use(s)**: It is used to treat brain cancer. It also possesses anticarcinogenic, antidiabetic, anti-atherogenic, anti-anorexic, antioxidant, antihyperlipidemic, and anti-inflammatory properties. It helps to lower cholesterol level and is used to treat skin disorders, ulcers on the skin, burns, scars, incisions, eczema, and abscess.			Graecunin E	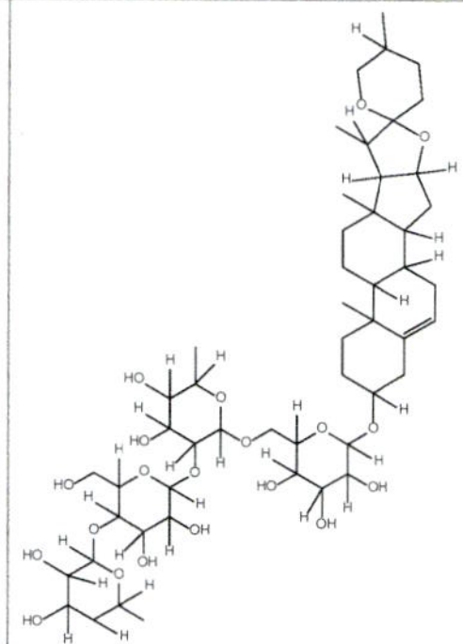	

(continued)

Table 4.59 (continued)

Family	Botanical name	Image	Plant part used	Bioactive compound(s)	Structure	Reference(s)
				Fenugreekine	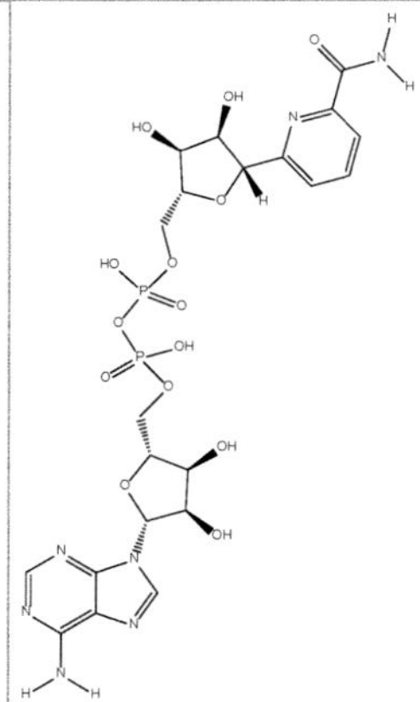	
				Phenolic compounds (salicylic acid)	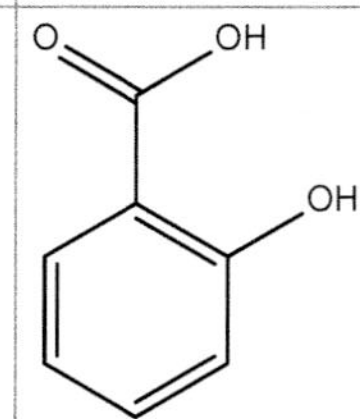	

	Vigna radiata (L.) R. Wilczek.		Entire plant	Cianidanol	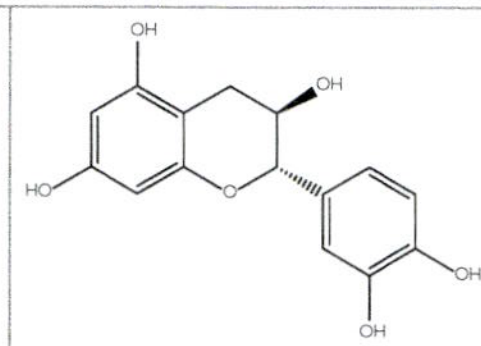	Duke and Ayensu (1985a, b) and Kumar and Singhal (2009)
	Synonym(s): *Azukia radiata* (L.) Ohwi, *Phaseolus abyssinicus* Savi, *Phaseolus aureus* Roxb., *Phaseolus chanetii* (H.Lev.) H.Lev., Phaseolus *hirtus* Retz., *Phaseolus radiatus* L., *Pueraria chanetii* H.Lev., *Rudua aurea* (Roxb.) F.Maek. **Common name(s)**: Green gram, Mung bean. **Botanical description**: It is an erect or semi-erect annual plant growing 13–130 cm tall from a well-developed taproot and is native to East Asia, China, India, Sri Lanka, Myanmar, Thailand, Indonesia, and New Guinea. It can be a much-branched, bushy plant or have somewhat twining stem and a climbing habit. It is widely cultivated in tropical and subtropical areas for its edible seeds. **Medicinal use(s)**: It is used to treat various types of cancer and has antitumor, antimutagenic, and anti-inflammatory properties.					

(continued)

Table 4.59 (continued)

Family	Botanical name	Image	Plant part used	Bioactive compound(s)	Structure	Reference(s)
	Vigna unguiculata (L.) Walp.		Stem	Gallic acid	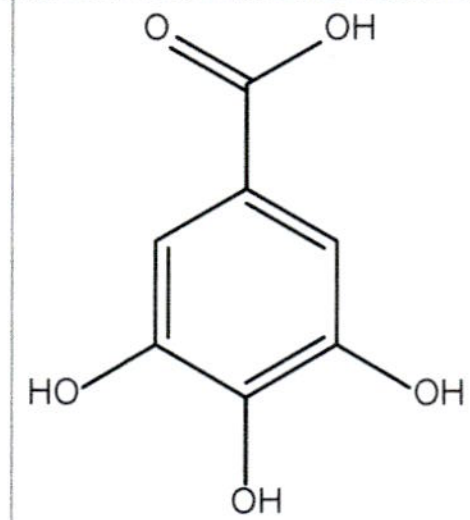	Gutierrez Uribe et al. (2011)
	Synonym(s): *Dolichos biflorus* L., *Dolichos catjang* Burm.f., *Dolichos hastifolius* Schnizl., *Dolichos lubia* Forssk., *Dolichos melanophtalmus* DC., *Dolichos melanophthalamus* DC., *Dolichos monachalis* Brot., *Dolichos obliquifolius* Schnizl., *Dolichos sinensis* L., *Dolichos tranquebaricus* Jacq., *Dolichos unguiculata* L., *Dolichos unguiculatus* L., *Liebrechtsia scabra* De Wild., *Phaseolus sphaerospermus* L., *Phaseolus unguiculatus* (L.) Piper, *Vigna brachycalyx* Baker f., *Vigna catjang* (Burm.f.) Walp., *Vigna catjang* Savi, *Vigna scabra* (De Wild.) T.Durand & H.Durand, *Vigna scabrida* Burtt Davy, *Vigna sinensis* (L.) Savi ex Hausskn. **Common name(s)**: Black-eyed pea. **Botanical description**: It is an annual climbing plant growing up to 4 m tall and is native to tropical Africa. The plant is often cultivated in the warm temperature and the tropical zones for its edible seeds. It can grow in a wide variety of soil from sandy loams of clays with a pH range up to 5.5–7.5 tolerating 4.3–8.8. **Medicinal use(s)**: It is used to treat prostate, breast, and colon cancer. It also possesses anti-inflammatory, antioxidant, antimicrobial, and antibacterial properties. The crushed leaves are used in a poultice to heal the broken bones. It is also used to treat swelling and skin infection, leaves are chewed to treat tooth ailments, powered carbonized seeds are applied on insect strings, and roots are used as an antidote for snakebites and to treat epilepsy, chest pain, constipation, and dysmenorrhea.			Protocatechuic acid	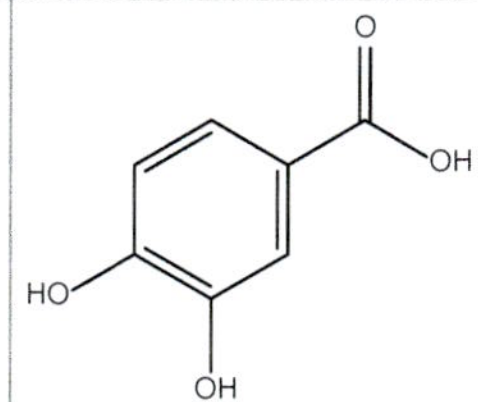	
				(p-Hydroxy benzoic acid)	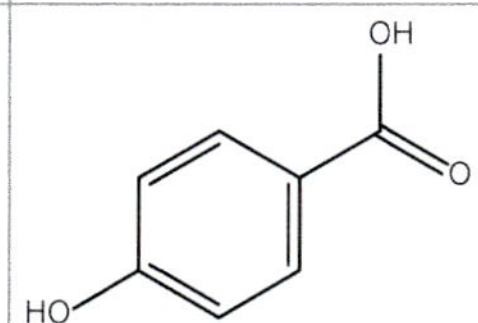	

Table 4.60 Anticancer plants of family Flacourtiaceae

Family	Botanical name	Image	Plant part used	Bioactive compound(s)	Structure	Reference(s)
Flacourtiaceae	*Casearia sylvestris* Sw		Leaf and stem	Casearin	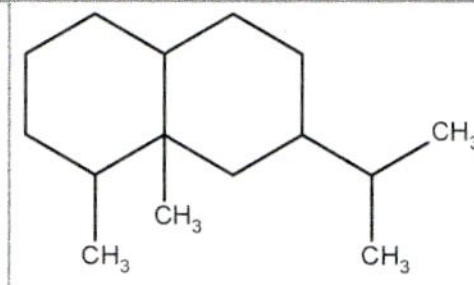	Garcia-Barriga (1975) and Silva et al. (2008)
	Synonym(s): *Anavinga parvifolia* Lam., *Anavinga samyda* C.F.Gaertn., *Casearia affinis* Gardner, *Casearia attenuata* Rusby, *Casearia benthamiana* Miq., *Casearia caudata* Uittien, *Casearia celtidifolia* de Vriese, *Casearia chlorophoroidea* Rusby, *Casearia herbert-smithii* Rusby, *Casearia integrifolia* Vahl ex DC., *Casearia lindeniana* Urb., *Casearia onacaensis* Rusby, *Casearia ovoidea* Sleumer, *Casearia parvifolia* Willd., *Casearia samyda* (P.Gaertn.) DC., *Casearia schulziana* O.C.Schmidt, *Casearia serrulata* Sw., *Casearia subsessiliflora* Lundell, *Chaetocrater capitatus* Ruiz & Pav., *Chaetocrater fasciculatus* Ruiz & Pav., *Crateria capitata* Pers., *Guidonia parvifolia* (Willd.) M.Gómez, *Guidonia sylvestris* (Sw.) M.Gómez, *Samyda sylvestris* (Sw.) Poir. **Common name(s)**: Wild sage. **Botanical description**: It is a native plant of South America – Argentina, Colombia, Mexico, and Brazil. It is an evergreen shrub or small tree with long, slender branches and a very dense, globule crown. It is usually 4–6 m tall but can sometimes grow as high as 20 m. It grows in the moist area. **Medicinal use(s)**: It is used to treat human cancer cell lines. It is also used as an antitumor agent and inhibits HIV replication. The oil is used for the treatment of leprosy. The bark and leaves are analgesic, anti-inflammatory, and anti-rheumatic, internally used for the treatment of gastric ulcers, inflammation, fever, and diarrhea. Externally used in the treatment of wounds.					

(continued)

Table 4.60 (continued)

Family	Botanical name	Image	Plant part used	Bioactive compound(s)	Structure	Reference(s)
	Hydnocarpus anthelmintica Pierre ex Laness.		Seed	Berberine	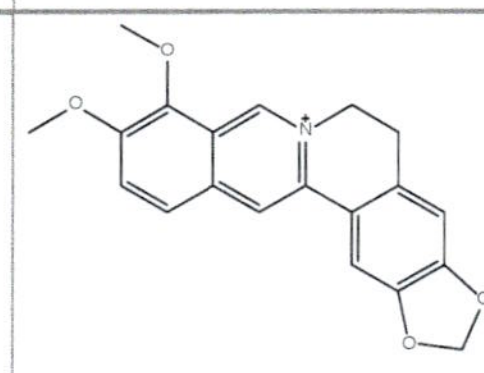	Wasuwat (1967), Duke and Ayensu (1985a, b), and Maffei (2003a, b, c, d, e, f, g, h, i, j, k, l, m)
	Synonym(s): No synonyms are recorded for this name. **Common name(s)**: Dafushi, Daepungja. **Botanical description**: It is a flowering deciduous tree (7 m) originated from Southeast Asia and found in China, Myanmar, Vietnam, India, and Thailand. It generally grows in tropical climate. **Medicinal use(s)**: It is used for cancer treatment. The extracted oil is used for leprosy. It also possesses antitumor, antibacterial, and anti-inflammatory properties.					

Table 4.61 Anticancer plants of family Fouquieriaceae

Family	Botanical name	Image	Plant part used	Bioactive compound(s)	Structure	Reference(s)
Fouquieriaceae	*Fragaria chiloensis* (L.) Mill.		Fruit	Quercetin	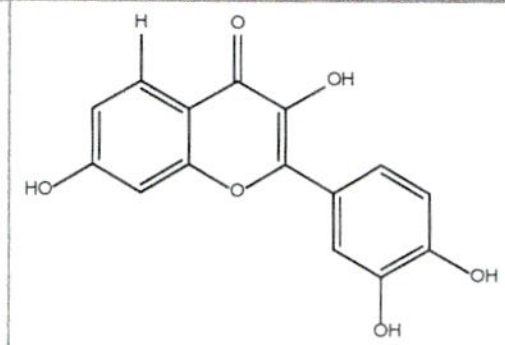	Wang and Lewers (2007)
	Synonym(s): No synonyms are recorded for this name. **Common name(s)**: Beach strawberry, Chilean strawberry, or Coastal strawberry. **Botanical description**: It is an evergreen plant growing up to 15–30 cm tall, with glossy green trifoliate leaves, and is a native plant of Canada. Its natural range is the Pacific Ocean coast. The flowers are white, produced in spring and early summer. The fruits are edible, red over the surface, and white inside. **Medicinal use(s)**: It is used as an antiseptic, astringent, and emmenagogue. It is also used to treat burns and to regulate menstrual cycle.					

(continued)

Table 4.61 (continued)

Family	Botanical name	Image	Plant part used	Bioactive compound(s)	Structure	Reference(s)
	Fouquieria fasciculata (Willd. ex Roem. & Schult.) Nash		Stem	Benzoic acid	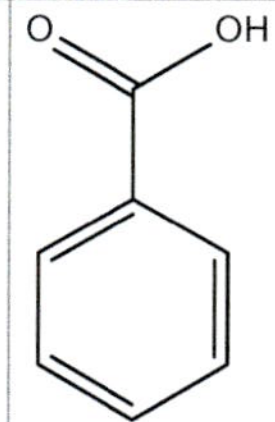	Moore (1979) and Maffei (2003a, b, c, d, e, f, g, h, i, j, k, l, m)
	Synonym(s): *Bronnia spinosa* Kunth. **Common name(s)**: Ocotillo. **Botanical description**: It is a shrub or small tree, reaching heights from about 4 m up to 50 m and is native to the United States. Plant has spines or sharp edges over the stem. It is drought tolerant and is suitable for xeriscaping. **Medicinal use(s)**: It is used to treat benign tumors and used in lotions for backache and hand sores. A tea made from the plant is used for the treatment of painful or irregular mensuration.			Kaempferol	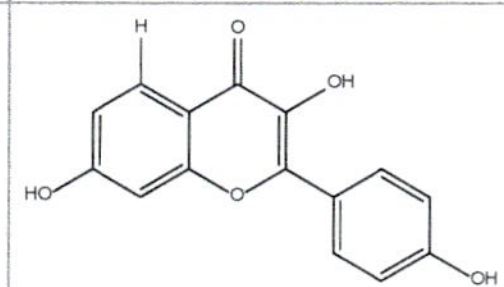	

Table 4.62 Anticancer plants of family Ganodermataceae

Family	Botanical name	Image	Plant part used	Bioactive compound(s)	Structure	Reference(s)
Ganodermataceae	*Ganoderma applanatum* (Pers.) Pat.		Fruiting body	β-Cymene Cis-2-methyl-4-penthyl-thian-5	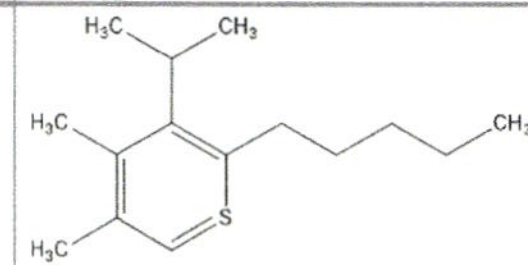	Kim et al. (1994) and Sun et al. (2015)
	Synonym(s): *Boletus applanatus* Pers., *Boletus fomentarius* var. *applanatus* (Pers.) Pers., *Polyporus applanatus* (Pers.) Wallr., *Fomes applanatus* (Pers.) Gillet, *Placodes applanatus* (Pers.) Quél, *Phaeoporus applanatus* (Pers.) J. Schröt., *Elfvingia applanata* (Pers.) P. Karst., *Friesia applanata* (Pers.) Lázaro Ibiza. **Common name(s)**: Artist's bracket. **Botanical description**: It is native to Britain and Ireland. It is found throughout Europe. It is a very common perennial bracket fungus. The underside is creamy white and can be scratched with a sharp point to leave brown marks and produce artistic images. **Medicinal use(s)**: It is used to treat breast and prostate cancer. It is also used for fatigue syndrome and hepatitis. It prevents the multiplication of cancer cell in a dose- and time-dependent manner. It is also used to boost the immune system for the people suffering from cancer and HIV/AIDS. It also possesses antioxidant and stress-reducing properties.					

(continued)

Table 4.62 (continued)

Family	Botanical name	Image	Plant part used	Bioactive compound(s)	Structure	Reference(s)
	Ganoderma lucidum (Leyss: Fr) Karst.		Fruiting body	Ganoderic acid	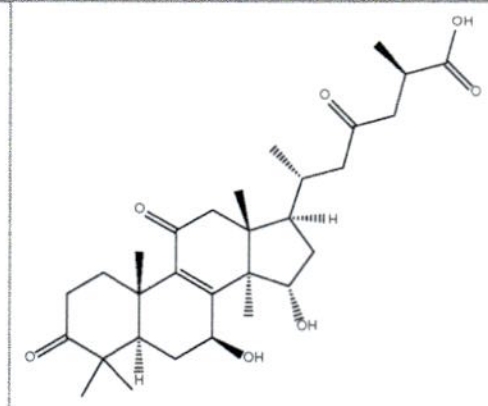	Huang et al. (1991), Furusawa et al. (1992), and Sliva (2003)
	Synonym(s): *Boletus rugosus* Jacq., *Boletus lucidus* Curtis, *Polyporus lucidus* (Curtis) Fr., *Grifola lucida* (Curtis) Gray, *Fomes lucidus* (Curtis) Cooke, *Placodes lucidus* (Curtis) Quél., *Phaeoporus lucidus* (Curtis) J. Schröt. **Common name(s)**: Reishi mushroom. **Botanical description**: It is native to North America, Japan, and China. It is also produced in other Asian countries. It is purple, brown fungus with a long thin stalk; spores are brown in color and shiny cap-shaped like a fan and tree stumps. **Medicinal use(s)**: It is used to treat prostate and breast cancer, bronchitis, asthma, allergies, herpetic conditions and HIV infection, and high blood pressure and reduce platelet aggregation.					

Table 4.63 Anticancer plants of family Gentianaceae

Family	Botanical name	Image	Plant part used	Bioactive compound(s)	Structure	Reference(s)
Gentianaceae	*Gentiana loureirii* (G. Don) Griseb.		Entire plant	Not available	Not available	Duke and Ayensu (1985a, b) and Maffei (2003a, b, c, d, e, f, g, h, i, j, k, l, m)
	Synonym(s): *Ericala loureirii* G. Don, *Gentiana indica* Steud, *Gentiana pedicellata* var. *rosulata* Kusn. **Common name(s)**: Hua nan long dan (Pinyin, China). **Botanical description**: It is a perennial plant 3–8 cm high, sometimes with stolons of 2.5 cm, and is native to South and Southeast Asia. Stalks are purple, erect, simple or slightly branched, slightly cespitose, and densely papillated. Petiole 3–7 mm, elliptic leaf blade, 1.5–3.0 mm × 3–5 mm, glabrous abaxially, adaxially densely and minutely papillated, apex obtuse to acute, often mucronate, corolla blue color a blue-purple or rarely pale greenish yellow. Seeds are of brown color, ellipsoids of 0.7–0.9 mm. **Medicinal use(s)**: It is used to treat tumors. It is also used for treatment of digestion problems such as loss of appetite, intestinal gas, diarrhea, gastritis, heartburn, and vomiting. It is also used for treatment of fever, hysteria, and high blood pressure. Some people use this to prevent muscle spasms, treat parasitic worms, and start menstrual periods and as a germ killer. It is also applied to the skin for treating wounds and cancer.					

(continued)

Table 4.63 (continued)

Family	Botanical name	Image	Plant part used	Bioactive compound(s)	Structure	Reference(s)
	Gentiana rhodantha Franch. ex Hemsl.		Entire plant	Mangiferin	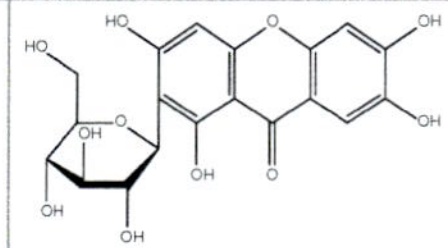	Duke and Ayensu (1985a, b) and Xu et al. (2011)
	Synonym(s): *Gentiana jankae* Kanitz. **Common name(s)**: Hong hua long dan (Pinyin, China). **Botanical description**: This perennial herb (50 cm) with erect stem grows in the grassland and forests of China. The branches are spreading and glabrous. The petiole is 0.5–1 cm long. The flowers are terminal, solitary, and sessile. The calyx lobes are lanceolate and 0.5 cm long. The corolla is funnel-shaped, is 4.5 cm long, and produces 5 lobes. The corolla lobes are ovate, deeply incised, and 0.9 cm long. The stamens are attached to the base of the corolla tube and are unequal. The filaments are 1.2 cm long. The stigma lobes are linear. **Medicinal use(s)**: It is used against five human cancer cell lines It is also used for the treatment of inflammation, cholecystitis, and tuberculosis by the local people of its growing areas.					

	Gentiana scabra Bunge		Root	Loganin	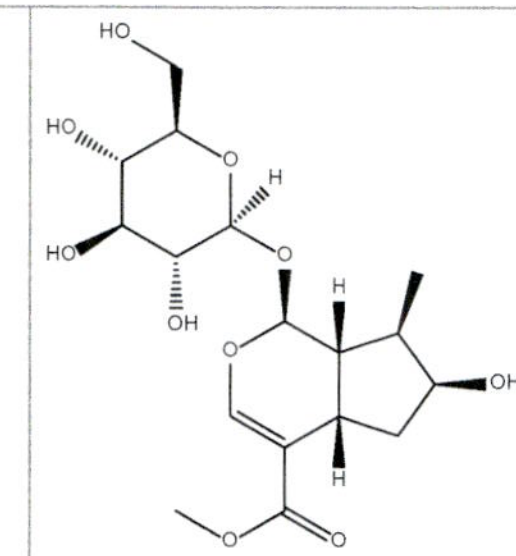	Duke and Ayensu (1985a, b) and Cheng et al. (2018)
	Synonym(s): *Dasystephana scabra* (Bunge) J. Sojak, *Dasystephana scabra* (Bunge) Zuev, *Gentiana fortunei* Hook. **Common name(s)**: Chinese gentian, Dancao, Japanese gentian, Kudancao, Longdan, Longdancao, Tourindou. **Botanical description**: It is a perennial herb with 10–15 cm long white roots with numerous short branches. Stem is 20–100 cm long with 10–20 pairs of leaves. Leaves are lanceolate to narrowly deltoid-ovate, 4–8 cm long, 1–3 cm wide, gradually acuminate, three-nerved, green above, paler beneath, usually sessile, margin of upper leaves papillose. Flowers are few to rather numerous, sessile, 4.5–6 cm long, purplish blue, calyx tube 12–18 mm long, the lobes rather unequal, linear-lanceolate, corolla toothed. Capsules stipulate not exerted, seeds broadly lanceolate, short-caudate at both ends. **Medicinal use(s)**: It possesses antitumor, anti-hepatotoxic, anti-inflammatory, anti-spasmodic, sedative, choleretic, and antimicrobial properties.					

(continued)

Table 4.63 (continued)

Family	Botanical name	Image	Plant part used	Bioactive compound(s)	Structure	Reference(s)
	Gentiana lutea L.		Root	Pyrocatechuic acid (2,3-dihydroxybenzoic acid	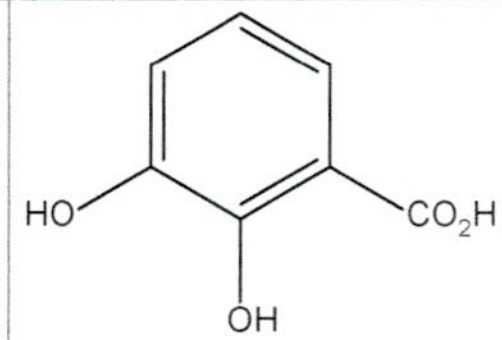	Khadem and Marles (2010)
	Synonym(s): *Asterias hybrid* G. Don, *Asterias lutea* (L.) Borkh., *Coilantha biloba* Bercht. & J. Presl, *Gentiana major* Bubani, *Gentianusa lutea* (L.) Pohl. **Common name(s)**: Bitter root. **Botanical description:** It is a perennial plant growing up to 1.2 m (4 ft.) by 0.6 m and is native to Central and South Europe. It is in flower from July to August. The species is hermaphrodite (has both male and female organs) and is pollinated by bees, flies, and beetles. It can grow in semi-shade (light woodland) or no shade. It prefers moist soil to grow. **Medicinal use(s)**: It is used to treat cancer and also possesses anti-inflammatory, antibacterial, and antimicrobial properties. It is taken internally for the treatment of liver complaints, indigestion, gastric infections, and anorexia.					

	Gentiana acaulis L.		Apical part	Gentiakochianin	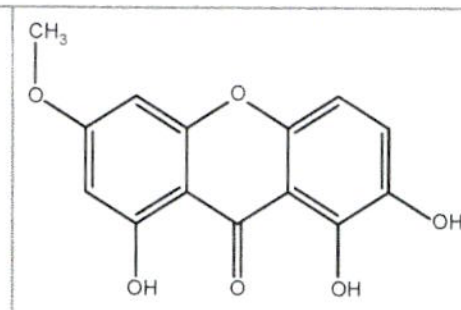	Leskovac et al. (2007)
	Synonym(s): *Ciminalis acaulis* Borkh., *Ciminalis acaulis* Moench, *Ciminalis grandiflora* Mayrh., *Ciminalis longiflora* Moench, *Ericala alpina* G. Don, *Gentiana excise* C. Presl, *Gentiana grandiflora* Lam., *Gentiana kochiana* Perr. & Songeon, *Gentiana vulgaris* (Neilr.) Beck, *Gentianusa acaulis* (L.) Pohl. **Common name(s)**: Bitter root, Bitterwort, Gall weed. **Botanical description**: Its large, trumpet-shaped flowers are held close to the mat of lanceolate (narrow and tapering to a point) or elliptic leaves, on a short stem. The glorious deep blue flowers can reach up to 7 cm long and have green spots on the throat (where the petal tube widens). The five-pointed corolla lobes (petals) are 6–9 mm long and are separated by small triangular lobe-like plicae (folds). **Medicinal use(s)**: It is used to treat cancers, high blood pressure, stomach disorders, diarrhea, heartburn, vomiting, fever, and menstrual disorders.			Gentiacaulein	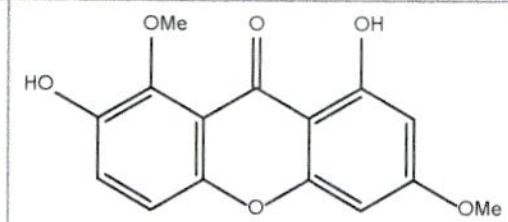	
				Decussatin	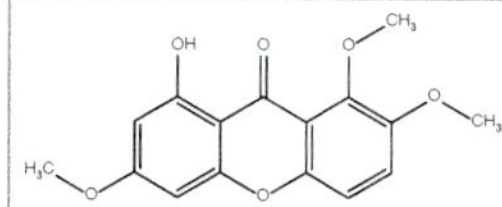	

(continued)

Table 4.63 (continued)

Family	Botanical name	Image	Plant part used	Bioactive compound(s)	Structure	Reference(s)
				Xanthone glycosides	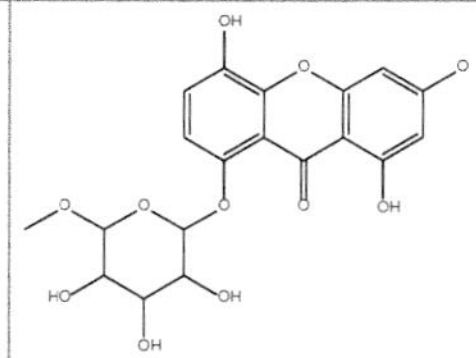	
				Isogentiakochianoside	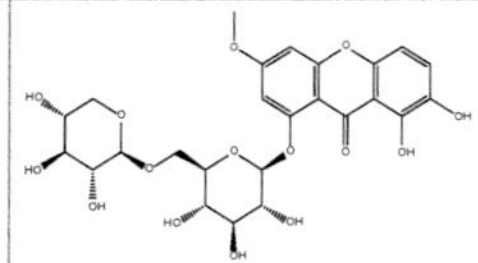	
				Gentia-caulein-1-O-glucoside	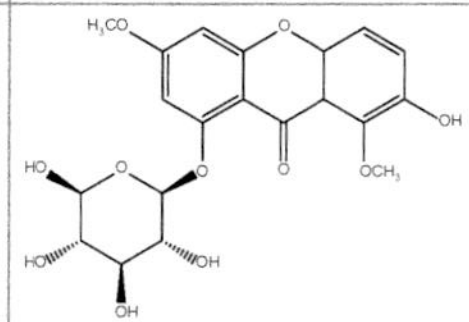	
				Bellidifolin	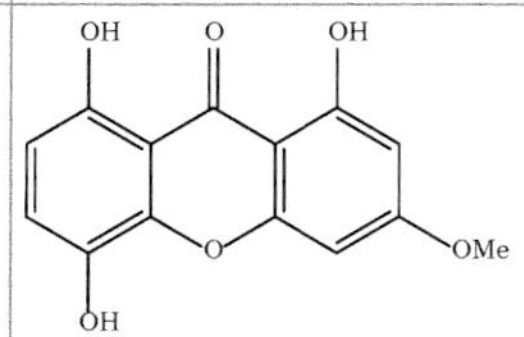	
				Dimethylbellidifolin	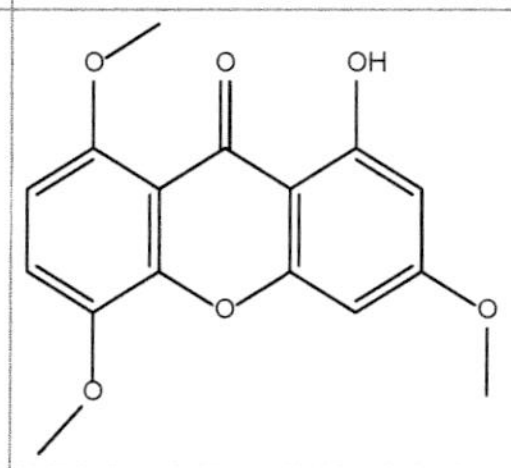	

Table 4.64 Anticancer plants of family Geraniaceae

Family	Botanical name	Image	Plant part used	Bioactive compound(s)	Structure	Reference(s)
Geraniaceae	*Pelargonium graveolens* L'Her		Essential oil	5-Flurouracil	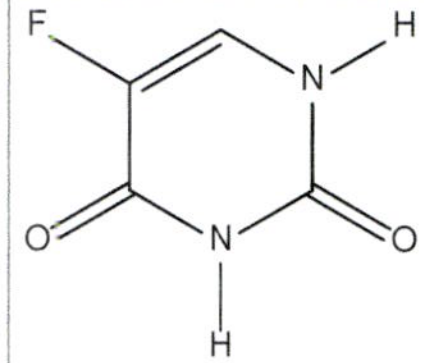	Duke and Ayensu (1985a, b), Bigos et al. (2012), and Ben Slima et al. (2013)
	Synonym(s): No synonyms are recorded for this name. **Common name(s)**: Rose geranium, Sweet-scented geranium. **Botanical description**: It is an evergreen shrub growing up to 1.2 m by 0.8 m at a medium rate and is native near the Hex river in South West Cape province. It is in flower from May to July. The flowers are hermaphrodite (have both male and female organs). It is suitable for light (sandy) and medium (loamy) soils and prefers well-drained soil, and suitable pH is acid, neutral, and basic (alkaline) soils. It cannot grow in the shade and prefers moist soil to grow. **Medicinal use(s)**: The whole plant has relaxant, anticancer, antidepressant, and antiseptic effects, reduces inflammation, and controls bleeding. It is used internally for the treatment of premenstrual and menopausal problems, nausea, tonsillitis, and poor circulation. Externally, it is used to treat acne, hemorrhoids, eczema, bruises, ringworm, and lice. The leaves can be used fresh at any time of the year. The essential oil from the leaves is used in aromatherapy and is also applied locally to treat cervical cancer.			Geraniol	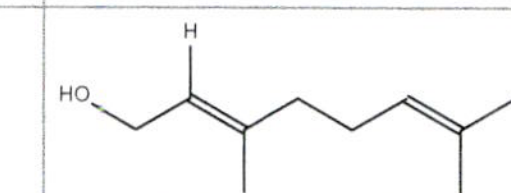	
				Citronellol	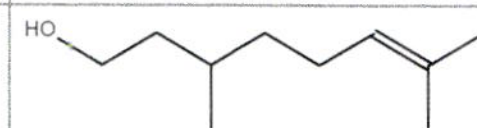	

Table 4.65 Anticancer plants of family Ginkgoaceae

Family	Botanical name	Image	Plant part used	Bioactive compound(s)	Structure	Reference(s)
Ginkgoaceae	*Ginkgo biloba* L.		Leaf	Ginkgolide A	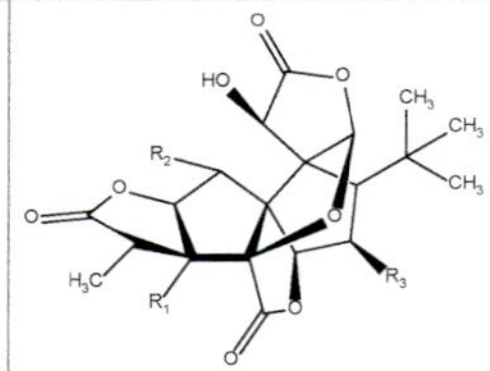	Chao and Chu (2004) and Biggs et al. (2010)
				Bilobalide	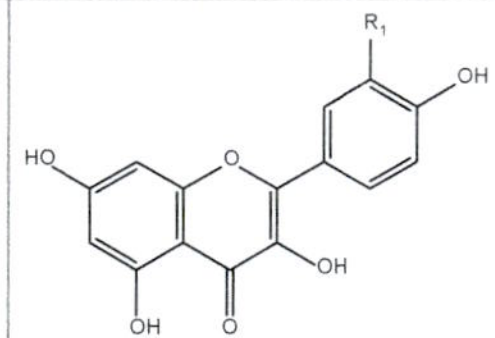	
	Synonym(s): *Ginkgo macrophylla* K.Koch, *Salisburia biloba* (L.) Hoffmanns., *Salisburia macrophylla* Reyn. **Common name(s)**: Maiden hair tree. **Botanical description**: The plants are native to China, Japan, and Korea. It is also cultivated in Europe. It is normally reaching a height of 20–35 m, with some specimens in China being over 50 m. The tree has an angular crown and long, somewhat erratic, branches and is usually deep rooted and resistant to wind and snow damage. Young trees are often tall and slender and sparsely branched; the crown becomes broader as the tree ages. **Medicinal use(s)**: It is used to treat breast cancer and ovarian cancer. It is also used for disorders related to Lyme disease, chemotherapy, and depression. Leaf is often taken orally for memory disorders including Alzheimer's disease. It is also used for conditions that seem to be due to reduced blood flow in the brain, especially in older people. These conditions include memory loss, headache, ringing in the ears, vertigo, dizziness, difficulty in concentration, mood disturbances, and hearing disorders.			Ginkgolide B	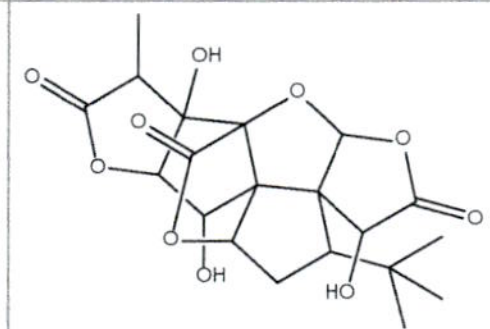	
				Ginkgetin	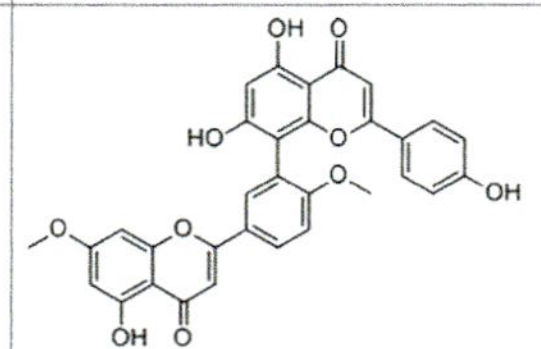	

Table 4.66 Anticancer plants of family Globulariaceae

Family	Botanical name	Image	Plant part used	Bioactive compound(s)	Structure	Reference(s)
Globulariaceae	*Globularia alypum* L.		Leaf	Coumarin(s)	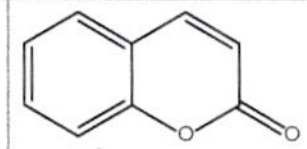	Boukef et al. (1982) and Harzallah et al. (2010)
	Synonym(s): *Alypum monspeliensium* Fourr., *Alypum salicifolium* Fisch, *Globularia alypa* St.-Lag., *Globularia turbith* Willk., *Globularia virgata* Salisb. **Common name(s)**: Sweet gum. **Botanical description**: It is native to North and South America. It is a perennial bushy plant that reaches up to 40–80 cm long. Juvenile branches and leaves are covered with minute calcareous dots. Leaves are 0.5–2 × 0.3–0.8 cm, alternate, or clustered at the nodes on old branches, glabrous, obovate to spathulate, with entire margins, mucronate apex. Flowers are small, blue, and arranged in terminal heads with a hairy calyx. **Medicinal use(s)**: It is known for its antimicrobial, antitumor, anti-inflammatory, antiseptic, antifungal, and antidiabetic properties, and it also treats cardiovascular diseases. It is also used to treat gynecological diseases, blood purification, and respiratory system diseases. This species is also considered to treat various cancerous lesions in the stomach, colon, rectum, and esophagus.			Yuccagenin	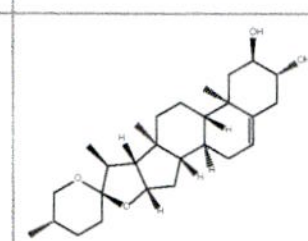	

Table 4.67 Anticancer plants of family Gramineae

Family	Botanical name	Image	Plant part used	Bioactive compound(s)	Structure	Reference(s)
Gramineae	*Sasa palmata* (hort. ex Burb.) E.G. Camus		Leaf	Paclitaxel	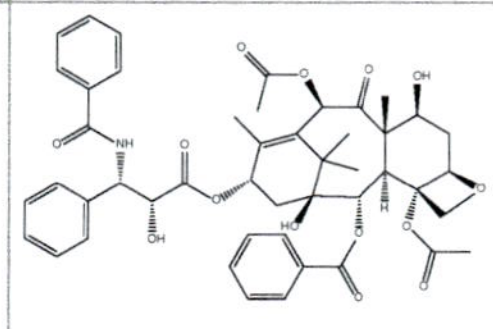	Namba and Bae (1982) and Woo et al. (2011)
	Synonym(s): *Arundinaria palmata* (Burbidge) Bean, *Bambusa ontakensis* (Franch. & Sav.) Makino, *Bambusa palmata* Burb., *Sasa amplissima* Koidz. **Common name(s)**: Broadleaf bamboo, Bamboo. **Botanical description**: It is an evergreen bamboo growing up to 2.5 m (8 ft.) by 5 m (16 ft.) at a fast rate native to East Asia and Japan and widely naturalized in Britain. The flowers are hermaphrodite (have both male and female organs) and are pollinated by wind. It is suitable for light (sandy), medium (loamy), and heavy (clay) soils, and suitable pH is acid, neutral, and basic (alkaline) soils. It can grow in semi-shade (light woodland) and prefers moist soil to grow. **Medicinal use(s)**: It is used to treat cancer and also possesses antitumor, antibacterial, and anti-inflammatory properties.					

	Sorghum bicolor (L.) Moench.		Entire plant	Prɔanthocyanidins	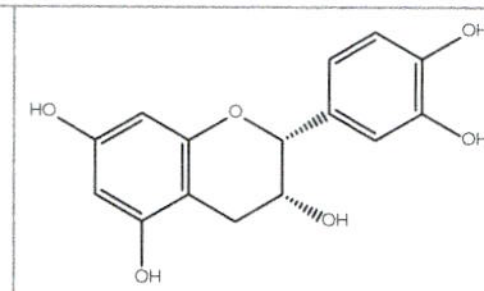	Han et al. (1984) and Suganyadevi et al. (2013)
	Synonym(s): *Agrostis nigricans* (Ruiz & Pav.) Poir., *Andropogon besseri* Kunth, *Andropogon bicolor* (L.) Roxb. **Common name(s)**: Jawar. **Botanical description**: It is a native plant of Northern Africa and now widely cultivated in tropical and subtropical regions. It is an erect, robust, herbaceous annual plant which is often 3–4 m in height. The *Sorghum* plant prefers tropical and subtropical climate. It can be grown successfully in a variety of soils ranging from sandy, loamy, to clay. **Medicinal use(s)**: It is a very effective medicinal herb to treat cancer including lung cancer, anemia, diabetes, high cholesterol, osteoporosis, body pain, headache, cataract, inflammation, sickle cell, as well as constipation.					

Table 4.68 Anticancer plants of family Gyrostemonaceae

Family	Botanical name	Image	Plant part used	Bioactive compound(s)	Structure	Reference(s)
Gyrostemonaceae	*Codonocarpus cotinifolius* (Desf.) F.Muell.		Leaf, root, bark	Myricadiol	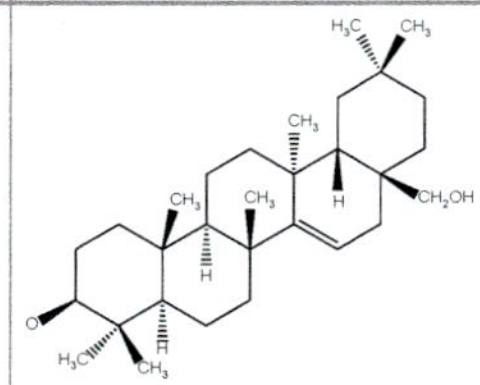	Reid and Betts (1979) and Johns and Kerr (2009)
	Synonym(s): *Gyrostemon cotinifolius* Desf., *Gyrostemon pungens* Lindl. **Common name(s)**: Bell fruit tree, Bellbush, Desert poplar, Western bell fruit, Bell-fruit tree, Horseradish tree. **Botanical description**: It is a short-lived shrub or tree growing up to 8 m tall, usually obconical in shape, and is native to South Australia. Leaves are obovate or oblanceolate, rarely elliptic or almost orbicular in young plants. Male flowers on pedicels are 4 mm long and 6 mm across at anthesis, calyx scarcely lobed to almost circular, female flowers often borne on branches below the male flowers or below the terminal vegetative growth, with pedicels to 20 mm long, calyx shallowly lobed to almost circular, with 30 or more carpels each with a terminal stigma arranged around the top of the central axis. **Medicinal use(s)**: Infusion used for the treatment of cancer.					

Table 4.69 Anticancer plants of family Hernandiaceae

Family	Botanical name	Image	Plant part used	Bioactive compound(s)	Structure	Reference(s)
Hernandiaceae	*Hernandia nymphaeifolia* (Presl.) Kubitzki		Whole plant	Deoxypodophyllotoxin	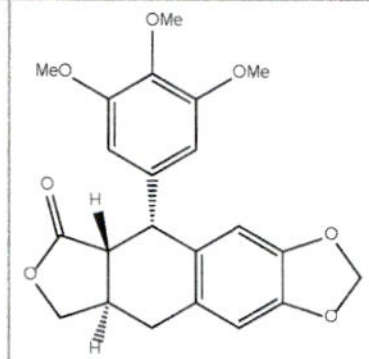	Chen et al. (2000), Chao et al. (2002), and Ain et al. (2016)
	Synonym(s): *Biasolettia nymphaeifolia* C.Presl, *Hernandia peltata* Meisn., *Hernandia vitiensis* Seem. ex Pax. **Common name(s)**: Sea hearse, Leampis. **Botanical description**: It is fairly fast-growing (10 m), beautiful, evergreen tree with an open, irregular crown and showy panicles of flowers and is native to Eastern Africa, tropical Asia to Australia, and the Pacific Islands. It is habituated to sparse forests near sea level in Southern China. The plant is also grown as an ornamental in gardens and along streets. **Medicinal use(s)**: It is used to treat human cancer cell lines, malarial fever, lymphocytic leukemia cell line, and human tumor cell lines. It also possesses antibacterial, anti-inflammatory, and antimicrobial properties.			Yatein	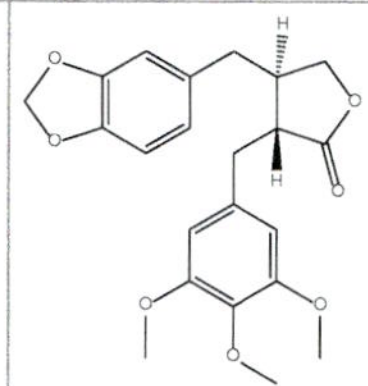	

(continued)

Table 4.69 (continued)

Family	Botanical name	Image	Plant part used	Bioactive compound(s)	Structure	Reference(s)
				Epiyangambin	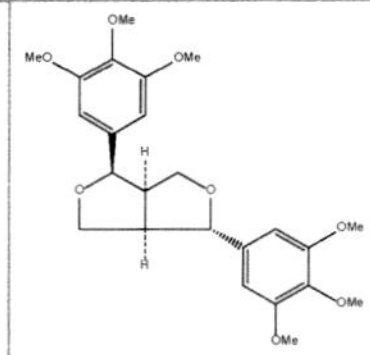	
				Epimagnolin	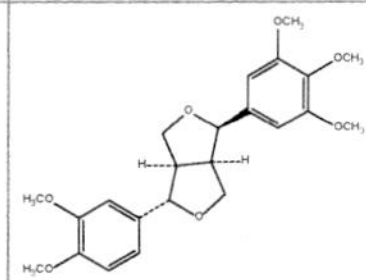	
				Epi-aschantin	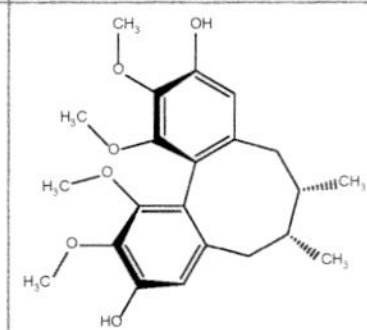	

	Hernandia nymphaeifolia (Presl) Kubitzki		Stem bark	(−)-deoxypodophyllotoxin	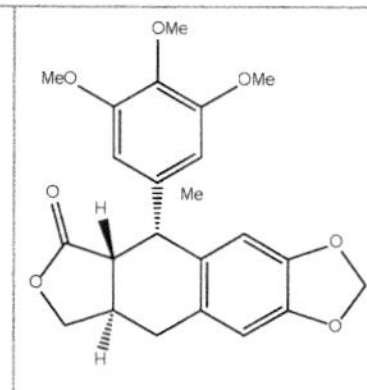	Pettit et al. (2004) and Kintzios (2006)
	Synonym(s): *Biasolettia nymphaeifolia* C.Presl, *Hernandia peltata* Meisn., *Hernandia vitiensis* Seem. ex Pax. **Common name(s)**: Sea hearse, Leampis. **Botanical description**: *Hernandia nymphaeifolia* is distributed in American Samoa, Chagos Archipelago, Gilbert Island, Marshall Island, Micronesia, N-Line Island, Northern Marianas, Palau Island, South-Cook Island, Society Island, Southern Marianas, Rotuma Island, Tokelau, Tuvalu, Western Samoa, New Caledonia, Wallis and Futuna Island, Fiji, Niue, Western Samoa, Bonin Island, Volcano Island, Marquesas Island, Tonga, Java, peninsular Malaysia, Indonesia, Southern Philippines, Australia, Kenya, North-East-Tanzania, South-Taiwan, China, Ryukyu Island, Seychelles, Comores, Mauritius, Madagascar, peninsular Malaysia, Sri Lanka, India, peninsular Thailand, Cambodia, Vietnam, Sulawesi, Moluccas, New Guinea, Cocos Island, North Keeling Island, Andamans, Nicobars Singapore, Lakshadweep Island, and Maldives. It is a tree of 40–50 ft. height with a clear trunk of 5–12 ft. in circumference; head wide spreading; wood soft and light, Gray in colour. Branchlets stout, glabrous, leaves broadly ovate, peltate, rounded-truncate, retuse or cordate at the base, shortly acuminate and subobtuse at the apex, more rarely obtuse or rounded, 4.5–12 in. long, 3–9 in. broad, subcoriaceous, glabrous, digitately 7-nerved at the insertion of the petiole, remaining nerves 3–4 on each side of the midrib; and petiole 2.5–8 in. long. **Medicinal use(s)**: It is used to treat human cancer cell lines, wounds, tumor, headache, epilepsy, and fits. It possesses anticancer, antitumor, antibacterial, anti-inflammatory, and antimicrobial properties.			Deoxypicropodophyllin	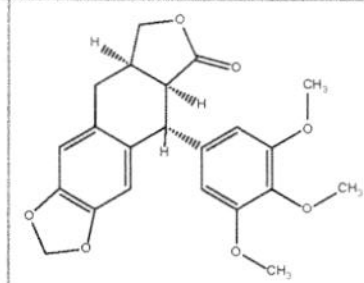	

(continued)

Table 4.69 (continued)

Family	Botanical name	Image	Plant part used	Bioactive compound(s)	Structure	Reference(s)
				(+)-epiaschantin	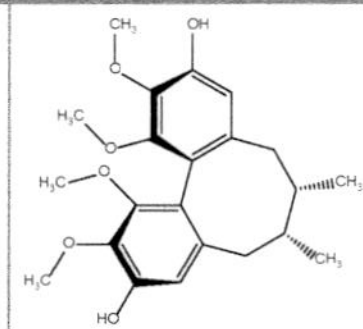	
				(+)-epieudesmin	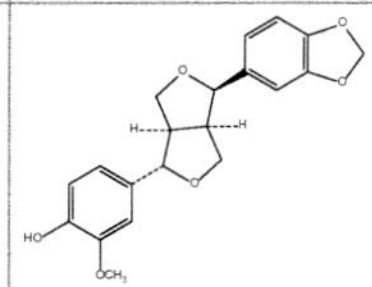	
				Praderin	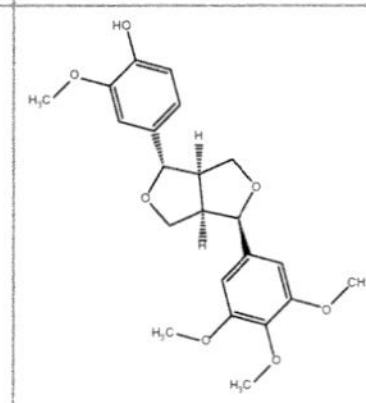	
				5′-Methoxyyatein	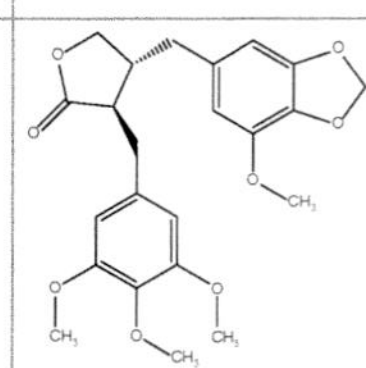	

	Gyrocarpus americanus Jacq.		Bark	Rooperol	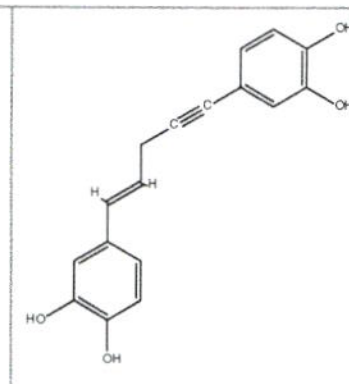	Singh et al. (1984) and Bradacs et al. (2009)
	Synonym(s): *Gyrocarpus acuminatus* Meisn., *Gyrocarpus asiaticus* Willd., *Gyrocarpus lobatus* Blanco, *Gyrocarpus rugosus* R.Br. **Common name(s)**: Helicopter tree, Propeller tree, Whirly whirly tree, Stinkwood or, Shitwood. **Botanical description**: It has a pantropical distribution. It is a slender, deciduous tree with smooth, gray bark. The tree grows to about 12 m in height. The leaves are spirally arranged, crowded near the ends of the branches, and grow up to 150 × 120 mm in size. They are ovate, often 3-lobed, dark green above, paler and grayer below, with velvety surfaces, 3-veined from the base. The veins are yellowish and the stalk up to 90 mm long. The cream yellowish green flowers grow in compact heads and have an unpleasant smell. The fruit is a woody nut with two long thin wings that help wind dispersal. The winged fruit and the smell of the flowers have given the tree its various common names. **Medicinal use(s)**: It is used to treat cervical cancer, diarrhea, kidney pain, and tumor. It possesses anticancer, antitumor, antibacterial, anti-inflammatory, and antimicrobial properties.					

Table 4.70 Anticancer plants of family Hyacinthaceae

Family	Botanical name	Image	Plant part used	Bioactive compound(s)	Structure	Reference(s)
Hyacinthaceae	*Merwilla plumbea* (Lindl.) Speta		Bulb	Allamcin	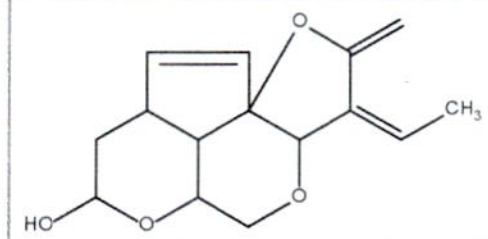	Mohammad (2006), Desai et al. (2008), and Madhuri and Pandey (2009)
				Benzoic acid	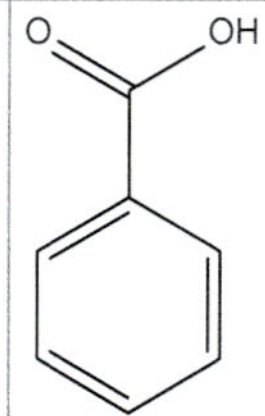	
	Synonym(s): *Merwilla kraussii* (Baker) Speta, *Merwilla natalensis* (Planch.) Speta, *Scilla kraussii* Baker, *Scilla natalensis* Planch., *Scilla plumbea* Lindl. **Common name(s)**: Blue squill, Wild squill, Blue hyacinth. **Botanical description**: It is a deciduous plant growing during summer and dormant in the winter and is native to South Africa. It produces a large bulb, 10–15 cm in diameter, covered with firm, hardened, papery brown or purplish tunics (bulb scales). A rosette of 6–9 broad, tapering leaves emerges from the top of the bulb in spring. The leaves are attractive with clearly distinct veins which give them a two-tone effect. The leaves of a well-grown plant can reach a height of 30–50 cm with about equal spread. **Medicinal use(s)**: It is used for the treatment of cancer, sprains, fractures, boils, and sores, to draw abscesses, and for female infertility. It also possesses antibacterial, antitumor, antioxidant, anti-inflammatory, and antimicrobial properties.					

Table 4.71 Anticancer plants of family Hypericaceae

Family	Botanical name	Image	Plant part used	Bioactive compound(s)	Structure	Reference(s)
Hypericaceae	*Hypericum caprifoliatum* Cham. & Schltdl.		Apical part	Hyperforin	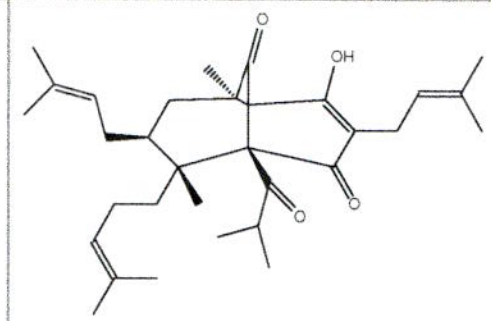	Ferraz et al. (2005) and Nürka and Crockettb (2012)
	Synonym(s): No synonyms are recorded for this name. **Common name(s)**: St. John's wort. **Botanical description**: It is a shrub (30–200 cm tall) native to East Asia, China, Northern India, Nepal, Myanmar, Thailand, and Indonesia (Sumatra). Its stems are erect, branched in the upper section, and can grow up to 1 m long. It has opposite, stalkless, narrow, oblong leaves that are 1–2 cm long. The flowers appear in broad cymes at the ends of the upper branches, between late spring and early summer to midsummer. The sepals are pointed, with black glandular dots. There are many stamens, which are united at the base into three bundles. The pollen grains are ellipsoidal. **Medicinal use(s)**: It is used for the treatment of colon cancer, depression, seasonal effective disorder, HIV and hepatitis C infection, and obsessive-compulsive disorder. It also possesses anticancer, antibacterial, anti-inflammatory, antimicrobial, antioxidant, and antitumor properties.			Hypericin	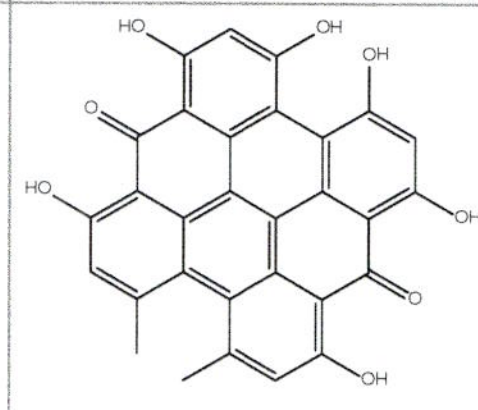	

(continued)

Table 4.71 (continued)

Family	Botanical name	Image	Plant part used	Bioactive compound(s)	Structure	Reference(s)
	Cratoxylum formosum (Jacq.) Benth. & Hook.f. ex Dyer		Root	Xanthone glycoside	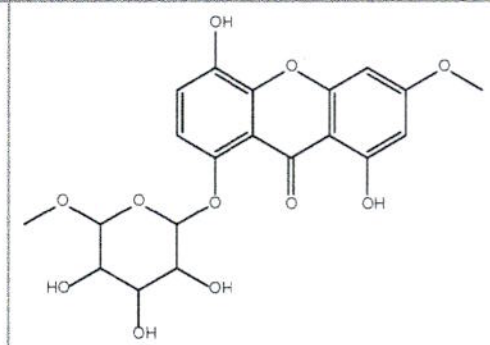	Boonsri et al. (2006), Promraksa et al. (2015), and Buranrat et al. (2017)
	Synonym(s): *Cratoxylum prunifolium* Dyer, *Elodes formosa* Jack. **Common name(s)**: Salinggogon, Pink mempat, Derum, Mempat, Entemu, Geronggang biabas, Gerunggung, Kemutul, Kemutun, Mempitis. **Botanical description**: It is a deciduous shrub or tree and is native to East Asia, Southern China, Myanmar, Thailand, Cambodia, Laos, and Vietnam. The bole has spreading long thorns on its lower part and can be straight or leaning up to 39 cm in diameter and slightly fluted at the base. **Medicinal use(s)**: It is used for the treatment of liver cancer, oral cancer cell lines, ovarian cancer, diarrhea, skin troubles, scabies, and leg wounds. It also possesses antibacterial, anti-inflammatory, antimicrobial, antioxidant, and antitumor properties.			1-Methoxy-2,4,5-trihydroxyxanthone	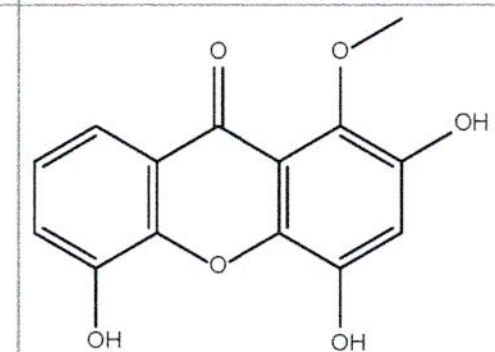	

	Hypericum annulatum Moris		Apical part	Hyperatomarin (a prenylated phloroglucinol)	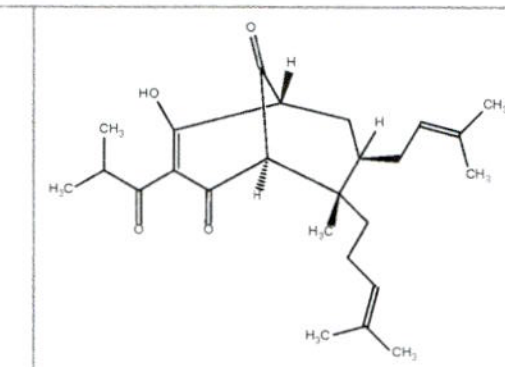	Momekov et al. (2008)
	Synonym(s): No synonyms are recorded for this name. **Common name(s)**: Not available. **Botanical description**: It is a bushy shrub (50–130 cm tall) with spreading branches and is native to East Asia, Southeast China, and Taiwan. The plant is harvested from the wild for local use as a medicine. It is also cultivated as an ornamental plant in many parts of the world, including Southeast Asia. **Medicinal use(s)**: It is used for the treatment of breast cancer cell lines, miasmatic disease, diarrhea, and vomiting. It also possesses antibacterial, anti-inflammatory, antimicrobial, antioxidant, and antitumor properties.					

(continued)

Table 4.71 (continued)

Family	Botanical name	Image	Plant part used	Bioactive compound(s)	Structure	Reference(s)
	Hypericum hookerianum Wight & Arn.		Stem	Labdane	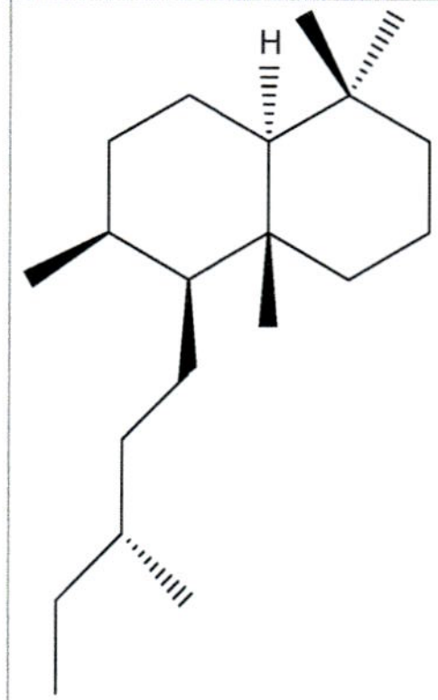	Dongre et al. (2007)
	Synonym(s): *Hypericum garrettii* Craib, *Norysca hookeriana* (Wight & Arn.) Wight. **Common name(s)**: Hooker's St. Johnswort. **Botanical description**: It is a perennial plant, often woody at the base, with stems from 23 to 100 cm tall and is native to South America, Argentina, Uruguay, Paraguay, Brazil, and Bolivia. The plant is sometimes gathered from the wild for local medicinal use. **Medicinal use(s)**: It is used for the treatment of cancer cell lines, angina, and mouth ulcers. It also possesses antibacterial, anti-inflammatory, antimicrobial, antioxidant, and antitumor properties.					

	Hypericum lysimachioides Boiss. & Noë		Flower	3-Hydroxytetradecanoic acid	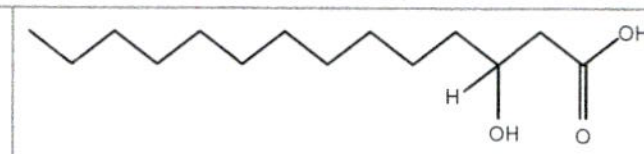	Özen et al. (2004)
	Synonym(s): *Hypericum callianthum* Boiss., *Hypericum leichtlinii* Stapf ex Stef. **Common name(s)**: Dyer's hypericum, Lysimachioides. **Botanical description**: It is a perennial, glabrous, erect herb (0.35–0.75 m tall) and is native to Turkey. Stem is eglandular with few or numerous amber or black gland dots or short streaks, not prominent, not glaucous, and glabrous or slightly scabrid above or on margins. Lamina (main stem) ovate or oblong to linear, apex acute or apiculate to rounded, margin rarely revolute, base cuneate to rounded; with petals spreading or becoming reflexed; buds ellipsoid to (rarely) globose, rounded to obtuse. **Medicinal use(s)**: It is used to treat breast cancer, burns in the skin, sciatica, shingles, and eczema. It also possesses antibacterial, anti-inflammatory, antimicrobial, antioxidant, and antitumor properties.			3-Hydroxy-octadecanoic acid	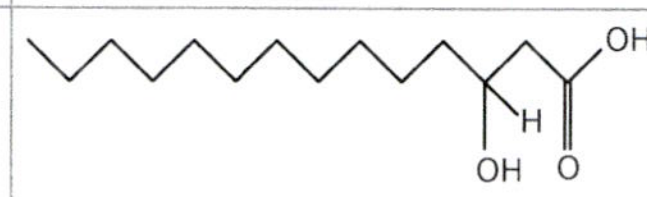	

(continued)

Table 4.71 (continued)

Family	Botanical name	Image	Plant part used	Bioactive compound(s)	Structure	Reference(s)
	Hypericum myrianthum Cham. & Schltdl.		Apical part	Hypericin	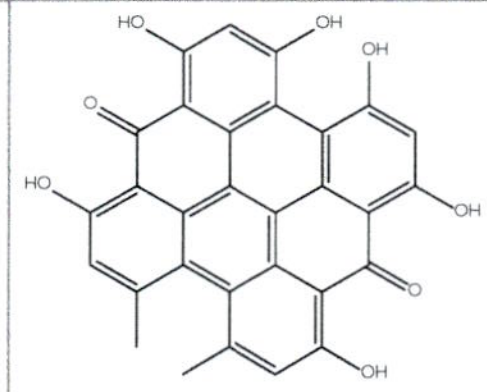	Ferraz et al. (2005) and Hashemian et al. (2017)
	Synonym(s): *Sarothra myriantha* (Cham. & Schltdl.) Y. Kimura. **Common name(s)**: St. John's wort, Rare flower. **Botanical description**: It is a shrub (30–200 cm tall) native to East Asia, China, Northern India, Nepal, Myanmar, Thailand, and Indonesia (Sumatra). Its stems are erect, branched in the upper section, and can grow up to 1 m long. It has opposite, stalkless, narrow, oblong leaves that are 1–2 cm long. The flowers appear in broad cymes at the ends of the upper branches, between late spring and early summer to midsummer. The sepals are pointed, with black glandular dots. There are many stamens, which are united at the base into three bundles. The pollen grains are ellipsoidal. **Medicinal use(s)**: It is used to treat human cancer cell lines. It also possesses antitumor, antibacterial, anti-inflammatory, antioxidant, antimicrobial, and antiproliferative properties.			Quercitrin	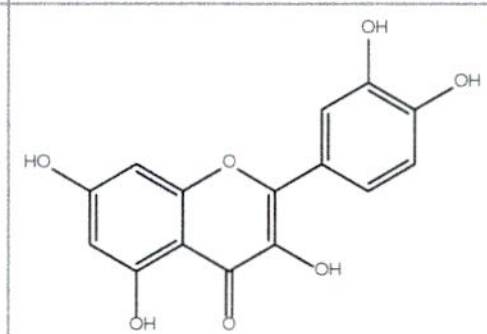	

	Hypericum mysorense F. Heyne.		Fruit, root, and stem	Isoastibin	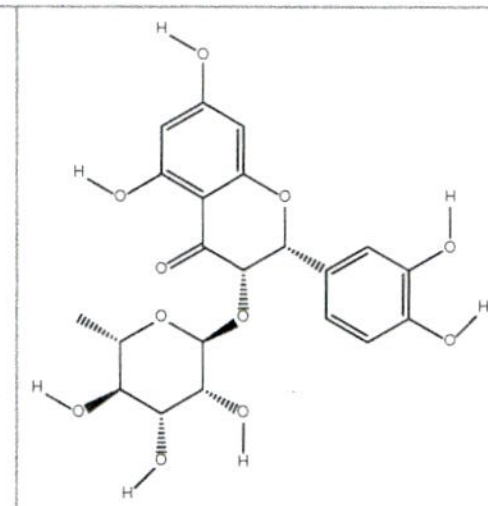	Chandrasekhar et al. (2009)
	Synonym(s): No synonyms are recorded for this name. **Common name(s)**: Mysore hypericum, Athlete foot. **Botanical description**: It is a perennial, glabrous, erect herb (0.35–0.75 m tall) and is native to Turkey. Stem is eglandular with few or numerous amber or black gland dots or short streaks, not prominent, not glaucous, and glabrous or slightly scabrid above or on margins. Lamina (main stem) ovate or oblong to linear, apex acute or apiculate to rounded, margin rarely revolute, base cuneate to rounded; with petals spreading or becoming reflexed; buds ellipsoid to (rarely) globose, rounded to obtuse. **Medicinal use(s):** It is used to treat cervical and penile cancers. It also possesses antitumor, antibacterial, anti-inflammatory, antioxidant, antimicrobial, and antiproliferative properties.					

(continued)

Table 4.71 (continued)

Family	Botanical name	Image	Plant part used	Bioactive compound(s)	Structure	Reference(s)
	Hypericum ternum A. St.-Hil.		Stem, leaf, flower	Hypericin	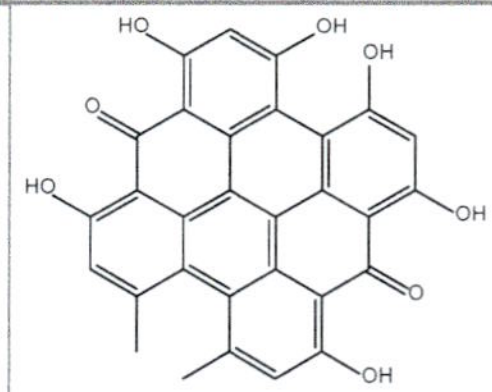	Ferraz et al. (2005) and Peron et al. (2013)
	Synonym(s): *Hypericum cordiforme* var. *glazioui* Briq., *Hypericum cordiforme* var. *hilairei* Briq. **Common name(s)**: Not available. **Botanical description**: It is a plant with ascending branches and is native to Argentina. Flowers about 2 cm in diameter, solitary and borne at the end of branches. Sepals are unequal in form and shape and narrowly lanceolate to oblong-lanceolate. **Medicinal use(s)**: It is used to treat jaundice and liver cancer. It also possesses antitumor, antibacterial, anti-inflammatory, antioxidant, antimicrobial, and antiproliferative properties.			Hyperforin	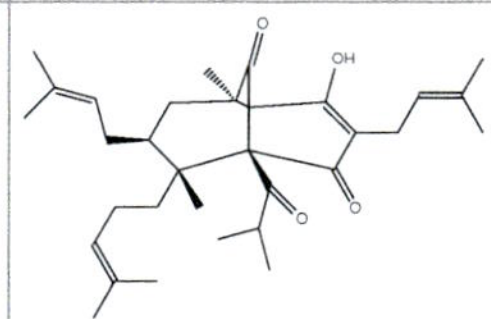	

Table 4.72 Anticancer plants of family Hypoxidaceae

Family	Botanical name	Image	Plant part used	Bioactive compound(s)	Structure	Reference(s)
Hypoxidaceae	*Hypoxis hemerocallidea* Fisch. Mey. & Ave-Lall.		Root	Hypoxoside	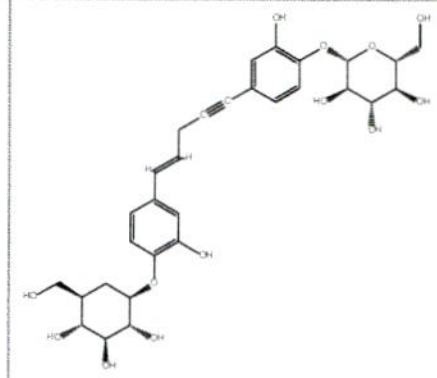	Mills et al. (2005), Steenkamp and Gouws (2006), and Drewes et al. (2008)
	Synonym(s): *Hypoxis obconica* Nel, *Hypoxis patula* Nel, *Hypoxis rooperi* T.Moore. **Common name(s)**: Star flower, Yellow star, Sterblom, Gifbol, molikharatsa. Wrongly referred to as African potato. **Botanical description**: It is native to Southern Africa, from South Africa as far north as Mozambique and Zimbabwe. The plant is tuberous perennial with strap-like leaves and yellow star-shaped flowers. The leaves are up to 400 mm long. The flowers are carried on 5 or 6 slender erect inflorescences, each carrying 5–13 bright yellow, star-shaped flowers with 6 tepals. **Medicinal use(s)**: It is used for the treatment of prostate cancer, HIV, diabetes, and bladder disorder. It possesses anticancer, antitumor, antibacterial, anti-inflammatory, and antioxidant properties.			Rooperol	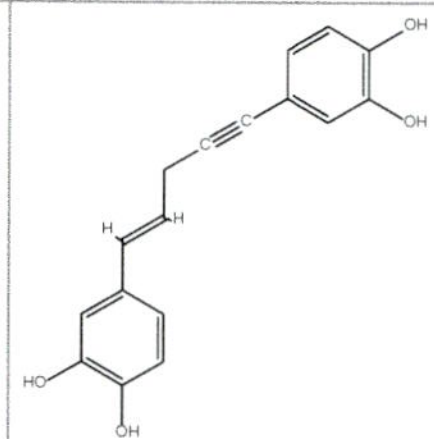	

(continued)

Table 4.72 (continued)

Family	Botanical name	Image	Plant part used	Bioactive compound(s)	Structure	Reference(s)
	Hypoxis nyasica Baker	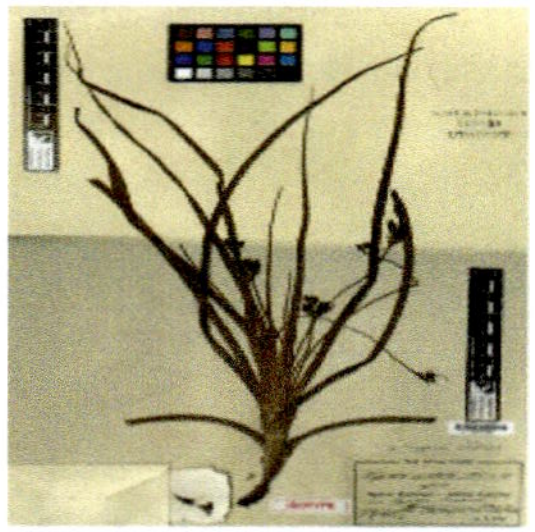	Rhizome, tuber	Campestrol	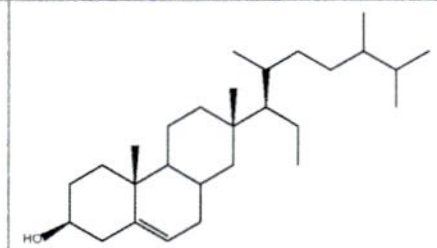	Kamwendo et al. (1985), Vinesi et al. (1990), and Maffei (2003a, b, c, d, e, f, g, h, i, j, k, l, m)
	Synonym(s): *Hypoxis campanulata* Nel, *Hypoxis engleriana* Nel, *Hypoxis ingrata* Nel, *Hypoxis probata* Nel, *Hypoxis retracta* Nel. **Common name(s)**: Star grass, Star lily, Yellow stars, African potato and stars. **Botanical description**: It is a perennial herb with corms or rhizomes and is native to Africa, the Americas, Asia, and Australia. The aboveground herbage is a layered cluster of lance-shaped, linear, or hairlike leaves, sometimes sheathed together at the bases. The blades are slightly hairy. **Medicinal use(s)**: It is used for the treatment of uterine and intestinal cancer and HIV. It also possesses antitumor, antibacterial, anti-inflammatory, and antimicrobial properties.					

	Hypoxis hemerocallidea Fisch., C.A.Mey. & Avé-Lall.		Tuber	Corynanthe	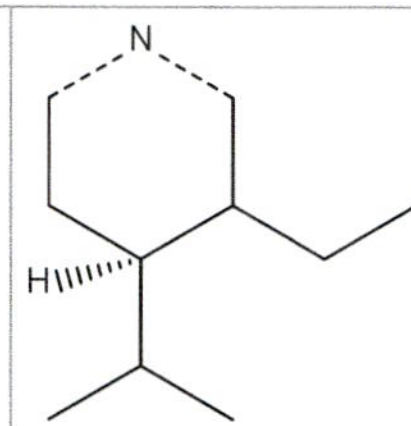	Amusan et al. (1995) and Mills et al. (2005)
	Synonym(s): *Hypoxis obconica* Nel, *Hypoxis patula* Nel, *Hypoxis rooperi* T.Moore. **Common name(s)**: Not available. **Botanical description**: It is native to Southern Africa, Mozambique, and Zimbabwe. Numerous fleshy contractile roots emerge from the bottom half of the corm, which are internally bright yellow in color and are mucilaginous. Yellow flowers are produced from September to February at the end of long peduncles. **Medicinal use(s)**: It is used to boost the immune system of patients having uterine cancer and HIV. It also possesses antibacterial, anti-inflammatory, antioxidant, and antimicrobial properties.					
	Hypoxis sobolifera Jacq.		Whole plant	Dregamine	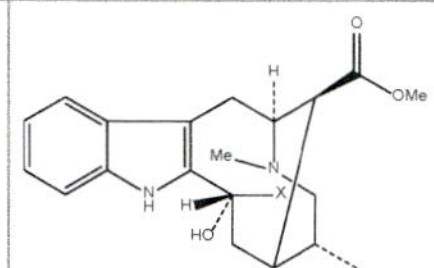	Boukes and Venter (2011)

(continued)

Table 4.72 (continued)

Family	Botanical name	Image	Plant part used	Bioactive compound(s)	Structure	Reference(s)
	Synonym(s): *Hypoxis canescens* Fisch. & C.A.Mey., *Hypoxis krebsii* Fisch., *Hypoxis pannosa* Baker. **Common name(s)**: African potato. **Botanical description**: It is a soft hairy herb 100–200 mm tall, growing singly or in tufts forming large clumps and is native to the Western Cape, the Eastern Cape, and Kwazulu-Natal, from Stellenbosch in the south to Zululand in the north. **Medicinal use(s)**: It is used for the treatment of uterine cancer and HIV/AIDS. It also possesses antibacterial, antioxidant, antimicrobial, anti-inflammatory, and antitumor properties.					
	Hypoxis stellipilis Ker Gawl.		Whole plant	Beta-sitosterol	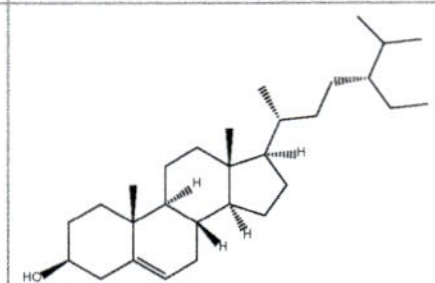	Nair and Kanfer (2008), Boukes and Venter (2011), and Boukes and Van de Venter (2016)
	Synonym(s): *Hypoxis lanata* Eckl. ex Baker. **Common name(s)**: Yellow star, Star grass. **Botanical description**: It is native to Australia and Africa. Its corms are globose 1–2 in. in diameter and crowned with a ring of bristles. Leaves are dozen or more, lanceolate, acuminate, neither rigid in texture nor conspicuously ribbed, 0.5–1 ft. long, 0.5–1 in. broad at the middle, green and glabrous above and covered with persistent white tomentum beneath. **Medicinal use(s)**: It is used for the treatment of cancer cell lines and AIDS. It also possesses antibacterial, antioxidant, antimicrobial, anti-inflammatory, and antitumor properties.			Beta-sitosterol glucoside	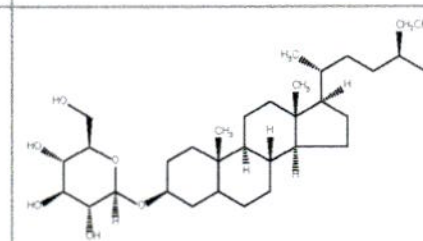	

	Psorospermum molluscum (Pers.) Hochr.		Aerial parts	Chloropsoroxanthin	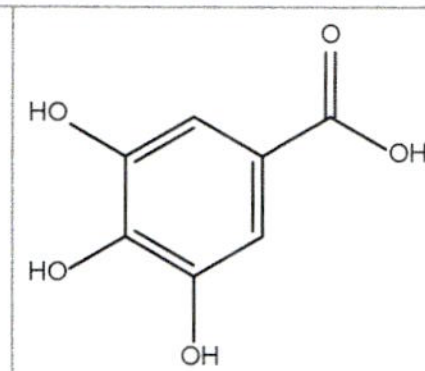	Li et al. (2010)
				p-Soroxanthin	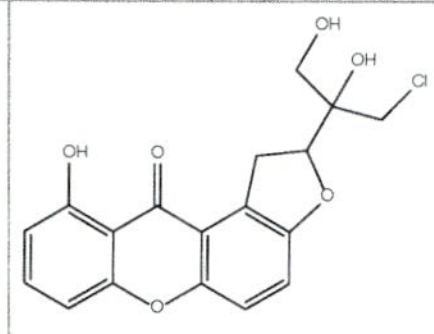	
	Synonym(s): *Haronga mollusca* (Pers.) Choisy, *Harungana mollusca* Pers., *Psorospermum bakeri* Hochr., *Psorospermum cuspidatum* Spach, *Psorospermum pauciflorum* Baker, *Psorospermum venulosum* Baker. **Common name(s)**: Fanterana. **Botanical description**: It is a much-branched shrub or small tree 0.3–6 m tall, occasionally 10 m, and is native to tropical Africa, Guinea to Ethiopia, south to Angola, Zimbabwe, and Mozambique. The plant is commonly harvested from the wild, for its medicinal purposes but also for its edible fruit and wood. It is also sold in local markets. **Medicinal use(s)**: It is used for the treatment of small-cell lung cancer, dysmenorrhea, dysentery, tuberculosis, and whooping cough. It also possesses antibacterial, antioxidant, antimicrobial, anti-inflammatory, and antitumor properties.					

(continued)

Table 4.72 (continued)

Family	Botanical name	Image	Plant part used	Bioactive compound(s)	Structure	Reference(s)
	Hypoxis argentea Harv. ex Baker		Corm	Lupeol	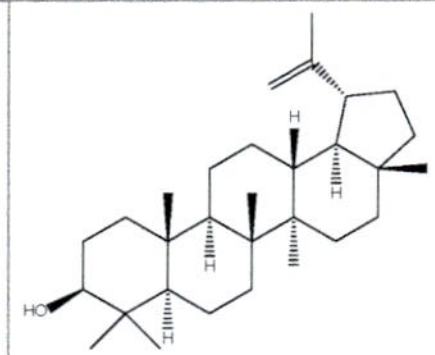	Madhuri and Pandey (2009) and Bradley et al. (2018)
	Synonym(s): *Hypoxis argentea* var. *argentea*. **Common name(s)**: Small yellow stars, Star grass, Star lily. **Botanical description**: It is native to grassland of Africa. Leaves are V-shaped and covered with silky yellowish hairs. The dainty little flowers make it an ideal plant for a small rockery where it can be interplanted with small grasses. Corms are ovoid, 1/2–3/4 in. diameter with a long neck and crowned with a dense ring of bristles. Leaves are 6–12, and anthers are lanceolate, deeply sagittate. Ovary is clavate, silky, 1/8–1/6 in. long, stigmas concrete, capsule turbinate, 1/4 in. long. **Medicinal use(s)**: It is used to boost the immunity of HIV/AIDS patients and to treat premalignant cancer cells. It also possesses antibacterial, antioxidant, antimicrobial, anti-inflammatory, and antitumor properties.			Proanthocyanidin A	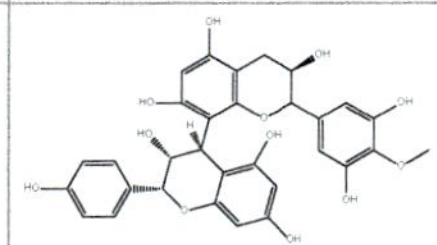	

Table 4.73 Anticancer plants of family Icacinaceae

Family	Botanical name	Image	Plant part used	Bioactive compound(s)	Structure	Reference(s)
Icacinaceae	*Icacina mannii* Oliv.		Root, tuber	Arctigenin	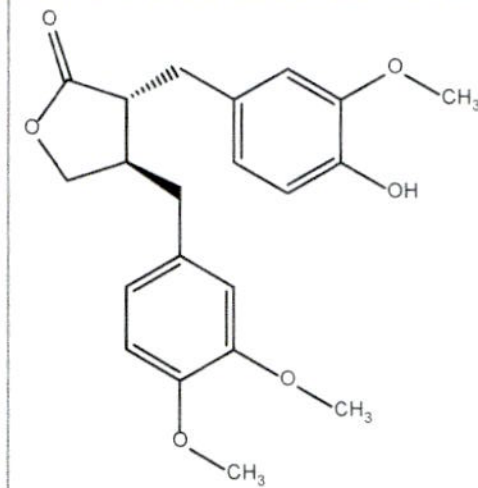	On'Okoko et al. (1985) and Zhao et al. (2015a, b)
	Synonym(s): No synonyms are recorded for this name. **Common name(s)**: Ohogha. **Botanical description**: It is native to Africa, Madagascar, and Australia. **Medicinal use(s)**: It is used for the treatment of ovarian and small-cell lung cancer. It also possesses antitumor, antibacterial, and anti-inflammatory properties. Palm wine is the prepared form of this plant in which the plant is boiled and drunk in the Congo (Brazzaville) for the treatment of blennorrhea. The plant is used as an analgesic for intestinal pain and for hernia.			(9βH)-17-Norpimaranes	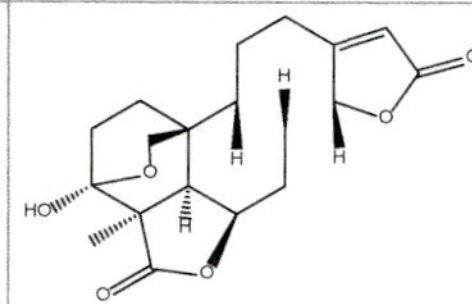	

(continued)

Table 4.73 (continued)

Family	Botanical name	Image	Plant part used	Bioactive compound(s)	Structure	Reference(s)
	Miquelia dentata Bedd.		Leaves	α-Amyrin	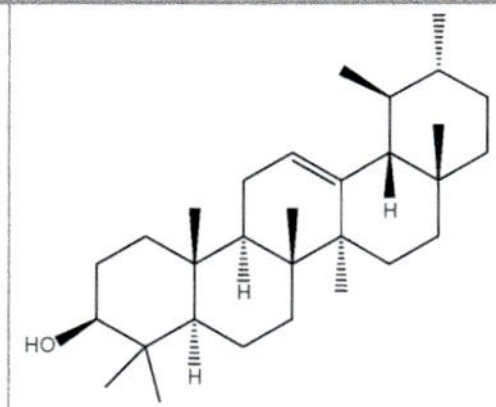	Thriveni et al. (2015)
	Synonym(s): *Miquelia assamica* (Griff.) D.G. Long. **Common name(s)**: Not available. **Botanical description**: It is a glabrous climber and is native to South India (Tamil Nadu, Kerala). Leaves are 16 × 7 cm, ovate, apex acute, base rounded or truncate, dentate, more sharply near the base; petiole 3–4 cm long. Flowers unisexual, peduncle 6 cm long, calyx cupular, 4-toothed, petals 4, 3 mm long, oblong, stamens 4 or 5, in male flowers, ovary 1-celled, 4-angled, style absent, stigma 4-angled, ovules 2, pendulous. **Medicinal use(s)**: It is used to treat ovarian cancer. It also possesses antitumor, antibacterial, anti-inflammatory, and antioxidant properties.			Camptothecin	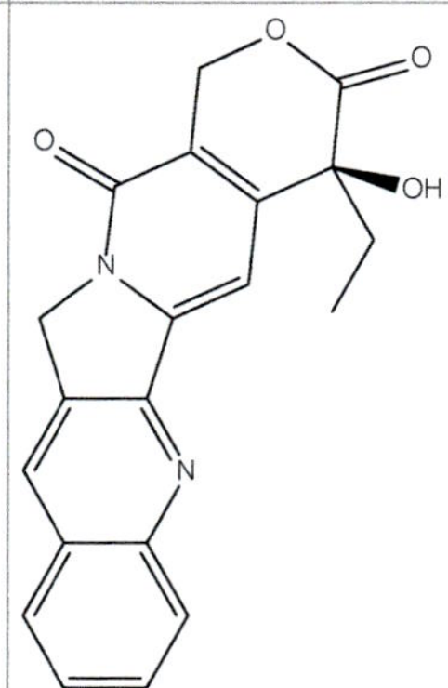	

Table 4.74 Anticancer plants of family Iridaceae

Family	Botanical name	Image	Plant part used	Bioactive compound(s)	Structure	Reference(s)
Iridaceae	*Crocosmia crocosmiiflora* (Lemoine) N.E.Br.		Corm	Berberine	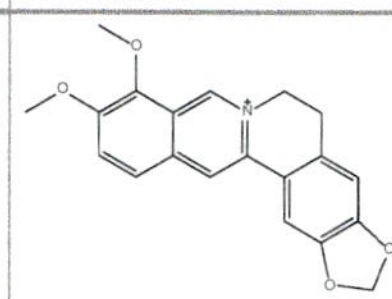	Asada et al. (1988) and Maffei (2003a, b, c, d, e, f, g, h, i, j, k, l, m)
	Synonym(s): *Crocosmia latifolia* N.E.Br., *Montbretia crocosmiiflora* Lemoine, *Tritonia crocosmiiflora* (Lemoine) G.Nicholson. **Common name(s)**: Crocosmia, Montbretia, Coppertip, Falling stars. **Botanical description**: It is commonly called montbretia and is native to grassland habitats in the Cape region of South Africa. In midsummer this makes a good background for the freesia-shaped sprays of flowers that are carried just above it. They are grown as ornamental pot plants. Leaves are mostly basal larger than cauline, blade lanceolate, 8–20 mm wide. **Medicinal use(s)**: It is used for the treatment of cancer, arthritis, and certain epidemic diseases.					

(continued)

Table 4.74 (continued)

Family	Botanical name	Image	Plant part used	Bioactive compound(s)	Structure	Reference(s)
	Iris pallasii Fisch. ex Trevir.		Leaf	Irisquinone	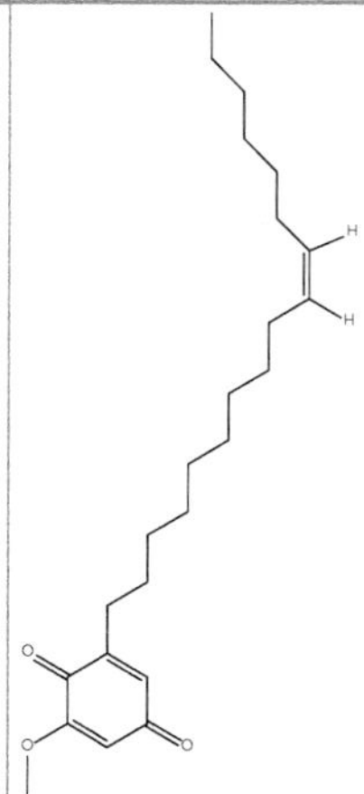	Asada et al. (1988), Hadfield et al. (2000), and Kaššák (2012)
	Synonym(s): *Iris iliensis* Poljakov, *Joniris pallasii* (Fisch. ex Trevir.) Klatt, *Xiphion pallasii* (Fisch. ex Trevir.) Alef. **Common name(s)**: Pallasii. **Botanical description**: It is native to temperate and middle Asia, Kazakhstan, Mongolia, Siberia Altay, Irkutsk, Tuva, and Krasnoyarsk. **Medicinal use(s)**: It is used for the treatment of breast cancer, ovarian cancer, and lymphoma. It also possesses antitumor, anti-inflammatory, antioxidant, and antibacterial properties.					

Table 4.75 Anticancer plants of family Juglandaceae

Family	Botanical name	Image	Plant part used	Bioactive compound(s)	Structure	Reference(s)
Juglandaceae	*Juglans mandshurica* Maxim		Cotyledons, Root	P-Hydroxymethoxy-benzobiju lone	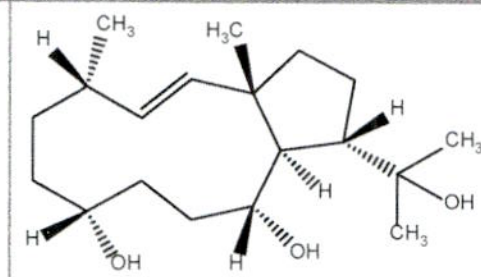	Duke and Ayensu (1985a, b), Son (1995), Joe et al. (1996), Li et al. (2007a, b, c), Xin et al. (2014), and Yao et al. (2017)
	Synonym(s): *Juglans cathayensis* Dode, *Juglans collapse* Dode, *Juglans draconis* Dode, *Juglans formosana* Hayata, *Juglans stenocarpa* Maxim. **Common name(s)**: Manchurian walnut. **Botanical description**: It is a tree or sometimes shrub (25 m tall) and is native to China, Korea, and South Asia. Leaves 40–90 cm long, petiole 5–23 cm, petiole and rachis sparsely to moderately glandular pubescent, leaflets (7–19), lateral ones sessile, blade elliptic to long elliptic or ovate-elliptic to long elliptic-lanceolate 6–17 × 2–7.5 cm, axially tomentose or occasionally slightly pubescent, generally without glandular hairs, slightly to densely glandular and glandular pubescent along midvein, base oblique, subcordate, margin serrate, rarely serrulate, apex acuminate, terminal petiolule 1–5 cm. **Medicinal use(s)**: It also possesses anticancer properties. It is used to treat liver cancer and ovarian cancer.			Coumarin(s)	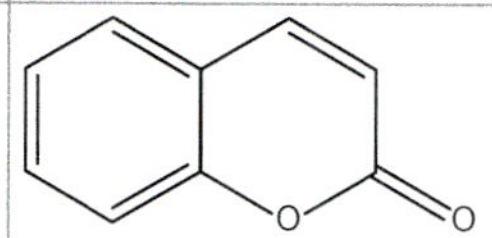	

(continued)

Table 4.75 (continued)

Family	Botanical name	Image	Plant part used	Bioactive compound(s)	Structure	Reference(s)
	Juglans regia L.		Stem, cotyledon, leaves	p-Hydroxymethoxy-benzobijuglone	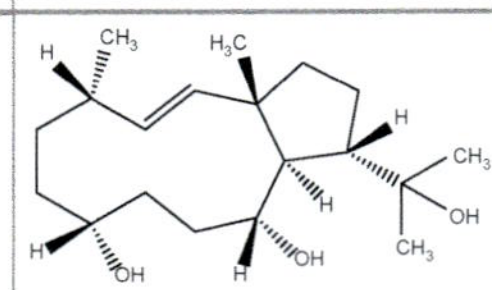	Kumudhavalli et al. (2010), Carvalho et al. (2010), and Salimi et al. (2012)
	Synonym(s): *Juglans duclouxiana* Dode, *Juglans fallax* Dode, *Juglans kamaonia* Dode, *Juglans orientis* Dode, *Juglans sinensis* (C.DC.) Dode. **Common name(s)**: Walnut, Persian walnut. **Botanical description**: It is a deciduous tree growing up to 20 m by 20 m at a medium rate and is native to the Balkans, the Himalayas, and Southwest China. It is in flower in June, and the seeds ripen in October. The flowers are monoecious (individual flowers are either male or female, but both sexes can be found on the same plant) and are pollinated by wind. The plant is self-fertile. It is suitable for light (sandy), medium (loamy), and heavy (clay) soils and prefers well-drained soil. It cannot grow in the shade. **Medicinal use(s)**: The cotyledons are used in the treatment of cancer. Walnut has a long history of folk use in the treatment of cancer. The leaves possess anthelmintic and anti-inflammatory properties and are also used as astringent. They are used internally for the treatment of constipation, chronic coughs, asthma, diarrhea, dyspepsia, etc. The leaves are also used to treat skin ailments and for blood purification. Male inflorescences are made into a broth and used in the treatment of coughs and vertigo.					

Table 4.76 Anticancer plants of family Juncaceae

Family	Botanical name	Image	Plant part used	Bioactive compound(s)	Structure	Reference(s)
Juncaceae	*Juncus acutus* L.		Leaf	Phenanthrenoid	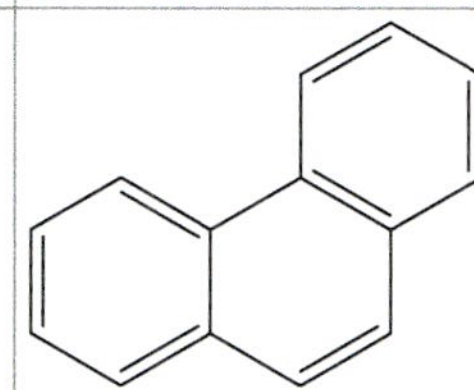	Martinez-Lirola et al. (1996), Bhanot et al. (2011), Rodrigues et al. (2014), and Oliveira et al. (2016)
	Synonym(s): *Juncus karelinii* Steud., *Juncus multibracteatus* Tineo, *Juncus spinosus* Forssk., *Juncus variegatus* Caruel. **Common name(s)**: Spiny rush, Sharp rush, Leopold's rush. **Botanical description**: It is a perennial plant growing up to 1.5 m (5 ft.) and is native to Southern Europe, Britain, South and East France, North Africa, and Macaronesia. The flowers are hermaphrodite (have both male and female organs) and are pollinated by wind. It is suitable to grow in light (sandy), medium (loamy), and heavy (clay) soils. It can grow in semi-shade (light woodland) or no shade. **Medicinal use(s)**: It is used for the treatment of cancer. It is used to treat human breast cancer. It is used for the treatment of diarrhea, and fruits are used in infusions to alleviate cold symptoms. It is also useful in the treatment of stomach disorders. It also possesses antioxidant, antibacterial, and anti-inflammatory properties.			8,8′-Bidehydrojuncusol	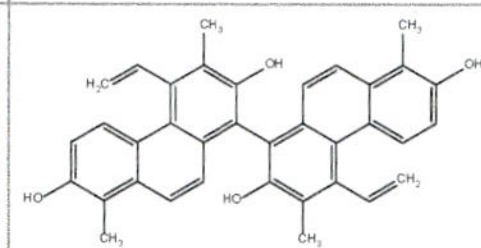	
	Juncus effusus L.		Whole plant	Tridecanone		Mohammad (2006), Desai et al. (2008), Madhuri and Pandey (2009), and Mukudai et al. (2013)

(continued)

Table 4.76 (continued)

Family	Botanical name	Image	Plant part used	Bioactive compound(s)	Structure	Reference(s)
	Synonym(s): *Juncus bogotensis* Kunth, *Juncus fistulosus* Guss., *Juncus lucens* Burnham, *Juncus luxurians* Colenso, *Juncus zebrinus* André. **Common name(s)**: Soft rush, common rush, lamp rush, Pacific rush. **Botanical description**: It is a perennial plant growing up to 1.5 m (5 ft.) and is native to Southern Europe, Britain, South and East France, North Africa, and Macaronesia. The flowers are hermaphrodite (have both male and female organs) and are pollinated by wind. It is suitable to grow in light (sandy), medium (loamy), and heavy (clay) soils. It can grow in semi-shade (light woodland) or no shade. **Medicinal use(s)**: It is used to treat human breast cancer and ovarian cancer. The pith of the stem is used as antiphlogistic, depurative, discutient, diuretic, febrifuge, lenitive, lithontripic, and sedative. It is also used in the treatment of sore throats, jaundice, edema, acute urinary tract infection, and morbid crying of babies.			Effusol	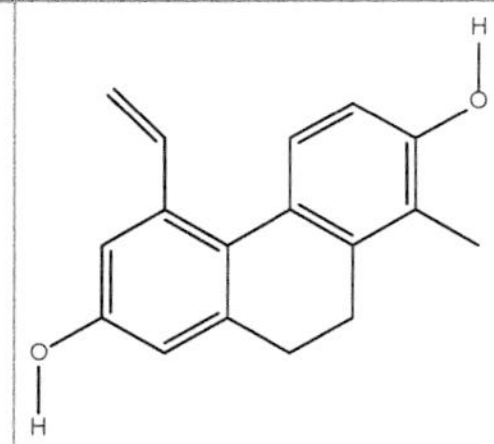	
				Juncanol	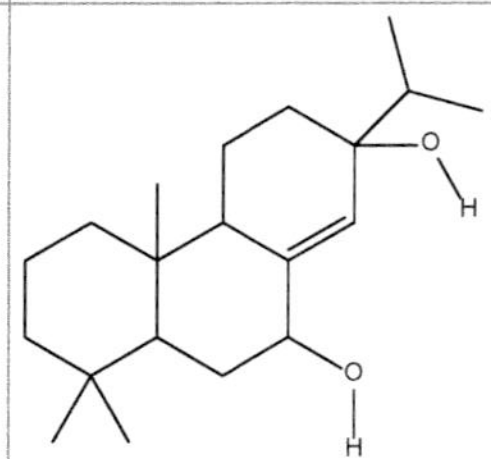	
				Phenylpropanoid	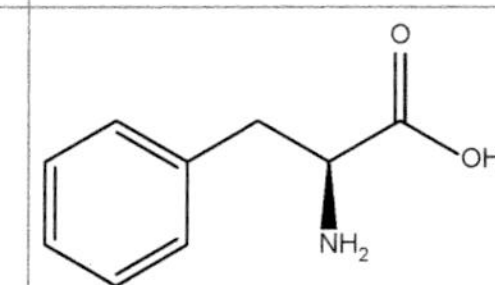	
				A-Tocopherol	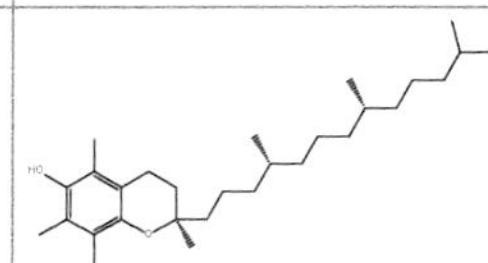	

Table 4.77 Anticancer plants of family Lamiaceae

Family	Botanical name	Image	Plant part used	Bioactive compound(s)	Structure	Reference(s)
Lamiaceae/ Labiatae	*Ajuga bracteosa* Wall.		Whole plant	Bractin A	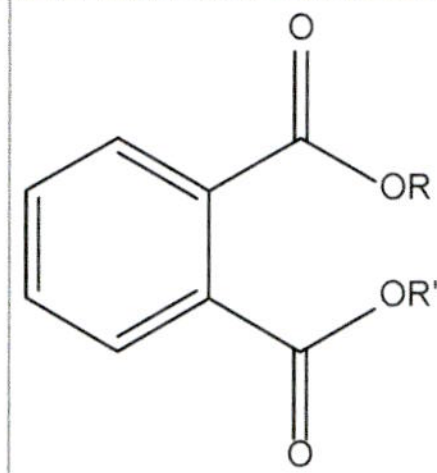	Riaz et al. (2007)
	Synonym(s): *Ajuga bracteosa* Wall. ex Benth., *Ajuga bracteosa* var. *alba* (Gürke) Engl. *Ajuga bracteosa* f. *alba* Gürke, *Ajuga bracteosa* var. *canescens* (Benth.) Engl. **Common name(s):** Bugleweed. **Botanical description:** *Ajuga bracteosa* is native to India and is another Indian herbal plant which has enormous traditional uses and used for Ayurvedic preparation. It is a perennial erect or ascending hairy herb, often prostrate with oblanceolate or subspathulate. Flowers are white or purplish violet tinged from lower surface in distant, axillary whorls in spike. **Medicinal use(s)**: *Ajuga bracteosa* is used as astringent and for hypoglycemia and gastrointestinal disorders. This plant is also used to treat brain tumor and liver cancer.					

(continued)

Table 4.77 (continued)

Family	Botanical name	Image	Plant part used	Bioactive compound(s)	Structure	Reference(s)
	Ajuga decumbens L.		Flower	12-O-Tetradecanoylphorbol-13-acetate (TPA)	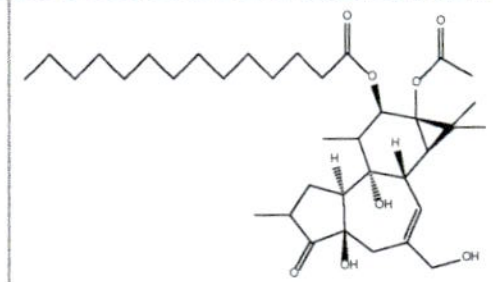	Tokuda et al. (1999)
	Synonym(s): *Ajuga decumbens* f. *condensata* Makino, *Ajuga decumbens* var. *decumbens*, *Ajuga decumbens* var. *glabrescens* Franch. & Sav. **Common name(s):** Bugle. **Botanical description:** *Ajuga decumbens* is an herbaceous flowering plant native to China, Japan, and Korea. It is commonly found in lightly forested sunny areas such as meadows and roadsides, growing at an altitude of 400 to 2300 m. This plant grows as a groundcover, and the leaf layer is usually no more than 100 mm (3.9 in.) tall. It flowers between April and June. **Medicinal use(s):** *Ajuga decumbens* is used as anthelmintic, diuretic, antifungal, anti-inflammatory, and antimycobacterial agents. The plant is famous for treating prostate and liver cancer.			Cyasterone	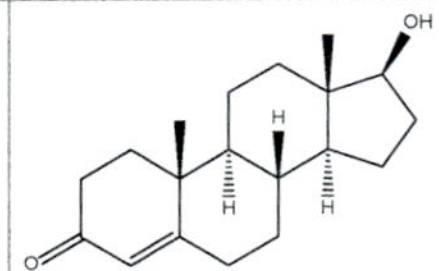	
				8-Acetylharpagide	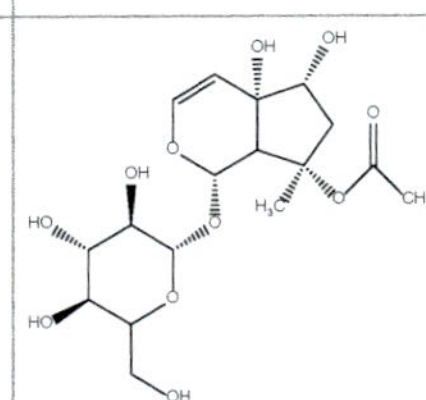	

	Ajuga genevensis L.		Twig	Flucrouracil	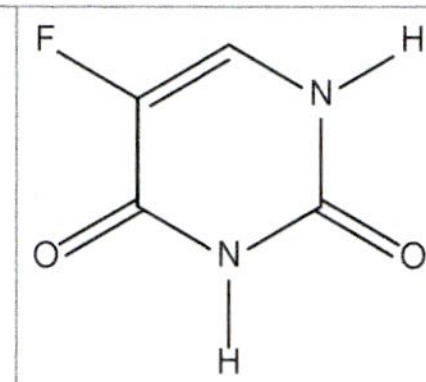	Pan (1960)
	Synonym(s): *Ajuga genevensis* var. *alpicola* Beck, *Ajuga genevensis* var. *alpina* (L.) Nyman, *Ajuga genevensis* var. *arida* Fr. **Common name(s):** Upright bugle, Blue bugle, Geneva bugleweed, Blue bugleweed. **Botanical description:** *Ajuga genevensis* is an herbaceous flowering plant native to Europe. It is a perennial plant (flowering between April and July) growing to a height of between 10 and 30 cm. Evergreen, it has long-stalked, obovate, basal leaves which are shallowly lobed or toothed. **Medicinal use(s):** *Ajuga genevensis* has been used in traditional Austrian medicine internally as tea for treatment of disorders related with the respiratory tract. The plant is used to treat vaginal cancer.					

(continued)

Table 4.77 (continued)

Family	Botanical name	Image	Plant part used	Bioactive compound(s)	Structure	Reference(s)
	Callicarpa americana L.		Fruit and leaf	Calliterpenone	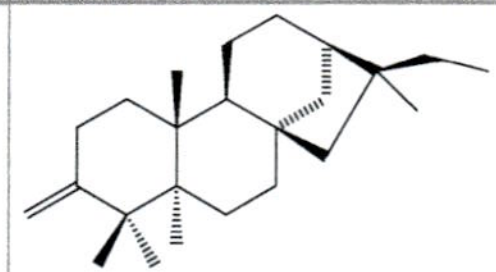	Jones et al. (2007)
				Euscaphic acid	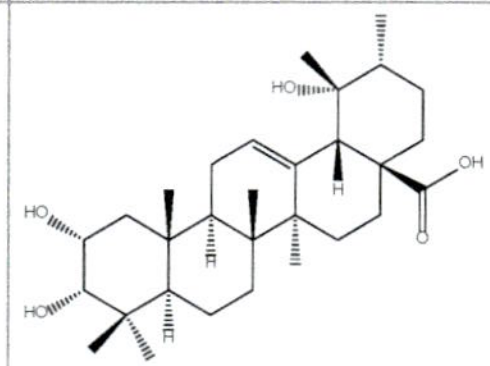	
	Synonym(s): *Callicarpa americana* var. *alba* Rehder, *Callicarpa americana* f. *lactea* (F.J.Müll.) Rehder, *Callicarpa americana* var. *lactea* F.J.Müll. **Common name(s):** American *beautyberry*. **Botanical description:** *Callicarpa americana* is an open-habit, native shrub of the Southern United States which is often grown as an ornamental in gardens and yards. American beautyberries produce large clusters of purple berries, which birds and deer eat, thus distributing the seeds. **Medicinal use(s):** The raw berries, while palatably sweet, are suitable for human consumption only in small amounts, because they are astringent. Some people have reported mild stomach cramps after consumption. The berries are also used in jellies and wine. The plant is also used to treat skin cancer.			Salvigenin (6,7-dimethoxy-4′-hydroxyflavone)	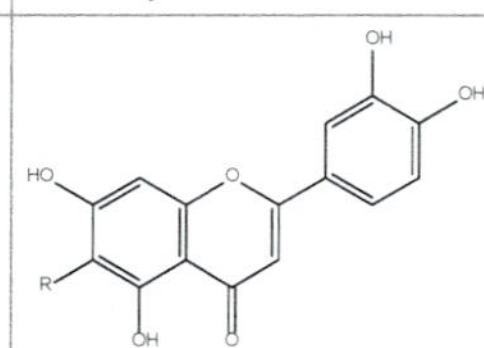	
				Genkwanin	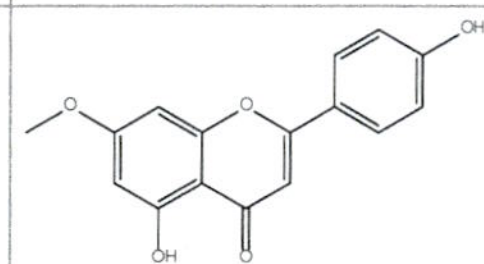	

	Glechoma hederacea L.		Fruit juice (unripe)	γ-Elemene	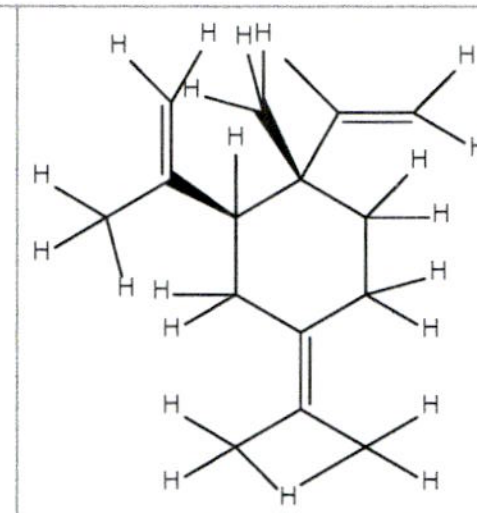	Culpeper (1950) and Mockute et al. (2007)
	Synonym(s): *Glechoma hederacea* f. *albiflorens* S.S.Ying, *Glechoma hederacea* var. *breviflora* Coss. & Germ., *Glechoma hederacea* var. *grandiflora* Hoffmanns. & Link. **Common name(s):** Ground ivy, Gill-over-the-ground, Creeping charlie, Alehoof, Tunhoof, Catsfoot, Field balm, and Run-away-robin. **Botanical description:** *Glechoma hederacea* is native to North America and can be identified by its round to reniform (kidney or fan shaped), crenate (with round toothed edges) opposed leaves 2–3 cm (0.79–1.18 in.) in diameter, on 3–6 cm (1.2–2.4 in.) long petioles attached to square stems which root at the nodes. Like crabgrass, creeping charlie's root has a ball. It is a variable species, its size being influenced by environmental conditions, from 5 to 50 cm (2.0–19.7 in.) tall. **Medicinal use(s):** *Glechoma hederacea* has been used in the traditional medicine of Europe going back thousands of years that recommends the plant to treat inflammation of the eyes. The herbalist recommended the plant to treat tinnitus, as well as a diuretic, astringent, tonic, and gentle stimulant. It is used to treat bile and liver cancer.			β-Elemene	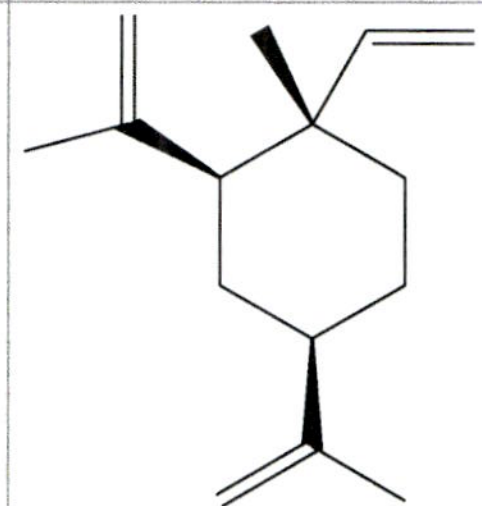	

(continued)

Table 4.77 (continued)

Family	Botanical name	Image	Plant part used	Bioactive compound(s)	Structure	Reference(s)
	Hyptis brevipes Jacq.		Whole plant	Brevipolide A	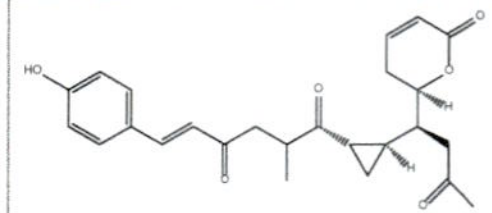	Ye (2011)
	Synonym(s): *Hyptis brevipes* var. *glabrior* Benth. *Hyptis brevipes* var. *remotidens* Briq., *Hyptis brevipes* var. *robusta* Loes. **Common name(s):** Vilayti tulsi. **Botanical description:** *Hyptis brevipes* is composed of approximately 350–400 species in the form of small herbs to large bushes, which are native to Mexico and West Indies, and is distributed in the tropics and warmer temperate regions all over the world. **Medicinal use(s):** This plant is traditionally used as an expectorant and astringent and to treat bronchitis and gastrointestinal cancer.					
	Isodon excisus (Maximowicz)		Whole plant	3-(3-Methoxy-phenyl)-N-(3, 4, 5-trimethoxyphenyl)-acrylamide	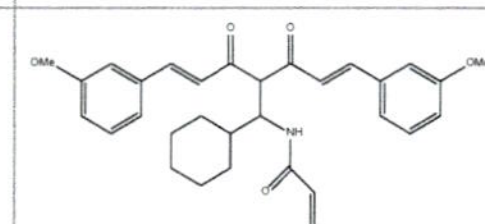	Sienkiewicz et al. (2007)

	Synonym(s): *Isodon excisus* var. *hakusanensis* Kudo, *Isodon excisus* var. *racemosus* (Hemsl.) Kudo. **Common name(s):** Kudo. **Botanical description:** *Isodon excisus* is a flowering plant native to tropical and subtropical parts of the Old World, primarily Asia, but two species are from Africa. Many of the species are endemic to China. **Medicinal use(s):** The essential oil of the plant has been used for centuries as a general tonic for colds and coughs and to relieve congestion of the mucous membranes. The plant is also used to cure gastrointestinal cancer.					
	Isodon henryi		Stem	Minkeryins A–G (1–7)	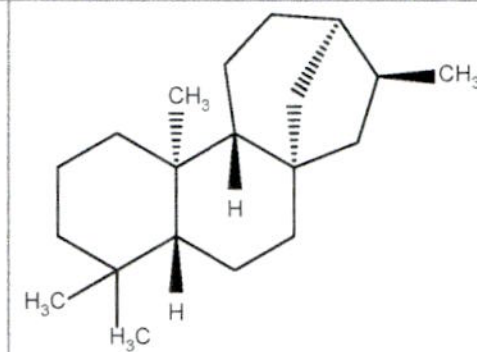	Zhao et al. (2009)
				Ent-kaurane diterpenoids	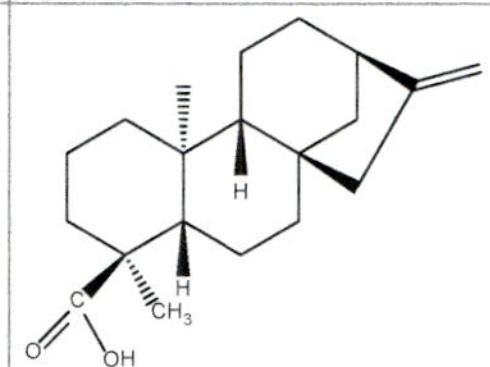	

(continued)

Table 4.77 (continued)

Family	Botanical name	Image	Plant part used	Bioactive compound(s)	Structure	Reference(s)
				Leukamenin F	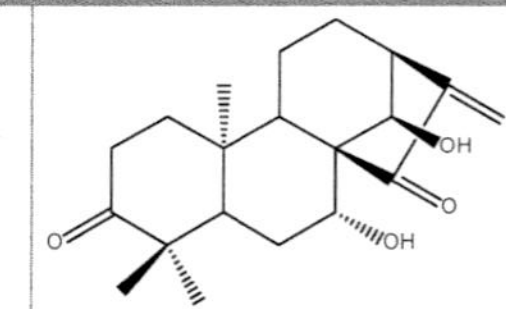	
	Synonym(s): *Isodon henryi* var. *dichromophyllus* (Diels) Kudô. **Common name(s):** Hemsley, Kudo, Henryi. **Botanical description:** *Isodon henryi* is native to China and is a perennial herb. Stems erect, (30–) 50–100 (–150) cm, base glabrescent, angles sparsely pubescent, much branched at apex, branches slender. Petiole to 4 cm; leaf blade rhombic-ovate to lanceolate, mid stem leaves ca. 6 × 4 cm. **Medicinal use(s):** This plant is used to treat disorders in the liver and bile and gastrointestinal cancer.			Leukamenin E	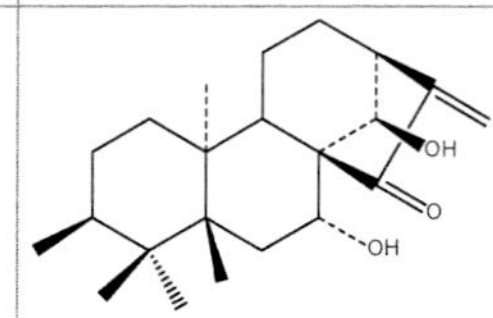	
				Wangzaozin A	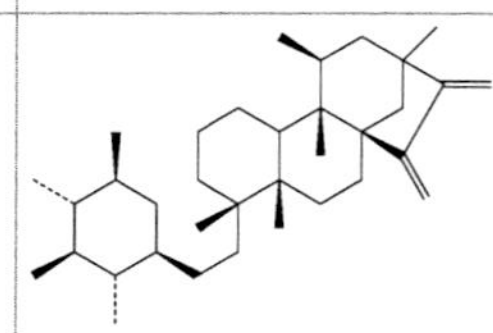	
				Racemosin A	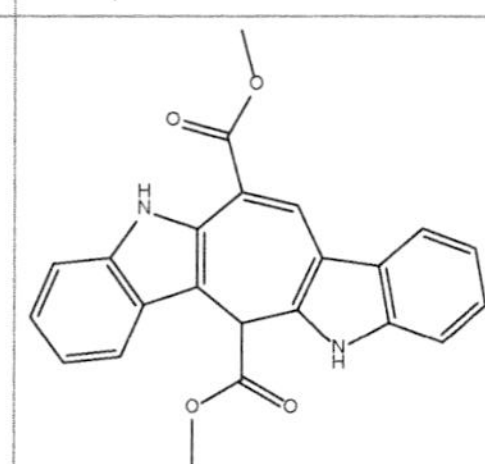	

	Isodon japonicus (Burm.f.) H.Hara		Leaf	Maoyecrystal 1	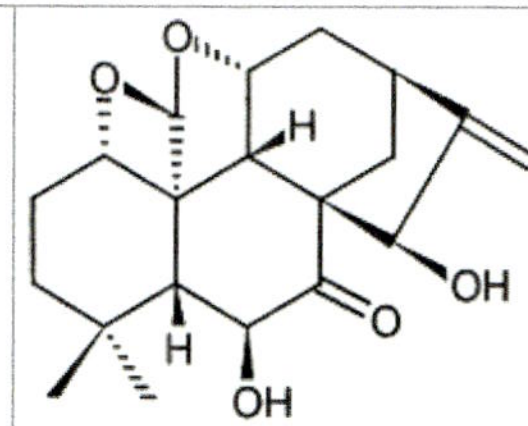	Zhang et al. (2002), Bai et al. (2005), and Liu et al. (2017b)
	Synonym(s): *Isodon alborubrus* (C.Y.Wu) H.Hara, *Isodon amethystoides* (Benth.) H.Hara, *Isodon angustifolius* (Dunn) Kudô. **Common name(s):** Japonica. **Botanical description:** *Isodon japonicus* is distributed in China North-Central, China South-Central Amur, India, West Himalaya, Japan, Korea, Manchuria, and Primorye (tropical and subtropical climates all over the world). **Medicinal use(s):** The plant is used to cure respiratory tract and kidney disorders. It is effective against human cancer cells, including ovarian, breast, colorectal, laryngeal, leukemia, liver, and prostate cancers.			Oridonin	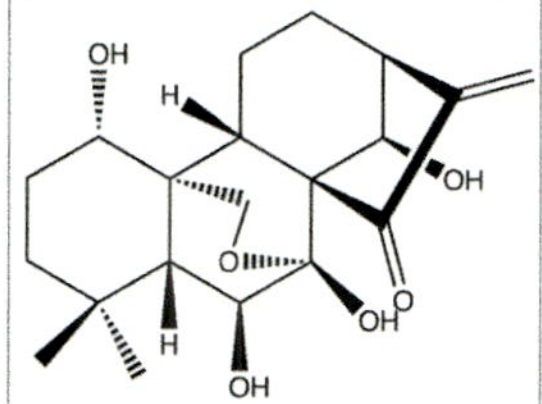	
				Rabdosin B	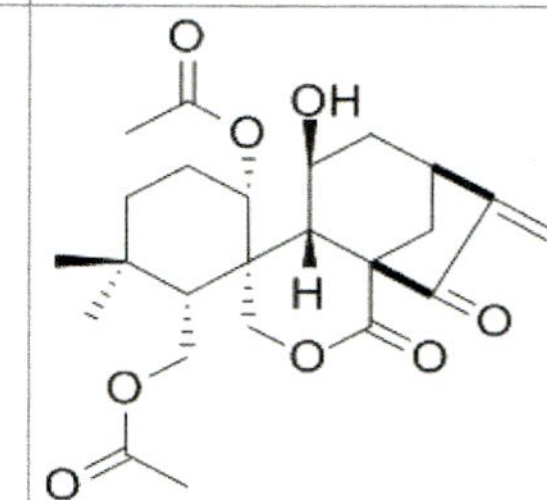	

(continued)

Table 4.77 (continued)

Family	Botanical name	Image	Plant part used	Bioactive compound(s)	Structure	Reference(s)
				Epinodosin	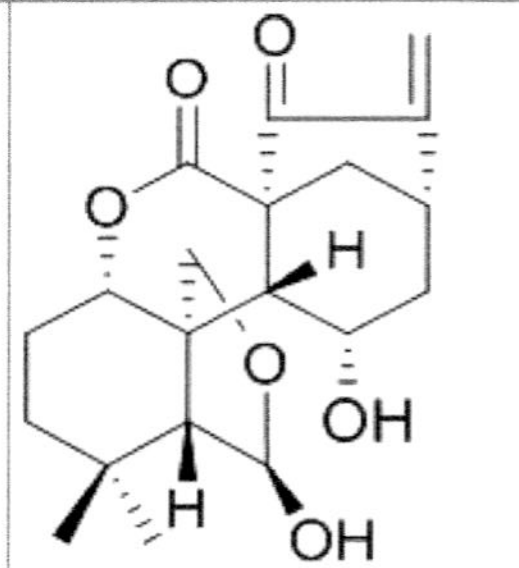	
				Lasiokaurin	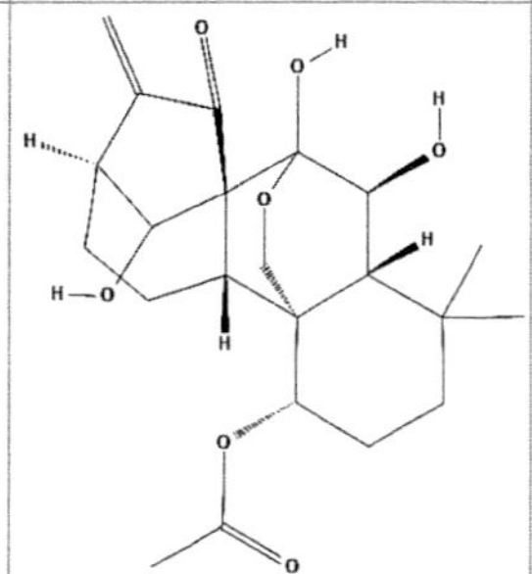	
				Ponicidin	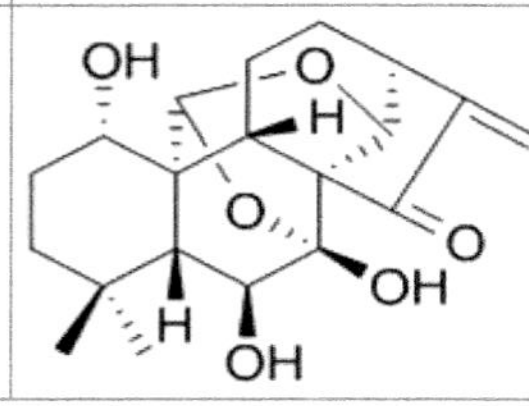	

				Shikokianin	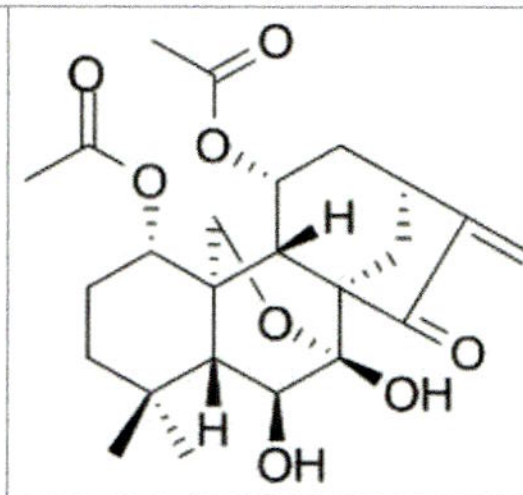	
	Isodon macrocalyx (Dunn) Kudo **Synonym(s):** *Amethystanthus taiwanensis* Masam., *Isodon bifidocalyx* (Dunn) H.Hara, *Isodon taiwanensis* (Masam.) S.S.Ying, *Plectranthus bifidocalyx* Dunn. **Common name(s):** Wrinkled leaf isodon. **Botanical description:** *Isodon macrocalyx* is a perennial herb native to China and America growing up to 1 m tall. Stems erect, unbranched or few branched, pubescent, glabrescent. Petiole 2–10 mm, obsolete in basal leaves, pubescent; leaf blade lanceolate, 3.5–13.5 × 1–2 cm, papery, adaxially pubescent along veins. **Medicinal use(s):** *Isodon macrocalyx* has been used in folk medicine as antibacterial, anti-inflammatory, stomachic, and anthelmintic. The plant is used to cure stomach cancer.		Rhizomes	Excisanin A	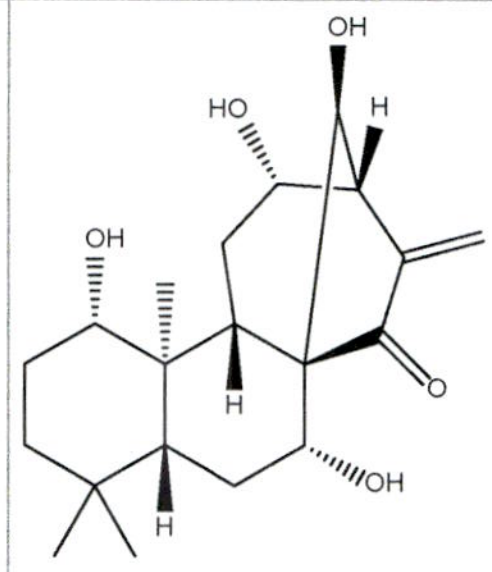	Deng et al. (2009a, b)

(continued)

Table 4.77 (continued)

Family	Botanical name	Image	Plant part used	Bioactive compound(s)	Structure	Reference(s)
	Isodon nervosus (Hemsl.) Kudo		Fruit	Ent-kaurane (diterpenoid)	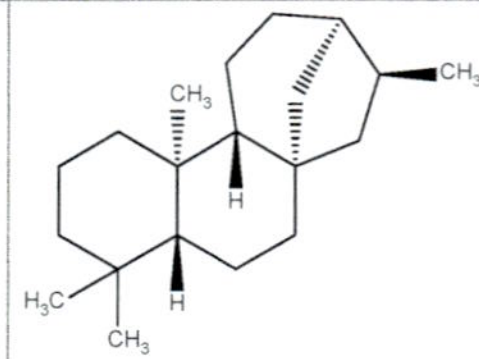	Wei et al. (2011) and Liu et al. (2017b)
	Synonym(s): *Plectranthus moslifolius* H.Lév., *Plectranthus nervosus* Hemsl., *Plectranthus salicarius* Hand.-Mazz., *Rabdosia nervosa* (Hemsl.) C.Y.Wu & H.W.Li. **Common name(s):** Nervosus. **Botanical description:** *Isodon nervosus* is a perennial herb native to China growing up to 1 m tall. Stems erect, unbranched or few branched, pubescent, glabrescent. Petiole 2–10 mm, obsolete in basal leaves, pubescent; leaf blade lanceolate, 3.5–13.5 × 1–2 cm, papery, adaxially pubescent along veins. **Medicinal use(s):** *Isodon nervosus* has many pharmacological properties including antitumor, antioxidant and antiplatelet. It has been used as a folk remedy for curing gastric cavity pain, sores, amenorrhea, wrestling injury, and brain tumor in China, Korea, and Japan.			Nervonin B	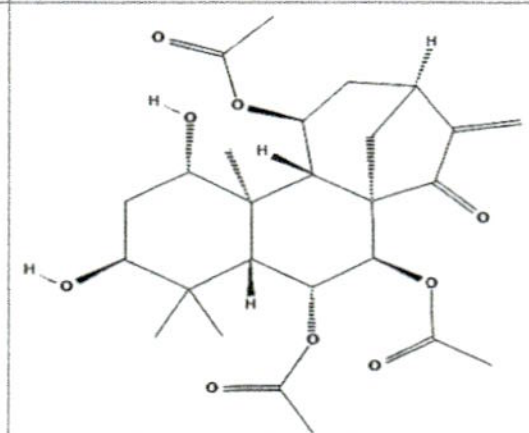	
				Weisiensin A	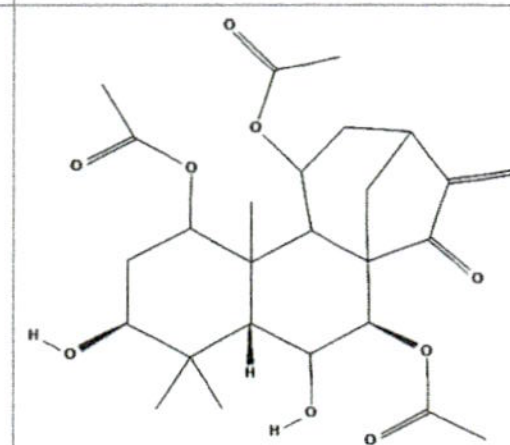	

		Calcicolin A	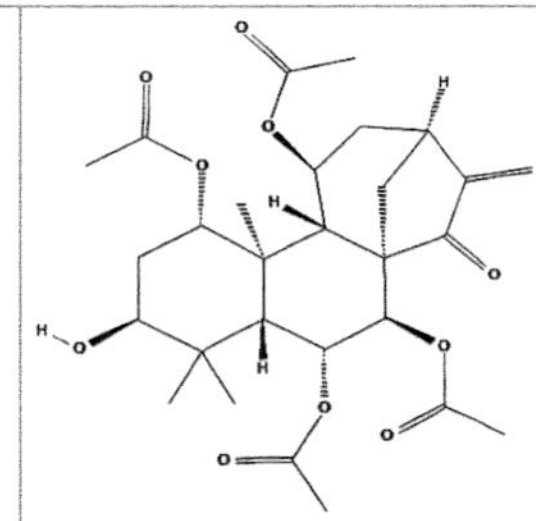	
		Calcicolin B	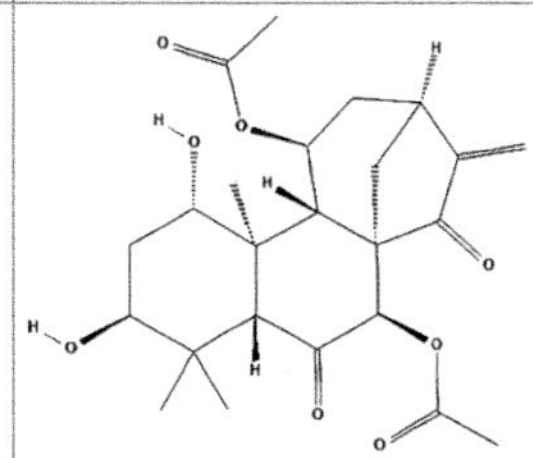	
		Adenanthin	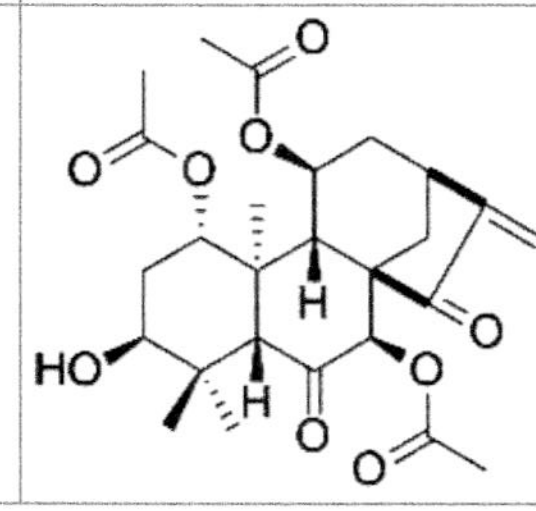	

(continued)

Table 4.77 (continued)

Family	Botanical name	Image	Plant part used	Bioactive compound(s)	Structure	Reference(s)
				Isodocarpin	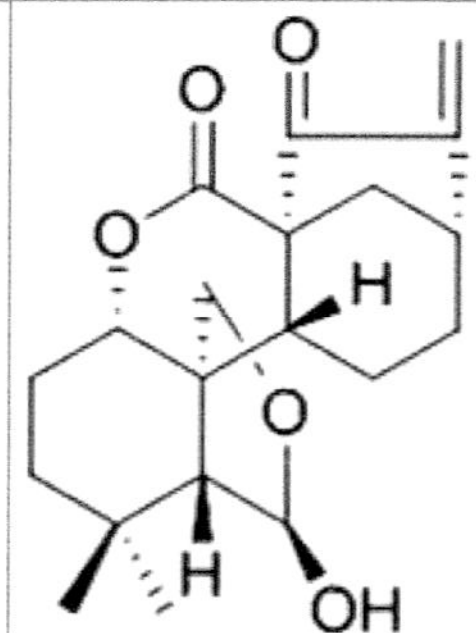	
				Effusanin A	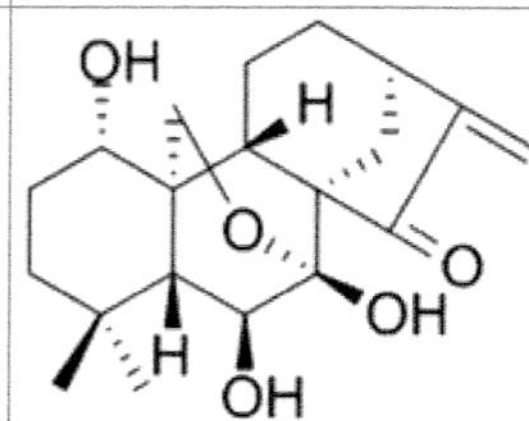	

		Effusanin E	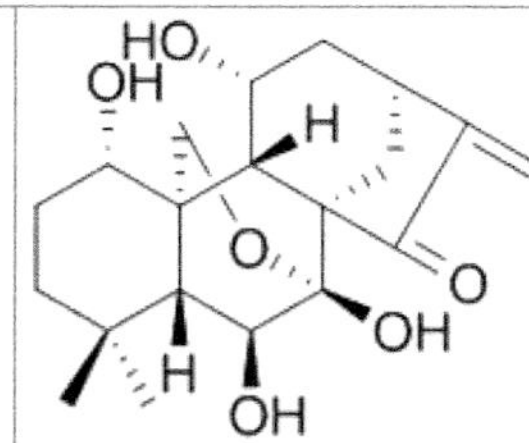	
		Rabdonervosin H	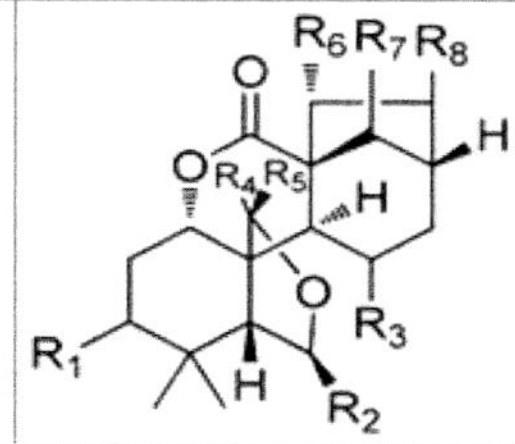	
		Rabdonervosin I	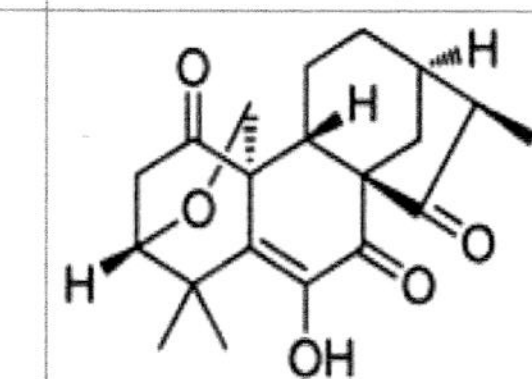	

(continued)

Table 4.77 (continued)

Family	Botanical name	Image	Plant part used	Bioactive compound(s)	Structure	Reference(s)
	Isodon phyllostachys (Diels) Kudo.		Aerial parts	Ent-kaurene (diterpenoid)	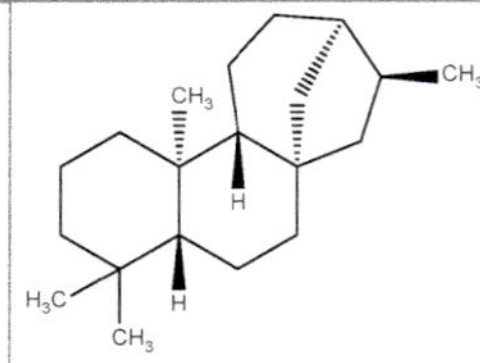	Li et al. (2006)
	Synonym(s): *Plectranthus phyllostachys* Diels, *Rabdosia phyllostachys* (Diels) H.Hara. **Common name(s):** *Phyllostachys.* **Botanical description:** *Isodon phyllostachys* native to China are shrubs or subshrubs growing up to 0.9–3 m. Branches erect, densely spreading pilose, glabrescent. Petiole 1–6 mm, pilose; leaf blade ovate, 1.5–5 × 0.8–4.5 cm, corrugate, adaxially puberulent or pilose, abaxially gray tomentose, base cordate, margin crenate, apex obtuse, sometimes acute. **Medicinal use(s):** It has been used as an antibiotic and antiphlogistic agent and is used to treat breast cancer in folk medicine.			Rabdoloxins A	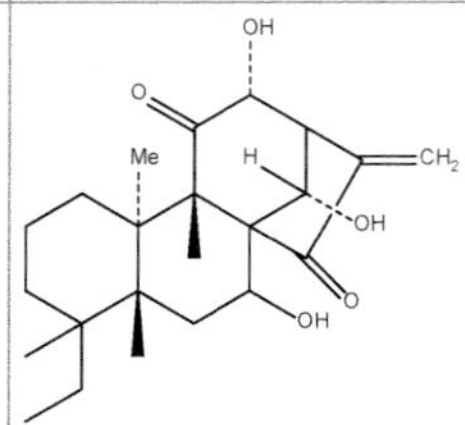	
				Rabdoinflexin B	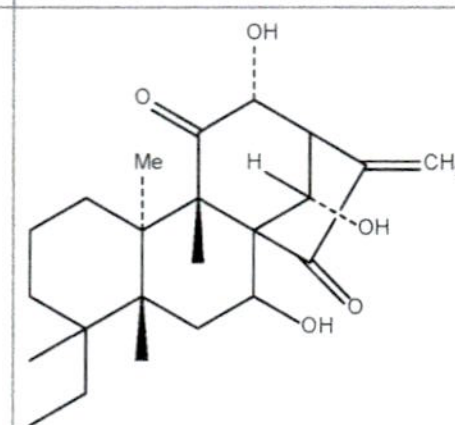	

		Amethystoidin A	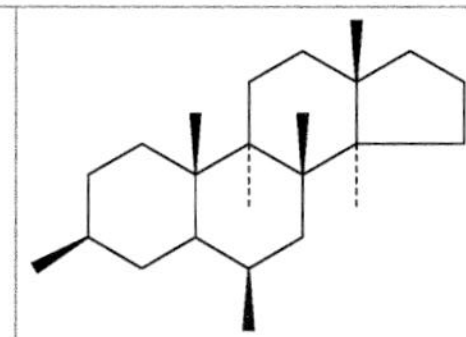	
		Oleanolic acid	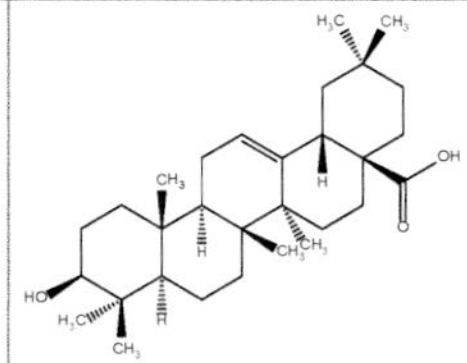	
		Daucosterol	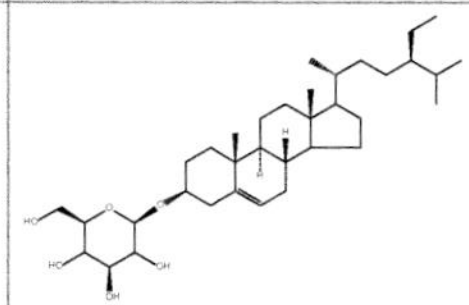	
		Phyllostachysin D	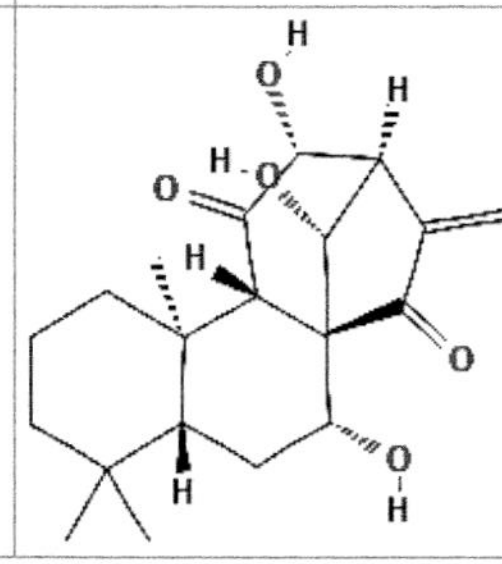	

(continued)

Table 4.77 (continued)

Family	Botanical name	Image	Plant part used	Bioactive compound(s)	Structure	Reference(s)
				Phyllostachysin F	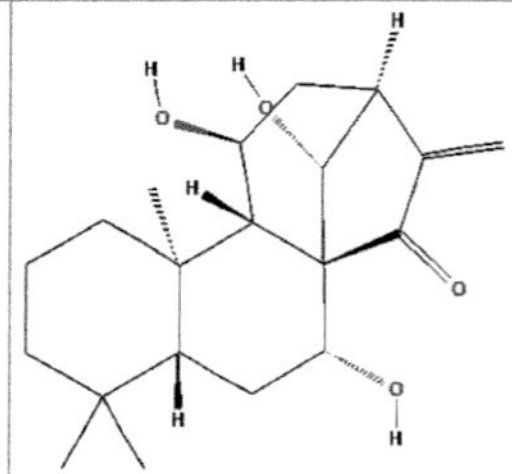	
				Rabdoloxin B	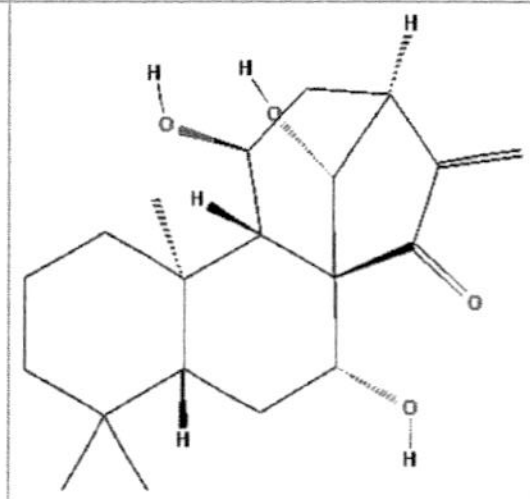	
				Xerophilusin B	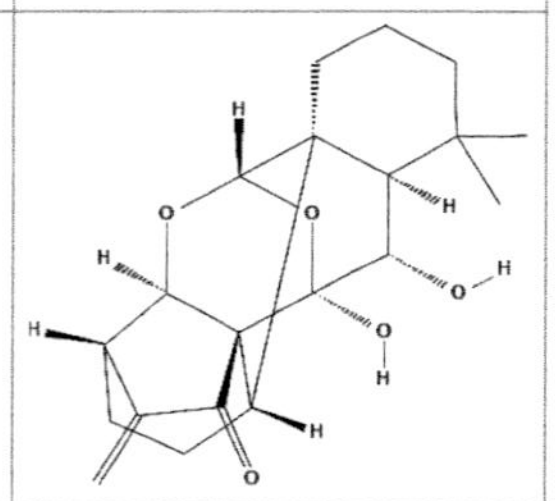	

	Isodon rubescens (Hemsley) H. Hara		Stem , bark, and fruit	Oridonin	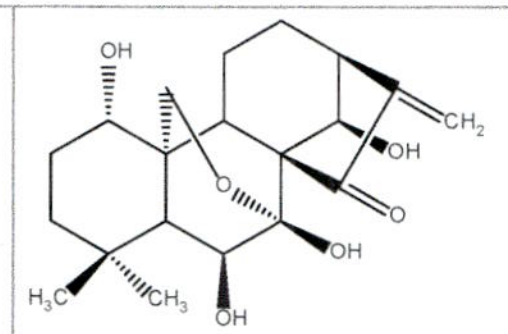	Li et al. (2011)
	Synonym(s): *Isodon rubescens* var. *eglandulosus* C.Chen. **Common name(s):** Chinese sage bush. **Botanical description:** *Isodon rubescens* is native to China, shrubs (30–) 50–100 (–120) cm. Stems numerous, erect, brownish gray, glabrous, much branched apically; branchlets reddish, densely tomentose. Petiole 1–3.5 cm; leaf blade ovate to rhombic-ovate, 2–6 × 1.3–3 cm, membranous to papery, adaxially sparsely pilose, glandular, or subglabrous, abaxially densely. **Medicinal use(s):** The plant extract is used for treating an enlarged prostate (benign prostatic hyperplasia, BPH) and cancer, including prostate cancer.			Bisrubescensin A	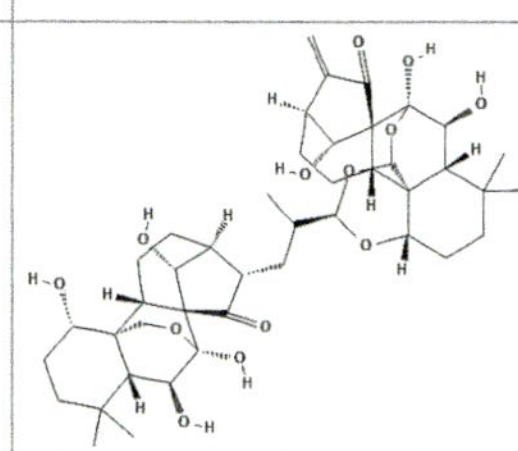	
				Hebeirubescensin H	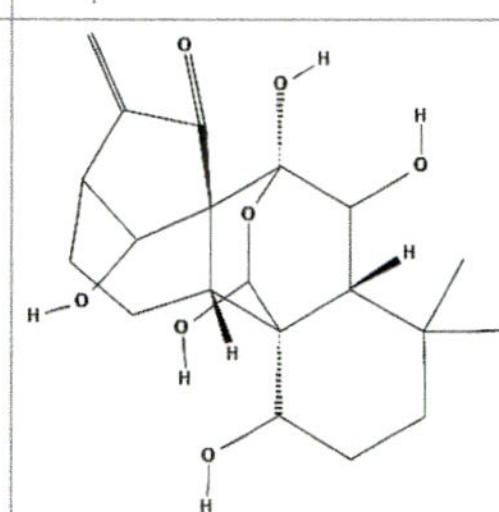	

(continued)

Table 4.77 (continued)

Family	Botanical name	Image	Plant part used	Bioactive compound(s)	Structure	Reference(s)
				Rabdoternin A	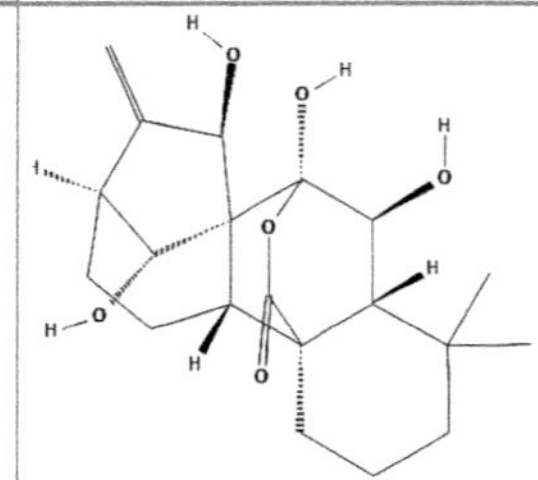	
				Rabdoternin F	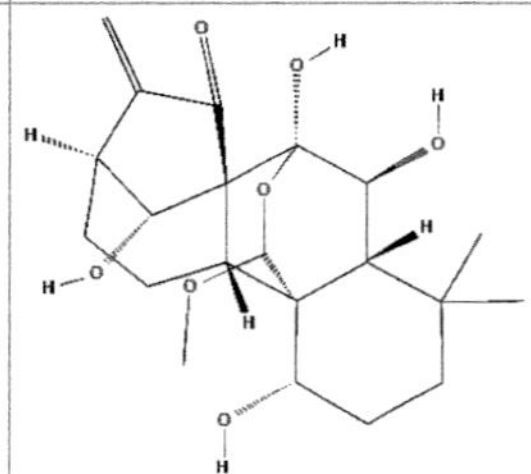	
				Rabdoternin E	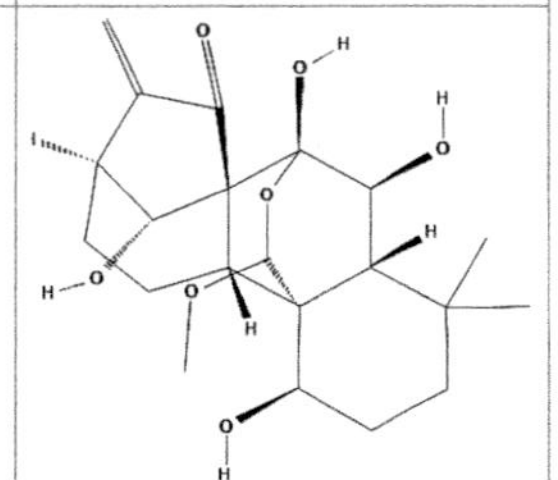	

		Lasiokaurin	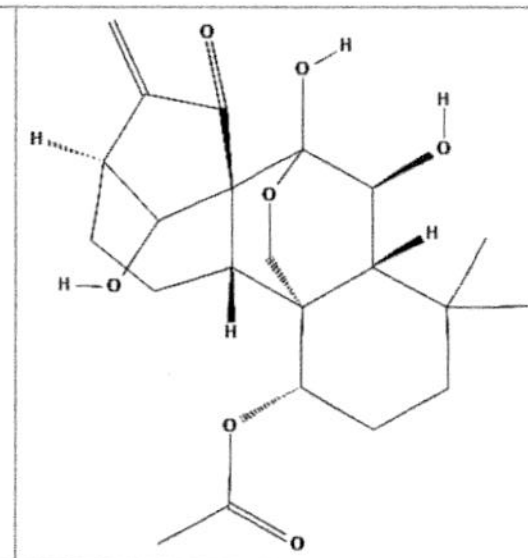	
		Ponicidin	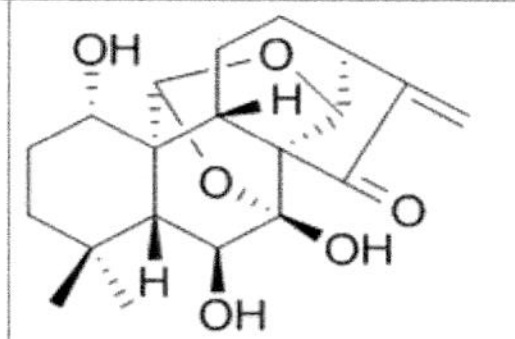	
		15α-Hydroxy-20-oxo-6,7-seco-ent-kaur-16-en-1,7α(6,11α)-Diolide	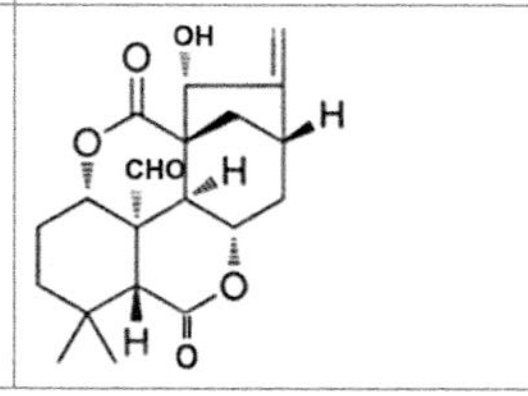	

(continued)

Table 4.77 (continued)

Family	Botanical name	Image	Plant part used	Bioactive compound(s)	Structure	Reference(s)
				Rabdosin A	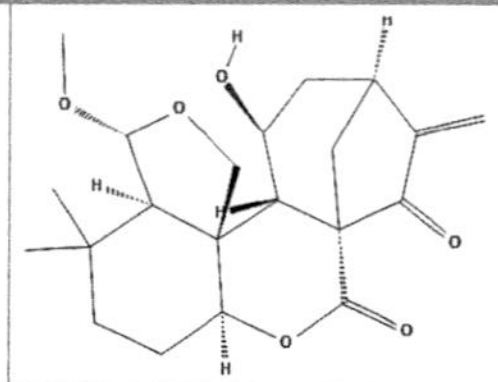	
				Isodocarpin	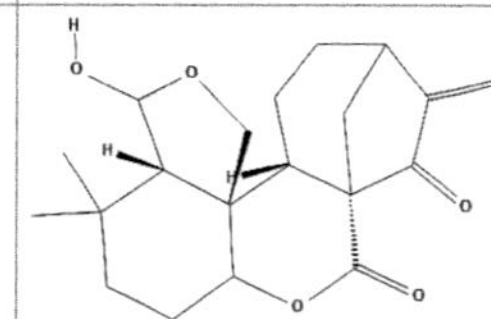	
				Jianshirubesin D	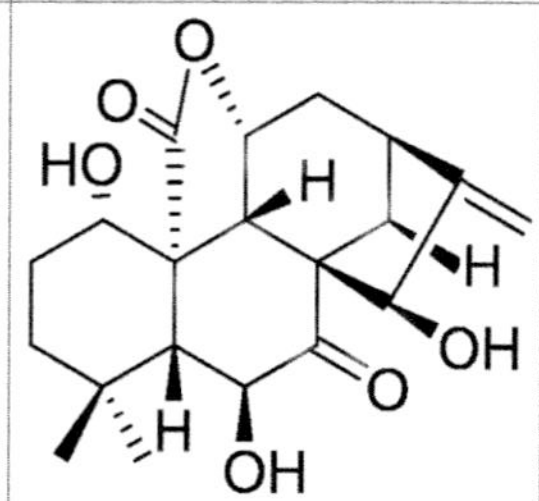	

	Isodon wikstroemioides (Hand.Mazz.) H.Hara		Leaf	Isowikstroemin A	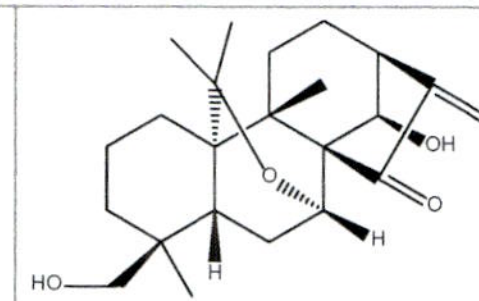	Wu et al. (1993a, b) and Wu et al. (2014)
	Synonym(s): *Plectranthus wikstroemioides* Hand.-Mazz., *Rabdosia wikstroemioides* (Hand.-Mazz.) H.Hara. **Common name(s):** Not available. **Botanical description:** *Isodon wikstroemioides* is a shrub, 0.6–1.5 m, much branched and native to China. Branches grayish brown, glabrescent; branchlets brown, densely glandular tomentulose. Petiole 1–4 mm, densely tomentulose, glandular; leaf blade lanceolate to oblanceolate, 0.8–1.5 cm × 5–7 mm, papery, adaxially densely tomentulose, papillate glandular, abaxially gray, densely floccose-tomentose. **Medicinal use(s):** Due to the presence of bioactive compounds, the plant possesses anticancer and anti-inflammatory properties and is used to cure breast cancer.					

(continued)

Table 4.77 (continued)

Family	Botanical name	Image	Plant part used	Bioactive compound(s)	Structure	Reference(s)
	Lamium purpureum L.		Flower	Flavonol 3-O-glucoside	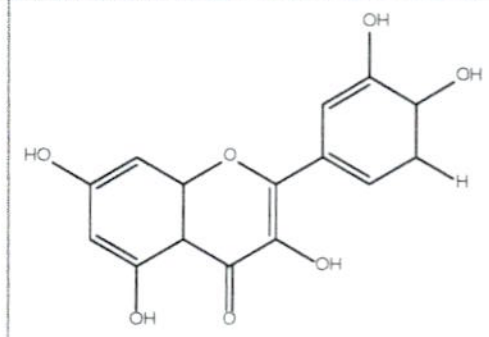	Suzuki et al. (2004)
	Synonym(s): *Lamium purpureum* var. *albiflorum* Dumort., *Lamium purpureum* subsp. *caucasicum* (Grossh.) Menitsky, *Lamium purpureum* var. *decipiens* Sond. ex W.D.J.Koch. **Common name(s):** Red deadnettle, Red dead-nettle, Purple dead-nettle, Red henbit, Purple archangel, or Velikdenche. **Botanical description:** *Lamium purpureum* is an herbaceous flowering plant native to Europe and Asia. It grows with square stems to 5–20 cm [3] (rarely 30 cm) in height. The leaves have fine hairs, are green at the bottom, and shade to purplish at the top; they are 2–4 cm long and broad, with a 1–2 cm petiole (leaf stalk), and wavy to serrated margins. **Medicinal use(s):** The whole plant is used as astringent, diaphoretic, diuretic, purgative, and styptic. The plant extract is used to treat liver and skin cancer.					
	Leonurus heterophyllus L.		Apical fruit	Leoheterin	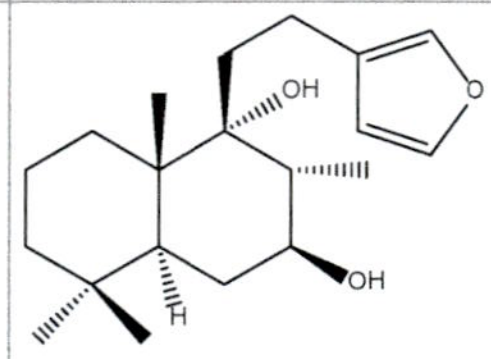	Morit et al. (1997) and Hung et al. (2011)

	Synonym(s): *Leonurus heterophyllus* Sweet, *Leonurus heterophyllus* f. *leucanthus* C.Y.Wu & H.W.Li. **Common name(s):** Oriental motherwort or Chinese motherwort. **Botanical description:** *Leonurus heterophyllus* is an herbaceous flowering plant native to Asia, including Korea and Japan and China to Cambodia. **Medicinal use(s):** The plant is a Chinese anticancer herbal medicine recommended for use in treating various cancers such as hepatoma, esophageal cancer, and stomach cancer.	Hispanone	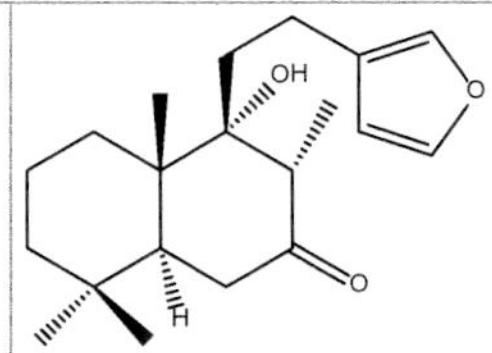	
		Galeopsin	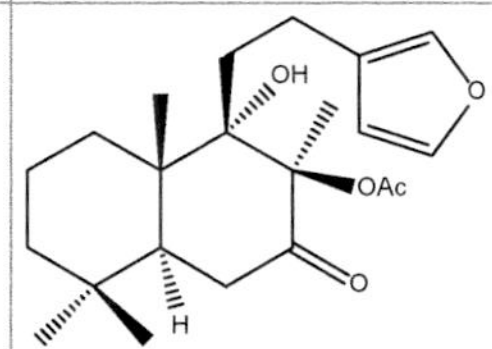	
		Leoheteronin A	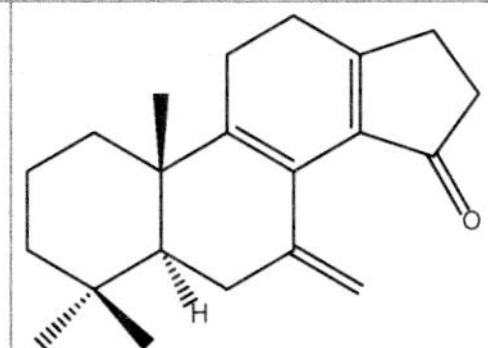	
		Leoheteronin D	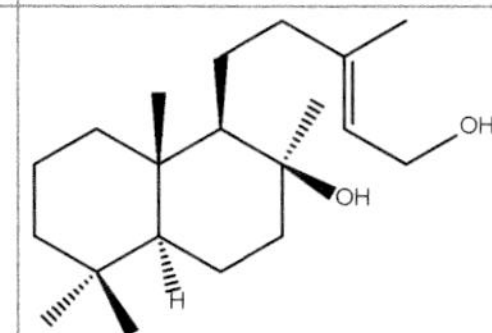	
		Leopersin G	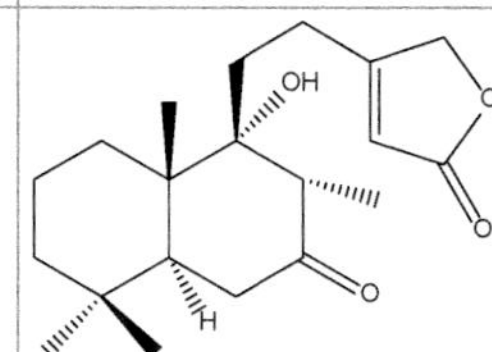	

(continued)

Table 4.77 (continued)

Family	Botanical name	Image	Plant part used	Bioactive compound(s)	Structure	Reference(s)
	Lepechinia caulescens (Ortega) Epling		Leaf	Ursolic acid	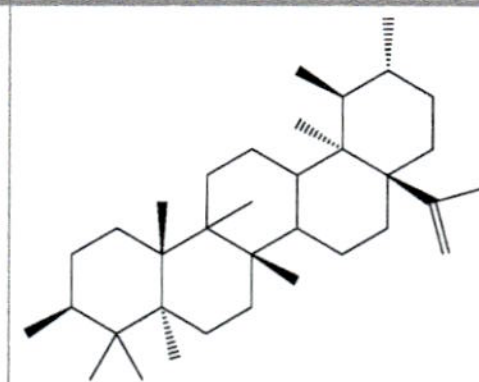	Moura et al. (2002) and Ramírez et al. (2018)
	Synonym(s): *Lepechinia spicata* Willd. **Common name(s):** Pitcher sage. **Botanical description:** *Lepechinia caulescens* can be found in Central and South America, Mexico, California, Hispaniola, and Hawaii, although the species in Hawaii is probably a human introduction. Many of them bear attractive pitcher-shaped flowers, often in shades of purple. **Medicinal use(s):** The dried leaves have been used as a poultice to stem hemorrhaging, while the fresh bruised leaves can be applied to external cuts and wounds. Plant extract is used to cure cervical, breast, and lung cancer.			Oleanolic acid	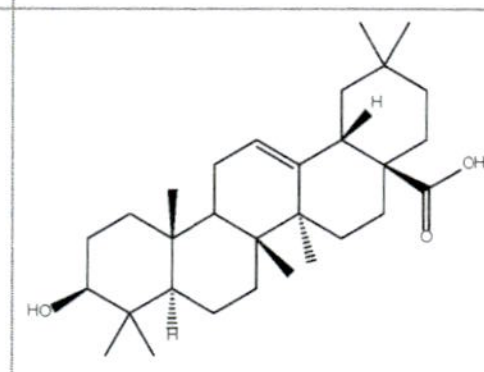	

Lamiaceae	*Glechoma hederacea* L.		Apical part, fruit juice	Germacrene D	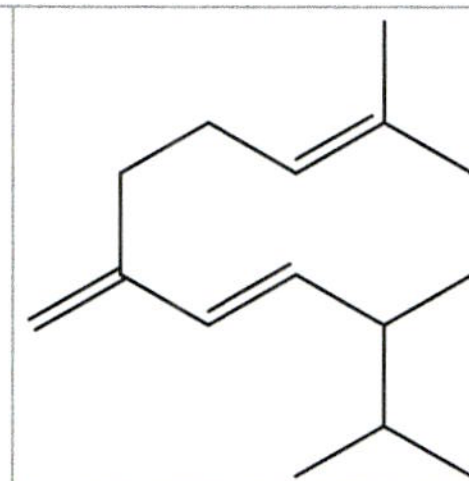	Culpeper (1950) and Mockute et al. (2005)
	Synonym(s): *Glechoma hederacea* var. *sardoa* Bég., *Glechoma hederacea* var. *micrantha* (Boenn. ex Rchb.) Nyman, *Glechoma hederacea* var. *magna* (Mérat) Lej., *Glechoma hederacea* var. *longituba* Nakai. **Common name(s):** *Creeping charlie, Ground ivy, Gill-over-the-ground, Alehoof, Tunhoof, Catsfoot, Field balm, Run-away-robin, Creeping jenny*. **Botanical description:** *Glechoma hederacea* is native to North America. It can be identified by its round to reniform (kidney or fan shaped), crenate (with round toothed edges) opposed leaves 2–3 cm (0.79–1.18 in.) diameter, on 3–6 cm (1.2–2.4 in.) long petioles attached to square stems which root at the nodes. Like crabgrass, creeping charlie's root has a ball. It is a variable species, its size being influenced by environmental conditions, from 5 to 50 cm (2.0–19.7 in.) tall. **Medicinal use(s):** The plant is used to treat inflammation of the eyes and tinnitus. It is also used as a diuretic, astringent, tonic, and gentle stimulant and to treat prostate cancer.			(Z)-β-ocimene	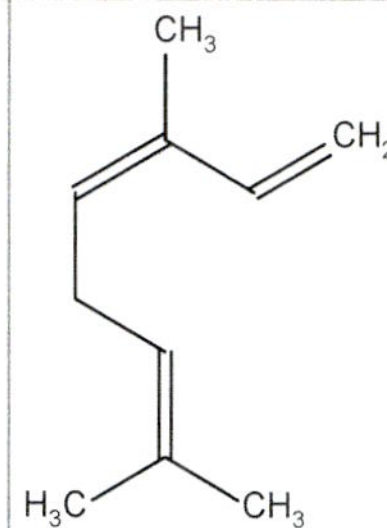	
				1.8-Cineole	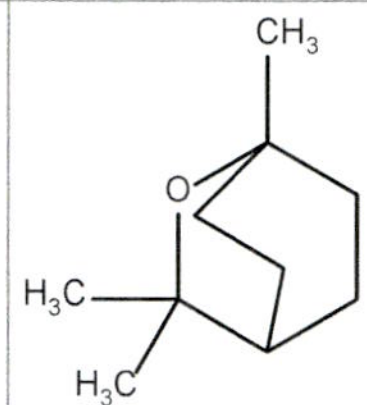	

(continued)

Table 4.77 (continued)

Family	Botanical name	Image	Plant part used	Bioactive compound(s)	Structure	Reference(s)
	Leonurus artemisia (Lour.) S.Y.Hu		Shoots	Leonurine	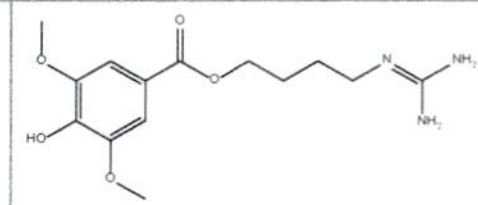	Hu (1976) and Tokuda et al. (1986)
	Synonym(s): *Leonurus artemisia* (Lour.) S.Y.Hu, *Leonurus artemisia* var. *albiflorus* (Migo) S.Y.Hu. **Common name(s):** Motherwort. **Botanical description:** Motherwort is a plant that is native to Asia but grows all around the world. Motherwort has been used most widely for women's health, including childbirth, afterbirth care, irregular menstrual cycles, and premenstrual anxiety, as well as to treat heart disorders. **Medicinal use(s):** *Leonurus artemisia* is used in traditional Chinese medicine to treat menstrual disorders and to facilitate parturition. In Eastern China, women consume a syrup made from motherwort to promote recovery of the uterus after childbirth. This plant is also used to treat uterine cancer.			Ursolic acid	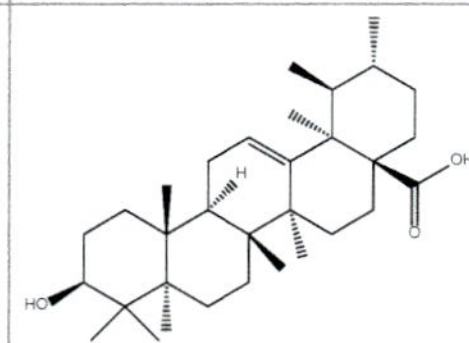	
	Lepechinia caulescens (Ortega) Epling		Leaves, stem	Ursolic acid	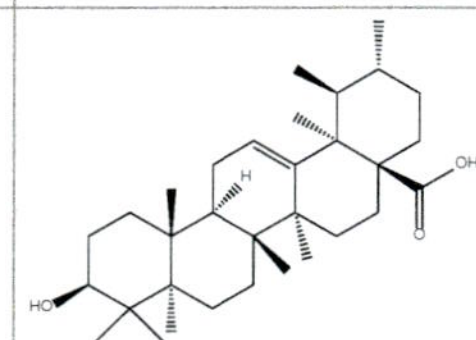	Ishikura (1982) and Ramírez et al. (2018)

	Synonym(s): *Lepechinia spicata* Willd., *Ulricia pyramidata* Jacq. ex Steud. **Common name(s):** Ortega Epling. **Botanical description:** *Lepechinia caulescens* is found in Central and South America, Mexico, California, Hispaniola, and Hawaii. It is a perennial herb (50 cm), rhizomatous and prostrate, often erect. Leaves are 4–12 × 2–5 cm, narrow to broadly oval, rough, glabrous to scatteredly hairy, truncated to obtuse base, jagged margins, obtuse or acute apex; petioles 0–3 cm. Inflorescences are 1.5–6 cm long, on peduncles of 1.5–3.5 cm, terminal and axillary. Corolla 5–8 mm, white, slightly 2-labiated, 5-lobed (2/3), short, sub-equal wolves; straight tube, with a trichome ring internally. Stamens are as long as the corolla. **Medicinal use(s):** *Lepechinia caulescens* has many functions such as promoting circulation, clearing clots, detoxification and treating breast pain and swelling. It has been used for breast cancer treatment in ancient China.			Oleanolic acid	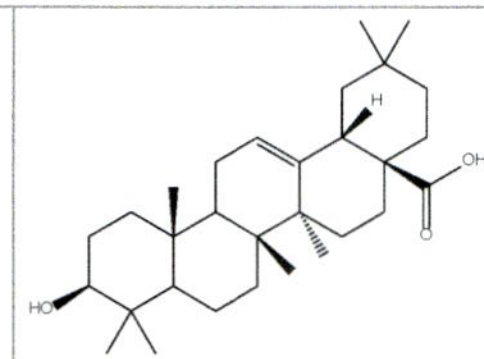	
	Leucas decemdentata var. decemdentata		Leaf	Not available		Holdsworth (1991)

(continued)

Table 4.77 (continued)

Family	Botanical name	Image	Plant part used	Bioactive compound(s)	Structure	Reference(s)
	Synonym(s): Isodeca flaccida (R.Br.) Raf., Leucas chinensis var. oliganthos Hassk., Leucas flaccida R.Br., Leucas flaccida var. petiolaris Benth., Leucas javanica Benth. **Common name(s):** Not available. **Botanical description:** *Leucas flaccida* is an annual herb native to India and mostly found in South India, the Himalayas, Myanmar, and Indochina, woody at base, up to 1.5 m tall, branches 4-gonous, covered with deflexed or deflexed and spreading hairs, glabrate or hairless to densely soft hairy. Leaves are opposite, $2–10 \times 0.7–5$ cm, ovate or ovate-lance-shaped, blunt or pointed, rounded, heart-shaped or wedge-shaped at base, rounded toothed-sawtoothed, membranous, hairless to bristly above, hairless or velvet hairy to thinly bristly beneath; leaf stalks up to 2.5 cm long. **Medicinal use(s):** *Leucas decemdentata* is used as anti-inflammatory, antioxidant, antibacterial, and anticancer agent. It is used for the treatment of abdominal tumors. The crushed leaves are used to cure skin diseases such as psoriasis and scabies. They are also used to treat rheumatism, snake bites, mild fevers and colds. A decoction of leaves is used against roundworm infection.					

	Lycopus lucidus Turcz. H	Europe Asia Australia North America 	Bark	Ferulic acid	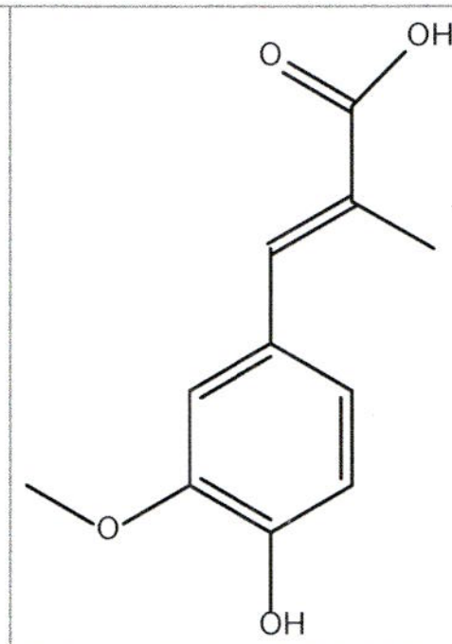	Yua et al. (2010) and Lu et al. (2018)
	Synonym(s): *Lycopus lucidus* var. *americanus* (Muhl. ex W.P.C.Barton) A.Gray, *Lycopus lucidus* subsp. *americanus* (Muhl. ex W.P.C.Barton) Hultén, *Lycopus lucidus* var. *formosanus* Hayata. **Common name(s):** Lycopi rhizoma or le zan, Gypsywort. **Botanical description:** *Lycopus lucidus* an herbaceous perennial traditional medicine in Eastern Asian countries. The root, as an edible and medicinal part from *L. lucidus*, containing high nutritional value and potential bioactivities, has recently become popular and been widely consumed in China as a typical plateau food. **Medicinal use(s):** The dried rhizomes of *Lycopus lucidus* are used in traditional Chinese medicine, having a reputation for promoting blood circulation and energy (qi) and preventing water retention. They are said to have a "warming" property and have a mild bittersweet flavor.			Caffeic acid	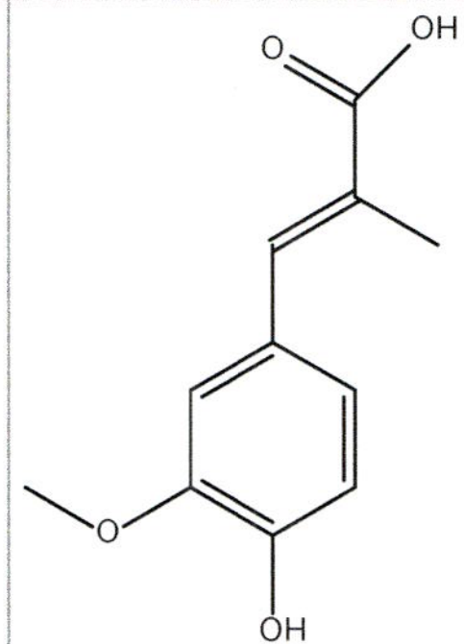	

(continued)

Table 4.77 (continued)

Family	Botanical name	Image	Plant part used	Bioactive compound(s)	Structure	Reference(s)
				Chlorogenic acid	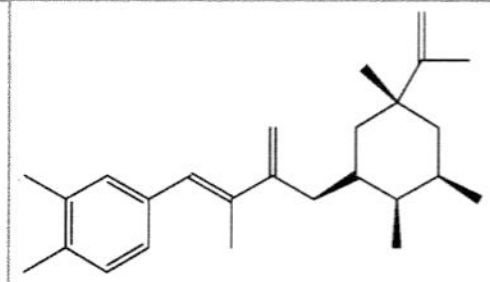	
				Vanillic acid	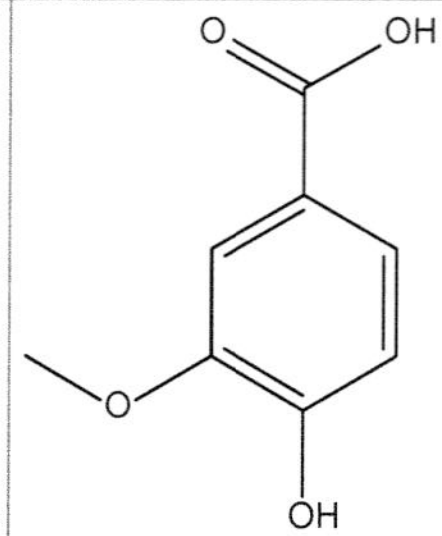 	
				p-Coumaric acid	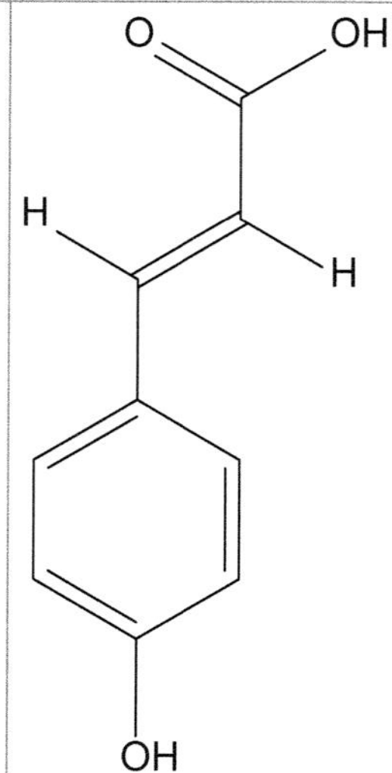 	

				Rosmarinic acid	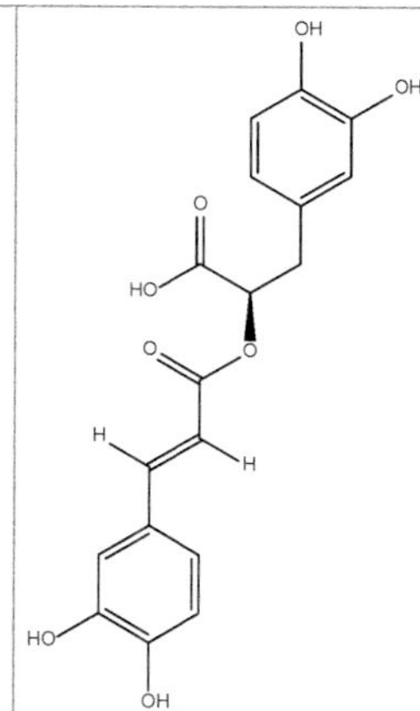	
	Marrubium parviflorum Fish et Mey.	 	Bark	Apigenin-7-O-glucoside	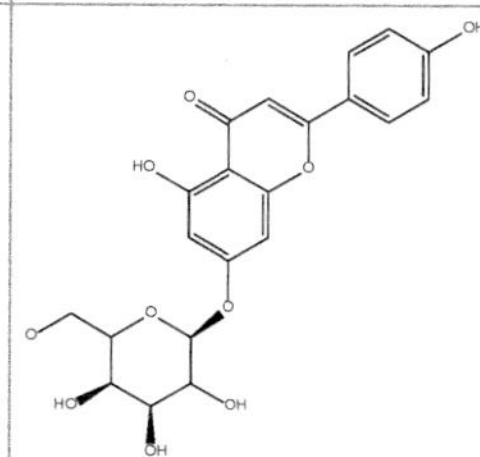	Yumrutas and Saygideg (2010)

(continued)

Table 4.77 (continued)

Family	Botanical name	Image	Plant part used	Bioactive compound(s)	Structure	Reference(s)
	Synonym(s): *Marrubium parviflorum* Fisch. & C.A.Mey., *Marrubium parviflorum* var. *oligodon* Boiss., *Marrubium parviflorum* subsp. *oligodon* (Boiss.) Seybold, *Marrubium parviflorum* subsp. *parviflorum*. **Common name(s):** Horehound. **Botanical description:** *Marrubium parviflorum* subsp. *oligodon* is a perennial herb that grows on hillsides, slopes, and steppes. It is endemic to Turkey. Fifty-one components were identified in the oil of this plant. **Medicinal use(s):** *Marrubium parviflorum* is associated with many pharmacological activities such as antioxidant, anti-inflammatory, antimicrobial activity, and anticancer activity against renal cancer. It also exhibits immunological, anti-allergic, hypoglycemic, and hypolipidemic effect.			β-Sitosterol	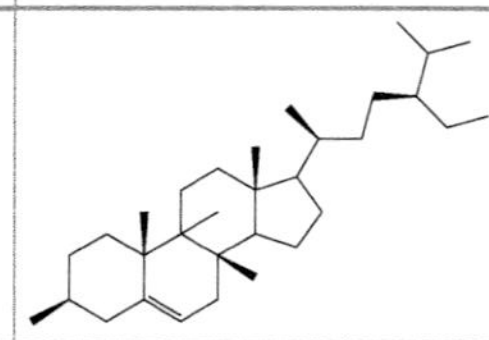	
				α-Amyrin	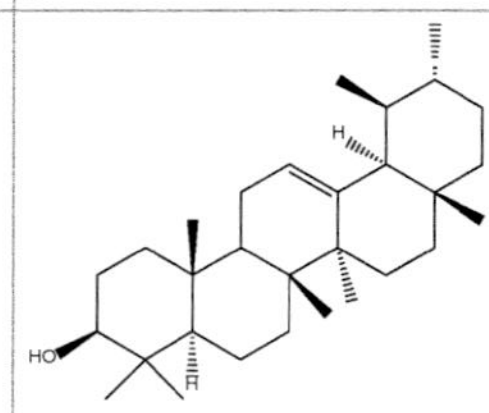	

	Melissa officinalis L.		Leaf	Rosmarinic acid	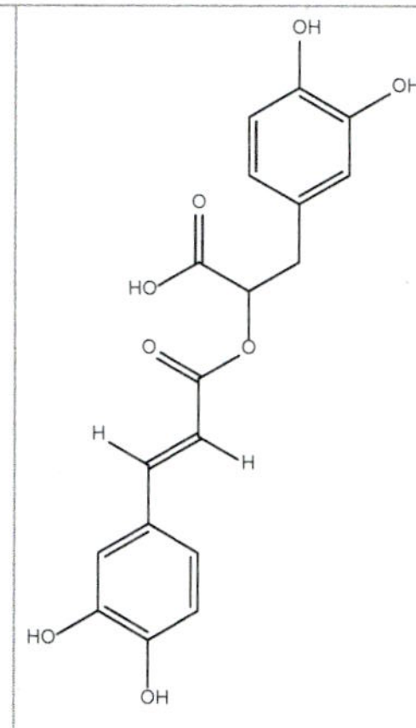	Gazola et al. (2004) and Ramanauskienė et al. (2015)
	Synonym(s): *Melissa officinalis* subsp. *altissima* (Sm.) Arcang., *Melissa officinalis* var. *altissima* (Sm.) K.Koch, *Melissa officinalis* var. *cordifolia* (Pers.) K.Koch, *Melissa officinalis* var. *foliosa* Briq. *Melissa officinalis* var. *graveolens* (Host) Nyman. **Common name(s):** Lemon balm, Balm, Common balm, or Balm mint. **Botanical description:** *Melissa officinalis* is a perennial herbaceous plant native to South-Central Europe, the Mediterranean Basin, Iran, and Central Asia but now naturalized in the Americas and elsewhere. It grows to a maximum height of 70–150 cm. **Medicinal use(s):** Lemon balm calms restlessness. It is used for treating some diseases such as attention deficit, hyperactivity disorder, swollen airways, rapid heartbeat due to nervousness, sores, tumors, uterine tumors, and breast cancer.					

(continued)

Table 4.77 (continued)

Family	Botanical name	Image	Plant part used	Bioactive compound(s)	Structure	Reference(s)
	Origanum majorana (L.) H. Karst		Leaf	Carvacrol	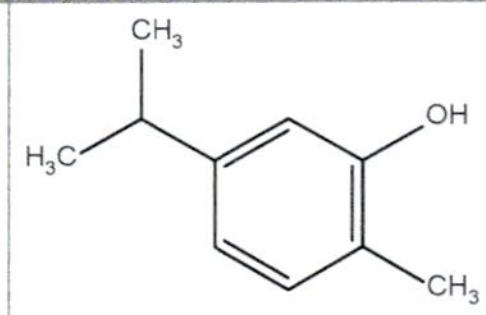 	Qari (2008), Romeilah (2009), and Farzaei et al. (2013)
	Synonym(s): *Origanum majorana* var. *majoranoides* (Willd.) Nyman, *Origanum majorana* var. *tenuifolium* Weston. **Common name(s):** Origanum vulgare, Majorana. **Botanical description:** *Origanum majorana* L. is native to Mediterranean region and cultivated in many countries of Asia, North Africa, and Europe, for example, Spain, Hungary, Portugal, Germany, Egypt, Poland, and France. It grows up to 30–60 cm. It is a perennial bushy plant. It has oblique rhizome, hairy shrub-like stalks, opposite dark green oval leaves, and white or red flowers in clustered bracts. The leaves are whole, larger ones being fragmented, oblate to broadly elliptical. This plant is widely used as a garnish and is used for different medicinal purposes in traditional and folklore medicine of different countries. **Medicinal use(s):** It is used as an antimicrobial, antiseptic, antidote, and antitussive and a carminative agent and for treating gastrointestinal disorder. The plant is also used to treat breast cancer.			α-Pinene	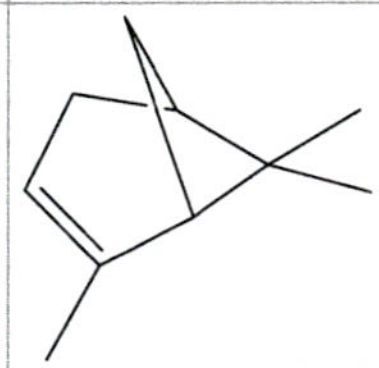	
				ρ-Cymene	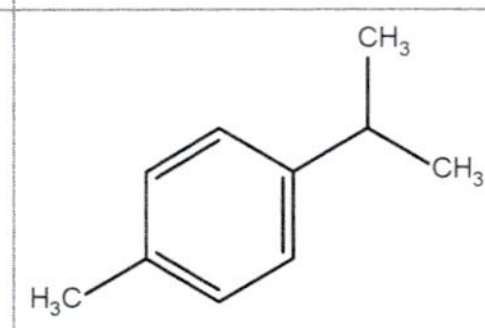 	
				Camphene	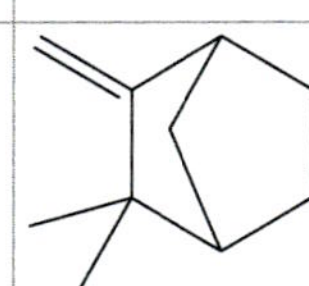	

	Origanum vulgare L.			Thymol	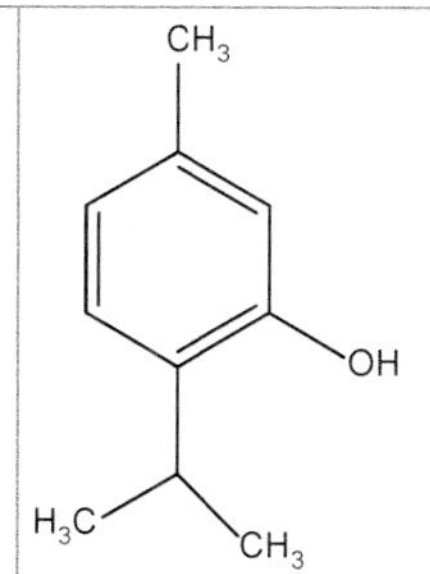 	Savini et al. (2009)
	Synonym(s): *Origanum vulgare* var. *albidum* Bellynck, *Origanum vulgare* var. *albiflorum* Lej. *Origanum vulgare* var. *album* Fraas. **Common name(s):** Oregano. **Botanical description:** *Origanum vulgare* is a flowering plant in the mint family. It is native to temperate Western and Southwestern Eurasia and the Mediterranean region. Oregano is a perennial herb, growing from 20 to 80 cm tall, with opposite leaves 1–4 cm long. The flowers are purple, 3–4 mm long, produced in erect spikes. **Medicinal use(s):** *Origanum vulgare* is used to treat head cool, sniffle, vision performance, otitis, melancholia accompanied by flatulence, and prostate, breast, and liver cancer.					

(continued)

Table 4.77 (continued)

Family	Botanical name	Image	Plant part used	Bioactive compound(s)	Structure	Reference(s)
	Isodon rubescens (Hemsl.) H.Hara		Aerial parts, leaf	Oridonin	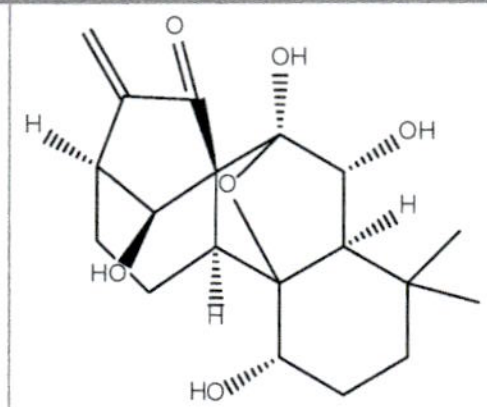	Sun et al. (1982, 1992), Li et al. (1987), and Chen et al. (2005)
	Synonym(s): *Isodon henryi* var. dichromophyllus (Diels) Kudô, *Isodon rubescens* var. eglandulosus C.Chen, *Rabdosia rubescens* (Hemsl.) H.Hara, *Plectranthus rubescens* Hemsl. **Common name(s):** Donglingcao, Blushred rabdosia leaf, Rabdosiae rubescentis herba. **Botanical description:** *Isodon rubescens* is a perennial herb native to China (Yellow river valley). Glandlular trichomes are distributed over the aerial and reproductive parts of the plant. **Medicinal use(s):** *Isodon rubescens* is used in Chinese folk medicine to treat stomachache, pharyngitis, sore throat, cough and as an antitumor medicine for the treatment of esophageal, cardiac, and prostate carcinoma. It is also used to treat unilateral facial paralysis, headache, ear pain, epilepsy, cataract, weakness of sight, respiratory and gastrointestinal bacterial infections and inflammation.					

	Rosmarinus officinalis		Whole plant	Rosmaricine	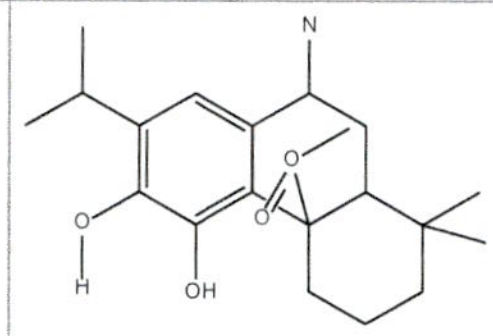	Madhuri and Pandey (2009)
	Synonym(s): *Rosmarinus angustifolius* Mill., *Rosmarinus chilensis* Molina. **Common name(s):** Rosemary. **Botanical description:** *Rosmarinus* is a perennial bush with simple linear to lanceolate leaves downward curving, and whole, often with a rough surface, either hairless or tomentose. It has erect or procumbent hairy stems and spherical gland as shown in Fig. 1. The plants have 5–15 flowers and are characterized by a lax inflorescence in axillary or cymose verticillaster. **Medicinal use(s):** *Rosmarinus* is used to treat many diseases like dyspnea, cardiac pain, dysrhythmia, and cramp and is very effective against breast and skin cancer.			Carnosic acid	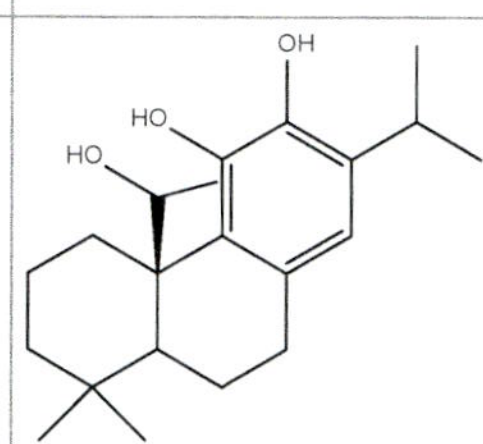	
				Carnosol	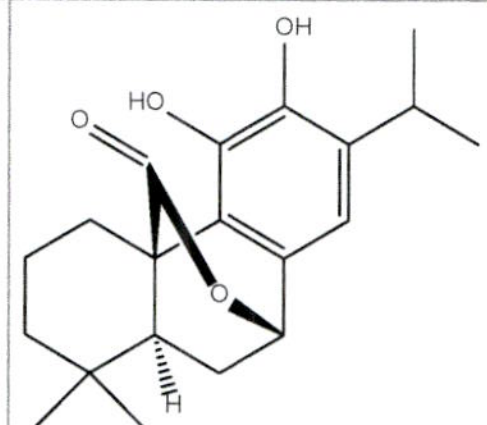	
				Rosmanol	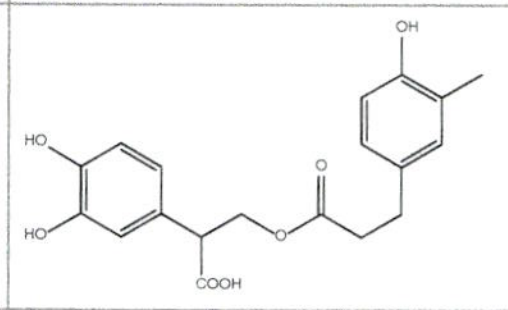	

(continued)

Table 4.77 (continued)

Family	Botanical name	Image	Plant part used	Bioactive compound(s)	Structure	Reference(s)
	Salvia chinensis Benth		Entire plant	Boswellic acid	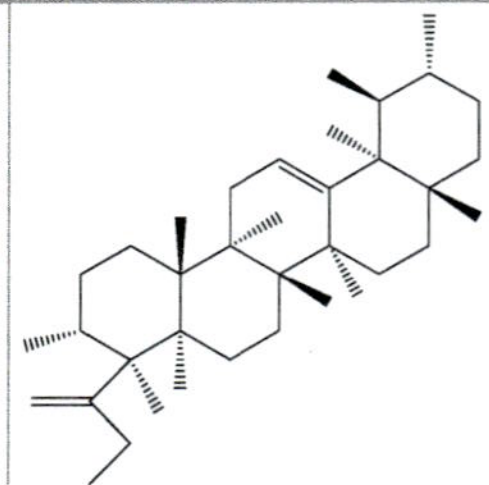	Qian and Li (1992) and Zhao et al. (2015a, b)
	Synonym(s): *Salvia chinensis* f. *alatopinnata* Matsum. & Kudô, *Salvia chinensis* var. *crenata* Makino, *Salvia chinensis* var. *intermedia* Makino. **Common name(s):** Hijianchuan Chinese sage. **Botanical description:** Chinensis is an annual plant that is native to several provinces in China, including Hubei, Sichuan, Guangxi, Guangdong, and Hunan, and grows in forests and in clusters of grass on hillsides or plains at 100 and 500 m elevation. *S. chinensis* grows on stems, which are erect or prostrate, up to a height of 20–60 cm. **Medicinal use(s):** The plant is used to treat obstruction of the large intestine and liver and kidney cancer and as an emmenagogue and a strangury agent.			Blumenol A	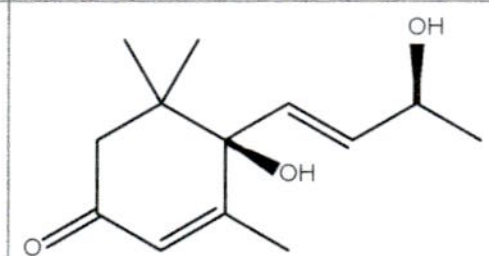	
				Pinafaenoic acid	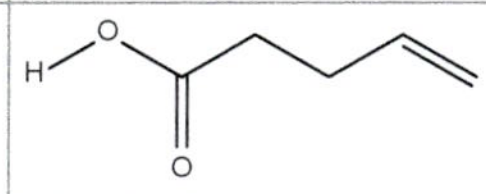	
				Salvianolic acid B	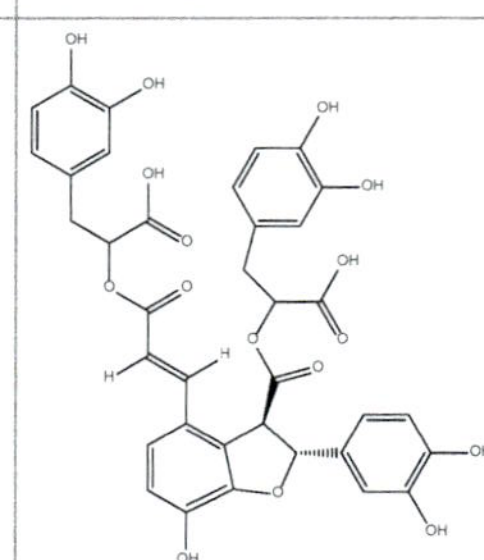	

				Salvianolic acid D	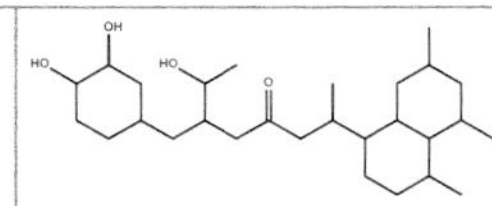	
				5,7,4′-Trihydroxydihydroflavonol	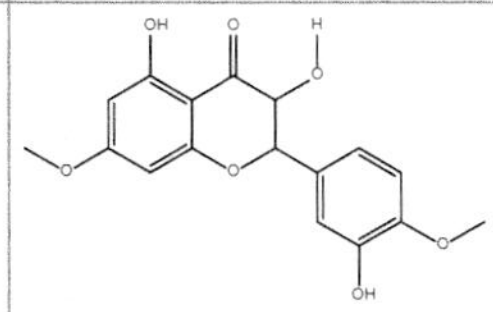	
				Protocatechuic acid	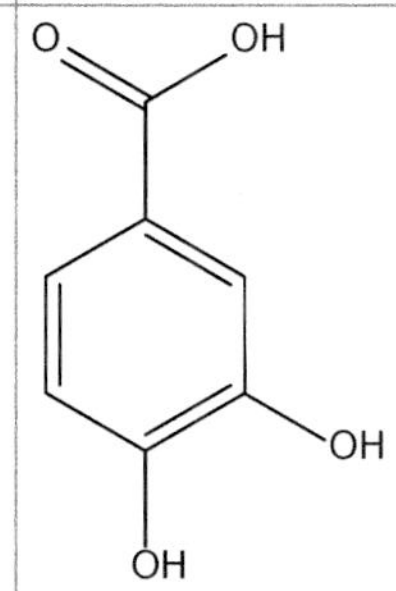	
				3,5,7-Trihydroxychromone	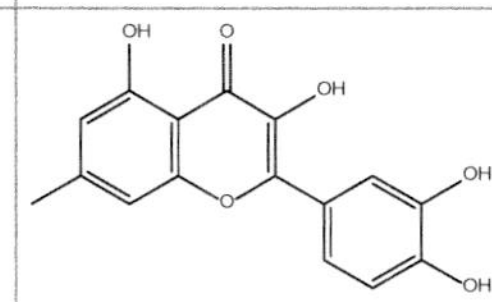	

(continued)

Table 4.77 (continued)

Family	Botanical name	Image	Plant part used	Bioactive compound(s)	Structure	Reference(s)
	Salvia hiptoides Mart. and Gal.		Entire plant	Tanshinone	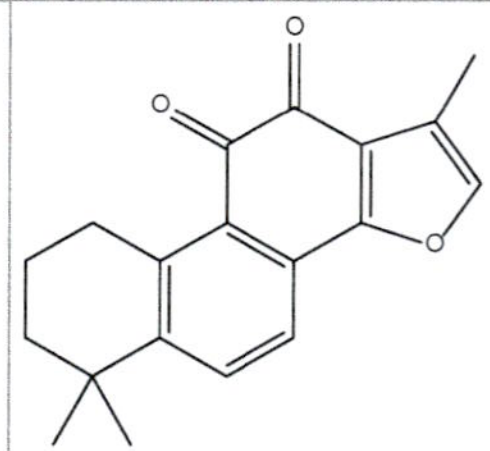	Hirschhorn (1982) and Hao et al. (2018)
	Synonym(s): *Salvia abbottii* Urb. **Common name(s):** Common sage. **Botanical description:** *Salvia hiptoides* is an herbaceous woody perennial growing up to 60 cm (2 ft.) tall, bears aromatic leaves that are the source of the culinary herb, and is native to the Old World and the Americas, with three distinct regions of diversity: Central and South America, Central Asia, and Mediterranean Eastern Asia. **Medicinal use(s):** *Salvia hiptoides* is used to cure dropsy, spondylolisthesis, groin pain, back pain, and liver cancer.			Cryptotanshinone	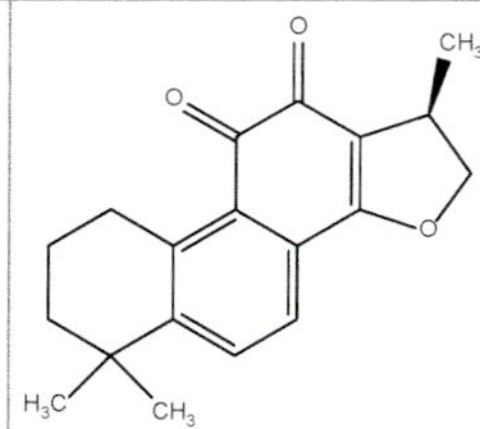	
	Salvia multicaulis Boiss.		Aerial parts	α-Humulene	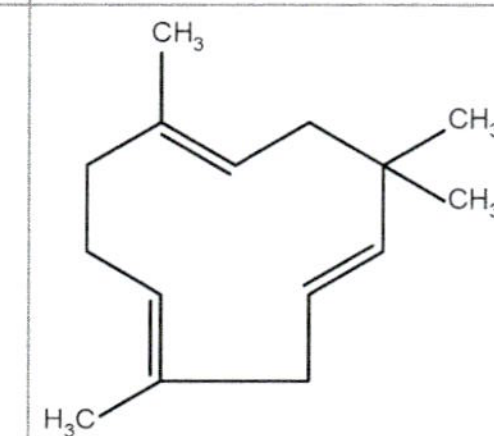	Loizzo et al. (2007)
	Synonym(s): *Salvia abyssinica* L.f, *Salvia pinardi*. **Common name(s):** Garden sage. **Botanical description:** *Salvia multicaulis* is a low-growing perennial shrub native to Turkey and bordering countries. Plants grow into mats up to 90 cm wide, with erect woody stems. The plant reaches 15 cm tall, though the flowering stems reach 30–45 cm. **Medicinal use(s):** The plant is very effective in the treatment of fatigue, freckle, migraine, and brain tumors.					

	Salvia plebeia Vahl.		Aerial parts	Salplebeone A	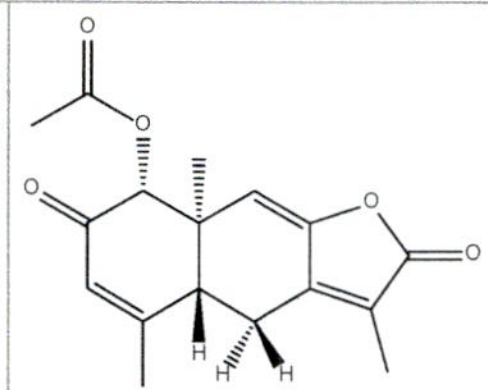	Um et al. (1996), Bae et al. (2007), and Ma et al. (2017)
	Synonym(s): *Salvia plebeia* var. *latifolia* E.Peter. **Common name(s):** Sage weed, Kamrkash, Samundarsok, Sathi. **Botanical description:** *Salvia plebeia* is an annual or biennial herb that is native to a wide region of Asia. It grows on hillsides, streamsides, and wet fields from sea level to 2800 m. *S. plebeia* grows on erect stems to a height of 15–90 cm tall, with elliptic-ovate to elliptic-lanceolate leaves. Inflorescences are 6-flowered verticillasters in racemes or panicles, with a distinctly small corolla (4.5 mm (0.18 in.)) that comes in a wide variety of colors: reddish, purplish, purple, blue purple, blue, and rarely white. **Medicinal use(s):** *Salvia plebeia* has been used to treat hepatitis, and breast, liver, and stomach cancer.			Salplebeone B	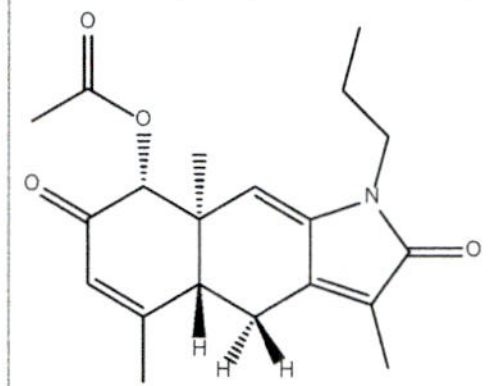	
				Salplebeone C	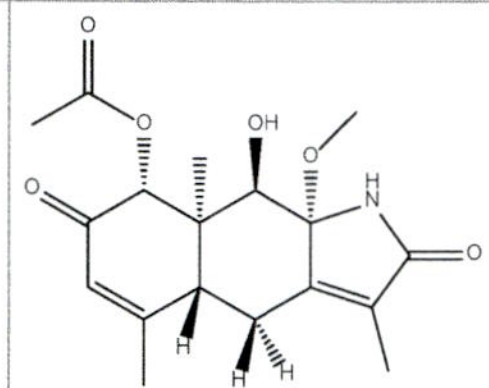	

(continued)

Table 4.77 (continued)

Family	Botanical name	Image	Plant part used	Bioactive compound(s)	Structure	Reference(s)
	Salvia sahendica Boiss. & Buhse		Apical part	Alpha-pinene	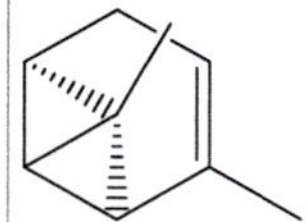	Jassbi et al. (2006)
	Synonym(s): *Salvia sahendica* subsp. *compar* Wissjul. **Common name(s):** Sage. **Botanical description:** *Salvia sahendica* is native of Iran. Aerial parts of the plant are aromatic and have been used in Iranian folk medicine. **Medicinal use(s):** The plant is used for the treatment of cervical cancer and colorectal adenocarcinoma.					
	Scutellaria barbata D.Don		Root Aerial parts	Scutellarein (6-hydroxyapigenin)	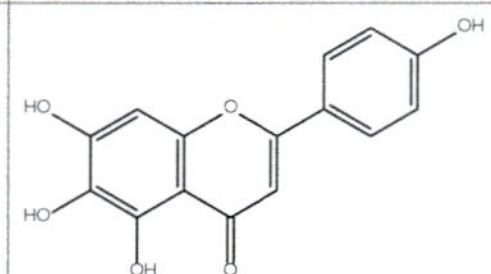	Wong et al. (1992a, b, 1993, 2009), Ducki et al. (1996), Wang et al. (1996), and Yin et al. (2004)

	Synonym(s): *Scutellaria adenophylla* Miq. *Scutellaria cavaleriei* H.Lév. & Vaniot, *Scutellaria komarovii* H.Lév. & Vaniot. **Common name(s):** Ban Zhi Lian or barbat skullcap, Barbed skullcap. **Botanical description:** *Scutellaria barbata* is a perennial herb of the Lamiaceae (mint) family that grows on moist flatlands in Southeastern China. Typically harvested in late summer when the plant is in full bloom, the aerial parts of these flowering herbs are dried to be used as medicine. **Medicinal use(s):** *Scutellaria adenophylla* is widely used to cure liver, lung, and rectal cancers.			Terpineol	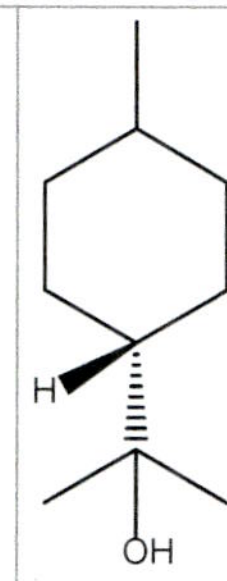	
	Scutellaria indica L.		Root	2(S)-5 , 7-Dihydroxy-8,2′-dimethoxyflavanone	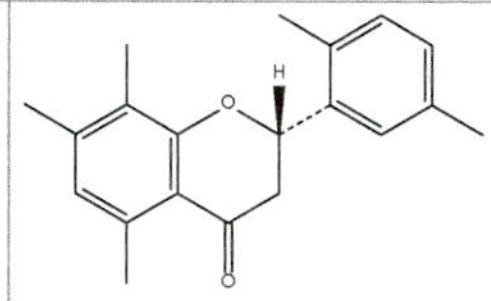	Bae et al. (1994), Tayarani-Najaran et al. (2009), and Bae et al. (1994)

(continued)

Table 4.77 (continued)

Family	Botanical name	Image	Plant part used	Bioactive compound(s)	Structure	Reference(s)
	Synonym(s): *Scutellaria indica* var. *alba* S.Kim & S.Lee, *Scutellaria indica* var. *ambigua* Hand.-Mazz. *Scutellaria indica* var. *coccinea* S.Kim & S.Lee. **Common name(s):** Han-xin- cao, Bagobo. **Botanical description:** *Scutellaria indica* is native to East Asia – China, Japan, and Korea – and is a perennial growing up to to 0.3 m (1 ft.) by 0.2 m (0 ft. 8 in.). It is hermaphrodite (has both male and female organs) and is pollinated by bees and flies. Suitable for light (sandy), medium (loamy), and heavy (clay) soils. Suitable pH: acid, neutral, and basic (alkaline) soils. It cannot grow in the shade. It prefers moist soil. **Medicinal use(s):** The plant is used to treat enlarged prostate (benign prostatic hyperplasia including prostate cancer).			Pyridine	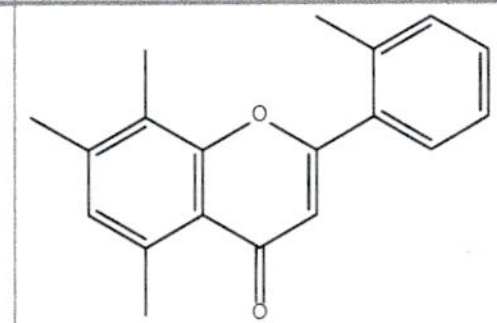	
				Wogonin (5,7-dihydroxy8-methoxyflavone)	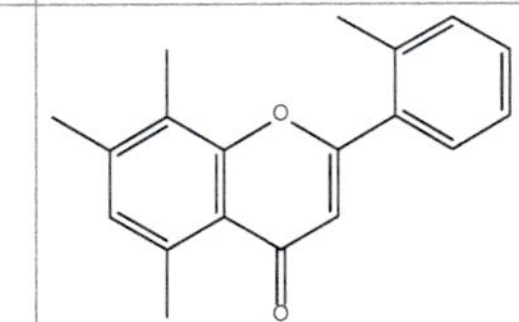	
				5, 7-Dihydroxy-8 , 2′-dimethoxyflavone	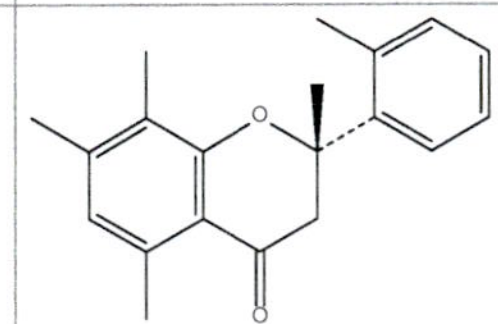	
				2(S)-S. 7,2′-Trihydroxy-8-methoxyflavanone	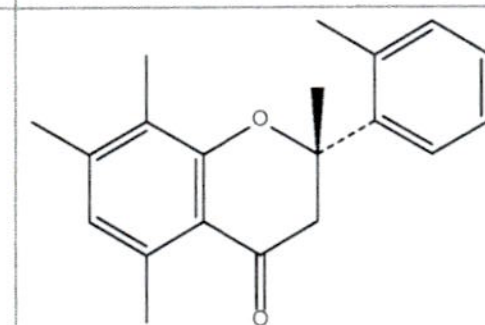	
	Scutellaria litwinowii Bornm. & Sint		Root	Baicalin	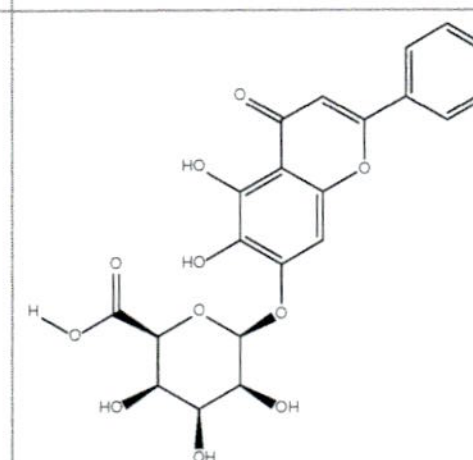	Tayarani-Najaran et al. (2009) and Tayarani-Najaran (2012)

	Synonym(s): No synonyms are recorded for this name. **Common name(s):** Skullcaps. **Botanical description:** *Scutellaria litwinowii* is one of the Iranian species of *Scutellaria*. It is erect or spreading annuals, rhizomatous and clump-forming herbaceous perennials, and more rarely, subshrubs, worldwide, excluding South Africa. **Medicinal use(s):** *Scutellaria litwinowii* inhibited the growth of malignant cells in a dose-dependent manner and is used to treat breast cancer.			Baicalein	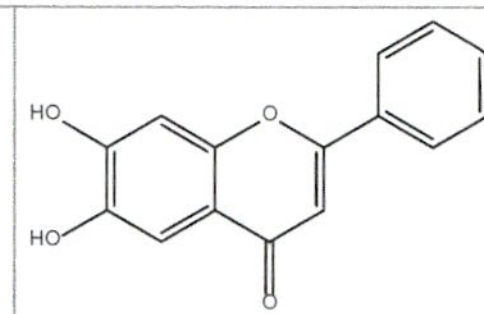	
				Wogonin	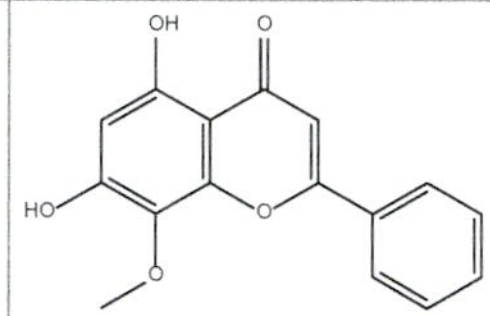	
				Ganhuangenin	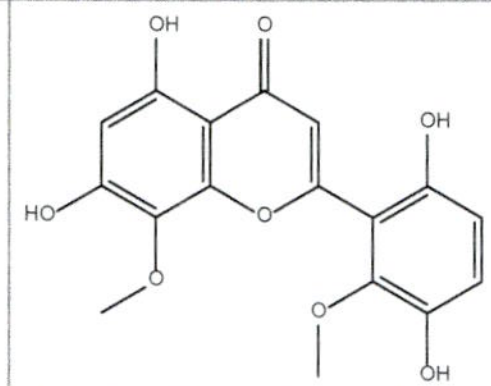	
	Scutellaria barbata D. Don		Root, aerial parts, entire plant	Hydroxyphenyl)-but-1 -en-3-one	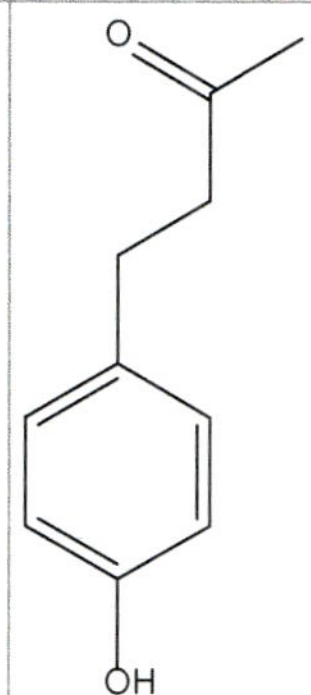	Lin (1987) and Ducki et al (1996)

(continued)

Table 4.77 (continued)

Family	Botanical name	Image	Plant part used	Bioactive compound(s)	Structure	Reference(s)
	Synonym(s): *Scutellaria rivularis* Wall. ex Benth. **Common name(s):** The barbed skullcap. **Botanical description:** *Scutellaria barbata* is native to Asia. It is a perennial herb generally reaching up to 35 cm tall, sometimes taller. The lightly toothed leaves are somewhat lance-shaped or triangular and up to about 3 cm long. **Medicinal use(s):** As an herb used in traditional Chinese medicine, it is known as Ban Zhi Lian. It has been tested in clinical trials for the treatment of metastatic breast cancer. Extracts induced apoptosis in prostate cancer cells in laboratory studies.					
	Teucrium polium L.		Leaf	Vincristine	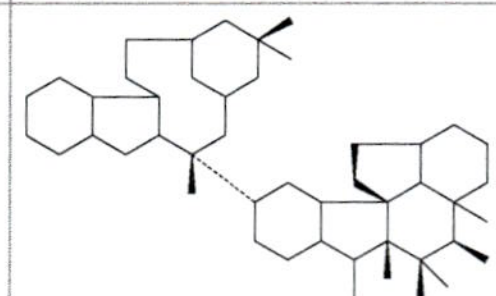	Rajabalian (2008)
	Synonym(s): *Teucrium polium* var. *achaemenis* (Schreb.) Nyman, *Teucrium polium* subsp. *aguilasense* S.Puech, *Teucrium polium* subsp. *album* (Poir.) Breistr. **Common name(s):** Felty germander. **Botanical description:** *Teucrium polium*, known popularly as felty germander, is a subshrub and herb native to the Western Mediterranean region. Its flowers are small and range from pink to white, and its leaves are used in cooking and for medicine. **Medicinal use(s):** The plant is used as an herbal remedy for inflammation, traumatic injury, toothache, soothing joints, and muscular pain and used to treat breast, lung, and liver cancer.			Vinblastine	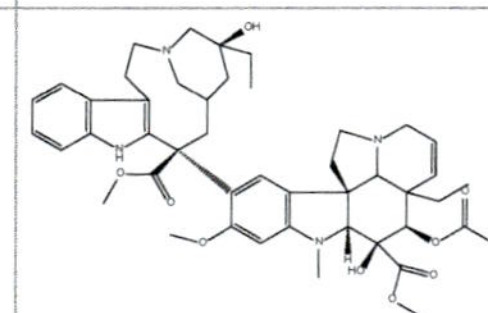	
				Doxorubicin	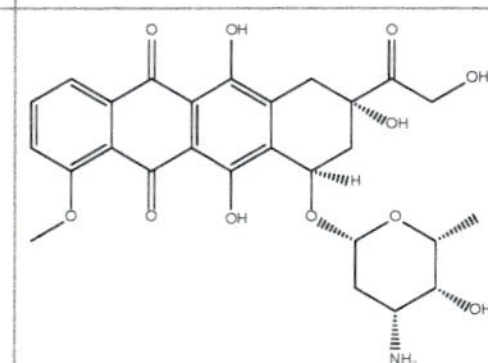	

	Thymus vulgaris		Whole plant	Thymol (2-isopropyl-5-methylphenol)	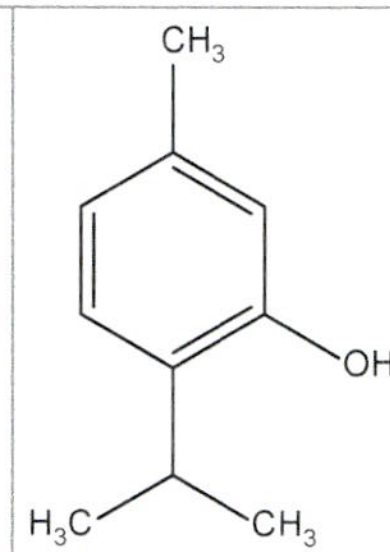	Mohammad (2006), Desai et al. (2008), Madhuri and Pandey (2009), and Javed et al. (2013)
	Synonym(s): *Thymus vulgaris* var. *aestivus* (Reut. ex Willk.) Font Quer, *Thymus vulgaris* var. *capitatus* Willk,, *Thymus vulgaris* var. *capitellatus* Font Quer. **Common name(s):** Garden thyme. **Botanical description:** *Thymus vulgaris* is native to Southern Europe from the Western Mediterranean to Southern Italy. Growing to 15–30 cm (6–12 in.) tall by 40 cm (16 in.) wide, it is a bushy, woody-based evergreen subshrub with small, highly aromatic, gray-green leaves and clusters of purple or pink flowers in early summer. **Medicinal use(s)**: It is useful in the garden as groundcover, where it can be short-lived, but is easily propagated from cuttings. It is also the main source of thyme as an ingredient in cooking and as an herbal medicine. It is slightly spicier than oregano and sweeter than sage and can cure cold, bronchial coughs, and asthmatic whooping. The plant extract is used to treat liver and prostate cancer.			Carvacrol (5-isopropyl-2-methylphenol	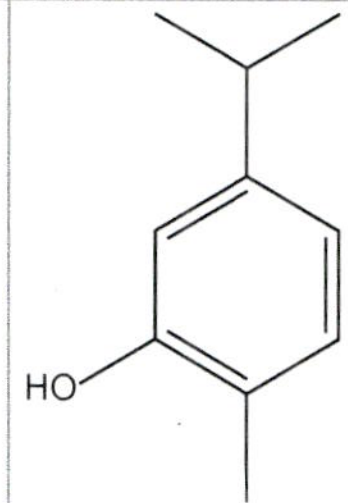	
				Eugenol (2-methoxy-4-(2-propenyl) phenol)	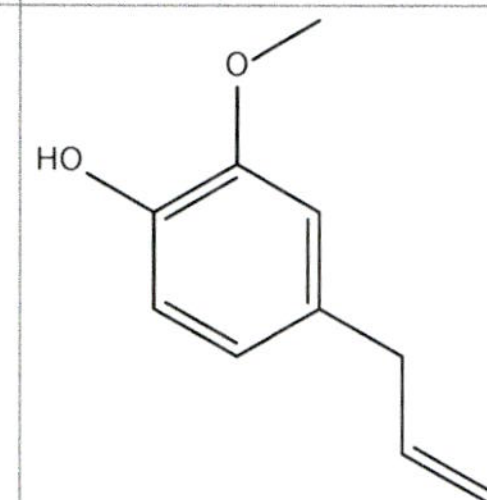	

(continued)

Table 4.77 (continued)

Family	Botanical name	Image	Plant part used	Bioactive compound(s)	Structure	Reference(s)
	Zhumeria majdae Rech. f. and Wendelbo		Entire plant	12,16-Dideoxy aegyptinone B	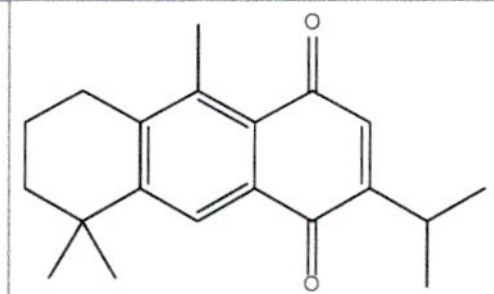	Zargari (1992) and Moein et al. (2008)
	Synonym(s): No synonyms are recorded for this name. **Common name(s):** Mehrkhosh. **Botanical description:** *Zhumeria majdae* grows in Southeastern Iran. This plant has a strong and pleasant odor. **Medicinal use(s):** *Z. majdae* is used for treating stomachache and dysmenorrhea and has been reported in traditional medicine. Its antinociceptive and anti-inflammatory activities have been recognized by researchers. *Z. majdae* is used for the treatment of various disorders including diarrhea, cold, acid reflux, and headache and is applied as a carminative for wound healing and anticancer to fight against stomach, bile, and liver cancer.					

Table 4.78 Anticancer plants of family Lardizabalaceae

Family	Botanical name	Image	Plant part used	Bioactive compound(s)	Structure	Reference(s)
Lardizabalaceae	*Akebia quinata* (Houtt.) Decne.		Fruit	Diosgenin	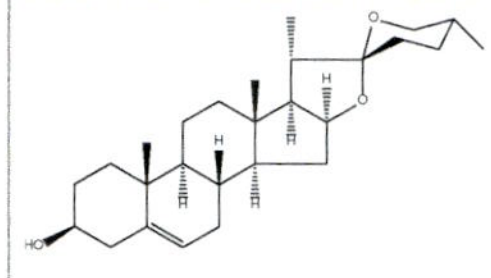	Mohammad (2006), Desai et al. (2008), and Madhuri and Pandey (2009)
	Synonym(s): *Akebia micrantha* Nakai, *Rajania quinata* Thunb. ex Houtt. **Common name(s)**: Chocolate vine. **Botanical description**: It is a deciduous, twining, woody vine that rapidly grows to 20–40 in. and is native to Japan, China, and Korea. Palmate leaves, each with 5 elliptic to oblong-obovate leaflets (1–3″ long) which are dark green above and glaucous below. Small chocolate-purple flowers bloom in drooping axillary racemes in spring. Flowers are monoecious, with both staminate (male) and pistillate (female) flowers appearing in the same raceme. Flowers are often hidden by the foliage but are quite interesting on close inspection and have a pleasant fragrance. **Medicinal use(s)**: It is used to treat breast cancer. The stem controls bacterial and fungal infections and is used for the treatment of urinary tract infections. It also possesses antitumor, antibacterial, anti-inflammatory, and antimicrobial properties.			Fenugreekine	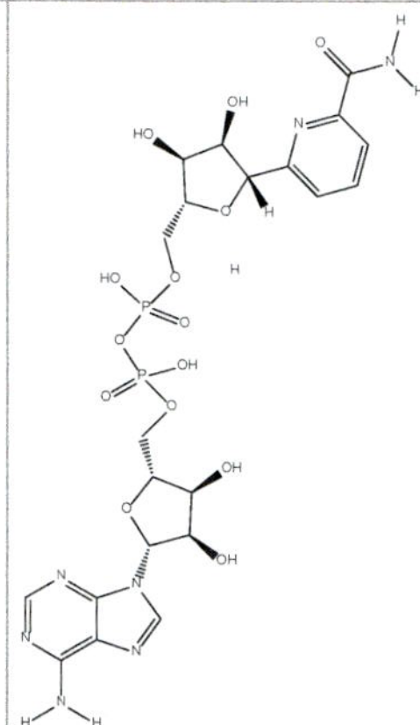	

Table 4.79 Anticancer plants of family Lauraceae

Family	Botanical name	Image	Plant part used	Bioactive compound(s)	Structure	Reference(s)
Lauraceae	*Ocotea acutifolia* (Nees) Mez		Leaf, stem, bark	(+)-6S-ocoteine N-oxide	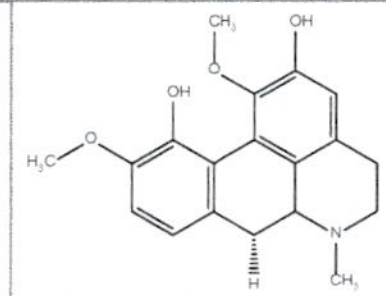	Garcez et al. (2011)
				(+)-norocoxylonine	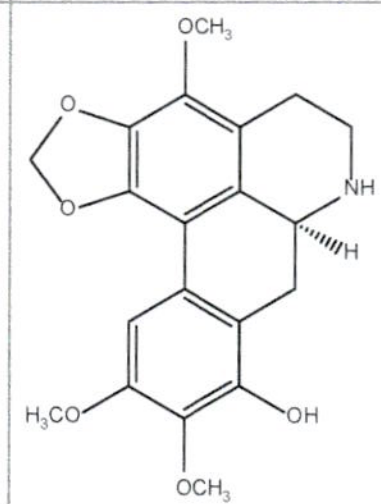	
	Synonym(s): *Oreodaphne acutifolia* Nees., *Oreodaphne acutifolia* var. *prolifera* Nees. **Common name(s)**: Black stinkwood. **Botanical description**: It is native to South America – Argentina; Uruguay; Paraguay; Southern, Central, and Eastern Brazil; and Bolivia. It is an evergreen tree with a low, dense, rounded crown; it can grow up to 5–10 m tall. The short cylindrical bole can be 20–30 cm in diameter. Seed is best sown as soon as it is ripe in a partially shaded position in a nursery seedbed. A low germination rate can be expected, with the seed sprouting within 40–60 days. The white wood is medium-textured, compact, light in weight, easy to cut, with moderate mechanical properties and a moderate resistance to wood-eating organisms. A low-value wood, it is used for making toys and boxes. The wood is used for fuel. It is habituated to Atlantic rainforest and savannah, always in more open and secondary growth areas. **Medicinal use(s)**: It is used to treat human cancer cell lines, gastric ulcers, and gastric cancers. It possesses antioxidant, antibacterial, anti-inflammatory, and antimicrobial properties.					

	Ocotea caparrapi Dugand		Oil, sap	Caparratriene	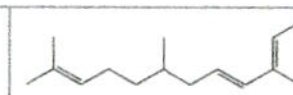	Garcia-Barriga (1974), Palomino et al. (1996), and Maffei (2003a, b, c, d, e, f, g, h, i, j, k, l, m)
	Synonym(s): No synonyms are recorded for this name. **Common name(s)**: Not available. **Botanical description**: It is native to Guyana, Surinam, French Guiana, Colombia, Costa Rica, North Brazil (Para, Amazonas), Central-West Brazil, and Venezuela. It is an evergreen tree with a low, dense, rounded crown; it can grow up to 5–10 m tall. The short cylindrical bole can be 20–30 cm in diameter. Seed is best sown as soon as it is ripe in a partially shaded position in a nursery seedbed. A low germination rate can be expected, with the seed sprouting within 40–60 days. **Medicinal use(s)**: It is used to treat vaginal cancer, bronchitis, laryngitis, and cancerous tumors. It possesses antibacterial, anti-inflammatory, antioxidant, and antimicrobial properties.					

(continued)

Table 4.79 (continued)

Family	Botanical name	Image	Plant part used	Bioactive compound(s)	Structure	Reference(s)
	Ocotea foetens (Aiton) Baill.		Branchlets	Burkinabin C	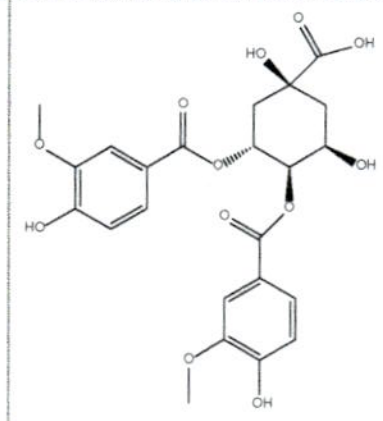	Rivera and Obon (1995) and Maffei (2003a, b, c, d, e, f, g, h, i, j, k, l, m)
	Synonym(s): *Borbonia foetens* (Aiton) J.Presl, *Keiria lutea* Bowdich, *Laurus foetens* Aiton, *Mespilodaphne foetens* (Aiton) Meisn., *Oreodaphne foetens* (Aiton) Nees, *Persea foetens* (Aiton) Spreng. **Common name(s)**: Stinkwood. **Botanical description**: It is native to Portugal and Spain. The leaves are about 9–12 cm (3.5–4.7 in.) long and 3–5 cm (1.2–2.0 in.) wide, oblong-lanceolate to almost elliptical, acuminate at the apex and slightly indented at the base. In adult plants, the leaves are leathery in texture, glossy on both sides, darker green on the upper surface, with 2 (–4) small gland-like depressions on the underside near the base. The petioles are short. The flowers of both sexes are white, with splashes of green and pale yellow, releasing a slight odor. **Medicinal use(s)**: It is used to treat stomach and vaginal cancer and tumors. It possesses antibacterial, anti-inflammatory, and antioxidant properties.					

	Ocotea leucoxylon (Sw.) Laness.		ND	Dicentrine	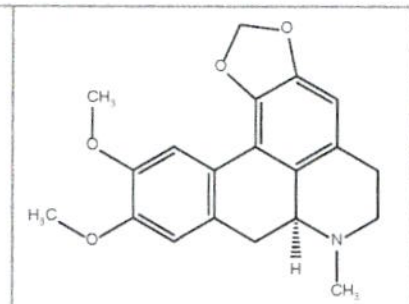	Zhou et al. (2000) and Wan Salleh and Ahmad (2017)
				Dicentrinone	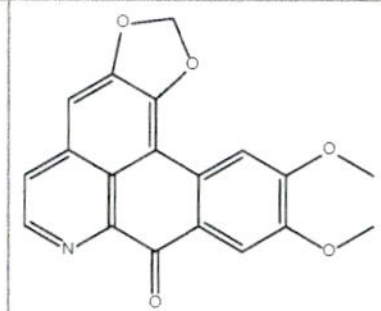	
	Synonym(s): *Laurus leucoxylon* Sw., *Ocotea duotincta* C.K.Allen, *Ocotea lasseriana* C.K.Allen, *Ocotea lenticellata* Lundell, *Ocotea persulcata* C.K.Allen, *Ocotea subsericea* Standl., *Oreodaphne leucoxylon* (Sw.) Nees, *Oreodaphne leucoxylon* (Sw.) Griseb., *Persea leucoxylon* (Sw.) Spreng. **Common name(s)**: Cigua boba. **Botanical description**: It is native to South America, Brazil, Peru, Ecuador, Colombia, Venezuela, and the Guianas; Central America, Nicaragua, and Belize; and the Caribbean, Trinidad to Cuba. It is an evergreen tree with a very dense, rounded crown; it can grow up to 15 m tall. The bole can grow up to 25 cm in diameter. Good-quality wood is harvested from the wild, mainly for local use. The plant flowers and produces fruit irregularly through the year. Dioecious species, both male and female forms, need to be grown if fruit and seed are required. **Medicinal use(s)**: It is used to treat colorectal cancer. It is used to treat headache and male urinary tract infections. It possesses antibacterial, antioxidant, anti-inflammatory, and antimicrobial properties.					

(continued)

Table 4.79 (continued)

Family	Botanical name	Image	Plant part used	Bioactive compound(s)	Structure	Reference(s)
	Ocotea usambarensis Engl.		Bark	Burkinabin C	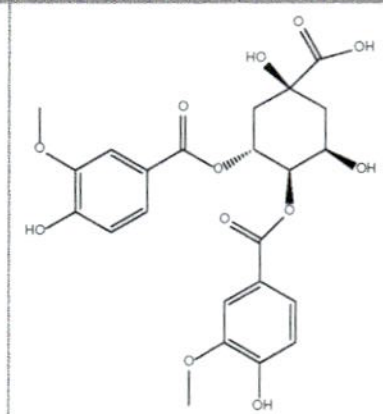 	Kokwaro (1976) and Maffei (2003a, b, c, d, e, f, g, h, i, j, k, l, m)
	Synonym(s): No synonyms are recorded for this name. **Common name(s)**: Maasi. **Botanical description**: It is native to east tropical Africa – Kenya, Tanzania, Zambia, and the Congo. It is a medium to large evergreen tree with a dense, spreading crown; usually growing 15–30 m tall but with occasional specimens to 40 m. It has a stout, straight bole that can be up to 10 m tall 2–3 m in diameter and buttressed at the base. A plant of the moist tropics, where it is found at elevations from 900 to 3000 m. It is found in areas where the mean annual rainfall ranges from 1200 to 2450 mm, often with 2–3 dry months. It prefers a mean annual temperature in the range of 15–20 °C but tolerates 8–0 °C. Grows best in a sunny position. It prefers a deep, fertile soil with good drainage. Prefers a pH in the range 5.5–6, tolerating 5–6.5. **Medicinal use(s)**: It is used for the treatment of human cancer cell lines. It is used to treat headache, malaria, back pain, stomachache, and whooping cough. It possesses antibacterial, anti-inflammatory, antioxidant, and antimicrobial properties.					

Table 4.80 Anticancer plants of family Leeaceae

Family	Botanical name	Image	Plant part used	Bioactive compound(s)	Structure	Reference(s)
Leeaceae	*Leea guineense* G.Don		Leaf	Campesterol	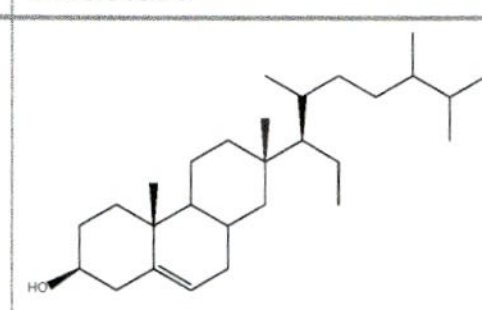	Vasileva (1969), Maffei (2003a, b, c, d, e, f, g, h, i, j, k, l, m)
	Synonym(s): No synonyms are recorded for this name. **Common name(s)**: West Indian holly. **Botanical description**: It is a tropical evergreen shrub to small tree that typically grows outdoors and is native to tropical Africa and tropical Asia. If grown indoors as a container plant, it typically grows 4–8 in. tall. This is an understory species that typically grows in its native habitat on forest floors in shady locations under the cover of taller trees where it typically adapts well to low light levels. It features 2–3 odd-pinnate compound leaves which emerge light green, sometimes with slight red tinges, but mature to a glossy green. **Medicinal use(s)**: It is used to treat cancer, urological problems, dermatological conditions, and skin tumors. It also possesses antioxidant, anti-inflammatory, and antibacterial properties.					

Table 4.81 Anticancer plants of family Liliaceae

Family	Botanical name	Image	Plant part used	Bioactive compound(s)	Structure	Reference(s)
Liliaceae	*Allium sativum* L.		Bud	Allicin	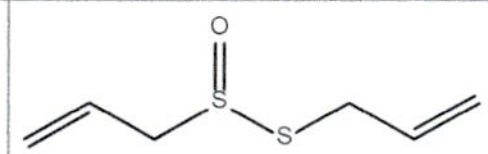	Ejaz et al. (2003)
	Synonym(s): *Allium controversum* Schrad. ex Willd., *Allium longicuspis* Regel, *Allium ophioscorodon* Link, *Allium pekinense* Prokh., *Porrum ophioscorodon* (Link) Rchb. **Common name(s)**: Garlic. **Botanical description**: It is a bulbous plant and is native to Central Asia and Northeastern Iran. It grows up to 1.2 m in height. It produces hermaphrodite flowers and pollinated by bees and other insects. It is easy to grow and can be grown year-round in mild climates. While sexual propagation of garlic is possible, nearly all of the garlic in cultivation is propagated asexually, by planting individual cloves in the ground. In colder climates, cloves are planted in the autumn, about 6 weeks before the soil freezes, and harvested in late spring or early summer. The cloves must be planted deep enough to prevent freeze/thaw, which causes mold or white rot. **Medicinal use(s)**: It is used to treat bladder cancer, tumor, gastric cancer, and dysentery. It also possesses anti-inflammatory, antioxidant, antibacterial, and antimicrobial properties.					

	Chlorophytum orchidastrum Lindl		Root	Chloromaloside D	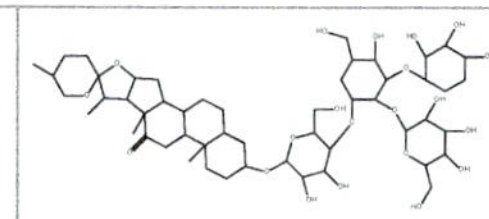	Acharya et al. (2009)
	Synonym(s): *Chlorophytum macropodum* K.Krause., *Chlorophytum mannii* Baker., *Chlorophytum petiolatum* Baker. **Common name(s)**: Spider plant, Fire flash. **Botanical description**: It is native to West Africa, particularly Sierra Leone. Its broad lance-shaped leaves are held in neat rosettes form. They are glossy and medium green and have lighter petioles or leaf stems. The roots are thick and fleshy and form small tubers. Mature plants produce long stems topped with branched spikes of small translucent white flowers. Unlike other spider plants, this one rarely forms plantlets on the flower spike; however it produces many fertile seeds. **Medicinal use(s)**: It is used to treat human cancer cell lines and tumor. It also possesses anti-inflammatory, antioxidant, antibacterial, and antitumor properties.			Galactopyranoside	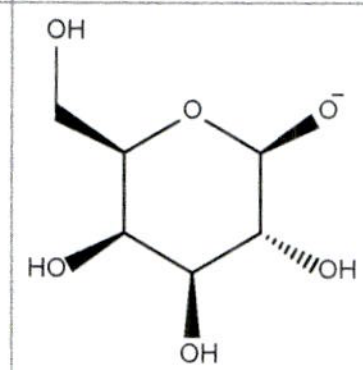	

(continued)

Table 4.81 (continued)

Family	Botanical name	Image	Plant part used	Bioactive compound(s)	Structure	Reference(s)
	Erythronium americanum Ker Gawl.		Whole plant	Alpha-methylenebu-tyrolactone	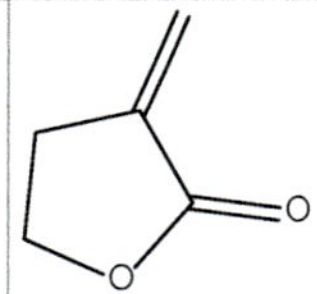	Jena et al. (2012) and Madhuri and Pandey (2009)
	Synonym(s): *Erythronium angustatum* Raf., *Erythronium aquatile* Salisb., *Erythronium aureum* Loisel., *Erythronium bracteatum* Bigelow, *Erythronium flavescens* Loisel., *Erythronium lanceolatum* Pursh, *Erythronium nuttallianum* Schult. & Schult.f. **Common name(s)**: Yellow trout lily, Yellow adder's tongue. **Botanical description**: It is native to North America and dwelling in woodland habitats. It does not flower for the first 4–7 years of its life. In any given trout lily colony, 99% plants will be nonflowering and only have one leaf and 0.5% will have paired leaves and flowers. Blooms in early spring before the trees growing above it develop leaves; this allows it to have unobstructed access to sunlight and also allows it to grow when soil nutrient levels are high. Each spring the plant will grow a single or pair of 8–23 cm long elliptic to lanceolate leaves. **Medicinal use(s)**: *Erythronium americanum* is used to treat cancer cells, tumors, gynecological diseases, general body malaise, suppressed menstruation, pruritus, ulcers, colds, skin inflammations, and intestinal worms. It also possesses anti-inflammatory, antibacterial, and antioxidant properties.					

	Fritillaria thunbergii Miq.		Fruit	Glutathione	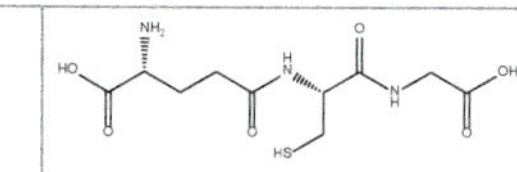	Wang et al. (2007)
	Synonym(s): *Fritillaria austroanhuiensis* Y.K.Yang & J.K.Wu, *Fritillaria chekiangensis* (P.K.Hsiao & K.C.Hsia) Y.K.Yang, Z.H.Lin & C.Lin, *Fritillaria collicola* Hance, *Fritillaria xiaobeimu* Y.K.Yang, J.Z.Shao & M.M.Fang, *Uvularia cirrhosa* Thunb. **Common name(s)**: Zhejiang mother. **Botanical description**: It is native to East Asia and East China. It is hardy to zone 8. It flowers from March to May. The flowers are hermaphrodite and are pollinated by insects. It is suitable for light and medium soils and prefers well-drained moist soil. It is habituated to bamboo forests, shady and moist places from near sea level to 600 m. **Medicinal use(s)**: It is used in the treatment of breast cancer, coughs, bronchitis, pneumonia, feverish illnesses, abscesses, and swelling of the throat, neck, and chest. It also possesses anti-inflammatory, antitumor, antibacterial, and antioxidant properties.					

(continued)

Table 4.81 (continued)

Family	Botanical name	Image	Plant part used	Bioactive compound(s)	Structure	Reference(s)
	Hosta plantaginea (Lam.) Asch.		Entire plant	Berberine	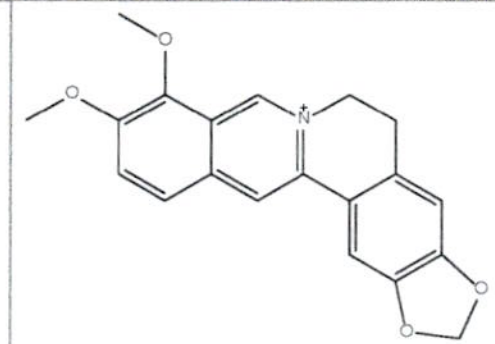	Han et al. (1984) and Maffei (2003a, b, c, d, e, f, g, h, i, j, k, l, m)
	Synonym(s): *Funkia alba* (Andrews) Sweet, *Funkia cordata* Siebold ex Steud., *Funkia grandiflora* Siebold & Zucc., *Funkia japonica* (Thunb.) Druce, *Funkia legendrei* H.Lév., *Funkia subcordata* Spreng., *Hemerocallis alba* Andrews, *Hemerocallis japonica* Thunb., *Hemerocallis plantaginea* Lam., *Hosta japonica* Tratt., *Niobe cordifolia* Salisb., *Niobe plantaginea* (Lam.) Nash, *Saussurea japonica* (Thunb.) Kuntze, *Saussurea plantaginea* (Lam.) Kuntze. **Common name(s)**: Plantain lily. **Botanical description**: It is erect, large, vigorously growing (26 in. tall and 46 in. diameter) and is native to China. It is mainly known for its beautiful foliage which provides color, contrast, and texture to the landscape. Dense foliage crowds out garden weeds. **Medicinal use(s)**: It is used to treat cancer cell lines, lymphoma, malignant ulcers, and tumors of the glands. It also possesses antibacterial, antioxidant, and antimicrobial properties.					

	Hosta longipes (Franch. & Sav.) Matsum.		Entire plant	Chlorogenic acid	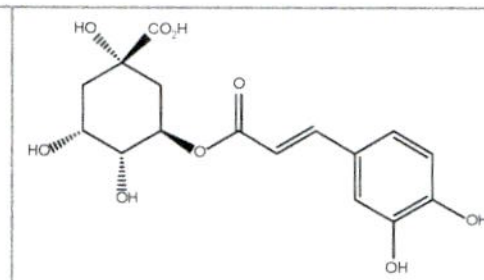	Han et al. (1984) and Maffei (2003a, b, c, d, e, f, g, h, i, j, k, l, m)
	Synonym(s): *Funkia longipes* Franch. & Sav., *Funkia tardiflora* W.Irving., *Hosta japonica* var. *tardiflora* (W.Irving) L.H.Bailey., *Hosta sparsa* Nakai, *Hosta tardiflora* (W.Irving) Stearn, *Hosta leptophylla* F.Maek. **Common name(s)**: Plantain lily. **Botanical description**: It is a perennial plant growing up to 0.3 m and is native to East Asia, Japan, and Korea. It is in flower from July to September, and the seeds ripen from September to October. The flowers are hermaphrodite and suitable to grow in light, medium, and heavy soils. It prefers moist soil and can tolerate drought conditions. **Medicinal use(s)**: It is used to treat malignant tumor, cough, sputum-retention, laryngopharyngitis, and burns. It also possesses anticancer, antioxidant, antibacterial, and anti-inflammatory properties.					

(continued)

Table 4.81 (continued)

Family	Botanical name	Image	Plant part used	Bioactive compound(s)	Structure	Reference(s)
	Colchicum trigynum (Steven ex Adam) Stearn		Bulb	Corynanthe	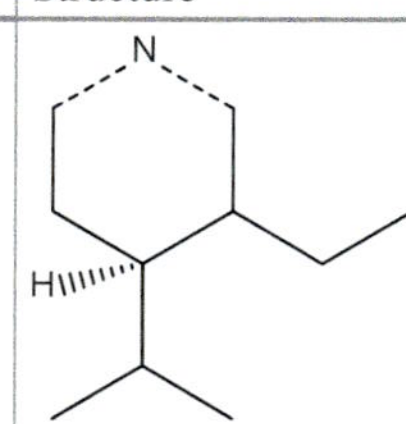	Ulubelen and Tanker (1975) and Hargittai (2015)
	Synonym(s): *Bulbocodium caucasicum* (M.Bieb.) Endl. ex Heynh., *Bulbocodium trigynum* Steven ex Adam, *Colchicum caucasicum* (M.Bieb.) Spreng., *Colchicum monogynum* M.Bieb. ex Schult. & Schult.f., *Merendera candidissima* Miscz. ex Grossh., *Merendera caucasica* M.Bieb., *Merendera ghalghana* Otsch., *Merendera trigyna* (Steven ex Adam) Stapf, *Sternbergia caucasica* (M.Bieb.) Willd. **Common name(s)**: Not available. **Botanical description**: It is native to Iran, the North Caucasus, Transcaucasus, and Turkey. Its corms are up to 3 cm long. Two to six leaves are produced at the same time as the flowers. Flowers are pinkish purple, linear, acute, ascending or spreading, glabrous or with scabrid-serrulate margins. Stamens with free parts of filaments up to 6 mm long, anthers 2–5 mm long. Ovary oblong, styles up to 30 mm long. Capsule 1.5 cm long, seeds 1.6 mm in size. **Medicinal use(s)**: It is used to treat breast and lung cancer. It also possesses antitumor, antibacterial, antioxidant, and anti-inflammatory properties.					

	Smilax menispermoidea A.DC.		Rhizome	Campesterol	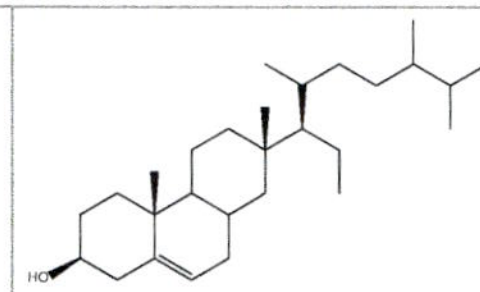	Ju and Jia (1992) and Maffei (2003a, b, c, d, e, f, g, h, i, j, k, l, m)
	Synonym(s): *Smilax luteocaulis* H.Lév., *Smilax rubriflora* Rehder. **Common name(s)**: Catbriers. **Botanical description**: It grows as shrubs, forming dense impenetrable thickets, and is native to tropics, subtropics worldwide, North America, and North Mexico. They will also grow over trees and other plants up to 10 m high, their hooked thorns allowing them to hang onto and scramble over the tree branches. The genus includes both deciduous and evergreen species. The leaves are heart-shaped and vary from 4 to 30 cm long in different species. It is dioecious. However, only about one in three colonies have plants of both sexes. Plants flower in May and June with white/green clustered flowers. If pollination occurs, the plant will produce a bright red to blue-black spherical berry fruit about 5–10 mm in diameter. **Medicinal use(s)**: It is used to treat breast cancer, tumor, sexual impotence, rheumatism, and skin ailments and used as a general tonic for weakness. It also possesses antibacterial, anti-inflammatory, and antioxidant properties.					

(continued)

Table 4.81 (continued)

Family	Botanical name	Image	Plant part used	Bioactive compound(s)	Structure	Reference(s)
	Smilax china L.		Rhizome	Gallic acid	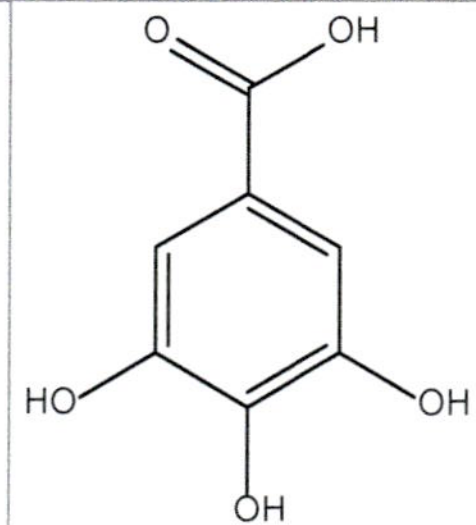	Xu et al. (2008) and Li et al. (2007b)
	Synonym(s): *Coprosmanthus japonicus* Kunth, *Smilax japonica* (Kunth) A.Gray, *Smilax pteropus* Miq., *Smilax taiheiensis* Hayata, *Smilax taquetii* H.Lév., *Smilax thomsoniana* A.DC. **Common name(s)**: China root. **Botanical description**: It is native to China, Korea, Taiwan, Japan (including Ryukyu and Bonin islands), the Philippines, Vietnam, Thailand, Myanmar, and Assam. The stem is woody, sparsely prickly, and 15 m long. Petiole is 0.5–1.5 cm long, and leaf blade is elliptic to orbicular 3–10 cm long and 1.5–6 cm wide, sometimes wider. Berries are red, globose, and 0.6–1.5 cm in diameter. It is habituated to forests, thickets, hillsides, grassy slopes, and shaded places along valleys or streams. **Medicinal use(s)**: It is used to treat ovarian cancer. It also possesses anti-inflammatory, antibacterial, antioxidant, and antimicrobial properties.			Apigenin	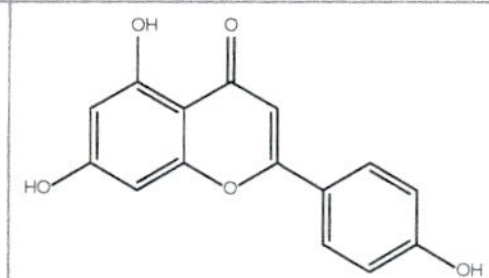	
				Kaempferol-7-O-β-d-glucoside	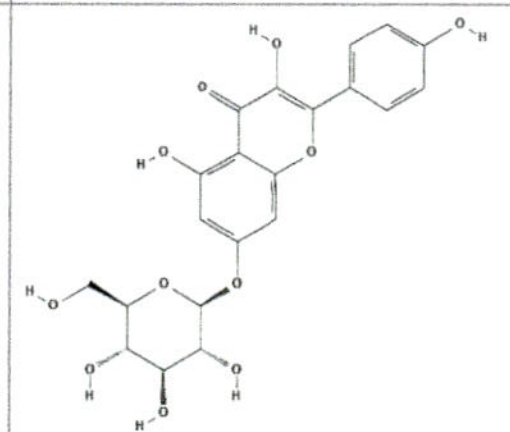	

	Smilax glabra		Rhizome	Thymol	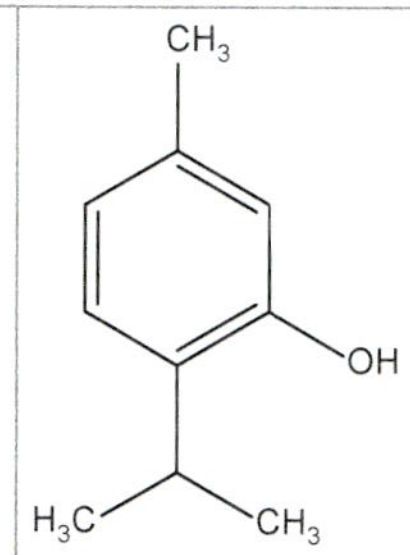	Madhuri and Pandey (2009)
	Synonym(s): *Smilax blinii* H.Lév., *Smilax calophylla* var. *concolor* C.H.Wright., *Smilax dunniana* H.Lév., *Smilax glabra* var. *maculata* H.Lév., *Smilax hookeri* Kunth, *Smilax mengmaensis* R.H.Miao, *Smilax trigona* Warb. **Common name(s)**: Tufuling. **Botanical description**: It is a deciduous climber growing up to 3 m and is native to East Asia, China, and the Himalayas. It is in flower from July to November, and the seeds ripen from November to April. The flowers are dioecious. It is suitable to grow in light, medium, and heavy soils. It can also grow in semi-shade or no shade and prefers moist soil. It is habituated to thickets in uplands and West China. **Medicinal use(s)**: It is used to treat cancer cell lines, diabetes, infection, and other skin and urinary disorders. It also possesses antitumor, antibacterial, antimicrobial, and anti-inflammatory properties.					

(continued)

Table 4.81 (continued)

Family	Botanical name	Image	Plant part used	Bioactive compound(s)	Structure	Reference(s)
	Smilax sieboldii Miq.		Entire plant, root	Phillyrin	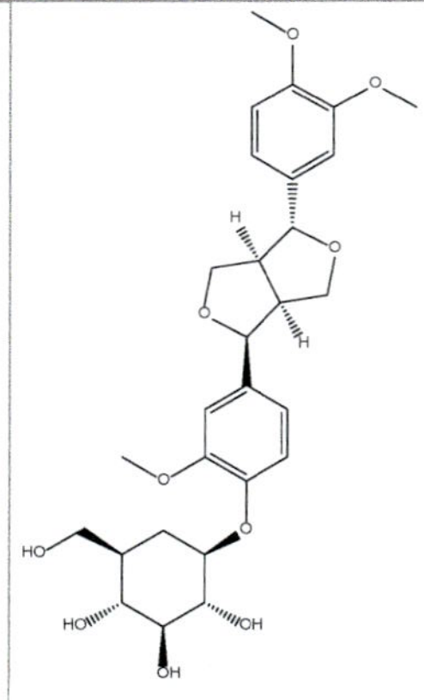	Woo et al. (1992), Suh et al. (1996) and Maffei (2003a, b, c, d, e, f, g, h, i, j, k, l, m)
	Synonym(s): *Coprosmanthus oldhamii* (Miq.) Masam., *Smilax formosana* (Hayata) Hayata., *Smilax nebelii* Gilg., *Smilax oldhamii* Miq. **Common name(s)**: Greenbriers. **Botanical description**: It is a deciduous climber growing up to 3 m and is native to East Asia, China, and the Himalayas. It is in flower from July to November, and the seeds ripen from November to April. The flowers are dioecious. It is suitable to grow in light, medium, and heavy soils. It can also grow in semi-shade or no shade and prefers moist soil. It is habituated to thickets in uplands and West China. **Medicinal use(s)**: It is used to treat ovarian cancer, breast cancer, lung cancer, nasopharyngeal cancer, HIV, malaria, and TB. It also possesses antitumor, antibacterial, anti-inflammatory, and antioxidant properties.					

	Yucca gloriosa L.		Roots, barks	Furostanol saponins	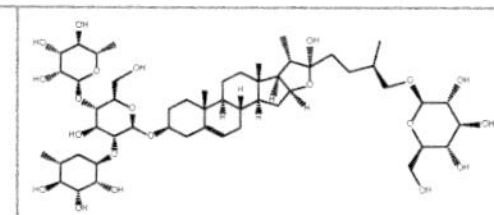	Skhirtladzea et al. (2006)
	Synonym(s): *Yucca acuminata* Sweet., *Yucca acutifolia* Truff., *Yucca ellacombei* Baker., *Yucca ensifolia* Groenl. **Common name(s)**: Moundlily yucca, Long-trunked Spanish dagger. **Botanical description**: It is an evergreen shrub growing up to 1.8 m at a slow rate and is native to North America, North Carolina, and Florida. It is in leaf on 12 January. It is in flower from July to September. The flowers are hermaphrodite and are pollinated by hand. It can grow in light and medium soils and nutritionally poor soil but prefers well-drained soil. It prefers dry or moist soil and can tolerate drought conditions. It is habituated to sand dunes and the borders of beaches of the seacoast. **Medicinal use(s)**: It is used to treat cancer cells, migraine headaches, colitis, ulcers, wounds, gout, bursitis, and hypertension. It also possesses antitumor, anti-inflammatory, antibacterial, and antimicrobial properties.			Gloriosacl C	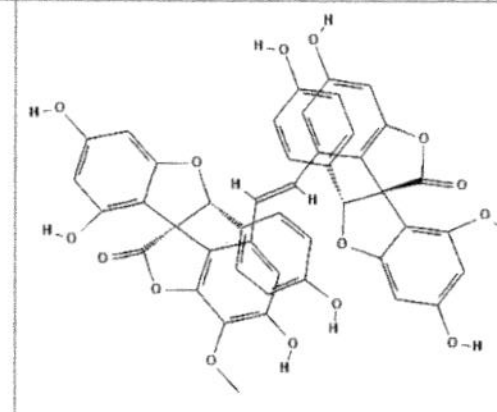	

(continued)

Table 4.81 (continued)

Family	Botanical name	Image	Plant part used	Bioactive compound(s)	Structure	Reference(s)
	Yucca schidigera Roezl ex Ortgies		Bark	Resveratrol	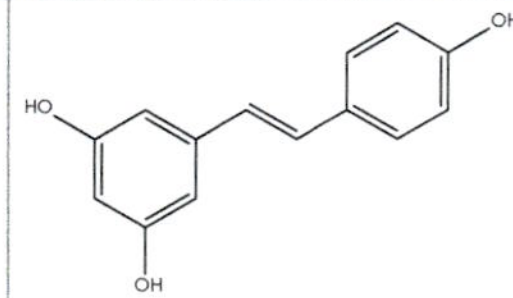	Piacente et al. (2004)
	Synonym(s): *Sarcoyucca mohavensis* (Sarg.) Linding., *Yucca californica* Nutt. ex Baker., *Yucca mohavensis* Sarg. **Common name(s)**: Mojave yucca, Spanish dagger. **Botanical description**: It is a small evergreen tree growing up to 5 m tall, with a dense crown of spirally arranged bayonet-like leaves on top of a conspicuous basal trunk and is native to the Mojave Desert, Chihuahuan Desert, and Sonoran Desert of Southeastern California, Baja California, New Mexico, Southern Nevada, and Arizona. The bark is gray-brown and is covered with brown dead leaves near the top, becoming irregularly rough and scaly-to-ridged closer to the ground. The leaves are 30–150 cm long and 4–11 cm broad at the base, concavo-convex, thick, very rigid, and yellow-green to blue-green in color. **Medicinal use(s)**: It is used to treat cancer cell lines, headaches, bleeding, gonorrhea, arthritis, and rheumatism. It also possesses antioxidant, anti-inflammatory, antibacterial, and antitumor properties.					

Table 4.82 Anticancer plants of family Linaceae

Linaceae	*Linum album* Kotschy ex Boiss.		Leaf	Podophyllotoxin	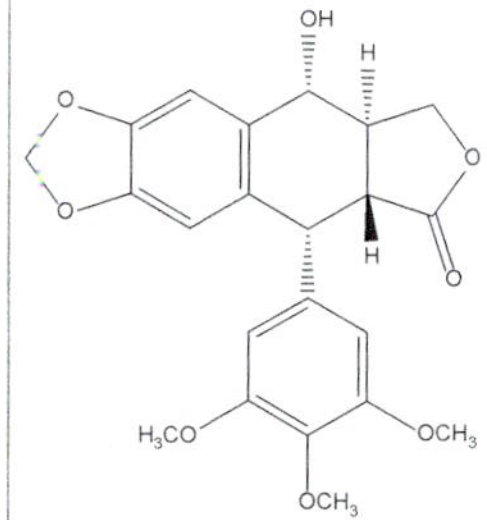	Arroo et al. (2002) and Wan Salleh and Ahmad (2017)
	Synonym(s): No synonyms are recorded for this name. **Common name(s)**: Not available. **Botanical description**: It is native to temperate and subtropical regions of the world. The flowers of most species are blue or yellow, rarely red, white, or pink, and some are heterostylous. There is an average of 6–10 seeds per boll. Flax plant grows up to 12–36 in. in height and has a typical life cycle of 45–60 days of vegetative period, followed by 15–25 days of flowering period, and 30–40 days of maturation period. The mature fruit of the flax plant is a dry boll or capsule with five segments. **Medicinal use(s)**: It is used to treat lung and testicular cancers. It also possesses antioxidant, antibacterial, anti-inflammatory, and antimicrobial properties.					

(continued)

Table 4.82 (continued)

	Linum campanulatum L.		Leaf	Podophyllotoxin	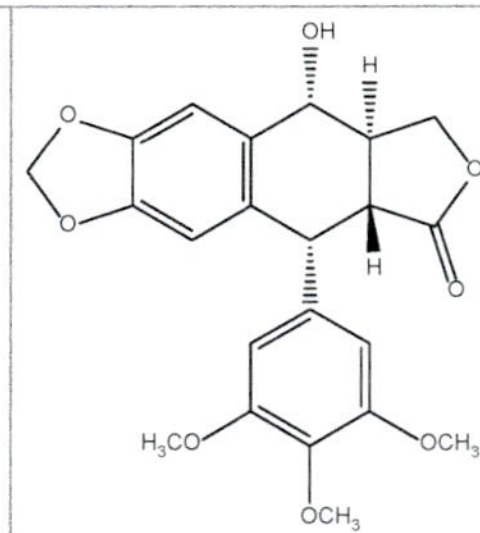	Arroo et al. (2002) and Ardalani et al. (2017)
	Synonym(s): *Xantholinum campanulatum* (L.) Rchb. **Common name(s)**: Yellow flax; Western flax. **Botanical description**: It is native to temperate and subtropical regions of the world. The flowers of most species are blue or yellow, rarely red, white, or pink, and some are heterostylous. There is an average of 6–10 seeds per boll. Flax plant grows to 12–36 in. in height and has a typical life cycle of 45–60 days of vegetative period, followed by 15–25 days of flowering period, and 30–40 days of maturation period. The mature fruit of the flax plant is a dry boll or capsule with five segments. **Medicinal use(s)**: It is used to treat lung cancer, epilepsy, respiratory and gastrointestinal inflammatory disorders, cold, flu, fever, rheumatism, eye infections, and gout. It also possesses antibacterial, antioxidant, anti-inflammatory, and antimicrobial properties.					

	Linum lewisii Pursh		Stem	Podophyllotoxin	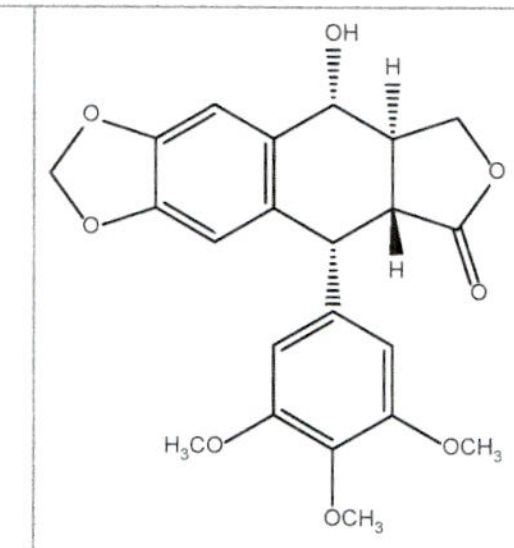	Arroo et al. (2002) and Wan Salleh and Ahmad (2017)
	Synonym(s): *Linum decurrens* Kellogg, *Linum lyallanum* Alef. **Common name(s)**: Prairie flax. **Botanical description**: It is native to western North America from Alaska south to Baja California and from the Pacific Coast east to the Mississippi River. The flowers are up to 2 in. long and homostylous (styles of all the same length) and have 5 broad, fan-shaped petals, styles longer than the stamens, 5 arrowhead-shaped anthers, and 5 spherical stigmas. The leaves are single-veined and linear to lanceolate in shape. The similar meadow flax (*Linum pratense*) grows at lower elevations and has narrower, very pale blue petals and shorter styles. **Medicinal use(s)**: It is used to treat breast and colon cancer, kidney diseases, psoriasis, and eczema. It also possesses antitumor, anti-inflammatory, antibacterial, antioxidant, antimicrobial, and antiviral properties.					

(continued)

Table 4.82 (continued)

	Linum persicum L.		Fruit	Carvacrol	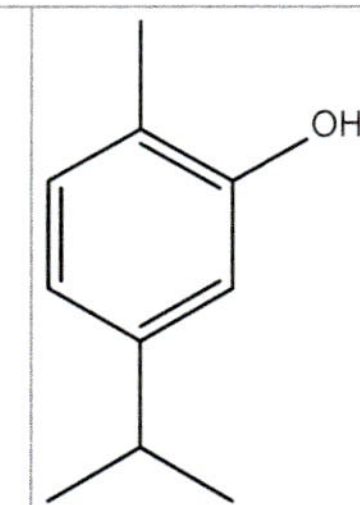 	Amirghofran et al. (2006)
	Synonym(s): *Linum abyssinicum* Hochst., *Linum acuminatum* Moench., *Linum acuminatum* E.Mey., *Linum adenophorum* Hausskn. & Bornm. **Common name(s)**: Flax. **Botanical description**: It is a perennial plant and is native to Iran. The flowers are up to 2 in. long and homostylous (styles of all the same length) and have 5 broad, fan-shaped petals, styles longer than the stamens, 5 arrowhead-shaped anthers, and 5 spherical stigmas. The leaves are single-veined and linear to lanceolate in shape. **Medicinal use(s)**: It is used to treat prostate cancer, infections, and inflammatory disorders. It also possesses antibacterial, antioxidant, anti-inflammatory, antimicrobial, and antitumor properties.					

	Linum virginianum L.		Seed	Benzcic acid	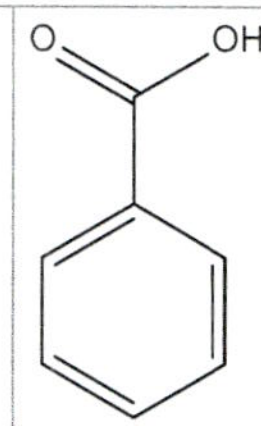	Hussey (1974) and Maffei (2003a, b, c, d, e, f, g, h, i, j, k, l, m)
	Synonym(s): *Linum virginianum* var. *microcarpum* Planch., *Linum virginianum* var. *virginianum*, *Nezera virginiana* (L.) Nieuwl. **Common name(s)**: Virginia yellow flax. **Botanical description**: It is a small native perennial plant of dry fields, roadsides, and clearings with small, yellow, solitary flowers and is native to Southern New England. Fruit length is 1.3–2 mm, dry and splits open when ripe. There are five petals, sepals, or tepals in the flower. The edge of the simple leaf blade is entire (has no teeth or lobes). **Medicinal use(s)**: It is used to treat skin and mouth cancer, cold, cough, lung and urinary ailments, fever, and gout and as laxative and poultice to relieve pain from burn and inflammation. It also possesses antitumor, anti-inflammatory, antibacterial, antimalarial, and antioxidant properties.					

(continued)

Table 4.82 (continued)

	Linum usitatissimum L.		Whole plant	Secoisolariciresinol (seco)	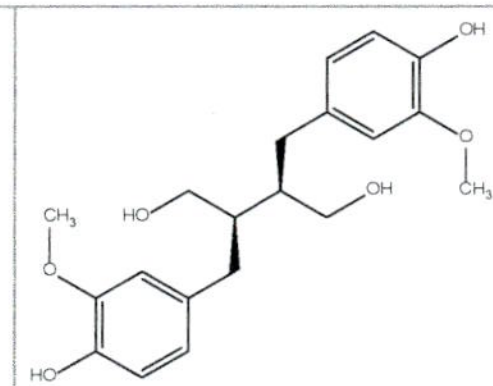	Zimmermann et al. (2006) and Marghescu et al. (2012)
				Matairesinol (mata)	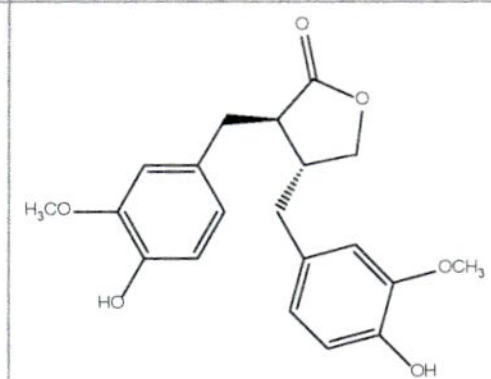	
	Synonym(s): *Linum crepitans* (Boenn.) Dumort., *Linum humile* Mill., *Linum indehiscens* (Neilr.) Vavilov & Elladi. **Common name(s)**: Common flax. **Botanical description**: It is an erect glabrous annual herb and is native to Europe. Stems 7.5–12 cm tall, corymbose-branched above, terete, grayish green; leaves gray-green, alternate, linear-lanceolate, numerous, flat, 2–3.5 cm long, acute to acuminate, gradually and slightly narrowed at base, glaucous; flowers borne in loose, terminal, leafy racemes or open cymes, on long erect pedicels, 2–4 cm long; sepals five, long acuminate, ovate-oblong, 5–7 mm long, entire, eglandular, 3-nerved, margins serrulate, about half as long as the petals; petals blue, or white or pale pink, 1 cm or more long, obovate. **Medicinal use(s):** It is used for the treatment of breast cancer, cold, cough, constipation, gravel rash, stones in the kidney and ureter, and burns and scalds. It also possesses antioxidant, anti-inflammatory, antibacterial, and antimicrobial properties.					

Table 4.83 Anticancer plants of family Loranthaceae

Family	Botanical name	Image	Plant part used	Bioactive compound(s)	Structure	Reference(s)
Loranthaceae	*Plicosepalus sagittifolius* (Engl.) Danser		Branches, leaf	Diosgenin	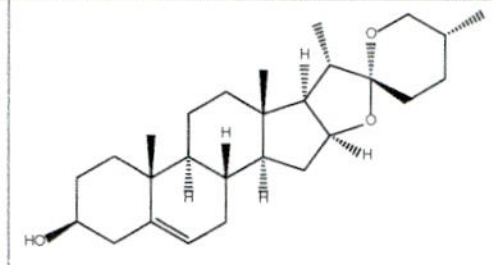	Chhabra et al. (1984, 1991) and Srinivasan et al. (2014)
	Synonym(s): *Loranthus sagittifolius* (Engl.) Sprague, *Loranthus undulatus* var. *sagittifolius* Engl. **Common name(s)**: Sprague. **Botanical description**: It is native to east of tropical Africa, from South Africa, Zambia, Malawi, coastal regions from Southern Somalia via Kenya and Tanzania to Northern Mozambique. It is shrub of around 1 m in diameter. Its ranches are grayish brown and carry opposite pairs. **Medicinal use(s)**: It is used to treat cancer. The plant is used to soften leather, while its branches are utilized as firewood. It possesses anticancer, antitumor, anti-inflammatory, and antimicrobial properties.					

(continued)

Table 4.83 (continued)

Family	Botanical name	Image	Plant part used	Bioactive compound(s)	Structure	Reference(s)
	Viscum album L.		Whole plant	Chlorogenic acid	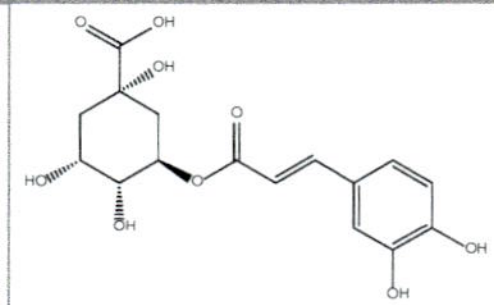	Tiwari (1995), Grossarth et al. (2001),Ostermann et al. (2009), Schoen-Angerer et al. (2015), and Bonamin et al. (2017)
	Synonym(s): *Viscum album* var. *album*. **Common name(s)**: Mistletoe. **Botanical description**: It is native to India. It is a small-sized herb. It is found in dense dry forest of slope on coarse limestone blocks. The flowers, in compact spikes, are bisexual, unisexual, or regular. **Medicinal use(s)**: It is used to treat cancer and possesses anticancer properties. It has been used to lower blood pressure and heart rate and to ease anxiety and is an herbal sleep aid. It has been used as a traditional herbal remedy for the treatment of cough, especially convulsive cough, bronchial asthma, and asthmatic attacks due to psychological tension.			Caffeic acid	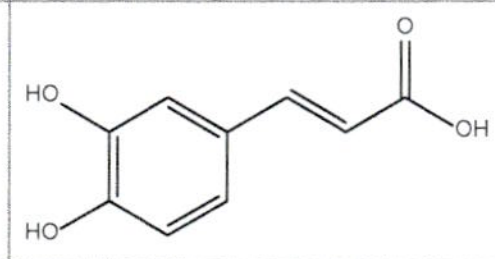	
				Sakuranetin	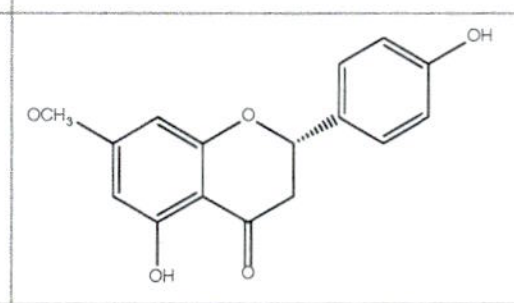	
				Naringenin 5-methyl ether	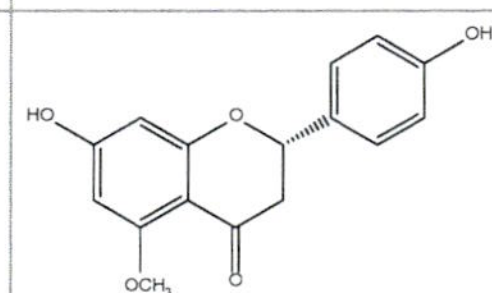	

		Syringenin 4-O-glucoside (eleutheroside B)	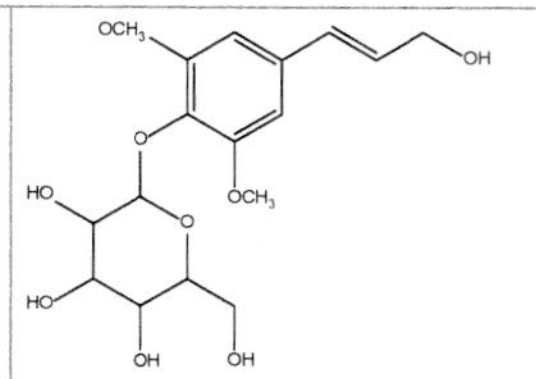	
		Syringenin 4-O-apiosyl-glucoside (polygalatenoside E)	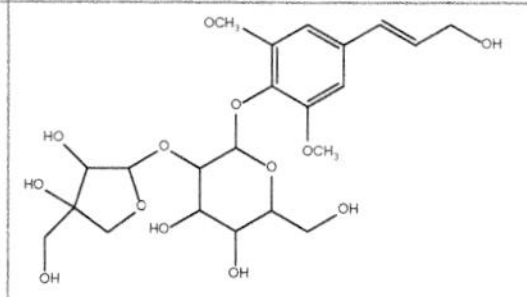	
		Alangilignoside C/ligalbumoside A	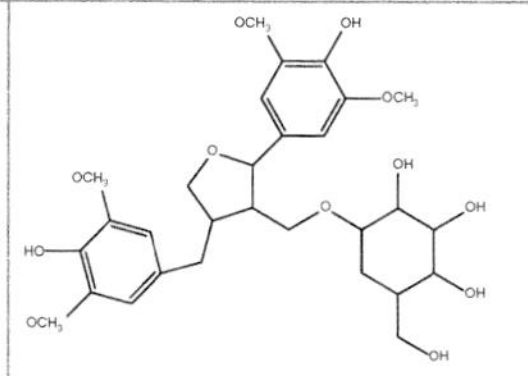	

Table 4.84 Anticancer plants of family *Lycoperdaceae*

Family	Botanical name	Image	Plant part used	Bioactive compound(s)	Structure	Reference(s)
Lycoperdaceae	*Lycoperdon bovista* L		Entire plant	(4-Carboxyphenyl)-Cyanoimino-Oxidoazanium	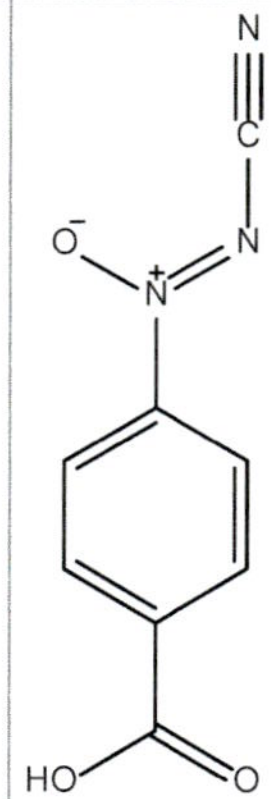	Burkill (1966a, b) and Burk (1983)
	Synonym(s): *Lycoperdon maximum* Schaeff., *Calvatia maxima* (Schaeff.) Morgan, *Calvatia bovista* (L.) T. Macbr., *Langermannia maxima* (Schaeff.) Pázmány, *Lycoperdon giganteum* Batsch, *Lycoperdon bovista* Pers. **Common name(s)**: Bovista, Puffballs, Bolita. **Botanical description**: It is native to Malaysia. It is a very small mushroom that grows to a diameter of only 3 cm. It is subglobose and a bit flattened. The surface is densely cobbled and light tan. The interior is straw-colored. The spores are spherical and brownish with a tinge of violet. They can be found on pastures, playing fields, edges of woods, roads, and paths. At low elevations, they are widely distributed and fruit during the summer in watered areas. **Medicinal use(s)**: It is used to treat lung, esophagus, breast, cervical, and ovarian cancer. It possesses antitumor, antimicrobial, antidote, antibacterial, and astringent properties. It is also used for several kinds of skin disorders and nose bleeding.					

Table 4.85 Anticancer plants of family Magnoliaceae

Family	Botanical name	Image	Plant part used	Bioactive compound(s)	Structure	Reference(s)
Magnoliaceae	*Magnolia officinalis* Rehd. Et Wils		Whole plant	Honokiol	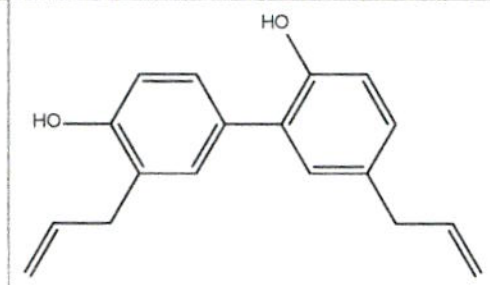	Lee et al. (2009), Seo et al. (2011), Jeong et al. (2012), McKeown (2015), and Arora et al. (2012)
	Synonym(s): No synonyms are recorded for this name. **Common name(s):** Houpu magnolia, Magnolia-bark magnolia. **Botanical description:** It is native to East Asia – West China. It is a deciduous tree growing up to 20 m (65 ft.) by 12 m (39 ft.). It is in flower from May to June, and the seeds ripen from August to October. The flowers are hermaphrodite (have both male and female organs) and are pollinated by beetles. It is suitable for light (sandy), medium (loamy), and heavy (clay) soils and prefers well-drained soil. Suitable pH, acid, neutral, and basic (alkaline) soils. It can grow in semi-shade (light woodland) or no shade. It prefers moist soil. It can tolerate atmospheric pollution. **Medicinal use(s):** *Magnolia officinalis* has antibacterial, anti-inflammatory, and anticancer properties and is also used in treating Alzheimer's disease, depression, diabetes, and menopause symptoms. The anticancer activity involves mitochondrial dysfunction and induced cell death mediated by increased endoplasmic reticulum stress that eventually results in apoptosis.			Magnolol	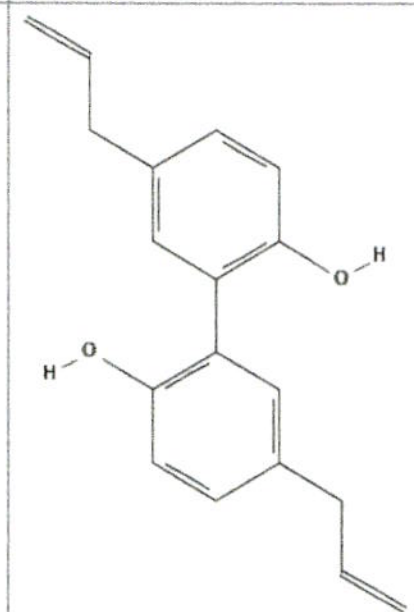	

Table 4.86 Anticancer plants of family Malvaceae

Family	Botanical name	Image	Plant part used	Bioactive compound(s)	Structure	Reference(s)
Malvaceae	*Bombax brevicuspe* Sprague		Root	Not available	Not available	Vasileva (1969)
	Synonym(s): No synonyms are recorded for this name. **Common name(s)**: Buma (Schultze). **Botanical description**: It is the native plant of India and also found in Guinea, Southeast Asia, and Africa. It is a deciduous tree and grows in tropical climate up to 20 m. Flowers are white and fragrant. **Medicinal use(s)**: It is used to treat breast cancer. It is also used as an antitumor agent. It is very effective as an aphrodisiac when combined with safed musli and black musli. It increases sexual stamina and takes care of sexual health. It also possesses anti-inflammatory, antioxidant, antibacterial, and antimicrobial properties.					

	Hibiscus sabdariffa L.		Apical part	Tiliroside	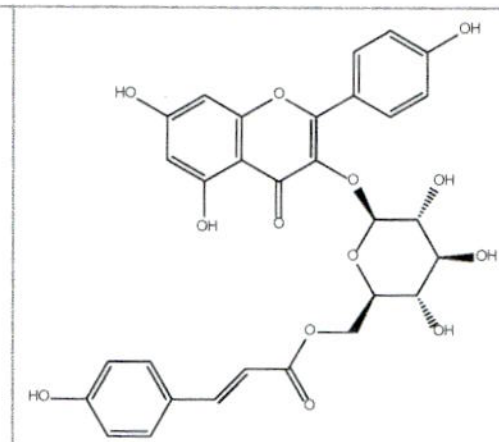	Olaleye (2007) and Malacrida et al. (2016)
	Synonym(s): *H. sabdariffa* var. *altissima*, *H. sabdariffa* var. *sabdariffa*. **Common name(s):** Patwa. **Botanical description:** It is native to the tropics – probably tropical Central or Western Africa. It is a perennial growing up to 3 m (9 ft.) by 2 m (6 ft.). It is in flower from August to October, and the seeds ripen from October to November. The flowers are hermaphrodite (have both male and female organs) and are pollinated by insects. It is suitable for light (sandy), medium (loamy), and heavy (clay) soils and prefers well-drained soil. Its suitable pH is acid, neutral, and basic (alkaline) soils. It cannot grow in the shade. It prefers moist soil. **Medicinal use(s):** It has anticancer properties. These may have diuretic and choleretic effects, decreasing the viscosity of the blood, reducing blood pressure, and stimulating intestinal peristalsis. It is used as a folk remedy in the treatment of abscesses, bilious conditions, cancer, cough, debility, dyspepsia, dysuria, fever, hangover, heart ailments, hypertension, neurosis, scurvy, and strangury.					

(continued)

Table 4.86 (continued)

Family	Botanical name	Image	Plant part used	Bioactive compound(s)	Structure	Reference(s)
	Sida cordifolia L.		Leaf, stem Root	Gallic acid equivalents	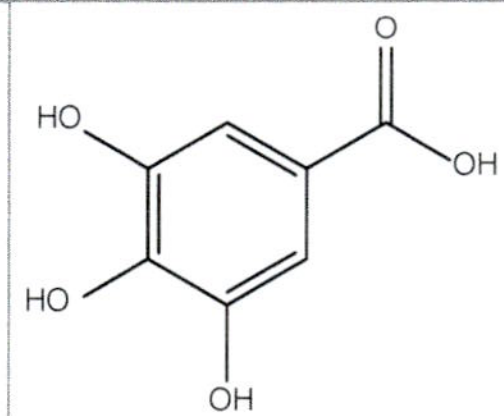	Dhalwal et al. (2005) and Silva et al. (2006)
	Synonym(s): *Sida rhombifolia.* **Common name(s):** Bala, Vatya, Bariyar, Kharethi. **Botanical description:** It is a shrub. It grows in a temperate climate. It grows up to 2 m. It is an erect perennial that reaches 50–200 cm (20–79 in.) tall, with the entire plant covered with soft white felt-like hair that is responsible for one of its common names, "flannel weed." The stems are yellow-green, hairy, long, and slender. The yellow-green leaves are oblong-ovate, covered with hairs, and 3.5–7.5 cm (1.4–3.0 in.) long by 2.5–6 cm (0.98–2.36 in.) wide. The flowers are dark yellow, sometimes with a darker orange center, with a hairy 5-lobed calyx and 5-lobed corolla. **Medicinal use(s):** It is applied directly to the skin for numbness, nerve pain, muscle cramps, skin disorders, tumors, joint diseases, wounds, ulcers, scorpion sting, and snakebite and as massage oil. The oil is used to massage persons suffering from paralysis, cervical spondylosis, and facial paralysis. It is potent in curing male infertility. It increases sperm motility and sperm count. It is an aphrodisiac herb and useful in treating erectile dysfunction and premature ejaculation.					

	Thespesia Populnea L.		Whole plant	Populene A	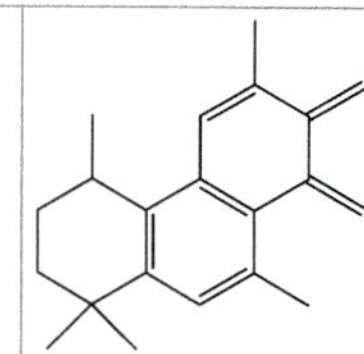	Boonsri et al. (2006) and Soysa and Silva (2016)
	Synonym(s): *Hibiscus populneus, Abelmoschus acuminatus, Hibiscus blumei.* **Common name(s):** Indian tulip tree, Aden apple, Portia tree. **Botanical description:** This is a good tree for small gardens or patios. Its name *Thespesia* means "divinely decreed" and was given by Daniel Solander who saw it in Tahiti as a member of Captain Cook's ship. Indian tulip tree is an evergreen bushy tree. It grows up to 40 ft. or more with a spread of 10–20 ft. It has heart-shaped leaves and cup-shaped yellow flowers that are produced intermittently throughout the year in warm climates. Each flower has a maroon eye that ages to purple. The flowers are followed by apple-shaped fruit. **Medicinal use(s):** The stem is employed in treating breast cancer. Juices from the pounded fruits, mixed with pounded leaves, are used in a poultice to treat headaches and itches. A decoction of the bark and fruit is mixed with oil and used to treat scabies. The inner bark is used to treat constipation and typhoid.					

(continued)

Table 4.86 (continued)

Family	Botanical name	Image	Plant part used	Bioactive compound(s)	Structure	Reference(s)
	Waltheria indica L.		Whole plant	Epicatechin	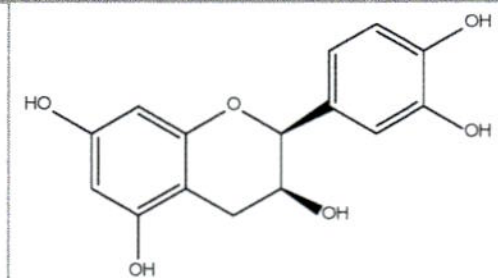	Koteswara et al. (2005) and Zailani et al. (2010)
				Quercetin	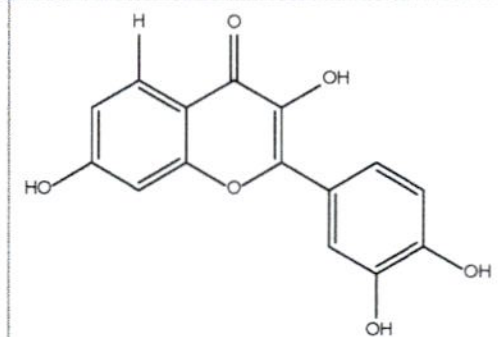	
	Synonym(s): *Waltheria americana*, *Waltheria pyrolaefolia*. **Common name(s):** Hala 'uhaloa, Hi'aloa, Kanakaloa, Kafaki. **Botanical description:** It is a short-lived, perennial plant producing several erect or ascending stems that that can be branched from the base. The stems become more or less woody and persist. The plant can grow from 0.5 to 2 m tall. It is widely spread through the tropics and subtropics. It grows in a wide range of soils. The plant is an ubiquitous weed and early colonizer in disturbed ground, tolerant of a wide spectrum of habitats and elevations. It is considered to be invasive in many areas. The plant can flower and produce fruit all year-round. **Medicinal use(s):** A decoction of various plant parts are used to treat fever and syphilis. It is used topically for wounds and skin eruptions. The decoction of leafy stem is used as a cure for fever, cough, cold, bladder ailments, vaginal infections, hypertension, ulcers, and hemoptysis. The decoction of root is used as antidiarrheal and tonic for children. It is used as a medicine for cough and for healing wounds.			Tiliroside	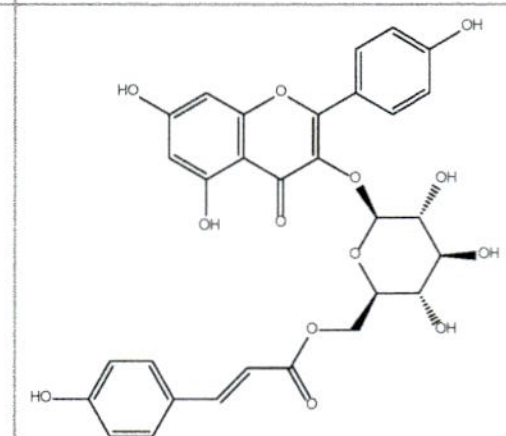	

Table 4.87 Anticancer plants of family Melastomataceae

Family	Botanical name	Image	Plant part used	Bioactive compound(s)	Structure	Reference(s)
Melastomataceae	*Medinilla crassinervia* Blume/*Medinilla macrocarpa* Blume		Entire plant	Berberine	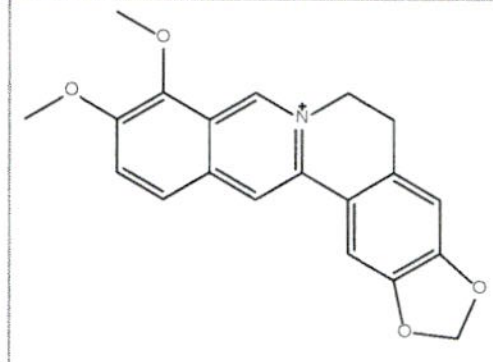	Holdsworth and Sakulas (1986) and Quattrocchi (2012)
	Synonym(s): *Medinilla acrochordonocarpa* Merr., *Medinilla acuminata* Merr., *Medinilla acutialata* P, *Medinilla acutissimifolia* H. Perrier. **Common name(s):** Not available. **Botanical description:** It is native to Indo-Malaysia, Japan, and Australia. It is an erect, free-flowering shrub that grows to a height of about 3 m. The plant is branched and has reddish stems that are covered with bristly scales and minute hairs. Its leaves are simple, elliptic-lanceolate with a rounded base, are up to 7 cm long, and have three distinct main veins running from the base to the apex. Its flowers are borne on short terminal cymes 2–8 cm across. Its fruit is a berry, which when ripe breaks irregularly to expose its soft, dark blue pulp and orange seeds. **Medicinal use(s):** It is used for the treatment of nose cancer, nasal cancer, and tumor. It possesses anticancer, antitumor, antioxidant, anti-inflammatory, antibacterial, and antimicrobial properties.					

(continued)

Table 4.87 (continued)

Family	Botanical name	Image	Plant part used	Bioactive compound(s)	Structure	Reference(s)
	Melastoma malabathricum L.		Flowers	Kaempferol	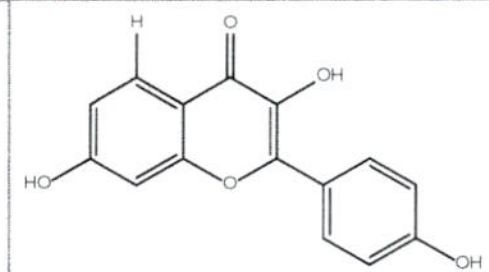	Mohandoss and Ravindran (1993) and David et al. (2017)
	Synonym(s): *Melastoma affine* D. Don., *Melastoma candidum* D. Don., *Melastoma cavaleriei* H. Lév. & Vaniot., *Melastoma esquirolii* H. Lév., *Melastoma malabathricum* subsp. *malabathricum*. **Common name(s):** Common sendudok, Singapore rhododendron, Indian rhododendron, Sesenduk, Malabar gooseberry, Straits rhododendron. **Botanical description:** It is native to Indomalaya, Japan, and Australia and is usually found between 100 and 2800 m on grasslands and sparse forests. It grows wild on a wide range of soils, from sea level up to an altitude of 3000 m. It is an erect, free-flowering shrub that grows to a height of about 3 m. The plant is branched and has reddish stems that are covered with bristly scales and minute hairs. Its leaves are simple, elliptic-lanceolate with a rounded base, are up to 7 cm long, and have three distinct main veins running from base to apex. Its flowers are borne on short terminal cymes 2–8 cm across. Its fruit is a berry, which when ripe breaks irregularly to expose its soft, dark blue pulp and orange seeds. **Medicinal use(s):** It is used for the treatment of lung cancer cell lines, diarrhea, dysentery, leukorrhea, wounds, infection during confinements, toothache, flatulence, and sore legs. It possesses anticancer, antioxidant, antibacterial, anti-inflammatory, and antimicrobial properties.					

Table 4.88 Anticancer plants of family Meliaceae

Family	Botanical name	Image	Plant part used	Bioactive compound(s)	Structure	Reference(s)
Meliaceae	*Aglaia elliptica* Blume		Bark	Quercetin	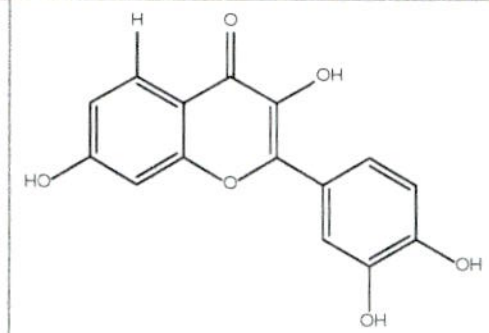	Pannell (1992) and Maffei (2003a, b, c, d, e, f, g, h, i, j, k, l, m)
	Synonym(s): *Aglaia apoana*, *Aglaia banahaensis* Elmer, *Aglaia havilandii* ridl. **Common name(s):** Not available. **Botanical description:** It is native to East Asia, Myanmar, Thailand, Malaysia, Indonesia, and the Philippines. It is an evergreen tree with an irregularly rounded crown; it usually grows from 2 to 20 cm tall, but with some specimens to 40 m. The bole is sometimes fluted throughout with L-shaped buttress up to 150 cm tall and 100 cm out from the bole. The bole can be free of branches for up to 15 m and 50 cm in diameter. **Medicinal use(s):** It is used to treat human oral, epidermal, prostate, colon, and breast cancer. It possesses anticancer, antifungal, antiviral, antibacterial, and antitumor properties. It is also used to treat asthma and jaundice.					

(continued)

Table 4.88 (continued)

Family	Botanical name	Image	Plant part used	Bioactive compound(s)	Structure	Reference(s)
	Aglaia elaeagnoidea (A. Juss.) Benth.		Fruit and apical part	Triterpenes roxburghiadiol A	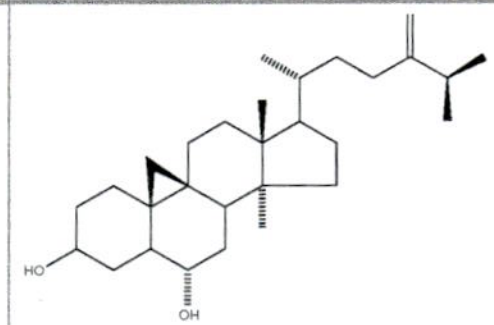	Janaki et al. (1999) and Singh and Peter (2018)
	Synonym(s): *Aglaia abbreviata* C. Y. Wu, *Aglaia canariifolia* Koord., *Aglaia elaeagnoidea* var. formosana Hayata, *Aglaia formosana* (Hayata) Hayata, *Aglaia grata* Wall. ex Voigt, *Aglaia hoanensis* Pierre, *Aglaia lepidota* Miq., *Aglaia parvifolia* Merr., *Aglaia roxburghiana* (Wight & Arn.) Miquel, *Aglaia spanoghei* Bl. ex Miq., *Aglaia talbotii* Sundara Raghavan, *Aglaia wallichii* Hiern. **Common name(s):** Droopy leaf, Priyangu. **Botanical description:** It is native to East Asia, India, Sri Lanka, Myanmar, Thailand, Cambodia, Malaysia, Indonesia, the Philippines, New Guinea, Australia, and Western Pacific. It is an evergreen shrub or small tree, usually grows 5–10 m tall but exceptionally to 20 m. The bole, which sometimes has small buttresses, is usually up to 25 cm in diameter, exceptionally to 50 cm. The flowers have ball-like appearance. **Medicinal use(s):** It is used to treat prostate, breast, oral, and colon cancer. It possesses anticancer, antipyretic, antidiarrheal, antidysenteric, anti-inflammatory, and astringent properties. The bark is used to treat tumors; tonic of leaves is used in the treatment of asthma and jaundice and after child birth.					

	Amoora ouangliensis (Levl.) C.Y. Wu		Bark	Diterpenoids	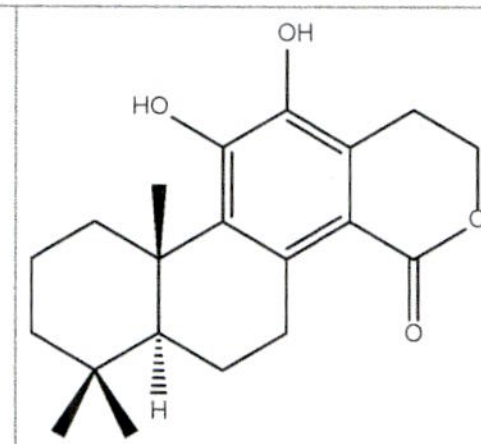	Yang et al. (2010a, b)
	Synonym(s): *Aglaia alternifoliola, Amoora dysoxyloides, Amoora tetrapetala, Amoora yunnanensis.* **Common name(s):** Not available. **Botanical description:** It is native to India, Thailand, Indochina, and throughout Malaysia towards the Solomon Islands. It is found in primary and secondary evergreen deciduous forest, sometimes in peat swamp or riverine forests, on sandy to clayey soil or limestone from sea level up to 1650 m in altitude. It is a tree which grows up to 40 m tall. The bole is branchless for up to 15 m and measures up to 200 cm in diameter. The dark surface is reddish brown to yellow brown or pale pinkish brown. **Medicinal use(s):** It is used to treat human lung and liver cancer. It is also used for the treatment of headaches. It possesses anticancer, antibacterial, antifungal, antiviral, and anthelmintic bioactivities.					

(continued)

Table 4.88 (continued)

Family	Botanical name	Image	Plant part used	Bioactive compound(s)	Structure	Reference(s)
	Aphanamixis polystachya (Wall.) R. N. Parker		Bark	Apigenin	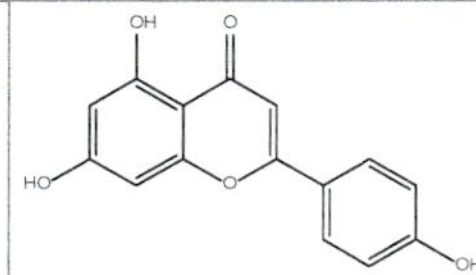	Chowdhury et al. (2003)
	Synonym(s): *Aglaia aphanamixis* (Schult. fil.) Pellegrin, *Aglaia gigantea* Pellegrin, *Amoora amboinensis* Miq., *Amoora gigantea* Pierre, *Amoora rohituka* (Roxb.) Wight & Arn., *Aphanamixis agusanensis* Elmer, *Buchanania spicata* Roxb. ex Wall., *Chuniodendron spicatum* Hu, *Guarea amaris* Buch.-Ham., *Meliacea wightiana* Wall., *Piper hyalinum* Reinw. ex Miq., *Ricinocarpodendron cumingianum* (C. DC.) D. J. Mabberley, *Sphaerosacme polystachya* (Wall.) Wall., *Sphaerosacme spicata* Wall., *Trichilia tripetala* Blanco. **Common name(s):** Pithraj tree, Shan lian. **Botanical description:** *Aphanamixis polystachya* is native to East Asia, South China, India, Sri Lanka, Thailand, Laos, Vietnam, Malaysia, Indonesia, the Philippines, New Guinea, and the Solomon Islands. It is an evergreen tree that attains a height of 20 m. The trunk bears bark that is grey and fissured; branchlets are terete and glabrous; Leaves are compound, imparipinnate, alternate, spiral, clustered at twig ends; rachis pulvinate, often lepidote scaly; leaflets opposite to subopposite, 4–8 pairs; lamina oblong-lanceolate, apex acuminate, base asymmetric, margin entire, coriaceous, glabrous. The inflorescence occurs as panicles and flowers are polygamous. The fruit is a capsule, subglobose, to 3 cm across, coriaceous, pale reddish, 3-celled; 1-seeded, orange red. **Medicinal use(s):** It is used to treat human breast, prostate, colon, cervical and pancreatic cancer, and leukemia cells. Stem bark extracts have antifungal property and is used to treat spleen and liver diseases, tumors, abdominal complaints, rheumatism, cold, and chest pains.					

	Aphanamixis polystachya (wall) Park		Leaf, root, stem bark	Meliacin	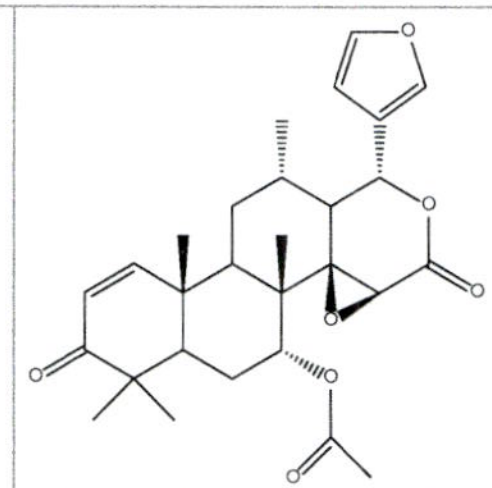	Jain and Srivastava (1985), Srivastava and Agnihotri (1985), Rabi and Gupta (1995), and Jagetia and Venkatesha (2006)
				Aphanamixinin	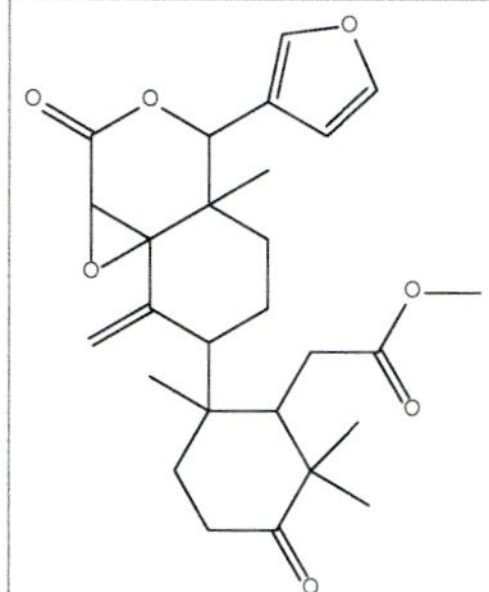	
	Synonym(s): *Amoora rohituka* Wight & Arn. **Common name(s):** Pithraj tree. **Botanical description:** It is native to East Asia, South China, India, Sri Lanka, Thailand, Laos, Vietnam, Malaysia, Indonesia, the Philippines, New Guinea, and the Solomon Islands. It is an evergreen plant that grows 2 m tall or more up to 32 m tall. The cylindrical bole can be free of branches for 15 m up to 70 cm in diameter, with buttresses up to 4 m high. **Medicinal use(s):** It is used to treat human breast, prostate, colon, and cervical cancers, pancreatic cancer cell, and leukemia cell panels. Stem bark is used in spleen and liver diseases, tumors, and abdominal complaints. The extracts also demonstrated mild antifungal effects. The bark is used to remedy rheumatism, cold, and chest pains.					

(continued)

Table 4.88 (continued)

Family	Botanical name	Image	Plant part used	Bioactive compound(s)	Structure	Reference(s)
	Carapa guianensis Aubl.		Seed oil	Epicatechin	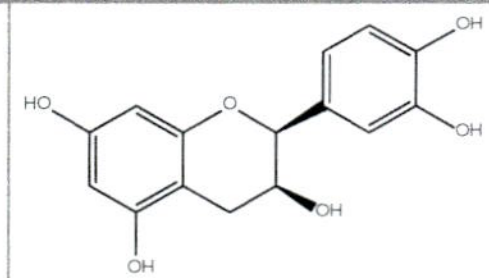	Hammer and Johns (1993), Henriques and Penido (2014)
	Synonym(s): *Amapa guianensis* (Aubl.) Steud., *Carapa latifolia Willd.* ex C. DC., *Carapa llanocarti* Kenfack, Carapa nicaraguensis C. DC., *Carapa slateri* Standl., *Granatum guianense* (Aubl.) Kuntze, *Granatum nicaraguense* (C. DC.) Kuntze, *Guarea mucronulata* C. DC., *Persoonia guareoides* Willd., *Xylocarpus carapa* Spreng. **Common name(s):** Andiroba. **Botanical description:** It is native to South America – Brazil, Peru, Ecuador, Colombia, Venezuela, the Guianas – and Central America, Panama to Guatemala. It is an evergreen tree, growing up to 35 m by 20 m at a fast rate. It is suitable for light (sandy), medium (loamy), and heavy (clay) soils, and suitable pH is basic (alkaline), acid, and neutral soils. It can grow in full shade, semi-shade, or no shade. It prefers moist soil. **Medicinal use(s):** It is used to treat uterine cancer and intestinal cancer. It possesses anticancer, anti-inflammatory, antibacterial, anticarcinogenic, anti-rheumatic, and antineoplastic properties. It is used internally in the treatment of malaria, stomachaches, diarrhea, dysentery, and rheumatism. The stem bark is used externally as a wash for treating a wide range of skin complaints including the spots of chicken pox and measles eczema, ulcers, burns, wounds, and sores.					

	Chukrasia tabularis A. Juss.		Leaf	Gallic acid	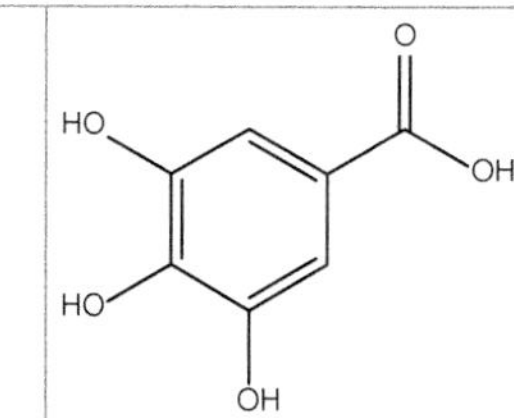	Kaur et al. (2011)
				Epicatechin	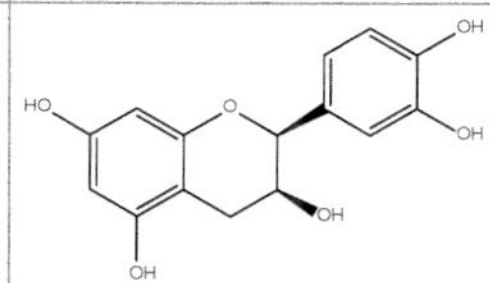	
	Synonym(s): *Chickrassia nimmonii* Graham ex Wight, *Chickrassia tabularis* Wi. **Common name(s)**: Chittagong wood, Indian red wood. **Botanical description**: It is native to East Asia, China, India, Bangladesh, Myanmar, Thailand, Malaysia, Indonesia, Laos, Cambodia, and Vietnam. It is a deciduous tree growing to up 30 m by 30 m at a medium rate, with a trunk diameter of up to 120 cm that may be branchless for up to 25 m. It has convex buttresses up to 150 cm tall. The bark is coarsely cracked and dark brown in color. The leaves are narrowly oval and tapes to the tip. The flowers are pale red and are located in branched clusters. **Medicinal use(s)**: It is used to treat cancer. It possesses anticancer, antimalarial, antifungal, antibacterial, and antimutagenic properties. An extract of the bark has powerful astringent properties and has been used to treat diarrhea and as a febrifuge.			7-Hydroxycoumarin	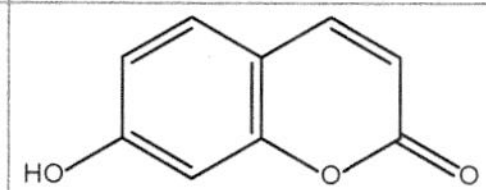	
				Rutin	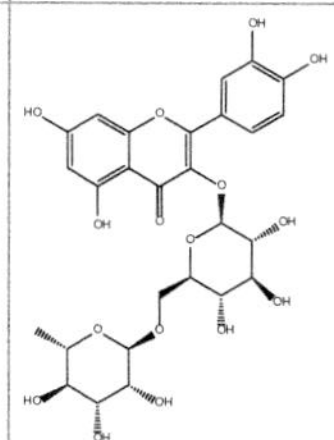	

(continued)

Table 4.88 (continued)

Family	Botanical name	Image	Plant part used	Bioactive compound(s)	Structure	Reference(s)
	Dysoxylum binectariferum Hook.f		Heart wood and gum	Rohitukine	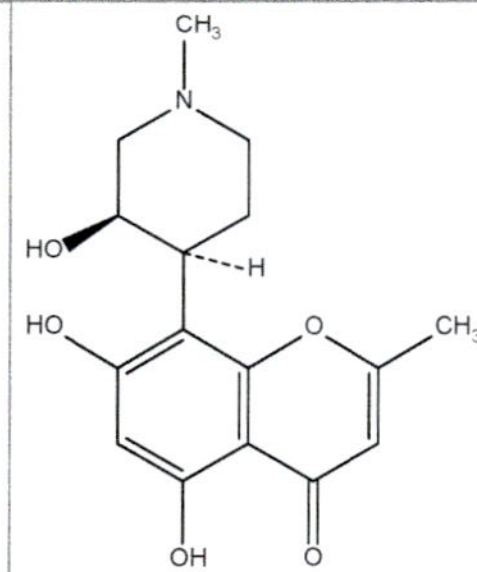	Mohanakumara et al. (2010)
	Synonym(s): *Dysoxylum ficiforme* (weight) Gamble. **Common name(s)**: Akil. **Botanical description**: It is native to India. It is an evergreen tree with leaflets cuneate at the base, entirely or obscurely dentate along the margins, acuminate at the apex. Flowers shortly pedicellate, calyx lip-shaped, entirely or obscurely 5-lobed, about half as along as petals. **Medicinal use(s)**: It is used to treat breast and ovarian cancer cell lines. It possesses anti-inflammatory, antibacterial, antimicrobial, and antifungal properties. It has a precursor, flavopiridol, which is a promising anticancer compound.					

	Melia azedarach L.		Bark	Toosendanin	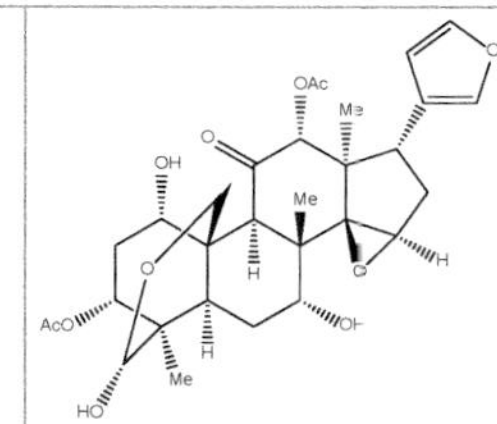	Nozaki et al. (2010)
	Synonym(s): *Antelaea javanica* Gaertn., *Azedara speciosa* Rafin., *Melia angustifolia* Schum. & Thonn., *Melia argentea* Buch.-Ham. ex Wall., *Melia toosendan* Sieb. & Zucc., *Melia robusta* Roxb. ex G. Don, *Melia japonica* G. Don, *Melia florida* Salisb., *Melia composita* Benth., *Melia cochinchinensis* M. Roem., *Melia bogoriensis* Koord. & Valet., *Melia azedarach* var. japonica (G. Don) Mak., *Melia azedarach* var. glandulosa Pierre. **Common name(s)**: Hog bush. **Botanical description**: It is native to East Asia, Central and Southern China, India, Sri Lanka, Nepal, Bhutan, Thailand, Laos, Vietnam, Indonesia, the Philippines, and Eastern Australia. It reaches a height of around 45 in. in closed moist forest, though it is much smaller, cooler 7 drier regions in which it is sometimes grown and where it is much more likely to be just 10–15 m tall. The straight, cylindrical bole can be free of branches up to 20 m. **Medicinal use(s)**: It is used to treat hepatocellular carcinoma. It possesses anticancer, anti-inflammatory, antifungal, antimicrobial, anti-parasitic, anti-leprotic, antiseptic, antipyretic, and antibacterial properties. It is also used in the treatment of hernia, abdominal, chest, and flank pain caused due to roundworms or tapeworms. This herb extract can stack the 'sphincter of Oddi' and contract the gallbladder to promote the release of bile.					

(continued)

Table 4.88 (continued)

Family	Botanical name	Image	Plant part used	Bioactive compound(s)	Structure	Reference(s)
	Swietenia humilis Zucc.		Seed	Rohitukine	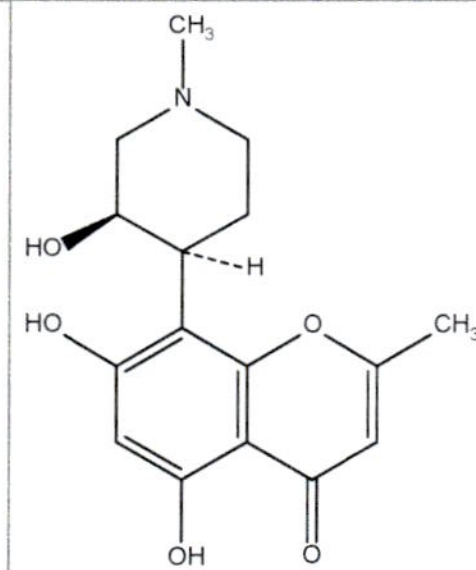	Segura-Correa et al. (1993)
	Synonym(s): No synonyms are recorded for this name. **Common name(s)**: Pacific coast mahogany. **Botanical description**: It is native to Central America – Costa Rica to Mexico. It is a small- to medium-sized deciduous tree growing up to 15–20 m tall. The unbuttressed bole is short, often crooked, and from 30 to 50 cm in diameter. It is found in dry deciduous forest, savannah, rough scrub, rocky hillsides, and cultivated fields. **Medicinal use(s)**: It is used to treat breast cancer. It possesses anticancer, antidiabetic, antimicrobial, anti-inflammatory, antioxidant, and antimutagenic properties. It is used for treating fever, diarrhea, chest pain, cough, and amebiasis.					

Table 4.89 Anticancer plants of family Menispermaceae

Family	Botanical name	Image	Plant part used	Bioactive compound(s)	Structure	Reference(s)
Menispermaceae	*Tinospora cordifolia* (Willd.) Miers ex Hook. F. & Thoms.		Whole plant	Quercetin	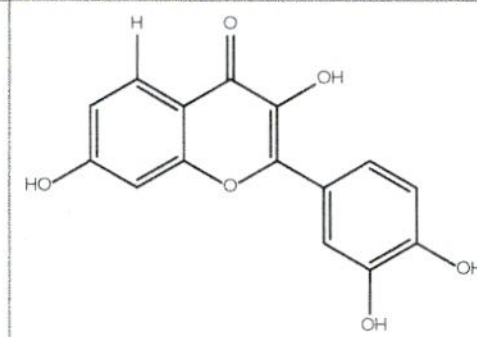	Jagetia and Rao (2006)
	Synonym(s): *Menispermum cordifolium* Willd. **Common name(s)**: Gulbel, Guduchi. **Botanical description**: It is native to the tropical areas of India, Myanmar, and Sri Lanka. It is a large, deciduous extensively spreading climbing shrub with several elongated twining branches. Leaves simple, alternate, exstipulate, long petioles up to 15 cm long, roundish, pulvinate, both at the base and apex with the basal one longer and twisted partially and half way around. Lamina broadly ovate or ovate-cordate, 10–20 cm long or 8–15 cm broad, 7-nerved and deeply cordate at base, membranous, pubescent above, whitish tomentose with a prominent reticulum beneath. Flowers unisexual, small on separate plants and appearing when plant is leafless, greenish yellow on axillary and terminal racemes. **Medicinal use(s)**: It is used to treat breast cancer, fever, jaundice, chronic diarrhea, cancer, dysentery, bone fracture, pain, asthma, skin disease, poisonous insect bite, snakebite, and eye disorders.			Berberine	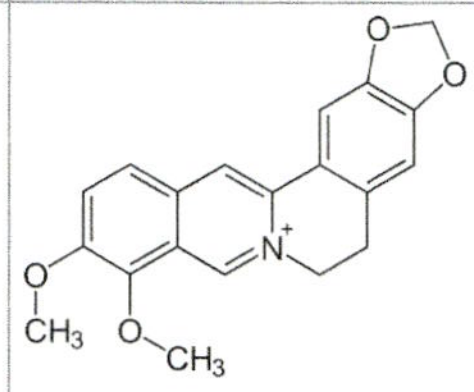	

Table 4.90 Anticancer plants of family Moraceae

Family	Botanical name	Image	Plant part used	Bioactive compound(s)	Structure	Reference(s)
Moraceae	*Antiaris africana* Engler		Stem and bark	3β-Acetoxy-1β,11α-dihydroxy-olean-12-ene	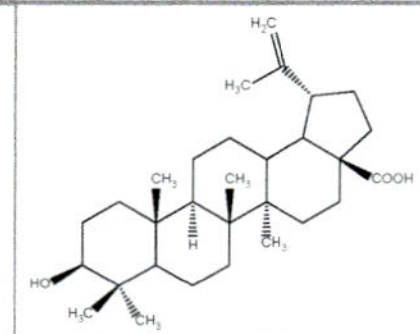	Kuete et al. (2009)
				Ursolic acid	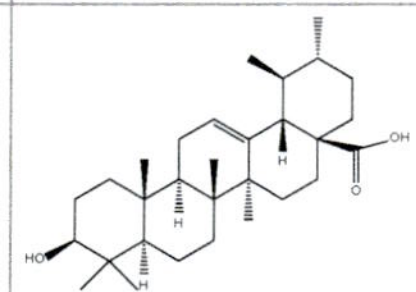	
	Synonym(s): *Antiaris toxicaria* subsp. *africana* (Engl.) C.C.Berg. **Common name(s)**: Bark cloth tree. **Botanical description**: It is native to tropical Africa, through tropical Asia to Northern Australia. It is a deciduous, evergreen, small to large tree with a crown of short spreading branches. It usually grows up to 45 m tall. The bole can be up to 150 cm and usually buttressed. **Medicinal use(s)**: It is used to treat prostate cancer cell lines and hepatocarcinoma HepG2 cells. It possesses antitumor, antioxidant, antimalarial, anti-inflammatory, antiproliferative, and anti-parasitic properties.			3,3′-Dimethoxy-4′-O-β-D-xylopyronosylellagic acid	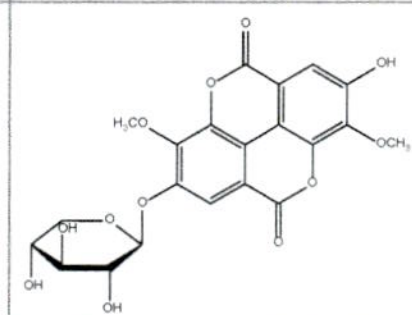	
				Methyl strophanthinate	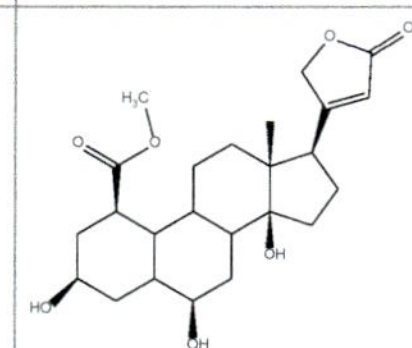	

		Stropha:thidinic acid	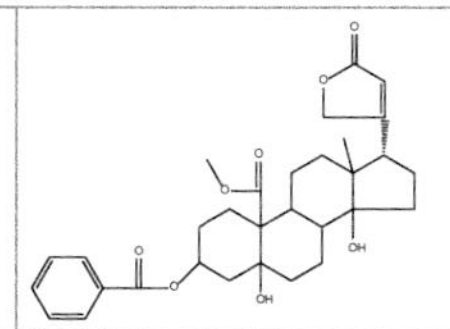	
		Convalla:oxin	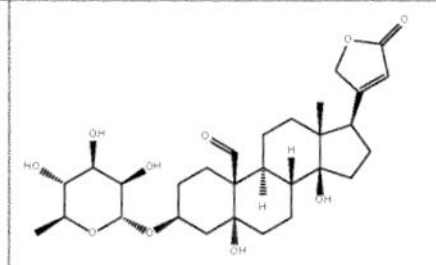	
		Periplogenin	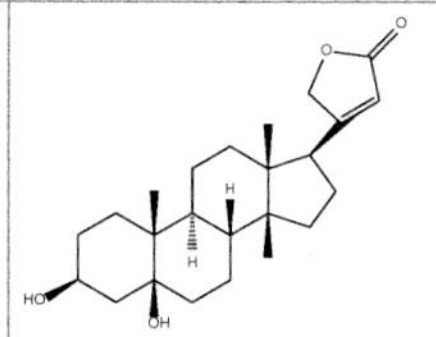	
		Strophanthido	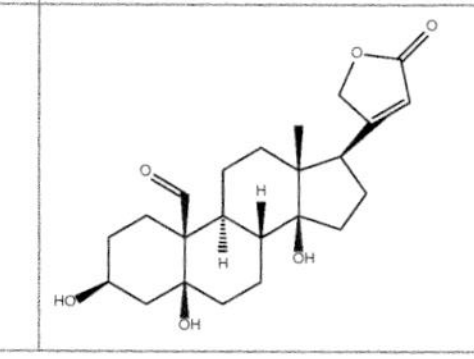	

(continued)

Table 4.90 (continued)

Family	Botanical name	Image	Plant part used	Bioactive compound(s)	Structure	Reference(s)
	Antiaris toxicaria Lesch.		Stem	Cardiac glycosides toxicarioside D	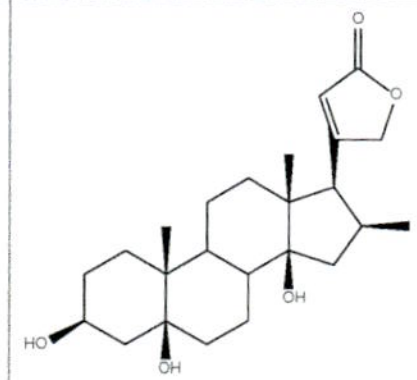	Jiang et al. (2008)
	Synonym(s): *Antiaris dubia* Span. ex Hook., *Antiaris innoxia* Blume, *Antiaris rufa* Miq., *Antiaris saccidora* Dalzell. **Common name(s)**: False iroko. **Botanical description**: It is native to tropical Africa, through tropical Asia to Northern Australia. It is a deciduous, evergreen, small to large tree with a crown of short spreading branches. It usually grows up to 45 m tall. The bole can be up to 150 cm and usually buttressed. **Medicinal use(s)**: It is used to treat prostate cancer. The bark is used as anodyne, astringent, febrifuge, and vermifuge. It is used in the treatment of hepatitis. The inner bark is chewed and the juice swallowed as a treatment for disorders of the spleen. An aqueous ethanol extract of the bark exhibited cytotoxic activity against tumor cell lines.					
	Ficus pretoriae Burtt Davy		Sap	Ellagic acid	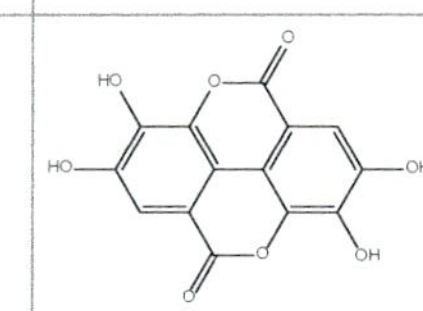	Ghazanfar and Al-Sabahi (1993) and Maffei (2003a, b, c, d, e, f, g, h, i, j, k, l, m)

	Synonym(s): *Ficus cordata* subsp. *salicifolia* (Vahl) C.C.Berg. **Common name(s)**: Wonderboom. **Botanical description**: It is native to Africa and Arabian peninsula, avoiding the regions of higher rainfall. It is a shrub growing up to 2–5 m tall or becoming graceful, sometimes drooping the trees with a dense, spreading crown; it usually grows 8–15 m tall, but in some parts, its range can reach up to 35 m. Often, the bole can be up to 100 cm in diameter. **Medicinal use(s)**: It is used to treat breast and prostate cancer. It is also used to treat diabetes, liver disorders, diarrhea, inflammatory conditions, hemorrhoids, and respiratory and urinary disorders.					
	Ficus benghalensis L.		Bark	Leucopelargonidin-3-0-α-Lrhamnoside	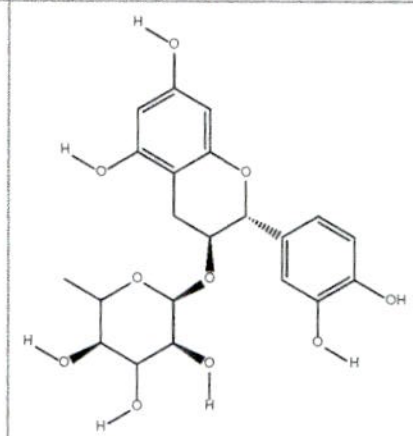	Josep and Raj (2011)
	Synonym(s): *Ficus banyana* Oken, *Ficus chauvieri* G.Nicholson, *Ficus cotoneifolia* Vahl, *Ficus cotonifolia* Stokes. **Common name(s)**: Banyan. **Botanical description**: It is native to tropical East Africa and India. It is a deciduous, evergreen, small to large tree with a crown of short spreading branches. It is usually grows up to 45 m tall. The bole can be up to 150 cm and usually buttressed. **Medicinal use(s)**: It is used to treat breast cancer. The leaf, bark, and seeds are used for treating diarrhea, polyuria, dental disorders, diabetes, and urine disorders. It is also used to treat painful joints, swelling, rashes, and intrinsic hemorrhage.			Leuco cynidin 3-0-α-D galactosyl cellobioside	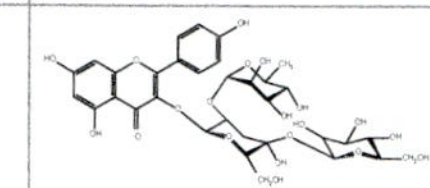	

(continued)

Table 4.90 (continued)

Family	Botanical name	Image	Plant part used	Bioactive compound(s)	Structure	Reference(s)
				Beta 19-20 sitosterolalpha-D-glucose	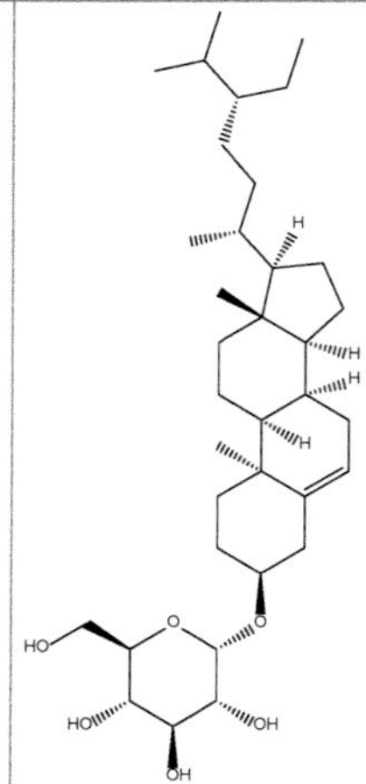	
				Pentatriacontan-5-one	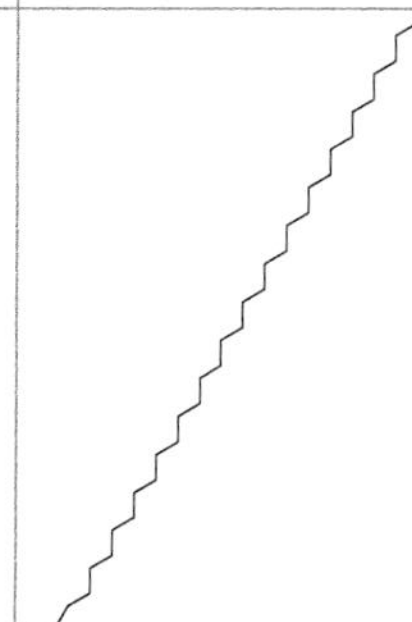	

<table>
<tr><td rowspan="4"></td><td rowspan="3">Ficus carica L.</td><td rowspan="3"></td><td rowspan="3">Leaf</td><td>Apigenin</td><td>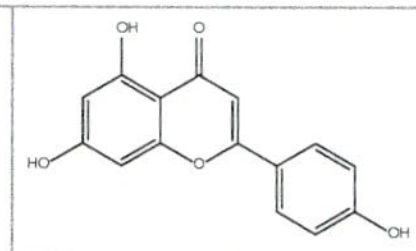</td><td rowspan="4">Rubnov et al. (2001)</td></tr>
<tr><td>Quercetin 3-O-glucoside</td><td>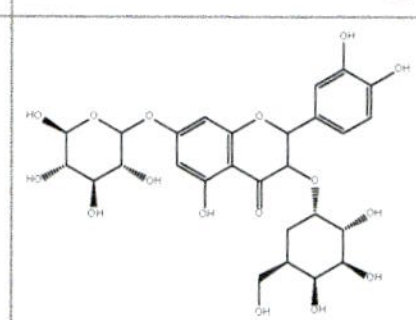</td></tr>
<tr><td>Kaempferol 3,7-di-O-rhamnoside</td><td>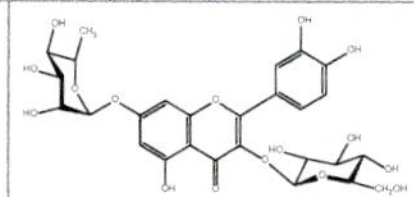</td></tr>
<tr><td colspan="3">Synonym(s): Ficus carica var. afghanica Popov, Ficus carica var. globosa Hausskn., Ficus carica var. longipes Bornm. ex Parsa, Ficus carica var. riparium Hausskn.
Common name(s): Fig, Edible fig.
Botanical description: It is native to West Asia – Turkey through Pakistan – and North Africa, Algeria, Morocco, and Tunisia. It is a deciduous tree growing up to 6 m at a medium rate. It is in flower from June to September, and the seeds ripen from August to September. The flowers are monoecious; the plants are self-fertile.
Medicinal use(s): It is used to treat esophageal and stomach cancer lines. It possesses anticancer, antioxidant, antibacterial, antifungal, antipyretic, anti-inflammatory, and antituberculosis properties. The latex and stem are used to treat corns, warts, and piles.</td><td>Apigenin 7-O-glucoside</td><td>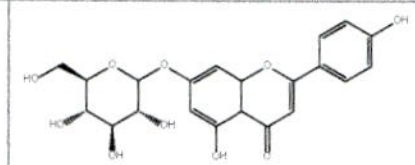</td></tr>
</table>

(continued)

Table 4.90 (continued)

Family	Botanical name	Image	Plant part used	Bioactive compound(s)	Structure	Reference(s)
	Ficus hispida L.		Bark	Quercetin 3-O-glucoside	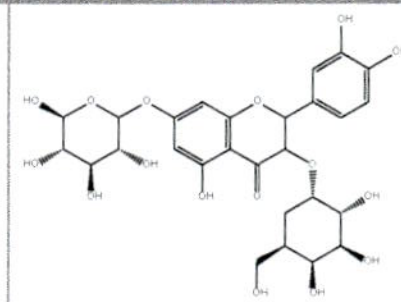	Josep and Raj (2011)
	Synonym(s): *Ficus hispida* var. *badiostrigosa* Corner, *Ficus hispida* f. *borneensis* Miq., *Ficus hispida* var. *hastata* Blanco. **Common name(s)**: Katgularia. **Botanical description**: It is native to East Asia – Southern China, Indian subcontinent, Myanmar, Indonesia to New Guinea, and Australia. It is a shrub or small tree with a spreading crown; it can grow up to 17 cm tall with a bole up to 25 cm in diameter. It is found in the base of foothills at elevations of 500–1100 m. **Medicinal use(s)**: It is used to treat human breast cancer cell lines. The juice of the root is used in the treatment of fever; bark is taken as an antipyretic, an emetic, and a tonic. The juice is also used in the treatment of liver problem.					

	Ficus indica L.		Leaf and flower	Psoralen	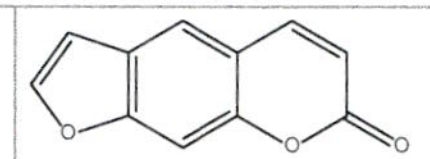	Rubnov et al. (2001)
				Bergapten	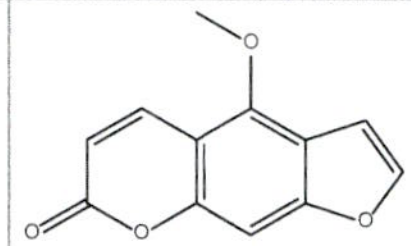	
	Synonym(s): *Ficus indica* Forssk., *Ficus indica* Willd., *Ficus indica* var. *sundaica* (Blume) Miq. **Common name(s)**: Chappathikkalli. **Botanical description**: It is native to Mexico. It is a pricky, evergreen, large, bushy cactus plant that can sometimes develop a definite woody trunk and is capable of growing up to 5 m tall. It has a shallow, fleshy root system with a horizontal spread. **Medicinal use(s)**: It is used to treat colon, ovarian, and cervical epithelial cancer cell lines. It possesses anticancer, antitumor, anti-inflammatory, and antioxidant properties. An infusion of the ground stem is used for treating stomach ulcers. The split stems have been bound around injured limbs as a first aid measure. The fruit is diuretic. The flowers are also used in the treatment of enlarged prostate gland.			Taraxasterol	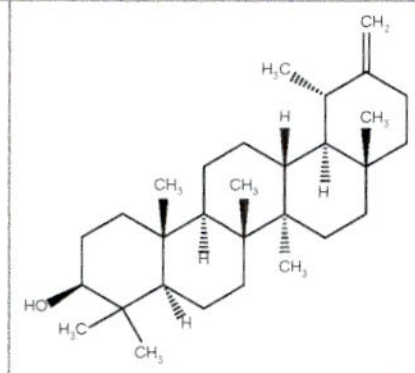	
				6-oacyl-β-D-Sitosterol	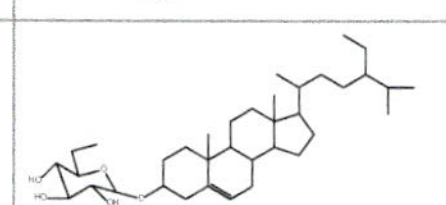	

(continued)

Table 4.90 (continued)

Family	Botanical name	Image	Plant part used	Bioactive compound(s)	Structure	Reference(s)
	Ficus retusa L.		Apical part	Retusaphenol	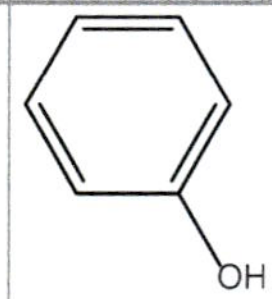	Sarg et al. (2011)
				Retusa afzelechin	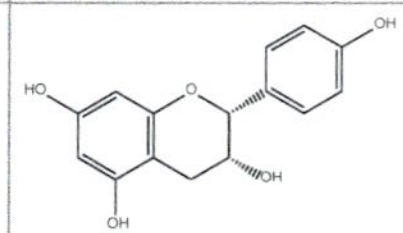	
	Synonym(s): *Ficus retusa* var. *borneensis* Corner, *Ficus retusa* var. *nitida* (Thunb.) Miq., *Ficus retusa* f. *nitida* (Thunb.) King, *Ficus retusa* var. *ovoidea* (Jack) Miq. **Common name(s)**: Indian laurel fig. **Botanical description**: It is native to India, Myanmar, and Bangladesh. It is distributed in forest at low and medium altitudes ascending to 1500 m; it is found in tropical and subtropical region; flowers are unisexual and numerous within the fig; leaves are short, elliptic oborate. The main lateral nerves are not very prominent, and stipules are lanceolate. **Medicinal use(s)**: It is used to treat lung cancer and leukemia. It possesses anticancer, antioxidant, anti-inflammatory, antibacterial, anti-ulcer, antidiabetic, antifungal, anti-rheumatic, and antihistamine properties. It is also used to treat ulcer, skin diseases, edema, and inflammations. The bark is given with buttermilk to cure liver diseases. The decoction of the bark is used as a cooling agent in case of abdominal pain, ulcers, liver problems, mouth ulcers, and leukorrhea.			Luteolin	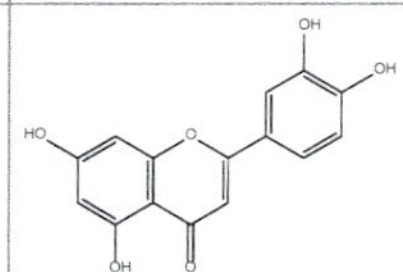	
				(+) -Afzelechin	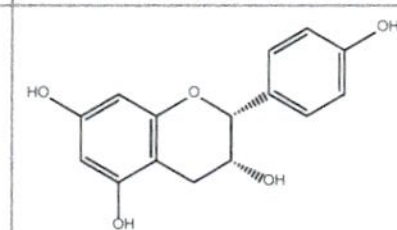	
				(+) – catechin	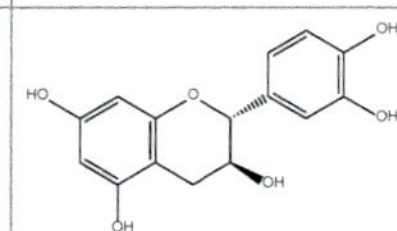	

		Vitexin	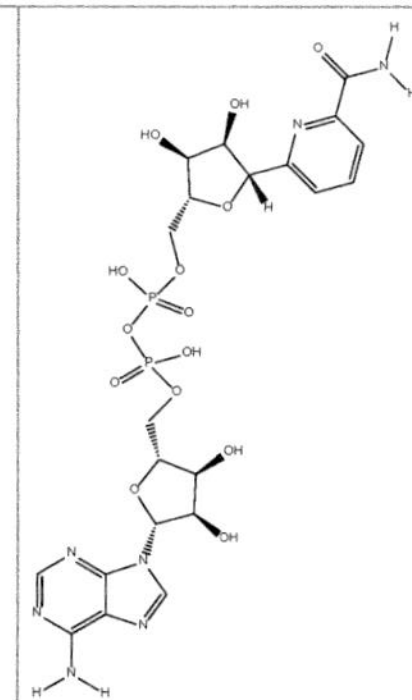	
		Amyrin acetate	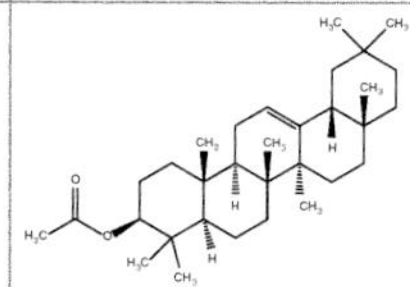	
		Moretenone	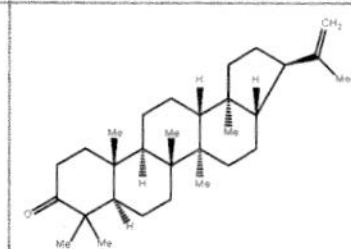	
		Friedelenol	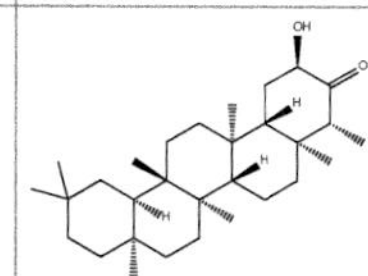	

(continued)

Table 4.90 (continued)

Family	Botanical name	Image	Plant part used	Bioactive compound(s)	Structure	Reference(s)
				Amyrin	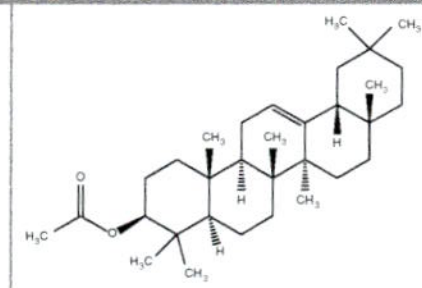	
				Sitosterol	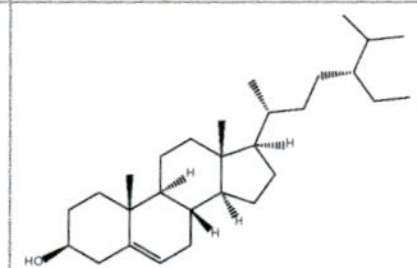	
	Maclura cochinchinensis (Lour.) Corner		Stemwood	Moretenone	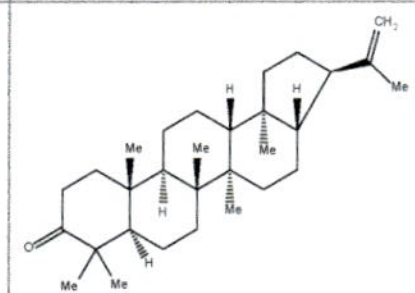	Karnjanapee and Natori (1966), Wasuwat (1967), and Bunyapraphatsara et al. (2000)
	Synonym(s): No synonyms are recorded for this name. **Common name(s)**: Cockspur thorn. **Botanical description**: It is native to East Asia – Indian subcontinent through Myanmar, Malaysia, Indonesia, and Indochina to Australia. It is a slow-growing thorny vine. The stem grows up to 10 m long and up to 15 cm in diameter. The thorns are present on the nodes. **Medicinal use(s):** It possesses anticancer properties. Wood is used to treat fevers. A decoction of the roots is used to treat alleviate coughing.					

	Maclura tinctoria L. (Gaud.)		Stem and bark	Luteolin	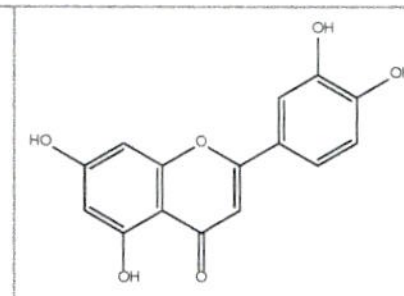	Kupeli et al. (2006)
	Synonym(s): *Broussonetia plumeri* Spreng., *Broussonetia tinctoria* (L.) Dum.Cours., *Broussonetia xanthoxylon* Mart., *Chlorophora tinctoria* (L.) Gaudich. **Common name(s)**: Old fustic. **Botanical description**: It is native to South America – Argentina and Paraguay north – to the Caribbean and through Central America to Mexico. It is a deciduous tree growing up to 20 m (65 ft.) by 15 m (49 ft.) at a medium rate. The plant is not self-fertile. It cannot grow in the shade. It prefers dry or moist soil. **Medicinal use(s)**: It possesses anticancer, antioxidant, antiproliferative, and anti-inflammatory properties. The bark is used as astringent, tonic, and vermifuge and as purgative in large doses.					

(continued)

Table 4.90 (continued)

Family	Botanical name	Image	Plant part used	Bioactive compound(s)	Structure	Reference(s)
	Maclura pomifera Rafin.		Fruit	Scandenone (isoflavone)	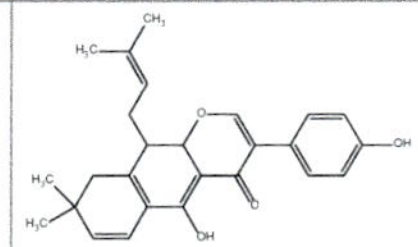	Kupeli et al. (2006)
				Auriculasin	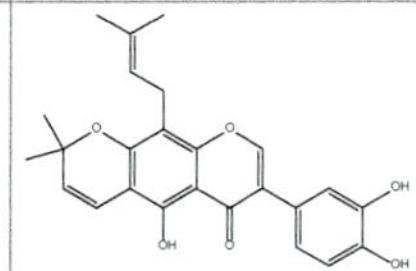	
	Synonym(s): *Ioxylon pomiferum* Raf., *Joxylon pomiferum* Raf., *Toxylon aurantiacum* (Nutt.) Raf., *Toxylon maclura* Raf. **Common name(s)**: Osage orange. **Botanical description**: It is native to southeastern North America – Arkansas to Texas. It is a deciduous tree growing up to 15 m (49 ft.) by 12 m (39 ft.) at a medium rate. It is in flower from May to June, and the seeds ripen from September to December. The species is dioecious (individual flowers are either male or female, but only one sex is to be found on any one plant, so both male and female plants must be grown if seed is required). The plant is not self-fertile. It cannot grow in the shade. It prefers dry or moist soil. The plant can tolerate maritime exposure. **Medicinal use(s)**: It possesses anticancer, antitumor, antioxidant, antiproliferative, and antimicrobial properties. The inedible fruits contain antioxidant and fungicidal compounds. A tea made from the roots has been used as a wash for sore eyes.					

Table 4.91 Anticancer plants of family Myristicaceae

Family	Botanical name	Image	Plant part used	Bioactive compound(s)	Structure	Reference(s)
Myristicaceae	*Knema angustifolia* (warb.)		Whole part	Norviburtinal	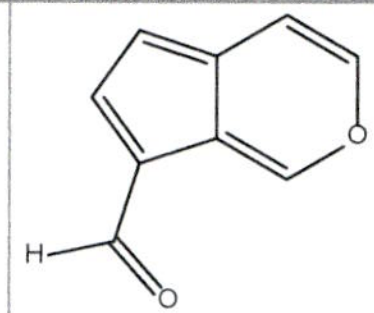	Phadungkit et al. (2010)
				β-Sitosterol	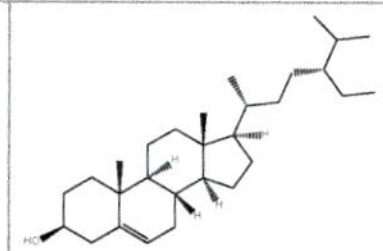	
	Synonym(s): *Myristica angustifolia* Roxb. **Common name(s)**: Wild nutmeg, Angustifolia. **Botanical description**: It is native to Asia – Southern China, India, Myanmar, Thailand, Laos, Cambodia, Vietnam, Malaysia, and Indonesia. It is an evergreen plant growing up to 5–20 m tall. The straight bole is fluted about 30–35 cm in diameter. It is found in the primary and secondary evergreen forest, along riverbank in hilly forest, at elevations from 150 to 1200 m. **Medicinal use(s)**: It is used to treat lung cancer. It possesses anticancer, antioxidant, antimicrobial, antifungal, anti-inflammatory, and antibacterial properties. The seeds are used in an externally applied medicine to cure scabies and other skin diseases. It is also used for virus infection such as the common cold and flu. It is used to strengthen the immune system. It has also been used for treating vaginal fungal infection (yeast infection).					

(continued)

Table 4.91 (continued)

Family	Botanical name	Image	Plant part used	Bioactive compound(s)	Structure	Reference(s)
	Knema glauca (Blume.)		Root	Myristinins A	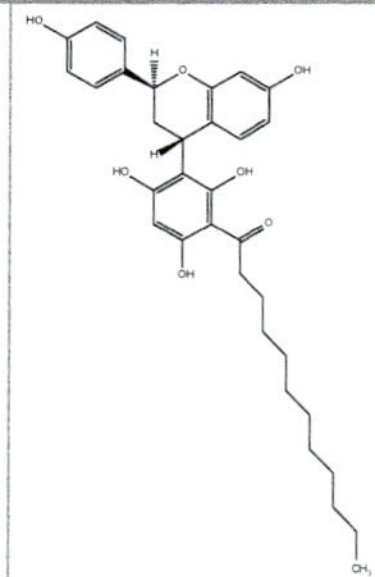	Rangkaew et al. (2009)
	Synonym(s): *Knema angustifolia* (Roxb.) Warb. **Common name(s)**: Dara-dara laki. **Botanical description**: It is native to Southeast Asia – Peninsular Thailand, Malaysia, and Indonesia. It is an evergreen tree with a light, open canopy; it can grow from 5 to 30 m tall. The bole can be free of branches for up to half its height; stilt roots up to 1 m tall are sometimes present. **Medicinal use(s)**: It is used to treat lung and breast cancer and leukemia. It has anti-inflammatory, anticancer, anti-rheumatic, antibacterial, anti-nematodal, cytotoxicity, and acetyl cholinesterase inhibitory activity. It is used to treat bone diseases.					

	Knema tenuinervia W.J.J.O.DeWilde		Stem bark	3-[(Z)-12-Phenyl-3-dodecenyl] phenol	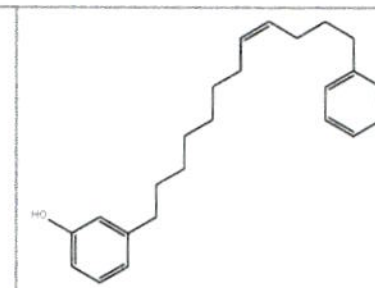	Kijjoa et al. (1991), Banerji et al. (1994), and Wangchuk et al. (2011)
	Synonym(s): *Knema elegans*. **Common name(s)**: Dara-dara laki. **Botanical description**: It is native to Southeast Asia – Peninsular, Thailand, Malaysia, and Indonesia. It is an evergreen tree with a light open canopy. It can grow from 5 to 30 m tall. The bole can be free of branches for up to half its height; stilt root up to 1m tall is sometimes present. It is found in primary and degraded evergreen forest, including mixed dipterocarp forest edges and riverine forest. **Medicinal use(s)**: It is used to treat lung and breast cancer and leukemia. It has anti-inflammatory, anticancer, anti-rheumatic, antibacterial, anti-nematodal, cytotoxic, and acetylcholinesterase inhibitory activity. It is used to treat bone diseases and skin tumors.					

(continued)

Table 4.91 (continued)

Family	Botanical name	Image	Plant part used	Bioactive compound(s)	Structure	Reference(s)
	Leptospermum scoparium J.R.Forst. & G.Forst.		Apical part	Glyoxal	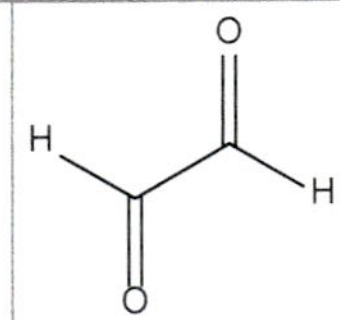	Mavric et al. (2008)
				Methylglyoxal	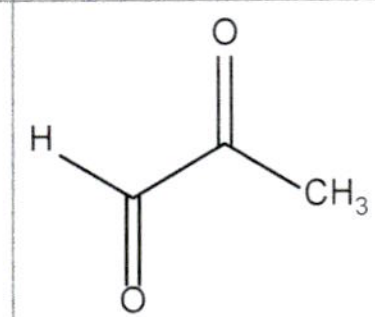	
	Synonym(s): *Knema elegans*. **Common name(s):** Manuka. **Botanical description:** It is native to New Zealand, Australia, the United States, Britain, South Africa, and Madeira. It is a prolific scrub-type tree and is often one of the first species to regenerate on cleared land. It is typically a shrub growing up to 2–5 m tall but can grow into a moderately sized tree, up to 15 m or so in height. It is evergreen, with dense branching and small leaves 7–20 mm long and 2–6 mm broad, with a short spine tip. The flowers are white, occasionally pink, 8–15 mm in diameter, with five petals. The wood is tough and hard. **Medicinal use(s):** It is used to treat lung and breast cancer and leukemia. It has anti-inflammatory, anticancer, anti-rheumatic, antibacterial, anti-nematodal, cytotoxic, and acetylcholinesterase inhibitory activity. It is used to treat bone diseases and skin tumors.					

	Pycnanthus angolensis (Welw.) Warb		Leaf	Pycnanolide A	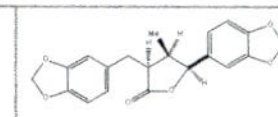	Onocha and Ali (2011)
	Synonym(s): *Pycnanthus kombo* Warb. **Common name(s)**: African nutmeg, Boxboard, Cardboard, False nutmeg, Pycnanthus, Akwa-mili, Oje, Lunaba, Munaba, Calabo, Ilomba, Akomu. **Botanical description**: It is native to tropical Africa – Senegal to Sudan and Uganda, south to Angola, Zambia, and Tanzania. It is a large, evergreen tree with a small crown of branches at right angles to the bole; it usually grows from 25 to 35 m tall, though trees up to 40 m have been recorded. The bole is usually straight and cylindrical, sometimes with a swollen base or root swelling; it can be unbranched for up to 18 m, with a bole 60–150 cm or more in diameter. **Medicinal use(s)**: It is used to treat human tumor cell lines. It is also used for the treatment of coughs and chest pains, malaria, anemia, ascites, and leprosy. It possesses anticancer, antitumor, anti-inflammatory, antibacterial, and antioxidant properties.			Pycnanolide B	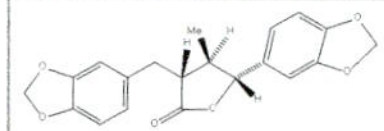	

(continued)

Table 4.91 (continued)

Family	Botanical name	Image	Plant part used	Bioactive compound(s)	Structure	Reference(s)
	Virola bicuhyba (Schott) Warb.		Seed	Kaempferol	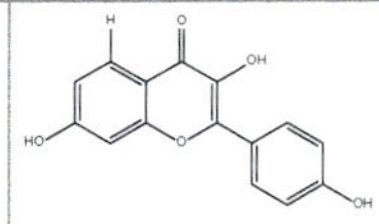	Plotkin and Schultes (1990)
	Synonym(s): *Myristica bicuhyba* Schott. **Common name(s)**: Red ucuuba. **Botanical description:** It is native to North and South America. It is a tall, thin tree which grows up to 5–30 m tall. The leaves are simple and can grow up to 30 cm long. The small flowers are single-sexed and are found in panicles. The fruit is reddish, oval-shaped, and about 10–15 mm long and 11 mm in diameter. The individual trees, which include 40–60 species, are difficult to differentiate from one another. **Medicinal use(s):** It is used to treat breast cancer. It possesses anti-inflammatory, antibacterial, antitumor, and antioxidant properties.					

Table 4.92 Anticancer plants of family Myrtaceae

Family	Botanical name	Image	Plant part used	Bioactive compound(s)	Structure	Reference(s)
Myrtaceae	*Eugenia jambos* L.		Leaf	1-O-galloyl castalagin	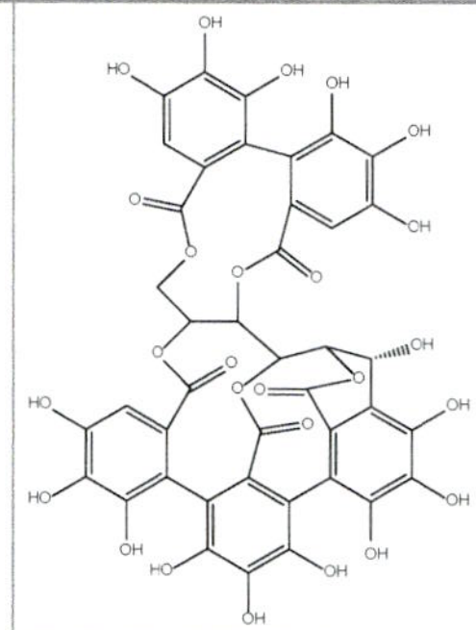	Yang et al. (2000) and Cock and Mohanty (2010)
	Synonym(s): *Eugenia jambos*, *Eugenia decora*, *Myrtus jambos*. **Common name(s)**: Rose apple. **Botanical description**: The rose apple is native to the East Indies and Malaya and is cultivated and naturalized in many parts of India, Ceylon and former Indochina, and the Pacific Islands. It is a shrub but is generally a tree reaching 25 or even 40 ft. (7.5–12 m) in height and has a dense crown of slender, wide-spreading branches, often the overall width exceeding the height. The evergreen leaves are opposite, lanceolate or narrow elliptic, tapering to a point; 4–9 in. (10–22 cm) long, and from 1 to 2 1/2 in. (2.5–6.25 cm) wide; somewhat leathery, glossy, dark green when mature, rosy when young. **Medicinal use(s)**: The active organic compounds in rose apples, combined with vitamin C and vitamin A, have become known as effective cancer prevention treatments. Early research and traditional medical texts claim that prostate and breast cancers are reduced. Rose apples are rich in vitamin C, dietary fiber, vitamin A, calcium, thiamin, niacin, iron, sulfur, and potassium. In terms of organic compounds, rose apples contain jambosine, betulinic acid, and friedelolactone.			Casuarinin	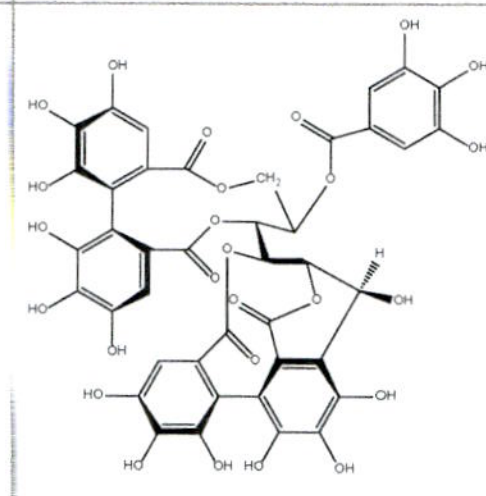	

(continued)

Table 4.92 (continued)

Family	Botanical name	Image	Plant part used	Bioactive compound(s)	Structure	Reference(s)
	Leptospermum scoparium Forst. And Forst.		Aerial parts	Quercetin	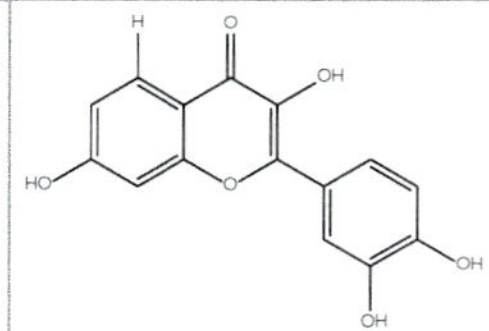	Mayer (1993) and Maddocks-Jennings et al. (2009)
	Synonym(s): *Leptospermum bullatum* Fitzh., *Leptospermum floribundum* Salisb., *Leptospermum humifusum* A.Cunn. ex Schauer, *Leptospermum linifolium* (Sol.) Dum.Cours., *Leptospermum multiflorum* Cav. **Common name(s)**: Tea tree, Broom tea tree, Manuka. **Botanical description**: *Leptospermum scoparium* is native to Australia and New Zealand. It is an evergreen shrub growing up to 5 m in height. The flowers are hermaphrodite and are pollinated by insects. The plant is suitable for light (sandy), medium (loamy), and heavy (clay) soils and prefers well-drained soil. Suitable pH, acid and neutral soils. It cannot grow in the shade. It prefers dry or moist soil. **Medicinal use(s)**: It is used for the treatment of cancers. It possesses antibacterial, antiviral, anti-inflammatory, antifungal, antiseptic, digestive, and antioxidant properties. It is also used for the treatment of flu, fever, colds, skin diseases, and ulcers.					

Table 4.93 Anticancer plants of family Nyssaceae

Family	Botanical name	Image	Plant part used	Bioactive compound(s)	Structure	Reference(s)
Nyssaceae	*Camptotheca acuminate* Decne.		Root	9-Methoxycamptothecin	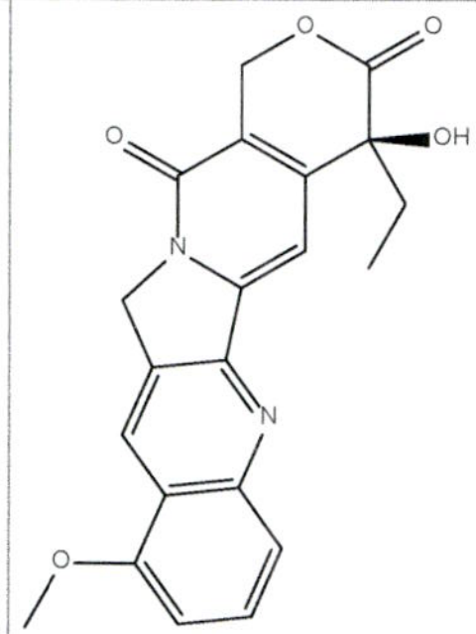	Wu et al. (1980) and Lin et al. (2014)
	Synonym(s): *Camptotheca yunnanensis* Dode, *Cephalanthus esquirolii* H.Lév. **Common name(s)**: The happy tree, Cancer tree, Tree of life. **Botanical description**: It is native to Southern China and Tibet. It grows to a height of 20–25 m tall and has a light gray bark and oval leaves with heavy pleated veining. The flowers are white to yellowish and are present in spherical clusters. **Medicinal use(s)**: The extract of tender is used to treat brain tumors, liver cancer, gastrointestinal tract cancer, leukemia, and other cancers. It is also used for the treatment of common colds, psoriasis, liver problems, and digestive problems.			10-Hydroxycamptothecin	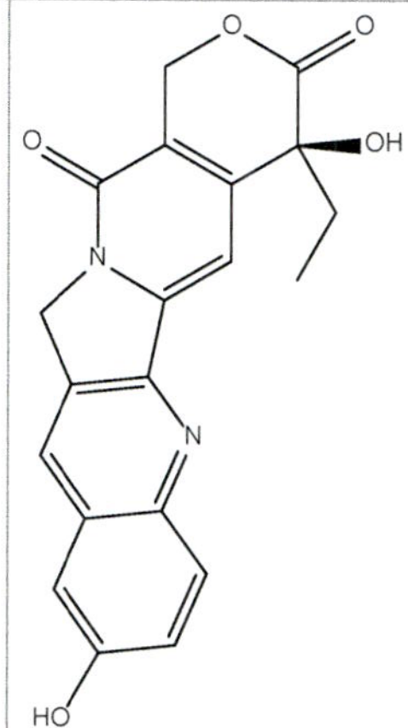	

(continued)

Table 4.93 (continued)

Family	Botanical name	Image	Plant part used	Bioactive compound(s)	Structure	Reference(s)
				Topotecan	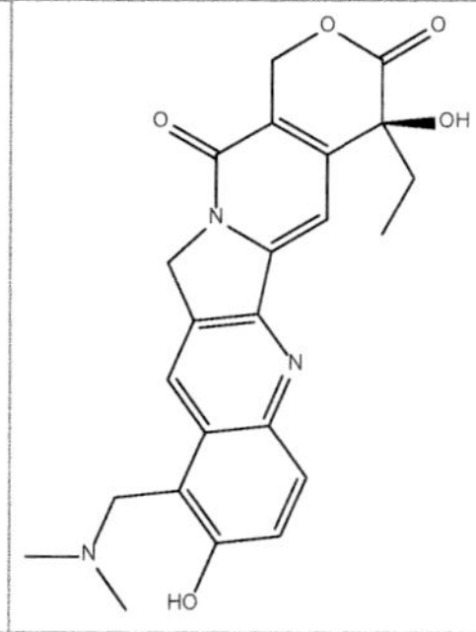	
				9-Aminocamptothecin	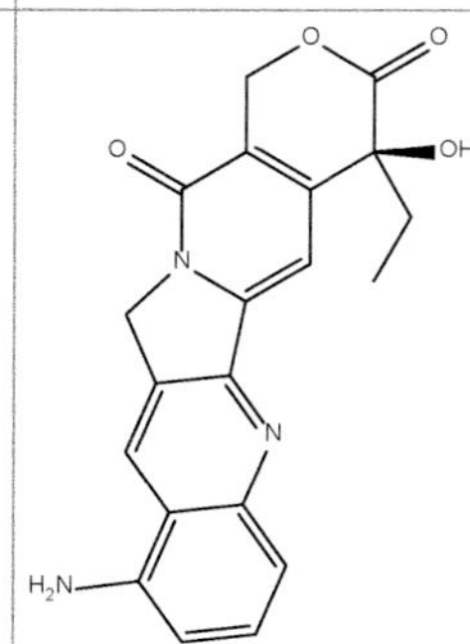	

		Irinotecan	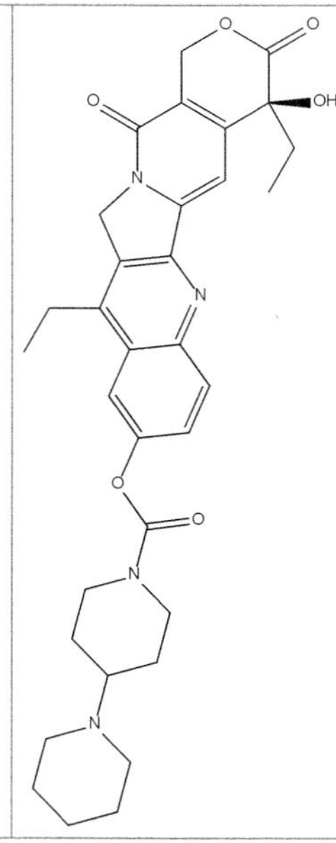	

(continued)

Table 4.93 (continued)

Family	Botanical name	Image	Plant part used	Bioactive compound(s)	Structure	Reference(s)
	Nyssa sinensis Oliv.		Rootbark	Etoposide	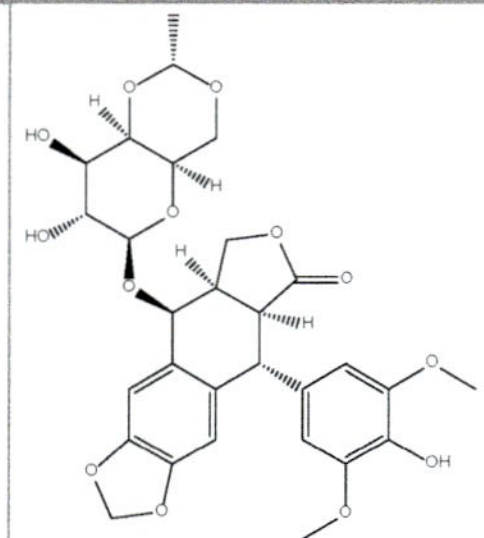	Luo and Xiong (1991) and Florence et al. (2011)
	Synonym(s): *Microrhamnus bodinieri* H. Lév., *Nyssa megacarpa* R.Parker. **Common name(s)**: Chinese tupelo. **Botanical description**: It is found in low acid woods, swamps, and shores in imperfectly drained soils and is native to China and Vietnam. It is a broadly conical deciduous tree and grows up to 10 m tall, with oval leaves 15–20 cm long. It turns into brilliant red, orange, and yellow color in autumn season. **Medicinal use(s)**: It is used for treatment of cancer. The bark is used as emetic, ophthalmic, and vermifuge. An infusion is given to children to get rid of the worms. A strong decoction is used to cause vomiting when unable to retain food. Strong ooze from the roots is used as eye drops.			Paclitaxel	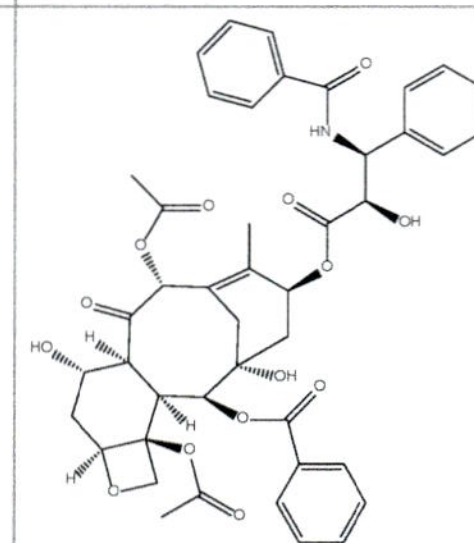	

		Camptothecin	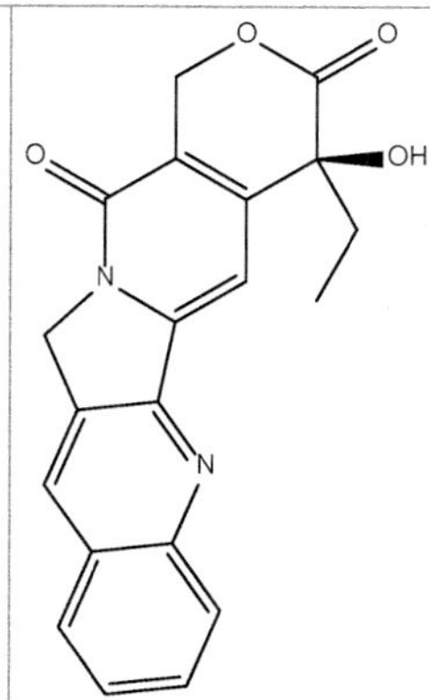	
		Topotecan	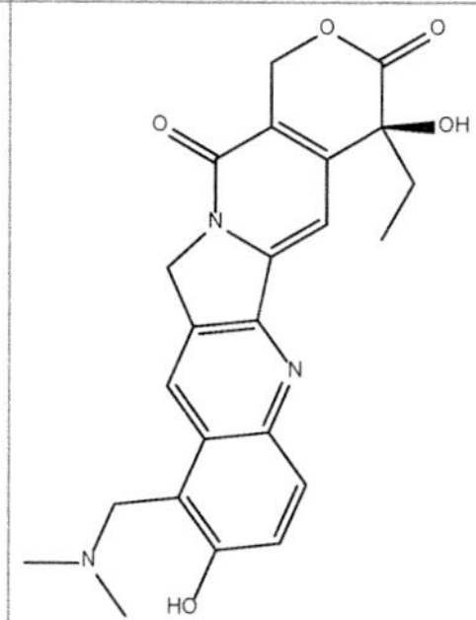	

Table 4.94 Anticancer plants of family Ochnaceae

Family	Botanical name	Image	Plant part used	Bioactive compound(s)	Structure	Reference(s)
Ochnaceae	*Ouratea semiserrata* (Mart. & Nees) Engl.		Leaf	Agathisflavone	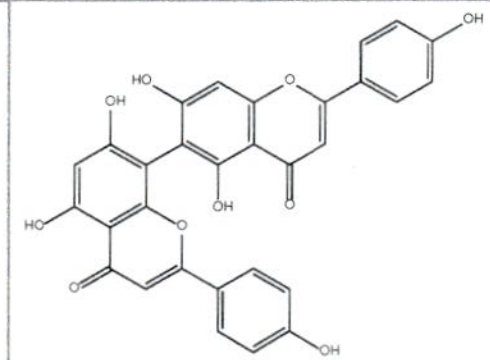	Grynberg et al. (2002)
	Synonym(s): No synonyms are recorded for this name. **Common name(s)**: Loro micuna, Araçá-do-campo. **Botanical description**: It is a shrub or small tree growing up to 3–5 m tall and is native to South America, Southern and Eastern Brazil. The plant is harvested from the wild for medicinal purposes. **Medicinal use(s)**: It is used for the treatment of stomach cancer, liver cancer, tumor, and leukemia. It also possesses antioxidant, antitumor, and antibacterial properties.					
	Luxemburgia octandra A. St.-Hil.		Root, stem, leaf	Luxenchalcone	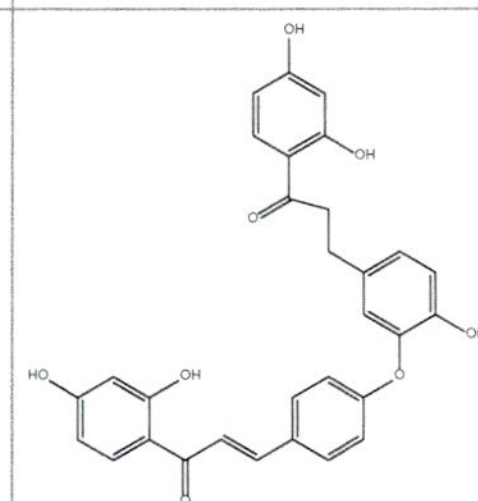	Daniel et al. (2007)

	Synonym(s): *Luxemburgia angustifolia* Planch., *Luxemburgia bracteata* Dwyer, *Luxemburgia ciliatibracteata* Sastre, *Luxemburgia ciliosa* (Mart.) Planch. **Common name(s)**: Rock osbeckia. **Botanical description**: It is native to Brazil, Sri Lanka, and peninsular region of India. It is a flowering plant. It grows up to 1 m tall; leaves are 1.5–3.3 cm in length and 0.5–1.8 cm in width. **Medicinal use(s)**: *Luxemburgia octandra* possesses anticancer, antiproliferative, anti-inflammatory, and antioxidant properties.					
	Ouratea hexasperma (A. St.-Hil.) Baill.		Leaf	7′-O-Methyl-agathisflavone	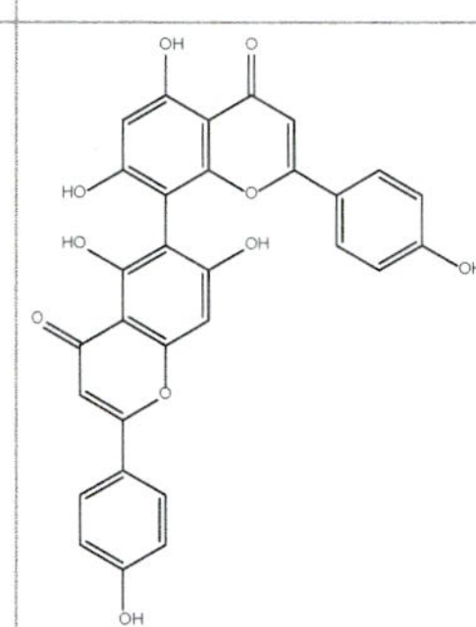	Grynberg et al. (2002)
	Synonym(s): No synonyms are recorded for this name. **Common name(s)**: Ponytail palm, Witches' broom. **Botanical description**: It is a shrub or small tree growing up to 3–5 m tall and is native to Brazil. The plant is harvested from the wild for medicinal purposes. **Medicinal use(s)**: It is used for the treatment of breast cancer and tumor. It also possesses anti-inflammatory, antioxidant, and antibacterial properties.					

(continued)

Table 4.94 (continued)

Family	Botanical name	Image	Plant part used	Bioactive compound(s)	Structure	Reference(s)
	Ouratea parvifolia Engl.		Leaf	Agathisflavone	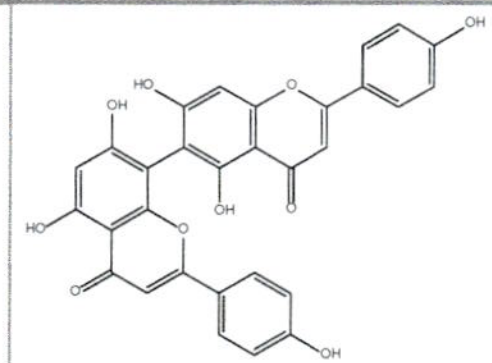	Araujo et al. (2011)
				7″-Methylagathisflavone	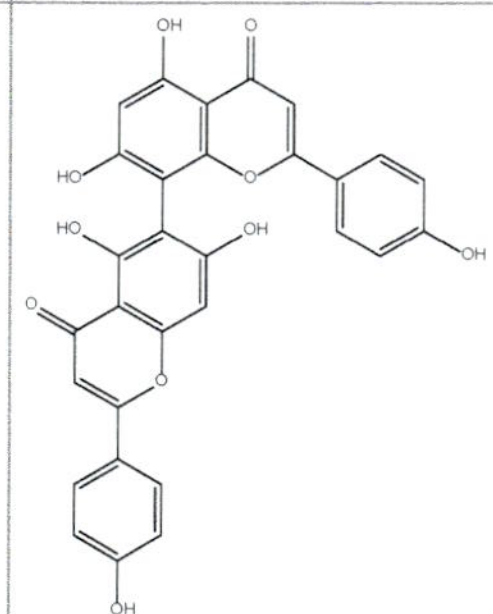	
	Synonym(s): No synonyms are recorded for this name. **Common name(s)**: Small flowered crape myrtle. **Botanical description**: It is a shrub or small tree growing up to 3–5 m tall and is native to Brazil. The plant is harvested from the wild for medicinal purposes. **Medicinal use(s)**: It is used for the treatment of breast cancer, inflammation, and gastric diseases. It also possesses anti-inflammatory, antioxidant, and antibacterial properties.			Amentoflavone	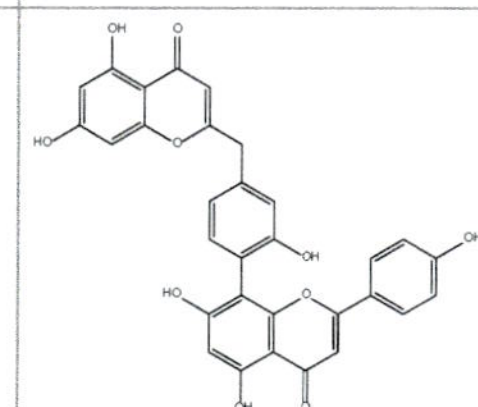	

		Apigenin	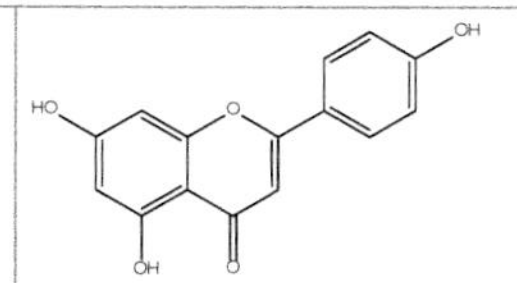	
		Stigmasterol	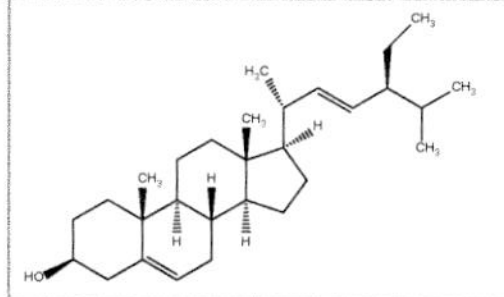	
		Campesterol	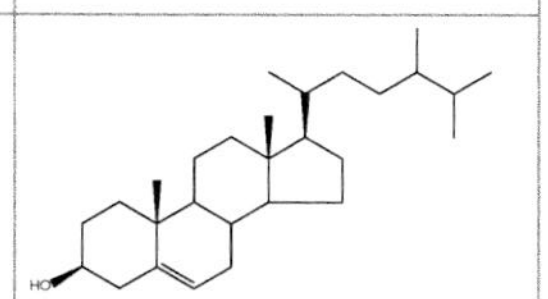	
		Lupeol	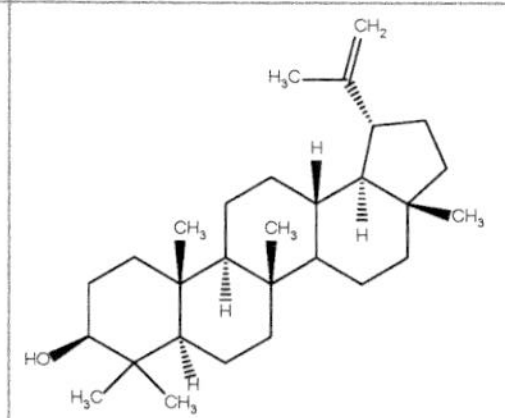	
		Friedelin	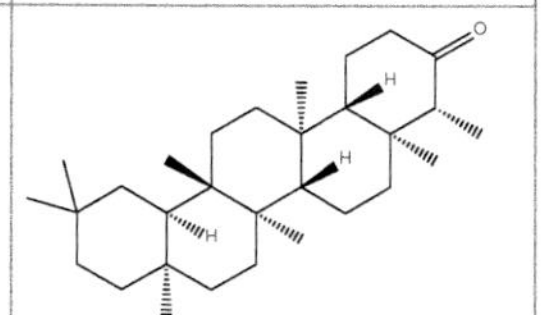	

(continued)

Table 4.94 (continued)

Family	Botanical name	Image	Plant part used	Bioactive compound(s)	Structure	Reference(s)
				α-Tocopherol	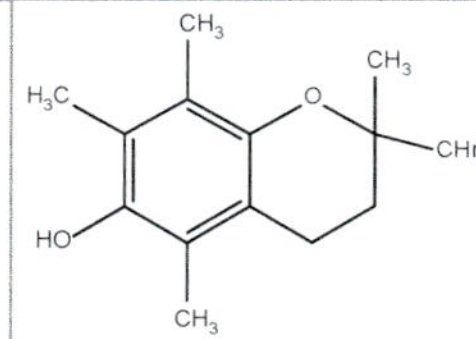	
	Minquartia guianensis Aubl.		Bark	Acalyphaser A	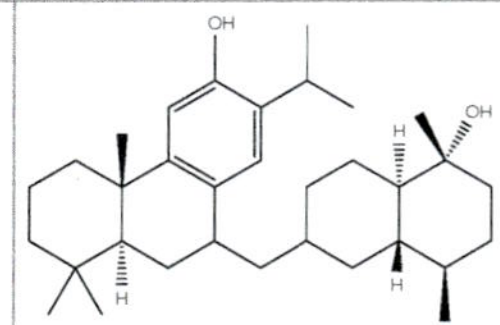	El-Seedi et al. (1994) and Reddy et al. (2010)
	Synonym(s): *Eganthus poeppigii* Tiegh., *Endusa punctata* Radlk., *Minquartia macrophylla* Ducke., *Minquartia parvifolia* A.C.Sm., *Minquartia punctata* (Radlk.) Sleumer, *Secretania loranthacea* Müll.Arg. **Common name(s)**: Urodibe, Platano, Pechiche, Minguar. **Botanical description**: It is an evergreen tree (10–25 m) with a short, oval crown and is native to South America, Brazil, Bolivia, Peru, Ecuador, Colombia, Venezuela, the Guianas, Central America, Panama, Costa Rica, and Nicaragua. The straight, cylindrical bole characteristically has many holes when it matures. It can be 40–120 cm in diameter, with buttresses at the base. A valuable and extremely durable timber, it is often harvested from the wild for local use and trade. Seed-producing trees are sometimes low in numbers, but regeneration does not appear to be inadequate. **Medicinal use(s)**: It is used for the treatment of lung cancer, malaria, and cold and to get rid of intestinal parasites. It also possesses anti-inflammatory, antitumor, antioxidant, and antibacterial properties.					

Table 4.95 Anticancer plants of family Olacaceae

Family	Botanical name	Image	Plant part used	Bioactive compound(s)	Structure	Reference(s)
Olacaceae	*Minquartia guianensis* Aubl.		Stem and bark	Minquartynoic acid		El-Seedi et al. (1994) and Sabitha et al. (2006)
	Synonym(s): *Eganthus poeppigii* Tiegh., *Endusa punctata* Radlk., *Minquartia macrophylla* Ducke, *Minquartia parvifolia* A.C.Sm., *Minquartia punctata* (Radlk.) Sleumer,*Secretania loranthacea* Müll.Arg. **Common name(s)**: Black manwood or Huambula. **Botanical description**: *Minquartia* is a monotypic genus of flowering plants in the Olacaceae family containing the single species *Minquartia guianensis*. It is an evergreen tree with a short, oval crown and can grow 10–25 m tall and is native to South America, Brazil, Bolivia, Peru, Ecuador, and Colombia. The straight, cylindrical bole characteristically has many holes when it is mature. It can be 40–120 cm in diameter, with buttresses at the base. It is found in the rainforests and can grow in both sandy and clayey soils at elevations from sea level to 1000 m. **Medicinal use(s)**: *Minquartia guianensis* are cytotoxic to diverse cancer cell lines including human lung cancer, ovarian, colon, and neuroblastoma. Decoction prepared from the bark is used as a remedy for herpes, lung cancer, hepatitis, and tuberculosis. It is also used to get rid of intestinal worms and parasites and to treat muscular pain and skin irritations. The pulverized bark of this plant is also used externally as a poultice for sore limbs, sore kidneys, and skin problems.					

(continued)

Table 4.95 (continued)

Family	Botanical name	Image	Plant part used	Bioactive compound(s)	Structure	Reference(s)
	Ximenia americana L.		Root	β-Sitosterol	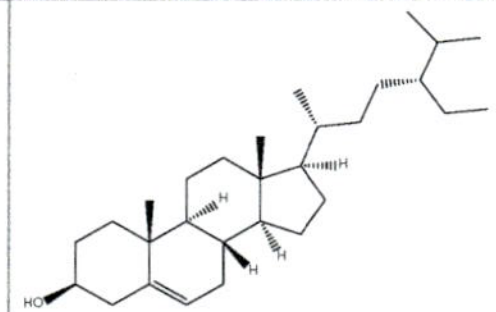	Suh et al. (1996) and da Silva et al. (2016)
	Synonym(s): *Amyris arborescens* P. Browne, *Heymassoli inermis* Aubl., *Heymassoli spinosa* Aubl., *Pimecaria odorata* Raf., *Ximenia aculeate* Crantz, *Ximenia elliptica* G. Forst., *Ximenia fluminensis* M. Roem., *Ximenia laurina* Delile, *Ximenia loranthifolia* Span., *Ximenia multiflora* Jacq., *Ximenia verrucosa* M. Roem. **Common name(s)**: Tallow wood, Yellow plum, Sea lemon, or Pi'ut (Chamorro). **Botanical description**: It is a small sprawling tree of woodlands native to the tropics. The leaves are oval-shaped and bright green and have a strong smell of almonds. Flowers are pale in color. Fruits are lemon-yellow or orange-red. **Medicinal use(s)**: It is used for treating cancer. It is effective against parasite that causes sleeping sickness and severe anemia in livestock. The leaf extract is active against *Escherichia coli*, *Pseudomonas aeruginosa*, and *Candida albicans*.					

Table 4.96 Anticancer plants of family Oleaceae

Family	Botanical name	Image	Plant part used	Bioactive compound(s)	Structure	Reference(s)
Oleaceae	*Forsythia koreana* (Rehder) Naka		Fruit	Myrcenol	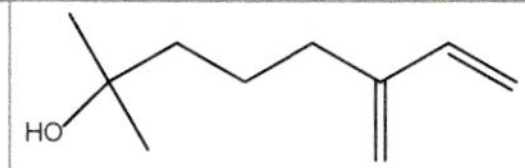	Moon et al. (1985), Nishibe (1995), and Yang et al. (2015)
	Synonym(s): *Forsythia koreana* var. *autumnalis* Uyeki, *Forsythia koreana* f. *autumnalis* (Uyeki) Nakai, *Forsythia koreana* var. *pilosa* (U.C.La & Chae G.Chen) U.C.La, *Forsythia pilosa* U.C.La & Chae G.Chen., *Forsythia viridissima* var. *koreana* Rehder, *Rangium koreanum* (Rehder) Ohwi, *Rangium koreanum* var. *autumnalis* (Uyeki) Uyeki. **Common name(s)**: Gaenari or Korean goldenbell tree, Forsythia. **Botanical description**: It is a deciduous shrub and is native to Eastern Asia. These plants typically grow to a height of 1–3 m and rarely up to 6 m, with rough gray-brown bark. The leaves are opposite, usually simple but sometimes trifoliate with a basal pair of small leaflets, and range from 2 to 10 cm in length and rarely up to 15 cm. The leaves may be either deciduous or evergreen, with evergreen species predominating in warm temperate and tropical regions and deciduous species predominating in colder regions. **Medicinal use(s)**: It possesses the potential to inhibit breast cancer-induced bone destruction through blocking invasion of breast cancer cells, osteoclastogenesis, and the activity of mature osteoclasts. It is also used in oriental traditional medicine to treat asthma, atopy, and allergic diseases.			Sylvatesmin	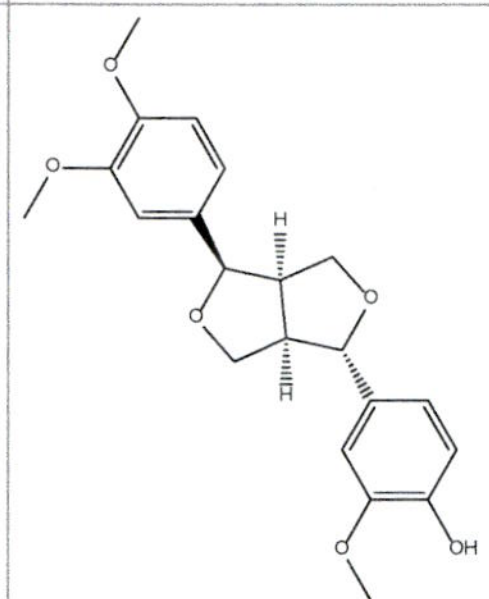	

(continued)

Table 4.96 (continued)

Family	Botanical name	Image	Plant part used	Bioactive compound(s)	Structure	Reference(s)
				Phillyrin	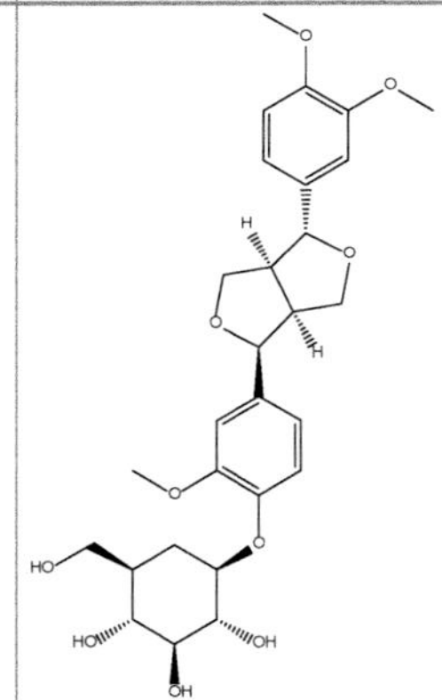	
				Salidroside	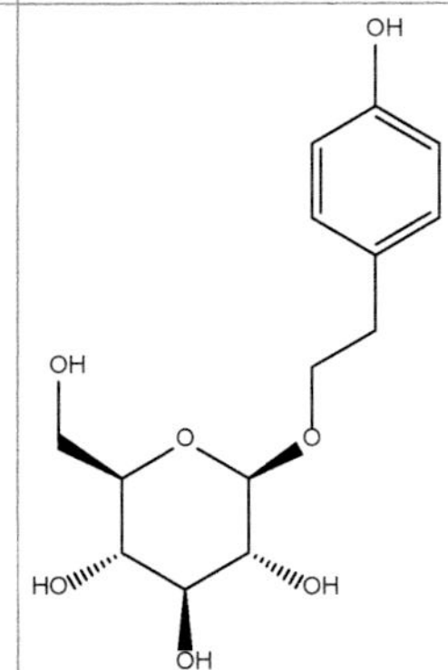	

<table>
<tr><td></td><td>Forsythia suspense (Thunb.) Vahl</td><td></td><td>Fruit</td><td>Rutin</td><td>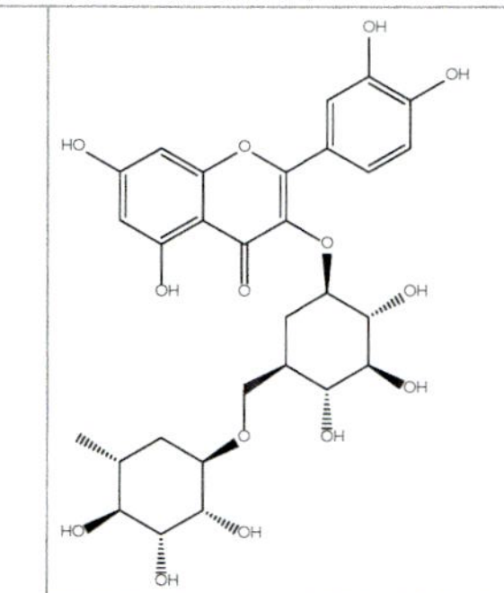</td><td>Zhang et al. (2002)</td></tr>
<tr><td></td><td colspan="3" rowspan="3">Synonym(s): Forsythia fortunei Lindl., Forsythia sieboldii (Zabel) Dippel, Ligustrum suspensum Thunb., Lilac perpensa Lam., Rangium suspensum (Thunb.) Ohwi, Syringa suspensa Thunb.
Common name(s): Weeping forsythia.
Botanical description: It is a straggling, deciduous shrub, with many spreading, pendulous branches. Weeping forsythia grows to around 3 m high as a free-standing shrub and higher with a support. The golden-yellow flowers are about 3 cm across and appear before the leaves, singly, or in small groups, in March to April. The opposite, broadly ovate, green leaves are usually simple (undivided), but are occasionally three-lobed, and have toothed margins, except at the base. The narrow capsules (fruits) appear from July to September.
Medicinal use(s): A decoction of the leaves and twigs is used to treat breast cancer, to get rid of intestinal worms, and to control menstruation pain. The roots are used to treat colds, fever, and jaundice, and decoction of the fruit is used to treat boils and other skin infections. It also possesses antibacterial and anti-inflammatory activities.</td><td>Oleanolic acid</td><td>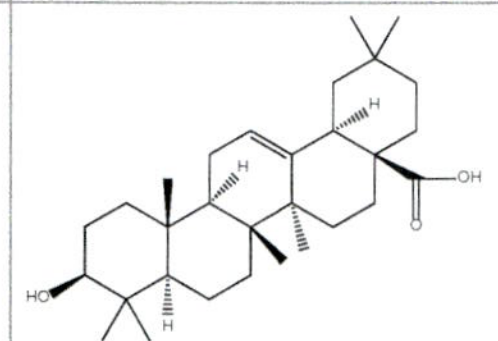</td><td></td></tr>
<tr><td></td><td>Isoforsythiaside</td><td>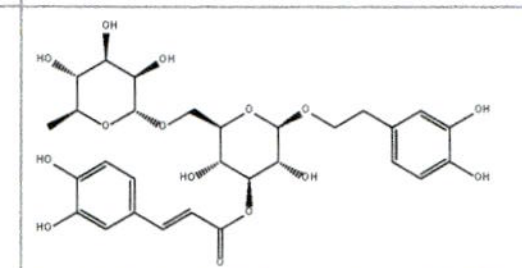</td><td></td></tr>
<tr><td></td><td>Betulinic acid</td><td>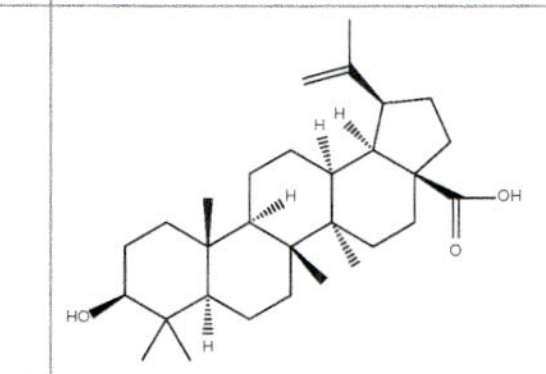</td><td></td></tr>
</table>

(continued)

Table 4.96 (continued)

Family	Botanical name	Image	Plant part used	Bioactive compound(s)	Structure	Reference(s)
	Forsythia viridissima Lindl.		Fruit	Isoquercetin	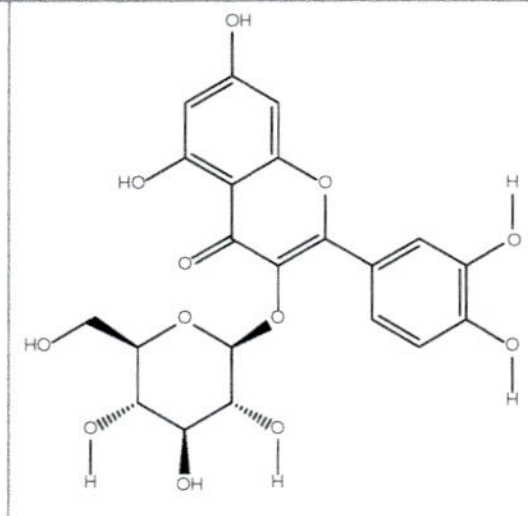	Nishibe (1995), Column et al. (2013), and Tokar and Klimek (2004)
	Synonym(s): *Forsythia viridissima* var. *incisa* Geerinck, *Rangium viridissimum* (Lindl.) Ohwi. **Common name(s)**: Green stem forsythia, Golden bells. **Botanical description**: It is a glabrous shrub with upright branches (3 m), branchlets green or yellow-green, pith lamellate. Leaves simple, petiole 6–12 mm, leaf blade long elliptic to lanceolate or long obovate-elliptic, base cuneate, margin serrate or entire along distal half, apex acute. **Medicinal use(s)**: A decoction of the leaves and twigs is used for the treatment of breast cancer. The root is used for the treatment of cancer, colds, fever, and jaundice. The fruit possess antiphlogistic and antitussive properties and also used as diuretic, emmenagogue, febrifuge, laxative, and tonic. It is used internally for the treatment of acute infectious diseases such as mumps and also for tonsillitis, urinary tract infections, allergic rashes, etc.			Beta-sitosterol	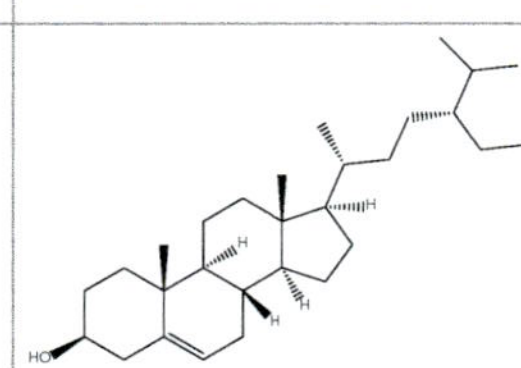	
				Ursolic acid	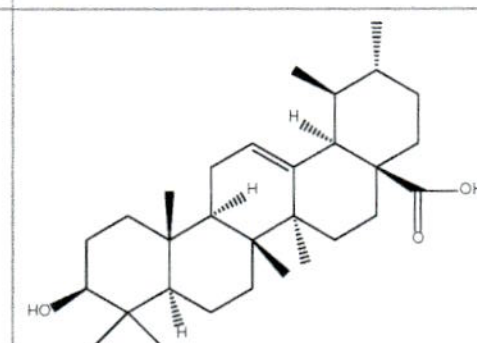	

	Jasminum sambac (L.) Aiton		Whole plant	Anthraquinone	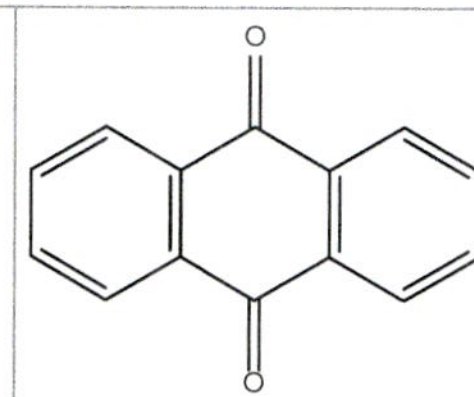	Kalaiselvi and Kalaivani (2011)
	Synonym(s): *Jasminum bicorollatum* Noronha, *Jasminum blancoi* Hassk., *Jasminum heyneanum* Wall. ex G.Don, *Jasminum odoratum* Noronha, *Jasminum quinqueflorum* B.Heyne ex G.Don, *Jasminum undulatum* (L.) Willd., *Mogorium gimea* Zuccagni, *Mogorium goaense* Zuccagni, *Mogorium sambac* (L.) Lam., *Mogorium undulatum* (L.) Lam., *Nyctanthes goa* Steud., *Nyctanthes sambac* L., *Nyctanthes undulata* L. **Common name(s)**: Arabian jasmine. **Botanical description**: It is a bushy vine or scrambling shrub with shiny dark green leaves and fragrant little white flowers. Some of the evergreen leaves are in whorls of three, and others are in opposite pairs. The long, angular shoots twist and twine as they clamber and sprawl over and through any support they can find. The waxy snow-white flowers are about 1 in., borne in clusters of 3–12, and intensely fragrant. They fade to pink as they age. Arabian jasmine blooms throughout the summer and almost continuously in warm climates. The fruits are small blackberries but are seldom formed in cultivation. **Medicinal use(s)**: It is used in the treatment of diarrhea, tumors, conjunctivitis, skin ulcers, dysentery, and fever; the roots are used in curing severe pain, headache, insomnia, pain, and inflammation due to broken bones or dislocated joints mainly attributed to the analgesic, anti-inflammatory, and anesthetic properties of this aromatic herb.			Tannic acid (tannin)	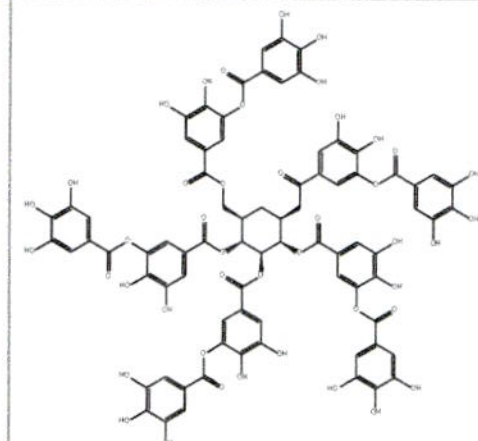	
				Coumarin(s)	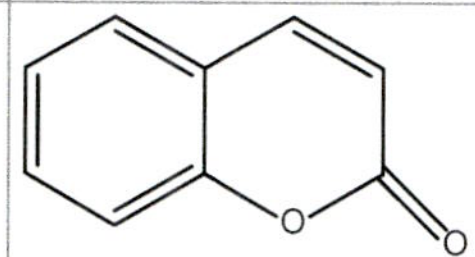	

(continued)

Table 4.96 (continued)

Family	Botanical name	Image	Plant part used	Bioactive compound(s)	Structure	Reference(s)
	Ligustrum lucidum W.T.Aiton		Seed	Palmitic acid	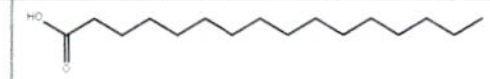	Wong et al. (1992a, b) and Hu et al. (2014)
	Synonym(s): *Esquirolia sinensis* H.Lév., *Ligustrum esquirolii* H.Lév., *Ligustrum magnoliifolium* Dippel, *Ligustrum roxburghii* Blume, *Ligustrum wallichii* Vis., *Olea chinensis* Sweet, *Olea clavata* G.Don, *Phillyrea paniculata* Roxb., *Visiania paniculata* (Roxb.) DC. **Common name(s)**: Chinese privet, Glossy privet, White wax tree, Tree ligustrum. **Botanical description**: It is a perennial shrub that can grow up to 16 ft. in height and is native to Europe, North Africa, and Asia. Ligustrum bark is tan to gray in color with a smooth texture. Leaves are elliptic to ovate in shape, oppositely arranged on twigs. Flowers have both male and female parts. Each flower has petals fused into a tube with four separate lobes. Flowers are borne on small panicles on short lateral branches at the end of the twigs. The oblong, blue/black fruit is a drupe containing 1–4 seeds. Fruit clusters persist through the winter. Mature trees can produce hundreds of fruits. **Medicinal use(s)**: The fruit possesses antibacterial, antiseptic, antitumor, cardiotonic, and diuretic properties. It is used internally for the treatment of complaints associated with weak kidney and liver functioning, menopausal problems (especially premature menopause), blurred vision, cataracts, tinnitus, rheumatic pains, palpitations, backache, and insomnia. Modern research has shown that the plant increases the white blood cell count and is used to prevent bone marrow loss in cancer chemotherapy patients.			Tormentic acid	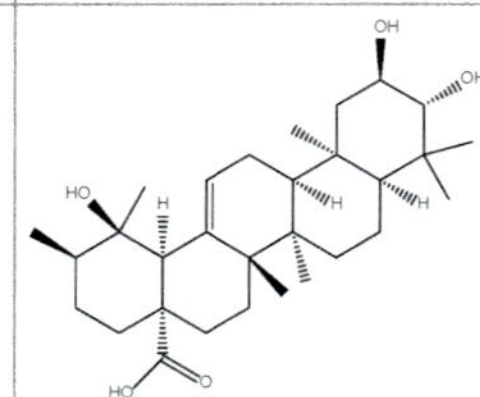	

	Ocimum gratissimum L.		Leaf	Eugenol	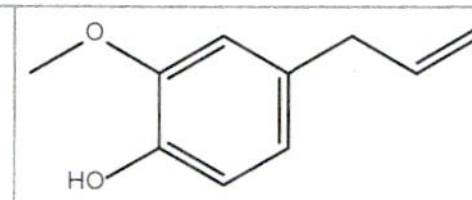	Ekunwe et al. (2010)
	Synonym(s): *Geniosporum discolor* Baker, *Ocimum arborescens* Bojer ex Benth., *Ocimum dalabaense* A.Chev., *Ocimum febrifugum* Lindl., *Ocimum guineense* Schumach. & Thonn., *Ocimum heptodon* P.Beauv., *Ocimum holosericeum* J.F.Gmel., *Ocimum petiolare* Lam., *Ocimum robustum* B.Heyne ex Hook.f., *Ocimum sericeum* Medik., *Ocimum suave* Willd., *Ocimum trichodon* Baker ex Gürke, *Ocimum urticifolium* Roth, *Ocimum viride* Willd., *Ocimum viridiflorum* Roth, *Ocimum* zeylanicum Medik. **Common name(s)**: Clove basil. **Botanical description**: It is an erect perennial herb or soft shrub, up to 2 m long, and is native to tropical Africa, East Asia, India, Sri Lanka, Nepal, Bangladesh, and Thailand. Leaves are opposite, ovate-lanceolate, variously pubescent on both surfaces, gland-dotted below. Margins are dentate in the upper half. Inflorescences terminal, simple, or sparingly branched. Calyx densely pubescent on the outside, lower lip often closing the mouth and obscuring the small white corolla. It cures coughs, colds, flu, and fever and helps in digestion. It has antiseptic properties, strengthens the immune system, and is considered to be a tonic and adaptogenic. It is an excellent wellness drink to help quit drinking coffee. **Medicinal use(s)**: The strong antioxidant activity slows the aging process and helps to prevent and treat cancer, heart disease, arthritis, diabetes, and dementia. It helps in normalizing both blood pressure and cholesterol levels.					

(continued)

Table 4.96 (continued)

Family	Botanical name	Image	Plant part used	Bioactive compound(s)	Structure	Reference(s)
	Olea europaea L.		Leaf, fruit	Maslinic acid	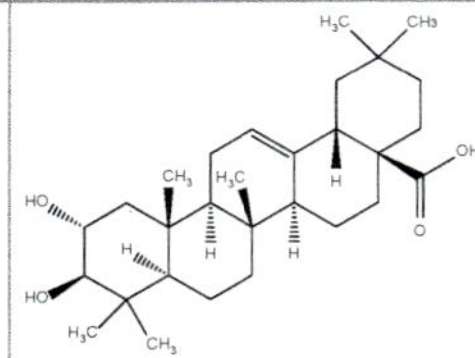	Reyes-Zurita et al. (2009)
	Synonym(s): *Olea europaea* subsp. *europaea*, *O. europaea* subsp. *cuspidata*, *O. europaea* subsp. *guanchica*, *O. europaea* subsp. *cerasiformis*, *O. europaea* subsp. *maroccana*. **Common name(s)**: Olive, African olive, European olive. **Botanical description**: It is an evergreen tree growing up to 10 m at a slow rate and is native to South Europe and Mediterranean region. The flowers are hermaphrodite (have both male and female organs) and are pollinated by wind. The plant is self-fertile. It is suitable to grow in light (sandy), medium (loamy), and heavy (clay) soils and prefers well-drained soil. It cannot grow in the shade and prefers dry or moist soil and can also tolerate drought conditions. **Medicinal use(s)**: Olive leaves contain many potentially bioactive compounds that have anticancer, antioxidant, antimicrobial, antihypertensive, antiviral, anti-inflammatory, hypoglycemic, and neuroprotective properties. Particularly, olive leaf has been reported to exhibit an antileukemic effect by inducing apoptosis in the acute myeloid leukemia HL-60 cells.					

	Olea ferrugenia Royal.		Whole plant	Butylated Hydroxytoluene	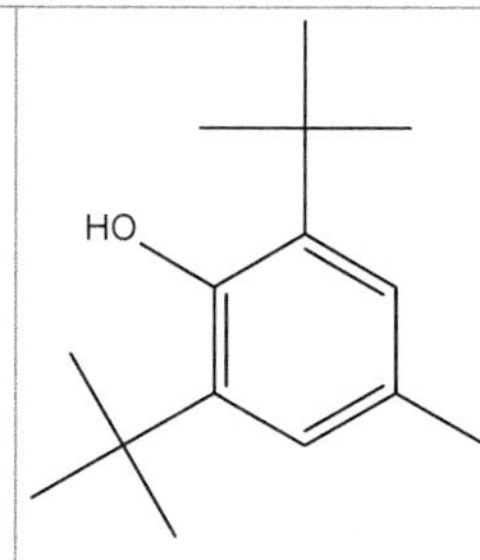	Siddiqui et al. (2011)
	Synonym(s): *Linociera lebrunii* Staner, *Olea africana* Mill., *Olea aucheri* A.Chev. ex Ehrend., *Olea chrysophylla* Lam., *Olea cuspidata* Wall. ex G.Don. **Common name(s)**: Wild olive. **Botanical description**: It is medium-sized tree up to 15 m long and is native to Afghanistan, Nepal, and India. Stem is erect, branched, woody, and hard, with dark gray bark. Leaves are simple, opposite, entire, petiolate, lanceolate, upper fleshy dark green and lower yellowish-green surface. Flowers are small, numerous, and creamish and are present in terminal or lateral cymes. Fruit drupe and oval-shaped. **Medicinal use(s)**: It is used to treat cancer. It is also used to treat primary bone tumors or invasion of bone by other cancers and diseases that increase bone turnover.					

(continued)

Table 4.96 (continued)

Family	Botanical name	Image	Plant part used	Bioactive compound(s)	Structure	Reference(s)
	Syringa pubescens subsp. patula (Palib.) M.C.Chang & X.L.Chen		Bud	Oleoside 11-methyl ester	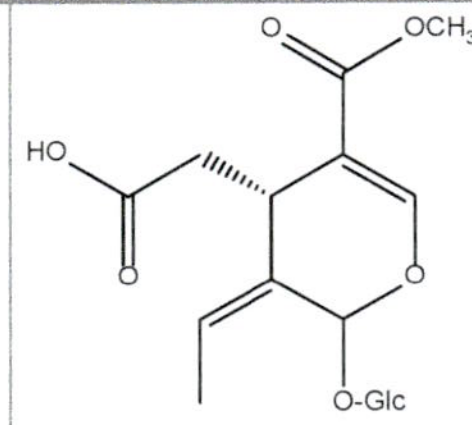	El-Desouky and Gamal-Eldeen (2009)
				Syringaresinol	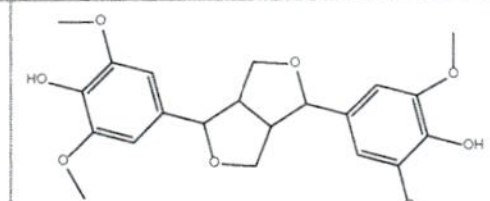	
	Synonym(s): *Syringa patula* (Palib.) Nakai, *Syringa pubescens* f. alba S.D.Zhao, *Syringa velutina* Kom., *Syringa venosa* Nakai, *Syringa villosa* var. lactea Nakai, *Ligustrum patulum* Palib., *Syringa debelderorum* J.L.Fiala, *Syringa palibiniana* Nakai, *Syringa fauriei* var. lactea (Nakai) Nakai, *Syringa kamibayashii* Nakai, *Syringa koehneana* C.K.Schneid., *Syringa micrantha* Nakai. **Common name(s):** Miss Kim. **Botanical description:** It is native to Korea and Northern China. This is a deciduous shrub, 6–8 ft. (1.8–2.7 m) tall with similar width, larger with age, dense. Leaves opposite, simple, elliptic to ovate-oblong, to 8 cm long, dull dark green. Flowers in paired clusters, 10–15 cm long, purple in bud than light blue, fragrant. **Medicinal use(s):** This plant has been extensively evaluated against various tumor cell lines. Aqueous extracts from the flowers and leaves inhibited the growth of L2215 (hepatitis B virus) cells. It also works against lung cancer cell lines.			Eugenol	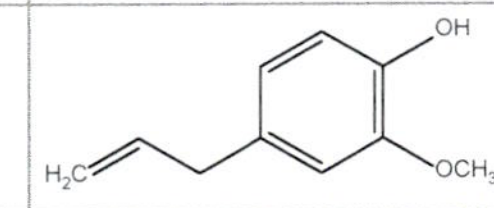	
				Nortropine	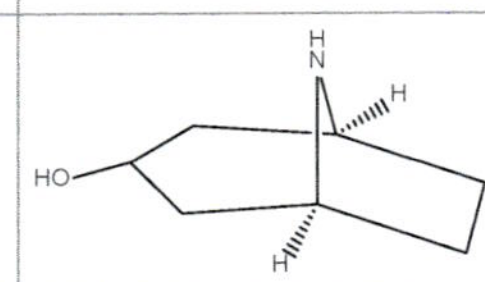	
				β-Sitosterol	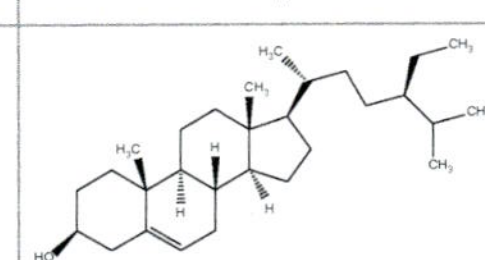	

Table 4.97 Anticancer plants of family Onagraceae

Family	Botanical name	Image	Plant part used	Bioactive compound(s)	Structure	Reference(s)
Onagraceae	*Epilobium hirsutum* L.		Entire plant	Gallic acid	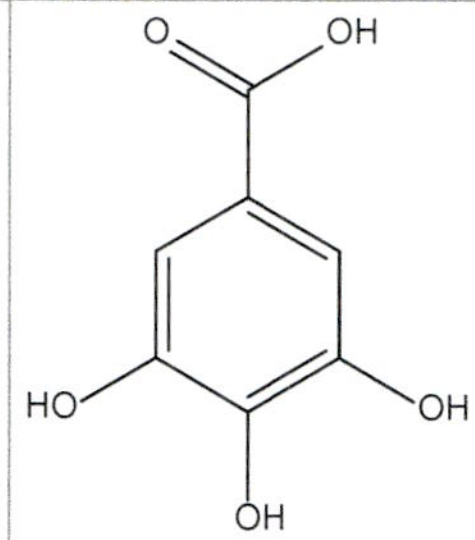	Barakat et al. (1997) and Stolarczyk et al. (2013)
	Synonym(s): *Chamaenerion grandiflorum* (Weber) Moench, *Chamaenerion hirsutum* (L.) Scop., *Epilobium amplexicaule* Lam., *Epilobium aquaticum* Thuill., *Epilobium dubium* Borbás., *Epilobium foliosum* Hochst., *Epilobium grandiflorum* Weber, *Epilobium himalense* Royle, *Epilobium incanum* Pers., *Epilobium mirei* Quézel, *Epilobium nassirelinulci* Stapf, *Epilobium ramosum* Huds., *Epilobium serratum* Jacquem. ex C.B.Clarke, *Epilobium tomentosum* Vent., *Epilobium villosum* Thunb. **Common name(s)**: Great willow herb, Great hairy willow herb, or Hairy willowherb. **Botanical description**: It is a perennial plant growing up to 2 m long and is native to Europe, including Britain, from South and East Sweden to Northeast and South Africa and temperate Asia. It is in flower from July to September. The flowers are hermaphrodite (have both male and female organs) and are pollinated by bees, hoverflies, and itself. **Medicinal use(s)**: It has anticancer and anti-inflammatory properties and also controls incontinence. The extracts are prostate cancer cells proliferation inhibitors. The tested extracts reduced prostate specific antigen (PSA) secretion inhibits arginase activity.			Quercetin	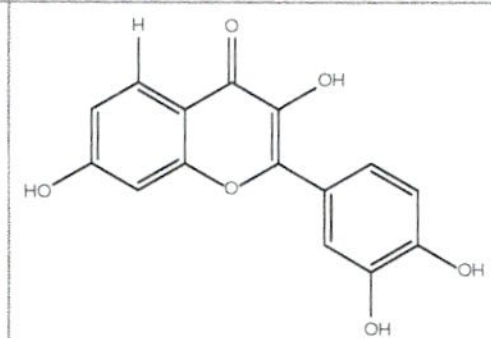	
				Kaempferol	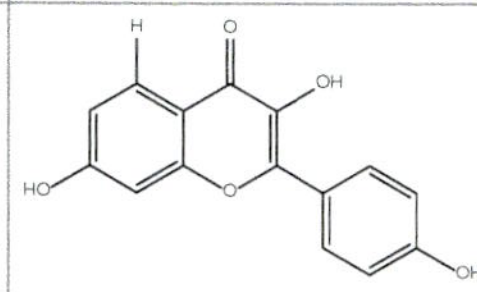	
				Beta-sitosterol	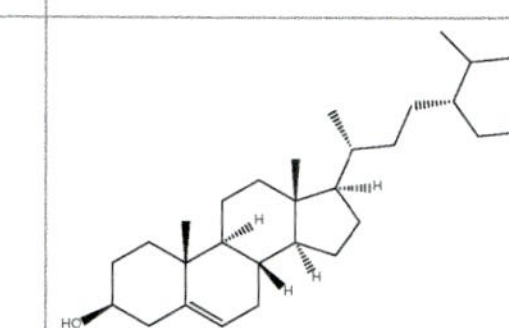	

(continued)

Table 4.97 (continued)

Family	Botanical name	Image	Plant part used	Bioactive compound(s)	Structure	Reference(s)
				Oenothein B	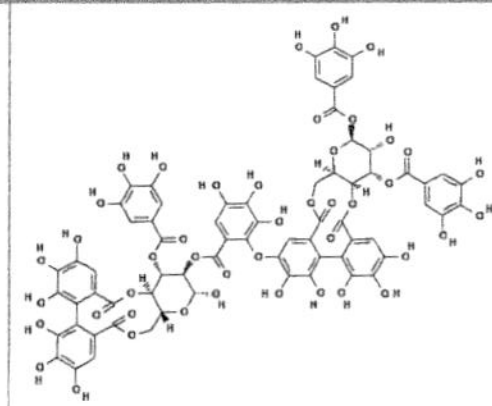	
				Quercetin-3-O-glucuronide	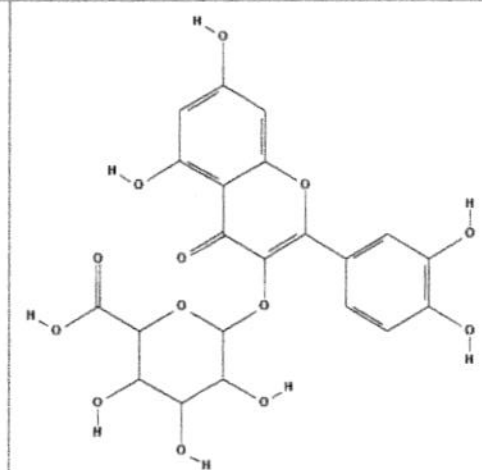	
				Myricetin-3-O-rhamnoside	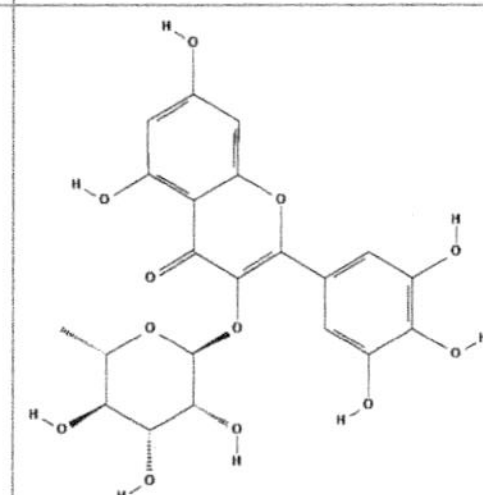	

Table 4.98 Anticancer plants of family Orchidaceae

Family	Botanical name	Image	Plant part used	Bioactive compound(s)	Structure	Reference(s)
Orchidaceae	*Pleione bulbocodioides* (Franch.) Rolfe		Tuber	Quercetin	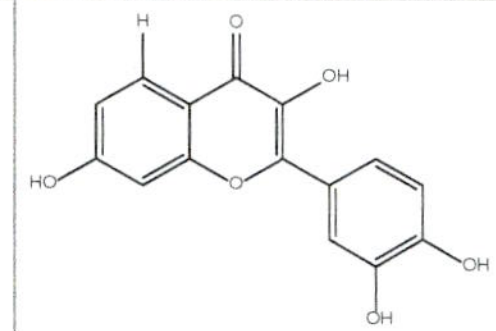	Bai et al. (1997) and Wang et al. (2013)
	Synonym(s): *Coelogyne bulbocodioides* Franch., *Coelogyne delavayi* Rolfe, *Coelogyne henryi* Rolfe, *Coelogyne pogonioides* Rolfe, *Pleione communis* Gagnep., *Pleione delavayi* (Rolfe) Rolfe, *Pleione fargesii* Gagnep., *Pleione ganchuenensis* Gagnep., *Pleione henryi* (Rolfe) Schltr., *Pleione mairei* Schltr., *Pleione pogonioides* (Rolfe) Rolfe, *Pleione rhombilabia Hand.*-Mazz., *Pleione smithii* Schltr. **Common name(s)**: The bulbocodium-like pleione. **Botanical description**: It is a species of *Pleione* found in East Asia. The plant is found growing on mossy rocks in Yunnan China at elevations of 900–3600 m as a small-sized, cool- to cold-growing terrestrial with clustered, narrowly ovoid pseudobulbs carrying 1–2, elliptic-lanceolate, acute, plicate leaf that blooms in the spring and early summer on a single flowered, 3–8, sheathed inflorescence that arises at the same time as the leaf. **Medicinal use(s)**: It is used to treat cancer. It is also used to treat tuberculosis and asthma, and the paste of this plant is applied on the boils and snakebite. It also possesses anticancer, anti-inflammatory, and antimicrobial properties.			Kaempferol	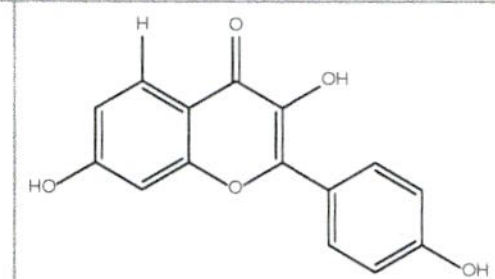	
				Phenylalanine	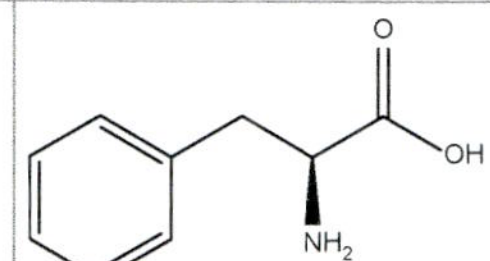	
				Stigmasterol	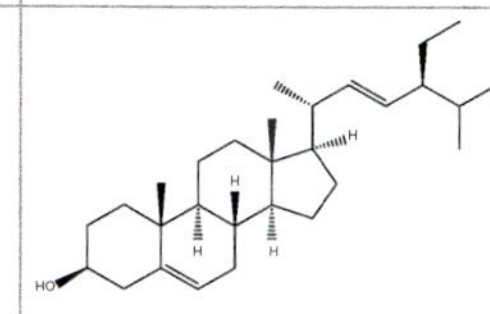	

Table 4.99 Anticancer plants of family Oxalidaceae

Family	Botanical name	Image	Plant part used	Bioactive compound(s)	Structure	Reference(s)
Oxalidaceae	*Oxalis corniculata* L.		Whole plant	Oleic acid	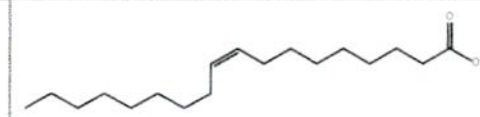	Kathiriya et al. (2010)
	Synonym(s): *Acetosella bakeriana* Kuntze, *Acetosella corniculata* (L.) Kuntze, *Acetosella fontana* (Bunge) Kuntze, *Acetosella herpestica* (Schltdl.) Kuntze, *Acetosella stricta* (L.) Kuntze, *Acetosella villosa* (Progel) Kuntze, *Oxalis albicans* Kunth, *Oxalis bradei R.* Knuth. **Common name(s)**: Woodsorrel, Yellow sorrel, Creeping woodsorrel. **Botanical description**: It is a perennial plant growing up to 0.1–0.3 m and is native to Britain. It is in flower from June to September. The flowers are hermaphrodite (have both male and female organs) and are pollinated by insects and itself. It is suitable for light (sandy), medium (loamy), and heavy (clay) soils and prefers well-drained soil. Suitable pH is acid, neutral, and basic (alkaline) soils. It cannot grow in the shade. It prefers dry or moist soil. **Medicinal use(s)**: It is used for the treatment of cancer, AIDS, malaria, liver diseases, kala-azar and other infectious diseases, hypertension, arthritis, and bronchial asthma. It also possesses antitumor, hepatoprotective, and anti-fibrotic activities.			Beta-carotene	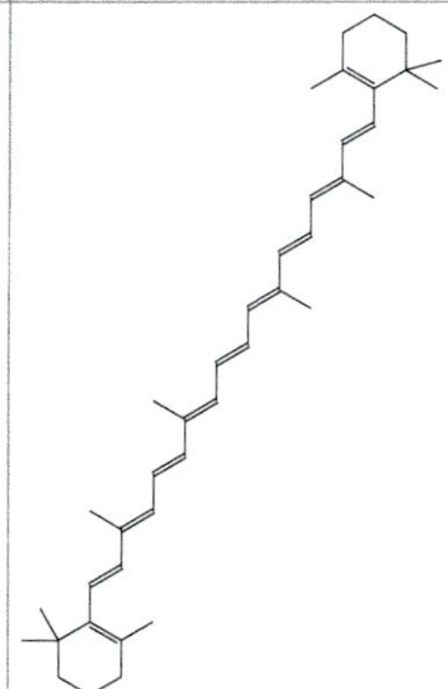	
				Oxalate	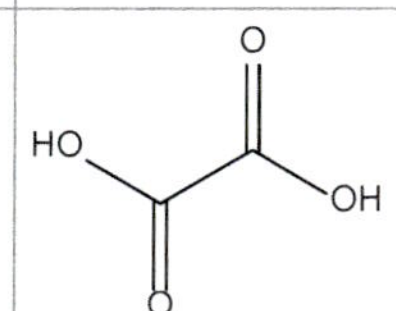	

Table 4.100 Anticancer plants of family Pandaceae

Family	Botanical name	Image	Plant part used	Bioactive compound(s)	Structure	Reference(s)
Pandaceae	*Panda oleosa* Pierre		Fruit	Lupeol	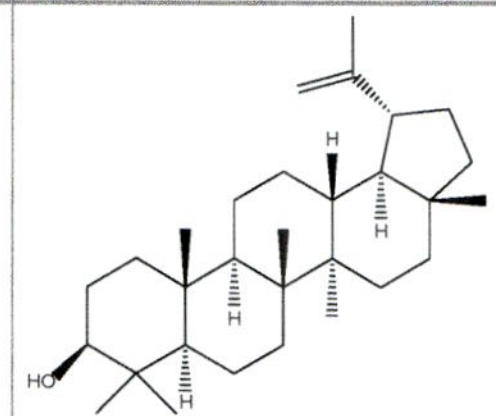	Katem et al. (2017)
	Synonym(s): *Porphyranthus zenkeri* Engl. *Sorindeia rubiflora* Engl. **Common name(s)**: Panda. **Botanical description**: It is native to west tropical Africa – Liberia to Central African Republic, south to Gabon and the Congo. It is a slow-growing, evergreen tree with a dense crown, usually growing from 10 to 20 m tall but with some specimens up to 35 m. **Medicinal use(s)**: It possesses anticancer properties. The bark possesses anti-inflammatory, analgesic, and aphrodisiac activities. The seed oil is applied to relieve ulcer pain. It is applied externally to treat rheumatism, wounds, yaws, sores, whitlow, swellings, and hemorrhoids.			Betulin	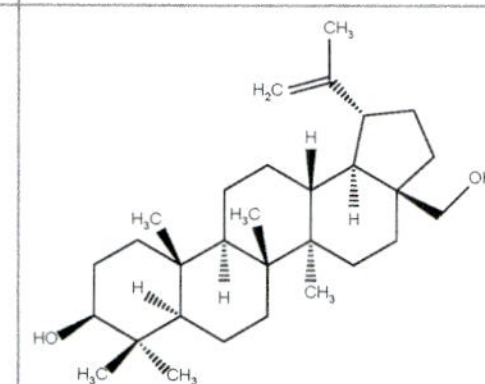	
				Betulic acid	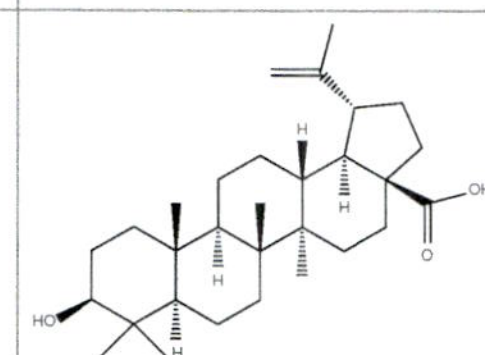	

Table 4.100 (continued)

Family	Botanical name	Image	Plant part used	Bioactive compound(s)	Structure	Reference(s)
	Microdesmis puberula Hook.f. exPlanch.		Seeds, leaves, fruits	Morphine	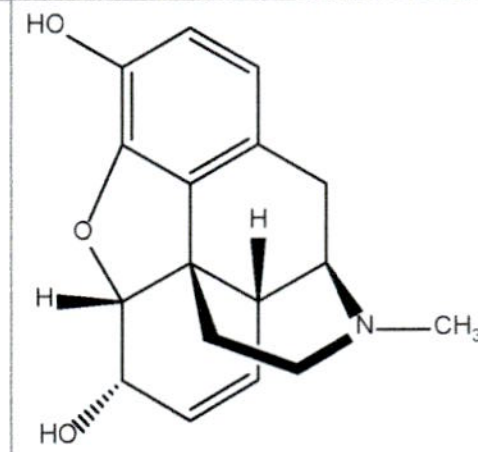	Vasileva (1969) and Lim (2016)
	Synonym(s): *Microdesmis zenkeri* Pax. **Common name(s)**: Fruiting branch. **Botanical description**: It is native to Guinea. It is usually a shrub growing up to 6 m tall, stem up to 8 cm in diameter; twigs usually densely short hairy. Leaves alternate, distichous, simple; stipules linear, up to 4 mm long, persistent; petiole 4–12 mm long; blade elliptical-oblong to ovate-lanceolate or ovate, base asymmetrical, cuneate to rounded, apex acute to acuminate, margin finely toothed to almost entire, shiny above, short hairy on the midrib. **Medicinal use(s)**: It is used to treat cancer. It also possesses antibacterial and anti-inflammatory properties. The roots, or pounded young leaves, mixed with the juice of sugar cane are considered to be aphrodisiac. The grated roots are mixed with corn flour and eaten to cure gonorrhea. Ash of burnt roots, mixed with palm oil, is rubbed into scarifications to treat renal pain and severe headache. The grated root is also applied on wounds and hernia. The pounded fruits, mixed with capsicum fruits, are eaten to treat cough.			Stigmasterol	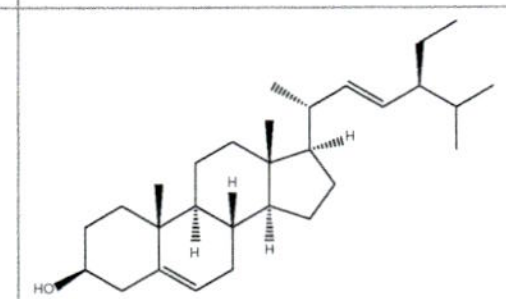	

Table 4.101 Anticancer plants of family Papaveraceae

Family	Botanical name	Image	Plant part used	Bioactive compound(s)	Structure	Reference(s)
Papaveraceae	*Chelidonium asiaticum* (Hara) Krahulc.		Whole plant	Chelidonine	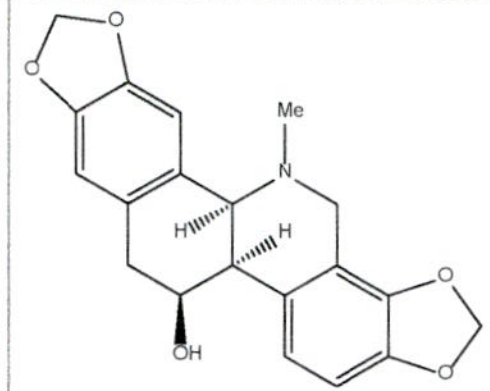 	Mohammad (2006), Desai et al. (2008), and Madhuri and Pandey (2009)
	Synonym(s): *Chelidonium majus* subsp. *asiaticum* H. Hara, *Chelidonium majus* var. *hirsutum* Trautv. & C. A. Mey. **Common name(s)**: Common celandine, Garden celandine, Tetterwort. **Botanical description**: This plant is grown in gardens and naturalized in a number of areas in North America including Quebec, Ontario, and much of the Eastern and Upper Midwestern United States plus several areas in the Pacific Northwest. It is native to rocky slopes, woodlands and waste areas and along roads in Europe and Western Asia. It is a short-lived perennial or biennial in the poppy family that bears yellow flowers and gray-green leaves. Plants typically grow in a clump to 12–24 in. tall. Loose umbels of 4-petaled, 1 in. wide, yellow flowers bloom from May to August. Flowers are followed by smooth, slender seed capsules that split open when ripe, to release tiny seeds. **Medicinal use(s)**: It possesses anticancer, antitumor, anti-inflammatory, and antimicrobial properties.			Stylopine	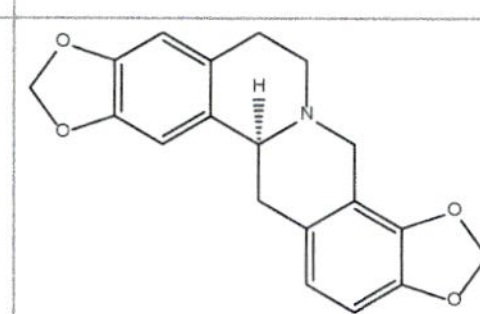 	

(continued)

Table 4.101 (continued)

Family	Botanical name	Image	Plant part used	Bioactive compound(s)	Structure	Reference(s)
	Chelidonium majus L.		Entire plant	Ukrain	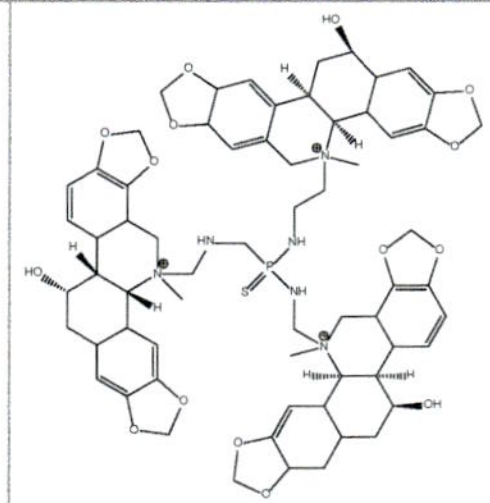	Han et al. (1984) and Gilca et al. (2010)
	Synonym(s): *Chelidonium haematodes* Moench, *Chelidonium laciniatum* Mill., *Chelidonium murale* P. Renault. **Common name(s)**: Greater celandine, Celandine, Chelandine, Common celandine, Garden celandine, Tetterwort. **Botanical description**: It is native to temperate regions of North Africa, Europe, and parts of Western Asia. It is a perennial herb with an erect stem and reaches 30–120 cm high. The blue-gray leaves are pinnate with lobed and wavy-edged margins, up to 30 cm long. The flowers appear from late spring to summer, May to September. The seeds are small and black, borne in a long, cylindrical capsule. **Medicinal use(s)**: It is used for stomach cancer, liver cancer, and gastrointestinal cancer. Plant extract is used to treat warts. It also possesses antibacterial and antifungal properties.					
	Sanguinaria canadensis L.		Entire plant	Benzo[c]phenanthrene	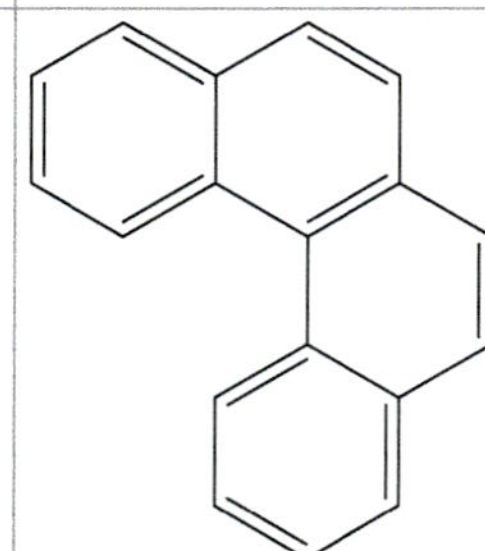	Chauret et al. (1990) and Croaker et al. (2016)

	Synonym(s): *Belharnosia canadensis* (L.) Nieuwl., *Sanguinaria australis* Greene, *Sanguinaria dilleniana* Greene. **Common name(s)**: Bloodworts. **Botanical description**: It is native to Eastern North America. It is found in moist to dry woods and thickets, often on flood plains and near shores or streams on slopes. They grow less frequently in meadows or on dunes and are rarely found in disturbed sites. **Medicinal use(s)**: It cures breast cancer and skin cancer. It also suppresses prostate tumor growth and inhibits survivin expression. It also inhibits proliferation and induces apoptosis in many malignant cell types.	Chelirubine	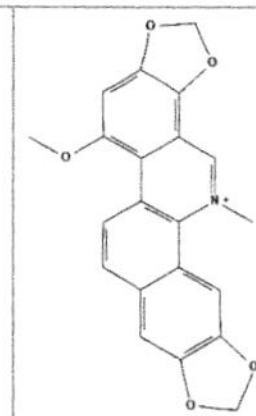	
		Sanguinarine	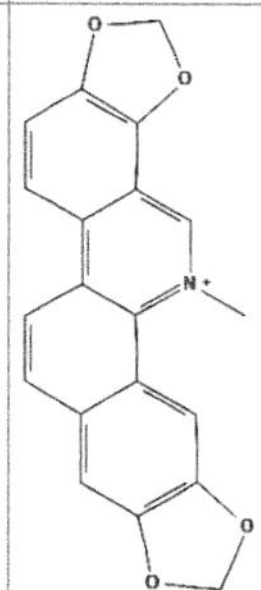	
		Chelerythrine	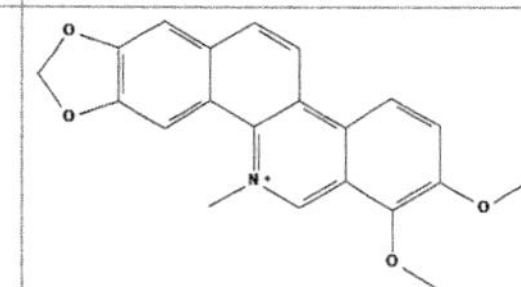	

Table 4.102 Anticancer plants of family Pedaliaceae

Family	Botanical name	Image	Plant part used	Bioactive compound(s)	Structure	Reference(s)
Pedaliaceae	*Rogeria adenophylla* J. Gay ex DC		Stem	Lectin	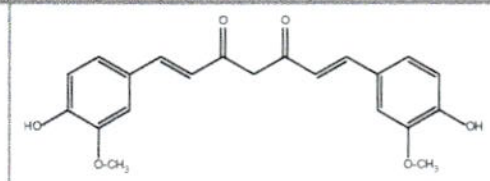	Yousif et al. (1983), West et al. (2000), and Bedigian (2013)
	Synonym(s): *Pedalium caillaudii* Hochst. ex A.DC., *Pedalium rogeria* Decne. **Common name(s)**: Niger. **Botanical description**: It is native to tropical Africa – savannah areas from Mauritania to Sudan and south to Angola. It is a robust, annual herb growing up to 2.5 m tall. The whole plant has an unpleasant acrid smell. **Medicinal use(s)**: It is used for the treatment of abscesses, tumors, adenoma, and prostate and stomach cancer. It is also used to cure digestive system disorders, infections/infestations, inflammation, injuries, pain, and skin/subcutaneous cellular tissue disorders.			Beta-sitosterol	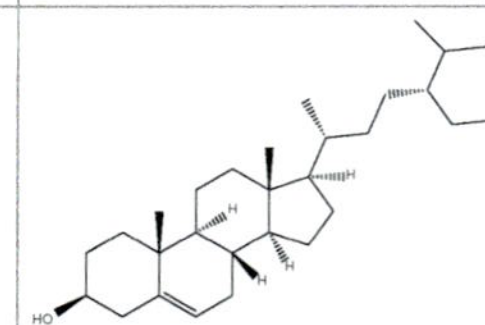	
				Leucine	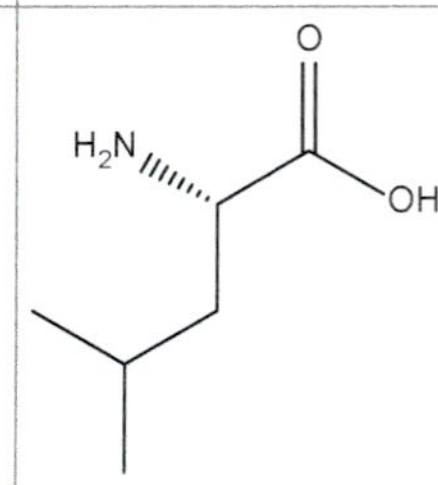	

	Juniperus phoenicea L.		Leaf	Widdrol	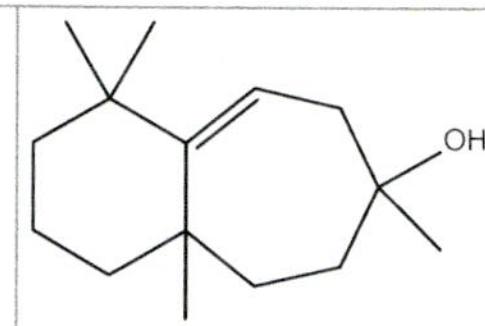	Cirses et al. (2011) and Hajjar et al. (2017)
	Synonym(s): *Cupressus devoniana* Beissn., *Cupressus tetragona* Humb. & Bonpl., *Juniperus bacciformis* Carrière. **Common name(s)**: Phoenicean juniper. **Botanical description**: It is native of Sinai Peninsula and Saudi Arabia along the Red Sea in the east. It is a shrub or small evergreen tree which can grow up to 5–8 m with a trunk up to 1–2 m in diameter. The crown is dense, first conical and then broadening and irregular in age, with ascending and often curved branches. The bark is dark gray in color. **Medicinal use(s)**: The berry oil works best in curing brain tumors and lung, liver, cervical, and breast carcinoma.			Camptothecin	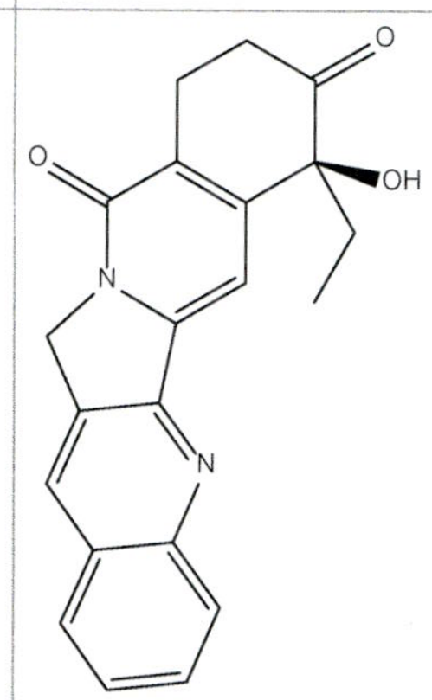	

(continued)

Table 4.102 (continued)

Family	Botanical name	Image	Plant part used	Bioactive compound(s)	Structure	Reference(s)
				Daunorubicin	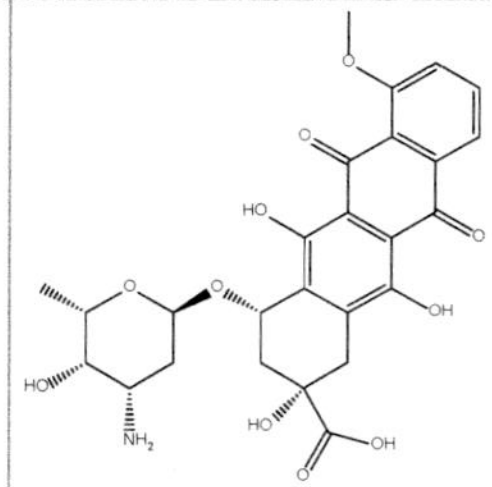	
				Doxorubicin	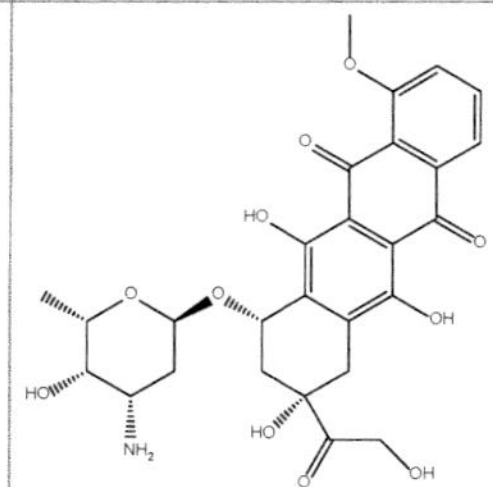	

Table 4.103 Anticancer plants of family Phyllanthaceae

Family	Botanical name	Image	Plant part used	Bioactive compound(s)	Structure	Reference(s)
Phyllanthaceae	*Glochidion eriocarpum* Champ. Ex Benth.		Apical part	Glochieriosides A	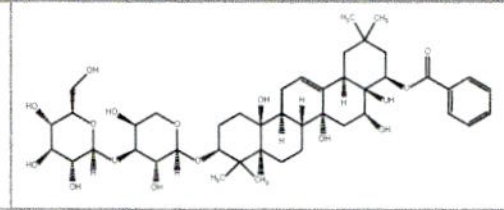	Kiem et al. (2009)
				Glochiericsides B	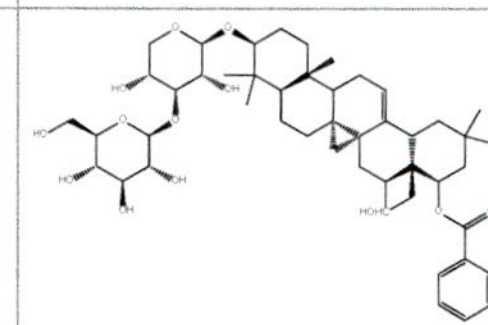	
	Synonym(s): *Diasperus anamiticus* Kuntze, *Diasperus eriocarpus*, *Diasperus villicaulis*, *Glochidion anamiticum* Kuntze. **Common name(s)**: Mao guosuan pan zi. **Botanical description**: It is a native plant of East Asia, Southern China, Myanmar, Thailand, and Vietnam. It is a deciduous shrub growing up to 4 m long. The flowers are monoecious and are suitable to grow in light (sandy), medium (loamy), and heavy (clay) soils. **Medicinal use(s)**: It is used to treat lung cancer, colon adenocarcinoma, and ovarian carcinoma. All parts are used as a medicine for urticaria, mastitis, toothache, menorrhagia, dysentery, skin eczema, and enteritis. The fresh leaves are crushed and applied topically to treat skin problems and rheumatic joints.			Lupeol	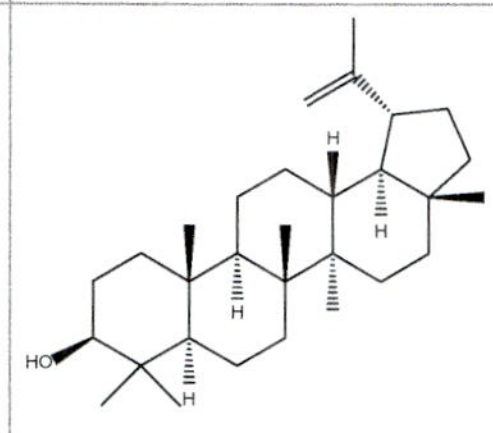	

(continued)

Table 4.103 (continued)

Family	Botanical name	Image	Plant part used	Bioactive compound(s)	Structure	Reference(s)
	Glochidion sphaerogynum (Mull.Arg.) Kurz.		Stem and root	5, 6 and lup-20(29) ene-3α,23-diol	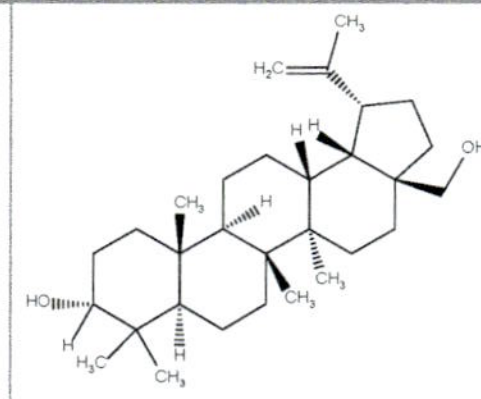	Puapairoj et al. (2005)
	Synonym(s): *Diasperus villicaulis*, *Glochidion anamiticum* Kuntze. **Common name(s)**: Buzo. **Botanical description**: It is a native plant of East Asia – Southern China, Myanmar, Thailand, and Vietnam. It is a deciduous shrub growing up to 4 m. The flowers are monoecious and suitable to grow in light (sandy), medium (loamy), and heavy (clay) soils. **Medicinal use(s)**: It is used to treat cancer. It also possesses antibacterial, anti-inflammatory, and antioxidant properties.					
	Phyllanthus niruri L.		Whole plant	Corilagin	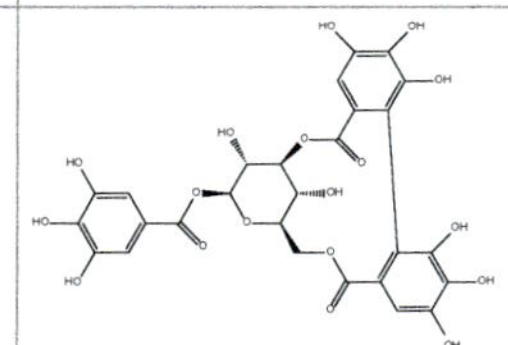	Jia et al. (2013)

	Synonym(s): *Diasperus chlorophaeus* (Baill.) Kuntze, *Diasperus lathyroides* (Kunth) Kuntze, *Diasperus microphyllus* (Mart.) Kuntze, *Diasperus niruri* (L.) Kuntze, *Diasperus rosellus* (Müll. Arg.) Kuntze, *Niruris annua* Raf., *Niruris indica* Raf. **Common name(s)**: Bhumyamalaki, Gale of the wind, Stonebreaker, or Seed-under-leaf. **Botanical description**: It is native of Central and Southern India. It grows throughout the world in tropical and subtropical regions in both the hemispheres. It grows easily as its seeds scatter around as they mature. It has mounting herbaceous branches. The bark is smooth and light greenish in color. It carries greenish flowers which are flushed with red color and also small fruits in a shape of a capsule which contains seeds. **Medicinal use(s)**: It prevents the metastasis of some lung and breast cancer. It also helps in hindering the growth of cancer cells, specifically sarcoma, lymphoma, and carcinoma.	Amarin	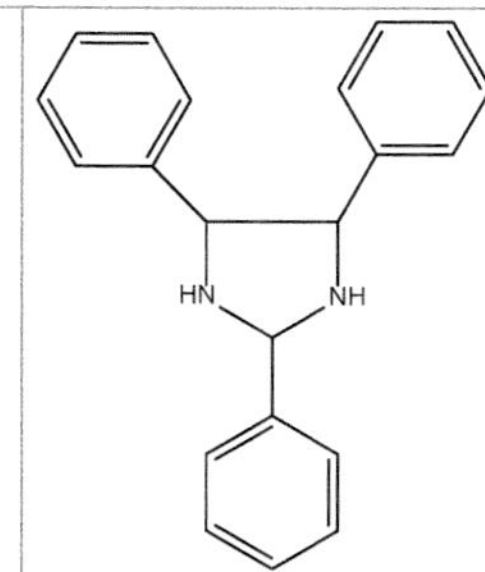	
		Corilagin	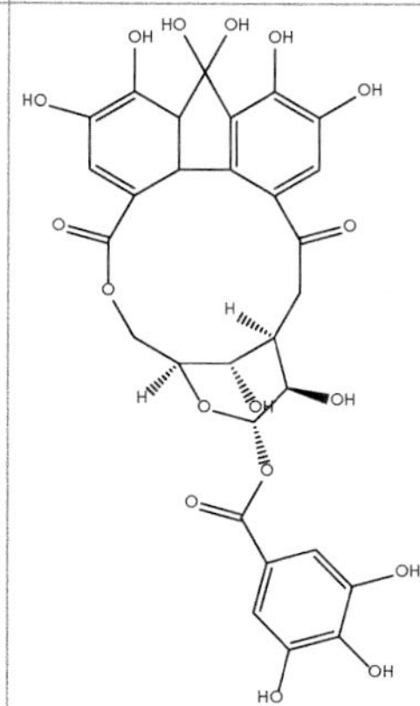	

Table 4.104 Anticancer plants of family Phytolaccaceae

Family	Botanical name	Image	Plant part used	Bioactive compound(s)	Structure	Reference(s)
Phytolaccaceae	*Phytolacca Americana* L.		Root, leaf, berry	Isoquinoline	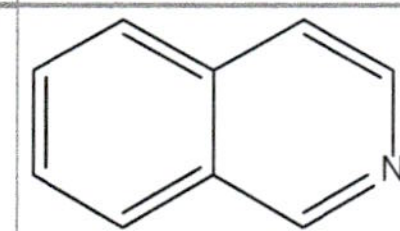	Wang et al. (2008c), Frenkel et al. (2010), and Maness et al. (2012, 2014)
	Synonym(s): *Phytolacca decandra* L., *Phytolacca americana* var. *americana*. **Common name(s)**: American pokeweed, Great pokeweed, Pokeweed, Pokeberry, Red ink plant, Pigeon berry. **Botanical description**: It is native to North America and commonly found in Mediterranean countries. The perennial root is large and fleshy, the stem is hollow, the leaves are alternate and ovate-lanceolate, and the flowers have a white calyx with no corolla. The fruit is a deep purple berry, covering the stem in clusters and resembling blackberries. **Medicinal use(s)**: The juice is used for the treatment of cancer, hemorrhoids, and tremors. A poultice made from the fruit is applied to sore breasts. A tea made from the fruit is used for the treatment of rheumatism, dysentery, etc. It contains potent anti-inflammatory agents, antiviral proteins, and substances that affect cell division.					

		Phytolaccagenic acid	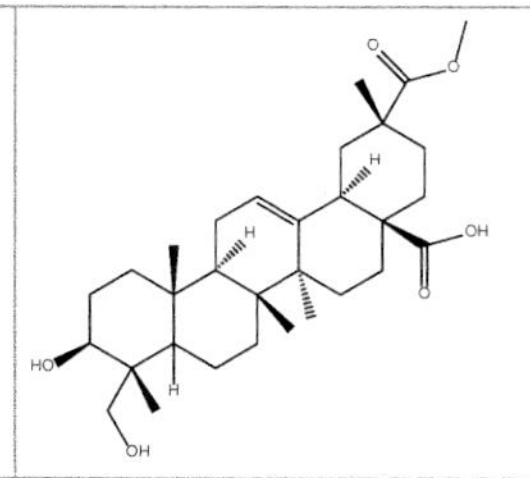	
		Alpha-spinasterol	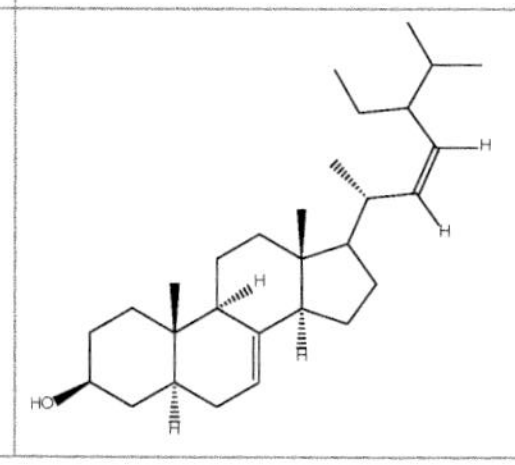	

(continued)

Table 4.104 (continued)

Family	Botanical name	Image	Plant part used	Bioactive compound(s)	Structure	Reference(s)
				Jaligonic acid	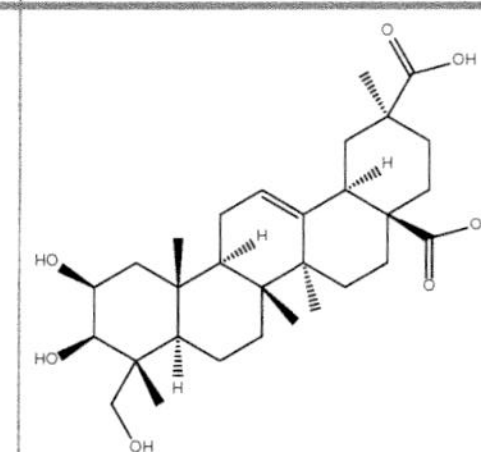	
				Oleanolic acid	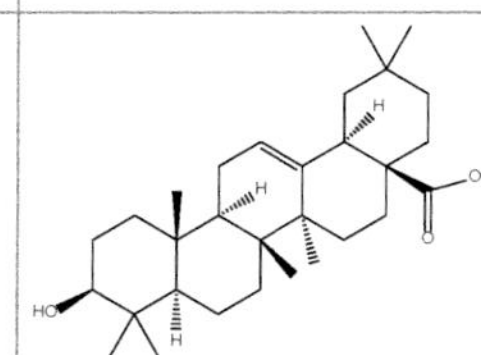	

	Phytolacca acinosa Roxb.		Bark	Acinospesigenin	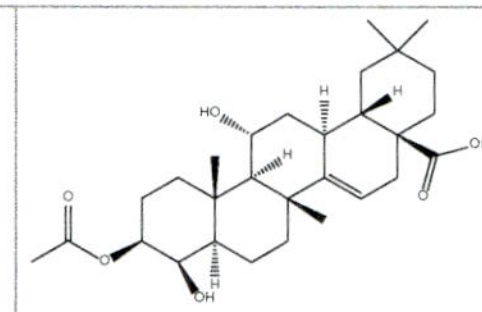	Koul et al. (2003)
	Synonym(s): *Phytolacca pekinensis* Hance. **Common name(s)**: Indian pokeweed, Himalayan pokeberry, Indian poke, Indian pokeberry. **Botanical description**: It is an herbaceous plant and is native to India. It has large lance-shaped leaves, 15–25 cm, which are long pointed, and narrowed at the base to a short stalk. Green-white flowers are borne in erect cylindrical clusters, opposite the leaves. These spikes are 5–15 cm long. Flowers are 7 mm across, with 5 obovate, spreading petals, with 8–10 stamens. Each fruit has about 8 fleshy dark purple carpels. The fruits are also crowded on erect cylindrical clusters. **Medicinal use(s)**: It is used to treat lung cancer, breast cancer, and cancer of lymph nodes. According to ancient medical documents, this plant is used to treat a wide variety of health problems, including hemorrhoids, rheumatoid arthritis, diphtheria, syphilis, adenitis (inflammation of any gland), conjunctivitis, and emesis (vomiting) and also used as a purgative.			Gamma-aminobutyric acid	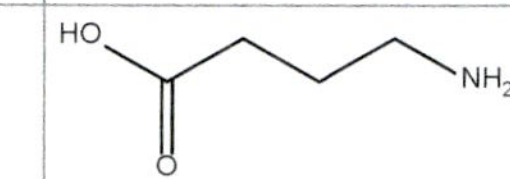	
				Myristyl palmitate		
				Esculentic acid	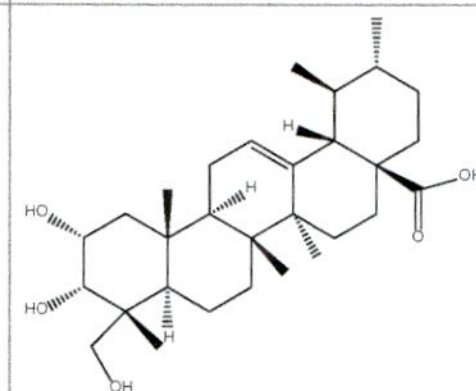	

(continued)

Table 4.104 (continued)

Family	Botanical name	Image	Plant part used	Bioactive compound(s)	Structure	Reference(s)
	Phytolacca bogotensis Kunth		Leaf	Oleanolic acid	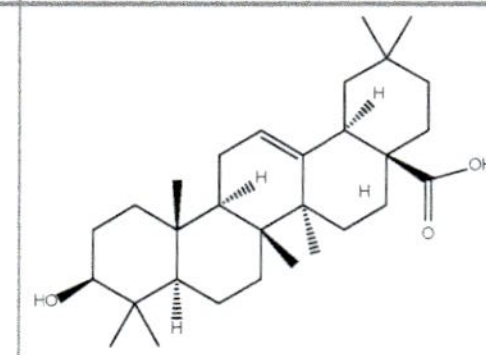	Garcia-Barriga (1974) and Montoya et al. (2009)
	Synonym(s): *Phytolacca micrantha* H. Walter, *Phytolacca australis* Phil., *Phytolacca parviflora* Hauman. **Common name(s)**: Pokeweeds. **Botanical description**: It is a scrambling or climbing perennial plant and is native to Colombia. It sends out shoots up to 5 m long. It is found in damp or wet thickets or forest, ascending from sea level to elevations of about 2600 m in Guatemala. **Medicinal use(s)**: It is used to treat tumors, breast cancer, cervical cancer, and endometrial cancer and is also used as chemoradiation adverse effect treatment. The root, poultices, and salves were used on skin ulcers, eczema, bunions, bruises, sprains, swollen joints, poison ivy rashes, and warts. Steam from a root decoction is used to ease hemorrhoids. The leaves of the plant were also used for skin problems and rheumatism. In addition, they are used as an emetic, as a laxative, and as a blood purifier tonic.					

	Phytolacca esculenta Van Houtte		Root	Lupeol	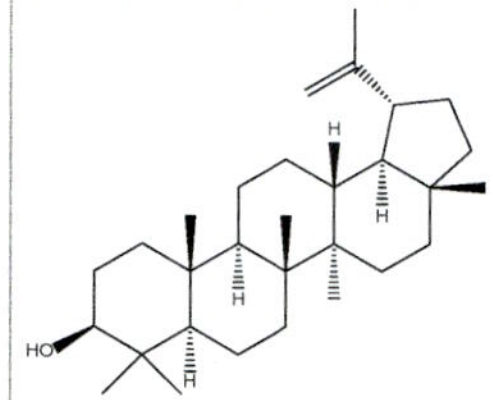	Yi and Dai (1991), Hua (1992), Yi (1992), and Wang et al. (2010a, b)
	Synonym(s): *Sarcoca esculenta* (Van Houtte) Skalický. **Common name(s)**: Pokeberry. **Botanical description**: It is a perennial and native plant of Asia and China. It is in flower from July to August, and the seeds ripen from August to September. The flowers are hermaphrodite (have both male and female organs), and it is suitable to grow in light (sandy), medium (loamy), and heavy (clay) soils. **Medicinal use(s)**: It is used to treat soft tissue sarcomas and prostate, breast, and ovarian cancers. It also possesses abortifacient, anticancer, anti-asthmatic, antibacterial, antifungal, anti-inflammatory, antiphlogistic, antitussive, diuretic, expectorant, hypotensive, and purgative properties. A decoction is used for the treatment of edema, beriberi, lumbago, rheumatism, abdominal distension, and numbness of the throat.			Aucubin	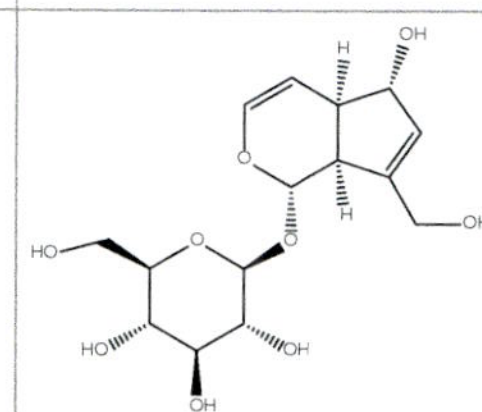	
				Riboflavin	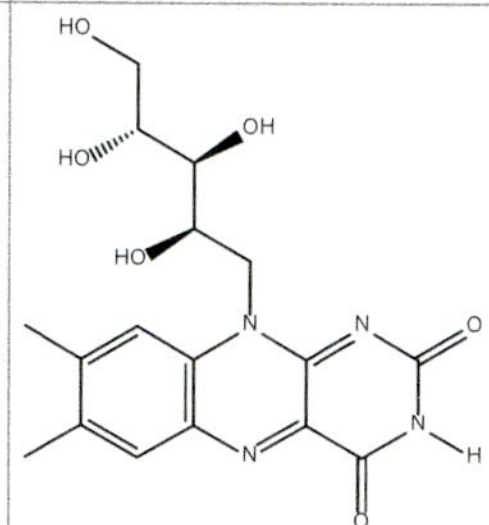	

(continued)

Table 4.104 (continued)

Family	Botanical name	Image	Plant part used	Bioactive compound(s)	Structure	Reference(s)
	Phytolacca rivinoides Kunth and Bouche		Leaf	Friedlin	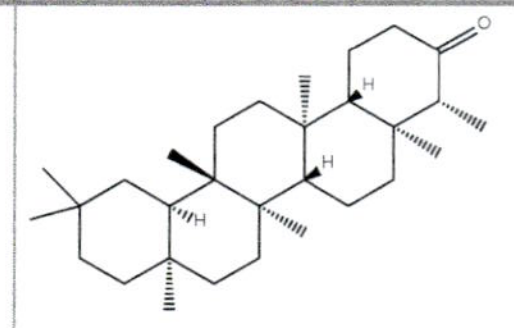	Garcia-Barriga (1974) and Maffei (2003a, b, c, d, e, f, g, h, i, j, k, l, m)
	Synonym(s): *Phytolacca acuminata* hort. ex Moq., *Phytolacca macrostachya* Willd. ex Moq., *Phytolacca polystigma* Benth. ex Moq. **Common name(s)**: Pokeroot. **Botanical description**: It is native to South America, Argentina, Brazil, Peru, North to the Caribbean, and through Central America to Mexico. It is a scrambling or climbing perennial plant that becomes woody at the base. It sends out shoots up to 5 m long. It is found in damp or wet thickets or forest, ascending from sea level to elevations of about 2600 m. **Medicinal use(s)**: It is used to treat breast, cervical, endometrial, and ovarian cancer. The leaves and young stem are eaten as a treatment for diabetes. Liquid from the leaves crushed in water is rubbed onto hair to kill lice and fleas and to treat dandruff. A decoction of the boiled leaves is used for treating fevers. The plant is used for treating rabies, lung diseases, and tumors.			Oleanolic acid	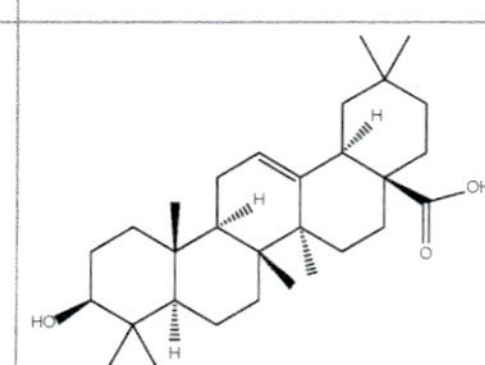	
				Oleic acid	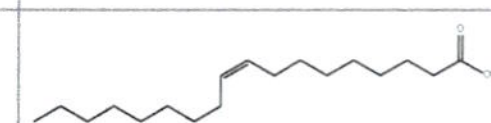	

	Phytolacca sanguinea H. Walter		Leaf	Isoquinoline	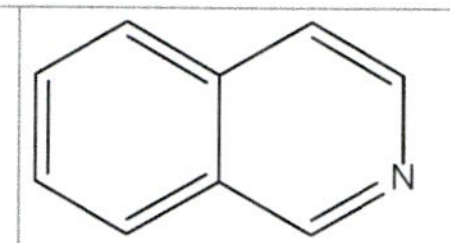	Garcia-Barriga (1974)
	Synonym(s): *Phytolacca esculenta*, *Phytolacca bogotensis*. **Common name(s)**: Pokeberry. **Botanical description**: It is a scrambling or climbing perennial plant that becomes woody at the base and is native to Colombia. It sends out shoots up to 5 m long. It is suitable to grow in light (sandy), medium (loamy), and heavy (clay) soils. **Medicinal use(s)**: *Phytolacca sanguine* is used to treat breast cancer, cervical cancer, and endometrial cancer and is also used as chemoradiation adverse effect treatment and integrative cancer treatment. It also possesses abortifacient, anti-asthmatic, antibacterial, antifungal, anti-inflammatory, antiphlogistic, antitussive, diuretic, expectorant, hypotensive, and purgative properties.			Lutein	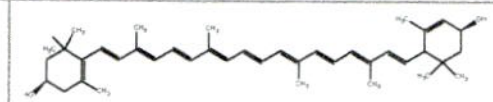	

Table 4.105 Anticancer plants of family Pinaceae

Family	Botanical name	Image	Plant part used	Bioactive compound(s)	Structure	Reference(s)
Pinaceae	*Pinus acutisleginum* L.		Whole plant	Aristolactams	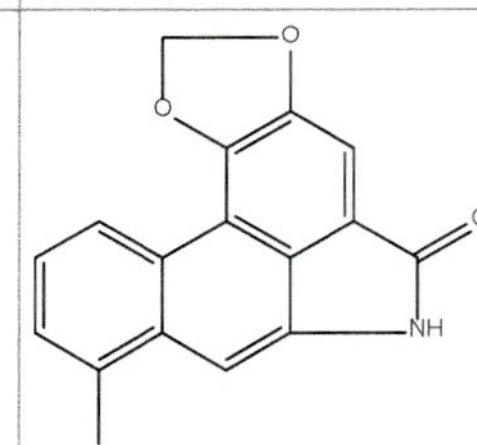	Kumar et al. (2003)
	Synonym(s): *Pinus longifolia* Salisb. **Common name(s)**: Digger pine. **Botanical description**: It is an evergreen plant and is native to Asia. The bark of the tree is 3–6 cm thick with brownish-red color sap. Wood is white to creamy white. Their leaves are needlelike, pointed, 15–30 in. long, bent towards the ground, and bright green in color. The flowers are yellow or rarely purple in color and 1–5 cm long. **Medicinal use(s)**: It is used to treat lung cancer, leukemia, typhoid fever, peptic ulcer, chronic bronchitis, and gangrene of the lungs. It is also very beneficial to the respiratory system and is useful in treating diseases of the mucous membranes and respiratory complaints such as coughs, colds, influenza, and TB. Externally it used for a variety of skin complaints, wounds, sores, burns, boils, etc. and is used in the form of liniment plasters, poultices, herbal steam baths, and inhalers. The wood is diaphoretic and stimulant.			4,5-Dioxoaporphines	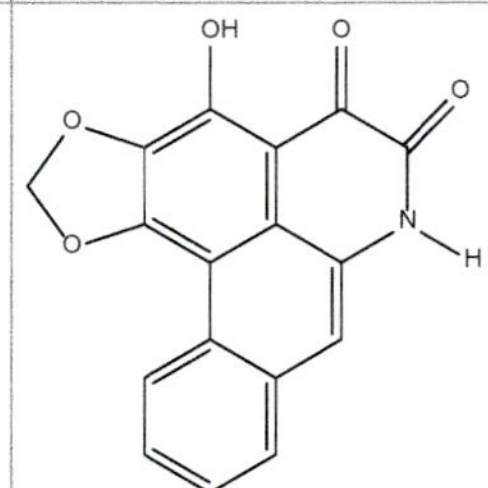	
				Abietic acid	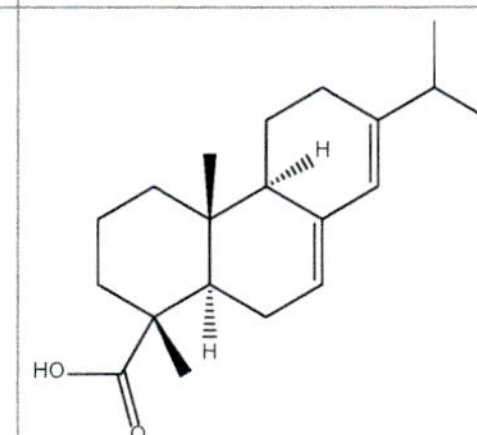	

<table>
<tr><td></td><td>Pinus bungeana Zucc. ex Endl.</td><td></td><td>Roots</td><td>Aristolactams</td><td>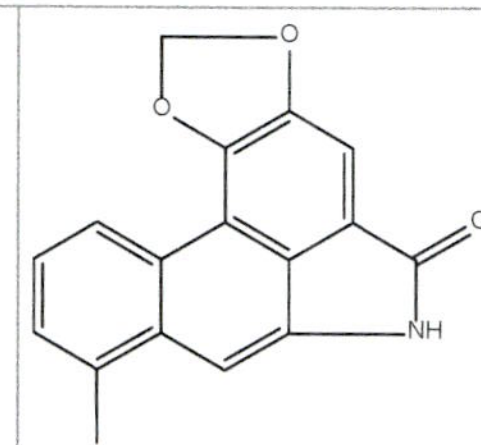</td><td rowspan="2">Kumar et al. (2003)</td></tr>
<tr><td></td><td colspan="3">Synonym(s): Pinus excorticata Lindl. & Gordon.
Common name(s): Chinese pinenut, Silver ghost, Chinese silver.
Botanical description: It is an evergreen tree and is native to East Asia and China. It is reaching up to 25–45 m in height, with a broad, rounded crown of long branches. The bark is thick, grayish brown, and scaly plated at the base of the trunk and orange-red, thin, and flaking higher on the trunk. The leaves are needlelike, dark green, with two per fascicle, 12–20 cm long and 0.8–1 mm wide, the persistent fascicle sheath 1.5–2 cm long. The cones are ovoid, 4–7 cm long, chestnut-brown, opening when mature in late winter. The seeds are winged, 4–6 mm long with a 10–15 mm wing. Pollination is in mid-spring.
Medicinal use(s): It possesses anticancer, antioxidant, anti-inflammatory, antimicrobial, antimutagenic, and antiproliferative effect. It is also used to treat breast and skin cancer. It is a valuable remedy which is used internally for the treatment of kidney and bladder complaints.</td><td>4,5-Dioxoaporphines</td><td>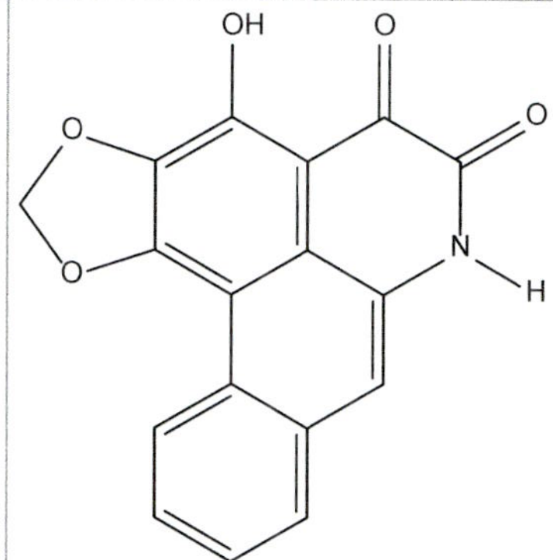</td></tr>
</table>

(continued)

Table 4.105 (continued)

Family	Botanical name	Image	Plant part used	Bioactive compound(s)	Structure	Reference(s)
	Pinus koraiensis Siebold & Zucc		Bark	Procyanidins	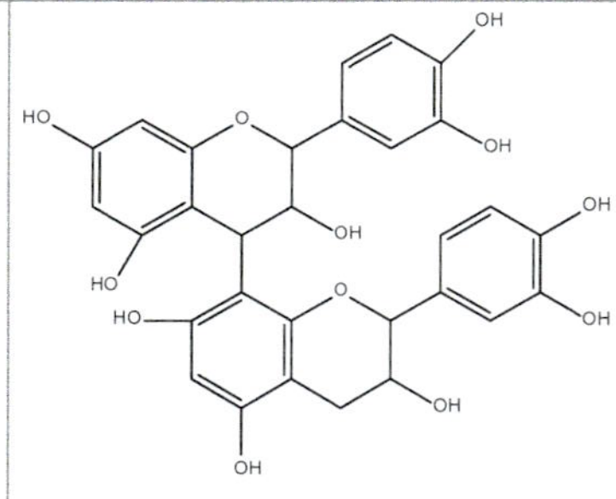	Li et al. (2011)
	Synonym(s): *Apinus koraiensis* (Siebold & Zucc.) Moldenke, *Pinus mandschurica* Rupr., *Pinus prokoraiensis* Y.T.Zhao, J.M.Lu & A.G.Gu. **Common name(s)**: Korean nut pine, Chinese pine nut. **Botanical description**: It is native to East Asia, China, Korea, Manchuria, and Siberia. It is an evergreen tree growing up to 20 m at a medium rate. Pollination occurs via wind. It is found in dry mountain slopes, especially those facing north, to 2600 m. **Medicinal use(s)**: It is used to treat prostate and cervical cancer. It also possesses anti-allergic, antibacterial, and anti-inflammatory properties. It is also very beneficial to the respiratory system and is useful in treating diseases of the mucous membranes and respiratory complaints such as coughs, colds, influenza, and TB.			Cianidanol	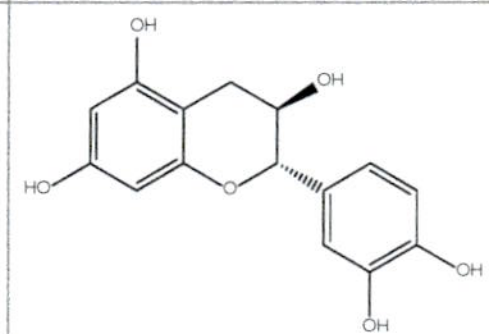	
				Taxifolin	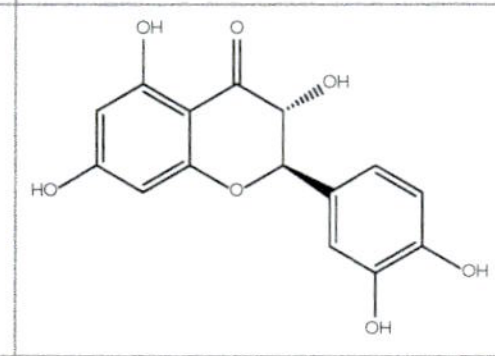	
				Pinolenic acid	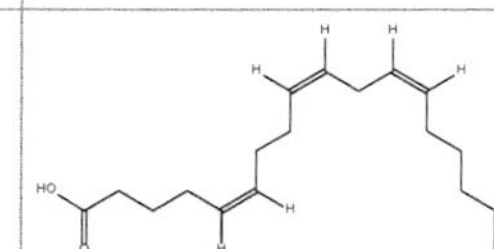	

	Pinus massoniana Lambert		Bark	Procyanidins A2	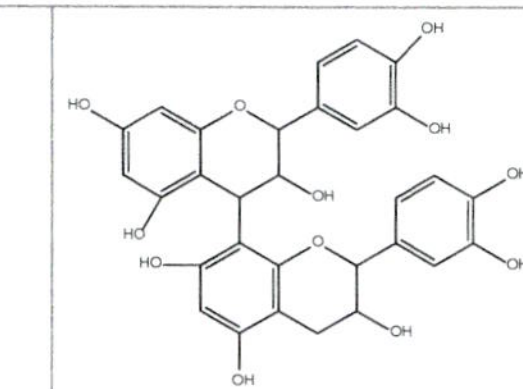	Cui et al. (2005)
	Synonym(s): *Pinea massoniana* (Lamb.) Opiz, *Pinus argyi* Lemée & H.Lév., *Pinus canaliculata* Miq. **Common name(s)**: Chinese red pine, Horsetail pine. **Botanical description**: It is native to East Asia and China. It is an evergreen tree reaching 25–45 m in height, with a broad, rounded crown of long branches. The bark is thick, grayish brown, and scaly plated at the base of the trunk and orange-red, thin, and flaking higher on the trunk. The leaves are needlelike, dark green, with two per fascicle, 12–20 cm long and 0.8–1 mm wide, the persistent fascicle sheath 1.5–2 cm long. The cones are ovoid, 4–7 cm long, chestnut-brown, opening when mature in late winter to 4–6 cm broad. The seeds are winged, 4–6 mm long with a 10–15 mm wing. Pollination is in mid-spring, with the cones maturing 18–20 months after. **Medicinal use(s)**: It is used to treat lung cancer, ovarian cancer, cervical cancer, and metastatic cancer. It also possesses anti-rheumatic, anti-inflammatory, and anti-apoptotic properties. It is also very beneficial to the respiratory system and is useful in treating diseases of the mucous membranes and respiratory complaints such as coughs, colds, influenza, and TB. Externally it is very beneficial for the treatment of skin complaints, wounds, sores, burns, boils, etc. and is used in the form of liniment plasters, poultices, herbal steam baths, and inhaler. The stem bark is used as astringent, parasiticide, and vulnerary. The root bark is used as astringent and tonic.			Procyanidin	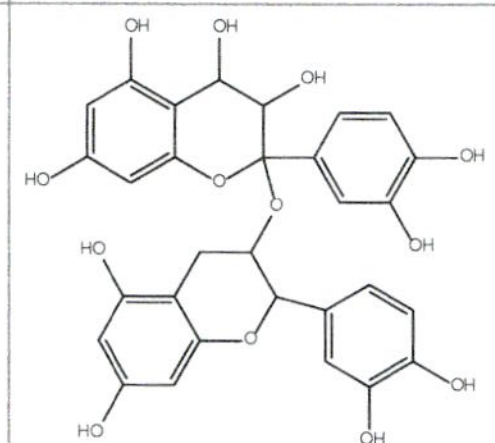	
				Antifade	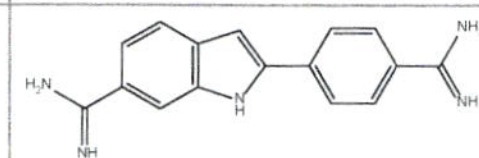	

(continued)

Table 4.105 (continued)

Family	Botanical name	Image	Plant part used	Bioactive compound(s)	Structure	Reference(s)
	Pinus halepensis Mill.		Bark	Phenolic acid	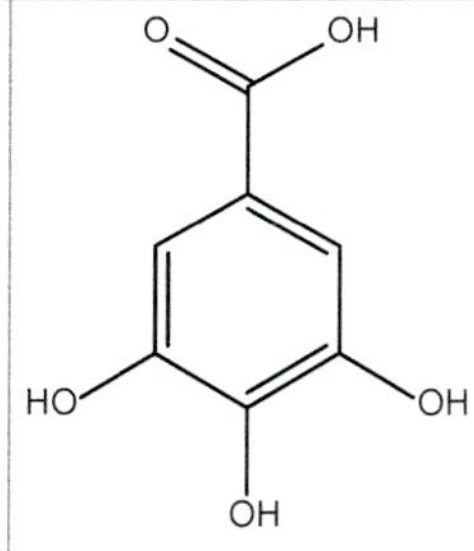	Kyriazia et al. (2006)
				Ferulic acid	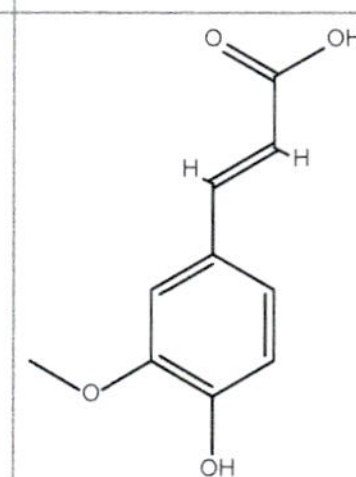	
	Synonym(s): *Pinus abasica* Carrière, *Pinus arabica* Sieber ex Spreng., *Pinus carica* D.Don, *Pinus ceciliae* Llorens & L.Llorens. **Common name(s)**: Maritime pine. **Botanical description**: It is native to Europe and West Mediterranean and naturalized in Britain in some parts of the New Forest. It is found in coastal sands and dunes and on mountains inland. It is an evergreen tree growing up to 20 m at a fast rate. **Medicinal use(s)**: It possesses anticancer, antioxidant, anti-inflammatory, antimicrobial, antimutagenic, and antiproliferative properties. It is used to treat breast cancer and skin cancer. It is a valuable remedy used internally for the treatment of kidney and bladder complaints.			Flavone(s)	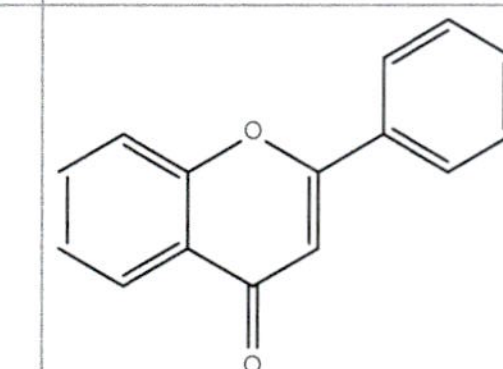	
				Isoquinoline	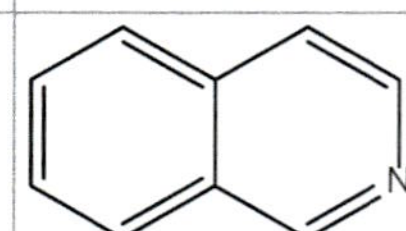	

		Compound	Structure	
		Beta-carotene	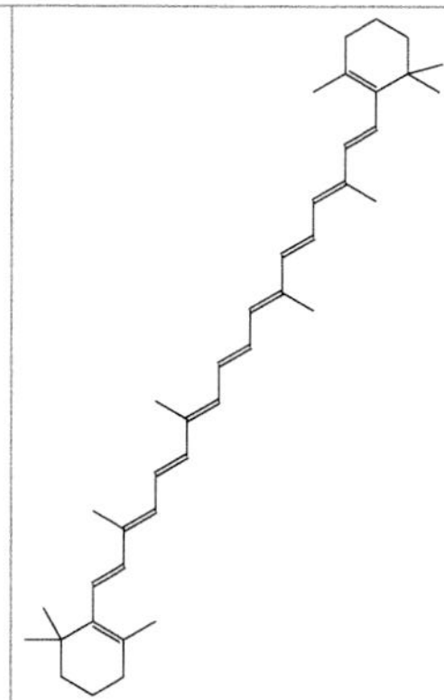	
		Taxifolin	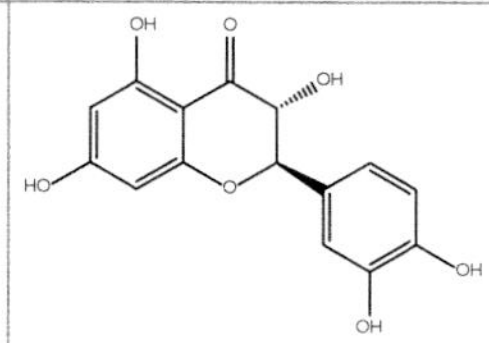	

(continued)

Table 4.105 (continued)

Family	Botanical name	Image	Plant part used	Bioactive compound(s)	Structure	Reference(s)
	Pinus morrisonicola Hay.		Bark	1,1-Diphenyl-2-picrylhydrazine	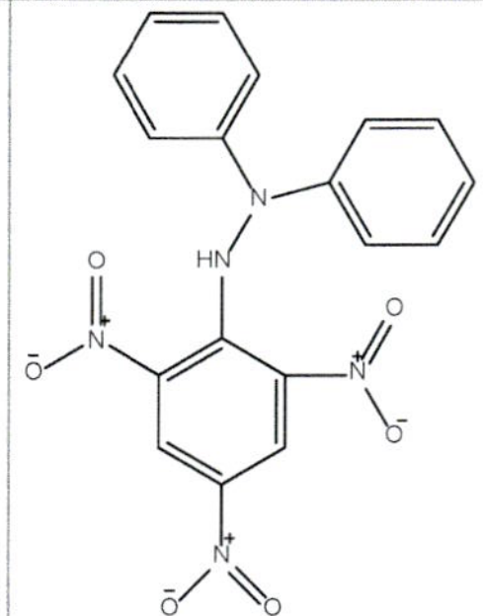	Hsu et al. (2005), Chen et al. (2011), and Qiu et al. (2018)
	Synonym(s): *Pinus hayatana* Businský, *Pinus uyematsui* Hayata. **Common name(s)**: Taiwan white pine. **Botanical description**: It is native to Taiwan. It is found in coastal sands and dunes and on mountains inland. It is an evergreen tree growing up to 20 m at a fast rate. The trunk is often crooked. Needles are in bundles of five. Mature cones are large, up to 10 cm long and 4–5 cm in diameter. **Medicinal use(s)**: It is used to treat lung cancer, breast cancer, and skin cancer. It also possesses antitumor, antimicrobial, antibacterial, and anti-inflammatory properties. It is also very beneficial to the respiratory system and is useful for treating diseases of the mucous membranes and respiratory complaints such as coughs, colds, influenza, and TB.			Lupeol	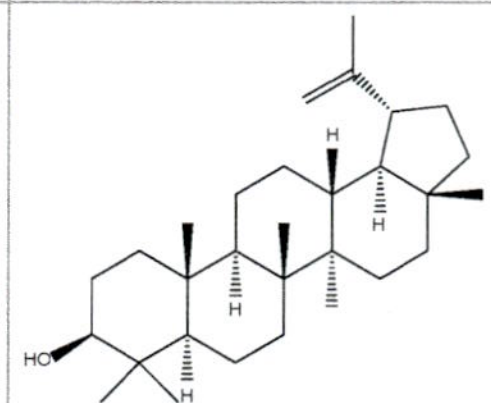	
				Lectin	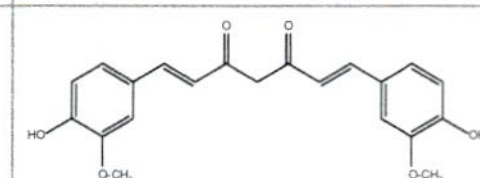	

	Pinus parviflora Siebold and Zucc.		Strobilus	L-Arabinose	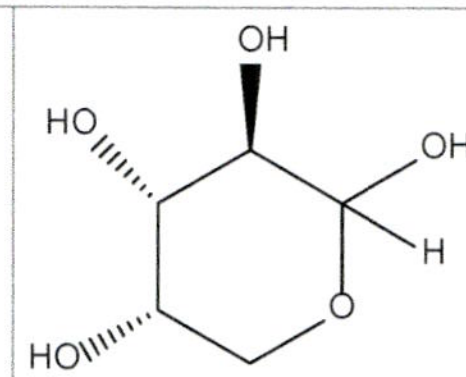	Sakagami et al. (1989), Lai et al. (1990), Sakagami et al. (1992), and Jimoh and Oladiji (2005)
	Synonym(s): *Pinus cembra* var. *japonica* J.Nelson, *Pinus himekomatsu* Miyabe & Kudô, *Pinus mayrii* Zabel. **Common name(s)**: Japanese white pine. **Botanical description**: It is native to Japan and Korea. It is an evergreen conifer that typically grows up to 30–50 m tall in cultivation but may reach 90 m tall in the wild. Trees have a dense pyramidal habit when young but with age develop more spreading branching and a flatter top. Often-twisted, green needles in bundles of five appear in dense tufts primarily at the branch ends. Needles are white on the interior surfaces. Oval reddish-brown cones appear solitary or in small clusters, remaining on the tree for up to 7 years. **Medicinal use(s)**: It is used to treat cancer. It also possesses anticancer, antitumor, anti-inflammatory, and antimicrobial properties. It is a valuable remedy used internally for the treatment of kidney and bladder complaints and is used externally as a rub and steam bath in the treatment of rheumatic affections.			D-Glucuronic acid	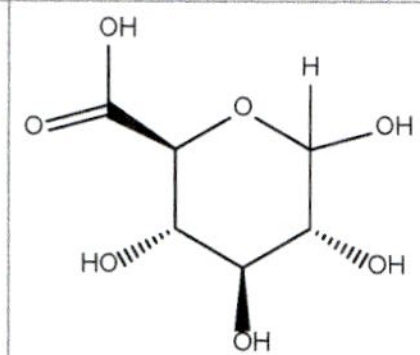	
				Galactose	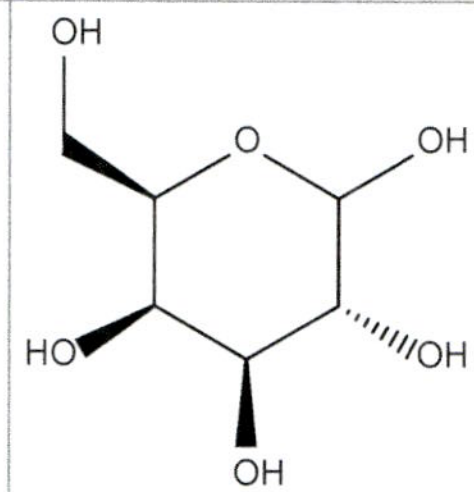	

Table 4.106 Anticancer plants of family Piperaceae

Family	Botanical name	Image	Plant part used	Bioactive compound(s)	Structure	Reference(s)
Piperaceae	*Piper betel* L		Leaf	Aristolactams	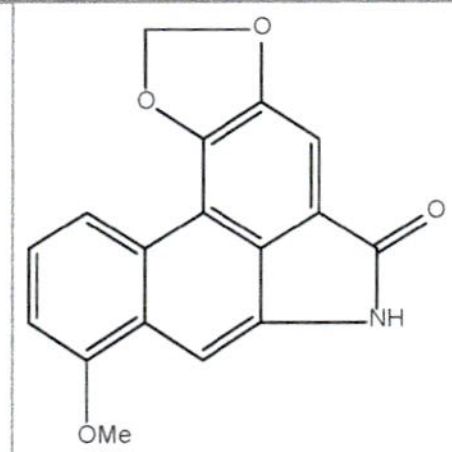	Kumar et al. (2003)
				4,5-Dioxoaporphines	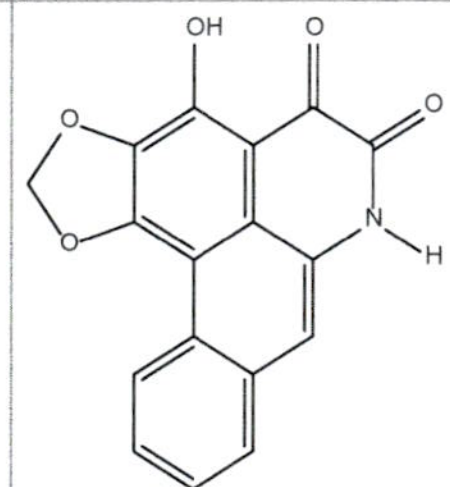	
	Synonym(s): *Artanthe hexagyna* Miq., *Betela mastica* Raf., *Chavica auriculata* Miq., *Chavica betle* (L.) Miq. **Common name(s)**: Pepper, Betel. **Botanical description**: It is native to India and Sri Lanka. The soil is plowed into furrows of 10–15 m in length, 75 cm in width, and 75 cm depth. Oil cakes, manure, and leaves are thoroughly incorporated with the topsoil of the furrows and wood ash. The cuttings are planted at the beginning of the monsoon season. Dried leaves and wood ash are applied to the furrows at fortnightly intervals, and cow dung slurry is sprinkled. In 3–6 months, the vines reach 150–180 cm in height. **Medicinal use(s)**: It is used for the treatment of colorectal cancer, oral cancer, and tumor. It also possesses anti-inflammatory, antioxidant, and antibacterial properties.					

	Piper boehmerifolium Wall.		Root	Carvacrol	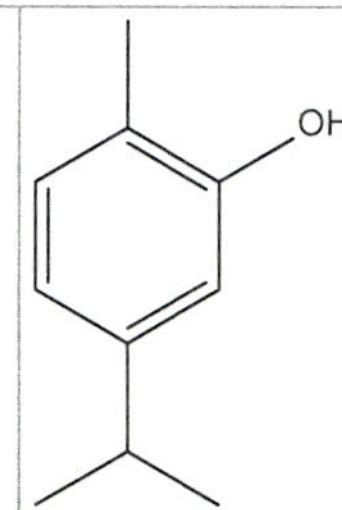	Mahanta et al. (1974) and Wang et al. (2014)
	Synonym(s): *Chavica boehmeriifolia* Miquel. **Common name(s)**: Kathali champa. **Botanical description**: It is native to India, Malaysia, Thailand, Myanmar, Sikkim, Bhutan, and North Vietnam. Subshrubs erect, 1–3 m high, glabrous to uniformly hairy, dioecious, most parts usually drying black. Stems terete to thickly ridged when dry, minutely papillate to smooth, usually glabrous. Petiole 3–10 mm, glabrous or sometimes sparsely pubescent; leaf blades towards base of stem elliptic, narrowly elliptic, oblong, oblong-lanceolate. **Medicinal use(s)**: It is used for the treatment of breast cancer, lung cancer, cervical cancer, prostate cancer, gastric cancer, uterine cancer, and colon cancer. It also possesses anti-inflammatory, antitumor, antioxidant, and antibacterial properties.					

(continued)

Table 4.106 (continued)

Family	Botanical name	Image	Plant part used	Bioactive compound(s)	Structure	Reference(s)
	Piper chaba L.		Fruit	Corynanthe	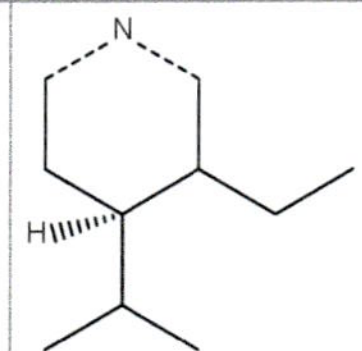	Sakpakdeejaron and Itharat (2009)
	Synonym(s): *Chavica longa* H.Karst., *Chavica roxburghii* Miq., *Chavica sarmentosa* Miq., *Piper roxburghii* (Miq.) J.Presl. **Common name(s)**: Long pepper, Indian long pepper, Lindi piper, Pipalli. **Botanical description**: It is native to East Asia, India, and Nepal. Long pepper is a climbing shrub with stems up to 3 m long. It may also grow around large trees. The leaves are oval-shaped and about 2–3 in. long. The flowers are monoecious and blossom during the monsoon. The fruit looks similar to other varieties of long pepper, with an elongated shape that can grow up to 3 in. long. The fruit is red when ripe, which turns dark brown or black when dry. **Medicinal use(s)**: It is used to treat breast cancer and tumor. It also possesses antioxidant, anti-inflammatory, and antibacterial properties.					

	Piper interruptum Opiz.		Stem	Corynanthe	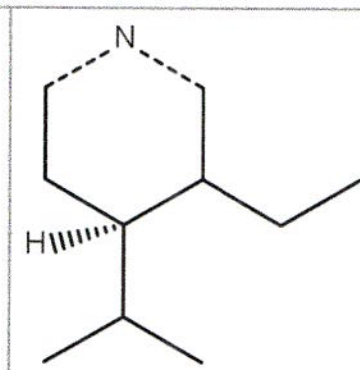	Sakpakdeejaron and Itharat (2009)
	Synonym(s): *Piper interruptum* var. *multinervum* C. DC. **Common name(s)**: Di-pli, Prik-hang. **Botanical description**: It is native to Southeast Asia, Taiwan, Indonesia, the Philippines, New Guinea to Northeast Australia and also to the Pacific Islands. It is a climbing shrub producing woody stems 20–45 mm thick. It is habituated in rainforest at elevations from near sea level to 750 m in Northeast Australia. A climber is dioecious. Female spikes 7–17 cm, flowers unevenly developed, sparse or interrupted in fruit; peduncle nearly as long as opposite leaves, glabrous; rachis and bracts as in male spikes. Ovary distinct, ovoid, apex acute; stigmas 4 or 5. Drupe ovoid or ovoid-globose, 3–6 × 2–4 mm, smooth. **Medicinal use(s):** It is used for the treatment of breast cancer and tumor. It also possesses antioxidant, anti-inflammatory, antibacterial, and antimicrobial properties.					

(continued)

Table 4.106 (continued)

Family	Botanical name	Image	Plant part used	Bioactive compound(s)	Structure	Reference(s)
	Piper latifolium Forst.f.		Leaf	D-Xylose	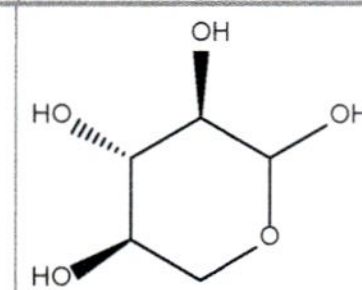	Holdsworth (1991) and Wang et al. (2014)
	Synonym(s): *Chavica longa* H.Karst., *Chavica roxburghii* Miq. **Common name(s)**: Dipli-chueak. **Botanical description:** It is native to East Asia – India and Nepal. It is a climbing shrub with stems up to 3 m long. The leaves are oval-shaped and about 2–3 in. long. The flowers are monoecious and blossom during the monsoon. The fruit looks similar to other varieties of long pepper, with an elongated shape that can grow up to 3 in. long. The fruit is red when ripe, which turns dark brown or black when dry. **Medicinal use(s):** It is used for the treatment of breast cancer and tumor. It also possesses antioxidant, anti-inflammatory, antibacterial, and antimicrobial properties.					

	Piper longum L.		Fruit	Beta-sitosterol	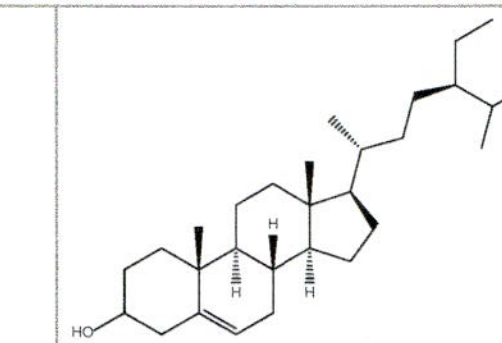	Hullatti and Murthy (2001)
	Synonym(s): *Chavica longa* H.Karst., *Chavica roxburghii* Miq., *Chavica sarmentosa* Miq., *Piper roxburghii* (Miq.) J.Presl. **Common name(s):** Long pepper. **Botanical description:** It is native to East Asia – India and Nepal. It is a climbing shrub with stems up to 3 m long. The leaves are oval-shaped and about 2–3 in. long. The flowers are monoecious and blossom during the monsoon. The fruit looks similar to other varieties of long pepper, with an elongated shape that can grow up to 3 in. long. The fruit is red when ripe, which turns dark brown or black when dry. **Medicinal use(s):** It is used for the treatment of breast cancer and tumor. It also possesses antibacterial, anti-inflammatory, antioxidant, and antimicrobial properties.					
	Piper methysticum G. Forster.		Whole plant, leaf	Flavokawain A	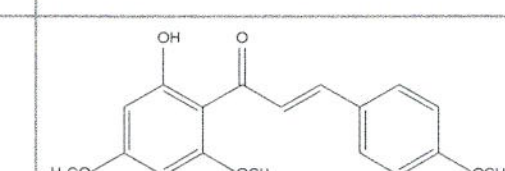	Tabudravu and Jaspar (2005)

(continued)

Table 4.106 (continued)

Family	Botanical name	Image	Plant part used	Bioactive compound(s)	Structure	Reference(s)
	Synonym(s): *Chavica sarmentosa* Miq., *Piper roxburghii* (Miq.) J.Presl. **Common name(s):** Kava pepper. **Botanical description:** It is native to Southeast Asia – Taiwan, Indochina, Indonesia, the Philippines, New Guinea to Northeast Australia and also to the Pacific Islands. It is a climbing shrub producing woody stems 20–45 mm thick. It is habituated in rainforest at elevations from near sea level to 750 m in northeast Australia. Stems 2–4.5 mm thick, ridged, glabrous. Petiole 1–2.5 cm, glabrous, sheathed at base only; leaf blade ovate to long ovate, 6–13 × 4–7 cm, membranous or papery, without evident glands, both surfaces glabrous, base rounded or shortly tapered, symmetric, apex acute or shortly acuminate; all basal; reticulate veins abaxially prominent, lax. Spikes leaf opposed. Male spikes 11–27 cm × 1.5–3 mm. **Medicinal use(s):** It is used for the treatment of breast cancer and tumor cells. It also possesses anti-inflammatory, antioxidant, and antibacterial properties.			Flavokawain B	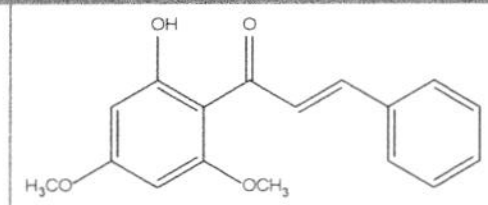	
				Dihydromethysticin	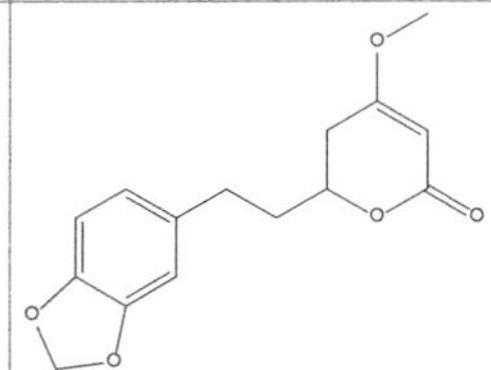	
				7,8 dihydrokawain	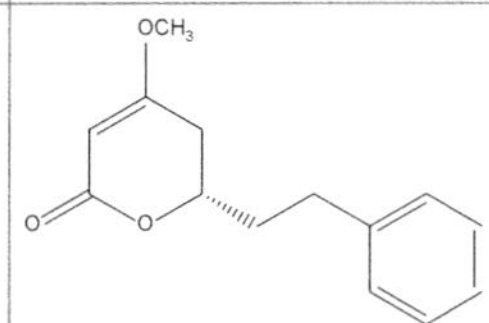	
				Demethoxyyangonin	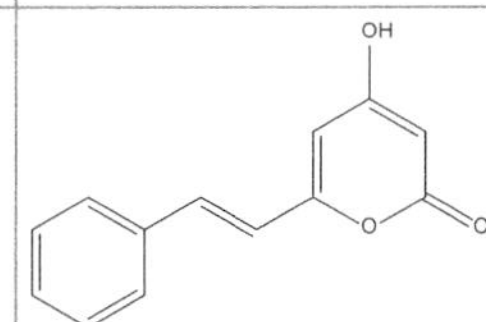	
				Trans-yangonin	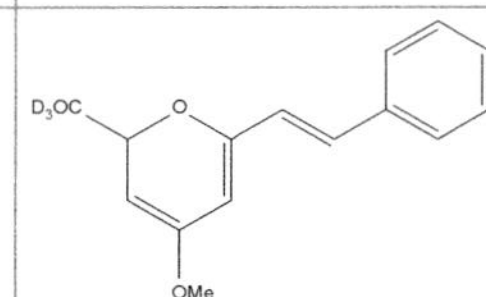	

	Piper sarmentosum Roxb.		Whole plant	D-Xylose	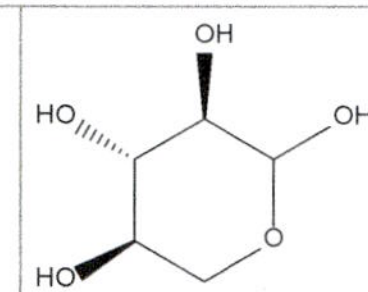 	Ariffin et al. (2009)
	Synonym(s): *Artanthe hexagyna* Miq., *Betela mastica* Raf. **Common name(s)**: Wild pepper, Kadok, Sirih tanah, Chabei. **Botanical description**: It is native to Southeast Asia and Vietnam. This plant is a perennial herb with creeping rhizomes and a striped stem that grows up to 40 cm high. Its leaves are thin, heart-shaped, and 8–10 cm long and 8–11 cm wide, with 5 main veins from the base of the blade, oil glands on the upper surface, and finely pubescent veins on its underside. Its petioles are 2.5–3 cm long. Erect white spikes of 1–2 cm long emerge at the axils. **Medicinal use(s)**: It is used for the treatment of human breast cancer cells. It also possesses antioxidant, anti-inflammatory, and antitumor properties.					

(continued)

Table 4.106 (continued)

Family	Botanical name	Image	Plant part used	Bioactive compound(s)	Structure	Reference(s)
	Piper schmidtii Hook. f.		Fruit	Aristolactams	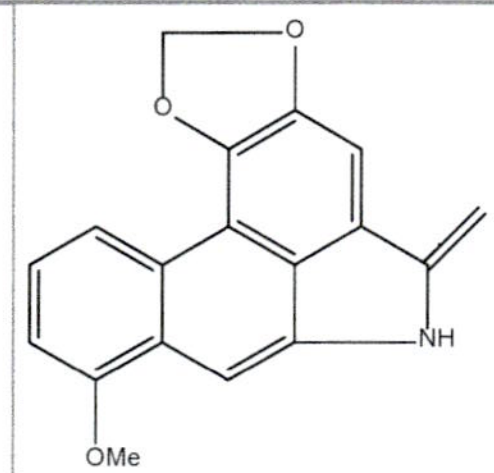	Kumar et al. (2003)
	Synonym(s): *Chavica auriculata* Miq., *Chavica betle* (L.) Miq. **Common name(s)**: Alajan pepper, Nilgiri pepper. **Botanical description**: It is native to Assam, Western Ghats, the Nilgiris, and Palani. Climbing shrubs; stem to 1.5 cm thick. Leaves alternate, each opposed by a spike or the spike scar, ovate-elliptic, often lop-sided, upper side dark green and glossy; lower very pale to white; coriaceous; main veins usually 5 with prominent reticulation between: stalk 1–2 cm, curved. Male spikes 1–3 cm, stalk 1–2 cm long, pendent; bracts circular with free margins. Stamens usually two, with thick filaments and anthers opening by vertical cross slits. Female spikes 5–7 cm, fruit longer, curved more or less and even spirally twisted; bracts square or roundish with raised margin. Berries red or yellow. **Medicinal use(s)**: It is used for the treatment of breast cancer cells and tumor. It also possesses anti-inflammatory, antioxidant, antibacterial, and antimicrobial properties.			4,5-Dioxoaporphines	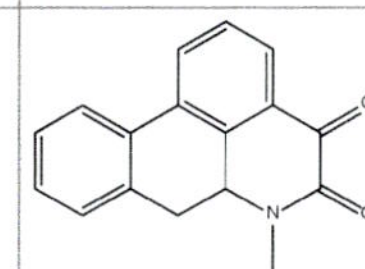	

	Piper sylvaticum Roxb.		Root	Aristolactams	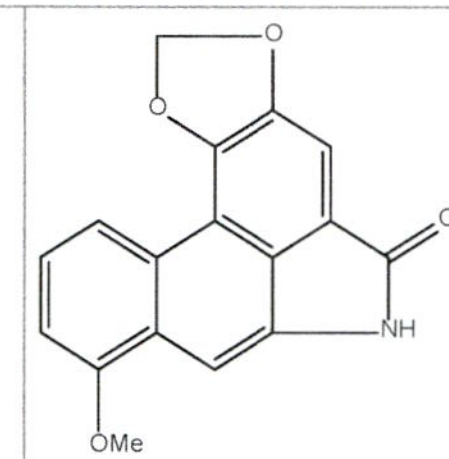	Mahanta et al. (1974) and Wang et al. (2014)
	Synonym(s): *Chavica sylvatica* (Roxb.) Miq. **Common name(s)**: Pahari pipul, Mountain long pepper. **Botanical description**: It is native to South Yunnan, Bangladesh, India, and Myanmar. Climbers herbaceous, dioecious. Stolons present. Stems ridged and furrowed when dry, very finely powdery pubescent when young. Petiole 1–7 cm, very finely powdery pubescent; prophylls 1/2 as long as petioles or slightly longer; leaf blade usually ovate, those at apex of stem ovate-lanceolate, 4–8.5 cm, papery, densely glandular, glabrous except for densely finely powdery pubescent veins abaxially and sometimes base of midvein adaxially, base cordate, symmetric, apex acuminate; veins 7, apical pair arising 0.7–1.5 cm above base, others basal; female spikes erect, 1.5–2.5 cm 3–4 mm; peduncle 0.5–2 cm, very finely powdery pubescent; bracts orbicular, sessile, adaxially pubescent; rachis pubescent. Ovary globose, distinct, stigmas 2 or 3, ovate, apex acuminate. **Medicinal use(s)**: It is used for the treatment of breast cancer, bronchitis, diseases of the spleen, and tumor. It also possesses anti-inflammatory, antitumor, antioxidant, and antibacterial properties.			Sylvatesmin	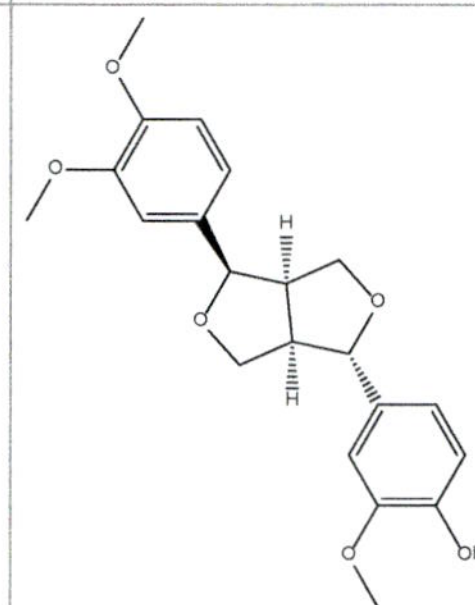	

(continued)

Table 4.106 (continued)

Family	Botanical name	Image	Plant part used	Bioactive compound(s)	Structure	Reference(s)
	Piper wightii Miq.		Leaf	Aristolactams	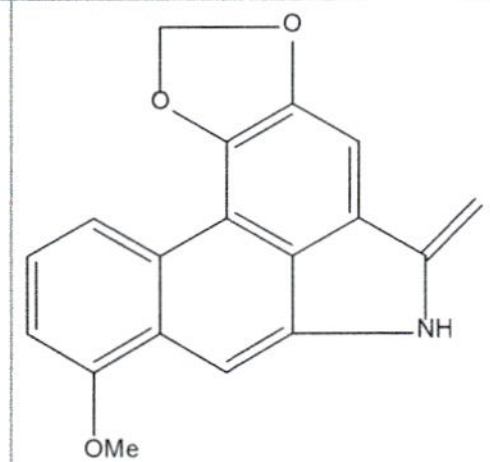	Kumar et al. (2003)
	Synonym(s): *Chavica sylvatica* (Roxb.) Miq. **Common name(s)**: Candela. **Botanical description**: It is native to southern Western Ghats and Kerala. Stout glabrous climbers. Leaves 14–17 × 6–8 cm, ovate, acuminate at apex, round to subcordate at base, 3–5-ribbed, coriaceous, midrib with 2–3 lateral nerves similar to ribs, nervules reticulate; petiole 2 cm long. Spike 8 cm long stout, drooping; peduncle 2 cm long; bracts curved and hooded at one side. Flowers distantly arranged; stamens 2. Berry reddish, to 7 mm across, globose, smooth; stigmatic lobes 3 or 4. **Medicinal use(s)**: It is used for the treatment of pancreatic cancer, skin cancer, colon cancer, breast cancer, tapeworm, burns, abscesses, and leukemia. It also possesses antioxidant, anti-inflammatory, antibacterial, and antimicrobial properties.					

Table 4.107 Anticancer plants of family Pittosporaceae

Family	Botanical name	Image	Plant part used	Bioactive compcund(s)	Structure	Reference(s)
Pittosporaceae	*Pittosporum viridiflorum* Sims.		Whole plant	Amylcdextrin	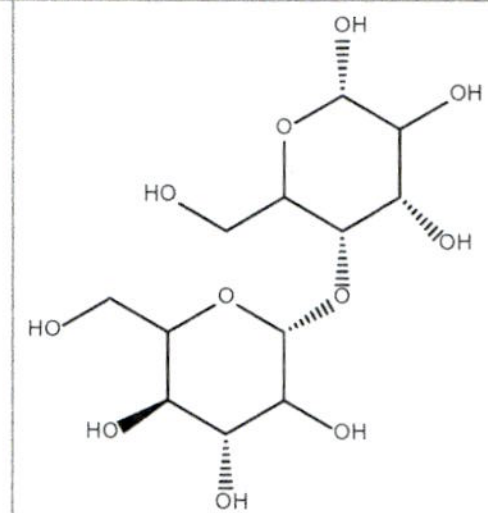	Seo et al. (2002), Ochwang'i et al. (2014), and Madikizela and McGaw (2017)
	Synonym(s): *Pittosporum antunesii* Engl., *Pittosporum commutatum* Putt., *Pittosporum dalzielii* Hutch., *Pittosporum kapiriense* Cufod. **Common name(s)**: Cheesewood, White cape beech. **Botanical description**: It is native to tropical Africa, mainly in the East from South Africa to Ethiopia, but also in the West, Yemen, and India. It is an evergreen plant, varying widely in size from a shrub about 4 m tall to a large forest tree up to 30 m tall, with a bole up to 50 cm. The bark is pale brown to grayish with distinctive white dots (lenticels). The leaves are usually wider above the middle, dark green, and glossy. This plant is very showy when the capsules split open to release numerous small, shiny, orange-red seeds, which are covered in a sticky, resinous exudate. **Medicinal use(s)**: It is used to treat prostate cancer. It possesses antimicrobial, antidiarrheal, antimalarial, anticancer, anti-inflammatory, antioxidant, and acaricidal properties. Decoctions or infusions are widely used to treat stomach complaints, abdominal pain, and fever. The roasted bark of young trees is used for the treatment of dysentery. It is said to release pain and have a calming effect. The dried, powdered root or bark is sometimes added to beer as an aphrodisiac.			Ruvosice	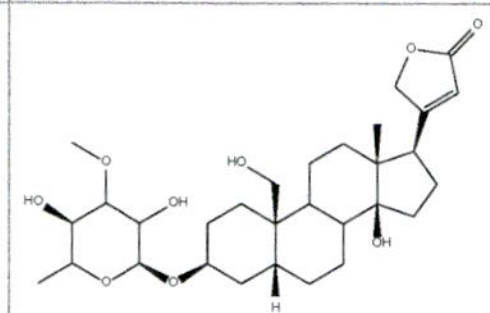	

Table 4.108 Anticancer plants of family Plantaginaceae

Family	Botanical name	Image	Plant part used	Bioactive compound(s)	Structure	Reference(s)
Plantaginaceae	*Plantago argentea* Chaix.		Apical part	Allamandin	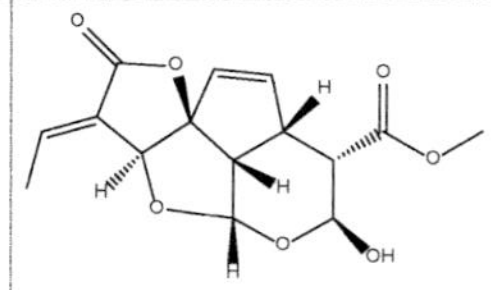	Beara et al. (2009)
	Synonym(s): No synonyms are recorded for this name. **Common name(s)**: Silver plantain. **Botanical description**: It is native to Europe and temperate parts of Asia. The plant has now become naturalized in all temperate regions of the world. It is a perennial plant; the leaves have 3–9 elastic veins that are thick and dark green. The flowers sit in a narrow cylindrical stem, which are approximately the same length as the flower stalk. Each flower has yellowish-brown petal lobes and purple anthers. **Medicinal use(s)**: It possesses anticancer, antitumor, antioxidant, antibacterial, and antidiabetic properties. The herb is used to treat eczema, psoriasis, and first-degree burns. The herb is also used to restore acid balance, to regulate gastric secretions, and to treat inflammation or irritations in the stomach and bowels.			Luteolin-7-O-glucoside	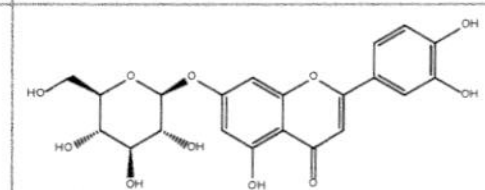	
				Apigenin-7-O-glucoside	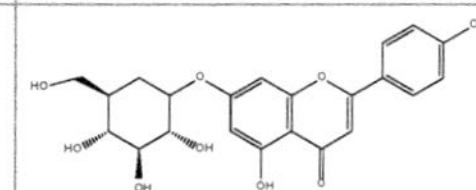	
				Luteolin	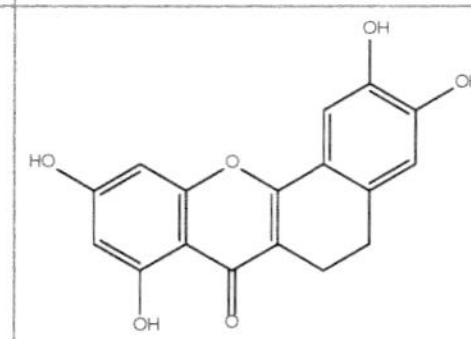	
				Apigenin	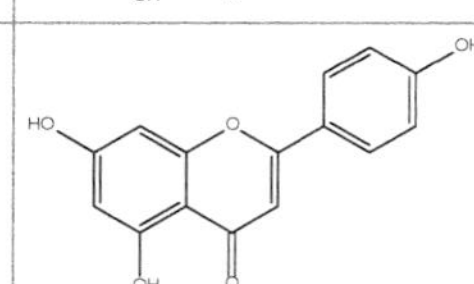	

				Rutin	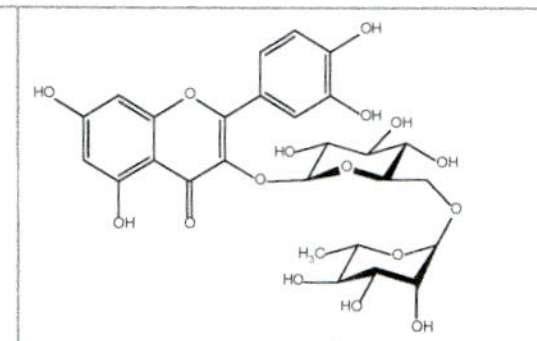	
				Quercetin	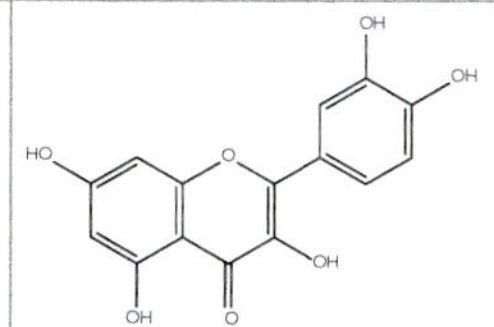	
	Plantago asiatica L.		Root, leaf	Acteoside	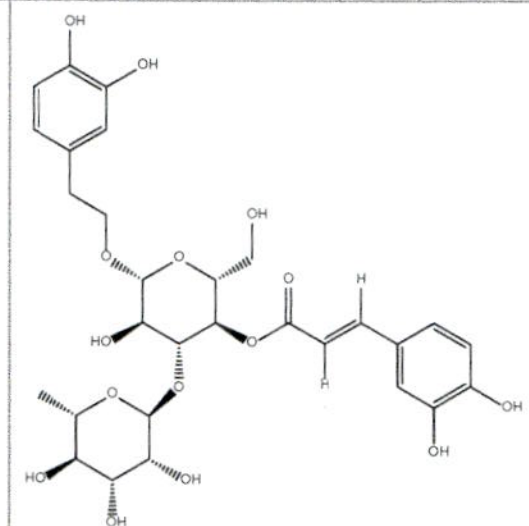	Holdsworth (1992) and Beara et al. (2009)

(continued)

Table 4.108 (continued)

Family	Botanical name	Image	Plant part used	Bioactive compound(s)	Structure	Reference(s)
	Synonym(s): *Plantago asiatica* subsp. *asiatica*, *Plantago hostifolia* Nakai & Kitag., *Plantago major* var. *folioscopa* T.Itô, *Plantago major* f. *paniculata* Makino. **Common name(s)**: Che Qian Zi, Ribwort plantain. **Botanical description**: It is native to East Asia, China, Japan, and Korea. It is found in woods and waste ground in lowland and mountains all over Japan. It is a perennial plant growing up to 0.5 m. It is in flower from July to August. The flowers are hermaphrodite (have both male and female organs) and are pollinated by wind. **Medicinal use(s)**: It is used to treat gastric cancer, breast cancer, and cervical cancer. It also possesses anticancer, antibacterial, anti-inflammatory, antiseptic, and antitussive effects. The seed is used internally for the treatment of a variety of ailments including asthma, bladder problems, bronchitis, fever, hypertension, and rheumatism. The leaves are also used for these purposes, though they are considered to be less powerful than the seed. The leaves can be harvested at any time in the growing season and are used fresh or dried. A decoction of the roots is used for the treatment of coughs.			Apigenin	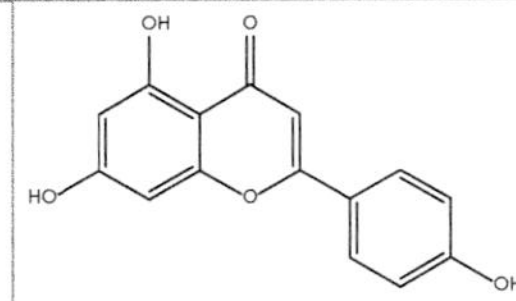	
				Plantainoside A	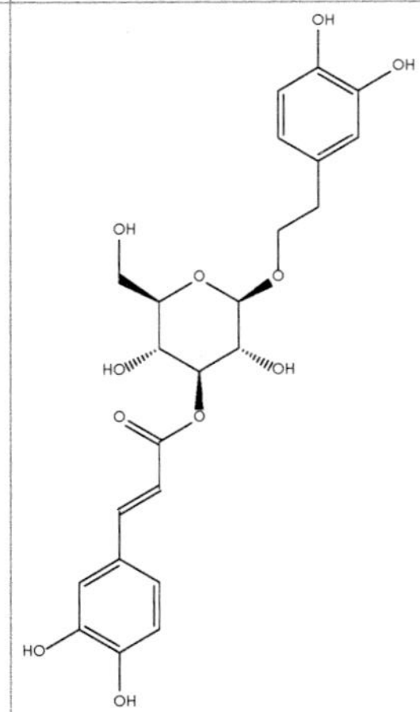	
				Plantainoside B	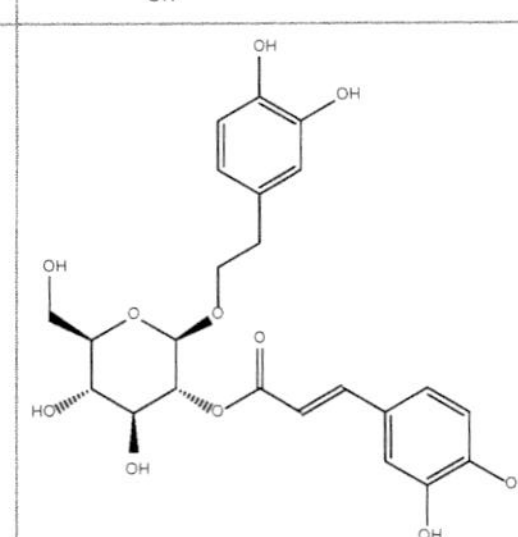	

		Plantainoside D	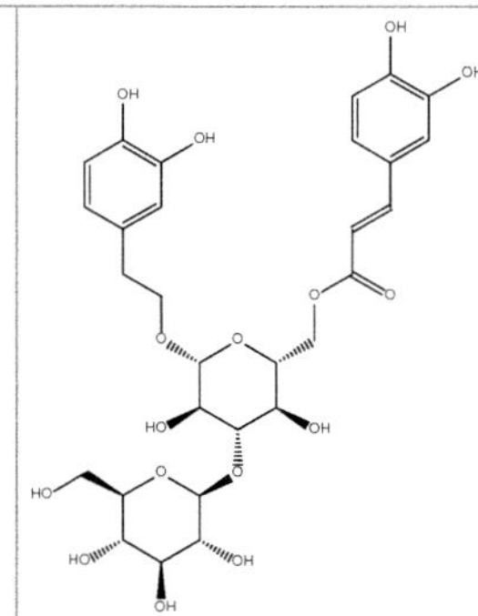	
		Plantamajoside	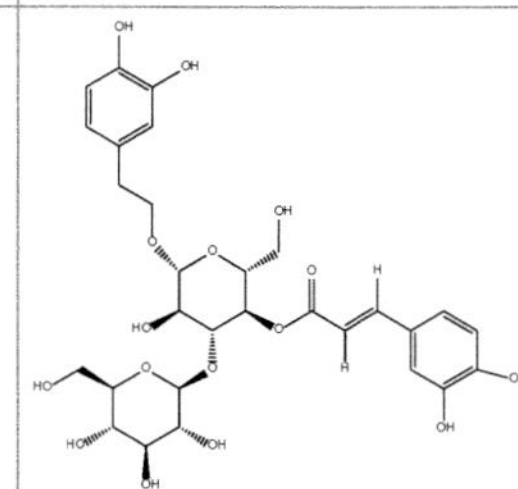	

(continued)

Table 4.108 (continued)

Family	Botanical name	Image	Plant part used	Bioactive compound(s)	Structure	Reference(s)
	Plantago cordata Lam.		Leaf	Luteolin-7-O-glucoside	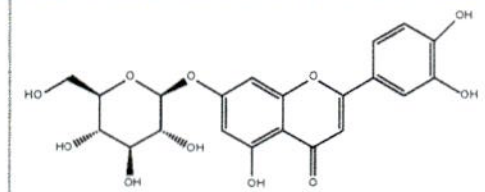	Galvez et al. (2003)
				Etoposide	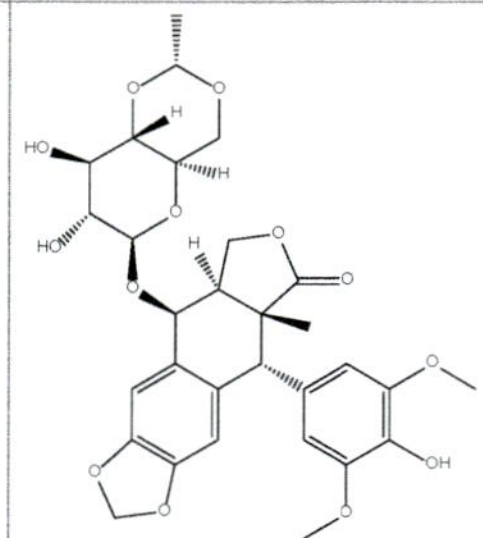	
	Synonym(s): No synonyms are recorded for this name. **Common name(s)**: Heartleaf plantain. **Botanical description**: It is native to Europe, including Britain, from Iceland south and east to Spain, Northern and Central Asia. It is a biennial to perennial plant producing a dense basal rosette of leaves from a thick rhizome. It can grow up to 10–60 cm tall. The plant can sometimes become annual and complete its life cycle in 6 weeks in the tropics. **Medicinal use(s)**: It is used to treat lung and bone cancer. It is used as antibacterial, antidote, astringent, anti-inflammatory, antiseptic, antitussive, cardiac, demulcent, diuretic, expectorant, hemostatic, laxative, ophthalmic, poultice, refrigerant, and vermifuge agents. It is also used as an alternative medicine for asthma, emphysema, bladder problems, bronchitis, fever, hypertension, rheumatism, and blood sugar control. A methanol extract of the plant showed no cytotoxic activity against a human tumor cell line of renal adenocarcinoma but did show total inhibition of breast adenocarcinoma and melanoma growth in culture.					

	Plantago australis Lam.		Entire plant	Plantamajoside	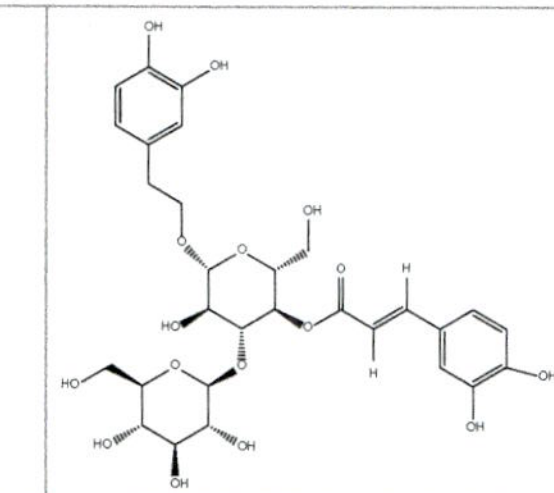	Jiu (1966) and Saeidnia (2015)
	Synonym(s): *Plantago hartwegii* Decne., *Plantago ovata* Phil., *Plantago orbignyana* var. *lasiantha* Pilg. **Common name(s)**: Broadleaf plantain. **Botanical description**: It is native to Europe and temperate parts of Asia. The plant has now become naturalized in all temperate regions of the world. It is a low-growing perennial plant. It could grow well near the roads, dirty muds, mountains, etc. But the preferred ideal habitat is a light, well-drained soil. The pH of the soil is not an issue as it could grow well in alkaline, neutral, as well as acid soil. The usual places it could be found are damped shades, plowed beds, and humid and sunny corners of the woods. **Medicinal use(s)**: It is used to treat cervical cancer, lung cancer, and breast cancer. It is used to possess strong actions against irritations and harmful microorganisms. The plant also possesses medicinal properties, like antibacterial, antiseptic, anti-inflammatory, and antitumor.			Flavone(s)	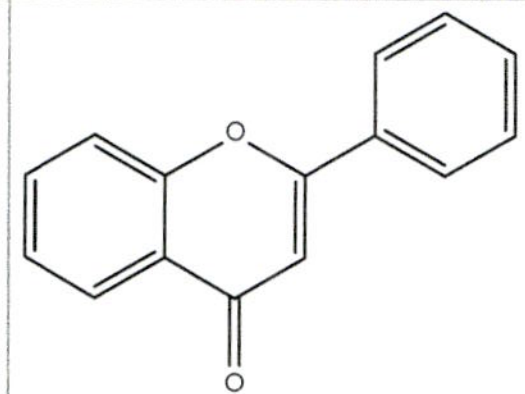	
				Gallic acid	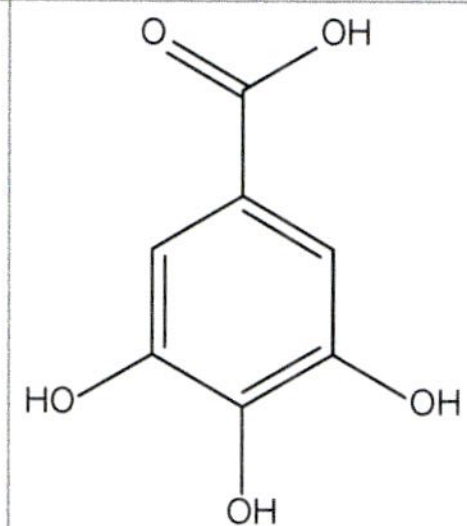	
				Oleanolic acid	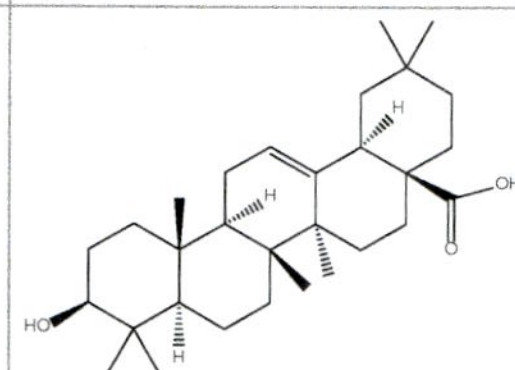	

(continued)

Table 4.108 (continued)

Family	Botanical name	Image	Plant part used	Bioactive compound(s)	Structure	Reference(s)
	Plantago subulata L.		Apical part	Etoposide	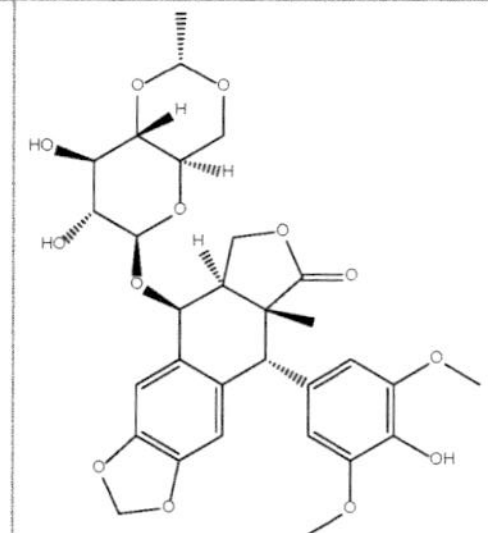	Galvez et al. (2003)
	Synonym(s): *Plantaginella carinata* Fourr., *Plantaginella subulata* (L.) Fourr., *Plantago acanthophylla* Decne., *Plantago holosteum* Scop. **Common name(s)**: Plantains or fleaworts**,** Pratimagri. **Botanical description**: It is native to America, Asia, Australia, New Zealand, Africa, and Europe. The leaves are sessile but have a narrow part near the stem which is a pseudo-petiole. They have three or five parallel veins that diverge in the wider part of the leaf. Leaves are broad or narrow, depending on the species. The inflorescences are borne on stalks typically 5–40 cm tall, and can be a short cone or a long spike, with numerous tiny wind-pollinated flowers. **Medicinal use(s)**: It is used to treat breast cancer, lung cancer, and leukemia. It also possesses anticancer, anti-inflammatory, antibacterial, antitumor, and antioxidant properties. It is used as a remedy for a variety of ailments such as mouth inflammation, throat infection, dropsy, asthma, earache, and epilepsy. It is also helpful in treating minor wounds, cuts, scrapes, and skin damage.			Luteolin-7-O-glucoside	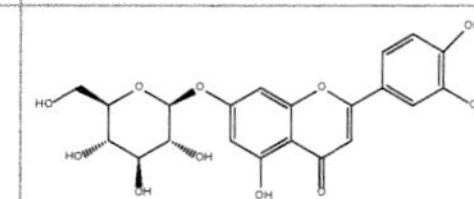	
				Apigenin-7-O-glucoside	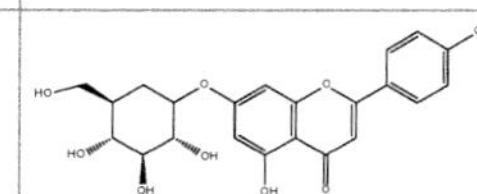	
				Luteolin	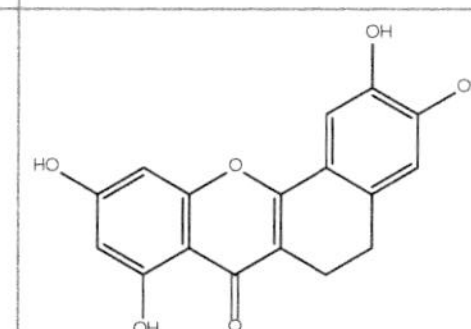	
				Quercetin	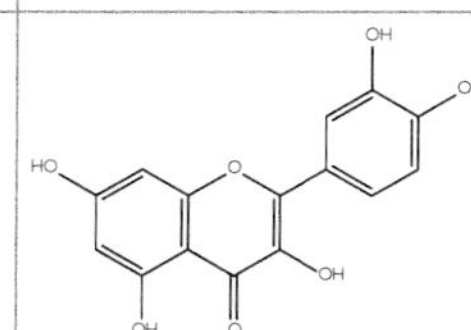	

<table>
<tr><td rowspan="3"></td><td>Plantago australis Lam.</td><td></td><td>Whole plant</td><td>Luteolin-7-O-glucoside</td><td>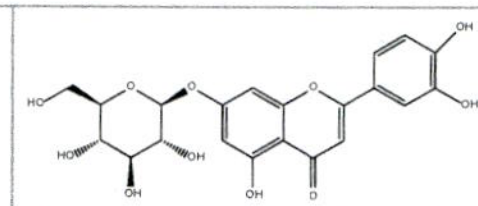</td><td rowspan="3">Galvez et al. (2003)</td></tr>
<tr><td></td><td></td><td></td><td>Luteolin</td><td>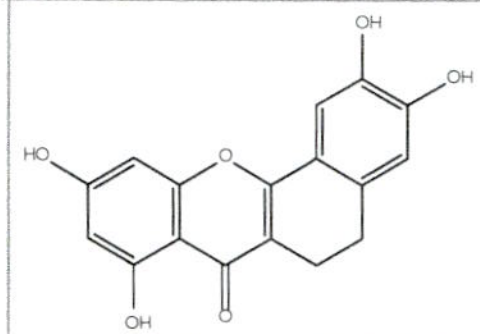</td></tr>
<tr><td colspan="3">Synonym(s): Plantago accrescens Pilg., Plantago asplundii Pilg.
Common name(s): American plantain, Blackseed plantain, Pale plantain, and Rugel's plantain, Whiteman's footprint.
Botanical description: It is native to most of Europe and Northern and Central Asia. It is an herbaceous perennial plant with a rosette of leaves 15–30 cm in diameter. Each leaf is oval-shaped, 5–20 cm long and 4–9 cm broad, rarely up to 30 cm long and 17 cm broad, with an acute apex and a smooth margin; there are five to nine conspicuous veins. The flowers are small, greenish brown with purple stamens, produced in a dense spike 5–15 cm long on top of a stem 13–15 cm tall.
Medicinal use(s): It is used to treat breast cancer, lung cancer, and leukemia. It has anticancer, anti-inflammatory, antibacterial, antitumor, and antioxidant properties. It is used as a remedy for a variety of ailments such as mouth inflammation, throat infection, dropsy, asthma, earache, and epilepsy. It is helpful in treating minor wounds, cuts, scrapes, and skin damage in order to hold bleeding.</td><td>Etoposide</td><td>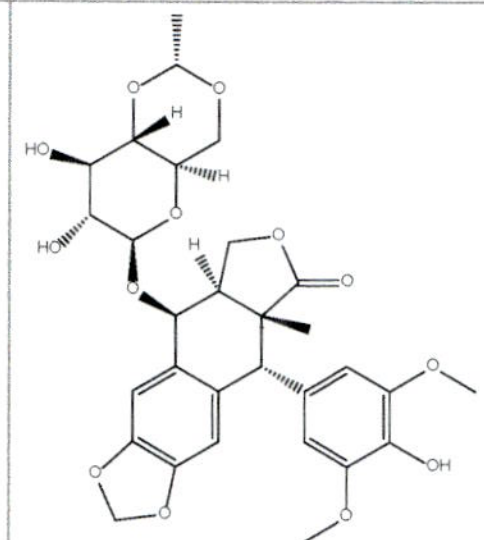</td></tr>
</table>

(continued)

Table 4.108 (continued)

Family	Botanical name	Image	Plant part used	Bioactive compound(s)	Structure	Reference(s)
	Plantago major L.		Leaf	Luteolin	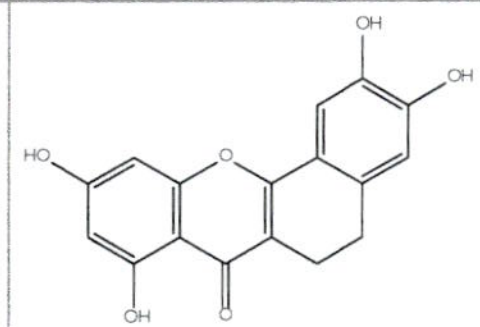	Ozaslan et al. (2009)
	Synonym(s): *Plantago borysthenica* Wissjul., *Plantago dregeana* Decne., *Plantago gigas* H. Lév., *Plantago jehohlensis* Koidz. **Common name(s)**: Blackseed plantain, Broadleaf plantain. **Botanical description**: It is native to America, Asia, Australia, New Zealand, Africa, and Europe. The leaves are sessile but have a narrow part near the stem which is a pseudo-petiole. They have three or five parallel veins that diverge in the wider part of the leaf. Leaves are broad or narrow, depending on the species. The inflorescences are borne on stalks typically 5–40 cm tall, and can be a short cone or a long spike, with numerous tiny wind-pollinated flowers. **Medicinal use(s)**: It is used to treat breast cancer and leukemia. It has anticancer, anti-inflammatory, antibacterial, antitumor, and antioxidant properties. It is used as a remedy for a variety of ailments such as mouth inflammation, throat infection, dropsy, asthma, earache, and epilepsy.						

	Plantago maritime L.		Apical part	Kaempferol	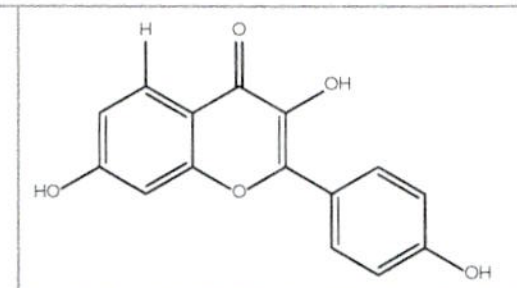	Beara et al. (2009)
				Luteolin-7-O-glucoside	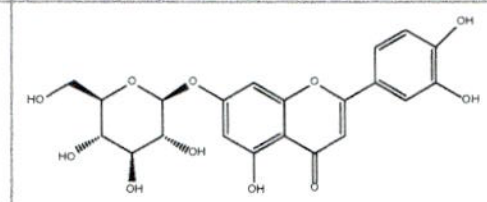	
	Synonym(s): *Plantaginella carinata* Fourr., *Plantaginella subulata* (L.) Fourr. **Common name(s)**: Pale plantain. **Botanical description**: It is native to most of Europe and Northern and Central Asia. It is an herbaceous perennial plant with a rosette of leaves 15–30 cm in diameter. Each leaf is oval-shaped, 5–20 cm long and 4–9 cm broad, rarely up to 30 cm long and 17 cm broad, with an acute apex and a smooth margin; there are five to nine conspicuous veins. The flowers are small, greenish brown with purple stamens, produced in a dense spike 5–15 cm long on top of a stem 13–15 cm tall. **Medicinal use(s)**: It is used to treat breast cancer and leukemia. It has anticancer property, anti-inflammatory, antibacterial, antitumor, and antioxidant properties. It is used as a remedy for a variety of ailments such as mouth inflammation, throat infection, dropsy, asthma, earache, and epilepsy.			Apigenin-7-O-glucoside	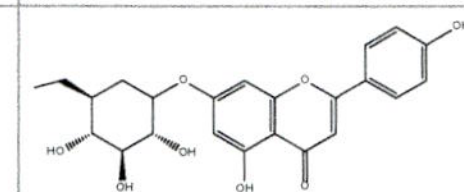	
				Apigenin	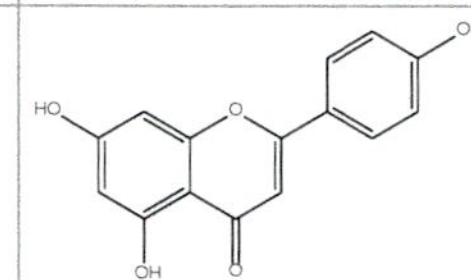	
				Rutin	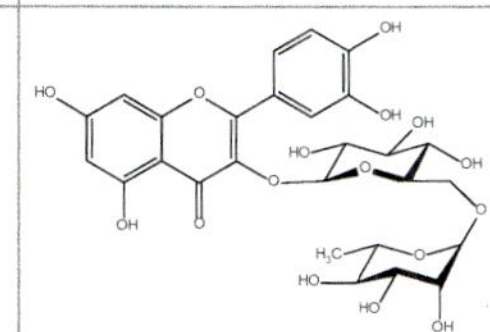	
				Quercetin	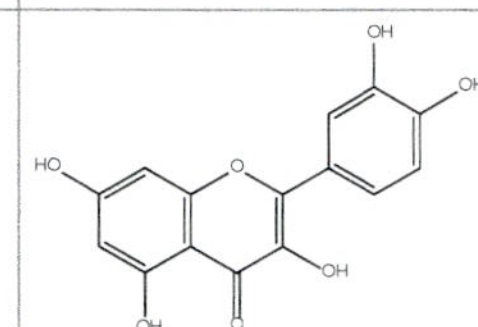	

(continued)

Table 4.108 (continued)

Family	Botanical name	Image	Plant part used	Bioactive compound(s)	Structure	Reference(s)
	Plantago mexicana (Link) Pilg		Apical part	Kaempferol	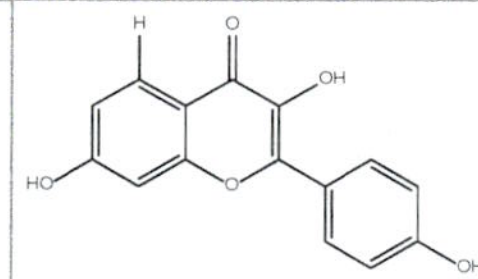	Beara et al. (2009)
	Synonym(s): *Plantaginella carinata* Fourr., *Plantaginella subulata* (L.) Fourr. **Common name(s)**: Broadleaf plantain, Tall coastal plantain. **Botanical description**: It is native to America, Asia, Australia, New Zealand, Africa, and Europe. The leaves are sessile but have a narrow part near the stem which is a pseudo-petiole. They have three or five parallel veins that diverge in the wider part of the leaf. Leaves are broad or narrow, depending on the species. The inflorescences are borne on stalks typically 5–40 cm tall, and can be a short cone or a long spike, with numerous tiny wind-pollinated flowers. **Medicinal use(s)**: It is used to treat breast cancer and leukemia. It has anticancer, anti-inflammatory, antibacterial, antitumor, and antioxidant properties. It is used as a remedy for a variety of ailments such as mouth inflammation, throat infection, dropsy, asthma, earache, and epilepsy.			Luteolin-7-O-glucoside	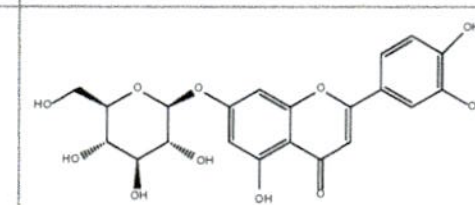	
				Apigenin-7-O-glucoside	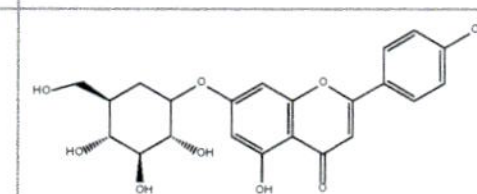	
				Luteolin	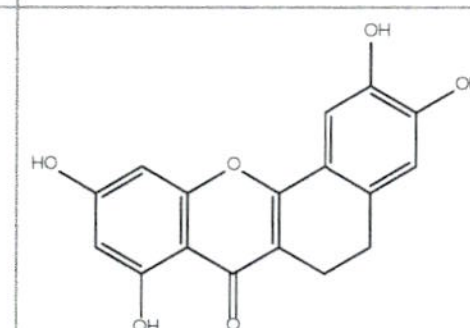	
				Quercetin	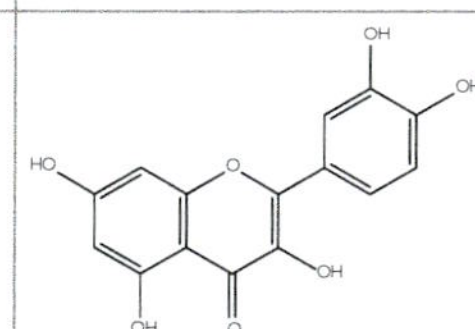	

				Luteolin-7-O-glucoside	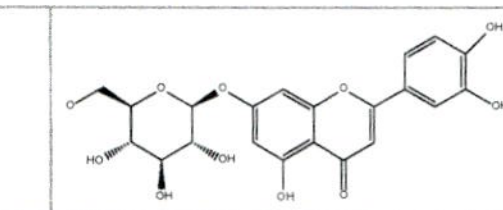	
	Plantago paralias Decne./*Plantago tomentosa* Lam.Tanchagem		Leaf	Quercetin	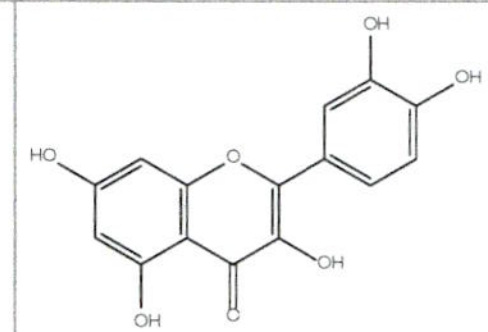	Bandoni et al. (1976) and de Melo et al. (2011)
	Synonym(s): *Plantago acanthophylla* Decne., *Plantago holosteum* Scop. **Common name(s)**: Rugel's plantain. **Botanical description**: It is native to Europe and temperate parts of Asia. The plant has now become naturalized in all temperate regions of the world. It is perennial plant; the leaves have 3–9 elastic veins that are thick and dark green. The flowers sit in a narrow cylindrical stem, which is approximately the same length as the flower stalk. Each flower has yellowish-brown petal lobes and purple anthers. **Medicinal use(s)**: It is used to treat breast cancer and leukemia. It has anticancer, anti-inflammatory, antibacterial, antitumor, and antioxidant properties. It is used as a remedy for a variety of ailments such as mouth inflammation.					

(continued)

Table 4.108 (continued)

Family	Botanical name	Image	Plant part used	Bioactive compound(s)	Structure	Reference(s)
	Pleione bulbocodioides (Franch.)		Tuber and stem	Luteolin	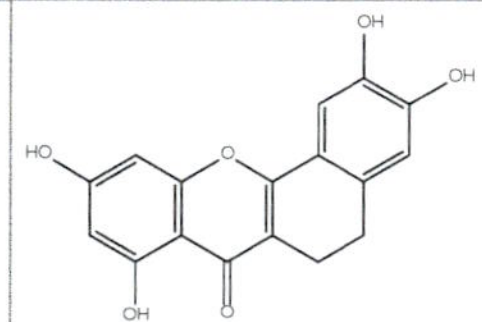	Liu et al. (2009)
	Synonym(s): *Coelogyne bulbocodioides* Franch., *Coelogyne delavayi* Rolfe., *Coelogyne henryi* Rolfe., *Coelogyne pogonioides* Rolfe., *Pleione communis* Gagnep. **Common name(s)**: Du suan lan, Bulbocodium-like Pleione. **Botanical description**: It is native to the Himalayas, India, Tibet, Nepal, Bhutan, Bangladesh, Burma, Thailand, Laos, Vietnam, Taiwan, and China. The species are well adapted to cold temperatures and even frost. A few, however, prefer warmer temperatures. They grow in well-drained habitats and on rocks covered with moss at altitudes of 600–4200 m. The pseudobulbs stay dormant, till a new one starts to develop. This one produces new roots and a short pedicel. **Medicinal use(s)**: It is used to treat gastric cancer. It also possesses anti-inflammatory and antioxidant properties.					

Table 4.109 Anticancer plants of family Plumbaginaceae

Family	Botanical name	Image	Plant part used	Bioactive compound(s)	Structure	Reference(s)
Plumbaginaceae	*Plumbago zeylanica* L.		Root	Etopcside	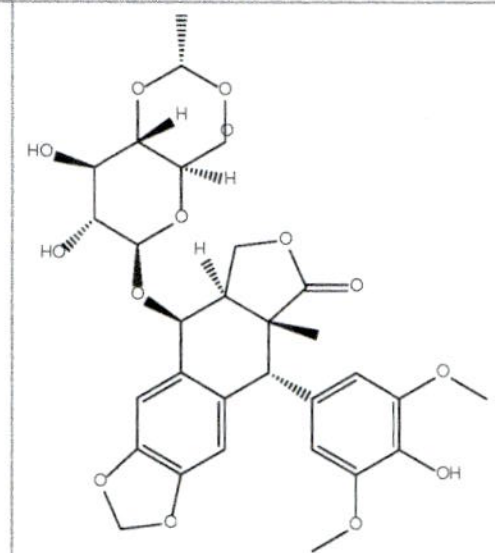	Sandur et al. (2006)
	Synonym(s): *Findlaya alba* Bowdich, *Molubda scandens* (L.) Raf., *Plumbagidium scandens* (L.) Spach. **Common name(s)**: Leadwort. **Botanical description**: It is native to Southeast Asia. Leaves are alternate, ovate, narrowed into the petiole. Flowers: borne in spikes, rachis of the spike pubescent or glandular. Corolla white, tube long and slender. Base of the style glabrous. The fruits are oblong and pointed capsules, enclosed by persistent viscid calyx. **Medicinal use(s)**: It is used to treat lung cancer and skin problems. It also possesses anticancer, anti-inflammatory, antioxidant, and antibacterial properties.					

(continued)

Table 4.109 (continued)

Family	Botanical name	Image	Plant part used	Bioactive compound(s)	Structure	Reference(s)
	Plumbago toxicaria Bertol.		Root	Kaempferol	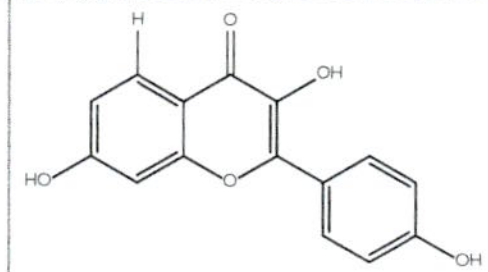	Ko (1933), Huai (2010), and Xu et al. (2013)
	Synonym(s): *Findlaya alba* Bowdich., *Molubda scandens* (L.) Raf., *Plumbagidium scandens* (L.) Spach., *Plumbago scandens* L. **Common name(s)**: Ceylon leadwort, Doctorbush, or Wild leadwort. **Botanical description**: It is native to Southeast Asia. Leaves are alternate, ovate, narrowed into the petiole. Flowers: borne in spikes, rachis of the spike pubescent or glandular. Corolla white, tube long and slender. Base of the style glabrous. The fruits are oblong and pointed capsules, enclosed by persistent viscid calyx. **Medicinal use(s)**: It is used to treat lung cancer and skin problems. It also possesses anticancer, anti-inflammatory, antioxidant, and antibacterial properties.					

Table 4.110 Anticancer plants of family Poaceae

Family	Botanical name	Image	Plant part used	Bioactive compound(s)	Structure	Reference(s)
Poaceae	*Elymus repens* (L.) Gould		Rhizome	Carvacrol	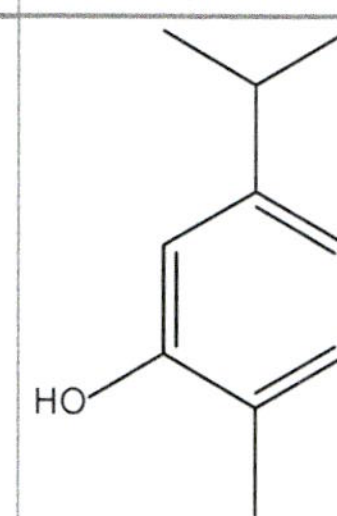	Mohammad (2006), Desai et al. (2008), and Madhuri and Pandey (2009)
	Synonym(s): *Agropyron arundinaceum* (Steud.) P.Candargy, *Agropyron caesium* J.Presl & C.Presl, *Agropyron caldesii* Goiran. **Common name(s)**: Quackgrass. **Botanical description**: It is native to Europe, Northern Asia, Australia, and America. Rhizome is long, stiff, pale yellow and smooth, about 1/10 in. in diameter, hollow except at the nodes, and strongly furrowed longitudinally. The rhizome is always free from rootlets, cut into short lengths of 1/8 to 1/4 in. and dried, straw-colored, many-edged tubular pieces, which are without odor, but have a sweet taste. **Medicinal use(s)**: It is used for the treatment of breast cancer. It also possesses anti-inflammatory, antioxidant, and antibacterial properties.					

(continued)

Table 4.110 (continued)

Family	Botanical name	Image	Plant part used	Bioactive compound(s)	Structure	Reference(s)
	Coix lacryma-jobi L.		Seed	Trans-feruloyl stigmasterol	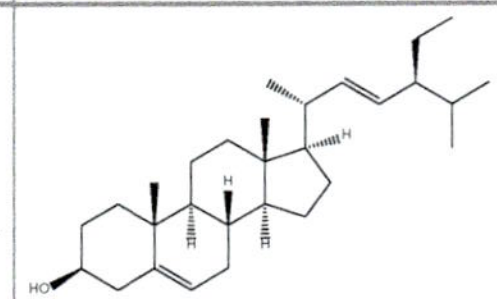	Mohammad (2006), Desai et al. (2008), and Madhuri and Pandey (2009)
	Synonym(s): *Coix agrestis* Lour., *Coix arundinacea* Lam., *Coix exaltata* Jacq. ex Spreng. **Common name(s)**: Job's tears. **Botanical description**: It is native to East Asia, China, Indian subcontinent, Myanmar, Thailand, Laos, Vietnam, Malaysia, Indonesia, the Philippines, and New Guinea. It is a tall grain-bearing perennial tropical plant. It is still occasionally cultivated and is a potentially very useful grain, having a higher protein to carbohydrate ratio than any other cereal. The plant is also grown as an ornamental in tropical to temperate areas. **Medicinal use(s)**: It is used to treat colorectal and lung cancer, lobar pneumonia, appendicitis, rheumatoid arthritis, beriberi, diarrhea, edema, and difficult urination. It also possesses anti-inflammatory, antipyretic, antiseptic, antitumor, and anti-spasmodic properties.					

	Sasa palmata (hort. ex Burb.) E.G. Camas P		Leaves, beans	α-Humulene	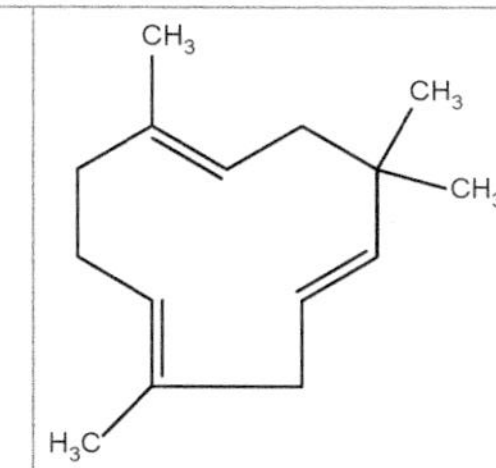	Loizzo et al. (2007)
	Synonym(s): *Arundinaria pygmaea*, *Bambusa pygmaea*, *Sasa pygmaea*. **Common name(s)**: Broadleaf bamboo or Broad-leaved bamboo. **Botanical description**: It is native to Japan. It is an evergreen bamboo growing up to 2.5 m at a fast rate. The flowers are hermaphrodite (have both male and female organs) and are pollinated by wind. It is habituated in woodland garden, sunny edge, and dappled shade. **Medicinal use(s)**: It is used to treat oral squamous cancer cells and tumor. It also possesses anti-inflammatory, antioxidant, antibacterial, antimicrobial, and antitumor properties.					

(continued)

Table 4.110 (continued)

Family	Botanical name	Image	Plant part used	Bioactive compound(s)	Structure	Reference(s)
	Zea mays L.		Whole plant	Camptothecin	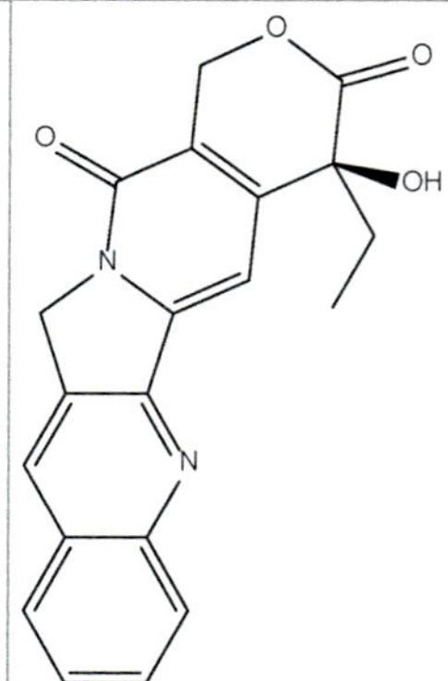	Habtemariam (1998) and Balasubramanian and Padma (2013)
	Synonym(s): *Mays vulgaris* Ser., *Mayzea vestita* Raf., *Zea alba* Mill. **Common name(s)**: Corn. **Botanical description**: It is native to North America including Canada. It is a tall, monecious annual grass with overlapping sheaths and broad conspicuously distichous blades. Plants have staminate spikelets in long spikelike racemes that form large spreading terminal panicles (tassels) and pistillate inflorescences in the leaf axils, in which the spikelets occur in 8–16 rows, approximately 30 long, on a thickened, almost woody, axis (cob). The whole structure (ear) is enclosed in numerous large foliaceous bracts and a mass of long styles (silks) protrude from the tip as a mass of silky threads. **Medicinal use(s)**: It is used to treat urinary and breast cancer, cardiovascular disease, obesity, diabetes, and chronic diseases. It also possesses anti-inflammatory, antitumor, and antioxidant properties.					

	Cymbopogon citratus (DC.) Stapf		Leaf	Cardiac glycosides	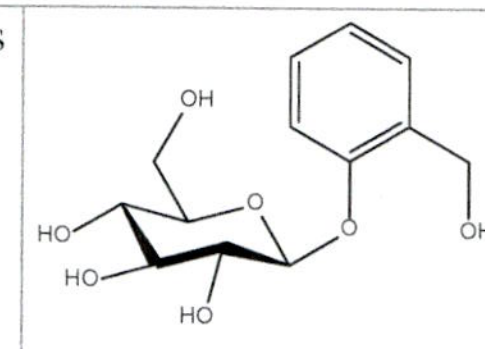	Gazola et al. (2004)
	Synonym(s): *Andropogon cerifer* Hack., *Andropogon ceriferus* Hack., *Andropogon ceriferus* Hack., *Andropogon citratus* DC., *Andropogon citratus* DC. ex Nees. **Common name(s)**: Lemon grass or Oil grass. **Botanical description**: It is native to South Asia and Southeast Asia. It is abundant in the Philippines and Indonesia where it is known as tanglad or sereh. Its fragrant leaves are traditionally used in cooking, particularly for lechon and roasted chicken. Fresh stalks of lemongrass (leaves and roots absent) can also be purchased from grocery stores specializing in Southeast Asian cuisine for rooting in a glass of water. Plant outside in spring after last frost date. **Medicinal use(s)**: It is used for the treatment of neck and head cancer and tumor. It also possesses antitumor, anti-inflammatory, and antioxidant properties.			Rhaponticin (flavonoids)	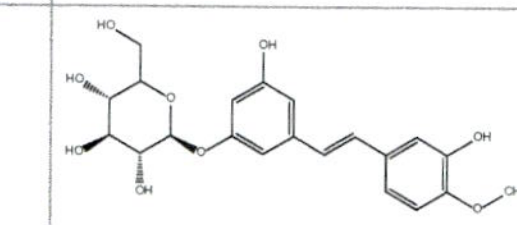	
				Ellagic acid	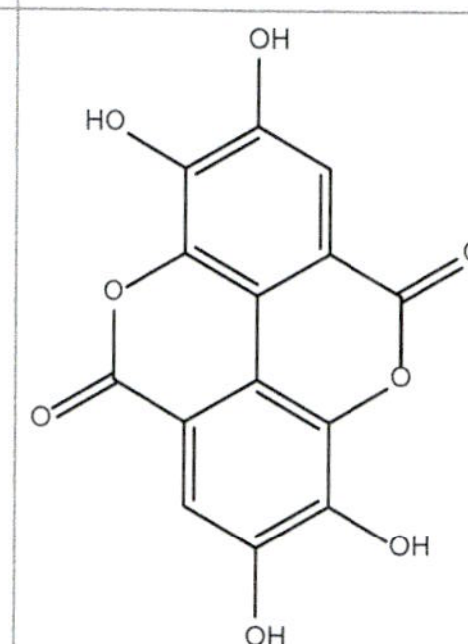	

(continued)

Table 4.110 (continued)

Family	Botanical name	Image	Plant part used	Bioactive compound(s)	Structure	Reference(s)
	Cymbopogon flexuosus (Nees ex Steud.) W.Watson		Whole plant	Geraniol	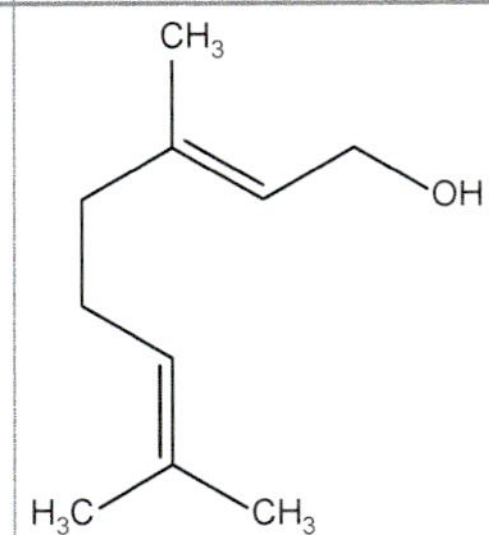	Sharma et al. (2009)
	Synonym(s): *Andropogon ampliflorus* Steud., *Andropogon flexuosus* Nees ex Steud., *Andropogon nardus* var. *flexuosus* (Steud.) Hack. **Common name(s)**: Oil grass, Fever grass. **Botanical description**: It is native to South Asia and Southeast Asia. It is abundant in the Philippines and Indonesia where it is known as tanglad or sereh. Its fragrant leaves are traditionally used in cooking, particularly for lechon and roasted chicken. Fresh stalks of lemongrass (leaves and roots absent) can also be purchased from grocery stores specializing in Southeast Asian cuisine for rooting in a glass of water. Plant outside in spring after last frost date. **Medicinal use(s)**: It is used for the treatment of cervical cancer. It also possesses antitumor, anti-inflammatory, and antioxidant properties.			Geranyl acetate	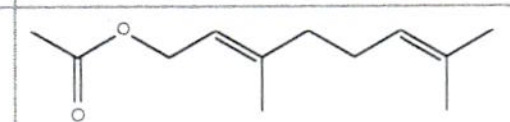	
				Alpha-bisabolol	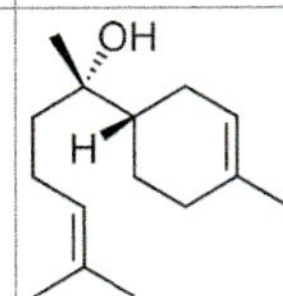	
				Isointermedeol	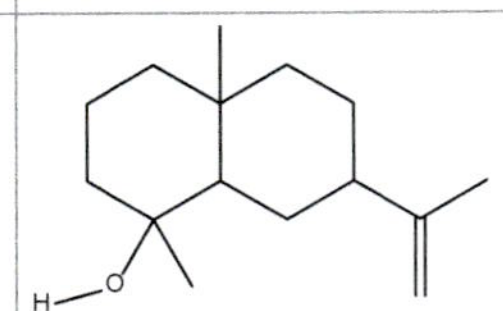	

	Dactyloctenium aegyptium (L.) P.B		Whole plant	Cianidanol	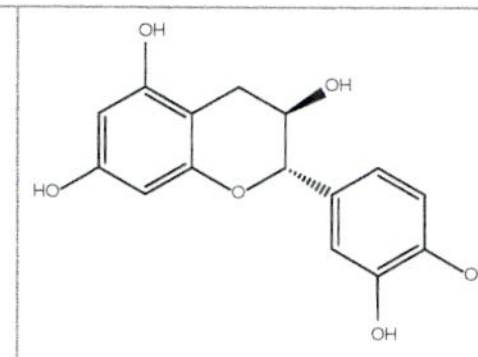 	Hansakul et al. (2009) and Hansakul et al. (2017)
	Synonym(s): *Aegilops saccharina* Walter, *Chloris guineensis* Schumach. & Thonn., *Chloris mucronata* Michx. **Common name(s)**: Crowfoot grass. **Botanical description**: It is native to Africa. It is still a traditional food plant used as a famine food in Africa, and this little-known grain has potential to improve nutrition, boost food security, foster rural development, and support sustainable land care. This grass creeps and has a straight shoot which is usually about 30 cm tall. The plant mostly grows in heavy soils at damp sites. **Medicinal use(s)**: It is used to treat lung and cervical cancer and typhoid. It also possesses anti-inflammatory, antioxidant, and antibacterial properties.					

(continued)

Table 4.110 (continued)

Family	Botanical name	Image	Plant part used	Bioactive compound(s)	Structure	Reference(s)
	Eleusine indica (L.) Gaerth.		Whole plant	Gallic acid	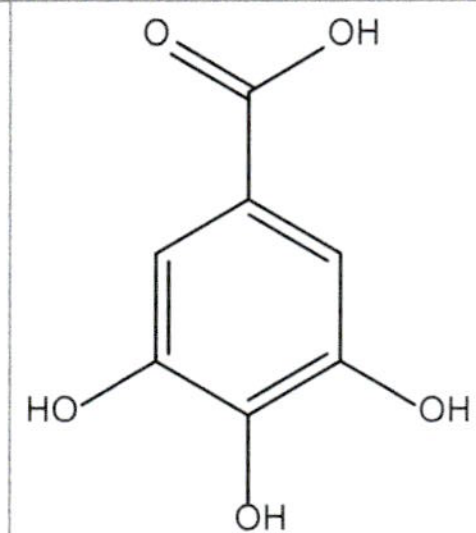	Hansakul et al. (2009), Sagnia et al. (2014), and Hansakul et al. (2017)
	Synonym(s): *Agropyron geminatum* Schult. & Schult.f., *Cynodon indicus* (L.) Raspail, *Cynosurus indicus* L. **Common name(s)**: Goosegrass, Wire grass. **Botanical description**: It is native to Asia and temperate zones. Its leaves are flat to V-shaped, up to 8 mm wide, 15 cm long, acute, boat-shaped tip. The leaves are glabrous and are usually fairly bright, fresh green in color. The ligule is a very short membranous rim up to 1 mm long, sparsely fringed with short hairs. The inflorescence consists of 3–8 racemes, each 5–10 cm long, about 5 mm (0.19 in.) wide, arranged more-or-less digitately, though one raceme may be inserted about 1 cm below the others. **Medicinal use(s)**: It is used to treat breast cancer, hypertension, influenza, oliguria, and urine retention. It also possesses anti-inflammatory, antioxidant, and antibacterial properties.					

	Hordeum vulgare L.		Stem	Gallic Protocatechuic	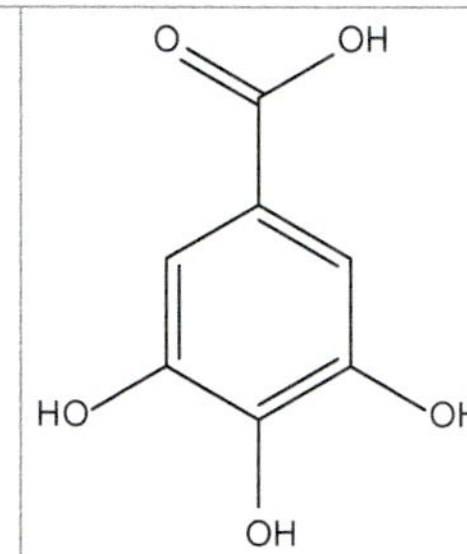	Dykes and Rooney (2007) and Woo et al. (2017)
	Synonym(s): *Frumentum hordeum* E.H.L.Krause., *Frumentum sativum* E.H.L.Krause., *Hordeum aestivum* R.E.Regel., *Hordeum americanum* R.E.Regel., *Hordeum bifarium* Roth. **Common name(s)**: Barley. **Botanical description**: It is native to Ukraine, Canada, and Australia. It is the main grain ingredient in a number of liquors, including beer. **Medicinal use(s)**: It is used to treat breast and prostate cancer. It also possesses anti-inflammatory, antioxidant, and antibacterial properties.					
				p-Hydroxybenzoic	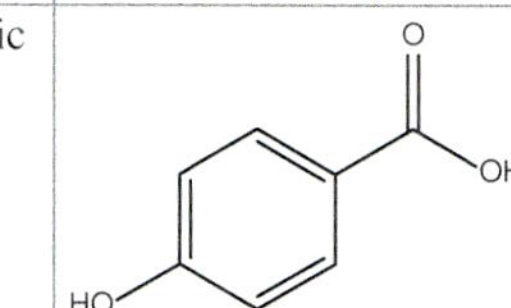	
				Gentisin	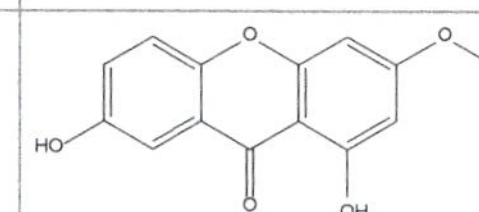	

(continued)

Table 4.110 (continued)

Family	Botanical name	Image	Plant part used	Bioactive compound(s)	Structure	Reference(s)
				Salicylic acid	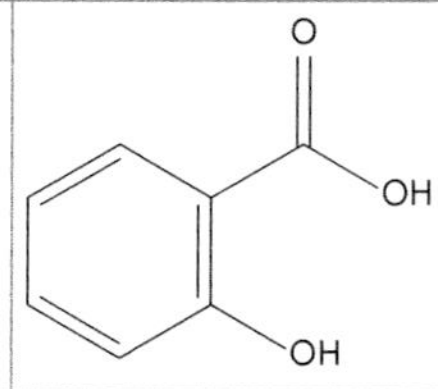	
				Vanillic acid	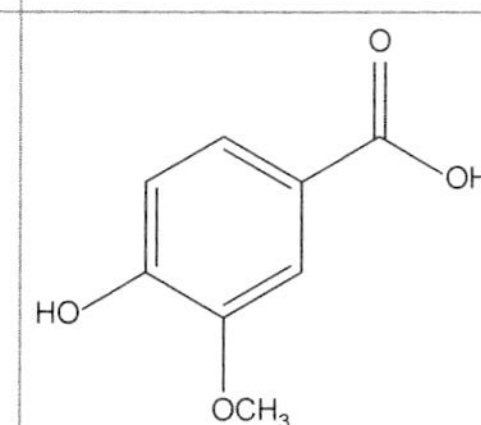	
				Hydroxycinnamic acid	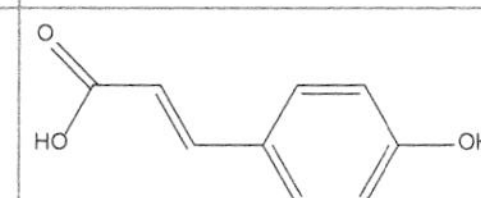	
				Caffeic acid derivatives	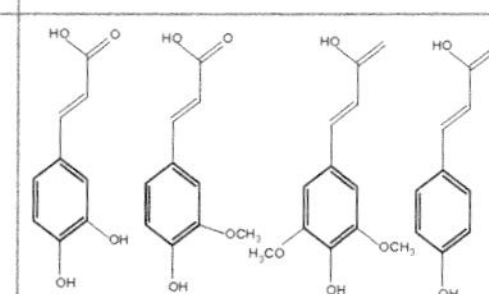	

	Oryza sativa L.		Seeds	Cyanidin	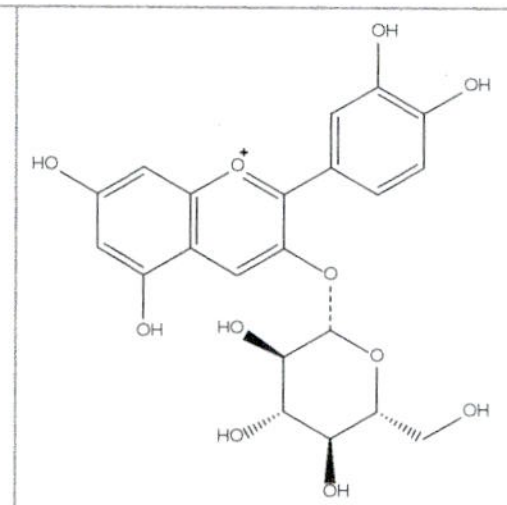	Hyun and Chung (2004), Tan and Norhaizan (2017), and Phannasorn et al. (2017)
	Synonym(s): *Oryza communissima* Lour., *Oryza denudata* Steud., *Oryza elongata* Steud., *Oryza emarginata* Desv. ex Steud. **Common name(s)**: Rice. **Botanical description**: It is native to China and India. It contains two major subspecies: the sticky, short-grained japonica or sinica variety and the nonsticky, long-grained indica variety. Japonica varieties are usually cultivated in dry fields, in temperate East Asia, upland areas of Southeast Asia, and high elevations in South Asia, while indica varieties are mainly lowland rice, grown mostly submerged, throughout tropical Asia. Rice occurs in a variety of colors, including white, brown, black, purple, and red. **Medicinal use(s)**: It is used to treat stomach, breast, and oral cancer. It also possesses antioxidant, anti-inflammatory, and antibacterial properties.			Malvidin	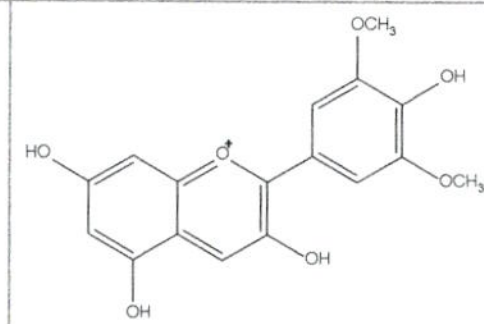	

Table 4.111 Anticancer plants of family Polygalaceae

Family	Botanical name	Image	Plant part used	Bioactive compound(s)	Structure	Reference(s)
Polygalaceae	*Monnina obtusifolia* H. B & K.		Aerial parts	Campesterol	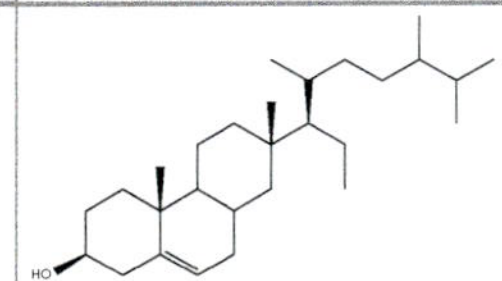	Pinto et al. (1994) and Bailon Moscoso et al. (2015)
	Synonym(s): *Monnina obtusifolia* var. *oblongifolia* Chodat. **Common name(s)**: Baby rubber plant. **Botanical description**: It is a shrub that grows up to 2–4 ft. and is native to Peru and Australia. **Medicinal use(s)**: It is used to treat cancer cell lines. It also possesses antitumor, anti-inflammatory, and antioxidant properties.					
	Polygala vulgaris L.		Whole plant	Coronaridine	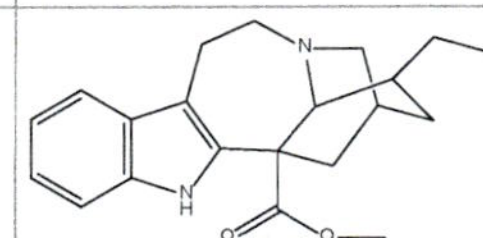	Cainelli et al. (2008)

	Synonym(s): *Polygala angustifolia* Gilib., *Polygala oxyptera* Rchb. **Common name(s)**: Common milkwort, Health milkwort. **Botanical description**: It is native to Europe, including Britain, from South and East Norway to North Africa and West Asia. It is a perennial plant growing up to 0.4 m. It is in flower from May to September. It is suitable to grow in light, medium, and heavy soils and prefers well-drained soil to grow. **Medicinal use(s)**: It is used for the treatment of cancer and respiratory problems such as chronic bronchitis, bronchial asthma, and convulsive coughs such as whooping cough. It also possesses antitumor, anti-inflammatory, and antioxidant properties.					
	Polygonatum multiflorum (L.) All.		Leaf	Isovitexin (flavonoids)	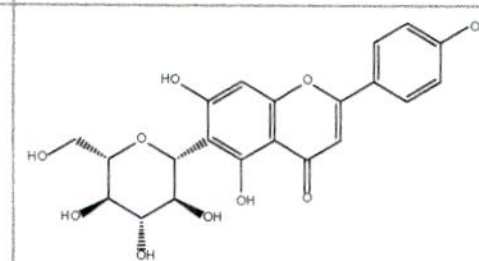	Damme et al. (1996) and Feng and Bounda (2015)
	Synonym(s): *Convallaria ambigua* Schrank, *Convallaria bracteata* B. Thomas, *Convallaria broteroi* Guss. **Common name(s)**: Solomon's seal, Eurasian Solomon's seal. **Botanical description**: It is native to Europe, including Britain, from South and East Scandinavia to Spain, and temperate Asia to Japan. It is a perennial plant growing to 1.2 m by 0.3 m. The species is hermaphrodite (has both male and female organs) and is pollinated by bees and itself. It is habituated to woodland, usually on limestone. **Medicinal use(s)**: It is used for the treatment of cancer; pulmonary problems, including tuberculosis; women's disorders; piles; bruises; and inflammation. It also possesses antitumor, anti-inflammatory, and antioxidant properties.					

(continued)

Table 4.111 (continued)

Family	Botanical name	Image	Plant part used	Bioactive compound(s)	Structure	Reference(s)
	Reynoutria japonica Houtt.		Root	Luteolin-7-O-glucoside	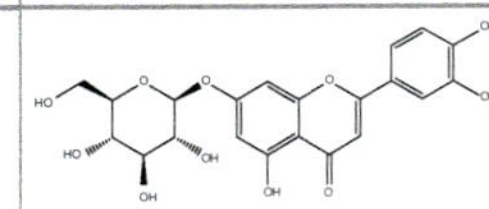	Kimura and Okuda (2001) and Eid et al. (2015)
	Synonym(s): *Fallopia compacta* (Hook.f.) G. H. Loos & P. Keil, *Fallopia japonica* (Houtt.) Ronse Decr., *Pleuropterus cuspidatus* (Siebold & Zucc.) H.Gross. **Common name(s)**: Japanese knotweed, Russian wine. **Botanical description**: It is native to Japan, China, and parts of Korea and Taiwan. It is a deciduous perennial, tough, vigorous, spreading rapidly from rhizomes to form very large clumps of red-brown jointed and hollow stems, 4–8 ft. tall. Leaves alternate, simple, 5–7.5 cm wide, heart-shaped with rounded lobes at the base, vein often reddish, upper surface dark green, paler green below; petioles ridged and 2.5 cm long. Mostly dioecious, i.e., male and female flowers mostly on separate plants. **Medicinal use(s)**: It is used to treat cancer, asthma, hypertension, cough, and tumor. It also possesses antitumor, anti-inflammatory, and antioxidant properties.					

	Persicaria tinctoria (Aiton) H.Gross		Leaf	Apigenin-7-O-glucoside	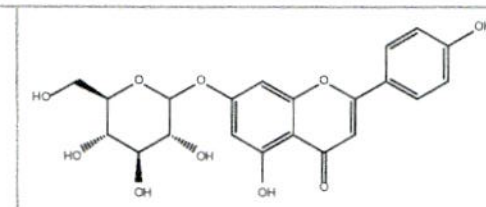	Kimoto et al. (2001) and Kim et al. (2015a, b, c)
	Synonym(s): *Ampelygonum tinctorium* (Aiton) Steud., *Persicaria tinctoria* (Aiton) Spach, *Pogalis tinctoria* (Aiton) Raf. **Common name(s)**: Chinese indigo. **Botanical description**: It is native to East Europe and to East Asia. It is annually growing up to 0.8 m. It is in flower from July to August, and the seeds ripen from September to November. The species is hermaphrodite and is pollinated by insects. It is habituated in weed infested places. **Medicinal use(s)**: It is used for the treatment of cancer, freckles, pimples, erysipelas, mumps, thrush, epidemic parotitis, infantile convulsions, and high febrile conditions of children. It also possesses antitumor, anti-inflammatory, and antioxidant properties.					
	Rheum emodi Wall. ex Meissn.		Rhizomes	D-Xylose	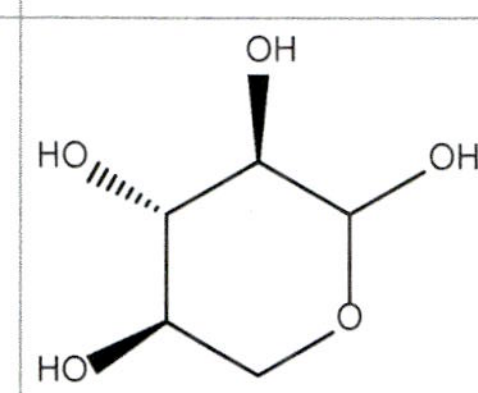	Rajkumar et al. (2010)

(continued)

Table 4.111 (continued)

Family	Botanical name	Image	Plant part used	Bioactive compound(s)	Structure	Reference(s)
	Synonym(s): No synonyms are recorded for this name. **Common name(s)**: Himalayan rhubarb. **Botanical description**: It is native to Southeastern Russia. **Medicinal use(s)**: It is used in various traditional systems to cure cancer, fever, cough, indigestion, and menstrual disorder since. It also possesses antitumor, anti-inflammatory, and antioxidant properties.					
	Rumex acetosa L.		Apical part	Chrysophanol	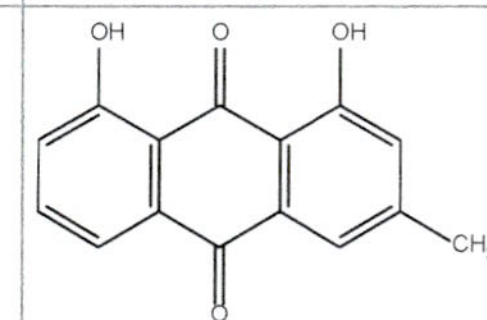	Lee et al. (2005) and Boon et al. (2010)
	Synonym(s): *Acetosa agrestis* Raf., *Acetosa amplexicaulis* Raf., *Acetosa angustata* Raf. **Common name(s)**: Common sorrel. **Botanical description**: It is native to northern temperate regions. It grows up to 1.50–2.00 ft. in height. It is a perennial herb that is primarily grown as a culinary herb for its leaves which are used in salads and cooking. Leaves have a tangy, acidic, sour-lemony flavor and are commonly used in salads, soups, omelets, and sauces. **Medicinal use(s)**: It is used to treat liver cancer. It also possesses antitumor, anti-inflammatory, and antioxidant properties.					

	Rumex japonicus Houtt		Leaf	Ellagic acid	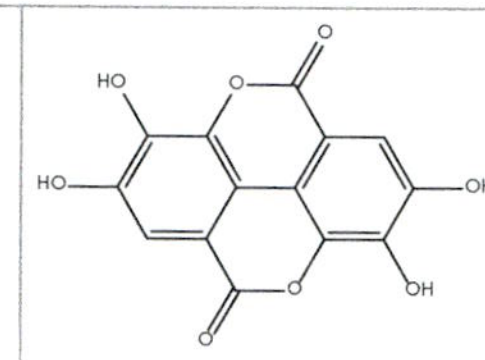	Kumar et al. (2011a, b), Xie and Yang (2014), and Özenver et al. (2018)
	Synonym(s): *Rumex cardiocarpus* Pamp., *Rumex crispus* subsp. *japonicus* (Houtt.) Kitam. **Common name(s)**: Not available. **Botanical description**: It is native to East Asia, China, Japan, and Korea. It is a perennial plant growing up to 1 m. It is in flower from May to June, and the seeds ripen from June to July. The species is hermaphrodite and is pollinated by wind. It is suitable to grow in light, medium, and heavy soils and prefers well-drained soil. **Medicinal use(s)**: It is used to treat cancer. It also possesses antitumor, anti-inflammatory, and antioxidant properties.			Gallic acid	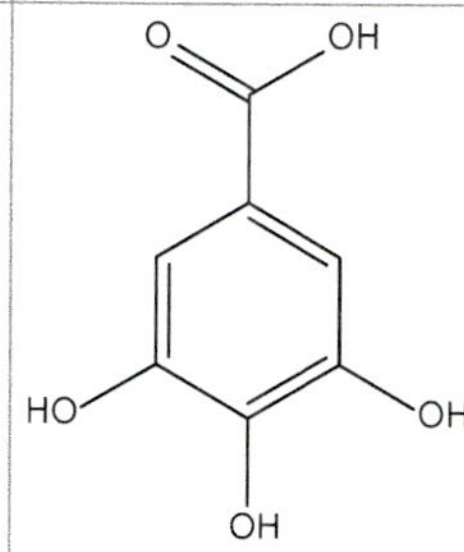	
				Cardiac glycosides	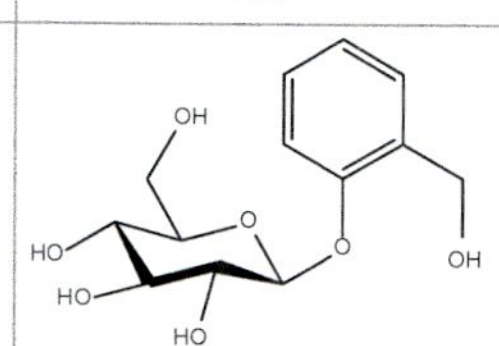	

(continued)

Table 4.111 (continued)

Family	Botanical name	Image	Plant part used	Bioactive compound(s)	Structure	Reference(s)
				Anthraquinone	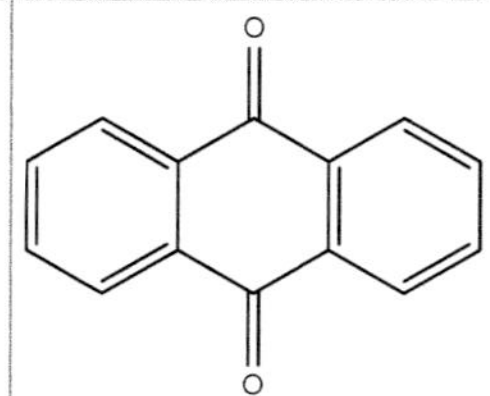	
				Vitexin	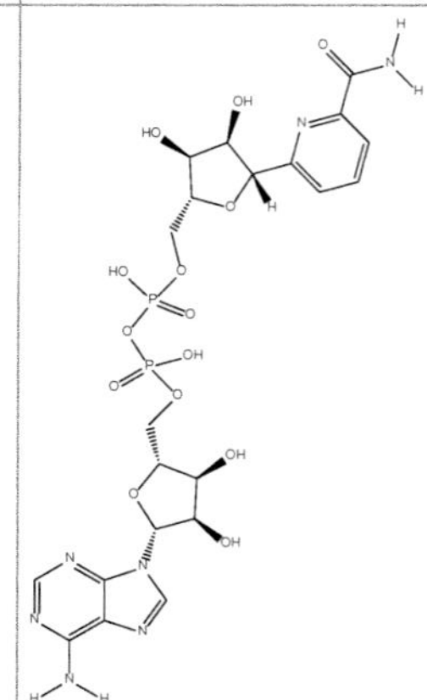	

	Rumex vesicarius L.		Stem	Isovitexin	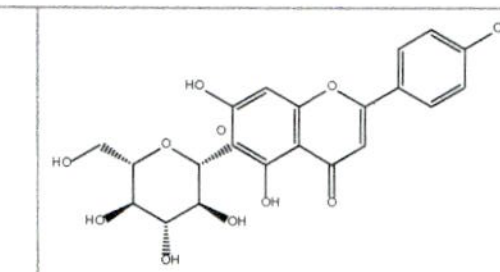	Mostafa et al. (2011) and Shahat et al. (2015)
	Synonym(s): *Lapathum vesicarium* Moench, *Acetosa vesicaria* (L.) Á. Löve. **Common name(s)**: Bladder dock. **Botanical description**: It is native to South Europe, North Africa, through Asia to India. It is an annual or perennial, rhizomatous herb growing up to 50 cm tall, strongly branched from the base. It is habituated to dry areas among loose stones, on grassy or gravelly slopes, at elevations from sea level up to 1150 m. **Medicinal use(s)**: It is used to treat cancer, dysentery, heat of the stomach, and toothache. It also possesses antitumor, anti-inflammatory, and antioxidant properties.					

Table 4.112 Anticancer plants of family Polypodiaceae

Family	Botanical name	Image	Plant part used	Bioactive compound(s)	Structure	Reference(s)
Polypodiaceae	*Dryopteris crassirhizoma*		Rhizome	Filicinic	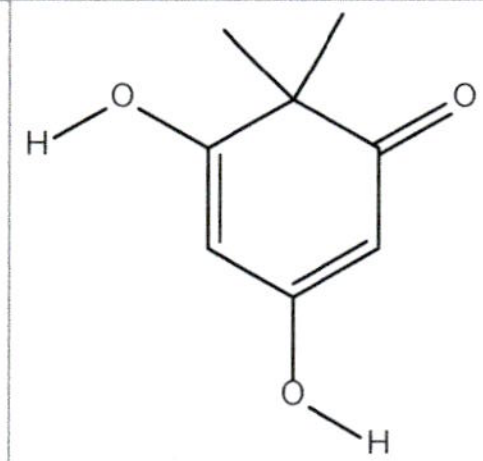	Mohammad (2006), Desai et al. (2008), and Madhuri and Pandey (2009)
	Synonym(s): *Dryopteris crassirhizoma* var. *crassirhizoma*. **Common name(s)**: Crown wood-fern, Thick stemmed wood fern. **Botanical description**: It is native to East Asia, Japan, Korea, and Manchuria. It is a fern growing up to 1 m on the humus-rich floor of temperate forests, often in colonies. It is habituated in mountains all over Japan. Larger clumps can be replanted directly into their permanent positions, though it is best to pot up smaller clumps and grow them in a cold frame until they are rooting well. **Medicinal use(s)**: It is used to treat prostate cancer. It also possesses anticancer, anti-inflammatory, antioxidant, and antibacterial properties.			Filicic acids	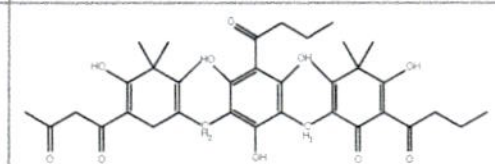	
				Aspidinol	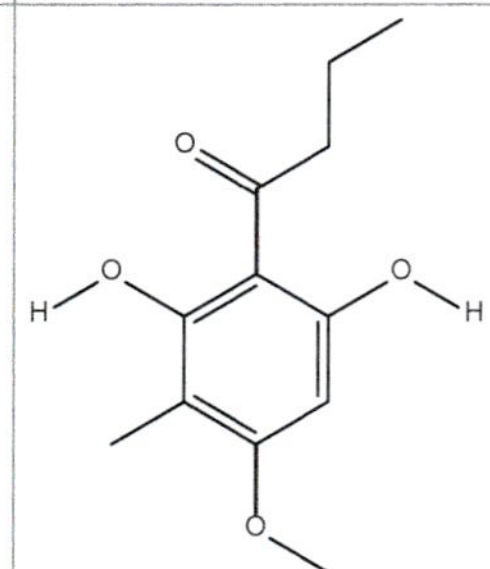	
				Aspidin	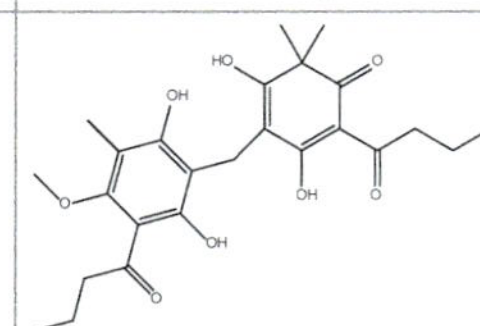	

	Phymatosorus diversifolium (Willd.) Pich.		Root	Campesterol	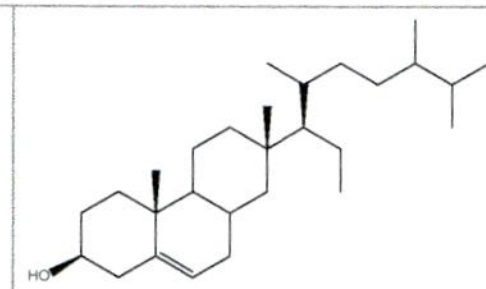	Holdsworth (1992)
	Synonym(s): *Polypodium diversifolium* Willd. **Common name(s)**: Paku wangi, Wart fern, Sakat hitam. **Botanical description**: It is native to Thailand and Cambodia. Rhizome long creeping, glaucous or not, about 5 mm diam., bearing fronds more than 1 cm apart, scaly; scales rather small, gradually narrowing from base towards long-acuminate apex, up to 3 by 0.7 mm, dark brown, clathrate, hairy at margin. Stipes up to 50 cm long, stramineous to pale castaneous. Laminae oblong-ovate to subdeltoid, simple, or pinnatifid with broadly winged rachis, up to 40 by 35 cm; lateral lobes up to 6 pairs coriaceous. **Medicinal use(s)**: It is used to treat breast cancer. It also possesses anticancer, anti-inflammatory, antioxidant, and antibacterial properties.					

(continued)

Table 4.112 (continued)

Family	Botanical name	Image	Plant part used	Bioactive compound(s)	Structure	Reference(s)
	Serpocaulon maritimum (Hieron.) A.R. Sm.		Rhizome	Carvacrol	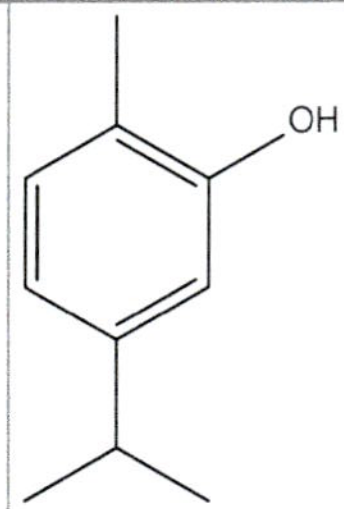	Gupta et al. (1979) and Maffei (2003a, b, c, d, e, f, g, h, i, j, k, l, m)
	Synonym(s): *Polypodium maritimum* Hieron. **Common name(s)**: Star fruit. **Botanical description**: It is native to Eastern Asia, Northern Asia, North America, and Europe. The fronds with triangular leaflets measure 10–50 cm. They are divided all the way back to the central stem into 10–18 pairs of segments or leaflets. The leaflets become much shorter at the end of the frond. The leaflets are generally whole or slightly denticulate and somewhat wider at their base, where they often touch each other. They have an alternating arrangement, those on one side being slightly offset from those on the other side. **Medicinal use(s)**: It is used to treat stomach cancer. It also possesses anticancer, anti-inflammatory, antioxidant, and antibacterial properties.					

	Polypodium subpetiolatum Hook.		Rhizome	D-Xylose	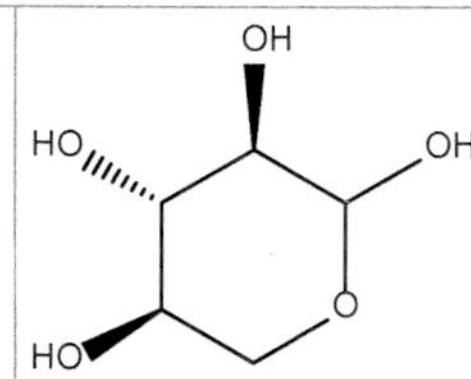	Anderson et al. (1979) and Maffei (2003a, b, c, d, e, f, g, h, i, j, k, l, m)
	Synonym(s): *Polypodium adelphum* Maxon., *Polypodium biserratum* M. Martens & Galeotti., *Polypodium firmulum* Baker., *Polypodium guilleminianum* E. Fourn. **Common name(s)**: Calaguala. **Botanical description**: It is native to Eastern Asia, Northern Asia, North America, and Europe. The fronds with triangular leaflets measure 10–50 cm. They are divided all the way back to the central stem into 10–18 pairs of segments or leaflets. The leaflets become much shorter at the end of the frond. The leaflets are generally whole or slightly denticulated and somewhat wider at their base, where they often touch each other. They have an alternating arrangement, those on one side being slightly offset from those on the other side. The petioles have no scales. **Medicinal use(s)**: It is used for the treatment of breast cancer and skin cancer. It also possesses anticancer, antioxidant, anti-inflammatory, and antibacterial properties.					

Table 4.113 Anticancer plants of family Polyporaceae

Family	Botanical name	Image	Plant part used	Bioactive compound(s)	Structure	Reference(s)
Polyporaceae	*Antrodia cinnamomea* Chang & Chou	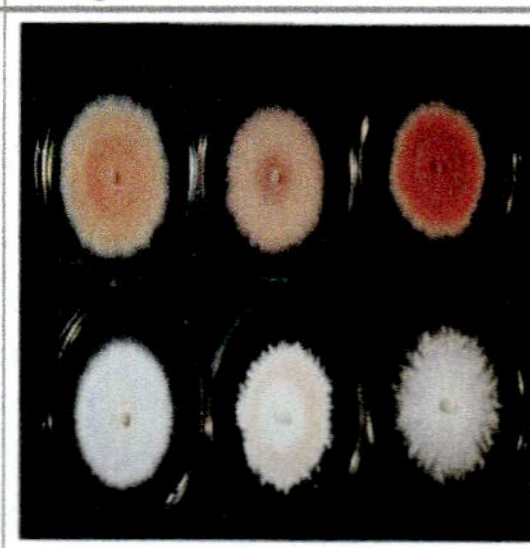	Fruit body	Coumaric acid	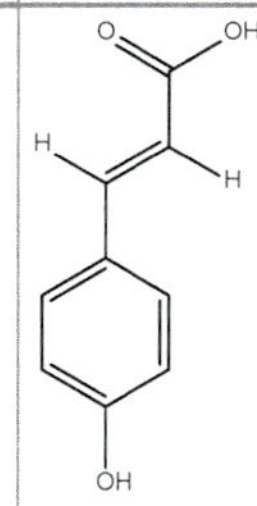	Cherng et al. (1996) and Liu et al. (2013)
	Synonym(s): *Ganoderma camphoratum* M., *Antrodia camphorata*. **Common name(s)**: Mushroom. **Botanical description**: It is native to Taiwan. It is growing in broad-leaved and coniferous forests at altitudes between 1300 and 2800 m in the central and northern parts of the island. A brown heart rot associated with resupinate, salmon-pink basidiomata in the empty rotten trunk is vernacularly called shiang-shan-chih. **Medicinal use(s)**: It is used for the treatment of liver and esophageal cancer, liver disease, food intoxication, drug intoxication, and hypertension. It also possesses anticancer, anti-inflammatory, antioxidant, and antibacterial properties.					

	Coriolus versicolor (L.:Fr.) Quelet/ *Trametes versicolor*		Fruit body	Apigenin	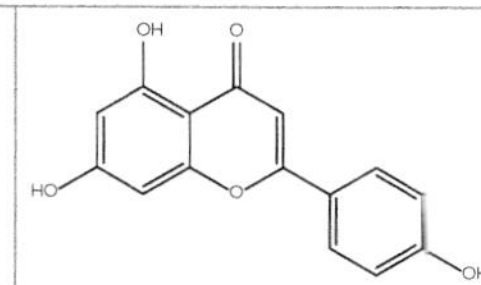	Liu et al. (1993) and Leanna et al. (2010)
	Synonym(s): *Boletus versicolor* L., *Poria versicolor* (L.) Scop., *Agaricus versicolor* (L.) Lam., *Boletus suberosus* Batsch. **Common name(s)**: Polypore mushroom. **Botanical description**: It is native to Asia and North America. The top surface of the cap shows typical concentric zones of different colors. The flesh is 1–3 mm thick and has leathery texture. Older specimens, such as the one pictured, can have zones with green algae growing on them, thus appearing green. It commonly grows in tiled layers. The cap is rust brown or darker brown, sometimes with blackish zones. **Medicinal use(s)**: It is used to treat breast cancer and tumor. It also possesses anticancer, antitumor, anti-inflammatory, antioxidant, antibacterial, and antimicrobial properties.					

(continued)

Table 4.113 (continued)

Family	Botanical name	Image	Plant part used	Bioactive compound(s)	Structure	Reference(s)
	Grifola frondosa (Dicks.: Fr.) Gray		Fruit body	Betulinic acid	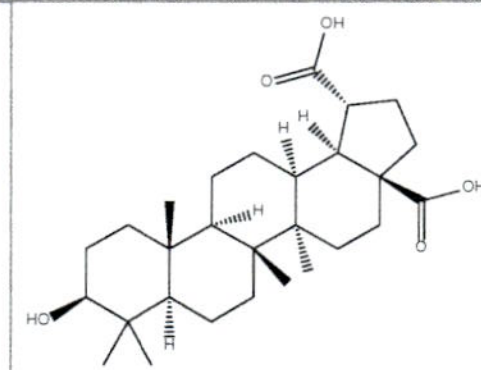	Ohno et al. (1985) and Deng et al. (2009a, b)
	Synonym(s): *Boletus frondosus* Dicks., *Polyporus frondosus* Fr. **Common name(s)**: Hen of the woods, Sheepshead, Maitake. **Botanical description**: It is native to Japan and North America. It grows from an underground tuber-like structure known as a sclerotium, about the size of a potato. The fruiting body, occurring as large as 100 cm, is a cluster consisting of multiple grayish-brown caps which are often curled or spoon-shaped, with wavy margins and 2–7 cm broad. The under surface of each cap bears about one to three pores per millimeter, with the tubes rarely deeper than 3 mm. The milky-white stipe (stalk) has a branchy structure and becomes tough as the mushroom matures. **Medicinal use(s)**: It is used for the treatment of breast cancer and tumor. It also possesses anticancer, antioxidant, antibacterial, and anti-inflammatory properties.					

Table 4.114 Anticancer plants of family Pteridaceae

Family	Botanical name	Image	Plant part used	Bioactive compound(s)	Structure	Reference(s)
Pteridaceae	*Adiantum macrophyllum* Sw.		Part not specified	Apigenin	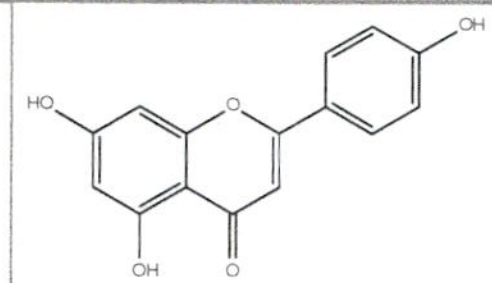 	Laferriere (1994) and Maffei (2003a, b, c, d, e, f, g, h, i, j, k, l, m)
	Synonym(s): *Adiantum aleuticum* (Rupr.) C.A. Paris., *Adiantum anceps* Maxon & C.V. Morton. **Common name(s)**: Black-stick maidenhair, Brittle maidenhair fern, Maidenhair fern. **Botanical description**: It is native to Eastern Asia, China, and Japan. They are distinctive in appearance, with dark, often black, stripes and rachises, and bright green, often delicately cut, leaf tissue. The sori are borne submarginally and are covered by reflexed flaps of leaf tissue which resemble indusia. Dimorphism between sterile and fertile fronds is generally subtle. They generally prefer humus-rich, moist, well-drained sites, ranging from bottomland soils to vertical rock walls. Many species are especially known for growing on rock walls around waterfalls and water seepage areas. **Medicinal use(s)**: It is used for the treatment of gastric cancer and gastric carcinoma. It also possesses anti-inflammatory, antioxidant, and antibacterial properties.					

(continued)

Table 4.114 (continued)

Family	Botanical name	Image	Plant part used	Bioactive compound(s)	Structure	Reference(s)
	Cheilanthes contracta (Kunze) Mett. ex Kuhn		Part not specified	Coumaric acid	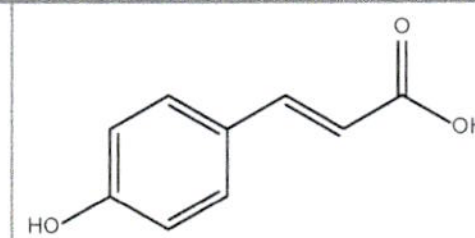	Charlson (1979) and Maffei (2003a, b, c, d, e, f, g, h, i, j, k, l, m)
	Synonym(s): *Cheilanthes acrostica* (Balb.) Tod., *Cheilanthes adiantoides* T.C. Chambers & P.A. Farrant., *Cheilanthes aemula* Maxon., *Cheilanthes afra* Kornaś., *Cheilanthes agetae* (Saiki) C.M. Kuo. **Common name(s)**: Lip ferns. **Botanical description**: It is native to Eastern and Southern Africa, Northern Cape, Western Cape, Eastern Cape, grassland, fynbos, and savannah. Plants are rupestral or terrestrial. Rhizomes compact to long-creeping, ascending or horizontal, scaly. Rhizome scales lanceolate to acicular, concolorous (tan to dark brown) or bicolorous (with dark central stripe and brown margins). Occasionally linear or pentagonal; adaxial surfaces glabrous or pubescent; abaxial surfaces scaly and/or pubescent or rarely glabrous. Ultimate segments round to oblong-ovate, minute to >1 cm long, the veins obscure and not ending in prominent hydathodes. Segment margins usually recurved, with a poorly differentiated false indusium. **Medicinal use(s)**: It is used for the treatment of pancreatic and gastric cancer. It also possesses anticancer, anti-inflammatory, antioxidant, and antibacterial properties.					

	Cheilanthes dalhousiae Hook		Leaf	Flavone-3-O-Glycosides	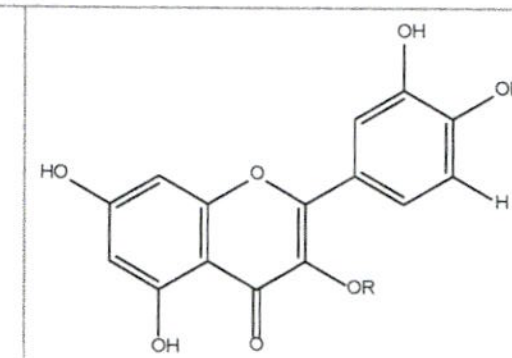	Mishra and Verma (2010)
	Synonym(s): *Cheilanthes acrostica* (Balb.) Tod., *Cheilanthes adiantoides* T.C. Chambers & P.A. Farrant. **Common name(s)**: Countess dalhousies spleenwort. **Botanical description**: It is native to Tibet, Nepal, India, Pakistan, Afghanistan, the United States, and Mexico. It is a genus of about 180 species of rock-dwelling ferns with a cosmopolitan distribution in warm, dry, rocky regions, often growing in small crevices high up on cliffs. Most are small, sturdy, and evergreen. The leaves are often densely covered in trichomes. Many of them are desert ferns, curling up during dry times and reviving with the coming of moisture. At the ends of veins, sporangia, or spore-bearing structures, are protected by leaf margins, which curl over them. **Medicinal use(s)**: It is used for the treatment of gastric cancer or tumor. It also possesses antioxidant, antitumor, antibacterial, and anti-inflammatory properties.					

(continued)

Table 4.114 (continued)

Family	Botanical name	Image	Plant part used	Bioactive compound(s)	Structure	Reference(s)
	Notholaena sulphurea (Cav.) J. Sm.		Rhizome	Quercetin	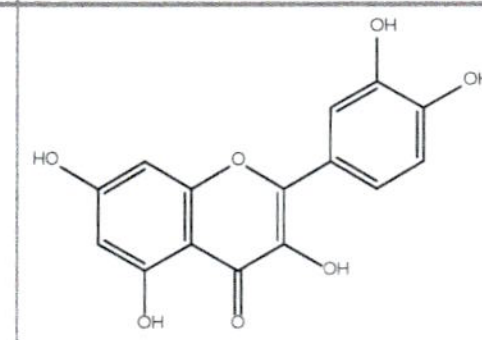	Ramirez et al. (1988) and Maffei (2003a, b, c, d, e, f, g, h, i, j, k, l, m)
	Synonym(s): *Aleuritopteris cretacea* (Liebm.) E. Fourn., *Aleuritopteris sulphurea* (Cav.) Fée, *Cheilanthes borsigniana* Rchb. f. & Warsz. ex W. Koch. **Common name(s)**: Countess dalhousies spleenwort. **Botanical description**: It is native to Tibet, Nepal, India, Pakistan, Afghanistan, the United States, and Mexico. It is a genus of about 180 species of rock-dwelling ferns with a cosmopolitan distribution in warm, dry, rocky regions, often growing in small crevices high up on cliffs. Most are small, sturdy, and evergreen. The leaves are often densely covered in trichomes. Many of them are desert ferns, curling up during dry times and reviving with the coming of moisture. At the ends of veins, sporangia, or spore-bearing structures, are protected by leaf margins, which curl over them. **Medicinal use(s)**: It is used for the treatment of gastric cancer, tumor, and gastric carcinoma. It also possesses antioxidant, antitumor, antibacterial, and anti-inflammatory properties.					

	Pteridium aquilinum (L.) Kuhn		Root	Apigenin	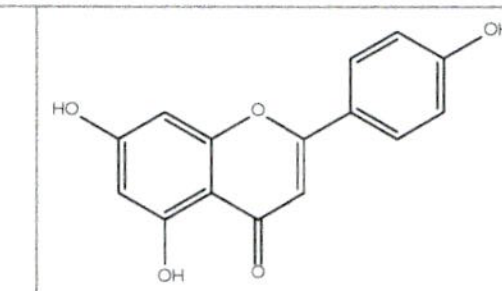	Latorre et al. (2009)
	Synonym(s): *Pteridium aquilinum* var. *aquilinum*., *Pteridium aquilinum* subsp. *aquilinum*., *Pteridium aquilinum* var. *lanuginosum* Henriq. **Common name(s)**: Eagle fern. **Botanical description**: It is native to China, Japan, Korea, and England. It is an herbaceous perennial plant and deciduous in winter. The large, roughly triangular fronds are produced singly, arising upwards from an underground rhizome, and grow up to 1–3 m tall; the main stem, or stipe, is up to 1 cm in diameter at the base. Bracken fern typically grows to 3 in. tall and 4 in. wide. Coarse, divided, triangular fronds rise directly from deep running rootstocks. Fronds typically tilt to being almost horizontal. Each frond (to 3′ long) is 2–3 times pinnately compound. **Medicinal use(s)**: It is used for the treatment of stomach cancer. The roots are used for the treatment of stomach cramps, chest pains, internal bleeding, diarrhea, and colds and also to expel worms. It also possesses antioxidant, anti-inflammatory, and antibacterial activities.					

(continued)

Table 4.114 (continued)

Family	Botanical name	Image	Plant part used	Bioactive compound(s)	Structure	Reference(s)
	Pteris ensiformis Burm.		Whole plant	Rutin	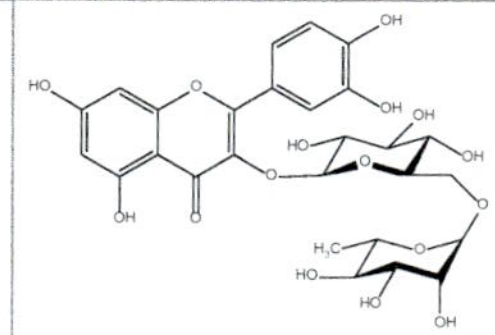	Chen et al. (2008a, b)
				Quercetin	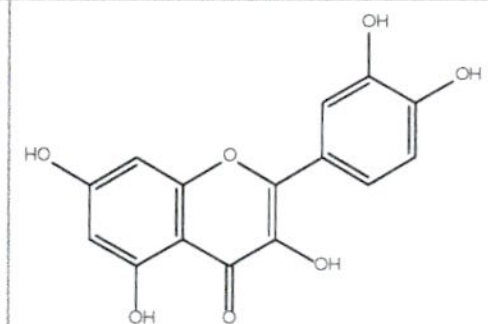	
	Synonym(s): *Pteris ensiformis* var. *ensiformis*, *Pteris serrulata* var. *obtusata* Christ. **Common name(s)**: Slender brake. **Botanical description**: It is native to China, Japan, Korea, and England. It is an herbaceous perennial plant and deciduous in winter. The large, roughly triangular fronds are produced singly, arising upwards from an underground rhizome, and grow up to 1–3 m tall; the main stem, or stipe, is up to 1 cm in diameter at the base. Bracken fern typically grows to 3 in. tall and 4 in. wide. **Medicinal use(s)**: It is used to treat breast cancer, diabetes, and snakebites. It also possesses anticancer, antioxidant, anti-inflammatory, and antibacterial activities.			Cyclolaudenol	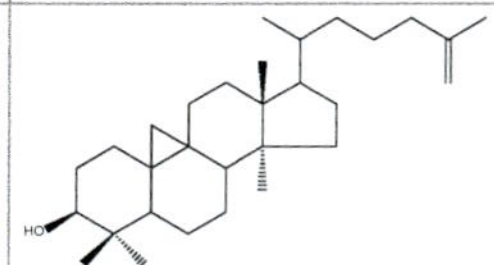	
				Coumaric acid	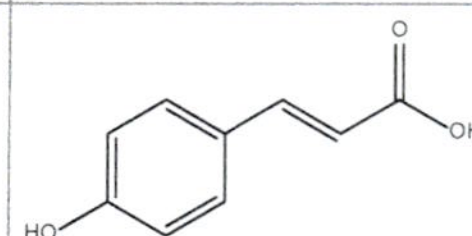	
				5-O-coumaroylquinic acid	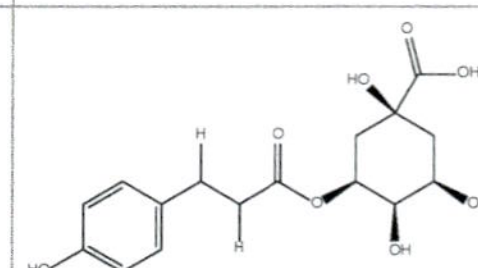	
				β-D-glucopyranosyl benzoic acid ester	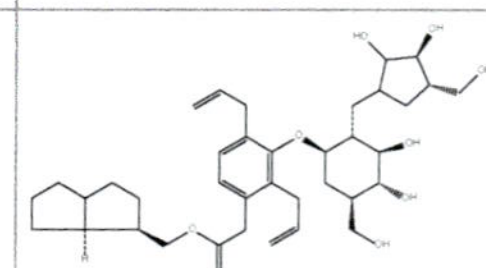	

				Pterosin	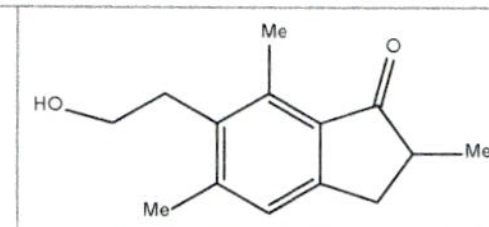	
	Pterocarpus marsupium Roxb.		Leaves Flowers	Pterostilbene	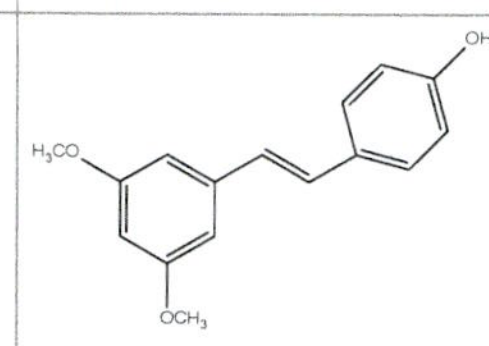	Chakraborty et al. (2010)
				Resveratrol	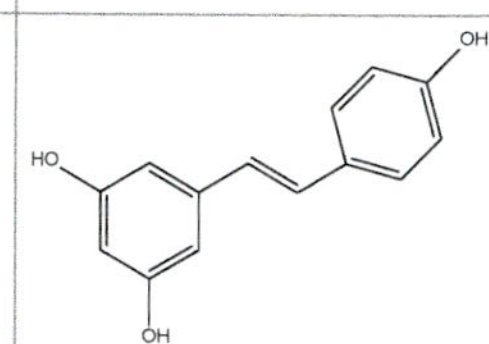	
	Synonym(s): *Lingoum marsupium* (Roxb.) Kuntze, *Pterocarpus bilobus* G.Don. **Common name(s)**: Malabar kino, Indian kino tree, or Vijayasar. **Botanical description**: It is native to central, western, and southern regions of India. It is a deciduous tree, up to 30 m tall, bark 10–15 mm, surface gray or grayish black, rough, deeply vertically cracked, exfoliations small, irregular, fibrous; blaze pink; exudation blood red. Leaves are compound, alternate; flowers are bisexual, yellow, at branch ends and in leaf axils, borne in panicles; 1.0–1.2 cm long; bracts small, dioecious; bracteoles 2, falling off; sepal tube bell-shaped, sepals short, the upper 2 often fused; Indian kino tree is found in Peninsular India and Sri Lanka. Flowering, September–October. **Medicinal use(s)**: It is used to treat breast and prostate cancer and diabetes. It also possesses antioxidant, antibacterial, and anti-inflammatory properties.					

(continued)

Table 4.114 (continued)

Family	Botanical name	Image	Plant part used	Bioactive compound(s)	Structure	Reference(s)
	Pterocarpus santalinus L.		Whole plant	Resveratrol	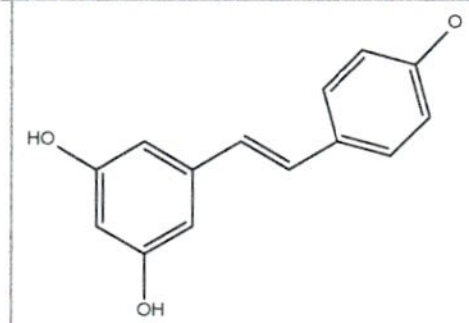	Kwon et al. (2006b)
	Synonym(s): *Lingoum santalinum* (L.f.) Kuntze. **Common name(s)**: Red sanders, Red sandalwood, Saunderswood. **Botanical description**: It is native to India. It is a light-demanding small tree, growing to 8 m tall with a trunk 50–150 cm in diameter. It is fast-growing when young, reaching 5 m tall in 3 years, even on degraded soils. The leaves are alternate, 3–9 cm long, trifoliate with three leaflets. The flowers are produced in short racemes. The fruit is a pod 6–9 cm long containing one or two seeds. **Medicinal use(s)**: It is used for the treatment of cancer cells, diseases of the skin, swollen limbs, ophthalmic disorders, sore eyes, and headache. It also possesses antioxidant, antidiabetic, antimicrobial, anticancer, and anti-inflammatory properties.					

Table 4.115 Anticancer plants of family Ranunculaceae

Family	Botanical name	Image	Plant part used	Bioactive compound(s)	Structure	Reference(s)
Ranunculaceae	*Actaea podocarpa* DC.		Root	Cycloartane	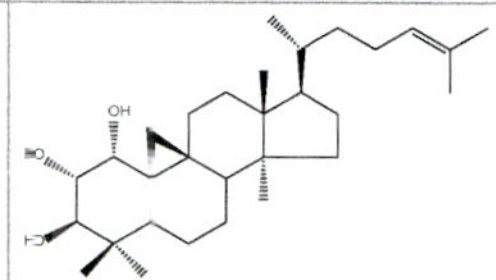	Ali et al. (2006) and Verbitski et al. (2008)
				Podocarpasides H	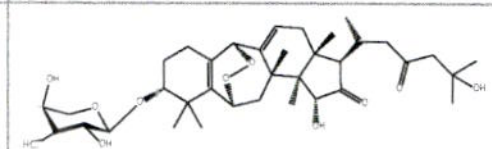	
	Synonym(s): *Actaea americana* Prantl, *Actaea pentacarpa* Michx. ex Torr. & A.Gray, *Actaea pentagyna* Walter. **Common name(s)**: Bugbane, Mountain bugbane. **Botanical description**: It is a perennial plant and is native to moist woods in the Appalachian Mountains from Pennsylvania to Tennessee and Georgia. It is in flower during October, and the seeds ripen from October to November. The flowers are hermaphrodite (have both male and female organs) and are pollinated by bees. **Medicinal use(s)**: It is used for the treatment of diabetes mellitus, nephritis, leukemia, and uterine cancer. It also possesses anti-inflammatory, hemolytic, and adjuvant properties.			Iodoazomycin arabinoside	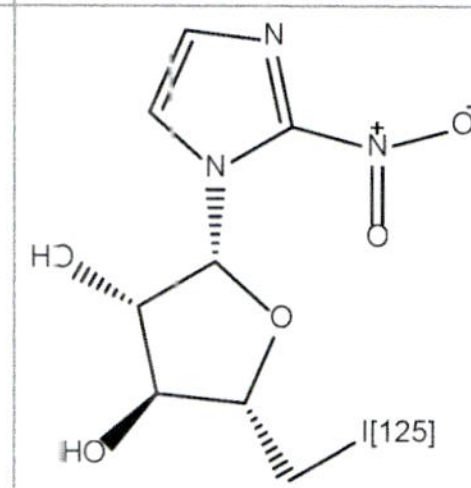	

(continued)

Table 4.115 (continued)

Family	Botanical name	Image	Plant part used	Bioactive compound(s)	Structure	Reference(s)
	Anemone rivularis Buch. Ham ex DC.		Root	Ranunculin	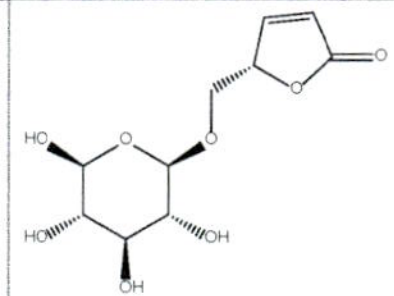	Jain and Puri (1984) and Chung et al. (2017)
	Synonym(s): *Anemone dubia* Wall. ex Wight & Arn., *Anemone esquirolii* H.Lév. & Vaniot, *Anemone geraniifolia* Wall. **Common name(s)**: River anemone, River windflower. **Botanical description**: It is an erect silky-haired perennial plant (30–90 cm) found on forest margins, grassy slopes, streamsides, and lakesides in the Himalayas and is native to India. Basal leaves are rounded, deeply 3-lobed, silky haired on both sides and long stalked. Lobes are further cut and shallowly toothed. Flowers are white, often flushed violet outside, 2–5 cm across; in lax umbel below which occur a ring of narrowly 3-lobed, sharply toothed leaves. The most interesting part of the plant is its unusual blue anthers. **Medicinal use(s)**: It possesses anticancer, anti-inflammatory, antimicrobial, and antioxidant properties. The seeds are used in Tibetan medicine. It is also used for the treatment of rotting tissues, snake poisoning, and stomach/intestinal pain from worm infestation.			Anemonin	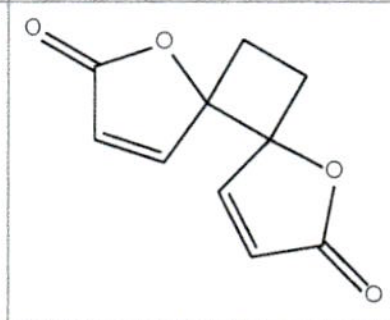	
				Protoanemonin	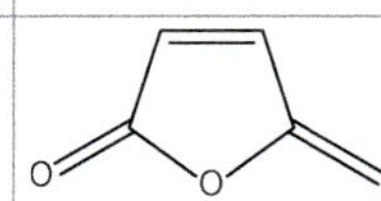	

	Plant	Image	Part	Compound	Structure	Reference
	Coptis japonica (Thunb.) Makino		Rhizome	4′,6-Diamidino-2-phenylindole	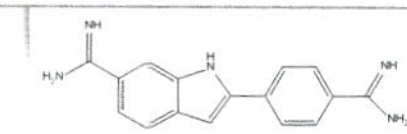	Cha (1977) and Park et al. (2005)
	Synonym(s): *Coptis anemonifolia* Siebold & Zucc., *Coptis orientalis* Maxim. **Common name(s)**: Japanese goldthread. **Botanical description**: It is an evergreen perennial plant and is native to East Asia and Japan. It is in flower in May. The flowers are hermaphrodite (have both male and female organs). It is suitable to grow in light (sandy), medium (loamy), and heavy (clay) soils and prefers moist soil. **Medicinal use(s)**: It is used to treat many types of cancer. The root is a pungent, very bitter, cooling herb that controls bacterial and viral infections, relaxes spasms, lowers fevers, and stimulates blood circulation. It is locally analgesic and an aesthetic and is also anti-inflammatory and stomachic. It is used for the treatment of intestinal catarrh, dysentery, enteritis, high fevers, inflamed mouth and tongue, conjunctivitis, etc. The root is harvested in the autumn and can be used fresh or dried.			Quinine	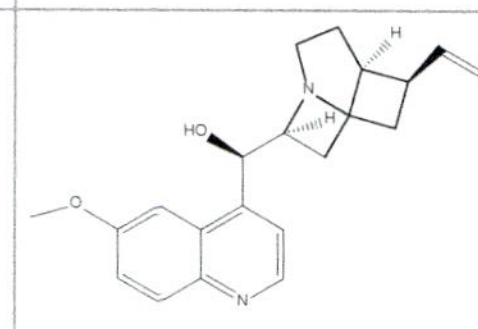	

(continued)

Table 4.115 (continued)

Family	Botanical name	Image	Plant part used	Bioactive compound(s)	Structure	Reference(s)
	Hydrastis canadensis L.		Rhizome	Berberine	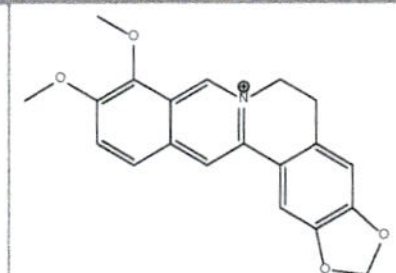	Jena et al. (2012) and Madhuri and Pandey (2009)
	Synonym(s): No synonyms are recorded for this name. **Common name(s)**: Goldenseal, Orange root. **Botanical description**: It is a perennial herb and is native to northeastern part of the United States. The stem bears one to three leaves which emerge from a creeping rhizome. The rhizome is twisted, yellow or brown in color, and wrinkled. It grows in rich shady soil. **Medicinal use(s)**: It is used for the treatment of constipation, hepatic origin, ulcers, cancer, and gastrointestinal diseases, such as gastric catarrh, accompanied by weakness, diarrhea, gas, and jaundice.			Canadine	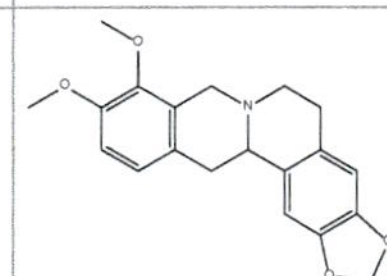	

	Knowltonia capensis (L.) Huth		Leaf	1,3-Dibenzylisoquinoline	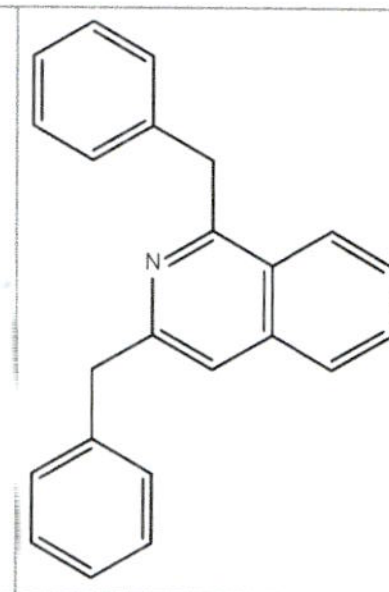	Madhuri and Pandey (2009)
	Synonym(s): *Knowltonia capensis* Huth, *Knowltonia bracteata* Harv. ex Zahlbr. **Common name(s)**: Blistering leaves, Brandblare, Katjiedrieblaar. **Botanical description**: It is a stemless perennial herb between 300 and 500 mm tall and is native to South Africa. Flowers consist of colored sepals and there are no petals. It has a slender flowering stem with a compound umbel of creamy-green flowers, from June to September. Fruits which are small, fleshy berries in dense clusters, green turning black when ripe. A short rhizome with fleshy roots is present. **Medicinal use(s)**: This plant is used for the treatment of cancer, toothache, colds, and influenza. These plants are widely used in traditional medicine to treat wounds, external cancers, and rheumatism.			Stigmasterol	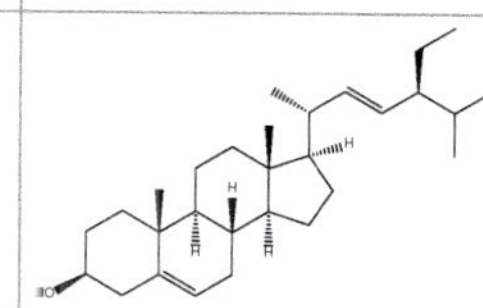	
				Beta-sitosterol	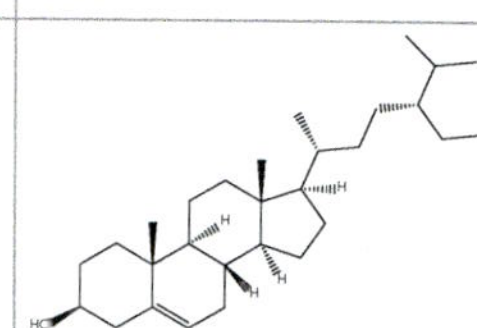	

(continued)

Table 4.115 (continued)

Family	Botanical name	Image	Plant part used	Bioactive compound(s)	Structure	Reference(s)
	Pulsatilla cernua (Thunb.) Bercht. ex J. Presl		Root	Cis-dichlorodiammineplatinum(II)	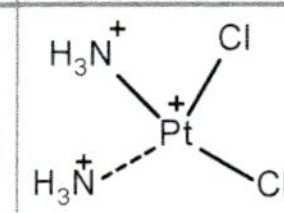	Shimizu et al. (1978), Park et al. (2012), Nho et al. (2013), Jain et al. (2014), and Shi et al. (2017)
	Synonym(s): *Pulsatilla cernua* var. *koreana* (Yabe ex Nakai) Y.N. Lee, *Pulsatilla koreana* (Yabe ex Nakai) Nakai ex T. Mori. **Common name(s)**: Wind flower, Prairie crocus, Easter flower, Meadow anemone. **Botanical description**: It is a perennial plant and is native to East Asia, China, Japan, and Korea. It is in flower from April to May, and the seeds ripen from May to June. The flowers are hermaphrodite (have both male and female organs) and are pollinated by insects. It is suitable to grow in light (sandy) and medium (loamy) soils and prefers well-drained soil. **Medicinal use(s)**: It has antitumor, antimicrobial, and anti-ulcer properties and is used to treat thyroid cancer, reproductive problems such as premenstrual syndrome, and epididymitis.					

<table>
<tr><td></td><td>Pulsatilla koreana (Yabe ex Nakai) Nakai ex T. Mori</td><td></td><td>Root</td><td rowspan="2">Beta-sitosterol</td><td rowspan="2">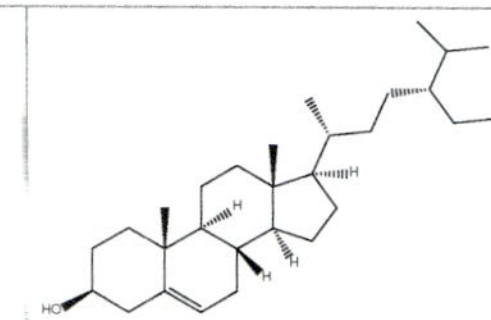</td><td rowspan="2">Kim et al. (2004), Hong (2011), and Park et al. (2012)</td></tr>
<tr><td></td><td colspan="3">Synonym(s): Pulsatilla cernua (Thunb.) Bercht. ex J. Presl.
Common name(s): Korean pasque flower.
Botanical description: It is a perennial plant and is native to East Asia and Korea. It is in flower in May. The flowers are hermaphrodite and are pollinated by insects. It is suitable to grow in light (sandy) and medium (loamy) soils and prefers well-drained soil.
Medicinal use(s): It has both cardiotoxic and cardiotonic properties and also possesses antitumor, anti-inflammatory, and anti-parasitic properties. It is used in Korea for the treatment of leukorrhea, dysentery, and scrofula and also used as contraceptive.</td></tr>
</table>

(continued)

Table 4.115 (continued)

Family	Botanical name	Image	Plant part used	Bioactive compound(s)	Structure	Reference(s)
	Ranunculus glaberrimus Hook.		Entire plant	Apigenin	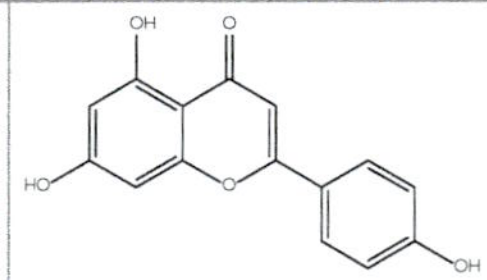	Turner (1984) and Maffei (2003a, b, c, d, e, f, g, h, i, j, k, l, m)
	Synonym(s): *Ranunculus glaberrimus* var. *glaberrimus*. **Common name(s)**: Bulbous buttercup, St. Anthony's turnip. **Botanical description**: It is native to Central British Columbia east to southern Saskatchewan. The stem is single or may be branched at the top. Basal leaves are narrow to spoon-shaped, entire to deeply 3-lobed, and tapered at the base to a petiole at least as long as the leaf. Flowers yellow, often turning white with age, petals 1/2 in. long. **Medicinal use(s)**: It is used to treat skin cancer. It also possesses antitumor properties, and the whole plant, and especially the sap, is used as acrid, anodyne, anti-spasmodic, diaphoretic, and rubefacient. The root is placed in a tooth cavity which acts as a painkiller.					

	Plant	Image	Part	Compound	Structure	Reference
	Semiaquilegia adoxoides (DC.) Makino		Root	Cyanogenic glycosides	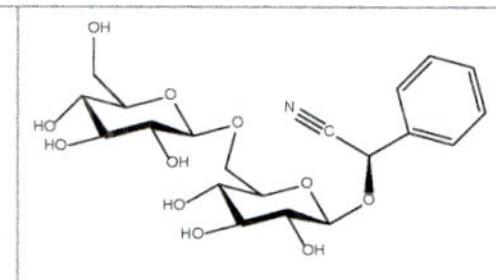	Duke and Ayensu (1985a, b) and Duan et al. (2013)
	Synonym(s): *Isopyrum adoxoides* DC, *Aquilegia adoxoides* (DC.) Ohwi, *Semiaquilegia adoxoides* var. *grandis* D.Q. Wang. **Common name(s)**: Tian kui. **Botanical description**: It is a perennial plant and is native to East Asia, China, Japan, and Korea. It is in flower in May and the flowers are hermaphrodite. It is suitable to grow in light (sandy), medium (loamy), and heavy (clay) soils. **Medicinal use(s)**: It is used to treat breast and lung cancer. The root is specifically recommended for the treatment of dry scabby skin and urinary disorders. The tubers are used as depurative, diuretic, and febrifuge.			Xeranthin	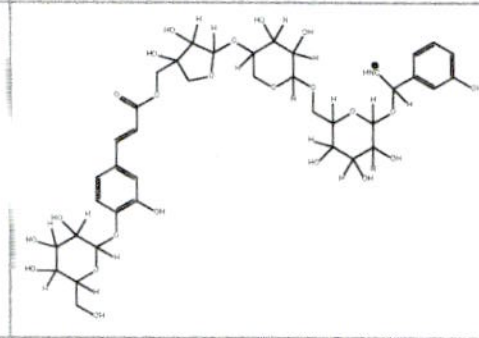	
				Beta-sitosterol	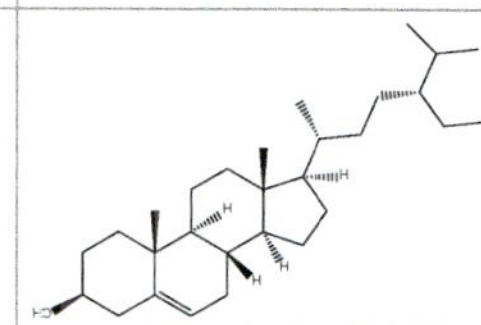	

(continued)

Table 4.115 (continued)

Family	Botanical name	Image	Plant part used	Bioactive compound(s)	Structure	Reference(s)
	Thalictrum faberi Ulbr.		Root	Thalifaberidine	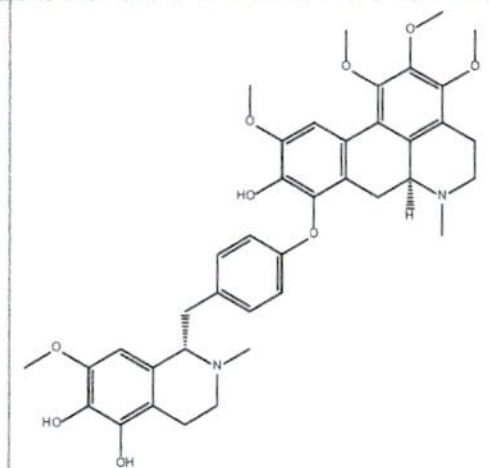	Lin et al. (1994a, b) and Maffei (2003a, b, c, d, e, f, g, h, i, j, k, l, m)
	Synonym(s): *Thalictrum macrophyllum* Migo. **Common name(s)**: Alpine meadow rue, Arctic meadow rue. **Botanical description**: It is native to Europe, Southwest Asia, and parts of Northwest Africa and South Africa. Plants are tufted, herbaceous (not woody) with colorless acrid (bitter) juice. The stem is 8–20 cm high, very thin, scapose (leafless), or rarely with one leaf. The elongated rhizomes, basal leaves withered at anthesis, stipules narrow, petiole 4.5–6 cm, leaf blade 2–3 ternate, leaflet blade broadly ovate or rhombic, papery, base rounded or subcordate, apex entire or 3-lobed, lobes obtuse or acute at apex, veins raised abaxially. **Medicinal use(s)**: It is used for treatment of stomach cancer and also shows antimalarial activity.			Thalifasine	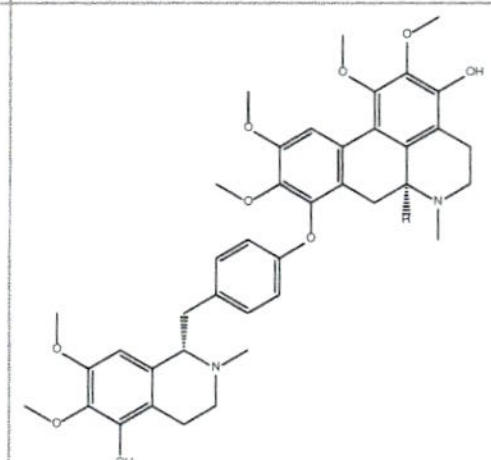	
				Aporphine	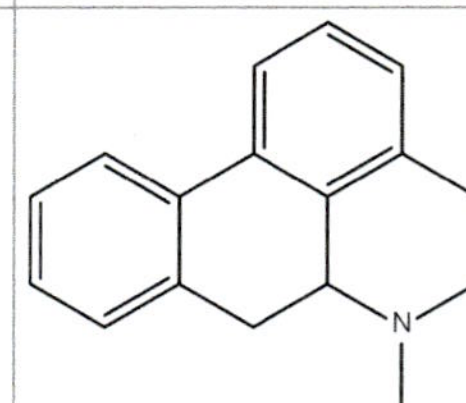	

		Isotetrandrine	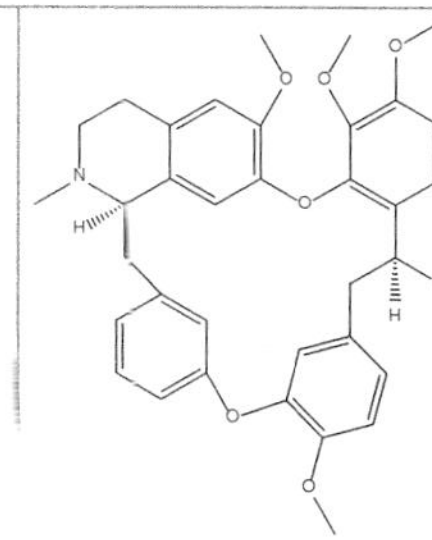	
		Thalicarpine	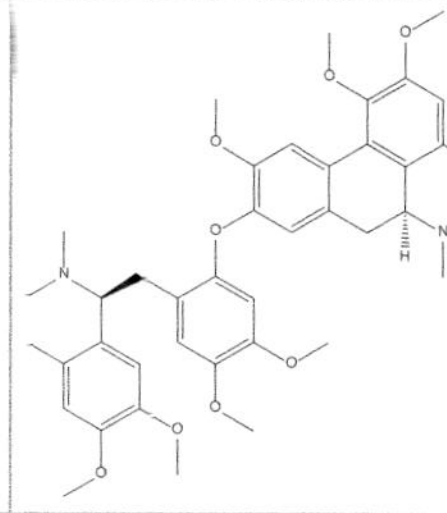	

(continued)

Table 4.115 (continued)

Family	Botanical name	Image	Plant part used	Bioactive compound(s)	Structure	Reference(s)
	Rubia cordifolia L.		Root	1-Hydroxytectoquinone	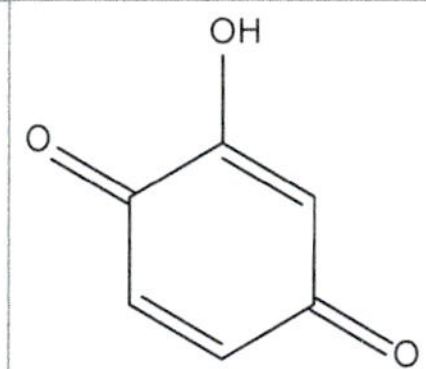	Ghosh et al. (2010), Bhatt and Kushwah (2015), and Verma et al. (2016)
	Synonym(s): *Dioscorea verticillata* Lam., *Galium cordifolium* (L.) Kuntze, *Rubia clematifolia* Reinw. ex Miq, *Rubia cordata* Thunb. **Common name(s)**: Indian madder. **Botanical description**: It is a climbing or creeping herbaceous perennial plant and is native to Asia, Afghanistan, India, China, Mongolia, Japan, Vietnam, Indonesia, and the Philippines. It supports itself upon other plants by means of its recurved prickles. The stems become woody at their base. The plant is mainly gathered from the wild for use as a dye plant, food, and medicine. Prior to the discovery of synthetic dyes, Indian madder was widely used as a dye plant in Asia and was exported in huge quantity to other areas of the globe. It is used much less nowadays but is still employed for high-quality traditional textiles. The plant can be used as an ornamental climber. **Medicinal use(s)**: It is used for cancer treatment. It also possesses anti-inflammatory, antimicrobial, and antidiabetic property.			1,3-Dihydroxyanthraquinone	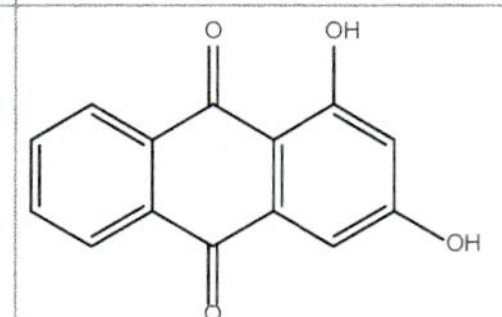	
				Munjistin	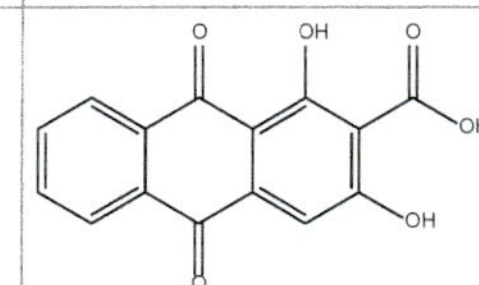	

		Compound	Structure	
		Purpurin	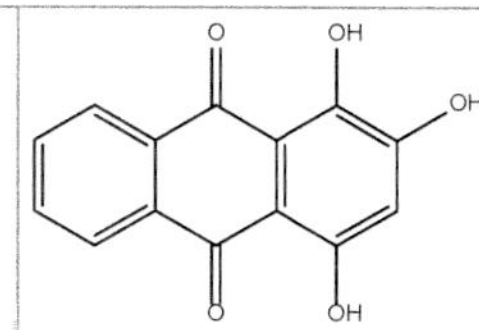	
		Rubicordifolin	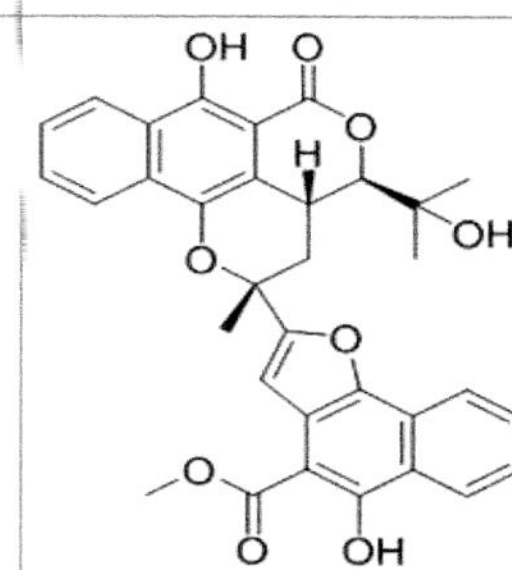	
		Mollugin	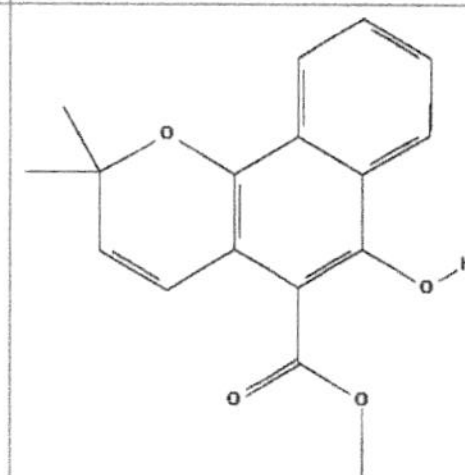	
		RA-XVII	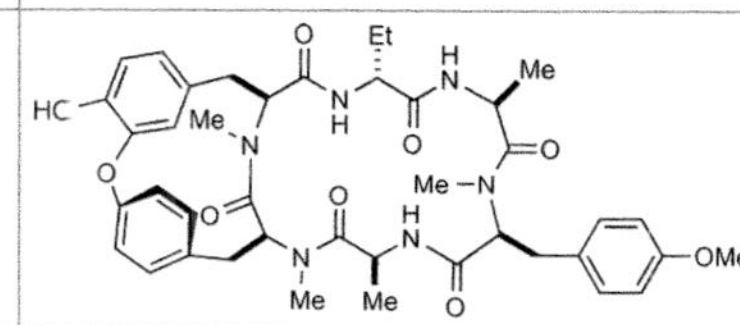	

(continued)

Table 4.115 (continued)

Family	Botanical name	Image	Plant part used	Bioactive compound(s)	Structure	Reference(s)
				RA-dimer A	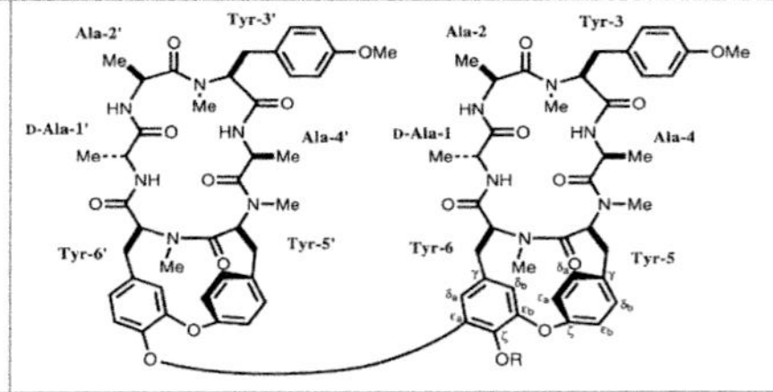	
				RA-V	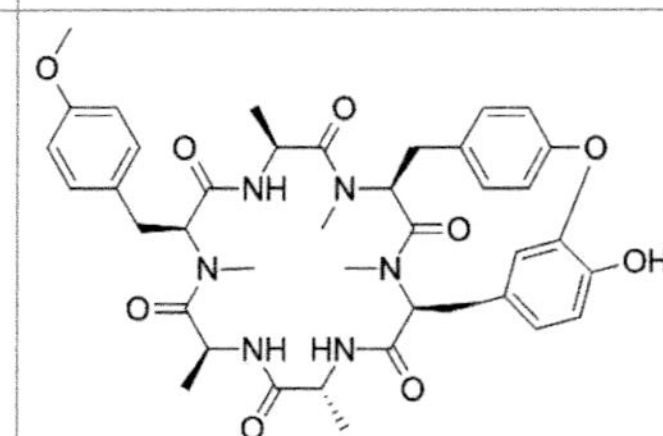	
				RA-VII	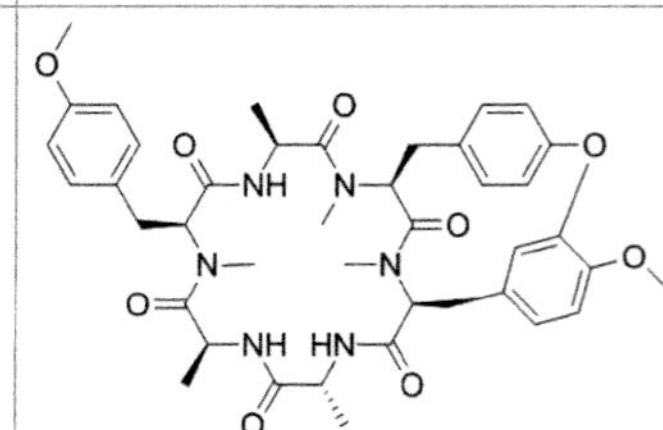	
				1,4-dihydroxy-2-naphthoic acid (DHNA)	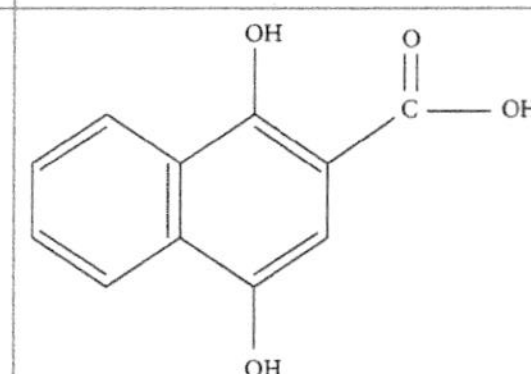	

Table 4.116 Anticancer plants of family Rhamnaceae

Family	Botanical name	Image	Plant part used	Bioactive compound(s)	Structure	Reference(s)
Rhamnaceae	*Pomaderris kumeraho* A.Cunn		Leaf	Kaempferol	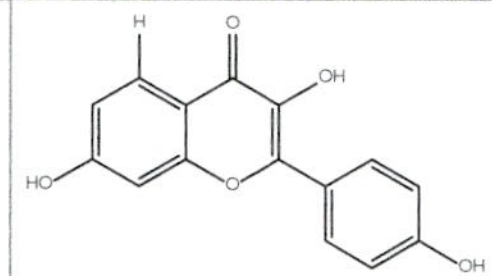	Brooker et al. (1989), Maffei (2003a, b, c, d, e, f, g, h, i, j, k, l, m)
	Synonym(s): *Pomaderris elliptica.* **Common name(s)**: Gum diggers soap, Kumerahou, New Zealand golden tainui, Papapa, Poverty weed. **Botanical description**: It is native to North Island of New Zealand. It grows up to 4 m in height and flowers in September, with yellow blossoms. It is found often on roadside banks and in gum land vegetation and occasionally seen in forested situations. **Medicinal use(s)**: It is used to treat skin and lung cancer. It has antitumor, anti-inflammatory, and antimicrobial effects. It is used for blood purification and to treat kidney, urinary, and all skin problems, bronchitis, pulmonary tuberculosis, coughs, colds, and asthma.			Quercetin	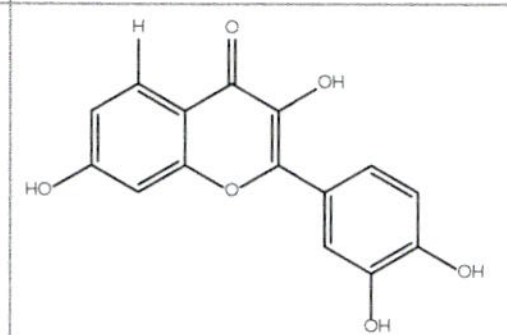	
				Taxifolin	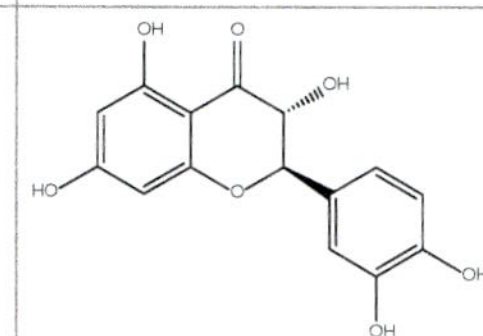	

(continued)

Table 4.116 (continued)

Family	Botanical name	Image	Plant part used	Bioactive compound(s)	Structure	Reference(s)
	Ziziphus jujuba Mill.		Leaf	Colubrinic acid	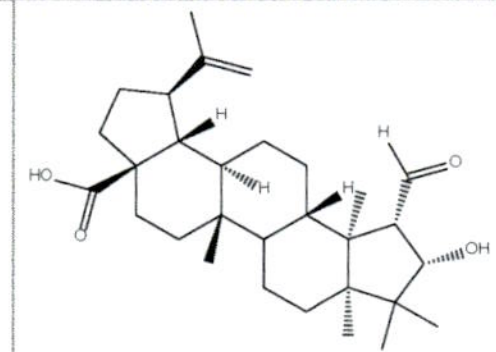	Lee et al. (2003) and Fard et al. (2015)
	Synonym(s): *Paliurus mairei* H. Lév., *Rhamnus jujube* L., *Rhamnus soporifera* Lour. **Common name(s)**: Jujube, Red date. **Botanical description**: It is a deciduous tree growing up to 10 m and is native to East Asia, China, and Japan. It is in flower from April to May, and the seeds ripen in October. The flowers are hermaphrodite (have both male and female organs) and are pollinated by insects. The plant is self-fertile. **Medicinal use(s)**: It is used to treat ovarian and breast cancer. It has anticancer, anti-inflammatory, anti-obesity, antioxidant, hepatoprotective, and gastrointestinal protective properties. A decoction of the root is used for the treatment of fevers. The root is made into a powder and applied to old wounds and ulcers. The leaves are used as astringent and febrifuge. They are said to promote the growth of hair. They are used to form a plaster in the treatment of strangury. The plant is a folk remedy for anemia, hypertonia, nephritis, and nerve diseases.			Alphitolic acid	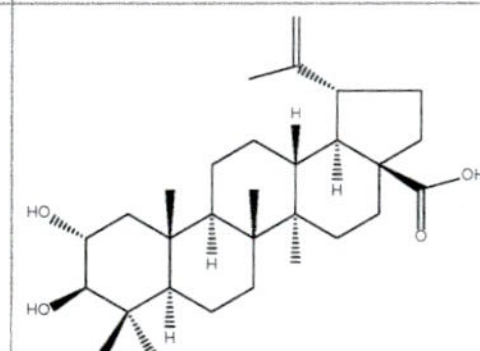	
				Zizyberenalic acid	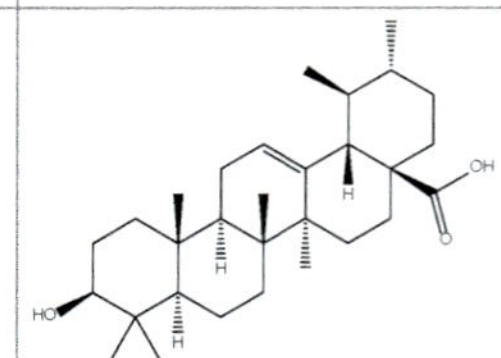	

Table 4.117 Anticancer plants of family Rhizophoraceae

Family	Botanical name	Image	Plant part used	Bioactive compound(s)	Structure	Reference(s)
Rhizophoraceae	*Bruguiera gymnorrhiza* (L.) Lam.		Flower	Bruguiesulfurol	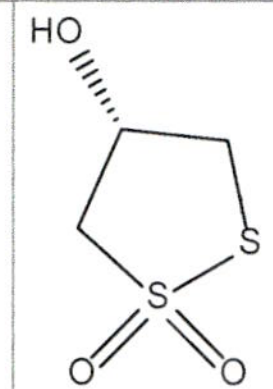	Huang et al. (2009a) and Sithranga Boopathy and Kathiresan (2010)
				Brugierol	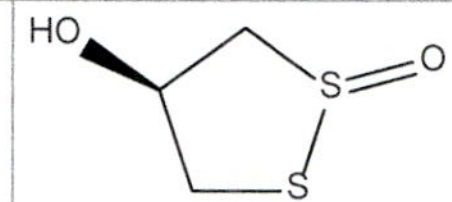	
	Synonym(s): *Bruguiera capensis* Blume, *Bruguiera conjugate* (L.) Merr. **Common name(s)**: Black mangrove, Swartwortelboom. **Botanical description**: It is an evergreen shrub or a tree that can grow up to 33 m tall and is native to East Asia, coastal areas from India and Sri Lanka through Southern Asia to New Guinea and New Britain. The bole can be up to 65 cm in diameter with buttresses up to 1 m high and pneumatophores (breathing roots) up to 45 cm. It is commonly found in the mangrove forests that are not frequently submerged and also grow along riverbanks and occasionally on sandy shores and in soils with water that is less saline than seawater. **Medicinal use(s)**: It is used to treat lung cancer, sarcoma, and carcinoma. It also possesses antitumor, antioxidant, anti-rheumatic, antimicrobial, and antiviral properties. Leaves and bark are used for the treatment of diarrhea, burns, liver disorders, fever, and diabetes and to get rid of intestinal worms, and the stems are used to cure viral fever.			Iscbrugierol	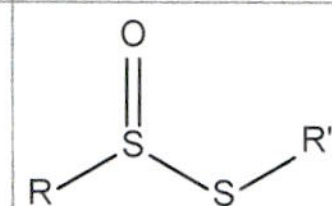	

(continued)

Table 4.117 (continued)

Family	Botanical name	Image	Plant part used	Bioactive compound(s)	Structure	Reference(s)
				Trans-3,30-dihydroxy1,5,10,50	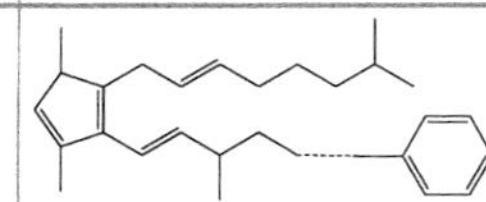	
				Tetrathiacyclodecane	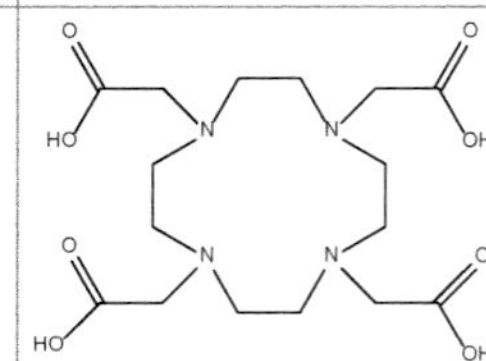	
				Cis-3,30-dihydroxy-1,5,10,50-tetrathiacyclodecane	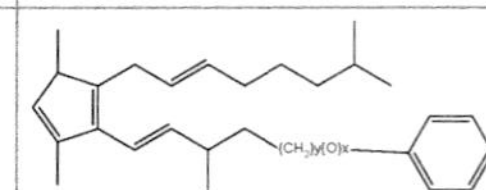	
				Gymnorrhizol	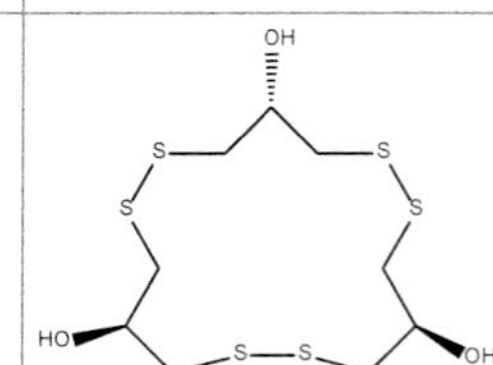	

	Rhizophora brevistyla Salvoza		Bark	Gallic acid	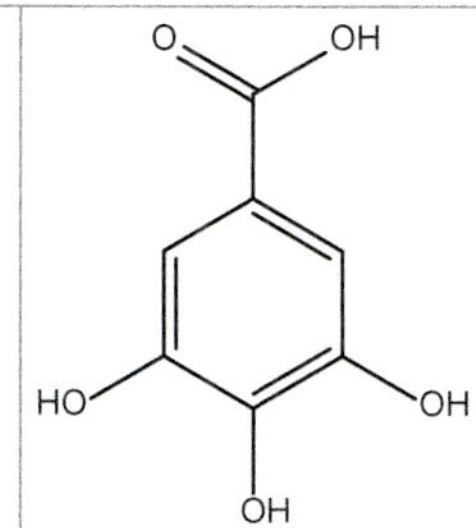	Garcia-Barriga (1975) and Maffei (2003a, b, c, d, e, f, g, h, i, j, k, l, m)
	Synonym(s): *Rhizophora harrisonii* Leechm. **Common name(s)**: Red mangrove. **Botanical description**: It is native to coastal areas of tropical America from Brazil and Peru north through the Caribbean and Central America to Mexico and Florida. It is a shrub or a small- to medium-sized, round-topped, bushy tree growing up to 5–20 m tall. The tree has branched, curved, and arching stilt roots 2–4.5 m tall when growing in salt water. It is found in mangrove swamp forests over large flat areas of silty or muddy shores in salt and brackish water. **Medicinal use(s)**: It is used to treat throat and esophageal cancer with gargles of mangrove bark. It also possesses antitumor, anti-inflammatory, and antimicrobial activities. The bark is used as an astringent and febrifuge and is used for the treatment of hemorrhages, dysentery, diarrhea, frequent micturition, and bladder diseases.			Oleanolic acid	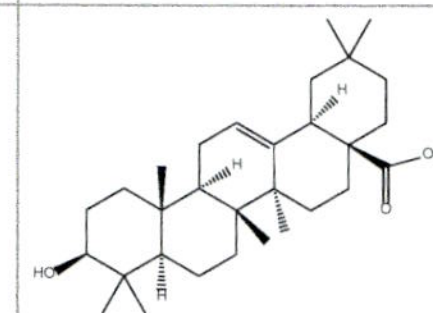	
				Fenugreekine	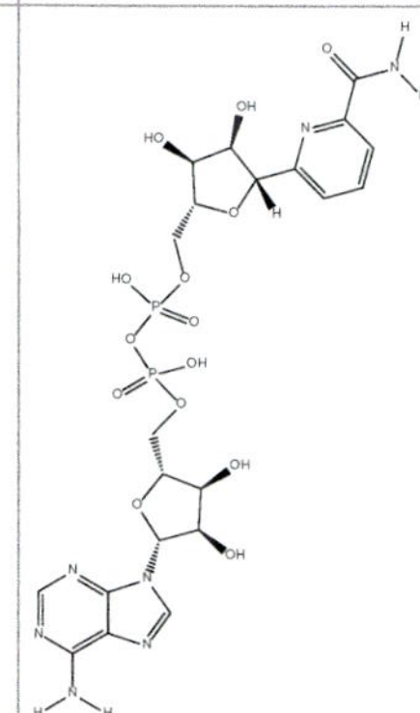	

Table 4.118 Anticancer plants of family Rosaceae

Family	Botanical name	Image	Plant part used	Bioactive compound(s)	Structure	Reference(s)
Rosaceae	*Agrimonia pilosa* Ledeb.		Whole plant	Agrimoniin	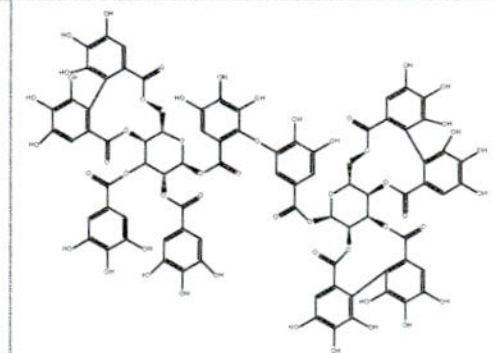	Murayama et al. (1992) and Nho et al. (2011)
	Synonym(s): *Agrimonia conopsea* Czern. ex C.A.Mey., *Agrimonia convergens* Czern. ex Juz., *Agrimonia dahurica* Willd. ex Ser., *Agrimonia davurica* (Link) Schltdl. ex Ledeb. **Common name(s)**: Hairy agrimony. **Botanical description**: It is a perennial plant and is native to East Europe and East Asia, China, and Japan. It is in flower from June to August, and the seeds ripen from August to September. The flowers are hermaphrodite (have both male and female organs) and are pollinated by bees, flies, and itself. It is suitable to grow in light (sandy), medium (loamy), and heavy (clay) soils. **Medicinal use(s)**: It is used to treat various types of cancer and also possesses antibacterial, anticancer, anticoagulant, anti-inflammatory, antidysenteric, and antipyretic properties. The plant is also used for the treatment of abdominal pain, coughs, colds, tuberculosis, diarrhea, peptic ulcer, sore throat, headaches, bloody and mucoid dysentery, bloody and white discharge, and heatstroke. The leaves are rich in vitamin K and are used to promote blood clotting and control bleeding.			Quercetin	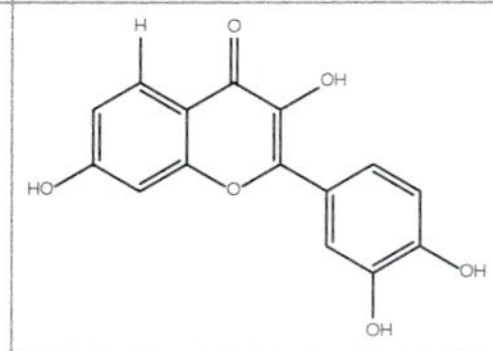	
				Sitosterol	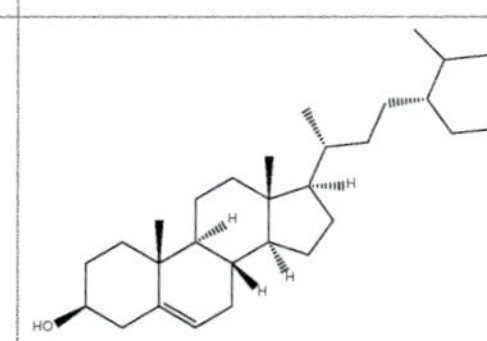	
				Apigenin	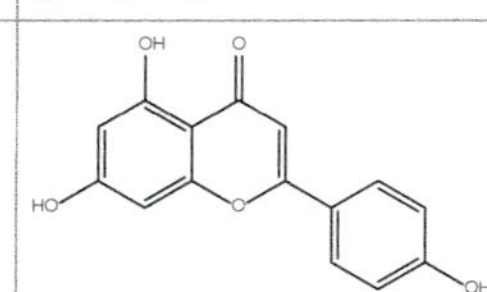	

	Aronia melanocarpa (Michx.) Elliott		Bark	Polyphenol-rich extract	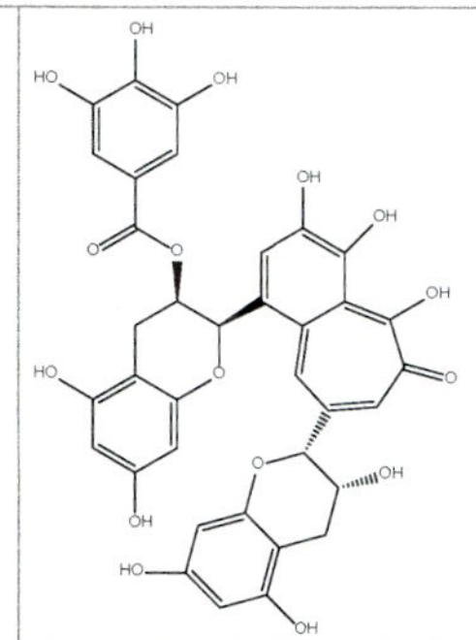	Kedzierska et al. (2009)
	Synonym(s): *Aronia arbutifolia* var. *nigra* (Willd.) F.Seym., *Aronia nigra* (Willd.) Britton. **Common name(s)**: Black chokeberry, Black berried aronia. **Botanical description**: It is a deciduous shrub and is native to east North America, Nova Scotia to Ontario, south to Florida and Michigan. It is in flower from July to August, and the seeds ripen from October to December. The flowers are hermaphrodite (have both male and female organs) and are pollinated by insects. It is suitable to grow in light (sandy), medium (loamy), and heavy (clay) soils and prefers well-drained soil to grow. **Medicinal use(s)**: It is used to treat brain, breast, and pancreatic cancer. It also possesses antitumor, antioxidant, antiaging, and anti-inflammatory properties. It is also used for the treatment of heart disease, diabetes, and high blood pressure.			Beta-carotene	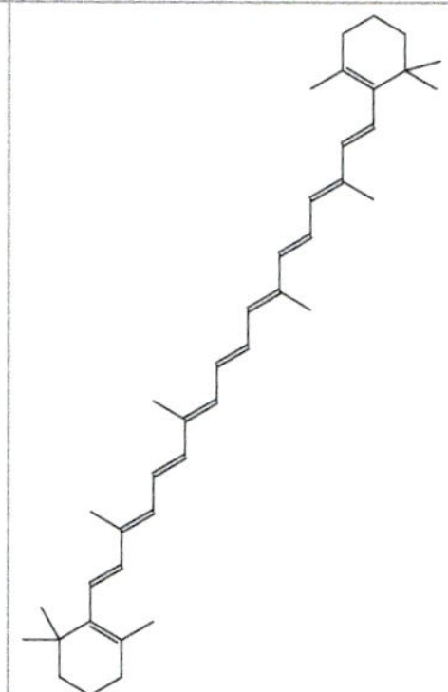	
				Lutein	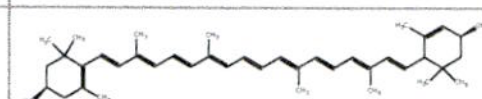	
				Quercetin	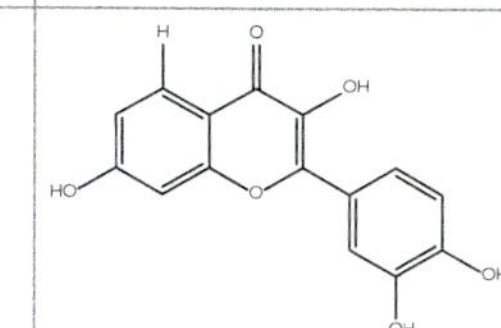	

(continued)

Table 4.118 (continued)

Family	Botanical name	Image	Plant part used	Bioactive compound(s)	Structure	Reference(s)
	Duchesnea indica (Jacks.) Focke		Entire plant	Flavone(s)	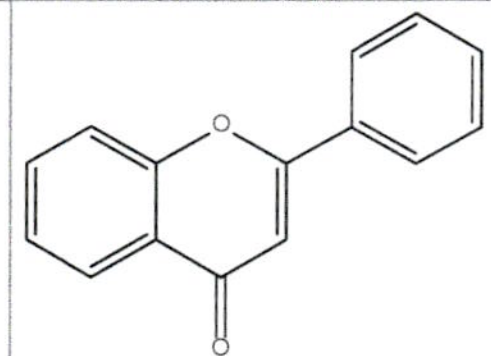	Peng et al. (1995a, b) and Peng et al. (2009)
	Synonym(s): *Duchesnea indica* var. *indica*. **Common name(s)**: Mock strawberry, Indian strawberry. **Botanical description**: It is an evergreen perennial plant and is native to East Asia, China, Japan, and the Himalayas. It is in flower from May to October, and the seeds ripen from July to October. The flowers are hermaphrodite (have both male and female organs) and are pollinated by insects. It is suitable to grow in light (sandy), medium (loamy), and heavy (clay) soils and prefers well-drained soil to grow. **Medicinal use(s)**: It is used to treat esophageal, cervical, ovarian, and skin cancer and various kinds of cancer cell lines. It also possesses antitumor, antioxidant, antimicrobial, antiseptic, and anticoagulant properties. It can be used in decoction, or the fresh leaves can be crushed and applied externally as a poultice. It is used for the treatment of boils and abscesses, weeping eczema, ringworm, stomatitis, laryngitis, acute tonsillitis, snake and insect bites, and traumatic injuries. A decoction of the leaves is used in the treatment of swellings. An infusion of the flowers is used to activate blood circulation.			Gallic acid	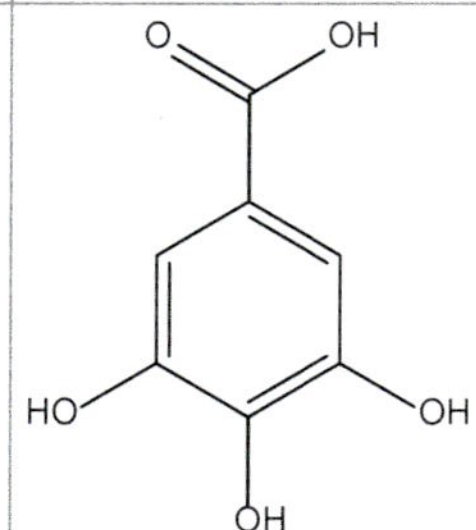	
				D-Xylose	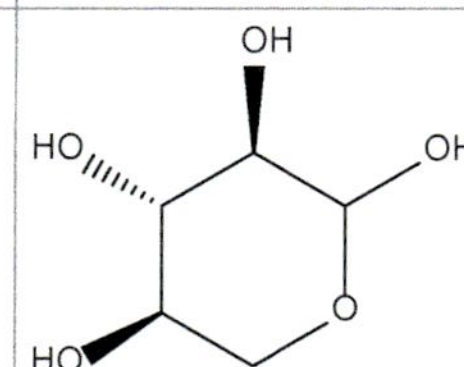	
				Lectin	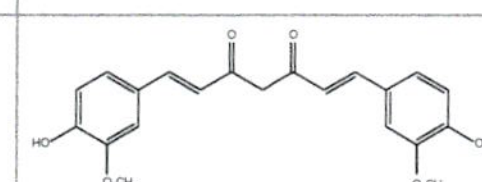	

	Fragaria vesca L.		Leaf, fruit	Ellag:c acid	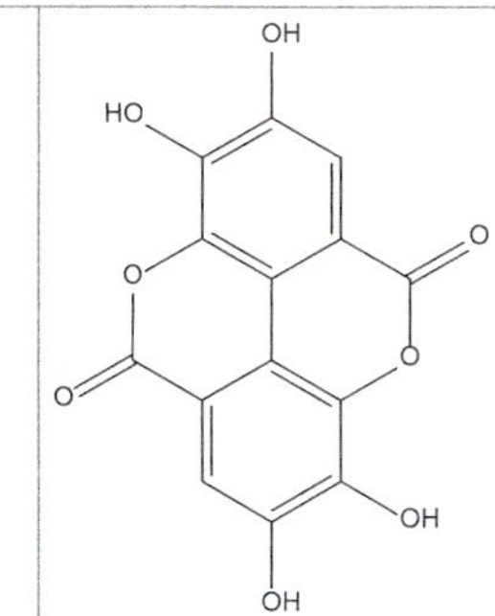	Mohammad (2006), Desai et al. (2008), and Madhuri and Pandey (2009)
	Synonym(s): *Dactylophyllum fragaria* Spenn., *Fragaria abnormis* Tratt., *Fragaria aliena* Weihe, *Fragaria alpina* (Weston) Steud. **Common name(s)**: Wood strawberry, Alpine strawberry, English strawberry. **Botanical description**: It is a perennial plant and is native to Europe, North Asia, Australia, and North America. It has short rhizome and long runners that grow on the soil surface and turn into roots to form new plants. The plants have light green trifoliate leaves with toothed margins that have silky hairs on the underside. The plant grows wild in areas with cool, temperate climate. The plant is in bloom from April to May and the seeds ripen from June to July. **Medicinal use(s)**: It is used to treat anal, breast, brain, liver, ovarian, uterine, colon, and cervical cancer. It also possesses antitumor, antioxidant, antibacterial, anti-rheumatic, antifungal, antiaging, antiviral, and antidysenteric effect. The berries are used as herbal remedy for gout, arthritis, throat inflammation, and kidney disease. They are also used to improve digestion and act as a general tonic. The berries are also used cosmetically in skin care creams to restore and strengthen the skin, reduce wrinkles, bleach freckles, and relieve sunburns.			Ursolic acid	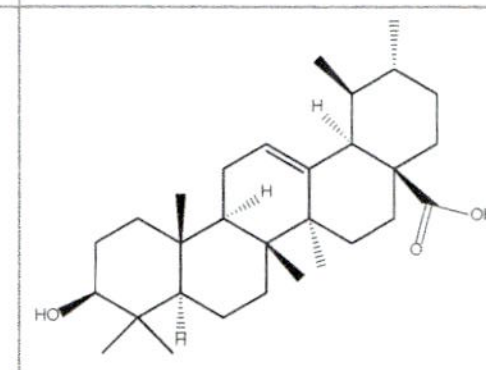	

(continued)

Table 4.118 (continued)

Family	Botanical name	Image	Plant part used	Bioactive compound(s)	Structure	Reference(s)
	Hagenia abyssinica (Bruce) J.F. Gmel.		Flower	Kosotoxin	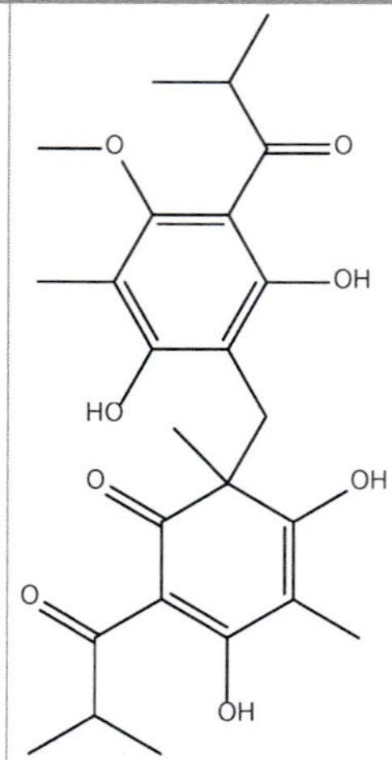	Woldemariam et al. (1992) and Tuasha et al. (2018)
	Synonym(s): *Hagenia abyssinica* (Bruce ex Steud.) J.F.Gmel. **Common name(s)**: Kousso, Hagenia kosso. **Botanical description**: It is native to East Africa, Central African Republic to Ethiopia, south to Tanzania and Zambia. The unbuttressed bole is usually short and crooked with the bark peeling in strips and sometimes very thick on old stems. It is found in the woodland zone just above the mountain at elevations of 2000–3000 m. **Medicinal use(s)**: It is used to treat cancer. It also possesses anticancer, antioxidant, antimicrobial, anti-inflammatory, and antibacterial properties. Roots are used to treat malaria and the bark is used as a remedy for diarrhea and stomachache.			Beta-kosin	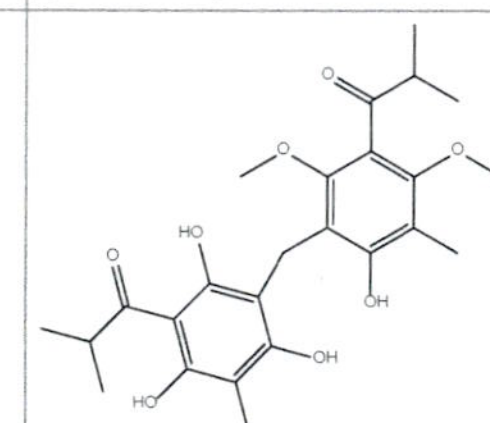	

		Alpha-kosin	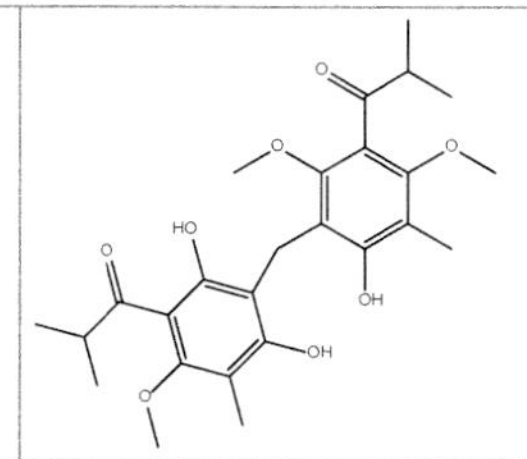	
		Praziquantel	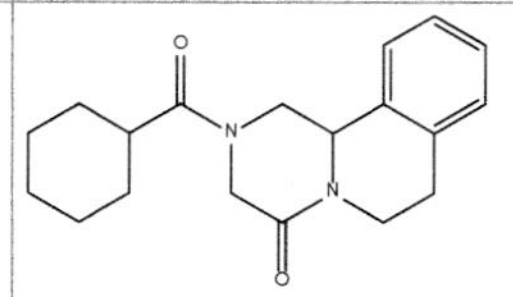	
		Niclosamide	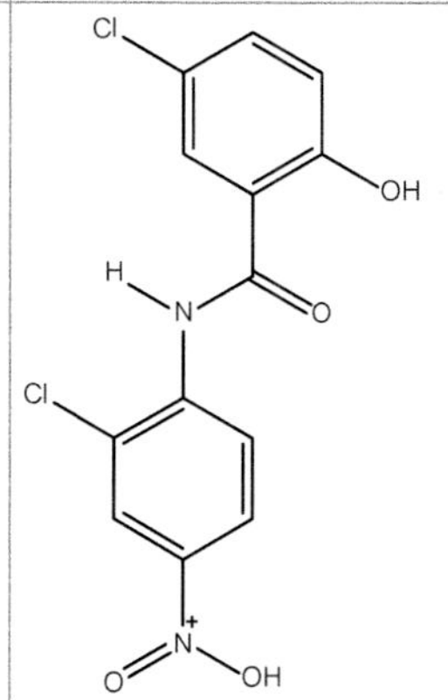	

(continued)

Table 4.118 (continued)

Family	Botanical name	Image	Plant part used	Bioactive compound(s)	Structure	Reference(s)
	Potentilla chinensis Ser.		Whole plant	Gallic acid	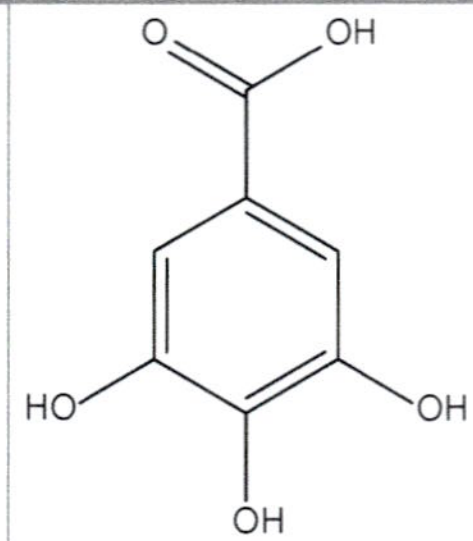	Mohammad (2006), Desai et al. (2008), and Madhuri and Pandey (2009)
	Synonym(s): *Potentilla exaltata* Bunge, *Potentilla chinensis* var. *chinensis*. **Common name(s)**: Chinese cinquefoil. **Botanical description**: It is native to East Asia, China, Japan, Korea, and Manchuria. It is in flower from May to October, and the seeds ripen from June to October. The flowers are hermaphrodite (have both male and female organs) and are pollinated by insects. It is suitable to grow in light (sandy), medium (loamy), and heavy (clay) soils and prefers well-drained soil to grow. **Medicinal use(s)**: It is used to treat osteosarcoma cancer. It is an astringent and is used for the treatment of diarrhea. It is used in Korea to treat fevers and is used as a women's tonic.			Oleanolic acid	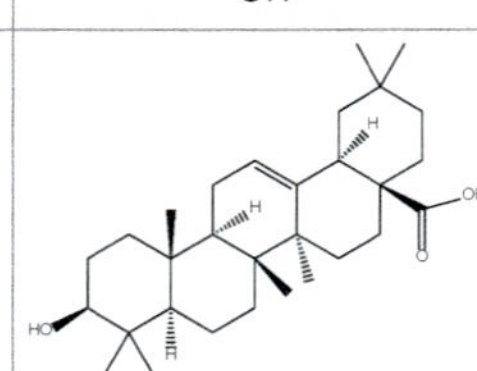	

	Potentilla discolor Bunge (PD)		Leaf, whole plant	Gallic acid	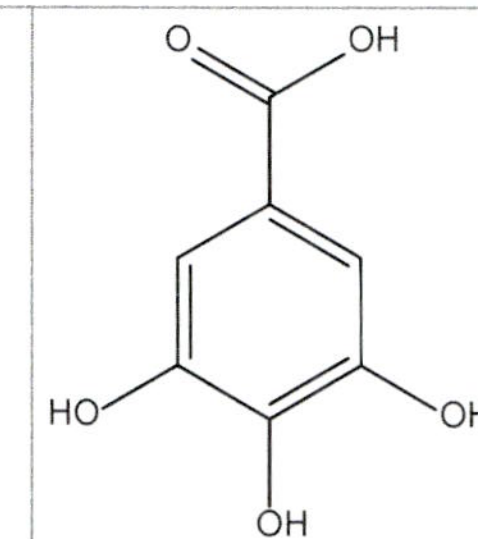	Jin et al. (2011), Xu et al. (2017), and Yu et al. (2018)
	Synonym(s): *Potentilla discolor* var. *formosana* (Hance) Franch., *Potentilla formosana* Hance. **Common name(s)**: Wei Ling Cai, Barren strawberry. **Botanical description**: It is native to East Asia, China, Japan, and Korea. It is in flower from May to September. The flowers are hermaphrodite (have both male and female organs) and are pollinated by insects. It is suitable to grow in light (sandy), medium (loamy), and heavy (clay) soils and prefers well-drained soil to grow. **Medicinal use(s)**: It is used to treat pancreatic B-cell carcinoma. It also possesses antitumor, antioxidant, and anti-inflammatory properties. Decoction is used to treat ulcers in the mouth, for relaxation of the uvula, for spongy gums, for fixing loose teeth, for toothache, and for preserving the gums from scurvy. It is also an excellent remedy for cramps in the stomach, heart, and abdomen.			Ellagic acid	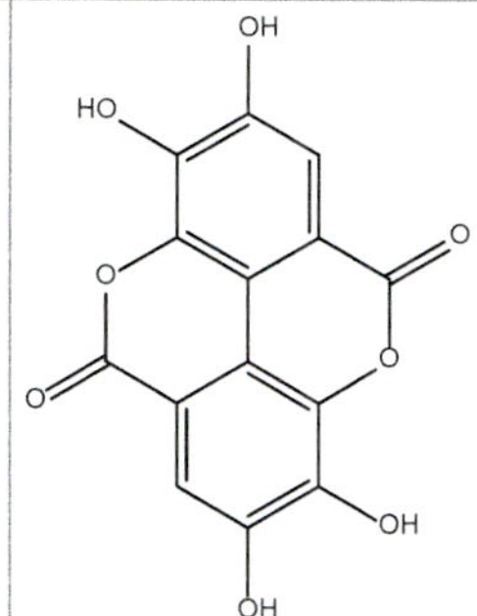	

(continued)

Table 4.118 (continued)

Family	Botanical name	Image	Plant part used	Bioactive compound(s)	Structure	Reference(s)
				Quercetin	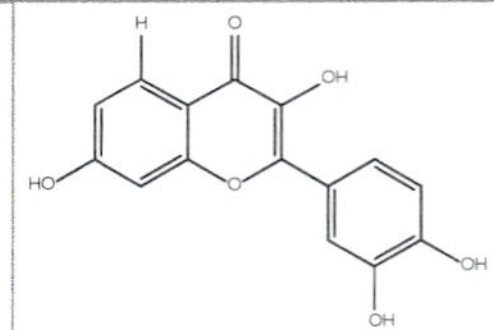	
	Prunus africana (Hook.f.) Kalkman		Aerial parts, bark	Beta-sitosterol	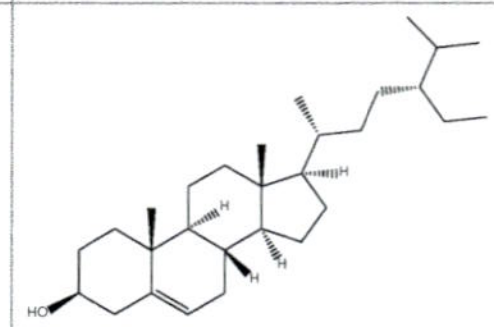	Shenouda et al. (2007), Mohammad (2006), Desai et al. (2008), Madhuri and Pandey (2009), and Komakech et al. (2017)
	Synonym(s): *Pygeum africanum* Hook.f. **Common name(s)**: African plum tree, African prune, Red stinkwood. **Botanical description**: It is found across the continent of Africa in the southern portion. It generally grows at altitudes of 3000 ft. or higher; it is becoming endangered due to the demand of bark. The fruit ripens in July and August. **Medicinal use(s)**: It is used to treat prostate cancer. It possesses antibacterial, antimicrobial, anti-inflammatory, antitumor, and anticancer properties. It is also used for pain caused by inflammation, kidney disease, urinary problems, malaria, stomachache, and fever and to increase sexual desire.					

		Carvacrol	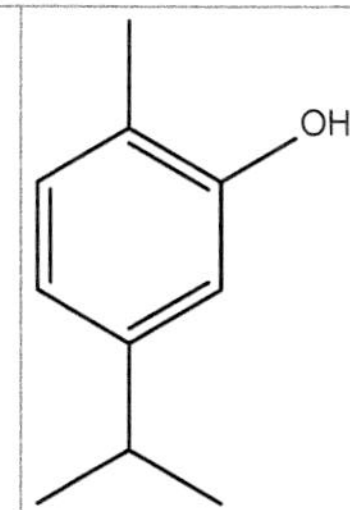	
		Oleanolic acid	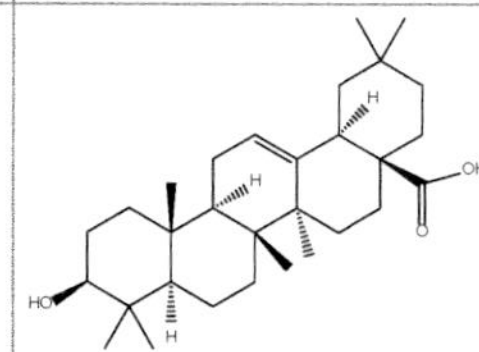	
		Ursolic acid	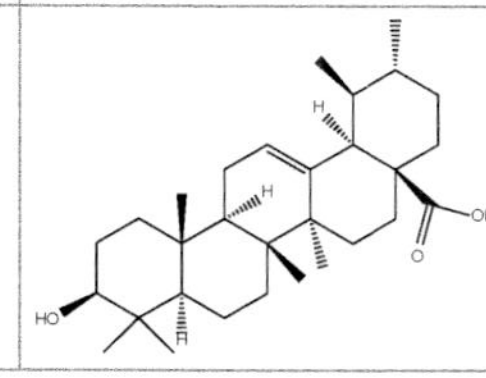	

(continued)

Table 4.118 (continued)

Family	Botanical name	Image	Plant part used	Bioactive compound(s)	Structure	Reference(s)
	Pyrus malus L.		Bark Fruit	Hydrogen Cyanide	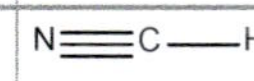	Madhuri and Pandey (2009)
	Synonym(s): *Malus domestica* Borkh. **Common name(s)**: Paradise apple, Common apple, Apple tree. **Botanical description**: It is native to Europe, including Britain, from Scandinavia south and east to Spain, Greece, and Southwest Asia. It is a deciduous tree growing up to 7 m at a medium rate. It is in flower in April. The flowers are hermaphrodite (have both male and female organs) and are pollinated by insects. It is suitable to grow in light (sandy), medium (loamy), and heavy (clay) soils and prefers well-drained soil. **Medicinal use(s)**: It is used to treat skin cancer. It also possesses antibacterial, antifungal, and antiseptic properties. The root bark is used as anthelmintic, refrigerant, and soporific; seed oil is used to treat skin ailments such as acne. It is rich in antioxidants that help to clean skin and soothe acne. It promotes hair growth, provides relief from stomach cramps, lowers blood sugar levels, and prevents diarrhea. It is also used to fight signs of aging, cure acne, and eliminate stretch marks.			Quercetin	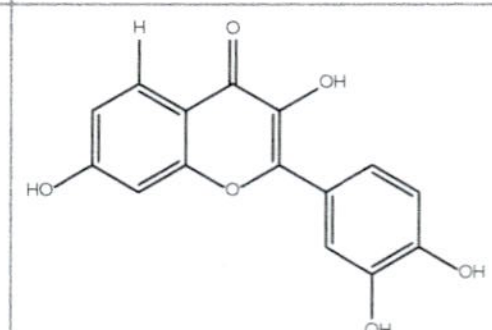	

		Gallic acid	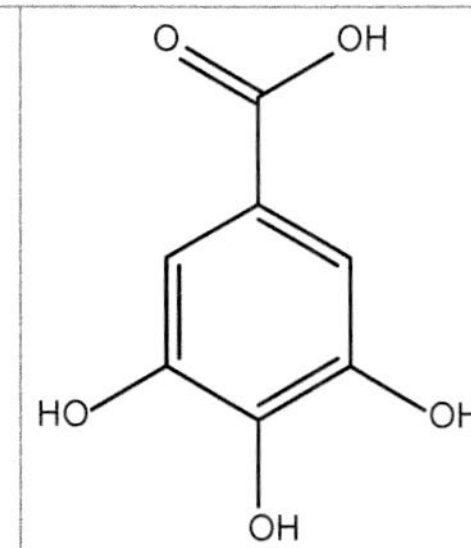	
		Cianidanol	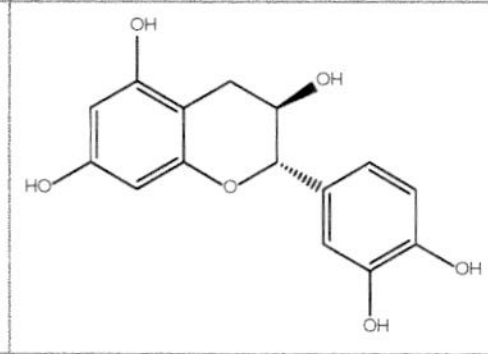	
		Flavone(s)	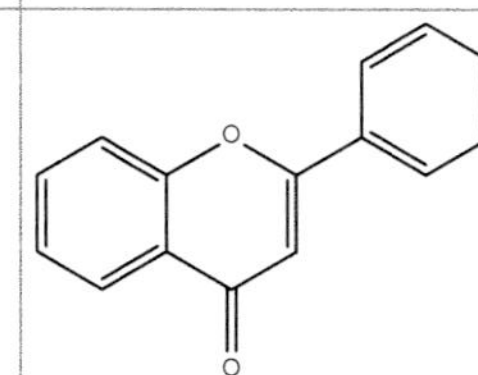	

(continued)

Table 4.118 (continued)

Family	Botanical name	Image	Plant part used	Bioactive compound(s)	Structure	Reference(s)
				Coumaric acid	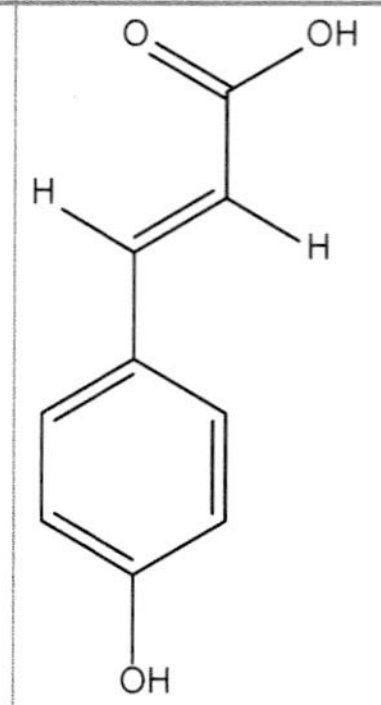	
				Phlorizin	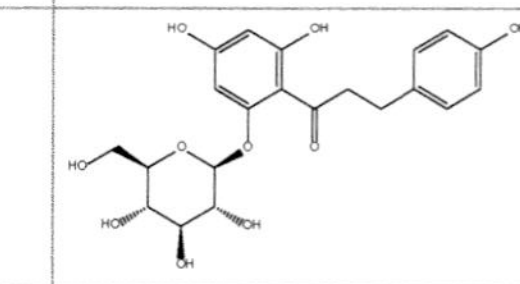	
				Procyanidin	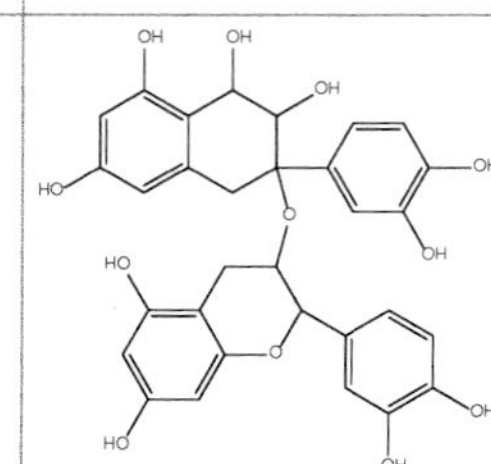	

	Rosa laevigata Michx.		Pericarp	Gallic acid	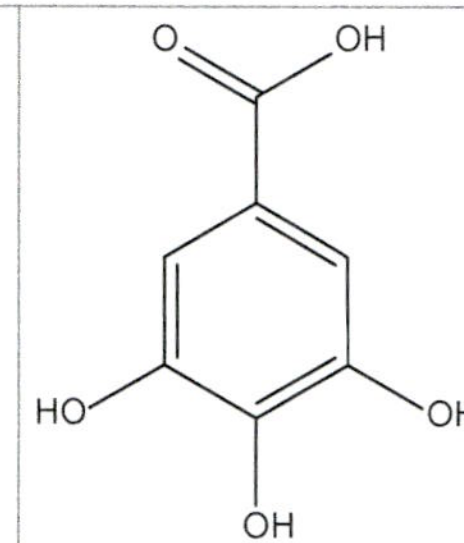	Yoshida et al. (1989) and Mehboob et al. (2017)
	Synonym(s): *Rosa amygdalifolia* Ser., *Rosa argyi* H.Lév., *Rosa* × *argyi* H.L,v. **Common name(s)**: Cherokee rose. **Botanical description**: It is an evergreen shrub and is native to East Asia, Southern China from Sichuan and Hubei to Taiwan. It is in flower from June to July, and the seeds ripen from August to November. The flowers are hermaphrodite (have both male and female organs) and are pollinated by bees. The plant can grow in light (sandy), medium (loamy), and heavy (clay) soils, but prefers well-drained soil. **Medicinal use(s)**: It is used to treat skin cancer. The dried fruits are used as astringent, carminative, diuretic, and stomachic. They are used internally for the treatment of urinary dysfunction, infertility, seminal emissions, urorrhea, leukorrhea, and chronic diarrhea. The root is used for the treatment of uterine prolapse. A decoction of the plant is used for the treatment of chronic dysentery, urinary tract infections, prolapse of the uterus, menstrual irregularities, and traumatic injuries.					

(continued)

Table 4.118 (continued)

Family	Botanical name	Image	Plant part used	Bioactive compound(s)	Structure	Reference(s)
	Rubus idaeus L.		Fruit	Beta-carotene	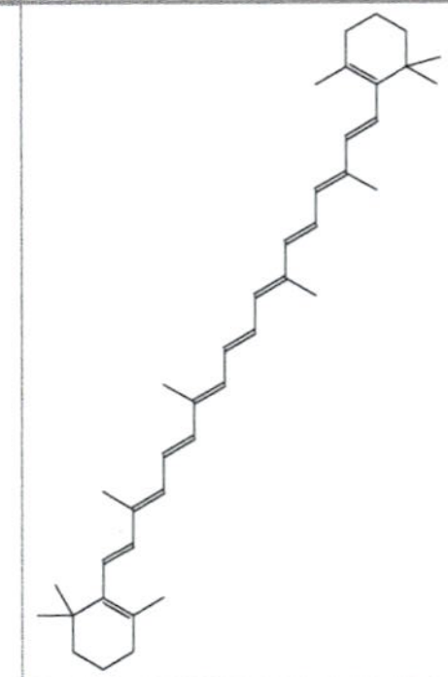	Bowen-Forbes et al. (2010), Chu et al. (2013), and Assad et al. (2015)
	Synonym(s): *Batidaea idea* (L.) Nieuwl., *Batidaea vulgaris* Nieuwl., *Rubus acanthocladus* Borb s. **Common name(s)**: Raspberry. **Botanical description**: It is a deciduous shrub and is native to Europe, Britain, from Iceland south and east to Spain and temperate Asia. It is in flower from June to August, and the seeds ripen from July to September. The flowers are hermaphrodite (have both male and female organs) and are pollinated by bees and flies. It is suitable to grow in light (sandy), medium (loamy), and heavy (clay) soils and prefers well-drained soil. **Medicinal use(s)**: It is used to treat lung cancer. It also possesses antifungal, antibacterial, antiviral, and antimutagenic properties. It strengthens the kidneys. The dried and fresh leaves contain tannin that is known to have astringent properties. The leaves have been used (usually as a tea) to treat diarrhea, heavy menstrual bleeding, menstrual cramps and foul vaginal discharge. They are also used for the treatment of fever, cold, and flu. The leaves can be used externally for treatment of ulcers, inflammation of the mouth and throat, tonsillitis, sore throat, varicose veins, burns, and slow-healing wounds.			Ellagic acid	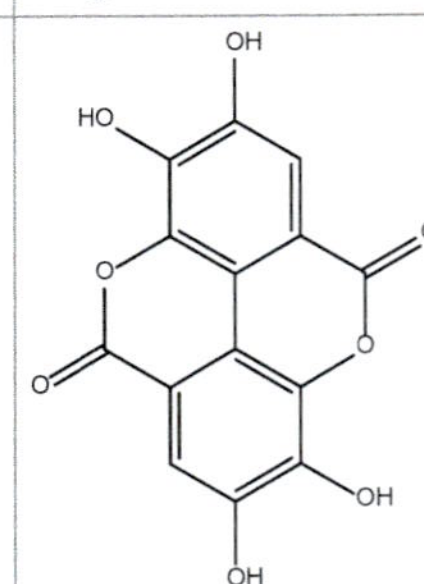	
				Gallic acid	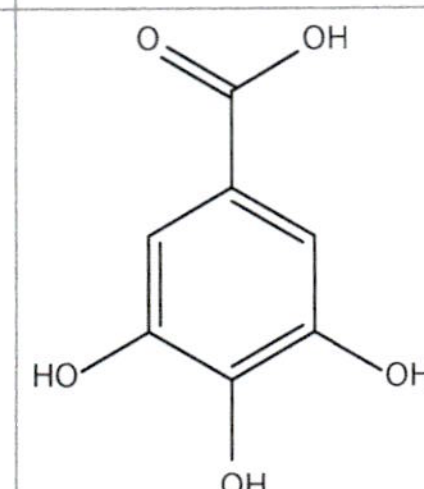	

Table 4.119 Anticancer plants of family Rubiaceae

Family	Botanical name	Image	Plant part used	Bioactive compound(s)	Structure	Reference(s)
Rubiaceae	*Chassalia kolly* (Schumach.) Hepper		Whole plant	Anthraquinone	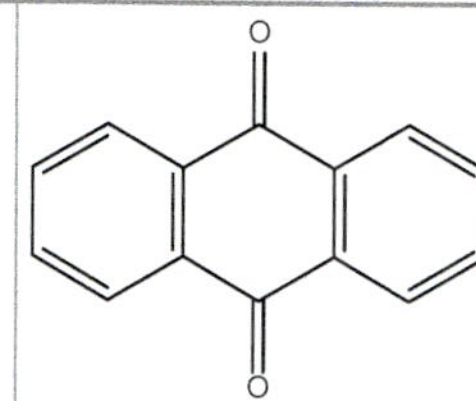	Onocha and Ali (2010)
	Synonym(s): *Chassalia parviflora* Benth., *Psychotria benthamiana* Hiern, *Psychotria kolly* Schumach. **Common name(s)**: Not available. **Botanical description**: It is native to west tropical Africa, Sierra Leone to Cameroon, south to Gabon and the Congo. It is a soft-stemmed shrub growing up to 3 m tall. **Medicinal use(s)**: It is used to treat cancer. It also possesses antifungal and antioxidant properties.			Isoquercetin	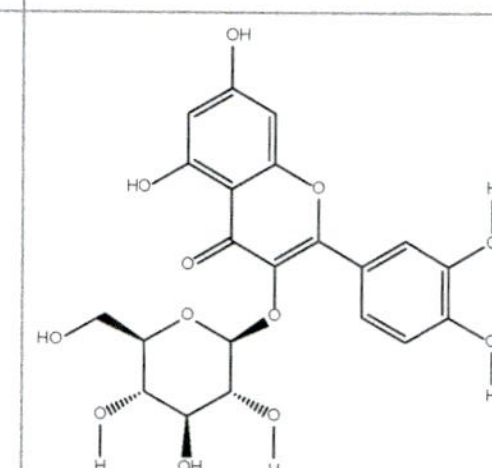	

(continued)

Table 4.119 (continued)

Family	Botanical name	Image	Plant part used	Bioactive compound(s)	Structure	Reference(s)
	Galium aparine L.		Cleaver	Anthraquinone	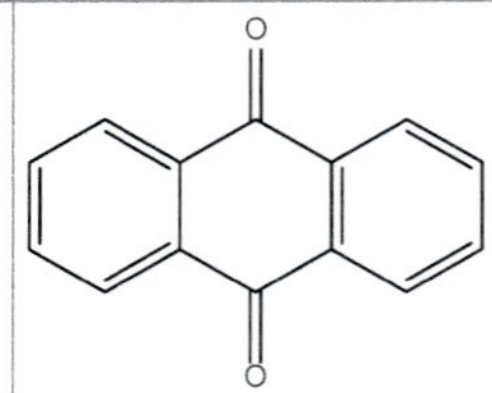	Mohammad (2006), Desai et al. (2008), and Madhuri and Pandey (2009)
	Synonym(s): *Aparine hispida* Moench. **Common name(s)**: Goosegrass, Coachweed, Catchweed, Stickywilly. **Botanical description**: It is native to Europe, including Britain, from Scandinavia south and east to Spain and North and West Asia. It is in flower from June to August, and the seeds ripen from August to September. The flowers are hermaphrodite (have both male and female organs) and are pollinated by flies and beetles. **Medicinal use(s)**: It is used to treat skin cancer, breast cancer, nipple cancer, tongue cancer, mouth cancer, and cancer of reproductive organs. It also possesses antioxidant, antimicrobial, anti-inflammatory, and anti-ulcer properties. It is also used for the treatment of glandular fever, tonsils, hepatitis, cystitis, etc. The plant is often used as part of a spring tonic drink with other herb. A tea made from the plant has traditionally been used internally and externally for the treatment of cancer.			Gallic acid	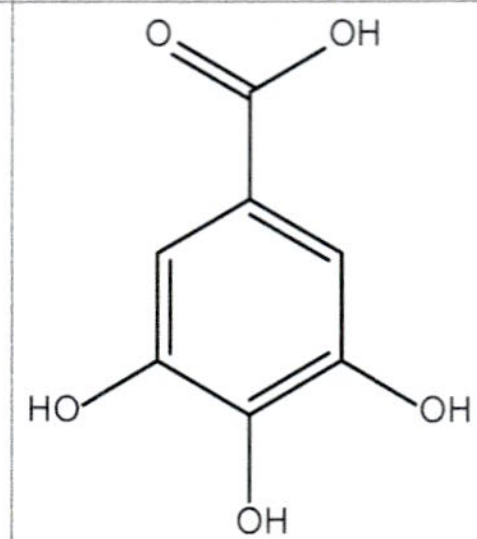	

<table>
<tr><td rowspan="3"></td><td>Gardenia obtusifolia Roxb. ex Hook.f.</td><td></td><td>Leaf</td><td>5,3′-Dihydrcxy-3,6,7,8,4′-pentamethoxyflavonea</td><td>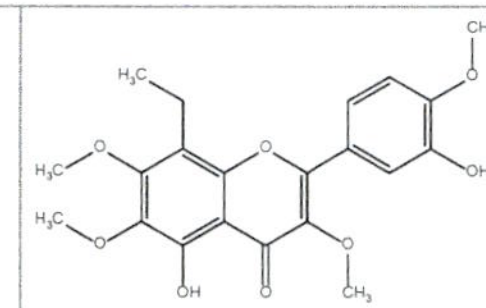</td><td rowspan="3">Phromnoi et al. (2010)</td></tr>
<tr><td colspan="3" rowspan="2">Synonym(s): Gardenia suavis Wall.
Common name(s): Gandhraj, Anant.
Botanical description: It is an evergreen shrub grows up to 12 m and is native to Asia. Most commonly it is found in tropical and subtropical climate. Leaves are 4–6 cm long. Flowers have a strong and sweet fragrance.
Medicinal use(s): It is used to treat breast cancer. It also possesses antitumor, antibacterial, antifungal, antiseptic, anti-spasmodic, and antidysenteric properties. It is also used for the treatment of heartburns and liver problems.</td><td>Beta-sitosterol</td><td>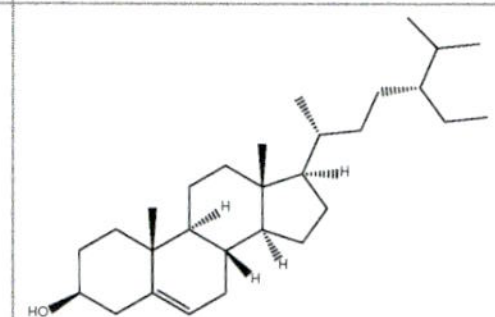</td></tr>
<tr><td>Ursolic acid</td><td>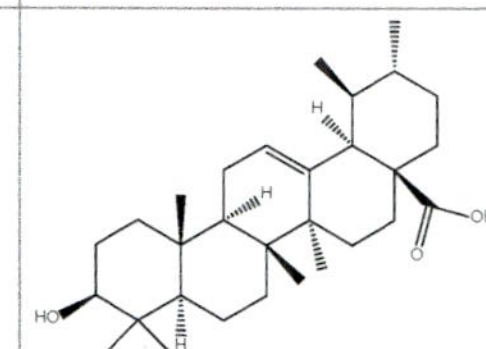</td></tr>
</table>

(continued)

Table 4.119 (continued)

Family	Botanical name	Image	Plant part used	Bioactive compound(s)	Structure	Reference(s)
	Gardenia tubifera Wall. ex Roxb.		Apical part	3,4-Secocycloartanes, gardenoins A-D (1-4)	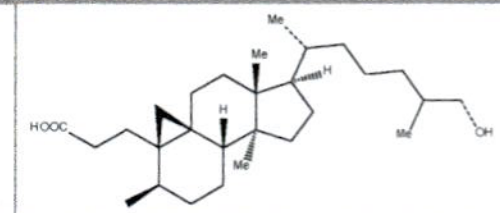	Nuanyai et al. (2010)
				Secaubryenol	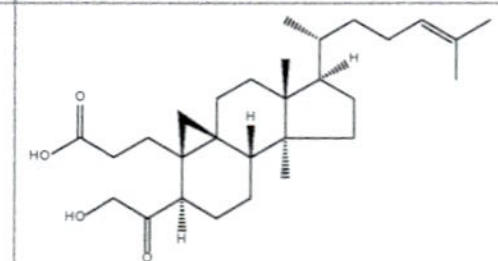	
	Synonym(s): *Gardenia tubifera* var. *tubifera*. **Common name(s)**: Pekanheran, Water gardenia, Chempakautan, Golden pinwheel. **Botanical description**: It is native to Thailand, Peninsular Malaysia, Singapore, Sumatra, and Borneo. It is a perennial plant. It is a shrub or tree up to 25 m tall. It grows in hill and lowland forests and often on banks of rivers. Its opposite, stalked leaves have leaf blades that are elliptic to reverse egg-shaped and 5–26 by 3–12 cm. Its flowers are white to pale yellow, but finally turning to intense orange, and are 0.8–3.5 cm long. Its round fruits are pale apple green, 2.5–5 cm wide, and split when ripe to expose many seeds in the orange to reddish pulp. **Medicinal use(s)**: It is used to treat lung cancer, prostate cancer, colorectal cancer, and malignant melanoma. It is used for the treatment of loss of appetite and anorexia, blood disorders, rhinitis, cough, asthma, sore throat and hiccup, sexual disorders, abdominal pain, worm infestation, pain, fever, inflammation, insomnia, urinary complications, cardiac weakness, etc.			Solasodine	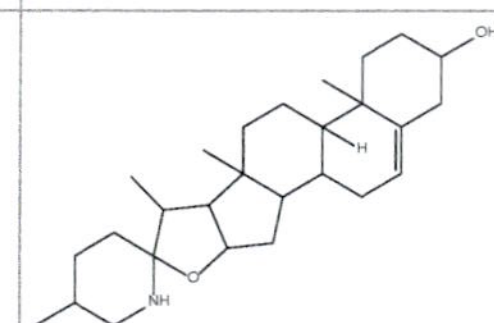	
				Lanosterol	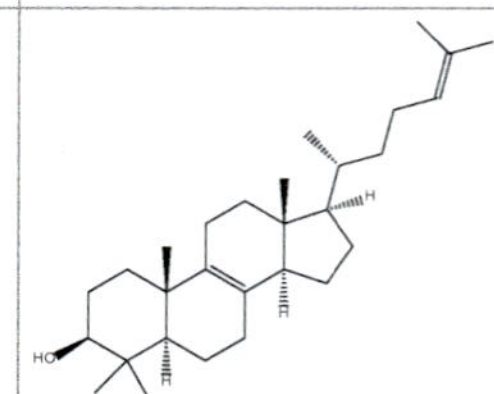	
				Diosgenin	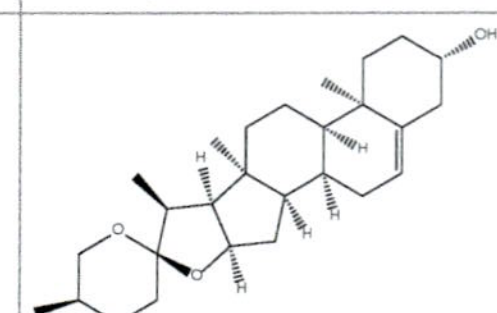	

	Oldenlandia corymbosa L.		Whole part	Ursolic acid	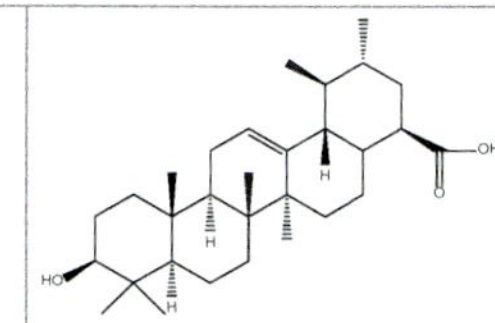	Haryanti et al. (2009)
	Synonym(s): *Gerontogea biflora* Cham. & Schltdl., *Gerontogea corymbosa* (L.) Cham. & Schltdl., *Gerontogea herbacea* Cham. & Schltdl. **Common name(s)**: Flat top mille grains, Old-world diamond flower, Five-leaved fumitory. **Botanical description**: It is widespread through tropical and subtropical Africa, Arabia, subtropical and tropical Asia to New Guinea. It is an erect or prostrate, sparsely branched annual plant with stems up to 40 cm long that sometimes root at the nodes. The leaves are 1.3–2 cm long; the lower leaves are often broader than upper ones, linear, acute, glabrous, usually with recurved margins. Flowers are white in pairs or in threes, usually on solitary axillary peduncles longer than the calyx. Fruits are loculicidal capsules, globose, and the seeds are minute, pale brown, angular. **Medicinal use(s)**: It is used to treat breast cancer. It also possesses antitumor, anti-inflammatory, anti-rheumatic, and antioxidant properties. The entire plant is used in decoction as depurative, diaphoretic, digestive, diuretic, febrifuge, pectoral, and stomachic. It is also used to treat jaundice and other liver disorders. In Chinese medicine, it is used to treat viral infections, cancer, acne, boils, appendicitis, hepatitis, eye problems, and bleeding.			Oleanolic acid	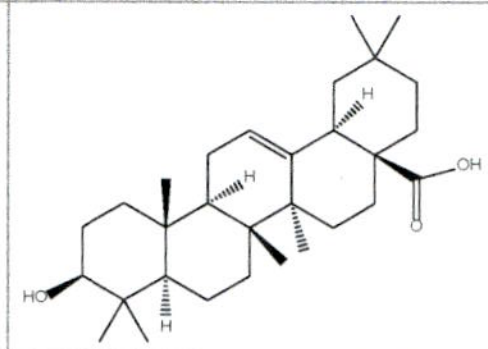	
				Rutin	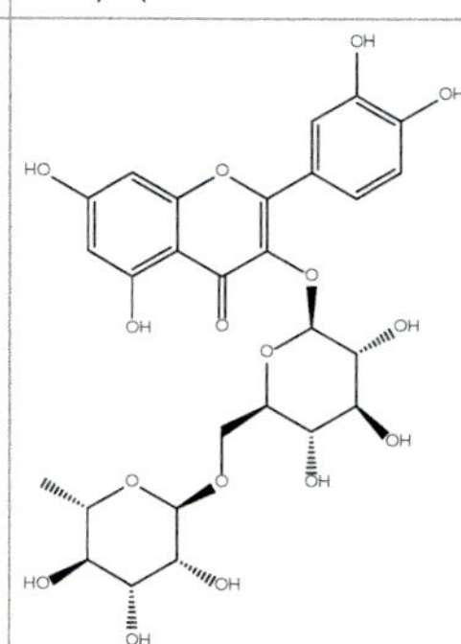	

(continued)

Table 4.119 (continued)

Family	Botanical name	Image	Plant part used	Bioactive compound(s)	Structure	Reference(s)
	Oldenlandia chrysotricha (Palib.) Chun			Ursolic acid	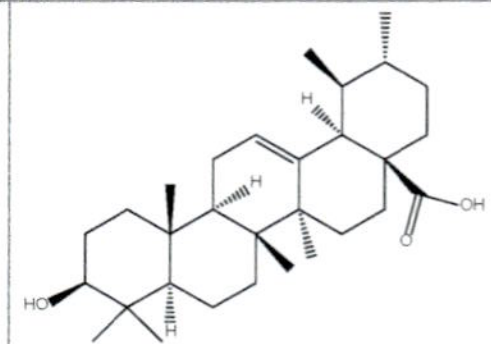	Peng et al. (1995a, b) and Ye et al. (2013)
	Synonym(s): *Anotis chrysotricha* Palib., *Oldenlandia kiusiana* Makino., *Hedyotis chrysotricha* (Palib.) Merr. **Common name(s)**: Bai hua she she cao. **Botanical description**: It is a perennial herb and is native to Asia and Southern China. It can grow in tropical and subtropical region and requires moist soil for proper growth; the leaves are subsessile to petiolate; petiole is 1–3 mm, elliptic or ovate. It has small white color flowers. Anther included 1–1.2 mm length, stigma is exerted and 1–1.2 mm in length. **Medicinal use(s)**: It is used to treat liver carcinoma, tumors of digestive tract, cervical carcinoma, and parenchymal, colorectal, and larynx carcinoma. It also possesses antibacterial, anti-inflammatory, antivirus, and antimicrobial properties.			Beta-carotene	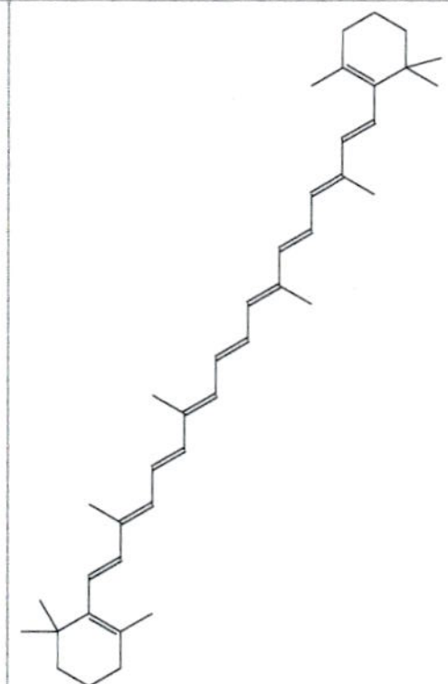	

	Oldenlandia diffusa (Willd.) Roxb		Whole plant	Methylanthraquinone	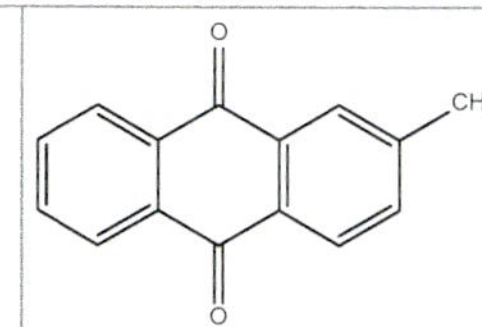	Hsu (1967), Islam et al. (2009), and Liu et al. (2010)
	Synonym(s): *Hedyotis diffusa* Willd. **Common name(s)**: Water hyssop. **Botanical description**: It is a perennial herb and is native to China, Nepal, and Japan. It grows in tropical and subtropical climate, grows up to 30 cm in length. Leaves are narrow, linear are stipules sheathing, flower are axillaries, solitary, or binate. **Medicinal use(s)**: It is used to treat lung cancer, cervical cancer, stomach cancer, ovarian cancer, colon cancer, and breast cancer. It also possesses anti-inflammatory, antibacterial, anti-ulcer, antidote, and antiviral properties. It is also used to make antidotes for snake venom and to treat sore throat, blood impurity, tonsillitis, and pneumonia.			Beta-sitosterol	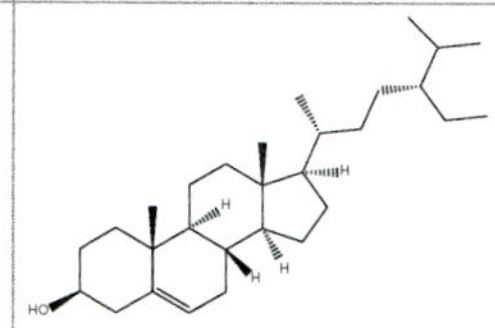	

(continued)

Table 4.119 (continued)

Family	Botanical name	Image	Plant part used	Bioactive compound(s)	Structure	Reference(s)
	Morinda citrifolia L.		Fruit, root	Gallic acid	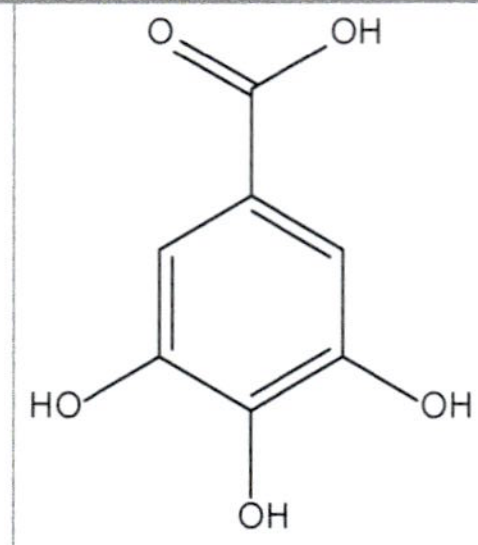	Holdsworth (1991) and Brown (2012)
	Synonym(s): *Belicea hoffimannioides* Lundell, *Morinda aspera* Wight &Arn., *Morinda asperula* Standl. **Common name(s)**: Indian mulberry, Brimstone tree. **Botanical description**: It is native to Asia, Australia, and the islands of Polynesia. It is a large shrub to medium tree varying from 3 to 12 m high. It has oval-shaped leaves of about 300 mm long by 150 mm wide. The white flowers occur in the leaf axils in clusters and occur mainly in summer and autumn. **Medicinal use(s)**: It is used to treat lung cancer and blood cancer. It also possesses antitumor, antioxidant, anti-inflammatory, antidiabetic, antimicrobial, and antiaging properties. The fruit juice is used for arthritis, diabetes, high blood pressure, muscle aches and pains, menstrual difficulties, headaches, heart disease, AIDS, cancers, gastric ulcers, sprains, depression, senility, poor digestion, atherosclerosis, circulation problems, and drug addiction.			Beta-sitosterol	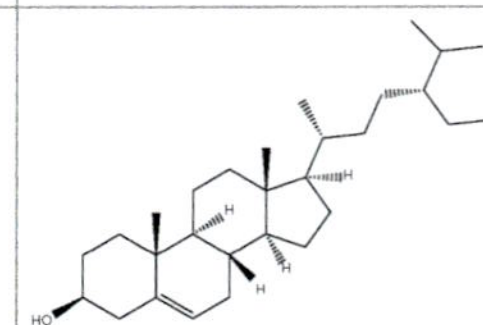	

Nauclea diderrichii (De Wild.) Merr.		Leaf	Quercetin	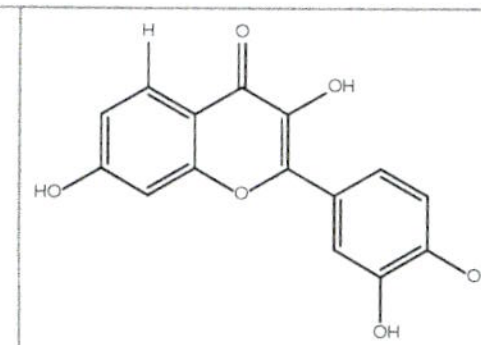	Vasileva (1969) and Ogbole et al. (2017)
Synonym(s): *Nauclea trillesii* (Pierre ex De Wild.) Merr., *Sarcocephalus diderrichii* De Wild. **Common name(s)**: Badi. **Botanical description**: It is native to western tropical Africa – Sierra Leone east to Central African Republic, south to Gabon. It is a very large evergreen tree with a dense, broad, spherical crown, usually growing 40 m tall, exceptionally to 50 m. The straight, cylindrical bole can be unbranched for up to 27 m. The tree is harvested from the wild for local use as a medicine and source of wood. The wood cultivated as a shade tree and nurse tree for more valuable timber species. **Medicinal use(s)**: It is used to cure cancer. It also possesses the properties of antitumor, antioxidant, and anti-inflammatory. A decoction of the bark is used for the treatment of anemia, stomachache, and indigestion. The bark is used in an infusion to treat gonorrhea and as part of an infusion for treating jaundice, piles, dyspepsia, rheumatism, and dysentery.			Rutin	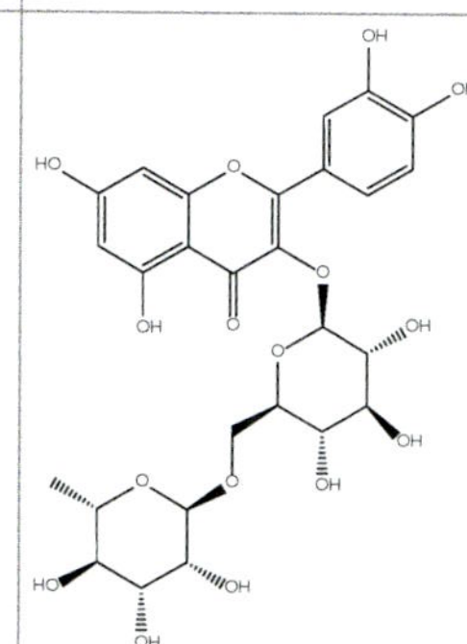	
			Apigenin	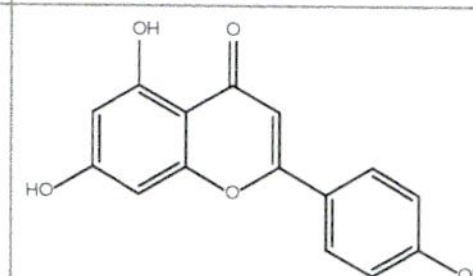	

(continued)

Table 4.119 (continued)

Family	Botanical name	Image	Plant part used	Bioactive compound(s)	Structure	Reference(s)
	Oldenlandia affinis (Roem. And Schult.) DC.		Entire plant	Rutin	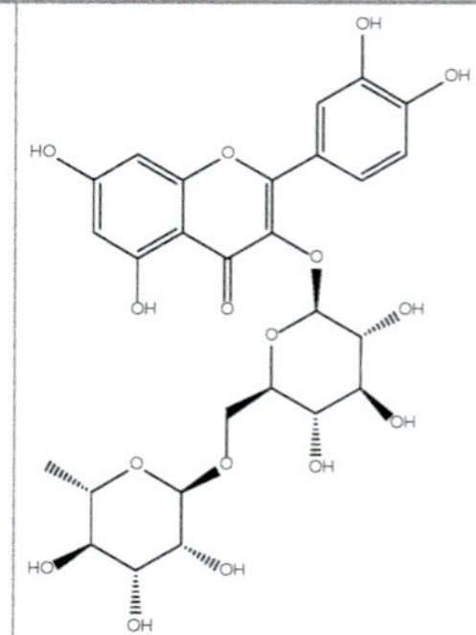	Hsu (1967) and Hu et al. (2015)
	Synonym(s): *Hedyotis affinis* Roem. & Schult., *Oldenlandia linearifolia* Wight &Arn. **Common name(s)**: Diamond flower, Daman pappar. **Botanical description**: It is native to Africa – Liberia to Sudan, south to South Africa, Madagascar, East Asia, India, Myanmar, Thailand, Vietnam, and Malaysia. It is a slender, erect or prostrate, much-branched annual or perennial herb that can grow up to 50 cm long, exceptionally to 120 cm. It is found in dry, often sandy locations, usually near the sea. In India it is commonly found on the slopes of hills, at elevations up to 1200 m. **Medicinal use(s)**: It is used to treat lung cancer, intestinal cancer, and breast cancer. It also possesses antitumor, antimicrobial, anti-inflammatory, antimalarial, and antioxidant effects. It is used to treat fever, indigestion, dehydration, malaria, and various types of skin problems.			Oleanolic acid	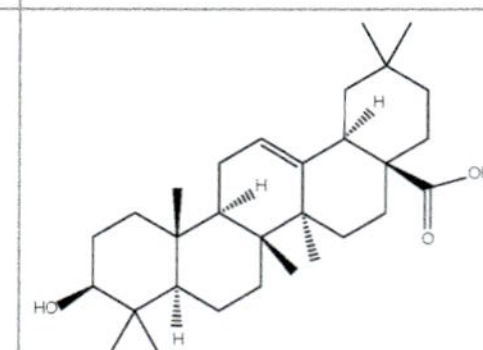	

				Sitosterol	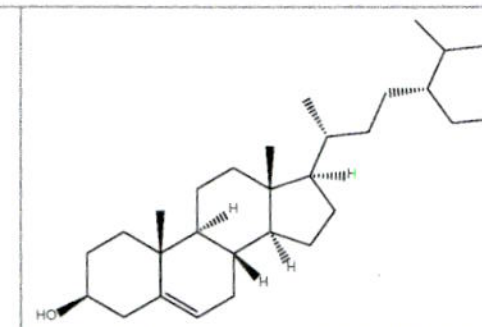	
	Ophiorrhiza mungos L.		Root and stem	CPT/10-methoxycamptothecin	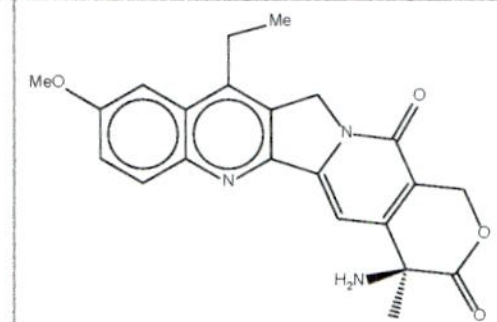	Beegum et al. (2007)
	Synonym(s): *Ophiorrhiza ostindica* Christm. **Common name(s)**: Indian snake root. **Botanical description**: It is native to East Asia – Southwest China, India, Sri Lanka, Bangladesh, Myanmar, Thailand, Vietnam, Malaysia, and Indonesia. It is an erect, herbaceous, evergreen perennial plant with stems that can become more or less woody and persist; it usually grows 10–50 cm tall but can reach 100 cm. It is an annual erect herb, rooting at lower nodes. Leaves are elliptic or elliptic-lance shaped, base narrowed, tip tapering, papery, hairy on veins below; leaf stalk up to 1.5 cm; stipules subulate, 3–5 mm long, 2-fid at tip. Flowers are borne in dense, branched, scorpioid cymes, at branch ends. **Medicinal use(s)**: It is used to treat lung cancer. It also possesses antiviral, antioxidant, and antitumor properties. A decoction of the leaves, roots, and bark is made into a bitter tonic and stomachic. The roots are used against snakebite, and the root bark is having sedative and laxative properties.			Camptothecin	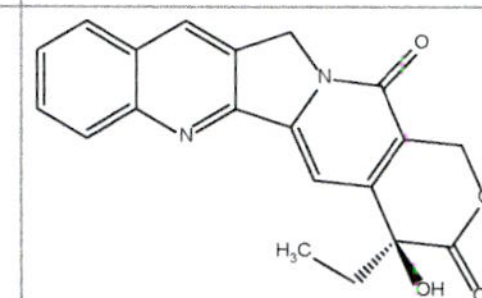	

(continued)

Table 4.119 (continued)

Family	Botanical name	Image	Plant part used	Bioactive compound(s)	Structure	Reference(s)
	Ophiorrhiza rugosa var. *prostrata* (D.Don) Deb &Mondal	 Ophiorrhiza rugosa var. prostrata (D.Don) Photo Courtesy : S.A.Punekar. Deb & Mondal	Leaf and stem	Camptothecin	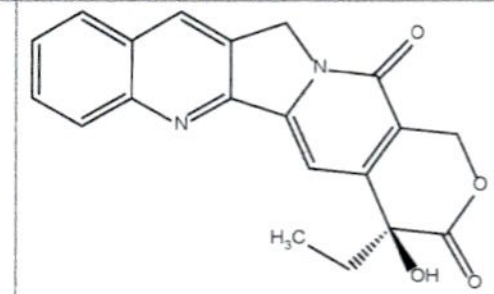	Beegum et al. (2007)
	Synonym(s): *Ophiorrhiza harrisiana* var. *condorensis* Pit., *Ophiorrhiza harrisonii* G. Don, *Ophiorrhiza prostrate* D. Don. **Common name(s)**: Bell weed. **Botanical description**: It is native to East Asia – Southwest China, India, Sri Lanka, Bangladesh, Myanmar, Thailand, Vietnam, Malaysia, and Indonesia. It is an erect, herbaceous, evergreen perennial plant with stems that can become more or less woody and persist; it usually grows 10–50 cm tall but can reach 100 cm. **Medicinal use(s)**: It is used to treat cervical cancer, lung cancer, ovarian cancer, and colon cancer, and it possesses anticancer, antitumor, anti-inflammatory, antiviral, antimicrobial, and antibacterial properties. A decoction of the leaves, roots, and bark make an agreeable, bitter tonic and stomachic. The plant is used as a poultice for keeping the skin moist and soft. The bark of the root is said to be laxative and sedative.			Quercetin	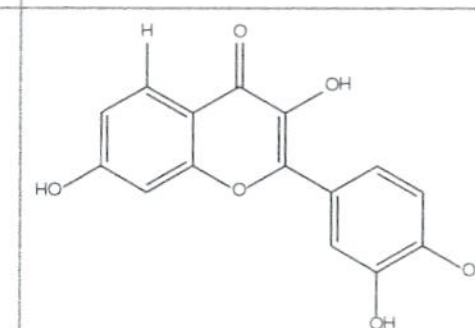	

	Rubia argyi (H.Lév. &Vaniot) Hara ex Lauener		Root	Anthraquinones	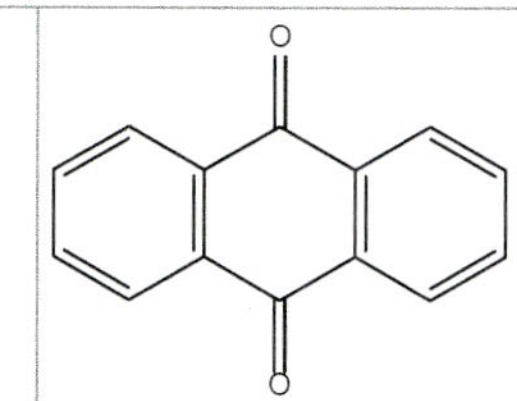	Park et al. (2009)
	Synonym(s): *Galium argyi* H. Lév. &Vaniot, *Rubia akane* Nakai. **Common name(s)**: Asian madder. **Botanical description**: It is native to East Asia, China, Japan, and Korea. It is a perennial climber growing to 1 m. The flowers are hermaphrodite (have both male and female organs). It is suitable to grow in light (sandy), medium (loamy), and heavy (clay) soils. **Medicinal use(s)**: It is used to treat lung cancer. It also possesses anticancer, antioxidant, antibacterial, antimicrobial, antitussive, anti-inflammatory, antiseptic, and anti-allergic properties. The roots are used internally for the treatment of abnormal uterine bleeding; internal and external hemorrhage; bronchitis; rheumatism; stones in the kidney, bladder, and gall; and dysentery. They are used as a poison antidote and to treat mouth sores and intestinal problems such as diarrhea.			Ursolic acid	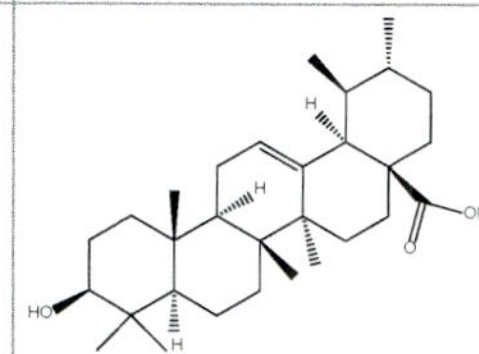	
				Rutin	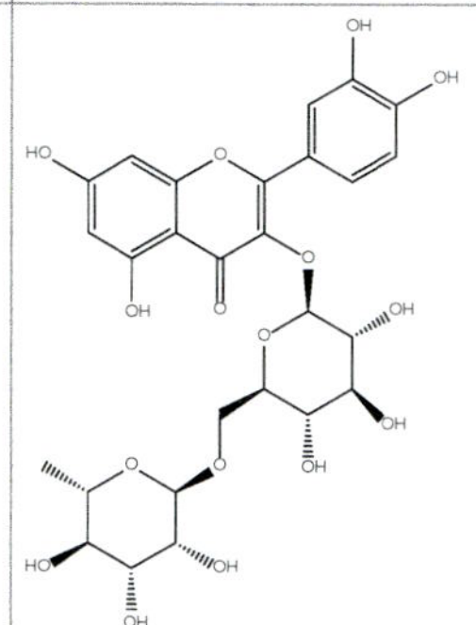	
				Alizarin	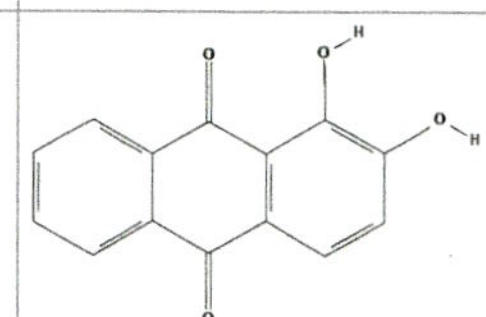	

(continued)

Table 4.119 (continued)

Family	Botanical name	Image	Plant part used	Bioactive compound(s)	Structure	Reference(s)
				Purpurin	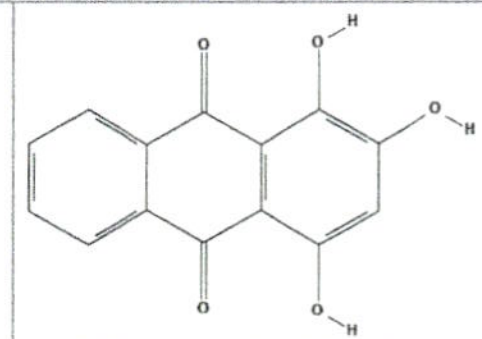	
	Rubus gothicus Frid. & Gelert ex E.H.L.Krause		Fruit	Apigenin	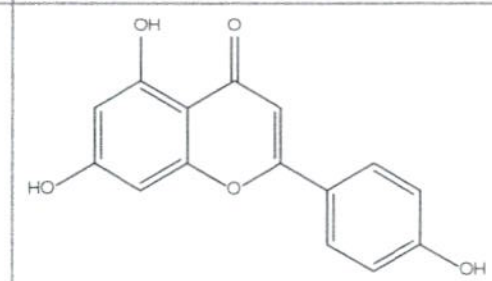	Bowen-Forbes et al. (2010)
	Synonym(s): *Rubus acuminatus* Lindeb., *Rubus acuminatus* (Lindblom) Gelert. **Common name(s)**: Cane fruit. **Botanical description**: It is native to Africa – Northern Nigeria to Ethiopia, South through Eastern Africa to South Africa. It is a very variable, scrambling shrub producing a cluster of stems from a woody rootstock. The plant can grow up to 3 m long. The stems are sparsely to moderately prickly, with prickles 3–5 mm long. Flowers are fragrant, minute, whitish, or greenish yellow. Fruit is minute, glabrous, 1–2 seeded, dark purplish or blackish when mature. **Medicinal use(s)**: It is used to treat cancer. It also possesses antitumor, anti-inflammatory, antibacterial, antidiabetic, and antiproliferative properties. A cooled and filtered decoction of raspberry leaves are used as eye drops for the treatment for conjunctivitis.			Beta-carotene	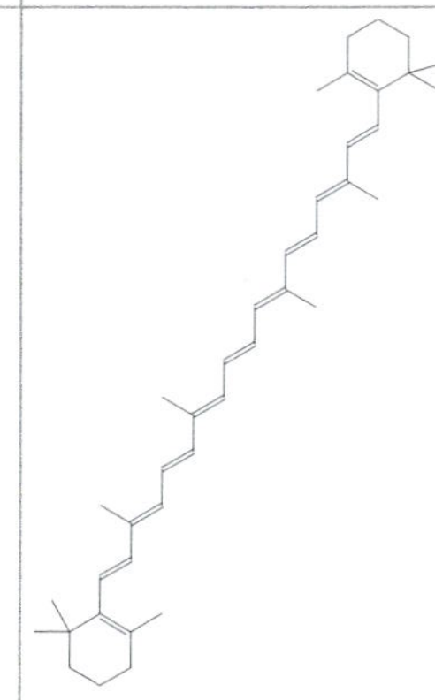	

	Rubia discolor Turcz.		Root	Apigenin	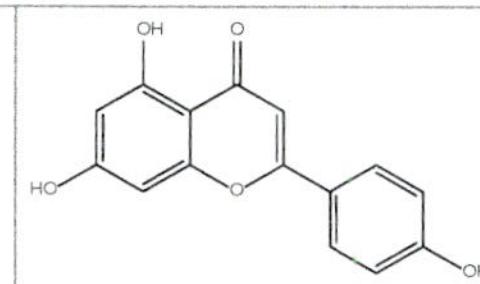	Kloos (1977), Kloos et al. (1978), and Maffei (2003a, b, c, d, e, f, g, h, i, j, k, l, m)
	Synonym(s): *Rubia cordifolia* var. *discolor* (Turcz.) K. Schum. **Common name(s)**: Indian madder. **Botanical description**: It is native to East Asia, China, Japan, and Korea. It is a perennial climber growing to 1 m tall. The flowers are hermaphrodite (have both male and female organs). It is suitable to grow in light (sandy), medium (loamy), and heavy (clay) soils. **Medicinal use(s)**: It is used to treat lung, larynx, breast, colon, cervical, and liver carcinoma. It also possesses antitumor, antimicrobial, antibacterial, anti-inflammatory, antioxidant, antiviral, anti-ulcer, anti-peroxidative, antiplatelet, and anti-arthritic properties. The roots are used internally for the treatment of abnormal uterine bleeding, internal and external hemorrhage, bronchitis, rheumatism stones in the kidney, bladder, and gall, and dysentery. The leaves are burnt, and the ash is applied externally to treat mastitis and itchy skin.			Beta-sitosterol	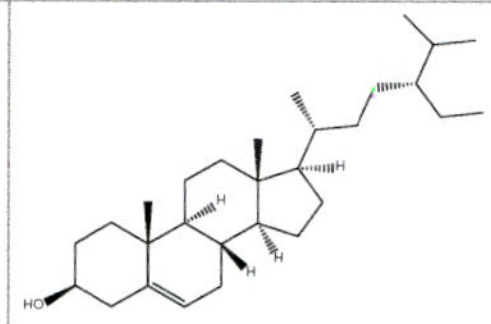	

(continued)

Table 4.119 (continued)

Family	Botanical name	Image	Plant part used	Bioactive compound(s)	Structure	Reference(s)
	Uncaria tomentosa (Willd. ex Roem. and Schult. DC.)		Root, bark, and stem	Ursolic acid	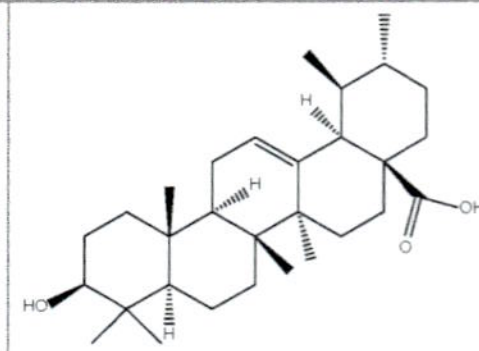	Stuppner et al. (1992, 1993), Wirth and Wagner (1997), and Sandoval et al. (2002a)
	Synonym(s): *Cinchona globifera* Pav. ex DC., *Nauclea aculeate* Kunth, *Nauclea polycephala* A. Rich. ex DC. **Common name(s)**: Cat's claw. **Botanical description**: It is native to South America, Brazil, Bolivia, Peru, Ecuador, Colombia, Venezuela, and the Guianas; the Caribbean, Trinidad; and Central America, Panama to Guatemala. It is a vigorous, evergreen, spiny climbing shrub scrambling into other plants and attaching itself by means of its stout, recurved spines. The stem can be up to 30 m long and 8–25 cm in diameter near the base. **Medicinal use(s)**: It is used to treat blood cancer, breast cancer, lung cancer, kidney cancer, and brain tumor. It also possesses anti-inflammatory, antitumor, anti-ulcer, and antibacterial properties. It is used as a wash for wounds in order to promote healing, to cleanse the skin of impurities. It is also used for treating chronic fatigue syndrome (CFS), Alzheimer's disease, asthma, hay fever, and various digestive system disorders including swelling and pain (inflammation) of the large intestine (diverticulitis), inflammation of the lower bowel (colitis), inflammation of the lining of the stomach (gastritis), stomach ulcers, hemorrhoids, and leaky bowel syndrome.			Oleanolic acid	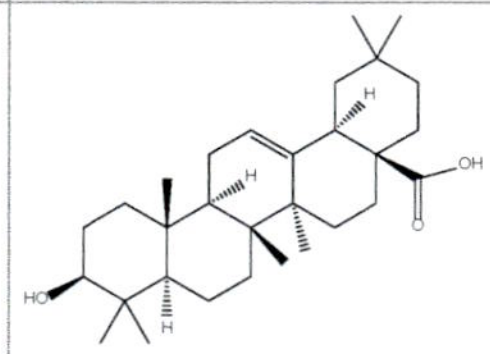	
				Quercetin	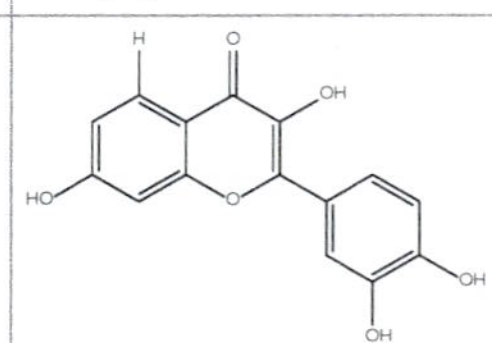	

Table 4.120 Anticancer plants of family Rutaceae

Family	Botanical name	Image	Plant part used	Bioactive compound(s)	Structure	Reference(s)
Rutaceae	*Acronychia pedunculata* (L.) Miq.		Fruit, leaf, root, Stem	1-[2,4-Dihydroxy-6-methoxy-3,5-bis (3-methylbut-2-en-1 yl) phenyl] ethanone	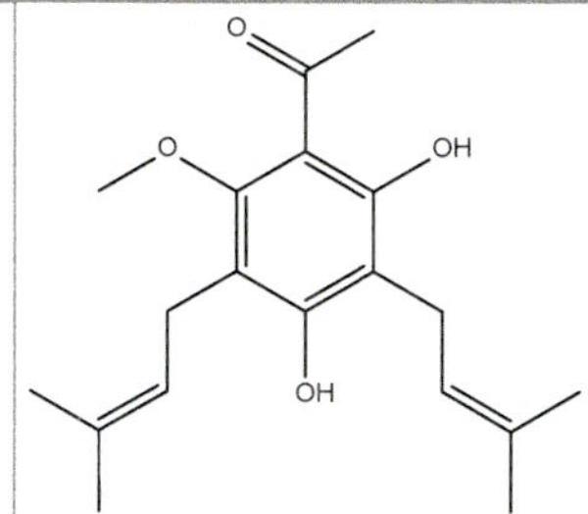	Pathmasiri et al. (2005)
				Acrovestenol	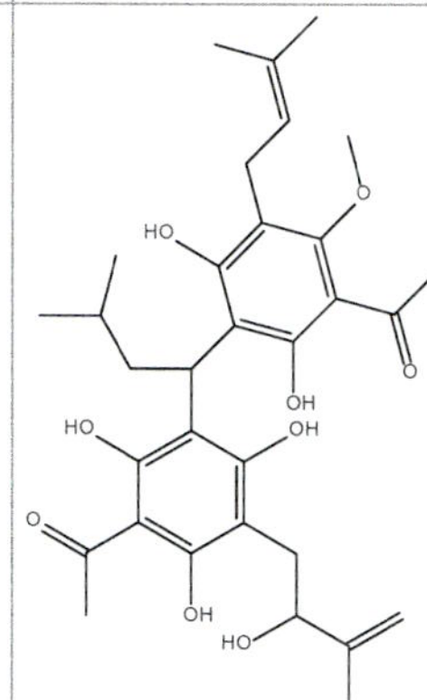	

(continued)

Table 4.120 (continued)

Family	Botanical name	Image	Plant part used	Bioactive compound(s)	Structure	Reference(s)
	Synonym(s): *Acronychia apiculata* Miq., *Acronychia arborea* Blume, *Acronychia barberi* Gamble. **Common name(s)**: Indian aspen, Chakkimaram. **Botanical description**: It is an evergreen plant that can range in size from a shrub to a large tree that can be up to 28 m tall and is native to East Asia, Southern China, Indian subcontinent, Myanmar, Thailand, Cambodia, Laos, Vietnam, Malaysia, Indonesia, the Philippines, and New Guinea. Leaves are elliptic to sub-long, often with tapered base. Twigs more or less angular, hairless. Flowers are greenish white, borne in axillary corymbose panicles, about 14 mm across in inflorescences of 4–24 cm wide. The fruits are cream to brownish yellow drupes. **Medicinal use(s)**: It is used to treat lung cancer, breast cancer, renal cancer, uterine cancer, ovarian cancer, and metastatic tumor in primary stage. It also possesses antitumor, anti-ulcer, anti-inflammatory, antiproliferative, antioxidant, and antimicrobial properties. A decoction of the roots, barks, and leaves is used to treat scabies, sores, ulcers, and a variety of stomach diseases.					
	Maclurodendron porteri (Hook. fil.) T. G. Hartley		Whole plant	5,3′-Dihydroxy-3,6,7,8,4′-pentamethoxyflavone	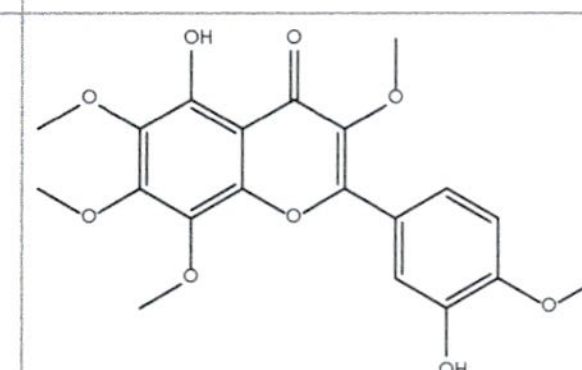	Lichius et al. (1994) and Kour (2014)

	Synonym(s): *Acronychia porteri* Hook. fil., *Jambolifera porteri* (Hook. fil.) Kuntze, *Melicope helferi* Hook. fil. **Common name(s)**: Tanaka. **Botanical description**: *Maclurodendron porteri* grows naturally in Burma, Sumatra, Thailand, Malaysia, Borneo, Singapore and the Philippines. It is an evergreen tree growing up to 25 m tall, though more commonly around 10 m, with a trunk diameter of up to 40 cm. It is a rainforest tree of small to medium height with a dense crown that bears useful and tasty fruits. It bears attractive and large glossy leaves. It flowers in summer and autumn, followed by fruiting. **Medicinal use(s)**: It is used to treat lung cancer, prostate cancer, breast cancer, and tumors. It also possesses antimicrobial, antiproliferative, and antibacterial properties.			Kaempferol	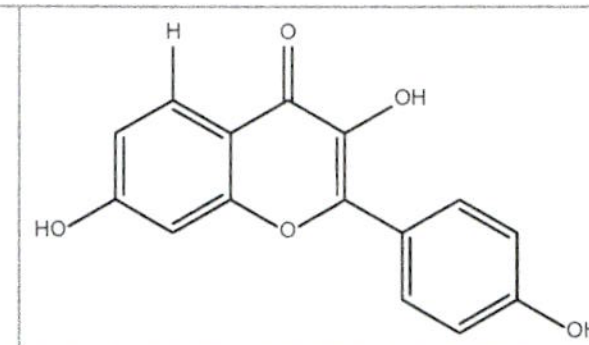	
				Anthraquninone	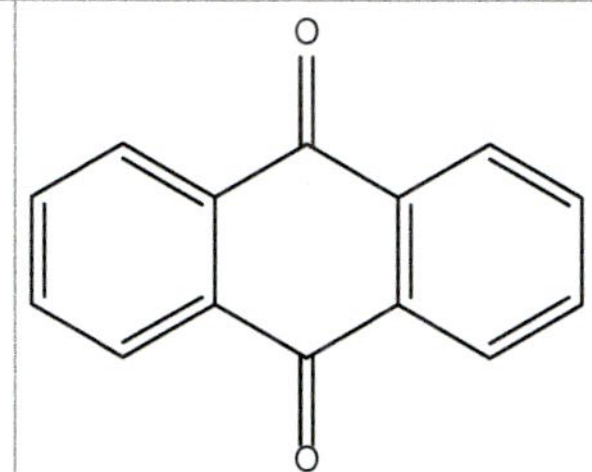	
	Aegle marmelos (L.) Correa		Fruit	Lupeol	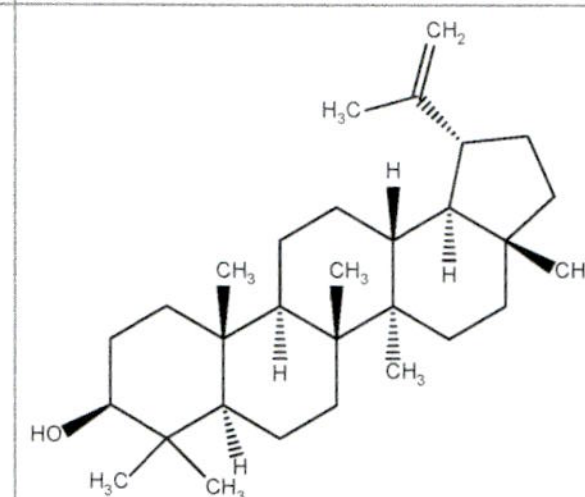	Kaur et al. (2009), Baliga et al. (2012), and Rahman and Parvin (2014)

(continued)

Table 4.120 (continued)

Family	Botanical name	Image	Plant part used	Bioactive compound(s)	Structure	Reference(s)
	Synonym(s): *Belou marmelos* (L.) Lyons, *Bilacus marmelos* (L.) Kuntze. **Common name(s)**: Bael fruit. **Botanical description**: It is a slow-growing deciduous shrub or tree that can reach an eventual height of 10–15 m and is native to East Asia, India, Nepal, Myanmar, and the Andaman and Nicobar Islands. The bole, which is usually fluted at the base, can be up to 50 cm in diameter. The tree is armed with 1–2 cm long spines on older branches and also produces very spiny basal suckers. **Medicinal use(s)**: It is used to treat breast cancer. It also possesses antimicrobial, hypoglycemic, astringent, antidiarrheal, antidysenteric, demulcent, analgesic, anti-inflammatory, antipyretic, wound-healing, insecticidal, and gastro-protective properties. The fruits are used for the treatment of diarrhea, dysentery, stomachache, and cardiac ailments. The leaves are astringent and are used for the treatment of peptic ulcers.			Eugenol	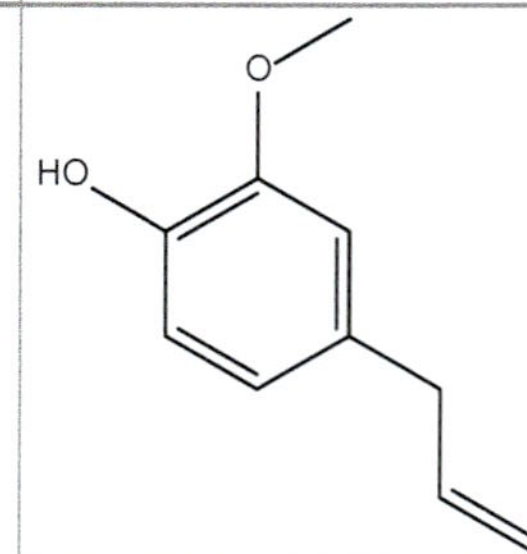	
				Eucalyptol	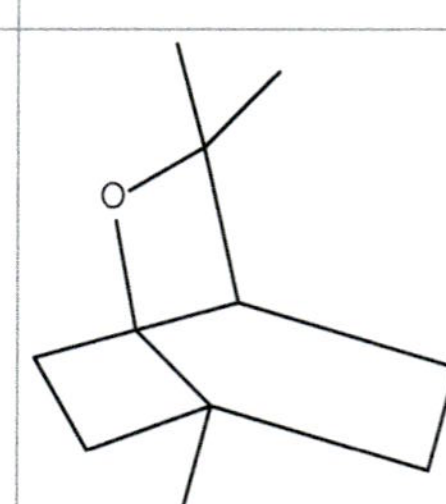	
	Clausena lansium (Lour.) Skeels		Peel	Carbazole	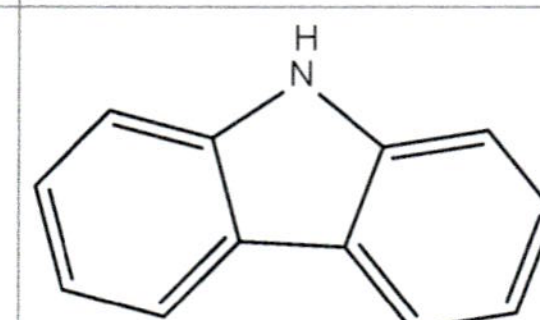	Prasad et al. (2010)

	Synonym(s): *Aulacia punctata* Raeusch., *Clausena wampi* (Blanco) Oliv. **Common name(s)**: Fool's curry leaf, Wampi. **Botanical description**: It is an attractive, fairly fast-growing evergreen shrub or small tree with a dense crown, reaching a height of 3–12 m, and is native to East Asia, China, and Vietnam. The short bole can be up to 40 cm in diameter. The fruits are pale yellow or brownish yellow which are available during May to July. The bark is rough and gray brown in color. The flowers are pentamerous and sweet-scented which are available in the late March. It prefers subtropical to tropical climate and grows well in rich, well-drained, and loamy soils. **Medicinal use(s)**: It is used to treat lung cancer, hepatocellular liver carcinoma, and gastric carcinoma. It also possesses anticancer, antioxidant, antimicrobial, and anti-inflammatory properties. The leaves, fruits, or seeds are all widely used in China and Vietnam for the treatment of gastrointestinal problems, including acute and chronic inflammation and ulcers. The fruit is said to have stomachic and cooling effects and act as a vermifuge.					
	Tetradium ruticarpum (A.Juss.) T.G.Hartley		Flower	Rutaecarpine	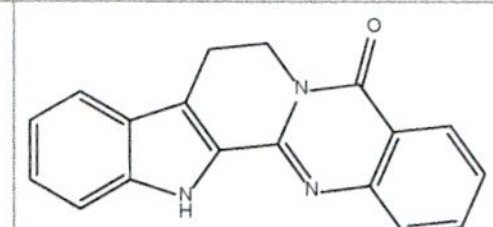	Lee et al. (2008)
				Pyridol	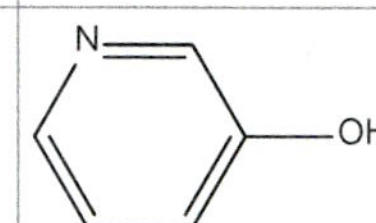	

(continued)

Table 4.120 (continued)

Family	Botanical name	Image	Plant part used	Bioactive compound(s)	Structure	Reference(s)
	Synonym(s): *Ampacus ruticarpa* (A.Juss.) Kuntze, *Cyclocarpus japonicus* Jungh., *Euodia bodinieri* Dode. **Common name(s)**: Evodia fruit. **Botanical description**: It is a shrub or small tree which can grow up to 4 m tall and is native to Australasia, New Guinea, and Southwest Pacific Islands. The bole is short, about 5 cm in diameter. It has dark purple twigs, and buds are covered with grayish-yellow or rust-colored hairs or sparse short pubescence. They commonly grow in the sunny slope of mountain woodland or bushes at an altitude of 1500 m. **Medicinal use(s)**: It is used to treat lung cancer and tumors. It also possesses anti-inflammatory, antitumor, antidiarrheal, anti-ulcer, and antiviral properties. The leaves are emmenagogue, febrifuge, and laxative. They are chewed as a remedy for toothache or stomach pain. An infusion is used to reduce fever. The bark is used to relieve thrushlike conditions, to retard menstruation, and to relieve pain during childbirth. Fluid from the bark is used to treat yellow eyes and yellow urine. The bark may be chewed with betel nuts and rubbed onto aching body parts.			Berberine	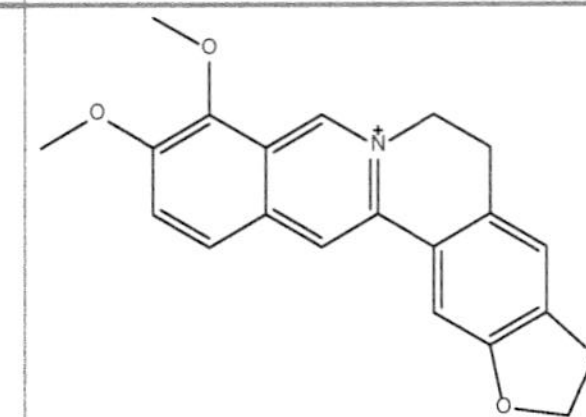	
				Aspartic acid	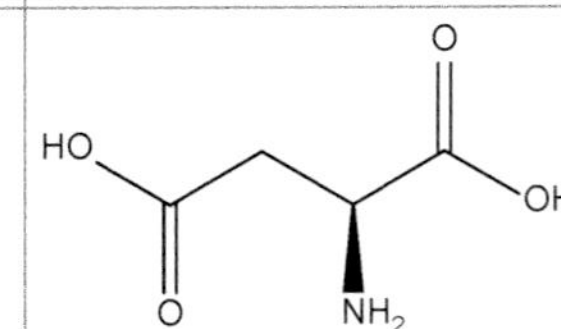	
				Tryptophan	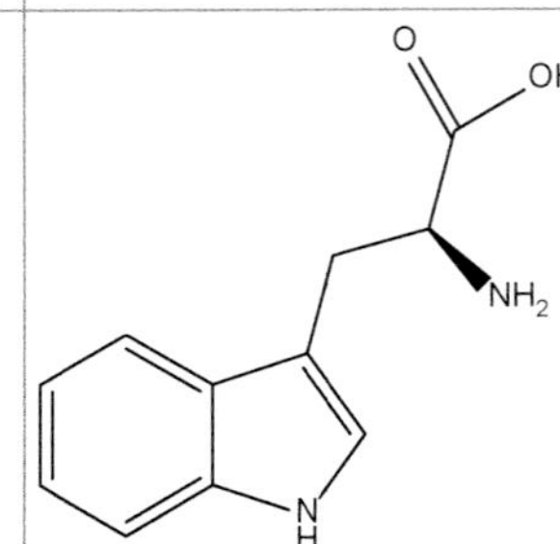	

	Zanthoxylum heitzii (Aubrév. & Pellegr.) P.G.Waterman		Stem and bark	Heitziamide A	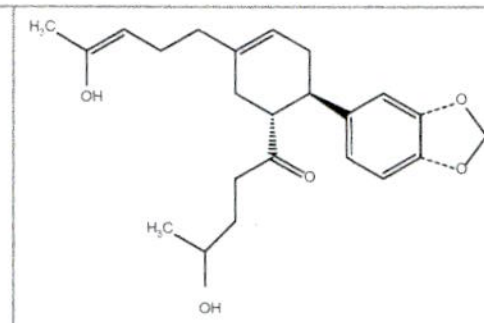	Mbaze et al. (2005)
	Synonym(s): *Fagara brieyi* Vermoesen ex Gilbert, *Fagara heitzii* Aubrév. & Pellegr. **Common name(s)**: Olon, Bongo. **Botanical description**: It is native to west tropical Africa, Cameroon to the Central African Republic, south to Gabon and the Western DR of the Congo. It is a spiny tree with a large crown of ascending branches, growing up to 35 m tall. The straight, cylindrical bole can be free of branches for up to 20 m and up to 150 cm in diameter. The bole is usually free of buttresses, though it is often slightly thickened at the base, and it is armed with spiny bosses 8–10 cm long, though these usually disappear in old trees. It is found in the evergreen and semi-deciduous forests at elevations up to 1200 m occurring most commonly in secondary forest. **Medicinal use(s)**: It is used to treat cervical, prostate, lung, and breast cancer and leukemia. It also possesses antitumor, anti-inflammatory, antifungal, antioxidant, antimicrobial, and anti-pathogenic properties. It is used to treat gonorrhea, abscesses, painful joints, and male sexual impotence. Scrapings from the stem bark are applied externally to treat malaria, rheumatism, and stiffness and to soothe toothache. A maceration of young twigs in lemon juice is used to treat heart complaints.			Heitziamide B	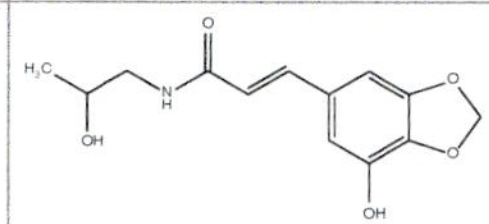	
				Phenylethanoids	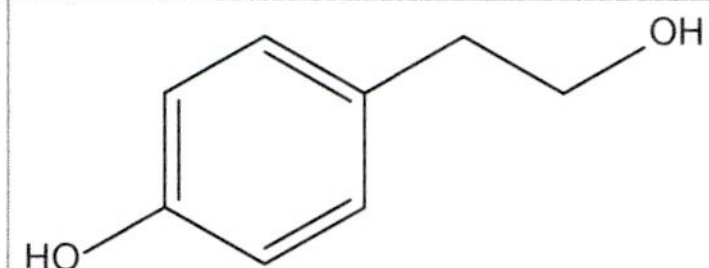	
				Heitziethanoid A	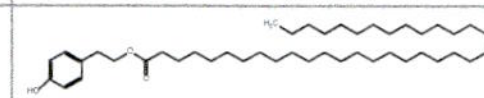	

(continued)

Table 4.120 (continued)

Family	Botanical name	Image	Plant part used	Bioactive compound(s)	Structure	Reference(s)
	Zanthoxylum zanthoxyloides (Lam.) Zepern. & Timler		Root	Apigenin	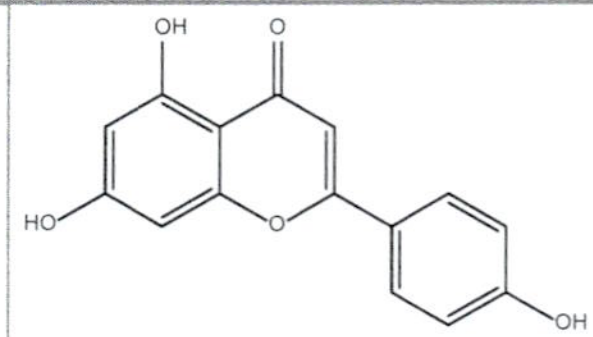	Ogwal-Okeng et al. (2003)
	Synonym(s): *Fagara senegalensis* (DC.) A.Chev., *Fagara zanthoxyloides* Lam. **Common name(s)**: Candle wood. **Botanical description(s)**: It is native to west tropical Africa, Senegal to Nigeria. It is a prickly shrub or low-branching shrub. The bole is armed with large, woody thorns that fall as the tree grows older. It is found in the savannah, thickets, dry and transitional forests, forest patches on dry ground, secondary dry forests, coastal dunes and thickets, and termite mounds and locally abundant in coastal areas. **Medicinal use(s)**: It is used to treat various types of cancer cell lines. It also possesses anticancer, antitumor, antifungal, antioxidant, antimalarial, and anti-inflammatory properties. Young shoots and twigs are used as chew sticks to clean the teeth and maintain oral hygiene. The wood is durable and termite-proof. The root bark extract is used for the treatment of elephantiasis, toothache, sexual impotence, gonorrhea, malaria, dysmenorrhea, and abdominal pain.			Burkinabin C	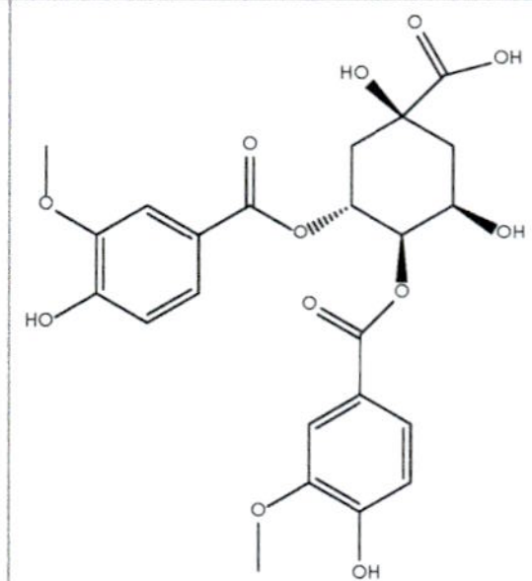	

	Zanthoxylum americanum (Mill)		Leaf	Pyranocoumarins	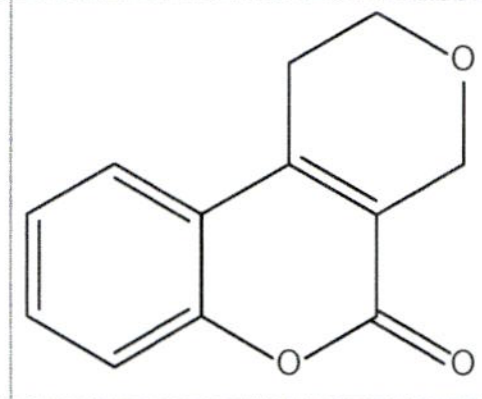	Ju et al. (2001)
				Alloxanthoxyletin	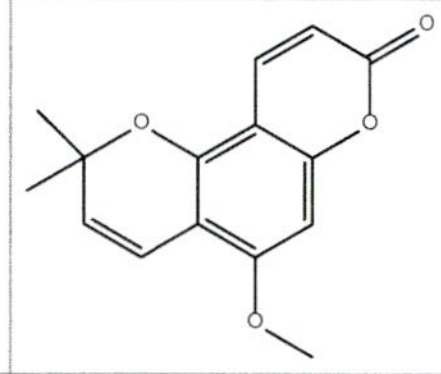	
	Synonym(s): *Thylax fraxineum* (Willd.) Raf., *Zanthoxylum americanum* f. *americanum*, *Zanthoxylum fraxineum* Willd. **Common name(s)**: Prickly ash-northern, Common pricklyash, Northern prickly ash. **Botanical description**: It is native to eastern North America – Quebec to Florida, west to Minnesota and Oklahoma. It is a deciduous shrub growing up to 4 m at a medium rate. It is in flower from May to June, and the seeds ripen from September to October. The flowers are dioecious (individual flowers are either male or female, but only one sex is to be found on any one plant, so both male and female plants must be grown if seed is required). It is suitable to grow on light (sandy), medium (loamy), and heavy (clay) soils and prefers well-drained soil. **Medicinal use(s)**: It is used to treat cervical, bone, lung, and prostate cancer. It also possesses anticancer, antitumor, antioxidant, antifungal, anti-inflammatory, and antimicrobial properties. It is used for the treatment of asthma, bronchitis, indigestion, varicose veins, diarrhea, rheumatism, dyspepsia, cholera, and toothache. Rubbing the fruit against the skin, especially on the lips or in the mouth, produces a numbing effect. A tea or tincture of the bark has been used in the treatment of rheumatism, dyspepsia, dysentery, heart and kidney troubles, etc. A tea made from the inner bark is used to treat itchy skin.			Xanthoxyletin	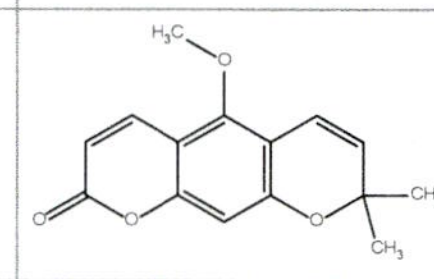	
				Lutein	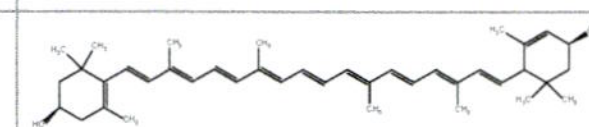	
				Asarinin	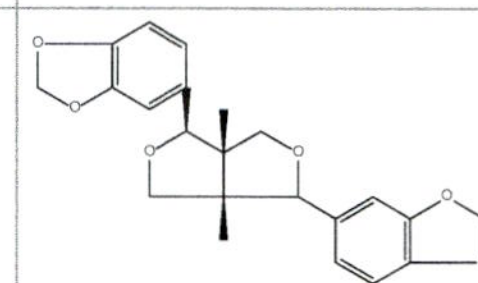	

(continued)

Table 4.120 (continued)

Family	Botanical name	Image	Plant part used	Bioactive compound(s)	Structure	Reference(s)
	Haplophyllum dauricum (L.) G.Don		Aerial parts	Lutein	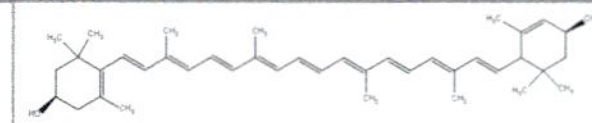	Bessonova et al. (1984) and Nho et al. (2007)
	Synonym(s): *Haplophyllum dahuricum* A.Juss., *Haplophyllum davuricum* Ledeb., *Harmala daurica* Crantz. **Common name(s)**: Harmal. **Botanical description**: It is a perennial plant which can grow to about 2.5 ft. tall and is native to China and Asia. The roots of the plant can reach a depth of up to 6.1 m. Leaves are stalkless, 4–8 cm long, irregularly and pinnately cut into 3–5 cm long, 2–5 mm broad, linear-lance shaped or subelliptic, pointed segments. The flowers are yellowish white and are about 2.5–3.8 cm in diameter. Petals are oblong-obovate, 1.5–2 cm long, and 6–9 mm wide. **Medicinal use(s)**: It is used to treat bladder, colorectal, ovarian, lung, and testicular cancer. It also possesses antitumor, anti-inflammatory, antioxidant, antidepressant, and antimicrobial properties. The root bark extract is used for the treatment of elephantiasis, toothache, sexual impotence, gonorrhea, malaria, dysmenorrhea, and abdominal pain.					

	Murraya koenigii (L.) Sprengel		Leaf and stem	Beta-sitosterol glucoside	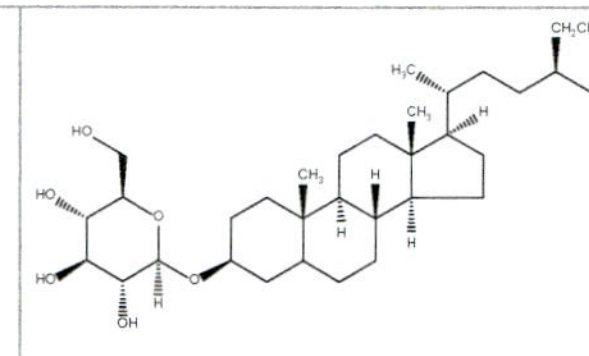	Philip (1981) and Iman et al. (2017)
	Synonym(s): *Bergera koenigii* L. **Common name(s)**: Curry leaf tree, Daun kari, Indian curry tree, Curry bush, Karapincha, Karwa pale, Garupillai, Kerupulai. **Botanical description**: It is native to East Asia, China, Indian subcontinent, Myanmar, Thailand, Laos, and Vietnam. It is found in the moist forests at elevations of 500–1600 m in Southern China. It is most attractive, small evergreen tree with wonderful sweet, curry-pungent foliage. Stem is dark green to brown, with numerous dots. Flowers are arranged in clusters at the terminal. **Medicinal use(s)**: It is used to treat lung, breast, and colorectal cancer. It also possesses antitumor, antioxidant, anti-inflammatory, and antidiabetic properties. The leaves, roots, and barks can be used internally for the treatment of digestive problems. The leaves can be used internally for treating constipation, diarrhea, and dysentery and can be applied externally as a poultice to treat burns and wounds.			Stigmasterol	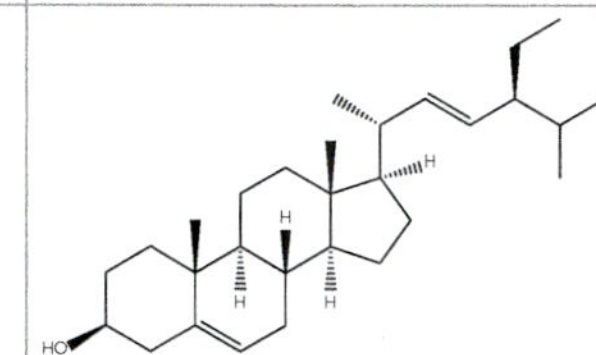	

(continued)

Table 4.120 (continued)

Family	Botanical name	Image	Plant part used	Bioactive compound(s)	Structure	Reference(s)
	Zanthoxylum acanthopodium DC.		Fruit	Burkinabin C	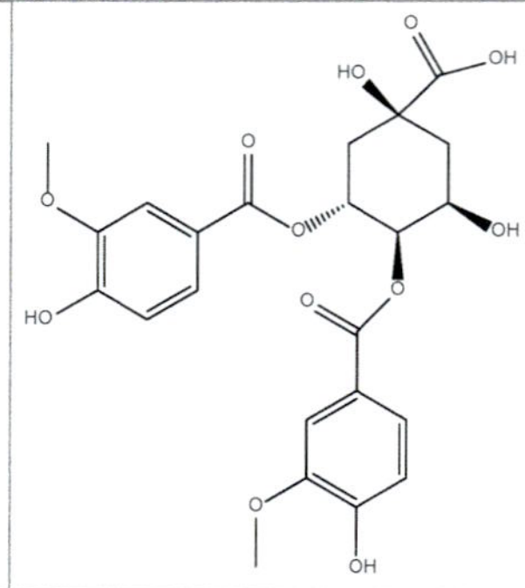	Bose and Bose (1939) and Harahap et al. (2018)
	Synonym(s): *Zanthoxylum acanthopodium* var. *acanthopodium*, *Zanthoxylum acanthopodium* var. *timbor* Hook. f. **Common name(s)**: Andaliman. **Botanical description**: It is native to East Asia, Southern China, India, Bhutan, Bangladesh, Myanmar, Thailand, Malaysia, and Indonesia. It is a prickly, evergreen plant that varies in habit from a shrub to a woody climber or a small tree. It can grow up to 6 m tall. It is found in open forests and thickets at elevations from 1400 to 3200 m. **Medicinal use(s)**: It is used to treat breast, lung, colorectal, and human cervical cancer. It also possesses anticancer, antitumor, anti-inflammatory, antimicrobial, antifungal, antioxidant, and antiseptic properties. The plant parts like leaves, stems, barks, fruits, seeds, and roots possess medicinal properties and are used in indigenous medicine preparation against various diseases like asthma, bronchitis, indigestion, varicose veins, diarrhea, rheumatism, dyspepsia, cholera, and toothache.			Lutein	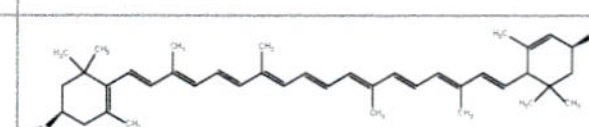	

	Species	Image	Part	Compound	Structure	Reference
	Zanthoxylum ailanthoides Sieb. & Zucc.		Leaf Fruit	Pheophorbide-a methyl ester	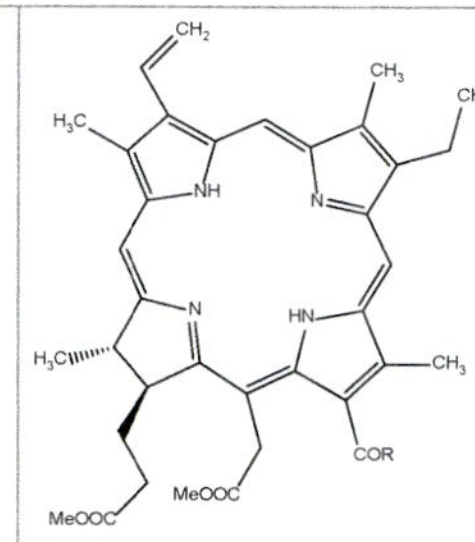	Chou et al. (2011)
				Pheophorbide-b methyl ester	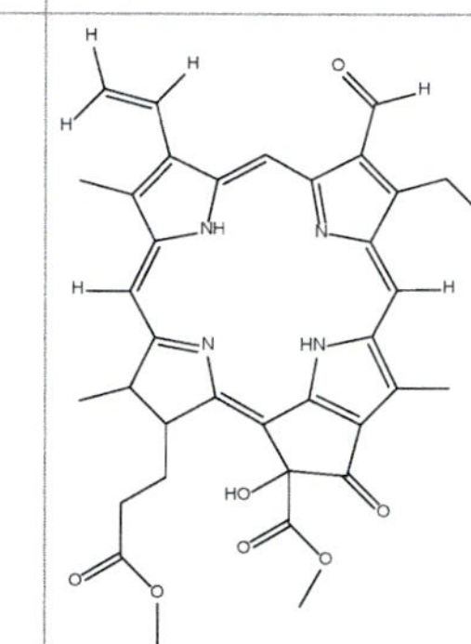	

(continued)

Table 4.120 (continued)

Family	Botanical name	Image	Plant part used	Bioactive compound(s)	Structure	Reference(s)
	Synonym(s): *Fagara boninshimae* Koidz., *Zanthoxylum emarginellum* Miq., *Zanthoxylum hemsleyanum* Makino. **Common name(s)**: Hercules club. **Botanical description**: It is native to East Asia, Southern China, Central and Southern Japan, Korea, and the Philippines. It is a deciduous tree growing up to 18 m tall. The flowers are dioecious (individual flowers are either male or female, but only one sex is to be found on any one plant, so both male and female plants must be grown if seed is required). **Medicinal use(s)**: It is used for the treatment of cervical, bone, lung, and prostate cancer. It also possesses anticancer, antitumor, antioxidant, antifungal, anti-inflammatory, and antimicrobial properties. It is also used to treat asthma, bronchitis, indigestion, varicose veins, diarrhea, rheumatism, dyspepsia, cholera, and toothache. Rubbing the fruit against the skin, especially on the lips or in the mouth, produces a numbing effect. A tea or tincture of the bark is used for the treatment of rheumatism, dyspepsia, dysentery, heart and kidney troubles, etc. A tea made from the inner bark is used to treat itchy skin.					
	Zanthoxylum armatum DC.		Fruit	Apigenin 7-O-glucoside	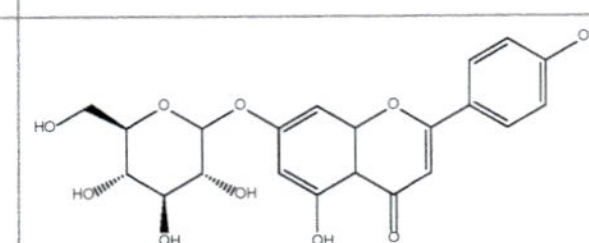	Chopra et al. (1960) and Alam et al. (2017a)

	Synonym(s): *Fagara armata* Thunb., *Zanthoxylum alatum* Roxb., *Zanthoxylum alatum* Hemsl. **Common name(s)**: Winged prickly ash. **Botanical description**: It is native to East Asia, China, Japan, Korea, India, Bhutan, Bangladesh, Myanmar, Thailand, Laos, Vietnam, Malaysia, Indonesia, and the Philippines. It is a deciduous shrub growing up to 4 m tall. The flowers are dioecious (individual flowers are either male or female, but only one sex is to be found on any one plant, so both male and female plants must be grown if seed is required). **Medicinal use(s)**: It is used to treat cervical, bone, lung, and prostate cancer. It also possesses anticancer, antitumor, antioxidant, antifungal, anti-inflammatory, and antimicrobial properties. The seeds and the barks are used as an aromatic tonic for the treatment of fevers, dyspepsia, and cholera. The fruits, branches, and thorns are considered to be carminative and stomachic. They are used as a remedy for toothache. It has a stimulating effect on the lymphatic system, circulation, and mucous membranes.			Stigmasterol	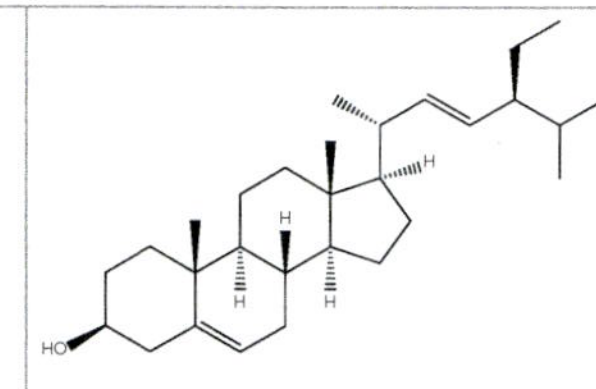	
	Zanthoxylum madagascariense Baker.		Leaf	Rutaceline	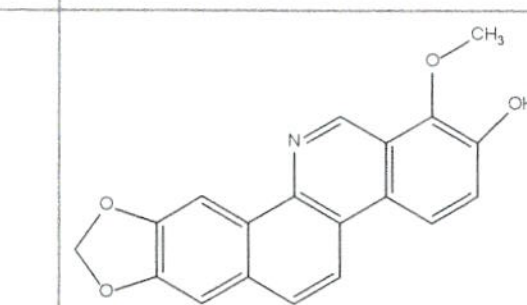	Pachon et al. (2005)

(continued)

Table 4.120 (continued)

Family	Botanical name	Image	Plant part used	Bioactive compound(s)	Structure	Reference(s)
	Synonym(s): *Zanthoxylum acanthopodium* DC., *Zanthoxylum aculeatissimum* Engl. **Common name(s)**: Spider plant. **Botanical description**: It is native to tropical Africa, Guinea to Ethiopia, south to Angola, Mozambique, and Madagascar. It is found in riverine, groundwater, lowland and submontane forests, rainforest, and other evergreen forests. It is an evergreen tree with drooping branches and a small, rounded crown. It usually grows 20–30 m tall. The straight, cylindrical, clean bole can be 50–100 cm in diameter. **Medicinal use(s)**: It is used to treat cervical, bone, lung, and prostate cancer. It also possesses anticancer, antitumor, antioxidant, antifungal, anti-inflammatory, and antimicrobial properties. The bitter stem bark decoction is taken to treat severe body pain, especially backache and arthritis, and also to get rid of intestinal worms. A bark decoction, together with the leaves, is used to relieve malarial symptoms, tiredness, muscular ache, and poisoning. Roots are pounded and soaked in cold water and the infusion mixed with porridge made out of finger millet flour. The porridge is drunk as a remedy for impotence.			Lutein	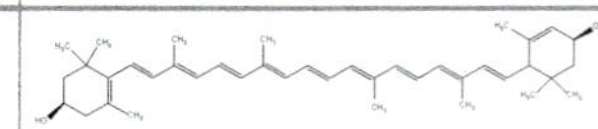	
	Zanthoxylum oxyphyllum Edgew.		Fruit	Rutaceline	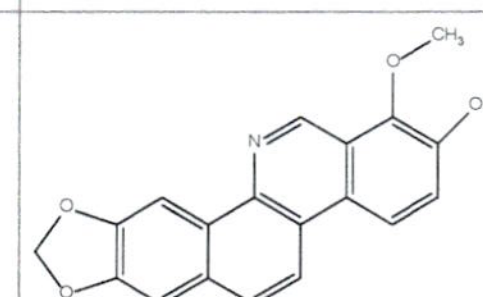	Suwal (1970) and Munda and Kakoti (2017)

	Synonym(s): *Fagara oxyphylla* (Edgew.) Engl., *Zanthoxylum alpinum* C.C. Huang, *Zanthoxylum taliense* C.C. Huang. **Common name(s)**: Prickly ash. **Botanical description**: It is native to East Asia, Southern China, India, Bhutan, Nepal, and Myanmar. It is a spiny shrub that sometimes adopts a climbing habit. It is found in open rocky slopes in the Himalayas at elevations of 2100–2800 m. **Medicinal use(s)**: It is used to treat cervical, bone, lung, and prostate cancer. It also possesses anticancer, antitumor, antioxidant, antifungal, anti-inflammatory, and antimicrobial properties. The bark, especially the root bark, is used as tonic. It can be used in the treatment of rheumatism and atonic dyspepsia. A paste made from the immature fruits is held between the teeth for about 10 min to relieve toothache. The fruits are said to be astringent, digestive, and stimulant.			Lutein	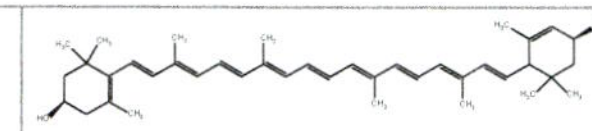	
	Zanthoxylum rhoifolium Lam.		Bark	α-Humulene	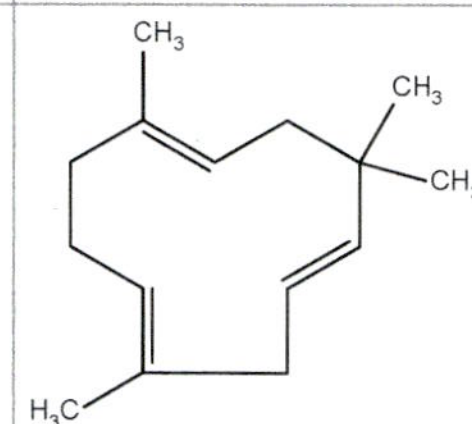	Silva et al. (2007)
				α-Pinene	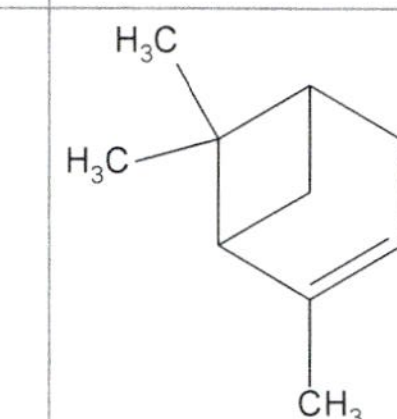	

(continued)

Table 4.120 (continued)

Family	Botanical name	Image	Plant part used	Bioactive compound(s)	Structure	Reference(s)
	Synonym(s): *Fagara astrigera* R.S.Cowan, *Fagara regnelliana* (Engl.) Chodat & Hassl., *Fagara rhoifolia* (Lam.) Engl. **Common name(s)**: Tambataru. **Botanical description**: It is native to South America; Central America, Panama to Mexico; the Caribbean, Trinidad, Grenada, and the Windward Isles. It is a semi-deciduous tree with a dense, spreading crown growing 6–12 m tall. The straight, cylindrical bole can be unbranched for 3–4 m and 30–40 cm in diameter. The bole is armed with stout spines about 6 mm long. It is found in secondary forest growth at elevations from 450 to 1050 m, slopes in the Atlantic rainforest and in semi-deciduous forests at higher elevations, usually in areas where the soil is well-drained. **Medicinal use(s)**: It is used to treat cervical, bone, lung, and prostate cancer. It also possesses anticancer, antitumor, antioxidant, antifungal, anti-inflammatory, and antimicrobial properties. The macerated bark, mixed with wine or rum, is used as a bitter tonic having antimalarial and anti-blennorrhagic properties. The inner bark is anti-spasmodic, bitter, febrifuge, and stomachic. It is used in the treatment of dyspepsia and blennorrhagia. It is boiled with the leaves of the plant, and the water is drunk as an antipyretic.			β-Pinene (inactive)	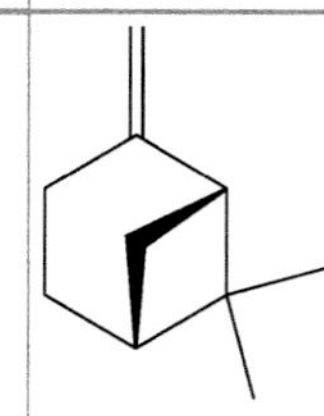	

	Picrella trifoliata Baillon/ *Zieridium pseudobtusifolium*		Fruit and aerial parts	Camptothecin	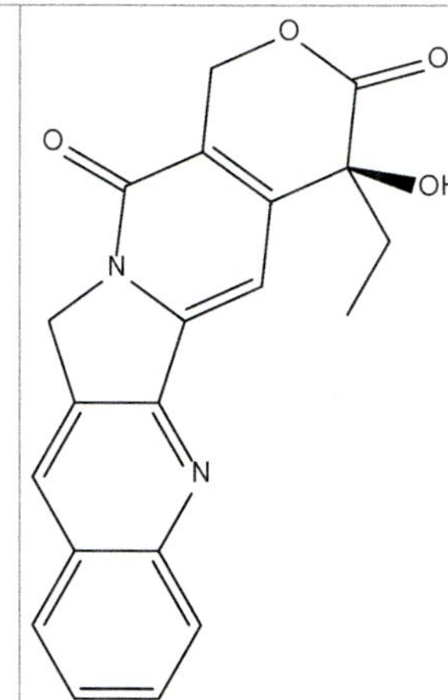	Lichius et al. (1994)
	Synonym(s): *Evodia pseudo-obtusifolia* Guillaumin. **Common name(s)**: Guillaum. **Botanical description**: It is native to South America, Peru, Northern Brazil, Colombia, Venezuela, north to Panama and Trinidad. It is a climbing plant with a perennial rootstock, producing annual stems that scramble over the ground or twine around other plants for support. **Medicinal use(s)**: It is used to treat cervical, bone, lung, and prostate cancer cell lines. It also possesses anticancer, antitumor, antioxidant, antifungal, anti-inflammatory, and antimicrobial properties. The macerated bark, mixed with wine or rum, is used as a bitter tonic having antimalarial and anti-blennorrhagic properties. The inner bark is anti-spasmodic, bitter, febrifuge, and stomachic. It is used in the treatment of dyspepsia and blennorrhagia. It is boiled with the leaves of the plant, and the water is drunk as an antipyretic.					

Table 4.121 Anticancer plants of family Salicaceae

Family	Botanical name	Image	Plant part used	Bioactive compound(s)	Structure	References
Salicaceae	*Casearia nigrescens* Tul.		Leaf	Casearlucin A/C	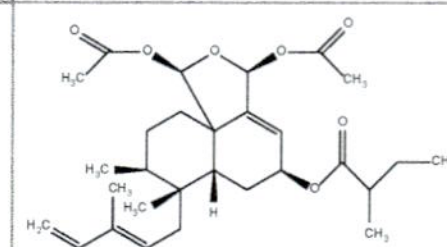	Fukamiya et al. (1986) and Williams et al. (2007),
	Synonym(s): *Guidonia elliptica* (Tul.) Baill., *Guidonia nigrescens* (Tul.) Baill. **Common name(s)**: Generic tree key. **Botanical description**: *Casearia nigrescens* is a large, evergreen shrub or tree growing up to 15 m tall. **Medicinal use(s)**: The wood is harvested from the wild for local use.			Caseanigrescens	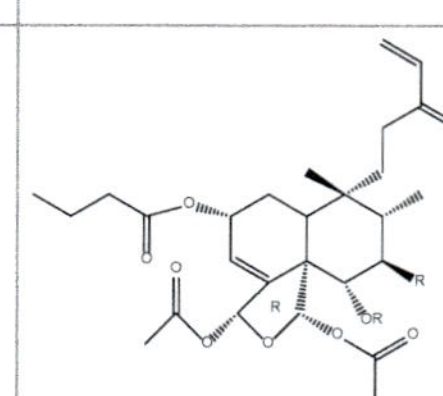	
	Casearia sylvestris Sw.		Leaf	Casearvestrin	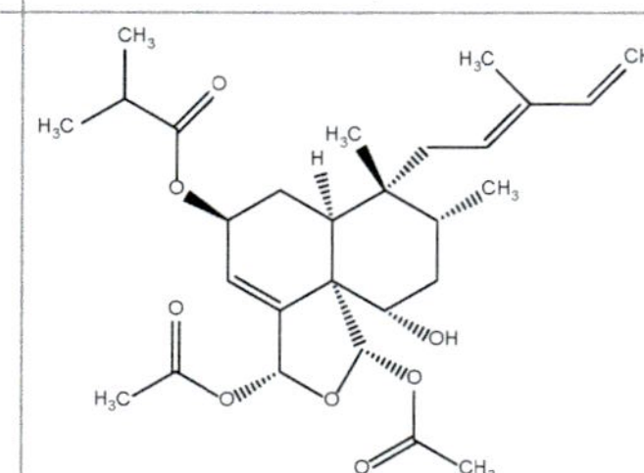	Guza (2007) and Felipe et al. (2014)

	Synonym(s): *Anavinga parvifolia* Lam., *Anavinga samyda* C.F.Gaertn. **Common name(s)**: Guaçatonga, Wild sage. **Botanical description**: Wild sage is an evergreen shrub or small tree with long, slender branches and a very dense, globose crown. It is usually 4–6 m tall but can sometimes grow as high as 20 m. The short, straight, cylindrical bole can be 20–30 cm in diameter. The plant is gathered from the wild for local use, mainly for its medicinal virtues. It can be used in reforestation schemes. **Medicinal use(s)**: The bark and leaves are analgesic, anti-inflammatory, anti-rheumatic, aphrodisiac, depurative, tonic, and vulnerary. They are used internally in the treatment of inflammation, fevers, gastric ulcers, and diarrhea. Externally, the leaves are used in the treatment of wounds. The bark and leaf are said to protect against snake venom. The macerated roots, or a decoction of them, are used in the treatment of wounds and leprosy. The oil from the seed is used in the treatment of leprosy. The leaves and twigs contain apache, chlordane intervenes (essential oils of leaves). They have antitumor action, inhibit HIV replication, and are used antibiotic.	Casearin C	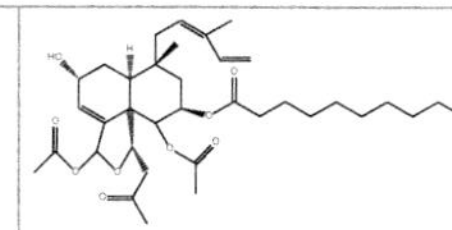	

Table 4.122 Anticancer plants of family Santalaceae

Family	Botanical name	Image	Plant part used	Bioactive compound(s)	Structure	References
Santalaceae	*Osyris lanceolata* Hochst. & Steud.		Leaf	(+)-catechin	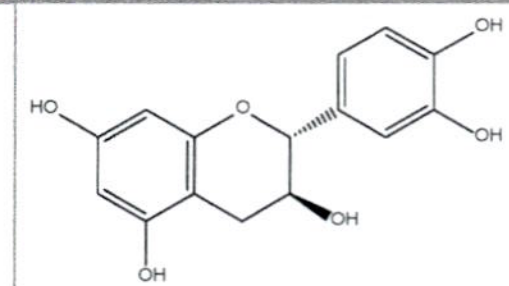	Abebaw et al. (2017)
	Synonym(s): *Osyris abyssinica* Hochst. ex A. Rich., *Osyris arborea* Wall. ex A. DC. **Common name(s)**: African sandal wood. **Botanical description**: *Osyris lanceolata* is an evergreen shrub or small tree, usually growing 1.2–9 m tall, with occasional specimens to 14 m. The bole is 60–90 cm in diameter at the base. The tree is harvested from the wild for local use as a food, medicine, and source of wood and materials. The wood is sold locally and also traded internationally for its essential oil which is used in making perfume. The plant is grown as an ornamental – its unusual bark and leaf combination makes it an interesting subject among the other trees in a garden; but due to its untidy growth form, it is not an attractive tree in a small garden. **Medicinal use(s)**: The roots and bark are used as a tonic in soups. The root is boiled in water for about 10 min and the liquid given to women after childbirth to control bleeding and boost energy. A decoction of the roots is used for treating diarrhea. The bark is boiled in water, strained, and then boiled again to form a gelatinous mass. This is used in the treatment of bloody dysentery. The mass is also applied externally to set dislocated bone and is put in the eye to relieve inflammation. The juice of the bark is used in the treatment of indigestion and is applied to foreheads to relieve headaches. A decoction of the bark and heartwood is used to treat sexually transmitted diseases and anemia. A paste of the fruit is applied to the forehead to relieve headaches. The leaves have emetic properties. A paste of the leaf buds is applied to cuts and wounds.			Procyanidin dimer B1	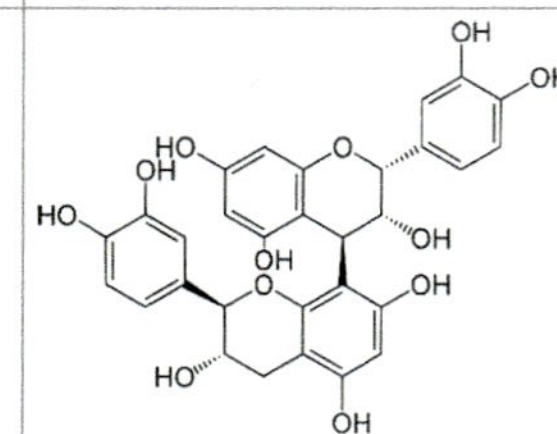	

Table 4.123 Anticancer plants of family Sapindaceae

Family	Botanical name	Image	Plant part used	Bioactive compound(s)	Structure	References
Sapindaceae	*Koelreuteria elegans* subsp. *formosana* (Hayata) F.G.Mey.		Fruit	Cyclolignans	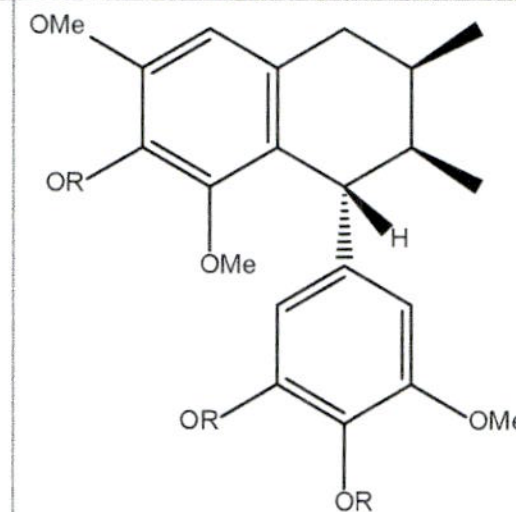	Song et al. (1994) and Miyagawa et al. (2009)
	Synonym(s): *Koelreuteria henryi* Dümmer. **Common name(s)**: Golden rain tree. **Botanical description**: It is a deciduous tree 15–17 m tall endemic to Taiwan. It is widely grown throughout the tropics and subtropical parts of the world as a street tree. It flowers in early to midsummer. Flowers are small and 20 mm in length and occur in branched clusters at the stem tips. They are butter yellow with five petals that vary in length until opening. Each flower contains seven to eight pale yellow stamens with hairy white filaments. The fruit is a brown-purplish three-lobed capsule that splits to reveal a number of black seeds. It is a declared weed in many parts of the world, particularly in Brisbane, Australia, and Hawaii. **Medicinal use(s)**: The flowers are ophthalmic. They are used in the treatment of conjunctivitis and epiphora.			Austrobailigans	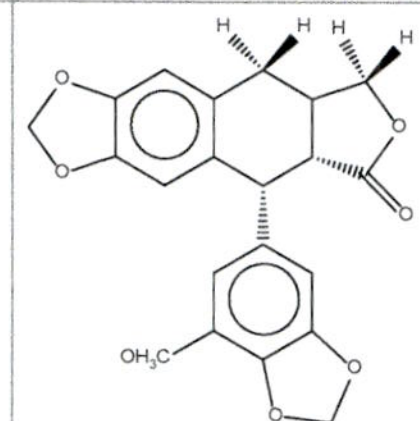	

(continued)

Table 4.123 (continued)

Family	Botanical name	Image	Plant part used	Bioactive compound(s)	Structure	References
	Sapindus emarginatus Vahl.		Fruit	Saponins	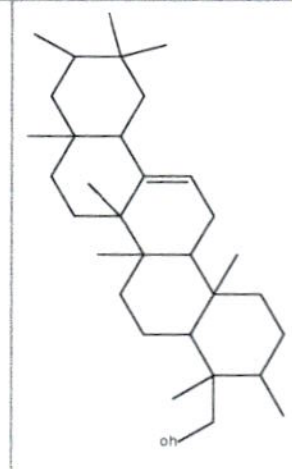	Rejinold et al. (2011)
	Synonym(s): No synonyms are recorded for this name. **Common name(s)**: Reetha. **Botanical description**: This tree is around 8–10 m high and has many branches with leaves and leaflets. Its flowers are white and fruits are round. **Medicinal use(s)**: Traditionally, it is used as anti-inflammatory and antipyretic. It is used to purify blood. The seed is an intoxicant and the fruit rind has oxytropic action. Its powder is used in nasal insufflations.					
	Litchi chinensis Sonn.		Leaf, pulp, seed	Procyanidin A2	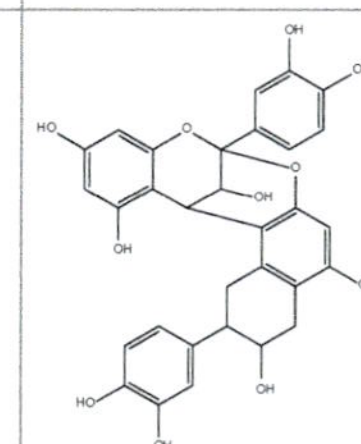	Wen et al. (2014)
	Synonym(s): *Corvinia litschi* Stadtm. ex Willemet, *Euphoria didyma* Blanco. **Common name(s)**: Lychee, Litchi. **Botanical description**: It is a non-climacteric subtropical fruit which originated from Southeast Asia. **Medicinal use(s)**: As a tissue of litchi plant, litchi leaf has been used in traditional Chinese medicine for the treatment of heart stroke, flatulence, and detoxification. Pharmacological studies conducted indicated that the petroleum ether extract of litchi leaf had significant anti-inflammatory, analgesic, and antipyretic activities.					

Table 4.124 Anticancer plants of family Sarcolaenaceae

Family	Botanical name	Image	Plant part used	Bioactive compound(s)	Structure	References
Sarcolaenaceae	*Schizolaena hystrix* Capuron		Aerial parts	Schizolaenone A/B	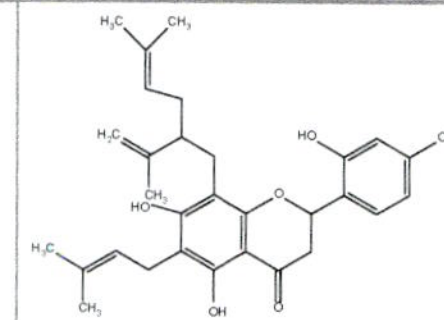	Murphy et al. (2005)
	Synonym(s): No synonyms are recorded for this name. **Common name(s)**: Wireweed. **Botanical description**: *Schizolaena hystrix* is known only from the northeastern regions of Sava, Alaotra-Mangoro, Analanjirofo, and Atsinanana. Its habitat is humid evergreen forests from sea level to 1500 m in altitude. It grows as a large tree up to 35 m tall. Its leaves are coriaceous. The spiny involucre is fleshy and is thought to attract lemurs, bats, and birds who in turn disperse the tree's seeds. **Medicinal use(s)**: Cytotoxic.			4′-*O*methylbonannione	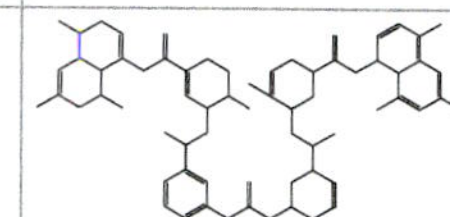	
				Nymphaeol A	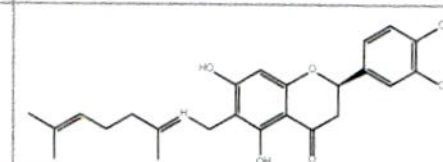	
				Bonannione A	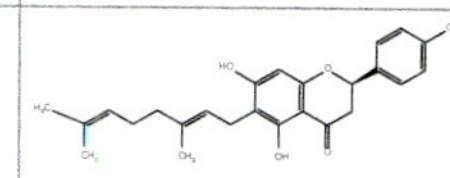	

(continued)

Table 4.124 (continued)

Family	Botanical name	Image	Plant part used	Bioactive compound(s)	Structure	References
				Macarangaflavanone B	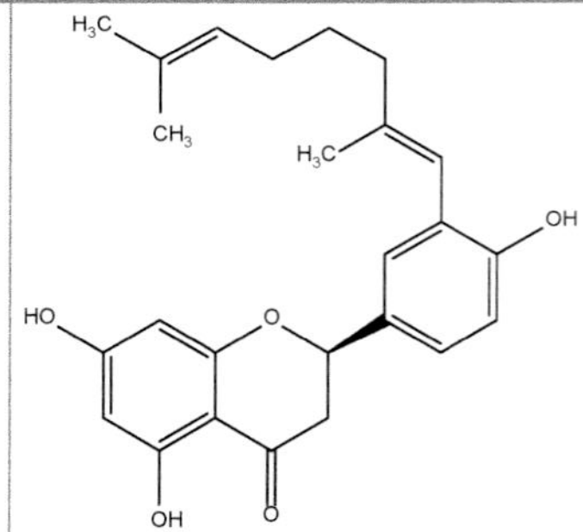	
				Flavanol bonanniol A	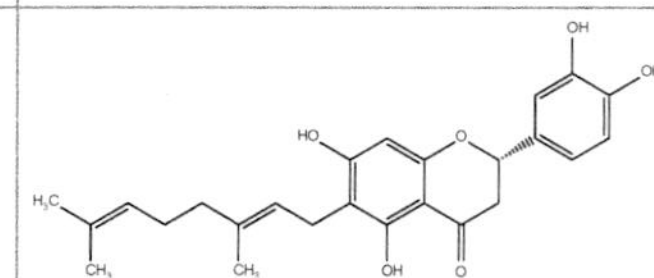	

Table 4.125 Anticancer plants of family Schisandraceae

Family	Botanical name	Image	Plant part used	Bioactive compound(s)	Structure	References
Schisandraceae	*Kadsura acsmithii* R.M.K.Saunders		Root, fruit	Arylnapthalene lignan elenoside(β-D glucoside)	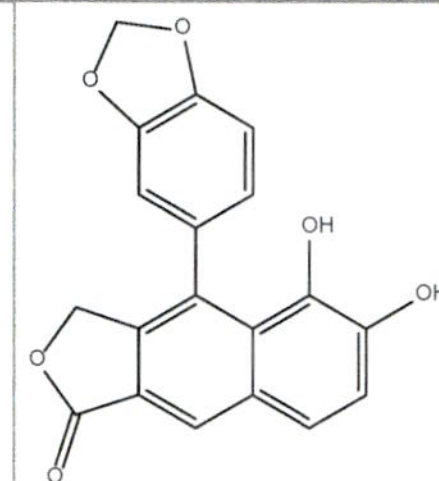	Yang et al. (2011a, b)
	Synonym(s): No synonyms are recorded for this name. **Common name(s)**: Not available. **Botanical description**: *Kadsura* is a genus of woody vines and is native to Eastern, Southern, and Southeastern Asia. **Medicinal use(s)**: Anticancer.			Aglycone	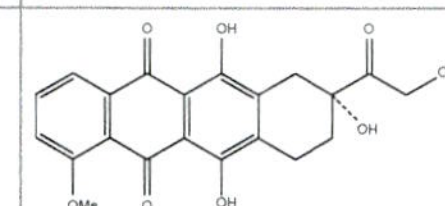	

Table 4.126 Anticancer plants of family Scrophulariaceae

Family	Botanical name	Image	Plant part used	Bioactive compound(s)	Structure	References
Scrophulariaceae	*Verbascum sinaiticum* Benth.		Leaf and flower	Verbascoside	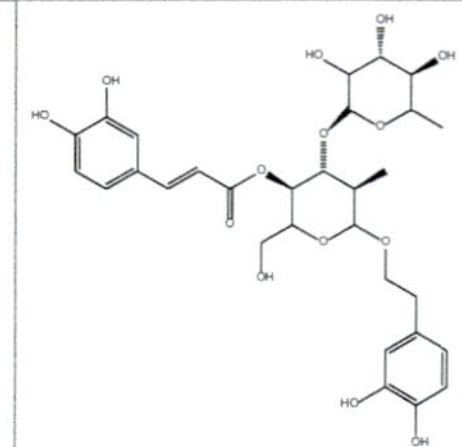	Senatore et al. (2007)
	Synonym(s): No synonyms are recorded for this name. **Common name(s)**: Velvet plant. **Botanical description**: It is a biennial plant growing up to 60–150 cm tall. **Medicinal use(s)**: It is traditionally used to treat wound, stomachache, viral infection, cancer, sunstroke, fever, abdominal colic, diarrhea, hemorrhage, anthrax, and hepatitis.			Aucubin	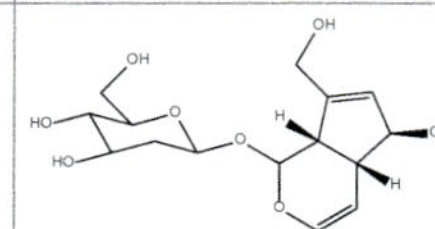	
				Luteolin-7-O-β-glucoside	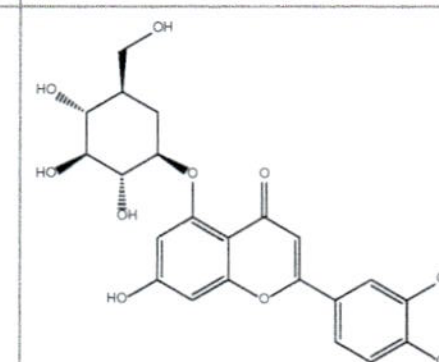	
				Sinuatol	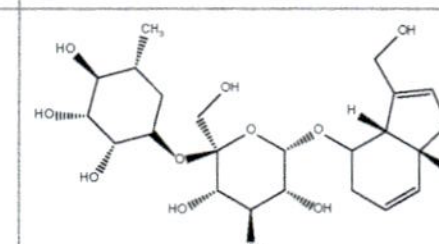	

				Ajugol	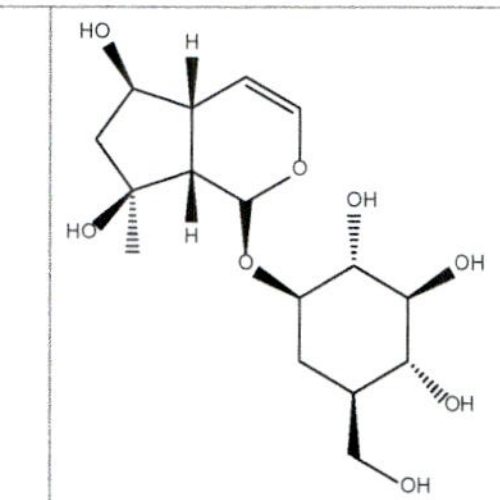	
	Scrophularia ningpoensis Hemsl		Root	Catalpol	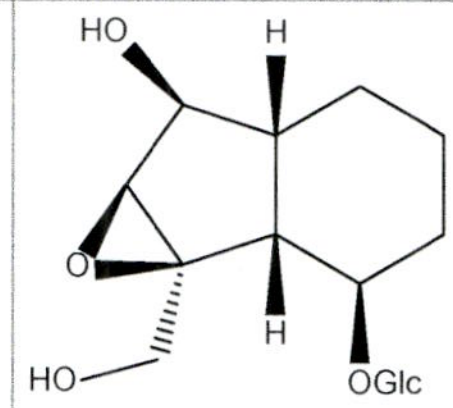	Nguyen et al. (2005), Shen et al. (2012)
	Synonym(s): *Scrophularia microdonta* Franch., *Scrophularia silvestrii* Bonati & Pamp. **Common name(s)**: Ningpo figwort. **Botanical description**: It reaches 1 m by 0.4 m. Its flowers are hermaphrodite and insect-pollinated, and the plant usually flowers in late spring. **Medicinal use(s)**: Known for its antipyretic and anti-inflammatory effects, RSN is widely used for the treatment of laryngitis, fever, swelling, constipation, and neuritis. Small doses of RSN act as a heart tonic, while higher doses rather suppress heart function.			Acetosice	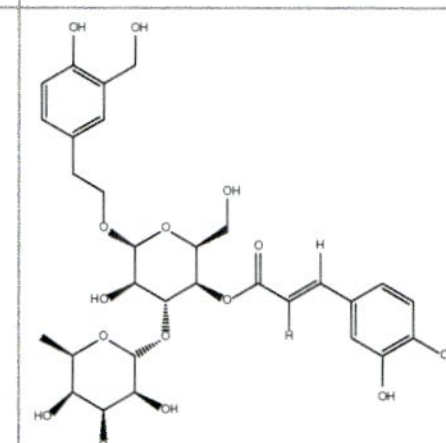	
				Ursolic acid	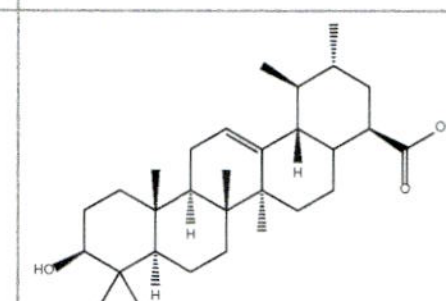	

(continued)

Table 4.126 (continued)

Family	Botanical name	Image	Plant part used	Bioactive compound(s)	Structure	References
				Cinnamic acid	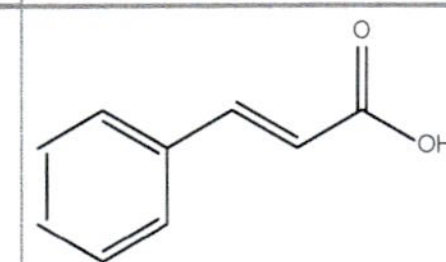	
				Oleanolic acid	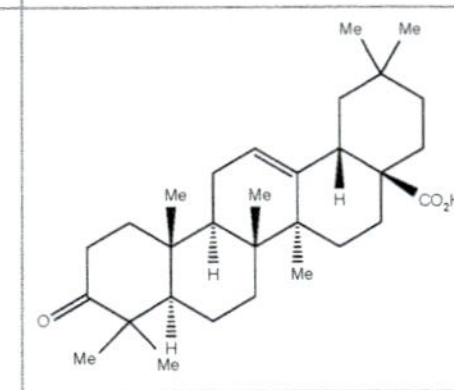	
	Bacopa monnieri (L.) Wettst.		Apical part	Bacobitacin	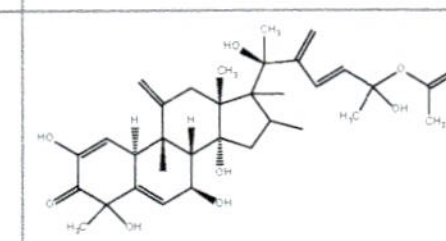	Bhandari et al. (2007) and Ghosh et al. (2011)

	Synonym(s): *Anisocalyx limnanthiflorus* (L.) Hance, *Bacopa micromonnieria* (Griseb.) B.L.Rob. **Common name(s)**: Water hyssop, Brahmi, Thyme-leafed gratiola, Water hyssop, Herb of grace, Indian pennywort. **Botanical description**: *Bacopa monnieri* is a nonaromatic herb. The leaves of this plant are succulent, oblong, and 4–6 mm (0.16–0.24 in.) thick. Leaves are oblanceolate and are arranged oppositely (opposite-decussate) on the stem. The flowers are small, actinomorphic, and white, with four to five petals. Its ability to grow in water makes it a popular aquarium plant. It can even grow in slightly brackish conditions. Propagation is often achieved through cuttings. **Medicinal use(s)**: *Bacopa monnieri* has been used in traditional Ayurveda for epilepsy, asthma, ulcers, tumors, ascites, enlarged spleen, inflammations, leprosy, anemia, gastroenteritis, and hair loss.			Stigmasterol	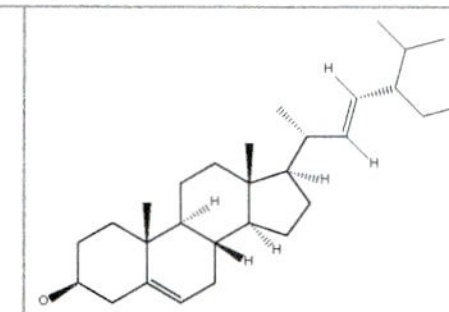	
	Pedicularis resupinata L.		Leaves, aerial parts	2,6-Dimethoxybenzoquinone	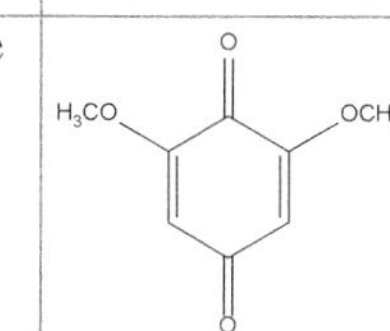	Yoo et al. (1993)
	Synonym(s): *Pedicularis resupinata* subsp. *resupinata*. **Common name(s)**: Lousewort. **Botanical description**: *Pedicularis resupinata* is a perennial up to 1.00 m tall. A semiparasitic plant, growing on grass roots. Rather difficult to establish in cultivation, it is best grown in conditions that approximate to its native habitat. It requires a moist peaty soil and the presence of host grasses. Requires a partially shaded to sunny site in a well-drained gritty but moist soil. **Medicinal use(s)**: Anti-rheumatic, diuretic, and febrifuge. The plant is used in the treatment of fevers, leukorrhea, rheumatism, sterility, and urinary difficulties. A decoction of the plant is used to wash foul ulcers.					

Table 4.127 Anticancer plants of family Selaginellaceae

Family	Botanical name	Image	Plant part used	Bioactive compound(s)	Structure	References
Selaginellaceae	*Selaginella doederleinii* Hieron.		Aerial parts	2″, 3″-Dihydro-3′, 3‴-biapigenin	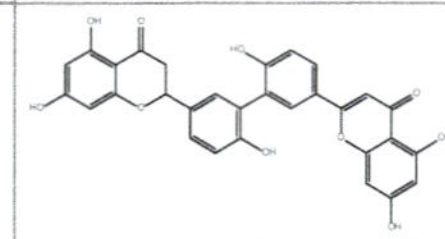	Lee et al. (2008), Li et al. (2014), and Wang et al. (2015)
				3′, 3‴-Binaringenin	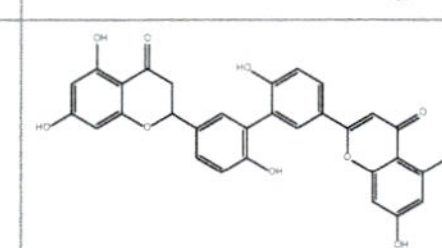	
				Amentoflavone	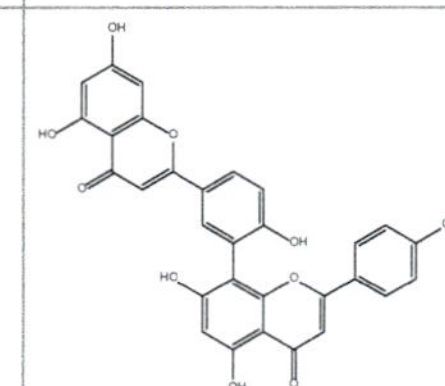	
	Synonym(s): *Lycopodioides doederleinii* (Hieron.) H.S. Kung, *Selaginella doederleinii* subsp. *doederleinii*. **Common name(s)**: Shi shang bai. **Botanical description**: It is a perennial tropical plant of Asian origin. It grows up to a height of 35 cm. **Medicinal use(s)**: It has been traditionally used as a folk medicine in China for the treatment of different tumors, especially for nasopharyngeal carcinoma, lung and liver cancer, and trophoblastic tumor.			2,2,3,3-Tetrahydrorobustaflavone 7,4,7-trimethyl ether	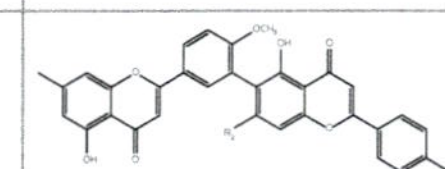	

	Selaginella tamariscina (P.Beauv.) Spring.		Leaf, entire plant	Selaginellins	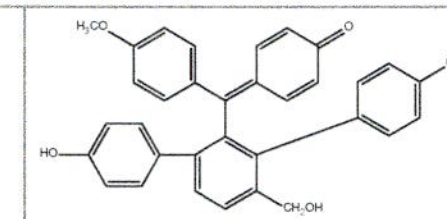	Zuo et al. (2012)
	Synonym(s): *Lycopodioides tamariscina* (P.Beauv.) H.S.Kung, *Lycopodium caulescens* Wall. ex Hook. & Grev. **Common name(s)**: Little club moss. **Botanical description**: *Selaginella tamariscina* is an evergreen, perennial plant that can grow in the soil or on rocks. The densely tufted stems are 4–15 cm long, branched nearly to the base. They form a rosette, the stems and roots entangled to form a treelike trunk that can exceptionally be 45 cm long. **Medicinal use(s)**: The whole plant is astringent and hemostatic. A decoction is used in the treatment of traumatic bleeding, hemoptysis in pulmonary disease, gastrointestinal bleeding, metrorrhagia, hematuria, persistence of postpartum lochial discharge, rectal prolapse, and leukorrhea. The plant is also used in the treatment of coughs, prolapse of the rectum, gravel, and old people complaints. Because the plant never seems to die, it is considered to have the property of prolonging life.					

(continued)

Table 4.127 (continued)

Family	Botanical name	Image	Plant part used	Bioactive compound(s)	Structure	References
	Selaginella willdenowii (Desv. ex Poir.) Baker Bif		Leaf	4′,7″-Di-O-methylamentoflavone	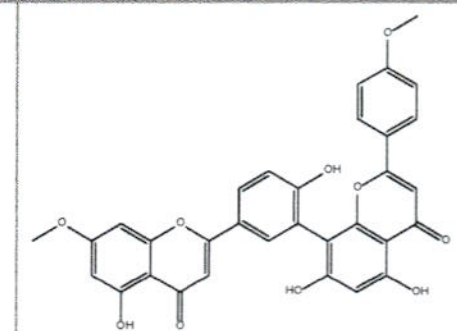	Silva et al. (1995)
				Isocryptomerin	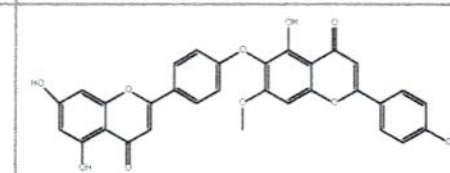	
	Synonym(s): *Selaginella willdenowii* (Desv. ex Poir.) Baker. **Common name(s)**: Spikemoss. **Botanical description**: Growth form: herbaceous, fernlike species with a scrambling growth habit. Foliage: iridescent fronds are mostly blue-green with pinkish hues depending on the angle at which they are viewed. They are composed of two leaf types, lateral and median leaves, which are both arranged in 2 rows along the stem. Lateral leaves are ovate to oblong (3–4 mm long and 1.5–2 mm wide), while median leaves are even smaller and vary in shape. Reproductive parts (nonflowering plant): cone-like strobili composed of sporophylls (spore-producing structures) are produced at the branch tips. The sporophylls are heart-shaped or egg-shaped to triangular. **Medicinal use(s)**: Antioxidant.					

Table 4.128 Anticancer plants of family Simaroubaceae

Family	Botanical name	Image	Plant part used	Bioactive compound(s)	Structure	References
Simaroubaceae	*Quassia undulata* (Guill. & Perr.) D.Dietr.		Stem and bark	Quassinoids	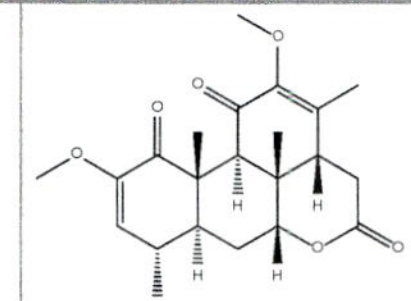	Francois et al. (1998)
	Synonym(s): *Hannoa chlorantha* Engl. & Gilg, *Hannoa undulata* (Guill. & Perr.) Planch. **Common name(s)**: Not available. **Botanical description**: *Harrisonia abyssinica* is a spiny, evergreen shrub that branches from the base or can become a spreading, much-branched tree; it usually grows up to 6 m tall, though some trees have been reported to reach 13 m. The bole and larger branches have thorns up to 2 cm long growing on conical corky outgrowths. The plant is widely used in traditional medicine in many parts of Africa, the roots being sold for medicinal use in many local markets. The tree also provides food and wood for the local population and is used as a living fence and provides shade around homestead. **Medicinal use(s)**: It is a popular traditional medicine in many parts of Africa, where the roots and the leaves are used in the treatment of a wide range of ailments. Considerable research has been carried out into the active compounds contained in the plant, much of which has supported the plant's traditional uses. Many pharmacologically active compounds have been isolated from the root and stem barks; these comprise mainly limonoids, which are highly oxygenated terpenoids, plus several other terpenoids, steroids, and chromone derivative. A leaf and twig decoction is used to treat a range of complaints including venereal diseases, fever and malaria, diarrhea, hemorrhoids, diabetes, urinary problems, and general body pain and to get rid of intestinal worms.			Chaparrinone	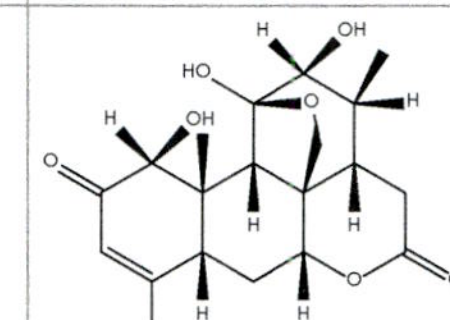	
				15-Desacetylundulatone	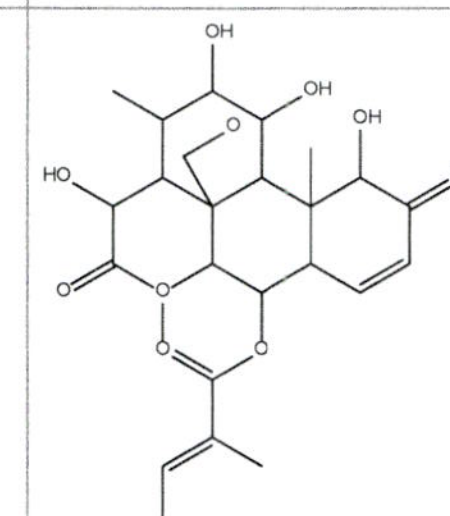	

(continued)

Table 4.128 (continued)

Family	Botanical name	Image	Plant part used	Bioactive compound(s)	Structure	References
	Quassia gabonensis Pierre		Stem and bark	Quassinoids	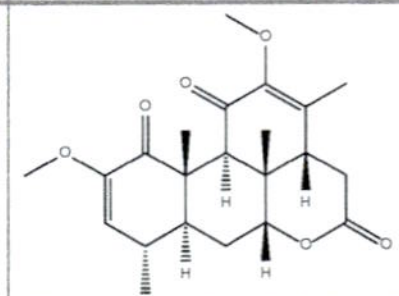	Francois et al. (1998)
				Chaparrinone	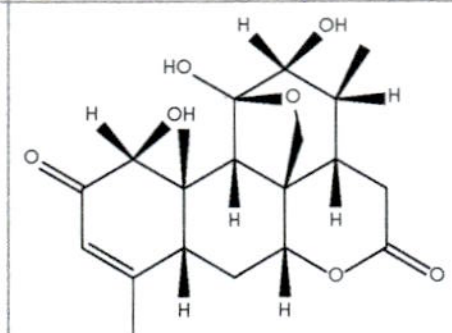	
	Synonym(s): *Hannoa klaineana* Pierre & Engl., *Odyendea gabunensis* (Pierre) Engl. **Common name(s)**: Not available. **Botanical description**: Shrub to high forest tree; branchlets glabrous. Leaves alternate, imparipinnate, 2–5 jugate, 8 cm long; leaflets opposite or subopposite, drying greenish brown above, paler beneath, terminal and basal leaflets usually smaller than the other laterals; leaflet blade narrowly to broadly elliptic or obovate, 2–20 cm long, 1–8 cm wide, emarginate or rounded to shortly acuminate at the apex, rounded to cuneate, often oblique at the base, coriaceous, glabrous, often with pitted glands on the upper surface, nervation indistinct, impressed, margin entire, sometimes slightly wavy; petiolule 1–40 mm, very variable within the species but fairly constant for each specimen; rachis terete. Flowers in terminal and axillary thyrses, inflorescence branches pubescent to glabrous. **Medicinal use(s)**: They are used in traditional medicine of Central African countries to treat fevers and malaria.			15-Desacetylundulatone	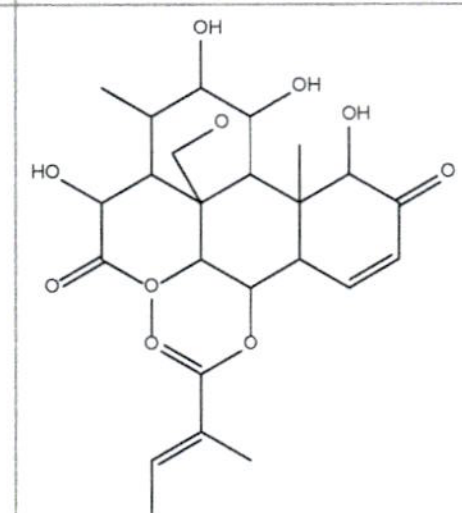	

	Ailanthus excelsa Roxb.		Root bark	AECHL-1	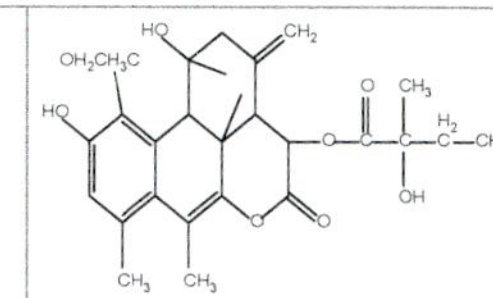	Ogura (1977) and Lavhale et al. (2009)
				Ailanthione	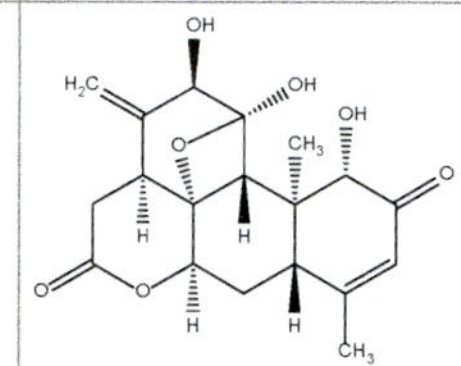	
				Glaucarubinone	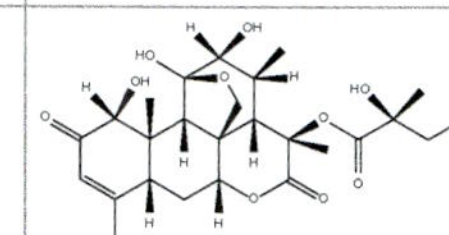	
	Synonym(s): *Ailanthus wightii* Tiegh., *Pongelion excelsum* (Roxb.) Pierre. **Common name(s)**: Tree-of-heaven. **Botanical description**: Indian tree of heaven is a large deciduous tree, 18–25 m tall; trunk straight, 60–80 cm in diameter; bark light gray and smooth, becoming gray brown and rough on large trees, aromatic, slightly bitter. Leaves alternate, pinnately compound, large, 30–60 cm or more in length; leaflets 8–14 or more pairs, long stalked, ovate or broadly lance-shaped from very unequal base, 6–10 cm long, 3–5 cm wide, often curved, long pointed, hairy gland; edges coarsely toothed and often lobed. Flower clusters droop at leaf bases, shorter than leaves, much branched; flowers many, mostly male and female on different trees, short stalked, greenish yellow. Five sepals, 5 narrow petals spreading 6 mm across. Fruit a 1-seeded samara, lance-shaped, flat, pointed at ends, 5 cm long, 1 cm wide, copper red, strongly veined, twisted at the base. **Medicinal use(s)**: Bark is used in India as a powerful fever cure and tonic. Leaves and bark in good repute as a tonic after labor, and the juice of the leaves and fresh bark employed by the Konkans as a remedy for afterpains.					

(continued)

Table 4.128 (continued)

Family	Botanical name	Image	Plant part used	Bioactive compound(s)	Structure	References
	Simarouba amara Aubl.		Tree	Canthin-6-one	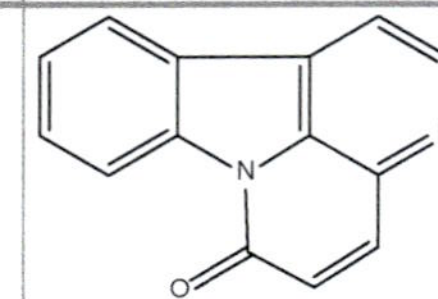	Rivero-Cruz et al. (2005)
				2-Methoxycanthin-6-one	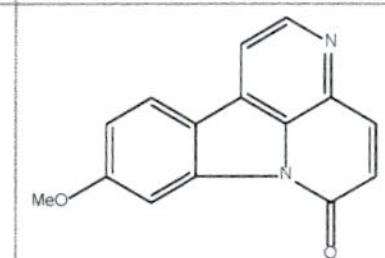	
	Synonym(s): *Quassia alatifolia* Stokes, *Quassia dioica* P.J.Bergius. **Common name(s)**: Paradise tree. **Botanical description**: An understory tree in moist or usually dry forest or thickets, often on dry open rocky hillsides, common in many regions along stream beds at elevations up to 900 m. **Medicinal use(s)**: The leaves and bark have a long history of medicinal use in the tropics, particularly in the treatment of malaria, fevers, and dysentery; as an astringent to stop bleeding; and as a tonic. They are also used as a digestive and emmenagogue and to kill parasites both within and on the body.			9-Hydroxycanthin-6-one	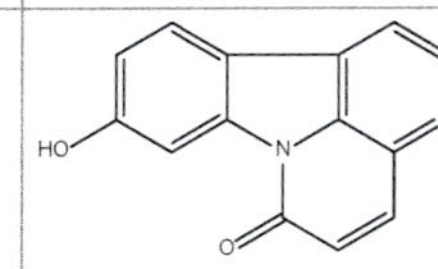	
				Melianodiol	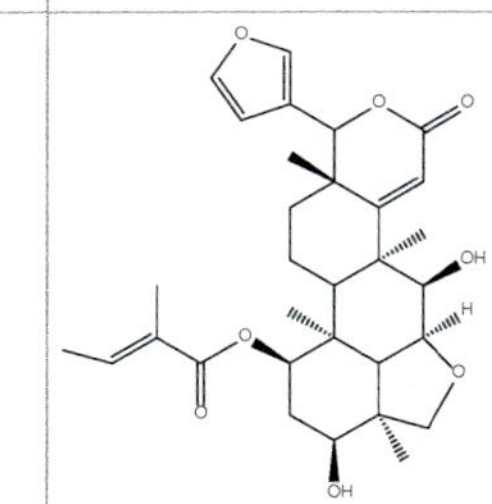	

		4,5-Dimethoxycanthin-6-one	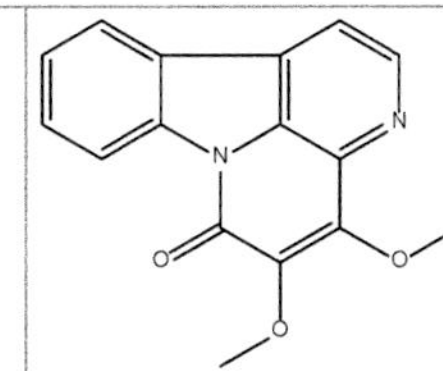	
		Scopolet:n	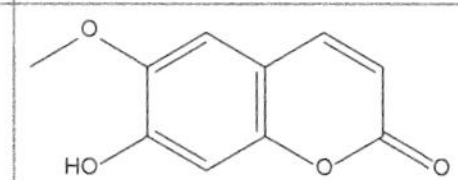	

Table 4.129 Anticancer plants of family Solanaceae

Family	Botanical name	Image	Plant part used	Bioactive compound(s)	Structure	References
Solanaceae	*Datura innoxia* P.Mill.		Leaf	Dinoxin B	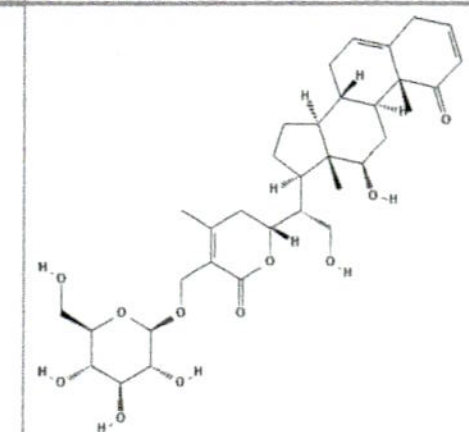	Arulvasu et al. (2010) and Vermillion et al. (2011)
	Synonym(s): *Datura guayaquilensis* Kunth, *Datura meteloides* DC. ex Dunal. **Common name(s)**: Jimsonweed, Thorn apple or Devil's trumpet, Pricklyburr, Recurved thorn-apple, Downy thorn-apple, Indian-apple, Lovache, Moonflower, Nacazcul, Toloatzin, Tolguache or Toloache. **Botanical description**: It is native to the Southwest United States and Central and South America and introduced in Africa, Asia, Australia, and Europe. It is an annual shrubby plant that typically reaches a height of 0.6–1.5 m. Its stems and leaves are covered with short and soft grayish hairs, giving the whole plant a grayish appearance. It has elliptic smooth-edged leaves with pinnate venation. All parts of the plant emit a foul odor similar to rancid peanut butter when crushed or bruised, although most people find the fragrance of the flowers to be quite pleasant when they bloom at night. The flowers are white, trumpet-shaped, 12–19 cm long. They first grow upright and later incline downwards. It flowers from early summer until late fall. **Medicinal use(s)**: *Datura innoxia* has many medicinal properties like anodyne, anti-spasmodic, hallucinogenic, hypnotic, narcotic, etc. *Datura innoxia* is used in the treatment of insanity, fevers with catarrh, diarrhea, scabies, piles, ulcers, colds, asthma, cardiac disorders, impotency, malaria, baldness, and skin diseases. The plant contains several alkaloids, the most active of which is scopolamine. It is also useful in respiratory ailments, rheumatism, elephantiasis, insanity, earache, and eye diseases.			Dinoxin B- Aglycone	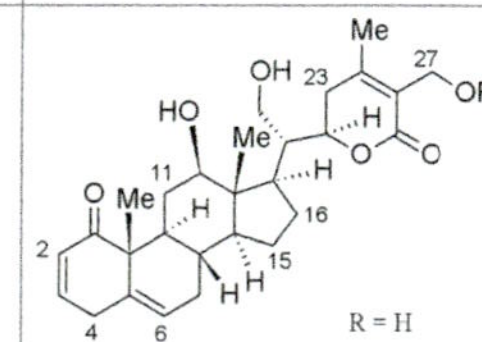	

	Withania somnifera (L.) Dunal		Leaf	Withanolides	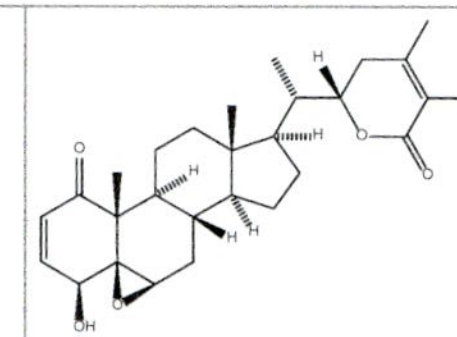	Jayaprakasam et al. (2003)
				Viscosalactone B	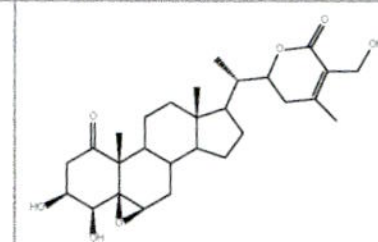	
	Synonym(s): *Physalis somnifera* L., *Withania kansuensis* Kuang & A. M. Lu. **Common name**(s): Ashwagandha, Indian ginseng, Poison gooseberry, Winter cherry. **Botanical description**: This species is a short, tender perennial shrub growing 35–75 cm tall. Tomentose branches extend radially from a central stem. Leaves are dull green, elliptic, usually up to 10–12 cm long. The flowers are small, green, and bell-shaped. The ripe fruit is orange-red. **Medicinal use(s)**: Ashwagandha is used for treating arthritis, anxiety, bipolar disorder, attention deficit hyperactivity disorder (ADHD), balance problems, obsessive-compulsive disorder (OCD), trouble sleeping (insomnia), tumors, tuberculosis, asthma, leukoderma (a skin condition marked by white patchiness), bronchitis, backache, fibromyalgia, menstrual problems, hiccups, Parkinson's disease, and chronic liver disease. It is also used to reduce side effects of medications and to treat cancer and schizophrenia. Ashwagandha is also used to reduce the levels of fat and sugar in the blood.					

(continued)

Table 4.129 (continued)

Family	Botanical name	Image	Plant part used	Bioactive compound(s)	Structure	References
	Solanum aculeastrum Dunal		Root, bark, leaf, fruit	Tomatidine	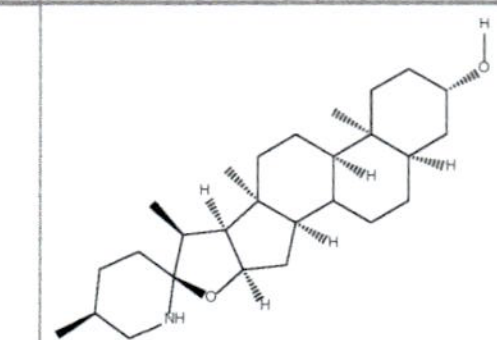	Koduru et al. (2007)
				Solasodine	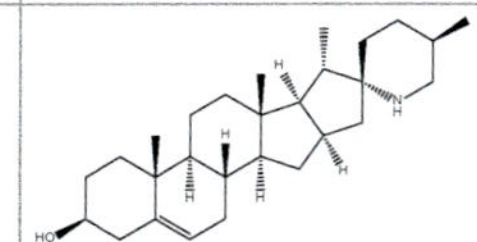	
	Synonym(s): *Solanum albifolium* C.H. Wright. **Common name(s)**: Goat bitter-apple, Poison apple. **Botanical description**: It is a shrub or small tree native to tropical Africa south to South Africa, in a wide range of soil, terrain, and climatic conditions. It is highly branched and reaches 1–5 m high, with numerous sharp, hooked, brown thorns. The leaves are ovate, up to 15 cm long and 13 cm broad, with lobed margins and downy undersides. It flowers (in South Africa) from September to July, peaking from November to March. The petals are white to pale violet surrounding the ovary; the flower also has a bitter, sour smell. These yield to fruit from April to January, peaking in June and November. **Medicinal use(s)**: Local healers use the extremely bitter berries for the treatment of various diseases in human beings and domestic animals. The fresh and boiled ripe berries are used to treat jigger wounds and gonorrhea.					

	Solanum dubium Dunal		Whole plant, fruit	β-Solamarine	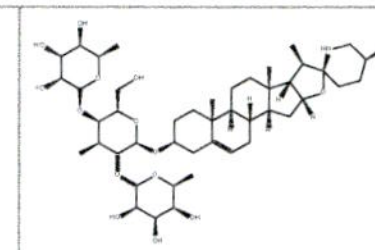	Kupchan et al. (1965) and El Kheir and Salih (1979)
	Synonym(s): No synonyms are recorded for this name. **Common name(s)**: Ground ivy. **Botanical description**: *Solanum dubium* Fresen, "Gubbain," is a well-kncwn wild plant that grows widely in Khartoum during the rainy season. It also grows in the west and east shores of the White Nile, South of the Blue Nile, Gezira, Kordofan, and Darfur regions. **Medicinal use(s)**: *S. dubium* seed in combination with honey is used for treatment of skin injuries and wounds.					
	Solanum schimperianum Hochst.		Aerial parts	Solanopubamine	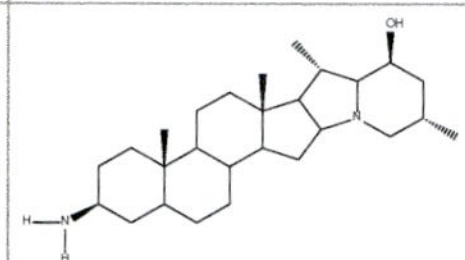	Al-Rehaily et al. (2013)
	Synonym(s): No synonyms are recorded for this name. **Common name(s)**: Climbing nightshade. **Botanical description**: NA. **Medicinal use(s)**: It is a medicinal plant that has long been used to treat various cancers and many other conditions in the Eastern Cape Province of South Africa.					

(continued)

Table 4.129 (continued)

Family	Botanical name	Image	Plant part used	Bioactive compound(s)	Structure	References
	Solanum torvum Swartz.		Fruit	Methyl caffeate	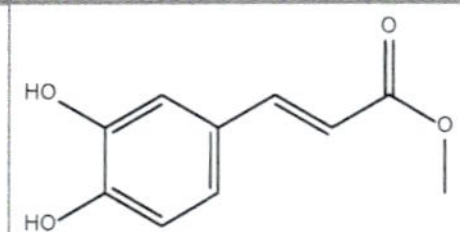	Kuo (2000), Panigrahi et al. (2014), and Balachandran et al. (2015)
				Solasodine	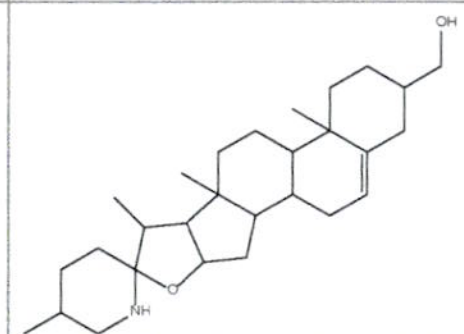	
	Synonym(s): *Solanum ficifolium* Ortega, *Solanum mayanum* Lundell. **Common name(s)**: Turkey berry, Prickly nightshade, Shoo-shoo bush, Wild eggplant, Pea eggplant, Pea aubergine. **Botanical description**: *Solanum torvum* is a shrub widely distributed in South India, Malaysia, China, the Philippines, Thailand, the West Indies, and tropical America. The plant is usually 2 or 3 m in height and 2 cm in basal diameter but may reach 5 m in height and 8 cm in basal diameter. The shrub usually has a single stem at ground level, but it may branch on the lower stem. The stem bark is gray and nearly smooth with raised lenticels. The inner bark has a green layer over an ivory color. The plants examined by the author, growing on firm soil, had weak taproots and well-developed laterals. The roots are white. Foliage is confined to the growing twigs. **Medicinal use(s)**: The fruits of *S. torvum* are edible and traditionally used for the treatment of abscesses, jigger wounds, skin infections, and athlete's foot. Pharmacological studies revealed antiviral, immunosecretory, antioxidant, analgesic, anti-inflammatory, and anti-ulcerogenic activities.					

	Solanum incanum L.		Fruit	Solamargine	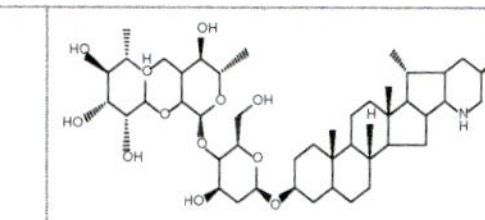	Kuo (2000)
	Synonym(s): *Solanum bojeri* Dunal, *Solanum sanctum* L. **Common name(s)**: Thorn apple. **Botanical description**: NA. **Medicinal use(s)**: NA.					
	Solanum americanum Mill.		Whole plant	Steroidal glycoside	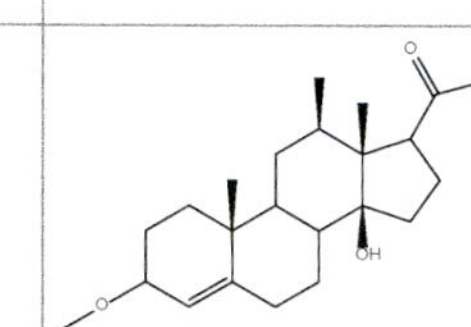	Ikeda et al. (2003) and Patel et al. (2009)
				Galactopyranoside	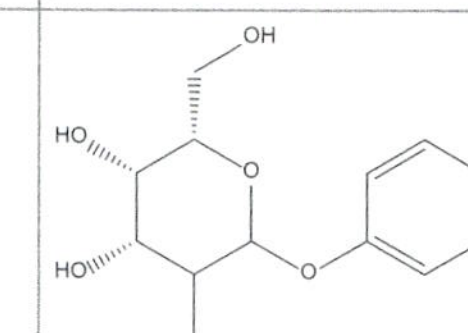	

(continued)

Table 4.129 (continued)

Family	Botanical name	Image	Plant part used	Bioactive compound(s)	Structure	References
	Synonym(s): *Solanum adventitium* Polgar, *Solanum amarantoides* Dunal. **Common name(s)**: Black night shade, Makoy, Deadly nightshade. **Botanical description**: Black nightshade is a common herb or short-lived perennial shrub, found in many wooded areas, as well as disturbed habitats. It reaches a height of 30–120 cm, leaves 4.0–7.5 cm long and 2–5 cm wide; ovate to heart-shaped, with wavy or large-toothed edges; both surfaces hairy or hairless; petiole 1–3 cm long with a winged upper portion. The flowers have petals greenish to whitish, recurved when aged and surround prominent bright yellow anthers. The berry is mostly 6–8 mm in diam., dull black or purple-black. In India, another strain is found with berries that turn red when ripe. **Medicinal use(s)**: It possesses medicinal properties like antimicrobial, antioxidant, cytotoxic, anti-ulcerogenic, and hepatoprotective activity.			Solamargine	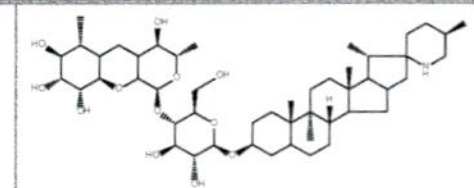	
				Solasonine	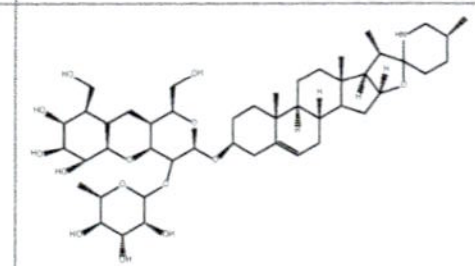	
	Solanum lyratum Thunb. in Murr.		Entire plant, aerial parts	Lyratol C	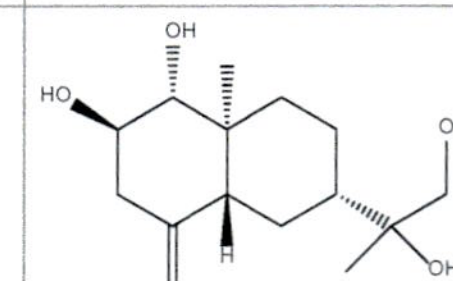	Hsu et al. (2008) and Ren et al. (2009)
				Lyratol D	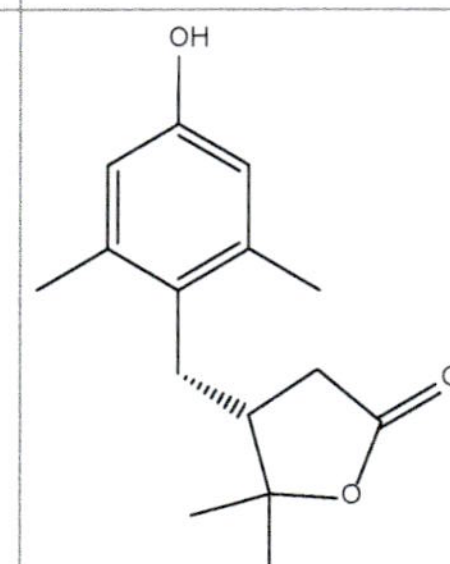	

				Dehydrovomifoliol	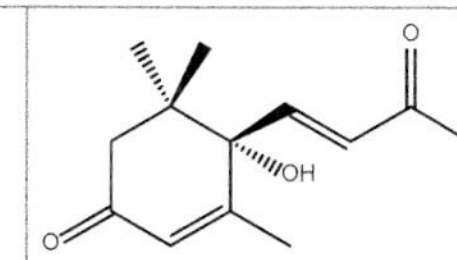	
				Blumenol A	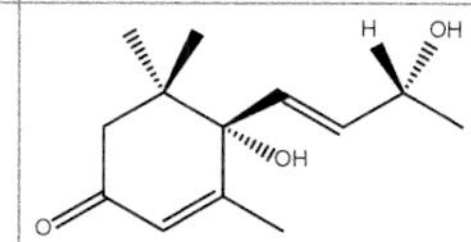	
	Synonym(s): *Solanum cathayanum* C.Y. Wu & S.C. Huang, *Solanum dulcamara* var. *chinense* Dunal. **Common name(s)**: Herba Solani Lyrati. **Botanical description**: *Solanum lyratum* is an outstretched shrub or herbaceous climber that lignifies at the base. It reaches stature heights of 0.5–3 m. The stems are pliable, not winged, and sparse to densely hairy. The hair is composed of translucent, glandular, and single-row trichomes of four to six cells with a length of 4 millimeters and a unicellular glandular tip, as well as shorter, glandular trichomes with a length of 0.5 mm. The trichomes are weak and confused; the hairiness of the young growth is similar to that of the stems. Older stems have a pale yellowish straw-colored bark and shed, as the longer trichomes usually break off and only the shorter trichomes remain. **Medicinal use(s)**: Plants of the species are used in traditional Chinese medicine under the name Baimaoteng among other things for the treatment of cancer, malaria, jaundice, edema, and hepatitis. Cytotoxic and apoptotic activity against different tumor cells could be detected. These effects are associated with various steroidal glucuronides and steroidal alkaloid glycosides that could be extracted from the plant.					

(continued)

Table 4.129 (continued)

Family	Botanical name	Image	Plant part used	Bioactive compound(s)	Structure	References
	Solanum indicum L.		Whole plant	β-Sitosterol	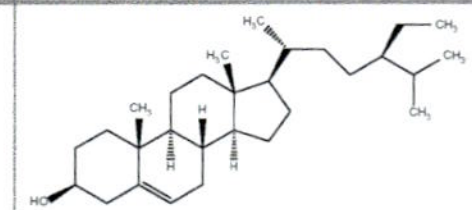	Chiang et al. (1991)
				Dioscin	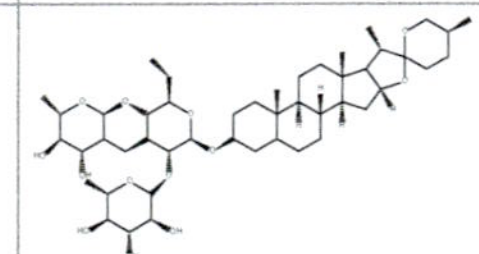	
	Synonym(s): No synonyms are recorded for this name. **Common name(s)**: Poison berry. **Botanical description**: It is an erect under-shrub, 0.3–1.5 m high. Stems are much branched, very prickly, and bearing compressed, stout, often recurved prickles. Leaves are ovate, 3.5–15 cm long, 2.5–8 cm wide, lobed or pinnatifid in the margins, blunt or pointed at the tip, pointed at the base, and stellately woolly beneath. Flowers are blue, borne in extraaxillary racemes. Fruit is yellow, rounded, about 1.5 cm in diameter. Root well developed, long, ribbed, woody, cylindrical, pale yellowish brown, 1–2.5 cm in dia., a number of secondary roots and their branches present, surface rough due to presence of longitudinal striations and root scars, fracture, short and splintery; no distinct odor and taste. **Medicinal use(s)**: Seeds, roots, leaves, and berries help in treating various diseases such as bronchitis, asthma, dry cough, rhinitis, dysuria, leucoderma, sexual disorders, insomnia, cardiac weakness, and pruritis.			Methyl protodioscin	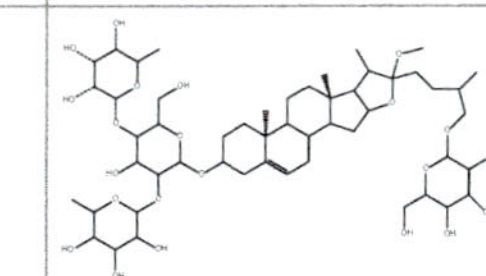	

Table 4.130 Anticancer plants of family Sparganiaceae

Family	Botanical name	Image	Plant part used	Bioactive compound(s)	Structure	References
Sparganiaceae	*Sparganium stoloniferum* (Buch.-Ham. ex Graebn.) Buch.-Ham. ex Juz.		Rhizome	Aqueous extract	Not available	Cho et al. (2006) and Sun et al. (2010a, b)
	Synonym(s): *Sparganium arcuscaulis* D.Yu & G.T.Yang, *Sparganium asiaticum* Graebn. **Common name(s)**: Bur-reed. **Botanical description**: It is widespread in wet areas in temperate regions of both the Northern and Southern Hemispheres. The plants are perennial marsh plants that can grow up to 3.5 m with epicene flowers. **Medicinal use(s)**: The herb in oriental medicine has been used for removing blood stasis, invigorating the flow of vital energy, dissolving lumps to relieve pain, etc.					

Table 4.131 Anticancer plants of family Stemonaceae

Family	Botanical name	Image	Plant part used	Bioactive compound(s)	Structure	References
Stemonaceae	*Stemona collinsiae* Craib.		Root	Dichloromethane-methanol (1:1) extract, 95% ethanol extract, aqueous extract	Not available	Akanitapichat et al. (2005)
	Synonym(s): No synonyms are recorded for this name. **Common name(s)**: Non tai yak. **Botanical description**: *Stemona collinsiae* is native to China, the Indian subcontinent, Southeast Asia, Papuasia, and Northern Australia. It is an erect plant, but as it grows older, it becomes a climbing or trailing plant up to 60 cm tall. The plant is a source of an insecticide. It is found in dry evergreen, dry deciduous and bamboo forest, mixed deciduous scrub and swampy thicket, and rugged limestone terrain at elevations from 25 to 450 m. **Medicinal use(s)**: An insecticide is obtained from the plant. The roots are collected when the plant withers, they are well bruised and soaked in water, and then the water is used for spraying on plants in order to kill any pests attacking them.					

Table 4.132 Anticancer plants of family Sterculiaceae

Family	Botanical name	Image	Plant part used	Bioactive compound(s)	Structure	References
Sterculiaceae	*Pterospermum acerifolium* (L.) Willd.		Leaf, root, branches	Ethanol and water extract	Not available	Dimayuga et al. (1987), Saboo et al. (2007), and Biswal and Tripathi (2019)
	Synonym(s): *Cavanilla acerifolia* (L.) J.F.Gmel., *Dombeya acerifolia* (L.) Gaertn. **Common name(s)**: Bayur tree, Maple-leafed bayur tree, Dinner plate tree. **Botanical description**: It is a flowering plant indigenous to Southeast Asia, from India to Burma. It is most likely to grow naturally along forested stream banks. The best growing conditions are a seasonally moist and then dry climate with access to full sunlight. **Medicinal use(s)**: The leaves can also serve as a primitive method of reenforcing roofs and preventing leaks. The pubescent under surface of the leaves is said to stop bleeding and can be used as tinder for a means of sparking fires. The flowers of the bayur tree can serve as a pleasant perfume and can even keep away insects.					

(continued)

Table 4.132 (continued)

Family	Botanical name	Image	Plant part used	Bioactive compound(s)	Structure	References
	Theobroma cacao L		Seeds	Pentameric procyanidin	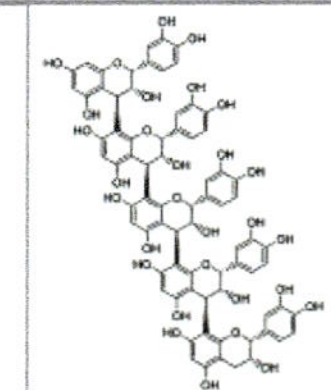	Preza et al. (2010) and Ramljak et al. (2005)
	Synonym(s): *Theobroma cacao* subsp. *cacao*, *Theobroma cacao* f. *leiocarpum* (Bernoulli) Ducke. **Common name(s)**: Cocoa tree. **Botanical description**: *Theobroma cacao* or cacao is a small, evergreen tree about 8 m in height and 30 cm in trunk diameter. It is native to Central and South America. The leaves are large, simple, and alternate, about 40 cm long and 5–20 cm broad. The flowers, small with pink calyx, occur in clusters on the trunk and older branches. It is pollinated by small flies. Fruits are ovoid, yellow to orange pods containing 20–60 seeds each. **Medicinal use(s)**: It is used to stimulate the nervous system, lower blood pressure, dilate the coronary arteries, and sooth and soften damaged skin. It is also used to treat anemia, angina, bruises, chapped skin and burns, diarrhea, and leprosy spots.					

Table 4.133 Anticancer plants of family Symplocaceae

Family	Botanical name	Image	Plant part used	Bioactive compound(s)	Structure	References
Symplocaceae	*Symplocos racemosa* Roxb.		Bark	Cyclophosphamide	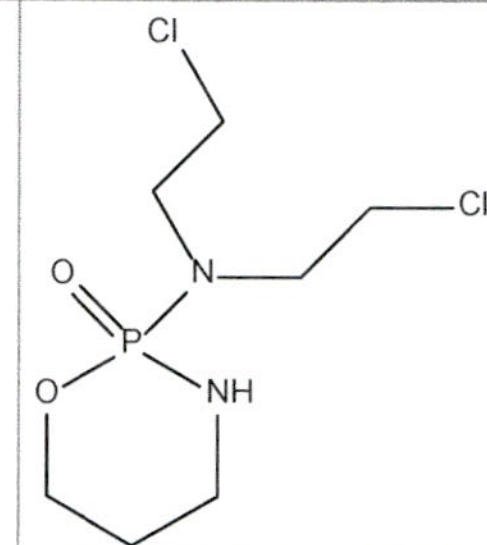	Bhuvan et al. (2009) and Raval et al. (2009)
	Synonym(s): *Symplocos intermedia* Brand, *Symplocos intermedia* var. *trichantha* Hand.-Mazz. **Common name(s)**: Lodh tree. **Botanical description**: *Symplocos racemosa* Roxb. is an evergreen Ayurvedic plant widely distributed in the tropics and subtropics of Asia, Australia, and America. Leaves simple, alternate, spiral; petiole up to 1.5 cm long, planoconvex in cross section, glabrous; lamina 6.5–12.5 × 3–4.3 cm, oblanceolate to narrow elliptic, apex narrowly acuminate, base acute to attenuate, margin serrate and slightly recurved, glabrous; midrib canaliculate above; secondary nerves 6–12 pairs; tertiary nerves obliquely distantly percurrent. **Medicinal use(s)**: This weed possesses a wide range of ethnomedicinal uses including treatment for dysentery, bowel complaints, inflammations, foul vaginal discharges, abortion and miscarriages, and snakebites. The bark is used as astringent, expectorant, anti-inflammatory, febrifuge, hemostatic, stomachic, suppurative and for constipation.					

(continued)

Table 4.133 (continued)

Family	Botanical name	Image	Plant part used	Bioactive compound(s)	Structure	References
	Symplocos cochinchinensis (Lour.) S. Moore		Bark	Methanol extract	Not available	Abida et al. (2016)
	Synonym(s): *Symplocos cochinchinensis* var. *cochinchinensis*, *Symplocos cochinchinensis* subsp. *cochinchinensis*. **Common name(s)**: Not available. **Botanical description**: *Symplocos cochinchinensis* is an evergreen tree that can grow up to 35 m tall. The straight, cylindrical bole is up to 50 cm in diameter. The tree is sometimes harvested from the wild for local use as a food, dye, and source of beads, tannins, and wood. The variety sessifolia is used as a pioneer species when reclaiming crater fields in volcanic areas. **Medicinal use(s)**: It is a medicinal plant with anti-inflammatory, antitumor, antimicrobial, and antidiabetic properties.					

Table 4.134 Anticancer plants of family Taxaceae

Family	Botanical name	Image	Plant part used	Bioactive compound(s)	Structure	References
Taxaceae	*Taxus brevifolia* Nutt.		Bark	Paclitaxel	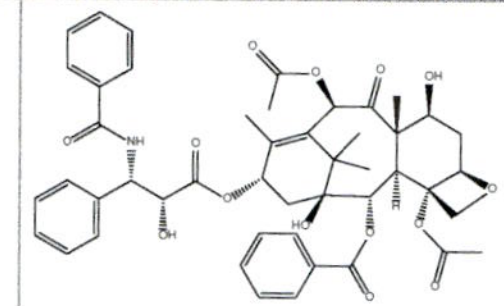	Forlines et al. (1992) and Dinda et al. (2013)
	Synonym(s): *Taxus baccata* var. *brevifolia* (Nutt.) Koehne, *Taxus bourcieri* Carrière, *Taxus lindleyana* A. Murray bis. **Common name(s)**: Pacific yew. **Botanical description**: It is native to western North America – Alaska to California. It is an evergreen tree growing up to 15 m at a slow rate. The flowers are dioecious (individual flowers are either male or female, but only one sex is to be found on any one plant, so both male and female plants must be grown if seed is required) and are pollinated by wind. It is suitable to grow in light (sandy), medium (loamy), and heavy (clay) soils and prefers well-drained soil. It can grow in full shade (deep woodland), semi-shade (light woodland), or no shade. **Medicinal use(s)**: It is used to treat breast, skin, lung, neck, and ovarian cancer, diphtheria, tapeworm infections, swollen tonsils (tonsillitis), seizures (epilepsy), muscle and joint pain (rheumatism), urinary tract conditions, and liver infections. Women use it for starting menstruation and causing abortions. A decoction of small woody pieces is used for the treatment of internal complaints including stomach pains and blood in the urine. The leaves are harvested in early autumn or spring and the bark from autumn to spring, for commercial extraction of taxol.			Docetaxel	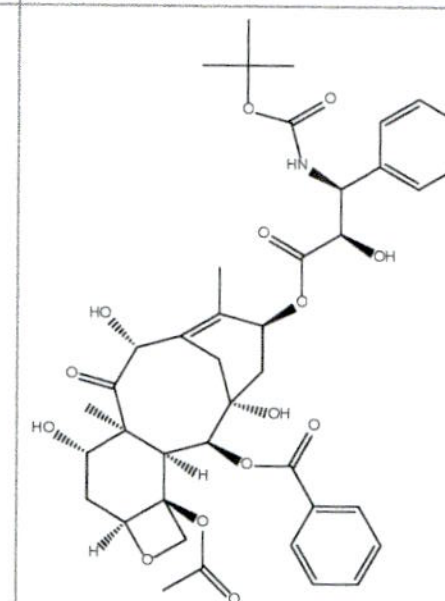	

Table 4.135 Anticancer plants of family Taxodiaceae

Family	Botanical name	Image	Plant part used	Bioactive compound(s)	Structure	References
Taxodiaceae	*Glyptostrobus pensilis* (Staunton ex D.Don) K.Koch		Stem	Quercetin	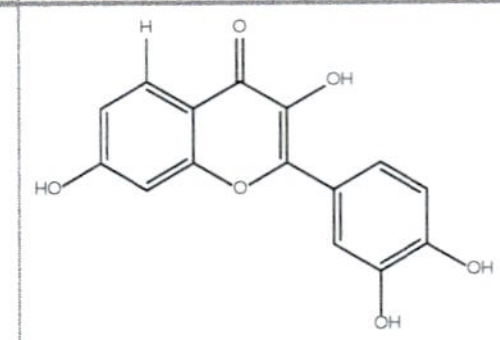	Duke and Ayensu (1985a, b) and Maffei (2003a, b, c, d, e, f, g, h, i, j, k, l, m)
	Synonym(s): *Cuprespinnata heterophylla* (Brongn.) J. Nelson, *Cuprespinnata sinensis* (J. Forbes) J. Nelson, *Cupressus nucifera* Carrière, *Juniperus aquatica* Roxb. **Common name(s):** Chinese swamp cypress. **Botanical description**: It is native to East Asia, Southern China, Vietnam, and Laos. It is a deciduous tree growing up to 10 m at a medium rate. It is in flower from January to March, and the seeds ripen from September to March. The flowers are monoecious (individual flowers are either male or female, but both sexes can be found on the same plant) and are pollinated by wind. **Medicinal use(s)**: It possesses anticancer properties. It is used as antidote and oxytocic. It is also used to cure ascites, treat animal bites, and treat dropsy of pregnant women. A decoction of the shoots is used for the treatment of fever, hepatitis, skin complaints, etc.			Quercetin-3-O-β-D-galactopyranoside	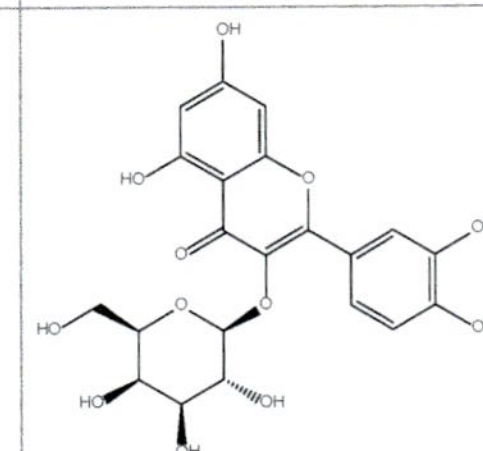	

Table 4.136 Anticancer plants of family Theaceae

Family	Botanical name	Image	Plant part used	Bioactive compound(s)	Structure	References
Theaceae	*Camellia sinensis* (L.) Kuntze		Leaf	Epicatechin	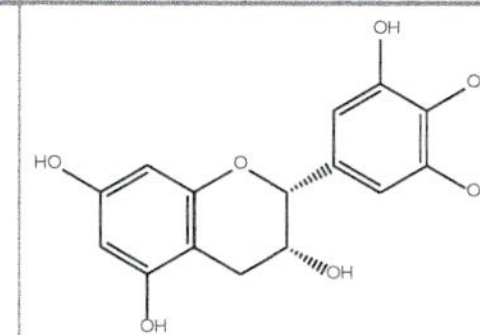	Ravindranath et al. (2006) and Rafieian-Kopaei and Movahedi (2017)
	Synonym(s): *Camellia sinensis* var. *kucha* H.T. Chang & S.S.Wang, *Camellia sinensis* f. *parvifolia* (Miq.) Sealy, *Camellia sinensis* f. *rosea* (Makino) Kitam., *Camellia sinensis* var. *sinensis*. **Common name(s)**: Mecha, Gyokura, Bancha, Kukicha, Asian tea, Senchu, Chinese tea, Japanese tea. **Botanical description**: It is native to East Asia – the exact origin is uncertain, but it is probably wild in Southern China, Northeast India, through Indochina, Southern Japan, and Korea. It is an evergreen, multi-stemmed shrub growing up to 3 m tall or a tree that can reach 10–15 m tall with one main stem. In cultivation the plants are usually kept pruned to a height of 1–1.5 m and trained as a low profusely branching and spreading bush. The leaves are from 3 to 14 cm in length and 2–5 cm in width, light green when young with short white hairs on the underside, become darker in maturity. **Medicinal use(s)**: It is used to treat anal, bladder, cervical, blood, liver, kidney, breast, ovarian, pancreatic, prostate, uterine, colon, lung, and skin cancer. It also possesses anticarcinogenic, antiaging, antimalarial, antioxidant, anti-inflammatory, anti-arthritic, antianemic, antithyroid, anti-rheumatic, anti-ulcer, anti-stress, antimicrobial, antibacterial, antiviral, and antidiabetic properties. It is effective in lowering blood sugar levels. It protects the liver from the damaging effects of toxic substances such as alcohol, including viral inflammation of the liver. It has less caffeine than coffee; it refreshes the body, particularly the central nervous system, skeletal muscles, heart, and liver.			Epigallocatechin (EGC)	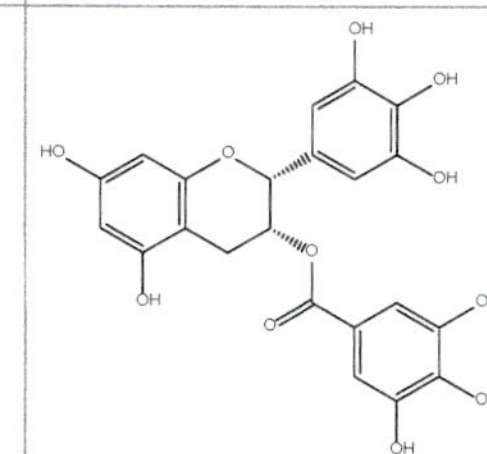	

Table 4.137 Anticancer plants of family Thymelaeaceae

Family	Botanical name	Image	Plant part used	Bioactive compound(s)	Structure	References
Thymelaeaceae	*Daphne mucronata* Royle		Stem	Allamandin	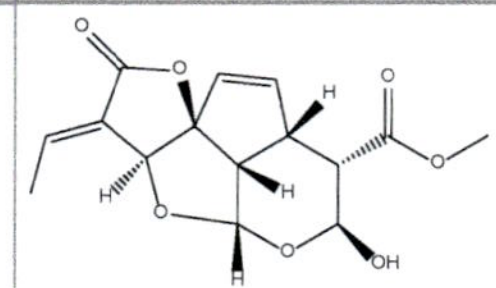	Sadeghi and Yazdanparast (2003)
	Synonym(s): *Daphne acuminata* Stocks, *Daphne angustifolia* K.Koch, *Daphne coriacea* Royle, *Daphne escalerae* Pau. **Common name(s)**: Kashmir daphne. **Botanical description**: It is native to the Himalayas, from Pakistan to Himachal Pradesh at altitudes of 1500–2700 m. It is common in Kashmir and Chenab valleys. It is an erect, much-branched evergreen wiry shrub, with narrowly inverted lance-shaped, gray leathery leaves, and creamy white or yellowish fragrant flowers borne in dense clusters at branch ends. Flowers have a tube of 1 cm long, densely woolly haired outside, petals spreading, ovate blunt. Leaves are 2–4 cm long, stalkless, alternately arranged on the branches. Stems are 2 m or more tall, with young branchlets hairy, reddish brown, older branches grayish. Fruit is ellipsoid, reddish orange, somewhat hairy, enclose in enlarged sepals. **Medicinal use(s)**: It is used to treat breast cancer, cervical cancer, and kidney cancer. It also possesses anti-analgesic, anti-inflammatory, antimicrobial, and antibacterial properties.					

	Phaleria macrocarpa (Scheff.) Boerl (Pm)		Fruit	Campesterol	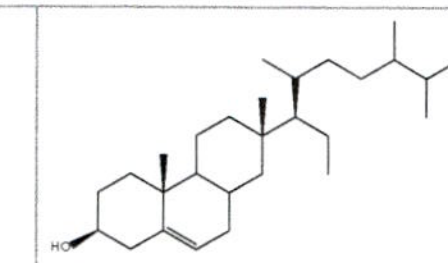	Chong et al. (2011)
	Synonym(s): *Phaleria calantha* Gilg, *Phaleria papuana* Warb. ex K. Schum. & Lauterb., *Phaleria wichmannii* Valeton. **Common name(s)**: The crown of Gods, Crown of the Lord, Makuto rojo, Makutadewa, Makuto mewo, Makuto queen. **Botanical description**: It is native to Asia including China and is very common in Indonesia and surrounding islands and Malaysia and New Guinea. Its height ranges from 1 to 18 m with greenish bark and white wood. It has green, tapered leaves. The flowers are made up to two to four petals and range from green to maroon in color. It grows 10–1200 m above sea level with a productive age that ranges from 10 to 20 years. The leaves are green and tapering with length and width ranging from 7 to 10 cm and 3 to 5 cm, respectively. Eclipse-shaped fruits are green when unripe and have a diameter of 3 cm. They are red when ripe. **Medicinal use(s)**: It is used for the treatment of bone cancer, breast cancer, cervical cancer, tumor, and lung, liver, and heart diseases. It also possesses anti-inflammatory, antibacterial, antifungal, antioxidant, and antihyperlipidemic properties.					

Table 4.138 Anticancer plants of family Tremellaceae

Family	Botanical name	Image	Plant part used	Bioactive compound(s)	Structure	References
Tremellaceae	*Tremella fuciformis* Berk.		Entire plant	Carvacrol	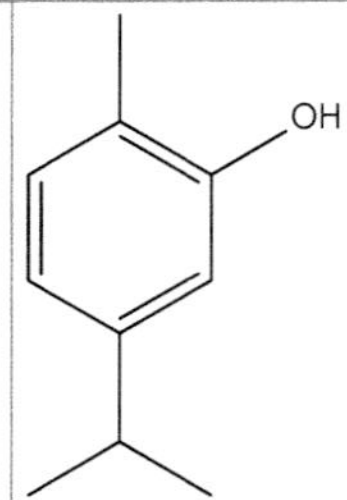	Lin et al. (1984a, b) and Shen et al. (2017)
	Synonym(s): *Nakaiomyces nipponicus* Kobayasi. **Common name(s)**: Snow fungus, Snow ear, Silver ear fungus, and White jelly mushroom. **Botanical description**: It is native to Asia, China and Japan, and Europe, Australia and New Zealand, and Southern and Eastern Asia. Fruit bodies are gelatinous, watery white, up to 7.5 cm across and composed of thin but erect, seaweed-like, branching fronds, often crisped at the edges. Microscopically, the hyphae are clamped and occur in a dense gelatinous matrix. Haustorial cells arise on the hyphae, producing filaments that attach to and penetrate the hyphae of the host. The basidia are tremelloid, 10–13 by 6.5–10 μm, sometimes stalked. The basidiospores are ellipsoid, smooth, 5–8 by 4–6 μm, and germinate by hyphal tube or by yeast cells. **Medicinal use(s)**: It is used for the treatment of lung cancer, cervical cancer, tumors, and lung diseases. It also possesses anticancer, antitumor, anti-inflammatory, antioxidant, antiaging, antihypertensive, and antimicrobial properties. It is used to treat dry cough and high blood pressure.					

Table 4.139 Anticancer plants of family Trilliaceae

Family	Plant	Image	Part	Compound	Structure	Reference
Trilliaceae	*Paris polyphylla* Sm. var. *yunnanensis* (Fr.)		Leaf	Polyphyllin D	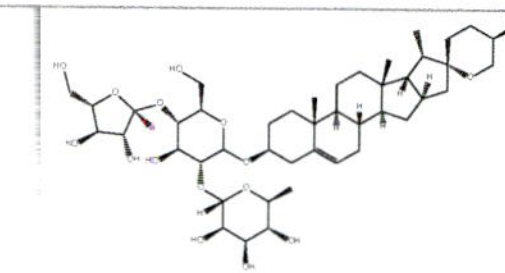	Wang and Liu (2010)
	Synonym(s): *Daiswa polyphylla* (Sm.) Raf., *Paris biondii* Pamp., *Paris daiswus* Buch.-Ham. ex D.Don., *Paris debeauxii* H.Lév., *Paris kwantungensis* R.H.Miao., *Paris polyphylla* var. *emeiensis* H.X.Yin, Hao Zhang & D.Xue. **Common name(s)**: Herb paris. **Botanical description**: It is native to China, the Indian subcontinent, and Indochina. It is a perennial plant growing up to 1 m by 0.3 m. It is in flower from July to August. The flowers are hermaphrodite. It is suitable to grow in light and medium soils. It can grow in full shade or semi-shade, broad-leaved and mixed woodlands to 3000 m in the Himalayan forests, bamboo forests, thickets, grassy or rocky slopes, and streamsides, 100–3500 m in Western China. **Medicinal use(s)**: It is used for the treatment of lung, liver, breast, esophageal, and brain cancers. It also possesses antibacterial, anti-inflammatory, anti-spasmodic, and analgesic properties. The leaves and berries are used as an allopathic medicine.			Diosgenin	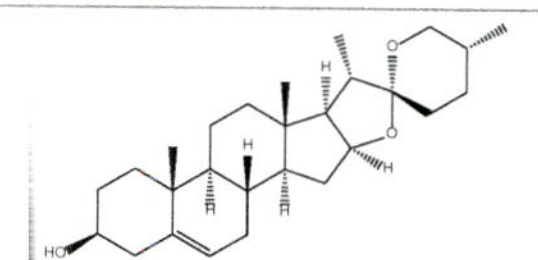	
				Dioscin	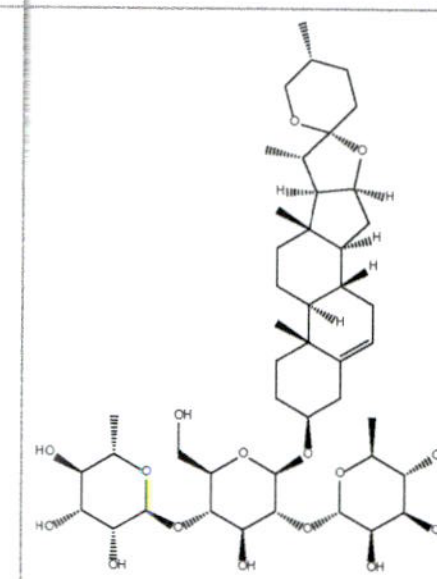	
				Triterpenoids (saponins)	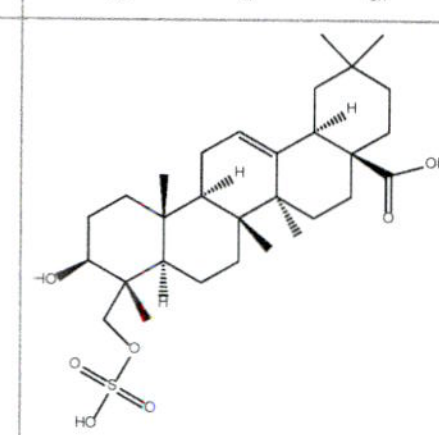	

Table 4.140 Anticancer plants of family Tropaeolaceae

Family	Botanical name	Image	Plant part used	Bioactive compound(s)	Structure	References
Tropaeolaceae	*Tropaeolum majus* L.		Leaf and stem	Kaempferol	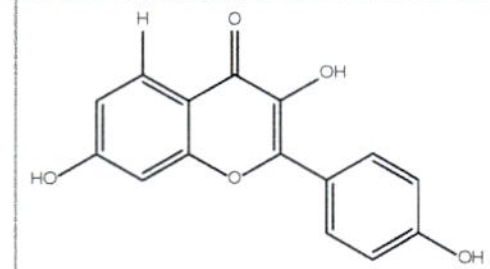	Morton (1977) and Kim et al. (2017)
	Synonym(s): *Cardamindum majus* Moench., *Nasturtium indicum* Garsault., *Tropaeolum elatum* Salisb., *Tropaeolum hortense* Sparre., *Tropaeolum hybridum* L., *Tropaeolum pinnatum* Andrews., *Tropaeolum quinquelobum* Bergius., *Trophaeum majus* (L.) Kuntze. **Common name(s)**: Indian cress. **Botanical description**: It is native to South and Central America. It can grow up to 2–3 m long during the summer. The stalked leaves have tissue that is touch-sensitive, causing it to bend around the bract and the plant climbs up trees, bushes or large herb-stemmed plants whenever it gets the chance. Its base can be very shady underneath other plants, but the shoots reach out towards the light. Garden nasturtium can also be grown on rocks, climbing up them or sprawling along the ground. Its leaf stalks are attached in a very distinctive fashion in the middle of the blade, and this kind of leaf type is called peltate. If the plant survives the spring frosts, it can flower quite well, ripen its seeds, sow itself, and adorn the same place with its colorful flowers year after year. **Medicinal use(s)**: It is used to treat lung cancer. It also possesses anticancer, antibiotic, antimicrobial, antifungal, and antiseptic properties. It is used for the treatment of bacterial infection affecting the skin, urinary tract, and respiratory and digestive system. It expels the mucus out of the chest and kills the bacteria. It is used to treat bones diseases and skin eruptions. A juice extracted by crushing the plant is used to improve blood circulation in the legs and toes.			Glucoside	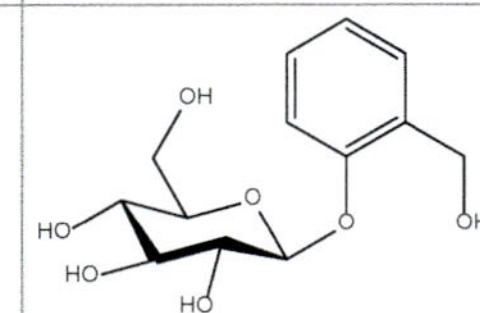	
				Quercetin-3-O-glucoside	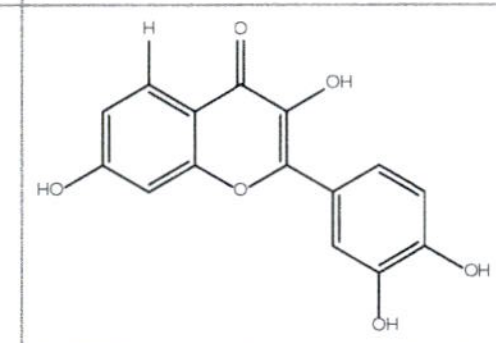	
				Glucotropaeolin	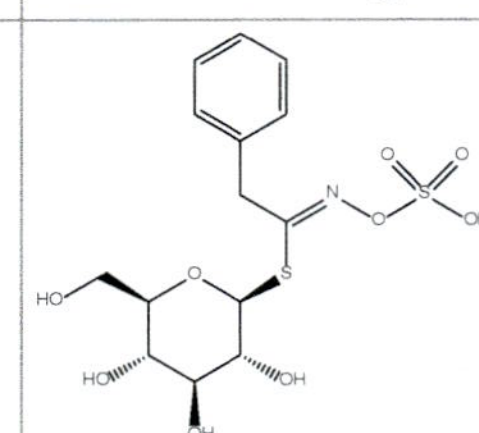	

Table 4.141 Anticancer plants of family Ulmaceae

Family	Botanical name	Image	Plant part used	Bioactive compound(s)	Structure	References
	Holoptelea integrifolia (Roxb.) Planch.		Bark	Beta-sitosterol	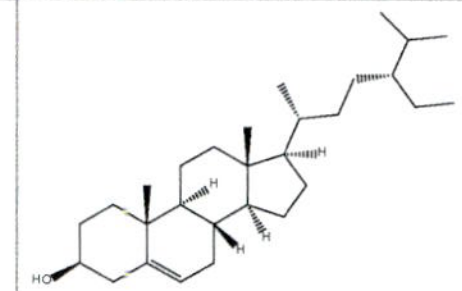	Sabnis and Bedi (1983) and Lakshmi et al. (2010)
	Synonym(s): *Ulmus integrifolia* Roxb. **Common name(s)**: Indian elm. **Botanical description**: It is native to East Asia – India, Sri Lanka, Nepal, Myanmar, Thailand, Cambodia, Laos, and Vietnam. It is a deciduous tree growing up to 22 m tall. The bole is often much twisted and branching. Bark is 6–8 mm thick, whitish gray, smooth with pubescent branchlets. Leaves are simple, alternate, stipulate, ovate or elliptic-ovate, and acuminate in shape. The bark when cut and leaves when crushed emit an unpleasant odor. Flowers and fruits are seen during the months of February–March. Seeds are small and whitish and are kidney-shaped. **Medicinal use(s)**: It is used for the treatment of intestinal cancer and bladder cancer. It is used traditionally for the treatment of inflammation, gastritis, dyspepsia, colic, intestinal worm infection, vomiting, leprosy, diabetes, hemorrhoids, dysmenorrhea, and rheumatism and for wound healing. Stem bark acts as an anti-inflammatory agent specifically for the eyes. It also possesses antibacterial, antifungal, antioxidant, anti-inflammatory, antidiabetic, and antitumor properties.			Friedlin	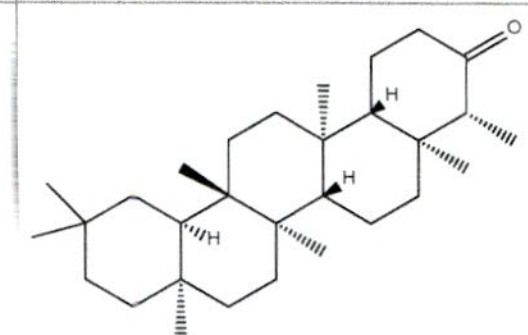	

(continued)

Table 4.141 (continued)

Family	Botanical name	Image	Plant part used	Bioactive compound(s)	Structure	References
				Betulinic acid	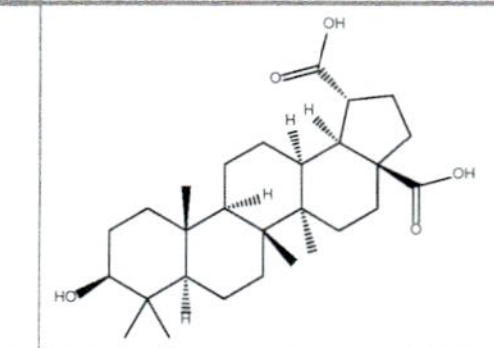	
	Ulmus davidiana Planch.		Root bark	Mansanone-E Camptothecin	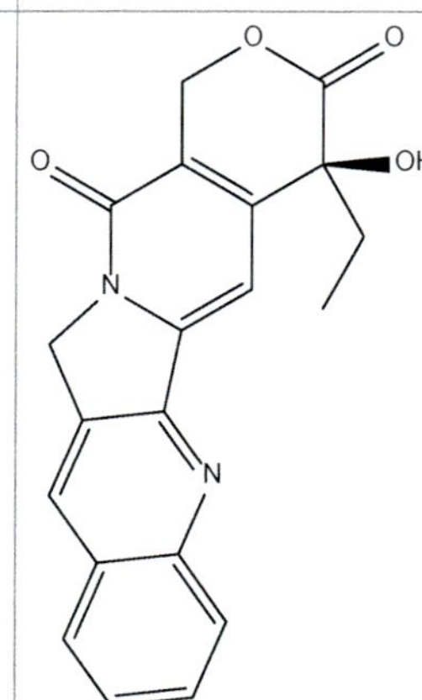	Kim et al. (1996a, b) and Lee et al. (2016a)
	Synonym(s): *Ulmus davidiana* var. *davidiana*, *Ulmus davidiana* var. *mandshurica* Skvortsov., *Ulmus davidiana* var. *pubescens* Skvortsov. **Common name(s)**: Japanese elm. **Botanical description**: It is native to East Asia – China, Japan, and Korea. It is a deciduous tree growing up to 15 m. The flowers are hermaphrodite and are pollinated by wind. It is suitable to grow in light, medium, and heavy soils and prefers well-drained soil. It is habituated to slopes, wetlands near streams, and valleys at elevations of 2000–2300 m. **Medicinal use(s)**: It is used for the treatment of lung cancer and human gastric cancer. It also possesses anti-inflammatory, anti-glycation, and anti-angiogenic properties.					

	Ulmus manschurica Nakai/*Ulmus manschurica* Nakai		Bark	Beta-sitosterol	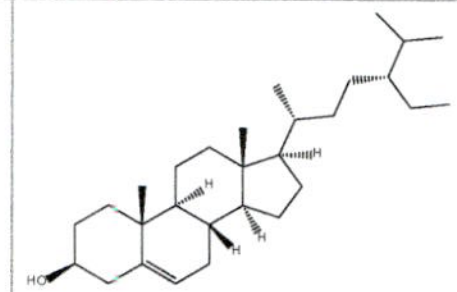	Rhan et al. (1992)
	Synonym(s): *Ulmus campestris* f. *pumila* Herder., *Ulmus campestris* var. *pumila* Maxim., *Ulmus manshurica* Nakai., *Ulmus pumila* var. *microphylla* Pers., *Ulmus turkestanica* Req., *Ulmus pinnato-ramosa* Dieck. **Common name(s)**: Chinese elm. **Botanical description**: It is native to Central Asia, Eastern Siberia, Mongolia, Xizang (Tibet), Northern China, India (Northern Kashmir), and Korea. It is usually a small- to medium-sized, often bushy, tree growing to 10–20 m tall, with a trunk up to 80 cm. The leaves are deciduous in cold areas, but semi-evergreen in warmer climates, 7 cm long and 3 cm broad, with an oblique base and a coarsely serrated margin, changing from dark green to yellow in autumn. The tree is short-lived in temperate climates, rarely reaching more than 60 years of age, but in its native environment may live to between 100 and 150 years. **Medicinal use(s)**: It is used for the treatment of lung cancer and gastric cancer cells. It also possesses anti-inflammatory, anti-glycation, and anti-angiogenic properties.					

Table 4.142 Anticancer plants of family Umbelliferae

Family	Botanical name	Image	Plant part used	Bioactive compound(s)	Structure	References
Umbelliferae	*Cnidium monnieri* (L.) Cusson		All plant parts	Osthol	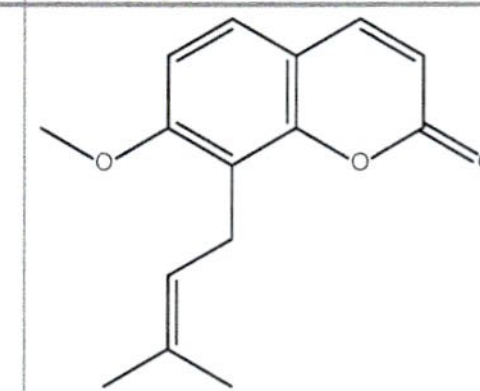	Chou et al. (2007), Pan et al. (2015), and Lim et al. (2017)
	Synonym(s): *Cicuta monnieri* (L.) Crantz., *Cicuta sinensis* Zuccagni., *Cnidium confertum* Moench., *Cnidium microcarpum* Turcz., *Cnidium mongolicum* H.Wolff., *Cnidium monnieri* var. *monnieri*. **Common name(s)**: Shechuangzi, Osthole, Jashoshi. **Botanical description**: It is native to China, Vietnam, Laos, Siberia, and Eastern Europe. Plants annual, 10–60 cm. Taproot 2–3 mm thick. Stem solitary, striate, scabrous. Lower petioles 3–8 cm, blade ovate-lanceolate, 3–8 × 2–5 cm, 2–3 pinnate; ultimate segments linear to linear-lanceolate, 3–10 × 1–1.5 mm, veins and margins scabrous. Umbels 2–3 cm across; bracts 6–10, linear to linear-lanceolate, 2–3 mm, persistent, margins narrowly white membranous, very finely ciliate; rays 8–20, 5–20 mm, unequal; bracteoles 5–9, linear, nearly equal pedicels, margins ciliate; umbellules 15–20 flowered; pedicels 3–5 mm. Calyx teeth obsolete or minute. Stylopodium conic; styles 3–4 times longer than stylopodium. Fruit ovoid, 1.5–3 × 1–2 mm; lateral ribs slightly broader than the dorsal. **Medicinal use(s)**: It is used for the treatment of breast cancer, colon cancer, and lung cancer. It also possesses anti-inflammatory, antibacterial, and antioxidant properties. It is also used to treat headache, back pain, stomach pain, indigestion, dizziness, or drops in blood pressure.					

	Hydrocotyle sibthorpioides Lam.		Whole plant	Carvacrol	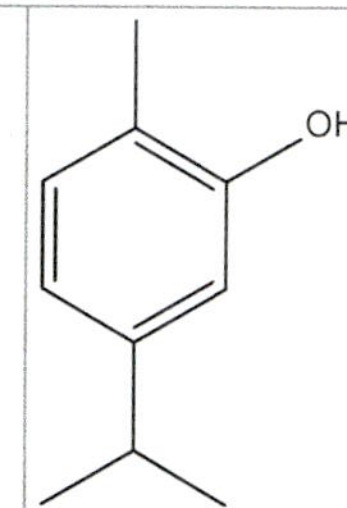	Desai et al. (2008) and Yu et al. (2007)
	Synonym(s): *Chondrocarpus sibthorpioides* Sweet., *Hydrocotyle americana* var. *monticola* (Hook. f.) Hiern., *Hydrocotyle keelungensis* Liu, Chao & Chuang., *Hydrocotyle monticola* Hook. f., *Hydrocotyle nitidula* A.Rich., *Hydrocotyle perexigua* Hance., *Hydrocotyle puncticulata* Miq., *Hydrocotyle rotundifolia* Roxb. **Common name(s)**: Carrot family, Lawn pennywort. **Botanical description**: It is native to tropical Africa, East Asia, China, Japan, India, Malaysia, and Indochina. It is a low-growing perennial plant that only reaches a height of 10 cm but spreads to form clumps 30 cm or more wide. **Medicinal use(s)**: It is used for the treatment of uterine cervical cancer, liver cancer, abscesses, boils, bruises, cirrhosis, colds, coughs, hepatitis, hepatoma, influenza, itch, jaundice, sinusitis, and sore throat. It also possesses antioxidant, antitussive, antitumor, anti-heptic, anti-hepatitis, and anti-inflammatory properties.			Fluorcuracil	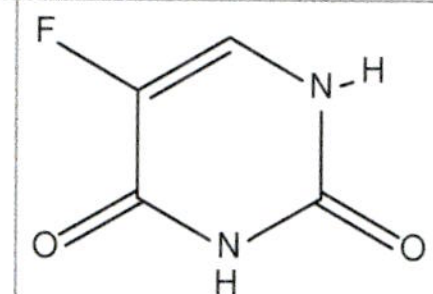	

(continued)

Table 4.142 (continued)

Family	Botanical name	Image	Plant part used	Bioactive compound(s)	Structure	References
	Oenanthe javanica (Blume) DC.		Leaf, stem, and roots	Corynanthe	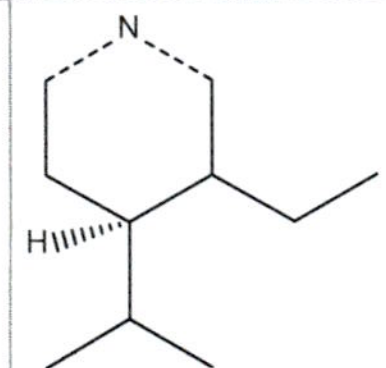	Kim et al. (2015a, b, c, 2016)
	Synonym(s): *Cyssopetalum javanicum* Turcz., *Dasyloma corticatum* Miq., *Dasyloma japonicum* Miq., *Dasyloma javanicum* (Blume) Miq., *Dasyloma laciniatum* (Blume) Miq., *Dasyloma latifolium* Lindl. **Common name(s)**: Water dropwort. **Botanical description**: It is native to Sakhalin Island, Japan, China, India, Thailand, the Philippines, Malaysia, Indonesia, New Guinea, and Northern Australia. It is a perennial plant growing up to 1 m. It is in flower from June to August, and the seeds ripen from August to October. The flowers are hermaphrodite and are pollinated by insects. It is habituated in ditches, ponds, and wet places in lowland areas all over Japan, marshlands, lakeshores, muddy stream banks, and shallow water at elevations of 600–3000 m in most parts of China. **Medicinal use(s)**: It is used for the treatment of colon cancer, epidemic influenza, fever and discomfort, jaundice, hematuria, and metrorrhagia. It also possesses antimutagenic, antioxidant, antimicrobial, and anti-inflammatory properties.			Methoxsalen	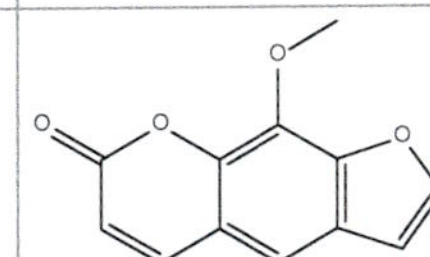	
				Isopimpinellin	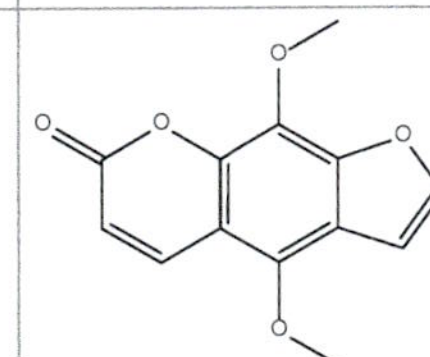	
				Bergapten	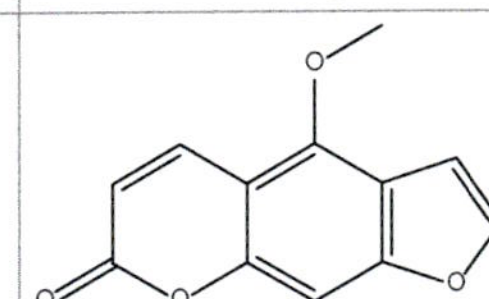	

	Sanicula caerulescens Franch.			Lute n	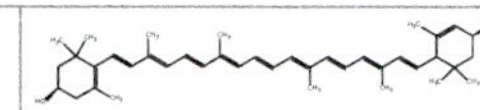	
	Synonym(s): *Sanicula dielsiana* H. Wolff., *Sanicula erythrophylla* Bobrov., *Sanicula stapfiana* H. Wolff. **Common name(s)**: Blue sanicle. **Botanical description**: It is native to China. Its height is up to 0.25–0.50 ft. It spread up to 0.50–1.50 ft. It grows in moist mountain woodlands. Plants form a low mound of divided bronzy-green leaves, with taller branching heads of small mauve-blue flowers appearing for weeks on end in the spring. **Medicinal use(s)**: It is used for the treatment of breast cancer, liver diseases, psoriasis, and tumor. It also possesses antioxidant, anti-inflammatory, and antibacterial properties.					

Table 4.143 Anticancer plants of family Urticaceae

Family	Botanical name	Image	Plant part used	Bioactive compound(s)	Structure	References
Urticaceae	*Laportea crenulata* Gaud		Roots	Berberine	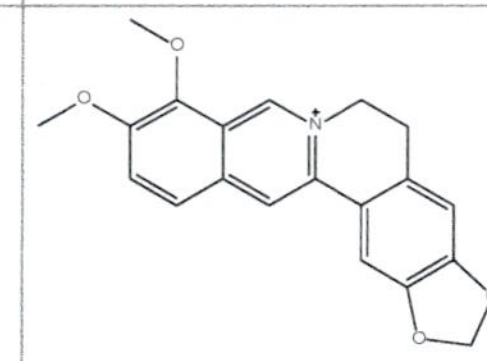	Rahman et al. (2009)
	Synonym(s): *Dendrocnide crenulata* Miq., *Dendrocnide microstigma* (Gaud. ex Wedd.) Chew., *Laportea crenulata* Gaud., *Laportea gigantea* Gaud., *Laportea integrifolia* C.Y. Wu., *Urera commersoniana* Wedd., *Urera crenulata* Wedd. **Common name(s)**: Batal. **Botanical description**: It is native to East Asia, China, and Japan. A shrub or small tree. Leaves ovate or elliptic, crenulate in the upper part or nearly entire. Flowers minute, green, dioecious, in axillary panicled cymes. It is habituated in woods in mountains all over Japan and in forest margins, thickets, roadsides, often partly shady, moist places at elevations of 700–3500 m in China. It is in flower from July to August, and the seeds ripen from August to December. The flowers are monoecious. It is suitable to grow in medium and heavy soils. **Medicinal use(s)**: It is used for the treatment of breast cancer, stomach cancer, inflammation, tumor, pain, heart diseases, and fever. It also possesses antioxidant, anti-inflammatory, and antibacterial activities.					

	Myrianthus arboreus Beauv.		Leaf, trunkwood	Dregamine	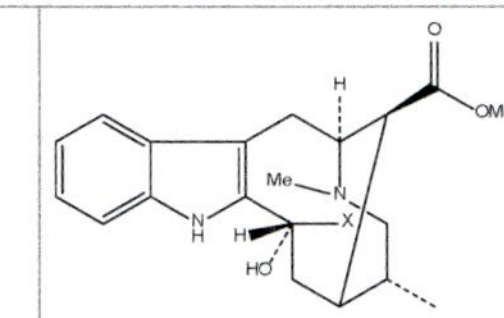	Vasileva (1969), Nqounou et al. (1990), and Kuete et al. (2015)
	Synonym(s): *Myrianthus talbotii* Rendle. **Common name(s)**: Giant yellow mulberry, Monkeyfruit. **Botanical description**: It is native to tropical Africa – Guinea to Central African Republic and Uganda, south to Angola, the DR of the Congo and Tanzania. It is habituated in secondary vegetation; more or less open places in rainforests; damp places in forest; forest gallery and clearings; stream or lagoon and lakesides; flooded ground; sometimes in villages at elevations from 700 to 1200 m in Tanzania. It is an evergreen tree growing up to 15 m by 10 m at a medium rate. **Medicinal use(s)**: It is used for the treatment of cancer, heart troubles, pregnancy complications, dysmenorrhea, incipient hernia, and dysentery. It also possesses antidysenteric, anti-inflammatory, antitussive, antidiarrheal, anti-plasmodial, antimycobacterial, and antitrypanosomal properties.					

(continued)

Table 4.143 (continued)

Family	Botanical name	Image	Plant part used	Bioactive compound(s)	Structure	References
	Urtica membranacea/ Urtica caudata Vahl, *Urtica dubia* Forssk.		Leaf	Ellagic acid	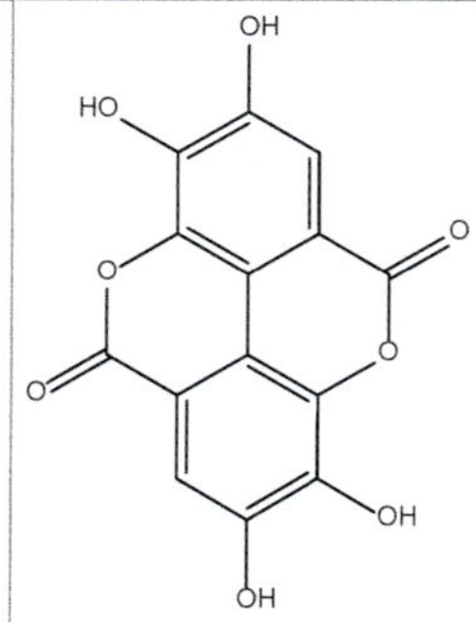	Solowey et al. (2014)
	Synonym(s): *Dubrueilia membranacea* Gaudich., *Urtica caudata* Vahl., *Urtica dubia* Forssk. **Common name(s)**: Large-leaved nettle, Membranous nettle. **Botanical description**: It is native to Asia, China, Japan, and Europe. It grows as annuals or perennial herbaceous plants and rarely shrubs. They can reach, depending on the type, location, and nutrient status, a height of 10–300 cm. The perennial species have underground rhizomes. The green parts have stinging hairs. Quadrangular stems are unbranched or branched, erect, ascending or spreading. Most leaves and stalks are arranged across opposite sides of the stem. The often-lasting bracts are free or fused to each other. The cystoliths are extended to more or less rounded. **Medicinal use(s)**: It is used for the treatment of lung, liver, bladder, and breast cancers and tumor. It also possesses antibacterial and anti-inflammatory properties.					

Table 4.144 Anticancer plants of family Valerianaceae

Family	Botanical name	Image	Plant part used	Bioactive compound(s)	Structure	References
Valerianaceae	*Clerodendrum infortunatum* L.		Leaf	Clerodin	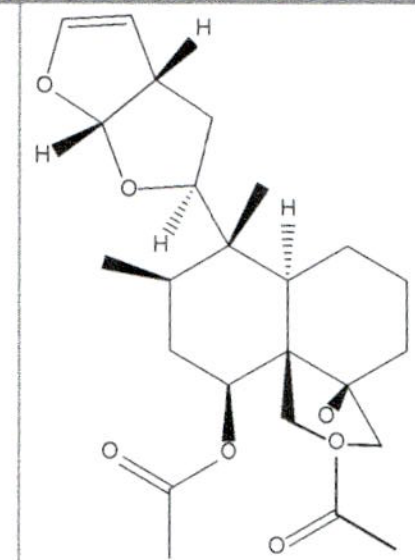	Gouthamchandra et al. (2010) and Debayan et al. (2011)
	Synonym(s): *Clerodendrum calycinum* Turcz., *Ovieda infortunata* (L.) Baill. **Common name(s)**: Hill glory bower. **Botanical description**: It is a flowering shrub or small tree and is native to tropical regions of Asia including Bangladesh, India, Myanmar, Pakistan, Thailand, Malaysia, the Andaman Islands, and Sri Lanka. The stem is erect, which is 0.5–4 m high, with no branches, and produces circular leaves, 15 cm in diameter. Leaves are simple, opposite, both surfaces sparsely villous-pubescent, elliptic, broadly elliptic, ovate or elongate ovate. The inflorescence is huge, consisting of many tubular snow-white flowers in a terminal cluster up to 0.6 m long. The tubes of the flowers are about 10 cm long and droop downwards, and the expanded corollas are about 5 cm across. The fruits are attractive dark metallic blue drupes, about 1 cm in diameter. **Medicinal use(s)**: It is used for the treatment of cancer, tumor, asthma, diabetes, jaundice, scabies, eczema, venereal diseases, snakebite, and gastric ulcer. It also possesses anti-inflammatory and antioxidant properties.					

(continued)

Table 4.144 (continued)

Family	Botanical name	Image	Plant part used	Bioactive compound(s)	Structure	References
	Lantana camara L.		Root	Caryophyllene	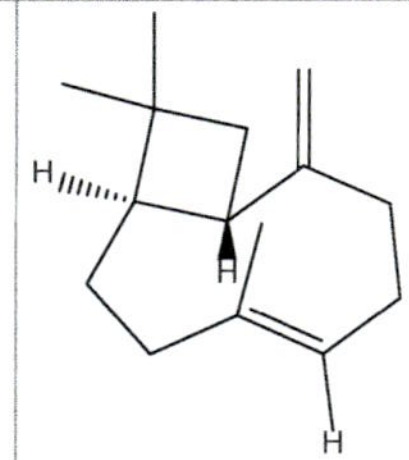	Badakashan et al. (2009) and Kurade et al. (2010)
	Synonym(s): *Camara vulgaris* Benth., *Lantana antillana* Raf., *Lantana asperata* Vis., *Lantana crocea* Jacq., *Lantana glandulosissima* Hayek, *Lantana mexicana* Turner, *Lantana mixta* Medik., *Lantana moritziana* Otto & A.Dietr., *Lantana sanguinea* Medik., *Lantana spinosa* L. ex Le Cointe, *Lantana undulata* Raf., *Lantana urticifolia* Mill. **Common name(s)**: Wild-sage, Red-sage, White-sage, Tickberry. **Botanical description**: It is a small perennial shrub which can grow to around 2 m tall and is native to Central and South America. It forms dense thickets in a variety of environments. Due to extensive selective breeding throughout the seventeenth and eighteenth centuries for use as an ornamental plant, there are now many different varieties of this plant. After pollination occurs, the color of the flowers changes (typically from yellow to orangish, pinkish, or reddish); this is believed to be a signal to pollinators that the pre-change color contains a reward as well as being sexually viable, thus increasing pollination efficiency. **Medicinal use(s)**: It is used for the treatment of skin cancer, chicken pox, measles, asthma, ulcers, swellings, eczema, tumors, high blood pressure, bilious fevers, catarrhal infections, and tetanus. It also possesses antioxidant, anti-inflammatory, and antibacterial properties.					

	Lippia alba Mill.		Stem and bark	Cardiac glycosides	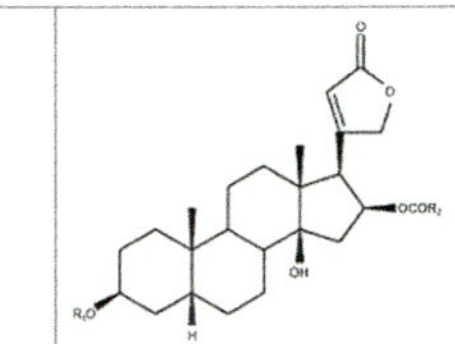	Gazola et al. (2004) and Glamočlija et al. (2011)
	Synonym(s): *Camara alba* (Mill.) Kuntze., *Lantana alba* Mill., *Lantana alba* f. *rubella* Moldenke., *Lantana cuneatifolia* Klotzsch ex Walp., *Lantana geminata* (Kunth) Spreng., *Lantana lippioides* Hook. & Arn., *Lantana malabarica* Hayek., *Lantana mollissima* Desf., *Lippia alba* f. *alba*., *Lippia alba* var. *carterae* Moldenke., *Lippia alba* f. *intermedia* Moldenke. **Common name(s)**: Bushy matgrass, Bushy lippia, Hierba negra, and Pitiona. **Botanical description**: It is native to Southern Texas in the United States, Mexico, the Caribbean, Central America, and South America. The species is also present in Australia and India. It is a multi-branched shrub, reaching a height of 1.5 m. Leaves are 1–3 cm in length and 0.9–2 cm in width. Flowers with white, pink, or light blue-purple corollas form on spikes 2 cm long. The plant is harvested from the wild for local use as food flavoring and for medicines. It is frequently cultivated in gardens throughout Central America as a medicinal plant. The plant is cultivated for its essential oil in Argentina. **Medicinal use(s)**: It is used to treat human cancer cell lines, colon cancer, gastric diseases, diarrhea, fever, asthma, and cough and possesses anti-spasmodic and emmenagogue activity. It also possesses anticancer, anti-inflammatory, antibacterial, and antioxidant properties.			Bifonazole	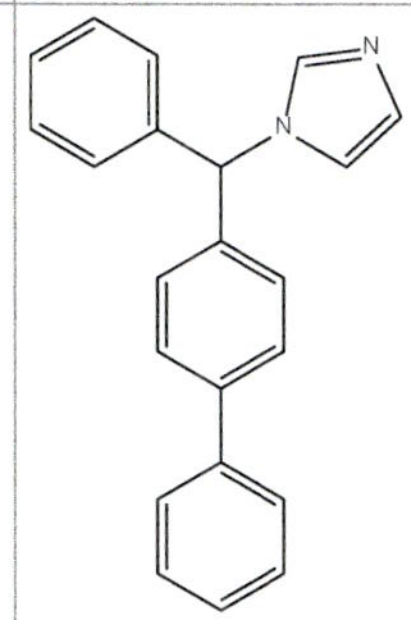	

(continued)

Table 4.144 (continued)

Family	Botanical name	Image	Plant part used	Bioactive compound(s)	Structure	References
	Lippia graveolens Kunth.		Leaf	Carvacrol	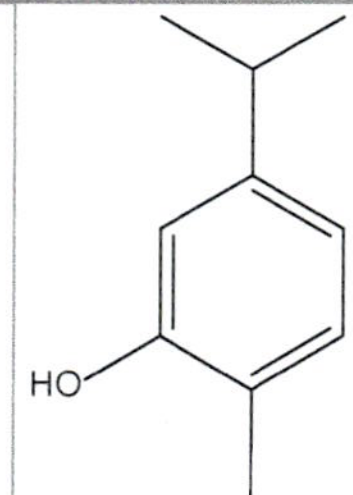	Martinez-Rocha et al. (2008) and Bueno-Durán et al. (2013)
	Synonym(s): *Camara alba* (Mill.) Kuntze., *Lantana alba* Mill., *Lantana alba* f. *rubella* Moldenke., *Lantana cuneatifolia* Klotzsch ex Walp., *Lantana geminata* (Kunth) Spreng., *Lantana lippioides* Hook. & Arn., *Lantana malabarica* Hayek. **Common name(s)**: Mexican oregano. **Botanical description**: It is native to Southern Texas in the United States, Mexico, the Caribbean, Central America, and South America. The species is also present in Australia and India. It is a multi-branched shrub, reaching a height of 1.5 m. Leaves are 1–3 cm in length and 0.9–2 cm in width. Flowers with white, pink, or light blue-purple corollas form on spikes 2 cm long. The plant is harvested from the wild for local use as food flavoring and for medicines. It is frequently cultivated in gardens throughout Central America as a medicinal plant. The plant is cultivated for its essential oil in Argentina. **Medicinal use(s)**: It is used to treat gastric diseases, diarrhea, fever, asthma, and cough and also possesses anti-spasmodic and emmenagogue activity.						

	Patrinia heterophylla Bunge		Whole plant	Penicillin G	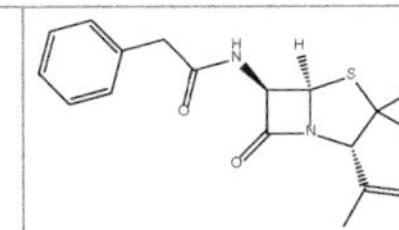	Lu et al. (2009)
	Synonym(s): *Patrinia angustifolia* Hemsl., *Patrinia heterophylla* subsp. *angustifolia* (Hemsl.) H.J. Wang., *Patrinia heterophylla* subsp. *heterophylla*. **Common name(s)**: Bitter herbs. **Botanical description**: It is a clump-forming perennial plant having thin, erect stems with few leaves and bearing a terminal inflorescence with yellow or white flowers and is native to grassy mountain habitats in China, Siberia, and Japan. **Medicinal use(s)**: It is used for the treatment of metrocarcinoma and cervical cancer, malaria, anthracia, stasis, typhoid, leukorrhea, metrorrhagia, and metrostaxis. It also possesses anticancer, anti-inflammatory, and antioxidant properties.					

(continued)

Table 4.144 (continued)

Family	Botanical name	Image	Plant part used	Bioactive compound(s)	Structure	References
	Patrinia scabra Bunge		Root	Lignans	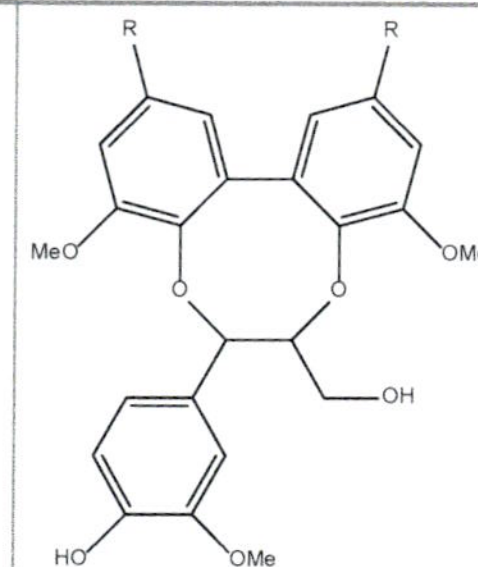	Kouno et al. (1994) and Sun et al. (2005)
	Synonym(s): *Patrinia angustifolia* Hemsl., *Patrinia heterophylla* subsp. *angustifolia* (Hemsl.) H.J. Wang., *Patrinia heterophylla* subsp. *heterophylla*. **Common name(s)**: Bitter herbs, Bunge. **Botanical description**: It is a clump-forming perennial plant having thin, erect stems with few leaves and bearing a terminal inflorescence with yellow or white flowers and is native to grassy mountain habitats in China, Siberia, and Japan. **Medicinal use(s)**: It is used for the treatment of cervical cancer, malaria, anthracia, stasis, typhoid, leukorrhea, metrorrhagia, and metrostaxis. It also possesses anticancer, anti-inflammatory, and antioxidant properties.			Azulenes	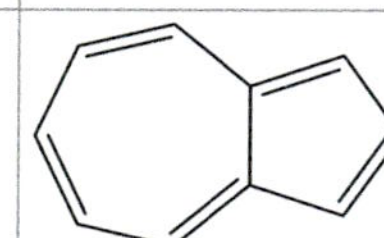	
				Caryophyllene oxide	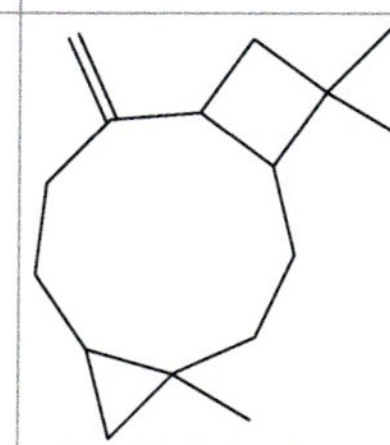	

	Patrinia scabiosaefolia		Whole plant	Isovalerate	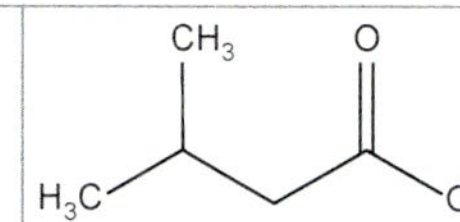	Madhuri and Pandey (2009) and Liu et al. (2017a)
	Synonym(s): *Patrinia angustifolia* Hemsl., *Patrinia heterophylla* subsp. *angustifolia* (Hemsl.) H.J. Wang., *Patrinia heterophylla* subsp. *heterophylla*. **Common name(s)**: Patrinia. **Botanical description**: It is a clump-forming perennial plant having thin, erect stems with few leaves and bearing a terminal inflorescence with yellow or white flowers and is native to grassy mountain habitats in China, Siberia, and Japan. **Medicinal use(s)**: It is used for the treatment of metrocarcinoma, cervical cancer, malaria, anthracia, stasis, typhoid, leukorrhea, and metrorrhagia. It also possesses anticancer, anti-inflammatory, and antioxidant properties.			3-Methylcrotonylglycine	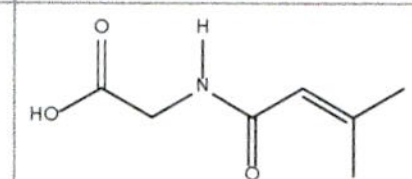	

(continued)

Table 4.144 (continued)

Family	Botanical name	Image	Plant part used	Bioactive compound(s)	Structure	References
	Valeriana officinalis L.		Roots	Valepotriate	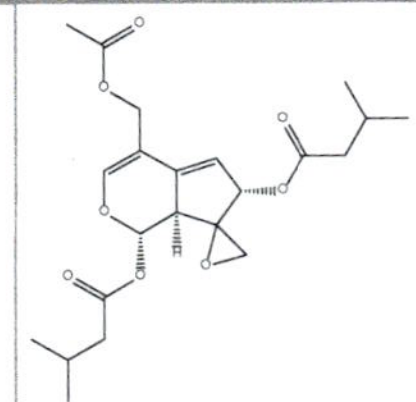	Pullela et al. (2005) and Patočka and Jakl (2010)
	Synonym(s): *Valeriana alternifolia* Bunge., *Valeriana alternifolia* var. *angustifolia* (Kom.) S.H. Li., *Valeriana alternifolia* f. *angustifolia* (Kom.) Kitag., *Valeriana baltica* Pleijel., *Valeriana chinensis* Kreyer ex Kom. **Common name(s)**: Garden heliotrope, Valerian. **Botanical description**: It is native to Europe and parts of Asia. The flowers are frequently visited by many fly species, especially hoverflies of the genus *Eristalis*. It is a perennial erect plant, which can be 4 ft. high. It has a short, conical, erect yellowish rootstock with multiple long thin roots. Only one stem usually grows from the root. It is round and hollow, with hairs near the base. The main stem then terminates in several flowering stems. It is consumed as food by the larvae of some Lepidoptera species including the grey pug. **Medicinal use(s)**: It is used to treat breast, gastrointestinal, gynecological, genitourinary, lung, and non-melanoma skin cancers. It is also used for the treatment of insomnia, leg syndrome, and anxiety. It possesses antioxidant and anti-inflammatory properties.					

	Valeriana sorbifolia Kunth.		Roots	Valepotriates (sorbifolivaltrates A-D (1–4))	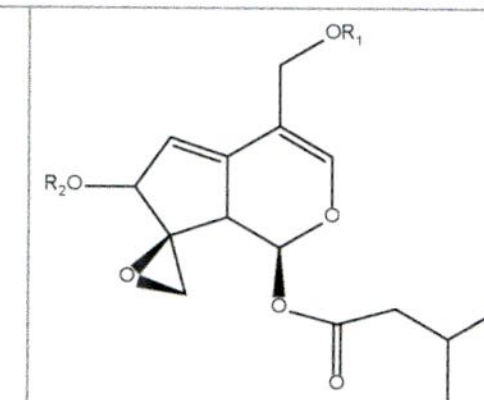	Xu et al. (2007)
				Tisovalerate	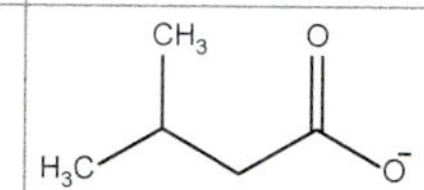	
	Synonym(s): *Valeriana alternifolia* Bunge., *Valeriana alternifolia* var. *angustifolia* (Kom.) S.H. Li., *Valeriana alternifolia* f. *angustifolia* (Kom.) Kitag., *Valeriana baltica* Pleijel., *Valeriana chinensis* Kreyer ex Kom. **Common name(s)**: Garden valerian, Pineland *valerian*. **Botanical description**: It is native to Europe and parts of Asia. The flowers are frequently visited by many fly species, especially hoverflies of the genus *Eristalis*. It is a perennial erect plant, which can be 4 ft. high. It has a short, conical, erect yellowish rootstock with multiple long thin roots. Only one stem usually grows from the root. It is round and hollow, with hairs near the base. It is consumed as food by the larvae of some Lepidoptera species including the grey pug. **Medicinal use(s)**: It is used to treat breast, gastrointestinal, gynecological, genitourinary, lung, and non-melanoma skin cancers. It is also used for the treatment of insomnia, leg syndrome, and anxiety. It also possesses antioxidant and anti-inflammatory properties.			Seneciovaltrate	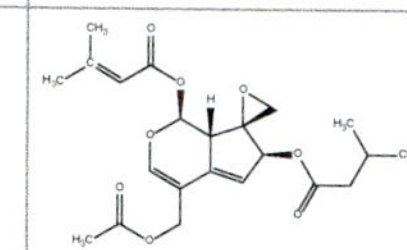	
				Valtrate hydrine B3	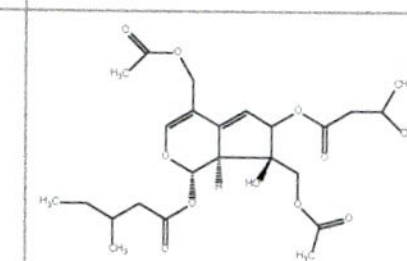	

Table 4.145 Anticancer plants of family Verbenaceae

Family	Botanical name	Image	Plant part used	Bioactive compound(s)	Structure	References
Verbenaceae	*Callicarpa rubella* Lindl.		Bark	Kaempferol	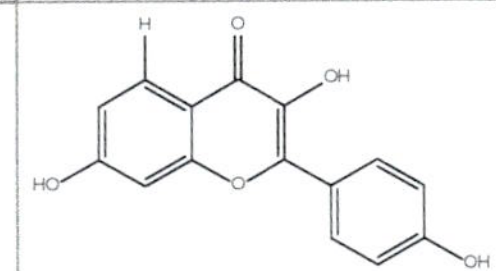	Jamir (1990) and Stephen and Peng (2018)
	Synonym(s): *Callicarpa dielsii* (H.Lév.) C.Pei., *Callicarpa lasiantha* Lem., *Callicarpa panduriformis* H.Lév., *Callicarpa purpurea* Lem., *Callicarpa rubella* f. *angustata* C.Pei., *Callicarpa rubella* f. *crenata* C.Pei., *Callicarpa rubella* var. *rubella*. **Common name(s)**: Dark purple berries. **Botanical description**: It is shrub about 4–8 ft. tall and is native to Southeast Asia. It is a shrub that produces pink or purple flowers followed by dark purple berries. The berries are actually drupes. It is grown in gardens as an ornamental plant. Stem is whitish, branches and branchlets obtusely 4 angular, grandular, stellate-pubescent, usually furrowed, nodes annulate. Leaves are opposite, simple, lanceolate-oblong, base broadly obtuse or subcordate, margin ciliate, serrate, entire at the apex and base, apex acuminate, coriaceous, chartaceous, sparsely pubescent above and paler stellate-pubescent beneath, stigma capitate. Fruit drupe, globose, 1.5–2 mm in diameter, reddish purple, glabrous, seeds 4. **Medicinal use(s)**: It is used for the treatment of blood cancer, human cancer cells, and tumor. It also possesses anti-inflammatory and antibacterial properties.					

	Clerodendrum serratum (L.) Moon		Leaf	Kaempferɔl	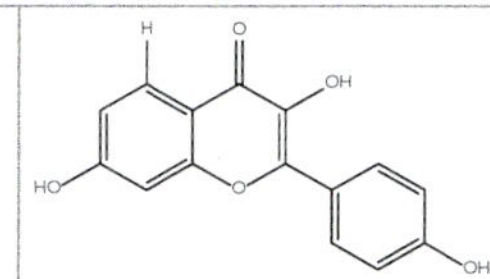 	Bhandary et al. (1995) and Patel et al. (2014)
	Synonym(s): *Clerodendrum ornatum* Wall., *Clerodendrum serratum* var. *amplexifolium* Moldenke., *Clerodendrum serratum* var. *glabrescens* Moldenke. **Common name(s)**: Glorybower, Bagflower, and Bleeding heart. **Botanical description**: It is a genus of small trees, shrubs, lianas, and subherbaceous perennials and is native to Sri Lanka and the Andaman Islands. Leaves are decussate or whorled. Inflorescence usually terminal. Sepals usually connate, often colored, usually accrescent. Corolla red to yellow, pink or white. Corolla tube 5-lobed, the lobes usually unequal. Stamens 4, usually in 2 pairs of unequal length and projecting well beyond the mouth of the corolla. Ovary incomplete 4-locular with 4 ovules. Fruit a drupe, usually with 4 grooves or lobes and 4-seeded. **Medicinal use(s)**: It is used for the treatment of lung cancer, tumor, diabetes, obesity, and hypertension. It also possesses anti-inflammatory, antitumor, and antibacterial activities.					

(continued)

Table 4.145 (continued)

Family	Botanical name	Image	Plant part used	Bioactive compound(s)	Structure	References
	Nyctanthes arbor-tristis L.		Part not specified	(1S, 4aS, 5R, 6S, 7S, 7aS)-Ethyl1,5,6-trihydroxy-7-(propionyloxy)-1,4a, 5,6, 7,7a-hexahydrocyclopenta[c] pyran-4-carboxylate	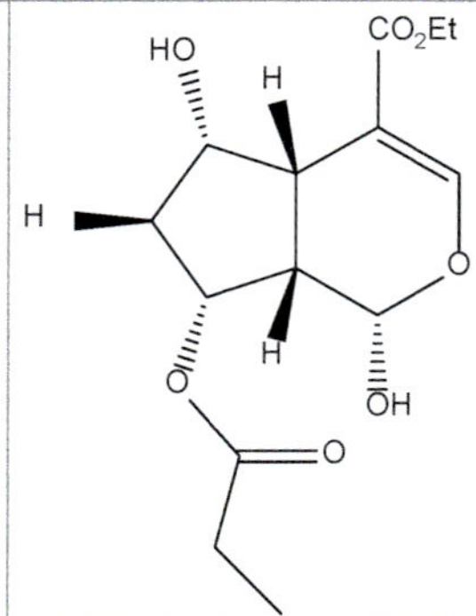	Jain (1970), Ramteke and Hussain (2012), and Kumari and Charya (2017)
	Synonym(s): *Bruschia macrocarpa* Bertol., *Nyctanthes dentata* Blume., *Nyctanthes tristis* Salisb., *Parilium arbor-tristis* (L.) Gaertn. **Common name(s)**: Night-flowering jasmine or Parijat. **Botanical description**: It is a shrub or a small tree with flaky gray bark growing up to 10 m tall and is native to South Asia and Southeast Asia. The leaves are opposite, simple, 6–12 cm long, and 2–6.5 cm broad with an entire margin. The flowers are fragrant, with 5–8-lobed white corolla with an orange-red center; they are produced in clusters of two to seven together, with individual flowers opening at dusk and finishing at dawn. The fruit is a flat brown heart-shaped to round capsule 2 cm diameter, with two sections each containing a single seed. **Medicinal use(s)**: It is used for the treatment of breast cancer. It is used to treat sickle cell anemia, atherosclerosis, Parkinson's disease, heart failure, myocardial infarction, Alzheimer's disease, and schizophrenia. It possesses anticancer, antioxidant, anti-inflammatory, and antibacterial activities.			Methylenenona-2,5-dien-4-yl)-4,4,10,13,14-pentamethy 2,3,4,7,8, 9, 10, 11, 12, 13,14,15,16,17-tetradecahydro-1H - cyclopenta [a] phenanthren-3-ol.	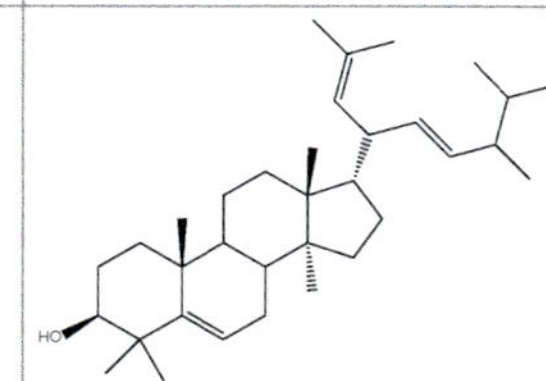	

	Pygmaeopremna herbacea (Roxb.) Moldenke *Premna herbacea* Roxb.		Root	Linoleic acid	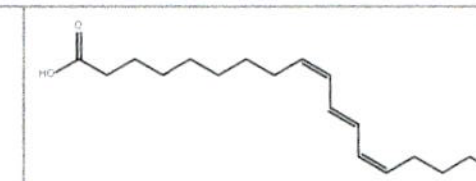	Sankaram and Rao (1978), Verma et al. (2012), Dhamija et al. (2013), and Chan et al. (2019)
	Synonym(s): *Gumira herbacea* (Roxb.) Kuntze., *Premna acaulis* (F.Muell.) Merr., *Premna humilis* Merr., *Premna nana* Collett & Hemsl., *Premna obovata* Merr., *Premna sessilifolia* H.J.Lam., *Pygmaeopremna acaulis* (F.Muell.) Moldenke., *Pygmaeopremna herbacea* (Roxb.) Moldenke., *Pygmaeopremna humilis* Merr. **Common name(s)**: Grass smelly yellow viper. **Botanical description**: It is a low-growing herbaceous perennial plant or a dwarf under-shrub with stems up to 15 cm tall, occasionally up to 30 cm, and is native to East Asia, Southern China, Indian subcontinent, Myanmar, Thailand, Cambodia, Laos, Vietnam, Malaysia, Indonesia, the Philippines, New Guinea, and Australia. Most of the stem is underground, the plant having a creeping woody rhizome, and the aboveground part is slender and simple or with single dichotomous branching. The plant is often harvested from the wild for local medicinal use and occasionally as a food. It is cultivated for medicinal use in India and Sri Lanka. **Medicinal use(s)**: It is used for the treatment of lung cancer and tumor. The juice from roots and rhizomes is used in India to treat dropsy, cough, asthma, fever, rheumatism, and cholera. It also possesses anti-inflammatory, antioxidant, antibacterial, and antimicrobial activities.			Scutellarein	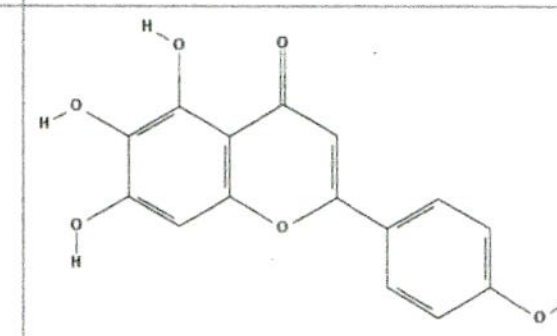	

(continued)

Table 4.145 (continued)

Family	Botanical name	Image	Plant part used	Bioactive compound(s)	Structure	References
	Verbena hybrid L.		Fruit	Flavonol 3-O-glucoside	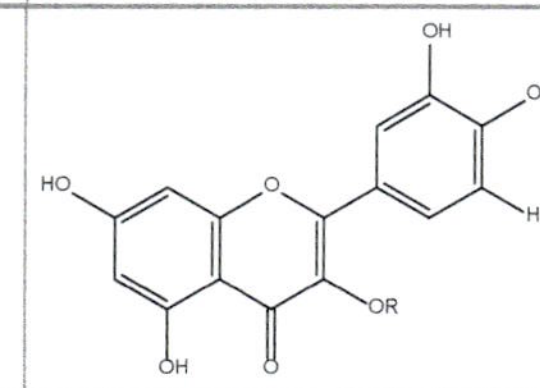	Suzuki et al. (2004)
				Malonyltransferases	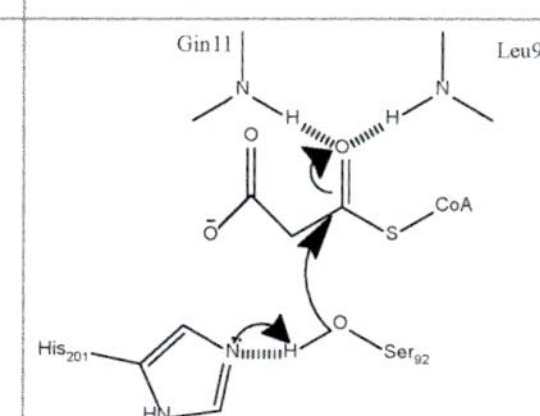	
	Synonym(s): *Verbena hortensis* Vilm. **Common name(s)**: Purpletop vervain, Lantanas. **Botanical description**: It is a short-lived perennial that is grown in St. Louis as an annual plant. It is noted for its profuse bloom of small five-petaled flowers in rounded clusters from late spring to fall. Flower colors include blue, violet, purple, rose, dark red, yellow, white, and bicolor. Some varieties are fragrant. Oblong toothed leaves are gray-green to dark green. It is grown as an annual in average, medium moisture, well-drained soils in full sun. Tolerates light shade, but best flowering usually occurs in full sun. Plants may also be purchased in six packs from local nurseries. **Medicinal use(s)**: It is used for the treatment of breast cancer. It also possesses anti-inflammatory and antioxidant activities.					

	Vitex negundo L.		Leaf	Leutic acid	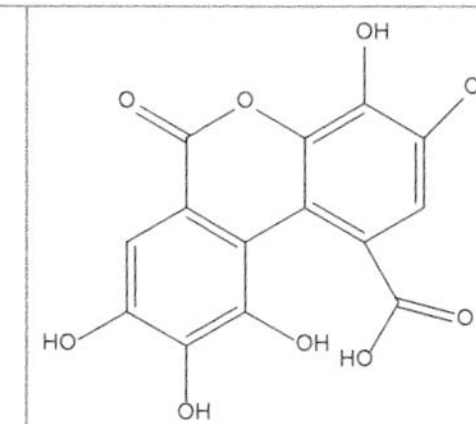	Masilungan et al. (1971) and Kadir et al. (2013)
	Synonym(s): *Agnus-castus incisa* (Lam.) Carrière., *Agnus-castus negundo* (L.) Carrière., *Vitex agnus-castus* var. *negundo* (L.) Kuntze., *Vitex agnus-castus* var. *negundoides* Kuntze. **Common name(s)**: Chaste tree. **Botanical description**: It is a large shrub or small tree of about 3 m in height with quadrangular branches and is native to tropical Eastern and Southern Africa and Asia. It can be described as a cross between a shrub and a tree with a single woody stem (trunk). Leaves opposite, 3–5 foliolate, leaflets lanceolate, entire or crenate, glabrate, dark green above, pale greenish-tomentose beneath, central leaflets larger. Flowers are small, bluish purple in lateral cymes, forming an elongated, terminal thyrsus. Calyx teeth triangular. Fruits, drupaceous, and black when ripe. Seeds, obovate or oblong. **Medicinal use(s)**: It is used to treat breast cancer, stomachache, eye disease, inflammation, enlargement of the spleen, bronchitis, and asthma. It also possesses anti-inflammatory, antioxidant, and antibacterial properties.					

(continued)

Table 4.145 (continued)

Family	Botanical name	Image	Plant part used	Bioactive compound(s)	Structure	References
	Vitex rotundifolia L.f.		Fruit	Luteolin	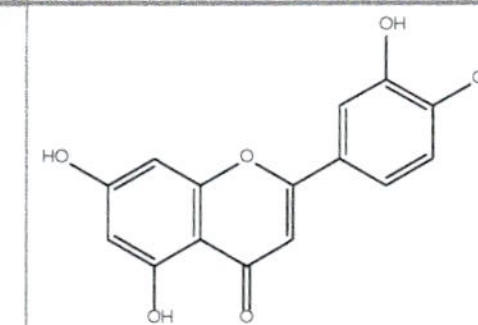	Ko et al. (2002)
	Synonym(s): *Vitex agnus-castus* var. *javanica* Kuntze., *Vitex agnus-castus* var. *subtrisecta* Kuntze., *Vitex agnus-castus* var. *trifolia* (L.) Kurz., *Vitex benthamiana* Domin., *Vitex bicolor* Willd., *Vitex indica* Mill., *Vitex integerrima* Mill. **Common name(s)**: Roundleaf chastetree. **Botanical description**: It is native to Asia, China, Japan, Oceania, Australia, and New Guinea. Stem is prostrate to creeping, rooting at the nodes. Young branches silky-tomentose. Branches terrate or obscurely 4-sided when mature. Leaves mostly 1-foliolate, rarely a few 2,3,5 or 7 foliolate, sessile or with short petioles. Blade obovate-oval, broadly oblong-elliptic, orbicular, or obovate-spathulate, 2.5–5 cm long, Flowers are zygomorphic with short pedicels; pedicels are densely glandular and tomentose, 0.5–2 mm long; bracteoles linear, 1–2.5 mm long. Calyx tube cylindrical, 3–4 mm long, 2–3 mm diam. at the top, 5-ribbed outside with 5 short teeth at the top, glandular and silky-tomentose outside, glabrous within. **Medicinal use(s)**: It is used for the treatment of breast cancer, lung cancer, headache, cold, migraine, eye pain, female hormonal disorders, asthma, chronic bronchitis, and gastrointestinal disorders. It also possesses anti-inflammatory, antioxidant, and antibacterial activities.					

	Vitex trifolia L.		Leaf	Luteolin	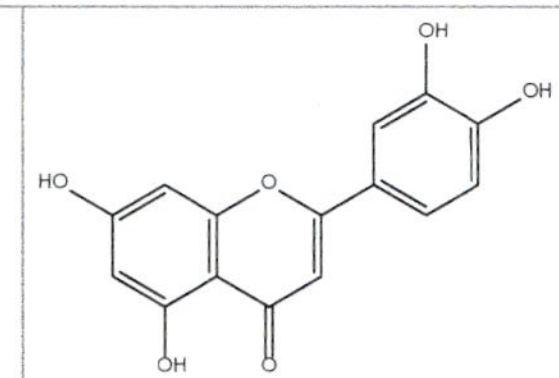	Hernandez et al. (1999), Li et al. (2005), and Garbi et al. (2015)
	Synonym(s): *Vitex agnus-castus* var. *javanica* Kuntze., *Vitex agnus-castus* var. *subtrisecta* Kuntze., *Vitex agnus-castus* var. *trifolia* (L.) Kurz., *Vitex benthamiana* Domin. **Common name(s)**: Simpleleaf chastetree. **Botanical description**: It is large coastal shrub or small tree, less than 5 m in height with the stems covered by soft hairs and is native to tropical east Africa and French Polynesia. The leaves are oppositely arranged along the stems and are usually compound, composed of 3 linear leaflets which range between 1 and 12 cm in length. The upper surface of the leaves is green and the lower surface is grayish green. The flowers are borne in panicles or clusters up to 18 cm in length. Individual flowers have purple to violet two-lipped corollas that are approximately 5 mm long. The stamens are in two pairs, and the ovary is superior or develops above the corolla. The fleshy fruits are about 6 mm in diameter and contain 4 small black seeds. **Medicinal use(s)**: It is used for the treatment of breast cancer. It also possesses anti-inflammatory, antioxidant, and antimicrobial properties.			Luteolin-3-O-β-D glucuronide	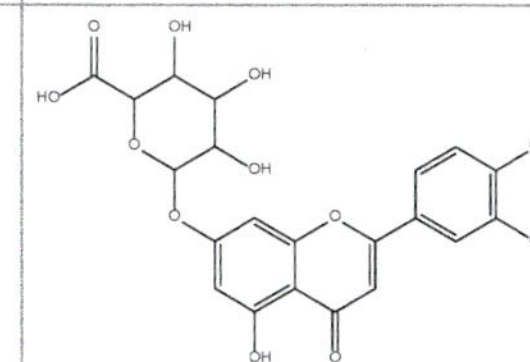	

(continued)

Table 4.145 (continued)

Family	Botanical name	Image	Plant part used	Bioactive compound(s)	Structure	References
	Viticis fructus P.E.		Leaf and stem	Casticin	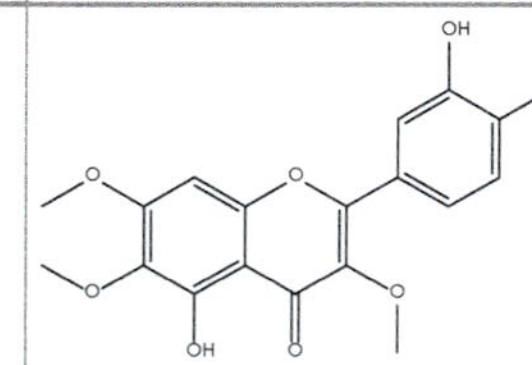	Hernandez et al. (1999) and Cao et al. (2016)
	Synonym(s): *Chaste tree berry* P.E, *Fructus Viticis* Simplicifoliae, *Chasteberry* Extract, *Fructus Vivicis* P.E, *Viticis fruits* Extract. **Common name(s)**: Chasteberry. **Botanical description**: It is native to the Mediterranean and Southern Europe. It is an ornamental plant in American gardens, where its attractive blue-violet flowers are appreciated in midsummer. They are deciduous shrub or small tree, which is about 3 m in height and with aroma. Young shoots are square and densely pubescent. **Medicinal use(s)**: It is used for the treatment of lung cancer. It is used to treat headaches, colds, migraines, and eye pain. It also possesses anti-inflammatory, antioxidant, and antibacterial activities.			3,6,7-Trimethyl quercetagetin	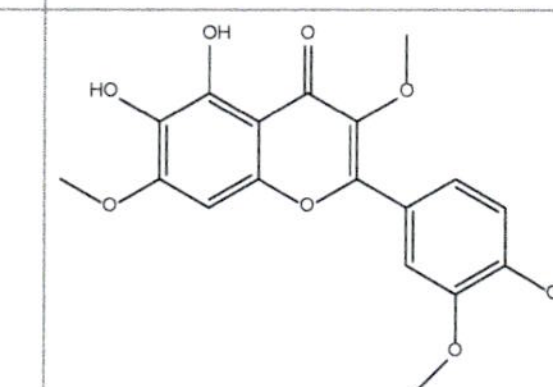	

				Vitexin	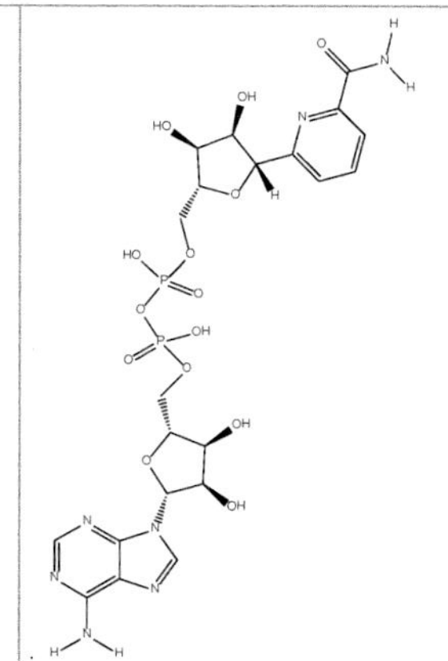	
				Artemetin	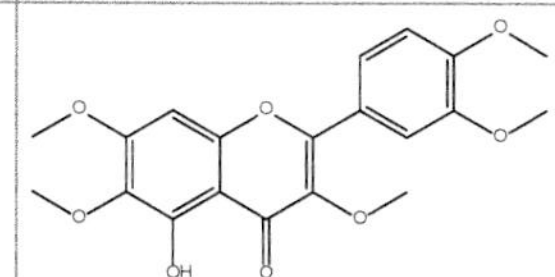	
				Luteolin	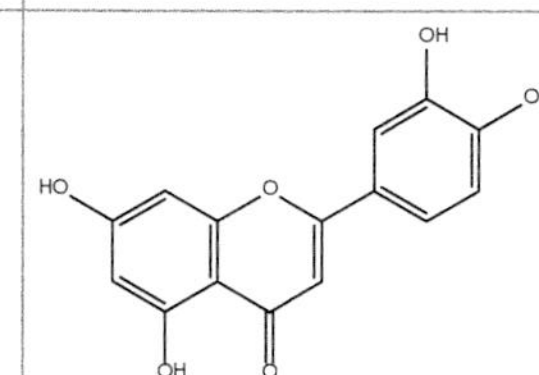	

(continued)

Table 4.145 (continued)

Family	Botanical name	Image	Plant part used	Bioactive compound(s)	Structure	References
				Luteolin-3-O-b-D glucuronide	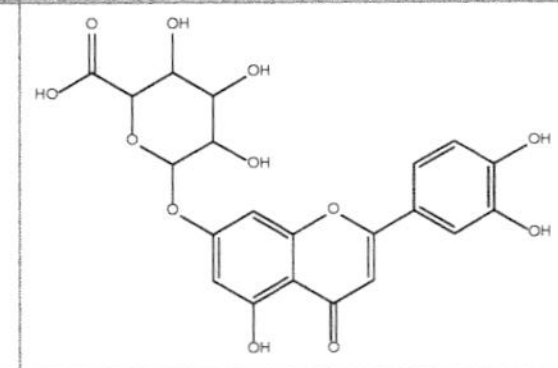	
				Isoorientin	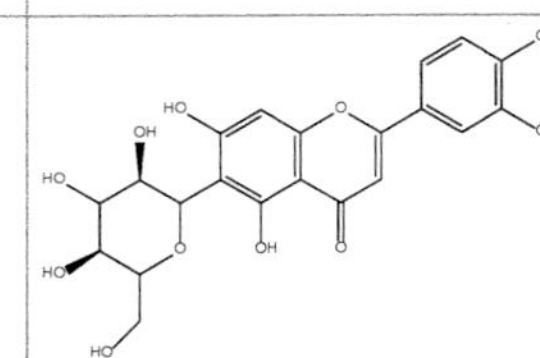	

Table 4.146 Anticancer plants of family Violaceae

Family	Botanical name	Image	Plant part used	Bioactive compound(s)	Structure	References
Violaceae	*Viola biflora* L.		Stem and bark	Resveratrol	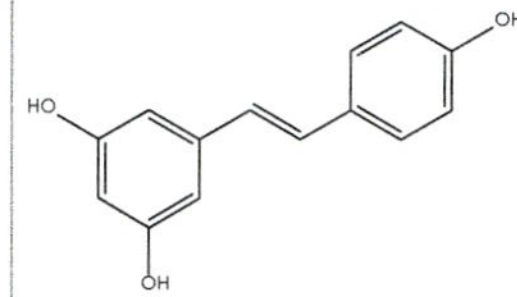	Herrmann et al. (2008) and Sadeghnia et al. (2014)
	Synonym(s): *Chrysion biflorum* (L.) Spach., *Dischidium biflorum* (L.) Opiz., *Viola biflora* var. *acuminata* Maxim., *Viola biflora* subsp. *biflora*, *Viola biflora* var. *biflora*, *Viola biflora* var. *hirsuta* W. Becker, *Viola biflora* var. *nudicaulis* W. Becker. **Common name(s)**: Alpine yellow-violet, Arctic yellow violet, Two-flower violet. **Botanical description**: It is an annual or perennial herb, is about 10–30 cm tall, and is native to Europe, Siberia, Central Asia, Pakistan, Western and Northern China, North Korea, Japan, and western North America. Roots are rhizomatous, slender, and stout with many thick root hairs. Stems are few to several, slender, erect or decumbent, 2–5 noded, glabrous or pubescent. Leaves are simple, orbicular-suborbicular, broadly ovate to reniform, sepals 5, subequal, linear lanceolate to oblong, base usually auriculate, margins scarious or ciliate at the lower parts, persistent, up to 7 mm across, petals 5, subequal, oblong-obovate, erect or spreading, about 7–15 × 3–6 mm long, lateral ones largest, spurred basally, spur erect, slightly curved or cylindrical, about 1–2.5 mm long. Stamens 5, hypogynous, filaments short, anthers 2 loculed, tipped with triangular nectariferous membranous appendages. **Medicinal use(s)**: It is used for the treatment of breast, lung, and digestive cancer. It is also used to treat inflammatory and skin disorders including scabs, itching, ulcers, eczema or psoriasis, and bronchitis or asthma. It possesses anti-inflammatory and antioxidant properties.			Violanthin	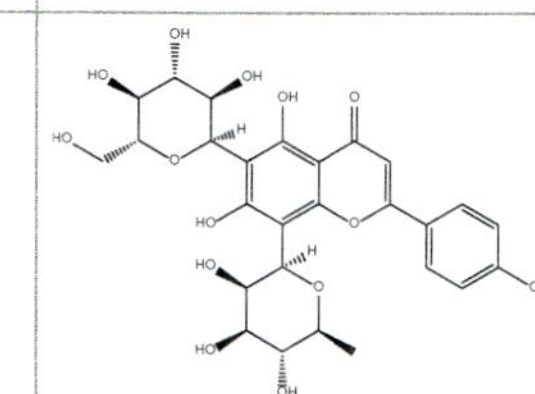	

(continued)

Table 4.146 (continued)

Family	Botanical name	Image	Plant part used	Bioactive compound(s)	Structure	References
	Viola diamantica Nakai		Aerial parts	Luteolin	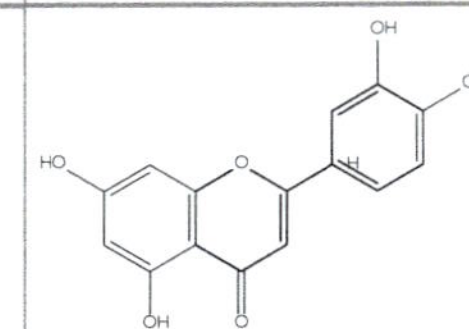	Yook et al. (1989) and Maffei (2003a, b, c, d, e, f, g, h, i, j, k, l, m)
	Synonym(s): *Chrysion biflorum* (L.) Spach., *Dischidium biflorum* (L.) Opiz., *Viola biflora* var. *acuminata* Maxim., *Viola biflora* subsp. *biflora*, *Viola biflora* var. *biflora*, *Viola biflora* var. *hirsuta* W. Becker, *Viola biflora* var. *nudicaulis* W. Becker. **Common name(s)**: Sweet violet. **Botanical description:** It is native to temperate Northern Hemisphere, Hawaii, Australia, and the Andes. They are perennial plants, some are annual plants, and few are small shrubs. A large number of species, varieties, and cultivars are grown in gardens for their ornamental flowers. In horticulture the term "pansy" is normally used for those multicolored, large-flowered cultivars which are raised annually or biennially from seed and used extensively in bedding. **Medicinal use(s)**: It is used for the treatment of gastric cancer. It also possesses anti-inflammatory, antioxidant, and antibacterial properties.			Violanthin	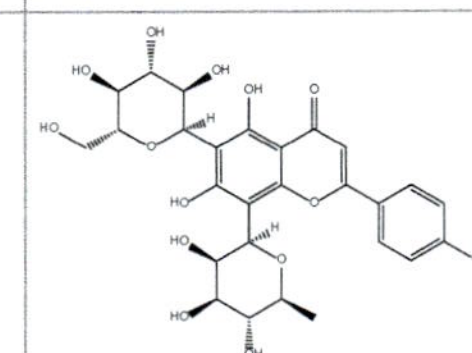	

	Viola odorata L.		Apical part	4-Ethylbenzaldehyde	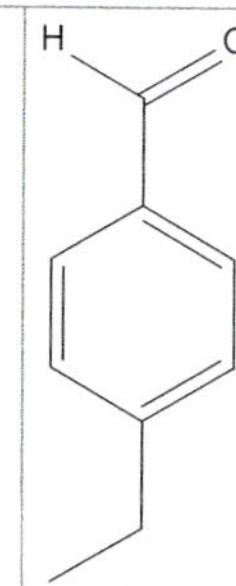	Gerlach et al. (2010) and Saint-Lary et al. (2014)
	Synonym(s): *Viola odorata* Neck., *Viola odorata* f. *odorata*, *Viola wiedemannii* Boiss. **Common name(s)**: Wood violet, Sweet violet, English violet, Common violet, Florist's violet, or Garden violet, Bayoleta. **Botanical description**: It is a perennial herb with stout creeping rootstock and is native to Pakistan, Iran, Kashmir, North America, and North and West Asia. Leaves all radical, up to 3 cm in diameter, petiolate, stipules elongated, leafy. Flowers pinkish blue or purplish, with long pedicels. Capsule 3-sided. **Medicinal use(s)**: It is used for the treatment of breast cancer, lung cancer, gastric cancer, cough, asthma, fever, and body ache. It also possesses anti-inflammatory, antioxidant, antibacterial, and antimicrobial properties.					

(continued)

Table 4.146 (continued)

Family	Botanical name	Image	Plant part used	Bioactive compound(s)	Structure	References
	Viola tricolor hortensis		Apical parts	Rutin	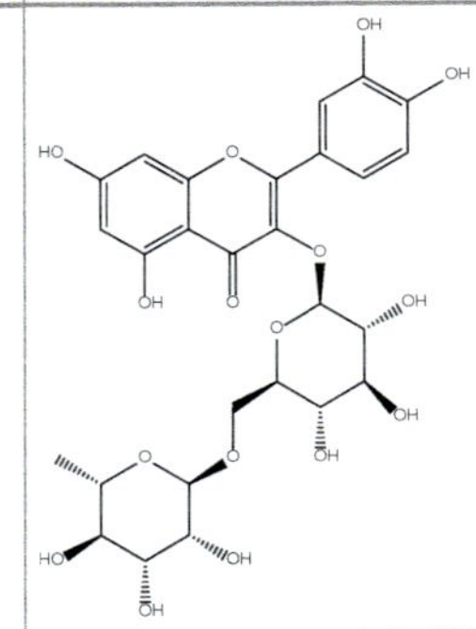	Svangard et al. (2004) and Vukis (2009)
	Synonym(s): *Jacea tricolor* (L.) Opiz., *Mnemion agreste* Fourr., *Mnemion alpestre* Fourr., *Mnemion contemptum* Fourr., *Mnemion elegans* Spach., *Mnemion gracilescens* Fourr., *Mnemion nemausense* Fourr., *Mnemion sagotii* Fourr., *Mnemion segetale* Fourr., *Mnemion sudeticum* Fourr. **Common name(s)**: Heartsease, Pansy. **Botanical description**: It is a small plant of creeping and ramping habit and is native to North America and Europe. It grows in short grassland on farms and wasteland, chiefly on acid or neutral soils. It is usually found in partial shade. The flowers can be purple, blue, yellow, or white. They are hermaphrodite and self-fertile and are pollinated by bees. **Medicinal use(s)**: It is used for the treatment of breast and lung cancer, tumor, and mastitis. It also possesses anti-inflammatory and antibacterial properties.					

Table 4.147 Anticancer plants of family Viscaceae

Family	Botanical name	Image	Plant part used	Bioactive compound(s)	Structure	References
Viscaceae	*Viscum cruciatum* Sieber ex Boiss.		Apical part	Diarylheptanoids	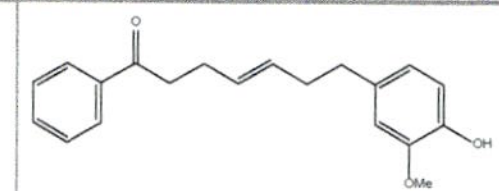	Martín-Cordero et al. (2001) and Assaf et al. (2013)
				Hirsutanone	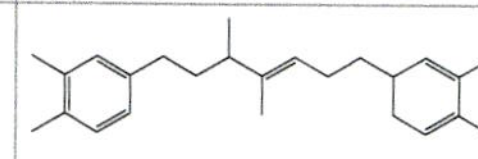	
	Synonym(s): *Viscum orientale* Willd. **Common name(s)**: Red-berry mistletoe. **Botanical description**: It is native to south West Spain, Southern Portugal, North Africa, Australia, and Asia. Leaves are opposite, 8 × 3 cm, spathulate, apex acute, entire to minutely crenate, basally 3–5 nerved, coriaceous. Flowers axillary in triads, lateral flowers male and central flowers female, bracts 1.5 mm, ovate, male flowers perianth lobes 4, 1 mm, ovate, stamens 4, female flowers, tepals 1 mm, ovate; ovary 0.6 mm, stigma rounded. Berry 0.6 × 0.5 mm, obovoid, acute at both ends. **Medicinal use(s)**: It is used for the treatment of larynx cancer. It also possesses anti-inflammatory, antioxidant, antimicrobial, and antibacterial properties.					

Table 4.148 Anticancer plants of family Zingiberaceae

Family	Botanical name	Image	Plant part used	Bioactive compound(s)	Structure	References
Zingiberaceae	*Alpinia galanga* (L.) Willd.		Rhizome	Kaempferide	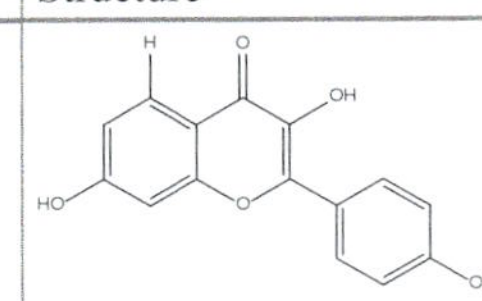	Mohammad (2006), Desai et al. (2008), Madhuri and Pandey (2009), and Samarghandian et al. (2014)
	Synonym(s): *Alpinia alba* (Retz.) Roscoe., *Alpinia bifida* Warb., *Alpinia carnea* Griff., *Alpinia pyramidata* Blume, *Alpinia rheedei* Wight, *Alpinia viridiflora* Griff., *Amomum galanga* (L.) Lour., *Amomum medium* Lour., *Galanga officinalis* Salisb., *Hellenia alba* (Retz.) Willd., *Heritiera alba* Retz., *Languas galanga* (L.) Stuntz, *Languas pyramidata* (Blume) Merr., *Languas vulgare* J.Koenig, *Maranta galanga* L., *Zingiber galanga* (L.) Stokes, *Zingiber medium* Stokes, *Zingiber sylvestre* Gaertn. **Common name(s)**: Blue ginger, Thai galangal. **Botanical description**: It is native to Indonesia, Thailand, Manipur, and Myanmar. The plant grows from rhizomes up to 2 m in height with abundant long leaves that bear red fruit. It is used most often in cookery. The robust rhizome has a sharp, sweet taste and smells like a blend of black pepper and pine needles. The red fruit is used in traditional Chinese medicine and has a flavor similar to cardamom. **Medicinal use(s)**: It is used for the treatment of breast and liver cancer. It also possesses antibacterial, anti-inflammatory, antioxidant, and antibacterial properties.			β-Sitosterol diglucosyl caprate	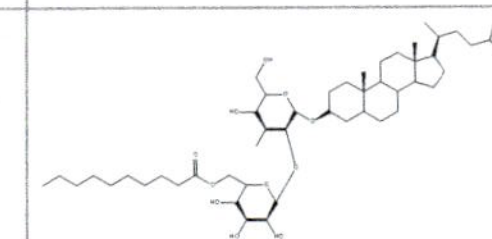	

	Alpinia hainanensis K.Schum.		Seed	Beta-selinene	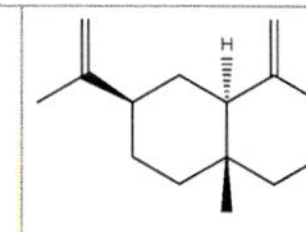	Saiki et al. (1978) and Ghosh and Rangan (2012)
	Synonym(s): *Alpinia henryi* K.Schum., *Alpinia kainantensis* Masam., *Alpinia katsumadae* Hayata, *Languas hainanensis* (K.Schum.) Merr., *Languas henryi* (K.Schum.) Merr., *Languas katsumadae* (Hayata) Merr. **Common name(s)**: Katsumadai seed, Katsumada's galangal seed. **Botanical description**: It is native to Asia, Australia, and the Pacific Islands, where they occur in tropical and subtropical climates. These herbs lack true stems but have pseudostems that usually grow up to 3 m long which are composed of overlapping leaf sheaths. A few species have been known to reach about 8 m. They grow from thick rhizomes. The leaves are lance-shaped to oblong. The inflorescence takes the form of a spike, a panicle, or a raceme. It may be hooded in bracts and bracteoles. The flower has a shallowly toothed calyx which is sometimes split on one side. The corolla is a cylindrical tube with three lobes at the mouth, the middle lobe is larger and hood like in some taxa. **Medicinal use(s)**: It is used for the treatment of stomach cancer, tumor, dysentery, diabetes mellitus, fever, dyspepsia, and urinary incontinence. It also possesses anti-inflammatory and antioxidant properties.					

(continued)

Table 4.148 (continued)

Family	Botanical name	Image	Plant part used	Bioactive compound(s)	Structure	References
	Alpinia purpurata (Vieill.) K.Schum.		Leaf	3-beta-trans-sinapoyloxylup-20(29)-en-28-ol	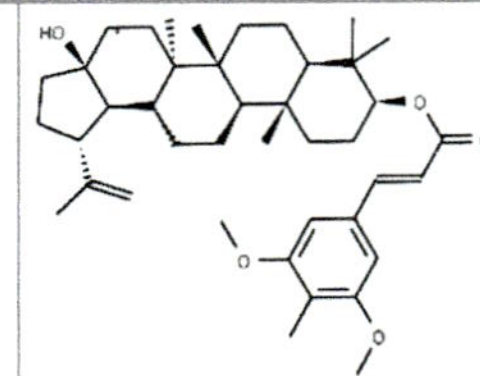	Arul et al. (2012), Chan and Wong (2015), and Zheng et al. (2016)
	Synonym(s): *Alpinia grandis* K.Schum., *Guillainia novo-ebudica* F.Muell., *Guillainia purpurata* Vieill., *Languas purpurata* (Vieill.) Kaneh. **Common name(s)**: Greater galangal. **Botanical description**: It is native to South Asia and Indonesia and cultivated in Malaysia, Laos, and Thailand. The robust rhizome has a sharp, sweet taste and smells like a blend of black pepper and pine needles. The red fruit is used in traditional Chinese medicine and has a flavor similar to cardamom. **Medicinal use(s)**: It is used to treat lung, liver, breast, stomach, colorectal, cervical, and prostate cancers. It also possesses antioxidant, anti-inflammatory, and antibacterial properties.			3-beta-trans-feruloyloxy-16-beta-hydroxylup-20(29)-ene	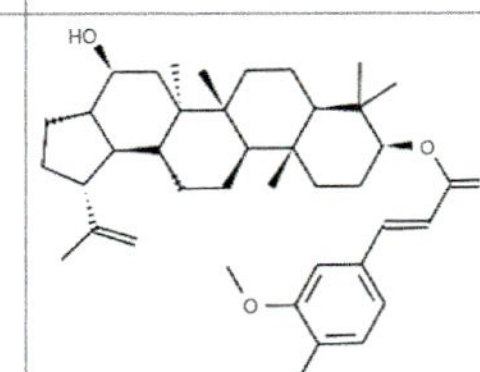	
				Ursolic acid	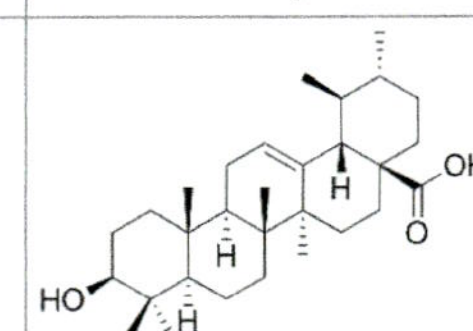	

	Alpinia pricei Hayata		Leaf and root	Caffeic acid	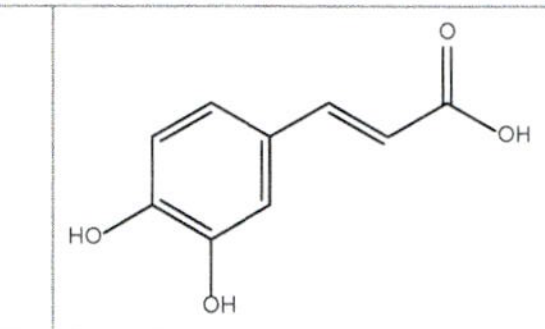	Hsu et al. (2010)
				Apigenin	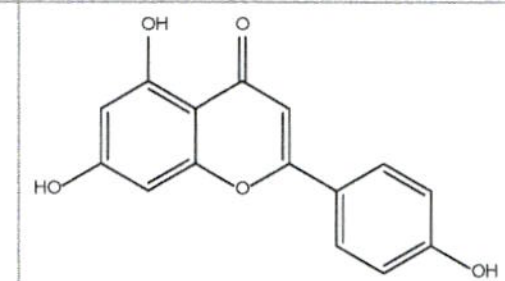	
				Curcumin	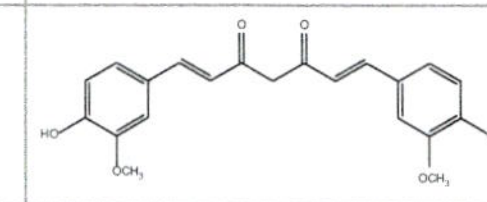	
	Synonym(s): *Alpinia sasakii* Hayata, *Languas pricei (Hayata)* Sasaki, *Languas sasakii (Hayata)* Sasaki, *Languas tarokoensis* Sasaki. **Common name(s)**: Shell ginger. **Botanical description**: It is native to Indonesia, Thailand, Manipur, and Myanmar. The plant grows from rhizomes up to 2 m in height with abundant long leaves that bear red fruit. It is used most often in cookery. The robust rhizome has a sharp, sweet taste and smells like a blend of black pepper and pine needles. The red fruit is used in traditional Chinese medicine and has a flavor similar to cardamom. **Medicinal use(s)**: It is used for the treatment of breast cancer and colorectal cancer. It possesses antitumor, antioxidant, and anti-inflammatory properties.			Pinocembrin	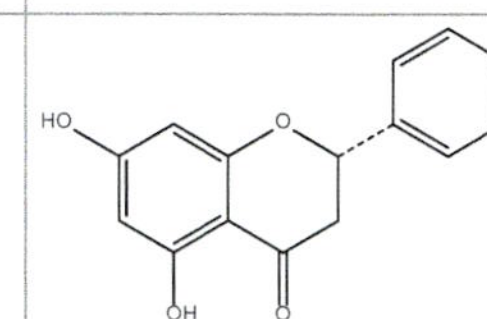	

(continued)

Table 4.148 (continued)

Family	Botanical name	Image	Plant part used	Bioactive compound(s)	Structure	References
	Curcuma longa L.		Rhizomes	Curcumin	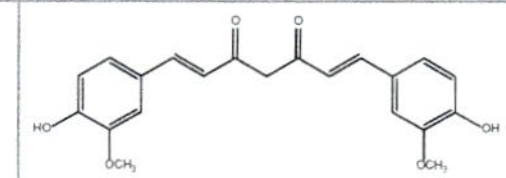	Sharma et al. (2005) and Vallianou et al. (2015)
	Synonym(s): *Amomum curcuma* Jacq., *Curcuma brog* Valeton, *Curcuma domestica* Valeton, *Curcuma ochrorhiza* Valeton, *Curcuma soloensis* Valeton, *Curcuma tinctoria* Guibourt, *Stissera curcuma* Giseke, *Stissera curcuma* Raeusch. **Common name(s)**: Curcuma, Tumeric, Jiang huang, Ukon, Renet, Rame, Temu kuning, Temu kunyit, Tius, and Haldi. **Botanical description**: It is native to South Asia and Southeast Asia. It is a perennial plant with oblong roots or tubers, palmate and deep orange inside, root leaves about 2 ft. long, lanceolate, long, petioled, tapering at each end, smooth, of a uniform green, petioles sheathing spike, erect, central, oblong, green, flowers dull yellow, three or five together surrounded by bracteolae. It is propagated by cuttings from the root. **Medicinal use(s)**: It is used for the treatment of breast cancer, liver ailments, and skin diseases and for wound healing. It is also used externally to heal sores and to treat dizziness, gonorrhea, and peptic ulcers and as an appetite stimulant, carminative, astringent, and antidiarrheal. The rhizome is also used to treat insect bites, ringworm, wounds, and bleeding from teeth and gums. It also possesses antioxidant, antitumor, anti-inflammatory, and antibacterial properties.					

	Kaempferia galanga L.		Leaf	Gemcitabine	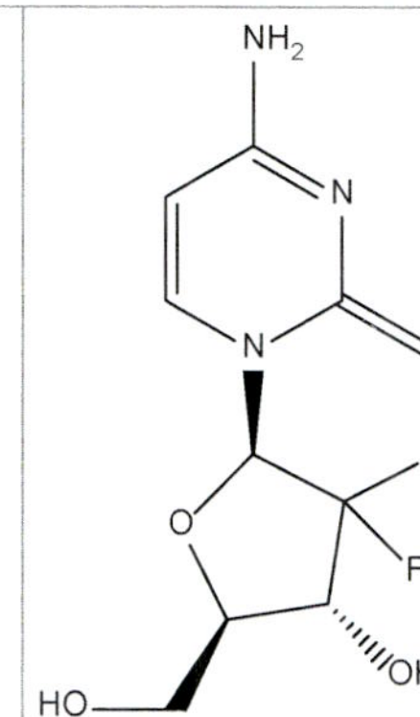	Jagdish et al. (2010) and Amuamuta et al. (2017)
	Synonym(s): *Alpinia sessilis* J.Koenig, *Kaempferia humilis* Salisb., *Kaempferia latifolia* Donn ex Hornem., *Kaempferia marginata* Carey ex Roscoe, *Kaempferia plantaginifolia* Salisb., *Kaempferia procumbens* Noronha. **Common name(s)**: Kencur, Aromatic ginger, Sand ginger, Cutcherry, or Resurrection lily. **Botanical description**: It is native to Indonesia, Southern China, Taiwan, Cambodia, and India but is also widely cultivated throughout Southeast Asia. The plant is used as an herb in cooking, especially in Thai cuisine. The plant has thick rounded leaves that lay flat on the ground. New leaves start growing in mid-spring from the small dormant rhizomes. **Medicinal use(s)**: It is used for the treatment of breast cancer and liver cancer. It is also used to treat cold, inflammation of the stomach, sprain, heartburn in the stomach, diarrhea and as a cough medicine. It possesses anticancer, antibacterial, antimicrobial, and anti-inflammatory properties.					

(continued)

Table 4.148 (continued)

Family	Botanical name	Image	Plant part used	Bioactive compound(s)	Structure	References
	Kaempferia rotunda L.		Rhizomes	Curcumin	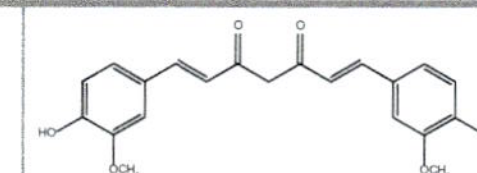	Lotulung et al. (2008)
	Synonym(s): *Kaempferia longa* Jacq., *Kaempferia versicolor* Salisb. **Common name(s)**: Blackhorm, Bhumi champa, Indian crocus, Peacock ginger, and Round-rooted galangale. **Botanical description**: It is a perennial plant that grows in warm places and is native to China, India, Nepal, Assam, Bangladesh, and Indochina. It is generally found in monsoon forests and open grassland at elevations around sea level to 2600 m. **Medicinal properties**: It is used for the treatment of breast and cervical cancer. It is also used to treat cough, nasal drop, indigestion, piles, skin diseases, jaundice, and wounds.					

	Zingiber zerumbet (L.) Roscoe ex Sm.		Fruit	Zerumbone	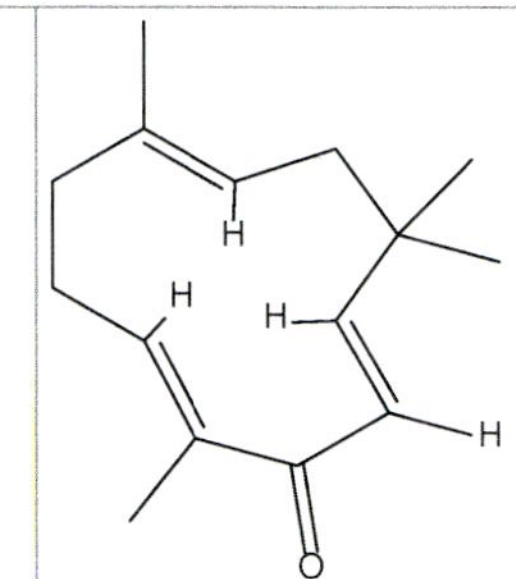 	Huang et al. (2005) and Rana et al. (2012)
	Synonym(s): *Amomum spurium* (J.Koenig) J.F.Gmel., *Amomum zerumbet* L., *Cardamomum spurium* (J.Koenig) Kuntze, *Dieterichia lampujang* Giseke, *Dieterichia lampuyang* Giseke, *Dieterichia major* Raeusch., *Dieterichia minor* Raeusch., *Dieterichia spuria* (J.Koenig) Giseke, *Zerumbet zingiber* T.Lestib., *Zingiber amaricans* Blume, *Zingiber aromaticum* Valeton, *Zingiber blancoi* Hassk., *Zingiber darceyi* H.J.Veitch, *Zingiber littorale* (Valeton) Valeton, *Zingiber ovoideum* Blume, *Zingiber spurium* J.Koenig, *Zingiber truncatum* Stokes. **Common name(s)**: Bitter ginger, Shampoo ginger, Pinecone ginger. **Botanical description**: It is a perennial plant and is native to India, Sri Lanka, and Malaysia. It goes dormant from autumn to spring, and then the leafy stems shrivel and die away. In the spring the plant grows up into a new plant. Leaves are blade-shaped, 15–20 cm long, and grow in an alternate arrangement. The stem is 1.2 m tall. The flowers are conical or club-shaped. The flower heads are initially green and are 3–10 cm long with overlapping scales, enclosing small yellowish-white flowers. **Medicinal use(s)**: It is used for the treatment of skin cancer and tumor. It also possesses anti-parasitic, anti-inflammatory, antinociceptive, antiproliferative, antihypertensive, and antitumor activities.					

Table 4.149 Anticancer plants of family Zygophyllaceae

Family	Botanical name	Image	Plant part used	Bioactive compound(s)	Structure	References
Zygophyllaceae	*Larrea divaricata* Cav		Leaf	Respirin	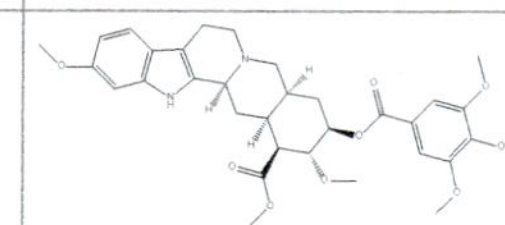	Bongiovanni et al. (2008)
	Synonym(s): *Neoschroetera divaricata* (Cav.) Briq, *Schroeterella divaricata* (Cav.) Briq. **Common name(s)**: Jarilla or Jarillo. **Botanical description**: It is native to South America. It attains a height of 3 m, with cylindrical stem, yellow five petals, and dark green small leaves. The bush is divided into stubby sections by dark glandular bands at the nodes that are leafy. The pinnate leaves are opposite, with two small, sessile, obliquely lance-ovate leaflets. **Medicinal use(s)**: It is used for the treatment of terminal cancer, acne, rheumatism, and diabetes. It also possesses anti-inflammatory and antioxidant effects and is used as a blood purifier.					

	Larrea mexicana Moric.		Leaf	Quercetin	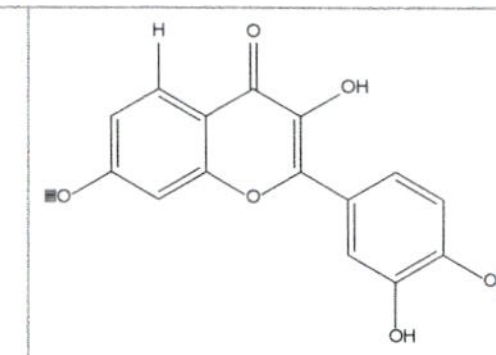	Gnabre et al. (2015)
	Synonym(s): No synonyms are recorded for this name. **Common name(s)**: Creosote bush, Créosotier, Greasewood chaparral, Gobernadora. **Botanical description**: It is native to South America, China, Japan, and Indonesia. The bush is an evergreen, with characteristic wide branches that droop when they have died off. It grows up to 3 m. Older branches are cylindrical, the younger branches square-edged. The bush is divided into stubby sections by dark glandular bands at the nodes that are leafy. The pinnate leaves are opposite, with two small, sessile, obliquely lance-ovate leaflets. **Medicinal use(s)**: It is used for the treatment of pancreatic cancer, bronchitis, coughs, colds, and chicken pox and used against viruses, bacteria, and parasites. It acts as analgesic and expectorant and is used to treat arthritis, bursitis, rheumatism, carpal tunnel syndrome, skin sores and wounds, urinary infections, and tuberculosis. Chaparral is useful for the treatment of vaginal yeast infections, such as trichomonas and *Candida albicans*. It also possesses antioxidant, antibacterial, and anti-inflammatory properties.					

(continued)

Table 4.149 (continued)

Family	Botanical name	Image	Plant part used	Bioactive compound(s)	Structure	References
	Larrea tridentata (Sessé & Moc. ex DC.) Coville		Whole plant	Nordihydro-guaiaretic acid	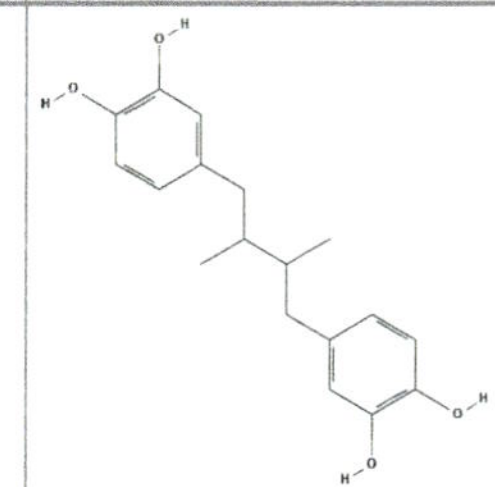	Wilkins et al. (1993), Lambert et al. (2005), Arteaga et al. (2005), and Martins et al. (2013)
	Synonym(s): *Larrea tridentata* var. *tridentata*. **Common name(s)**: Creosote bush or greasewood, Black bush, Chaparral, Dwarf evergreen oak, Gobernadora. **Botanical description**: It is a drought-tolerant, evergreen shrub growing up to 4 m tall and is native to the Southwestern United States and Mexico. Its numerous branches are brittle and densely leafy at the tips. Because of leaf and stem alignment, creosote bush provides little shade during the full desert sunshine. **Medicinal use(s)**: It is used for the treatment of pancreatic cancer. Decoction of the dried root is used as an abortive and for diabetes. It also acts as a blood purifier and is used to treat kidney problems, urinary tract infections, frigidity, fore gallstones, rheumatism and arthritis, wounds, skin injuries, displacement of the womb, and paralysis and to dissolve tumors. It possesses anti-inflammatory and antioxidant properties.			Lignans: 7S,8S,7′S, 8′S)-3,3′, 4′-trihydroxy-4-methoxy-7, 7′-epoxylignan, meso-(rel 7S,8S,7′R, 8′R)-3,4,3′, 4′-tetrahydroxy-7,7′-epoxylignan, and (E)-4,4′-dihydroxy-7, 7′-dioxolign-8(8′)-ene	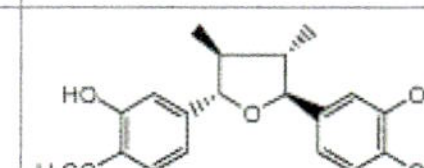 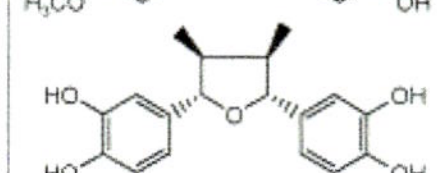 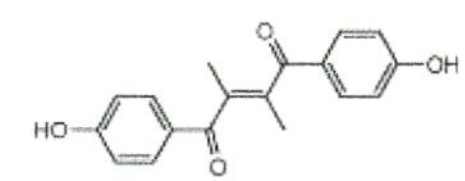	

	Fagonia cretica L.		Apical part	Apigenin	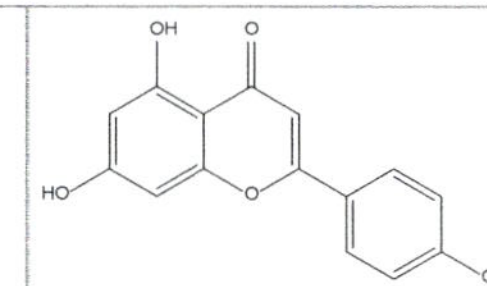	Hussain et al. (2007)
	Synonym(s): *Fagonia deflexa* Moench, *Fagonia elongata* Salisb., *Fagonia erecta* Mill., *Fagonia forskalii Pfund*, *Fagonia hispanica* L., *Fagonia virens* Coss. ex T.Anderson, *Fagonia viscosa* Hochst. ex Boiss. **Common name(s)**: Dhamasa, Virgin's mantle. **Botanical description**: It is small spiny shrub present in dry calcareous rocks of Pakistan and is native to Nigeria, Ghana, Zambia, Mali, Alicante province, Spain, and the Balearic Islands. Plant community analysis, using the phytosociological method, shows that it grows in wintergreen perennial dominated dry grasslands. **Medicinal use(s)**: It is used for the treatment of cancer, fever, thirst, vomiting, dysentery, asthma, urinary discharges, liver trouble, typhoid, toothache, stomach troubles, and skin diseases. Boiled residue of the plant in water is used to induce abortion. It is externally applied as a paste on tumors and other swellings of the neck. It also possesses anti-inflammatory activity and antioxidant activity.					

(continued)

Table 4.149 (continued)

Family	Botanical name	Image	Plant part used	Bioactive compound(s)	Structure	References
	Fagonia indica Burm. fil.		Aerial parts, leaf, and twigs	Gallic acid	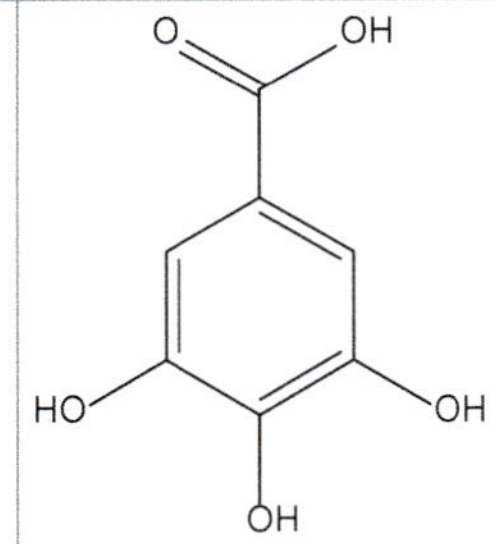	Ansari and Kenne (1982) and Urruticoechea et al. (2010)
	Synonym(s): *Fagonia aucheri* Boiss, *Fagonia elliptica* Tabassum, Omer & Qaiser, *Fagonia jolyi* Batt., *Fagonia mysorensis* B.Heyne ex Roth, *Fagonia ovalifolia* Hadidi, *Fagonia parviflora* Boiss. **Common name(s)**: Mushikka or White spine. **Botanical description**: It is a perennial herb and is native to Asia and Africa. Leaves are opposite, trifoliate with spinescent stipules, a pink corolla and smooth fruits. **Medicinal use(s)**: *Fagonia indica* is used for the treatment of breast cancer. It also possesses anti-inflammatory, wound healing, androgenic, anti-allergic, neuroprotective, endocrinological, antimicrobial, analgesic, cytotoxic, and antitumor properties.			Apigenin	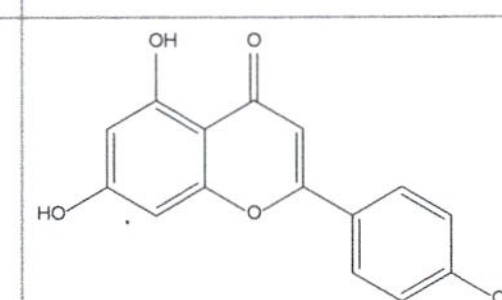	
	Tribulus terrestris L.		Root and fruit	Terrestrinins A	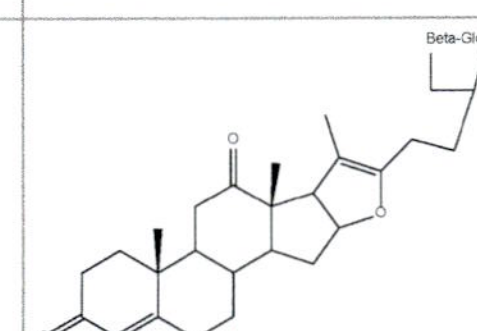	Chen et al. (2010) and Huang et al. (2003)

				Hitogenin	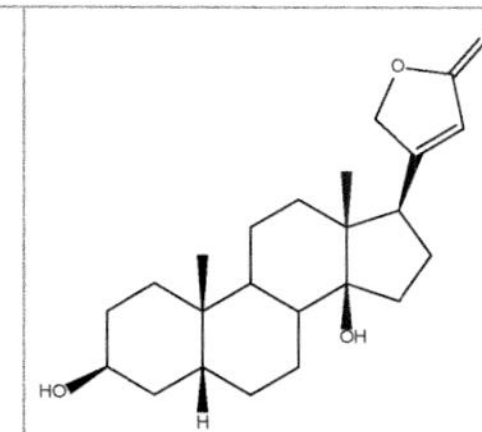	
				Ruscogenin	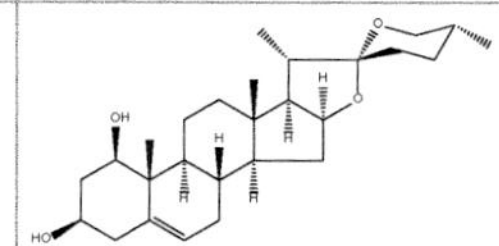	
	Synonym(s): *Tribulus lanuginosus* L., *Tribulus saharae* A. Chev. **Common name(s)**: Goat's head, Bindii, Bullhead, Burragokharu, Bhakhdi, Caltrop, Small caltrops, Cat's head, Devil's eyelashes, Devil's thorn, Devil's weed, Puncturevine, and Tackweed. **Botanical description**: It is native to Southern Europe, Southern Asia, throughout Africa and Australia. It can grow even in desert climates and poor soil too. The flowers are 4–10 mm wide, with five lemon-yellow petals, five sepal, and ten stamens. Leaves are opposite and pinnately compound. Densely hairy leaflets are opposite and up to 3 mm long. **Medicinal use(s)**: It is used for the treatment of cancer. It balances the nervous system, promotes good mood, reduces blood pressure, and is effective for treating cardiovascular disease, diabetes, and cholesterol. It also possesses antioxidant and anti-inflammatory properties.			Hecogenin	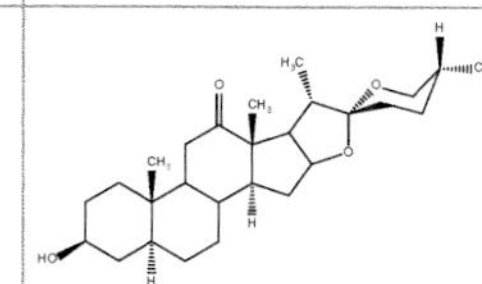	

References

Abdel-Hameed ESS, Bazaid SA, Al Zahrani O, El-Halmouch Y, El-Sayed MM, El-Wakil E (2014) Chemical composition of volatile components, antimicrobial and anticancer activity of n-hexane extract and essential oil from *Trachyspermum ammi* L. seeds. Orient J Chem 30(4):1653–1662

Abdel-Kader MS, Wisse J, Evans R, van der Werff H, Kingston DG (1997) Bioactive Iridoids and a New Lignan from *Allamanda cathartica* and *Himatanthus f allax* from the Suriname Rainforest. J Nat Prod 60(12):1294–1297

Abdel-Kader MS, Bahler BD, Malone S, Werkhoven MCM, Troon FV, David WJH, Bursuker I, Neddermann KM, Mamber SW, Kingston DGI (1998) DNA-damaging steroidal alkaloids from *Eclipta alba* from the Suriname rainforest. J Nat Prod 61(10):1202–1208

Abdel-Monem AR, Abdel-Sattar E, Harraz FM, Petereit F (2008) Chemical investigation of *Euphorbia schimperi*. J Nat Prod 2(2):39–45

Abebaw M, Mishra B, Gelayee DA (2017) Evaluation of anti-ulcer activity of the leaf extract of *Osyris quadripartita* Decne. (Santalaceae) in rats. J Exp Pharmacol 9:1

Abela CA, Santos AC (1974) Cyclotetracontane from *Cyanotis* vega. Acta Manilana Series A 12:80–82

Abida P, De Britto AJ, Antoney J, Raj TLS (2016) Evaluation of in vitro anticancer activity of *Symplocos cochinchinensis* (Lour.) S. Moore bark. Int J Herb Med 4:117–119

Acharya D, Mitaine-Offer AC, Kaushik N, Miyamoto T, Paululat T, Mirjolet JF, Duchamp O, Lacaille-Dubois MA (2009) Cytotoxic spirostane-type saponins from the roots of *Chlorophytum borivilianum*. J Nat Prod 72(1):177–181

Adesogan EF, Durodola JI (1976) Antitumor and antibiotic principles of *Annona senegalensis*. Phytochemistry 15:1311–1312

Adinarayana G, Rao R, Syamasundar K, Rajput D, Nagaraju G (2012) Biodiversity, conservation and cultivation of medicinal plants. J Pharmacogn 3(2):59–62

Ahmad FB, Holdsworth DK (1995) Traditional medicinal plants of Sabah, Malaysia part III. The Rungus people of Kudat. Int J Pharm 33:262–264

Ahmad Y, Le Quesne PW, Neuss N (1971) Sewarine: a new phenolic akuammicine alkaloid from *Rhazya stricta*. J Pharm Sci 60:1581–1583

Ahmad S, Sukari MA, Ismail N, Ismail IS, Abdul AB, Abu Bakar MF, Kifli N, Gwendoline CLE (2015) Phytochemicals from *Mangifera pajang* Kosterm and their biological activities. BMC Complement Altern Med 15(83):1–8

Ahn BT, Oh KJ, Ro JS, Lee KS (1996) A new flavonoid from *Euphorbia ebracteolata*. Planta Med 62:383–384

Ain Q, Khan H, Mubarak M, Pervaiz A (2016) Plant alkaloids as antiplatelet agent: drugs of the future in the light of recent developments. Front Pharmacol 7(1):292

Aixia X (2014) Chinese medicinal preparation for treating Chong-and-Ren channel disorder type mammary gland hyperplasia. CN103961488A

Akanitapichat P, Tongngok P, Wangmaneerat A, Sripanidkulchai B (2005) Antiviral and anticancer activities of *Stemona collinsae*. Thai J Pharm Sci 29(4):125–136

Akter R, Uddin Shaikh J, Darren Grice I, Tiralongo E (2014) Cytotoxic activity screening of Bangladeshi medicinal plant extracts. J Nat Med 68:246–252

Alam A, Rashid M, Abdur Rahman M, Baki A, Sadik M (2002) Chemical constituents of *Hemigraphis hirta* T. Anders (Acanthaceae). Pak J Biol Sci 5:1264–1266

Alam F, Najum us Saqib Q, Waheed A (2017a) Cytotoxic activity of extracts and crude saponins from *Zanthoxylum armatum* DC. against human breast (MCF-7, MDA-MB468) and colorectal (Caco-2) cancer cell lines. Complemen Altern Med 17:368

Alam M, Kwon K, Lee S, Lee S (2017b) *Lannea coromandelica* (Houtt.) Merr. induces heme oxygenase 1 (HO-1) expression and reduces oxidative stress via the p38/c-Jun N-terminal kinase–nuclear factor erythroid 2-related factor 2 (p38/JNK–NRF2)-mediated antioxidant pathway. Int J Mol Sci 18:266

Alayande K, Pohl C, Ashafa A (2016) Assessment of antidiarrhoea properties of *Euclea crispa* (Thunb.) leaf extract and fractions. S Afr J Bot 103:306

Ali Z, Khan SI, Khan IA (2006) Phytochemical study of *Actaea rubra* and biological screenings of isolates. Planta Med 72(14):1350–1352

Al-Jassabi S, Al-Omari A, Saad A, Azirun MS (2011) Tamsulosin-Induced hepatotoxicity and nephrotoxicity and its prevention by potato peel extract. Am-Euro J Toxicol Sci 3(2):52–58

Allen A (1995) Purification and characterization of an N-acetyllactosamine-specific lectin from tubers of Arum maculatum. Biochim Biophys Acta 1244(1):129–132

Alonso-Castro A, Villarreal M, Salazar-Olivo L, Gomez-Sanchez M, Dominguez F, Garcia-Carranca A (2011) Mexican medicinal plants used for cancer treatment: pharmacological, phytochemical and ethnobotanical studies. J Ethnopharmacol 133:945–972

Al-Rehaily AJ, Ahmad MS, Mustafa J, Al-Oqail MM, Hassan WH, Khan SI, Khan IA (2013) Solanopubamine, a rare steroidal alkaloid from *Solanum schimperianum*: synthesis of some new alkyl and acyl derivatives, their anticancer and antimicrobial evaluation. J Saudi Chem Soc 17(1):67–76

Amico A (1977) Medicinal plants of Southern Zambesia. Fitoterapia 48:101–139

Amin A, Mousa M (2007) Merits of anti-cancer plants from the Arabian Gulf region. J Cancer Ther 5:55–66

Amirghofran Z, Bahmani H, Azadmehr A, Javidnia K (2006) Induction of apoptosis in leukemia cell lines by *Linum persicum* and *Euphorbia cheiradenia*. J Cancer Res Clin Oncol 132(7):427–432

Amoo S, Ndhlala A, Finnie J, Van Staden J (2011) Antifungal, acetylcholinesterase inhibition, antioxidant and phytochemical properties of three *Barleria* species. S Afr J Bot 77:435–445

Amuamuta A, Plengsuriyakarn T, Na-Bangchang K (2017) Anticholangiocarcinoma activity and toxicity of the *Kaempferia galanga* Linn. Rhizome ethanolic extract. BMC Complement Altern Med 17(1):213

Amusan OOG, Msonthi JD, Makhubu LP (1995) Molluscicidal activity of *Spathodea campanulata*, Andrachne ovalis, *Phytolacca dodecandra* and *Hypoxis rooperi*. Fitoterapia 66:113–116

Anajwala CC, Patel RM, Dakhara SL, Jariwala JK (2010) In vitro cytotoxicity study of *Agave americana*, *Strychnos nuxvomica* and *Areca catechu* extracts using MCF-7 cell line. J Adv Pharm Technol Res 1(2):245–252

Anderson C, Fuller F, Epstein WW (1979) Nonpolarpentacyclic triterpenes of the medicinal fern *Polypodium subpetiolatum*. J Nat Prod 42:168–173

Angayarkanni J, Ramkumar KM, Poorinam T, Priyadarshani U (2007) Cytotoxic activity of *Amorphophallus paeoniifolius* tuber extracts in vitro. Am-Eur J Agric Environ Sci 2(4):395–398

Anon (1976) Cephalotaxine esters in the treatment of acuteleukemia. Chung-Hua I Hsueh Tsa Chih (New Series) 2:263–272

Anon (1985) Ayurvedic drug to fight cancer. Probe 24:234

Ansari AA, Kenne L (1982) Isolationprocedure and structure of new saponins from *Fagoniaindica*. Planta Med 45:143

Ansil P, Wills P, Varun R, Latha M (2014) Cytotoxic and apoptotic activities of *Amorphophallus campanulatus* (Roxb.) Bl. tuber extracts against human colon carcinoma cell line HCT-15. Saudi J Biol Sci 21(6):524–531

Appendino G, Porta DC, Conseil G, Sterner O, Mercalli E, Dumontet C, Pietro AD (2003) A new P-glycoprotein inhibitor from the caper spurge (*Euphorbia lathyris*). J Nat Prod 66(1):140–142

Aquino R, Pizza C, De Tommasi N, De Simone F (1995) New polyoxypregnane ester derivatives from *Leptadeniahastata*. J Nat Prod 58:672–679

Araujo MF, Santos CB, Cavalcanti JF, Pereira FS, Mendes GS, Werle AA, Romanos MTV, Carvalho MG (2011) Proposed active compounds from *Ouratea parviflora*. J Med Plant Res 5(12):2489–2493

Ardalani H, Mobarhan M, Avan A (2017) Podophyllotoxin: a novel potential natural anticancer agent. Avicenna J Phytomed 7(23):285–294

Arfan M, Amin H, Karamac M, Kosinska A, Shahidi F, Wiczkowski W, Amarowicz R (2007) Antioxidant activity of extracts of *Mallotus philippinensis* fruit and bark. J Food Lipids 14(3):280–297

Arı S, Temel M, Kargıoğlu M, Konuk M (2015) Ethnobotanical survey of plants used in Afyonkarahisar-Turkey. J Ethnobiol Ethnomed 11:84

Ariffin SHZ, Omar WHW, Ariffin ZZ, Safian MF, Senafi S, Wahab RMA (2009) Intrinsic anticarcinogenic effects of *Piper sarmentosum* ethanolic extract on a human hepatoma cell line. Int J Cancer Cell 9(6):1–9

Arisawa M, Fujita A, Morita N, Okuyama T, Nishino H (1991) Inhibition of tumorpromoter-enhanced 3h-choline incorporation into cellular phospholipids by phloroglucinol derivatives from *Mallotus japonicas*. J Nat Prod 54(5):1409–1412

Arora S, Tandon S (2014) *Achyranthes aspera* root extracts induce human colon cancer cell (COLO-205) death by triggering the mitochondrial apoptosis pathway and S phase cell cycle arrest. Sci World J:1–15

Arora S, Singh S, Piazza G, Contreras C, Panyam J, Singh A (2012) Honokiol: a novel natural agent for cancer prevention and therapy. Curr Mol Med 12(10):1244–1252

Arroo RRJ, Alfermann AW, Medarde M, Petersen M, Pras N, Woolley JG (2002) Plant cell factories as a source for anti-cancer lignans. Phytochem Rev 1:22–35

Arteaga S, Andrade-Cetto A, Cardenas R (2005) *Larrea tridentata* (Creosote bush), an abundant plant of Mexican and US-American deserts and its metabolite nordihy-droguaiaretic acid. J Ethnopharmacol 98:231–239

Arul CR, Sophia D, Ragavendran P, Starlin T, Rathi MA, Gopalakrishnan VK (2012) Leaf extract of *Alpinia purpurata* (Vieill.) K. Schum screened for its phytochemical constituents and antibacterial and anticancer activities. Zhong xi yi jie he xue bao= J Chin Integr Med 10(12):1460–1464

Arulvasu C, Babu G, Manikandan R, Srinivasan P, Sellamuthu S, Prabhu D, Dinesh D (2010) Anticancer effect of *Datura innoxia* P. Mill. Leaf extract in vitro through induction of apoptosis in human Colon Adenocarcinoma and larynx cancer cell lines. J Pharm Res 3(7):1485–1488

Arulvasu C, Vasantha Suppriya S, Gajendran B (2012) Anticancer activity of *Pongamia glabra* V. seed oil extract against selected human cancer cell lines. Int Res J Pharm 3(8):131–134

Arunjyothi B, Venkatesh K, Chakrapani P, Anupalli RR (2011) Phytochemical and Pharmacological potential of *Annona cherimola*-A review. Int J Phytomed 3(4):439

Asada Y, Hirayama Y, Furuya T (1988) Acylatedflavonols from *Crocosmia crocosmiiflora*. Phytochemistry 27:1497–1501

Ashidi JS, Houghton PJ, Hylands PJ, Efferth T (2010) Ethnobotanical survey and cytotoxicity testing of plants of south-western Nigeria used to treat cancer, with isolation of cytotoxic constituents from *Cajanus cajan* Millsp leaves. J Ethnopharmacol 128:501–512

Assad N, Dheeba B, Aljubouri F, Hamad A (2015) Anti-cancer activity of the *Rubus idaeus* extracts against HepG2 and L20B cell lines using tissue culture technique. Egypt Acad J Biol Sci 7(2):17–23

Assaf A, Haddadin R, Aldouri N, Alabbassi R, Mashallah S, Mohammad M, Bustanji Y (2013) Anti-cancer, anti-inflammatory and anti-microbial activities of plant extracts used against hematological tumors in traditional medicine of Jordan. J Ethnopharmacol 145(3):72–736

Ata A, Kalhari KS, Samarasekera R (2009) Chemical constituents of Barleria prionitis and their enzyme inhibitory and free radical scavenging activities. Phytochem Lett 2:7–40

Atta-ur-Rahman, Choudhary M (2017) Frontiers in drug design and discovery 8:222

Aumsuwan P, Khan SI, Khan IA, Ali Z, Avula B, Walker LA, Shariat-Madar Z, Helferich WG, Katzenellenbogen BS, Dasmahapatra AK (2016) The anticancer potential of steroidal saponin, dioscin, isolated from wild yam (Dioscoreavillosa) root extract in invasive human breast cancer cell line MDA-MB-231 in vitro. Arch Biochem Biophys 591:98–110

Auyeung KKW, Cho CH, Ko JKS (2009) A novel anticancer effect of *Astragalus saponins*: Transcriptional activation of NSAID-activated gene. Int J Cancer 125(5):1082–1091

Ayatollahi AM, Ghanadian M, Afsharypuor S, Siddiq S, Pour-Hosseini M (2010) Biological screening of *Euphorbia aellenii*. Iran J Pharm Res 9(4):429–436

Ayensu ES (1978) Medicinal plants of West Africa. Reference Publications Inc, Algonac

Ayob Z, Samad A, Bohari S (2013) Cytotoxicity activities in local *Justicia gendarussa* crude extracts against human cancer cell lines. J Tek 64(2):45–52

Aziz A, Taha H, Ismail N, Mohd Yusof F, Bakar M (2016) Therapeutic potential of plant species derived from some annonaceae genus. Int J Agric Forestry 6:214–221

Badakashan MP, Sreenivasan S, Jegathambigai RN, Surash R (2009) Anti-leukemia activity of methanolic extracts of *Lantana camara*. Pharm Res 1(5):274–279

Bae KH, Min BS, Park KL, Ahn BZ (1994) Cytotoxic flavonoids from *Scutellaria indica*. Planta Med 60(3):280–281

Bae MJ, Ye EJ, Kim SJ, Kim JM, Yee ST, Park EM (2007) The effects of *Plebeiae herba* (*Salvia plebeia* R. Br.) on the anticancer (in vitro) and activation of immune cells. J Korean Soc Food Sci Nutr 36(4):377–382

Bai L, Yamaki M, Takagi S (1997) Lignans and a bichromanfrom *Pleione bulbocodioides*. Phytochemistry 44:341–343

Bai SP, Wei QY, Jin XL, Wu QX, Yang L (2005) Two novel ent-kauranoid diterpenoids from Isodon japonica leaves. Planta Med 71(8):764–769

Bailon-Moscoso N, Romero-Benavides J, Tinitana-Imaicela F, Ostrosky-Wegman P (2015) Medicinal plants of Ecuador: a review of plants with anticancer potential and their chemical composition. Med Chem Res 24:2283–2296

Bailon-Moscoso N, Romero Benavides JC, Ramirez Orellana MI, Ojeda K, Granda G, Ratoviski EA, Ostrosky-Wegman P (2016) Cytotoxic and genotoxic effects of extracts from *Annona Montana* M. Fruit. Food Agric Immunol 27(4):559–569

Balachandran C, Emi N, Arun Y, Yamamoto Y, Ahilan B, Sangeetha B, Al Dhabi NA (2015) In vitro anticancer activity of methyl caffeate isolated from *Solanum torvum* Swartz. fruit. Chem Biol Interact 242:81–90

Balaji R, Suba V, Rekha N, Deecaraman M (2009) Hepatoprotective activity of methanolic fraction of *Jatropha curcas* on aflatoxin b1 induced hepatic carcinoma. Int J Pharm Sci 1(2):287–296

Balasubramanian K, Padma P (2013) Anticancer activity of Zea mays leaf extracts on oxidative stress-induced Hep2 Cells. J Acupunct Meridian Stud 6(3):149–158

Balick M, Arvigo R (2015) Messages from the Gods: a guide to the useful plants of Belize, p 295

Baliga MS (2010) *Alstonia scholaris* L. in the treatment and prevention of cancer: past, present, and future. Integr Cancer Ther 9(3):261–269

Baliga MS (2012) Review of the phytochemical, pharmacological and toxicological properties of *Alstonia Scholaris* Linn. R. In: Br (Saptaparna) Chin J Integr Med

Baliga M, Thilakchand K, Rai M, Rao S, Venkatesh P (2012) *Aegle marmelos* (L.) Correa (Bael) and its phytochemicals in the treatment and prevention of cancer. Integr Cancer Ther 12(3):187–196

Bandoni AL, Mendiondo ME, Rondina RVD, Coussio JD (1976) Survey of Argentine medicinal plants. Folkloreand phytochemical screening. II. Econ Bot 30:161–185

Banerji A, Deshpande AD, Prabhu BR, Pradhan P (1994) Tomentonone: a new xanthonoid from the stem bark of *Calophyllum tomentosum*. J Nat Prod 57:396–399

Barakat HH, Hussein SAM, Marzour MS, Merfort I, Linscheid M, Nawwar MAM (1997) Polyphenolicmetabolites of *Epilobium hersutum*. Phytochemistry 46:935–941

Barton D, Soori G, Bauer B, Sloan J, Johnson P, Figueras C, Duane S, Mattar B, Liu H, Atherton P, Christensen B, Loprinzi C (2009) Pilot study of *Panax quinquefolius* (American ginseng) to improve cancer-related fatigue: a randomized, double-blind, dose-finding evaluation: NCCTG trial N03CA. Support Care Cancer 18:179–187

Bastos ML, Lima MR, Conserva LM, Andrade VS, Rocha EM, Lemos RP (2009) Studies on the antimicrobial activity and brine shrimp toxicity of Zeyheria tuberculosa (Vell.) bur. (Bignoniaceae) extracts and their main constituents. Ann Clin Microbiol Antimicrob 8(1):16

Beara IN, Lesjak MM, Jovin ED, Balog KJ, Anackov GT, Orcic DZ, Mimica-Dukic NM (2009) Plantain (*Plantago*) species as novel sources of flavonoid antioxidants. J Agric Food Chem 57:9268–9273

Bedigian D (2013) Ecogeography and taxonomy of Rogeria J. Gay ex Delile (Pedaliaceae). Webb 68(2):103–126

Beegum AS, Martin KP, Zhang C, Nishitha IK, Ligimol Slater A, Madhusoodanan PV (2007) Organogenesis from leaf and internode explants of *Ophiorrhiza prostrata*, an anticancer drug (camptothecin) producing plant. Electron J Biotechnol 10(1):114–123

Ben Slima A, Ali M, Barkallah M, Traore A, Boudawara T, Allouche N, Gdoura R (2013) Antioxidant properties of *Pelargonium graveolens* L'Her essential oil on the reproductive damage induced by deltamethrin in mice as compared to alpha-tocopherol. Lipids Health Dis 12(1):30

Bendaoud H, Romdhane M, Souchard JP, Cazaux S, Bouajila J (2010) Chemical composition and anticancer and antioxidant activities of Schinus molle L. and *Schinus terebinthifolius* Raddi berries essential oil. J Food Sci 75(6):466–472

Benosman A, Oger JM, Richomme P, Bruneton J (1997) New terpenylated dihydrochalcone derivatives isolated from Mitrella kentia. J Nat Prod 60(9):921–924

Bentley R (2010) Medicinal plants. Domville-Fife Press, London, pp 23–46

Bernabé-Antonio A, Álvarez-Berber LP, Cruz-Sosa F (2014) Biological importance of phytochemicals from *Calophyllum brasiliense* Cambess. Ann Res Rev Biol 30:1502–1517

Bessonova IA, Batsuren D, Yunusov SY (1984) Alkaloids of *Haplophyllium dauricum*. Khimia Priroidin Soedin 20:73–76

Bezerra DP, Marinho-Filho JD, Alves AP, Pessoa C, de Moraes MO, Pessoa OD, Torres MC, Silveira ER, Viana FA, Costa-Lotufo LV (2009) Antitumor activity of the essential oil from the leaves of *Croton regelianus* and its component ascaridole. Chem Biodivers 6(8):1224–1123

Bhagat M, Arora JS, Saxena AK (2007) In vitro cytotoxicity of extracts and fractions of *Calotropis procera* (Ait.) roots against human cancer cell lines. Int J Green Pharm 4(1):36–40

Bhandari P, Kumar N, Singh B, Kaul VK (2007) Cucurbitacins from *Bacopa monnieri*. Phytochemistry 68:1248–1254

Bhandary MJ, Chandrashekar KR, Kaveriappa KM (1995) Medical ethnobotany of the Siddis of Uttara Kannadadistrict, Karnataka, India. J Ethnopharmacol 47:149–158

Bhanot A, Sharma R, Noolvi MN (2011) Natural sources as potential anti-cancer agents: a review. Int J Phytomed 3:9–26

Bhat KS, Sharma A (2013) Therapeutic potential of cardiac glycosides of *Calotropis gigantea* for breast cancer. Int Res J Pharm 4(6):164–167

Bhat PR, Prajna PS, Kumar V, Hegde MA, Singh L (2011) Antimicrobial properties of leaves of *Calycopteris floribunda* Lam. J Med Plant Res 5(12):2448–2451

Bhatia K, Kaur M, Atif F, Ali M, Rehman H, Rahman S, Raisuddin S (2006) Aqueous extract of *Trigonella foenum-graecum* L. ameliorates additive urotoxicity of buthionine sulfoximine and cyclophosphamide in mice. Food Chem Toxicol 44(10):1744–1750

Bhatt P, Kushwah A (2015) *Rubia cordifolia* overview: a new approach to treat cardiac disorders. Int J Drug Dev Res 5(2)

Bhuvan RP, Jignesh PD, Bhavik PA, Ashok GL (2009) Potent in vitro anticancer activity of *Symplocos racemosa* bark. J Plant Biol 54(2):135–140

Bi W, Gao Y, Shen J, He C, Liu H, Peng Y, Zhang C, Xiao P (2016) Traditional uses, phytochemistry, and pharmacology of the genus Acer (maple): a review. J Ethnopharmacol 189:31–60

Bibi G, Ihsan-ul-Haq UN, Mannan A, Mirza B (2011) Antitumor, cytotoxic and antioxidant potential of *Asterthomsonii* extracts. Afr J Pharm Pharmacol 5(2):252–258

Biggs M, Sorkin B, Nahin R, Kuller L, Fitzpatrick A (2010) *Ginkgo biloba* and risk of cancer: secondary analysis of the Ginkgo Evaluation of Memory (GEM) Study. Pharmacoepidemiol Drug Saf 19(7):694–698

Bigos M, Wasiela M, Kalemba D, Sienkiewicz M (2012) Antimicrobial activity of Geranium oil against clinical strains of *Staphylococcus aureus*. Molecules 17(9):10276–10291

Bijak M (2017) Silybin, a major bioactive component of milk thistle (*Silybum marianum* L. Gaernt.) Chemistry, bioavailability, and metabolism. Molecules 22(11):1942

Biswal BK, Tripathi SK (2019) Phytotherapeutic potential of *Pterospermum acerifolium* L. in human cancer cell lines through mitochondrial mediated ROS generation

Biswas R, Mandal SK, Dutta S, Bhattacharyya SS, Boujedaini N, Khuda-Bukhsh AR (2010) Thujone-rich fraction of *Thuja occidentalis* demonstrates major anti-cancer potentials: evidences from in vitro studies on A375 cells. Evi Based Comp Alt Med Adv Acc 42:1–15

Bonamin L, de Carvalho A, Waisse S (2017) *Viscum album* (L.) in experimental animal tumors: a meta-analysis. Exp Ther Med 13(6):2723–2740

Bongiovanni G, Cantero J, Eynard A, Goleniowski M (2008) Organic extracts of *Larrea divaricata* cav. induced apoptosis on tumoral MCF-7 cells with an higher cytotoxicity than nordihydroguaiaretic acid or paclitaxel. J Exp Ther Oncol 7(1):1–7

Bonigala B, Kiranmayi M, Viswanath K, Poda S (2017) A comprehensive report on cancer and cytotoxic activity of plants on cancer cell lines. J Pharm Res 11(8):967

Boon H, Zick S, Green J, Olatunde S (2010) Trial of Essiac to ascertain its effect in women with breast cancer (TEA-BC). J Altern Complement Med 12(10):6–6

Boonsri S, Karalai C, Ponglimanont C, Kanjana-opas A, Chantraprommа K (2006) Antibacterial and cytotoxic xanthones from the roots of *Cratoxylumformosum*. Phytochemistry 67(7):723–727

Bose PK, Bose J (1939) Natural Flavones. Part III. On theconstitution of Tambulin. J Indian Chem Soc 16:183

Boukef K, Souissi HR, Balansard G (1982) Contributionto the study on plants used in traditional medicine inTunisia. Plant Med Phytother 162:60–279

Boukes G, Van de Venter M (2016) In vitro modulation of the innate immune response and phagocytosis by three Hypoxis spp. and their phytosterols. S Afr J Bot 102(2016):120–126

Boukes GJ, Venter MVD (2011) Cytotoxicity and mechanism(s) of action of Hypoxis spp. (African potato) against HeLa, HT-29 and MCF-7 cancer cell lines. J Med Plant Res 5:2766–2774

Bousserouel S, Le Grandois J, Gosse F, Werner D, Barth S, Marchioni E, Marescaux J, Raul F (2013) Methanolic extract of white asparagus shoots activates trail apoptotic death pathway in human cancer cells and inhibits colon carcinogenesis in a preclinical model. Int J Oncol 43:394–404

Bowen-Forbes CS, Zhan Y, Nair MG (2010) Anthocyanin content, antioxidant, anti-inflammatory and anticancer properties of blackberry and raspberry fruits. J Food Compos Anal 23(6):554–560

Bradacs G, Maes L, Heilmann J (2009) In vitro cytotoxic, antiprotozoal and antimicrobial activities of medicinal plants from Vanuatu. Phytother Res 24(6):800–809

Bradley G, Akinrinde A, Afolayan A (2018) Phytochemical composition and antioxidant activities of *Dianthus Thunbergii* hooper and *Hypoxis Argentea* harv ex baker: plants used for the management of diabetes mellitus in Eastern Cape, South Africa. Pharmacogn Mag 14(54):195–202

Brisibe EA, Umoren UE, Brisibe F, Magalhaes PM, Ferreira JFS, Luthria D, Wu X, Prior RN (2009) Nutritional characterisation and antioxidant capacity of different tissues of Artemisia annua L. Food Chem 115(4):1240–1246

Brooker SG, Cambie RC, Cooper RC (1989) Economicnative plants of New Zealand. Econ Bot 43:79–106

Brown A (2012) Anticancer activity of Morinda citrifolia (Noni) fruit: a review. Phytother Res 26(12):1427–1440

Bueno-Durán A, Cervantes-Martínez J, Obledo-Vázquez E (2013) Composition of essential oil from *Lippia graveolens*. Relationship between spectral light quality and thymol and carvacrol content. J Essent Oil Res 26(3):153–160

Bunyapraphatsara N, Dechsree S, Yoosook C, Herunsalee A, Panpisutchai Y (2000) Anti-herpes simplex virus component isolated from *Maclura cochinchinensis*. Phytomedicine 6(6):421–424

Buranrat B, Mairuae N, Kanchanarach W (2017) Cytotoxic and antimigratory effects of *Cratoxy formosum* extract against HepG2 liver cancer cells. Biomed Rep 6(4):441–448

Burk W (1983) Puffball usages among North American Indians. J Ethnobiol 3(1):55–62

Burkill IH (1966a) Dictionary of the economic products of the Malay peninsula, vol II. Ministry of agriculture and cooperatives, Kuala Lumpur

Burkill IH (1966b) Dictionary of the economic products of the Malay peninsula, vol I. Ministry of agriculture and cooperatives, Kuala Lumpur

Cai Y, Chen ZP, Phllipson J (1993) Diterpenes from *Croton lechleri*. Phytochemistry 32:755–760

Cai Y, Luo Q, Sun M, Corke H (2003) Antioxidant activity and phenolic compounds of 112 traditional Chinese medicinal plants associated with anticancer. Life Sci 74:2157–2184

Cainelli G, Angeloni C, Cervellati R, Galletti P, Giacomini D, Hrelia S, Sinisi R (2008) New polyphenolic β-lactams with antioxidant activity. Chem Biodivers 5(5):811–829

Çalişkan UK, Aka C, Bor E (2017) Melationin in edible and non-edible plants. Turk J Pharm Sci 14(1):75–83

Calixto JB, Sant'ana EG (1987) Pharmacological analysisof the inhibitory effect of jatrophone, a diterpene isolated from *Jatropha elliptica*, on smooth and cardiac muscles. Phytother Res 1:122–126

Campagnuolo C, Fattorusso E, Petrucci F, Taglialatela-Scafati O, Appendino G, Marquez N, Munoz E (2004) Study on the diterpenes and their bioactivities in Croton plant. J Nat Prod 67(4):532–536

Cao S, Sng VHL, Wu X, Sim K, Tan BHK, Pereira JT, Goh SH (1998) Novel cytotoxic polyprenylated xanthonoids from *Garcinia gaudichaudii* (Guttiferae). Tetrahedron 54:10915–10924

Cao S, Norris A, Wisse JH, Miller JS, Evans R, Kingsto DGI, Ipomoeassins AE (2005) Cytotoxic macrocyclic glycoresins from the leaves of *Ipomoea squamosa* from the Suriname Rainforest. J Nat Prod 68(4):487–492

Cao X, Zou H, Cao J, Cui Y, Sun S, Ren K, Song Z, Li D, Quan M (2016) A candidate Chinese medicine preparation-Fructus Viticis Total Flavonoids inhibits stem-like characteristics of lung cancer stem-like cells. BMC Complement Altern Med 16(1):364

Carmona ER, Escobar B, Obando V (2016) Antimutagenic activity of *Buddleja globosa* extracts in the Drosophila wingspot test. Fresenius Environ Bull 25:5758–5764

Cartagena E, Bardon A, Catalan CA, de Hernandez ZN, Hernandez LR, Joseph-Nathan P (2000) Germacranolides and a new type of guaianolide from *Acanthospermum hispidum*. J Nat Prod 63(10):1323–1332

Cartea ME, Francisco M, Soengas P, Velasco P (2011) Phenolic compounds in Brassica vegetables. Molecules 16(1):251–280

Carvalho M, Ferreira P, Mendes V, Silva R, Pereira J, Jerónimo C, Silva B (2010) Human cancer cell antiproliferative and antioxidant activities of *Juglans regia* L. Food Chem Toxicol 48(1):441–447

Cernakova M, Kostalova D, Kettmann V, Plodova M, Toth J, Drimal J (2002) Potential antimutagenic activity of berberine, a constituent of *Mahonia aquifolium*. BMC Complement Altern Med 2(2):1–6

Cha S (1977) Potential anticancer medicinal plants. A statistical evaluation of their frequencies of appearance in oriental medicine formularies. Korean J Pharma 8:1–14

Chakraborty A, Guptaa N, Ghosh K, Roy P (2010) In vitro evaluation of the cytotoxic, anti-proliferative and anti-oxidant properties of pterostilbene isolated from *Pterocarpus marsupium*. Toxicol in Vitro 24(4):1215–1228

Chan EW, Wong SK (2015) Phytochemistry and pharmacology of ornamental gingers, *Hedychium coronarium* and *Alpinia purpurata*: a review. J Integr Med 13(6):368–379

Chan EW, Lim CS, Lim WY, Loong ZJ, Wong CW (2019) Role of scutellarin in human cancer—a review. J Appl Pharm Sci 9(01):142–146

Chandrasekhar HR, Raj PV, Rao JV, Udupa N (2009) Anticancer activity of *Hypericum mysorense*. J Biomed Pharm Eng 2(4):1–3

Chandrasekaran R, Seetharaman P, Krishnan M, Gnanasekar S, Sivaperumal S (2018) Carica papaya (Papaya) latex: a new paradigm to combat against dengue and filariasis vectors *Aedes aegypti* and *Culex quinquefasciatus* (Diptera: Culicidae). 3 Biotech 8(2):83

Chang Y, Seo E, Gyllenhaal C, Block K (2003) Panax ginseng: a role in cancer therapy. Integr Cancer Ther 2:13–33

Chang H, Huang G, Agrawal D, Kuo C, Wu C, Tsay H (2007) Antioxidant activities and polyphenol contents of six folk medicinal ferns used as "Gusuibu". Bot Stud 48:397–406

Chao JC, Chu CC (2004) Effects of *Ginkgo biloba* extract on cell proliferation and cytotoxicity in human hepatocellular carcinoma cells. World J Gastroenterol 10(1):37–41

Chao WW, Lin BF (2010) Isolation and identification of bioactive compounds in *Andrographis paniculata* (Chuanxinlian). Chin Med 5(1):17

Chao YY, Jan CR, Ko YC, Chen JJ, Jiann BP, Lu YC, Chen WC, Su W, Chen IS (2002) Effect of lignans isolated from *Hernandia nymphaeifolia* on estrogenic compounds induced calcium mobilization in human neutrophil. J Life Sci 70:3109–3121

Charlson AJ (1979) Antineoplastic constituents in some southern African Plants. Newsl Reg Netw Chem Nat Prod Southeast Asia 3:22–23

Charlson AJ (1980) Antineoplastic constituents of some southern African plants. J Ethnopharmacol 2:323–335

Chaturvedi P, Bhui K, Shukla Y (2008) Lupeol: connotations for chemoprevention. Cancer Lett 263(1):1–13

Chauhan NS, Dixit VK (2010) *Asteracantha longifolia* (L.) Nees, Acanthaceae: chemistry, traditional, medicinal uses and its pharmacological activities-a review. Rev Bras 20(5):812–817

Chauret N, Rho D, Archambault J (1990) Solid-phase extraction and fluorimetric detection of benzophenanthridine alkaloids from *Sanguinaria canadensis* cell cultures. J Chromatogr A 519(1):99–107

Chavez H, Estevez-Braun A, Ravelo AG, González AG (1998) Friedelane triterpenoids from *Maytenus macrocarpa*. J Nat Prod 61(1):82–85

Cheenpracha S, Karalai C, Ponglimanont C, Chantrapromma K (2004) New cytotoxic cardenolide glycoside from the seeds of Cerbera manghas. Chem Pharm Bull 52(8):1023–1025

Cheenpracha S, Karalai C, Ponglimanont C, Kanjana-Opas A (2009) Candenatenins a– F, Phenolic compounds from the heartwood of *Dalbergia candenatensis*. J Nat Prod 72(8):1395–1398

Chen J, Chang Y, Teng C, Chen I (2000) Anti-platelet aggregation alkaloids and lignans from *Hernandia nymphaeifolia*. Planta Med 66(3):251–256

Chen X, Krakauer T, Oppenheim JJ, Howard OMZ, Huang YZ (2004) An injectable multicomponent Chinese herbal medicine, is a potent inhibitor of T-Cell activation. J Altern Complement Med 10(3):519–526

Chen S, Gao J, Halicka HD, Huang X, Traganos F, Darzynkiewicz Z (2005) The cytostatic and cytotoxic effects of oridonin (Rubescenin), a diterpenoid from *Rabdosia rubescens*, on tumor cells of different lineage. Int J Oncol 26(3):579–588

Chen FA, Wu AB, Shieh P, Kuo DH, Hsieh CY (2006) Evaluation of the antioxidant activity of *Ruellia tuberosa*. Food Chem 94(1):14–18

Chen YH, Chang FR, Lin YJ, Hsieh PW, Wu MJ, Wu YC (2008a) Identification of antioxidants from rhizome of *Davallia solida*. Food Chem 107(2):684–691

Chen Y, Chang F, Lu M, Hsieh P, Wu M, Du Y, Wu Y (2008b) New benzoyl glucosides and cytotoxic pterosin sesquiterpenes from *Pteris ensiformis* Burm. Molecules 13:255–266

Chen C, Kuo P, Chen Y, Huang J, Ho M, Lin R, Chang J, Wang H (2010) Tyrosinase inhibition, free radical scavenging, antimicroorganism and anticancer proliferation activities of *Sapindus mukorossi* extracts. J Taiwan Inst Chem Eng 41(2):129–135

Chen Y, Hsieh P, Mau J, Sheu S (2011) Antioxidant properties and mutagenicity of *Pinus morrisonicola* and its vinegar preparation. LWT Food Sci Technol 44(6):1477–1481

Chen G, Tian Y, Wu J, Li N, Guo M (2016a) Antiproliferative activities of Amaryllidaceae alkaloids from Lycoris radiata targeting DNA topoisomerase I. Sci Rep 6:38284

Chen X, Liu X, Wink M, Ma Y, Guo Y (2016b) A myrsinol diterpene isolated from *Euphorbia prolifera* reverses multidrug resistance in breast cancer cells. Pharmazie 71(9):537–539

Chen S, Yu H, Luo H, Wu Q, Li C, Steinmetz A (2016c) Conservation and sustainable use of medicinal plants: problems, progress, and prospects. Chin Med 11(1):37

Chen Y, Lien H, Kao M, Lo U, Lin L, Lin C, Chang S, Chen C, Hsieh J, Lin H, Tang C, Lai C (2017) Sensitization of radioresistant prostate cancer cells by resveratrol isolated from arachis hypogaea stems. PLoS One 12(1):0169204

Cheng H, Lin T, Yu K, Yang C, Lin CA (2003) Antioxidant and free radical scavenging activities of *Terminalia chebula*. Biol Pharm Bull 26(9):1331–1335

Cheng G, Liu H, Wang Y, Liu F, Cao J (2009) Phenolic compounds with anticancer activity from the buds of *Vaccinium dunalianum*. Mod Pharm Res 2(5):75–82

Cheng Y, Hsieh C, Tsai T (2018) Concurrent administration of anticancer chemotherapy drug and herbal medicine on the perspective of pharmacokinetics. J Food Drug Anal 26:S88–S95

Cherdshewasart W (2009) Extracts derived from pueraria mirifica, butea superba and/or mucuna collettii and extraction thereof. EP1242103A1, 7 January 2009

Cherng IH, Wu DP, Chiang HC (1996) Triterpenoids from *Antrodia cinnamomea*. Phytochemistry 41:263–267

Chhabra SC, Uiso FC, Mshiu EN (1984) Phytochemicalscreening of Tanzanian medicinal plants. I. J Ethnopharmacol 11:157–179

Chhabra SC, Mahunnah RLA, Mshiu EN (1987) Plantsused in traditional medicine in eastern Tanzania. I. Pteridophytes and Angiosperms (Acanthaceae to Canellaceae). J Ethnopharmacol 21:253–277

Chhabra SC, Mahunnah RLA, Mshiu EN (1991) Plantsused in traditional medicine in Eastern Tanzania. V. Angiosperms (Passifloraceae to Sapindaceae). J Ethnopharmacol 33:143–157

Chiang HC, Tseng TH, Wang CJ, Chen CF, Kan WS (1991) Experimental antitumor agents from *Solanum indicum* L. Anticancer Res 11(5):1911–1917

Chiang HC, Juilo Y, Lu FJ (1994) Xanthine oxidase inhibitors from the leaves of *Alsophila spinulosa* (Hook)Tryon. J Enzym Inhib 8:61–71

Chiang LH, Chen SH, Lo YC, Yeh AI (2018) Estrogenic activity of yam via a yeast model and its effects on two cancer cell lines. J Funct Foods 40:86–92

Chinkwo KA (2005) Sutherlandia frutescens extracts can induce apoptosis in cultured carcinoma cells. J Ethnopharmacol 98:163–170

Cho SI, Cho SJ, Kim BY, Dhayal M (2006) Anticancer activities of *Sparganium stoloniferum* on the proliferation of MCF-7 Cells. J Biomed Nanotechnol 2(2):125–128

Choi HJ, Seon MR, Lim SS, Kim S, Chun HS, Park JHY (2008) Hexane/ethanol extract of *Glycyrrhiza uralensis* licorice suppresses doxorubicin-induced apoptosis in H9C2 rat cardiac myoblasts. Exp Biol Med 233:1554–1560

Choi H, Kim M, Choi Y, Shin Y, Cho S, Ko S (2016) *Rhus verniciflua* Stokes (RVS) and butein induce apoptosis of paclitaxel-resistant SKOV-3/PAX ovarian cancer cells through inhibition of AKT phosphorylation. BMC Complement Altern Med 16:122

Chong SC, Dollah MA, Chong PP, Maha A (2011) *Phaleria macrocarpa* (Scheff.) Boerl fruit aqueous extract enhances LDL receptor and PCSK9 expression in vivo and in vitro. J Ethnopharmacol 137(1):817–827

Chopra RN, Nayar SL, Chopra IC (1956) *Barleria prionitis* Linn. In: Glossary of Indian medicinal plants. Council of Scientific and Industrial Research Publication, New Delhi, pp 33–34

Chopra CL, Bhatia MC, Chopra IC (1960) In vitro antibacterial activity of oils from Indian medicinal plants. J Am Pharm Assoc 49:780. Scientific Edition

Chou S, Hsu C, Wang K, Wang M, Wang C (2007) Antitumor effects of Osthol from Cnidium monnieri: an in vitro and in vivo study. Phytother Res 21(3):226–230

Chou ST, Chan HH, Peng HY, Liou MJ, Wu TS (2011) Isolation of substances with antiproliferative and apoptosis-inducing activities against leukemia cells from the leaves of *Zanthoxylum ailanthoides* Sieb. and Zucc. Phytomedicine 18(5):344–348

Choudhary M, Kumar V, Singh S (2014) Phytochemical and Pharmacological activity of Genus Plumeria: An updated review. Int J Biomed Adv Res 5(6):266–271

Chowdhury R, Hasan CM, Rashid MA (2003) Guaiane sesquiterpenes from *Amoora rohituka*. J Phytochem 62(8):1213–1216

Chu S, Hsieh Y, Hsu L, Chen K, Chiang C, Chen P (2013) *Rubus idaeus* L inhibits invasion potential of human a549 lung cancer cells by suppression epithelial-to-mesenchymal transition and Akt pathway in vitro and reduces tumor growth in vivo. Integr Cancer Ther 13(3):259–273

Chung T, Lee J, Choi H, Park M, Kim E, Han J, Jang S, Lee S, Lee S, Hang J, Yi L, Ha K (2017) *Anemone rivularis* inhibits pyruvate dehydrogenase kinase activity and tumor growth. J Ethnopharmacol 203:47–54

Cirses D, Ollxegvx A, Kixgsto EDI (2011) Plant anticancer agents. X. lignans from *Juniperus phoenicea*. J Nat Prod 43(4):495–497

Cock I, Mohanty S (2010) Bioactivity of *Syzygium jambosmethanolic* extracts: antibacterial activity and toxicity. Pharmacognosy Res 2(1):4–9

Column M, Aoyoung K, Nam-hee L, Chung Ju L, Kim T, Young-pil J (2013) Pharmaceutical composition and functional health food for preventing or treating cancer containing natural extract as active ingredient. WO2014168447A1, 11 Apr 2013

Cong Y, Sun K, He X, Li J, Dong Y, Zheng B, Tan X, Song X (2015) A traditional chinese medicine xiao-ai-tong suppresses pain through modulation of cytokines and prevents adverse reactions of morphine treatment in bone cancer pain patients. Mediat Inflamm 2015:1–8

Cornelius M (2015) Climate change: what would nature do. Biomimicry Institute

Correa GM, Alcantara AF (2011) Chemical constituents and biological activities of species of Justicia – a review. Braz J Pharm 11:1–19

Corrêa GM, Alcântara AF (2012) Chemical constituents and biological activities of species of Justicia: a review. Rev Bras Farmacogn 22(1):220–238

Cragg GM, Newman DJ (2005) Plants as a source of anti-cancer agents. J Ethnopharmacol 100(1–2):72–79

Croaker A, King G, Pyne J, Anoopkumar-Dukie S, Liu L (2016) *Sanguinaria canadensis*: traditional medicine, phytochemical composition, biological activities and current uses. Int J Mol Sci 17(4):1414

Csupor-Loffler B, Hajdu Z, Zupko I, Rethy B, Falkay G, Forgo P, Hohmann J (2009) Antiproliferative effect of flavonoids and sesquiterpenoids from *Achillea millefolium* on cultured human tumour cell lines. Physiol Res 23(5):672–676

Cui CB, Tezuka Y, Kikuchi T, Nakano H, Tamaoki T, Park JH (1990) Constituents of Fern, *Davallia mariesii* Moore.I. Isolation and structures of Davallialactone and a new flavanone glucuronide. Chem Pharm Bull 38:3218–3225

Cui YY, Xie H, Qi KB, He YM, Wang JF (2005) Effects of *Pinus massoniana* bark extract on cell proliferation and apoptosis of human hepatoma BEL-7402 cells. World J Gastroenterol 11(4):5277–5282

Culpeper N (1950) Culpeper's complete herbal. W. Foulshamand Co. Ltd., London, p 430

da Costa PM, Ferreira PM, da Silva Bolzani V, Furlan M, Macedo VA, Corsino J, de Moraes MO, Costa-Lotufo LV, Montenegro RC, Pessoa C (2008) Antiproliferative activity of pristimerin isolated from *Maytenus ilicifolia* (Celastraceae) in human HL-60 cells. Toxicol in Vitro 22(4):854–863

da Silva R, de Lemos T, Ferreira D, Monte F (2016) *Ximenia americana*: chemical and spectral studies of extracts of seeds. Analysis of trimethylsilyl derivatives by gas chromatography and mass spectrometry. Am J Anal Chem 7(2):192–202

Dagang W, Sorg B, Hecker E (1994) Oligo- and macrocyclicditerpenes in Thymeleaceae and Euphorbiaceae occurring and utilized in Yunnan (Southwest China). 6. Tigliane type diterpene esters from latex of *Euphorbia prolifera*. Phytother Res 8:95–99

Damme EJ, Barre A, Rouge P, Leuven FV, Balzarini J, Peumans WJ (1996) Molecular cloning of the lectin and a lectin-related protein from common Solomon's seal (*Polygonatum multiflorum*). Plant Mol Biol 31(3):657–672

Daniel JFDS, Alves CCF, Grivicich I, Rocha ABD, Carvalho MGD (2007) Antitumor activity of biflavonoids from *Ouratea* and *Luxemburgia* on human cancer cell lines. J Indian Pharm 39(4):184–186

Darias V, Bravo L, Rabanal R, Sanchez Mateo C, GonzalezLuis RM, Hernandez Perez AM (1989) New contribution to the ethnopharmacological study of the Canary Islands. J Ethnopharmacol 25:77–92

Das S, Das MK, Mazumder PM, Das S, Basu SP (2009) Cytotoxic activity of methanolic extract of *Berberis aristata* DC on colon cancer. Int J Green Pharm 3(3):137–140

Dashtdar M, Dashtdar M, Dashtdar B, Kardi K, Shirazi M (2016) The concept of wind in traditional chinese medicine. J Pharmacopuncture 19:293–302

Davey MS, Atlee CW (2011) Inotropic and cardioprotective effect of *Terminalia paniculata* bark extract in doxorubicin induced cardiotoxicity in rats. Int J Res Ayur Pharm 2(3):869–875

David S, Idris A, Zulkipli I, Zulhilmi N, Lee H, Rajabalaya R, Chee L, Majid M (2017) Melastoma malabathricum ethyl acetate fraction induces secondary necrosis in human breast and lung cancer cell lines. Pharmacogn Mag 13(51):688

Davydov M, Krikorian A (2000) *Eleutherococcus senticosus* (Rupr. & Maxim.) Maxim. (Araliaceae) as an adaptogen: a closer look. J Ethnopharmacol 72:345–393

De Feo V, Aquino R, Menghini A, Ramundo E, Senatoare F (1992) Traditional phytotherapy in the peninsula Sorrentina, Campania, Southern Italy. J Ethnopharmacol 36:113–125

de Melo J, Santos A, de Amorim E, Nascimento S, de Albuquerque U (2011) Medicinal plants used as antitumor agents in brazil: an ethnobotanical approach. Evid Based Complement Alternat Med 2011:1–14 de Souza PO, Bianchi SE, Figueiró F, Heimfarth L, Moresco KS, Gonçalves RM, Hoppe JB, Klein CP, Salbego CG, Gelain DP, Bassani VL (2018) Anticancer activity of flavonoids isolated from *Achyrocline satureioides* in gliomas cell lines. Toxicol in Vitro 51:23–33

De L, Duarte Weinberg M, Gottlieb OR, De Oliveira GG (1976) Chemistry of Brazilian Bignoniaceae. Part 2. Naphthoquinones from *Zeyhera tuberculosa*. Phytochemistry 15:570

Debayan B, Arup D, Sreemoy Kanti D, Guno Sindhu C (2011) *Clerodendrum Infortunatum* Linn.: a review. JAPHR 1(3):82–85

Decosterd L, Gustafson K, Cardellina J, Cragg G, Boyd M (1994) The differential cytotoxicity of cardenolides from Thevetia ahouia. Phytother Res 8(2):74–77

Deiab S, Mazzio E, Eyunni S, McTier O, Mateeva N, Elshami F, Soliman K (2015) 1,2,3,4,6-penta-o-galloylglucose within galla chinensis inhibits human LDH-a and attenuates cell proliferation in MDA-MB-231 breast cancer cells. Evid Based Complement Alternat Med 2015:1–11

De-Leon L, Lopez MR, Moujir L (2009) Antibacterial properties of zeylasterone, a triterpenoid isolated from *Maytenus blepharodes*, against Staphylococcus aureus. Microbiol Res 165(8):617–626

Deng R, Tang J, Xia LP, Li DD, Zhou WJ, Wang LL, Feng GK, Zeng YX, Gao YH, Zhu XF, Excisanin A (2009a) A diterpenoid compound purified from *Isodon macrocalyxin* D, induces tumor cells apoptosis and suppresses tumor growth through inhibition of PKB/AKT kinase activity and blockade of its signal pathway. J Mol Cancer Ther 8:873–882

Deng G, Lin H, Seidman A, Fornier M, D'Andrea G, Wesa K, Yeung S, Cunningham-Rundles S, Vickers A, Cassileth B (2009b) A phase I/II trial of a polysaccharide extract from Grifola frondosa (Maitake mushroom) in breast cancer patients: immunological effects. J Cancer Res Clin Oncol 135(9):1215–1221

Desai A, Qazi G, Ganju R, El-Tamer M, Singh J, Saxena A, Bedi Y, Taneja S, Bhat H (2008) Medicinal plants and cancer chemoprevention. Curr Drug Metab 9(7):581–591

Devappa RK, Makkar HPS, Becker K (2011) Jatropha Diterpenes: a review. J Am Oil Chem Soc 88:301–322

Dey Y, Srikanth N, Wanjari M, Ota S, Jamal M (2012) A phytopharmacological review on an important medicinal plant – *Amorphophallus paeoniifolius*. AYU 33(1):27–32

Dey SA, Roy SU, Deb NI, Sen KK, Besra SE (2013) Anti-carcinogenic activity of *Ruellia tuberosa* L. (Acanthaceae) leaf extract on hepatoma cell line & increased superoxide dismutase activity on macrophage cell lysate. Int J Pharm Pharm Sci 5:854–861

Dhalwal K, Deshpande YS, Purohit AP, Kadam SS (2005) Evaluation of the antioxidant activity of *Sida cordifolia*. Pharm Biol 43(9):754–761

Dhamija I, Kumar N, Manjula S, Parihar V, Setty M, Pai K (2013) Preliminary evaluation of in vitro cytotoxicity and in vivo antitumor activity of Premna herbacea Roxb. in Ehrlich ascites carcinoma model and Dalton's lymphoma ascites model. Exp Toxicol Pathol 65(3):235–242

Dhuna V, Bains JS, Kamboj SJ, Singh J, Kamboj S, Saxena AK (2005) Purification and characterization of a lectin from *Arisaema tortuosum* schott. having in-vitro anticancer against human cancer cell lines. J Biochem Microbiol 38(5):526–532

Díaz C, Quesada S, Brenes O, Aguilar G, Cicció J (2008) Chemical composition of *Schinus mole* essential oil and its cytotoxic activity on tumour cell lines. Nat Prod Res 22:1521–1534

Dimayuga RE, Murillo RF, Pantoja ML (1987) Traditionalmedicine of *Baja California* Sur (Mexico) II. J Ethnopharmacol 20:209–222

Dinda A, Biswal I, Chowdhury P, Mohapatra R (2013) Formulation development and evaluation of paclitaxel loaded solid lipid nanoparticles using glyceryl monostearate. J Appl Pharm Sci 3(08):133–138

Ding YL, Jia ZJ (1992) Two phenolic derivatives from *Euphorbia kansui*. Phytochemistry 31:1435–1436

Djakpo O, Yao W (2010) *Rhus chinensis* and *Galla chinensis*-Folklore to modern evidence. Phytother Res 24:1739–1747

Dongre SH, Badami S, Natesan SHRC (2007) Antitumor activity of the methanol extract of *Hypericum hookerianum* stem against Ehrlich ascites carcinoma in Swiss albino mice. J Pharm Sci 103(4):354–359

Dos Santos AF, Sant'Ana AEG (1999) Molluscicidal activity of the diterpenoids jatrophone and jatropholones a and B isolated from *Jatropha elliptica* (Pohl) Muell. Arg Phytother Res 13:660–664

Drewes S, Elliot E, Khan F, Dhlamini J, Gcumisa M (2008) *Hypoxis hemerocallidea*—not merely a cure for benign prostate hyperplasia. J Ethnopharmacol 119(3):593–598

Driggins SN, Whalen M, Myles EL (2017) The inhibitory effect of *Echinacea purpurea* and *Echinacea pallida* on BT-549 and natural killer cells. MOJ Cell Sci Rep 4(3):00091

Du W, He T, Yuan C, Mehendale S, Aung H, Fishbein A, Ni M, Zhang B, Xie J, Wang C (2007) Chemopreventive effects of *Panax notoginseng* and its major constituents on SW480 human colorectal cancer cells. Int J Oncol 31:1149–1156

Duan S, Jin C, Hao J, Hou L, Li A, Zhu Z, Gao J, Pan Y (2013) A study on the inhibitory effect of radix semiaquilegiae extract on human hepatoma HEPG-2 AND SMMC-7721 cells. Afr J Tradit Complement Altern Med 10(5):336–340

Ducki S, Hadfield JA, Lawrence NJ, Zhang XG, McGowen AT (1995) Isolation of paeonol from *Arisaemaeubescens*. Planta Med 61:586–587

Ducki S, Hadfield JA, Lawrence NJ, Liu CY, McGown AT, Zhang X (1996) Isolation of E-1-(4′-Hydroxyphenyl)-but-1-en-2-one from *Scutellaria barbata*. Planta Med 62(2):185–186

Dudai N, Weinstein Y, Krup M, Rabinski T, Ofir R (2005) Citral is a new inducer of caspase-3 in tumor cell lines. Planta Med 71(5):484–488

Dufoura D, Pichettea A, Mshvildadzea V, Bradette-Heberta M, Lavoiea S, Longtina A, Lapriseb C, Legault J (2007) Antioxidant, anti-inflammatory and anticancer activities of methanolic extracts from *Ledum groenlandicum* Retzius. J Ethnopharmacol 111(1):22–28

Duke JA, Ayensu ES (1985a) Medicinal plants of China, vol 1. Reference Publications, Inc, Algonac, 52, p 361

Duke JA, Ayensu ES (1985b) Medicinal plants of China, vol 2. Reference Publications, Inc, Algonac, pp 363–705

Duraipandiyan V, Al-Dhabi N, Balachandran C, Ignacimuthu S, Sankar C, Balakrishna K (2015) Antimicrobial, antioxidant, and cytotoxic properties of vasicine acetate synthesized from vasicine isolated from *Adhatoda vasica* L. Biomed Res Int 2015:1–7

Durodola JI (1975a) Viability and transplantability of developedtumour cells treated in vitro with antitumour agent C:M2 isolated from a herbal cancer remedy, of *Annona senegalensis*. Planta Med 28:359

Durodola JI (1975b) Antitumour effects against sarcoma 180 ascites of fractions from *Annona senegalensis*. Planta Med 28:32

Dykes L, Rooney LW (2007) Phenolic compounds in cereal grains and their health benefits. Cereal Foods World 52(3):105–111

Eid S, El-Readi M, Ashour M, Wink M (2015) *Fallopia japonica*, a natural modulator, can overcome multidrug resistance in cancer cells. Evid Based Complement Alternat Med 2015(1):1–8

Ejaz S, Woong LC, Ejaz A (2003) Extract of Garlic (*Allium sativum*) in cancer chemoprevention. Exp Oncol 25:93–97

Ekunwe SIN, Thomas MS, Luo MS, Wang H, Chen Y, Zhang X, Begonia GB (2010) Potential cancer-fighting *Ocimum gratissimum* leaf extracts: increased antiproliferation activity of partially purified fractions and their spectral fingerprints. Ethn Dis 20:1–5

El Kheir YM, Salih MH (1979) Investigation of the alkaloidalcontent of *Solanum dubium* L. growing in Sudan. Fitoterapia 50:255–258

El-Desouky SK, Gamal-Eldeen AM (2009) Cytotoxic and anti-inflammatory activities of some constituents from the floral buds of *Syringa patula*. J Pharm Biol 47(9):872–877

Elingold I, Isollabella MP, Casanova MB, Celentano AM, Perez C, Cabrera JL, Diez RA, Dubin M (2008) Mitochondrial toxicity and antioxidant activity of a prenylated flavonoid isolated from *Dalea elegans*. Chem-Bio Interact 171:294–305

Elkady A (2013) Crude alkaloid extract of *Rhazya stricta* inhibits cell growth and sensitizes human lung cancer cells to cisplatin through induction of apoptosis. Genet Mol Biol 36:12–21

El-Seedi HR, Hazell AC, Torssell KBG (1994) Triterpenes, Lichexanthone and an acetylenic acid from *Minquartiaguianensis*. Phytochemistry 35:1297–1299

Eowyndbh (2012) *Asclepias speciosa* wild edible and medicinal plants. https://keys2liberty.wordpress.com/2011/12/14/wild-edible-and-medicinal-plants-67-68-milkweedvalerian-root/

Esposito-Avella M, Brown P, Tejeira I, Buitrago R, Barrios L, Sanchez C, Gupta MP, Cedeno J (1985) Pharmacological screening of Panamanian medicinalplants. Part 1. Int J Crude Drug Res 23:17–25

Evans FJ, Kinghorn AD (1975) The succulent euphorbiasof Nigeria. Part I. Lloydia 38:363–365

Fadeyi S, Fadeyi O, Adejumo A, Okoro C, Myles E (2013) In vitro anticancer screening of 24 locally used Nigerian medicinal plants. BMC Complement Altern Med 13:79

Falodun A, Sheng-Xiang Q, Parkinson G, Gibbons S (2012) Isolation and characterization of a new anticancer diterpenoid from *Jatropha gossypifolia*. Pharm Chem J 45(10):636–639

Fan Y, Lu H, An L, Wang C, Zhou Z, Feng F, Ma H, Xu Y, Zhao Q (2016) Effect of active fraction of *Eriocaulon sieboldianum* on human leukemia K562 cells via proliferation inhibition, cell cycle arrest and apoptosis induction. Environ Toxicol Pharmacol 43:13–20

Fard M, Beydokhti H, Tahergorabi Z, Abedini M, Mitra M (2015) "*Ziziphus jujuba*": a red fruit with promising anticancer activities. Pharmacogn Rev 9(18):99–106

Farias F, Williamson J, Rodriguez S, Angeles G, Portugal V (2009) Bark anatomy in Croton draco var. draco (Euphorbiaceae). Am J Bot 96:2155–2167

Farzaei MH, Abbasabadi Z, Ardekani MR, Rahimi R, Farzaei F (2013) Parsley: a review of ethnopharmacology, phytochemistry and biological activities. J Tradit Chin Med 33(6):815–826

Fatope MO, Zeng L, Ohayaga JE, Shi G, McLaughlin JL (1996) Selectively cytotoxic diterpenes from *Euphorbia poisonii*. J Med Chem 39(4):1005–1008

Felipe KB, Kviecinski MR, da Silva FO, Buecker NF, Farias MS, Castro LS, Pedrosa RC (2014) Inhibition of tumor proliferation associated with cell cycle arrest caused by extract and fraction from *Casearia sylvestris* (Salicaceae). J Ethnopharmacol 155(3):1492–1499

Félix-Silva J, Giordani R, Silva-Jr A, Zucolotto S, Fernandes-Pedrosa M (2014) *Jatropha gossypiifolia* L. (Euphorbiaceae): a review of traditional uses, phytochemistry, pharmacology, and toxicology of this medicinal plant. Evid Based Complement Alternat Med 2014:1–32

Feng Y, Bounda G (2015) Review of clinical studies of Polygonum multiflorum Thunb. and its isolated bioactive compounds. Pharmacogn Res 7(3):225–236

Feng T, Wang YY, Su J, Li Y, Cai XH, Luo XD (2011) Amaryllidaceae alkaloids from *Lycoris radiata*. Helvetica Chim Acta 94(1):178–183

Ferraz A, Faria DH, Benneti MN, da Rocha AB, Schwartsmann G, Henriques A, Von Poser GL (2005) Screening for antiproliferative activity of six southern Brazilian species of *Hypericum*. Int J Phytother Phytopharm 12(1):112–115

Filly C, Chui CH, Albert SCC, Lau FY, Gregory YMC, Raymond SMW, Stanton HLK, Ivy TNT, Cheng CHA, Johnny COT (2005) Inhibition of proteasome activity in *Gleditsia sinensis* fruit extract-mediated apoptosis on human carcinoma cells. Int J Mol Med 16(5):925–929

Flavin DF (2007) A lipoxygenase inhibitor in breast cancer brain metastases. J Neuro-Oncol 82(1):91–93

Florence T, Hines M, Gan D (2011) Topical skin care formulations comprising plant extracts. US8454943B2, 6 April 2011

Flores JS, Ricalde RV (1996) The secretions and exudatesof plants used in mayan traditional medicine. J Herbs Spices Med Plants 4:53–59

Fongod AG, Modjenpa NB, Veranso MC (2013) Ethnobotany of Acanthaceae in the Mount Cameroon region. J Med Plants Res 7(38):2859–2866

Forlines DR, Tavenner T, Malan JCS, Karchesy JJ (1992) Plants of the Olympic coastal forests, ancient knowledgeof materials and medicines and future heritage. Basic Life Sci 59:767–782

Francois G, Diakanamwa C, Timperman G, Bringmann G, Steenackers T, Atassi G, Van Looveren M, Holenz J, Tassin JP, Assi LA, Vanhaelen-Fastre R, Vanhaelen M (1998) Antimalarial and cytotoxic potential of four quassinoids from *Hannoa chlorantha* and *Hannoa klaineana*, and their structure-activity relationships. Int J Parasitol 28(4):635–640

Freitas A, Ximenes N, Aguiar J, Nascimento S, Lins T, Magalhães L, Coelho L, Carneiro-da-Cunha M, Gonçalves-Silva T, Correia M (2012) Biological activities of *Libidibia* (Caesalpinia) *ferrea* var. Parvifolia (Mart. ex Tul.) L. P. Queiroz Pod Preparations. Evid Based Complement Alternat Med 2012:1–7

Frenkel M, Mishra BM, Sen S, Yang P, Pawlus A, Vence L, Leblanc A, Cohen L, Banerji P, Banerji P (2010) Cytotoxic effects of ultra-diluted remedies on breast cancer cells. Int J Oncol 36(2):395–403

Fukamiya N, Okano M, Aratani T, Negoro K, McPhail AT, Ju-ichi M, Lee KH (1986) Antitumor agents, cytotoxic antileukemic alkaloids from *Brucea antidysenterica*. J Nat Prod 49(3):428–434

Furusawa E, Chou SC, Furusawa S, Hirazumi A, Dang Y (1992) Antitumour activity of *Ganoderma lucidum*, anedible mushroom, on intraperitoneally implanted lewislung carcinoma in synergenic mice. Phytother Res 6:300–304

Gabriel OA, Olubunmi A (2009) Comprehensive scientific demystification of *Kigelia africana*: a review. Afr J Pure Appl Chem 3(9):158–164

Galvez M, Martin-Cordero C, Lopez-Lazaro M, Cortes F, Ayuso MJ (2003) Cytotoxic effect of *Plantago*spp. on cancer cell lines. J Ethnopharmacol 88:125–130

Gamal-Eldeen AM, Djemgou PC, Tchuendem M, Ngadjui BT, Tane P, Toshifumi H (2007) Anticancer and immunostimulatory activity of chromones and other, constituents from *Cassia petersiana*. Z. Z Naturforsch C 62:331–338

Gangadharan A, Nishath TK, Benny PJ (2013) Evaluation of phytochemical analysis, anti oxidant and anti-elastase activity of *Hemigraphis colorata*. Int J Pharmacogn Phytochem Res 5(4):292–298

Gangwar M, Goel R, Nath G (2014) *Mallotus philippinensis* Muell. Arg (Euphorbiaceae): ethnopharmacology and phytochemistry review. Biomed Res Int 2014:1–13

Garbi MI, Osman FE, Dahab MM, Koko WS, Kabbashi AS, Elegami AA, Hamed SY (2015) Antigiardial, antiamoebic and cytotoxic activity of the leaves extracts of *Vitex trifolia*. Advancement Med Plant Res 3(1):1–7

Garcez FR, Francisca da Silva AG, Garcez WS, Linck G, de Fatima Matos MC, Santos EC, Queiroz LM (2011) Cytotoxic aporphine alkaloids from *Ocotea acutifolia*. Planta Med 77(4):383–387

Garcia-Barriga H (1974) Flora Medicinal De Colombia 1. Universidad Nacional, Bogota

Garcia-Barriga H (1975) Flora Medicinal De Colombia 2:3. Universidad Nacional, Bogota

Gazola R, Machado D, Ruggiero C, Singi G, Alexandre MM (2004) *Lippia alba Melissa officinalis* and *Cymbopogon citratus*: effects of the aqueous extracts on the isolated hearts of rats. Pharm Res 50(5):477–480

Ge YW, Zhu S, Shang MY, Zang XY, Wang X, Bai YJ, Li L, Komatsu K, Cai SQ (2013) Aristololactams and aporphines from the stems of *Fissistigma oldhamii* (Annonaceae). Phytochemistry 86:201–208

Gentry AH (1980) Bignoniaceae-Part I (Crescentieae and Tourrettieae). Flora Neotropica Monogr 25:1–130

George S, Bhalerao SV, Lidstone EA, Ahmad IA, Abbasi A, Cunningham BT, Watkin KL (2010) Cytotoxicity screening of Bangladeshi medicinal plant extracts on pancreatic cancer cells. BMC Complement Altern Med 10:1–11

Gerlach SL, Rathinakumar R, Chakravarty G, Goransson U, Wimley WC, Darwin SP, Mondal D (2010) Anticancer and chemosensitizing abilities of cycloviolacin oxygen from Viola odorata and psyle cyclotides from *Psychotria leptothyrsa*. J Pept Sci 94:617–625

Ghazanfar SA, Al-Sabahi MA (1993) Medicinal plants of Northern and Central Oman (Arabia). Econ Bot 47:89–98

Ghosh S, Rangan L (2012) Alpinia: the gold mine of future therapeutics. 3 Biotech 3(3):173–185

Ghosh S, Sarma MD, Patra A, Hazra B (2010) Anti-inflammatory and anticancer compounds isolated from *Ventilago madraspatana* Gaertn, *Rubia cordifolia* L and *Lantana camara* L. J Pharm Pharmacol 62(9):1158–1166

Ghosh T, Maity TK, Singh J (2011) Evaluation of antitumor activity of stigmasterol, a constituent isolated from *Bacopa monnieri* Linn aerial parts against Ehrlich Ascites Carcinoma in mice. Orient Pharm Exp Med 11(1):41–49

Giang PM, Jin HZ, Son PT, Lee JH, Hong YS, Lee JJ (2003) Ent-kaurane diterpenoids from *Croton tonkinensis* inhibit lps-induced Nf-Kb activation and no production. J Nat Prod 66:1217–1220

Gilca M, Gaman L, Panait E, Stoian I, Atanasiu V (2010) *Chelidonium majus* – an integrative review: traditional knowledge versus modern findings. Res Comp Med 17(5):241–248

Giri A, Narasu ML (2000) Production of podophyllotoxin from *Podophyllum hexandrum*: a potential natural product for clinically useful anticancer drugs. J Cytotechnol 34(2):17–26

Glamočlija J, Soković M, Teević V, Linde G, Colauto N (2011) Chemical characterization of essential oil: an alternative to control green molds. Braz J Microbiol 42(4):1537–1546

Gnabre J, Bates R, Huang R (2015) Creosote bush lignans for human disease treatment and prevention: perspectives on combination therapy. J Tradit Complement Med 5(3):119–126

Goey AKL, Meijerman I, Rosing H, Burgers JA, Mergui-Roelvink M, Keessen M, Marchetti S, Beijnen JH, Schellens JH (2013) The effect of *Echinacea purpurea* on the pharmacokinetics of docetaxe. Br J Clin Pharmacol 76(3):467–474

Gonzalez AG, Hernandez JC, Leon F, Padron JI, Estevez F, Quintana J, Bermejo J (2003) Steroidal saponins from the bark of *Dracaena draco*and their cytotoxic activities. J Nat Prod 66:793–798

Gonzalez-Tejero MR, Molero-Mesa J, Casares-Porcel M, Lirola MJM (1995) New contributions to the ethnopharmacology of Spain. J Ethnopharmacol 45:157–163

Gouthamchandra K, Mahmood R, Manjunatha H (2010) Free radical scavenging, antioxidant enzymes and wound healing activities of leaves extracts from *Clerodendrum infortunatum* L. Environ Toxicol Pharmacol 30:11–18

Govindan G, Sambandan TG, Govindan M, Sinskey A, Vanessendelft J, Adenan I, Rha CK (2007) A bioactive polyacetylene compound isolated from *Centella asiatica*. Planta Med 73(6):597–599

Graham J, Quinn M, Fabricant D, Farnsworth N (2000) Plants used against cancer – an extension of the work of Jonathan Hartwell. J Ethnopharmacol 73:347–377

Graziano MN, Ferrari F, Delle Monache F, Coussio JD (1970) Polyphenols of *Schinus lentiscifolius* March. Annali dell Instituto Superiore de Sanita 6:310–311

Greenwell M, Rahman PK (2015) Medicinal plants: their use in anticancer treatment. Int J Pharm Sci Res 6(10):4103–4112

Grossarth M, Baumgartner S, Kiene H, Ziegler R (2001) Use of Iscador, an extract of European mistletoe (*Viscum album*), in cancer treatment: prospective nonrandomized and randomized matched-pair studies nested within a cohort study. Altern Ther Health Med 7(3):57–66

Grynberg NF, Carvalho MG, Velandia JR, Oliveira MC, Moreira IC, Braz-Filho R, Echevarria A (2002) DNA topoisomerase inhibitors: biflavonoids from *Ouratea* species. Braz J Med Biol Res 35:819–822

Gupta SK (1979) Apocynaceous plants of Varanasi with notes on their medicinal importance. J Res Indian Med Yoga Homeopath 14:140–142

Gupta MP (1995) Panamanian flora: Source of bioactive molecules. In: Hostettmann K, Marston A, Maillard M, Hamburger M (eds) Phytochemistry of plants used in traditional medicine. Oxford University Press, New York, pp 359–398

Gupta MP, Arias TD, Correa M, Lamba SS (1979) Ethnopharmacognositc observations on Panamanianmedicinal plants. Part I. Q J Crude Drug Res 17:115–130

Gupta MP, Monge A, Karikas GA, Lopez De Cerain A, Solis PN, De Leon E, Trujillo M, Suarez O, Wilson F (1996) Screening of Panamanian medicinal plants for brine shrimp toxicity, crown gall tumor inhibition, cytotoxicity and DNA intercalation. Int J Pharm 34:19–27

Gupta M, Mazumder UKA, Haldar PKA, Kandar CC, Manikandan L, Senthil GP (2007) Anticancer activity of *Indigofera aspalathoides* and *Wedeliacalendulaceae* in Swiss albino mice. Iran J Pharma Res 6(2):141–145

Gupta D, Bleakley B, Gupta R (2008) Dragon's blood: botany, chemistry and therapeutic uses. J Ethnopharmacol 115(3):361–380

Gutierrez-Uribe JA, Romo-Lopez I, Serna-Saldıvar SO (2011) Phenolic composition and mammary cancer cell inhibition of extracts of whole cowpeas (*Vigna unguiculata*) and its anatomical parts. J Funct Foods 3:290–297

Guza RC (2007) Isolation of natural products from *Casearia nigrescens*. Master's thesis, Virginia Polytechnic Institute and State University, pp 12–44

Guzman M (2003) Cannabinoids: potential anticancer agents. Nat Rev Cancer 3:744–755

Habtemariam S (1998) Extract of corn silk (Stigma of *Zea mays*) inhibits tumour necrosis Factor-α- and bacterial lipopolysaccharide-induced cell adhesion and ICAM-1 expression. Planta Med 64(4):314–318

Hadfield JA, McGown AT, Butler J (2000) A high-yielding synthesis of the naturally occurring antitumour agent irisquinone. Molecules 5:82–88

Hajjar D, Kremb S, Sioud S, Emwas A, Voolstra C, Ravasi T (2017) Anti-cancer agents in Saudi Arabian herbals revealed by automated high-content imaging. PLoS One 12(6):017731

Hamburger M (2002) *Isatis tinctoria* – From the rediscovery of an ancient medicinal plant towards a novel anti-inflammatory phytopharmaceutica. Bio Life Sci 1(3):333–344

Hamburger M, Cordell G (1987) Traditional medicinal plants of Thailand, viii. Isoflavonoids of *Dalbergia candenatensis*. J Nat Prod 50:696–699

Hammer MLA, Johns EA (1993) Tapping an Amazonianplethora, four medicinal plants of Marajo Island, Para (Brazil). J Ethnopharmacol 40:53–75

Han DS, Lee SJ, Lee HK (1984) Ethnobotanical surveyin Korea. In: Han BH, Han DS, Han YN, Woo WS (eds) Proceedings of the fifth Asian symposium on medicinal plants and spices, Seoul, Korea, 20–24 August, pp 125–144

Han S, Zhao W, Sun H, Zhou N, Zhou F, An G, Li P (2015) *Marsdenia tenacissima* extract enhances gefitinib efficacy in non-small cell lung cancer xenografts. Phytomedicine 22:560–567

Hansakul P, Ngamkitidechakul C, Ingkaninan K, Sireeratawong S, Panunto W (2009) Apoptotic induction activity of *Dactyloctenium aegyptium* (L.) P.B. and *Eleusine indica* (L.) Gaerth. Extracts on human lung and cervical cancer cell lines. Songklanakarin J Sci Technol 31:273–279

Hansakul P, Wongnoppavich A, Ingkaninan K, Sireeratawong S, Panunto W (2017) Apoptotic induction activity of *Dactyloctenium aegyptium* (L.) P.B. and *Eleusine indica* (L.) Gaerth. extracts on human lung and cervical cancer cell lines. Songklanakarin J Sci Technol 31(3):273–279

Hao DC, Ge GB, Xiao PG (2018) Anticancer drug targets of *Salvia* phytometabolites: chemistry, Biology and omics. Curr Drug Targets 19(1):1–20

Hapsari A, Asti D, Hidayati R, Kumalla N, Suhendi A (2016) The Potency of Kitolod (*Isotoma longiflora* (L) Presl) Herb Extract as a Cure for Cervical Cancer an in Vitro Study of Hela Cells

Harahap U, Hasibuan P, Sitorus P, Arfian N, Satria S (2018) Antimigration activity of an ethylacetate fraction of Zanthoxylum acanthopodium DC. Fruits in 4T1 breast cancer cells. Asian Pac J Cancer Prev 19(2):565–569

Hargittai M (2015) Women scientists: reflections, challenges, and breaking boundaries. Oxford University Press, Oxford, p 254

Haridas V, Higuchi M, Jayatilake GS, Bailey D, Mujoo K, Blake ME, Arntzen CJ, Gutterman JU (2001) Anticancer and antimutagenic properties of *Acacia nilotica* L. on 7,12-Dimethylbenz(a) anthracene-induhuced skin papillomagenesis in Swiss albino mice. Asian Pac J Cancer 7:627–632

Haryanti S, Junedi S, Meiyanto E (2009) Ethanolic extract of *Hedyotis corymbosa* L. increases cytotoxic activity of doxorubicin on MCF-7 breast cancer cell. Indonesian J Biotech 14(1):1146–1154

Harzallah HJ, Neffati A, Skandrani I, Maaloul E, Chekir-Ghedira L, Mahjoub T (2010) Antioxidant and antigenotoxic activities of *Globularia alypum* leaves extracts. J Med Plant Res 4(19):2048–2053

Hashemian F, Mansouri P, Mirafzal S, Najafizadeh P, Safaei-Naraghi Z, Salehi-Surmaghi M (2017) The impact of topical Saint John's Wort (*Hypericum perforatum*) treatment on tissue tumor necrosis factor-alpha levels in plaque-type psoriasis: a pilot study. J Postgrad Med 63(4):215–220

Hastings RB (1990) Medicinal legumes of Mexico, Fabaceae, Papilionoidea, Part One. Econ Bot 44:336–348

He K, Valcic S, Timmerman BN, Montenegro G (1997) Indole alkaloids from *Aristotelia chilensis* (Mol.) Stuntz. Int J Pharmacol 35:215–217

He M, Hou J, Wang L, Zheng M, Fang T, Wang X, Xia J (2017) *Actinidia chinensis* Planch root extract inhibits cholesterol metabolism in hepatocellular carcinoma through upregulation of PCSK9. Oncotarget 8(26):42136

Hegazy MF, Mohamed AH, Aoki N, Ikeuchi T, Ohta E, Ohta S (2007) Bioactive jatrophane diterpenes from *Euphorbia guyoniana*. Z Naturforsch C 63(5):403–408

Henriques M, Penido C (2014) The therapeutic properties of *Carapa guianensis*. Curr Pharm Des 20(6):850–856

Hernandez MM, Heraso C, Villarreal ML, Vargas-Arispuro I, Aranda E (1999) Biological activities of crude plant extracts from *Vitex trifolia* L. (Verbenaceae). J Ethnopharmacol 67:37–44

Herrmann A, Burman R, Mylne JS, Karlsson G, Gullbo J, Craik DJ, Clark RJ, Goransson U (2008) The Alpine violet, *Viola biflora*, is a rich source of cyclotides with potent cytotoxicity. Phytochemistry 69(4):939–952

Hirschhorn HH (1982) Botanical remedies of south, central America, and the Caribbean, an archival analysis. Part 2. Conclusion. J Ethnopharmacol 5:163–180

Ho LK, Ou JC, Sun ML, Sun CM (1995) Two rare alkaloids from *Pratia nummularia*. Planta Med 61:567–568

Holdsworth DK (1991) Traditional medicinal plants of Rarotonga, Cook Islands. Part II. Int J Pharm 29:71–79

Holdsworth DK (1992) A preliminary study of medicinal plants of Easter Island, South Pacific. Int J Pharm 30:27–32

Holdsworth D, Sakulas H (1986) Medicinal plants of the Morobe province, Part II. The Aseki Valley. Int J Crude Drug Res 24:31–40

Holdsworth D, Wamoi B (1982) Medicinal plants of the Admiralty Islands, Papua New Guinea. Part I. Int J Crude Drug Res 20:169–181

Holdsworth D, Pilokos B, Lambes P (1983) Traditional medicinal plants of New Ireland, Papua New Guinea. Int J Crude Drug Res 21:161–168

Hong S (2011) Apoptotic and anti-angiogenic effects of *Pulsatillaï koreana* extract on hepatocellular carcinoma. Int J Oncol 40(2):452–460

Hossain MS, Urbi Z, Sule A, Hafizur Rahman KM (2014) *Andrographis paniculata* (Burm. f.) Wall. ex Nees: a review of ethnobotany, phytochemistry and pharmacology. Sci World J:1–28

Houghton PJ (1984) Ethnopharmacology of some *Buddleja* species. J Ethnopharmacol 11:293–308

Hsu YT (1967) Study on the Chinese drugs used as cancer remedy. J Southeast Asian Res 3:63

Hsu TY, Sheu SC, Liaw ET, Wang TC, Lin CC (2005) Anti-oxidant activity and effect of *Pinus morrisonicola* Hay. on the survival of leukemia cell line U937. Phytomedicine 12(9):663–669

Hsu SC, Lu JH, KUO CL, Yang JS, Lin MW, Chen GW, Chung JG (2008) Crude extracts of *Solanum lyratum* induced cytotoxicity and apoptosis in a human colon adenocarcinoma cell line (colo 205). Anticancer Res 28(2A):1045–1054

Hsu CL, Yu YS, Yen GC (2010) Anticancer effects of *Alpinia pricei* Hayata roots. J Agric Food Chem 58(4):2201–2208

Hu SY (1976) A contribution to our knowledge of Leonurus, I-Mu-Ts'ao, the Chinese Motherwort. Am J Chin Med 4:219

Hu K, Yao X (2003) The cytotoxicity of methyl protoneogracillin (NSC-698793) and gracillin (NSC-698787), two steroidal saponins from the rhizomes of *Dioscorea collettii* var. hypoglauca, against human cancer cells in vitro. Phytother Res 17(6):620–626

Hu CC, Lin JT, Liu SC, Yang DJ (2007) A spirostanol glycoside from wild yam (*Dioscorea villosa*) extract and its cytostatic activity on three cancer cells. J Food Drug Anal 15(3):310–315

Hu B, Du Q, Deng S, An H, Pan C, Shen K, Xu L, Wei M, Wang S (2014) *Ligustrum lucidum* Ait. fruit extract induces apoptosis and cell senescence in human hepatocellular carcinoma cells through upregulation of p21. Oncol Rep 32(3):1037–1042

Hu E, Wang D, Chen J, Tao X (2015) Novel cyclotides from Hedyotis diffusa induce apoptosis and inhibit proliferation and migration of prostate cancer cells. Int J Clin Exp Med 8(3):4059–4065

Hua YY (1992) Two new saponins from the roots of *Phytolacca esculenta*. Planta Med 58:99–101

Hua W, Yu L, Wang MH (2010) Antioxidant and antiproliferative properties of water extract from *Mahonia bealei* (Fort.) Carr leaves. Food Chem Toxicol 131(2):449–455

Huai H (2010) Ethnomedicinal analysis of TOXIC plants from five ethnic groups in China. Ethnobot Res Appl 8:169–179

Huang CC, Han CS, Yue XF, She CM, Wang SW, Wu FG, Xu B (1983) Cytotoxicity and sister chromatid exchanges induced in vitro by six anticancer drugs developed in the People's Republic of China. J Natl Cancer Inst 71:841–847

Huang SL, Hada S, Kakiuchi N, Hattori M, Kikuchi T, Namba T (1991) Effects of *Ganoderma lucidum* extractsand their constituents on the spontaneous beating of myocardial cell sheet in culture. Shoyakugaku Zasshi 45:132–136

Huang J, Tan C, Jiang S, Zhu D (2003) Terrestrinins A and B, two new steroid saponins from Tribulus terrestris. J Asian Nat Prod Res 5(4):285–290

Huang GC, Chien TY, Chen LG, Wang CC (2005) Antitumor effects of zerumbone from *Zingiber zerumbet* in P-388D1 cells in vitro and in vivo. Planta Med 71:219–224

Huang XY, Wang Q, Liu HL, Zhang Y, Xinb GR, Shen X, Dong L (2009a) Diastereoisomeric macrocyclic polydisulfides from the mangrove *Bruguiera gymnorrhiza*. Phytochemistry 70:2096–2100

Huang WW, Meng SS, Pan Q, Wu JF, Hu MZ, Fan J (2009b) Mechanisms of Shikonin-induced apoptosis in JEG-3 cells [J]. Carcinogenesis Teratogenesis Mutagen 6

Hullatti KK, Murthy UD (2001) Activity guided isolation of cytotoxic compounds from Indian medicinal plants using bsl bioassay. J Curr Pharma Res 1:16–18

Hung TM, Luan TC, Vinh BT, Cuong TD, Min BS (2011) Labdane-type diterpenoids from *Leonurus heterophyllus* and their cholinesterase inhibitory activity. Phytother Res 25(4):611–614

Hussain SJ, Alvi AB, Jahan M (1993) A study on unanimedicinal plants, *Asteratiqus*. J Res Edu Indian Med 2:35–39

Hussain A, Zia M, Mirza B (2007) Cytotoxic and antitumor potential of *Fagonia cretica* L. Turk J Biol 31:19–24

Hussey JS (1974) Some useful plants of early New England. Econ Bot 28:311

Hwang BY, Chai HB, Kardono LB, Riswan S, Farnsworth NR, Cordell GA, Pezzuto JM, Kinghorn AD (2003) Cytotoxic triterpenes from the twigs of *Celtis philippinensis*. Phytochemistry 62(2):197–201

Hyun JW, Chung HS (2004) Cyanidin and malvidin from *Oryza sativa* cv. Heugjinjubyeo mediate cytotoxicity against human monocytic leukemia cells by arrest of G2/M phase and induction of apoptosis. J Agric Food Chem 52:2213–2217

Ibrahim LA, Kawashty SA, El-Hagrassy AM, Nassar MA, Mabry TJ (2008) a new Kaempferol triglycoside from *Fagonia taeckholmiana*: cytotoxic activity of its extracts. Carbohydr Res 343(1):155–158

Ikeda T, Tsumagari H, Honbu T, Nohara T (2003) Cytotoxic activity of steroidal glycosides from *Solanum* plants. Bio Pharma Bull 26(8):1198–1201

Ikewuchi J, Onyeike E, Uwakwe A, Ikewuchi C (2011) Effect of aqueous extract of the leaves of *Acalypha wilkesiana* 'Godseffiana' Muell Arg (Euphorbiaceae) on the hematology, plasma biochemistry and ocular indices of oxidative stress in alloxan induced diabetic rats. J Ethnopharmacol 137(3):1415–1424

Iman V, Mohan S, Abdelwahab S, Karimian H, Karimian H, Fadaeinasab M, Noordin M, Noor S (2017) Anticancer and anti-inflammatory activities of girinimbine isolated from Murraya koenigii. Drug Des Devel Ther 11:103–121

Iqbal J, Abbasi B, Mahmood T, Kanwal S, Ali B, Shah S, Khalil A (2017) Plant-derived anticancer agents: a green anticancer approach. Asian Pac J Trop Biomed 7(12):1129–1150

Ishiguro K, Yamaki M, Kashihara M, Takagi S (1987) Saroaspidin A, B and C, additional antibiotic compoundsfrom *Hypericum japonicum*. Planta Med 53:415–417

Ishiguro K, Nagata S, Fukumoto H, Yamaki M, Isoland K, Yamagata Y (1994a) A 2-pyrone derivative from*Hypericum japonicum*. Phytochemistry 37:283–284

Ishiguro K, Nagata S, Fukumoto H, Yamakland M, Isoi K (1994b) Phloroglucinol derivatives from *Hypericum japonicum*. Phytochemistry 35:469–471

Ishikura N (1982) Flavonol glycosides in the flowers of *Hibiscus mutabilis* F. *versicolor*. Agric Biol Chem 46:1705–1706

Islam I, Shaikh AU, Shahidul IM (2006) Antioxidative and antimutagenic potentials of phytochemicals from *Ipomoea batatas* (L.) Lam. Int J Cancer Res 5:83–94

Islam MS, Akhtar MM, Rahman MM, Rahman MA, Sarker KK, Alam MA (2009) Antitumor and phytotoxic activities of leaf methanol extract of *Oldenlandia diffusa* (Willd.) Roxb. Glob J Pharmacol 3(2):99–106

Islam MS, Rahman MM, Rahman MA, Qayum MA, Alam MF (2011) In vitro evaluation of *Croton bonplandianum* Baill. As potential antitumor properties using Agrobacterium tumefaciens. J Agric Sci Technol 7(3):711–719

Itharat A, Houghton P, Eno-Amooquaye E, Burke P, Sampson J, Raman A (2004) In vitro cytotoxic activity of Thai medicinal plants used traditionally to treat cancer. J Ethnopharmacol 90:33–38

Jafarian A, Ghannadi A, Mohebi B (2014) Cytotoxic effects of chloroform and hydroalcoholic extracts of aerial parts of *Cuscuta chinensis* and *Cuscuta epithymum* on Hela, HT29 and MDA-MB-468 tumor cells. Res Pharm Sci 9:115–122

Jafarian A, Aliomrani M, Zolfaghari B (2017) Phytochemical screening and cytotoxic evaluation of *Euphorbia turcomanica* on Hela and HT-29 tumor cell lines. Adv Biomed Res 6(1):68

Jagdish PC, Chandrashekhar HR, Kumar SV (2010) Potent selective cytotoxic activity of *Kaempferia galanga* rhizome against cancer cell cultures. Int J Pharm Biosci 1(2):1–5

Jagetia GC, Rao SK (2006) Evaluation of the antineoplastic activity of guduchi *Tinospora cordifolia* in Ehrlich ascites carcinoma bearing mice. Bio Pharma Bull 29(3):460–466

Jagetia GC, Venkatesha VA (2006) Treatment of mice with stem bark extract of *Aphanamixis polystachya* reduces radiation-induced chromosome damage. Int J Radiat Biol 82(3):197–209

Jahan S, Chaudhary R, Goyal PK (2009) Anticancer activity of an Indian medicinal plant, *Alstonia scholaris*, on skin carcinogenesis in mice. Integr Cancer Ther 8(3):273–279

Jain SK (1970) Medicinal plant-lore of the Santals. Econ Bot 24:241–278

Jain AK, Barton MC (2018) p53: emerging roles in stem cells, development and beyond. Development 145(8):Dev158360

Jain SP, Puri HS (1984) Ethnomedicinal plants of Jaunsar-Bawar Hills, Uttar Pradesh, India. J Ethnopharmacol 12:213–222

Jain A, Srivastava SK (1985) 8-C-Methyl-Quercetin-3-OBeta-D-Xylopyranoside, a new flavone glycoside from the roots of *Amoora rohituka*. J Nat Prod 48:299–301

Jain V, Kaur R, Jaiswal M (2013) Protective effect of *Lannea coromandelica* Houtt. Merrill. against three common pathogens. J Ayur Integr Med 4:224

Jain P, Pandey R, Shukla S (2014) Inflammation: natural resources and its applications. Springer, New Delhi, p 73

Jamil SS, Nizami Q, Salam M (2007) *Centella asiatica* (L.) urban a review. Nat Prod Radiat 6(2):158–170

Jamir NS (1990) Some interesting medicinal plants used by Nagas. J Res Edu Indian Med 9:81–87

Janaki S, Vijayasekaran V, Viswanathan S, Balakrishna K (1999) Anti-inflammatory activity of *Aglaia roxburghiana* var. *beddomei* extract and triterpenes roxburghiadiol A and B. J Ethnopharmacol 67(1):45–51

Janakiraman N, Jansi J, Johnson M, Hussain M, Jeeva S (2014) Antibacterial efficacy of *Abrus precatorius* L. and *Asystasia gangetica* (L.) T. Anderson. Anti-Infective Agents 12:165–170

Jassbi AR, Mehrdad M, Eghtesadi F, Ebrahimi SN, Baldwin IT (2006) Novel rearranged abietane diterpenoids from the roots of *Salvia sahendica*. Chem Biodivers 3:916–922

Javed H, Erum S, Tabassum S, Ameen F (2013) An overview on medicinal importance of *thymus vulgaris*. J Asian Sci Res 3(10):974

Jayaprakasam B, Zhang Y, Seeram NP, Nair MG (2003) Growth inhibition of human tumor cell lines by withanolides from *Withania somnifera* leaves. J Life Sci 74:125–132

Jemal A, Bray F, Center M, Ferlay J, Ward E (2011) Global cancer statistics. C A Cancer J Clin 61(1):69–90

Jena J, Ranjan R, Ranjan P, Sarangi MK (2012) A study on natural anticancer plants. Int J Pharm Chem Sci 1(1):365–368

Jeong SI, Zhou B, Bae JB, Kim NS, Kim SG, Kwon J, Kim DK, Shin TY, Jeon H, Lim JP, Kim H, Kim HK, Oh CH (1999) Apoptosis-inducing effect of Akebia saponin D from the roots of *Dipsacus asper* Wall in U937 cells. Arch Pharm Res 31(11):1399–1404

Jeong S, Zhou B, Bae J, Kim N, Kim S, Kwon J, Kim D, Shin T, Jeon H, Lim J, Kim H, Kim H, Oh C (2008) Apoptosis-inducing effect of Akebia saponin D from the roots of *Dipsacus asper* Wall in U937 cells. Arch Pharm Res 31(11):1399–1404

Jeong EJ, Yang H, Kim SH, Kang SY, Sung SH, Kim YC (2011) Inhibitory constituents of *Euonymus alatus* leaves and twigs on nitric oxide production in BV2 microglia cells. Food Chem Toxicol 49(6):1394–1398

Jeong JJ, Lee JH, Chang KC, Kim HJ (2012) Honokiol exerts an anticancer effect in T98G human glioblastoma cells through the induction of apoptosis and the regulation of adhesion molecules. Int J Oncol 41(4):1358–1364

Jetawattana S, Boonsirichai K, Charoen S, Martin S (2015) Radical intermediate generation and cell cycle arrest by an aqueous extract of *Thunbergia Laurifolia* Linn. in human breast cancer cells. Asian Pac J Cancer Prev 16:4357–4361

Jewers K, Pusey DFG, Sharma SR, Ahmad Y (1980) 13-C-NMR spectral analysis of Rhazine, Quebrachamine and Rhazinilam. Planta Med 38:359–362

Jia L, Jin H, Zhou J, Chen L, Lu Y, Ming Y, Yu Y (2013) A potential anti-tumor herbal medicine, Corilagin, inhibits ovarian cancer cell growth throughblocking the TGF-β signaling pathways. BMC Complement Altern Med 13(33):1–11

Jiang MM, Dai Y, Gao H, Zhang X, Wang GH, Jia-You HE, Qiong-Ying HU, Zeng JZ, Zhang XK, Yao XH (2008) Cardenolides from *Antiaris toxicaria* as potent selectivenur77 modulators. Chem Pharm Bull 56(7):1005–1008

Jiang Z, Qian J, Dong H, Yang J, Yu X, Chen J, Chen H, Shi Q, Jia L (2017) The traditional chinese medicine *Achyranthes bidentata* and our de novo conception of its metastatic chemoprevention: from phytochemistry to pharmacology. Sci Rep 7:3888

Jimoh FO, Oladiji AT (2005) Preliminary Studies on *Piliostigma thonningii* seeds: proximate analysis, mineral composition and phytochemical screening. Afr J Biotechnol 4(12):1439–1442

Jin W, Shi Q, Hong C, Cheng Y, Ma Z, Qu H (2008) Cytotoxic properties of thiophenes from *Echinops grijissi* Hance. Int J Phytotherp Phytopharm 15(9):768–774

Jin Q, Nan J, Lian L (2011) Antitumor activity of leaves from *Potentilla discolor* on human hepatocellular carcinoma cell line HepG-2. Chin J Nat Med 9(1):61–64

Jing L, Yua NJ, Lib YS, Fua L, Zhaoa YM (2011) Novel lignans from the stems and leaves of *Trachelospermum jasminoides*. Chin Chem Lett 22:1075–1077

Jiu J (1966) A survey of some medicinal plants of Mexico for selected biological activities. Lloydia 29:250–259

Joe YK, Son JK, Park SH, Lee IJ, Moon DC (1996) New naphthalenyl glucosides from the roots of *Juglansmandshurica*. J Nat Prod 59:159–160

Johns N, Kerr P (2009) West Australian desert plants: Australian aboriginal medicines with anticancer potential. Afr J Tradit Complement Altern Med 6:316

Jones WP, Lobo-Echeverri T, Mi Q, Chai H, Lee D, Soejarto DD, Cordell GA, Pezzuto JM, Swanson SM, Kinghorn AD (2005) Antitumour activity of 3-chlorodeoxylapachol, a naphthoquinone from *Avicennia germinans* collected from an experimental plot in Southern Florida. Pharmacologia 57(9):1101–1108

Jones WP, Lobo-Echeverri T, Mi Q, Chai HB, Soejarto DD, Cordell GA, Swanson SM, Kinghorn AD (2007) Cytotoxic constituents from the fruiting branches of *Callicarpa americana* collected in Southern Florida. J Nat Prod 70:372–377

Jongsuwat Y (1981) Antileukemic activity of *Acanthus illicifolius*. Master thesis, Chulalunkhorn University, Bangkok, Thailand, p 151

Josep B, Raj SJ (2011) An overview *Ficus bengalensis*L. Int J Pharm Sci Rev Res 6(1):21–24

Ju Y, Jia ZJ (1992) Steroidal saponins from the rhizomes of *Smilax menispermoidea*. Phytochemistry 31:1349–1351

Ju Y, Still CC, Sacalis JN, Li J, Ho CT (2001) Cytotoxic coumarins and lignans from extracts of the northern prickly ash (*Zanthoxylum americanum*). Phytother Res 15(5):441–443

Ju EM, Lee SE, Hwang HJ, Kim JH (2004) Antioxidant and anticancer activity of extract from *Betula platyphylla* var. japonica. Life Sci 74(8):1013–1026

Jung H, Roy A, Abdul Q, Kim H, Park H, Choi J (2017) Luteolin 5-O-glucoside from Korean Milk Thistle, *Cirsium maackii*, exhibits anti-inflammatory activity via activation of the Nrf2/HO-1 Pathway. Nat Prod Sci 23:183

Jura-Morawiec J, Tulik M (2016) Dragon's blood secretion and its ecological significance. Chemoecology 26:10–105

Kadil FE, Nassar MI (1998) A tannin anti-cancer promotor from *Terminalia arjuna*. Phytochemistry 47(8):1567–1568

Kadir F, Kassim N, Abdulla M, Yehye W (2013) PASS-predicted Vitex negundo activity: antioxidant and antiproliferative properties on human hepatoma cells-an in vitro study. BMC Complement Altern Med 13(1):343

Kainsa S, Kumar P, Rani P (2012) Medicinal plants of asian origin having anticancer potential: short review. Asian J Biomed Pharm Sci 2(10):1–11

Kalaiselvi M, Kalaivani K (2011) Phytochemcial analysis and anti lipid peroxidative effect of *Jasminum sambac* (L.) Ait oleaceae. Pharmacology 1:38–43

Kambara H, YamadaT TM, Matsunaga S, Tanaka R, Ali HI, Wiart C, Yusof M, Hassan H, Hanifah A, Fauzi ZM, Mazlan NH, Jay M, Kunishima M, Akaho E (2006) A study on medicinal plants from Malaysia focused on *Acalypha siamensis* oliv. Exgage Isolation and structure of a new tetraterpene. Chem Biodivers 3(12):1301–1306

Kamwendo WY, Chiotha SS, Msonthi JD (1985) Screeningof plants used traditionally to control schistosomiasisin Malawi. Fitoterapia 56:229–232

Kanchanapoom T, Kasai R, Yamasaki K (2002) Iridoid glucosides from *Thunbergia laurifolia*. Phytochemistry 60(8):769–771

Kang M, Lee C, Yun J, Oh S, Park S, Lee K, Kim H, Han S, Kim H, Kang J (2012) Methanolic extract isolated from root of *Lycoris aurea* inhibits cancer cell growth and endothelial cell tube formation in vitro. Toxicol Res 28:33–38

Kao WJ, Hung DZ, Tsai WJ, Lin KP, Deng JF (1992) Podophyllotoxin intoxication. Toxic effect of Bajiaolianin herbal therapeutics. Hum Exp Toxicol 11:480–487

Kariuki A, Njoroge G (2011) Ethnobotanical and antimicrobial studies of some plants used in kibwezi (Kenya) for management of lower respiratory tract infections. Afr J Tradit Complement Altern Med 8(2):144–149

Karnjanapee P, Natori S (1966) Studies of *Cudrania javanensis*Trec (Moraceae). Bull Dept Med Sci Mahidol Univ 8:96–106

Karuppaiya P, Tsay HS (2016) Therapeutic values, chemical constituents and toxicity of Taiwanese *Dysosma pleiantha*–A review. Toxicol Lett 236(2):90–97

Karuppannan K, Subramanian D, Venugopal S (2013) Phytopharmacological properties of Albiza species: a review. Int J Pharm Pharm Sci 5(3):70–73

Kaššák P (2012) Secondary metabolites of the choosen genus iris species. Acta Univ Agric Silvic Mendel Brun 60(8):269–280

Katagi A, Sui L, Kamitori K, Suzuki T, Katayama T, Hossain A, Noguchi C, Dong Y, Yamaguchi F, Tokuda M (2016) Inhibitory effect of isoamericanol A from *Jatropha curcas* seeds on the growth of MCF-7 human breast cancer cell line by G2/M cell cycle arrest. Heliyon 2(1):00055

Katem F, Kadima J, Marini R (2017) Comparative phytochemical composition and hypoglycemic activity of some plants used by traditional healers to treat diabetes in kisangani. Int J Pharm Pharma Res 9(4):312–321

Kathiriya A, Das K, Kumar EP, Mathai KB (2010) Evaluation of antitumor and antioxidant activity of *Oxalis corniculata* linn. against Ehrlich ascites carcinoma on mice. Iran J Cancer Prev 3(4):157–165

Katsube N, Iwashita K, Tsushida T, Yamaki K, Kobori M (2003) Induction of apoptosis in cancer cells by bilberry (*Vaccinium myrtillus*) and the anthocyanins. J Agric Food Chem 51:68–75

Kaur M, Singh K, Rup PJ, Kamboj SS, Saxena AK, Sharma M, Bhagat M, Sood SK, Singh J (2006) A tuber lectin from *Arisaema jacquemontii* blume with anti-insect and antiproliferative properties. Biochem Mol Biol Rep 39(4):432–440

Kaur P, Walia A, Kumar S, Kaur S (2009) Antigenotoxic activity of polyphenolic rich extracts from *Aegle marmelos*. J Nat Prod 3(1):68–75

Kaur R, Sharma U, Singh B, Arora S (2011) Antimutagenic and antioxidant characteristics of *Chukrasia tabularis* A. Juss extracts. Int J Toxicol 30(1):21–34

Keawpradub N, Kirby GC, Steele JC, Houghton PJ (1999) Antiplasmodial activity of extracts and alkaloids of three Alstonia species from Thailand. Planta Med 65(08):690–694

Kedzierska M, Olas B, Wachowicz B, Stochmal A, Oleszek W, Jeziorski A, Piekarski J, Glowacki R (2009) An extract from berries of *Aronia melanocarpa* modulates the generation of superoxide anion radicals in blood platelets from breast cancer patients. Planta Med 75:1405–1409

Khade KV, Dubey H, Tenpe CR, Yeole PG, Patole AM (2011) Anticancer activity of the ethanolicextracts of *Agave americana* leaves. Pharmacol Online 2:53–68

Khadem S, Marles RJ (2010) Monocyclic phenolic acids; hydroxy- and polyhydroxybenzoic acids: occurrence and recent bioactivity studies. Molecules 15:7985–8005

Khajure PV, Rathod JL (2011) Potential anticancer activity of *Acanthus ilicifolius* extracted from the mangroves forest of karwar. World J Sci Technol 1(1):1–6

Khamis SB, Brown JE, Bibby MC, Saleh K, Wright KC (1997) Evaluation of two plants species used in Malay traditional medicine for the treatment of breast cancer. J Pharm Pharmacol 49:113

Khamis S, Bibby M, Brown J, Cooper P, Scowen I, Wright C (2004) Phytochemistry and preliminary biological evaluation of *Cyathostemma argenteum*, a Malaysian plant used traditionally for the treatment of breast cancer. Phytother Res 18:507–510

Khan A (2017) Flowering plants: structure and industrial products, p 216

Khan AL, Hussain J, Hamayun M, Gilani SA, Ahmad S, Rehman G, Kim Y, Kang S, Lee I (2010) Secondary metabolites from *Inula britannica* L. and their biological activities. Molecules 15(3):1562–1577

Khan M, Harun N, Rehman AHNA, Elhussein SAA (2013) In vitro antioxidant evaluation of extracts of three wild Malaysian plants. Procedia Eng 53:29–36

Kiem PV, Thu VK, Yen PH, Nhiem NX, Tung NH, Cuong NX, Minh CV, Huong HT, Hyun JH, Kang HK, Kim YH (2009) New triterpenoid saponins from *Glochidion eriocarpum* and their cytotoxic activity. Chem Pharm Bull 57:102–105

Kijjoa A, Gonzalez MJTG, Pinto MMM, Monanondra IO, Herz W (1991) Constituents of *Knema laurina* and *Knema tenuinervia* subsp. *setosa*. Planta Med 57:575–577

Kim YS, Ryu KH, Mo YK, Lee CK, Han SS (1994) Effects of the antitumor component, f-d-p, isolated from *Efvingia applanata* on the immune response. Korean J Pharmacol 25:348–355

Kim JP, Kim WG, Koshino H, Jung J, Yoo ID (1996a) Sesquiterpene o-naphthoquinones from the rootbark of *Ulmus davidiana*. Phytochemistry 43(2):425–430

Kim JP, Kim WG, Koshino H, Park JH, Jung J, Yoo ID (1996b) Isolation and identification of sesquiterpeneo-naphthoquinones, mansonones E, F and H, from theroot bark of *Ulmus davidiana* Planch. Hanguk Nonghwa Hakhoe Chi 39:89–94

Kim SE, Kim YH, Lee JJ (1998) A new sesquiterpene ester from *Celastrus orbiculatus* reversing multidrug resistance in cancer cells. J Nat Prod 61:108–111

Kim Y, Bang SC, Lee JH, Ahn BZ (2004) Pulsatilla saponin D: the antitumor principle from *Pulsatilla koreana*. Arch Pharm Res 27(9):915–918

Kim HJ, Yeom SH, Kim MK, Shim JG, Paek IN, Lee MW (2005) Nitric oxide and prostaglandin E2 synthesis inhibitory activities of diarylheptanoids from the barks of *Alnus japonica* steudel. Arch Pharm Res 28(2):177–179

Kim JS, Lee SY, Park SU (2009) An efficient protocol for Arachis hypogaea L. transformation mediated by *Agrobacterium rhizogenes*. Rom Biotech Lett 14(5):4641–4647

Kim J, Lee C, Park J, Cho J, Kim I, Ahn J, Lee J, Chen B, Shin B, Tae H, Bae E, Kang I, Won M (2015a) Effect of *Oenanthe Javanica* extract on antioxidant enzyme in the rat liver. Chin Med J 128(12):1649–1654

Kim K, Moon E, Choi S, Pang C, Kim S, Lee K (2015b) Identification of cytotoxic and anti-inflammatory constituents from the bark of *Toxicodendron vernicifluum* (Stokes) F.A. Barkley. J Enthopharmacol 162:231–237

Kim S, Jang T, Kim D, Park J (2015c) Study on antioxidant and anti-inflammatory activities of *Persicaria tinctoria*. Korea J Herb 30(6):17–24

Kim KH, Moon E, Kim HK, Oh JY, Kim SY, Choi SU, Lee KR (2015d) Phenolic constituents from the rhizomes of *Acorus gramineus* and their biological evaluation on antitumor and anti-inflammatory activities. Bioorg Med Chem Lett 22:6155–6159

Kim J, Shin E, Park G, Kim Y, Shin D (2016) Root extract of water dropwort, *Oenanthe javanica* (Blume) DC, induces protein and gene expression of phase I carcinogen-metabolizing enzymes in HepG2 cells. Springerplus 5(1):413

Kim G, Kim J, Kim G, Choi S (2017) Anti-adipogenic effects of *Tropaeolum majus* (nasturtium) ethanol extract on 3T3-L1 cells. Food Nutr Res 61(1):1339555

Kimoto T, Hino K, Koya-Miyata S, Yamamoto Y, Takeuchi M, Nishizaki Y, Micallef MJ, Ushio S, Iwaki K, Ikeda M, Kurimoto M (2001) Cell differentiation and apoptosis of monocytic and promyelocytic leukemia cells (U-937 and HL-60) by tryptanthrin, an active ingredient of *Polygonum tinctorium* Lour. Pathol Int 51(5):315–325

Kimura Y, Okuda H (2001) Resveratrol isolated from *Polygonum cuspidatum* root prevents tumor growth and metastasis to lung and tumor-induced neovascularization in Lewis lung carcinoma-bearing. J Nutr 131:1844–1849

Kimura S, Ito C, Jyoko N, Segawa H, Kuroda J, Okada M, Adachi S, Nakahata T, Yuasa T, Filho VC, Furukawa H, Maekawa T (2005) Inhibition of leukemic cell growth by anovel anti-cancer drug (GUT-70) from *Calophyllum brasiliense* that acts by induction of apoptosis. Int J Cancer 113(1):158–165

Kintzios S (2006) Terrestrial plant-derived anticancer agents and plant species used in anticancer research. Crit Rev Plant Sci 25(2):79–113

Kite GC, Veitch NC, Boalch ME, Lewis GP, Leon CJ, Simmonds MS (2009) Flavonol tetraglycosides from fruits of Styphnolobium japonicum (Leguminosae) and the authentication of *Fructus Sophorae* and *Flos Sophorae*. Phytochemistry 70(6):785–794

Kloos H (1977) Preliminary studies of medicinal plants and plant products in markets of central Ethiopia. Ethnomedicine 4:63–104

Kloos H, Tekle A, Yohannes LW, Yosef A, Lemma A (1978) Preliminary studies of traditional medicinal plants in nineteen markets in Ethiopia, use patterns and public health aspects. Ethiop Med J 16:33–35

Kma L, Sharan R, Hadem K (2014) Inhibitory potential of methanolic extracts of *Aristolochia tagala* and *Curcuma caesiaon* hepatocellular carcinoma induced by diethylnitrosamine in BALB/c mice. J Carcinog 13:7

Ko K (1933) On the pharmacological action of Plumbagin. Jpn J Med Sci Pharm 6:259–261

Ko WG, Kang TH, Lee SJ, Kim YC, Lee BH (2002) Effects of luteolin on the inhibition of proliferation and induction of apoptosis in human myeloid leukaemia cells. Physiol Res 16:295–298

Ko S, Koh SH, Jun CY, Nam CG, Bae HS, Shin MK (2004) Induction of apoptosis by *Saussurea lappa* and Pharbitis nil on AGS gastric cancer Cells. Bio Pharm Bull 27(10):1604–1610

Koduru S, Grierson DS, Van de Venter M, Afolayan AJ (2007) Anticancer activity of steroid alkaloids isolated from *Solanum aculeastrum*. Pharm Biol 45(8):613–618

Kokwaro JO (1976) Medicinal plants of East Africa. East African Literature Bureau, Nairobi

Komakech R, Kang Y, Lee J, Omujal F (2017) A review of the potential of phytochemicals from *Prunus africana* (Hook f.) Kalkman stem bark for chemoprevention and chemotherapy of prostate cancer. Evid Based Complement Alternat Med 2017:1–10

Konda Y, Iguchi M, Harigaya Y, Takayanagi H, Ogura H, Li X, Lou HX, Onda M (1990) Hancokinol, a novel triterpene, from *Cynanchum hancokianum*. Tetrahedron Lett 31:5315–5318

Koteswara RY, Hua FS, Nd-Min TY (2005) Inhibitory effects of the flavonoids isolated from *Waltheria indica* on the production of NO, TNF-ALPHA. And IL-12 in activated macrophages. Bio Pharm Bull 28(5):912–915

Koul S, Razdan TK, Andotra CS (2003) Acinospesigenin-A, -B, and -C: three new triterpenoids from *Phytolacca acinosa*. J Nat Prod 66:1121–1123

Kouno I, Yasuda I, Mizoshiri H, Tanaka T, Marubayashi N, Yang DM (1994) Two new iridolactones and their glycosides from the roots of *Patrinia scabra*. Phytochemistry 37:467–472

Kour A (2014) Plants exhibiting potential for cancer treatment. Int J Pharm Sci Rev Res 27(2):23–53

Kuduk-Jaworska J, Szpunar J, Gasiorowski K, Brokos B (2004) Immunomodulating polysaccharide fractions of *Menyanthes trifoliata* L. Z Naturforsch C 59(7):485–493

Kuete V, Vouffo B, Mbaveng AT, Vouffo EY, Siagat RM, Dongo E (2009) Evaluation of *Antiaris africana* methanol extract and compounds for antioxidant and antitumor activities. Pharm Biol 47(11):1042–1049

Kuete V, Sandjo L, Mbaveng A, Seukep J, Ngadjui B, Efferth T (2015) Cytotoxicity of selected Cameroonian medicinal plants and *Nauclea pobeguinii* towards multi-factorial drug-resistant cancer cells. BMC Complement Altern Med 15(1):309

Kulathuran P, Narayanan N, Chidambaranathan N, Mohamed H, Jayaprakash S (2012) Antitumor activity of ethanolic extract of Cnidoscolus chayamansa McVaugh against Dalton's Ascitic Lymphoma in mice. Int J Pharm Pharm Sci 4:647–652

Kumar VL, Singhal A (2009) Germinating seeds of the mung bean, *Vigna radiate* (Fabaceae), as a model for the preliminary evaluation of cytotoxic effects of drugs. Biocell 33(1):19–24

Kumar V, Poonam, Prasad AK, Parmar VS (2003) Naturally occurring aristolactams, aristolochic acids and dioxoaporphines and their biological activities. J Nat Prod 20:565–583

Kumar RA, Sridevi K, Kumar NV, Nanduri S, Rajagopal S (2004) Anticancer and immunostimulatory compounds from *Andrographis paniculata*. J Ethnopharmacol 92(2):291–295

Kumar P, Misra CS, James J, Dev L, AKT V, Thankamani V (2011a) Ethnobotanical and pharmacological study of *Alstonia* (Apocynaceae) – a review. J Pharm Sci Res 3(8):1394–1403

Kumar RS, Rajkapoor B, Perumal P (2011b) Antioxidant activities of *Indigofera cassioides* Rottl. Ex. DC. using various in vitro assay models. Asian Pac J Trop Biomed 2(4):1–6

Kumar S, Fathima S, Alexander P, Jeniffa R, Praseetha P (2015) *Typhonium divarictum*: a promising endangered anti-cancer plant. Acad Edu 8:125

Kumar A, Shanthi V, Ramanathan K (2016) Management of crizotinib resistance in lung cancer using traditional plant source: an in-silico strategies. Biomed Res 27(3):794–800

Kumar H, Agrawal R, Kumar V (2018) *Barleria cristata*: perspective towards phytopharmacological aspects. J Pharm Pharmacol 70:475–487

Kumari TS, Charya MS (2017) Phytochemistry, Anti-cancer and anti-inflammatory activities of solvent leaf extracts of *Nyctanthes arbor-tristis*. Int J Pharm Sci Res 8(4):1654–1663

Kumari V, Kaushal K, Sharma A, Chandra Mishra R, Soni P (2018) Some phytochemicals found in medicinal plants used in cancer – a review. Med Chem 08:18–20

Kumoro AC, Hasan M, Singh H (2018) Extraction of andrographolide from *Andrographis paniculata* dried leaves using supercritical CO2 and ethanol mixture. Ind Eng Chem Res 58(2):742–751

Kumudhavalli MV, Jayakar B, Kumar GA (2010) Phytochemical and pharmacological evaluation of the dried fruit of the plant *Juglans regia* L. oil. Drug Invention Today 2(7):362–365

Kuo KW (2000) Anticancer activity evaluation of the Solanum glycoalkaloid solamargine triggering apoptosis in human hepatoma cells. Biochem Pharmacol 60:1865–1873

Kupchan SM, Barboutis SJ, Knox JR, Cam CAL (1965) Beta-Solamarine: tumor inhibitor isolated from *Solanum dulcamara*. J Sci 150:1827–1828

Kupeli E, Orhan I, Toker G, Yesilada E (2006) Anti-inflammatory and antinociceptive potential of *Maclura pomifera* (Rafin.) Schneider fruit extracts and its major isoflavonoids, scandenone and auriculasin. J Ethnopharmacol 107(2):169–174

Kurade N, Jaitak V, Kaul V, Sharma O (2010) Chemical composition and antibacterial activity of essential oils of *Lantana camara*, *Ageratum houstonianum* and *Eupatorium adenophorum*. Pharm Biol 48(5):539–544

Kusari S, Zühlke S, Spiteller M (2011) Chemometric evaluation of the anti-cancer pro-drug podophyllotoxin and potential therapeutic analogues in *Juniperus* and *Podophyllum* species. Phytochem Anal 22:128–143

Kviecinski MR, Felipe KB, Schoenfelder T, de Lemos Wiese LP, Rossi MH, Goncalez E, Felicio JD, Filho DW, Pedrosa RC (2008) Study of the antitumor potential of Bidens pilosa (Asteraceae) used in Brazilian folk medicine. J Ethnopharmacol 117:69–75

Kwon D, Yoon S, Carter O, Bailey GS, Dashwoodb RH (2006a) Antioxidant and antigenotoxic activities of Angelica keiskei, *Oenanthe javanica* and *Brassica oleracea* in the Salmonella mutagenicity assay and in HCT116 human colon cancer cells. Biofactors 26(4):231–244

Kwon HJ, Hong YK, Kim KH, Han CH, Cho SH, Choi JS, Kim BW (2006b) Methanolic extract of *Pterocarpus santalinus* induces apoptosis in HeLa cells. J Ethnopharmacol 105:229–234

Kyogoku K, Hatayama K, Yokomori S, Seki T, Tanaka I (1975) Synthesis of isoprenyl chalcone "Sophoradin"isolated from *Sophora subprostrata*. Agric Biol Chem 39:133

Kyriazia M, Yova D, Rallis M, Lima A (2006) Cancer chemopreventive effects of *Pinus maritima* bark extract on ultraviolet radiation and ultraviolet radiation- 7,12,dimethylbenz(a) anthracene induced skin carcinogenesis of hairless mice. Cancer Lett 237:234–241

Laferriere JE (1994) Medicinal plants of the lowland Ingapeople of Colombia. Int J Pharm 32:90–94

Lai PK, Donovan J, Takayama H, Sakagami H, Tanaka A, Konno K, Nonoyama M (1990) Modification of human immunodeficiency viral replication by pinecone extracts. AIDS Res Hum Retrovir 6:205–217

Lai CS, Mas RH, Nair NK, Mansor SM, Navaratnam V (2010) Chemical constituents and in vitro anticancer activity of *Typhonium flagelliforme* (Araceae). J Ethnopharmacol 127(2):486–494

Lakshmi KS, Sharma SS, Rajesh T, Chitra V (2010) Antitumour activity of ethanolic extract of leaves of *Holoptelea integrifolia* on Dalton's ascitic lymphoma in Swiss albino mice. Int J Green Pharm 4(1):44–47

Lambert JD, Sang S, Dougherty A, Caldwell CG, Meyers RO, Dorr RT, Timmermann BN (2005) Cytotoxic lignans from *Larrea tridentate*. Phytochemistry 66:811–815

Lan T, Sulaiman S, Najimudin N, Muhammad T (2007) Anticancer medicinal plant, *Epipremnum pinnatum* (L.) Engl. chloroform extracts elicited both apoptotic and non-apoptotic cell deaths in T-47D mammary carcinoma cells. KMITL Sci Tech J 7:24–30

Lao A, Fujimoto Y, Tatsuno T (1984) Studies on the constituents of *Artemisia argyi* Levl et Vant. Chem Pharm Bull 32:723–727

Latorre AO, Furlan MS, Sakai M, Fukumasu H, Hueza IM, Haraguchi M, Górniak SL (2009) Immunomodulatory effects of *Pteridium aquilinum* on natural killer cell activity and select aspects of the cellular immune response of mice. J Immunotoxicol 6(2):104–114

Lavhale MS, Kumar S, Mishra SH, Sitasawad SL (2009) A novel triterpenoid isolated from the root bark of *Ailanthus excelsa* Roxb (Tree of Heaven), AECHL-1 as a potential anti-cancer agent. PLoS One 4(4):5365

Leanna S, Cynthia W, Erin S, Bridge C, Nelson A, Martzen M, Novack J, Torkelson C (2010) Trametes versicolor mushroom immune therapy in breast cancer. J Soc Integr Oncol 6(3):122–128

Lechner-Knecht S (1982) Sacred healing plants in Nepal. Deutsche Apotheker Zeitung 122:2122–2129

Lee KH, Furukawa H, Huang HC, Huang ES (1972) Antitumor agents II: eupatolide, a new cytotoxic principle from *Eupatorium formosanum* hay. J Pharm Sci 61(4):629–631

Lee IR, Song JY, Lee YS (1992) Cytotoxicity of folkloric medicines in murine and human cancer cells. Korean J Pharm 23:132–136

Lee SM, Min BS, Lee CG, Kim KS, Kho YH (2003) Cytotoxic triterpenoids from the fruits of *Zizyphus jujube*. Planta Med 69(11):1051–1054

Lee N, Choi J, Koo B, Ryu S, Han Y, Lee S, Lee D (2005) Antimutagenicity and cytotoxicity of the constituents from the aerial parts of *Rumex acetosa*. Bio Pharma Bull 28(11):2158–2161

Lee NY, Min HY, Lee J, Nam JW, Lee YJ, Han AR, Wiryawan A, Suprapto W, Lee SK, Seo EK (2008) Identification of a new cytotoxic biflavanone from *Selaginella doederleinii*. Chem Pharm Bull 56(9):1360–1361

Lee SJ, Cho YH, Park K, Kim EJ, Kang BS, Jung KH, Kim CH, Kim WJ, Moon SK (2009) Inhibitory effects of the aqueous extract of *Magnolia officinalis* on the responses of human urinary bladder cancer 5637 cells in vitro and mouse urinary bladder tumors induced by N-Butyl-N-(4-hydroxybutyl) nitrosamine in vivo. Phytother Res 23(1):20–27

Lee MS, Lee CM, Cha EY, Thuong PT, Bae K, Song IS, Noh SM, Sul JY (2010) Activation of AMP-activated protein kinase on human gastric cancer cells by apoptosis induced by corosolic acid isolated from *Weigela subsessilis*. Phytother Res 24(12):1857–1861

Lee YM, Jung HY, Bae JY, Lee JY, Jang HH, Park DS, Kim J (2011a) Screening of Korean edible wild plants for estrogenic activity using MCF-7 cell proliferation assay. Food Sci Biotechnol 20(3):829–833

Lee H, Guo J, Wu C, Ti M, Lan K, Hu C, Lo C, Yang C, Lin Y, Wang T (2011b) Anti-inflammatory effects of*Cynanchum taiwanianum rhizome* aqueous extract in IL-1β-induced NRK-52E cells. Pharma Bio 49:437–444

Lee MY, Shin IS, Kyoung H, Seo CS, Son JK, Shin HK (2014) α-spinasterol from *Melandrium firmum* attenuates benign prostatic hyperplasia in a rat model. Mol Med Rep 9(6):2362–2366

Lee K, Joo H, Lee Y, Park M, Kang G, Choi S, Lee K, Jeon B (2016a) *Ulmus davidiana* ethanol extract inhibits monocyte adhesion to tumor necrosis factor-alpha-stimulated endothelial cells. Integr Med Res 5(2):131–139

Lee HJ, Shin JS, Lee WS, Shim HY, Park JM, Jang DS, Lee KT (2016b) Chikusetsusaponin IVa methyl ester isolated from the roots of Achyranthes japonica suppresses LPS-induced iNOS, TNF-α, IL-6, and IL-1β expression by NF-κB and AP-1 inactivation. Biol Pharm Bull 39(5):657–664

Leskovac A, Joksic G, Jankovic T, Savikin K, Menkovic N (2007) Radioprotective properties of the phytochemically characterized extracts of *Crataegusmonogyna*, cornus mas and *Gentianella austriaca* on human lymphocytes in vitro. Planta Med 73(11):1169–1175

Levings C (2006) A monograph of the genus Maackia. Degree of Master of Science, Miami University, pp 1–24

Li JC, Sun HD, Lin ZW (1987) The structure of Lushanrubescensin E. Yunnan Zhiwu Yanjiu 9:485–488

Li W, Cui C, Cai B, Wang H, Yao X (2005) Flavonoids from *Vitex trifolia* L. inhibit cell cycle progression at G2/M phase and induce apoptosis in mammalian cancer cells. J Asian Nat Prod Res 7(4):615–626

Li X, Xiao W, Pu J, Ban L, Shen Y, Li ZWS, Sun H (2006) Cytotoxic ent-kaurene diterpenoids from *Isodon phyllostachys*. Phytochemistry 67:1336–1340

Li Z, Wang J, Jiang B, Zhang X, An L, Bao Y (2007a) Benzobijuglone, a novel cytotoxic compound from *Juglans mandshurica*, induced apoptosis in HeLa cervical cancer cells. Phytomedicine 14:846–852

Li YL, Gan GP, Zhang HZ, Wu HZ, Li CL, Huang YP, Liu YW, Liu JW (2007b) A flavonoid glycoside isolated from *Smilax china* L. rhizome *in vitro* anticancer effects on human cancer cell lines. J Ethnolpharmacol 113(1):115–124

Li W, Chiu L, Lam W, Wong W, Chan Y, Ho Y, Wong E, Wong Y, Ooi V (2007c) Ethyl acetate extract of Chinese medicinal herb *Sarcandra glabra* induces growth inhibition on human leukemic HL-60 cells, associated with cell cycle arrest and up-regulation of pro-apoptotic Bax/Bcl-2 ratio. Oncol Rep 17:425–431

Li S, Shi Y, Shang XY, Cui BS, Yuan Y, Chen XG, Yang YC, Shi JG (2010) Cytotoxic xanthones from *Psorospermum molluscum* from the Madagascar rain Forest. J Nat Prod 71(3):460–463

Li K, Li Q, Zhang T, Han Z, Li J, Liu Z, Zheng F (2011) Procyanidins from *Pinus koraiensis* bark inhibits HeLa cell growth by inducing apoptosis and reducing survivin protein expression. Afr J Biotechnol 10(4):7766–7771

Li S, Zhao M, Li Y, Sui Y, Yao H, Huang L, Lin X (2014) Preparative isolation of six anti-tumour biflavonoids from *Selaginella doederleinii* Hieron by high-speed counter-current chromatography. Phytochem Anal 25(2):127–133

Li W, Xu Q, He Y, Liu Y, Yang S, Wang Z, Zhang J, Zhao L (2015) Anti-tumor effect of steamed *Codonopsis lanceolata* in H22 tumor-bearing mice and its possible mechanism. Nutrients 7:8294–8307

Liang HR, Vuorela P, Vuorela H, Hiltunen R (1997) Isolation and immunomodulatory effect of flavonol glycosides from *Epimedium hunanense*. Planta Med 63:316–319

Liang QL, Dai CC, Jiang JH, Tang YP, Duan JA (2009) A new cytotoxic casbane diterpene from *Euphorbia pekinensis*. Fitoterapia 80(8):514–516

Liaw CC, Chang FR, Chen SL, Wu CC, Lee KH, Wu YC (2005) Novel cytotoxic monotetrahydrofuranic Annonaceous acetogenins from *Annona montana*. Bioorg Med Chem 13(15):4767–4776

Lichius JJ, Thoison O, Montagnac A, Pais M, Gueritte-Voegelein F, Sevenet T, Cosson JP, Hadi AH (1994) Antimitotic and cytotoxic flavonols from *Zieridium pseudobtusifolium* and *Acronychia porter*. J Nat Prod 7(7):1012–1016

Lim TK (2016) Edible medicinal and non-medicinal plants, 3rd edn. Springer, Dordrecht, p 181

Lim E, Kim E, Kim B, Kim S, Ha S, Kim Y (2017) Ethanol extract from *Cnidium monnieri* (L.) cusson induces G1 cell cycle arrest by regulating Akt/GSK-3β/p53 signaling pathways in AGS gastric cancer cells. J Korean Soc Food Sci Nutr 46(4):417–425

Lin YL (1987) Flavonoids and coumarins from *Achillea schischkinii*. Planta Med 53:507–509

Lin C, Lin C (1993) Pharmacological and pathological studies on Taiwan folk medicine (IX): the hepatoprotective effect of the methanolic extract from *Echinops Grijisii*. Am J Chin Med 21(1):33–44

Lin Z, Qin Z, Xia H (1984a) Effect of *Tremella polysaccharide* on clearance rate of charcoal particles and hemolysinreaction in mice. In: American Society for pharmaceutical and experimental therapeutics annual meeting, August 19–23, Indianapolis, p 184

Lin ZB, Qin ZL, Xia HL (1984b) Effect of tremellapolysaccharide on clearance rate of charcoal particles and hemolysin reaction in mice. Pharmacology 26:189

Lin CC, Lin CH, Chiu HF, Hu MF (1992) The Pharmacologicaland pathological studies on Taiwan folkmedicine (VII). The anti-inflammatory effect of *Echinops grjiisii*. Am J Chin Med 22:127–134

Lin LZ, Hu SF, Zaw K, Angerhoffer CK, Chai HY, Pezzuto JM, Cordell GA, Lin J, Zheing DM (1994a) Thalifaberidine, a cytotoxic aporphine-benzylisoquinoline alkaloid from *Thalictrum faberi*. J Nat Prod 57:1430–1436

Lin RC, Skaltsounis AL, Sequin E, Tilleguin F, Koch M (1994b) Phenolic constituents of *Selaginella doederleinii*. Planta Med 60:168–170

Lin YL, Huang RL, Chang CM, Kuo YH (1997a) Twonew puriniums and three new pyrimidines from *Heterostemma brownii*. J Nat Prod 60:982–985

Lin YL, Wu YM, Kuo YH (1997b) Revised structures for four acetophenones from *Cynanchum tawanianum*. Phytochemistry 45:1057–1061

Lin RJ, Chen CY, Lo WL (2008) Cytotoxic activity of Ipomoea cairica. Nat Prod Res 22(9):747–753

Lin C, Chen P, Wang C, Wang C, Chang Y, Tai C, Tai C (2014) Antitumor effects and biological mechanism of action of the aqueous extract of the *Camptotheca acuminata* fruit in human endometrial carcinoma cells. Evid Based Complement Alternat Med 2014:1–10

Lin M, Tang S, Zhang C, Chen H, Huang W, Liu Y, Zhang J (2017) Euphorbia factor L2 induces apoptosis in A549 cells through the mitochondrial pathway. Act Pharm Sin B 7(1):59–64

Lirdprapamongkol K, Kramb JP, Chokchaichamnankit D, Srisomsap C, Surarit R, Sila-Asna M, Bunyaratvej M, Dannhardt G, Svasti J (2008) Juice of *Eclipta prostrata* inhibits cell migration in vitro and exhibits anti-angiogenic activity in vivo. In vivo 22(3):363–368

Liu WK, Ng TB, Sze SF, Tsui KW (1993) Activation of peritoneal macrophages by polysaccharopeptide from the mushroom, *Coriolus versicolor*. Immunopharmacology 26:139–146

Liu RX, Wang Q, Guo HZ, Lia L, Bib KS, Guo DE (2005) Simultaneous determination of nine flavonoids in *Dalbergia odorifera* by LC. Chromatographia 61:409–413

Liu XQ, Yuan QY, Guo YQ (2009) Two new stilbenoids from *Pleione bulbocodioides*. Asian J Nat Prod Res 11(1):75–78

Liu Z, Liu M, Liu M, Li J (2010) Methylanthraquinone from *Hedyotis diffusa* Willd induces Ca^{2+}-mediated apoptosis in human breast cancer cells. Toxicol in Vitro 24(1):42–147

Liu QM, Zhao HY, Zhong XK, Jiang JG (2012) *Eclipta prostrata* L. phytochemicals: isolation, structure elucidation, and their antitumor activity. Food Chem Toxicol 50(11):4016–4022

Liu Y, Liu Y, Lan K, Lee Y, Tsai T, Chen Y (2013) Medicinal fungus *Antrodia cinnamomeaIn-hibits* growth and cancer stem cell characteristics of hepatocellular carcinoma. Evid Based Complement Alternat Med 2013:1–8

Liu Z, Ma R, Yang L, Li J, Hou B, Hu J, Zhou J (2017a) Triterpenoids and iridoids from *Patrinia scabiosaefolia*. Fitoterapia 119:130–135

Liu M, Wang WG, Sun HD, Pu JX (2017b) Diterpenoids from Isodon species: an update. Nat Prod Rep 23:673–698

Loizzo MR, Tundis R, Menichini F, Saab AM, Statti GA, Menichini F (2007) Cytotoxic activity of essential oils from Labiatae and Lauraceae families against in vitro human tumor models. Anticancer Res 27:3293–3300

Lopes M, Saffi J, Echeverrigaray S, Henriques J, Salvador M (2004) Mutagenic and antioxidant activities of *Croton lechleri* sap in biological systems. J Ethnopharmacol 95:437–445

Lotulung PDN, Minarti, Kardono LBS, Kawanishi K (2008) Antioxidant compound from the rhizomes of *Kaempferia rotunda*. Pak J Biol Sci 11(20):2447–2450

Lou HX, Li X, Onda MY, Konda Y, Urano M, Harigaya Y, Takayanagi H, Ogura H (1991) Stereochemistry of novel triterpenes from *Cynanchum hancokianum*. Chem Pharm Bull 39:2271–2276

Lu G, Liao J (1990) Detection of the anti-cancer biological effect of naphthoquinone pigment-LIII. Zhong xi yi jie he za zhi. Chinese J Modern Develop Trad Med 10(7):422–425

Lu W, Li Q, Li J, Liu F, Yang X (2009) Polysaccharide from *Patrinia heterophylla* Bunge inhibits hela cell proliferation through induction of apoptosis and cell cycle arrest. Am J Clin Pathol 40(3):161–167

Lu K, Chang Y, Yin P, Chen T, Ho Y, Chang Y, Chi C (2010) In vitro and in vivo apoptosis-inducing antileukemic effects of *Mucuna macrocarpa* stem extract on HL-60 human leukemia cells. Integr Cancer Ther 9(3):298–308

Lu K, Lee H, Huang M, Lai S, Ho Y, Chang Y, Chi C (2012) Synergistic apoptosis-inducing antileukemic effects of arsenic trioxide and *Mucuna macrocarpa* stem extract in human leukemic cells via a reactive oxygen species-dependent mechanism. Evid Based Complement Alternat Med 2012(1):1–14

Lu YH, Tian CR, Gao CY, Wang WJ, Yang WY, Kong X, Chen YX, Liu ZZ (2018) Protective effect of free phenolics from *Lycopus lucidus* Turcz. root on carbon tetrachloride-induced liver injury in vivo and in vitro. Food Nutr Res 2018:62

Luo YM, Xiong WS (1991) Chemical constituents of anti-cancer plant *Nyssa sinensis*. Oliv. Zhongguo ZhongyaoZazhi 16:424–425

Luo Y, Xu X, Liu J, Li J, Sun Y, Liu Z, Liu J, Van DE, Balzarini J, Bao J (2007) A novel mannose-binding tuber lectin from *Typhonium divaricatum* (L.) Decne (family Araceae) with antiviral activity against HSV-II and anti-proliferative effect on human cancer cell lines. Biochem Mol Biol 40(3):358–367

Luo ZF, Yang Y, Wang LS, Li HP (2017) Isolation of three cyanins from *Lonicera caerulea* L. fruits and its anticancer activity. J Cent South Univ 24(7):1573–1581

Lv J, Wang L, Shen H, Wang X (2018) Regulatory roles of OASL in lung cancer cell sensitivity to *Actinidia chinensis* Planch root extract (acRoots). Cell Biol Toxicol 34(3):207–218

Ma H (2014) Phytochemical and biological investigation of gallotannins from red maple (*Acer rubrum*) species. Thesis, University of Rhode Island

Ma B, Shi YP, Jia ZJ (1997) A seco-oleane-type triterpenoid acids from *Legularia intermedia*. Planta Med 63:973–974

Ma YX, Fu HZ, Li M, Sun W, Xu B, Cui JR (2007) An anticancer effect of a new saponin component from *Gymnocladus chinensis* Baillon through inactivation of nuclear factor-kappa B. Anti-Cancer Drugs 18(1):41–46

Ma LF, Wang PF, Wang JD, Tong XM, Shan WG, Zhang H, Zhan ZJ (2017) New eudesmane sesquiterpenoids from *Salvia plebeia* R. Br. Chem Biodivers 14(8):1700127

Maddocks-Jennings W, Wilkinson J, Cavanagh H, Shillington D (2009) Evaluating the effects of the essential oils *Leptospermum scoparium* (manuka) and Kunzea ericoides (kanuka) on radiotherapy induced mucositis: a randomized, placebo-controlled feasibility study. Eur J Oncol Nurs 13(2):87–93

Madhuri S, Pandey G (2009) Some anticancer medicinal plants of foreign origin. Curr Sci 96(6):778–779

Madikizela B, McGaw L (2017) *Pittosporum viridiflorum* Sims (Pittosporaceae): a review on a useful medicinal plant native to South Africa and tropical Africa. J Ethnopharmacol 205(1):217–230

Madziga HA, Sanni S, Sandabe UK (2010) Phytochemical and elememtal analysis of*Acalypha wilkesiana* leaf. Am J Sci 6(11):510–514

Maffei M (2003a) Dietary supplements of plant origin, p 212

Maffei M (2003b) Dietary supplements of plant origin, p 213

Maffei M (2003c) Dietary supplements of plant origin, p 215

Maffei M (2003d) Dietary supplements of plant origin, p 216

Maffei M (2003e) Dietary supplements of plant origin: a nutrition and health approach. CRC Press, Taylor & Francis, Boca Raton, p 209

Maffei M (2003f) Dietary supplements of plant origin: a nutrition and health approach. Taylor & Francis, Boca Raton, p 217

Maffei M (2003g) Dietary supplements of plant origin: a nutrition and health approach. Taylor & Francis, Boca Raton, p 221

Maffei M (2003h) Dietary supplements of plant origin: a nutrition and health approach. Taylor & Francis, Boca Raton, p 218

Maffei M (2003i) Dietary supplements of plant origin: a nutrition and health approach. Taylor & Francis, Boca Raton, p 230

Maffei M (2003j) Dietary supplements of plant origin: a nutrition and health approach. Taylor & Francis, Boca Raton, p 223

Maffei M (2003k) Dietary supplements of plant origin: a nutrition and health approach. Taylor & Francis, Boca Raton, p 224

Maffei M (2003l) Dietary supplements of plant origin: a nutrition and health approach. Taylor & Francis, Boca Raton, p 226

Maffei M (2003m) Dietary supplements of plant origin: a nutrition and health approach. Taylor & Francis, Boca Raton, p 225

Magesh V, Yuvaraj G, Deecaraman M, Kalaichelvan PT (2009) Methanolic extract of *Indigofera tinctoria* induces apoptosis in HCT116 cells. J Appl Biosci 14:768–774

Mahanta PK, Ghanim A, Gopinath KW (1974) Chemical constituents of *Piper sylvaticum* (Roxb) and *Piper boehmerifolium* (Wall). J Pharm Sci 63:1160–1161

Malacrida A, Maggioni D, Cassetti A, Nicolini G, Cavaletti G, Miloso M (2016) Antitumoral effect of *Hibiscus sabdariffa* on human squamous cell carcinoma and multiple myeloma cells. Nutr Cancer 68(7):1161–1170

Mallmann R, Ethur E, Bianchetti P, Faleiro D, Hoehne L, Goettert M (2018) Effectiveness of aqueous and hydroalcoholic extracts of *Acanthospermum australe* (Loefl.) Kuntze against diarrhea-inducing bacteria. Braz J Biol 78(4):1–6

Mandal SM, Migliolo L, Das S, Mandal M, Franco OL, Hazra TK (2012) Identification and characterization of a bactericidal and proapoptotic peptide from *Cycas revoluta* seeds with DNA binding properties. J Cell Biochem 113:184–193

Mandal N, Ghate N, Hazra B, Sarkar R (2014) Heartwood extract of *Acacia catechu* induces apoptosis in human breast carcinoma by altering bax/bcl-2 ratio. Pharmacogn Mag 10(37):27–33

Maness L, Goktepe I, Hardy B, Yu J, Ahmedna M (2012) Antiproliferative and apoptotic effects of *Phytolacca americana* extracts and their fractions on breast and colon cancer cells. Res J Med Plant 6:17–26

Maness L, Goktepe I, Chen H, Ahmedna M, Sang S (2014) Impact of *Phytolacca Americana* extracts on gene expression of colon cancer cells. Phytother Res 28(2):219–223

Mansoor TA, Ramalho RM, Mulhovo S, Rodrigues CMP, Ferreir MJU (2009) Induction of apoptosis in HuH-7 cancer cells by monoterpene and b-carboline indole alkaloids isolated from the leaves of *Tabernaemontana elegans*. Bioorganic Med Chem Lett 19:4255–4258

Marghescu F, Teodorescu M, Radu D (2012) The positive impact of flaxseed (*Linum Usitatissimum*) on breast cancer. J Agroaliment Process Technol 18:161–168

Marius M, Jabeen A, Gilbert A, Simjee S, Desire B, Dastagir N (2018) NO-cGMP-K channel-dependent anti-nociceptive activities of methanol stem bark extract of *Piptadeniastrum africanum* (Mimosaceae) on rats. Asian Pac J Trop Biomed 8(3):150–159

Maroyi A (2017) *Euclea undulata* Thunb.: review of its botany, ethnomedicinal uses, phytochemistry and biological activities. Asian Pac J Trop Med 10(11):1030–1036

Martin-Cordero C, Saenz MT, Ayuso MJ, Caviedes MA (1995) Cytotoxic activity of *Retama sphaerocarpa*. Fitoterapia 66:495–498

Martín-Cordero C, López-Lázaro M, Agudo MA, Navarro E, Trujillo J, Ayuso MJ (2001) A cytotoxicdiarylheptanoid from *Viscum cruciatum*. Phytochemistry 58(4):567–569

Martinez-Lirola MJ, Gonzalez-Tejero MR, Molero-Mesa J (1996) Ethnobotanical resources in the province of Almeria, Spain, Campos De Nijar. Econ Bot 50:40–56

Martinez-Rocha A, Puga R, Hernandez-Sandoval L, Loarca-Pina G, Mendoza S (2008) Antioxidant and antimutagenic activities of Mexican oregano (*Lippiagraveolens* Kunth). Plant Foods Human Nutr 63(1):1–5

Martins S, Teixeira J, Mussatto S (2013) Solid-state fermentation as a strategy to improve the bioactive compounds recovery from *Larrea tridentata* leaves. Appl Biochem Biotechnol 171(5):1227–1239

Martucciello S, Balestrieri ML, Felice F, Estevam Cdos S, Sant-Ana AE, Pizza C, Piacente S (2009) Effects of triterpene derivatives from *Maytenus rigida* on VEGF-induced Kaposi's sarcoma cell proliferation. Chem-Biol Interact 183(3):450–445

Masilungan VA, Vadlamudi S, Goldin A (1971) Screeningof Philippine medicinal plants for anticancer agents using CCNSC protocols. Cancer Chemother Rep Part 2:135–140

Maslovskaya LA, Savchenko AI, Gordon VA, Reddell PW, Pierce CJ, Parsons PJ, Williams CM (2011) Isolation and confirmation of the proposed cleistanthol biogentic link from *Croton insularis*. Org Lett 13(5):1032–1035

Massimo M (2003) Dietary supplements of plant origin. CRC Press, p 108

Mathew A (2014) Cytotoxicity and antimicrobial effects of *Pistia stratiotes* leaves. Int J Drug Dev Res 6(4):208–217

Matsunaga K, Saitoh M, Ohizumi Y (1996) Acanthostral, a novel antineoplastic cis,cis,cis-germacranolide from *Acanthospermum australe*. Tetrahedron Lett 37(9):1455–1456

Mavric E, Wittmann S, Barth G, Henle T (2008) Identification and quantification of methylglyoxal as the dominant antibacterial constituent of Manuka (*Leptospermum scoparium*) honeys from New Zealand. Mol Nutr Food Res 52(4):483–489

Mayer R (1993) A beta-hydroxychalcone from *Leptospermum acoparium*. Planta Med 59:269–271

Mazzio E, Soliman K (2009) In vitro screening for the tumoricidal properties of international medicinal herbs. Phytother Res 23:385–398

Mbaze LM, Lado JA, Wansi JD, Shiao TC, Chiozem DD, Mesaik MA, Choudhary MI, Lacaille-Dubois MA, Wandji J, Roy R, Sewald N (2005) Oxidative burst inhibitory and cytotoxic amides and lignans from the stem bark of *Fagara heitzii* (Rutaceae). Phytochemistry 70(11):1442–1447

Mbwambo ZH, Moshi MJ, Masimba PJ, Kapingu MC, Nondo RS (2007) Antimicrobial activity and brine shrimp toxicity of extracts of *Terminalia brownii* roots and stem. BMC Complement Alt Med 7:1–9

McCuddin CR (1974) Samoan medicinal plants and their usage. Office of comprehensive health planning, department of medical services, Government of American Samoa

McKeown BT (2015) Magnolol affects cellular growth and proliferation in human prostate cancer cells in vitro (Doctoral dissertation, University of Prince Edward Island)

Medeiros R, Otuki M, Avellar M, Calixto J (2007) Mechanisms underlying the inhibitory actions of the pentacyclic triterpene α-amyrin in the mouse skin inflammation induced by phorbol ester 12-O-tetradecanoylphorbol-13-acetate. Eur J Pharmacol 559(2–3):227–235

Meena PD, Kaushik P, Shukla S, Soni AK, Kumar M, Kumar A (2006) Anticancer and antimutagenic properties of *Acacia nilotica* (Linn.) on 7,12-dimethylbenz(a)anthracene-induced skin papillomagenesis in Swiss albino mice. Life Sci 7(4):627–632

Mehboob H, Iqbal M, Ejaz M, Bibi G, Sarwar U, Iftikhar S, Shaheen S, Safdar I (2017) A review on secondary metabolites of *Rosa laevigata* Michaux: an important medicinal plant. Biochem Anal Biochem 6(3):2161–1009

Melchor-Martínez E, Silva-Mares D, Torres-López E, Waksman-Minsky N, Pauli G, Chen S, Niemitz M, Sánchez-Castellanos M, Toscano A, Cuevas G, Rivas-Galindo V (2017) Stereochemistry of a second riolozane and other diterpenoids from *Jatropha dioica*. J Nat Prod 80(8):2252–2262

Mills E, Cooper C, Seely D, Kanfer I (2005) African herbal medicines in the treatment of HIV: Hypoxis and Sutherlandia. An overview evidence. Pharm J Nutr 4(1):1–19

Mishra R, Verma DL (2010) Flavone-5-O-Glycosides from*Cheilanthes dalhousiae* (Hook). Nat Sci 8(5):139–143

Miyagawa T, Ohtsuki T, Koyano T, Kowithayakorn T, Ishibashi M (2009) Cardenolide glycosides of *Thewetia peruwiana* and triterpenoid saponins of *Sapindus emarginatus* as trail resistance-overcoming compounds. J Nat Prod 72:1507–1511

Miyase T, Melek FR, Warashina T, Selim MA, Kassem IAA (2009) Bisdesmosidic saponins from fruits of *Gleditsia caspica* desf. Rev Latinoam Quim 37(3):218–229

Mockute D, Bernotiene G, Judþentiene A (2005) Chemical composition of essential oils of *Glechoma hederacea* L. growing wild in Vilnius district. J Ecol 16(3):47–50

Mockute D, Bernotiene G, Judzentiene A (2007) The essential oil of Ground Ivy (*Glechoma hederacea* L) growing wild in Eastern Lithuania. J Essent Oil Res 19(5):449–451

Moein MR, Pawar RS, Khan SI, Tekwani BL, Khan IA (2008) Antileishmanial, antiplasmodial and cytotoxic activities of 12, 16-dideoxy aegyptinone B from *Zhumeria majdae* Rech. f. & Wendelbo. Phytother Res 22(3):283–285

Mohammad S (2006) Anticancer agents from medicinal plants. Bangladesh J Pharm 1:35

Mohanakumara P, Sreejayan N, Priti V, Ramesha BT, Ravikanth G, Ganeshaiah KN, Vasudeva R, Mohan J, Santhoshkumar TR, Mishra PD, Ram V, Shaanker RU (2010) *Dysoxylum binectariferum* Hook.f (Meliaceae), a rich source of rohitukine. Fitoterapia 81(2):145–148

Mohandoss S, Ravindran P (1993) Flavonoids from *Melastoma melabathricum*. Fitoterapia 64:277–278

Momekov G, Ferdinandov D, Zheleva-Dimitrova D, Nedialkov P, Girreser U, Kitanov G (2008) Cytotoxic effects of hyperatomarin, a prenylated phloroglucinol from *Hypericum annulatum* Moris subsp. annulatum, in a panel of malignant cell lines. Int J Phytother Phytopharm 15(11):1010–1015

Monteiro L, Bastos K, Barbosa-Filho J, de Athayde-Filho P, Diniz M, Sobral M (2014) Medicinal plants and other living organisms with anti-tumor potential against lung cancer. Evid Based Complement Alternat Med 2014:1–15

Montoya G, Arango G, Unger M, Holzgrabe U (2009) O-glycoside sequence of pentacyclic triterpene saponins from *Phytolacca bogotensis* using HPLC-ESI/multi-stage tandem mass spectrometry. Phytochem Anal 20(6):465–474

Moon HI, Zee O (2010) Anticancer activity of sesquiterpene lactone from plant food (*Carpesium rosulatum*) in human cancer cell lines. Int J Food Sci Nutr 62(2):102–105

Moon CK, Park KS, Lee SH, Ha BJ, Lee BG (1985) Effects of antitumor polysaccharides from *Forsythia corea*on the immune function. Arch Pharm Res 8:31–39

Moon SS, Rahman AA, Manir M, Ahamed VJ (2010) Kaempferol glycosides and cardenolide glycosides, cytotoxic constituents from the seeds of *Draba nemorosa* (Brassicaceae). Arch Pharm Res 33(8):1169–1173

Moore M (1979) Los Remedios De La Gente, published by the author

Morit H, Gond A, Takey K, Itokaw H, Iitaka Y (1997) Cycloleonuripeptide D, a new proline-rich cyclic decapeptide from *Leonurus heterophyllus*. Tetrahedron 53(5):1617–1626

Morit H, Hirasawa Y, Muto A, Yoshida T, Sekita S, Shirota O (2008) Antimitotic quinoid triterpenes from*Maytenus chuchuhuasca*. Bioorg J Med Chem Lett 18(3):1050–1052

Morton JF (1975) Current folk remedies of Northern Venezuela. Q J Crude Drug Res 13:97–121

Morton JF (1977) Some folk-medicine plants of Central American markets. Q J Crude Drug Res 15:165–169

Moshi M, Otieno D, Mbabazi P, Weisheit A (2010) Ethnomedicine of the Kagera Region, north western Tanzania. Part 2: the medicinal plants used in Katoro Ward, Bukoba District. J Ethnobiol Ethnomed 6:19

Mostafa HAM, Elbakry AA, Eman AA (2011) Evaluation of antibacterial and antioxidant activities of different plant parts of *Rumex vesicarius* l. (Polygonaceae). Int J Pharm Pharm Sci 3(2):109–118

Moura MDD, Silva JDS, Oliveira RAGD, Diniz MDFFM, Barbosa-Filho JM (2002) Natural products reported as potential inhibitors of uterine cervical neoplasia. Acta Farm Bonaer 21(1):67–74

Mousinho KC, Oliveira CD, Ferreira JR, Carvalho AA, Magalhães HI, Bezerra DP, Alves AP, Costa-Lotufo LV, Pessoa C, de Matos MP, Ramos MV (2011) Antitumor effect of laticifer proteins of *Himatanthus drasticus* (Mart.) Plumel–Apocynaceae. J Ethnopharmacol 137(1):421–426

Msonthi JD, Magombo D (1983) Medicinal herbs in Malawi and their uses. Hamdard 26:94–100

Muhammad I, Dunbar DC, Takamatsu S, Walker LA, Clark AM (2001) Antimalarial, cytotoxic, and antifungal alkaloids from *Duguetia hadrantha*. J Nat Prod 64(5):559–562

Mukherjee P, Rai S, Kumar V, Mukherjee K, Hylands P, Hider R (2007) Plants of Indian origin in drug discovery. Expert Opin Drug Discovery 2(5):633–657

Mukudai Y, Kondo S, Shiogama S, Koyama T, Li C, Yazawa K, Shintani S (2013) Root bark extracts of *Juncus effusus* and *Paeonia suffruticosa* protect salivary gland acinar cells from apoptotic cell death induced by cis-platinum (II) diammine dichloride. Oncol Rep 30(8):2665–2671

Munda S, Kakoti B (2017) Pharmacological evaluation of methanolic extract of *Zanthoxylum Oxyphyllum* Edgw. with special reference to anti-inflammatory and analgesic activity. Int J Appl Biol Pharma Tech 8:5–10

Murayama T, Kishi N, Koshiura R, Takagi K, Furukawa T, Miyamoto K (1992) Agrimoniin, an antitumor tannin in *Agrimonia pilosa* Ledeb. Induces interleukin-1. J Anticancer Res 12(5):1471–1474

Murphy BT, MacKinnon SL, Yan X, Hammond GB, Vaisberg AJ, Neto CC (2003) Identification of triterpene hydroxycinnamates with in vitro antitumor activity from whole cranberry fruit (*Vaccinium macrocarpon*). J Agric Food Chem 51:3541–3545

Murphy BT, Cao S, Norris A, Miller JS, Ratovoson F, Andriantsiferana R, Rasamison VE, Kingston DG (2005) Cytotoxic flavanones of *Schizolaena hystrix*from the Madagascar rainforest. J Nat Prod 68(3):417–419

Mustapha AA (2013) *Annona senegalensis* Persoon: a multipurpose shrub, its phytotherapic, phytopharmacological and phytomedicinal uses. Int J Sci Technol 2:862–865

Nagamine M, da Silva T, Matsuzaki P, Pinello K, Cogliati B, Pizzo C, Akisue G, Haraguchi M, Górniak S, Sinhorini I, Rao K, Barbuto J, Dagli M (2009) Cytotoxic effects of butanolic extract from *Pfaffia paniculata* (Brazilian Ginseng) on cultured human breast cancer cell line MCF-7. Exp Toxicol Pathol 61:75–82

Nair V, Kanfer I (2008) Sterols and sterolins in *Hypoxis hemerocallidea* (African potato). S Afr J Sci 104(6–7):322–324

Nair PK, Melnick SJ, Wnuk SF, Rapp M, Escalon E, Ramachandran C (2009) Isolation and characterization of an anticancer catechol compound from *Semecarpus anacardium*. J Ethnopharmacol 122:450–456

Nakao H, Senokuchi K, Umebayashi C, Kanemaru K, Masuda T, Oyama Y, Yonemori S (2004) Cytotoxic activity of maytanprine isolated from *Maytenus diversifolia* in human leukemia K562 cells. Biol Pharm Bull 27(8):1236–1240

Namba T, Bae KH (1982) Pharmacognostical studies on the crude drug "Zhu-Ye" and the Rambusaceous plants the botanical origins of the folk medicine "Kuma-Zasa" on the Japanese markets and the comparative anatomical studies of the leaves of the genus Sasa (Sections S). Shoyakugaku Zasshi 36:43–54

Narayan C, Kumar A (2013) Identification and characterization of phenolic compounds in hydro methanolic extract of *Achyranthes aspera* (HMEA) by UPLC and MALDI-TOF-MS and in vivo antioxidant activity. Orient Pharm Exp Med 13:51–59

Nasir NA, Bohari SP (2015) Cytotoxicity effects of *Typhonium flagelliforme* and *Clinacanthus nutans* on breast cancer cells. Jurnal Teknologi 77(31)

Nath R, Roy S, De B, Choudhury M (2013) Anticancer and antioxidant activity of Croton: a review. Int J Pharm Pharm Sci 5(2):63–70

Navarro E, Alonso SJ, Trujillo J, Jorge E, Perez C (2001) General behavior, toxicity and cytotoxic activity of elenoside, a lignan from *Justicia hyssopifolia*. J Nat Prod 64:134–135

Navarro E, Alonso SJ, Trujillo J, Jorge E, Pérez C (2004) Central nervous activity of elenoside. Phytomedicine 11:498–503

Nguyen AT, Fontaine J, Malonne H, Claeys M, Luhmer M, Duez P (2005) A sugar ester and an iridoid glycoside from *Scrophularia ningpoensis*. Phytochemistry 66:1186–1191

Nho C, Kang K, Lee S, Dulamjav B, Jigjidsuren T (2007) Pharmaceutical composition containing daurinol for the prevention and treatment of cancers. US8227512B2, 30 May 2007

Nho K, Chun J, Kim H (2011) Agrimonia pilosa ethanol extract induces apoptotic cell death in HepG2 cells. J Ethnopharmacol 138(2):358–363

Nho D, Lee H, Keum D, Lee G, Lee C, Lee J, Bae H (2013) Renoprotective effect of *Pulsatillae Radix* on cisplatin-induced nephrotoxicity in mice. Mol Cell Toxicol 9(4):393–399

Ni L, Zhong X, Chen X, Zhang B, Bao M, Cai X (2018) Bioactive norditerpenoids from *Cephalotaxus fortunei* var. alpina and *C. lanceolata*. Phytochemistry 151:50–60

Nibret E, Sporer F, Asres K, Wink M (2009) Antitrypanosomal and cytotoxic activities of pyrrolizidine alkaloid-producing plants of Ethiopia. J Pharm Pharmacol 61(6):801–808

Nile S, Park S (2013) HPTLC analysis, antioxidant, anti-inflammatory and antiproliferative activities of *Arisaema tortuosum*tuber extract. Pharma Bio 52:221–227

Nille G, Reddy K (2015) A phyto pharmacological review of plant – *Cassia auriculata*. Int J Pharm Bio Arch 6(6):1–9

Nirmala MJ, Samundeeswari A, Sankar PD (2011) Natural plant resources in anti-cancer therapy-A review. Res Plant Biol 1(3)

Nisa M, Akbar S, Tariq M (1985) Anti-inflammatory activity of *Cuscuta chinensis*. Fitoterapia 56:315–317

Nishibe S (1995) Bioactive phenolic compounds in traditional medicines. Pure Appl Chem 66:2263–2266

Noolvi M, Sharma R, Bhanot A (2011) Natural sources as potential anti-cancer agents: a review. Int J Phytomed 3:09–26

Norizan AH, Said IM, Latip J, Din LB, Syah YM, Hakim EH (2007) Styryldehydropyrone and clerodane-type diterpene from *Croton argyratus* the Malayasian. J Anal Sci 11(1):189–192

Noufou O, Wamtinga S, André T, Christine B, Marius L, Emmanuelle H, Jean K, Marie-Geneviève D, Pierre G (2012) Pharmacological properties and related constituents of stem bark of *Pterocarpus erinaceus* Poir. (Fabaceae). Asian Pac J Trop Med 5(1):46–51

Nozaki H, Hall IH, Kasai R, Hirayama T, Suzuki H, Wu RY (2010) Toosendanin inhibits hepatocellular carcinoma cells by inducing mitochondria-dependent apoptosis. Planta Med 76(13):1447–1453

Nozakia H, Hayashia K, Kidob M, Kakumotoa K, Ikedaa S, Matsuurac N, Tani H, Takaokad D, Iinumae M, AkaofY, Pauferrol A (2007) a novel chalcone trimer with a cyclobutane ring from *Caesalpinia ferrea* Mart exhibiting DNA topoisomerase II inhibition and apoptosis-inducing activity. Tetrahedron Lett 48(47):8290–8292

Nqounou FN, Lontsi D, Sandengam BL, Tsoupras MG, Tabacchi R (1990) Myrianthiphyllin, the first LignanCinnamate from *Myrianthus arboreus*. Planta Med 56:227–229

Nuanyai T, Chokpaiboon S, Vilaivan T, Pudhom K (2010) Cytotoxic 3,4-secocycloartane triterpenes from the exudate of *Gardenia tubifera*. J Nat Prod 73(1):51–54

Nürka N, Crockettb S (2012) Morphological and phytochemical diversity among Hypericum species of the Mediterranean basin. Med Aromat Plant Sci Biotechnol 5(1):14–28

Nzowa LK, Teponno RB, Tapondjou LA, Verotta L, Liao Z, Graham D, Zink MC, Barboni L (2013) Two new tryptophan derivatives from the seed kernels of *Entada rheedii*: effects on cell viability and HIV infectivity. Fitoterapia 87:37–42

Obiora Nwonuma C (2015) Effect of aqueous leaf extract of *Annona senegalensis* on selected testicular function indices of Wistar rats. Am J Life Sci 3:203–212

Ochwang'i D, Kimwele C, Oduma J, Gathumbi P, Mbaria J, Kiama S (2014) Medicinal plants used in treatment and management of cancer in Kakamega County, Kenya. J Ethnopharmacol 151(3):1040–1055

Ogbole O, Segun P, Adeniji A (2017) In vitro cytotoxic activity of medicinal plants from Nigeria ethnomedicine on Rhabdomyosarcoma cancer cell line and HPLC analysis of active extracts. BMC Complement Altern Med 17(1):494

Ogila K, Aduol O (2017) Protection of cyclophosphamide induced myelosuppression by extracts of *Asparagus setaceous* kunth and *Caesalpinia volkensii* harm in albino rats. Int J Pharm Pharmacol 1(3):114

Ogura M (1977) Potential anticancer agents vi. constituents of *Ailanthus excels* (*Simaroubaceae*). Lloydia 40(6):579–584

Ogwal-Okeng JW, Obua C, Anokbonggo WW (2003) Acute toxicity effects of the methanolic extract of *Fagara zanthoxyloides* (Lam.) root-bark. Afr Health Sci 3(3):124–126

Oh DR, Yoo JS, Kim Y, Kang H, Lee H, Lm SJ, Choi EJ, Jung MA, Bae D, Oh KN, Hong J (2018) *Vaccinium bracteatum* leaf extract reverses chronic restraint stress-induced depression-like behavior in mice: regulation of hypothalamic-pituitary-adrenal axis, serotonin turnover systems, and ERK/Akt phosphorylation. Front Pharmacol 9:604

Ohno N, Iino K, Suzuki I, Oikawa S, Sato K, Miyazaki T, Yadomae T (1985) Neutral and acidic antitumor polysaccharides extracted from cultured fruit bodies of *Grifola frondosa*. Chem Pharm Bull 33:1181–1186

Okoli CO, Akah PA, Onuoha NJ, Okoye TC, Nwoye AC, Nworu CS (2008) *Acanthus montanus*: an experimental evaluation of the antimicrobial, anti-inflammatory and immunological properties of a traditional remedy for furuncles. BMC Complement Altern Med 8(1):1–11

Olaleye MT (2007) Cytotoxicity and antibacterial activity of methanolic extract of *Hibiscus sabdariffa*. J Med Plant Res 1(1):9–13

Olas B, Skalski B, Ulanowska K (2018) The anticancer activity of Sea Buckthorn [*Elaeagnus rhamnoides* (L.) A. Nelson]. Front Pharmacol 9(2018):232

Oliveira S, Nascente P, Escareño J, Carvalho R, Piva E, Lund R (2013) Evaluation of anti-Candida activity and cytoxicity of *Jatropha dioica* Cerv. extracts. Afr J Microbiol Res 7(30):3903–3907

Oliveira M, Sales Junior P, Rodrigues M, DellaGreca M, Barreira L, Murta S, Romanha A, Custódio L (2016) Unlocking the in vitro anti-*Trypanosoma cruzi* activity of halophyte plants from the southern Portugal. Asia Pac J Trop Med 9(8):735–741

Olorunnisola OS, Adetutu A, Owoade AO, Adesina BT, Adegbola P (2017) Toxicity evaluation and protective effect of *Rhus longipes* Engl. leaf extract in paracetamol induced oxidative stress in wister rats. J Phytopharmacol 6(2):73–77

Omar R (2013) Compounds from A. *Platanoides* Bark, V. Corymbosum Roots & Topical Formulations using Maple Syrup. Open Access Dissertations. Paper 46. https://digitalcommons.uri.edu/oa_diss/46

Omino EA, Kokwaro JO (1993) Ethnobotany of Apocynaceae species in Kenya. J Ethnopharmacol 40:167–180

Omoregie EH, Oluyemisi KF, Koma OS, Ibumeh OJ (2012) Chemical constituents of the essential oil of *Laggera pterodonta* (DC.) sch. bip. from north-central Nigeria. J Appl Pharm Sci 2(8):198

On'Okoko P, Vanhaelen M, Vanhaelen-Fastre R, Declercq JP, Van Meerssche M (1985) Icacenone, a furanoditerpene with a pimarane skeleton from *Icacina mannii*. Phytochemistry 24:2452–2453

Onocha PA, Ali SM (2010) Insecticidal, antimicrobial, phyto and cytotoxicity of *Chassalia kolly* plant extract. Arch Appl Sci Res 2(4):151–156

Ostermann T, Raak C, Büssing A (2009) Survival of cancer patients treated with mistletoe extract (Iscador): a systematic literature review. BMC Cancer 9(1):451

Osunderu O (2017) Medicinal and economic values of forest products in the treatment of cancer in southwest Nigeria. Nat Prod Chem Res 5:6

Otsuki N, Dang NH, Kumagai E, Kondo A, Iwata S, Morimoto C (2010) Aqueous extract of *Carica papaya* leaves exhibits anti-tumor activity and immunomodulatory effects. J Ethnopharmacol 127:760–767

Ozaslan M, Karagoz ID, Kiliç IB, Cengiz B, Kalender MN, Guldur ME, Karagoz A, Zumrutdal ME (2009) Effect of *Plantago major* sap on Ehrlich ascites tumours in mice. Afr J Biotechnol 8(6):955–959

Özen HÇ, Başhan M, Toker Z, Keskin C (2004) 3-Hydroxy fatty acids from the flowers of *Hypericum lysimachioides* var. lysimachioides. Turk J Chem 28(2):223–226

Özenver N, Saeed M, Demirezer L, Efferth T (2018) Aloe-emodin as drug candidate for cancer therapy. Oncotarget 9(25):17770–17796

Pachon G, Rasoanaivo H, Azqueta A, Rakotozafy JC, Raharisololalao A, Lopez De Cerain A, De Lapuente J, Borra V, Moukha S, Centelles JJ, Creppy EE, Cascante M (2005) Anticancer effect of a new benzophenanthridine isolated from *Zanthoxylum madagascariense* (*Rutaceline*). In vivo 21:417–422

Palomino E, Maldonado C, Kempff MB, Ksebati MB (1996) Caparratriene, an active sesquiterpene hydrocarbon from *Ocotea caparrapi*. J Nat Prod 59:77–79

Pan PC (1960) The application of Chinese traditional medicine in the treatment of vaginal cancer. Chin J Obstet Gynaecol 8:156–162

Pan S, Litscher G, Gao S, Zhou S, Yu Z, Chen H, Zhang S, Tang M, Sun J, Ko K (2014) Historical perspective of traditional indigenous medical practices: the current renaissance and conservation of herbal resources. Evid Based Complement Alternat Med 2014:1–20

Pan Z, Fang Z, Lu W, Liu X, Zhang Y (2015) Osthole, a coumadin analog from *Cnidium monnieri* (L.) Cusson, stimulates corticosterone secretion by increasing steroidogenic enzyme expression in mouse Y1 adrenocortical tumor cells. J Ethnopharmacol 175(2015):456–462

Pandey S (2017) In vivo antitumor potential of extracts from different parts of *Bauhinia variegata* linn. Against b16f10 melanoma tumour model in c57bl/6 mice. Appl Cancer Res 37(1):33

Pandey K, Gupta H, Kamble B (2018) *Barleria prionitis*: journey from Ayurveda to modern medicine. Nat Prod J 8(2):109–130

Panigrahi SW, Sundaram Muthuraman M, Natesan RA, Pemiah BR (2014) Anticancer activity of ethanolic extract of *Solanum torvum* sw. Int J Pharm Pharm Sci 6:93–98

Pannell CM (1992) A taxonomic monograph of the genus *Aglaia* Lour (Meliaceae). Kew Bull 273(S16):S273–S283

Park H, Kim Y, Leem K, Park S, Seo J, Kim H, Chung J (2005) *Coptis japonica* root extract induces apoptosis through caspase 3 activation in SNU-668 human gastric cancer cells. Phytother Res 19(3):189–192

Park SU, Kim YK, Lee SY (2009) Establishment of hairy root culture of *Rubia akane* Nakai for alizarin and purpurin production. Sci Res Essays 4(2):94–97

Park B, Jung K, Son M, Seo J, Lee H, Lee J, Hong S (2012) Antitumor activity of *Pulsatilla koreana* extract in anaplastic thyroid cancer via apoptosis and anti-angiogenesis. Mol Med Rep 7(1):26–30

Park JY, Kim HY, Shibamoto T, Jang TS, Lee SC, Shim JS, Hahm DH, Lee HJ, Lee S, Kang KS (2017) Beneficial effects of a medicinal herb, *Cirsium japonicum* var. maackii, extract and its major component, cirsimaritin on breast cancer metastasis in MDA-MB-231 breast cancer cells. Bioorg Med Chem Lett 27(17):3968–3973

Pasqua G, Monacelli B, Valletta A (2004) Cellular localisation of the anti-cancer drug camptothecin in *Camptotheca acuminata* Decne (Nyssaceae). Eur J Histochem 48(3):321–327

Patel NM, Nozaki S, Shortle NH, Bhat-Nakshatri P, Newton TR, Rice S, Gelfanov V, Boswell SH, Goulet RJ Jr, Sledge GW Jr, Nakshatri H (2000) Paclitaxel sensitivity of breast cancer cells with constitutively active NF-kB is enhanced by IkBa super-repressor and parthenolide. Oncogene 19:4159–4169

Patel S, Gheewala N, Suthar A, Shah A (2009) In-vitro cytotoxicity activity of *Solanum nigrum* extract against Hela cell line and Vero cell line. Int J Pharm Sci 1(1):38–46

Patel C, Divakar K, Santani D, Solanki H, Thakkar J (2012) Remedial prospective of *Hippophae rhamnoides* Linn. (Sea Buckthorn). ISRN Pharmacol 2012:1–6

Patel J, Acharya S, Acharya N (2014) Clerodendrum serratum (L.) Moon. – a review on traditional uses, phytochemistry and pharmacological activities. J Ethnopharmacol 154(2):268–285

Paterna A, Kincses A, Spengler G, Mulhovo S, Molnár J, Ferreira M (2017) Dregamine and tabernaemontanine derivatives as ABCB1 modulators on resistant cancer cells. Eur J Med Chem 128:247–257

Pathmasiri W, El-Seedi HR, Han X, Janson JC, Huss U, Bohlin L (2005) Aryl Ketones from *Acronychia pedunculata* with cyclooxygenase-2 Inhibitory effects. Chem Biodivers 2(2):463–469

Patil SB, Naikwade MRSNS, Magdum CS (2009) Review on phytochemistry and pharmacological aspects of *Euphorbia hirta* linn. Asian J Pharm Res Health Care 1(1):113–133

Pattanayak S, Maity D, Mitra S, Debnath PK, Mandal TK, Bandyopadhyay SK (2013) Use of fresh parts of medicinal plants for health and production in livestock–a new concept of farming. Explor Anim Med Res 3(1):7–16

Patočka J, Jakl J (2010) Biomedically relevant chemical constituents of *Valeriana officinalis*. J Appl Biomed 8(1):11–18

Peng J, Lu Y, Chen D (1995a) Chemical constituents of Indian Mock Strawberry (*Duchesnea indica*). Chung TsaoYao 26:339–341

Peng JN, Feng XZ, Liang XT (1995b) A new Iridoid from *Hedyotis crysotrichae*. Chin Chem Lett 6:965–966

Peng YR, Li YB, Liu XD, Zhang JF, Duan JA (2008) Antitumor activity of C-21 steroidal glycosides from *Cynanchum auriculatum* Royle ex Wight. Phytomedicine 15:1016–1020

Peng B, Hu Q, Liu X, Wang L, Chang Q, Li J, Tang J, Wang N, Wang Y (2009) Duchesnea phenolic fraction inhibits *in vitro* and *in vivo* growth of cervical cancer through induction of apoptosis and cell cycle arrest. Exp Biol Med 234(1):74–83

Pereira CG, Carvalho JE, Meireles MA (2006) Anticancer activity of *Tabernaemontana catharinensis* extract obtained by supercritical fluid extraction. Braz J Med Biol Res 8(4):144–149

Peron A, Mariucci R, de Almeida I, Düsman E, Mantovani M, Vicentini V (2013) Evaluation of the cytotoxicity, mutagenicity and antimutagenicity of a natural antidepressant, *Hypericum perforatum* L. (St. John's wort), on vegetal and animal test systems. BMC Complement Altern Med 13(1):97

Pettit GR, Cragg GM, Herald DL, Schmidt JM, Lohavanuaya P (1982) Isolation and structure of Combretastatin. Can J Chem 60:1374–1376

Pettit GR, Meng Y, Gearing RP, Herald DL, Pettit RK, Doubek DL, Chapuis JC, Tackett LP (2004) Antineoplastic agents. 522. *Hernandia peltata* (Malaysia) and *Hernandia nymphaeifolia* (Republic of Maldives). J Nat Prod 2:214–220

Phadungkit M, Rattarom R, Rattana S (2010) Phytochemical screening, antioxidant, antibacterial and cytotoxic activities of *Knema angustifolia* extracts. J Med Plant Res 4(13):1269–1272

Phannasorn W, Khanaree C, Wongnoppavich A, Chewonarin T (2017) The effect of purple rice (*Oryza sativa* L. indica) extract on the inflammatory response in a colon cancer cell line and dextran sulfate-induced tumor promotion in the rat colon. Mol Cell Toxicol 13(4):433–442

Philip J (1981) Curry leaf, and its uses. World Crops:125–126

Phromnoi K, Reuter S, Sung B, Limtrakul P, Aggarwal BB (2010) A dihydroxy-pentamethoxy-flavone from *Gardenia obtusifolia* suppresses proliferation and promotes apoptosis of tumor cells through modulation of multiple cell signaling pathways. Anticancer Res 30:3599–3610

Piacente S, Belisario MA, Del Castillo H, Pizza C, DeFeo V (1998) Croton ruizianus: platelet proaggregating activity of two new prgnane glycoides. J Nat Prod 61:318–322

Piacente S, Montoro P, Oleszek W, Pizza C (2004) *Yucca schidigera* bark: phenolic constituents and antioxidant activity. J Nat Prod 67(5):882–885

Pieters L, De Bruyne T, Mei G, Lemiere G, Van denBerg D, Vlietinck AJ (1992) In vitro and in vivo biological activity of South American Dragon's Blood and its constituents. Planta Med 58:A582–A583

Pimm SL, Russell GJ, Gittleman JL, Brooks TM (1995) The future of biodiversity. Science 269(5222):347–350

Pinto DCG, Fuzzati N, Pazmino XC, Hostettmann K (1994) Xanthone and antifungal constituents from *Monnina obtusifolia*. Phytochemistry 37:875–878

Plotkin MJ, Schultes RE (1990) Virola, a promising genusfor ethnopharmacological investigation. J Psychoactive Drugs 22:357–361

Poornima P, Efferth T (2016) Ayurveda for cancer treatment. Med Aromat Plants (Los Angel) 5:e178

Potvin S, Bédard V, Béland F, Béland G, Abe K (2011) Nutriprotective diet. US20150140143A1, 9 Sept 2011

Prachayasittikul S, Suphapong S, Worachartcheewan A, Lawung R, Ruchirawat S, Prachayasittikul V (2009) Bioactive metabolites from *Spilanthes acmella* Murr. Molecules 14(4):850–867

Pradhan D, Panda PK, Tripathy G, Nayak JR, Pattanayak A (2009) Anticancer activity of biflavonoids from *Lonicera japonica* and *Benincasa hispida* on human cancer cell lines. J Pharm Res 2(5):983–985

Prajakta JP, Ghosh JS (2010) Antimicrobial activity of *Catharanthus roseus* – a detailed study. Br J Pharm Toxicol 1(1):40–44

Prameela Rani A, Nagarjuna Babu E, Prem Madhur S, Ravi Chandra Sekhara Reddy D, Phani Kumar K (2018) Evaluation of anticancer activity of *Parkinsonia Aculeata* leaves extract on Ehrlich's Ascites Carcinoma-Induced Mice. Asian J Pharm Clin Res 11(1):390–393

Prasad KN, Xie H, Hao J, Yang B, Qiu S, Wei X, Chen F, Jiang Y (2010) Antioxidant and anticancer activities of 8-hydroxypsoralen isolated from wampee [*Clausena lansium* (Lour.) Skeels] peel. Food Chem 118(1):62–66

Prasanna R, Harish CC, Pichai R, Sakthisekaran D, Gunasekaran P (2009) Anti-cancer effect of *Cassia auriculata* leaf extract in vitro through cell cycle arrest and induction of apoptosis in human breast and larynx cancer cell lines. Int J Biochem Cell Biol 33(2):127–134

Pratibha P, Sophia D, Perumal PC, Gopalakrishnan VK (2014) In-silico docking analysis of *Emilia sonchifolia* (l.) dc. gas chromatography-mass spectroscopy derived terpenoid compounds against pancreatic cancer targets (AKT and BRCA2). World J Pharm Pharm Sci 3:1844–1855

Pratyush K, Misra CS, James J, Dev LM, Veettil AK, Thankamani V (2011) Ethnobotanical and pharmacological study of Alstonia (Apocynaceae)-A review. J Pharm Sci Res 3(8):1394

Preza AM, Jaramillo ME, Puebla AM, Mateos JC, Hernández R, Lugo E (2010) Antitumor activity against murine lymphoma L5178Y model of proteins from cacao (*Theobroma cacao* L.) seeds in relation with in vitro antioxidant activity. BMC Complement Altern Med 10(1):61

Priya L, Priya KB, Subbaiah Kotakadi V, Josthna P (2015) Herbal and medicinal plants molecules towards treatment of cancer: a mini review. Am J Ethnomed 2:136–142

Promraksa B, Daduang J, Chaiyarit P, Tavichakorntrakool R, Khampitak T, Rattanata N, Tangrassameeprasert R, Boonsiri P (2015) Cytotoxicity of *Cratoxylum Formosum* Subsp. Pruniflorum gogel extracts in oral cancer cell lines. Asian Pac J Can Prev 16(16):7155–7159

Puapairoj P, Naengchomnong W, Kijjoa A, Pinto MM, Pedro M, Nascimento MS, Silva AM, Herz W (2005) Cytotoxic activity of lupane-type triterpenes from *Glochidion sphaerogynum* and *Glochidion eriocarpum* two of which induce apoptosis. Planta Med 71(3):208–213

Pullela SV, Choi YW, Khan SI, Khan IA (2005) New acylated clionasterol glycosides from*Valeriana officinalis*. Planta Med 71(10):960–961

Puthur S, Anoopkumar A, Rebello S, Aneesh E (2018) *Hypericum japonicum:* a double-headed sword to combat vector control and cancer. Appl Biochem Biotech:1–11

Qari SH (2008) Assessment of antimutagenic and genotoxic potential of *Origanum majorana* aqueous extract using in vitro assays. Saudi J Biol Sci 15(2):207–212

Qian TX, Li LN (1992) Isosalvianolic acid C, a depsidepossessing a dibenzooxepin skeleton. Phytochemistry 31:1068–1070

Qiu B, Jiang W, Qiu W, Mu W, Qin Y, Zhu Y, Zhang J, Wang Q, Liu D, Qu Z (2018) Pine needle oil induces G2/M arrest of HepG2 cells by activating the ATM pathway. Exp Ther Med 15(2):1975–1981

Quattrocchi U (2012) CRC world dictionary of medicinal and poisonous plants: common names, scientific names, eponyms, synonyms, and etymology. M-Q, Volume 4. CRC Press/Taylor & Francis, p 2446

Quílez AM, Fernández-Arche MA, García-Giménez MD, De la Puerta R (2018) Potential therapeutic applications of the genus Annona: Local and traditional uses and pharmacology. J Ethnopharmacol 225:244–270

Quintans JDSS, Soares BM, Ferraz RPC, Oliveira ACA, da Silva TB, Menezes LRA, Sampaio MFC, Prata APDN, Moraes MO, Pessoa C, Antoniolli AR, Costa EV, Bezerra DP (2013) Chemical constituents and anticancer effects of the essential oil from leaves of Xylopia laevigata. Planta Med 79:123–130

Rabi T, Gupta RC (1995) Antitumor and cytotoxic investigation of *Amoora rohituka*. Int J Pharm 33:359–361

Rachana SB, Mamta P, Priyanka KM, Sonam S (2011) Review and future perspectives of using vasicine and related compounds. J Pharm Sci 1(1):85–98

Rafieian-Kopaei M, Movahedi M (2017) Breast cancer chemopreventive and chemotherapeutic effects of Camellia Sinensis (green tea): an updated review. Electron Physician 9(2):3838–3844

Rahman S, Parvin R (2014) Therapeutic potential of *Aegle marmelos* (L.) – an overview. Asian Pac J Trop Dis 4(1):71–77

Rahman MM, Khan A, Haque ME, Rahman MM (2009) Antimicrobial and cytotoxic activities of *Laportea crenulata*. Fitoterapia 79(7):584–586

Rahmatullah M, Samarrai W, Jahan R, Rahman S, Sharmin N, Miajee EU, Chowdhury MH, Bari S, Jamal F, Bashar AM, Azad AK (2010) An ethnomedicinal, pharmacological and phytochemical review of some Bignoniaceae family plants and a description of Bignoniaceae plants in folk medicinal uses in Bangladesh. Adv Nat Appl Sci 4(3):236–253

Rai K, Rasal V, Chanu M (2015) Efficacy of chemopreventive agent, *Dipteracanthus prostratus*, on 1,2-dimethylhydrazine induced colon carcinogenesis. J Sci Soc 42:144

Rajabalian S (2008) Methanolic extract of *Teucrium polium* L. potentiates the cytotoxic and apoptotic effects of anticancer drugs of vincristine, vinblastine and doxorubicin against a panel of cancerous cell lines. Exp Oncol 30(2):133–138

Rajagopal S, Kumar RA, Deevi DS, Satyanarayana C, Rajagopalan R (2003) Andrographolide, a potential cancer therapeutic agent isolated from *Andrographis paniculata*. J Exp Ther Oncol 3(3):147–158

Rajchal R (2009) Seabuckthorn (*Hippophae salicifolia*) management guide. The Rufford small grants for nature conservation 1:1–45

Rajkapoor B, Jayakar B, Murugesh N (2003) Antitumour activity of *Bauhinia variegata* against Ehrlich ascites carcinoma induced mice. J Ethnopharmacol 41(8):604–607

Rajkumar V, Guha G, Kumar RA (2010) Antioxidant and anti-cancer potentials of Rheum emodi rhizome extracts. Evid Based Complement Alternat Med 36(7):1–7

Raju S, Kavimani S, Uma M, Sreeramulu R (2011) Kyllinga nemoralis (Hutch & Dalz) (Cyperaceae): ethnobotany, phytochemistry and pharmacology. Pharmacogn J 3:7–10

Ramalakshmi S, Muthuchelian K (2011) Analysis of bioactive constituents from the leavs of *Mallotus tetracoccus* (roxb.) Kurz, by gas chromatography – mass spectrometry. Int J Pharm Sci Res 2(6):1449–1454

Ramanauskienė K, Stelmakiene A, Majienė D (2015) Assessment of lemon balm (*Melissa officinalis* L.) hydrogels: quality and bioactivity in skin cells. Evid Based Complement Alternat Med 2015

Ramirez VR, Mostacero LJ, Garcia AE, Mejia CF, Pelaez PF, Medina CD, Miranda CH (1988) Vegetalesempleados en medicina tradicional Norperuana. Banco Agrario del Peru and Universidad Nacional deTrujillo, Trujillo, Peru, p 54

Ramírez J, Gilardoni G, Ramón E, Tosi S, Picco AM, Bicchi C, Vidari G (2018) Phytochemical study of the ecuadorian species *Lepechinia mutica* (Benth.) Epling and high antifungal activity of carnosol against *Pyricularia oryzae*. Pharmaceuticals 11(2):33

Ramírez-Erosa I, Huang Y, Hickie RA, Sutherland RG, Barl B (2007) Xanthatin and xanthinosin from the burs of *Xanthium strumarium* L. as potential anticancer agents. Can J Physiol Pharmacol 85(11):1160–1172

Ramljak D, Romanczyk LJ, Metheny-Barlow LJ, Thompson N, Knezevic V, Galperin M, Ramesh A, Dickson RB (2005) Pentameric procyanidin from *Theobroma cacao* selectively inhibits growth of human breast cancer cells. Mol Cancer Ther 4(4):537–546

RamosSilva A (2017) Anticancer potential of thevetia peruviana fruit methanolic extract. BMC Complement Altern Med 17(1):241

Ramteke A, Hussain A (2012) Flower extract of *Nyctanthes arbor*-tristis modulates glutathione level in hydrogen peroxide treated lymphocytes. Pharmacogn Res 4(4):230

Rana V, Verdeguer M, Blazquez M (2012) Chemical composition of the essential oil of *Zingiber zerumbet* var. darcyi. Nat Prod Commun 7(10):1369–1370

Rangkadilok N, Worasuttayangkurn L, Bennett RN, Satayavivad J (2005) Identification and quantification of polyphenolic compounds in Longan (*Euphoria longana* Lam.) fruit. J Agric Food Chem 53(5):1387–1392

Rangkaew N, Suttisri R, Moriyasu M, Kawanishi K (2009) A new acyclic diterpene acid and bioactive compounds from *Knema glauca*. Arch Pharm Res 32(5):685–692

Ranjbarnejad T, Saidijam M, Moradkhani S, Najafi R (2017) Methanolic extract of *Boswellia serrata* exhibits anti-cancer activities by targeting microsomal prostaglandin E synthase-1 in human colon cancer cells. Prostaglandins Other Lipid Mediat 131:1–8

Rao KV, McBride TJ, Oleson JJ (1968) Recognition and evaluation of lapachol as an antitumor agent. Cancer Res 28:1952–1954

Rascón-Valenzuela L, Jiménez-Estrada M, Velázquez-Contreras C, Garibay-Escobar A, Vilegas W, Campaner L, Coqueiro A, Robles-Zepeda R (2015) Chemical composition and antiproliferative activity of *Acalypha californica*. Ind Crop Prod 69:48–54

Raval BP, Suthar MP, Patel RK (2009) Potent in vitro anti-tumor activity of *Symplocos racemosa* against leukemia and cervical cancer. Electron J Biol 5(4):89–91

Ravikanth V, Reddy LNV, Reddy VA, Ravinder K, Rao PT, Ram ST, Kumar AK, Vamanarao PD, Venkateswarlu Y (2003) Three new ingol diterpenes from *Euphorbia nivulia*: evaluation of cytotoxic activity. Chem Pharm Bull 51:431–434

Ravindranath MH, Saravanan TS, Monteclaro CC, Presser N, Ye X, Selvan SR, Brosman SE (2006) Epicatechins purified from green tea (*Camellia sinensis*) differentially suppress growth of gender-dependent human cancer cell lines. Evid Based Complement Alternat Med 3(2):237–247

Reboucas SDO, Grivicich I, Santos MSD, Rodriguez P, Gomes MD, Oliveira SQD, Silva JD, Ferraza ADBF (2011) Antiproliferative effect of a traditional remedy, *Himatanthus articulatus* bark, on human cancer cell lines. J Ethnopharmacol 137(1):926–929

Reddy D (2016) Phytochemical analysis of active constituents of *Alstonia scholaris* and their cytotoxicity in vitro. Int J Pharm Sci Res 7(8):3262–3273

Reddy S, Mani R (2010) G1 arrest and caspase-mediated apoptosis in HL-60 cells by Dichloromethane extract of *Centrosema pubescens*. Am J Chin Med 38(6):61–163

Reddy VVS, Sirsi M (1969) Effect of *Abrus precatorius* L. on experimental tumors. Cancer Res 29:1447–1451

Reddy MB, Reddy KR, Reddy MN (1991) Ethnobotany of Cuddapah district, Andhra Pradesh, India. Int J Pharm 29:273–280

Reddy K, Kumar A, Lakshmi S, Angothu S (2010) Antioxidant properties of methanolic extract of *Oxalis corniculata*. Int J Phytopharmacol 1(1):43–46

Rehecho S, Uriarte-Pueyo I, Calvo J, Vivas LA, Calvo MI (2011) Ethnopharmacological survey of medicinal plants in Nor-Yauyos, a part of the Landscape Reserve Nor-Yauyos-Cochas, Peru. J Ethnopharmacol 133(1):75–85

Reid EJ, Betts TJ (1979) Records of Western Australian plants used by aboriginals as medicinal agents. Planta Med 36:164–173

Rejinold NS, Muthunarayanan M, Muthuchelian K, Chennazhi KP, Nair SV, Jayakumar R (2011) Saponin-loaded chitosan nanoparticles and their cytotoxicity to cancer cell lines in vitro. Carbohydr Polym 84(1):407–416

Rejiya CS, Cibin TR, Abraham A (2009) Leaves of *Cassia tora* as a novel cancer therapeutic – an in vitro study. Toxicol in Vitro 23:1034–1038

Rekha JB, Jayakar B (2011) Anti cancer activity of ethanolic extract of leaves of *Plumeria rubra* (Linn). Current Pharm Res 1(2):175

Ren Y, Shen L, Zhang DW, Dai SJ (2009) Two new sesquiterpenoids from *Solanum lyratum* with cytotoxic activities. Chem Pharm Bull 57(4):408–410

Renouard S, Lopez T, Hendrawati O (2011) Podophyllotoxin and deoxypodophyllotoxin in *Juniperus bermudiana* and 12 other *Juniperus* species: optimization of extraction, method validation, and quantification. J Agric Food Chem 59:8101–8107

Reyes-Zurita FJ, Rufino-Palomares EE, Lupianez JA, Cascante M (2009) Maslinic acid, a natural triterpene from *Olea europaea* L., induces apoptosis in HT29 human coloncancer cells via the mitochondrial apoptotic pathway. Cancer Lett 273(1):44–54

Rhan LI, Yong SJ, Sil LY (1992) Cytotoxicity of folkloric medicine in murine and human cancer cell. Korean J Pharma 23:62

Riaz N, Nawaz SA, Mukhtar N, Malik A, Afza N, Ali S, Ullah S, Choudhary MP (2007) Isolation and enzyme-inhibition studies of the chemical constituents from *Ajuga bracteosa*. Chem Biodivers 4(1):72–83

Riaz A, Rasul A, Hussain G, et al (2018) Astragalin: a bioactive Phytochemical with potential therapeutic activities. Adv Pharmacol Sci 2018, Article ID: 9794625, 1–15

Ribeiro-Varandas E, Ressurreição F, Viegas W, Delgado M (2014) Cytotoxicity of *Eupatorium cannabinum* L. ethanolic extract against colon cancer cells and interactions with Bisphenol A and Doxorubicin. BMC Complement Altern Med 14(1):264

Rivera D, Obon C (1995) The ethnopharmacology of Madeira and Porto Santo islands, a review. J Ethnopharmacol 46:73–93

Rivero-Cruz JF, Lezutekong R, Lobo-Echeverri T, Ito A, Mi Q, Chai HB, Soejarto DD, Cordell GA, Pezzuto JM, Swanson SM, Morelli I, Kinghorn AD (2005) Cytotoxic constituents of the twigs of *Simarouba glauca* collected from a plot in Southern Florida. Phytother Res 19(2):136–140

Robinson A, Kumar TV, Sreedhar E, Naidu VG, Krishna SR, Babu KS, Srinivas PV, Rao JM (2008) A new sesquiterpene lactone from the roots of *Saussurea lappa*: structure-anticancer activity study. Bioorg Med Chem Lett 18(14):401–457

Rodrigues M, Gangadhar K, Vizetto-Duarte C, Wubshet S, Nyberg N, Barreira L, Varela J, Custódio L (2014) Maritime halophyte species from Southern Portugal as sources of bioactive molecules. Mar Drugs 12(4):2228–2244

Roeder E, Liu K (1991) Pyrrolizidine alkaloids from *Senecio integrifolius varfauriri*. Phytochemistry 30:1734–1737

Romanucci V, D'Alonzo D, Guaragna A, Di Marino C, Davinelli S, Scapagnini G, Di Fabio G, Zarrelli A (2016) Bioactive compounds of Aristotelia chilensis Stuntz and their pharmacological effects. Curr Pharm Biotech 17:513–523

Romeilah RM (2009) Anticancer and antioxidant activities of *Matricaria chamomilla* and *Marjorana hortensis* essential oils. Res J Med Sci 4(2):332–339

Rossini C, Menéndez P, Dellacassa E, Moyna P (1996) Essential oils from leaves of Schinus molle and S. lentiscifolius of Uruguayan origin. J Essent Oil Res 8(1):71–73

Roy MC, Chang FR, Huang HC, Chiang MYN, Wu YC (2005) Cytotoxic principles from the formosan milkweed, *Asclepias curassavica*. J Nat Prod 68(10):1494–1499

Rubnov S, Kashman Y, Rabinowit R, Schlesinger M, Mechoulam R (2001) Suppressors of cancer cell proliferation from fig (*Ficus carica*) resin: isolation and structure elucidation. J Nat Prod 64:993–996

Sabandar CW, Ahmat N, Mohd Jaafar F, Sahidin I (2012) Medicinal property, phytochemistry and pharmacology of several Jatropha species (Euphorbiaceae): a review. Phytochemistry. (Article in press)

Sabitha G, Reddy CS, Yadav JS (2006) Total synthesis of the highly potent anti-cancer polyacetylenes, (S)-18-hydroxyminquartynoic acid, (S)-minquartynoic acid and (E)-15,16-dihydrominquartynoic acid. Tetrahedron Lett 47(2):4513–4516

Sabnis SD, Bedi SJ (1983) Ethnobotanical studies in Dadra-Nagar Haveli and Daman. Indian J Forestry 6:65–69

Saboo S, Deore SL, Khadabadi SS, Deokate UA (2007) Evaluation of antimitotic and anticancer activity of the crude extracts of *Pterospermum acerifolium* wild leaves (sterculiaceae). Niger J Nat Prod Med 11(1):76–79

Sadeghi H, Yazdanparast R (2003) Anti-tumor activity and cell cycle arrest of a new diterpene ester from *Daphne mucronata* using K562 cells. Arch SID 7:127–131

Sadeghi-Aliabadi A, Sajjadib SE, Khodamoradic M (2009) Cytotoxicity of *Euphorbia macroclada* on MDA-MB-468 breast cancer cell line. Iran J Pharm Sci 5(2):103–108

Sadeghnia H, Ghorbani Hesari T, Mortazavian S, Mousavi S, Tayarani-Najaran Z, Ghorbani A (2014) Viola tricolor Induces apoptosis in cancer cells and exhibits antiangiogenic activity on chicken chorioallantoic membrane. BioMed Res Inter 2014(1):1–8

Saeidnia S (2015) New approaches to natural anticancer drugs. Springer, Cham, pp 18–20

Sagnia B (2014) Correction: antioxidant and anti-inflammatory activities of extracts from *Cassia alata*, *Eleusine indica*, *Eremomastax speciosa*, *Carica papaya* and *Polyscias fulva* medicinal plants collected in Cameroon. PLoS ONE 9:e112573

Sagnia B, Fedeli D, Casetti R, Montesano C, Falcioni G, Colizzi V (2014) Antioxidant and anti-inflammatory activities of extracts from *Cassia alata*, *Eleusine indica*, *Eremomastax speciosa*, *Carica papaya* and *Polyscias fulva* medicinal plants collected in Cameroon. PLoS One 9(8):e103999

Sagwan S, Rao DV, Sharma RA (2011) In-vitro antioxidant activity of *Maytenus emarginata* (Willd.) using DPPH assay. J Pharm Res 4(7):2276–2278

Saha S, Paul S (2017) Potential of *Hygrophila auriculata* (Schumach.) heine as a source of future anti-cancer drugs: a comprehensive review. J Pharmacogn Phytother 6(4):1725–1740

Saied S, Batool F, Naz S (2011) Two new diarylesters from *Ipomoea cairica*. J Basic Appl Sci 7(2):97–99

Saiki Y, Ishikawa Y, Uchida M, Fukushima S (1978) Essential oil from Chinese drug Caodoukou, the seeds of *Alpinia katsumadai*. Phytochemistry 17:808–809

Saint-Lary L, Roy C, Paris J, Tournayre P, Berdagué J, Thomas O, Fernandez X (2014) Volatile compounds of *Viola odorata* absolutes: identification of odorant active markers to distinguish plants originating from France and Egypt. Chem Biodivers 11(6):843–860

Sakagami H, Oh-Hara T, Kaiya T, Kawazoe Y, Nonoyama M, Konno K (1989) Molecular species of the antitumor and antiviral fraction from pine cone extract. Anticancer Res 9:1593–1598

Sakagami H, Konno K, Kawazoe Y, Lai P, Nonoyama M (1992) Multiple immunological functions of extractfrom the cone of Japanese white pine, *Pinus parviflora* Sieb.et Zucc. Adv Exp Med Biol 319:331–335

Sakamoto S, Kuroyanagi M, Ueno A, Sekita S (1992) Triterpenoid saponins from *Sophora subprostrata*. Phytochemistry 3:1339–1342

Sakpakdeejaron I, Itharat A (2009) Cytotoxic compounds against breast adenocarcinoma cells (MCF-7) from pikutbenjakul. J Health Res 23(2):71–76

Sakthive K, Kannan N, Angeline A, Guruvayoorappan C (2012) Anticancer activity of *Acacia nilotica* (L.) Wild. Ex. Delile Subsp. indica against Dalton's ascitic lymphoma induced solid and ascitic tumor model. Asian Pacif J Can Prev 13(8):3989–3995

Salama M, El-Hawary S, Mousa O, El-Askary N, Esmat A (2015) In vivo TNF-and IL-1 inhibitory activity of phenolics isolated from *Trachelospermum jasminoides* (Lindl.) Lem. J Med Plants Res 9(2):30–41

Salie F, Eagles PFK, Leng HMJ (1996) Preliminary antimicrobial screening of four South African *Asteraceae* species. J Ethnopharmacol 52:27–33

Salimi M, Majd A, Sepahdar Z, Azadmanesh K, Irian S, Ardestaniyan M, Hedayati M, Rastkari N (2012) Cytotoxicity effects of various *Juglans regia* (walnut) leaf extracts in human cancer cell lines. Pharm Biol 50(11):1416–1422

Samal J (2016) Medicinal plants and related developments in India: a peep into 5-year plans of India. Indian J Med Sci 9(1):14–19

Samarghandian S, Hadjzadeh M, Afshari J, Hosseini M (2014) Antiproliferative activity and induction of apoptotic by ethanolic extract of Alpinia galanga rhizhome in human breast carcinoma cell line. BMC Complement Altern Med 14(192):1–9

San Martin AJ (1983) Medicinal plants in Central Chile. Econ Bot 37:216–227

Sánchez-Carranza J, Alvarez L, Marquina-Bahena S, Salas-Vidal E, Cuevas V, Jiménez E, Veloz GR, Carraz M, González-Maya L (2017) Phenolic compounds isolated from *Caesalpinia coriaria* induce S and G2/M phase cell cycle arrest differentially and trigger cell death by interfering with microtubule dynamics in cancer cell lines. Mole 22(4):666

Sanchez-Hernandez I, Barragan-Alvarez C, Torres Gonzalez O, Padilla-Camberos E (2017) Nutraceutical potential of *Cnidoscolus aconitifolius*. ARC J Nutr Growth 3(2):27–30

Sandoval M, Okuhama NN, Zhang XJ, Condezo LA, Lao J, Angeles FM, Musah RA, Bobrowski P, Miller MJ (2002a) Anti-inflammatory and antioxidant activities of cat's claw (*Uncaria tomentosa* and *Uncaria guianensis*) are independent of their alkaloidcontent. Phytomedicine 9:325–337

Sandoval M, Okuhama NN, Clark M, Angeles FM, Lao J, Bustamante S, Miller MJ (2002b) Sangre de Grado *Croton palanostigma* induces apoptosis in human gastrointestinal cancer cells. J Ethnopharmacol 80(2–3):121–129

Sandur SK, Ichikawa H, Sethi G, Ahn KS, Aggarwal BB (2006) Plumbagin (5-Hydroxy-2- methyl-1,4-naphthoquinone) suppresses NF-B Activation and NF-B-regulated Gene products through modulation of p65 and I-B Kinase activation, leading to potentiation of apoptosis induced by cytokine and chemotherapeutic agents. J Biol Chem 281(2):17023–17033

Sankaram AVB, Rao GLN (1978) Bharangin. a novelditerpenoid quinonemethide from *Pygmacopremnaherbaceae*. In: Proceedings of IUPAC 11th international symposium of chemical natural products, vol 2, pp 97–100

Santos LP, Boaventura MA, Sun NJ, Cassady JM, De Oliveira AB (1996) Araticulin, a bis-tetrahydrofuran polyketide from Annona itcrassiflora seeds. Phytochemistry 42(3):705–707

Sarg TM (1975) Phytochemical investigation of *Pulicariacrispa* growing in Saudi Arabia. Egypt J Pharm Sci 16:421–426

Sarg TM, Abbas FA, El-Sayed ZI, Mustafa AM (2011) Two new polyphenolic compounds from *Ficus retusa* L. "variegata" and the biological activity of the different plant extracts. J Pharmacogn Phytother 3(7):346–348

Sati OP, Bahuguna S, Uniyal S, Sakakibara J, Kaiya T, Nakamura A (1990) A new saponin from *Deeringia amaranthoides*. J Nat Prod 53:466–469

Savigni D (2010) Evaluation of the anticancer potential of plants used as traditional medicines by Aboriginal people, pp 82–108

Savini I, Arnone R, Catani MV, Avigliano L (2009) *Origanum vulgare* induces apoptosis in human colon cancer caco2 cells. Nutr Cancer 61(3):381–389

Saxena VK, Dixit G (1993) Costunolide and dihydrocostunolide from *Saussurea lappa*. Fitoterapia 64:84–88

Schacter L (1996) Etoposide phosphate: what, why, where, and how? Semin Oncol 6(13):1–7

Schmelzer GH (2012) *Sarcophyte sanguinea* Sparrm. In: Schmelzer GH, Gurib-Fakim A (eds) Prota 11(2): Medicinal plants/Plantes médicinales 2. PROTA, Wageningen

Schoen-Angerer T, Wilkens J, Kienle G, Kiene H, Vagedes J (2015) High-dose *Viscum album* extract treatment in the prevention of recurrent bladder cancer: a retrospective case series. Perm J 19(4):76–83

Schultes RE (1979) De plantis toxicariie e mundo novotropicale commentationes. XIX. Biodynamic apocynaceous plants of the Northwest Amazon. J Ethnopharmacol 1:165–192

Seangphakdeea P, Pompimona W, Meepowpanb P, Panthongc A, Chiranthanutd N, Banjerdpongchaie R, Wudtiwaie B, Nuntasaenf N, Pitchuanchoma S (2013) Anti-inflammatory and anti-cancer activities of (-)-zeylenol from stems of *Uvaria grandiflora*. Sci Asia 39:610–614

Seca A, Pinto D (2018) Plant secondary metabolites as anticancer agents: successes in clinical trials and therapeutic application. Int J Mol Sci 19(1):263

Segura-Correa R, Mata R, Anaya AL, Hernandez-Bautista B, Villena R, Soriano-Garcis M, Bye R, Linares E (1993) New tetranortriterpenoids from *Swietenia humilis*. J Nat Prod 56:1567–1574

Senatore F, Rigano D, Formisano C, Grassia A, Basile A, Sorbo S (2007) Phyto growth inhibitory and antibacterial activity of *Verbascum sinuatum*. Fitoterapia 78:244–247

Seo EK, Kim NC, Mi Q, Chai H, Wall ME, Wani MC, Navarro HA, Burgess JP, Graham JG, Cabieses F, Tan GT, Farnsworth NR, Pezzuto JM, Kinghorn AD, Macharistol (2001) A new cytotoxic cinnamylphenol from the stems of *Machaerium aristulatum*. J Nat Prod 64(11):1483–1485

Seo Y, Berger JM, Hoch J, Neddermann KM, Bursuker I, Mamber SW, Kingston DG (2002) A new triterpene saponin from *Pittosporum viridiflorum*from the Madagascar rainforest. J Nat Prod 65(1):65–68

Seo JM, Kang HM, Son KH, Kim JH, Lee CW, Kim HM, Chang SI, Kwon BM (2003) Antitumor activity of flavones isolated from *Artemisia argyi*. Planta Med 69(3):218–222

Seo JU, Kim MH, Kim HM, Jeong HJ (2011) Anticancer potential of magnolol for lung cancer treatment. Arch Pharm Res 34(4):625–633

Sermakkani M, Thangapandian V (2010) In vitro cytotoxicity of *Cassia italica* Miller. Int J Pharma Res Dev 2(9):99–106

Shabani A (2016) A review of anticancer properties of herbal medicines. J Pharm Care Health Syst 3(2):1–4

Shafaat-Al-Mehedi M, Hasan CM, Haque MR (2015) Isolation of flavonoids from the bark of *Entada rheedii* Spreng. Orient Pharm Exp Med 15(4):347–351

Shahat A, Alsaid M, Kotob S, Ahmed H (2015) Significance of *Rumex Vesicarius* as anticancer remedy against hepatocellular carcinoma: a proposal-based on experimental animal studies. Asian Pac J Cancer Prev 16(10):4303–4310

Shamona LA (1997) Evaluation of the mutagenic, cytotoxic, and antitumor potential of triptolide, a highly oxygenated diterpene isolated from *Tripterygium wilfovdii*. Cancer Lett 112:113–117

Shangary S, Singh J, Kamboj S, Kamboj K, Sandhu R (1995) Purification and properties of four monocot lectins from the family araceae. Phytochemistry 40(2):449–455

Sharma T, Jana S (2019) Boswellic acids as natural anticancer medicine: precious gift to humankind. J Herbal Med. (In press)

Sharma MP, Ahmad J, Hussain A, Khan S (1992) Folklore medicinal plants of Mewat (Gurgaon district), Haryana, India. Int J Pharm 30:135–137

Sharma AR, Gescher JA, Steward PW (2005) Curcumin: the story so far. Eur J Cancer 41:1955–1968

Sharma P, Mondhe D, Muthiah S, Pal H, Shahi A, Saxena A, Qazi G (2009) Anticancer activity of an essential oil from *Cymbopogon flexuosus*. Chem Biol Interact 179(2–3):160–168

She GM, Xiao X, Ba YY, Shi RB, Zhang LZ (2013) Chemical Constituents of *Dioscorea collettii* var. hyplauca. Chem Nat Compd 49(5):983–984

Shen JH, Zhou BN (1992) Studies on diterpene-quinonesof *Tripterygium regelii* Sprague. Chin Chem Lett 3:113–116

Shen T, Xie WD, Jia ZJ (2005) Two new sesquiterpenes from *Ligularia macrophylla*. Chin Chem Lett 16(9):1220–1222

Shen X, Eichhorn T, Greten HJ, Efferth T (2012) Effects of *Scrophularia ningpoensis* Hemsl. on inhibition of proliferation, apoptosis induction and NF-κB signaling of immortalized and cancer cell lines. Pharmaceuticals 5(2):189–208

Shen T, Duan C, Chen B, Li M, Ruan Y, Xu D, Shi D, Yu D, Li J, Wang C (2017) *Tremella fuciformis* polysaccharide suppresses hydrogen peroxide-triggered injury of human skin fibroblasts via upregulation of SIRT1. Mol Med Rep 16(2):1340–1346

Shenouda NS, Sakla MS, Newton LG, Besch-Williford C, Greenberg NM, MacDonald RS, Lubahn DB (2007) Phytosterol *Pygeum africanum* regulates prostate cancer in vitro and in vivo. Endocrine 31(1):72–81

Shepherd R (2012) Ginseng for cancer patients says Mayo Clinic. Med News Today

Shi JG, Kua ZJ, Yang L (1994) Lathyrane and ingenane diterpenoids from *Euphorbia micractina*. Planta Med 60:588–589

Shi Y, Zhao M, Yao H, Yang P, Xin T, Li B, Sun W, Chen S (2017) Rapidly discriminate commercial medicinal *Pulsatilla chinensis* (Bge.) Regel from its adulterants using ITS2 barcoding and specific PCR-RFLP assay. Sci Rep 7(2017):40000

Shim S, Lee M, Lee K (2016) *Achyranthes japonica* Nakai water extract suppresses binding of IgE antibody to cell surface FcεRI. Prev Nutr Food Sci 21:323–329

Shimizu M, Shingyouchi K, Morita N, Kizu H, Tomimori T (1978) Triterpenoid saponins from *Pulsatilla cernua*.I. Chem Pharm Bull 26:1666–1668

Shirataki S, Wakae M, Yamamoto Y, Hashimoto K, Satoh K, Ishihara M, Kikuchi H, Nishikawa H, Minagawa K, Motohashi N, Sakagami H (2004) Cytotoxicty and radical modulating activity of isoflavones and isoflavanones from *Sophora species*. Anticancer Res 24:1481–1488

Shirota O, Morita H, Takeya K, Itokawa H (1995) Structures of xuxuarines, stereoisomeric triterpene dimmers from *Maytenus chuchuhuasca*. Tetrahedron 51:1107–1120

Shirota O, Tamemura T, Morita H, Takeya K, Itokawa H (1996) Triterpenes from Brazilian medicinal plant "Chuchuhuasi"(*Maytenus krukovii*). J Nat Prod 59(11):1072–1075

Shirota O, Morita H, Takeya K, Itokawa H (1998) Isolation of antitumor substance, dulcitol, from *Maytenus ebenifolia*. Nat Med 52(2):184–186

Shoeb M, Celik S, Jaspar M, Kumarasamy Y, MacManus SM, Nahar L, Thoo-Lind PK, Sarker SD (2005) Isolation, structure elucidation and bioactivity of schischkiniin, a unique indole alkaloid from the seeds of Centaurea *schischkinii*. Tetrahedron Lett 61:9001–9006

Shoeb M, Jaspars M, MacManus SM, Celik S, Nahar L, Kong-Thoo-Lin PP, Sarker SD (2007a) Anti-colon cancer potential of phenolic compounds from the aerial parts of *Centaurea gigantea* (Asteraceae). J Nat Med 61:164–169

Shoeb M, Jaspars M, MacManus SM, Celik S, Nahar L, Kong-Thoo-Lin PP, Sarker SD (2007b) Montamine, a unique dimeric indole alkaloid, from the seeds of *Centaurea montana* (Asteraceae), and its *in vitro* cytotoxic activity against theCaCo2 colon cancer cells. Tetrahedron Lett 62:11172–11177

Shoji N, Umeyama A, Takemoto T, Ohizumi Y (1984) Serotonergic receptor antagonist from *Nandina domestica* Thunberg. J Pharm Sci 73:568–570

Shylesh B, Nair SA, Subramoniam A (2005) Induction of cell-specific apoptosis and protection from Dalton's lymphoma challenge in mice by an active fraction from *Emilia sonchifolia*. Indian J Pharm 37(4):232–237

Sibanda S, Mebe PP, Multari G (1992) Pentacyclic triterpenoids from *Euclea crispa*. Fitoterapia 63:274–277

Siddiqui SZ, Abbasi MA, Aziz-ur-Rehman RT, Shahzadi T, Ajaib M, Khan KM (2011) *Olea ferrugenia*: a potential natural source of protection from oxidative stress. J Med Plant Res 5(17):4080–4086

Siegel R, Miller K, Jemal A (2017) Cancer statistics. C A Cancer J Clin 67(1):7–30

Sienkiewicz P, Ciolino HP, Leslie BJ, Hergenrother PJ, Singletary K, Yeh GC (2007) A novel synthetic analogue of a constituent of *Isodon excisus* inhibits transcription of CYP1A1, -1A2 and -1B1 by preventing activation of the aryl hydrocarbon receptor. J Carcinogenet 28(5):1052–1057

Sigstedt SC, Hooten CJ, Callewaert MC, Jenkins AR, Romero AE, Pullin MJ, Kornienko A, Lowrey TK, Slambrouck SV, Steelant WF (2008) Evaluation of aqueous extracts of *Taraxacum officinale* on growth and invasion of breast and prostate cancer cells. Int J Oncol 32:1085–1090

Sihotang V (2011) Ethnomedicinal study of the Sundanese people at the Bodogol area, Gede Pangrango Mountain National Park, West Java 63(4):519–526

Silva GL, Chai H, Gupta MP, Farnsworth NR, Cordell GA, Pezzuto JM, Beecher CW, Kinghorn AD (1995) Cytotoxic biflavonoids from *Selaginella willdenowii*. Phytochemistry 40(1):129–134

Silva R, Melo G, Melo V, Antoniolli Â, Michellone P, Zucoloto S, Picinato M, Franco C, Mota G, Castro e Silva O (2006) Effect of the aqueous extract of *Sida cordifolia* on liver regeneration after partial hepatectomy. Acta Cirurgica Brasileira 21(1):37–39

Silva SLD, Figueiredo PMS, Yano T (2007) Chemotherapeutic potential of the volatile oils from *Zanthoxylum rhoifolium* Lam leaves. Eur J Pharmacol 576:180–188

Silva S, Chaar J, Figueiredo P, Yano T (2008) Cytotoxic evaluation of essential oil from Casearia sylvestris Sw on human cancer cells and erythrocytes. Acta Amazon 38(1):107–112

Silva AM, Paiva SR, Figueiredo MR, Kaplan MA (2012) Atividade biológica de naftoquinonas de espécies de Bignoniaceae. Farmacologia 7:207–2015

Silva VA, Alves AL, Rosa MN, Silva LR, Melendez ME, Cury FP, Gomes IN, Tansini A, Longato GB, Martinho O, Oliveira BG (2019) Hexane partition from Annona crassiflora Mart. promotes cytotoxity and apoptosis on human cervical cancer cell lines. Investig New Drugs 37(4):602–615

Simon S (2018) Facts & figures 2018: rate of deaths from cancer continues decline. American Cancer Society

Simpore J, Bayala B (2014) Anticancer activity of essential oils and their chemical components – a review. Am J Cancer Res 4(6):591–607

Singh B, Peter K (2018) New age herbals: resource, quality and pharmacognosy. Springer, Singapore, p 242

Singh MP, Malla SB, Rajbhandari SB, Manandhar A (1979) Medicinal plants of Nepal-retrospects and prospects. Econ Bot 33:185–198

Singh YN, Ikahihifo T, Panuve M, Slatter C (1984) Folk medicine in Tonga. A study on the use of herbal medicines for obstetric and gynaecological conditions and disorders. J Ethnopharmacol 12:305–329

Singh V, Kapahi BK, Srivastava TN (1996a) Medicinal herbs of Ladakh especially used in home remedies. Fitoterapia 67:38–48

Singh VK, Ali ZA, Zaidi STH, Siddiqui MK (1996b) Ethnomedicinal uses of plants of Gonda district forests of Uttar Pradesh, India. Fitoterapia 67:129–139

Singh B, Sharma RA, Vyas GK (2011) Antimicrobial, antineoplastic and cytotoxic activities of indole alkaloids from *Tabernaemontana divaricata* (L.) R.Br. Curr Pharm Anal 7:125–132

Singh G, Passsari AK, Singh P, Leo VV, Subbarayan S, Kumar B, Singh BP, Kumar NS (2017) Pharmacological potential of *Bidens pilosa* L. and determination of bioactive compounds using UHPLC-QqQ LIT-MS/MS and GC/MS. BMC Complement Altern Med 17(1):492

Sithranga Boopathy N, Kathiresan K (2010) Anticancer drugs from marine flora: an overview. J Oncol 2010:1–18

Skhirtladzea A, Plazab A, Montorob P, Benidzea M, Kemertelidzea E, Pizzab C, Piacente S (2006) Furostanol saponins from *Yucca gloriosa* L. rhizomes. Biochem Syst Ecol 34(11):809–814

Sliva D (2003) *Ganoderma Lucidum* (Reishi) in cancer treatment. Integr Cancer Ther 2(4):358–364

Smith-Kielland I, Dornish JM, Malterud KE, Hvistendahl G, Romming C, Bockman OC, Kolsaker P, Stenstrom Y, Nordal A (1996) Cytotoxic triterpenoids from the leaves of *Euphorbia pulcherrima*. Planta Med 62(4):322–325

Soares F, Cavalcante L, Romero N, Bandeira M (2016) *Himatanthus Willd.* ex Schult. (Apocynaceae): review. Pharmacogn Rev 10(19):6–10

Sofi MS, Sateesh MK, Bashir M, Harish G, Lakshmeesha TR, Vedashree S, Vedamurthy AB (2013) Cytotoxic and pro-apoptotic effects of *Abrus precatorius* L. on human metastatic breast cancer cell line, MDA-MB-231. Cytotechnology 65(3):407–417

Solis-Quispe L, Tomaylla-Cruz C, Callo-Choquelvica Y, Solís-Quispe A, Rodeiro I, Hernández I, Fernández M, Pino J (2015) Chemical composition, antioxidant and antiproliferative activities of essential oil from *Schinus areira* L. and *Minthostachys spicata* (Benth.) Epl. grown in Cuzco, Peru. J Essent Oil Res 28:234–240

Solowey E, Lichtenstein M, Sallon S, Paavilainen H, Solowey E, Lorberboum-Galski H (2014) Evaluating medicinal plants for anticancer activity. Sci World J

Son JK (1995) Isolation and structure determination of a newtetralone glucoside from the roots of *Juglans mandshurica*. Arch Pharm Res 18:203–205

Song YN, Zang HL, Chang CZ (1994) Cytotoxic cyclolignans from *Koelreuteria henryi*. J Nat Prod 57(12):1670–1674

Soysa P, Silva I (2016) Evaluation of phytochemical composition and antioxidant capacity of a decoction containing *Adenanthera pavonina* L. and *Thespesia populnea* L. BMC Complement Altern Med 16(27):70

Srinivasan R, Lakshmana G, Anjinayalu B (2014) Nature evolution effective against cancer therapy-review (Treating cancer diseases using plant products). Int J Pharm Bio Sci 5(2):357–365

Srivastava SK, Agnihotri VK (1985) New glycosides fromthe stem bark of *Aphanamixis polystachya*. Curr Sci 54:38–40

Steenkamp V, Gouws M (2006) Cytotoxicity of six South African medicinal plant extracts used in the treatment of cancer. S Afr J Bot 72(4):630–633

Stephen J, Peng K (2018) Measles virus for cancer therapy. Curr Top Microbiol Immunol 330:213–241

Stolarczyk M, Piwowarski J, Granica S, Stefańska J, Naruszewicz M, Kiss A (2013) Extracts from Epilobium sp. Herbs, their components and gut microbiota metabolites of *Epilobium Ellagitannins*, Urolithins, inhibit hormone-dependent prostate cancer cells-(LNCaP) proliferation and PSA secretion. Phytother Res 27(12):1842–1848

Stuppner H, Sturm S, Konwalinka G (1992) Capillary electrophoretic analysis of oxindole alkaloids from *Uncaria tomentosa*. J Chromatogr 609:375–380

Stuppner H, Sturm S, Geisen G, Zillian U, Konwalinka G (1993) A differential sensitivity of oxindole alkaloids tonormal and leukemic cell lines. Planta Med 59:A583

Subbarayan PR, Sarkar M, Impellizzeri S, Raymo F, Lokeshwar BL, Kumar P, Agarwal RP, Ardalan B (2010) Anti-proliferative and anti-cancer properties of *Achyranthes aspera*: specific inhibitory activity against pancreatic cancer cells. J Ethnopharmacol 131(1):78–82

Subramonian A, Rajasekharan S, Latha PG, Evans DA, Pushpangadan P (1996) Immunomodulatory and antitumour activities of *Janakia arayalpathra* root. Fitoterapia 67:140–144

Subramonium A, Rajasekharan S, Puphangadan P, George V, Nair G (2010) A process for preparation of a novel herbal medicinal composition for cancer treatment from *Janakia arayalpathra* root and *Triphcopus zeylanicus* leaf 193609, 10 September 2010

Suganyadevi P, Saravanakumar K, Mohandas S (2013) The antiproliferative activity of 3–deoxyanthocyanins extracted from red sorghum (*Sorghum bicolor*) bran through P53–dependent and Bcl-2 gene expression in breast cancer cell line. Life Sci 92(6–7):379–382

Sugata M, Lin CY, Shih YC (2015) Anti-inflammatory and anticancer activities of Taiwanese purple-fleshed sweet potatoes (*Ipomoea batatas* L. Lam) extracts. BioMed Res Int Article ID 768093:1–10

Suh HW, Song DK, Son KH, Woo MH, Do J, Choi YS, Lee KH, Kim YH (1996) Antinociceptive effect of Smilaxin B administered intracerebroventricularly in the mouse. Planta Med 62:141–145

Suksamrarn S, Komutiban O, Ratananukul P, Chimnoi N, Lartpornmatulee N, Suksamrarn A (2006) Cytotoxic prenylated xanthones from the young fruit of *Garcinia mangostana*. Chem Pharm Bull 54:301–305

Sultan A, Raza A (2015) Steroids: a diverse class of secondary metabolites. Sultan Rauf Raza 5:7

Sun HD, Chao JH, Lin ZW, Marunaka T, Minami Y, Fujita T (1982) The structure of Rubescensin C., A new minor diterpenoid isolated from *Rabdosia rubescens*. Chem Pharm Bull 30:341–343

Sun HD, Zhou QZ, Fujita T, Takeda Y, Minami Y, Maronaka T, Lin ZW, Shen XY (1992) RubescensinD, a diterpenoid from *Rabdosia rubescens*. Phytochemistry 31:1418–1419

Sun H, Sun C, Pan Y (2005) Cytotoxic activity and constituents of the volatile oil from the roots of *Patrinia scabra* Bunge. Chem Biodivers 2:1351–1357

Sun S, Wang C, Tong R, Li X, Fishbein A, Wang Q, He T, Du W, Yuan C (2010a) Effects of steaming the root of *Panax notoginseng* on chemical composition and anticancer activities. Food Chem 118:307–314

Sun J, Wang S, Ma D, Li H, Guo B, Wei Y (2010b) *Sparganium stoloniferum* flavonoids extraction and HPLC analysis of ingredients against HeLa cervical cancer. Acta Botan Boreali-Occiden Sin 30(12):2530–2535

Sun X, Zhao C, Pan W, Wang J, Wang W (2015) Carboxylate groups play a major role in antitumor activity of *Ganoderma applanatum* polysaccharide. Carbohydr Polym 123(6):283–287

Sundaram MM, Karthikeyan K, Sudarsanam D, Brindha P (2010) Antimicrobial and anticancer studies on *Euphorbia heterophylla*. J Pharm Res 3(3):685–688

Supratman U, Fujita T, Akiyama K, Hayashi H, Murakami A, Sakai H et al (2001) Anti-tumor promoting activity of bufadienolides from *Kalanchoe pinnata* and K. daigremontiana × butiflora. Biosci Biotechnol Biochem 65(4):947–949

Suwal PN (1970) Medicinal plants of Nepal. Ministry offorests, Department of medicinal plants, Thapathali, Kathmandu, Nepal

Suzuki H, Nakayama T, Nagae S, Yamaguchi M, IwashitaT FY, Nishino T (2004) cDNA cloning and functional characterization of flavonol 3-O-glucoside-6-Omalonyltransferases from flowers of *Verbena hybrida* and *Lamium purpureum*. J Mol Catal B Enzym 28(2):87–93

Svangard E, Goransson U, Hocaoglu Z, Gullbo J, Larsson R, Claeson P, Bohlin L (2004) Cytotoxic cyclotides from *Viola tricolor*. J Nat Prod 67(2):144–147

Svarcova I, Heinrich J, Valentova K (2007) Berry fruits as a source of biologically active compounds: the case of *Lonicera caerulea*. J Agric Food Chem 151(2):163–174

Swain S, Sinha B, Murthy P (2008) Antiinflammatory, diuretic and antimicrobial activities of *Rungia pectinata* Linn. and *Rungia repens* Nees. Indian J Pharm Sci 70(5):679–683

Sylvestre M, Pichette A, Longtin A, Nagau F, Legault J (2006) Essential oil analysis and anticancer activity of leaf essential oil of *Croton flavens* L. from Guadeloupe. J Ethnopharmacol 103(1):99–102

Tabudravu JN, Jaspar M (2005) Anticancer activities of constituents of kava (*Piper methysticum*). South Pac J Nat Sci 23(1):26–29

Tai PS, Hung YF (1981) Some antitumor constituents of *Chloranthus glaber*. Chung Tsao Yao 12:9–13

Taira J, Tsuchida E, Uehara M, Ohhama N, Ohmine W, Ogi T (2015) The leaf extract of *Mallotus japonicus* and its major active constituent, rutin, suppressed on melanin production in murine B16F1 melanoma. Asian Pac J Trop Biomed 5(10):819–823

Takaishi Y, Tokura K, Noguchi H, Nakano K, Murakami K, Tomimatsu T (1991) Sesquiterpene esters from *Tripterygium wilfordii*. Phytochemistry 30:1561–1566

Takeda Y, Yamashita H, Matsumoto T, Terao H (1993) Chloranthalactone F, a sesquiterpenoid from the leaves of *Chloranthus glabra*. Phytochemistry 33:713–715

Takemoto T, Odashima S, Hayashi S, Nishimoto N, Nakia S, Takagi N, Watanabe Y (1983) Pfaffosides. Patent-Japan-(Sho-Wa)-58-57547:21

Talib WH, Mahasneh AM (2010) Antiproliferative activity of plant extracts used against cancer in traditional medicine. Sci Pharm 78(1):33–45

Talib WH, Mahasneh AM (2012) Combination of *Ononis hirta* and *Bifidobacterium longum* decreases syngeneic mouse mammary tumor burden and enhances immune response. J Cancer Res Ther 8(3):417

Tamilselvan V, Rajeswari M, Velayutham P (2014) GC-MS analysis and "in vitro" anticancer activity of methanolic root extract of *Asystasia Gangetica* (L.) T. Anderson. World J Pharm Pharm Sci (WJPPS) 3(12):957–967

Tan B, Norhaizan M (2017) Scientific evidence of rice by-products for cancer prevention: chemopreventive properties of waste products from rice milling on carcinogenesisin vitro and in vivo. Biomed Res Int 2017:1–18

Tangah J, Bajau F, Jilimin W, Chan H, Wong S, Chan E (2017) Phytochemistry and pharmacology of *Mangifera pajang*: an iconic fruit of Sabah, Malaysia. Syst Rev Pharm 8:86–91

Tao HW, Hao XJ, Liu PP, Zhu WM (2003) Cytotoxic macrocyclic diterpenoids from *Euphorbia helioscopia*. Arch Pharm Res 31(12):1547–1551

Tao SJ, Guan SH, Wang W, Lu ZQ, Chen GT, Sha N, Yue QX, Liu X, Guo DA (2009) Cytotoxic polyprenylated xanthones from the resin of *Garciniahanburyi*. J Nat Prod 72(1):117–124

Tapondjou AL, Pursaethosides AE (2005) Triterpene Saponins from *Entada pursaetha*. J Nat Prod 68(8):1185–1190

Tayarani-Najaran Z, Emami SA, Asili J, Mirzaei A, Mousavi SH (2009) Analyzing cytotoxic and apoptogenic properties of *Scutellaria litwinowii* root extract on cancer cell lines. Int J Mol Med 16:1–9

Tayarani-Najaran Z, Mousavi SH, Vahdati-Mashhadian N, Emami SA, Parsaee H (2012) *Scutellaria litwinowii* induces apoptosis through both extrinsic and intrinsic apoptotic pathways in human promyelocytic leukemia cells. Nutr Cancer 64(1):80–88

Teklehaymanot T (2017) An ethnobotanical survey of medicinal and edible plants of Yalo Woreda in Afar regional state, Ethiopia. J Ethnobiol Ethnomed 13:40

Thoison O, Fahy J, Dumontet V, Chiaroni A, Riche C, Tri MV, Sevenet T (2000) Cytotoxic prenylxanthones from *Garcinia bracteata*. J Nat Prod 63(4):441–446

Thomas S (2012) *Leptadenia hastata*: a review of its traditional uses and its pharmacological activity. Med Chem 02:148–150

Thomas CM, Wood RC III, Wyatt JE, Pendleton MH, Torrenegra RD, Rodriguez OE, Harirforoosh S, Ballester M, Lightner J, Krishnan K, Ramsauer VP (2012) Anti-neoplastic activity of two flavone isomers derived from *Gnaphalium elegans* and *Achyrocline bogotensis*. PLoS One 7(6):39806

Thriveni H, Gudasalamani R, Vasudeva R, Shaanker R (2015) Camptothecine and methoxy camptothecine from callus cultures of Miquelia dentata Bedd.- A rare plant of the Western Ghats of India. Indian J Biotechnol 14(1):123–126

Tiwari VJ (1995) Ethnobotanical survey of Halbi tribe of Chandrapur and Gadchiroli districts of Maharashtra State, India. Fitoterapia 66:346–350

Tiwari TN, Pandey VB, Dubey NK (2002) Plumieride from *Allamanda cathartica* as an antidermatophytic agent. Physiol Res 16(4):393–394

Tokar M, Klimek B (2004) Isolation and identification of biologically active compounds from *Forsythia viridissima* flowers. Acta Pol Pharm 61(3):191–197

Tokuda H, Ohigashi H, Koshimizu K (1986) Inhibitory effects of ursolic and oleanolic acid on skin tumor promotion by 12-O-tetradecanoylphorbol-13-acetate. Cancer Lett 33:279–285

Tokuda H, Nishino H, Konoshima T (1999) Cancer chemopreventive agents (antitumorpromoters) from *Ajuga decumbens*. J Nat Prod 62(7):972–975

Tor YS, Yazan LS, Foo JB, Armania N, Cheah YK, Abdullah R, Imam MU, Ismail N, Ismail M (2014) Induction of apoptosis through oxidative stress-related pathways in MCF-7, human breast cancer cells, by ethyl acetate extract of *Dillenia suffruticosa*. BMC Complement Altern Med 14(1):55

Torrenegra RD, Escarria S, Raffelsberger B, Achenbach H (1980) Colombian plants of the genus Gnaphalium. Part E. 5,7-Dihydroxy-3, 6, 8-Trimethoxyflavone from the flowers of *Gnaphalium elegans*. Phytochemistry 19:2795–2796

Train P, Henrichs JR, Archer WA (1957) Medicinal uses of medicinal plants by Indian tribes of Nevada. Contrib Flora Nevada 45:1–139

Tsai CH, Lin FM, Yang YC, Lee MT, Cha TL, Wu GJ, Hsieh SC, Hsiao PW (2009) Herbal extract of *Wedelia chinensis* attenuates androgen receptor activity and orthotopic growth of prostate cancer in nude mice. Clin Cancer Res 15(17):543–544

Tshikalange TE, Hussein A (2010) Cytotoxicity activity of isolated compounds from *Elaeodendron transvaalense* ethanol extract. J Med Plant Res 4(16):1695–1697

Tsui WY, Brown GD (1996) Cycloeudesmanolides from *Sarcandra glabra*. Phytochemistry 43:819–821

Tuasha N, Petros B, Asfaw Z (2018) Medicinal plants used by traditional healers to treat malignancies and other human ailments in Dalle District, Sidama Zone, Ethiopia. J Ethnobiol Ethnomed 14(1):15

Turner NJ (1984) Counter-irritant and other medicinal uses of plants in Ranunculaceae by native peoples in British Columbia and neighbouring areas. J Ethnopharmacol 11:181–201

Uddin MJ, Patari P, Nag SK, Sil SK (2018) A comparative study on ethno medicinal and nutraceutical potentials of the leaves of three species of genus Ipomea from Tripura, NE-India. J Pharmacogn Phytochem 7(1):2073–2082

Ueda J-Y, Athikomkulchai S, Miyatake R, Saiki I, Esumi H, Awale S (2014) (+)-Grandifloracin, an antiausterity agent, induces autophagic PANC -1 pancreatic cancer cell death. Drug Design Dev Ther 8:39–47

Ulubelen A, Cole JR (1965) Preliminary phytochemical investigation of *Maytenus trichotomus* (Celastraceae). J Pharm Sci 54:1763–1765

Ulubelen A, Tanker M (1975) Polypeptides and amino acids in the bulbs of *Merendera caucasica*. Planta Med 28:379–381

Um SH, Lee KR, Zee OP, Pyo S (1996) The effect of *Salvia plebeia* on murine macrophage-mediated cytotoxicity. Nat Prod Sci 2:43–47

Uniyal MR (1990) Utility of hitherto unknown medicinal plants traditionally used in Ladakh. J Res Edu Indian Med 9:89–95

Upadhyay S, Jeena GS, Shukla RK (2018) Recent advances in steroidal saponins biosynthesis and in vitro production. Planta 248(3):519–544

Urruticoechea A, Alemany R, Balart J, Villanueva A, Vinals F, Capella G (2010) Recent advances in cancer therapy: an overview. Curr Pharm Des 16(1):3–10

Uto T, Tung NH, Appiah-Opong R, Aning A, Morinaga O, Edoh D, Nyarko AK, Shoyama Y (2015) Antiproliferative and pro-apoptotic activity of diarylheptanoids isolated from the bark of *Alnus japonica* in human leukemia cell lines. Am J Chin Med 43(04):757–767

Vaidya A, Devasagayam T (2007) Current status of herbal drugs in India: an overview. J Clin Biochem Nutr 41(1):1–11

Vallianou NG, Evangelopoulos A, Schizas N, Kazazis C (2015) Potential anticancer properties and mechanisms of action of curcumin. Anticancer Res 35(2):645–651

Van Erp NP, Baker SD, Zhao M, Rudek MA, Guchelaar HJ, Nortier JW, Sparreboom A, Gelderblom H (2005) Effect of milk thistle (*Silybum marianum*) on the pharmacokinetics of irinotecan. Clin Cancer Res 11(21):7800–7806

Varshney AK, Mah T, Khan NU, Rahman W, Hwa CW, Okigawa M, Kawano N (1973) Biflavones from *Cycas revoluta*, *C. circinalis*, and *C. rumphii* Indian. J Chemother 11:1209–1214

Vashi IG, Patel HC (1989) Chemical analysis and antibiotic potential of Alstonia scholaris R.Br. bark. Comp Physiol Ecol 14:1–3

Vasileva B (1969) Plantes Medicinales De Guinee. Republique De Guinee, Conakry

Vega MRJ, Esteves-Souza A, Vieira IJC, Mathias LD, Braz-Filho R, Echevarria A (2007) Flavonoids from Annona dioica leaves and their effects in Ehrlich carcinoma cells, DNA-topoisomerase I and II. J Braz Chem Soc 18(8):1554–1559

Velasco G, Hernández-Tiedra S, Dávila D, Lorente M (2016) The use of cannabinoids as anticancer agents. Prog Neuro-Psychopharmacol Biol Psychiatry 64:259–266

Vennapusa N, Pkm N, Madanu D (2013) Evaluation of anti-cancer activity of *Ruellia tuberosa* on EAC induced mammary tumor. J Pharmacol Toxicol 1:36–42

Verbitski S, Gourdin G, Ikenouye L, McChesney J, Hildreth J (2008) Detection of *Actaea racemosa* adulteration by thin-layer chromatography and combined thin-layer chromatography-bioluminescence. J AOAC Int 91(2):268–275

Verma VK, Siddiqui NU, Aslam M (2012) Isolation of scutellarein from *Pygmaeopremna herbacea* Roxb. J Appl Pharm Sci 2(6):241

Verma A, Rizvi SM, Shaikh S, Ansari MA, Shakil S, Ghazal F, Siddiqui MH, Haneef M, Rehman A (2014) Compounds isolated from *Ageratum houstonianum* inhibit the activity of matrix metalloproteinases (MMP-2 and MMP-9): an oncoinformatics study. Pharmacogn Mag 10(37):18

Verma A, Kumar B, Alam P, Singh V, Gupta S (2016) *Rubia cordifolia*- a review on pharmaconosy and phytochemistry. Int J Pharm Sci Res 7(7):2720

Vermillion K, Holguin FO, Berhow MA, Richins RD, Redhouse T, O'Connell MA, Posakony J, Mahajan SS, Kelly SM, Simon JA (2011) Dinoxin b, a withanolide from datura inoxia leaves with specific cytotoxic activities. J Nat Prod 74(2):267–271

Versiani MA, Ahmed SK, Ikram A, Ali ST, Yasmeen K, Faizi S (2014) Chemical constituents and biological activities of Adenium obesum (Forsk.) Roem. et Schult. Chem Biodivers 11(2):171

Vieira JRC, Souza IAD, Nascimento SCD, Leite SP (2007) *Indigofera suffruticosa*: an alternative anticancer therapy. Evid Based Complement Alternat Med 4(3):355–359

Vinesi P, Serafini M, Nicoletti M, Spano L, Betto P (1990) Plant regeneration and hypoxoside content in *Hypoxisobtusa*. J Nat Prod 53:196–199

Vukis V (2009) Antioxidant flavanoid glycosides in *Viola Tricolor* L. Graduate, Semmelweis University, p 222

Wan Salleh W, Ahmad F (2017) Phytochemistry and biological activities of the genus Ocotea (Lauraceae): a review on recent research results (2000–2016). J Appl Pharm Sci 7(5):204–218

Wang SY, Ballington JR (2007) Free radical scavenging capacity and antioxidant enzyme activity in deerberry (*Vaccinium stamineum* L.). LWT-Food Sci Technol 40(8):1352–1361

Wang SY, Lewers KS (2007) Antioxidant activities and anticancer cell proliferation properties of wild strawberries. J Am Soc Hortic Sci 132(5):647–658

Wang HW, Liu YQ (2010) Evaluation of trace and toxic element concentrations in *Paris polyphylla* from China with empirical and chemometric approaches. Food Chem 121:887–892

Wang ZQ, Xu FM, Yan XZ, Zhu Y (1996) ScutebarbatineA, a new neoclerodane-type diterpenoid alkoid from *Sculleteria barbata*. Chin Chem Lett 7:333–334

Wang H, Ng TB, Ooi VEC, Liu WK (2000) Effects of lectins with different carbohydratebinding specificities on hepatoma, choriocarcinoma, melanoma and osteosarcoma cell lines. Int J Biochem Cell Biol 32(3):365–372

Wang Y, Li X, Li LH, Meng DL, Li ZL, Li N (2007) Two new thiophenes from Echinops latifolius and their phototoxic activities. Planta Med 73(7):696–698

Wang CZ, Aung HH, Zhang B, Sun S, Li XL, He H, Xie JT, He TC, Du W, Yuan CS (2008a) Chemopreventive effects of heat-processed *Panax quinquefolius* root on human breast cancer cells. J Anticancer Res 28:2545–2552

Wang WS, Dai X, Hong LY, Lu P, Feng JC, Jiao YG (2008b) New sesquiterpenoids from *Ligularia duciformis*. Helvetica Chimica Acta 91(6):1118–1123

Wang L, Bai L, Nagasawa T, Hasegawa T, Yang X, Sakai JI, Bai Y, Kataoka T, Oka S, Hirose K, Tomida A (2008c) Bioactive triterpene saponins from the roots of *Phytolacca americana*. J Nat Prod 271(1):35–40

Wang GY, Lv QH, Dong Q, Xu RZ, Dong QH (2009) Berbamine induces Fas-mediated apoptosis in human hepatocellular carcinoma HepG2 cells and inhibits its tumor growth in nude mice. J Asian Nat Prod Res 11(3):219–228

Wang Q, Chen T, Bastow KF, Lee K, Chen D, Altaicalarins AD (2010a) Cytotoxic Bisabolane Sesquiterpenes from *Ligularia altaica*. J Nat Prod 73(2):139–142

Wang P, Deng G, Yuan W, Li S (2010b) Cytotoxity of saponin-rich extracts from *Phytolacca americana* L. (Phytolaccaceae). Planta Med 76(5):87

Wang GF, Guo YW, Feng B, Li L, Huang CG, Jiao BH (2010c) Tanghinigenin from seeds of *Cerbera manghas* L. induces apoptosis in human promyelocytic leukemia HL-60 cells. Environ Toxicol Pharmacol 30(1):31–36

Wang Y, Hong C, Zhou C, Xu D, Qu H (2011a) Screening antitumor compounds psoralen and isopsoralen from *Psoralea corylifolia* L. Seeds. Evid Based Complement Alternat Med 34(1):31–36

Wang Z, Liu H, Zhang Y, Guo L, Li Z, Shi X (2011b) Anticancer potential of *Euphorbia helioscopia* L extracts against human cancer cells. Anat Rec Adv Integr Anat Evol Biol 295(2):223–233

Wang Y, Guan S, Feng R, Zhang J, Li Q, Chen X, Bi K, Guo D (2013) Elution-extrusion counter-current chromatography separation of two new benzyl ester glucosides and three other high-polarity compounds from the tubers of *Pleione bulbocodioid* (Stolarczyk et al. 2013) es. Phytochem Anal 24(6):671–676

Wang Y, Morris-Natschke S, Yang J, Niu H, Long C, Lee K (2014) Anticancer principles from medicinal piper (*Piper boehmerifolium*) plants. J Tradit Complement Med 4(1):8–16

Wang JZ, Li J, Zhao P, Ma WT, Feng XH, Chen KL (2015) Antitumor activities of ethyl acetate extracts from *Selaginella doederleinii* Hieron in vitro and in vivo and its possible mechanism. Evid Based Complement Alternat Med 2015:865714

Wangchuk P, Keller P, Pyne S, Taweechotipatr M, Tonsomboon A, Rattanajak R, Kamchonwongpaisan S (2011) Evaluation of an ethnopharmacologically selected Bhutanese medicinal plants for their major classes of Phytochemicals and biological activities. J Ethnopharmacol 137(1):730–742

Ward PF, Hall RJ, Peters RA (1964) Fluoro-fatty acids in the seeds of *Dichapetalum toxicarium*. Nature 201(4919):611–612

Wasuwat S (1967) A list of Thai medicinal plants, A.S.R.C.T, Bangkok. Report No.1 on Research Project 17

Watt JM, Breyer-Brandwijk MG (1962) The medicinal and poisonous plants of southern and eastern Africa, 2nd edn. E. and S. Livingstone Ltd., London

Wei Z, Gao Y, Yang F, Lu H, Ni L, Zhou X (2011) Three new diterpenoids from *Isodon nervosus*. Helvetica Chimica Acta 94(7):1320–1325

Wen L, Wu D, Jiang Y, Prasad KN, Lin S, Jiang G, Yang B (2014) Identification of flavonoids in litchi (*Litchi chinensis* Sonn.) leaf and evaluation of anticancer activities. J Funct Foods 6:555–563

West B, Jensen C, Palu A, Deng S, Wasden J (2000) *Morinda citrifolia* and iridoid based formulations. US8790727B2, 12 May 2000

Whitson EL, Damayanthi SMV, Mala VCA, Bugni TS, Dilip de Silva CM (2006) Ireland Oppositines A and B, sesquiterpenes pyridine alkaloids from a Sri Lankan *Pleurostylia opposita*. J Nat Prod 69:1833–1835

Wijisekera ROB (1978) An ethnobotanical list of the Guyanese Flora, unpublished manuscript

Wilkins AL, Lu Y, Tan S (1993) Extractives from New Zealand Honeys. 4. Linalool derivatives and other components from nodding thistle (*Carduus nutans*) honey. J Agric Food Chem 41:873–878

Williams R, Norris A, Miller J, Razafitsalama L, Andriantsiferana R, Rasamison V, Kingston D (2006) Two new cytotoxic naphthoquinones from *Mendoncia cowanii* from the rainforest of Madagascar. Planta Med 72:564–566

Williams RB, Norris A, Miller JS, Birkinshaw C, Ratovoson F, Andriantsiferana R, Rasamison VE, Kingston DG (2007) Cytotoxic clerodane diterpenoids and their hydrolysis products from *Casearia nigrescens* from the rainforest of Madagascar. J Nat Prod 70(2):206–209

Wilson RT, Mariam WG (1979) Medicine and magic in central Tigre. A contribution to the ethnobotany of the Ethiopian Plateau. Econ Bot 33:29–34

Wirth C, Wagner H (1997) Pharmacologically active procyanidines from the bark of *Uncaria tomentosa*. Phytomedicine 4:265–266

Woldemariam TZ, Fell AF, Linley PA, Bibby MC, Phillips RM (1992) Evaluation of the antitumour action and acute toxicity of kosins from *Hagenia abyssinica*. J Pharm Biomed Anal 10(8):555–560

Wong KT, Tan BKH (1996) In vitro cytotoxicity and immunomodulating property of *Rhaphidophora korthalsii*. J Ethnopharmarmacol 52:53–57

Wong BYY, Lau BHS, Teel RW (1992a) Chinese medicinal herbs modulated mutagenesis, DNA binding and metabolism of Benzo[A] Pyrene. Phytother Res 6:10–14

Wong BYY, Lau BHS, Tadi PP, Teel RW (1992b) Chinese medicinal herbs modulate mutagenesis, DNA binding and metabolism of Aflatoxin B1. Mutat Res 279:209–216

Wong BYY, Lau BHS, Yamasaki T, Teel RW (1993) Modulation of cytochrome P-450ia1-mediated mutagenicity, DNA binding and metabolism of Benzo[A]Pyrene by Chinese medicinal herbs. Cancer Lett 68:75–82

Wong BY, Nguyen DL, Lin T, Wong HH, Cavalcante A, Greenberg NM, Hausted RP, Zheng J (2009) Chinese medicinal herb *Scutellaria barbata* modulates apoptosis and cell survival in murine and human prostate cancer cells and tumor development in TRAMP mice. Eur J Cancer Prev 18(4):331–341

Wonkchalee O, Boonmars T, Aromdee C, Laummaunwai P, Khunkitti W, Vaeteewoottacharn K, Sriraj P, Aukkanimart R, Loilome W, Chamgramol Y, Pairojkul C (2012) Anti-inflammatory, antioxidant and hepatoprotective effects of Thunbergia laurifolia Linn. on experimental opisthorchiasis. Parasitol Res 111(1):353–359

Woo WS (1985) Phytochemical study on the stembark of *Albizia julibrissin*. Korean J Pharm 16:39–43

Woo EH, Woo WS (1989) Flavonoid glycosides from *Melandrium firmum*. Arch Pharm Res 12:223–225

Woo MH, Do JC, Son KH (1992) Five new spirostanol glycosides from the subterranean parts of *Smilax sieboldii*. J Nat Prod 55:1129–1135

Woo S, Jo T, Kim D, Sung S, Do S, Lee Y, Nam J, Ryu J, Lee K, Sung H, Heo Y, Oh M, Cho J, Sung S, Lee J, Kim T, Song J, Lee S, Kim M (2011) Composition comprising bamboo extract and the compounds isolated therefrom showing treating and preventing activity for inflammatory and blood circulation disease. US20110117224A1, 19 May 2011

Woo S, Kwon S, Ko S, Cho S (2017) Barley grass extract causes apoptosis of cancer cells by increasing intracellular reactive oxygen species production. Biomed Rep 6(6):681–685

Wu TS, Tien HJ, Yeh MY (1980) Studies on the constituents of Formosan Folk Medicine. VII. Constituents of the flowers of *Vanilla somai* hayata and the roots of *Camptotheca acuminata* Dcne. Cheng-Kung Ta HsuenHsuen Pao 15:65–67

Wu TS, Lin YM, Haruna M, Pan DJ, Shingu T, Chen YP, Hsu HY, Nakano T, Lee KH (1991) Antitumor agents, 119. Kansuiphorins A and B, two novel antileukemic diterpene esters from *Euphorbia kansui*. J Nat Prod 54(3):823–829

Wu JB, Cheng YD, Chiu NY, Huang SC, Kuo SC (1993a) A novel morphinandienone alkaloid from *Fissistigma oldhamii*. Planta Med 59:179–181

Wu SH, Zhang HJ, Chen YP, Lin ZW, Sun HD (1993b) Diterpenoids from *Isodon wikstroemoides*. Phytochemistry 34:1099–1102

Wu YC, Chang GY, Chang-Yih D, Shang-Kwei W (1993c) Cytotoxic alkaloids of *Annona montana*. Phytochemistry 33(2):497–500

Wu JB, Cheng YD, Kuo SC, Wu TS, Iitaka Y, Ebizuka Y, Sankawa U (1994) Fissoldhimine, a novel skeleton alkaloid from *Fissistigma oldhamii*. Chem Pharm Bull 42:2202–2204

Wu DG, Sorg B, Hecker E (1995) New myrsinol-related polyfunctional pentacyclic diterpene esters from roots of *Euphorbia prolifera*. J Nat Prod 58:408–413

Wu JH, Chang FR, Hayashi K, Shiraki H, Liaw CC, Nakanishi Y, Bastow KF, Yu D, Chen IS, Lee KH (2003) Antitumor agents. Part 218: Cappamensin A, a new in vitro anticancer principle, from *Capparis sikkimensis*. Bioorg Med Chem Lett 13:223–225

Wu QH, Liu CM, Chen YJ, Gao K (2006) Terpenoids from the roots of *Ligularia muliensis*. Helvetica Chimica Acta 89(5):985–992

Wu SB, Ji YP, Zhu JJ, Zhao Y, Xia G, Hu YH, Hu JF (2009) Steroids from the leaves of *Chinese Meliaazadirachta* and their cytotoxic effects on human cancer cell lines. Steroids 74(9):761–765

Wu HY, Wang WG, Jiang HY, Du X, Li XN, Pu JX, Sun HD (2014) Cytotoxic and anti-inflammatory ent-kaurane diterpenoids from *Isodon wikstroemioides*. Fitoterapia 98:192–198

Wu Q, Patocka J, Nepovimova E, Kuca K (2019) *Jatropha gossypiifolia* L. and its biologically active metabolites: a mini review. J Ethnopharmacol 34:197–203

Xia EQ, Wang BW, Xu XR, Zhu L, Song Y, Li H (2011) Microwave-assisted extraction of oleanolic acid and ursolic acid from *Ligustrum lucidum* Ait. Int J Mol Sci 12:5319–5329

Xie Q, Yang Y (2014) Anti-proliferative of physcion 8-O-β-glucopyranoside isolated from *Rumex japonicus* Houtt. on A549 cell lines via inducing apoptosis and cell cycle arrest. BMC Complement Altern Med 14(1):377

Xin N, Hasan M, Li W, Li Y (2014) *Juglans mandshurica* Maxim extracts exhibit antitumor activity on HeLa cells in vitro. Mol Med Rep 9(4):1313–1318

Xiong W, Luo G, Zhou L, Zeng Y, Yang W (2009) In vitro and in vivo antitumor effects of acetyl-shikonin isolated from *Arnebia euchroma* (Royle) Johnst (Ruanzicao) cell suspension cultures. Chin Med 4(1):14

Xu J (2016) Cancer inhibitors from Chinese natural medicines, p 252

Xu Y, McLaughlin SP, Gunatilaka AAL (2007) Sorbifolivaltrates A–D, Diene Valepotriates from *Valeriana sorbifolia*. Nat Prod 70(12):2045–2048

Xu W, Liu J, Li C, Wu HZ, Liu YW (2008) Kaempferol-7-O-beta-D-glucoside (KG) isolated from *Smilax china* L. rhizome induces G2/M phase arrest and apoptosis on HeLa cells in a p53- independent manner. Cancer Lett 264(2):229

Xu G, Kan WLT, Zhou Y, Song J, Han Q, Qiao C, Cho C, Rudd JA, Lin GL, Xu H (2010) Cytotoxic acylphloroglucinol derivatives from the twigs of *Garcinia cowa*. J Nat Prod 73(2):104–108

Xu M, Zhang M, Wang D, Yang C, Zhang Y (2011) Phenolic compounds from the whole plants of *Gentiana rhodantha* (Gentianaceae). Chem Biodivers 8(10):1891–1900

Xu T, Shen H, Liu L, Shu Y (2013) Plumbagin from *Plumbago Zeylanica* L induces apoptosis in human non-small cell lung cancer cell lines through NF-κB inactivation. Asian Pac J Cancer Prev 14(4):2325–2331

Xu X, Zheng Y, Liu L, Huang W, Huang J (2017) Extracts of *Potentilla discolor* bunge protect islet β-cells against streptozotocin induced injury. Biomed Res 28(13):5713–5716

Yahya AL, Sayed EL, Mossa JS (1988) Potential cancer chemopreventive and cytotoxic agents from *pulzcaria*. J Nat Prod 51(3):621–624

Yang RZ, Yang TR, Zhou J (1981) The structures of Tenacigenin A, B and C. Yun-Nan Chih Wu Yen Chiu 3:271–278

Yang LL, Lee CY, Yen KY (2000) Induction of apoptosis by hydrolyzable tannins from *Eugenia jambos* L. on human leukemia cells. Cancer Lett 157:65–75

Yang G, Zhang L, Chen G (2010a) Determination of four phenolic compounds in *Scirpus yagara* Ohwi by CE with amperometric detection. Chromatographia 71:143–147

Yang SM, Wu DG, Liu XK (2010b) Anticancer activity of diterpenoids from *Amoora ouangliensis* and *Amoora stellato-squamosa*. Z Naturforsch C 65(1):39–42

Yang H, Kim J, Lee J, Lee B, Park K, Shim K, Lee M, Seo K (2011a) Celastrol isolated from *Tripterygium regelii* induces apoptosis through both caspase-dependent and – independent pathways in human breast cancer cells. Food Chem Toxicol 49(2):527–532

Yang JH, Zhang HY, Wen J, Du X, Chen JH, Zhang HB, Xiao WL, Pu JX, Tang XC, Sun HD (2011b) Dibenzocyclooctadiene lignans with anti-neurodegenerative potential from *Kadsura ananosma*. J Nat Prod 74(5):1028–1035

Yang H, Chen P, Wang J (2014) Chinese herbs containing aristolochic acid associated with renal failure and urothelial carcinoma: a review from epidemiologic observations to causal inference. Biomed Res Int 2014:1–9

Yang X, Khan I, Kang S (2015) Chemical composition, mechanism of antibacterial action and antioxidant activity of leaf essential oil of *Forsythia koreana* deciduous shrub. Asian Pac J Trop Med 8(9):694–700

Yao G, Cheng Z, Shang X, Gao P, Huang X, Song S (2017) Coumarins from the bark of *Juglans mandshurica* exhibited anti-hepatoma activities via inducing apoptosis. J Asian Nat Prod Res 19(11):1134–1142

Yazan L, Ong Y, Zaaba N, Ali R, Foo J, Tor Y (2015) Anti-breast cancer properties and toxicity of Dillenia suffruticosa root aqueous extract in BALB/c mice. Asian Pac J Trop Biomed 5:1018–1026

Ye D (2011) Bioactive constituents of two medicinal plants from Indonesia. Doctor of Philosophy, Ohio State University. Pharmacy, p 198

Ye M, Su J, Liu S, Cao L, Xiong J, Zhao Y, Fan H, Yang G, Xia G, Hu J (2013) (24S)-Ergostane-3β,5α,6β-triol from Hedyotis chrysotrichawith inhibitory activity on migration of SK-HEP-1 human hepatocarcinoma cells. Nat Prod Res 27(12):1136–1140

Yeap S, Omar A, Ho W, Beh B, Ali A, Alitheen N (2013) *Rhaphidophora korthalsii* modulates peripheral blood natural killer cell proliferation, cytokine secretion and cytotoxicity. BMC Complement Altern Med 13:145

Yi YH (1992) A triterpenoid saponin from *Phytolacca esculenta*. Phytochemistry 31:2552–2554

Yi YH, Dai FB (1991) A new triterpenoid and its glycoside from *Phytolacca esculenta*. Planta Med 57:162–164

Yim N, Gu M, Hwang Y, Cho W, Ma J (2016) Water extract of Galla Rhois with steaming process enhances apoptotic cell death in human colon cancer cells. Integr Med Res 5:284–292

Yin X, Zhou J, Jie C, Xing D, Zhang Y (2004) Anticancer activity and mechanism of *Scutellaria barbata* extract on human lung cancer cell line A 549. Life Sci 75(18):2233

Yin H, Zhang S, Wu J, Nan H, Long L, Yang J, Li Q (2006) Pongaflavanol: a prenylated flavonoid from *Pongamia pinnata* with a modified ring A. J Mol 11:786–791

Yin S, Wei W, Jian F, Yang N (2013) Therapeutic applications of herbal medicines for cancer patients. Evid Based Complement Alternat Med 2013:1–15

Yokosuka A, Mimaki Y, Kuroda M, Sashida Y (2000) A new steroidal saponin from the leaves of Agave americana. Planta Med 66(4):393–396

Yoo SJ, Yim DS, Lee SY (1993) Biological activities of *Pedicularis resupinata* var. *oppositifolia*. Korean J Pharma 24:258–262

Yook CS, Lee WT, Moon CK (1989) Studies on flavonoid glycoside of the leaves of *Viola diamantica*. Yakhak Hoe Chi 33:124–128

Yoshida T, Tanaka K, Chen XM, Okuda T (1989) *Dimeric ellagitannins*, laevigatins e, f and g, from *Rosa laevigata*. Phytochemistry 28:2451–2454

You Y, Wu Y, Mao J, Zou L, Liu S (2009) Screening of Chinese Brassica species for anticancer sulforaphane and erucin. Afr J Biotechnol 7(2):147–152

Young R (2016) The use of alkaline herbs in the prevention and treatment of any cancerous condition, p 3

Youngsa-ad W, Ngamrojanavanich N, Mahidol C, Ruchirawat S, Prawat H, Kittakoop P (2007) Diterpenoids from the roots of *Croton oblongifolius*. Lett Planta Medica 73:1491–1494

Yousif G, Iskander GM, Eisa EB (1983) Investigation of the alkaloidal components in the Sudan Flora Pt. 1. Fitoterapia 54:81–85

Yousif G, Iskander GM, Eisa EB (1984) Alkaloids of *Cadaba farinosa* and *C. rotundifolia*. Fitoterapia 55:117–118

Yu LJ, Ma R, Wang YQ, Nishino H (1994a) Potent anti-tumor activity and low toxicity of tubeimoside 1 isolated from *Bolbostemma paniculatum*. Planta Med 60:204–208

Yu SM, Kim HJ, Woo ER, Park HK (1994b) Some sesquiterpenoids and 5-alpha, 8-alpha-epidioxysterols from *Solanum lyratum*. Arch Pharm Res 17:1–4

Yu ZL, Liu XR, McCulloch M, Gao J (2004) Anticancer effects of various fractions extracted from *Dioscorea bulbifera* on mice bearing HepA. Zhongguo Zhong Yao Za Zhi 29(6):563–567

Yu F, Yu F, McGuire P, Li R, Wang R (2007) Effects of *Hydrocotyle sibthorpioides* extract on transplanted tumors and immune function in mice. Phytomedicine 14(2–3):166–171

Yu C, Xu F, Ming W, Xiaodong J, Youyi Z, Yunfa D (2009) Triterpene glycosides from Lonicera. II. Isolation and structural determination of glycosides from flower buds of*Lonicera macranthoides*. Chem Nat Compd 45(4):514–518

Yu J, Ma Y, Drisko J, Chen Q (2013a) Antitumor activities of *Rauwolfia vomitoria* extract and potentiation of carboplatin effects against ovarian cancer. Curr Ther Res 75:8–14

Yu B, Luo J, Wang J, Zhang D, Yu S, Kong L (2013b) Pentasaccharide resin glycosides from Ipomoea cairica and their cytotoxic activities. Phytochemistry 95:421–427

Yu H, Ahn C, Yang I, Won D, Jin B, Cho N, Hong S, Shin J, Cho S (2018) Apoptosis induced by methanol extract of *Potentillaï discolor* in human mucoepidermoid carcinoma cells through STAT3/PUMA signaling axis. Mol Med Rep 17(4):5258–5264

Yua J, Leib J, Zhangc X, Yud H, Tianc D, Liaoa Z, Zoud G (2010) Anticancer, antioxidant and antimicrobial activities of the essential oil of *Lycopus lucidus* Turcz, var. hirtus Regel. Food Chem 126(4):1593–1595

Yuan YL, Fang LZ, Bai SP, Liang HJ, Ye DD (2013) A novel eremophilane sesquiterpenoid from *Ligularia intermedia* roots and rhizomes. Chem Nat Compd 49(2):258–260

Yumrutas O, Saygideg SD (2010) Determination of in vitro antioxidant activities of different extracts of *Marrubium parviflorum* Fish et Mey. and*Lamium amplexicaule* L. from South east of Turkey. J Med Plant Res 4(20):2164–2172

Zahra A, Masoud B, Abbas A, Katayoun J (2006) Induction of apoptosis in leukemia cell lines by *Linum persicum*and *Euphorbia cheiradenia*. J Cancer Res Clin Oncol 132(7):427–432

Zailani A, Jada S, Wurochekke Usman A (2010) Antimicrobial activity of *Waltheria indica*. J Am Sci 6:1591–1594

Zani CL, Chaves PPG, Queiroz R, De Oliveira AB, Cardoso JE, Anjos AMG, Grandi TSM (1995) Brineshrimp lethality assay as a prescreening system for anti-trypanosomacruzi activity. Phytomedicine 2:47–50

Zargari A (1992) Medicinal plants, vol 3, 5th edn. Tehran University Publications, No. 1810:3, Tehran, Iran, p 889

Zeng L, Fagotto F, Zhang T, Hsu W, Vasicek TJ, Perry WL, Lee JJ, Tilghman SM, Gumbiner BM, Costantini F (1997a) The mouse fused locus encodes axin, an inhibitor of the Wnt signaling pathway that regulates embryonic axis formation. Cell 90(1):181–192

Zeng JF, Li GL, Shen JK, Zhu DY, Chen K, Lee KH (1997b) Manuifolins D, E, and F., new isoflavonoids from*Maackia tenuifolia*. J Nat Prod 60:918–920

Zhang JX, Han QB, Zhao QS, Li SH, Sun HD (2002) Two New ent-Kaurane Diterpenoids from *Isodon japonica*. Chin Chem Lett 13(11):1075–1078

Zhang XQ, Ye WC, Jiang RW, Yin ZQ, Zhao SX, Mak TC, Yao XS (2006) Two new eremophilanolides from *Xanthium sibiricum*. Nat Prod Res 20(13):1265–1270

Zhang ZX, Lin CJ, Li PL, Jia ZJ (2007a) New weakly cytotoxic eremophilane sesquiterpenes from the roots of *Ligularia virgaurea*. J Planta Medica 73(6):585–590

Zhang L, Guo H, Tian L, Cao S, Du C (2007b) Study of inhibitory effect of extracts from *Actinidia arguta* on human carcinoma of esophagus cells. Zhong Yao Cai 30:564–566

Zhang P, Liang D, Jin W, Qu H, Cheng Y, Li X, Ma Z (2009) Cytotoxic thiophenes from the root of *Echinops grijisii* Hance. Z Naturforsch C 64(3–4):193–196

Zhang N, Cai H, Cai B, Yang H, Li J, Yang G (2010) Two new cytotoxic acetophenone derivatives from*Euphorbia ebracteolata* Hayata. Fitoterapia 81(5):385–388

Zhang H, Lu P, Cao W, Zhang Z, Meng X (2012a) The effect-enhancing and toxicity-reducing activity of *Hypericum japonicum* Thunb. extract in murine liver cancer chemotherapy. Mol Clin Oncol 1(2):395–399

Zhang GG, Jing Y, Zhang HM, Ma EL, Guan J, Xue FN, Liu HX, Sun XY (2012b) Isolation and cytotoxic activity of selaginellin derivatives and biflavonoids from *Selaginella tamariscina*. Planta Med 78(4):390–392

Zhang Z, Yu C, Zhang C (2014) Chemopreventive effects of oplopantriol A, a novel compound isolated from *Oplopanax horridus*, on colorectal cancer. Nutrients 6:2668–2680

Zhang W, Shan L, Su J, Huizi JI, Li H, Yunheng SH, Xu X, Liu R (2016) Line leaf inula flower lactone A and methods for preparing and using the same for treating myocarditis. 8(22):143–155

Zhao Y, Huang SX, Yang LB, Pu JX, Xiao WL, Li LM, Lei C, Weng ZY, Han QB, Sun HD (2009) Cytotoxic ent-kaurane diterpenoids from *Isodon henryi*. Planta Med 75(01):65–69

Zhao Q, Guo Y, Feng B, Li L, Huang C, Jiao B (2011) Neriifolin from seeds of *Cerbera manghas* L. induces cell cycle arrest and apoptosis in human hepatocellular carcinoma HepG2 cells. Fitoterapia 82(5):735–741

Zhao M, Onakpa M, Chen W, Santarsiero B, Swanson S, Burdette J, Asuzu I, Che C (2015a) 17-Norpimaranes and (9βH)-17-Norpimaranes from the tuber of *Icacina trichantha*. J Nat Prod 78(4):789–796

Zhao Q, Huo X, Sun F, Dong R (2015b) Polyphenol-rich extract of *Salvia chinensis* exhibits anticancer activity in different cancer cell lines and induces cell cycle arrest at the G0/G1-phase, apoptosis and loss of mitochondrial membrane potential in pancreatic cancer cells. Mol Med Rep 12(4):4843–4850

Zheng J, Zhou Y, Li Y, Xu D, Li S, Li H (2016) Spices for prevention and treatment of cancers. Nutrients 8(8):495

Zhou BN, Johnson RK, Mattern MR, Wang X, Hecht SM, Beck HT, Ortiz A, Kingston DG (2000) Isolation and biochemical characterization of a new topoisomerase I inhibitor from *Ocotea leucoxylon*. J Nat Prod 63:217–221

Zhou WB, Tao JY, Xu HM, Chen KL, Zeng GZ, Ji CJ, Zhang YM, Tan NH (2010) Three new antiviral triterpenes from *Aster tataricus*. Z Naturforsch 65b:1393–1396

Zhou J, Han N, Lv G, Jia L, Liu Z, Yin J (2017) Two new β dihydroagarofuran sesquiterpenes from *Celastrus orbiculatus* Thunb and their anti-proliferative activity. Molecules 22:948–954

Zhuang S, Chen S, Tsai J, Huang C, Wu T, Liu W, Tseng H, Lee H, Huang M, Shane G, Yang C, Shen Y, Yan Y, Wang C (2009) Effect of citronellol and the Chinese medical herb complex on cellular immunity of cancer patients receiving chemotherapy/radiotherapy. Phytother Res 23:785–790

Zhuang Q, Li J, Chen Y, Lin J, Lai F, Chen X, Lin X, Peng J (2015) Ethyl acetate extract of *Hypericum japonicum* induces apoptosis via the mitochondria-dependent pathway in vivo and in vitro. Mol Med Rep 12(4):4851–4485

Zimmermann R, Bauermann U, Morales F (2006) Effects of growing site and nitrogen fertilization on biomass production and lignan content of linseed (*Linum usitatissimum* L.). J Sci Food Agric 86:415–419

Zuo GY, An J, Han J, Zhang YL, Wang GC, Hao XY, Bian ZQ (2012) Isojacareubin from the Chinese herb *Hypericum japonicum*: potent antibacterial and synergistic effects on clinical methicillin-resistant *Staphylococcus aureus* (MRSA). Int J Mol Sci 13(7):8210–8218